Biofluid Mechanics

Condensing 40 years of teaching experience, this unique textbook will provide students with an unrivalled understanding of the fundamentals of fluid mechanics, and enable them to place that understanding firmly within a biological context. Each chapter introduces, explains, and expands a core concept in biofluid mechanics, establishing a firm theoretical framework for students to build upon in further study. Practical biofluid applications, clinical correlations, and worked examples throughout the book provide real-world scenarios to help students quickly master key theoretical topics. Examples are drawn from biology, medicine, and biotechnology with applications to normal function, disease, and devices, accompanied by over 500 figures to reinforce student understanding. Featuring over 120 multicomponent end-of-chapter problems, flexible teaching pathways to enable tailor-made course structures, and extensive Matlab and Maple code examples, this is the definitive textbook for advanced undergraduate and graduate students studying a biologically grounded course in fluid mechanics.

James B. Grotberg is a PhD and MD Professor of Biomedical Engineering, and a Professor in the Department of Surgery, at the University of Michigan, with many years of research, teaching and clinical experience focused on biofluid mechanics. He is a Fellow of the American Physical Society, a Fellow of the American Institute for Medical and Biological Engineering, an Inaugural Fellow of the Biomedical Engineering Society, and a Fellow of the American Society of Mechanical Engineers.

Cambridge Texts in Biomedical Engineering

Cambridge Texts in Biomedical Engineering provide a forum for high-quality textbooks targeted at undergraduate and graduate courses in biomedical engineering. They cover a broad range of biomedical engineering topics from introductory texts to advanced topics, including biomechanics, physiology, biomedical instrumentation, imaging, signals and systems, cell engineering, and bioinformatics, as well as other relevant subjects, with a blending of theory and practice. While aiming primarily at biomedical engineering students, this series is also suitable for courses in broader disciplines in engineering, the life sciences and medicine.

Biofluid Mechanics

Analysis and Applications

JAMES B. GROTBERG

University of Michigan

CAMBRIDGE
UNIVERSITY PRESS

University Printing House, Cambridge CB2 8BS, United Kingdom

One Liberty Plaza, 20th Floor, New York, NY 10006, USA

477 Williamstown Road, Port Melbourne, VIC 3207, Australia

314–321, 3rd Floor, Plot 3, Splendor Forum, Jasola District Centre, New Delhi – 110025, India

103 Penang Road, #05–06/07, Visioncrest Commercial, Singapore 238467

Cambridge University Press is part of the University of Cambridge.

It furthers the University's mission by disseminating knowledge in the pursuit of education, learning, and research at the highest international levels of excellence.

www.cambridge.org
Information on this title: www.cambridge.org/9781107003118
DOI: 10.1017/9781139051590

First published 2021

Printed in the United Kingdom by TJ Books Limited, Padstow Cornwall

A catalogue record for this publication is available from the British Library.

ISBN 978-1-107-00311-8 Hardback

Additional resources for this publication at www.cambridge.org/grotberg.

CONTENTS

PREFACE

Biofluid mechanics embraces the flow of life. It is intimate and personal and situates at the intersection of engineering, chemistry, physics, mathematics and biology. The author has dual expertise in fluid mechanics and medicine, and shares this unique perspective for the book's two major goals: (1) provide the student with an understanding of fundamental fluid mechanics and (2) put that understanding into a biological context. The level of the book is aimed at the advanced undergraduate and graduate student. So for most it would be a second course in fluid mechanics, though a novice with strong mathematical experience can benefit as well. The content is derived from class notes developed over many years of teaching the subject at Northwestern University and the University of Michigan. Students from many backgrounds have been taught from these notes, including engineering, mathematics, physics and medicine. With 650 references, the book can also serve as a general reference for biofluid mechanics.

To serve this diverse group, the book is organized to cover fundamentals of fluid mechanical principles in Chapters 1–6 which are common to the general field but with biofluid highlights. This includes how we describe fluid motion, characterize mechanical systems with dimensionless variables and parameters, conserve mass and momentum, and relate stresses to rates of deformation. Then biofluid applications are presented in Chapters 7–17 in the context of specialized areas of fluid mechanics. Within each chapter, essential information appears in the early sections, while the more advanced material is in the later sections. Worked problems, biofluid applications and clinical correlations are included in the text proper.

This layout allows the course director to choose a format to fit the needs of the students. Either focus on the essentials in the early chapter sections and then move to applications while covering more topics, or go more in depth, further into each chapter, while sampling the applications as desired. The fluid mechanics curriculum in some departments often consists of a second level, standalone, general course that delves more deeply into the derivation and understanding of fluid flow. Then they provide a separate biofluid mechanics course. This text can combine those goals into one course, or act as a separate course emphasizing the biofluid applications of Chapters 7–17. A mix of balances can be struck as well, but the flexibility is meant to be provided.

Good preparation for this book would include a first course in fluid mechanics, mathematical skills including differential equations, and experience with computational and problem solving software like MATLAB®, Maple, Mathematica, Python or others. There are 122 multicomponent homework problems with 659 separate parts that guide the student through the relevant flow, physical interpretation, and application. They will employ these skills and software for solving and plotting various features of the many flows covered in the chapters. Examples in MATLAB® and Maple are provided to orient the student. Plotting gives the student a chance to learn by creating and sorting out the key issues: where is the fluid, what is it doing, and why?

Acknowledgements: I am grateful to my family for their encouragement, support, and patience. As well many thanks to the hundreds of students whose labors in my course created the texture of this book. Special recognition goes to the trainees in my lab who interacted with the course as students, teaching assistants and enthusiastic advisors. Of the 64 so far, 32 have become professors around the world, reflecting on their valuable contributions.

1 Introduction to Biofluid Mechanics

Introduction

What are biofluids? There is a long list to answer this question, but a modest attempt would include: air, water, blood, mucus, urine, gastrointestinal (saliva, bile, chyme, stool), lymph, ocular (aqueous humor, vitreous humor), synovial, reproductive, cerebrospinal, milk, auditory and vestibular, intracellular, interstitial, venom, odorants, sap, nectar. These are fluids produced or used by living organisms. Of course, if we introduce a foreign fluid we need to consider it as a biofluid since it interacts with the organism. For example, perfluorocarbon liquids have been used as an artificial blood due to their ability to carry relatively large amounts of oxygen (Leach, 1996; Bull et al., 2009). They do not derive from a life form, but are inserted into one. So, one can think of a biofluid as a fluid involved in a living system.

Medical devices and drug delivery often involve biofluid mechanics. An intravenous drip, artificial heart, coronary stent, mechanical ventilator, barium swallow, magnetic resonance imaging with contrast, and inhaled bronchodilator aerosols, are all examples of therapies and diagnostics involving biofluid mechanics and transport processes. The analysis of cells and molecules in microfluidic and array devices for high throughput also are dependent on the flow of fluids within the construct which carry the target objects (Huh et al., 2005; Whitesides, 2006; Tavana et al., 2009). Then there are fluids that are man-made from biological materials. This is most evident in the field of food processing, where biofluids such as mayonnaise, ketchup, apple sauce, peanut butter, melted chocolate and soft serve ice cream flow in pipes and devices to end up in containers or cones.

What are some functions of biofluids? Life depends inherently on fluid mechanics. Land animals and plants are all immersed in the atmospheric air, for example. So we rely on it for oxygen and carbon dioxide exchange. The same is true for aquatic life forms, though their fluid is much more dense and viscous. Animals and plants also have internal fluids. In general, biofluids maintain internal milieu (chemicals, temperature); transport of metabolic gases, nutrients and waste; provide means of locomotion for finding food or prey, avoiding predators, mating, temperature regulation, migration; reproduction; protection; communication (mechanical, acoustic, chemical, visual).
And one cannot view a frisky dolphin without acknowledging that biofluids can provide an outlet for "play."

Topics Covered

In this introductory chapter we explore what is a fluid and the general concepts of what makes a fluid move with examples from a variety of situations. The last two sections are more in depth. They analyze the transition from mass point mechanics to continuum mechanics for wave propagation in an array of springs and masses, and how a system of ordinary differential equations (ODEs) describing that array becomes a single partial differential equation (PDE) in the continuum limit of an elastic bar. Finally the continuum limit in gases is shown by introducing the Knudsen number, which is the dimensionless ratio of the mean free path of a gas molecule to the characteristic length scale of the problem being analyzed.

1.1 What is a Fluid?

In Figure 1.1 there are three examples of fluids shown: first is a liquid (glass of beer), which is a pending biofluid; second is a gas (Hurricane Katrina); and last is a glacier. We generally think of ice as a solid, so why are we calling a glacier a fluid? The answer is that it depends on how long one observes the glacier. Over short times (seconds, minutes, hours), it appears to behave like a solid. But over long time scales, years, it clearly flows like a fluid. An important dimensionless ratio related to the glacier in particular is the internal time scale of a material, say its stress relaxation time, divided by the observation time. That ratio is called the Deborah number. The history of this name (Reiner, 1964) is Biblical (Judges 5:5), when the prophet Deborah said "The mountains flowed before the Lord." Her insight was significant, that mountains also can flow like a fluid if observed long enough, and who else can observe that long but a deity? There is more about her in Section 14.3.

(a)

Figure 1.1 Three examples of fluids: (a) liquid (mug of beer); (b) gas (Hurricane Katrina, August 28, 2005); (c) the Kaskawulsh Glacier, which flows from snowy mountains in Kluane National Park, Yukon, Canada.

A more precise definition comes from the response of a material to an applied shear stress, see Figure 1.2. When the shear stress, τ, is applied to the top of a material rectangle, which is fixed to a rigid plate on its bottom, the rectangle deforms by the angle ϕ. For a

(b)

(c)

Figure 1.1 (*continued*)

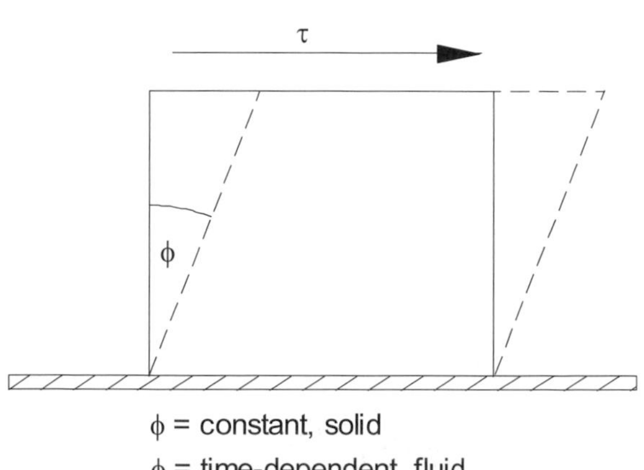

Figure 1.2 Shearing of a continuum material element.

solid, ϕ reaches a constant value; however, for a fluid it is time-dependent and will continue to increase. The fluid flows in response to a shear stress. Another way to say it is that a fluid cannot sustain a shear stress and also remain at rest.

1.2 What Makes a Fluid Move?

Without going into the details as of yet, motion of a fluid obeys Newton's second law in appropriate form. Motion results from imposed forces, or the forces result from imposed motion, or some combination. For our purposes we will restrict ourselves to fluid motion that results from the following:

(i) Pressure gradients: a good example in human biofluid mechanics is blood flow in the heart, see Figure 1.3. The arrows indicate the direction of flow into and out of the four main chambers.
(ii) Forces from a distance: gravity and magnetism act from a distance without contacting the fluid. A surfer enjoys the benefits of gravity-induced wave motion, as shown in Figure 1.4.
(iii) Motion at a boundary: a shark on the attack propels through the water by moving its tail

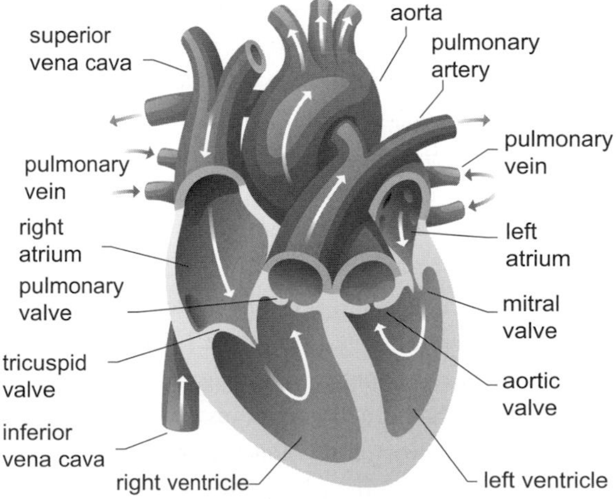

Figure 1.3 Blood flow in the heart showing chambers and major vessels.

Figure 1.4 Tyler Wright surfing at Lululemon Maui Pro – Women's WSL Championship 2019 Maui, Hawaii – December 2.

Figure 1.5 A shark finds a seal meal.

and fins. Going airborn takes strong muscular effort to catch prey, see Figure 1.5.

(iv) Force at a boundary: creating a squeal from a deflating balloon by stretching the outlet applies an elastic force on the boundary that interacts in oscillation with the exiting gas. Vocalization and respiratory wheezing also depend on this type of boundary force, see Figure 1.6.

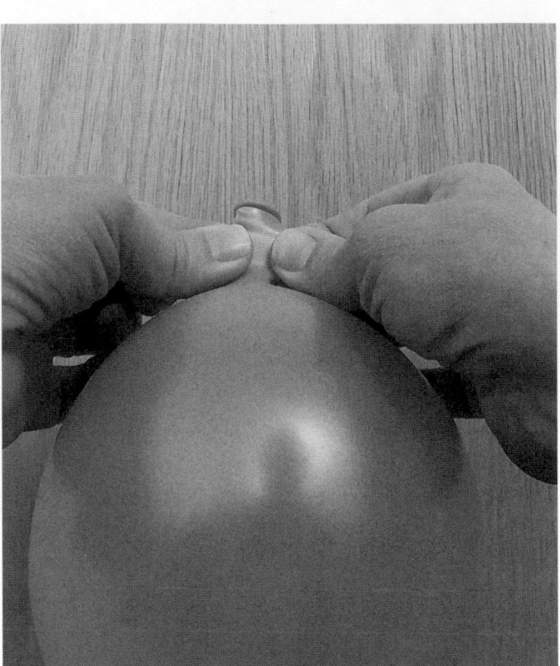

Figure 1.6 Balloon squeals need fine tuning of the force stretching the elastic boundary.

(v) Osmosis (diffusion): red blood cells placed in a hypertonic solution lose fluid by osmosis and shrink in size, placed in hypotonic solution they gain fluid and can burst, see Figure 1.7.

1.3 Properties of Fluids

This is a second course in fluid mechanics and the reader will already have familiarity with fluid properties. The main ones that concern this book are fluid density, ρ, g/cm^3, viscosity, μ, g/cm·s, kinematic viscosity, $\nu = \mu/\rho, cm^2/s$, and surface tension (force/length), σ, g/s^2. A significant feature of some biofluids like mucus or blood is that their viscosity is not constant, as in a Newtonian fluid, but instead depends on the local strain rate of the fluid, a non-Newtonian behavior. Properties such as thermal conductivity, compressibility, vapor pressure and others are important in contexts not covered in this book.

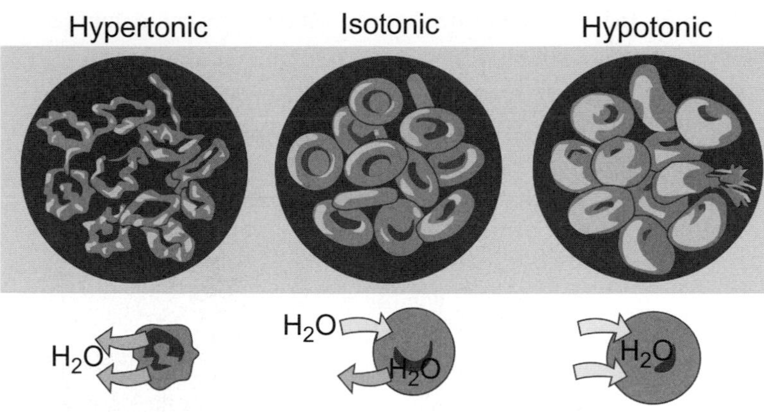

Figure 1.7 Water flow in or out of red blood cells due to osmotic differences between the internal and external fluid.

1.4 Concept of a Continuum: Transition from Point Mass Mechanics to Continuum Mechanics

Fluid mechanics is a branch of the general field called continuum mechanics, the other branch being solid mechanics. Materials that have both fluid and solid behavior, depending on the physical circumstances, fit into continuum mechanics as well. Here we develop the concept of a continuum by starting with a familiar point mass mechanics entity: the mass and spring oscillator. Consider a set of masses and springs as shown in Figure 1.8. Each spring has a spring constant, *k*, and resting length L_0. Each mass point has the same mass, *m*, and the position of the i-th mass point is a function of time, $X^{(i)}(t)$. Its initial resting position is $X_0^{(i)} = X^{(i)}(t = 0)$. We are interested in a disturbance to the system; say we move one of the masses off of its equilibrium position. Then the spring lengths attached to that mass will change leading to an oscillation that can propagate along the sequence of masses and springs. As scientists and engineers, we may want to know, for example, at what speed does the disturbance propagate?

Let's apply Newton's Second Law to the i-th mass point. The force from the spring on the right is proportional to that spring's length, $X^{(i+1)} - X^{(i)}$, relative to the resting length, L_0, that is

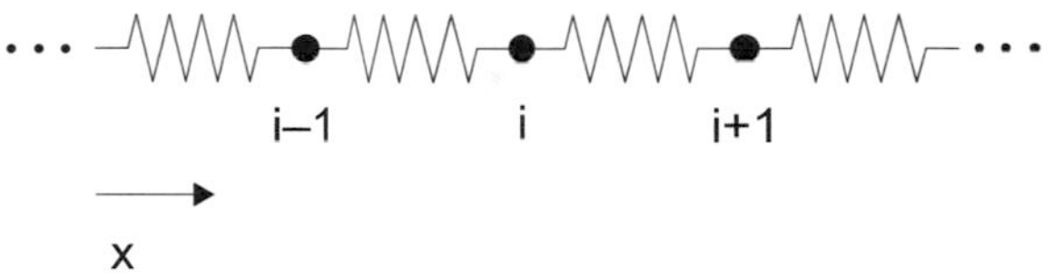

Figure 1.8 A linear array of identical springs and masses.

$$F_R^{(i)} = k\left[\left(X^{(i+1)} - X^{(i)}\right) - L_o\right] \tag{1.4.1}$$

Equation (1.4.1) tells us that when the spring on the right is longer (shorter) than L_0, then it applies a force in the positive (negative) x-direction, since it is under tension (compression). The force on the i-th mass from the spring on its left is

$$F_L^{(i)} = -k\left[\left(X^{(i)} - X^{(i-1)}\right) - L_o\right] \tag{1.4.2}$$

Here we have introduced the negative sign so that when the spring on the left is longer (shorter) than L_0, then it applies a force in the negative (positive) x-direction, since it is under tension (compression). Adding the forces together, they must equal the mass × acceleration of the i-th mass point

$$\begin{aligned} m\frac{d^2X^{(i)}}{dt^2} &= k\left[\left(X^{(i+1)} - X^{(i)}\right) - L_o\right] - k\left[\left(X^{(i)} - X^{(i-1)}\right) - L_o\right] \\ &= k\left[X^{(i+1)} - 2X^{(i)} + X^{(i-1)}\right] \end{aligned} \tag{1.4.3}$$

Note that the resting length, L_0, has cancelled out. The right hand side is a second-order central difference,

$$\Delta^2 X^{(i)} = X^{(i+1)} - 2X^{(i)} + X^{(i-1)} \tag{1.4.4}$$

In view of Eq. (1.4.4) we see Eq. (1.4.3) can be represented as

$$m\frac{d^2X^{(i)}}{dt^2} = k\Delta^2 X^{(i)} \tag{1.4.5}$$

Now let's take the limit of Eq. (1.4.5) as the resting spring length approach zero, $L_0 \to 0$. First of all, the masses will be crowded closer and closer, so in a continuum limit we can describe the mass distribution in terms of a density = mass/length along the x-direction

$$\lim_{L_0 \to 0} \frac{m}{L_0} = \rho \tag{1.4.6}$$

where ρ is the constant density. In addition, as a spring's resting length is shortened, it behaves more stiffly. For example, let k increase like $1/L_0$ as $L_0 \to 0$. Then

$$\lim_{L_0 \to 0} kL_0 = E \tag{1.4.7}$$

where E is a constant elasticity modulus. Inserting Eq. (1.4.6) and Eq. (1.4.7) into Eq. (1.4.5) and taking the limit as $L_0 \to 0$ gives us

$$\rho\frac{d^2X^{(i)}}{dt^2} = E \lim_{L_0 \to 0} \frac{\Delta^2 X^{(i)}}{L_0^2} \tag{1.4.8}$$

The second-order difference operator, divided by the measure of the distance between the difference points, is just the second spatial derivative in L_0 in the limit of $L_0 \to 0$. Keeping in mind that

$$L_0 = X_0^{(i)} - X_0^{(i-1)} = X_0^{(i+1)} - X_0^{(i)} \tag{1.4.9}$$

In this limit, the discrete initial positions $X_0^{(i)}$ form a continuum, X_0. That is, we can switch from identifying the point masses by their index, and instead identify them by where there are located in a continuum at t = 0. Since $X^{(i)}$ is also a function of time, the derivative we seek is a partial derivative with respect to the initial positions

$$\lim_{L_0 \to 0} \frac{\Delta^2 X^{(i)}}{L_0^2} = \frac{\partial^2 X}{\partial X_0^2} \tag{1.4.10}$$

Finally, then, we can represent the continuum version of Eq. (1.4.8):

$$\rho\frac{\partial^2 X}{\partial t^2} = E\frac{\partial^2 X}{\partial X_0^2} \tag{1.4.11}$$

which is a "wave equation" for $X(X_0, t)$.

A traveling wave solution to Eq. (1.4.11) is

$$X = A\sin\left(\frac{2\pi}{\lambda}(X_0 - ct)\right) \tag{1.4.12}$$

where λ is the wavelength and c is the wave speed. This particular wave is traveling in the positive x-direction.

When Eq. (1.4.12) is substituted into Eq. (1.4.11) we find

$$c^2 = \frac{E}{\rho} \tag{1.4.13}$$

The wave travels faster for a higher elastic modulus or a lower density material.

During this process of moving from point mass mechanics to continuum mechanics, the mathematical limits of the particles being ever closer together forced us to change from a discrete form of particle identification to a continuous one, initial position. Other reference positions are also candidates for particle identification. Also, notice that we started with an ordinary differential equation for the i-th mass point. Its solution depends on the motion of its immediate neighbors, as does theirs and so on. So the discrete system is an infinite set of coupled ordinary differential equations which must be solved simultaneously. Upon our transition to the continuum approximation, however, the system became a single partial differential equation for the dependent variable, X, which is a function of the two independent variables, X_0 and t.

1.5 Continuum Approximation for Gases

Continuum mechanics is the study of the mechanics of materials that can be treated as continuous, i.e. filling all of the space where it resides. Solids and fluids are typical examples, and fluids are further divided into liquids and gases. This approach ignores the space between atoms or molecules of the material, which is justifiable if the length scale of interest in the problem is much larger than the distance between them.

For liquids the continuum approach seems reasonable for many applications since the molecules are so close. Nanofluidics is an area where even these ideas can break down, however. For a gas the molecules can be much farther apart. A dimensionless ratio that helps us to understand the limits of the continuum approximation for a gas is the Knudsen Number, Kn, which is defined as the ratio of the mean free path, λ, to the characteristic length scale of the problem, L,

$$\mathrm{Kn} = \frac{\lambda}{L} \tag{1.5.1}$$

The mean free path for an ideal gas can be calculated

$$\lambda = \frac{k_B T}{\sqrt{2}\pi\sigma^2 p} \tag{1.5.2}$$

where k_B is the Boltzmann constant 1.380650 4 $\times 10^{-23}$ J/K in SI units), T is the thermodynamic temperature in Kelvin, σ is the particle hard shell diameter, and p is the total pressure. For particle dynamics in the atmosphere, and assuming standard temperature and pressure, i.e. 25 °C and 1 atm, we have $\lambda \approx 8 \times 10^{-8}\,\mathrm{m} = 80$ nm. We expect that the continuum assumption is not valid for $\mathrm{Kn} \geq 1$.

Summary

We have introduced the basic format of the textbook and some early ideas about defining a fluid, its properties, and what causes a fluid to move. We also focused on biofluids and their many functions in the biological world. The last sections dealt with the general concept of a continuum, since fluid mechanics is a branch of the more general field of continuum mechanics. In that discussion, the transition from thinking in the realm of mass point mechanics to a continuous distribution of a material was explored conceptually and mathematically. As the topics in each of the following chapters are covered, this introduction will take on more meaning so can be reviewed from time to time as a touchstone for the student.

Image Credits

Figure 1.1(a) Source: Burazin/ Getty
Figure 1.1(b) Source: StocktrekImages/ Getty
Figure 1.1(c) Source: NASA/Christy Hansen, 2013, www.nasa.gov/mission_pages/icebridge/multimedia/spr13/DSCN3044.html#.Xqwt5kmWyLg
Figure 1.3 Source: AlonzoDesign/ Getty
Figure 1.4 Source: EdSloane/ Getty
Figure 1.5 Source: USO/ Getty
Figure 1.7 Source: Osmotic pressure on blood cells diagram, LadyofHats, 2007 (public domain) https://commons.wikimedia.org/wiki/File:Osmotic_pressure_on_blood_cells_diagram.svg

2 Scalings, Parameters and Variables

Introduction

Biofluid mechanics, as a field of study, tends to attract students from a wide variety of backgrounds such as Mechanical Engineering, Biomedical Engineering, Chemical Engineering, Aerospace Engineering, Physics, Mathematics, Physiology and Medicine. Each of these disciplines has its own unique way of envisioning and solving problems, and its own operative language of analysis. With so many entry doorways into this house, it is meaningful to present a common way to address biofluid mechanical phenomena for purposes of organizing our thinking and solution activities. In this chapter, the goal is to present some problems that are already well known to the student, from as early as a high school physics course, but using an approach that will allow us to build a pathway of understanding for more complex problems as we advance in this textbook.

In much of what you have likely learned from mass point mechanics or introductory fluid mechanics, dealing with variables and parameters was usually done with dimensional variables (e.g. space and time) and dimensional parameters (e.g. diameter, density, viscosity). Distance in meters or miles, time in seconds or hours, force in dynes or pounds, etc. However, a more advanced view that helps to simplify complex physics, which is often instructive and insightful, is to represent quantities as dimensionless ratios. This approach allows us to compare the importance of specific competing effects in a conservation law, like momentum conservation.

For example, dropping your pencil to the floor is a simple process to analyze, if we initially ignore the fluid interaction from the air on it. The same is not true for dropping your paper, which can catch the air and zig-zag down to the floor, getting both lift and drag from the fluid. The conservation of momentum equation is the same for both, but the fluid drag force term, F_d, for the pencil will be much smaller than the gravitational force on it, i.e. its weight W. That is, $F_d/W<<1$ for the pencil and to first approximation we do not have to deal with the fluid drag in air. This simplification does not occur for the paper, where the two forces may be comparable in size. Using some formality to compare the relative size of terms in a force balance permits us to simplify the system by ignoring some terms and solving approximate equations which can be far more manageable and just as insightful. It also allows us to reduce the number of independent parameters by combining the dimensional ones into dimensionless groups, which is a saving in both experiments and computations, as well as revealing for the main competing phenomena in the system.

Thinking non-dimensionally is a common event, we just may not be aware that we are doing it. The process starts even in our childhood. For example, a seesaw holds person A with body weight W_A and person B with body weight W_B. The seesaw works for small, medium, and large people, so it is not the dimensional body weights, W_A and W_B, that matter. Instead, it is the dimensionless ratio W_A/W_B that is the important consideration. Children learn early on to choose a partner so that the dimensionless ratio, $W_A/W_B \sim 1$. If not, then they learn to adjust the distance of their seat from the fulcrum.

Likewise, a child learns that in order to blow up bubble gum, a viscoelastic material, they have to chew it for a while so that its consistency is just right, see Figure 2.1. If it is too hard, then they cannot create enough air pressure from their lungs to inflate it. It needs more chewing and saliva, an important biofluid, mixed in to soften it. If it is too soft, then the bubble can rupture prematurely. The optimal chewing allows the expired air to form a big bubble. The lung air pressure they create, P_L, relative to the elastic modulus of the gum, E, is a dimensionless ratio, P_L/E, that children learn, through trial and error experiments, to tune for success. Diving into a pool requires that the height from which the dive begins, H_D, be compared to the depth of the water, H_W, so that the stopping distance upon entering the water is achieved by fluid drag, rather than striking the bottom of the pool. The ratio H_W/H_D must be safe. Finally, in a pickup game of basketball, each player tends to guard an opponent who is approximately their height, $H_{me}/H_{you} \sim 1$.

Figure 2.1 Successful bubble gum blowing is a dimensionless endeavor.

Topics Covered

In this chapter we will introduce some concepts of dimensionless quantities by starting with simple and familiar mechanics settings. We first focus on the point mass mechanics of a ball in a gravity field but without the influence of a surrounding fluid, i.e. in a vacuum. This is a well-studied problem in high school physics. The ball is dropped from a height and we discuss the effect of the governing parameters on the time to impact. The solutions to Newton's second law yield the ball's position as a function of time. These two dimensional variables, position vs time, are scaled into dimensionless form and the resulting simplifications for the number of experiments needed to characterize the system as well as the physical interpretations are discussed. Then the ball is thrown and the analysis leads to a single dimensionless parameter, the Froude number, which is proportional to the ratio of the initial kinetic energy to the initial potential energy. The entire system is governed by the value of this ratio. The fundamental approach to finding the dimensionless parameters governing a mechanical system is then discussed in terms of the Buckingham Pi Theorem. We next revisit the thrown ball problem, but this time the governing equations are cast in dimensionless form first, and then solved. This is the normal procedure when approaching fluid mechanical problems, since the resulting dimensionless parameters appear in the governing equations and may suggest simplified approaches if they are very big or very small, for example. It allows us to ignore some terms in the equations which can make the solutions easier to find. Computational software like MATLAB®, Maple, Mathematica and others are useful for transforming equations into dimensionless form and solving them, and the thrown ball problem is a presented example. The ball is considered to be in a fluid which has a drag force on it and a biofluid mechanics application to the erythrocyte sedimentation rate is explored. Finally, a cell sorting microfluidic device, based on sedimentation and cross flow, is presented as a means of learning about design parameters using fluid mechanics.

2.1 The Dropped Ball

Consider a ball dropped from a height, h, in the absence of any fluid, i.e. in a vacuum. This is a typical problem studied in a first course in physics. The ball has mass m, and is under the influence of gravity acceleration, g, see Figure 2.2. There is a floor at $y = 0$. The ball position, call it y_p, is a function of time, t, i.e. $y_p(t)$. The dependent variable, y_p, "depends" on the independent variable, t. The time derivative of the position is the velocity, $v_p(t) = dy_p/dt$. There are three parameters in the problem: h, m and g.

In order to characterize this mechanical system, what would be a good scalar output measurement that captures the nature of the falling process? One choice is the impact time, t_i, such that $y_p(t = t_i) = 0$. To investigate how t_i depends on the three parameters, write the functional relationship as

$$t_i = f_0(m, h, g) \tag{2.1.1}$$

where the function f_0 depends on m, h and g and we seek its specific form.

When you approached the falling ball problem in Physics class, you began with Newton's second law, which is the conservation of momentum (linear momentum) equation in the y-direction. For this example, it is

$$F = \frac{d}{dt}(mv_p) \tag{2.1.2}$$

where F is the force. In words, the equation says that the force, F, is equal to the time rate of change of the momentum, mv_p. Equation (2.1.2) is a "differential equation," since it involves derivatives of the unknown dependent quantity, v_p.

The force is the constant weight, $F = -mg$, in the negative y-direction. For constant m, Eq. (2.1.2) simplifies to

$$-mg = m\frac{dv_p}{dt} \quad \rightarrow \quad -g = \frac{dv_p}{dt} \qquad (2.1.3)$$

The first thing we see is that the motion does not depend on m, so Eq. (2.1.1) becomes

$$t_i = f_0(h, g) \qquad (2.1.4)$$

It makes sense to vary each of the input parameters (h and g) multiple times while measuring t_i for each combination of parameter values. So we may consider 10 values of h and two for g (Earth and Moon). That is 20 combinations. For good experimental methods, it would be typical to perform, say, five trials for each combination so that variability in the experiments can be averaged. Now we are up to 100 experiments, half of them on the Moon. Can we simplify this system to reduce the number of experiments and their location?

We solve Eq. (2.1.3) showing the individual steps for clarity in the early going. It can be integrated for constant g

$$\int dv_p = -g\int dt + c_1 \quad \rightarrow \quad v_p = -gt + c_1 \qquad (2.1.5)$$

where the constant of integration, c_1, may be determined by imposing an initial velocity on the ball at a reference time, say $t = 0$. Let that initial velocity be zero, so the ball starts at rest. Then

$$v_p(t = 0) = c_1 = 0 \qquad (2.1.6)$$

So Eq. (2.1.5) becomes

$$v_p = -gt \qquad (2.1.7)$$

The velocity is just the time derivative of the y-position, so that Eq. (2.1.7) may be written as

$$\frac{dy_p}{dt} = -gt \quad \rightarrow \quad dy_p = -gt\,dt \qquad (2.1.8)$$

Again showing the steps, we integrate Eq. (2.1.8) to find

$$\int dy_p = -g\int t\,dt + c_2 \quad \rightarrow \quad y_p = -g\frac{t^2}{2} + c_2 \qquad (2.1.9)$$

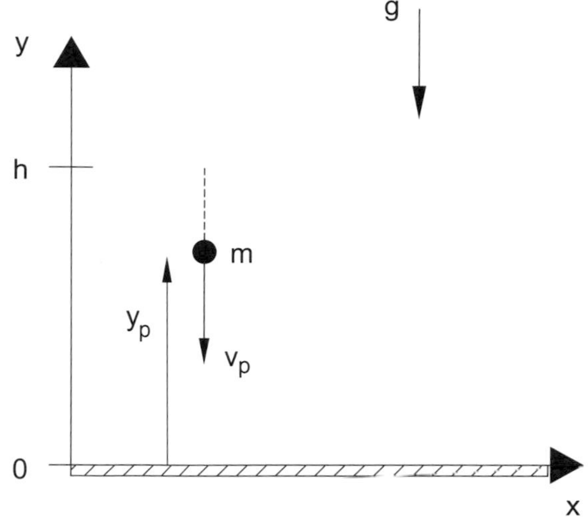

Figure 2.2 Ball of mass, m, dropped from height h with gravitational acceleration, g.

and the second constant of integration, c_2, may be determined by imposing an initial position at $t = 0$, call it h, as shown in Figure 2.2

$$y_p(t = 0) = c_2 = h \qquad (2.1.10)$$

Now the form is the well-known solution

$$y_p = h - g\frac{t^2}{2} \qquad (2.1.11)$$

Let's restrict ourselves to $h > 0$ and follow the solution until the particle hits at $y = 0$. Some sample solutions for choices of the two remaining parameters, h, g, are graphed as follows in Figure 2.3. We have chosen two values of the initial position, $h = 10,\ 20$ cm, and two values of the gravitational constant, $g = 980, 163\ \text{cm/s}^2$, which are the values for gravity on Earth and on the Moon, respectively. The goal would be to calculate and plot 20 combinations. The impact time is when $y_p = 0$ on this plot. Clearly the time to impact increases with larger initial height, h, and small gravity, g.

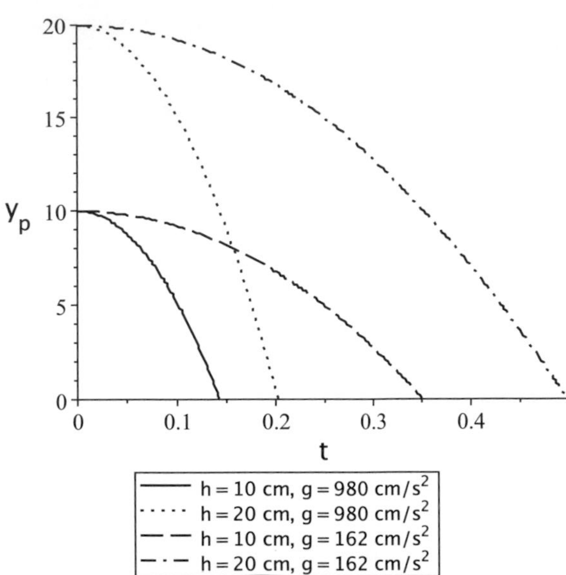

Figure 2.3 Vertical position, y_p, versus time, t, for four choices of the parameters h and g. The impact time, t_i, is when $y_p = 0$.

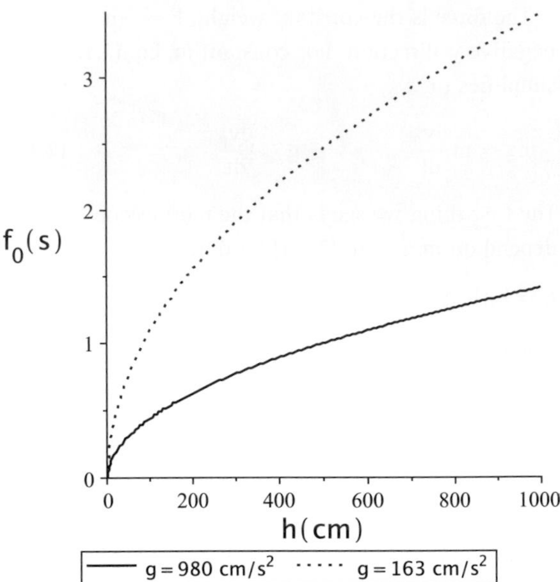

Figure 2.4 f_0 vs h for $g = 980\,\text{cm/s}^2$ (Earth) and $g = 163\,\text{cm/s}^2$ (Moon).

2.2 Impact Time: Theory and Experiments

How can we characterize the dropped ball system? That is, what do we want to measure or calculate that will let us compare the phenomenon as it varies from one set of parameter choices to another? What about characterizing the system by the average speed? What about the impact time? Since we have already paid attention to the impact time, let's choose it for the output of interest. The equation for the impact time, t_i, comes from setting $y_p = 0$ in Equation (2.1.11) and letting $t = t_i$, then solve for t_i

$$y_p(t = t_i) = 0 = -g\frac{t_i^2}{2} + h \quad \rightarrow \quad t_i = \left(\frac{2h}{g}\right)^{1/2} \tag{2.2.1}$$

Equation (2.1.4) can now be given its mathematical form for f_0

$$t_i = f_0(h, g) = \left(\frac{2h}{g}\right)^{1/2} \quad \text{dropped ball} \tag{2.2.2}$$

So for fixed $g = 980\,\text{cm/s}^2$ (Earth) or $g = 163\,\text{cm/s}^2$ (Moon), the dependence of f_0 on h is a square root, as graphed in Figure 2.4.

For our experimental results, hopefully the method would have been good enough to fit well to the theory. For example, if we set $g = 9.8\,\text{m/s}^2$ and choose 10 values of h in meters with five trials at each value, then a plot of the data could look like Figure 2.5 where the theoretical curve is also shown. There is reasonably good correlation between theory and experiment in Figure 2.5.

Another approach is to assume a functional form of the dependence of f_0 on g, h and m and test that form against the data to fill in the detailed relationship. The functional form can be guided in some way by knowledge of the physical issues, or it can be an *ad hoc* guess. It is common, for example, to propose the following type of relationship for a measurable outcome like

$$f_0 = k_0 g^a h^b m^c \tag{2.2.3}$$

where k_0, a, b and c are unknown constants. Equation (2.2.3) may fit the data well over some range of the parameters to be discovered. A regression analysis of the data, including trials with varying m, hopefully will show that the constants are close to the values

$$k_0 = \sqrt{2}, a = -\frac{1}{2}, b = \frac{1}{2}, c = 0 \tag{2.2.4}$$

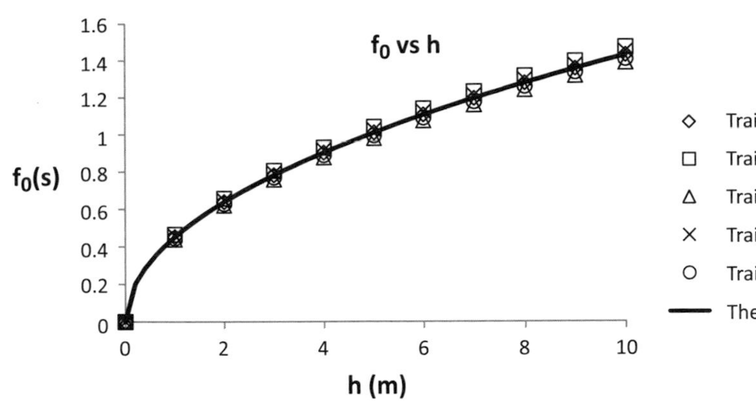

Figure 2.5 Data from 50 experiments measuring f_0 vs h for $g = 9.8\,m/s^2$. Comparison to theory is shown in the solid line.

with relatively small standard deviations. Again we see that the impact time does not depend on the mass, m, since c is zero. Then we would have good experimental correlation to the theoretical predictions of Eq. (2.2.2).

2.3 Scalings and Dimensionless Variables

We are going to put f_0 to use. Just for a first step in simplifications, more will be discussed later, we can rearrange Eq. (2.1.11) by dividing through by h as follows

$$\frac{y_p}{h} = -\frac{g}{2h}t^2 + 1 \qquad (2.3.1)$$

But we see that the coefficient of t^2 is f_0^{-2}, so Eq. (2.3.1) may be re-written as

$$\frac{y_p}{h} = -\frac{t^2}{f_0^2} + 1 \qquad (2.3.2)$$

Note that $0 \leq y_p/h \leq 1$ and $0 \leq t/f_0 \leq 1$, so we have the convenience of limited ranges for the variables. In fact, if we plot y_p /h vs t/f_0, the several curves of Figure 2.3 collapse to a master curve shown in Figure 2.6.

So we have found a way to simplify the problem, compare Figure 2.6 to Figure 2.3. We now have only one curve instead of four (or 20). For the experiments we would only need five trials using one set of parameters, instead of 100 experiments using 20 sets and five trials each. This is a huge savings of time and

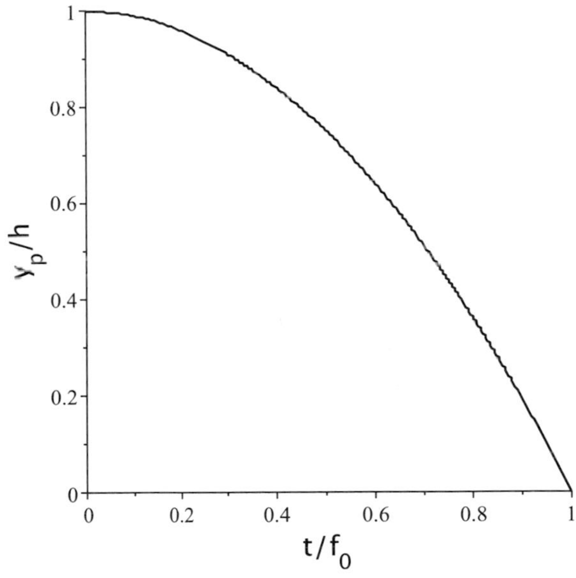

Figure 2.6 Master curve for a dropped ball with dimensionless variables, y_p/h vs t/f_0.

expense (no need to go to the Moon) and we have gained some physical insight into the problem.

Our approach was to make a "dimensionless" version of the dependent variable, y_p, by forming y_p/h, and a dimensionless version of the independent variable, t, by forming t/f_0 We can give these forms their own name, Y_p and T, respectively

$$Y_p = \frac{y_p}{h}, \quad T = \frac{t}{\left(\dfrac{2h}{g}\right)^{1/2}} \qquad (2.3.3)$$

Now the dimensionless solution of Eq. (2.3.2) is

$$Y_p = 1 - T^2 \tag{2.3.4}$$

Clearly the dimensionless impact time occurs for $T = 1$ when $Y_p = 0$, which makes sense since our time scale for forming T is the dimensional impact time, f_0, for this problem. In fact, we see that with the scalings, the range of the dimensionless variables is $0 \leq Y_p \leq 1, 0 \leq T \leq 1$.

In the process of defining the dimensionless variables, Y_p and T, we used the "time scale," $\sqrt{2h/g}$, which has units of time, and the "length scale," h, which has units of length. Of course, $\sqrt{h/g}$ or any multiple is an equally reasonable choice. It is routine in fluid mechanics to find the characteristic scales of the problem at hand for the independent variables, which include time (t) and space (x, y, z), as well as dependent variables like pressure, velocity, position, temperature, concentration and others.

In terms of understanding the mechanics, we have also shown that this approach has given us important physical interpretations of the dimensionless variables. Y_p is the fraction of the initial height, while T is the fraction of the final time, and the mechanics of the system only "knows" these relative values. If we choose feet or centimeters for y_p, or seconds or hours for t, the ratios Y_p or T don't change. The mechanics does not know what units we are using, the phenomena only know relative magnitudes of forces and motion.

2.4 The Thrown Ball: A Dimensionless Parameter

Now let's add an initial velocity, v_0, to the problem, so that the ball is thrown rather than dropped. The analysis is the same, except now Eq. (2.1.6) is modified to include a non-zero initial velocity

$$v_p(t = 0) = c_1 = v_0 \tag{2.4.1}$$

and that carries through to the final form following the steps leading to Eq. (2.1.11)

$$y_p(t) = -g\frac{t^2}{2} + v_0 t + h \tag{2.4.2}$$

We have three parameters now: h, v_0, g. Let's, again, try to simplify the result. Divide Eq. (2.4.2) by h to obtain the form

$$\frac{y_p}{h} = -\frac{g}{2h}t^2 + \frac{v_0}{h}t + 1 \tag{2.4.3}$$

Let's choose scales for dimensionless variables as we did in the previous section, Eq. (2.3.3),

$$Y_p = \frac{y_p}{h}, \quad T = \left(\frac{g}{h}\right)^{1/2} t \tag{2.4.4}$$

but without the $\sqrt{2}$ factor in the time scale, which had a special significance for the dropped ball. Inserting Eqs. (2.4.4) into Eq. (2.4.3) gives us

$$Y_p = -\frac{1}{2}T^2 + \frac{v_0}{\sqrt{gh}}T + 1 \tag{2.4.5}$$

Now we have one dimensionless parameter in the solution, $v_0/\sqrt{gh}$. This combination arises in fluid mechanics when v_0 is a characteristic fluid velocity, often with regard to flow of rivers or waves in the ocean. It is called the "Froude number," Fr. At this point, we have not introduced a fluid into the problem, but we can use this terminology anyway. So $Fr = v_0/\sqrt{gh}$ and it can be viewed as a ratio of a measure of the initial kinetic energy to the initial potential energy in this problem, as shown

$$Fr = \frac{v_0}{\sqrt{gh}} \quad \rightarrow \quad Fr^2 = \frac{v_0^2}{gh}$$

$$= \frac{mv_0^2}{mgh} \sim \frac{2 \times \text{initial kinetic energy}}{\text{initial potential energy}} \tag{2.4.6}$$

Equation (2.4.5) has the more compact form

$$Y_p = -\frac{1}{2}T^2 + Fr\,T + 1 \tag{2.4.7}$$

Instead of a single, master plot for $Y_p(T)$ as in Figure 2.6, for the dropped ball, a plot of $Y_p(T)$ for the thrown ball requires us to choose values for Fr. Four choices of Fr are shown in the $Y_p(T)$ plots of Figure 2.7. $Fr < 0$ occurs when the ball is thrown downward, while $Fr > 0$ corresponds to the ball being thrown upward.

As before, the dimensionless impact time, call it T_i, is found from setting $Y_p = 0$, which gives us

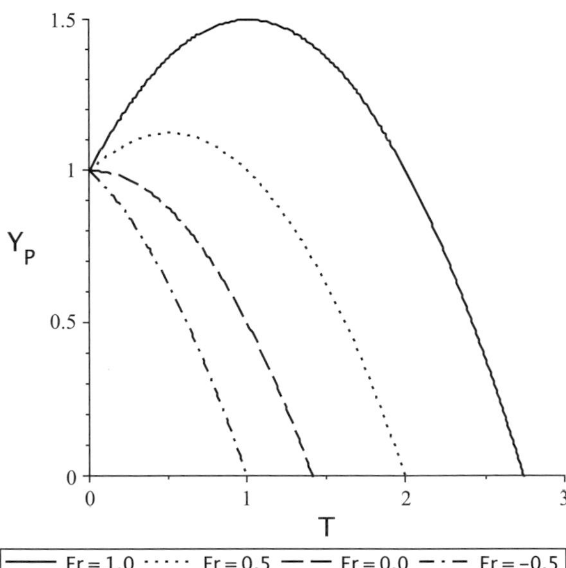

Figure 2.7 Dimensionless vertical position, Y_p, vs dimensionless time, T, for a thrown ball with four different values of the Froude number, $Fr = 1.0, 0.5, 0.0, -0.5$.

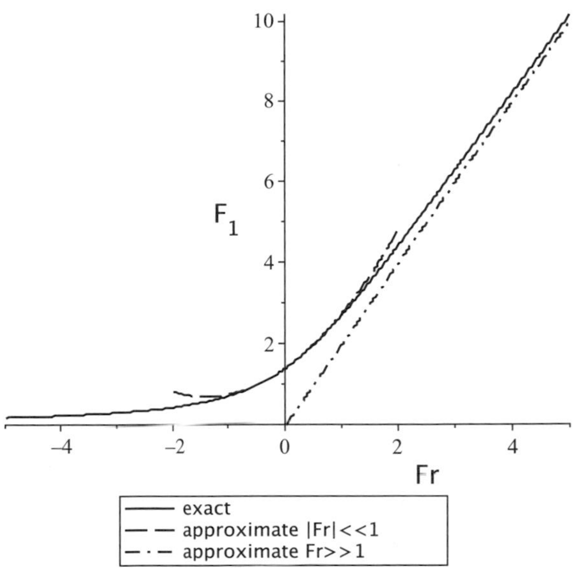

Figure 2.8 Dimensionless impact time F_1 vs the Froude number, Fr, for the total solution and an approximate solution for $|Fr|<<1$ and for $Fr>>2$.

$$\frac{1}{2}T_i^2 - Fr\,T_i - 1 = 0 \tag{2.4.8}$$

Equation (2.4.8) is a quadratic equation for T_i where T_i is now a function of the dimensionless parameter, Fr, i.e. $T_i = F_1(Fr)$

$$F_1 = Fr \pm \left(2 + Fr^2\right)^{1/2} \quad \rightarrow \quad F_1 = Fr + \left(2 + Fr^2\right)^{1/2} \tag{2.4.9}$$

Since the square root term is always greater than Fr, we choose the positive root so that F_1 is always positive. A plot of $F_1(Fr)$ is shown in Figure 2.8 for the total solution and for an approximate, or asymptotic, solution for $|Fr|<<1$.

Note that for $Fr = 0$ there is no initial velocity, i.e. $v_0 = 0$, so the problem should reduce to that of the dropped ball. To check, from Eq. (2.4.9) $F_1(Fr = 0) = \sqrt{2}$. According to our scaling, then, $F_1(Fr = 0) = t_0/\sqrt{h/g} = \sqrt{2}$, which is consistent. With only one dimensionless parameter, we are free to do experiments which change Fr by any means. We could change g or h or v_0. For negative values of Fr, we must have a negative v_0, when the ball is thrown downward. Clearly $F_1 \rightarrow 0$ as $Fr \rightarrow -\infty$, as can be seen in both Figure 2.8 and Eq. (2.4.9). The dimensional impact time for this new situation for a thrown ball is obtained from multiplying F_1 by the time scale, $(h/g)^{1/2}$.

Box 2.1 Greek Alphabet

The Latin alphabet, which you are currently reading, runs out of options in mathematics and physics, so the Greek alphabet often is called into play. Figure 2.9 is a primer on the Greek alphabet which is called upon throughout this book. In this chapter we have already used ϕ, π, α. You can see from the first two entries the origin of the word "alpha-bet(a)."

Greek Letter		Name	Equivalent	Pronunciation
Α	α	Alpha	A	al-fah
Β	β	Beta	B	bay-tah
Γ	γ	Gamma	G	gam-ah
Δ	δ	Delta	D	del-tah
Ε	ε	Epsilon	E	ep-si-lon
Ζ	ζ	Zeta	Z	Zay-tah
Η	η	Eta	E	ay-tah
Θ	θ	Theta	Th	thay-tah
Ι	ι	Iota	I	eye-o-tah
Κ	κ	Kappa	K	cap-ah
Λ	λ	Lambda	L	lamb-dah
Μ	μ	Mu	M	mew
Ν	ν	Nu	N	new
Ξ	ξ	Xi	X	zzEye
Ο	ο	Omicron	O	om-ah-cron
Π	π	Pi	P	pie
Ρ	ρ	Rho	R	row
Σ	σ	Sigma	S	sig-ma
Τ	τ	Tau	T	tawh
Υ	υ	Upsilon	U	oop-si-lon
Φ	ϕ	Phi	Ph	figh
Χ	χ	Chi	Ch	kigh
ψ	ψ	Psi	Ps	sigh
Ω	ω	Omega	O	o-may-gah

Figure 2.9 Greek alphabet with upper case, lower case, Latin alphabet equivalent and pronunciation.

There are many occasions in fluid mechanics where solutions to governing equations are achieved in limits of small and large parameters. In the current context, it is instructive to obtain an approximation to the exact solution, $F_1(Fr)$ in Eq. (2.4.9), for $|Fr| << 1$. Use a Taylor series

$$F_1 = Fr + (2)^{1/2}(1+\varepsilon)^{1/2}$$
$$= Fr + (2)^{1/2}\left(1 + \frac{1}{2}\varepsilon - \frac{1}{8}\varepsilon^2 + \frac{1}{16}\varepsilon^3 + \cdots\right) \tag{2.4.10}$$

where ε (Greek epsilon, lower case see Figure 2.9) is defined as $\varepsilon = Fr^2/2 << 1$. In Eq. (2.4.10) the Taylor series $\sim \varepsilon^n, n \geq 4$ are not shown. Rather than using the ellipsis punctuation "$\cdots$" for all of the absent terms, it is standard to use the notation $O(\varepsilon^4)$, which reads "order epsilon to the fourth," to signify the size of the largest of those neglected terms. It is a measure of the expected error magnitude in truncating the series. More about expansions of solutions and equations for small parameters is discussed in Chapter 17.

For small enough $|Fr| << 1$, try keeping just the first two terms,

$$F_1 = Fr + (2)^{1/2}\left(1 + \frac{1}{2}\varepsilon + O(\varepsilon^2)\right)$$
$$= (2)^{1/2} + Fr + \frac{(2)^{1/2}}{4}Fr^2 + O(Fr^4) \tag{2.4.11}$$

where now we use $O(Fr^4)$ to indicate the size of the error in terms of Fr. Figure 2.8 shows this approximate solution from Eq. (2.4.11) in comparison to the exact solution, revealing its accuracy for $|Fr| << 1$.

From this discussion, in the event we did not know the exact solution to the thrown ball problem, one can see that a polynomial approach to fitting experimental data of $F_1(Fr)$ would do well in this range of $|Fr| << 1$. Indeed a proposed quadratic relationship

$$F_1(Fr) = a_2 + b_2 Fr + c_2 Fr^2 \tag{2.4.12}$$

would hopefully lead to good data fit with $a_2 = 1.41, b_2 = 1, c_2 = 0.35$, which corresponds to Eq. (2.4.11). See Figure 2.14 for an example of a quadratic regression for the dimensionless data.

A different polynomial also does well for large $Fr >> 2$. Let a proposed form to fit data be

$$F_1 = k_3 Fr^{a_3} \tag{2.4.13}$$

For $Fr >> 2$, Eq. (2.4.9) takes the form

$$F_1 \sim Fr + \left(Fr^2\right)^{1/2} = 2Fr \quad Fr >> 2 \tag{2.4.14}$$

Data would yield fitting values for $k_3 = 2$ and $a_3 = 1$, which approximates the exact solution as seen in Figure 2.8.

2.5 Scaling a Mechanical System: The Buckingham Pi Theorem

If we did not know the governing equations, or had trouble solving them for some reason, we still could have determined that a dimensionless F_1 would only depend on one dimensionless parameter, in this case Fr, which we could define. The Buckingham Pi Theorem is useful for this inquiry and is due to Edgar Buckingham in 1914/1915 (Buckingham, 1914; 1915) who formalized earlier work by Lord Rayleigh from 1892 (Rayleigh, 1892a). First we count the number of independent dimensional parameters in the problem that contribute to the result (m doesn't contribute, for example). So in this case of the thrown ball, the dimensional impact time, call it $t_i = f_1$, will depend on g, h and v_O,

$$t_i = f_1(g, h, v_O) \tag{2.5.1}$$

which gives us four dimensional parameters. Three of them are input parameters (h, g, v_O) which are independent of one another and the fourth is an output value, f_1, which depends on the other three. Then we count the number of reference dimensions. Usually these are chosen from mass (M), length (L), time (T) and temperature (Θ). Note, this use of "T" means only units of time, and is not the T symbol used in Eq. (2.4.4) as an independent dimensionless variable. For our system we check for the reference dimensions by listing the parameters and stating their dimensions

$$[h] = L, [g] = \frac{L}{T^2}, [v_O] = \frac{L}{T}, [f_1] = T \tag{2.5.2}$$

The total number of reference dimensions is two, i.e. length and time (not mass or temperature). The number of dimensionless parameters that we can derive is found by subtracting the number of parameters minus the number of measures, $4 - 2 = 2$. So our goal is to discover the dimensionless form of the impact time, call it π_2, on the dimensionless form of the remaining parameters, call it π_1, so that Eq. (2.5.1) looks like

$$\pi_2 = \phi(\pi_1) \tag{2.5.3}$$

To discover these dimensionless groups, also called Pi groups, we first choose from h, g, v_0, f_1 two so-called "repeating parameters," so that L and T are represented by them: h does well for L and g brings in T, so that pair should do nicely. Importantly, they do not, themselves, form a dimensionless group.

For the first dimensionless group (Pi group), we look for the one involving v_0, which we multiply by the repeating parameters raised to unknown exponents to be solved

$$\pi_1 = h^a g^b v_0 \tag{2.5.4}$$

Substituting the dimensions of the parameters gives us

$$[\pi_1] = L^a \left(\frac{L}{T^2}\right)^b \frac{L}{T} \rightarrow L^{a+b+1} T^{-2b-1} = L^0 T^0 \tag{2.5.5}$$

Since the result must be dimensionless, the exponents of L and T must be zero, as indicated. Equation (2.5.5) tells us that we need to solve for a and b from the algebraic equations

$$a + b + 1 = 0, \; -2b - 1 = 0 \rightarrow a = -1/2, \; b = -1/2 \tag{2.5.6}$$

Using the solutions for a and b yields the Froude number

$$\pi_1 = h^{-1/2} g^{-1/2} v_0 = Fr \tag{2.5.7}$$

For the remaining dimensionless group, use the same repeating parameters but now we seek the dimensionless version of f_1

$$\pi_2 = h^c g^d f_1 \quad \rightarrow \quad L^{c+d} T^{1-2d} = L^0 T^0 \tag{2.5.8}$$

The solutions for c and d, then, are derived from the algebraic equations

$$c + d = 0, \; -2d + 1 = 0 \quad \rightarrow \quad c = -1/2, \; d = 1/2 \tag{2.5.9}$$

which yields a dimensionless impact time

$$\pi_2 = h^{-1/2} g^{1/2} f_1 = F_1 \tag{2.5.10}$$

which we found in Eq. (2.4.4).

Worked Example 2.1

Let's try the Buckingham Pi Theorem for the fluid drag on the wing of a bird. The bird wing shown in Figure 2.10, and in idealized form in Figure 2.11, has two important length scales that we will consider: the distance from the bird's body to wing tip called the wing length, L_W, or wing semi-span; and the distance from the leading front edge to trailing rear edge, called the chord. Since the chord varies along the length, an average value, $\bar{c}$, is used, defined as the ratio of the wing area, A, to the wing length, $\bar{c} = A/L_W$. The angle of attack, α (Greek alpha, lower case, see Figure 2.9), is the angle between the wind direction and the wing. The drag

Figure 2.10 The drag on a bald eagle's wing can be analyzed in dimensionless terms.

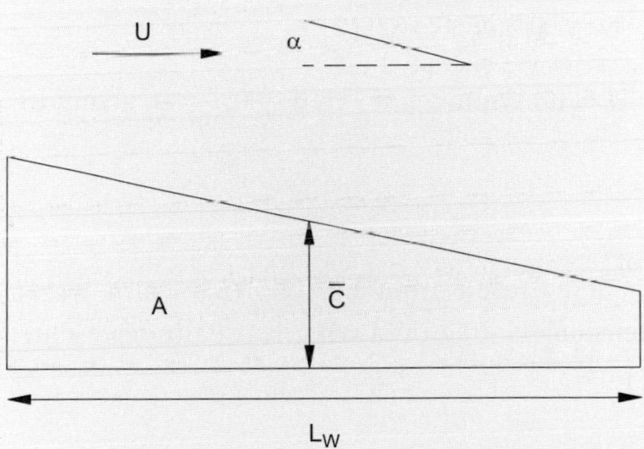

Figure 2.11 Angle of attack, α, wing area, A, wing length, L_W, and average chord length, $\bar{c}$, for an idealized bird wing.

force on the wing is an output quantity, we will call F_D. The fluid has properties of density, ρ, viscosity, μ (Greek mu, lower case), and relative velocity to the bird, U. The dimensions of these six parameters are

$$[F_D] = MLT^{-2}, [L_W] = L, \ [\bar{c}] = L \ [\rho] = ML^{-3}, [\mu] = ML^{-1}T^{-1}, [U] = LT^{-1} \tag{2.5.11}$$

while we note that α is already dimensionless. Choosing $\bar{c}$ for our length scale, ρ for mass, and μ for time, the first Pi group is

$$\pi_1 = \bar{c}^a \rho^b \mu^c U = L^{a-3b-c+1} M^{b+c} T^{-c-1} = M^0 L^0 T^0 \tag{2.5.12}$$

From Eq. (2.5.12) we readily find $a = 1, b = 1, c = -1$, so that π_1 becomes

$$\pi_1 = \frac{\rho U \bar{c}}{\mu} = \text{Re} \tag{2.5.13}$$

which is Re, the Reynolds number, the ratio of inertial to viscous forces in the fluid. Likewise for the second Pi group, the dimensionless drag, we have

$$\pi_2 = \bar{c}^d \rho^e \mu^f F_D = L^{d-3e-f+1} M^{e+f+1} T^{-f-2} = M^0 L^0 T^0 \tag{2.5.14}$$

The equations for the exponents derived from Eqs. (2.5.14) yield $d = 0, e = 1, f = -2$ so that

$$\pi_2 = \rho \mu^{-2} F_D \tag{2.5.15}$$

which is certainly dimensionless but a bit unorthodox. Usually we want the dimensionless form to be a ratio of meaningful quantities with a recognizable physical interpretation. We will address this issue shortly. The obvious third dimensionless parameter is the aspect ratio, $AR = L_W/\bar{c}$, which we can derive as π_3

$$\pi_3 = \bar{c}^g \rho^h \mu^i L_W = L^{g-3h-i+1} M^{h+i} T^{-i} = M^0 L^0 T^0 \tag{2.5.16}$$

The exponents of Eqs. (2.5.16) require $g = -1, h = 0, i = 0$, giving us

$$\pi_3 = AR = \frac{L_W}{\bar{c}} \tag{2.5.17}$$

as expected.

Now we return to our π_2 result from Eq. (2.5.15), which was a bit disappointing. In aerodynamics the dimensionless drag on a wing is usually represented as the drag coefficient, C_D,

$$C_D = \frac{F_D}{\frac{1}{2}\rho U^2 A} = 2\frac{\pi_2}{\pi_1^2 \pi_3} \tag{2.5.18}$$

which is simply a combination of our three Pi groups as shown. Combinations of the dimensionless parameters are, of course, also dimensionless as are numbers like 2. Now the drag force is in a ratio with a dynamic fluid pressure term that arises in the Bernoulli equation, $\frac{1}{2}\rho U^2$, multiplying the wing area, A. The denominator has a readily interpreted and measured physical meaning. Pulling together our three computed dimensionless parameters plus the angle of attack, the output drag coefficient C_D depends on the Reynolds number and the aspect ratio in a dimensionless relationship:

$$C_D = \phi(\alpha, \text{Re}, AR) \tag{2.5.19}$$

where ϕ(Greek phi, lower case, see Figure 2.9) here means function. The student will certainly be somewhat uncomfortable that there is no single, unique solution to the dimensionless parameters, but this is part of the art form of fluid dynamics. It was important to stumble a bit on π_2 and pick ourselves up. Although π_2 is perfectly fine as a dimensionless parameter, it does not exhibit a ratio of readily interpreted quantities. C_D on the other hand does, and that is why it is widely used.

Examples of drag (and lift) measurements on bird wings is found in Withers (1981) for several species and gives us a form of Eq. (2.5.19). The data there shows that a hawk wing has an aspect ratio $AR = 3.0$ and flies in a Reynolds number range of $1 \times 10^4 < Re < 5 \times 10^4$. Under those conditions, a quadratic fit to the drag coefficient data for varying the angle of attack, α, is

$$C_D = a + b\alpha + c\alpha^2 \tag{2.5.20}$$

where the values for a,b,c and their standard error after multiple trials are given as $a = 0.0944 \pm 0.0028$, $b = 0.0018 \pm 0.0005$, $c = 0.0006 \pm 0.0001$.

Worked Example 2.2 | Thrown Ball Data and the Master Curve

Suppose we perform thrown ball experiments to measure the impact time, $t_i = f_1$, for 11 values of v_0 and four values of h on Earth where $g = 9.8\,m/s^2$. For these 44 sets of parameter values we perform five trials each totaling 220 experiments. The average f_1 values of the five trials for the 44 parameter pairs, v_0 and h, are shown in spreadsheet of Figure 2.12(a). The rows have 11 different values of v_0, $-10\,m/s \leq v_0 \leq 10 m/s$ and the columns have four different values of $h = 10, 20, 30, 40\,m$. Plotting the data in Figure 2.12 (b), f_1 vs v_0 is shown for each of the four values of h, resulting in four labeled plots or curves.

As we saw in Section 2.5, we can use the Buckingham Pi Theorem to find two dimensionless groups, $F_1 = f_1/\sqrt{h/g}$ and $Fr = v_0/\sqrt{gh}$ for the thrown ball problem. For each pair of v_0 and h in Figure 2.12(a), calculate the corresponding value of the Froude number, $Fr = v_0/\sqrt{gh}$. The result is shown in the table of Figure 2.13(a). Then for each value of the dimensional impact time, f_1, in Figure 2.12(b), calculate the corresponding dimensionless impact time $F_1 = f_1/\sqrt{h/g}$, which is shown in the table of Figure 2.13(b). Finally, plot F_1vs Fr for each of the four h values, e.g. plot the $h = 10$ m/s column in Figure 2.13(b) vs the $h = 10\,m/s$ column in Figure 2.13(a). Repeat for the other columns.

The resulting four dimensionless plots are shown in Figure 2.14 and we see that they now collapse to a single, master curve F_1 (Fr). Often the first attempt to curve-fitting data is using a

(a)

	f_l (s)			
h (m)	10	20	30	40
v_0 (m/s)				
−10	0.76	1.29	1.72	2.09
−8	0.79	1.29	1.70	2.05
−6	0.97	1.54	1.99	2.38
−4	1.06	1.62	2.06	2.43
−2	1.26	1.86	2.32	2.71
0	1.39	1.96	2.40	2.77
2	1.68	2.28	2.74	3.13
4	1.82	2.37	2.80	3.16
6	2.25	2.83	3.29	3.68
8	2.41	2.94	3.35	3.71
10	2.89	3.42	3.84	4.22

(b)

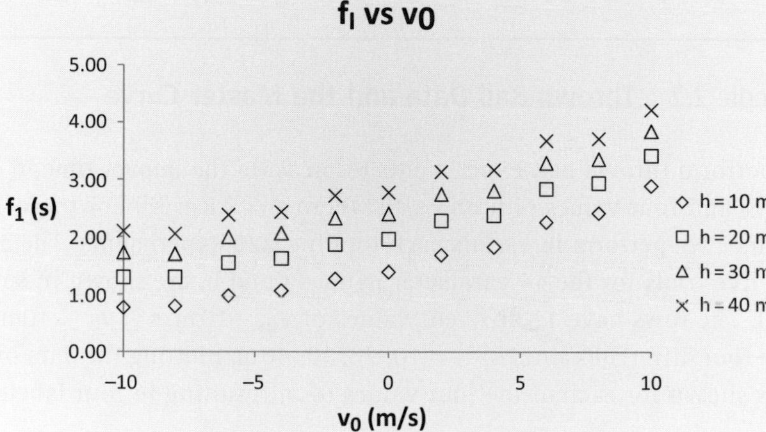

Figure 2.12 (a) Raw averaged data of measured impact time, f_1, for 11 values of v_0 and four values of h. (b) Graph of the data in (a).

straight line, i.e. linear regression. Performing this process on the data yields $F_1 = 1.49 + 1.02\,Fr$ with a coefficient of determination $r^2 = 0.95$. By comparison, fitting a quadratic polynomial yields $F_1 = 1.41 + 1.02\,Fr + 0.38\,Fr^2$ with a coefficient of determination $r^2 = 0.99$, which is much better and looks that way. These coefficients are in good agreement with Eq. (2.4.12). This is one of the advantages of using dimensionless parameters, since such an equation could cover a large number of combinations of v_0 and h in compact form.

(a)

	Fr			
h (m)	10	20	30	40
v_0 (m/s)				
−10	−1.01	−0.71	−0.58	−0.51
−8	−0.81	−0.57	−0.47	−0.40
−6	−0.61	−0.43	−0.35	−0.30
−4	−0.40	−0.29	−0.23	−0.20
−2	−0.20	−0.14	−0.12	−0.10
0	0.00	0.00	0.00	0.00
2	0.20	0.14	0.12	0.10
4	0.40	0.29	0.23	0.20
6	0.61	0.43	0.35	0.30
8	0.81	0.57	0.47	0.40
10	1.01	0.71	0.58	0.51

(b)

	F_1			
h (m)	10	20	30	40
v_0 (m/s)				
−10	0.76	0.90	0.98	1.04
−8	0.78	0.91	0.97	1.01
−6	0.96	1.08	1.14	1.18
−4	1.05	1.13	1.18	1.20
−2	1.25	1.30	1.33	1.34
0	1.37	1.37	1.37	1.37
2	1.66	1.60	1.57	1.55
4	1.80	1.66	1.60	1.57
6	2.23	1.98	1.88	1.82
8	2.39	2.05	1.92	1.84
10	2.86	2.39	2.20	2.09

Figure 2.13 (a) Calculation of $\mathrm{Fr} = v_0/\sqrt{gh}$ for each combination of v_0 and h in Figure 2.12 with $g = 9.8\,\mathrm{m/s^2}$. (b) Calculation of $F_1 = f_1/\sqrt{h/g}$ for each f_1 in Figure 2.12.

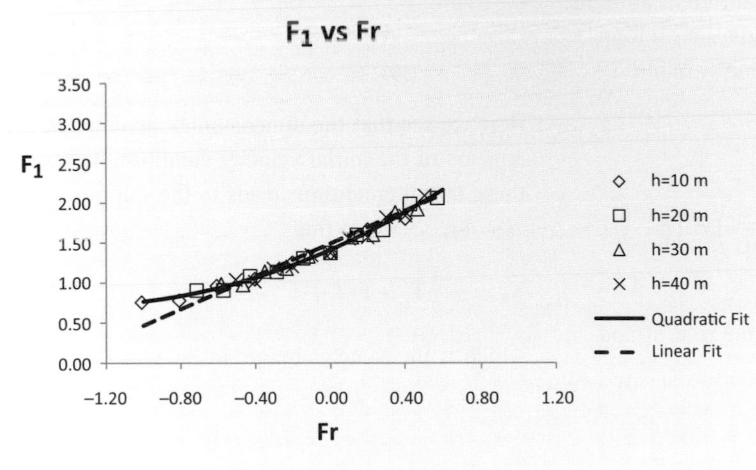

Figure 2.14 Graph of F_1 vs Fr showing the data collapsing onto a single (master) curve. Both linear and quadratic regression curves are presented.

2.6 Thrown Ball Revisited: Scaling the Governing Equations First

In Section 2.4 we solved the dimensional equations and then recast the solution using dimensionless variables and parameters. In fluid mechanics we normally make the governing equations dimensionless first, and then solve the system. This assists us in gauging the size of different terms in case we can neglect them to make the problem easier to solve while retaining the essential physics. Starting with the momentum balance we have the dimensional version

$$\frac{d^2y_p}{dt^2} = -g \tag{2.6.1}$$

and the dimensional initial conditions

$$y_p(t=0) = h, \quad \frac{dy_p}{dt}(t=0) = v_0 \tag{2.6.2}$$

Recall our dimensionless variables are $Y_p = y_p/h$, $T = (g/h)^{1/2}t$. As an example of how to proceed, we use the relationship $y_p = hY_p$ and insert that in Eqs. (2.6.1) and (2.6.2). For the derivative terms treat h as a constant which can be brought out in front of the derivative,

$$\frac{d}{dt}\left(y_p\right) = \frac{d}{dt}\left(hY_p\right) = h\frac{dY_p}{dt} \quad \text{and} \quad \frac{d^2}{dt^2}\left(y_p\right)$$
$$= \frac{d^2}{dt^2}\left(hY_p\right) = h\frac{d^2Y_p}{dt^2} \tag{2.6.3}$$

Now we need to use the chain rule of differentiation to express the dimensional time derivatives with respect to t in terms of the dimensionless time variable, T

$$\frac{d}{dt} = \frac{dT}{dt}\frac{d}{dT} = \left(\frac{g}{h}\right)^{1/2}\frac{d}{dT}$$
$$\frac{d^2}{dt^2} = \frac{d}{dt}\left(\frac{d}{dt}\right) = \left(\frac{g}{h}\right)^{1/2}\frac{d}{dT}\left(\left(\frac{g}{h}\right)^{1/2}\frac{d}{dT}\right) = \left(\frac{g}{h}\right)\frac{d^2}{dT^2} \tag{2.6.4}$$

Now apply the results of the chain rule to d/dt and d^2/dt^2 in Eq. (2.6.3)

$$\frac{dy_p}{dt} = h\frac{dY_p}{dt} = h\left(\frac{g}{h}\right)^{1/2}\frac{dY_p}{dT} = (gh)^{1/2}\frac{dY_p}{dT}$$
$$\frac{d^2y_p}{dt^2} = h\frac{d^2Y_p}{dt^2} = h\left(\frac{g}{h}\right)\frac{d^2Y_p}{dT^2} = g\frac{d^2Y_p}{dT^2} \tag{2.6.5}$$

The governing equation (2.6.1) can now be recast in dimensionless terms

$$g\frac{d^2Y_p}{dT^2} = -g \quad \rightarrow \quad \frac{d^2Y_p}{dT^2} = -1 \tag{2.6.6}$$

The solution to Eq. (2.6.6) is

$$Y_p = -\frac{1}{2}T^2 + c_0T + c_1 \tag{2.6.7}$$

where c_0 and c_1 are determined from the dimensionless initial conditions. Those are obtained from Eqs. (2.6.2) and found to be

$$y_p(t=0) = hY_p(T=0) = h \rightarrow Y_p(T=0) = 1$$
$$\left.\frac{dy_p}{dt}\right|_{t=0} = h\left(\frac{g}{h}\right)^{1/2}\left.\frac{dY_p}{dT}\right|_{T=0} = v_0 \rightarrow \left.\frac{dY_p}{dT}\right|_{T=0} \tag{2.6.8}$$
$$= \frac{v_0}{(gh)^{1/2}} = Fr$$

Here we see that the dimensionless parameter, Fr, appears in the initial velocity condition. Imposing these initial conditions leads to the solution of $c_0 = Fr$, $c_1 = 1$ so that

$$Y_p = -\frac{1}{2}T^2 + FrT + 1 \tag{2.6.9}$$

which is the same as before in Eq. (2.4.7).

Box 2.2 | Using Computing Software to Scale and Solve Equations

Students often find scaling equations a frustrating process. Fortunately, computing software can provide some assistance. This is a transformation of variables involving symbolic calculations. Code 2.1(a) is a Maple file which scales the differential equation and its initial conditions of the thrown ball problem, solves the resulting system, and plots the solutions.

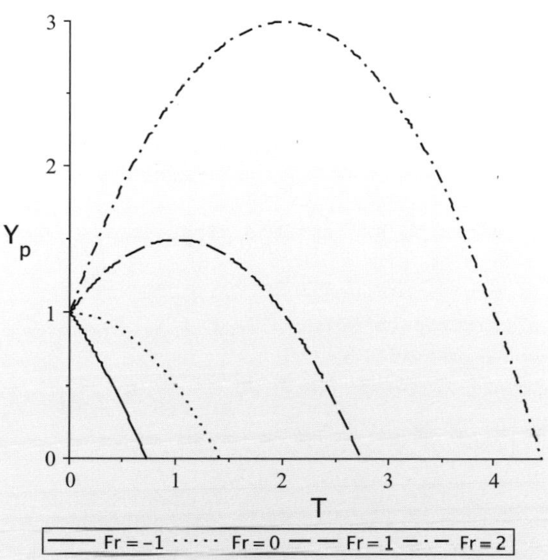

Figure 2.15 $Y_p(T)$ for four values of $Fr = -1, 0, 1, 2$.

Figure 2.15 shows the results of the plotting "display" command, $Y_p(T)$ for four values of the Froude number, Fr = −1, 0, 1, 2. The software was able to find an analytic solution to the ODE with initial conditions. The corresponding MATLAB® file is in Code 2.1(b).*

```
(a)
restart;
#First define the ode to solve
ode := diff(yp(t), t, t) = -g;
```

$$ode := \frac{d^2}{dt^2} yp(t) = -g \qquad (1)$$

```
#Next load PDEtools which has the "dchange" command that transforms
variables with (PDEtools):
#Define the variable transformation as tr
```

$$tr := \left\{ t = sqrt\left(\frac{h}{g}\right) \cdot T, yp(t) = Yp(T) \cdot h \right\};$$

$$tr := \left\{ t = \sqrt{\frac{h}{g}}\, T, yp(t) = Yp(T)\, h \right\} \qquad (2)$$

```
#Then use the "dchange" command to change variables of ode, call the new
equation ODE  ODE := dchange(tr, ode, [T, Yp]);
```

$$ODE := \left(\frac{d^2}{dT^2} Yp(T)\right) g = -g \qquad (3)$$

```
#There are many ways to use Maple to simplify expressions like ODE to
cancel g, including manually.
```

Code 2.1 Code examples: (a) Maple, (b) MATLAB®.

* Note: Dr. Francesco Romanò has graciously supplied all of the MATLAB® codes.

Using the "simplify" command and dividing through by g leads to the form we seek

```
ODE := simplify(ODE/g);
```

$$ODE := \frac{d^2}{dT^2} Yp(T) = -1 \tag{4}$$

```
#Next, use the "eval" command to express the first initial condition
ic1 := eval(yp(t), t = 0) = h;
```

$$ic1 := yp(0) = h \tag{5}$$

```
#and perform the transformation on ic1, call it IC1, followed by simplification
IC1 := eval(dchange(tr, ic1,[T, Yp]), T=0);
```

$$IC1 := Yp(0)\, h = h \tag{6}$$

```
IC1 := simplify(IC1/h);
```

$$IC1 := Yp(0) = 1 \tag{7}$$

```
#The second initial condition using "eval" is given by
ic2 := eval(diff(yp(t), t), t=0)= v0;
```

$$ic2 := \left(\frac{d}{dt} yp(t)\right)\bigg|_{\{t=0\}} = v0 \tag{8}$$

```
#Applying the variable transformation to ic2, as IC2, we obtain
IC2 := eval(dchange(tr, ic2, [T, Yp]), T=0);
```

$$IC2 := \frac{\left(\left(\frac{d}{dT} Yp(T)\right)\bigg|_{\{T=0\}} h\right)}{\sqrt{\frac{h}{g}}} = v0 \tag{9}$$

```
#Because g and h appear in square roots, to simplify IC2 we need to state they are
positive valued using the "assuming" command with "simplify"

IC2 := simplify(IC2/h . sqrt(h/g)) assuming h >0, g>0;
```

$$IC2 := \left(\frac{d}{dT} Yp(T)\right)\bigg|_{\{T=0\}} = \frac{v0}{\sqrt{h}\,\sqrt{g}} \tag{10}$$

```
#Define the Froude number, Fr, through v0 and check IC2 to find the form we want
v0 == Fr·sqrt(h)·sqrt(g);
```

$$v0 := Fr\,\sqrt{h}\,\sqrt{g} \tag{11}$$

```
IC2;
```

$$\left(\frac{d}{dT} Yp(T)\right)\bigg|_{\{T=0\}} = Fr \tag{12}$$

Code 2.1 (*continued*)

```
#The system is ready to be solved first by defining the initial conditions in proper
format, here as ICS, and then using "dsolve" where we "assign" the solution
ICS := Yp(0) = 1, D(Yp) (0) = Fr;
                          ICS := Yp(0) = 1, D(Yp) (0) = Fr                    (13)
sol := dsolve( { ODE, ICS}) :
assign( sol);
#Check the solution
Yp(T);
```

$$-\frac{1}{2}T^2 + FrT + 1 \qquad (14)$$

```
#The impact time is given by setting Yp(T)=0

 solve(Yp(T) = 0, T);
```

$$Fr+\sqrt{Fr^2+2},\ Fr-\sqrt{Fr^2+2} \qquad (15)$$

```
#Choose the positive root.
Ti := Fr+(Fr^2 + 2)^(1/2);
```

$$Ti := Fr+\sqrt{Fr^2+2} \qquad (16)$$

```
#Call "with(plots)" to use "display" for multiple curve plotting
with (plots) :
#Define 4 plots of Yp(T) with 4 different values of Fr where 0≤T≤Ti using the "subs"
command for Yp
    (T) and Ti
PL1 := plot(subs(Fr=-1,Yp(T)), T=0 .. subs(Fr=-1,Ti), linestyle=1, legend="Fr=-1"):
PL2 := plot(subs(Fr=0, Yp(T)), T=0 .. subs(Fr=0,Ti), linestyle=2, legend="Fr=0"):
PL3 := plot(subs(Fr=1, Yp(T)), T=0 .. subs(Fr=1,Ti), linestyle=3, legend="Fr=1"):
PL4 := plot(subs(Fr=2, Yp(T)), T=0 .. subs(Fr=2,Ti), linestyle=4, legend="Fr=2"):
display(PL1, PL2, PL3, PL4, labels=["T",Y[p]], labelfont=[HELVETICA, 12]);

(b)
clear all, close all, clc
syms h g v0 Fr t y(t)T Y(T)
dy_dt(t)     = diff(y,t)
d2y_dt2(t)   = diff(dy_dt,t)
DY_DT(T)     = diff(Y,T)
D2Y_DT2(T)   = diff(DY_DT,T)
ode(t)       = d2y_dt2 == -g
ODE(T)       = subs(ode(t),[d2y_dt2,t],[D2Y_DT2*h/(h/g),T])
ODE(T)       = ODE(T)/g
ic1          = y(0)==h
IC1          = subs(ic1,y,Y*h)
IC1          = IC1/h
ic2          = dy_dt(0)==v0;
IC2          = subs(ic2,[dy_dt,t],[DY_DT*h/((h/g)^(1/2)),T])
IC2          = simplify(IC2)
IC2          = subs(IC2,v0,Fr*h^(1/2)*g^(1/2))
```

Code 2.1 *(continued)*

```
assume(h>0)
assume(g>0)
IC2     = simplify(IC2)
Y(T)    = dsolve(ODE,[IC1 IC2])
T       = linspace(0,5,1000);
Fr      = -1;
plot(T,eval(Y),'-k'), hold on;
Fr      = 0;
plot(T,eval(Y),':k')
Fr      = 1;
plot(T,eval(Y),'--k')
Fr      = 2;
plot(T,eval(Y),'-.k')
axis([0 5 0 3.5])
legend('Fr=-1','Fr=0','Fr=1','Fr=2','location','northeast')
xlabel('T')
ylabel('Y_p')
```

Code 2.1 *(continued)*

Worked Example 2.3

Figure 2.16 shows another familiar physics problem of a projectile shot at angle θ with initial speed v_0. There are two momentum balances governing the position of the projectile (x_p, y_p), one in the x-direction and the other in the y-direction

$$m\frac{d^2x_p}{dt^2} = 0,\ m\frac{d^2y_p}{dt^2} = -mg \tag{2.6.10}$$

subject to initial conditions

$$x_p(0) = 0, y_p(0) = 0, \frac{dx_p}{dt}_{t=0} = v_0\cos\theta, \frac{dy_p}{dt}_{t=0} = v_0\sin\theta \tag{2.6.11}$$

From the two-dimensional parameters, g and v_0 we can form a length scale, v_0^2/g, and a time scale, v_0/g. Use these to define the dimensionless variables

$$X_p = \frac{g}{v_0^2}x_p, Y_p = \frac{g}{v_0^2}y_p,\ T = \frac{g}{v_0}t \tag{2.6.12}$$

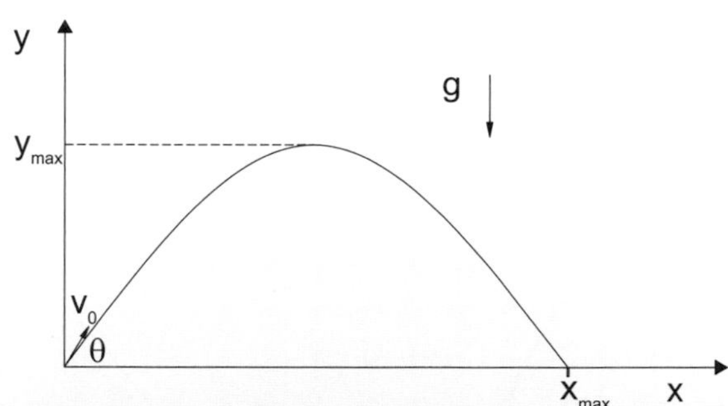

Figure 2.16 Projectile problem for initial speed v_0 at angle θ with gravity g.

By pencil or software, transforming the variables leads to the governing equations in dimensionless form

$$\begin{aligned} &\frac{d^2X_p}{dT^2} = 0, \quad \frac{d^2Y_p}{dT^2} = -1 \\ &X_p(0) = 0, Y_p(0) = 0, \frac{dX_p}{dT}_{T=0} = \cos\theta, \frac{dY_p}{dT}_{T=0} = \sin\theta \end{aligned} \tag{2.6.13}$$

Integrating the differential equations and imposing the initial conditions gives the solutions

$$X_p = T\cos\theta, \; Y_p = -\frac{1}{2}T^2 + T\sin\theta \tag{2.6.14}$$

which depend on T and the dimensionless parameter, θ. From the Buckingham Pi Theorem we knew that the two dimensional parameters, g and v_0 were only in two measures, length and time, so neither appears in a dimensionless group.

Clearly, $Y_p = 0$ at $T = 0$, which is its initial condition, but also at $T = T_i = 2\sin\theta$, which is the impact time. The value of $X_p(T_i) = 2\sin\theta\cos\theta = \sin(2\theta)$ is the maximum of X_p or the range, X_i. The maximum range occurs for the value of θ which satisfies $dX_i/d\theta = 2\cos(2\theta) = 0$, which is $\theta = \pi/4$. The height reached is where $dY_p/dT = 0$, which occurs when $T = \sin\theta$, half T_i. The corresponding value of $Y_p(T = \sin\theta) = (\sin^2\theta)/2 = Y_{max}$.

Figure 2.17(a) shows seven trajectories for seven values of $\theta = n\pi/16, n = 1, 2, 3, 4, 5, 6, 7$, while Figure 2.17(b) shows T_i, X_i, Y_{max} vs θ. As θ increases, the height, Y_{max}, and impact time, T_i, both increase monotonically. However, the range, X_i, passes through a maximum at $\theta = \pi/4$ yielding the same value for two values of θ. Examples are shown in Figure 2.17(a) where the pairs $\theta = \pi/16, 7\pi/16$, $\theta = \pi/8, 3\pi/8$ and $\theta = 3\pi/16, 5\pi/16$ have the same range.

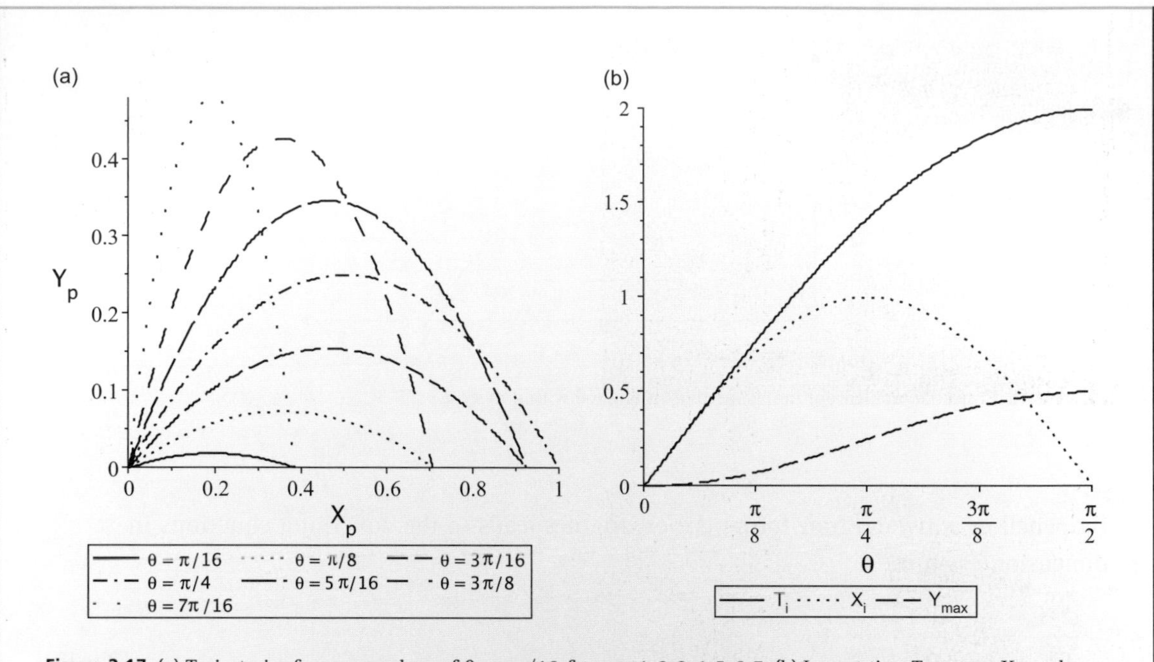

Figure 2.17 (a) Trajectories for seven values of $\theta = n\pi/16$, for $n = 1, 2, 3, 4, 5, 6, 7$. (b) Impact time, T_i, range, X_i, and height, Y_{max}, vs θ.

2.7 Viscous and Buoyancy Effects of a Dropped Particle: Erythrocyte Sedimentation Rate

Whole blood is made of serum, which is a watery liquid, and cellular constituents such as: red blood cells (RBCs), which are also called erythrocytes; white blood cells (WBCs) also called leukocytes; platelets, also called thrombocytes; and other components of lesser amounts. The RBCs carry oxygen on their hemoglobin molecules, WBCs respond to inflammation and fight infections, and platelets are the building block for making blood clots. Figure 2.18 (a) shows an example of an RBC on the left and a WBC on the right with a platelet in between, while Figure 2.18(b) shows several RBCs and some WBCs (types include lymphocytes, a monocyte, a neutrophil), and many small disc-shaped platelets. The mature RBC has no nucleus and is shaped like a bi-concave disc, as shown in Figure 2.18(c) with a diameter of 7.5 μm and a thickness at the edge of 2.0 μm. The bi-concave disc shape endows the RBC with more relative surface area than if it were spherical, which is important for gas exchange across its membrane. A WBC is irregularly shaped, but more spherical than an RBC, and has a diameter in the range 12–15 μm (depending on which type), and has a nucleus. A platelet has a diameter of 2–4 μm.

How numerous is each blood cell type? An average RBC count is 5.4 million per μL, or microliter (a microliter is 10^{-6} liters or one cubic millimeter), of whole blood in an adult male with a range of 4.6–6.2 million. For an adult female an average value is 4.8 million per microliter with a range of 4.2–5.4 million. There are far fewer WBCs, by a factor of ~1/1000, in whole blood with a typical normal range of 4,500 to 10,000 per microliter. Platelets number in the range 150,000 to 400,000 per microliter which is about 1/20th of the RBC count. The volume of a single RBC is 80–100 femtoliters (a femtoliter is 10^{-15} liters), while a

(a)

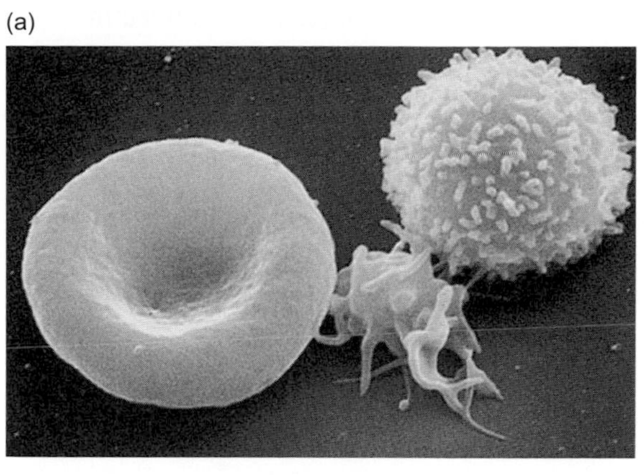

(b)

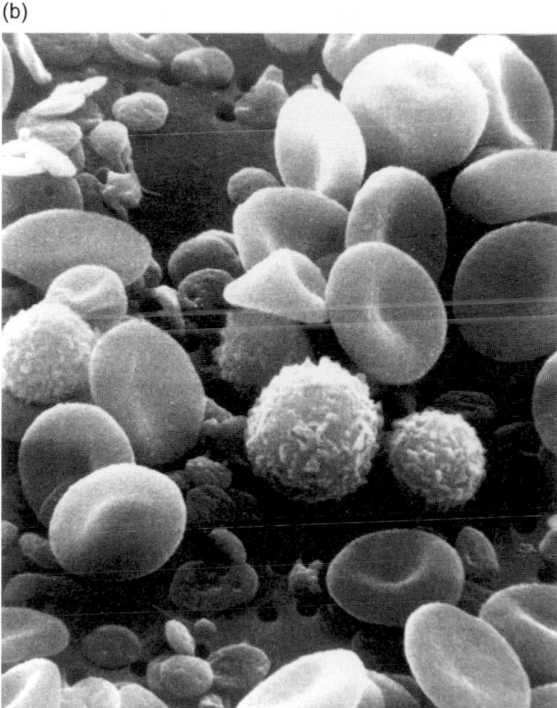

(c)

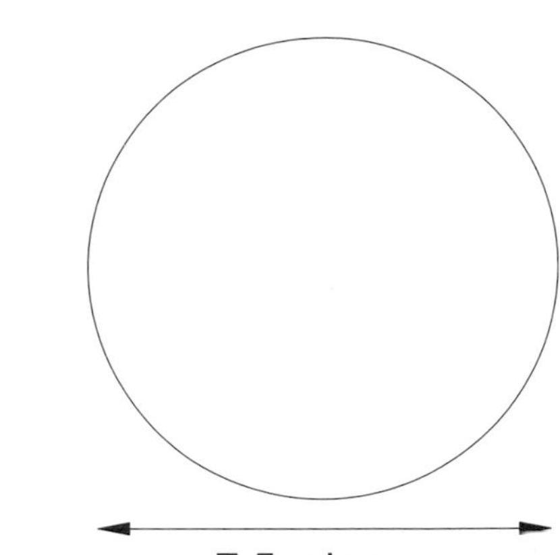

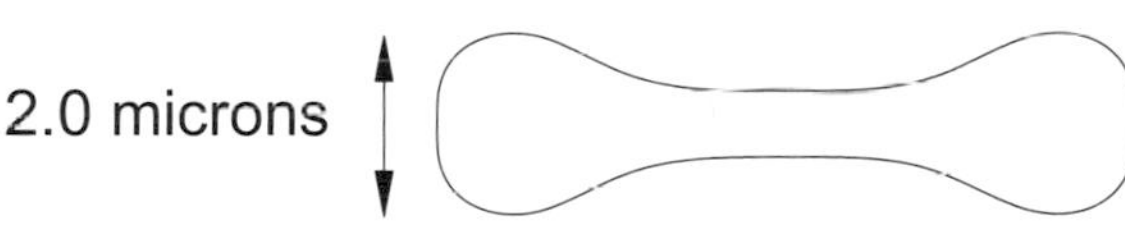

Figure 2.18 (a) A scanning electron micrograph (SEM) of an individual RBC (left) and WBC (right) and a platelet in between. (b) An SEM of RBCs, WBCs (lymphocytes, a monocyte, a neutrophil) and many small disc-shaped platelets. (c) Dimensions of an RBC with a top view and side view of the biconcave disc.

Figure 2.18 (*continued*)

WBC volume depends on the type, but are larger than the RBC volume. A sphere with diameter 10 μm has a volume of ~500 femtoliters, for example. The much smaller platelet has a volume of 9.7–12.8 femtoliters.

Given the number of cells and their individual sizes, a volume of whole blood with serum and cells is approximately 40–45% due to RBC volume. That is the definition of the hematocrit (HCT)

$$HCT = \frac{\text{RBC volume}}{\text{whole blood volume}} \tag{2.7.1}$$

The WBC and platelet volumes are ~1–2% of the total. Clinically the HCT is measured by putting a whole blood sample into a centrifuge to force the heavier cells to the bottom of the test tube.

Consider the biofluid mechanics problem of gravitational settling of RBCs in their own serum. The speed of the RBC falling is used in a clinical lab test called the erythrocyte sedimentation rate (ESR). The value of the ESR can vary with several forms of inflammatory disease (e.g. rheumatoid arthritis, systemic lupus erythematosus) that, in addition to their systemic manifestations, also make the RBC membrane surfaces sticky, leading to clumping of RBCs into bigger groups.

The RBC has mass, m_p, and we may treat it as an equivalent sphere of diameter, d_p, so that

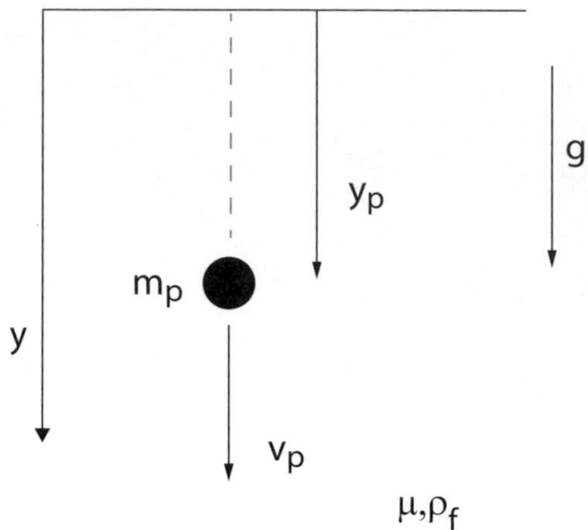

Figure 2.19 Falling particle with buoyancy force.

$$m_p = \frac{4}{3}\pi\left(\frac{d_p}{2}\right)^3 \rho_p \tag{2.7.2}$$

where ρ_p is the particle density. The weight of the particle is simply

$$W = m_p g = \frac{4}{3}\pi\left(\frac{d_p}{2}\right)^3 \rho_p g \tag{2.7.3}$$

The fluid has viscosity, μ, and density, ρ_f. Gravity, g, is in the positive y-direction, see Figure 2.19, where y is positive in the downward direction

The particle will fall in the positive y-direction due to its weight. The fluid exerts both a drag force on the falling particle, which is linearly related to the particle velocity, and a buoyancy force. Define the Reynolds number as

$$Re = \frac{\rho_f U d_p}{\mu} \sim \frac{\text{fluid inertia}}{\text{viscous forces}} \tag{2.7.4}$$

where U is a characteristic velocity and μ is the serum viscosity. It is appropriate for $Re << 1$ to use Stokes drag given by

$$F_d = -3\pi\mu d_p(v_p - v_f) \tag{2.7.5}$$

where v_p is the particle velocity, which is positive downward, and v_f is the fluid velocity ($v_f = 0$ for this example). Equation (2.7.5) is derived in Section 7.8. The negative sign tells us that the drag is always in the opposite direction of the relative particle to fluid velocity. In this case the drag is upwards in the negative y direction. The buoyancy force, F_b, from Archimedes' Principle, is that the upward force of the fluid on the particle is equivalent to the weight of the fluid displaced by the particle's volume,

$$F_b = -\frac{4}{3}\pi\left(\frac{d_p}{2}\right)^3 \rho_f g \tag{2.7.6}$$

Newton's second law for the particle is

$$m_p \frac{dv_p}{dt} = W + F_b + F_d \tag{2.7.7}$$

or

$$\frac{4}{3}\pi\left(\frac{d_p}{2}\right)^3 \rho_p \frac{dv_p}{dt} = \frac{4}{3}\pi\left(\frac{d_p}{2}\right)^3 \left(\rho_p - \rho_f\right) g - 3\pi\mu d_p v_p \tag{2.7.8}$$

Equation (2.7.8) has a steady state solution, $v_p = v_t$, where v_t is the "terminal velocity" or "Stokes settling velocity." We can solve for v_t by setting the time derivative to zero, $dv_p/dt = 0$,

$$0 = \frac{1}{6}\pi d_p^3\left(\rho_p - \rho_f\right) g - 3\pi\mu d_p v_t \tag{2.7.9}$$

or rearranging, we find

$$v_t = \frac{\left(\rho_p - \rho_f\right) d_p^2 g}{18\mu} \tag{2.7.10}$$

It makes sense that v_t is proportional to the density difference, $(\rho_p - \rho_f)$. Cells in the blood generally have a higher density than the surrounding serum, because of their various internal constituents (RBCs have hemoglobin, for example). So this density difference is positive and $v_t > 0$. If the two densities are the same, then $v_t = 0$ and the particle is neutrally buoyant. Neutral buoyancy is important for man-made or natural particles used to image flows. Likewise, if the fluid density is greater than the particle density, then the particle rises since $v_t < 0$. It is also sensible that v_t is inversely proportional to the serum fluid viscosity, μ. The more viscous the fluid, the greater the drag, and the slower the motion.

What about particle size? There are competing effects. The particle mass, which increases the magnitude of v_t, is proportional to d_p^3, while the drag force, which decreases the magnitude of v_t, is proportional to d_p. The overall effect of increasing particle size is to increase the magnitude v_t by d_p^2 dependence. This is one of the important results of ESR. RBCs tend to clump together in certain diseases because the surface of the RBC membrane becomes sticky. So the clump has a larger apparent diameter and hence falls faster. A linear clump of RBCs looks like a roll or cylinder, and the French equivalent word "rouleau" (plural, "rouleaux") is used (see Figure 2.20). At this stage, we are not taking into account the details of this cylindrical shape, but as a first approximation treat the entire clump as an equivalent sphere which is larger than a single RBC.

What about the unsteady equation, Eq. (2.7.8)? Let's simplify it by dividing through by the Stokes drag coefficient, $3\pi\mu d_p$, to find

$$\begin{aligned} T_S \frac{dv_p}{dt} &= v_t - v_p \\ T_S &= \frac{\rho_p d_p^2}{18\mu} \end{aligned} \tag{2.7.11}$$

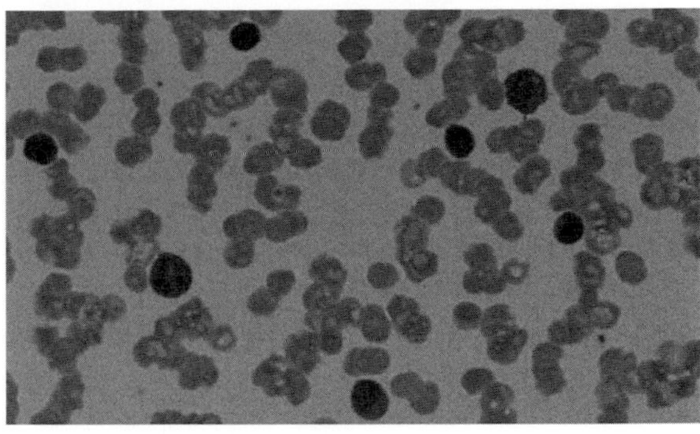

Figure 2.20 Rouleaux formation of RBCs in a peripheral blood smear. Note there are several linear clumps of RBCs stuck together as well as individual RBCs.

where T_S has the units of time. So Eq. (2.7.11) has two independent dimensional parameters, v_t, and T_S, even though they are made of combinations from the five parameters. We can make Eq. (2.7.11) dimensionless by selecting the following scales

$$V_p = \frac{v_p}{v_t}, \quad T = \frac{t}{T_S} \tag{2.7.12}$$

Inserting these definitions into Eq. (2.7.11) yields the dimensionless force balance

$$\frac{dV_p}{dT} = 1 - V_p \tag{2.7.13}$$

Worked Example 2.4

Let's solve Eq. (2.7.13) and analyze the results using physiological data. Integrating Eq. (2.7.13) leads to

$$\ln(1 - V_p) = -T + c \tag{2.7.14}$$

where c is a constant of integration. Exponentiate both sides of the equation

$$e^{\ln(1-V_p)} = e^{(-T+c)} \quad \rightarrow \quad (1 - V_p) = e^{(-T+c)} = c_1 e^{-T} \quad c_1 = e^c \tag{2.7.15}$$

and rearrange, now with c_1 as the constant

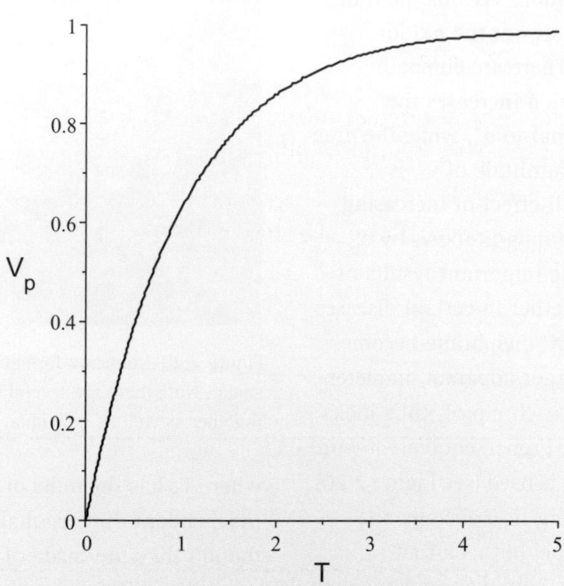

Figure 2.21 RBC sedimentation velocity vs time in dimensionless variables.

$$V_p = 1 - c_1 e^{-T} \tag{2.7.16}$$

Using the initial condition that the particle velocity starts from zero, that solves for c_1

$$V_p(T = 0) = 0 \quad \rightarrow \quad c_1 = 1 \tag{2.7.17}$$

and leads to the final form

$$V_p = 1 - e^{-T} \tag{2.7.18}$$

The two dimensional groups, v_t and T_S, have two units of measure, length and time. So there are no dimensionless parameters in Eq. (2.7.18). There is only one curve to plot, see Figure 2.21. We see that the terminal velocity is reached asymptotically as $t/T_S >> 1$. For example, when $t/T_S = 5$, then $e^{-5} \sim 0.01$, so $V_p = 0.99$ and v_p is 99% of the terminal velocity, see Figure 2.21.

Some physiologic values of the five parameters are

$$\begin{aligned} \rho_p &= 1.092\,\text{g/cm}^3 \\ \rho_f &= 1.022\,\text{g/cm}^3 \\ d_p &= 8\,\mu\text{m} = 8 \times 10^{-4}\,\text{cm} \\ \mu &= 0.01\,235\,\text{gm/(cm·s)} \\ g &= 980\,\text{cm/s}^2 \end{aligned} \tag{2.7.19}$$

Using these estimates, the dimensional terminal velocity is

$$v_t = 1.97 \times 10^{-4}\text{cm/s} = 7.1\ \text{mm/h} \tag{2.7.20}$$

which falls in the normal clinical range. Our time scale, T_S, from Eq. (2.7.11) calculates to be $T_S = \rho_p d_p^2/(18\mu) = 3.14 \times 10^{-6}$ s. So the equilibration to steady state which occurs within $5T_S$ as we saw from the unsteady analysis is essentially instantaneous.

Let's check the size of the Reynolds number based on the terminal velocity, setting its velocity scale $U = v_t$

$$Re = 1.3 \times 10^{-5} << 1 \tag{2.7.21}$$

so using the Stokes drag is justified.

Clinical Correlation 2.1

The erythrocyte sedimentation rate is used as a non-specific screening test for various inflammatory processes such as rheumatoid arthritis, systemic lupus erythematosus, and multiple myeloma. In the hematology lab, the blood sample from a patient is placed upright in a tube next to a ruler in millimeters. The RBCs settle, leaving a cell-free plasma as the top layer and an RBC layer at the bottom. The line of demarcation between these two layers moves downward and its speed is calculated simply by the distance it covers in one hour, see Figure 2.22.

Figure 2.22 Sedimentation of blood samples in several pipettes. The lower dark portion of each column contains the RBCs and the light portion above is the remaining serum.

The ESR test was invented in 1897 by a Polish physician, doctor Edmund Faustyn Biernacki (1866–1911). Among his discoveries was that ESR increased in febrile diseases, like rheumatic fever, and that decreased plasma fibrinogen (i.e. lower plasma viscosity) resulted in higher ESR (Grzybowski and Sak, 2011). In 1917 another Polish scientist, Ludwik Hirszfeld (1884–1954), was working on blood groups (e.g. A, B, AB, O) and noted ESR differences for patients with malaria. In 1918 the Swedish pathologist Robert Sanno Fåhræus (1888–1968), apparently unaware of the earlier discoveries, found a similar phenomenon of ESR studying the difference of blood from pregnant and non-pregnant women, showing that the ESR is higher in pregnancy. In fact, Fahraeus envisioned the ESR as a possible pregnancy test. Another Swedish physician and scientist, Alf Vilhelm Albertsson Westergren (1891–1968), in 1921 made a similar observation of ESR differences studying the blood of patients with pulmonary tuberculosis.

Typical normal upper limits for ESR vary with age and gender. Below the age of 50: men 15 mm/hr, women 25 mm/hr. Above the age of 50: men 20 mm/hr, women 30 mm/hr (Böttiger and Svedberg, 1967). Our simplistic approach ignores important aspects of the real test. For example, cell aggregation develops during the test as the RBCs fall and that can increase the local viscosity while decreasing the local density difference, both events slowing the ESR. As well, the hematocrit itself influences the aggregation, and since women tend to have lower hematocrits they have faster ESRs than men.

2.8 Microfluidic Cell Sorting

A fluid sample containing different cells occurs frequently in clinical and research settings, and there often is a need to separate these cells from one another in an efficient manner. A microfluidic flow chamber can provide a platform to achieve this goal. Consider a two-dimensional channel whose half-width is b, see Figure 2.23. The channel conveys a flow in the x-direction which has a parabolic profile. A cell dropped into the top of the channel falls in the negative y-direction at terminal velocity, v_t, while being convected in the x-direction by the parabolic x-velocity profile. Eventually it reaches the lower wall at an impact distance, x_i, as indicated. In our scheme, the goal is to have x_i be significantly distinct between cell types, so that they land at different places along the wall, can attach there, and later be collected. The linear dependence of v_t on cell density is one way to sort, but since v_t is proportional to the square of particle diameter, size can be a more sensitive way to differentiate cells in this flow.

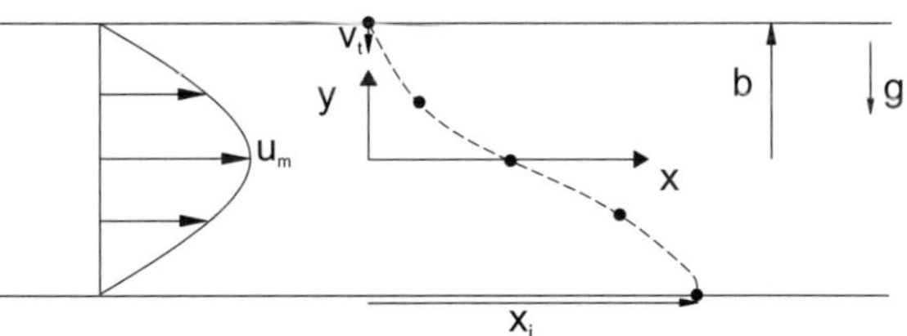

Figure 2.23 Channel flow and cell sedimentation contribute to a cell sorting scheme where the impact distance x_i can distinguish between cells of different size.

The channel conveys a flow in the x-direction, which has a parabolic profile with respect to y,

$$u(y) = u_m\left(1 - \left(\frac{y}{b}\right)^2\right) \tag{2.8.1}$$

We note that the velocity of Eq. (2.8.1) is the maximum, u_m, when $y = 0$, and that $u = 0$ for $y = \pm b$ at the walls to satisfy the no-slip condition there. The system is fed cells of different sizes that enter at the top $(x = 0, y = b)$. We can assume that the x-direction velocity of the cell is equal to the fluid velocity, u, at the particle's y-position, i.e. $u(y = y_p)$. The particle's x_p and y_p positions are then given by

$$\frac{dx_p}{dt} = u_m\left(1 - \left(\frac{y_p}{b}\right)^2\right), \quad \frac{dy_p}{dt} = -v_t \tag{2.8.2}$$

while the initial conditions are

$$x_p(0) = 0, y_p(0) = b \tag{2.8.3}$$

The dimensional input parameters for this system are b, v_t and u_m. The particle density, ρ_p, fluid density, ρ_f, fluid viscosity, μ, and gravity, g, are also dimensional parameters, but they only appear in the one combination, v_t. So they are not independent parameters. Then we expect the output parameter, impact distance x_i, will have a functional dependence on the three input parameters in a form $x_i = x_i(b, v_t, u_m)$. Because we have two units of measure, length and time, and four dimensional parameters, we will have two dimensionless groups. Without too much effort, clearly these will turn out to be $X_i = x_i/b$ and $\beta = u_m/v_t$, and we seek $X_i(\beta)$.

Natural time and position scales for the problem are

$$X_p = \frac{x_p}{b}, \quad Y_p = \frac{y_p}{b}, \quad T = \frac{v_t t}{b} \tag{2.8.4}$$

To scale Eqs. (2.8.2) and (2.8.3), we insert these definitions, applying the chain rule for the time derivatives that $d/dt = (dT/dt)d/dT = (v_t/b)d/dT$. After rearranging and simplifying, the dimensionless velocity equations become

$$\frac{dX_p}{dT} = \beta\left(1 - Y_p^2\right), \quad \frac{dY_p}{dT} = -1 \tag{2.8.5}$$

and corresponding dimensionless initial conditions are

$$X_p(0) = 0, Y_p(0) = 1 \tag{2.8.6}$$

Note that in Eqs. (2.8.5), dX_p/dT depends on Y_p, but dY_p/dT does not depend on X_p. So we can solve the dY_p/dT equation first, and then substitute the result into the dX_p/dT equation. Integrating dY_p/dT of Eq. (2.8.5) yields $Y_p = -T + c_1$. The constant of integration, c_1, is solved from the initial condition $Y_p(0) = 1$, in Eqs. (2.8.6), which gives us $c_1 = 1$. So the final form of Y_p is

$$Y_p = -T + 1 \tag{2.8.7}$$

The particle impact is when $Y_p = -1$, so the time to impact is $T = 2$, according to Eq. (2.8.7).

Inserting the result of Eq. (2.8.7) into the dX_p/dT equation of Eq. (2.8.5) gives us

$$\frac{dX_p}{dT} = \beta\left(1 - (1 - T)^2\right) = \beta\left(2T - T^2\right) \tag{2.8.8}$$

Integrating Eq. (2.8.8) with respect to T results in $X_p = \beta(T^2 - T^3/3) + c_2$ and the constant of integration, c_2, is determined from the initial condition in Eq. (2.8.6), $X_p(0) = 0$, which makes $c_2 = 0$. Then X_p is given by

$$X_p = \beta\left(T^2 - \frac{T^3}{3}\right) \tag{2.8.9}$$

Figure 2.24 shows plots of $X_p(T)$ vs $Y_p(T)$ parameterized by T in the range $0 \leq T \leq 2$ since $T = 2$ corresponds to impact. Three values of $\beta = 10, 20, 30$ are chosen for the trajectories. As β increases, the x-direction velocity, u_m, increases compared to the y-direction velocity, v_t. Consequently, the particle lands farther downstream. The corresponding impact range, X_I, is found at $X_p(T - 2)$, or

$$X_i = X_p(T = 2) = \frac{4}{3}\beta \tag{2.8.10}$$

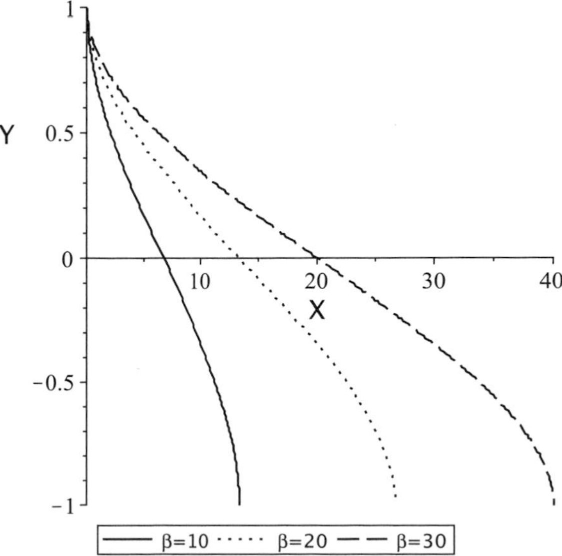

Figure 2.24 Three particle trajectories for $\beta = 10, 20, 30$.

To designing a useful cell separator that employs this technique, we would want a reasonable difference in X_i for, say, two different cell sizes. Let's say that cell type 1 is twice the diameter of cell type 2. That makes the corresponding ratio of v_t for these two cell types $v_{t1}/v_{t2} = 4$, because v_t depends on the square of the diameter. Then the ratio of the β values is $\beta_1 = \beta_2/4$ since u_m is common to both cell trajectories in the same flow chamber. From Eq. (2.8.10) this makes the impact distances related by

$$\frac{X_{i\,1}}{X_{i\,2}} = \frac{\frac{4}{3}\beta_1}{\frac{4}{3}\beta_2} = \frac{1}{4} \tag{2.8.11}$$

In terms of designing a useful flow chamber, note that the dimensionless impact range for the smaller cell, which will take a further distance to land, is $X_{i2} = (4/3)(\beta_2)$ from Eq. (2.8.10). Since $X_i = x_i/b$ from the scalings of Eq. (2.8.4), the dimensional impact range $x_{i2} = (4/3)(\beta_2 b) = (4/3)(bu_m/v_{t2})$. Then rearranging we find an expression for the flow maximum

$$u_m = \frac{3}{4}\frac{x_{i2}}{b}v_{t2} \tag{2.8.12}$$

In Eq. (2.8.12) the value of v_{t2} is characteristic of cell type 2, given a constant fluid viscosity of the serum. Suppose v_{t2} is measured separately to be 5×10^{-4} cm/s. We are free to choose the remaining parameters to make an efficient cell sorting device. Suppose our fabrication technology allows $b = 100\ \mu m = 0.01$ cm, and our separation process works well for $x_{i\,2} = 4\,cm$. Inserting those values into Eq. (2.8.12), we need to impose a flow with $u_m = 0.15$ cm/s. Cell type 1 will be located at $x_{i1} = x_{i2}/4 = 1$ cm according to Eq. (2.8.11). Because there is variability in the cell sizes of the two populations, having a distance of 3 cm between their landing location allows for some scatter in the impact positions.

Summary

We have visited some mechanics problems well known to the reader from earlier experiences. A ball dropped in a gravity field with no surrounding fluid was examined for its impact time as a function of the problems dimensional input parameters (g, h). It was shown that using scalings for the independent and dependent variables to make them dimensionless simplified the solution to one with no free parameters, thus reducing the number of experiments and computations. The dimensionless position, Y_p, and the dimensionless time, T, both varied between 0 and 1. An important aspect of well-chosen scales is to keep the range of their values within a finite limit like 0 to 1, for example. Then it is more reliable to gauge the size of other terms in the problem.

Throwing the ball added a new dimensional parameter, the ball's initial velocity, v_o, and that led to the appearance of our first dimensionless parameter, Fr, the Froude number, which we interpret as reflecting the ratio of initial kinetic energy to initial potential energy. The more general approach of identifying the possible dimensionless groups in a problem was explored via the Buckingham Pi Theorem. We then addressed the scaling of variables for the governing equations prior to solving them, since that is what is typically done in fluid mechanics. Finally, we added real fluid effects to particle dynamics. First we did this in the form of drag and buoyancy on a particle falling in a viscous fluid. The application to the erythrocyte sedimentation rate was explored as the fluid mechanics behind an important clinical measurement, and an extension to a cell sorting microfluidic device was presented. These approaches are standard for advanced fluid mechanics, and in particular are important in biofluid mechanics where additional features of flow phenomena interact with a living system and the problems become complex. Scaling the problem reduces that complexity. Students are usually not familiar with the dimensionless approach, so this chapter is meant to be transitional for using well-acquainted mechanics situations seen in a new way.

Problems

2.1 In rectangular (Cartesian) coordinates, consider the vectors

$$\underline{A} = xy^2\underline{e}_x + x\sin(y)\underline{e}_y + e^{5z}\tan(y)\underline{e}_z\underline{B}$$
$$= xe^{3z}\sin(y)\underline{e}_x + x^2\tan(y)\underline{e}_y + e^{3z}\underline{e}_z \quad (1)$$

and the scalar function

$$f(x, y, z) = x^2\sin(y)\,e^{5z} \quad (2)$$

The del operator in Cartesian coordinates is

$$\underline{\nabla} = \underline{e}_x\frac{\partial}{\partial x} + \underline{e}_y\frac{\partial}{\partial y} + \underline{e}_z\frac{\partial}{\partial z} \quad (3)$$

a. Calculate the gradient of f, i.e. $\underline{\nabla} f$, and evaluate it at $\underline{x} = \underline{x}_0 = (1, \frac{\pi}{3}, 0)$.
b. Calculate the dot product $\underline{A}\cdot\underline{\nabla} f$ and evaluate it at $\underline{x}_0$.
c. Calculate the magnitude of $\underline{B}$, i.e. $|\underline{B}| = (\underline{B}\cdot\underline{B})^{1/2}$ and evaluate it at $\underline{x}_0$.
d. Calculate the dot product $\underline{A}\cdot\underline{B}$ and evaluate it at $\underline{x}_0$.
e. Calculate the cross product $\underline{A}\times\underline{B}$ and evaluate it at $\underline{x}_0$.
f. Calculate the divergence of $\underline{A}$, i.e. $\underline{\nabla}\cdot\underline{A}$, and evaluate it at $\underline{x}_0$.
g. Calculate the curl of $\underline{A}$, i.e. $\underline{\nabla}\times\underline{A}$ and evaluate it at $\underline{x}_0$.

2.2 Suppose you perform experiments on a dropped ball starting at three different heights, $h = 100, 200, 300$ cm in earth's gravity $g = 980\,\text{cm/s}^2$. Your data acquisition system allows you to measure the vertical position of the ball as a function of time at intervals of 0.05 seconds. For each value of h, you do five experiments and average your data. Here is your averaged data organized into a table:

h cm	100		200		300
t (s)	y_p (cm)	t (s)	y_p (cm)	t (s)	y_p (cm)
0	100	0	200	0	300
0.05	99	0.05	197	0.05	299
0.10	97	0.10	189	0.10	301
0.15	93	0.15	178	0.15	283
0.20	82	0.20	175	0.20	286
0.25	66	0.25	169	0.25	264
0.30	57	0.30	162	0.30	261
0.35	38	0.35	136	0.35	235
0.40	23	0.40	124	0.40	226
0.45	1	0.45	96	0.45	197
		0.50	80	0.50	181
		0.55	50	0.55	149
		0.60	22	0.60	129
				0.65	88
				0.70	57
				0.75	26

a. Plot these three data sets as y_p vs t on a single graph using different symbols for the three different values of h. Hint: you might be able to copy and paste the columns from this document.
b. From dimensional analysis you determine that a good length scale is h, so define $Y_p = y_p/h$ and, without knowledge of any theory, a good time scale is $(h/g)^{1/2}$, so define $T = t/(h/g)^{1/2}$. Now plot the three curves in dimensionless form on the same graph Y_p vs T.
c. Form two columns, T and Y_p, for all of the data in part (b) and perform a nonlinear regression for a quadratic curve fit $Y = a_1 + b_1T + c_1T^2$. What are your values of a_1, b_1, c_1 and what is the r^2 value?

Add a fourth entity to your plot in part (b), which is this quadratic curve fit, show it as a line.

Hint: don't struggle with software. Here is a website where you can paste the two columns and select a three-parameter nonlinear regression: http://www.xuru.org/rt/nlr.asp.

d. Now perform the same nonlinear regression procedure, but just on the $h = 300$ cm data. What are your values of a_1, b_1, c_1 and r^2? Compare to part (c). What do you think about it?
e. You have already done 15 experiments. Your professor asks you to perform five each for $h = 400, 500, 600, 700, 800, 900, 1000$ cm: another 35 experiments over the weekend. Recast your dimensionless curve fit from part (c) as a dimensional equation y_p vs t explicitly showing g and h. Do you need to do more experiments, or can you tell your professor what the outcomes will be without doing them?

2.3

s_i (m/s)				
h (m)	10	20	30	40
v_0 (m/s)				
−20	24.41	28.14	31.43	34.41
−16	21.26	25.46	29.05	32.25
−12	18.44	23.15	27.06	30.46
−8	16.12	21.35	25.53	29.12
−4	14.56	20.20	24.58	28.28
0	14.00	19.80	24.25	28.00
4	14.56	20.20	24.58	28.28
8	16.12	21.35	25.53	29.12
12	18.44	23.15	27.06	30.46
16	21.26	25.46	29.05	32.25
20	24.41	28.14	31.43	34.41

a. For the thrown ball the table lists the average measured impact speed, s_i, for 11 values of $v_0(m/s)$ and four values of $h(m)$. Plot s_i vs v_0 for each column of constant h. What do you notice about the effect of throwing the ball downward vs throwing it upward?
b. Create a corresponding table of dimensionless impact speed $S_i = s_i/(gh)^{1/2}$ and a corresponding table of Fr values. Let $g = 9.8\,m/s^2$. Then plot S_i vs Fr for each column of constant h. What has happened to the data?
c. For the dimensionless impact speed let $S_i = \left|dY_p/dT\right|_{T=T_i}$. Derive this expression from the text which uses the same scalings. What does it say about the sign of v_0? Is there an underlying physical principle?
d. Plot S_i vs Fr and compare to your data plot in part (b).
e. What would be a good curve fit $S_i = kFr^a$ for $Fr << 1$? What about $Fr >> 1$? Determine k and a for both.

2.4 Consider the ejection flow from the heart, through the aortic valve, and into the aorta, see Figure 1.3. You propose an investigation of possible damage to red blood cells as they pass through the aortic valve where the fluid velocity, U, is high and the shear stress magnitude on the red blood cells is also high. Of interest is the critical shear stress magnitude, call it $\tau_c \sim g/(cm{\cdot}s^2)$, that causes damage to the cells. The other dimensional parameters are the fluid density, $\rho \sim g/cm^3$, the fluid viscosity, $\mu \sim g/(cm{\cdot}s)$, the red blood cell diameter, $d \sim cm$, and the valve opening cross-sectional area, $A \sim cm^2$. The red blood cell is elastic, and you lump all of those properties into an overall elastic modulus, $E \sim g/(cm{\cdot}s^2)$. So there are six input dimensional parameters affecting the output parameter τ_c,

$$\tau_c = f(\mu, \rho, U, d, A, E)$$

Using the Buckingham Pi Theorem, form four dimensionless groups from these seven dimensional parameters. Hint: one of them is a Reynolds number.

For your repeating parameters try length (L) units using d, time (T) units using μ, and mass units (M) using ρ.

2.5

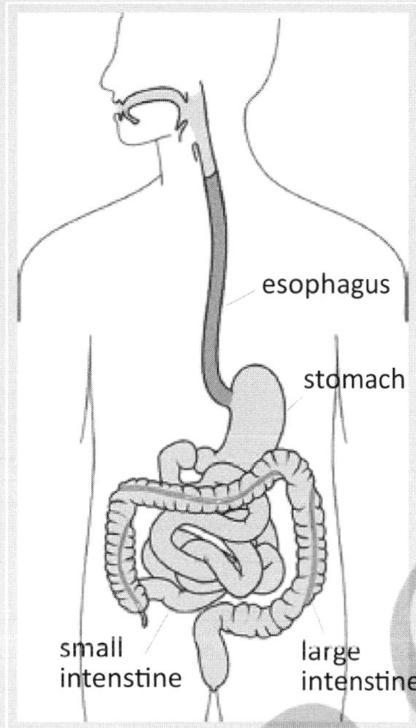

As a gastroenterology bioengineer, you want to study the speed of the material that is being swallowed into the esophagus which connects the mouth to the stomach as shown in the figure. Call that material speed, U_0. Peristaltic waves in the esophagus travel at speed, C, and have wavelength, λ. The material swallowed for this example is a non-Newtonian fluid that has a yield stress, τ_y. When the local shear stress in the material is less than the yield stress, that part of the material behaves like an elastic solid with bulk elastic modulus, E, and shear elastic modulus, G. When the local shear stress in the material is above the yield stress, it flows like a fluid with viscosity, μ, and density, ρ. The esophagus has finite length and diameter as well.

a. What are the dimensional parameters that will influence U_0 ? Hint: read the above description carefully.

b. Form the related dimensionless parameters, including U_0. Choose your own repeating parameters for L, T and M.

2.6

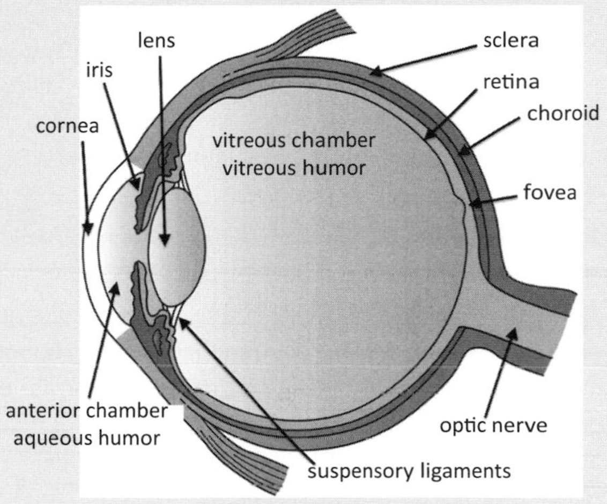

Anatomy of the eye.

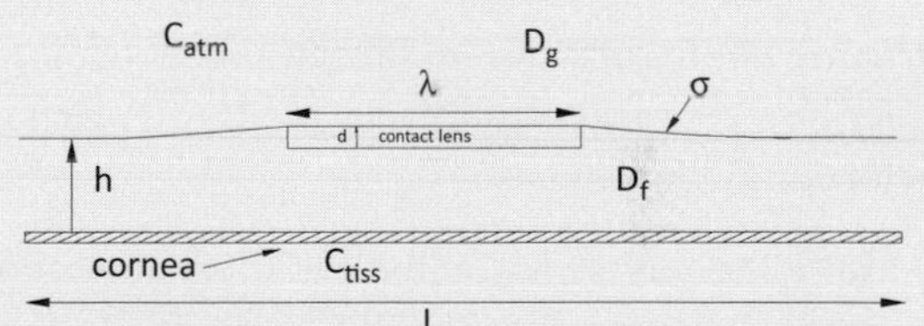

Model of oxygen transport to the cornea with a contact lens.

The cells of the eye's cornea rely on oxygen delivery through the liquid tear film covering it, see the anatomy figure. You are hired by a contact lens company to study the influence of a contact lens on the oxygen mass transport to the cornea.

You create a simplified analytical model, as shown in the figure, where C is an oxygen concentration in the atmosphere or tissue, D, is the molecular diffusivity of oxygen in the gas or tear fluid film, h is the tear film thickness, L is its length. The contact lens has length λ and thickness d. The tear film has surface tension, σ, with the atmosphere. The liquid and gas are stationary.

a. Define the dimensionless parameters of this system which will influence the mass flow rate of oxygen to the cornea
b. Improvements in contact lens technology have led to the lens having flexibility and its own gas diffusivity. What additional dimensionless parameters do they create?
c. Why is the company interested in this study?

2.7 For the case of the thrown ball, we had other choices to scale the dependent variable of position, y_p, and the independent variable of time, t. We can let the scale for position remain as h, so that $Y_p = y_p/h$, but how about choosing a different time scale? From the parameters h, g, and v_0 it is not too difficult to arrive at three combinations that have the units of time: $(h/g)^{1/2}, v_0/g, h/v_0$. Let's try the third option for a time scale, so that the new dimensionless time is $T = v_0 t/h$. We can restrict ourselves to $v_0 > 0$ so there are no singularities in the time definition. If it is easier, use your software (Maple, MATLAB®, etc.) to perform the calculations.

a. Make the governing equation, $d^2y_p/dt^2 = -g$, dimensionless using the definitions of T and Y_p.
b. Make the initial conditions, $y_p(t = 0) = h, dy_p/dt|_{t=0} = v_0$, dimensionless too.
c. Where does Fr appear now in the problem? Is that different from Section 2.6?
d. Solve the system for $Y_p(T)$.
e. Solve for the impact time, $T_i(Fr)$, and plot it for $0.1 \leq Fr \leq 4$. Describe your results.
f. Expand your solution in a Taylor series for $0 < Fr << 1$ keeping the first two terms. Use your software if it helps.

2.8

a. Make the following equation dimensionless using the scaled variables as shown

$$\frac{d^2y}{dx^2} + \frac{1}{\lambda}\frac{dy}{dx} = K$$

where the variables y and x are measured in units of length and K and λ are constants. Define dimensionless variables X,Y as $X = x/\lambda, Y - y/\lambda$. Identify any dimensionless parameters.

b. Make the following equation dimensionless using the scaled variables as shown

$$\frac{d^4f}{dz^4} + \beta\frac{d^2f}{dz^2} + kf = 1$$

where z has units of length. Let the dimensionless variables be $Z = \beta^{1/2}z, F = \beta^2 f$. Identify any dimensionless parameters.

c. Make the following equation dimensionless using the scaled variables as shown

$$a\frac{d^3g}{dt^3} + \frac{d^2g}{dt^2} + \alpha g = 1$$

where t has units of time. Let the dimensionless variables be $T = \alpha^{1/2}t, G = \alpha g$. Identify any dimensionless parameters.

2.9

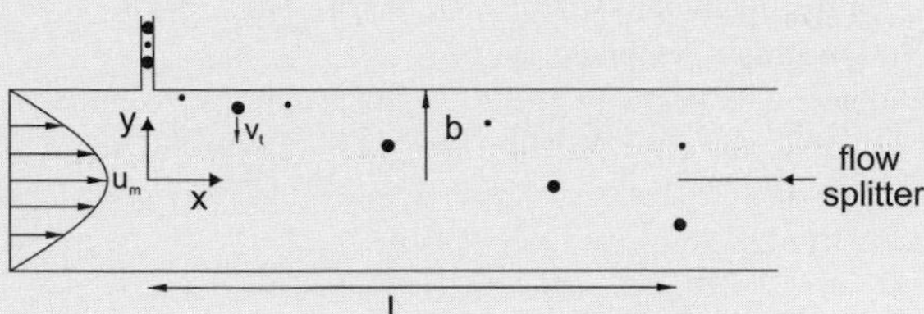

Designing a cell sorting device with a split channel.

Building on Section 2.8, you are designing a microfluidic, cell sorting device as shown in Figure 2.23, which is a two-dimensional channel of width 2b and length L. Cells drop into a liquid flow chamber at their terminal velocity, v_t. At $x = L$ there is a splitter plate that divides the flow into two downstream channels each of width b. Our goal will be to design a device so that two cell populations of different size can be separated. One type, the larger cell, should end up in the lower channel and the other, the smaller cell, should flow into the upper channel.

As before, the x-velocity profile of the liquid is parabolic in y of the form

$$u = u_m\left[1 - \left(\frac{y}{h}\right)^2\right] \tag{1}$$

where u_m is the maximum velocity at the centerline, $y = 0$, and the fluid has zero velocity on the boundaries, $y = \pm b$. We can assume for this example that u does not change near $x = L$. We assumed that the cell maintains its constant y-velocity, v_t, but that its x-velocity matches the fluid's x-velocity. Then the governing equations for the cell position (x_p, y_p) are

$$\begin{aligned} \frac{dx_p}{dt} &= u_m\left[1 - \left(\frac{y_p}{b}\right)^2\right] \\ \frac{dy_p}{dt} &= -v_t \end{aligned} \tag{2}$$

with initial conditions $x_p(t = 0) = 0, y_p(t = 0) = b$.

a. Use the scaled variables

$$X_p = \frac{x_p}{L},\ Y_p = \frac{y_p}{b},\ T = \frac{v_t t}{b} \tag{3}$$

and express Eqs. (1), (2), and the initial conditions in dimensionless form. Feel free to use your software.

b. Solve for $X_p(T)$ and $Y_p(T)$ and let the dimensionless parameter be $\beta = bu_m/Lv_t$. β can be viewed as a ratio of two time scales. What are they and what do they represent?

c. Consider two cell types, 1 and 2, where type 2 is twice the diameter of type 1, so the ratio of terminal velocities is $v_{t2} = 4v_{t1}$ and then $\beta_2 = \beta_1/4$. Plot the two cell trajectories, X_{p1} vs Y_{p1} (type 1) and X_{p2} vs Y_{p2} (type 2) for $\beta_1 = 1$ and place them on the same graph. Make sure in your plots you can "see" the entire larger channel.

d. Now adjust β_1 to get the best separation at $x = L$, so that type 1 enters the upper channel and type 2 the lower channel with a comfortable margin of error.

e. Repeat this process for type 2 diameter three times that of type 1. What is a good value for β_1?

Image Credits

Figure 2.1 Source: pkline/ Getty
Figure 2.9 Source: Jason Davey (public domain). https://en.wikipedia.org/wiki/File:Greek_ext.png
Figure 2.10 Source: Adam Jones/ Getty
Figure 2.18(a) Source: Bruce Wetzel & Harry Schaefer/ National Cancer Institute. https://visualsonline.cancer.gov/details.cfm?imageid=2129
Figure 2.18(b) Source: Bruce Wetzel & Harry Schaefer/ National Cancer Institute. https://visualsonline.cancer.gov/details.cfm?imageid=2129

Figure 2.20 Source: Spontaneous splenic rupture in Waldenstrom's macroglobulinemia: a case report. *Journal of Medical Case Reports*. 2010; 4: 300.

Figure 2.22 Source: MechESR. Public domain. https://en.wikipedia.org/wiki/Erythrocyte_sedimentation_rate#/media/File:StaRRsed_pipet_array.jpg

Problem 2.5 Source: Source: Olek Remesz (CC BY SA); https://commons.wikimedia.org/wiki/File:Tractus_intestinalis_esophagus.svg

Problem 2.6, Anatomy of the eye Source: https://commons.wikimedia.org/wiki/File:Three_Main_Layers_of_the_Eye.png

3 Kinematics, Lagrangian and Eulerian Variables

Introduction

Any discipline, no matter how modest, makes headway by forming a set of guiding principles on how to describe events. Fluid mechanics is no exception, and, in fact, we have some choices in this regard. Let's start with some simple examples. Suppose you are backed up at a toll booth on the Interstate. From your viewpoint you see how many vehicles are "flowing" in front of you, get a general sense of when you will reach the booth, and you fumble for pocket change to pay the toll in anticipation of when you actually get there. Your view is moving with the flow of traffic from the perspective of your car. The toll booth attendant, on the other hand, just sees cars and trucks pull up to the booth, regardless of who is driving. The attendant's view is fixed and not moving. Nevertheless, the two views coincide when you arrive at the booth and pay the cash.

In a biological flow we can think of a similar transaction. Blood flows through a capillary in the tissues and delivers oxygen while picking up carbon dioxide. If you are traveling with the blood and measuring oxygen levels with a tiny meter, for example, the reading progressively drops with time as you pass through the systemic circulation from the arterial tree, through the capillaries, and into the venous system. You are recording oxygen levels as a function of time while attached to the fluid, forming an oxygen time history. However, if you take a second tiny meter and hold it in the arterial flow at a fixed location, its reading essentially is not changing with time. You are recording oxygen levels as a function of time while fixed in space, as a continuous stream of fluid passes by. The two meters give the same reading when the one moving with the fluid passes the one that is fixed, since at that moment they are in the same place at the same time.

We can think of these two views of events as each being a reference frame. The reference frame which is fixed in space and measuring (or predicting) fluid behavior as it flows by is what we call the Eulerian frame. It is a natural one for fluid mechanical investigations since a probe measuring velocity or oxygen level, for example, can be held in place in the lab while the fluid passes by. The frame of reference which moves with the fluid, however, is called the Lagrangian frame, precisely what was used to describe the motion of the mass points in Chapter 2. In the Lagrangian frame the behavior of a particular fluid particle is followed as it flows along. Each frame has its own set of variables (position, velocity, pressure, temperature, etc.) which coincide when the fluid particle being followed in the Lagrangian frame passes through the spatial point of interest fixed in the Eulerian frame.

Many treatises on fluid mechanics focus entirely on the Eulerian description which is fully adequate for air flow over an airplane wing, for example, since the air is not "living." In biofluid mechanics, however, the fluid is often alive in that it potentially carries nutrients, signaling molecules, cells, toxins, etc. In addition, the boundaries of the flow are alive since they are formed by living tissue and can move, like the heart as it contracts and pumps blood. Thus knowing what happens to a particular fluid particle, or boundary point, can be essential to an investigation of normal and abnormal physiology.

Topics Covered

The topics progress by introducing the concepts of Lagrangian and Eulerian variables, or reference frames, and how they are related through the fluid mapping. The mapping is the function describing the position of each fluid particle as it moves in space and time, also known as its pathline or trajectory. From that concept of the mapping we compare how a function, like fluid temperature or speed, would be viewed in the two reference frames. In order to use this information to develop conservation equations in the next chapter, the relationship between partial derivatives in the two frames is derived using the chain rule. Specific examples are given for inviscid stagnation point flow. In addition to the fluid pathline, three additional important fluid lines are discussed: streamlines, streaklines and timelines. These lines all have important uses in analyzing and measuring fluid mechanical phenomena. Finally, an application is shown for aerosol particle deposition in the lung using the simple stagnation point flow. Included in the aerosol analysis is an introduction to using computational software to find analytic solutions to linear ordinary differential equations. Many fluid mechanics courses do not study streaklines or timelines, for example, so Sections 3.7 and 3.8 can be skipped without loss of continuity in the book. Finally, the general concepts of numerical methods are introduced since this is often a very new topic for students. Since computation software like MATLAB®, Maple, Mathematica and others have user friendly interfaces to solve problems numerically, from integrations to differential equations, this introduction will help to put things into perspective. Some homework assignments In this book will require using numerical methods.

3.1 Lagrangian and Eulerian Variables

The emphasis in Chapter 2 was to revisit many of the concepts you learned in your Introductory Physics course about describing the motion of a mass point from where it starts at an initial time to where it is located for later times and its velocity history along the way. Likewise, we can also think of the fluid itself as made up of fluid particles and describe its motion in the same way as the mass point. It is a bit more complicated, since the fluid is a continuum. Instead of one initial position, there are an infinite number, i.e. a continuum, of starting locations for the fluid particles. Each mathematical point in the continuum is a fluid point.

As discussed in the Introduction, there are two approaches to studying fluid phenomena and we outline their basic concepts here. Suppose we are interested in the fluid motion including its kinematics like velocity and acceleration. One approach is to follow an individual fluid particle as it moves, similar to the analysis of the solid particle motion in Chapter 2. If we identify a fluid particle as being at some initial position, say $\underline{X}_0 = (X_0, Y_0, Z_0)$ when $t = 0$, we can, in principle, keep a running history of its spatial position, the vector $\underline{X} = (X, Y, Z)$, i.e.

$$\underline{X} = \underline{X}(\underline{X}_0, t) \qquad \text{fluid mapping} \tag{3.1.1}$$

There is a continuum of particles, each with its own starting position, the vector $\underline{X}_0$. For the solid particle motion in Chapter 2, we found $\underline{X}_p = \underline{X}_p(\underline{X}_0, t)$ as that particle's trajectory. The same concept is used for following a fluid particle, where the relationship of Eq. (3.1.1) is often called the mapping or fluid particle path, pathline or trajectory. With careful enough records of $\underline{X}(\underline{X}_0, t)$ we could calculate the velocity, $\underline{U}$, of the particle by calculating the partial time derivative holding particle identity, $\underline{X}_0$, fixed

$$\underline{U}(\underline{X}_0,t) = \left(\frac{\partial \underline{X}}{\partial t}\right)_{\underline{X}_0} \qquad \text{Lagrangian fluid velocity vector} \tag{3.1.2}$$

Alternatively, we could, in principle, measure the velocity of that particular particle as we track it. This approach is called the Lagrangian description or reference frame since we determined the velocity following a particular fluid particle which we identified by its initial position. It is marked by the independent variables being $(\underline{X}_0,t)$. Part of the notation here is purposeful to use upper case variables for the Lagrangian frame. One can appreciate, however, that keeping track of an individual fluid particle as it flows in a continuum of fluid particles is a challenge.

We have just been studying particle motion based on their initial position and time. The alternative approach to describing the fluid velocity is not to follow individual fluid particles on their journey in the flow, but rather to let them come to us, so to speak. To do that we focus on a fixed position in space, call it $\underline{x}$, and then keep a running history of the fluid particle velocities that pass through that point for each value of time, t. Let that velocity be called the Eulerian fluid velocity vector,

$$\underline{u}(\underline{x},t) \qquad \text{Eulerian fluid velocity vector} \tag{3.1.3}$$

This is the Eulerian approach or frame, sometimes also called the "lab" frame, since most laboratories are stationary, not moving with a fluid. It is marked by the independent variables being $(\underline{x}, t)$, where the lower case denotes the Eulerian frame.

The two approaches are connected, because the Eulerian position, $\underline{x}$, though fixed in the fluid space, always has fluid particles going through it whose Lagrangian position, $\underline{X}$, is coincident when

$$\underline{x} = \underline{X}(\underline{X}_0, t) \tag{3.1.4}$$

Equation (3.1.4) seems obvious, but it is the key to relating the two frames and we will employ it frequently.

3.2 Mappings and Inverse Mappings

Let's consider a general deformation of a continuum (i.e. fluid or solid) as shown in Figure 3.1. A body is seen at two times, $t = 0$ and $t = t$, and has moved and deformed in between. A material point in the body starts at position $\underline{X}_0$ when $t = 0$, and is then located at

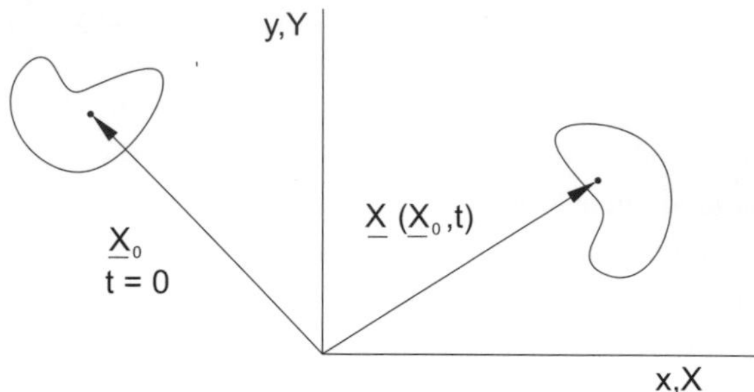

Figure 3.1 General deformation of a continuum. A body starts at $t = 0$ then flows and deforms until a later time, t. A point in the continuum starts at $\underline{X}_0$, and later is at position $\underline{X}(\underline{X}_0, t)$, which is the mapping.

position $\underline{X}$ at time t. The function $\underline{X}(\underline{X}_0, t)$ is the mapping discussed in Eq. (3.1.1). It defines the time-dependent position of a particle which is at $\underline{X} = \underline{X}_0$ when $t = 0$. Reference positions other than initial positions may also be used.

A simple example of a mapping is uniform flow in the x-direction at speed *q*. All of the fluid particles are now represented by mathematical points, a continuum. Their initial positions are (X_0, Y_0, Z_0) and only the x-component is time-dependent. The mapping is

$$X = X_0 + qt, \quad Y = Y_0, \quad Z = Z_0 \tag{3.2.1}$$

Given the initial position (X_0, Y_0, Z_0) of any particle, the mapping tells us where that particle is at any future time. We can perform a partial derivative with respect to time on the mapping, holding the initial position variables fixed. This is the Lagrangian velocity as defined in Eq. (3.1.2), and the components are

$$U = \left(\frac{\partial X}{\partial t}\right)_{\underline{X}_0} = q, V = \left(\frac{\partial Y}{\partial t}\right)_{\underline{X}_0} = 0, W = \left(\frac{\partial Z}{\partial t}\right)_{\underline{X}_0} = 0 \tag{3.2.2}$$

When dealing with a single mass point, it was natural to differentiate with respect to time and view it as an ordinary derivative since we knew the identity of the single object under consideration. Here we hold particle identity, $\underline{X}_0$, fixed in a demonstrable way, using the partial derivative for time.

Now let's move on to the inverse mapping. The inverse mapping for this simple flow is symbolized by $\underline{X}_0 = \underline{X}^{-1}(\underline{x}, t)$, which tells us the initial positions of the particles that are passing through the Eulerian position $\underline{x}$ at time t. Here, $\underline{X}^{-1}$ means inverse function, not $1/\underline{X}$. It is easy to solve Eq. (3.2.1) for the initial positions

$$X_0 = X - qt, \quad Y_0 = Y, \quad Z_0 = Z \tag{3.2.3}$$

Employing Eq. (3.1.4) we can replace $\underline{X}$ with $\underline{x}$ to have the independent variables be Eulerian

$$X_0 = x - qt, \quad Y_0 = y, \quad Z_0 = z \tag{3.2.4}$$

Equation (3.2.4) is the inverse mapping we seek. It can be represented by the following vector equation

$$\underline{X}_0 = \underline{X}^{-1}(\underline{x}, t) = \begin{pmatrix} x - qt \\ y \\ z \end{pmatrix} \tag{3.2.5}$$

Let's examine what the inverse mapping really says. For $q > 0$ and flow to the right, the x-component of the inverse mapping states that given a fixed Eulerian value of x, as t increases the initial position of the particle passing through x starts further and further to the left, that is, it starts further upstream, since X_0 decreases with time. This makes sense since it takes more time for the particles starting farther to the left to arrive at the position x where we are observing. Figure 3.2 shows both the mapping and the inverse mapping at four values of time $t = 0, 1, 2, 3$ and for $q = 1$. For the mapping, the fluid particle starting at $X(t = 0) = X_0 = 1$ moves to the position $1 + qt$ in the positive X-direction. For the inverse mapping, the initial position of the particle observed at $x = 1$ starts farther and farther to the left at position $1 - qt$.

3.3 Functions in Lagrangian and Eulerian Reference Frames

In this section we study an example of how to view a flow and, in this case, a temperature field in either the Lagrangian or Eulerian frame of reference. We begin with some mathematical statements that, while

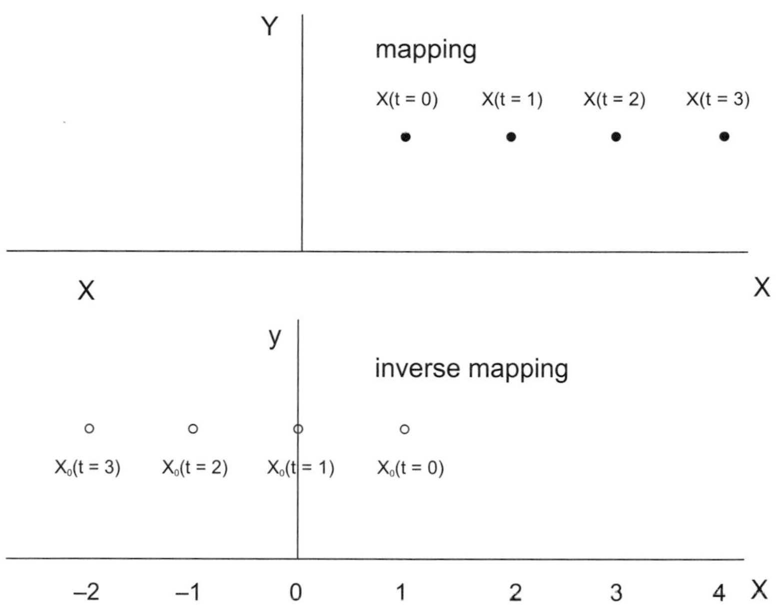

Figure 3.2 Mapping and inverse mapping of uniform flow in the positive x-direction. Four time values are shown. Note that the mapping defines the fluid particle starting at $X(t = 0) = X_0 = 1$ moves to the right, but the inverse mapping shows the initial position of the particle at $x = 1$ starts farther and farther to the left with increasing time.

precise, are not particularly demonstrative yet. It is the example later in the section that will bring more enlightenment to the student's curiosity. An Eulerian function $f(\underline{x}, t)$ can be written in the Lagrangian frame as $F(\underline{X}_0, t)$, by using the mapping, Eq. (3.1.4), where we replace $\underline{x}$ with $\underline{X}(\underline{X}_0, t)$

$$f(\underline{x}, t) = f(\underline{X}(\underline{X}_0, t), t) = F(\underline{X}_0, t) \tag{3.3.1}$$

Likewise the Lagrangian function $F(\underline{X}_0, t)$ can be written in the Eulerian frame by using the inverse mapping, $\underline{X}_0 = \underline{X}^{-1}(\underline{x}, t)$

$$F(\underline{X}_0, t) = F(\underline{X}^{-1}(\underline{x}, t), t) = f(\underline{x}, t) \tag{3.3.2}$$

As an example, consider the dimensionless Eulerian temperature field

$$\theta(x, t) = px + kt \tag{3.3.3}$$

where p and k are constants. We can think of $p = \partial\theta/\partial x$ as a constant spatial gradient and $k = \partial\theta/\partial t$ as a constant time rate. This field is possible when there is a uniformly distributed heat source or sink, $S = k + qp$.

Let's consider $p = 1/2$ and $k = \pm 1$. A graph of the Eulerian temperature field is shown in Figure 3.3 for

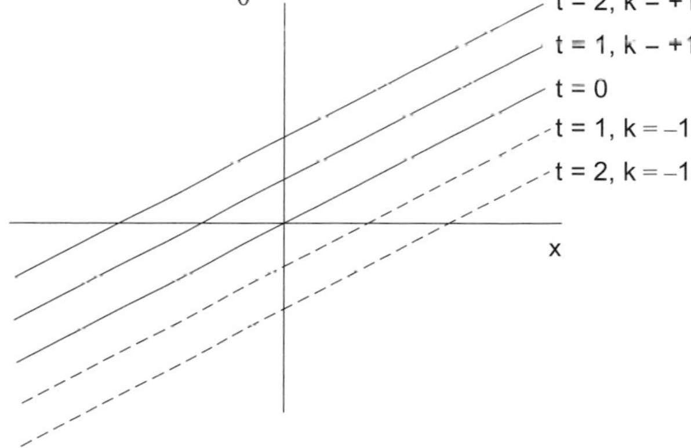

Figure 3.3 Eulerian temperature field at three values of time for $p = 1/2$, $k = 1$, solid, and $k = -1$, dashed.

three values of time. The $t = 0$ line passes through the origin and has positive slope, $p = 1/2$. The temperature increases from left to right. The next value of time is $t = 1$ which shifts the line upward for $k = +1$ (solid lines) and downward for $k = -1$ (dashed lines), with the same slope. Temperatures are warming everywhere with time for positive k, and cooling for negative k. That pattern continues for $t = 2$.

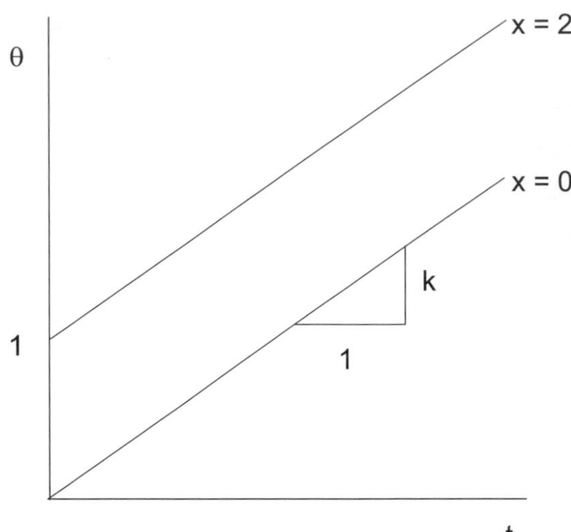

Figure 3.4 Time history of the Eulerian temperature at two values of x. The slope of both curves is k.

For an example of Eulerian measurements of temperature, let $k = 1$ in Figure 3.3. If we place a thermometer at $x = 0$ and one at $x = 2$, then the readouts will be those shown in Figure 3.4. Both lines have the same positive slope, k, as temperatures increase with time at that rate. However, the thermometer at $x = 2$ starts out at a hotter location and always has higher temperatures than at $x = 0$.

What is the Lagrangian temperature, $\Theta(\underline{X}_0, t)$, following a fluid point? To show the Lagrangian temperature we need to use the mapping given in Eq. (3.2.1). Replacing $\underline{x} = \underline{X}(\underline{X}_0, t)$ in Eq. (3.3.3) yields the Lagrangian temperature

$$\begin{aligned}\theta(\underline{x}, t) &= px + kt = pX(\underline{X}_0, t) + kt \\ &= p(X_0 + qt) + kt = \Theta(\underline{X}_0, t)\end{aligned} \quad (3.3.4)$$

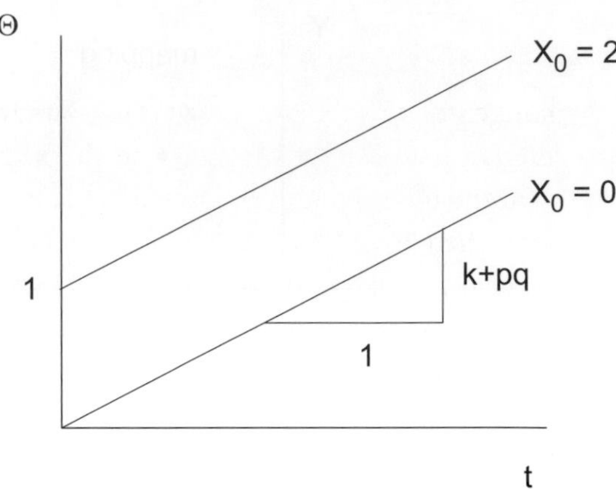

Figure 3.5 Lagrangian temperature of two fluid particles. The slope of both curves is $k + pq$.

or

$$\Theta(\underline{X}_0, t) = pX_0 + (k + pq)t \quad (3.3.5)$$

Now $\Theta(\underline{X}_0, t)$ is the Lagrangian temperature history of the particle which starts at $\underline{X}_0$ when $t = 0$. This Lagrangian temperature field is also unsteady, in general, and the fluid particle temperature depends on time in two different ways. First, because of local unsteadiness, i.e. kt, temperatures are increasing everywhere in the field with time. Second, because of the fluid velocity, q, the particle is being convected in the positive x-direction where conditions are hotter.

Figure 3.5 shows the Lagrangian temperature of two fluid particles identified by their initial positions $X_0 = 0, 2$. Again, we let $p = 1/2$. The plots are a time history of the temperature experienced by each particle. The slope of the straight lines is $k + pq$ and the particle starting at $X_0 = 2$ begins at a higher temperature than the $X_0 = 0$ particle.

Worked Example 3.1 | **Steady and Unsteady Processes**

Processes that are steady are independent of time. However, that definition depends on which reference frame is involved. For example, if we choose $k = 0$, then the Eulerian temperature field given by Eq. (3.3.3) is independent of time, i.e. steady $\theta(x) = px$. However, from Eq. (3.3.5) we see that the Lagrangian temperature following a fluid particle is still unsteady if

$k = 0$, since p and q are non-zero, $\Theta(\underline{X}_0, t) = pX_0 + pqt$. The particle experiences a time change in its temperature because it is flowing to warmer conditions. So in this case the Eulerian temperature is steady but the Lagrangian temperature is unsteady.

On the other hand, suppose we let $k = -1$ in Figure 3.3, so the temperature is cooling everywhere with time, but still increasing spatially in the positive x-direction. For the special choice of, $k + pq = 0$, that is, the fluid speed $q = 2$, we find the interesting result that the Lagrangian temperature field from Eq. (3.3.5) becomes the steady value $\Theta = pX_0$, while the Eulerian temperature field is unsteady, $\theta(x, t) = x/2 - t$, since k is not zero. That is because the time rate at which temperatures are dropping in the Eulerian frame is just balanced by the spatial rate of increase in temperature seen by the flowing particle which is moving in the +x-direction, where conditions are warmer, at this specially selected speed. So this is an example of a steady Lagrangian temperature but an unsteady Eulerian temperature.

3.4 Partial Derivatives in Eulerian and Lagrangian frames

The equations of motion depend on time rates of change and spatial rates of change, so it is essential that we can relate partial derivatives in both frames. First let us review some multivariable calculus. Consider the function b(r, s) where the variables (r, s) depend on two other variables, (m, n), i.e.

$$\begin{aligned} r &= r(m, n) \\ s &= s(m, n) \end{aligned} \tag{3.4.1}$$

so that in the m, n variables the function is B(m, n). What is $(\partial B/\partial n)_m$? The chain rule of partial differentiation says that

$$\left(\frac{\partial B}{\partial n}\right)_m = \left(\frac{\partial b}{\partial r}\right)_s \left(\frac{\partial r}{\partial n}\right)_m + \left(\frac{\partial b}{\partial s}\right)_r \left(\frac{\partial s}{\partial n}\right)_m \tag{3.4.2}$$

For our purposes consider the definitions

$$\begin{aligned} r &= x, \quad s = t \\ m &= X_0, n = t \end{aligned} \tag{3.4.3}$$

so that

$$\left(\frac{\partial B}{\partial t}\right)_{X_0} = \left(\frac{\partial b}{\partial x}\right)_t \left(\frac{\partial x}{\partial t}\right)_{X_0} + \left(\frac{\partial b}{\partial t}\right)_x \left(\frac{\partial t}{\partial t}\right)_{X_0} \tag{3.4.4}$$

Clearly,

$$\left(\frac{\partial t}{\partial t}\right)_{X_0} = 1 \tag{3.4.5}$$

and we can transform the convective term as

$$\underbrace{\left(\frac{\partial x}{\partial t}\right)_{X_0} = \left(\frac{\partial X}{\partial t}\right)_{X_0}}_{\downarrow\rightarrow \text{ mapping } \rightarrow\uparrow} = \underbrace{U(\underline{X}_0, t) = u(\underline{x}, t)}_{\downarrow\rightarrow \text{ inverse mapping } \rightarrow\uparrow} \tag{3.4.6}$$

by using the mapping and the inverse mapping as shown in Eq. (3.4.6). The result is the relationship between the partial time derivative in the Lagrangian frame, following a fluid particle, and the partial time and space derivatives in the Eulerian frame,

$$\left(\frac{\partial B}{\partial t}\right)_{X_0} = \frac{\partial b}{\partial t} + u\frac{\partial b}{\partial x} \tag{3.4.7}$$

where we again see that there is both a local unsteady term and a convective term in the Eulerian frame. The transformations of Eq. (3.4.6) were essential in simplifying the result, yet we did not need to know the details of the mapping or the inverse mapping to do it.

Let's check the result by using an example from our temperature field and take its time derivative. Let $B = \Theta$, the Lagrangian temperature, and $b = \theta$, the Eulerian temperature. Substituting the Lagrangian temperature from Eq. (3.3.5) into the left hand side of Eq. (3.4.7) gives us

$$\left(\frac{\partial \Theta}{\partial t}\right)_{X_0} = k + pu \tag{3.4.8}$$

for $q = u$. Likewise, if we substitute the Eulerian temperature, Eq. (3.3.3), into the right hand side of Eq. (3.4.7), the result is

$$\frac{\partial \theta}{\partial t} + u\frac{\partial \theta}{\partial x} = k + qp \tag{3.4.9}$$

which is the same as Eq. (3.4.8) as expected.

We may generalize Eq. (3.4.7) for three-dimensional systems

$$\left(\frac{\partial B}{\partial t}\right)_{\underline{X}_0} = \frac{\partial b}{\partial t} + u\frac{\partial b}{\partial x} + v\frac{\partial b}{\partial y} + w\frac{\partial b}{\partial z} = \frac{\partial b}{\partial t} + \left(\underline{u}\cdot\underline{\nabla}\right)b \tag{3.4.10}$$

In Eq. (3.4.10) we have used the compact notation for the scalar operator

$$\underline{u}\cdot\underline{\nabla} = u\frac{\partial}{\partial x} + v\frac{\partial}{\partial y} + w\frac{\partial}{\partial z} \tag{3.4.11}$$

The right hand side of Eq. (3.4.10), which is in the Eulerian frame, is equivalent to the Lagrangian derivative on the left hand side. Sometimes the right hand side is called the material derivative (also the substantial derivative, the Stokes derivative, the total derivative, the convective derivative, the derivative following a fluid point). It is common to use a different symbol for the Lagrangian time derivative when the arguments are cast in Eulerian variables

$$\left(\frac{\partial B}{\partial t}\right)_{\underline{X}_0} = \frac{Db}{Dt} = \frac{\partial b}{\partial t} + \left(\underline{u}\cdot\underline{\nabla}\right)b \tag{3.4.12}$$

So D/Dt is the symbol for the material derivative. This derivative definition can also apply for operations on vectors. For example, the Lagrangian acceleration vector is given by the equivalent Eulerian version

$$\left(\frac{\partial \underline{U}}{\partial t}\right)_{\underline{X}_0} = \frac{D\underline{u}}{Dt} = \frac{\partial \underline{u}}{\partial t} + \left(\underline{u}\cdot\underline{\nabla}\right)\underline{u} \tag{3.4.13}$$

Equation (3.4.13) will be a valuable identity when we deal with conservation of momentum for a fluid.

Clinical Correlation 3.1

The distinction between an Eulerian velocity field and the corresponding Lagrangian velocity field can be the insight needed to understand important fluid motion effects in biological systems. An interesting argument in the early biofluid mechanics literature concerned peristaltic pumping, particularly that of the ureter, which is a tube connecting the kidney to the bladder. The peristaltic wave propagates on the ureter, from the kidney to the bladder, squeezing urine along and thus filling the bladder. However, under the right conditions of pressures and peristaltic wave dynamics, a reflux flow can occur in the opposite direction, with the potential to carry bacteria from the bladder to the kidney. A bladder infection is an uncomfortable but not particularly dangerous condition; however, spreading it to the kidney can lead to a much more serious illness (Shapiro and Jaffrin, 1971).

3.5 Mapping Example: Stagnation Point Flow

Given an Eulerian velocity field, in general we can find the corresponding mapping. For example, inviscid stagnation point flow in two-dimensions is defined by the velocity vector $\underline{u} = (u, v)$, where

$$u = Ax, \quad v = -Ay \tag{3.5.1}$$

where A is a constant with units of reciprocal time that sets the speed.

The flow is directed at a wall and impinges at a right angle, as shown in Figure 3.6, and the stagnation point is at $\underline{x} = (0, 0)$. Figure 3.6 is a representation of the Eulerian velocity vector field with selected Eulerian points in a grid and the velocity vector at each point shown. For unsteady Eulerian flows, such vector fields are time-dependent, so the value of time needs to be specified to create the instantaneous velocity vector field.

The Maple file for Figure 3.6 is given by Code 3.1(a) and the MATLAB® file is shown in Code 3.1(b).

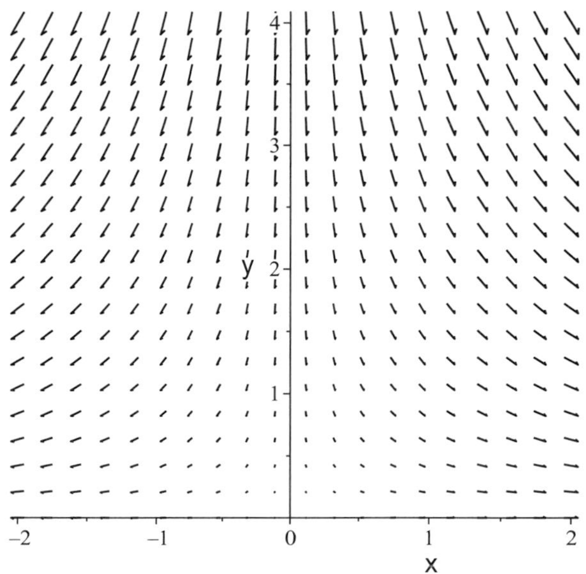

Figure 3.6 Eulerian velocity vector field for inviscid stagnation point flow, A = 1.

This flow is a two-dimensional approximation to the local stagnation point (really a line) that occurs

(a)

```
restart;
u := A•x:
v := -A•y:
A := 1:
with(plots):
fieldplot([u,v],x=-2..2, y = 0..4);
```

(b)

```
clear all, close all, clc
x     =linspace(-2, 2, 11);
y     =linspace(0, 4, 10);
[X, Y]= meshgrid(x, y);
A     =1;
U     =A*X;
V     =-A*Y;
quiver(X, Y, U, V, '-k'),hold on;
plot(x, 0*x, 'k')
plot(0*y, y, 'k')
xlabel('x')
ylabel('y')
axis([-2204])
```

Code 3.1 Code examples: (a) Maple, (b) MATLAB®.

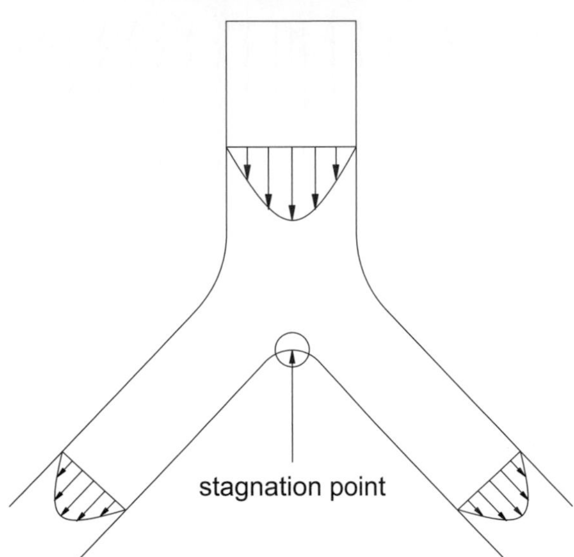

Figure 3.7 Stagnation point flow at a branching artery or airway.

whenever blood flow encounters an arterial branch or air flow in the lung during inspiration encounters an airway bifurcation, see Figure 3.7. There the fluid starts in the parent tube then splits at its bifurcation into two daughter tubes.

To solve for the mapping, we need to rewrite Eulerian velocities in the Lagrangian frame. Using Eq. (3.3.1) the general form is

$$\underline{u}(\underline{x}, t) = \underline{u}(\underline{X}(\underline{X}_0, t), t) = \underline{U}(\underline{X}_0, t) \tag{3.5.2}$$

and we note the universal definition of Lagrangian velocity is the partial time-derivative of the position for a fixed initial position identification in Eq. (3.1.2),

$$\underline{U}(\underline{X}_0, t) = \left(\frac{\partial \underline{X}}{\partial t}\right)_{\underline{X}_0} \tag{3.5.3}$$

Applying Eq. (3.5.3) to stagnation point flow yields

$$\frac{\partial X}{\partial t} = AX, \quad \frac{\partial Y}{\partial t} = -AY \tag{3.5.4}$$

Solutions to Eqs. (3.5.4) are derived by separating the variables and integrating. For example, the X component separated is

$$\frac{\partial X}{X} = A\partial t \tag{3.5.5}$$

and integrating both sides gives

$$\ln(X) = At + c_1(X_0, Y_0) \tag{3.5.6}$$

where c_1 is an integration constant which depends, at most, on (X_0, Y_0). Let both sides of Eq. (3.5.6) be an exponent of e

$$e^{\ln(X)} = e^{At + c_1} \quad \Rightarrow \quad X = ce^{At} \tag{3.5.7}$$

The new constant is just $c = e^{c_1}$. Imposing the initial condition that $X(t = 0) = X_0$ yields $c = X_0$. Performing the same steps for the Y equations yields the mapping

$$\begin{aligned} X &= X_0 e^{At} \\ Y &= Y_0 e^{-At} \end{aligned} \tag{3.5.8}$$

Equations (3.5.8) are the mapping we seek. The fluid particle pathline is parameterized by time, t. We can choose a value of A, then start t from zero and increment its value. Equations (3.5.8) yield the X and Y positions, which can be plotted against one another, giving the trajectory, or Lagrangian particle pathline. However, in this case we can eliminate t between Eqs. (3.5.8) to find the trajectory or pathline

$$e^{At} = \frac{X}{X_0} = \frac{Y_0}{Y} \tag{3.5.9}$$

so that

$$Y = \frac{X_0 Y_0}{X} \tag{3.5.10}$$

For a choice of initial position(X_0, Y_0), Eq. (3.5.10) represents a hyperbola that starts at (X_0, Y_0) and extends in the flow direction of increasing X and decreasing Y with increasing t, as

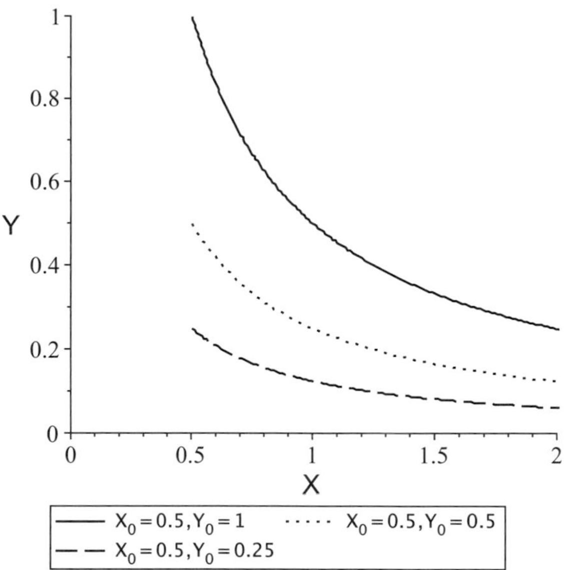

Figure 3.8 Three fluid particle trajectories or pathlines for inviscid stagnation point flow for $(X_0, Y_0) = (0.5, 1), (0.5, 0.5), (0.5, 0.25)$.

shown in Figure 3.8 for three initial positions $(X_0,\ Y_0) = (0.5, 1), (0.5, 0.5), (0.5, 0.25)$.

Equations (3.5.1) are the Eulerian velocities, but what are the Lagrangian velocities? We can determine them either by taking the Lagrangian time derivative of Eqs. (3.5.8), or by inserting $\underline{X}$ for $\underline{x}$ and $\underline{U}$ for $\underline{u}$ into Eqs. (3.5.1),

$$U = AX_0e^{At}, \quad V = -AY_0e^{-At} \tag{3.5.11}$$

Figure 3.9 shows the Lagrangian particle position and its velocity vector at three values of time. Figure 3.9(a) shows the velocity vectors attached to the particle in the X, Y axes. Figure 3.9(b) shows the same velocity vectors for the particle, but attached to its initial position in the X_0, Y_0 axes.

Note that the inverse mapping is

$$X_0 = Xe^{-At} = xe^{-At}, \quad Y_0 = Ye^{At} = ye^{At} \tag{3.5.12}$$

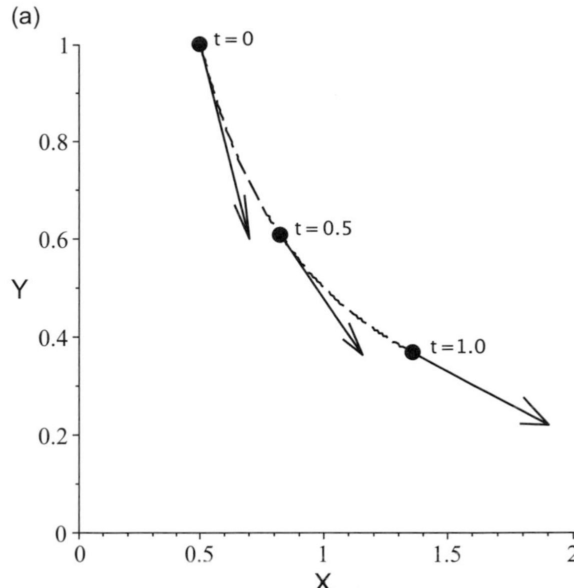

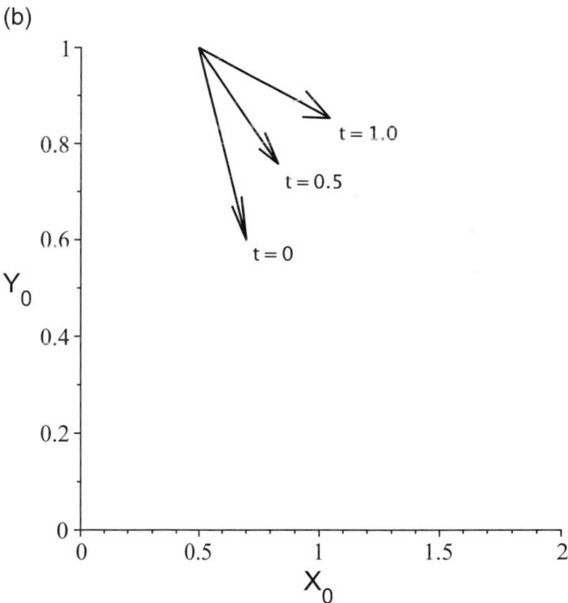

Figure 3.9 Lagrangian velocity vectors for a particle starting at $\underline{X}_0 = (0.5, 1.0)$. Three times are shown, $t = 0, 0.5, 1$ for $A = 1$. (a) Velocity vectors attached to particle, (b) velocity vectors attached to initial position.

Consider a fixed Eulerian position (x, y) in the upper right quadrant where x and y are positive. The initial position of the particle passing through that point starts at progressively smaller values of x and larger values of y for increasing t. The initial positions are farther and farther upstream as t increases. Eliminating t between the two equations in Eqs. (3.5.12) gives us

$$Y_0 = \frac{xy}{X_0} \tag{3.5.13}$$

Equation (3.5.13) is the same as Eq. (3.5.10) with (x, y) replacing (X, Y). For three choices of (x, y) the initial positions are shown in Figure 3.10. Initial positions start at (x, y) and progress upstream with increasing t for smaller X_0 and larger Y_0.

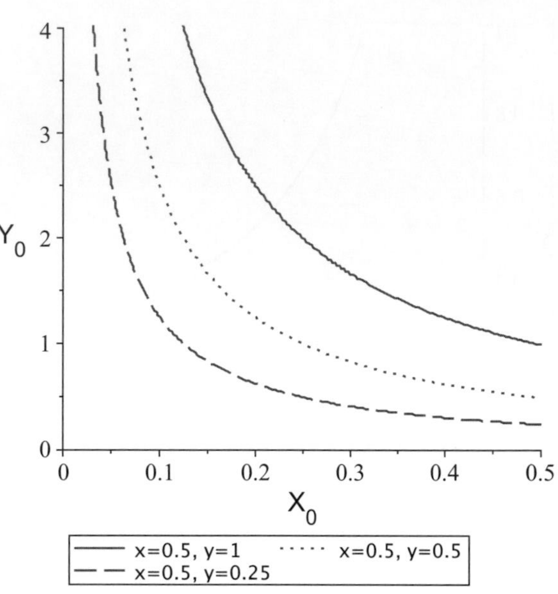

Figure 3.10 The inverse mapping for three choices of (x, y).

Worked Example 3.2 **Eulerian and Lagrangian Speeds**

Let's compare the fluid particle speed in both the Eulerian and Lagrangian frames. In the Eulerian frame, the speed, $s(\underline{x}, t)$, is the amplitude of the Eulerian velocity vector

$$s = (\underline{u}\cdot\underline{u})^{1/2} = A(x^2 + y^2)^{1/2} \tag{3.5.14}$$

which is independent of time. Curves in x–y space where the speed is constant consist of circles, since that is where $x^2 + y^2$ is constant.

In the Lagrangian frame the speed is $S(\underline{X}_0, t)$ where

$$S = (\underline{U}\cdot\underline{U})^{1/2} = A(X_0^2 e^{2At} + Y_0^2 e^{-2At})^{1/2} \tag{3.5.15}$$

where we have used Eqs. (3.5.11).

Choosing $A = 1$, Figure 3.11 shows the Lagrangian speed history of fluid particles starting from four different initial positions. Notice that there is a local minimum speed for $t > 0$ for some of the particles. As time progresses the negative exponential of Eq. (3.5.15) decreases exponentially so that the Lagrangian speed is dominated by the X component of velocity, $S \sim AX_0 e^{At}$ for $At >> 1$.

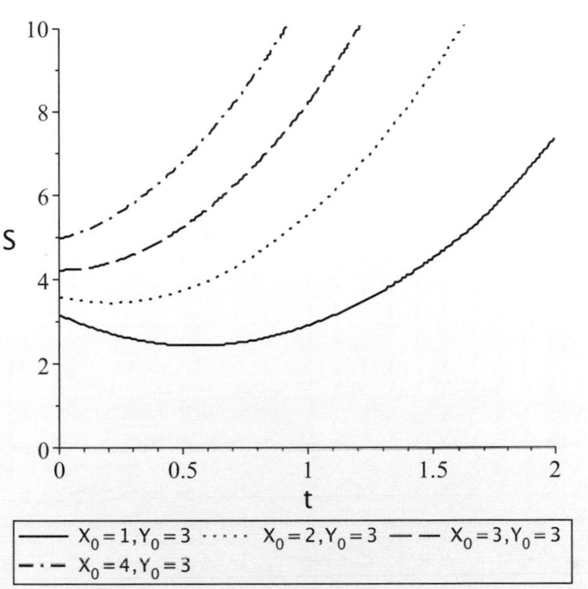

Figure 3.11 The Lagrangian speed of fluid particles from four different initial positions.

Worked Example 3.3 | Mapping from an Unsteady Eulerian Velocity Field

Consider the Eulerian velocity field

$$\begin{aligned} u &= yt/T_0^2 \\ v &= V_0 \end{aligned} \tag{3.5.16}$$

where T_0 is a parameter with units of time and V_0 is a constant velocity in the y-direction. With a time-dependent flow, the velocity vector field will depend on the choice of t.

Figure 3.12 shows examples of the velocity field for $V_0 = 1, T_0 = 1$ at two different time values, $t = 1, 5$ on the domain $-1 \le x \le 1, \; -1 \le y \le 1$. Notice the change in direction of the vectors between Figure 3.12(a) and Figure 3.12(b) as the x-component increases with time giving a more horizontal tilt. To facilitate presentation the vector lengths or speed, $s = (u^2 + v^2)^{1/2} = (y^2t^2 + 1)^{1/2}$, within each figure are scaled on the longest. Within Figure 3.12(a), $t = 1$, the range of s is $1 \le s \le 1.41$ while in Figure 3.12(b), $t = 5$, the range is $1 \le s \le 5.1$. It is also instructive to animate the flow and watch the vectors change magnitude and direction continuously.

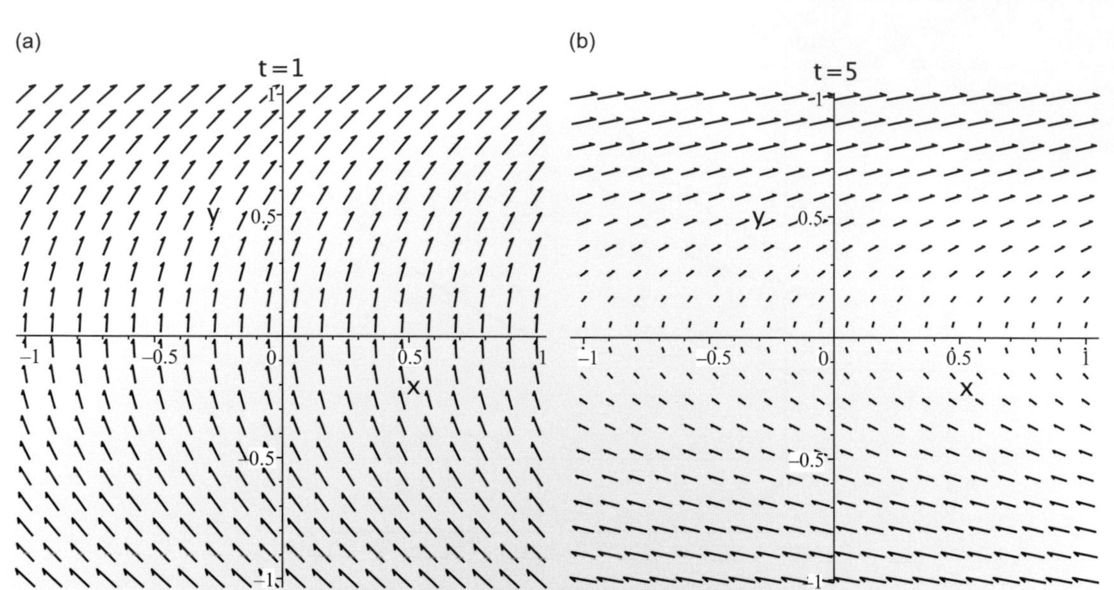

Figure 3.12 Unsteady velocity field of Eq. (3.5.16) for $V_0 = 1, T_0 = 1$: (a) $t = 1$ (b) $t = 5$.

The Maple file to create Figure 3.12(a,b) and animate the flow over the time range $0 \leq t \leq 5$ is shown in Code 3.2(a) and the MATLAB® file is in Code 3.2(b).

```
(a)
restart
u := y·t;
                            u := yt                                        (1)
v := 1;
                            v := 1                                         (2)
with(plots):
fieldplot([subs(t=1,u),subs(t=1,v)],x=-1..1,y=-1..1,labels=["x","y"],labelfont
     =[HELVEtiCA, 14],title = "t=1",titlefont=[HELVETICA, 14], thickness=1):

fieldplot([subs(t=5,u),subs(t=5,v)],x=-1..1,y=-1..1,labels=["x","y"],labelfont
     =[HELVEtiCA, 14],title="t=5",titlefont=[HELVETICA, 14], thickness=1):
animate(fieldplot, [[u,v],x=-1..1,y=-1..1],t=0..5):
```

Code 3.2 Code examples: (a) Maple, (b) MATLAB®.

```
(b)
clear all, close all, clc
u = @(y,t) y*t;
v = @(y) ones(size(y));
x = linspace(-1,1,20);
y = linspace(-1,1,20);
t = linspace(0,5,500);
[X,Y] = meshgrid(x,y);
figure(1)
U = u(Y,1);
V = v(Y);
quiver(X,Y,U,V,'K');
axis([-11-11])
xlabel('x')
ylabel('y')
ax = gca;
ax.FontSize = 11;
ax.FontName = 'Arial';
title('t-1')
figure(2)
U = u(Y,5);
V = v(Y);
quiver(X,Y,U,V,'k');
axis([-11-11])
xlabel('x')
ylabel('y')
ax = gca;
ax.FontSize = 11;
ax.FontName = 'Arial';
title('t=5')
figure(3)
for T = t
        U = u(Y,T);
        V = v(Y);
        quiver(X,Y,U,V,'k');
        axis([-11-11])
        xlabel('x')
        ylabel('y')
        ax = gca;
        ax.FontSize = 11;
        ax.FontName = 'Arial';
        drawnow;
end
```

Code 3.2 (*continued*)

The the corresponding Lagrangian equations are

$$\frac{\partial X}{\partial t} = \frac{Yt}{T_0^2}$$
$$\frac{\partial Y}{\partial t} = V_0 \tag{3.5.17}$$

We cannot solve the equation for $X(\underline{X}_0, t)$ until we know $Y(\underline{X}_0, t)$ since Y appears in the X equation. It is a common error for students to treat Y in this situation as a constant in the X equation here, so be cautious.

Solving for Y, first, integrate and impose the initial condition, $Y(t = 0) = Y_0$, to find

$$Y = V_0 t + Y_0 \tag{3.5.18}$$

Inserting Eq. (3.5.18) into Eq. (3.5.17) for X gives us

$$\frac{\partial X}{\partial t} = (V_0 t + Y_0)\frac{t}{T_0^2} \tag{3.5.19}$$

Integrating Eq. (3.5.19) and using the initial condition $X(t = 0) = X_0$ gives us the answer

$$X = \frac{1}{T_0^2}\left(V_0\frac{t^3}{3} + \frac{Y_0 t^2}{2}\right) + X_0 \tag{3.5.20}$$

Equations (3.5.20) and (3.5.18) are the mapping for this flow.

If we scale time on T_0 and lengths on $V_0 T_0$, such that

$$\xi = \frac{X}{V_0 T_0},\ \eta = \frac{Y}{V_0 T_0}, \xi_0 = \frac{X_0}{V_0 T_0}, \eta_0 = \frac{Y_0}{V_0 T_0}, T = \frac{t}{T_0} \tag{3.5.21}$$

then the pathlines are defined by the equations

$$\xi = \frac{T^3}{3} + \frac{\eta_0 T^2}{2} + \xi_0,\quad \eta = T + \eta_0 \tag{3.5.22}$$

which are parameterized by dimensionless time, T. Eliminating T between them yields the pathlines directly

$$\xi = \frac{(\eta - \eta_0)^3}{3} + \frac{\eta_0(\eta - \eta_0)^2}{2} + \xi_0 \tag{3.5.23}$$

Plotting ξ vs η gives us four trajectories for four initial positions, $(\xi_0, \eta_0) = (1, 1)$, $(-1, 1), (1, -1), (-1, -1)$, as seen in Figure 3.13.

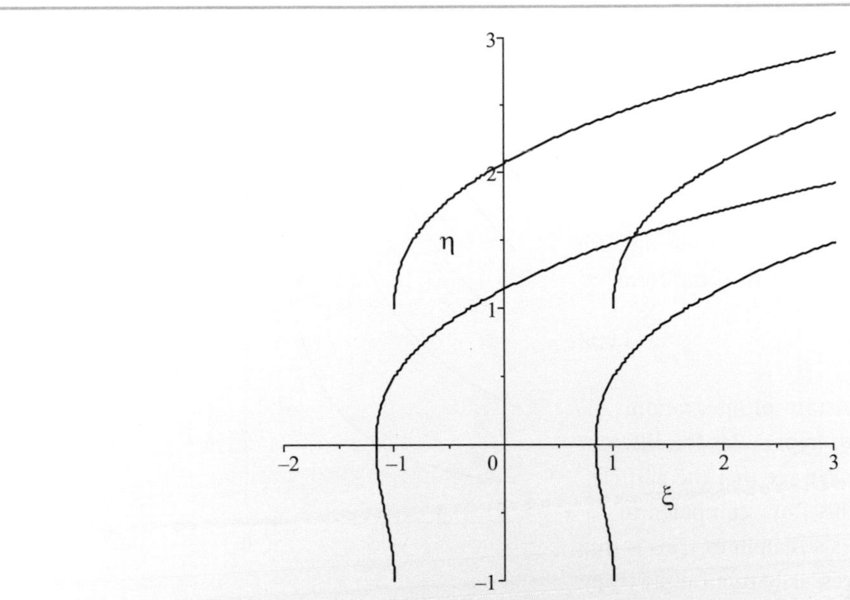

Figure 3.13 Pathlines for four initial positions in an unsteady flow: $(\xi_0, \eta_0) = (1, 1), (-1, 1), (1, -1), (-1, -1)$.

3.6 Streamlines

A streamline is defined at an instant of time as a line everywhere parallel to the Eulerian velocity field. Let the line element along a streamline be designated as $\underline{dx}_s$, see Figure 3.14. The mathematical statement that $\underline{dx}_s$ is parallel to $\underline{u}$ is that their vector cross product is zero,

$$\underline{dx}_s \times \underline{u} = \det \begin{vmatrix} \underline{i} & \underline{j} & \underline{k} \\ dx_s & dy_s & dz_s \\ u & v & w \end{vmatrix} = \underline{0} \tag{3.6.1}$$

In a two-dimensional flow, $\underline{u} = u\underline{i} + v\underline{j}$, and $\underline{dx}_s = dx_s\underline{i} + dy_s\underline{j}$, where the unit vectors in the x-, y-directions are $\underline{i}, \underline{j}$ respectively. The $\underline{k}$ component of Eq. (3.6.1) becomes

$$v\,dx_s - u\,dy_s = 0 \tag{3.6.2}$$

or equivalently

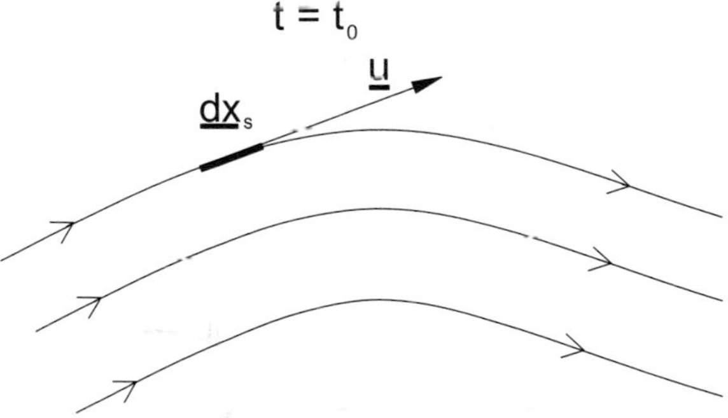

Figure 3.14 Streamlines, velocity vector and differential line segment at a fixed value of time, $t = t_0$.

$$\frac{dx_s}{u} = \frac{dy_s}{v} \tag{3.6.3}$$

For example, stagnation point flow has the Eulerian velocity components given in Eq. (3.5.1).

Inserting these into Eq. (3.6.3) results in

$$\frac{dy_s}{y_s} = -\frac{dx_s}{x_s} \tag{3.6.4}$$

Equation (3.6.4) may be integrated to yield

$$\ln y_s = \ln\left(\frac{1}{x_s}\right) + c_0 \tag{3.6.5}$$

where c_0 is the constant of integration. Inserting both sides as the exponent of e, leads to the final form,

$$y_s = \frac{c}{x_s} \tag{3.6.6}$$

where $c = e^{c_0}$ is the new constant of integration. Equation (3.6.6) is a family of hyperbolas for different values of c, see Figure 3.15. We see that the particle trajectories or pathlines of this flow, compared to Figure 3.8, are the same as its streamlines. This is true for steady flows, but not necessarily true for unsteady flows, as we now demonstrate.

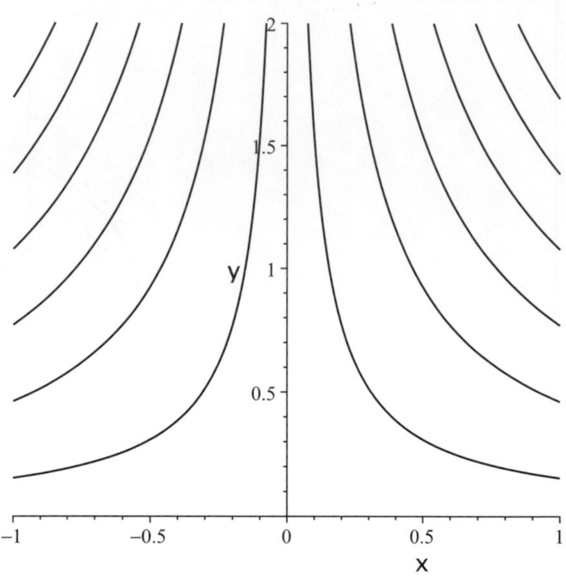

Figure 3.15 Streamlines for stagnation point flow.

Worked Example 3.4 | Streamlines in Unsteady Eulerian Velocity Field

Using the unsteady Eulerian velocity field of Eq. (3.5.16), $u = yt/T_0^2, v = V_0$, the streamlines are defined from Eq. (3.6.3). Using similar scales as in Eqs. (3.5.21), the definition of the streamlines is

$$\frac{dx_s}{u} = \frac{dy_s}{v} \quad \rightarrow \quad \frac{d\xi_s}{\eta_s T} = d\eta_s \tag{3.6.7}$$

We see right away that a value of T needs to be imposed so we can solve for the streamline, reinforcing the concept that a streamline is defined as a line everywhere parallel to the velocity field at an instant of time. So carrying T along as a constant, we solve Eq. (3.6.7) by separating variables to find

$$\xi_s = T\frac{\eta_s^2}{2} + c_0 \tag{3.6.8}$$

The equation for the streamline at an instant of time, $T = 1$ for example, and then different choices of c_0 yields a family of parabolas, quite different from the particle paths, Eq. (3.5.20), solved earlier. **This is generally true for unsteady flows, that the particle paths and the streamlines are different.** If we choose $c_0 = 1/2$ and $c_0 = -3/2$, the two streamlines pass through the initial positions of the pathlines as shown in Figure 3.16.

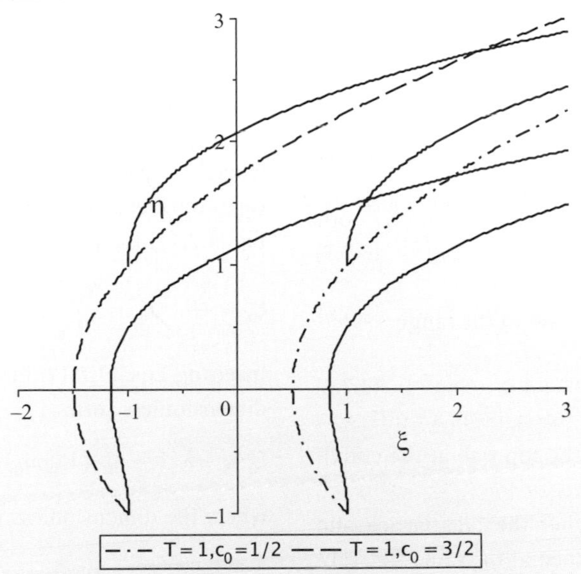

Figure 3.16 Two streamlines at T = 1 selected to intersect the pathlines at their initial positions. Pathlines and streamlines are different, in general, for unsteady flows.

3.7 Streaklines

In addition to streamlines and fluid particle pathlines (trajectories, mapping) is the line created by tagging fluid particles called a "streakline." Suppose we have flow with a parabolic profile in a two-dimensional channel (Hagen–Poiseuille flow), whose width is 2b, see Figure 3.17. The parabolic fluid velocity u(y) has a maximum value at $y = 0$ indicated as U_m, while the tagger starts at the lower wall with constant velocity V_T.

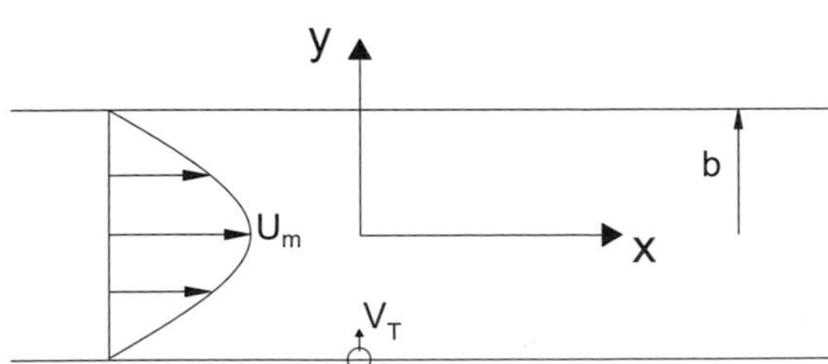

Figure 3.17 Hagen–Poiseuille flow in a channel with a tagger starting on the lower wall.

The corresponding Eulerian velocity field is

$$u(x,y) = U_m\left(1 - \left(\frac{y}{b}\right)^2\right) \qquad (3.7.1)$$
$$v(x,y) = 0$$

which is the familiar parabolic profile for u. U_m is the maximum velocity, which occurs at $y = 0$, and b is the channel half-width. The mapping for this flow is $\underline{X}(\underline{X}_0, t)$, and we find the solution by substituting $\underline{x} = \underline{X}, \underline{u} = \underline{U} = (\partial \underline{X}/\partial t)_{\underline{X}_0}$ as follows

$$\frac{\partial X}{\partial t} = U_m\left(1 - \left(\frac{Y}{b}\right)^2\right) \qquad (3.7.2)$$
$$\frac{\partial Y}{\partial t} = 0$$

First we need to solve the Y equation, with the initial condition that $Y(t = 0) = Y_0$. The result is

$$Y = Y_0 \qquad (3.7.3)$$

Now we can substitute for Y in the X equation and integrate, imposing that $X(t = 0) = X_0$, and the result is

$$X = U_m\left(1-\left(\frac{Y_0}{b}\right)^2\right)t + X_0 \tag{3.7.4}$$

Equations (3.7.3) and (3.7.4) constitute the mapping of the fluid, $\underline{X}(\underline{X}_0, t)$. Let the tagger move vertically through the fluid as a function of the time variable "s." For a constant speed, V_T, the position of the tagger is

$$\begin{aligned} x_T &= 0 \\ y_T &= -b + V_T s \end{aligned} \tag{3.7.5}$$

and we consider the s-time to be in the range

$$0 \le s \le \frac{2b}{V_T} \tag{3.7.6}$$

so that the tagger arrives at the top wall at the end of the s-time interval.

A fluid particle is tagged when the fluid particle and the tagger are at the same place at the same time

$$\begin{aligned} X &= x_T \\ Y &= y_T \\ t &= s \end{aligned} \tag{3.7.7}$$

The corresponding equations are

$$\begin{aligned} &U_m\left(1-\left(\frac{Y_0^*}{b}\right)^2\right)s + X_0^* = 0 \\ &Y_0^* = -b + V_T s \end{aligned} \tag{3.7.8}$$

where the * indicates that these are the special set of initial positions of the fluid particles which get tagged. We can substitute for Y_0^* in the first equation and solve for these initial positions, i.e. the inverse mapping

$$\begin{aligned} X_0^* &= -U_m\left(1-\left(\frac{-b+V_T s}{b}\right)^2\right)s \\ Y_0^* &= -b + V_T s \end{aligned} \tag{3.7.9}$$

Equations (3.7.9) describe a curve, parameterized by s, formed by the initial positions of the tagged fluid particles. Let's eliminate s between the two equations by solving for s in the second equation, $s = (Y_0^* + b)/V_T$ and substituting into the first

$$X_0^* = -\frac{U_m}{V_T}\left(1-\left(\frac{Y_0^*}{b}\right)^2\right)(Y_0^* + b) \tag{3.7.10}$$

Let's make this analysis dimensionless. An obvious length scale is the channel half-width, b, and a time scale can be the transit time of the tagger across the distance b. So define the following scaled variables

$$\xi = \frac{X}{b}, \quad \eta = \frac{Y}{b}, \quad S = \frac{sV_T}{b} \tag{3.7.11}$$

where ξ is the lower case Greek xi and η is the lower case Greek eta, see Figure 2.9. With the definitions in Eq. (3.7.11) the tagged initial positions are

$$\xi_0 = \frac{X_0^*}{b}, \quad \eta_0 = \frac{Y_0^*}{b} \tag{3.7.12}$$

Inserting Eqs. (3.7.12) into Eq. (3.7.10) results in the dimensionless form

$$\xi_0 = -\lambda(1-\eta_0^2)(1+\eta_0) \tag{3.7.13}$$

where the dimensionless parameter is

$$\lambda = U_m/V_T \tag{3.7.14}$$

which is the ratio of the maximal fluid speed in the x-direction to the tagger speed in the y-direction. Some plots of the results are shown in Figure 3.18 for three values of λ. As λ increases, the initial positions of the fluid particles that become tagged start farther upstream since the relative fluid speed is higher. λ is the Greek lower case lambda.

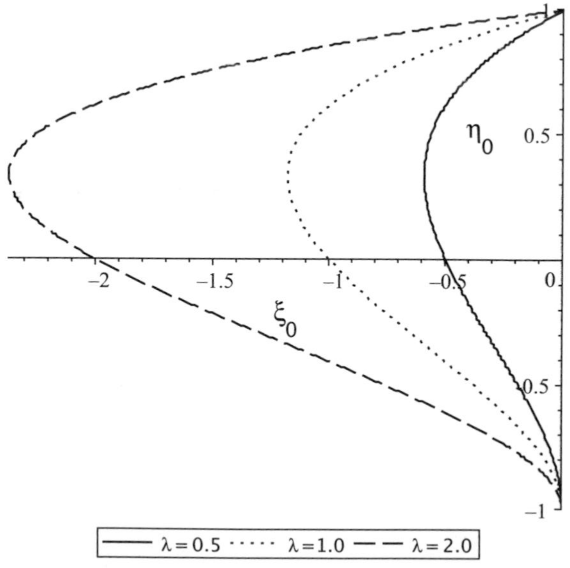

Figure 3.18 Initial positions of the fluid particles that will be tagged.

We would like to know where these tagged initial positions end up when the tagger just reaches the top. So substitute them into the mapping

$$X = U_m\left(1 - \left(\frac{-b + V_T s}{b}\right)^2\right)(t - s) \qquad Y = -b + V_T s \tag{3.7.15}$$

and let $t = 2b/V_T$. The result is

$$X = U_m\left(1 - \left(\frac{-b + V_T s}{b}\right)^2\right)\left(\frac{2b}{V_T} - s\right) \qquad Y = -b + V_T s \qquad 0 \le s \le \frac{2b}{V_T} \tag{3.7.16}$$

and this is the curve we seek, parameterized by s. We can make Eqs. (3.7.16) dimensionless by using the scales in Eq. (3.7.11), which gives us

$$\xi = \lambda(1 - (-1 + S)^2)(2 - S), \quad \lambda = \frac{U_m}{V_T} \qquad \eta = -1 + S \qquad 0 \le S \le 2 \tag{3.7.17}$$

In this case, we can eliminate S between the ξ (S) and η(S) in Eqs. (3.7.17), to obtain

$$\xi = \lambda(1 - \eta^2)(1 - \eta) \tag{3.7.18}$$

Figure 3.19 shows plots of the streakline for $\lambda = 0.5, 1.0, 2.0$. As λ increases, the streakline shape stretches farther downstream since the characteristic fluid speed is larger compared to the tagger speed. Note that the maximum x-displacement of the streaklines is well below the midline of $y = 0$, since the fluid there has had a longer time to flow before the tagger reaches the upper wall.

3.8 Timelines

Another fluid line is called a "timeline," and it is a set of initial positions which are tagged at the same time and then followed in the flow via the mapping. Suppose we have Poiseuille flow in a two-dimensional channel, whose width is 2b, as in Figure 3.17. We found the mapping for this flow as shown in Eqs. (3.7.3) and (3.7.4),

$$X = U_m\left(1 - \left(\frac{Y_0}{b}\right)^2\right)t + X_0, \quad Y = Y_0 \tag{3.8.1}$$

Suppose we create a timeline across the channel at $x = 0$ when $t = 0$. This is shown in Figure 3.20 as a

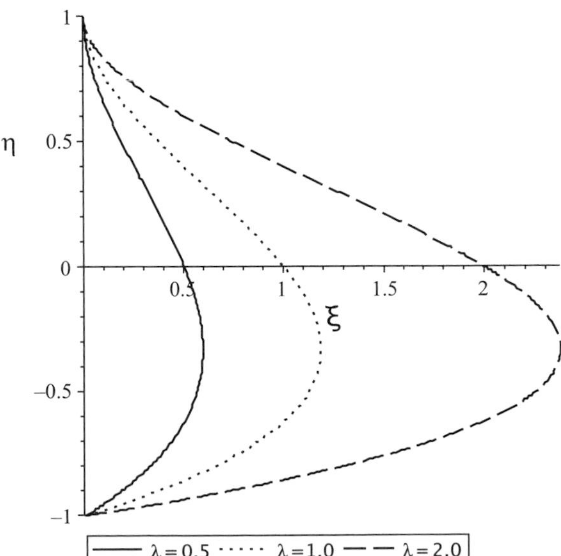

Figure 3.19 Streaklines for $\lambda = 0.5, 1.0, 2.0$.

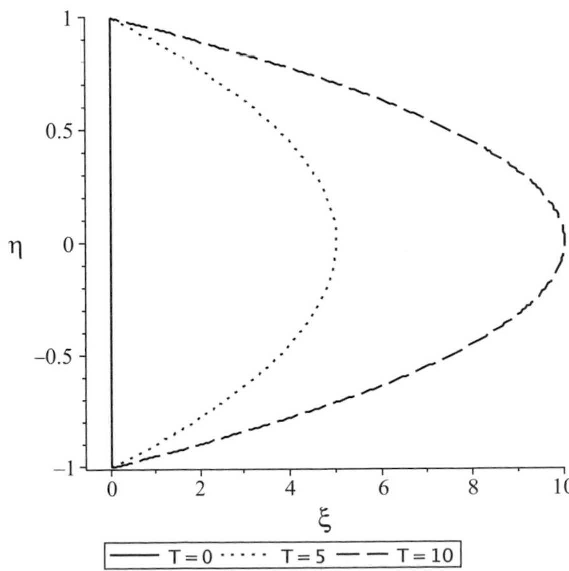

Figure 3.20 Sequence of timelines for Hagen–Poiseuille flow.

solid line. These initial positions can be parameterized using the dummy variable, h, as follows

$$\begin{aligned} X_0 &= 0 \\ Y_0 &= hb \end{aligned} \tag{3.8.2}$$

where the limits of h are $-1 \leq h \leq 1$. The mapping of the timeline using these initial positions is

$$\begin{aligned} X &= U_m\left(1 - \left(\frac{hb}{b}\right)^2\right)t \\ Y &= hb \end{aligned} \tag{3.8.3}$$

We can make this mapping non-dimensional with the definitions

$$\xi = \frac{X}{b}, \quad \eta = \frac{Y}{b}, \quad T = \frac{U_m t}{b} \tag{3.8.4}$$

Using these scales, the mapping becomes

$$\begin{aligned} \xi &= (1 - h^2)T \\ \eta &= h \end{aligned} \tag{3.8.5}$$

Some examples of the timeline are shown for three values of time in Figure 3.20. Note that the timeline has its maximal displacement in the center of the channel.

3.9 Aerosol Particle Deposition in the Lung

Consider the inviscid stagnation point flow and its Eulerian velocity field of Eqs. (3.5.1). This flow is a two-dimensional approximation to the local stagnation region of flow at a bifurcation that occurs in the lung and the vasculature, as shown in Figure 3.7. For the lung, inhaled aerosol particles may be medications, toxins, ambient dust, products of combustion or smoking, carriers of infective agents like coronavirus, etc. Some of these particles can cause cancer, and the occurrence of pulmonary tumors from such exposures tend to dominate at the airway bifurcations where the particles preferentially impact due to the stagnation flow (Balashazy et al., 2003). Understanding the deposition of particles in this flow is a first step in developing a model of aerosol deposition in the lung.

The equations of motion for a particle, ignoring gravity, are

$$\begin{aligned} m\frac{d^2x_p}{dt^2} + 3\pi\mu d_p\left(\frac{dx_p}{dt} - u\left(\underline{x} = \underline{x}_p\right)\right) &= 0 \\ m\frac{d^2y_p}{dt^2} + 3\pi\mu d_p\left(\frac{dy_p}{dt} - v\left(\underline{x} = \underline{x}_p\right)\right) &= 0 \end{aligned} \tag{3.9.1}$$

using Stokes drag from Section 2.7 for the particle velocity relative to the fluid velocity; d_p is the particle diameter and μ is the fluid viscosity. Substituting $u = Ax$, $v = -Ay$ for $x = x_p$, $y = y_p$ yields

$$\begin{aligned} m\frac{d^2x_p}{dt^2} + 3\pi\mu d_p\left(\frac{dx_p}{dt} - Ax_p\right) &= 0 \\ m\frac{d^2y_p}{dt^2} + 3\pi\mu d_p\left(\frac{dy_p}{dt} + Ay_p\right) &= 0 \end{aligned} \tag{3.9.2}$$

We note that this system of ODEs is uncoupled, that is, x_p only appears in one equation and y_p only in the other. Initial conditions for the particle will be its starting position and its starting velocity. For simplicity, we can assume that the aerosol particle starts at the dimensional position $x_p(0) = L, y_p(0) = L$, and its initial velocity is the same as the local fluid velocity there, which is $u(x = L) = AL$, $v(y = L) = -AL$, for example. We have, then, the initial conditions

$$x_p(0) = L, \quad \frac{dx_p}{dt}(0) = AL, \quad y_p(0) = L, \quad \frac{dy_p}{dt}(0) = -AL \tag{3.9.3}$$

Choosing scales as follows

$$X_p = \frac{x_p}{L}, \quad Y_p = \frac{y_p}{L}, \quad T = At \tag{3.9.4}$$

we make the governing equation dimensionless by inserting Eqs. (3.9.4) into Eqs. (3.9.2)

$$\text{Stk}\frac{d^2X_p}{dT^2}+\frac{dX_p}{dT}-X_p=0$$
$$\text{Stk}\frac{d^2Y_p}{dT^2}+\frac{dY_p}{dT}+Y_p=0 \tag{3.9.5}$$

where Stk is the dimensionless Stokes number, which is the ratio of inertial force to viscous drag force

$$\text{Stk}=\frac{mA}{3\pi\mu d_p}=\frac{\rho_p d_p^2 A}{18\mu} \tag{3.9.6}$$

In Eq. (3.9.6) we have replaced the particle mass with the product of its density, ρ_f and its volume $m=\rho_p(4/3)\pi(d_p/2)^3$. The dimensionless versions of the initial conditions of Eqs. (3.9.3) are

$$X_p(0)=1,\quad \frac{dX_p}{dT}(0)=1,\quad Y_p(0)=1,\quad \frac{dY_p}{dT}(0)=-1 \tag{3.9.7}$$

Since impact occurs when $Y_p=0$, and Eqs. (3.9.5) are uncoupled, we can examine that component first. The governing equation for Y_p in Eqs. (3.9.5) is a linear ODE with constant coefficients. So, for example, we propose that $Y_p\sim e^{rT}$, insert into the second of Eqs. (3.9.5) and find the characteristic equation for r as

$$r=\frac{-1\pm(1-4\text{Stk})^{1/2}}{2\text{Stk}}=r^+,r^- \tag{3.9.8}$$

so that

$$Y_p=c_1e^{r^+T}+c_2e^{r^-T} \tag{3.9.9}$$

The constants c_1 and c_2 are solved from the initial conditions Eqs. (3.9.7) and the resulting form is

$$Y_p=-\left(\frac{1+r^-}{r^+-r^-}\right)e^{r^+T}+\left(\frac{1+r^+}{r^+-r^-}\right)e^{r^-T} \tag{3.9.10}$$

We are interested in particle impact (i.e. $Y_p=0$) within a finite time. The Y_p solution gives us some important information about the criteria for impact. The roots of r as given in Eq. (3.9.8) yield purely exponential decay with time if Stk < 1/4. Therefore, the particle cannot impact the wall in finite time. However, if Stk > 1/4, then the term in the square root is negative so a complex solution is obtained with real and imaginary parts

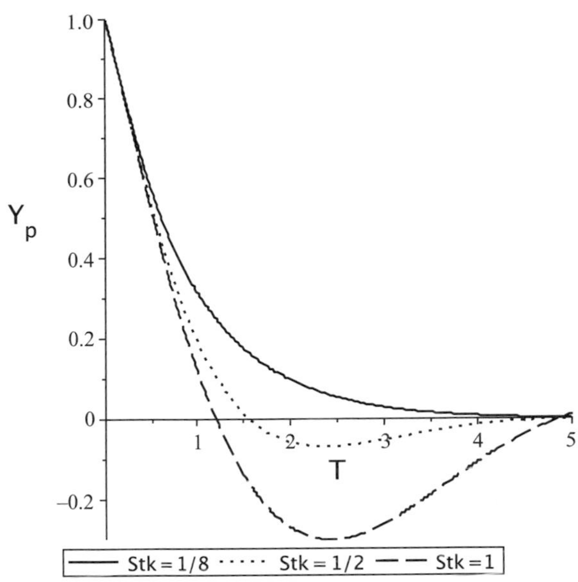

Figure 3.21 $Y_p(T)$ for three values of the Stokes number, Stk = 1/8, 1/2, 1. Note that impact occurs when $Y_p=0$ for the first time.

$$r=\frac{-1}{2\text{Stk}}\pm\frac{i(4\text{Stk}-1)^{1/2}}{2\text{Stk}},\quad \text{Stk}>\frac{1}{4} \tag{3.9.11}$$

For Stk > 1/4, $Y_p(T)$ is a decaying sinusoidal wave, since the equivalent form of the complex exponential term is

$$\begin{aligned}e^{rT}&=e^{-\frac{1}{2\text{Stk}}T}e^{\pm i\frac{(4\text{Stk}-1)^{1/2}}{2\text{Stk}}T}\\&=e^{-\frac{1}{2\text{Stk}}T}\left(\cos\left(\frac{(4\text{Stk}-1)^{1/2}}{2\text{Stk}}T\right)\pm i\sin\left(\frac{(4\text{Stk}-1)^{1/2}}{2\text{Stk}}T\right)\right)\end{aligned} \tag{3.9.12}$$

Impact occurs at the first $Y_p=0$ crossing. A sample plot is shown in Figure 3.21 for three values of

Stk = 1/8, 1/2, 1. For Stk = 1/8 there is no impact in finite time since it is smaller than 1/4. The other two values of Stk are greater than 1/4 and lead to impact as shown.

Using a similar approach, the solution for X_p is

$$X_p = \left(\frac{1-q^-}{q^+-q^-}\right)e^{q^+T} - \left(\frac{1-q^+}{q^+-q^-}\right)e^{q^-T} \tag{3.9.13}$$

where q is given by

$$q = \frac{-1 \pm (1+4\mathrm{Stk})^{1/2}}{2\mathrm{Stk}} = q^+, q^- \tag{3.9.14}$$

Note that the term inside the square root is always positive and greater than unity, unlike the Y_p case. Plotting some solutions for the aerosol particle trajectory (X_p, Y_p), we see in Figure 3.22 how they differ from the fluid streamlines, crossing them more readily as the relative particle inertia, Stk, is increased.

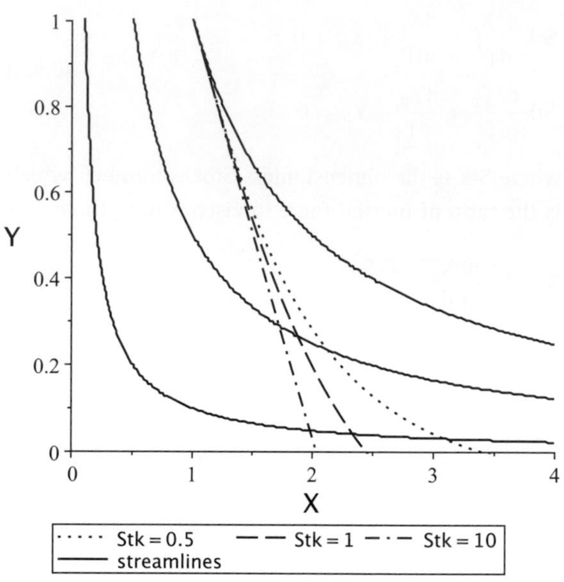

Figure 3.22 Aerosol particle trajectories and impact in stagnation point flow for three values of the Stokes number, Stk, and $(X_p(0), Y_p(0)) = (1, 1)$. Sample streamlines are also shown.

Worked Example 3.5 Using Computational Software

Let's solve the ODE system above for the aerosol deposition problem using computational software. Since it is linear with constant coefficients, one could expect an analytic solution as we have found by hand. However, computational software is capable of finding analytic solutions as well. This is sometimes easier and less fraught with error. An example of a Maple code which solves and plots aerosol trajectories is shown in Code 3.3(a). The output is shown in Figure 3.23. The corresponding MATLAB® code is shown in Code 3.3(b).

```
(a)
restart;
#first we define the system of 2 ODEs already in dimensionless form
odesys := Stk.diff(Xp(T),T,T) + diff(Xp(T),T) - Xp(T), Stk.diff(Yp(T),T,T) + diff(Yp(T),T)
    +Yp(T)
```

$$odesys := Stk\left(\frac{d^2}{dT^2}Xp(T)\right) + \frac{d}{dT}Xp(T) - Xp(T),\ Stk\left(\frac{d^2}{dT^2}Yp(T)\right) + \frac{d}{dT}Yp(T) + Yp(T) \quad (1)$$

```
#and then the initial conditions in appropriate format
ics := Xp(0) = 1, Yp(0) = 1, D(Xp)(0) = 1, D(Yp)(0)=-1;
```

$$ics := Xp(0) = 1,\ Yp(0) = 1,\ D(Xp)(0) = 1,\ D(Yp)(0) = -1 \quad (2)$$

Code 3.3 Code examples: (a) Maple, (b) MATLAB®.

```
#Use "dsolve" to find the analytic solution and assign it
sol := dsolve({odesys, ics}):
assign(sol):
#Check the solution
Xp(T);
```

$$\frac{\left(2\sqrt{1+4\ Stk}\ Stk+\sqrt{1+4\ Stk}+4\ Stk+1\right)\,e^{\frac{(-1+\sqrt{1+4\ Stk})\,T}{2\ Stk}}}{2\ (1+4\ Stk)} - \frac{\left(2\ \sqrt{1+4\ Stk}\ Stk-1+\sqrt{1+4\ Stk}-4\ Stk\right)\,e^{-\frac{(1+\sqrt{1+4\ Stk})\,T}{2\ Stk}}}{2\ (1+4\ Stk)} \qquad (3)$$

```
Yp(T);
```

$$\frac{\left(2\ \sqrt{1-4\ Stk}\ Stk-\sqrt{1-4\ Stk}+4\ Stk-1\right)\,e^{\frac{(-1+\sqrt{1-4\ Stk})\,T}{2\ Stk}}}{2\ (-1+4\ Stk)} - \frac{\left(2\sqrt{1-4\ Stk}\ Stk+1-\sqrt{1-4\ Stk}-4\ Stk\right)e^{\frac{(1+\sqrt{1-4\ Stk})\,T}{2\ Stk}}}{2\ (-1+4\ Stk)} \qquad (4)$$

```
#Call "with(pots)" to allow use of "display" for plotting multiple curves
with(plots):
#Choose Stk=0.1 which does not impact the wall
PL1 := plot([subs(Stk=0.1,Xp(T)),subs(Stk = 0.1, Yp(T)), T = 0..1.5],linestyle=2,legend
     = "Stk=0.1"):

#Choose Stk=0.5 which impact the wall at Tf2 where Yp(Tf2)=0. Slove for Tf2 using "fsolve" and limit
   the solution to  the range 0≤Tf2≤5
Tf2 := Re(fsolve(subs(Stk = 0.5, Yp(T)) = 0,T =0..3));
                                            Tf2 := 1.570796327                                    (5)

PL2 := plot([subs(Stk = 0.5, Xp(T)),subs(Stk=0.5,Yp(T)), T=0..Tf2], linestyle=3, legend
     = "Stk=0.5"):

#Choose Stk=1
Tf3 := fsolve((subs(Stk=1, Yp(T)), T = 0..5));
                                            Tf3 := 1.209199576                                    (6)

PL3 := plot([subs(Stk=1, Xp(T)),subs(Stk=1,Fr=0.2,Yp(T)), T=0..Tf3], linestyle=, legend
     = "Stk=1"):
#Finally, for comparison consider the streamline passing through (1,1). Recall it is also the fluid particle
    pathline
```

$$PL4 := plot\left(\frac{1}{x}\ x=1..4,\ linestyle=1,\ legend="streamline"\right):$$

```
#Display all 4 plots on one graph
display(PL1, PL2, PL3, PL4, labels=[X_p, Y_p]);
```

Code 3.3 (*continued*)

```
(b)
Clear all, close all, clc
syms x(t) y(t) Stk
Dx =diff(x,1);
Dy =diff(y,1);
odex=Stk*diff(x,2) + diff(x,1)- x== 0;
ic1  = x(0) == 1;
ic2  = Dx(0) == 1;
icsx = [ic1,ic2];
x    = dsolve(odex,icsx);

odey = Stk*diff(y,2) + diff(y,1) + y == 0;
ic3  = y(0)   == 1;
ic4  = Dy(0) ==-1;
icsy = [ic3, ic4];
y    = dsolve(odey,icsy);
Y    = matlabFunction(y);
t    = linspace(0,5,1000);
Stk  = 0.1;
plot(eval(x),eval(y),':k'), hold on;
Stk  = 0.5;
Yf   = @(t) Y(Stk,t);
Tf2  = fzero(Yf,1)
plot(eval(x),eval(y),'--k'), hold on;
Stk  = 1;
plot(eval(x),eval(y),'-.k')
xStr = linspace(1,4,1000);
plot(xStr,1./xStr,'-k')
axis([0 4 0 1])
legend('Stk=0.1','Stk=0.5','Stk=1','streamline','location','northeast')
xlabel('X_p')
ylabel('Y_p')
```

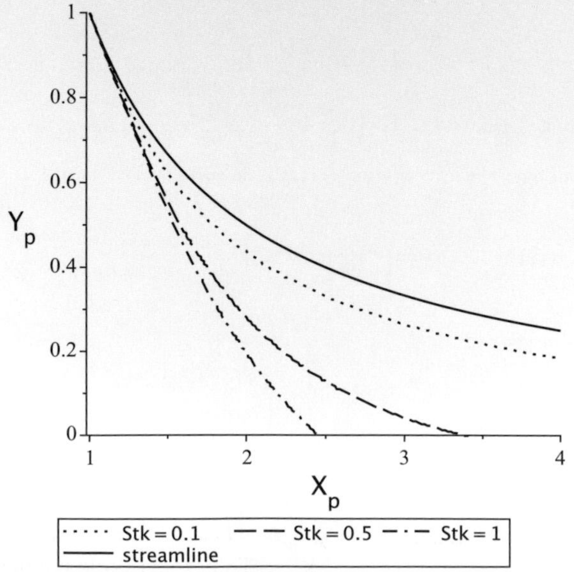

Figure 3.23 Aerosol particle deposition in stagnation point flow for three values of Stk $= 0.1, 0.5, 1.0$ and the streamline passing through (1, 1).

3.10 Numerical Methods

Many students are not quite sure what the term "numerical methods" means, since it is often new territory for them. In this section we take more time to discuss numerical methods in a broad sense. Nearly everyone has had an introduction to numerical methods when studying calculus. For example, a numerical approximation to the integral of $f(x)$ on the interval $a \leq x \leq b$,

$$\int_a^b f(x)\,dx \tag{3.10.1}$$

was likely explored using the Trapezoidal Rule or Simpson's Rule. To approximate the area under $f(x)$ using the Trapezoidal Rule, we first divide the x-interval into n equal segments of length Δx where $\Delta x = (b - a)/n$.

In Figure 3.24 a function, $f(x)$, is shown with the x-axis divided into five equal segments, $n = 5$, forming five trapezoids. The continuous x variable is now represented by a finite number of discrete values given by $x_j = x_1 + (j - 1)\Delta x$ where the integer j is in the range $1 \leq j \leq n + 1$ and $x_1 = a$, $x_{n+1} = x_6 = b$. The area of the first trapezoid, shown in Figure 3.24, is

$$\Delta x\left[\frac{f(x_1 = a) + f(x_2 = x_1 + \Delta x)}{2}\right] \tag{3.10.2}$$

Adding to that the area of the other trapezoids yields an approximation to the total area under $f(x)$

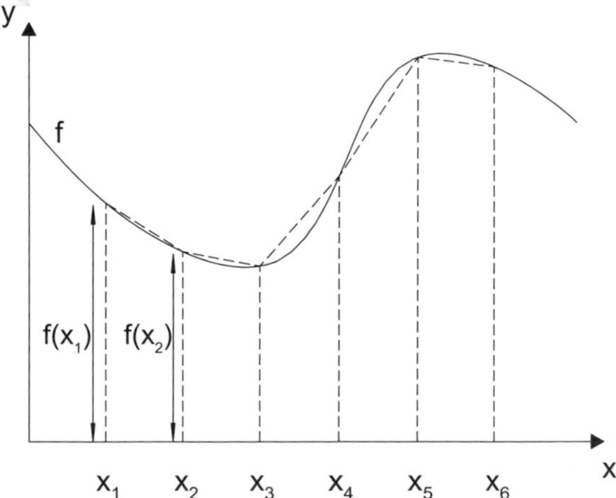

Figure 3.24 Example of the Trapezoidal Rule.

$$\begin{aligned}\text{area} &= \Delta x\left[\left[\frac{f(x_1)+f(x_2)}{2}\right]+\left[\frac{f(x_2)+f(x_3)}{2}\right]+\cdots\right.\\ &\quad\left.+\left[\frac{f(x_n)+f(x_{n+1})}{2}\right]\right]\\ &= \frac{\Delta x}{2}[f(x_1)+2f(x_2)+2f(x_3)+\cdots+2f(x_n)\\ &\quad+f(x_{n+1})]\end{aligned} \tag{3.10.3}$$

As a specific example, let $f(x) = \sin x$ and integrate over the interval $a = 0, b = \pi$ for $n = 4$ so that $\Delta x = \pi/4$. We will need to calculate $f(x)$ at the five values of $j = 1, 2, 3, 4, 5$, which are $f(x_j) = \sin((j - 1)\Delta x)$. Substituting those values of $f(x_j)$ into Eq. (3.10.3) gives us an approximation to the area

$$\begin{aligned}\text{area} &= \frac{\pi}{8}[0 + 2(0.707) + 2(1.000) + 2(0.707) + 0]\\ &= 1.896\end{aligned} \tag{3.10.4}$$

The analytic, or closed form, solution is

$$\int_0^\pi \sin x\,dx = -(\cos\pi - \cos 0) = 2 \tag{3.10.5}$$

Analytic solutions are ones that can be written in terms of known functions, in this case the cosine function. Our numerical solution for the area, 1.896, represents approximately a 5% error. Increasing the number of steps to $n = 100$, for example, yields area = 1.9998, which is a 0.01% error. We point out that it is more typical in discussions of numerical methods to start the index at $j = 0$ rather than $j = 1$. However, some computational software programs do not allow indexing of vectors to include 0.

If our interest were to calculate the integral of $f(x)$ to a variable upper limit, x', we would have the form

$$g(x') = \int_0^{x'} \sin x\,dx \tag{3.10.6}$$

where $g(x')$ is, itself, a function. Using our Trapezoidal Rule method for $n = 4$, we would keep a running total of the trapezoid areas as we add each subsequent segment

$$
\begin{aligned}
g(x' = x_1) &= g_1 = 0.000 \\
g(x' = x_2) &= g_2 = \frac{\Delta x}{2}[f(x_1) + f(x_2)] = 0.278 \\
g(x' = x_3) &= g_3 = \frac{\Delta x}{2}[f(x_1)+2f(x_2)+ f(x_3)]= 0.948 \\
g(x' = x_4) &= g_4 = \frac{\Delta x}{2}[f(x_1)+2f(x_2)+2f(x_3)+ f(x_4)] \\
&= 1.618 \\
g(x' = x_5) &= g_5 = \frac{\Delta x}{2}[f(x_1) + 2f(x_2) + 2f(x_3) \\
&\quad + 2f(x_4) + f(x_5)] = 1.896
\end{aligned}
\tag{3.10.7}
$$

where we note that last entry, 1.896, is the total area already discussed. Now the numerical solution to Eq. (3.10.6) can be represented by two, paired, 1×5 vectors

$$
\begin{aligned}
(x_i') &= (0,\ 0.785, 1.571, 2.356, 3.142) \\
(g_i) &= (0.000, 0.2777, 0.9481, 1.6184, 1.8961)
\end{aligned}
\tag{3.10.8}
$$

or, equivalently, one 2×5 matrix where the first column is (x_i') and the second column is (g_i)

x_i'	g_i
0.000	0.000
0.785	0.278
1.571	0.948
2.356	1.618
3.142	1.896

(3.10.9)

The analytic solution is

$$
g(x') = \int_0^{x'} \sin x\, dx = 1 - \cos x' \tag{3.10.10}
$$

which are known functions. By comparison, the numerical solution is a list of numbers.

Figure 3.25 shows the comparison between the analytic solution to Eq. (3.10.6) and two numerical solutions, one for $n = 4$ using the results of Eq. (3.10.9) and one for $n = 20$. The $n = 20$ numerical solution is a much better fit to the analytic solution which is typical for smaller step sizes. Letting $n = 500$ soon makes it clear why computers are essential for numerical methods.

At this stage of their education it is far less pervasive for students to have any experience, or even awareness, that differential equations can be solved numerically. With the availability of computational software like MATLAB®, Maple, Mathematica, Python and others, however, students have this capability at their fingertips. The methods are much more complicated than approximating an integral like the Trapezoidal Rule, but the broad concept is similar, that numerical solutions give us lists of numbers as compared to analytic functions.

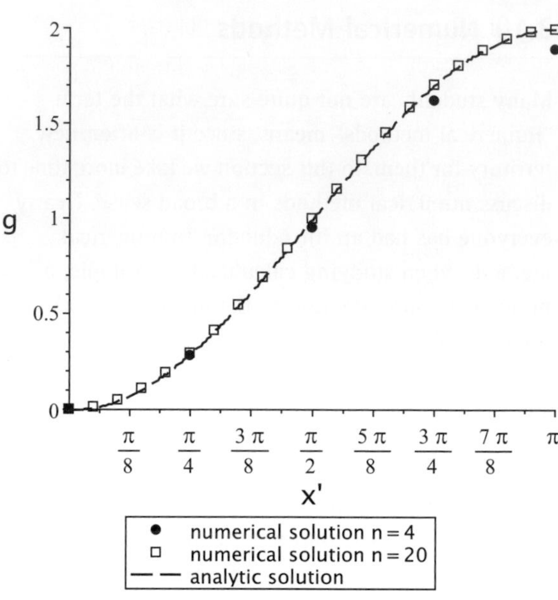

Figure 3.25 Comparison of the numerical solution for $n = 4$ and $n = 20$ to the analytic solution for the integral in Eq. (3.10.6).

This section is not meant to teach numerical methods for differential equations, or to review, exhaustively, the many popular strategies. Instead, it is to give a context so that the student has some bearings with using their software by examining a simple example.

Let's solve the following differential equation numerically

$$
\frac{dh}{dx} = -h, \quad h(x = 0) = 1 \tag{3.10.11}
$$

First, we know that the analytic solution to Eq. (3.10.11) is $h = e^{-x}$, which satisfies the initial condition $h(x = 0) = 1$. Of the many methods available, let's use the Euler Method to solve

Eq. (3.10.11) numerically on the interval $0 \leq x \leq 1$, since it is rather straightforward for this simple example.

Let's use the notation that $x_1 = 0$ and $h(x_1) = h_1 = 1$. We can extrapolate the value of $h_2(x_2 = x_1 + \Delta x)$, where the increment in x is $\Delta x = (1 - 0)/n$, by using the slope at x_1

$$h_2 = h_1 + \left.\frac{dh}{dx}\right|_{x_1} \Delta x \tag{3.10.12}$$

However, from Eq. (3.10.11) we know that $(dh/dx)_{x_1} = -h_1$, so Eq. (3.10.12) becomes

$$h_2 = h_1 - h_1 \Delta x \tag{3.10.13}$$

In general from Eq. (3.10.13) we see that

$$h_{i+1} - h_i - h_i \Delta x \tag{3.10.14}$$

where the subscript $1 \leq i \leq n$, where n is the number of equally spaced intervals. As an example, let $n = 5$, so $\Delta x = 0.2$. Now we can employ the recursive formula of Eq. (3.10.14) to solve the differential equation over the interval $0 \leq x \leq 1$

$$\begin{aligned}
h_1 &= 1 \\
h_2 &= h_1 - h_1 \Delta x = 1 - 1(0.2) = 0.800 \\
h_3 &= h_2 - h_2 \Delta x = 0.8 - 0.8(0.2) = 0.640 \\
h_4 &= h_3 - h_3 \Delta x = 0.64 - 0.64(0.2) = 0.512 \\
h_5 &= h_4 - h_4 \Delta x = 0.512 - 0.512(0.2) = 0.410 \\
h_6 &= h_5 - h_5 \Delta x = 0.410 - 0.410(0.2) = 0.328
\end{aligned} \tag{3.10.15}$$

As in the numerical integration example, we now have lists of numbers for our numerical solution to a differential equation, here in matrix form of paired values

x_i	h_i
0.0	1.000
0.2	0.800
0.4	0.640
0.6	0.512
0.8	0.410
1.0	0.328

(3.10.16)

The numerical solution from Eq. (3.10.16) is plotted along with the exact solution in Figure 3.26. The error in the numerical solution increases with x; however, we can do better by decreasing the step size, $\Delta x = 0.2$ to $\Delta x = 0.02$ by increasing n from 5 to 50. The $n = 50$ numerical solution is also shown in Figure 3.26 and provides a much better approximation to the analytic solution.

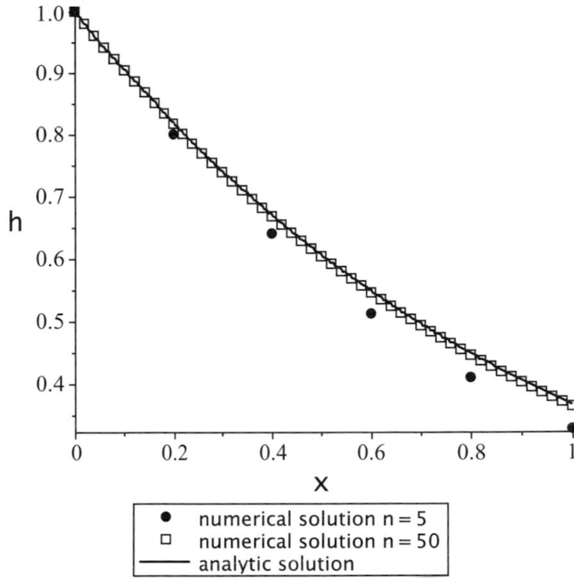

Figure 3.26 Comparison of analytic vs numerical solution for Eq. (3.10.11).

This example hopefully takes away some of the mystery for students, enough at least to allow them to try using their software. Whichever software is used, it will have help pages for numerical solutions to differential equations and systems of differential equations. While several named methods are available, often the default method is the Runge–Kutta Method which, itself, has various forms. It is a robust and accurate algorithm which is very popular, more so than the Euler Method. The ease of computational software is that it allows the user to interface with the numerical method, and employ or plot its output, without having to do much programming. Generally speaking, entering the equations and boundary or initial conditions in the particular required format gets you started. If values of the solution are needed at points in between the calculated points, then the software interpolates a solution for the user from the stored data.

Worked Example 3.6 | Numerical Solution of an ODE

As an example, consider a nonlinear ordinary differential equation which does not have an analytic solution

$$\frac{d^2x}{dt^2} + 3\frac{dx}{dt} - Bx^2 = 1 \quad x(0) = 1, \left.\frac{dx}{dt}\right|_{t=0} = 1 \tag{3.10.17}$$

subject to initial conditions as shown and with a parameter, B. The Maple code to solve x(t) numerically is shown in Code 3.4(a) with a plot of x(t) for $0 \le t \le 1$ for $B = 0.5, 1.0, 2.0$ and the equivalent MATLAB® code is in Code 3.4(b).

```
(a)
restart;
#Define the ODE with a parameter, B, and its initial conditions
ODE :=diff(x(t),t,t) + 3·x(t)·diff(x(t),t) -B·x(t)^2 = 0;
```

$$ODE := \frac{d^2}{dt^2}x(t) + 3\,x(t)\left(\frac{d}{dt}x(t)\right) - B\,x(t)^2 = 0 \tag{1}$$

```
ICS := x(0) = 1, D(x) (0) = 1;
                    ICS := x(0) = 1, D(x) (0) = 1                      (2)
#Choose 3 values of B and solve ODE for each one.
sol1 := dsolve({subs(B = 0.5, ODE), ICS}, type = numeric);
                    sol1 := proc(x_rkf45) ... end proc                 (3)
sol2 := dsolve({subs(B = 1, ODE), ICS}, type = numeric);
                    sol2 := proc(x_rkf45) ... end proc                 (4)
sol3 := dsolve({subs(B = 2, ODE), ICS}, type = numeric);
                    sol3 := proc(x_rkf45) ... end proc                 (5)
#Create 3 plots on a single graph for each value of B
with (plots) :
PL1 := odeplot(sol1, t = 0..1, linestyle = 1, legend = "B=0.5") :
PL2 := odeplot(sol2, t = 0..1, linestyle = 2, legend = "B=1.0") :
PL3 := odeplot(sol3, t = 0..1, linestyle = 3, legend = "B=2.0") :
display(PL1, PL2, PL3);
```

```
(b)
clear all; close all; clc
tspan    = [0 1];
x0       = 1;
Dx0      = 1;
ICS      = [x0 Dx0];
B        = 0.5;
[t, sol] = ode45(@(t, sol) Chapter3_odeEquation3_10_17(t, sol, B), tspan, ICS);
x        = sol(:, 1);
Dx       = sol(:, 2);
```

Code 3.4 Code examples: (a) Maple, (b) MATLAB®.

```
plot(t,x,'-k'), hold on;
B       = 1;
[t,sol] = ode45(@(t,sol) Chapter3_odeEquation3_10_17(t,sol,B),tspan,ICS);
x       = sol(:,1);
Dx      = sol(:,2);
plot(t,x,':k')
B       = 2;
[t,sol] = ode45(@(t,sol) Chapter3_odeEquation3_10_17(t,sol,B),tspan,ICS);
x       = sol(:,1);
Dx      = sol(:,2);
plot(t,x,'--k')
axis([0 1 1 2])
legend('B=0.5','B=1','B=2','location','northwest')
xlabel('t')
ylabel('x')
```

The following MATLAB function must be saved with the name "Chapter3_odeEquation3_10_17.m"

```
function dxdt = Chapter3_odeEquation3_10_17(t,x,B)
  dxdt = zeros(2,1);
  dxdt(1) = x(2);
  dxdt(2) = -3.*x(1).*x(2)+B.*x(1).^2;
end
```

Code 3.4 *(continued)*

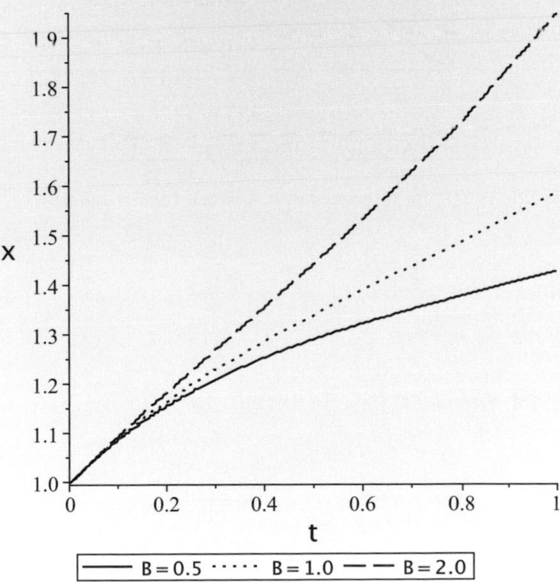

Figure 3.27 Numerical solution to sample ODE for three values of B = 0.5, 1.0, 2.0.

Now consider a system of coupled, nonlinear ODEs with a parameter A

$$A\frac{d^2X}{dT^2}+2\frac{dX}{dT}-XY=0$$
$$\frac{d^2Y}{dT^2}+A\frac{dY}{dT}-\frac{Y}{X}=0 \tag{3.10.18}$$

subject to the initial conditions

$$X(0)=1, Y(0)=1, \left.\frac{dX}{dT}\right|_{T=0}=0, \left.\frac{dY}{dT}\right|_{T=0}=0 \tag{3.10.19}$$

An example of Maple code which solves this system is shown in Code 3.5(a). The default method is a Fehlberg fourth–fifth-order Runge–Kutta with degree four interpolant.

Figure 3.28 shows plots of the solution to this system of coupled, nonlinear ODEs for three values of $A = 0.1, 1.0, 5.0$. The equivalent code for MATLAB® is shown in Code 3.5(b).

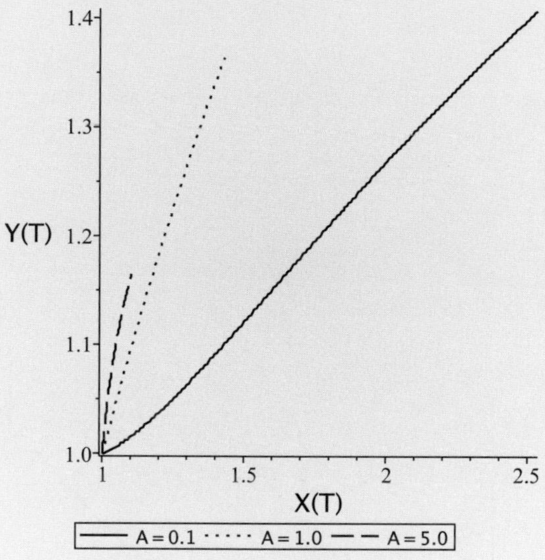

Figure 3.28 Parametric plots of X(T) vs Y(T) for three values of $A = 0.1, 1.0, 5.0$ and $0 \leq T \leq 1$.

```
(a)
restart;
#Define two coupled ODEs with parameter A and their initial conditions.

EQ1 := A·diff(X(T), T, T) + diff(X(T), T)  - X(T)·Y(T) = 0;
```

$$EQ1 := A\left(\frac{d^2}{dT^2}X(T)\right)+\frac{d}{dT}X(T)-X(T)\,Y(T)=0 \tag{1}$$

```
EQ2 := diff(Y(T), T, T) + A·diff(Y(T), T)  - Y(T)/X(T);
```

$$EQ2 := \frac{d^2}{dT^2}Y(T)+A\left(\frac{d}{dT}Y(T)\right)-\frac{Y(T)}{X(T)} \tag{2}$$

```
ICS := X(0) = 1, Y(0) = 1, D(X)(0) = 0, D(Y)(0) = 0;
```

$$ICS := X(0)=1,\ Y(0)=1,\ D(X)(0)=0,\ D(Y)(0)=0 \tag{3}$$

```
#substitute A for 3 different values and compute solutions
```

Code 3.5 Code examples: (a) Maple, (b) MATLAB®.

```
sol1 := dsolve( { subs(A = 0.1, EQ1), subs(A = 0.1, EQ2), ICS}, {X(T), Y(T)}, numeric);
                    sol1 := proc(x_rkf45) ... end proc                              (4)
sol2 := dsolve( { subs(A = 1, EQ1), subs(A = 1, EQ2), ICS}, {X(T), Y(T)}, numeric);
                    sol2 := proc(x_rkf45) ... end proc                              (5)
sol3 := dsolve( { subs(A = 5, EQ1), subs(A = 5, EQ2), ICS}, {X(T), Y(T)}, numeric);
                    sol3 := proc(x_rkf45) ... end proc                              (6)
Tf := 1 :
#Define parametric plots of X(T) vs Y(T) for 0≤T≤Tf for each value of
A and display on one graph
with (plots) :
PL1 := odeplot(sol1, [X(T), Y(T)], 0..Tf, linestyle = 1, legend = "A=0.1") :
PL2 := odeplot(sol2, [X(T), Y(T)], 0..Tf, linestyle = 2, legend = "A=1.0") :
PL3 := odeplot(sol3, [X(T), Y(T)], 0..Tf, linestyle = 3, legend = "A=5.0") :
display(PL1, PL2, PL3);
```

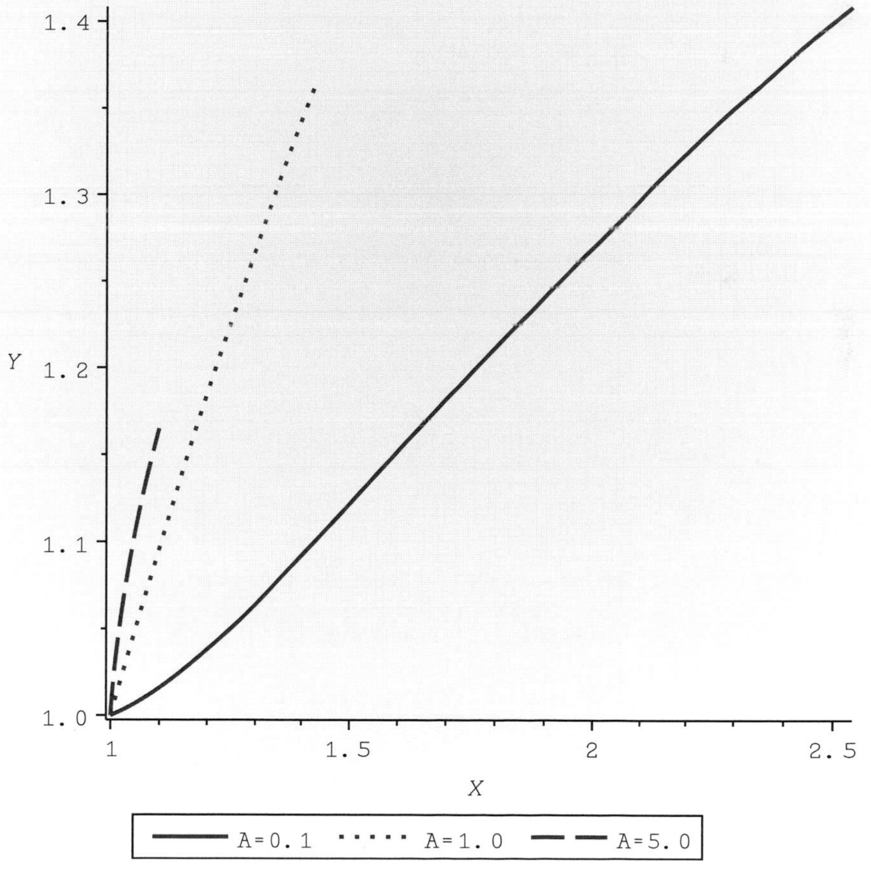

Code 3.5 (*continued*)

```
(b)
clear all; close all; clc
tspan    = [0 1];
x0       = 1;
Dx0      = 0;
y0       = 1;
Dy0      = 0;
ICS      = [x0 Dx0 y0 Dy0];
A        = 0.1;

[t,sol] = ode45(@(t,sol) Chapter3_odeSystem3_10_18(t,sol,A),tspan,ICS);
x        = sol(:,1);
Dx       = sol(:,2);
y        = sol(:,3);
Dy       = sol(:,4);
plot(x,y,'-k'), hold on;
A        = 1;
[t,sol] = ode45(@(t,sol) Chapter3_odeSystem3_10_18(t,sol,A),tspan,ICS);
x        = sol(:,1);
Dx       = sol(:,2);
y        = sol(:,3);
Dy       = sol(:,4);
plot(x,y,':k')
A        = 5;
[t,sol] = ode45(@(t,sol) Chapter3_odeSystem3_10_18(t,sol,A),tspan,ICS);
x        = sol(:,1);
Dx       = sol(:,2);
y        = sol(:,3);
Dy       = sol(:,4);
plot(x,y,'--k')
axis([1 2.5 1 1.4])
legend('A=0.1','A=1','A=5','location','southeast')
xlabel('X(T)')
ylabel('Y(T)')
```

The following MATLAB function must be saved with the name "Chapter3_odeSystem3_10_18.m"

```
function dsdt = Chapter3_odeSystem3_10_18(t,s,A)
  dsdt = zeros(4,1);
  dsdt(1)  = s(2);
  dsdt(2)  =(-s(2)+s(1).*s(3))./A;
  dsdt(3)  = s(4);
  dsdt(4)  =-A.*s(4)+s(3)./s(1);
end
```

Code 3.5 (*continued*)

Summary

The concepts of Eulerian and Lagrangian frames, and how they are related, can be challenging. This chapter is designed to introduce the two ideas in concert, and to provide several examples that are easy enough for the student to handle mathematically, while focusing on the main objectives. In addition, these examples have been analyzed in both dimensional and dimensionless terms to build on what was learned in Chapter 2 about scalings and parameters.

Connected to the two descriptions we found the definitions and use of important fluid mechanical lines associated with the flow. The trajectory or pathline is a Lagrangian concept and the mathematical description of the time history of a fluid particle's spatial position, starting with its initial position, is called the fluid "mapping." The concept of the mapping is important for both theoretical and practical bases, since it allows us to manipulate relationships between Lagrangian and Eulerian variables as well as their time derivatives, a building block of formulating the conservation of mass and momentum in Chapters 4 and 5, respectively. Special use of the mapping gave us two other important fluid mechanical lines. The streakline is a set of fluid particles that are tagged as they pass by a particular spatial location. Flow visualization techniques can involve injecting a dye at a point (or a moving point) in a fluid to characterize the behavior and understanding the streakline relationship to the motion is essential for interpretation of what is recorded visually. For a biological flow, the tagger could be a point in the flow where a signaling molecule or infectious agent is released. The timeline is a set of fluid particles tagged simultaneously so that wherever they go they indicate the same time instant. In the Eulerian frame the important fluid mechanical line is the streamline, defined at an instant of time and always tangent to the Eulerian velocity vectors. Streamlines and their related velocity vector fields are a routine way of showing flow information from experimental measurements and from theoretical/computational analysis of the governing equations. An example of particle trajectory from Chapter 2, as a balance of its inertia and drag, is investigated in stagnation point flow to draw together our tools so far and use them to approach an important biofluid mechanical phenomenon of aerosol deposition in the lung. In addition, introducing the concepts of numerical methods opens up further development later in the book for numerical solutions as well.

Problems

3.1 Consider a 1D flow field with the dimensionless Eulerian velocity given by

$$u(x, t) = \frac{x}{1+t}.$$

a. Find the mapping $X(X_0, t)$ with the initial condition $X(T = 0) = X_0$. Do this by replacing $u \sim U = dX/dt$ and $x \sim X$, then separate variables and integrate.
b. What is the inverse mapping $X_0(x, t)$?
c. Calculate the Lagrangian velocity $U(X_0, t)$. Is the Lagrangian velocity steady? If so, why? Is the Eulerian velocity steady?
d. Determine the Jacobian $J = \partial X/\partial X_0$. Is this flow incompressible?

3.2 Consider the Eulerian velocity field, $\underline{u}(\underline{x}, t)$, given by

$$\begin{aligned} u &= y \\ v &= -x + 0.2 \sin(t/2) \end{aligned} \tag{1}$$

a. Use your computational software program to plot this Eulerian velocity vector field for $t = \pi$. Describe this flow. Maple users should try "fieldplot," MATLAB® users should try "quiver." Let the domain be $-1 \le x \le 1, -1 \le y \le 1$.
b. The Lagrangian acceleration is given in Eulerian variables using the material derivative given in Eq. (3.4.13).
Using the definition of $\underline{u}$ in Eq. (1) form the Lagrangian acceleration vector, $\underline{A}(\underline{x}, t)$ from the x and y components of $D\underline{u}/Dt$ in Eq. (2). Show your answer.
c. Plot this vector field over the interval $-1 \le x \le 1, -1 \le y \le 1$ for $t = \pi$.
d. Find the streamlines for this flow where t is considered a constant. What is the shape of the streamlines and how does it change with t?
e. Find the mapping $\underline{X}(\underline{X}_0, t)$ for this flow in an analytical expression with initial positions (X_0, Y_0). Hint: solve for Y first in the v equation, or use your software if needed.
f. Plot the mapping, also called the particle pathline, for three fluid particles whose initial positions are $(0.2, 0.2)$, $(0.2, 0.4)$, $(0.2, 0.6)$ all on the same graph with a legend identifying the three curves. Let the range of t be $0 \le t \le 2\pi$. Describe the pathlines.

3.3 For uniform flow the Eulerian velocity field in 2D is $\underline{u} = (q, 0)$ with constant velocity $q > 0$ in the positive x-direction. The corresponding mapping is given in Eq. (3.2.3), $X = X_0 + qt$, $Y = Y_0$. Consider a tagger starting at the point $(0, 0)$ which travels at constant speed, v_T, in the y-direction to the point $(0, L)$.

a. What is the time-dependent position of the tagger $x_T(s), y_T(s)$, where s is a time variable? What is the range of s?
b. Solve for the initial positions of the fluid particles that get tagged when $t = s$. Call them $(X_0^*(s), Y_0^*(s))$.
c. Using $(X_0^*(s), Y_0^*(s))$ from part (b) as the initial positions in the mapping, solve for the streakline when the tagger first arrives at (0, L), which fixes the value of t. Now your solution depends on the s time variable.
d. Scale your results in part (c) as follows $\xi = X/L, \eta = Y/L, S = v_T s/L$ to make your solution dimensionless. Let the dimensionless parameter be $\lambda = q/v_T$. What is the range of S?
e. Make a parametric plot of your solutions for the streakline using three values of the dimensionless parameter $\lambda = 0.5, 1.0, 2.0$
f. Now represent the initial positions (X_0^*, Y_0^*) in dimensionless form using the same scales as in part (d). Call them $\xi_0(S), \eta_0(S)$ and make a parametric plot for the same values of λ as in part (e). Explain your answer.

3.4 For stagnation point flow the Eulerian velocity field is given in Eq. (3.5.1) and the mapping in Eq. (3.5.8). Consider a tagger starting at the point $(-L, L)$ which travels at constant speed, v_T, in the x-direction to the point (L, L).

a. What is the time-dependent position of the tagger $x_T(s), y_T(s)$ where s is a time variable?
b. Solve for the initial positions of the fluid particles that get tagged when $t = s$. Call them $(X_0^*(s), Y_0^*(s))$.
c. Using $(X_0^*(s), Y_0^*(s))$ from part (b) as the initial positions in the mapping, solve for the streakline when the tagger first arrives at (L, L), which fixes the value of t. Now your solution depends on the s time variable.
d. Scale your results in part (c) as follows $\xi = X/L, \eta = Y/L, S = v_T s/L$ to make your solution dimensionless.
e. Plot solutions for the streakline using four values of the dimensionless parameter $AL/v_T = 0.05, 0.10, 0.15, 0.20$.

f. Explain the shapes of your results.

3.5 Consider a dimensionless Eulerian velocity field for an incompressible fluid, $u = 1 + \lambda y, v = \cos(t)$.

a. Solve the mapping, $\underline{X}(\underline{X}_0, t)$, analytically, where the initial positions are $\underline{X}_0 = (X_0, Y_0)$ at $t = 0$. Hint: solve for Y first.
b. Plot the trajectory (pathline) from (a) for $(X_0, Y_0) = (0, 0)$ over one oscillation, $0 \le t \le 2\pi$, for $\lambda = 0,\ 1$.
c. A dye tagger is stationary at the origin, $(x_T(s),\ y_T(s)) = (0, 0)$ where s is the time variable for the tagger. Set $(X(X_0,\ Y_0,\ t),\ Y(X_0,\ Y_0,\ t)) = (x_T(s),\ y_T(s))$for $t = s$ and solve for the tagged initial positions, call them $(X_0^*(s), Y_0^*(s))$.
d. Insert (X_0^*, Y_0^*) into the mapping and plot the resulting streakline after one cycle $0 \le s \le 2\pi, t = 2\pi$, for $\lambda = 0, 1$.
e. Find the streamline equation. Set $t = 2\pi$ and plot the streamline that passes through (0, 0) for $\lambda = 0, 1$ and indicate the flow direction.
f. Discuss a comparison of your results for streamline, streakline, and pathline.

3.6 For stagnation point flow the Eulerian velocity field is given in Eq. (3.5.1) and the mapping in Eq. (3.5.8). Let the timeline start as a straight line between the points $(-L_0, L_0/2)$ and $(+L_0, 3L_0/2)$. That means the initial positions are given by $X_0 = -L_0 + 2hL_0,\ Y_0 = (L_0 + 2hL_0)/2$ where h is the parameterization of the timeline and has the limits $0 \le h \le 1$.

a. Solve for the mapping of this timeline.
b. Make your solution in part (a) dimensionless by using the following scales: $\xi = X/L_0$, $\eta = Y/L_0,\ T = At$.
c. Plot your solution at times $T = 0.0, 0.3, 0.6, 1.0$.
d. Discuss your results in terms of the length and angle of the timeline.
e. Repeat your analysis for the initial timeline, which is a semicircle of radius L_0 centered at the origin. In addition, show that the shape at any time is an ellipse. What are the major and minor axes as functions of T? Plot your results for $T = 0, 0.3, 0.6, 1.0$.

3.7 Equation (3.9.1) ignored gravitational effects on aerosol deposition. Let's include it now by adding the term $-mg$ to the right hand side of the y-component equation. We can assume that the initial particle position and velocity are given by Eq. (3.9.3) and choose the same scales as in Eq. (3.9.4).

a. Make the governing equations dimensionless, including the initial conditions.
b. You should have two dimensionless parameters already defined in class, i.e. the Stokes Number, $Stk = \rho_p d_p^2 A/18\mu$, and the Froude number, $Fr^2 = A^2L/g$. Go back and derive these same two parameters using the Buckingham Pi Theorem.
c. Solve this system. Hint: you will need a homogeneous solution for the x and y components, and, in addition, a particular solution for the y component.
d. Plot aerosol particle trajectories from their starting point at (1, 1) to impact for four combinations of the parameter values, $Fr = 0.2, 0.5$ and $Stk = 0.05, 0.1$. Also plot the streamline that passes through (1, 1).
e. Discuss the results of part (d). How do your results compare to zero gravity?
f. Standard values for air viscosity and kinematic viscosity are $\mu = 0.00018\,g/cm{\cdot}s,\ \ \nu = 0.15\,cm^2/s$ and we can assume the particle density is $\rho_p = 1.0\ g/cm^3$ and the gravitational constant is $g = 980\ cm/s^2$. Let $L = 1$ cm. For the four combinations of Stk and Fr, calculate A and dp. Then calculate the Reynolds number $Re = UL/\nu = AL^2/\nu$ for each combination. Our stagnation point theory assumes inviscid flow, which is reasonable for $Re >> 1$. Is that so?
g. In order to justify our use of Stokes' drag we must have the Reynolds number based on the particle diameter be small, i.e. $Re_p = Ud_p/\nu = ALd_p/\nu << 1$. Is that assumption correct for the cases in part (d)?

3.8

a. Use your computational software to solve the following differential equation numerically with the three choices of x(0) and plot your answers on the same graph for

$$\frac{d^2x}{dt^2} - \cos(t)\frac{dx}{dt} + \sin(2t)\,x = 0 \qquad (1)$$
$$x(0) = 0,\ 0.5,\ 1,\ \frac{dx}{dt}(0) = 1$$

b. Solve the system of coupled differential equations numerically

$$\frac{d^2x}{dt^2} + \sin(y)\,x = 0$$
$$\frac{d^2y}{dt^2} + \sin(x)\frac{dy}{dt} + y = 0$$
$$x(0) = 0,\ \frac{dx}{dt}(0) = 1$$
$$y(0) = 0,\ \frac{dy}{dt}(0) = 1$$

Plot x(t) and y(t) on the same graph. Then plot y(x) in a parametric plot on a separate graph, both graphs for $0 \le t \le 2\pi$.

3.9 Let an Eulerian velocity field be the following

$$u = ye^{-x} + \sin y,\quad v = \frac{1}{2}y^2e^{-x} + \sin x \qquad (1)$$

a. Plot the Eulerian velocity vector field for $-1 \le x \le 1,\ -1 \le y \le 1$. Hint: use "fieldplot" in Maple, "quiver" in MATLAB®.
b. Plot contours of constant speed $s(x, y) = (u^2 + v^2)^{1/2}$ on the same interval as in part (a), use color or shaded filled regions if you like. Overlay the two plots from parts (a) and (b).
c. Determine the Eulerian acceleration field and show your answer. Plot the Eulerian acceleration vector field for $-1 \le x \le 1$, $-1 \le y \le 1$.
d. Solve the mapping X, Y using numerical methods. On the same graph plot X(t) and Y(t) for initial position $\underline{X}_0 = (0.5, 0.5)$ for $0 \le t \le 4$.
e. On a new graph, plot two pathlines, X vs Y, parameterized by t, for the initial position $\underline{X}_0 = (0.5, 0.5)$ and $\underline{X}_0 = (0.5, 1.0)$ for $0 \le t \le 4$. What is the difference in the results?

4 Conservation of Mass

Introduction

Conservation of mass is a basic concept spanning many branches of science. We need to know how much mass of a particular material that there is in a physical situation, and how it is changing, and of course why it is changing. In point mass mechanics we could easily deal with the mass, m, of each, say, spherical body. For a group of identical spheres, this process reduces to just a nose count if you will, i.e. how many bodies are there? In any description of the group motion, we would have to insure that two spheres do not occupy the same space at the same time. For a continuous media like a fluid, instead we speak of its mass per unit volume, i.e. the density, $\rho(\underline{x}, t)$, which has units of mass per unit volume ML^{-3}. Mass conservation of the fluid is a constraint on its motion that couples with the force balance, i.e. its momentum conservation, which is discussed in the next chapter. It guarantees that fluid particle A, which is a mathematical point in a continuum, does not pass through fluid particle B. Particle A has to go around B, or push it out of the way. B is not allowed to disappear and leave a hole, i.e. the fluid is continuous. Often, in fact, the equation for fluid mass conservation is called the "continuity equation." A simple example is blowing up a balloon. The mass of air exhaled from your lung ends up in the balloon. The air flows out of your lung due to imposed forces, but the overall fluid motion is constrained to be an equal mass exchange from the lung to the balloon. Mouth to mouth resuscitation is an essentially pure biofluid mechanics example of the same process, though more complicated and more urgent. Every occurrence of fluid flow must satisfy conservation of mass for the fluid; however, the balloon and resuscitation examples are easy ways to appreciate its presence.

Another mass conservation principle is that applied to dissolved (i.e. soluble) species or constituents in the fluid. The mass of the species per unit volume of fluid is its concentration, $c(\underline{x}, t)$, units of ML^{-3}. For example in blood serum there are many soluble species that include respiratory gases (e.g. oxygen, carbon dioxide), nutrients (e.g. glucose, proteins, fatty acids), waste products (e.g. blood urea nitrogen, lactic acid), electrolytes (e.g. sodium, potassium) amongst others. One can add to those naturally occurring substances anything that is infused into the blood such as medications, contrast agents for radiographic imaging, as well as more blood itself from a transfusion. The oxygen level of the atmosphere needs to be above a critical value to sustain life, and other agents ranging from environmental pollutants such as ozone to anesthetic gases that put you to sleep are all soluble gases in the inspired air. Notable expired gases may include ethanol vapor – which can land you in jail for failing a breathalyzer test. Unlike conservation of the bulk fluid mass, which depends primarily on its convection, dissolved species are transported both by convection of the fluid in which they are dissolved and also by diffusion through the fluid.

Topics Covered

In Chapter 4 we will introduce the concept of an Eulerian control volume to generate a ledger sheet of mass conservation. We will keep track of the net inflow of mass across its boundaries, any sources from the boundaries, and sources within the volume but not on the boundaries. The resulting equation of integrals and ordinary time derivative that holds for control volume is then transformed to give us a partial differential equation, with partial derivatives in space and time, that holds at an arbitrary point in the continuum. From the resulting form we then explore simplifications for incompressible fluids and introduce the concept of the stream function, ψ, which exists for certain flows that are restricted to two dimensions. From the stream function we can derive the streamlines which were introduced in Chapter 3, and an important clinical example of streamlines is presented for flow through the nose. Another mass conservation concept involves a diffusible solute in the fluid, like a dissolved chemical species. The corresponding convection–diffusion equation is derived and applied to stagnation point flow for an example of solute transport at a branch bifurcation in the vascular or lung airway system. The form of the conservation of mass equation and the possible stream function definitions are then explored in other coordinate systems (cylindrical, spherical). Conservation of mass has an important consequence in the fluid mapping discussed in Chapter 3, and that is explored for the case of an incompressible fluid. Finally a more complete analysis of fluid deformation is used to derive the same conservation of mass equation, but from the standpoint of a Lagrangian control volume, i.e. one that moves with the fluid. The important Reynolds Transport Theorem is developed and then used to this end. Although it would appear that formulating conservation of mass by two different methods is not necessary, it does allow the student and teacher to explore further the kinematics of the fluid deformation required to achieve the Lagrangian method via the Reynolds Transport Theorem. So there is a richness there not otherwise appreciated unless addressed, and conservation laws are a good motivation for discussing them. On the other hand, if the course goal is more applied then one can certainly skip Sections 4.6, 4.7 and 4.8.

4.1 Mass Conservation from an Eulerian Control Volume

Figure 4.1 shows an arbitrary Eulerian control volume, V, in space whose surface is S. The outward pointing unit normal to the surface is the vector $\underline{n}$, and fluid flows through the surface, into and out of V, as indicated by the streamlines.

For conservation of mass in V, the overall bookkeeping can be written as a balance sheet where the time rate of increase of the mass in V is equal to the net influx through or from the boundary S plus any sources or sinks in V, see Eq. (4.1.1)

$$\begin{array}{l}\text{Time rate of} \\ \text{mass increase in V}\end{array} = \begin{array}{l}\text{Net influx of mass} \\ \text{through or from S}\end{array} + \begin{array}{l}\text{Sources} \\ \text{of mass in V}\end{array} \tag{4.1.1}$$

Since the density is mass per unit volume, then in continuum mechanics the mass, m(t) inside V is the volume integral of the Eulerian density field, $\rho(\underline{x}, t)$, in V

$$m(t) = \iiint_V \rho(\underline{x}, t)\, dV \tag{4.1.2}$$

The limits of integration in Eq. (4.1.2) are fixed spatial coordinates that are not time-dependent. However, those fixed spatial coordinates are arbitrary. For example, if the control volume were a unit cube centered at the origin with edges that are 2b units long, then Eq. (4.1.2) would look like

$$m(t) = \int_{-b}^{b}\int_{-b}^{b}\int_{-b}^{b} \rho(x, y, z, t)\, dx\, dy\, dz \tag{4.1.3}$$

The time rate of increase of m(t) is just its time derivative, a positive value indicating increase and a negative value indicating decrease. Taking the derivative of Eq. (4.1.2),

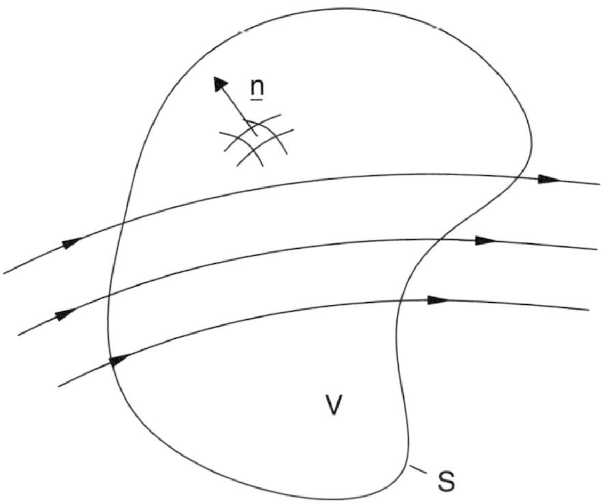

Figure 4.1 An Eulerian control volume, V, whose boundary is S and outward unit normal $\underline{n}$.

$$\frac{dm}{dt} = \frac{d}{dt}\iiint_V \rho(\underline{x}, t)\, dV \tag{4.1.4}$$

Mathematically, the balance of Eq. (4.1.1) is stated as follows

$$\frac{d}{dt}\iiint_V \rho(\underline{x}, t)\, dV = \iint_S \Sigma_\rho\, dS + \iiint_V \sigma_\rho\, dV \tag{4.1.5}$$

where Σ_ρ is the surface contribution and σ_ρ is any mass source (or sink) inside V. Equation (4.1.5) is an integro-differential equation, with an ordinary time derivative, that is a balance for the entire control volume. Eventually we will transform it into a partial differential equation, with partial derivatives in space and time, and no integrals or ordinary time derivative. This final form will hold for an arbitrary point in space at an arbitrary time. The required steps to that end follow.

The surface contribution, Σ_ρ, can be divided into two parts: one from fluid convection across the boundary, and the other from any non-convective processes at the boundary. Then Σ_ρ may be written as the sum

$$\Sigma_\rho = (\rho\underline{u})\cdot(-\underline{n}) + \Sigma'_\rho \tag{4.1.6}$$

The first term in Eq. (4.1.6) is the convective contribution where $\underline{u}\cdot(-\underline{n})$ is the surface fluid velocity component in the $-\underline{n}$-direction, which is the component that crosses the boundary; Σ'_ρ is the non-convective contribution, such as surface sources on S. Inserting Eq. (4.1.6) into Eq. (4.1.5), and neglecting surface sources, i.e. $\Sigma'_\rho = 0$, gives us

$$\frac{d}{dt}\iiint_V \rho(\underline{x}, t)\, dV = -\iint_S (\rho\underline{u})\cdot\underline{n}\, dS + \iiint_V \sigma_\rho\, dV \tag{4.1.7}$$

An Eulerian control volume is fixed in space, so the limits of integration over the volume, V, are time independent, see Eq. (4.1.3) for an example. That allows us to represent the first term of Eq. (4.1.7) as

$$\frac{d}{dt}\iiint_V \rho dV = \iiint_V \frac{\partial \rho}{\partial t}\, dV \tag{4.1.8}$$

where we have moved the time derivative inside the integral, becoming a partial derivative. The surface integral term of Eq. (4.1.7) has the form

$$\iint_S \underline{\beta}\cdot\underline{n}\, dS, \quad \underline{\beta} = -\rho\underline{u} \tag{4.1.9}$$

Recall that the divergence theorem states that we can recast Eq. (4.1.9), the surface integral of the scalar product of the vector $\underline{\beta}$ with the unit normal, $\underline{n}$, as a volume integral of the divergence of the vector $\underline{\beta}$

$$\iint_S \underline{\beta}\cdot\underline{n}\, dS = \iiint_V \underline{\nabla}\cdot\underline{\beta}\, dV \tag{4.1.10}$$

Introducing index notation in Cartesian coordinates, the scalar product in Eq. (4.1.10) is

$$\underline{\beta}\cdot\underline{n} = \beta_1 n_1 + \beta_2 n_2 + \beta_3 n_3 = \beta_k n_k \tag{4.1.11}$$

where the indices 1, 2, 3 correspond to the x, y, z components, respectively. It is convenient to use the Einstein summation notation for repeated indices as shown in Eq. (4.1.11). The repeated index, k, signifies that we evaluate the expression $\beta_k n_k$ for $k = 1$, then for $k = 2$, and then for $k = 3$, and sum the results. The Cartesian divergence operator can similarly be given in index notation

$$\underline{\nabla}\cdot\underline{\beta} = \frac{\partial \beta_k}{\partial x_k} = \frac{\partial \beta_1}{\partial x_1} + \frac{\partial \beta_2}{\partial x_2} + \frac{\partial \beta_3}{\partial x_3} \tag{4.1.12}$$

Substituting Eqs. (4.1.11) and (4.1.12) into Eq. (4.1.10), the divergence theorem in index notation becomes

$$\iint_S \beta_k n_k\, dS = \iiint_V \frac{\partial \beta_k}{\partial x_k}\, dV \tag{4.1.13}$$

We will use this approach later when we derive conservation of momentum and β is a tensor (matrix) with two indexes.

We can now rewrite Eq. (4.1.7) as the volume integral and collect all of the terms in the integrand

$$\iiint_V \left(\frac{\partial \rho}{\partial t} + \underline{\nabla}\cdot(\rho\underline{u}) - \sigma_\rho\right) dV = 0$$
$$\Leftrightarrow \quad \iiint_V \left(\frac{\partial \rho}{\partial t} + \frac{\partial}{\partial x_k}(\rho u_k) - \sigma_\rho\right) dV = 0 \tag{4.1.14}$$

Equation (4.1.14) gives both vector and index notation forms.

Equation (4.1.14) is an integral over the arbitrary volume, V, of the bracketed term. Since the control volume we chose is arbitrary, it must be that this integrand itself is zero. An example of this logic in one dimension is as follows. Suppose we have an integral of a function, f(x), with specific limits of integration $\int_0^{2\pi} f(x)\,dx = 0$, What can we say about f(x), if anything? Certainly $f(x) = 0$ is a solution, but that is not the only possibility. Some other examples of solutions are $f(x) = \sin x$, $f(x) = \cos x$, etc. So we cannot reach any useful conclusions about f(x). However, if the limits of integration are arbitrary as in the form $\int_{x_1}^{x_2} f(x)\,dx = 0$, where x_1 and x_2 can have any value, then it must be that $f(x) = 0$. That is the conclusion in Eq. (4.1.14), which gives us the conservation law for mass in either vector or index form

$$\frac{\partial \rho}{\partial t} + \underline{\nabla} \cdot (\rho \underline{u}) - \sigma_\rho = 0 \quad \Leftrightarrow \quad \frac{\partial \rho}{\partial t} + \frac{\partial}{\partial x_k}(\rho u_k) - \sigma_\rho = 0 \tag{4.1.15}$$

In Eq. (4.1.15), the divergence of the product of a scalar and a vector can be expanded using the product rule for differentiation. In vector or index form we have

$$\begin{aligned} \underline{\nabla} \cdot (\rho \underline{u}) &= \underline{u} \cdot (\underline{\nabla} \rho) + \rho (\underline{\nabla} \cdot \underline{u}) \Leftrightarrow \frac{\partial}{\partial x_k}(\rho u_k) \\ &= u_k \frac{\partial \rho}{\partial x_k} + \rho \frac{\partial u_k}{\partial x_k} \end{aligned} \tag{4.1.16}$$

In the absence of any sources, $\sigma_\rho = 0$, we apply the product rules of Eq. (4.1.16) to the divergence term of Eq. (4.1.15) to find the final form of the conservation of mass equation, also called the continuity equation

$$\frac{\partial \rho}{\partial t} + \underline{u} \cdot \underline{\nabla} \rho = -\rho \underline{\nabla} \cdot \underline{u} \quad \Leftrightarrow \quad \frac{\partial \rho}{\partial t} + u_k \frac{\partial \rho}{\partial x_k} = -\rho \frac{\partial u_j}{\partial x_j} \tag{4.1.17}$$

Notice that the operator on the left hand side of Eq. (4.1.17) is the time derivative following a fluid particle, i.e. the Lagrangian time derivative or material derivative

$$\frac{\partial \rho}{\partial t} + \underline{u} \cdot \underline{\nabla} \rho = \left(\frac{\partial R(\underline{X}_0, t)}{\partial t} \right)_{\underline{X}_0} = \frac{D\rho}{Dt} = -\rho \underline{\nabla} \cdot \underline{u} \tag{4.1.18}$$

Recall that the material derivative in Eulerian variables is simply the Lagrangian time derivative

$$\frac{D\rho}{Dt} = \left(\frac{\partial R(\underline{X}_0, t)}{\partial t} \right)_{\underline{X}_0} \tag{4.1.19}$$

where $R(\underline{X}_0, t)$ is the Lagrangian density following a particle.

To recount the events of this section, we began with an integro-differential equation for the mass balance of an Eulerian control volume. By using the divergence theorem to represent surface integral terms as volume integral terms, and moving the time derivative into the volume integrand, since the boundaries are fixed, we imposed that the Eulerian control volume was arbitrary, which made the integrand zero. From these steps, then, we derived a partial differential equation for mass balance that is valid at every point of the fluid and for all values of time.

4.2 Incompressible Fluids

An incompressible fluid can be viewed as one where the density of a fluid particle does not change as it moves about in the flow field; i.e. the Lagrangian density is constant. While neighboring particles may have a different density, all particles retain their initial value as they flow. i.e. $R(\underline{X}_0, t) = R(\underline{X}_0)$. For example, the density of sea water depends, in part, on its salt concentration, so it is stratified vertically with heavier sea water the deeper one goes. The fluid flows as an incompressible material, nevertheless. So this incompressible fluid is one in which

$$\left(\frac{\partial R(\underline{X}_0, t)}{\partial t} \right)_{\underline{X}_0} = \frac{D\rho}{Dt} = 0 \tag{4.2.1}$$

From Eq. (4.1.18), the restriction of incompressibility is equivalent to

$$\underline{\nabla} \cdot \underline{u} = \frac{\partial u_k}{\partial x_k} = 0 \tag{4.2.2}$$

In the conservation of momentum equations, discussed in the next chapter, the density would, in general, be non-constant in the Eulerian frame, i.e. $\rho(\underline{x}, t)$.

Another simplification is to consider the density as a constant in space in the Eulerian frame, so that spatial derivatives of ρ are zero. From Eq. (4.1.17) this implies

$$\frac{\partial \rho}{\partial t} = -\rho \underline{\nabla} \cdot \underline{u} \quad \Leftrightarrow \quad \frac{\partial \rho}{\partial t} = -\rho \frac{\partial u_j}{\partial x_j} \tag{4.2.3}$$

Yet another simplification is to consider the density as a constant in time and space in the Eulerian frame, so derivatives of ρ with respect to time and Eulerian space are zero. From Eq. (4.1.17) this also gives us Eq. (4.2.2). However, now the density is constant in the Eulerian frame, and that will be true in the momentum conservation equations, $\rho(\underline{x}, t) = \rho$.

Worked Example 4.1

Let's check the inviscid stagnation point flow discussed in Chapter 3, $(u, v) = (Ax, -Ay)$, to see if it is incompressible

$$\underline{\nabla} \cdot \underline{u} = \frac{\partial u}{\partial x} + \frac{\partial v}{\partial y} = A - A = 0 \tag{4.2.4}$$

It is incompressible. How about two-dimensional (2D) Poiseuille flow? Since $u = u(y)$, $v = 0$, it is easy to see that it is also incompressible.

The partial differential equation for conservation of mass gives us an opportunity to determine a component of the velocity field given the others. For example, in 2D, if we are given the x-component of velocity u, then we solve for v for an incompressible fluid as

$$\frac{\partial u}{\partial x} + \frac{\partial v}{\partial y} = 0 \quad \Rightarrow \quad v = -\int \frac{\partial u}{\partial x} dy + f(x, t) \tag{4.2.5}$$

where f(x, t) is a function of integration.

Worked Example 4.2

For example, assume a steady flow with u given by

$$u = e^{-kx} \cos(ky) \tag{4.2.6}$$

then we can determine the y-component for an incompressible fluid,

$$v = -\int \frac{\partial u}{\partial x} dy + f(x) = -\int -ke^{-kx} \cos(ky)\, dy + f(x) = e^{-kx} \sin(ky) + f(x) \tag{4.2.7}$$

The integration function f(x) could be determined from some boundary condition on v at $y = \text{constant}$. For example, $v(y = 0) = 0$ makes $f = 0$.

4.3 The Stream Function

For incompressible fluids, we learned that the conservation of mass equation for two dimensions in rectangular coordinates is

$$\frac{\partial u}{\partial x}+\frac{\partial v}{\partial y}=0 \tag{4.3.1}$$

Here we have assumed that the z-direction fluid velocity is zero, i.e. $w = 0$, and that u and v do not depend on z. Under these circumstances, one can define a stream function ψ which is related to the velocity components (u, v) as

$$u=\frac{\partial \psi}{\partial y}, \quad v=-\frac{\partial \psi}{\partial x} \tag{4.3.2}$$

This definition of ψ in rectangular coordinates is called the Lagrange stream function, named for its discoverer Joseph-Louis Lagrange (1736–1813), born Giuseppe Lodovico Lagrangia in Turin, Italy. It solves the continuity equation, Eq. (4.3.1), identically

$$\frac{\partial^2 \psi}{\partial x \partial y}-\frac{\partial^2 \psi}{\partial y \partial x}\equiv 0 \tag{4.3.3}$$

since we can interchange the order of differentiation with respect to x and y assuming that real flow is sufficiently smooth.

Worked Example 4.3

What is the stream function for 2D stagnation point flow? The velocity components are $(u, v) = (Ax, -Ay)$. Starting with the x-component of Eq. (4.3.2) we have

$$\frac{\partial \psi}{\partial y}=Ax \tag{4.3.4}$$

Integrating partially with respect to y results in

$$\psi = Axy + f(x) \tag{4.3.5}$$

where $f(x)$ is an unknown function of integration. The reason it is a function of integration, and not a constant of integration, is because of the partial derivative with respect to y that we integrated. To find more about $f(x)$, we insert v into the y component of Eq. (4.3.2) to find

$$v=-Ay=-\frac{\partial \psi}{\partial x} \tag{4.3.6}$$

and then substitute for ψ our result from Eq. (4.3.5)

$$Ay+\frac{df}{dx}=Ay \quad \rightarrow \quad \frac{df}{dx}=0 \tag{4.3.7}$$

Since $df/dx = 0$, f must be a constant. The definition of the stream function is often simplified by setting $\psi = 0$ on a boundary. For this flow, we know that $y = 0$ can be viewed as a boundary, so

$$\psi(y=0)=0 \quad \rightarrow \quad f=0 \tag{4.3.8}$$

and the stream function is determined to be

$$\psi = Axy \tag{4.3.9}$$

and we see that $x = 0$ is also a boundary.

4.4 Streamlines and Stream Tubes: Vascular and Respiratory Branching Networks

The stream function is so named because streamlines in the flow are along lines of constant ψ. To prove this, consider the differential of ψ and keep the linear terms

$$d\psi = \frac{\partial \psi}{\partial x}dx + \frac{\partial \psi}{\partial y}dy \tag{4.4.1}$$

Substituting Eqs. (4.3.2) into Eq. (4.4.1) results in

$$d\psi = -vdx + udy \tag{4.4.2}$$

A line along which ψ is a constant is a line where $d\psi = 0$, and then Eq. (4.4.2) simplifies to

$$vdx = udy \quad \text{or} \quad \frac{dx}{u} = \frac{dy}{v} \tag{4.4.3}$$

which is the equation for the streamlines from Chapter 3. Different streamlines are identified as different values of $\psi(x, y) = \text{constant}$.

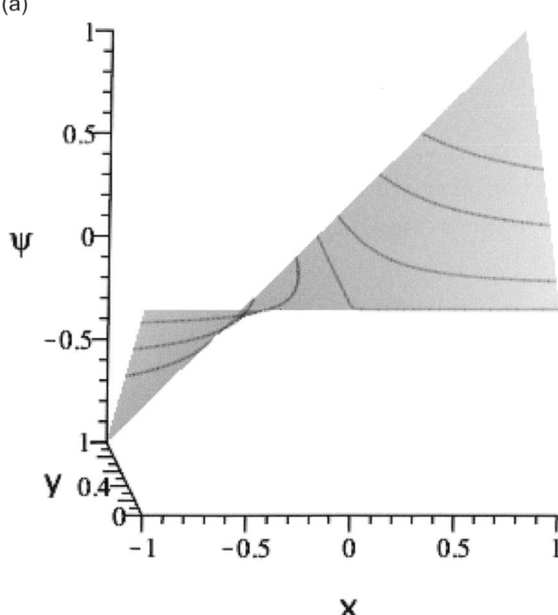

Figure 4.2 (a) The surface $\psi(x, y) = Axy$, $A = 1$ showing contours of constant ψ; (b) streamlines associated with (a) are the contours projected onto the x–y plane; (c) streamlines from (b) and the associated Eulerian vector field $u = Ax$, $v = -Ay$, $A = 1$.

A three-dimensional plot of the stream function $\psi(x, y) = Axy$, $A = 1$ for stagnation point flow is shown in Figure 4.2(a), where contour lines of $\psi(x, y) = \text{constant}$ are indicated. Figure 4.2(b) shows the streamlines which are the contours of Figure 4.2(a) projected onto the x–y plane. Since streamlines do not show the direction of the flow, it is useful to plot the associated Eulerian velocity vector field $\underline{u} = (u, v) = (Ax, -Ay)$ to understand where the fluid is going. Figure 4.2(c) shows the streamlines and also the Eulerian velocity vector field of stagnation point flow. The Maple code for Figure 4.2 is shown in Code 4.1(a). The corresponding MATLAB® code is in Code 4.1(b).

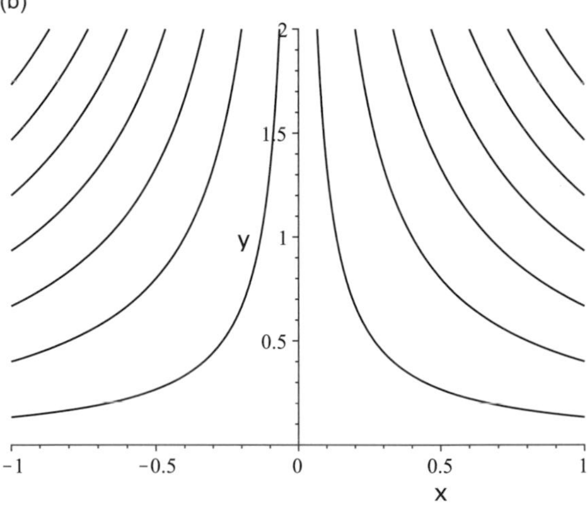

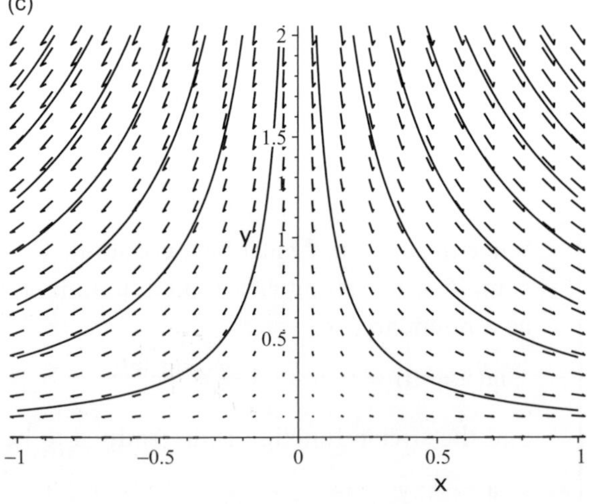

Figure 4.2 *(continued)*

(a)

```
restart;
#Define U, V, Ψ and set A=1
u := A·x:
v := -A·y:
psi := A·x·y:
A := 1:
#Plot Ψ(x, y) in 3D
plot3d(psi, x=-1..1, y=0..1, labels=[x, y, "Ψ"]);
with(plots):
#Use "contourplot" to plot lines of constant Ψ which are streamlines
PL1 := contourplot(psi, x=-1..1, y=0..1):
#Use "fieldplot" to plot the Eulerian velocity field u, v
PL2 := fieldplot([u, v]), x=-1..1, y=0..1):
#Display the plots on the same graph
display(PL1, PL2);
```

(b)

```
clear all, close all, clc
x      = linspace(-1, 1, 110);
y      = linspace(0, 1, 100);
[X, Y] = meshgrid(x, y);
A      = 1;
U      = A*X;
V      =-A*Y;
PSI    = A*X.*Y;
surf(X, Y, PSI), colormap gray, shading interp, caxis([-4 2])
hold on
contour3(X, Y, PSI, '-k')
view([-7.5 8.4])
xlabel('x')
ylabel('y')
zlabel('\psi')
figure()
contour(X, Y, PSI, 20, '-k'), hold on, plot(0*y, y, 'k')
xlabel('x')
ylabel('y')
figure()
contour(X, Y, PSI, 20, '-k'), hold on, plot(0*y, y, 'k'), plot(x, 0*x, 'k')
quiver(X(1:11:end, 1:10:end), Y(1:11:end, 1:10:end), U(1:11:end, 1:10:end), V(1:11:end, 1:10:end), '-k')
xlabel('x')
ylabel('y')
axis([-1 1 -0.05 1])
```

Code 4.1 Code examples: (a) Maple, (b) MATLAB®.

The mass flow rate between streamlines can be shown to be related to the difference between stream function values of those streamlines. To show this, consider the flow shown in Figure 4.3 with two neighboring streamlines defined stream function values $\psi = \psi_1$ and $\psi = \psi_2$.

The curve BD is described by the equation $y = f(x)$ or, equivalently, $F = y - f(x) = 0$.

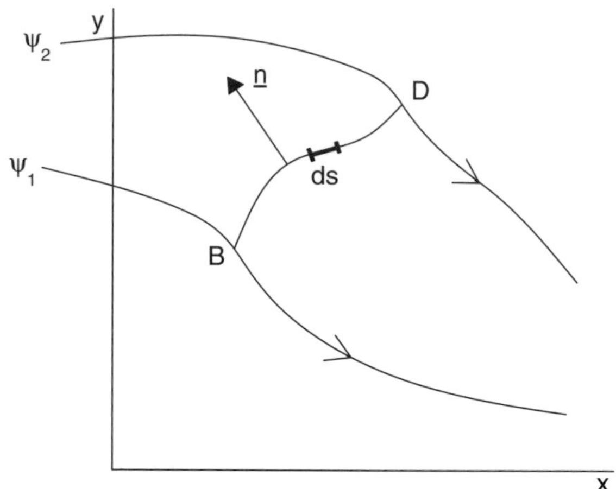

Figure 4.3 The volumetric flow rate between streamlines.

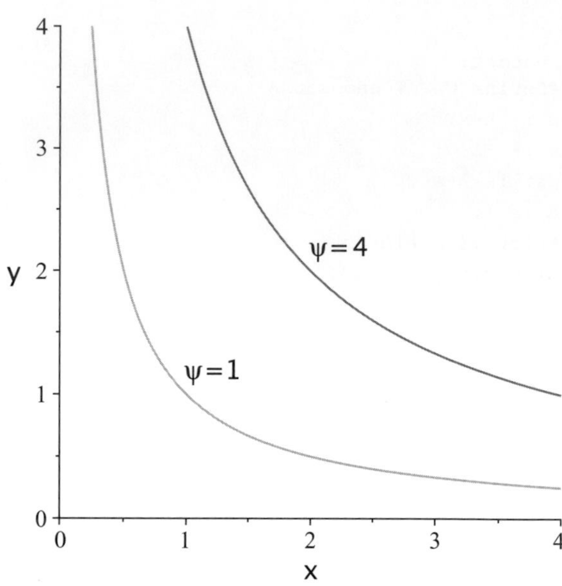

Figure 4.4 Two streamlines in stagnation point flow for $\psi = 1$ and $\psi = 4$ with $A = 1$.

The curve has a differential arc length $ds = \left(1 + (df/dx)^2\right)^{1/2} dx$. The unit normal to the curve is $\underline{n}$, which can be given in term of F as

$$\underline{n} = \frac{\nabla F}{|\nabla F|} = \frac{\left(-\dfrac{df}{dx}, 1\right)}{\left(1 + \left(\dfrac{df}{dx}\right)^2\right)^{1/2}} \tag{4.4.4}$$

The volumetric flow rate across BD is the fluid velocity component in the $-\underline{n}$-direction integrated along the curve BD

$$Q = -\int_B^D \underline{u} \cdot \underline{n}\, ds \tag{4.4.5}$$

Using the stream function to define the velocities, and substituting for ds and $\underline{n}$ from Eq. (4.4.4), the flow rate in Eq. (4.4.5) becomes

$$Q = \int_B^D \left(\frac{\partial \psi}{\partial y}\frac{df}{dx} + \frac{\partial \psi}{\partial x}\right) dx = \int_B^D \left(\frac{\partial \psi}{\partial y} dy + \frac{\partial \psi}{\partial x} dx\right)$$
$$= \int_B^D d\psi = \psi_2 - \psi_1 \tag{4.4.6}$$

For the first term in the integrand of Eq. (4.4.6) we used the relationship that $(df/dx)dx = df$ and then that $df = dy$. The units of ψ in rectangular coordinates are L^2/T, and the difference in the two values gives the two-dimensional volumetric flow rate between the corresponding streamlines. The average fluid velocity increases in regions where the two streamlines are closer to one another and slower where they are farther apart since Q is the same between them at every location.

For example, the stream function for stagnation point flow is $\psi = Axy$. For $A = 1$, choosing two values of ψ to be 1 and 4 yields the two streamlines shown in Figure 4.4. If x and y are in units of cm, and A in units of s^{-1}, then the flow rate between the two streamlines is $4 - 1 = 3 cm^2/s$. Note that the distance between the streamlines is largest along the line of identity, $y = x$. How does that relate to Figure 3.11, where the Lagrangian velocity passes through a local minimum depending on the initial position?

The Maple code for Figure 4.4 is given by Code 4.2(a) and the corresponding MATLAB® code is Code 4.2(b).

(a)
```
restart;
#Define Ψ and choose A=1
psi := A·x·y:
A := 1:
with(plots):
#Use "contourplot" to plot the streamlines for Ψ=1,4
PL1 := contourplot(psi, x=0..4, y=0..4, contours=[1,4]):
display(PL1, view={0..4, 0..4]);
```

(b)
```
clear all, close all, clc
x      = linspace(0,4,110);
y      = linspace(0,4,100);
[X,Y]  = meshgrid(x,y);
A      = 1;
PSI    = A*X.*Y;
contour(X,Y,PSI,[1,4],'-k')
xlabel('x')
ylabel('y')
```

Code 4.2 Code examples: (a) Maple, (b) MATLAB®.

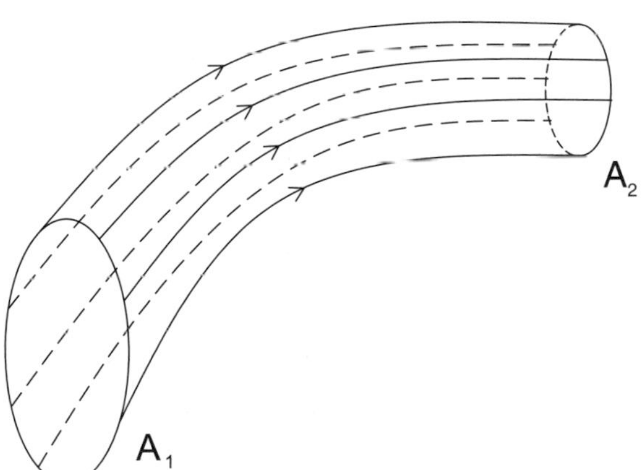

Figure 4.5 An example of a stream tube.

A bundle of streamlines is called a stream tube, as shown in Figure 4.5. The tube is defined by its surface which is formed from streamlines, so no flow crosses that surface. In the same sense as the planar case, the average velocity in a stream tube at a particular station is inversely proportional to its cross-sectional area for constant flow rate Q. The two stations shown in the stream tube, where the cross-sectional areas are A_1 and A_2, and the average velocities are $\bar{u}_1$ and $\bar{u}_2$, respectively, have the relationship

$$Q = A_1\bar{u}_1 = A_2\bar{u}_2 \tag{4.4.7}$$

The simple mass conservation for a stream tube, Eq. (4.4.7), is used often to analyze gross features of flow in a branching network like the vascular system or the airways of the lung. The basic branch geometry of a simple bifurcating system is shown in Figure 4.6. Each parent tube splits into two daughter tubes, and three generations of tubes are shown, $n = 0, 1, 2$. For a system that only bifurcates, there are 2^n tubes at each generation. Each cross-sectional area is labeled in the figure with a subscript to indicate the generation (first number) and tube identification within the generation (second number).

The entire system is a stream tube, so that the total flow in the parent must equal the total flow in the daughters at each generation. That means that the following equation holds

$$\begin{aligned} Q &= A_0\bar{u}_0 = A_{11}\bar{u}_{11} + A_{12}\bar{u}_{12} \\ &= A_{21}\bar{u}_{21} + A_{22}\bar{u}_{22} + A_{23}\bar{u}_{23} + A_{24}\bar{u}_{24} \end{aligned} \tag{4.4.8}$$

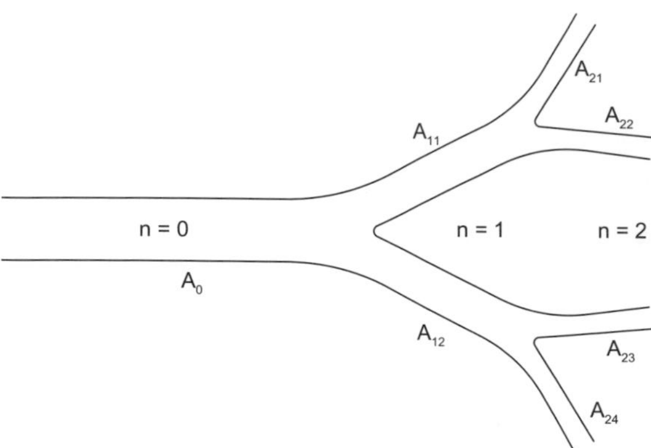

Figure 4.6 A branching network of bifurcating tubes representative of respiratory airways or vascular blood vessels. Three generations are shown, n = 0, 1, 2.

In the event that the system is symmetric, then $A_{11} = A_{12}$ and the total cross-sectional area at generation $n = 1$ is given by $A_1 = A_{11} + A_{12} = 2A_{11}$. Likewise at generation $n = 2$ we have equal cross-sectional areas for each daughter, $A_{21} = A_{22} = A_{23} = A_{24}$. Then the total cross-sectional area at $n = 2$ is $A_2 = 4A_{21}$. Under these conditions, the relationships in Eq. (4.4.8) simplify to

$$Q = A_0\bar{u}_0 = A_1\bar{u}_1 = A_2\bar{u}_2 \tag{4.4.9}$$

Clinical Correlation 4.1 | Flow Through the Nose

An excellent biofluid mechanics example of mass conservation, with related velocities and streamlines, appears in the following coupled experimental (Hahn et al., 1993) and computational (Keyhani et al., 1995) studies of flow through the nose. The nose serves many functions regarding respiration. It rapidly warms and humidifies inspired air prior to its reaching the lungs, and also filters particles that land on its mucus coating. And, of course, the nose has the sense of smell or "olfaction" where odorant molecules come in contact with the olfactory mucosa. A fluid dynamical analysis of nasal airflow can lead to better understanding of normal and abnormal physiology and disease processes of the nose. In addition, better therapeutic methods such as nasal drug delivery or surgical procedures, which modify its architecture, can first be done in a model to predict the impact on function.

Figure 4.7 shows the anatomy of a nasal passage: (a) is a lateral view with the olfactory region indicated by hatching; (b) is a computerized axial tomography CAT scan of an adult nasal passage at the level of the dashed line in (a); and (c) points out the general structures for nasal cross-section including turbinates and olfactory region. A common nasal surgery is to trim the turbinates to relieve flow obstruction.

Figure 4.8 is the experimental setup: (a) an enlarged (×20), anatomically correct, physical model of the right nasal passage for an adult male human was fabricated from Styrofoam and sits, upside down, on a table top. A fan provides either inspiration (shown) or expiration. Five dotted lines across the entire model indicate planes of measurements with the hot-film anemometer probe, starting with plane 1, inflow, at the naris (top right) and ending with plane 5, outflow, at the nasopharynx (left). The much bigger size of this nose model made it easier to use (b) the hot-film anemometer probe which measures velocity. To keep the system

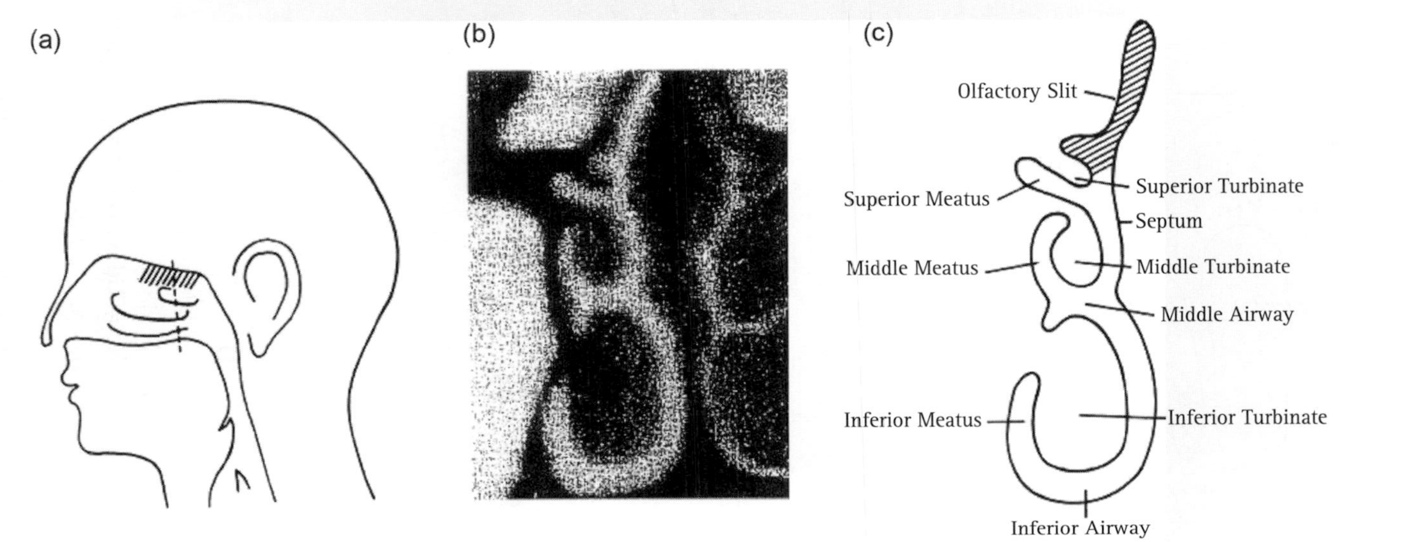

Figure 4.7 (a) Human nasal anatomy from lateral view and hashed lines indicating the olfactory region; (b) CAT scan of cross-section at level of dashed line in (a); (c) structures for nasal cross-section.

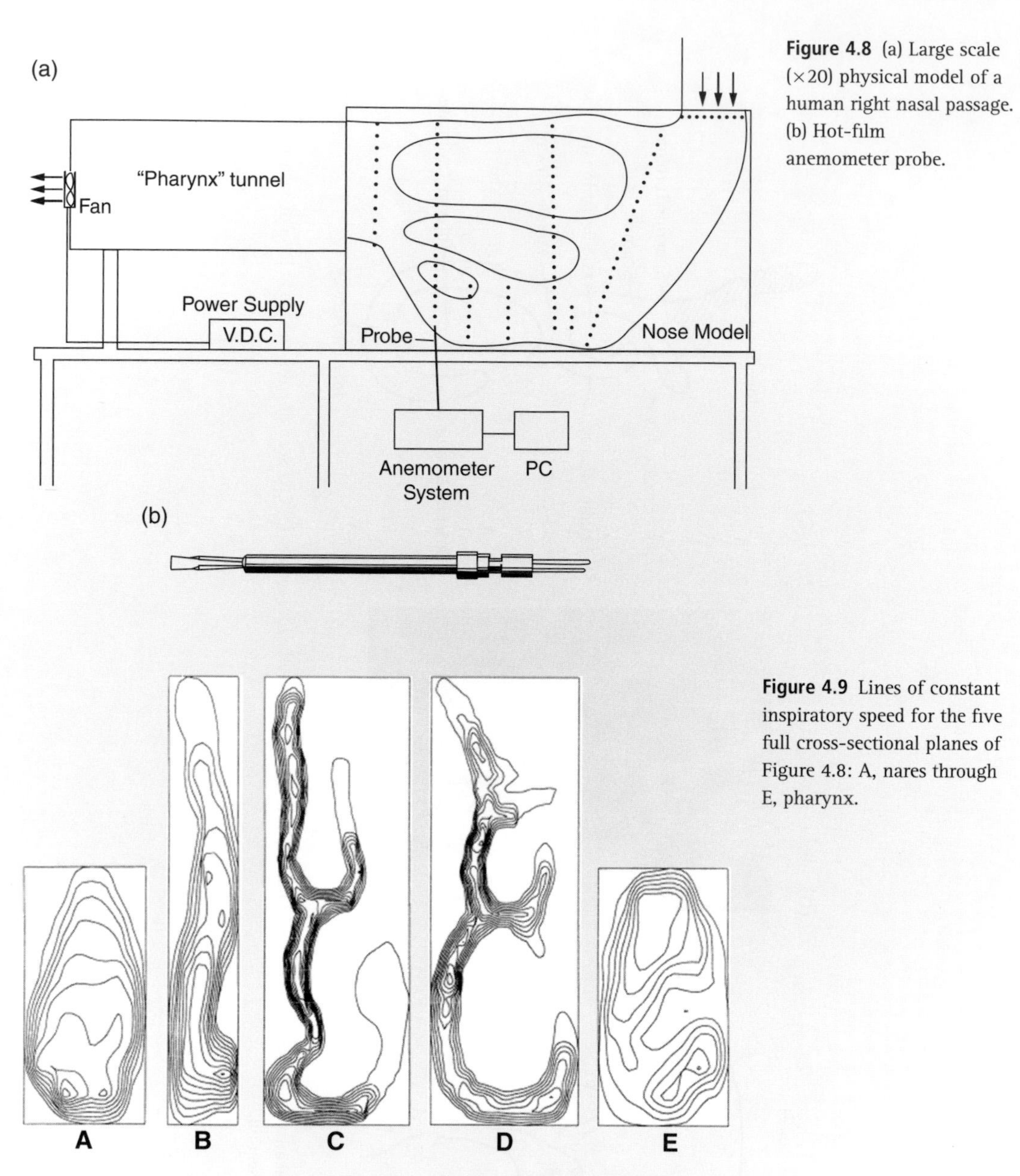

Figure 4.8 (a) Large scale (×20) physical model of a human right nasal passage. (b) Hot-film anemometer probe.

Figure 4.9 Lines of constant inspiratory speed for the five full cross-sectional planes of Figure 4.8: A, nares through E, pharynx.

dynamically similar, the Reynolds number, $Re = Ud/\nu$, was matched to the normal human nose. Since the characteristic length scale, d, was ×20 in the model, U in the model was reduced ×1/20 from real nose velocities.

Figure 4.9 shows lines of constant speed (isovelocity) during inspiration measured at the five planes of Figure 4.8, starting with plane 1 (A) at the entrance to the nasal passage.

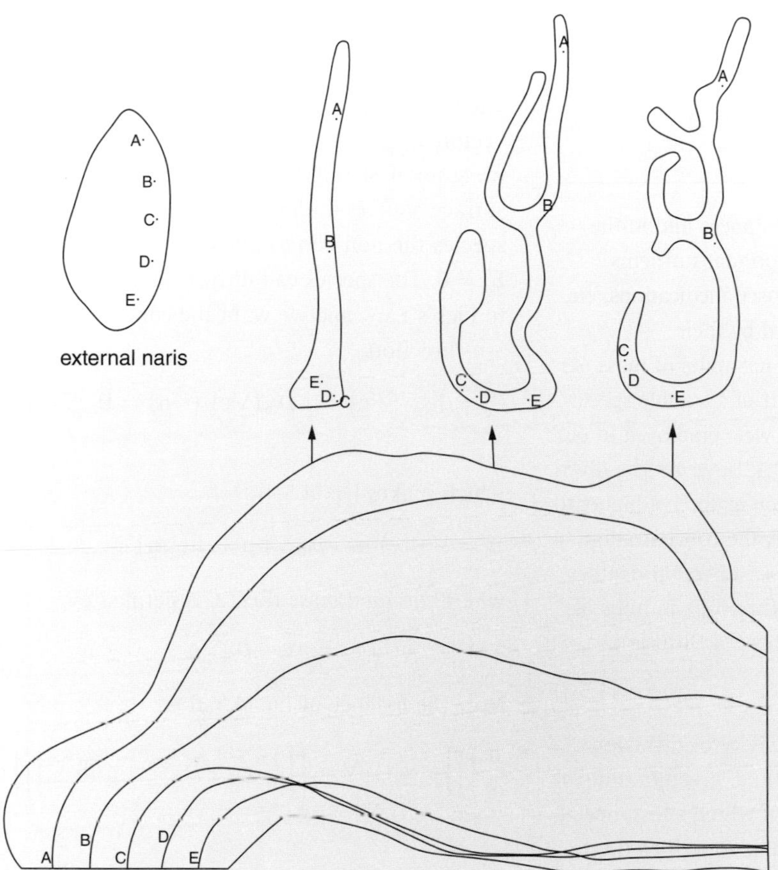

Figure 4.10 Streamlines calculated from computational solutions of the flow equations. Seven different streamlines are shown A–G with cross-sectional views and the larger lateral view.

All speeds are normalized to $U_0 = 8.88\,\text{m/s}$, which was measured at (A) and each family of curves are at 10% intervals. Crowding of the streamlines indicates large gradients of velocity, and hence shear, across them.

Computational fluid dynamics (CFD) was also performed on flow through this nasal model, where solutions to conservation of mass and momentum (see Chapter 5) are performed using numerical analysis and algorithms constrained by the boundary shape (Keyhani et al., 1995). Part of the analysis leads to the streamline pattern shown in Figure 4.10, which actually was derived from the trajectories of neutrally buoyant mass point particles released from locations A–G at the entrance. Both cross-sectional views (above) and a lateral view (below) show where the streamlines are located. Note that streamline A ends up in the olfactory region.

where $\bar{u}_1$ and $\bar{u}_2$ are the average fluid velocities at generations 1 and 2, respectively.

4.5 Conservation of Mass for a Soluble Species

Fluids often convey dissolved substances and in the biological realm these include important nutrients, blood gases, waste products, imposed medications, etc. These soluble species are described by their concentration, c(x, y, z, t), which has units of mass per unit volume, ML^{-3}. Mass transport of a soluble species includes convective processes, as were employed in our balance of fluid mass in Section 4.1. However, dissolved species also move by diffusion from regions of higher to lower concentration. For example, the concentration gradient for diffusive flux in the +x-direction requires $dc/dx < 0$, so that the diffusive flux, J_{diff}, in units of $ML^{-2}T^{-1}$, is given by Fick's First Law of Diffusion,

$$J_{diff} = -D_m(dc/dx) \tag{4.5.1}$$

where D_m is the molecular diffusivity or diffusion coefficient of the species in the fluid, given in units of L^2T^{-1}. The diffusivity depends on which substance is diffusing and through which fluid, as well as the temperature. At 25 °C, oxygen, for example, has $D_m = 0.176176\,cm^2/s$ for diffusion through air, but $D_m = 2.1\,x \times 10 - 55\,cm^2/s$ for diffusion through water. In more than one dimension, Eq. (4.5.1) generalizes to the diffusive vector flux, $\underline{J}_{diff}$

$$\underline{J}_{diff} = -D_m \underline{\nabla} c \tag{4.5.2}$$

The balance equation of a soluble species is similar to Eq. (4.1.5) with c replacing ρ to indicate a species concentration

$$\frac{d}{dt}\iiint_V c(\underline{x},t)\,dV = \iint_S \Sigma_c\,dS + \iiint_V \sigma_c\,dV \tag{4.5.3}$$

The surface term, Σ_c, in Eq. (4.5.3) again has two possible components as described in Eq. (4.1.6) for just the fluid itself. There is a convective contribution proportional to the product $\underline{u}c$ and a non-convective contribution, Σ_c',

$$\Sigma_c = (\underline{u}c)\cdot(-\underline{n}) + \Sigma_c' \tag{4.5.4}$$

The term $\underline{u}c$ is also known as the convective flux vector, $\underline{J}_{conv} = \underline{u}c$. For conservation of the fluid mass we ignored, setting $\Sigma_\rho' = 0$, by assuming there were no surface sources on S of fluid mass. For the soluble species the non-convective contribution is not zero, $\Sigma_c' \neq 0$. The species can diffuse across S according to Fick's Law, and we want the component in the $-\underline{n}$-direction,

$$\Sigma_c' = \underline{J}_{diff}\cdot(-\underline{n}) = -D_m(\underline{\nabla}c)\cdot(-\underline{n}) = D_m\underline{\nabla}c\cdot\underline{n} \tag{4.5.5}$$

which makes Eq. (4.5.4)

$$\Sigma_c = -(\underline{u}c)\cdot\underline{n} + D_m\underline{\nabla}c\cdot\underline{n} = \underline{J}\cdot(-\underline{n}) \tag{4.5.6}$$

where the total mass flux, $\underline{J}$, is defined by

$$\underline{J} = \underline{J}_{conv} + \underline{J}_{diff} = \underline{u}c - D_m\underline{\nabla}c \tag{4.5.7}$$

Now the balance of Eq. (4.5.3) is

$$\frac{d}{dt}\iiint_V c(\underline{x},t)\,dV = \iint_S (-c\underline{u} + D_m\underline{\nabla}c)\cdot\underline{n}\,dS + \iiint_V \sigma_c\,dV \tag{4.5.8}$$

Again, we employ the divergence theorem for the surface integral to turn it into a volume integral

$$\iint_S (-c\underline{u} + D_m\underline{\nabla}c)\cdot\underline{n}\,dS = \iiint_V \underline{\nabla}\cdot(-c\underline{u} + D_m\underline{\nabla}c)\,dV \tag{4.5.9}$$

Substituting Eq. (4.5.9) into Eq. (4.5.8) and bringing the time derivative of Eq. (4.5.8) into the integral as a partial derivative, we can again put all the terms together under one volume integral

$$\iiint_V \left(\frac{\partial c}{\partial t} - \underline{\nabla}\cdot(-c\underline{u} + D_m\underline{\nabla}c) - \sigma_c\right) dV = 0 \tag{4.5.10}$$

Since the Eulerian volume is arbitrary, then the integrand is zero and the resulting partial differential equation valid at a point in space is

$$\frac{\partial c}{\partial t} + \underline{\nabla} \cdot (c\underline{u}) = \underline{\nabla} \cdot (D_m \underline{\nabla} c) + \sigma_c \qquad (4.5.11)$$

For constant molecular diffusivity, D_m, the divergence of the diffusive flux yields the Laplacian operator, i.e.

$$\underline{\nabla} \cdot (D_m \underline{\nabla} c) = D_m \underline{\nabla} \cdot (\underline{\nabla} c) = D_m \nabla^2 c \qquad (4.5.12)$$

where the Laplacian in Cartesian coordinates is $\nabla^2 = \partial^2/\partial x^2 + \partial^2/\partial y^2 + \partial^2/\partial z^2$, for example. Cylindrical and spherical coordinates are treated in the next section. The chain rule of partial differentiation can be used for the divergence of the convective transport term

$$\underline{\nabla} \cdot (c\underline{u}) = \underline{u} \cdot \underline{\nabla} c + c \underline{\nabla} \cdot \underline{u} \qquad (4.5.13)$$

For an incompressible fluid, $\underline{\nabla} \cdot \underline{u} = 0$. Gathering all the terms now we have the conservation of mass for a soluble species, also called the transport equation, the convection–diffusion equation, the advection–dispersion equation, amongst others

$$\frac{\partial c}{\partial t} + \underline{u} \cdot \underline{\nabla} c = D_m \nabla^2 c + \sigma_c \qquad (4.5.14)$$

The source (sink) term σ_c is generally due to chemical reactions. Additionally, the concentration field,

Clinical Correlation 4.2

For stagnation point flow, suppose there is a diffusible constituent in the fluid which is being transported by convection and diffusion. This situation can be a model for transport to the wall of an airway or artery in the region of the bifurcation where the flow splits and there is a stagnation point. Such transported substances can be a gas, drug, nutrient, toxin, etc. The conservation of mass for a dissolved species in a fluid is the transport equation

$$\frac{\partial c}{\partial t} + u\frac{\partial c}{\partial x} + v\frac{\partial c}{\partial y} = D_m \nabla^2 c \qquad (4.5.15)$$

where c is the concentration, u is the Eulerian velocity field and D_m is the molecular diffusion coefficient. For two-dimensional stagnation point flow, the Eulerian velocity field is

$$\underline{u} = (u, v) = (Ax, -Ay) \qquad (4.5.16)$$

Inserting Eq. (4.5.16) into Eq. (4.5.15) and seeking a steady concentration solution which is independent of x, i.e. $c = c(y)$ only, we have

$$-Ay\frac{dc}{dy} = D_m \frac{d^2 c}{dy^2} \qquad (4.5.17)$$

Let the concentration boundary conditions be $c(y = 0) = 0$, and $c(y \to \infty) = C_0$. Equation (4.5.17) is a first-order ODE for dc/dy. Let $f = dc/dy$, then Eq. (4.5.17) becomes

$$D_m \frac{df}{dy} = -Ayf \qquad (4.5.18)$$

Rearranging Eq. (4.5.18) to separate variables and integrate gives us

$$\int \frac{df}{f} = -\frac{A}{D_m}\int y\,dy + a_1 \tag{4.5.19}$$

where a_1 is a constant of integration. The solution to Eq. (4.5.19) is $\ln(f) = -(A/2D_m)y^2 + a_1$, which can be exponentiated to give us

$$f = \frac{dc}{dy} = a_2 e^{-(A/2D_m)y^2} \tag{4.5.20}$$

where $a_2 = e^{a_1}$ is the modified form of the integration constant.

Now we can integrate (4.5.20) to solve for the concentration, c, setting the lower limit of integration to $y = 0$ and the upper limit to a variable y

$$c = a_2 \int_0^y e^{-(A/2D_m)\hat{y}}\,2d\hat{y} + a_3 \tag{4.5.21}$$

Here, a_3 is another integration constant and $\hat{y}$ is the dummy y variable of integration. There is no analytic, closed form solution to the integral in Eq. (4.5.21). Instead, it is related to the error function, which is a tabulated special function, erf(z), where

$$\operatorname{erf}(z) = \frac{2}{\sqrt{\pi}}\int_0^z e^{-\hat{z}}\,2d\hat{z} \tag{4.5.22}$$

For our purposes, we can rescale y to be $\eta = (A/2D_m)^{1/2}y$, so Eq. (4.5.21) has an integral similar to the error function

$$c = a_2\left(\frac{2D_m}{A}\right)^{1/2}\int_0^\eta e^{-\hat{\eta}^2}\,d\hat{\eta} + a_3 = a_2\left(\frac{2D_m}{A}\right)^{1/2}\frac{\sqrt{\pi}}{2}\operatorname{erf}(\eta) + a_3 \tag{4.5.23}$$

A plot of the error function, erf(z), is shown in Figure 4.11(a). Imposing the boundary condition $c(\eta = 0) = 0$ on Eq. (4.5.23) gives us $a_3 = 0$; however, the condition at infinity gives us

$$c(\eta \to \infty) = a_2\left(\frac{2D_m}{A}\right)^{1/2}\frac{\sqrt{\pi}}{2}\operatorname{erf}(\infty) = C_0 \tag{4.5.24}$$

One of the properties of the error function is that $\operatorname{erf}(\infty) = 1$, as appreciated in Figure 4.11(a), which inserted into Eq. (4.5.24) solves for $a_2 = (2/\sqrt{\pi})(A/2D_m)^{1/2}C_0$. Using that result and rescaling the concentration to be dimensionless $C = c/C_0$ finally gives us

$$C = \frac{2}{\sqrt{\pi}}\int_0^\eta e^{-\hat{\eta}^2}\,d\hat{\eta} = \operatorname{erf}(\eta) \tag{4.5.25}$$

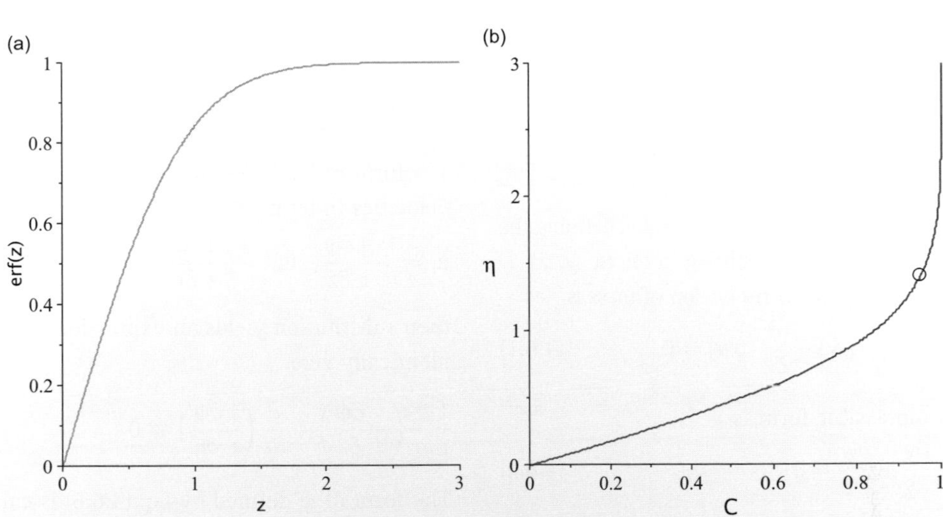

Figure 4.11 (a) The error function erf(z), (b) solution C(η) for stagnation point transport.

which is plotted in Figure 4.11(b). Let's choose a finite distance, $y = L_0$, from the boundary where the concentration is 95% of the value at infinity, i.e. $C = 0.95$, the circled point in Figure 4.11(b). From the error function $\text{erf}(\eta = 1.4) = 0.95$ so $L_0(A/2D_m)^{1/2} = 1.4$. The combination $L_0(A/2D_m)^{1/2}$ is proportional to the square root of the Péclet number, Pe, given by

$$\text{Pe} = \frac{UL}{D_m} = \frac{AL_0^2}{D_m} \sim \frac{\text{convective transport}}{\text{diffusive transport}} \tag{4.5.26}$$

The Péclet number involves a characteristic velocity, U, a characteristic length, L, and the molecular diffusivity, D_m. Another way to interpret Eq. (4.5.26) is $\text{Pe} = (L^2/D_m)/(L/U)$, which is the ratio of a diffusive time scale L^2/D_m to a convective time scale L/U. Pe>>1 signifies mass transport dominated by convection, while Pe<<1 is a diffusion dominated process. For $\sqrt{\text{Pe}} = 1.4$, increasing A, the fluid velocity, will decrease L_0 to be closer to the wall, for example. Slower diffusion, smaller D_m, also reduces L_0.

$c(x, y, z, t)$, is subject to boundary conditions on surfaces in contact with the fluid. Typical situations include setting $c = \text{constant}$ on the boundary or $\underline{n}\cdot\nabla c = 0$ for an insulated boundary.

4.6 Other Coordinate Systems

For a rectangular Cartesian coordinate system, defining the velocity components (u, v, w) as being in the (x, y, z)-directions, respectively, the conservation of mass is

$$\frac{\partial \rho}{\partial t} + \frac{\partial}{\partial x}(\rho u) + \frac{\partial}{\partial y}(\rho v) + \frac{\partial}{\partial z}(\rho w) = 0 \qquad (4.6.1)$$

and the incompressible form is

$$\underline{\nabla}\cdot\underline{u} = \frac{\partial u}{\partial x} + \frac{\partial v}{\partial y} + \frac{\partial w}{\partial z} = 0 \qquad (4.6.2)$$

For a cylindrical coordinate system, see Figure 4.12, we define the velocity components (u_r, u_θ, u_z) as being in the (r, θ, z)-directions, respectively, then the conservation of mass is given by

$$\frac{\partial \rho}{\partial t} + \frac{1}{r}\frac{\partial}{\partial r}(\rho r u_r) + \frac{1}{r}\frac{\partial}{\partial \theta}(\rho u_\theta) + \frac{\partial}{\partial z}(\rho u_z) = 0 \qquad (4.6.3)$$

The incompressible form is

$$\frac{1}{r}\frac{\partial}{\partial r}(r u_r) + \frac{1}{r}\frac{\partial}{\partial \theta}(u_\theta) + \frac{\partial}{\partial z}(u_z) = 0 \qquad (4.6.4)$$

In the case of axisymmetric flow for cylindrical coordinates, $u_\theta = 0$ and $\partial/\partial\theta = 0$, then Eq. (4.6.4) simplifies to

$$\frac{1}{r}\frac{\partial}{\partial r}(r u_r) + \frac{\partial}{\partial z}(u_z) = 0 \qquad (4.6.5)$$

The form of Eq. (4.6.5) tells us that if we define the velocities in terms of a stream function as

$$u_r = -\frac{1}{r}\frac{\partial \psi}{\partial z}, \quad u_z = +\frac{1}{r}\frac{\partial \psi}{\partial r} \qquad (4.6.6)$$

then substitution yields an expression which is identically zero,

$$\frac{1}{r}\frac{\partial}{\partial r}\left(-\frac{\partial \psi}{\partial z}\right) + \frac{\partial}{\partial z}\left(\frac{1}{r}\frac{\partial \psi}{\partial r}\right) \equiv 0 \qquad (4.6.7)$$

The form of ψ defined by Eq. (4.6.6) is called the Stokes stream function. Setting ψ equal to a constant forms a surface which is an axisymmetric stream tube enclosing the flow. The units for ψ in this case are L^3T^{-1}, fully three-dimensional volumetric flow rate.

For a spherical coordinate system, see Figure 4.12, we define the velocity components (u_r, u_θ, u_ϕ) as being in the (r, θ, ϕ)-directions, respectively,

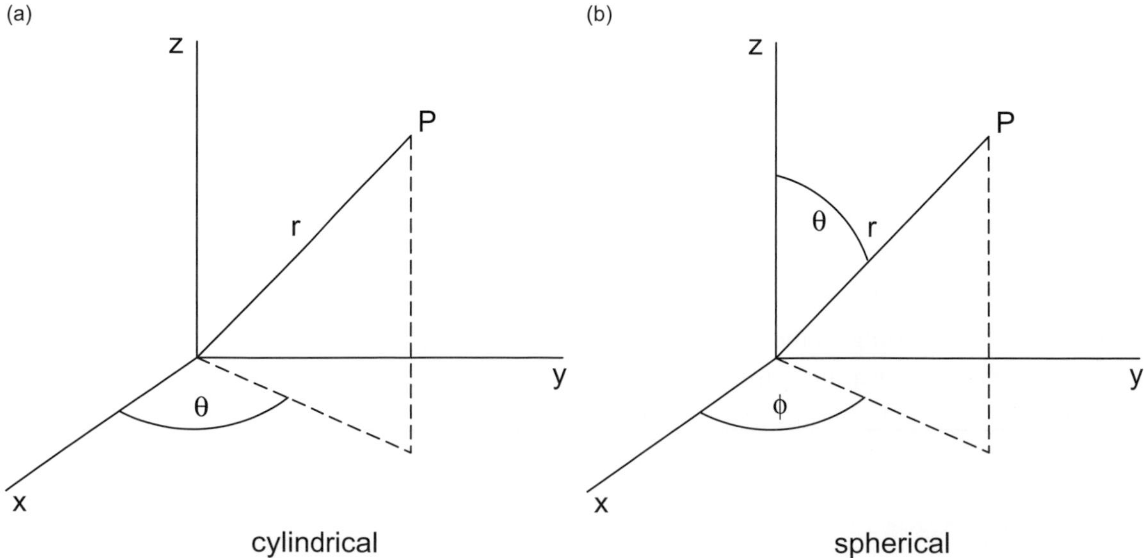

Figure 4.12 Cylindrical coordinates (r, θ, z), where $r \geq 0, 0 \leq \theta \leq 2\pi$, and spherical coordinates (r, θ, ϕ), where $r \geq 0,\ 0 \leq \theta \leq \pi,\ 0 \leq \phi \leq 2\pi$.

$$\frac{\partial \rho}{\partial t}+\frac{1}{r^2}\frac{\partial}{\partial r}\left(\rho r^2 u_r\right)+\frac{1}{r\sin\theta}\frac{\partial}{\partial \theta}\left(\rho u_\theta \sin\theta\right) +\frac{1}{r\sin\theta}\frac{\partial}{\partial \phi}\left(\rho u_\phi\right)=0 \tag{4.6.8}$$

The incompressible form is

$$\frac{1}{r^2}\frac{\partial}{\partial r}\left(r^2 u_r\right)+\frac{1}{r\sin\theta}\frac{\partial}{\partial \theta}\left(u_\theta \sin\theta\right)+\frac{1}{r\sin\theta}\frac{\partial}{\partial \phi}\left(u_\phi\right)=0 \tag{4.6.9}$$

For axisymmetric flow in spherical coordinates, where $u_\phi = 0$ and $\partial/\partial\phi = 0$, Eq. (4.6.9) simplifies to

$$\frac{1}{r^2}\frac{\partial}{\partial r}\left(r^2 u_r\right)+\frac{1}{r\sin\theta}\frac{\partial}{\partial \theta}\left(u_\theta \sin\theta\right)=0 \tag{4.6.10}$$

Defining the relationship between the velocities and the stream function as

$$u_r = +\frac{1}{r^2\sin\theta}\frac{\partial \psi}{\partial \theta},\quad u_\theta = -\frac{1}{r\sin\theta}\frac{\partial \psi}{\partial r} \tag{4.6.11}$$

results in Eq. (4.6.11) being identically zero

$$\frac{1}{r^2}\frac{\partial}{\partial r}\left(\frac{1}{\sin\theta}\frac{\partial \psi}{\partial \theta}\right)+\frac{1}{r\sin\theta}\frac{\partial}{\partial \theta}\left(-\frac{1}{r}\frac{\partial \psi}{\partial r}\right)\equiv 0 \tag{4.6.12}$$

The axisymmetric stream function in spherical coordinates is also called the Stokes stream function.

Likewise, the three forms of the transport equation for a soluble species are:

Rectangular

$$\frac{\partial c}{\partial t}+u\frac{\partial c}{\partial x}+v\frac{\partial c}{\partial y}+w\frac{\partial c}{\partial z}=D\left(\frac{\partial^2 c}{\partial x^2}+\frac{\partial^2 c}{\partial y^2}+\frac{\partial^2 c}{\partial z^2}\right) \tag{4.6.13}$$

Cylindrical

$$\frac{\partial c}{\partial t}+u_r\frac{\partial c}{\partial r}+\frac{u_\theta}{r}\frac{\partial c}{\partial \theta}+u_z\frac{\partial c}{\partial z} =D\left(\frac{1}{r}\frac{\partial}{\partial r}\left(r\frac{\partial c}{\partial r}\right)+\frac{1}{r^2}\frac{\partial^2 c}{\partial \theta^2}+\frac{\partial^2 c}{\partial z^2}\right) \tag{4.6.14}$$

Spherical

$$\frac{\partial c}{\partial t}+u_r\frac{\partial c}{\partial r}+\frac{u_\theta}{r}\frac{\partial c}{\partial \theta}+\frac{u_\phi}{r\sin\theta}\frac{\partial c}{\partial \phi} = D\left(\frac{1}{r^2}\frac{\partial}{\partial r}\left(r^2\frac{\partial c}{\partial r}\right)+\frac{1}{r^2\sin\theta}\frac{\partial}{\partial \theta}\left(\sin\theta\frac{\partial c}{\partial \theta}\right) +\frac{1}{r^2\sin^2\theta}\frac{\partial^2 c}{\partial \phi^2}\right) \tag{4.6.15}$$

Box 4.1 | 2D Planar Flow in Polar Coordinates

Two-dimensional flow in polar coordinates obey the simplified conservation of mass equation by dropping the u_z term in Eq. (4.6.4). The result is

$$\frac{\partial}{\partial r}\left(r u_r\right)+\frac{\partial}{\partial \theta}\left(u_\theta\right)=0 \tag{4.6.16}$$

We can define a stream function, ψ, for this flow

$$u_r = \frac{1}{r}\frac{\partial \psi}{\partial \theta},\quad u_\theta = -\frac{\partial \psi}{\partial r} \tag{4.6.17}$$

which satisfies Eq. (4.6.16) identically.

Two well-known flows in polar coordinates are a point source (or sink) at $r = 0$ and the point vortex circulating around $r = 0$. The point source has only radial flow,

$$u_r = \frac{Q}{2\pi r},\ u_\theta = 0 \tag{4.6.18}$$

where Q is the strength of the source and is equal to the total flow across a circle of radius r,

$$\int_0^{2\pi} u_r \, r \, d\theta = \int_0^{2\pi} \frac{Q}{2\pi r} r \, d\theta = Q \tag{4.6.19}$$

Note that for $Q > 0$, $Q < 0$, the flow is radially outward, inward, modeling a point source, sink, respectively.

The corresponding stream function is

$$\psi = \frac{Q}{2\pi} \theta \tag{6.20}$$

Figure 4.13(a) shows the stream function plotted in 3D and Figure 4.13(b) has the corresponding streamlines with arrows indicating outward flow with $Q > 0$.

The point vortex has only azimuthal flow

$$u_r = 0, \; u_\theta = \frac{\Gamma}{2\pi r} \tag{4.6.21}$$

where Γ is the vortex strength or circulation strength. The definition of circulation in such flows is

$$\int_0^{2\pi} u_\theta \, r \, d\theta = \int_0^{2\pi} \frac{\Gamma}{2\pi r} r \, d\theta = \Gamma \tag{4.6.22}$$

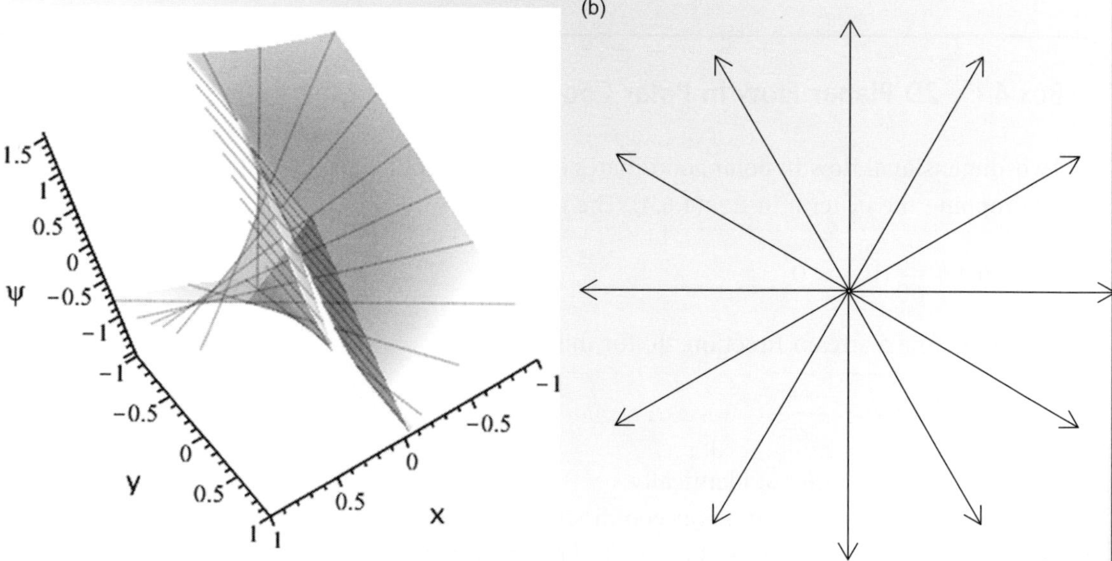

Figure 4.13 Stream function (a) and streamlines (b) for a point source with $Q > 0$.

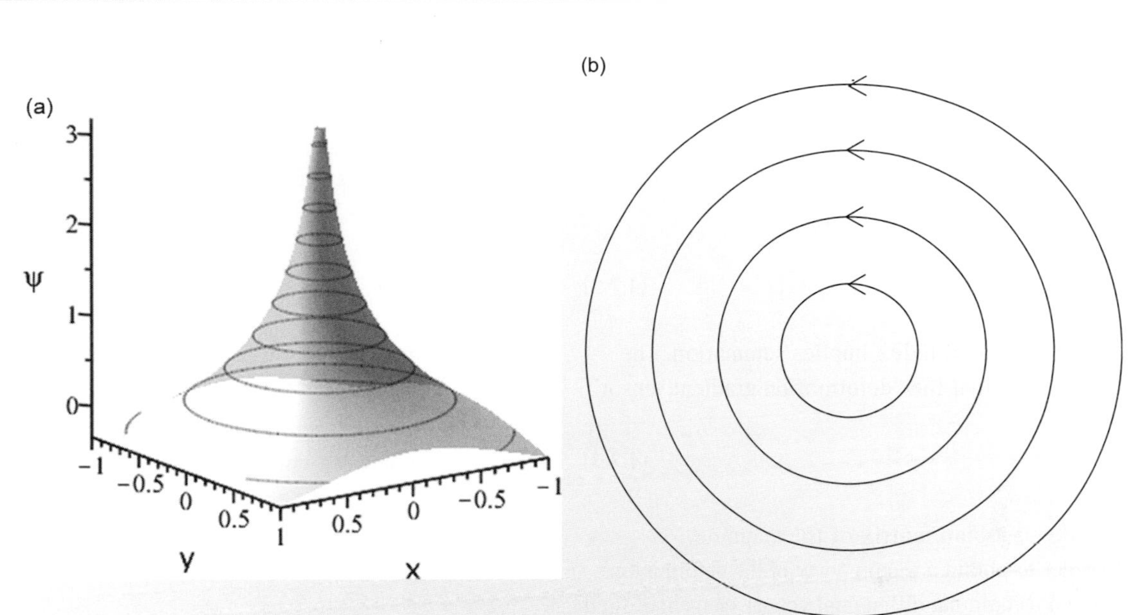

Figure 4.14 Stream function (a) and streamlines (b) for a point vortex with $\Gamma > 0$.

For $\Gamma > 0$, $\Gamma < 0$, the flow is in the positive, negative, θ-direction respectively.

The corresponding stream function is

$$\psi = \frac{\Gamma}{2\pi} \ln\left(\frac{1}{r}\right) \tag{4.6.23}$$

Figure 4.14(a) shows the stream function plotted in 3D for a point vortex, while Figure 4.14(b) has the corresponding streamlines with arrows indicating $\Gamma > 0$.

4.7 Kinematics of Fluid Deformation

A differential material line element of a continuum at starts at $t = 0$, in its initial state, call it $d\underline{X}_0 = (dX_0, dY_0, dZ_0) = \left(dX_0^1, dX_0^2, dX_0^3\right)$. After deformation over a small time interval, the new state of the line element is $d\underline{X} = (dX, dY, dZ) = (dX_1, dX_2, dX_3)$, see Figure 4.15. By a linearized Taylor series for small deformation, the two states are related by

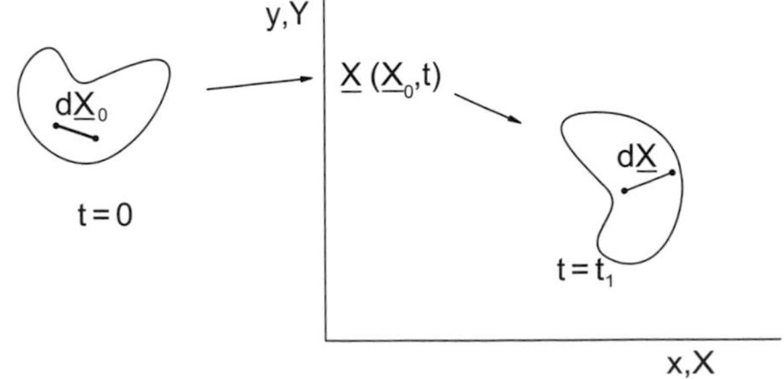

Figure 4.15 Deformation of a differential material line element.

$$\begin{pmatrix} dX \\ dY \\ dZ \end{pmatrix} = \begin{pmatrix} \frac{\partial X}{\partial X_0} & \frac{\partial X}{\partial Y_0} & \frac{\partial X}{\partial Z_0} \\ \frac{\partial Y}{\partial X_0} & \frac{\partial Y}{\partial Y_0} & \frac{\partial Y}{\partial Z_0} \\ \frac{\partial Z}{\partial X_0} & \frac{\partial Z}{\partial Y_0} & \frac{\partial Z}{\partial Z_0} \end{pmatrix} \begin{pmatrix} dX_0 \\ dY_0 \\ dX_0 \end{pmatrix} \quad (4.7.1)$$

The index form of Eq. (4.7.1) is

$$dX_i = \frac{\partial X_i}{\partial X_0^j} dX_0^j = F_{ij} dX_0^j \quad (4.7.2)$$

where the repeated index implies summation. The tensor, $\underline{\underline{F}}$, is called the "deformation gradient tensor"

$$F_{ij} = \frac{\partial X_i}{\partial X_0^j} \quad (4.7.3)$$

and is the Jacobian matrix of the mapping.

In order to obtain a picture view of the deformation, consider a rectangular differential volume element of fluid at time $t = 0$ whose edges are the differential vectors $d\underline{X}_0$, $d\underline{Y}_0$, $d\underline{Z}_0$ that are aligned with their respective axes, as shown in Figure 4.16. The fluid then undergoes a deformation from the mapping of fluid points,

$$\underline{X} = \underline{X}(\underline{X}_0, t) \quad (4.7.4)$$

such that at a later time, t_1, the volume element is a parallelepiped, also shown in Figure 4.16.

The initial volume of the element may be derived from the vector relationship

$$dV_0 = d\underline{X}_0 \cdot (d\underline{Y}_0 \times d\underline{Z}_0) = dX_0 dY_0 dZ_0 \quad (4.7.5)$$

where dX_0, dY_0, dZ_0 are the lengths of the respective differential line elements. The volume after the deformation may be derived similarly

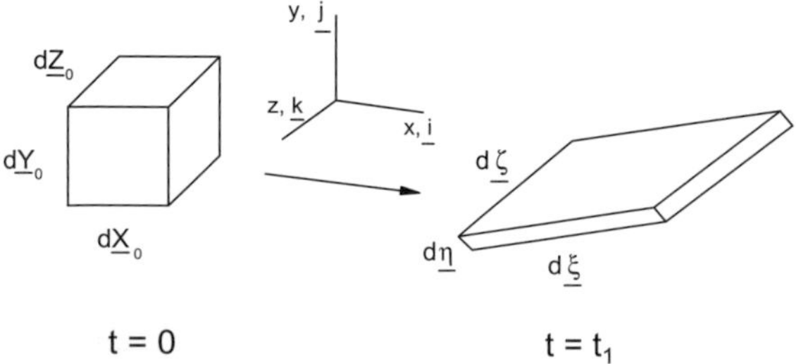

Figure 4.16 A differential volume element undergoing deformation.

$$dV = d\underline{\xi} \cdot \left(d\underline{\eta} \times d\underline{\zeta}\right) \quad (4.7.6)$$

where the new edges are related to the old edges by the differential vector relationship, which is simply a Taylor series truncated after the linear term

$$d\underline{\xi} = \underline{i}\frac{\partial X}{\partial X_0} dX_0 + \underline{j}\frac{\partial X}{\partial Y_0} dY_0 + \underline{k}\frac{\partial X}{\partial Z_0} dZ_0$$

$$d\underline{\eta} = \underline{i}\frac{\partial Y}{\partial X_0} dX_0 + \underline{j}\frac{\partial Y}{\partial Y_0} dY_0 + \underline{k}\frac{\partial Y}{\partial Z_0} dZ_0$$

$$d\underline{\zeta} = \underline{i}\frac{\partial Z}{\partial X_0} dX_0 + \underline{j}\frac{\partial Z}{\partial Y_0} dY_0 + \underline{k}\frac{\partial Z}{\partial Z_0} dZ_0 \quad (4.7.7)$$

Inserting Eq. (4.7.7) into Eq. (4.7.6) leads to the relationship between volume elements

$$dV = J dV_0 \quad (4.7.8)$$

where J is the determinant of the Jacobian matrix,

$$J = \det \begin{vmatrix} \frac{\partial X}{\partial X_0} & \frac{\partial X}{\partial Y_0} & \frac{\partial X}{\partial Z_0} \\ \frac{\partial Y}{\partial X_0} & \frac{\partial Y}{\partial Y_0} & \frac{\partial Y}{\partial Z_0} \\ \frac{\partial Z}{\partial X_0} & \frac{\partial Z}{\partial Y_0} & \frac{\partial Z}{\partial Z_0} \end{vmatrix} \quad (4.7.9)$$

The determinant of the Jacobian matrix, J, is often just called the Jacobian after Carl Gustav Jacob Jacobi (1804–1851), and we will follow that convention. Equation (4.7.8) tells us that the Jacobian of the mapping has a physical interpretation as the local ratio of the deformed volume to the initial volume. If there are no sources or sinks of fluid mass, then we also have that the mass in the initial volume, $R_0 dV_0$ equals the mass in the deformed volume, R dV so that the relationship to J is

$$J = \frac{dV}{dV_0} = \frac{R_0}{R} \quad (4.7.10)$$

where R_0 is the initial Lagrangian density and R is the density in the deformed state.

If a fluid is incompressible, then the differential volume elements are equal, and that occurs when

$$J = 1 \quad (4.7.11)$$

which is the Lagrangian constraint on the mapping for an incompressible fluid. Let us check to see if stagnation point flow satisfies the $J = 1$ criterion. Recalling that the mapping for stagnation point flow is $X = X_0 e^{At}, Y = Y_0 e^{-At}$, then the determinant of the Jacobian is given by

$$\det \begin{vmatrix} \frac{\partial X}{\partial X_0} & \frac{\partial X}{\partial Y_0} \\ \frac{\partial Y}{\partial X_0} & \frac{\partial Y}{\partial Y_0} \end{vmatrix} = \det \begin{vmatrix} e^{At} & 0 \\ 0 & e^{-At} \end{vmatrix} = 1 \tag{4.7.12}$$

and we see again that stagnation point flow satisfies incompressibility.

To develop the conservation laws in the Lagrangian frame, we first need to describe the time rate of change of a differential fluid volume element. The Lagrangian time derivative of the differential Lagrangian volume, dV, from Eq. (4.7.8), is

$$\begin{aligned} \frac{D}{Dt}(dV) &= \frac{D}{Dt}(J dV_0) = J\frac{D}{Dt}(dV_0) + dV_0\frac{D}{Dt}(J) \\ &= dV_0\frac{D}{Dt}(J) \end{aligned} \tag{4.7.13}$$

since dV_0 is independent of time. To determine the Lagrangian time derivative of J, let us do a simple example in 1-dimension where the mapping is $X(X_0, t)$. Consider the Jacobian

$$J = \det\left|\frac{\partial X}{\partial X_0}\right| = \frac{\partial X}{\partial X_0} \tag{4.7.14}$$

Now take the time derivative of J, and interchange the order of differentiation noting that the material time derivative of X is simply the Lagrangian velocity component, U,

$$\frac{DJ}{Dt} = \frac{D}{Dt}\left(\frac{\partial X}{\partial X_0}\right) = \frac{\partial}{\partial X_0}\left(\frac{DX}{Dt}\right) = \frac{\partial U}{\partial X_0} \tag{4.7.15}$$

To obtain a final result useful in Eulerian variables, we anticipate needing to express the velocity components in Eq. (4.7.15), $U(X_0, t)$, in the form $U(X, t)$ by using the inverse mapping. However, we do not need the inverse mapping explicitly, we can simply use the chain rule of partial differentiation for the spatial derivative of the velocity assuming the variable change

$$\frac{\partial U}{\partial X_0} = \frac{\partial X}{\partial X_0}\frac{\partial U}{\partial X} = J\underline{\nabla}\cdot\underline{U} \tag{4.7.16}$$

Then translating to Eulerian variables, $x = X$ and $u = U$ gives us the general result, which holds for two and three dimensions as well

$$\frac{DJ}{Dt} = J\underline{\nabla}\cdot\underline{u} \tag{4.7.17}$$

where the gradient operator and the velocity are both in Eulerian variables.

Solving for the divergence of the velocity in Eq. (4.7.17) and inserting into Eq. (4.1.18) we have

$$\frac{D\rho}{Dt} + \rho\underline{\nabla}\cdot\underline{u} = \frac{D\rho}{Dt} + \rho\frac{1}{J}\frac{DJ}{Dt} = \frac{1}{J}\frac{D(\rho J)}{Dt} = 0 \tag{4.7.18}$$

which simplifies to

$$\frac{D(\rho J)}{Dt} = 0 \tag{4.7.19}$$

Equation (4.7.18) and its equivalence in (4.7.19) gives us the full relationship between Eulerian and Lagrangian conservation of mass.

Worked Example 4.4

Given the Eulerian velocity field in one dimension

$$u(x, t) = \frac{2xt}{1 + t^2} \tag{4.7.20}$$

find the corresponding Eulerian density field $\rho(x, t)$, assuming its initial value is

$$\rho(x, t = 0) = a_0 x \tag{4.7.21}$$

where a_0 is a constant. First we need to find the mapping, so rewriting Eq. (4.7.20) in the equivalent Lagrangian form we have

$$\left(\frac{\partial X}{\partial t}\right)_{X_0} = \frac{2Xt}{1+t^2} \tag{4.7.22}$$

Equation (4.7.22) can be separated and integrated

$$\int \frac{\partial X}{X} = \int \frac{2t}{1+t^2}\, \partial t + c_0 \quad \rightarrow \quad \ln X = \ln\left(1+t^2\right) + c_0 \tag{4.7.23}$$

where c_0 is an integration constant that depends in general on X_0 which was being held fixed in both the differentiation and the integration. Exponentiating both sides yields

$$X = c_1\left(1+t^2\right) \tag{4.7.24}$$

and $c_1 = e^{c_0}$ is just a different version of the integration constant. The initial condition for the mapping, $X(t = 0) = X_0$, determines that $c_1 = X_0$, so that the mapping is

$$X = X_0\left(1+t^2\right) \tag{4.7.25}$$

while the corresponding inverse mapping, $X_0(x, t)$, is easily found to be

$$X_0 = x\left(1+t^2\right)^{-1} \tag{4.7.26}$$

The Jacobian of the mapping is simply

$$J = \frac{\partial X}{\partial X_0} = 1 + t^2 \tag{4.7.27}$$

Now to discover the density field, recast Eq. (4.7.19) in Lagrangian variables as

$$\left(\frac{\partial (RJ)}{\partial t}\right)_{X_0} = 0 \tag{4.7.28}$$

where $R(X_0, t)$ is the Lagrangian density field. Equation (4.7.28) states that the product RJ is at most a function of X_0,

$$RJ = R\left(1+t^2\right) = f(X_0) \tag{4.7.29}$$

which gives us that

$$R = \frac{f(X_0)}{(1+t^2)} \tag{4.7.30}$$

We determine f from the initial density condition translated into Lagrangian variables replacing ρ with R and x with $X(X_0, t)$ in Eq. (4.7.21)

$$R(X_0, t=0) = a_0 X(t=0) = a_0 X_0 = f(X_0) \quad \rightarrow \quad R = \frac{a_0 X_0}{(1+t^2)} \tag{4.7.31}$$

To determine the Eulerian density field, $\rho(x, t)$ we only need to insert the inverse mapping for X_0, Eq. (4.7.26), into Eq. (4.7.31), which gives the final result

$$\rho(x,t) = \frac{a_0 x}{(1+t^2)^2} \tag{4.7.32}$$

4.8 The Reynolds Transport Theorem

This simple formula, Eq. (4.7.17), forms the backbone of an important equation used in fluid mechanics, the Reynolds Transport Theorem. That theorem deals with derivatives of integrals over Lagrangian volumes and surfaces which are moving with the fluid. Consider an arbitrary Lagrangian, or material, volume V(t) shown in Figure 4.17. The surface, S(t), consists of fluid material points and moves with the fluid. Consequently, no fluid crosses the surface, S(t).

The Lagrangian time derivative of the integral over the material volume, V(t), of an Eulerian function $f(\underline{x}, t)$, is

$$\frac{D}{Dt}\iiint_{V(t)} f(\underline{x}, t)\ dV \tag{4.8.1}$$

where we note that the limits of integration for this moving volume of fluid points are time-dependent. The same fluid particles are always inside this volume and none may be convected across its surface S(t). We utilize the mapping and represent the integral in initial position space where the limits of integration over the volume at $t = 0$, call it V_0, are no longer dependent on time. Using $dV = J dV_0$ from Eq. (4.7.8) we have

$$\frac{D}{Dt}\iiint_{V(t)} f(\underline{x}, t)\ dV = \frac{D}{Dt}\iiint_{V_0} F(\underline{X}_0, t) J\ dV_0 \tag{4.8.2}$$

In Eq. (4.8.2) $f(\underline{x}, t)$ has been transformed to its Lagrangian representation, $F(\underline{X}_0, t)$ through the mapping and the volume has been transformed to the initial position space. The limits of integration are independent of time so we may move the time derivative inside of the integral signs of Eq. (4.8.2) as follows, using the product rule for differentiation

$$\begin{aligned}\iiint_{V_0}\left(\frac{DF}{Dt}J + F\frac{DJ}{Dt}\right) dV_0 &= \iiint_{V_0}\left(\frac{DF}{Dt}J + FJ\underline{\nabla}\cdot\underline{U}\right) dV_0 \\ &= \iiint_{V_0}\left(\frac{DF}{Dt} + F\underline{\nabla}\cdot\underline{U}\right) J dV_0\end{aligned} \tag{4.8.3}$$

where the second step inserts our formula for the Lagrangian time derivative of J, Eq. (4.7.17), and we again note that dV_0 is time-independent. Having collected the common term, $J\ dV_0$ in the last step of Eq. (4.8.3), we may relate the integral back to the moving volume and surface since $J\ dV_0 = dV$, and use Eulerian forms of $\underline{U}$ and F, i.e. $\underline{u}$ and f,

$$\iiint\limits_{V_0} \left(\frac{DF}{Dt} + F\underline{\nabla}\cdot\underline{U}\right) J dV_0 = \iiint\limits_{V(t)} \left(\frac{\partial f}{\partial t} + \underline{u}\cdot\underline{\nabla} f + f\underline{\nabla}\cdot\underline{u}\right) dV = \iiint\limits_{V(t)} \left(\frac{\partial f}{\partial t} + \underline{\nabla}\cdot(f\underline{u})\right) dV \tag{4.8.4}$$

The middle step in Eq. (4.8.4) has used the definition of the material derivative. Now we have shown that

$$\frac{D}{Dt}\iiint\limits_{V(t)} f(\underline{x}, t)\ dV = \iiint\limits_{V(t)} \left(\frac{\partial f}{\partial t} + \underline{\nabla}\cdot(f\underline{u})\right) dV = \iiint\limits_{V(t)} \left(\frac{\partial f}{\partial t} + \frac{\partial}{\partial x_k}(f u_k)\right) dV \tag{4.8.5}$$

We can use the divergence theorem to change the last volume integral of $\nabla\cdot(f\underline{u})$ to a surface integral of $(f\underline{u})\cdot\underline{n}$

$$\frac{D}{Dt}\iiint\limits_{V(t)} f(\underline{x},\ t)\ dV = \iiint\limits_{V(t)} \left(\frac{\partial f}{\partial t}\right) dV + \iint\limits_{S(t)} ((f\underline{u})\cdot\underline{n})\ dS \tag{4.8.6}$$

Equation (4.8.5) and its alternate form, Eq. (4.8.6), are known as the Reynolds Transport Theorem and named for Osborne Reynolds (1842–1912). This theorem is used to derive the conservation laws using a Lagrangian control volume. It is a three-dimensional extension of Leibniz's rule for derivatives of integrals when the limits of integration depend on the independent variable

$$\frac{d}{dt}\int_{x_1(t)}^{x_2(t)} f(x, t)\ dx = \int_{x_1(t)}^{x_2(t)} \frac{\partial f}{\partial t} dx + f(x_2, t)\frac{dx_2}{dt} - f(x_1, t)\frac{dx_1}{dt} \tag{4.8.7}$$

due to the German mathematician Gottfried Wilhelm Leibniz (1646–1716).

4.9 Mass Conservation from a Lagrangian Control Volume

Let $f = \rho$ in Eq. (4.8.5) and recall that because V(t) is a Lagrangian volume, its surface S(t) is made of fluid material points, see Figure 4.17, where the volume V(t) is shown at two different times. The fluid inside the volume V(t), and bounded by the surface V(t), always remains in V(t) and does not cross S(t), i.e. S(t) has no holes. Assuming there are no other surface sources of mass, then the mass inside V(t) is constant, i.e.

$$\frac{D}{Dt}\iiint\limits_{V(t)} \rho(\underline{x}, t)\ dV = 0 = \iiint\limits_{V(t)} \left(\frac{\partial \rho}{\partial t} + \underline{\nabla}\cdot(\rho\underline{u})\right) dV \tag{4.9.1}$$

from Eq. (4.8.5). Then, because V(t) is an arbitrary Lagrangian fluid volume, the integrand must be zero and we have

$$\frac{\partial \rho}{\partial t} + \underline{\nabla}\cdot(\rho\underline{u}) = 0 \tag{4.9.2}$$

which is identical to Eq. (4.1.15) for $\sigma_\rho = 0$.

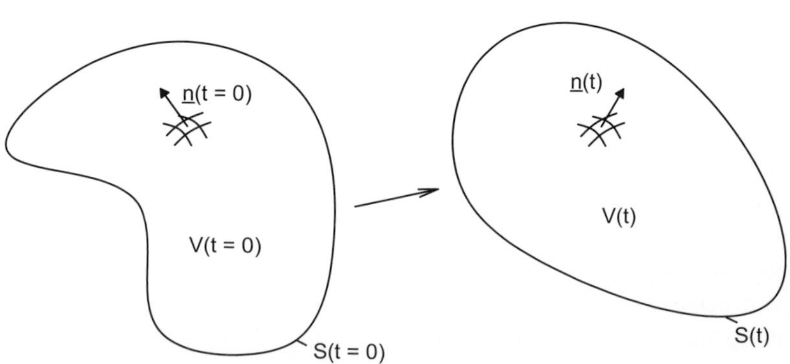

Figure 4.17 Lagrangian control volume V(t) and its surface, S(t), and outward pointing unit normal, $\underline{n}(t)$, shown at two times, t = 0 and t = t.

Summary

Chapter 4 is our first introduction to the control volume concept, approached either as an Eulerian control volume fixed in space, or a Lagrangian control volume moving with the fluid. Either through the divergence theorem for an Eulerian approach, or the Reynolds Transport Theorem for a Lagrangian approach, the goal was to start with an integro-differential equation for mass balance of the volume and then derive from that a partial differential equation which holds at any spatial point and time in the fluid. The definition of a stream function, ψ, is a way to simplify certain flows which are restricted to two velocity components and no variations in the remaining third direction. Flow through the nose gave us a chance to learn about real streamlines and flow analysis in hardware models which assist medical and surgical interventions. Rectangular, cylindrical and spherical coordinate systems each have their version of a stream function. In addition, we shall discover in conservation of momentum that the stream function allows us to represent the governing balance as a single equation in the scalar ψ, rather than the two vector component equations for (u, v), to use rectangular coordinates as an example. The conservation of a soluble species is an essential study in biofluid mechanics since life depends on this interplay of transport by convection and diffusion at different scales.

Problems

4.1 Flow in the x–y plane has the unsteady Eulerian velocity field

$$u = y + t, v = -x \qquad (1)$$

where the variables are already dimensionless.

a. Is this flow incompressible?
b. What is the stream function, $\psi(x, y, t)$, for this flow? To determine any integration constants you can assume $\psi(x = 0, y = 0) = 0$
c. Plot ψ in 3D for $t = 0.1$ and $-1 \leq x \leq 1$, $-1 \leq y \leq 1$. What shape is the surface and where is its minimum located?
d. Plot both the streamlines, using ψ, and the Eulerian velocity vector field on the same graph for the domain $-1 \leq x \leq 1$, $-1 \leq y \leq 1$ at time $t = 0.1$. Describe your results.
e. Plot two sets of streamlines using ψ for the domain $-1 \leq x \leq 1$, $-1 \leq y \leq 1$ at two times, $t = 0, 0.1$, on the same graph. What shapes do you see? If the shapes look familiar, try to recast your mathematical form of ψ to reflect that observation.
f. What is the volume flow rate across $x = 0$ from $-0.5 \leq y \leq 0$ at $t = 0.1$? Calculate your answer in two ways, one from a direct integration of the Eulerian velocity and the other from the difference in stream function values.
g. Solve for the mapping, either manually, which involves taking the time derivative of one component and substituting from the other, or using computational software to find an analytic solution. Use the initial conditions $(X, Y)_{t=0} = (X_0, Y_0)$. Plot pathlines for two fluid particles that start at $(X_0, Y_0) = (0, 2), (0, 5)$ for $0 \leq t \leq 10$. Describe these trajectories and relate to the streamlines and velocity vector field.

4.2 For stagnation point flow where $\underline{u} = (u, v) = (Ax, -Ay)$ our solution of the transport equation in dimensionless form is given in Eq. (4.5.25) and the mass flux in Eq. (4.5.7).

a. What is the unit normal pointing into the fluid from the wall $y = 0$? Call it $\underline{n}_w$.
b. Let $A = 1\,s^{-1}$ and $D_m = 0.18\ cm^2/s$, which is typical for diffusion through a gas. Keep C_0 arbitrary. Calculate the total wall flux $F_w = \underline{J}\cdot\underline{n}_w$ and integrate it over the wall segment $0 \leq x \leq 1$ cm to obtain the total mass flow rate, $\dot{M}_w$, there.
c. What is the unit normal along the line $y = x$? Call it $\underline{n}_{45}$. Hint: make sure $\underline{n}_{45}$ has unit length and choose $\underline{n}_{45}\cdot\underline{e} > 0$.
d. For the parameter values in part (b), calculate the total flux across this line $F_{45} = \underline{J}\cdot\underline{n}_{45}$ and integrate it over the line segment joining the two points $(0, 0)$ and $(1, 1)$cm to obtain the total mass flow rate, $\dot{M}_{45}$, there. Hint: do your integration with respect to y letting $x = y$ and use your computational software as needed.
e. Using the mapping for the flow, plot the Lagrangian concentration history for a fluid particle which starts at $\underline{X}_0 = (1, 1)$cm for $A = 0.1, 0.3$ and $D_m = 0.18\,cm^2/s$. Discuss your results.

4.3 Consider the Eulerian velocity field for translating cellular flow

$$u = U_0 \sin(ky), \quad v = V_0 \sin(kx - \omega t) \tag{1}$$

a. Is this flow incompressible?
b. Let $\xi = kx, \eta = ky, T = \omega t, \hat{u} = u/U_0, \hat{v} = v/V_0$ and rewrite Eq. (1) in these dimensionless variables so that you have the velocities $\hat{u}, \hat{v}$ in terms of the spatial and time independent variables ξ, η, T.
c. Plot the velocity vector field from part (b) for $T = 0$ over the interval $-2\pi \leq \xi \leq 2\pi, -2\pi \leq \eta \leq 2\pi$.
d. Determine the stream function $\psi(\xi, \eta, T)$ from your results in part (b).
e. For the same independent variable values in part (c), plot the streamlines from constant values, contours, of ψ.
f. Animate the vector field plot and the stream function plot together in one graph over one cycle, $0 \leq T \leq 2\pi$, and describe your results.

4.4

a. For point source flow the Eulerian velocity field in polar coordinates is $u_r = Q/(2\pi r), u_\theta = 0$. Find the mapping of the fluid such that Lagrangian velocities are given by $U_R = \partial R/\partial t$ and $U_\Theta = R\partial\Theta/\partial t$. It should be in terms of initial positions R_0, Θ_0, the source strength, Q, and time, t.
b. Consider a tagger which moves at constant speed, v_T, in a circle of constant radius r_1 starting at $\theta = 0$ and ending at $\theta = 2\pi$. What is the tagger position if $\underline{r}_T = (r_T, \theta_T(s))$ in terms of the parameters and the time variable, s? What is the range of s?
c. Solve for the initial positions of the fluid particles that get tagged when $t = s$. Call them $(R_0^*(s), \theta_0^*(s))$.
d. Using $(R_0^*(s), \theta_0^*(s))$ from part (c) as the initial positions in the mapping in part (a), solve for the streakline when the tagger first arrives at $\theta_T = 2\pi$ which fixes the value of t. Now your solution depends on the s time variable, not t.
e. Scale your results in part (d) as $\eta = R/r_1, S = v_T s/2\pi r_1$ to make your solution dimensionless, using $\beta = Q/(2\pi r_1 v_T)$. What is the range of S? What is the physical interpretation of the dimensionless parameter β?
f. Plot solutions for the streakline using four values of $\beta = 0.5, 2.0, 5.0, 10.0$ and discuss your results.

4.5

a. For vortical flow the Eulerian velocity field in polar coordinates is $u_r = 0, u_\theta = K/2\pi r$ where K is the vortex strength. Find the mapping of the fluid such that Lagrangian velocities are given by $U_R = \partial R/\partial t$ and $U_\Theta = R\partial\Theta/\partial t$. It should be in terms of initial positions R_0, Θ_0, the vortex strength, K, and time, t.
b. Consider a tagger which moves at constant speed, v_T, in the outward radial direction along $\theta = 0$ starting at r_1 and ending at r_2. What is the tagger position if $\underline{r}_T = (r_T(s), \theta_T)$ in terms of the parameters and the time variable, s? What is the range of s?

c. Solve for the initial positions of the fluid particles that get tagged when $t = s$. Call them $(R_0^*(s), \Theta_0^*(s))$.
d. Using $(R_0^*(s), \theta_0^*(s))$ from part (c) as the initial positions in the mapping in part (a), solve for the streakline when the tagger first arrives at $r_T = r_2$ which fixes the value of t. Now your solution depends on the s time variable, not t.
e. Scale your results in part (d) as $\eta = R/r_1$, $S = v_T s/(r_2 - r_1)$ to make your solution dimensionless. Note that Θ is already dimensionless. What is the range of S? Formulate your solution in terms of two dimensionless parameters: $\hat{K} = K/2\pi r_1 v_T$ and $\lambda = (r_2 - r_1)/r_1$. If it is easier, substitute for K in terms of $\hat{K}$, s in terms of S and $r_2 - r_1$ in terms of λ and simplify. What are the physical interpretations of $\hat{K}, \lambda$?
f. Plot streakline solutions $(\eta(S), \Theta(S))$ over the range of S using parametric plotting in polar coordinates. Do this for four values of $\hat{K} = 2.0, 4.0, 8.0, 12.0$ for $\lambda = 0.5$ and discuss your results.

4.6

a. For vertical flow the Eulerian velocity field in polar coordinates is $u_r = 0, u_\theta = K/2\pi r$ where K is the vortex strength. Find the mapping of the fluid such that Lagrangian velocities are given by $U_R = \partial R/\partial t$ and $U_\Theta = R\partial\Theta/\partial t$. It should be in terms of initial positions R_0, Θ_0, the vortex strength, K, and time, t.
b. Let a timeline at $t = 0$ be the line segment $r_1 \leq r \leq r_2$ for $\theta = 0$. You can parameterize it as the set of initial points $R_0 = r_1 + h(r_2 - r_1)$ and $\Theta_0 = 0$, where the parameter has the range $0 \leq h \leq 1$. Insert this set of initial positions into the mapping and scale your results as $\eta = R/r_1, T = kt/2\pi r_1^2$ to make your solution dimensionless.
c. What is the physical interpretation of the dimensionless parameter $\lambda = (r_2 - r_1)/r_1$?
d. Plot solutions for the timeline in polar coordinates using four values of $T = 1.0, 2.0, 4.0, 6.0$ for $\lambda = 0.5$.
e. Discuss your results.

4.7 Let's consider a one-dimensional, compressible flow that must obey conservation of mass as given by

$$\frac{D(\rho J)}{Dt} = \rho\frac{DJ}{Dt} + J\frac{D\rho}{Dt} = 0 \tag{1}$$

Let the mapping of a fluid motion be

$$X = X_0(1 + \varepsilon \sin(\omega t)) \tag{2}$$

According to Eq. (2), each fluid particle is oscillating around its initial position, $X(t = 0) = X_0$, with an amplitude that is proportional to the initial position, εX_0, and angular frequency ω. Assume that the amplitude is small, $\varepsilon << 1$, and we are limited to the strip $0 \leq X_0 \leq L$.

a. What is the Jacobian, J, of this mapping?
b. Insert J from part (a) into Eq. (1) and use Lagrangian notation $R(X_0, t) = \rho(x, t)$ to arrive at an equation involving only R and $\partial R/\partial t$.
c. Rearrange your result in part (b) to have R and ∂R on one side and ∂t on the other. Integrate that form and solve for R(t) to within a function of integration which can depend onX_0.
d. Suppose we let the initial condition be $R(X_0, t = 0) = R_0(1 + a_0 X_0)$ where R_0 and a_0 are constants, so that the initial Lagrangian density depends on the initial fluid particle position. Use this initial condition to solve for your function of integration in part (c). Now you have the unsteady Lagrangian density field $R(X_0, t)$.
e. Solve the inverse mapping of Eq. (2) and insert X_0 in terms of x and t into $R(X_0, t)$from part (d) to obtain the Eulerian density field $\rho(x, t)$.
f. The corresponding Lagrangian velocity is the time derivative of Eq. (2), $U(X_0, t) = \partial X/\partial t$. Substitute the inverse mapping into U to find the Eulerian velocity field $u(x, t)$
g. Scale your Eulerian equations for velocity, from part (f), and density, from part (e), using $\hat{\rho} = \rho/R_0, \hat{u} = u/\omega L, T = \omega t, \xi = x/L$. Also, let $\beta = a_0 L$. Choose $\varepsilon = 0.2$, $\beta = 0.2$ and display animated plots of $\hat{u}(\xi, T)$ and $\hat{\rho}(\xi, T)$ on the same graph for the spatial interval $0 \leq \xi \leq 1$ over one cycle, $0 \leq T \leq 2\pi$. Choose continuous replay and describe your results.

Image Credits

Figure 4.7 Source: Figure 1 from Hahn I, Scherer PW, Mozell MM. Velocity profiles measured for air-flow through a large-scale model of the human nasal cavity. *Journal of Applied Physiology.* 1993; 75(5):2273–87.

Figure 4.8 Source: Figure 3 from Hahn I, Scherer PW, Mozell MM. Velocity profiles measured for air-flow through a large-scale model of the human nasal cavity. *Journal of Applied Physiology.* 1993; 75(5):2273–87.

Figure 4.9 Source: Figure 4 from Hahn I, Scherer PW, Mozell MM. Velocity profiles measured for air-flow through a large-scale model of the human nasal cavity. *Journal of Applied Physiology.* 1993; 75(5):2273–87.

Figure 4.10 Source: Figure 12 from Keyhani K, Scherer PW, Mozell MM. Numerical simulation of airflow in the human nasal cavity. *Journal of Biomechanical Engineering – Transactions of the ASME.* 1995; 117(4):429–41.

5 Conservation of Momentum

Introduction

Conservation of momentum is Newton's second law of motion, which, for a point mass, states that the time rate of change of its momentum, $m\underline{v}$, is equal to the sum of the forces on the mass. We need to adapt that law to a continuum, such as a fluid, so we are interested in the time rate of change of momentum per unit volume, $\rho\underline{u}$. In addition, the forces on a fluid may be from contact (traction) between fluid particles, like pressure and viscous effects, or from forces at a distance, like gravity or electromagnetism, or from a boundary which may be formed by a solid or another fluid. The linear momentum conservation law we derive at this stage, Cauchy's momentum equation, holds for any continuum, solid or fluid. It is when we relate the local stresses to measures of the local continuum deformation that eventually we distinguish the two. Generally speaking, a solid would have stresses depending on the local strain, while fluid stresses depend on the local time rate of strain. More complicated behaviors exist, however, particularly in biological fluids (e.g. mucus, blood, synovial fluid), and we explore those in later chapters. For many traditional applications of fluid mechanics in engineering, the flows and stresses are not potentially damaging to either the fluid or the boundaries confining its motion. For biofluid mechanics the situation is significantly different. For example, regions of high shear stress in blood flow can tear open the red blood cells which make up 40% of its volume. On the other hand, regions of low shear stress can provoke the formation of blood clots from a biochemical cascade resulting in platelet activation. In addition, cells which make up the inner lining of blood vessels and airways are subject to fluid stresses and may secrete bioactive molecules if provoked sufficiently, as well as become damaged. In biofluid mechanics, the fluid and its boundary can be living.

Topics Covered

Having introduced the concept of a control volume in Chapter 4, it is possible for us to start more directly in this chapter to get to the main results. Using an Eulerian control volume, we derive the linear momentum balance, also known as Cauchy's momentum equation. It is derived for a general continuum, either fluid or solid, the approach is indifferent to that specification in the early going. Part of the derivation is, again, to change surface integrals to volume integrals and to postulate that the surface traction stress vector, $\underline{t}$, depends on the local surface unit normal, $\underline{n}$. This postulate is due to Cauchy, as is its relation to the stress tensor, $\underline{\underline{T}}$, which is examined in detail. It is the dependence of $\underline{\underline{T}}$ on measures of the continuum deformation that distinguish between fluids and solids. One of the key features of the stress tensor is that it is symmetric, $T_{ij} = T_{ji}$, and we show that this property implies conservation of angular momentum within the fluid. Considering a static fluid, the concept of pressure is explored and shown to be isotropic, the same in all directions at a point in the fluid. In the next chapter we build on this notion when extra stresses are derived from the viscous effects of fluid motion.

5.1 Conservation of Linear Momentum

The Eulerian control volume, V, that we used for conservation of mass in Figure 4.1, is shown here again in Figure 5.1. The balance of terms for conservation of linear momentum is given in Eq. (5.1.1)

$$\begin{array}{l}\text{Time rate of} \\ \text{momentum increase} \\ \text{in V}\end{array} = \begin{array}{l}\text{Net influx of momentum} \\ \text{through or from S}\end{array} + \begin{array}{l}\text{Sources of} \\ \text{momentum in V}\end{array} + \begin{array}{l}\text{Forces} \\ \text{on V}\end{array} \tag{5.1.1}$$

which states that the time rate of increase of linear momentum in V is equal to the net influx of momentum through S, or from it, plus any source of momentum inside V, plus any forces on V. The momentum inside the Eulerian control volume, V, is the volume integral of the Eulerian density field, $\rho(\underline{x}, t)$, multiplying the velocity vector $\underline{u}(\underline{x}, t)$. So its time derivative is

$$\frac{d}{dt}\iiint_V \rho\underline{u}\,dV \tag{5.1.2}$$

Mathematically, the balance of Eq. (5.1.1) is stated as follows

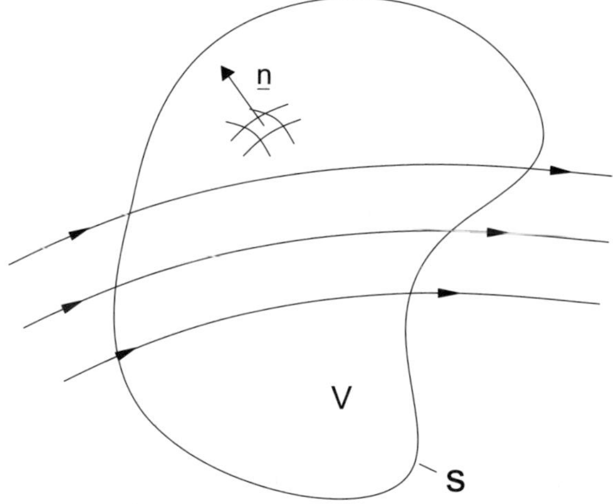

Figure 5.1 An Eulerian control volume, V, whose boundary is S and outward unit normal $\underline{n}$.

$$\frac{d}{dt}\iiint_V \rho\underline{u}\,dV = \iint_S \underline{\Sigma}_{mom}\,dS + \iiint_V \underline{\sigma}_{mom}\,dV + \underline{B} \tag{5.1.3}$$

where $\underline{\Sigma}_{mom}$ is the surface contribution, $\underline{\sigma}_{mom}$ is any momentum source (or sink) inside V, and $\underline{B}$ is the force on V. The force $\underline{B}$ can act at a distance, as in a gravity or magnetic field, and does not have a counterpart in mass conservation. Another major difference in momentum balance, as opposed to mass balance, is that momentum is a vector quantity, so the resulting equation, Eq. (5.1.3), is a vector equation with three components in the x_1-, x_2- and x_3-directions. As we did with mass conservation, $\underline{\Sigma}_{mom}$ can be divided into two parts; one from fluid convection across the boundary, and the other from any non-convective processes at the boundary. Then $\underline{\Sigma}_{mom}$ may be written as the sum

$$\underline{\Sigma}_{mom} = (\rho \underline{u}) \left(-\underline{u} \cdot \underline{n} \right) + \underline{\Sigma}'_{mom} \tag{5.1.4}$$

The first term in Eq. (5.1.4) is the convective contribution where $-\underline{u} \cdot \underline{n}$ is the surface fluid velocity component in the $-\underline{n}$-direction. The fluid carries the momentum $\rho \underline{u}$ and it is only the component that crosses the boundary that we seek. $\underline{\Sigma}'_{mom}$ is the non-convective contribution, such as surface stresses or traction on S. We ignored this term in mass conservation, but for momentum conservation it is very important and we will keep it in the balance, calling it $\underline{t}$, the surface stress vector, so $\underline{\Sigma}'_{mom} = \underline{t}$. Inserting Eq. (5.1.4) into Eq. (5.1.3), gives us the vector or index equation

$$\begin{aligned} \frac{d}{dt} \iiint_V \rho \underline{u} \, dV &= -\iint_S (\rho \underline{u}) (\underline{u} \cdot \underline{n}) \, dS + \iint_S \underline{t} \, dS + \\ &+ \iiint_V \underline{\sigma}_{mom} \, dV + \underline{B} \\ \frac{d}{dt} \iiint_V \rho u_i \, dV &= -\iint_S (\rho u_i)(u_k n_k) \, dS + \iint_S t_i \, dS \\ &+ \iiint_V \sigma_{i_{mom}} \, dV + B_i \quad i = 1, 2, 3 \end{aligned} \tag{5.1.5}$$

The value of the free index, i, in Eq. (5.1.5) identifies the vector component. As we did with mass balance, the Eulerian control volume is fixed in space, so the limits of integration over the volume, V, are time independent. That allows us to represent the first term of Eq. (5.1.5) as

$$\frac{d}{dt} \iiint_V \rho \underline{u} \, dV = \iiint_V \frac{\partial}{\partial t} (\rho \underline{u}) \, dV \tag{5.1.6}$$

where we have moved the time derivative inside the integral, becoming a partial derivative. The surface integral term of Eq. (5.1.5) for convective contributions of momentum has the form, in index notation,

$$-\iint_S \beta_{ik} n_k \, dS \quad \beta_{ik} = \rho u_i u_k \tag{5.1.7}$$

Notice the difference from conservation of mass, where we had $\beta_k n_k$ in the integrand of the surface integral, Eq. (4.1.13). $\beta_k n_k$ is a scalar, i.e. there is no free index, only the dummy summation index, and it represents the dot product, or inner product, of the vector β_k with the unit normal n_k. Now in Eq. (5.1.7) we have β_{ik}, which has two indices, so we can view it as a second-order tensor (matrix). This particular tensor, $\beta_{ik} = \rho u_i u_k$, is called a dyadic. The divergence theorem is handled the same way as before. In index notation it becomes

$$\iint_S \beta_{ik} n_k \, dS = \iiint_V \frac{\partial \beta_{ik}}{\partial x_k} \, dV \tag{5.1.8}$$

where the dummy index of summation is still the one denoting the components n_k as was done for the divergence theorem applied to the scalar $\beta_k n_k$. The form of Eq. (5.1.8) assumes a Cartesian coordinate system. When applied to the convective momentum term we get

$$-\iint_S (\rho u_i u_k) n_k \, dS = -\iiint_V \frac{\partial}{\partial x_k} (\rho u_i u_k) \, dV \tag{5.1.9}$$

Turning to the stress vector term, $\underline{t}$, we again would like to use the divergence theorem to change the surface integral to a volume integral. In order to do that, it is necessary to have the unit normal $\underline{n}$ appear in the surface integral term. The successful approach is to propose, or postulate, that the surface stress vector, $\underline{t}$, is linearly proportional to $\underline{n}$ through the product

$$\underline{t} = \underline{\underline{T}} \cdot \underline{n} \qquad t_i = T_{ij} n_j \tag{5.1.10}$$

where T_{ij} is called the stress tensor, due to Cauchy. Now the divergence theorem applied to this term becomes, in index form,

$$\iint\limits_S t_i\, dS = \iint\limits_S T_{ij} n_j\, dS = \iiint\limits_V \frac{\partial T_{ij}}{\partial x_j}\, dV \quad (5.1.11)$$

where the last term of Eq. (5.1.11) is the integral of the divergence of the stress tensor. Finally, the body force term of Eq. (5.1.5) can be represented in the volume integral as well. For example, if the body force is gravity, then B_i is the weight of the fluid in V, which must be the integrated density over V multiplied by gravity and given the direction of the gravity vector,

$$B_i = mg_i = \iiint b_i\, dV \quad \text{where} \quad b_i = \rho g_i \quad (5.1.12)$$

We can now rewrite Eq. (5.1.5) as the volume integral

$$\iiint\limits_V \left(\frac{\partial}{\partial t}(\rho u_i) + \frac{\partial}{\partial x_k}(\rho u_i u_k) - \frac{\partial T_{ij}}{\partial x_j} - \sigma_{i_{mom}} - b_i \right) dV = 0 \quad (5.1.13)$$

As we observed with mass conservation, Eq. (5.1.13) is an integral over the arbitrary volume, V, of the bracketed term. Since the control volume we chose is arbitrary, it must be that this integrand itself is zero

$$\frac{\partial}{\partial t}(\rho u_i) + \frac{\partial}{\partial x_k}(\rho u_i u_k) = \frac{\partial T_{ij}}{\partial x_j} + b_i + \sigma_{i_{mom}} \quad (5.1.14)$$

5.2 Cauchy Momentum Equation

The left hand side of Eq. (5.1.14) can be expanded using the rule for differentiation of products, and then pieced back together strategically to become

$$u_i \left(\cancel{\frac{\partial}{\partial t}(\rho) + \frac{\partial}{\partial x_k}(\rho u_k)} \right) + \rho \left(\frac{\partial}{\partial t}(u_i) + u_k \frac{\partial}{\partial x_k}(u_i) \right) \quad (5.2.1)$$

However, the first bracketed term in Eq. (5.2.1) is the conservation of mass expression, which we know is zero. So the final form of the conservation of linear momentum equation in index form is

$$\rho \left(\frac{\partial u_i}{\partial t} + u_k \frac{\partial u_i}{\partial x_k} \right) = \frac{\partial T_{ij}}{\partial x_j} + b_i \quad (5.2.2)$$

where we have ignored the source term. In vector notation the balance in Eq. (5.2.2) is

$$\rho \left(\frac{\partial \underline{u}}{\partial t} + (\underline{u} \cdot \nabla)\underline{u} \right) = \nabla \cdot \underline{\underline{T}} + \underline{b} \quad (5.2.3)$$

Equation (5.2.2) or Eq. (5.2.3) are called the Cauchy momentum equation, named for the French mathematician and physicist Augustin-Louis Cauchy (1789–1857), who was a professor at Ecole Polytechnique in Paris, but fled France in exile, following the 1830 revolution, then returned in 1838. It represents three equations for the three component directions. In Eq. (5.2.2) for rectangular coordinates we have the $x(i = 1)$, $y(i = 2)$ and $z(i = 3)$ components. The dummy indices j and k are summed over and do not carry information about which component is under consideration.

For the index form, it is the free index "i" which signifies which component we are considering. The three component equations in Cartesian coordinates are

$$\begin{aligned}
&\rho \left(\frac{\partial u_1}{\partial t} + \left(u_1 \frac{\partial u_1}{\partial x_1} + u_2 \frac{\partial u_1}{\partial x_2} + u_3 \frac{\partial u_1}{\partial x_3} \right) \right) = \frac{\partial T_{11}}{\partial x_1} + \frac{\partial T_{12}}{\partial x_2} + \frac{\partial T_{13}}{\partial x_3} + b_1 \\
&\rho \left(\frac{\partial u_2}{\partial t} + \left(u_1 \frac{\partial u_2}{\partial x_1} + u_2 \frac{\partial u_2}{\partial x_2} + u_3 \frac{\partial u_2}{\partial x_3} \right) \right) = \frac{\partial T_{21}}{\partial x_1} + \frac{\partial T_{22}}{\partial x_2} + \frac{\partial T_{23}}{\partial x_3} + b_2 \\
&\rho \left(\frac{\partial u_3}{\partial t} + \left(u_1 \frac{\partial u_3}{\partial x_1} + u_2 \frac{\partial u_3}{\partial x_2} + u_3 \frac{\partial u_3}{\partial x_3} \right) \right) = \frac{\partial T_{31}}{\partial x_1} + \frac{\partial T_{32}}{\partial x_2} + \frac{\partial T_{33}}{\partial x_3} + b_3
\end{aligned} \quad (5.2.4)$$

The bracketed term on the left hand side of Eq. (5.2.3) we recognize to be the time derivative of the velocity following a fluid particle, i.e. the material derivative of the velocity

$$\rho \left(\frac{\partial \underline{u}}{\partial t} + (\underline{u} \cdot \underline{\nabla})\underline{u} \right) = \rho \frac{D\underline{u}}{Dt} \quad (5.2.5)$$

So another representation of the Cauchy momentum equation is

$$\rho \frac{D\underline{u}}{Dt} = \nabla \cdot \underline{\underline{T}} + \underline{b} \quad (5.2.6)$$

which is more compact. In Section 5.6 we will discuss the corresponding forms of the stress tensor and the Cauchy momentum equation in cylindrical and spherical coordinate systems.

Our derivation of Eq. (5.2.2) has been general for a continuum, so would be appropriate in fluids or solids. It is the stress tensor, T_{ij}, that tells us the type of continuum and its properties. For a linearly elastic solid, for example, we expect the stress to be proportional to strain, as in Hooke's law. For a simple viscous fluid, we expect stress to be proportional to the time rate of strain, in contrast.

5.3 The Stress Tensor

The stress tensor T_{ij} has nine components, since both indexes run 1, 2, 3. As a matrix it appears as

$$\underline{\underline{T}} = T_{ij} = \begin{bmatrix} T_{11} & T_{12} & T_{13} \\ T_{21} & T_{22} & T_{23} \\ T_{31} & T_{32} & T_{33} \end{bmatrix} \quad (5.3.1)$$

A picture of a cube of the continuum, as shown in Figure 5.2, is helpful to understand the components of the stress tensor T_{ij} in rectangular coordinates. The first subscript identifies the face of the cube by signifying the direction of its unit normal. The second subscript identifies the direction of the stress. For example, T_{31}, as shown in Figure 5.2, is the stress on the face of the cube whose unit normal is in the x_3-direction, and the direction of the stress is in the x_1-direction.

The stresses where $i = j$ are seen to be the normal stresses, i.e. perpendicular (normal) to the cube face. The stresses where $i \neq j$ are the shear or tangential stresses, since they are tangent to the cube face.

An important property of the stress tensor, for the fluids of interest to us, is that it is symmetric, i.e.

$$T_{ij} = T_{ji}, \quad \text{symmetric} \quad (5.3.2)$$

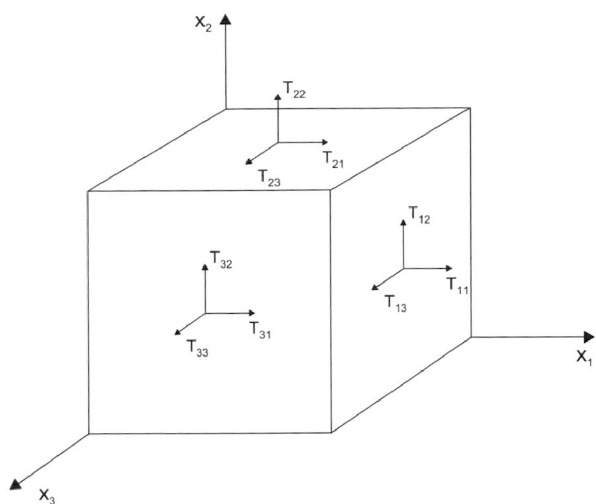

Figure 5.2 A differential volume element illustrating the stress tensor components.

so let's prove it by imposing conservation of angular momentum.

5.4 Angular Momentum Conservation

Consider a two-dimensional fluid element as shown in Figure 5.3, where the stress tensor components T_{12} and T_{21} are shown. The conservation of angular momentum is

$$\tau = I\alpha \quad (5.4.1)$$

where τ is the torque, I is the moment of inertia of the fluid element and α is the angular acceleration. The torque is the sum of moments around a chosen point, in this case the origin.

The moment produced by T_{21} at the top of the fluid element is the force it generates, F_2, multiplied by the lever arm from the origin, dx_2. Since stress is force per unit area, F_2 is derived from multiplying T_{21} times the surface area of the face where T_{21} is applied. We can consider a differential length in the x_3-direction as dx_3. The force then is

$$F_2 = T_{21} dx_1 dx_3 \quad (5.4.2)$$

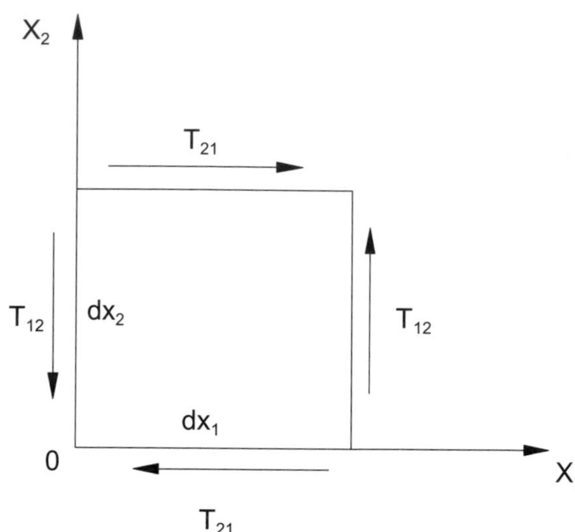

Figure 5.3 A two-dimensional differential volume element undergoing shear stresses on its surface.

and the moment is

$$M_{O_2} = -(T_{21}dx_1dx_3)dx_2 \tag{5.4.3}$$

The same process for the other stress component, which is in the opposite rotational direction is

$$M_{O_1} = (T_{12}dx_2dx_3)dx_1 \tag{5.4.4}$$

Of course there is no contribution from the two forces that pass through the origin and have no lever arm.

Now the conservation of angular momentum, Eq. (5.4.1), becomes

$$\begin{aligned} I\alpha &= (T_{21}dx_1dx_3)dx_2 - (T_{12}dx_2dx_3)dx_1 \\ &= (T_{21} - T_{12})dx_1dx_2dx_3 \end{aligned} \tag{5.4.5}$$

For simplicity, let $dx_1 = dx_2 = dx_3 = \delta$, so

$$I\alpha = (T_{21} - T_{12})\delta^3 \tag{5.4.6}$$

The moment of inertia, I, for this body has the form $I \sim m\delta^2$ and we can replace m with the following equation which uses the constant fluid density, ρ, $m = \rho\delta^3$. Inserting these into Eq. (5.4.6) gives us

$$\rho\delta^5\alpha \sim (T_{21} - T_{12})\delta^3 \tag{5.4.7}$$

which simplifies to

$$\alpha \sim \frac{(T_{21} - T_{12})}{\rho\delta^2} \tag{5.4.8}$$

Equation (5.4.8) must be satisfied in the limit of $\delta \to 0$. However, we cannot have an unbounded angular acceleration in that limit. So this implies that the numerator of Eq. (5.4.8) must be zero, or

$$T_{12} = T_{21} \tag{5.4.9}$$

and the local fluid angular acceleration is zero.

This reasoning can be applied to all of the stress components, so we have proved that the stress tensor is symmetric, $T_{ij} = T_{ji}$, on the basis of conservation of angular momentum. This is a very powerful result, since it says we do not have to solve for conservation of angular momentum within the fluid. That conservation law is imbedded in the symmetry of the stress tensor. Part of our argument assumed that the fluid element does not have or generate internal moments which are proportional to its volume. That assumption does not always hold true, for example in an electrostatic field with a conducting fluid. The symmetry means the nine components of the stress tensor reduce to six, since

$$T_{12} = T_{21}, T_{23} = T_{32}, T_{13} = T_{31} \tag{5.4.10}$$

For some complex fluids, however, the stress tensor is not symmetric and constitutive models can include six different viscosity coefficients depending on the component (Ericksen, 1960; Leslie, 1968). Familiar examples include liquid crystals where the alignment of the crystals strongly influences its behavior. They are used in flat panel electronic displays, but were originally discovered as a biofluid by Austrian botanist Friedrich Richard Reinitzer (1857–1927) through his studies of cholesteryl benzoate.

5.5 Fluid Pressure

Let's first consider a fluid at rest in static equilibrium. This means that there are no tangential stresses in the fluid. Figure 5.4 shows a static force balance on a two-dimensional triangle of fluid. Each of the three faces has a force given by the product of the fluid stress normal to the face and the face length. A unit length into the paper can be assumed.

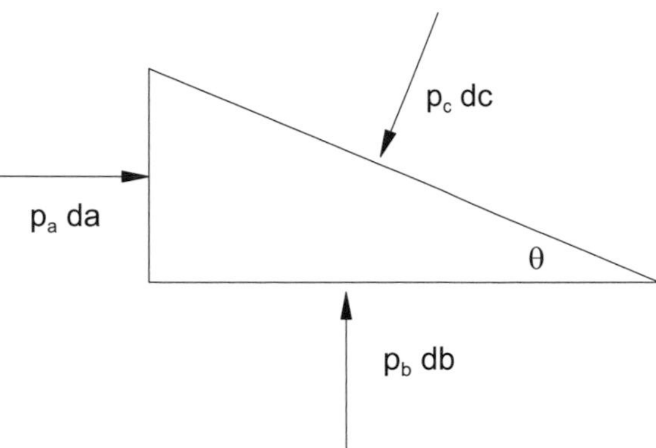

Figure 5.4 Static force balance on a fluid triangle.

Summing the forces in the horizontal and vertical directions yield the two equations

$$\begin{aligned} p_a\, da &= p_c\, dc \sin\theta \\ p_b\, db &= p_c\, dc \cos\theta \end{aligned} \tag{5.5.1}$$

However, we can replace the $\sin\theta$, $\cos\theta$ terms with

$$\sin\theta = da/dc, \quad \cos\theta = db/dc \tag{5.5.2}$$

Inserting Eq. (5.5.2) into Eq. (5.5.1) results in $p_a = p_c$, $p_b = p_c$. So the normal stress on each face of the triangle is the same. We call this normal stress the pressure, p. Since θ is arbitrary, we see that fluid pressure, p, is the same in all directions at a point, that is, after we let the triangle shrink to a point. Thus, fluid pressure is isotropic.

The stress tensor for a static fluid, then, is just the pressure component. To represent the pressure, which is a scalar, in a form useful for the stress tensor, we need only multiply it by the isotropic tensor also known as the identity matrix

$$\underline{\underline{T}} = -p\underline{\underline{I}} \qquad \underline{\underline{I}} = \begin{bmatrix} 1 & 0 & 0 \\ 0 & 1 & 0 \\ 0 & 0 & 1 \end{bmatrix} \tag{5.5.3}$$

or its equivalent index form known as the Kronecker delta, δ_{ij},

$$T_{ij} = -p\delta_{ij} \qquad \delta_{ij} = \begin{cases} 0, & i \neq j \\ 1, & i = j \end{cases} \tag{5.5.4}$$

The Kronecker delta is named for the German mathematician, Leopold Kronecker (1823–1891). It is an isotropic, second-order tensor. The convention of using the minus sign gives us the correct sense of pressure in terms of the stress vector it forms. Recall the definition of the stress vector in Eq. (5.1.10) and use Eq. (5.5.4) to obtain

$$t_i = T_{ij}n_j = -p\delta_{ij}n_j = -pn_i \tag{5.5.5}$$

So a positive pressure has the sense of pushing in the opposite direction to the normal, i.e. it "presses" against the surface, which is physically intuitive. In Eq. (5.5.5) we see that δ_{ij} behaves like a substitution tensor, since it is zero unless you replace j with i in the expression.

5.6 Cauchy Momentum Equation in Other Coordinate Systems

Cylindrical Coordinates

The stress tensor components in cylindrical coordinates are

$$\underline{\underline{T}} = \begin{bmatrix} T_{rr} & T_{r\theta} & T_{rz} \\ T_{\theta r} & T_{\theta\theta} & T_{\theta z} \\ T_{zr} & T_{z\theta} & T_{zz} \end{bmatrix} \tag{5.6.1}$$

The Cauchy momentum equation in cylindrical coordinates has components as follows:

$$\begin{aligned}
r:\ & \rho\left(\frac{\partial u_r}{\partial t} + u_r\frac{\partial u_r}{\partial r} + \frac{u_\theta}{r}\frac{\partial u_r}{\partial \theta} + u_z\frac{\partial u_r}{\partial z} - \frac{u_\theta^2}{r}\right) \\
&= \left\{\frac{1}{r}\left[\frac{\partial}{\partial r}(rT_{rr}) + \frac{\partial T_{\theta r}}{\partial \theta} + \frac{\partial}{\partial z}(rT_{zr})\right] - \frac{T_{\theta\theta}}{r}\right\} + b_r \\
\theta:\ & \rho\left(\frac{\partial u_\theta}{\partial t} + u_r\frac{\partial u_\theta}{\partial r} + \frac{u_\theta}{r}\frac{\partial u_\theta}{\partial \theta} + u_z\frac{\partial u_\theta}{\partial z} + \frac{u_r u_\theta}{r}\right) \\
&= \left\{\frac{1}{r}\left[\frac{\partial}{\partial r}(rT_{r\theta}) + \frac{\partial T_{\theta\theta}}{\partial \theta} + \frac{\partial}{\partial z}(rT_{z\theta})\right] + \frac{T_{r\theta}}{r}\right\} + b_\theta \\
z:\ & \rho\left(\frac{\partial u_z}{\partial t} + u_r\frac{\partial u_z}{\partial r} + \frac{u_\theta}{r}\frac{\partial u_z}{\partial \theta} + u_z\frac{\partial u_z}{\partial z}\right) \\
&= \left\{\frac{1}{r}\left[\frac{\partial}{\partial r}(rT_{rz}) + \frac{\partial T_{\theta z}}{\partial \theta} + \frac{\partial}{\partial z}(rT_{zz})\right]\right\} + b_z
\end{aligned} \tag{5.6.2}$$

Note that the left hand side brackets contain the acceleration terms, and the curly brackets on the right hand side contain the $\underline{\nabla} \cdot \underline{\underline{T}}$ contribution to the momentum balance. We can see right away that the divergence of a tensor in curvilinear coordinates involves extra terms compared to Cartesian tensors, such as $-T_{\theta\theta}/r$ and $+T_{r\theta}/r$ in the r and θ components, respectively, of Eqs. (5.6.2). The body force vector is $\underline{b} = (b_r, b_\theta, b_z)$, which for gravity translates to $\underline{b} = \rho\underline{g} = (\rho g_r, \rho g_\theta, \rho g_z)$.

The stress tensor in spherical coordinates has the following components

$$\underline{\underline{T}} = \begin{bmatrix} T_{rr} & T_{r\theta} & T_{r\phi} \\ T_{\theta r} & T_{\theta\theta} & T_{\theta\phi} \\ T_{\phi r} & T_{\phi\theta} & T_{\phi\phi} \end{bmatrix} \tag{5.6.3}$$

The Cauchy momentum equation in spherical coordinates has the following components

$$r:\rho\left(\frac{\partial u_r}{\partial t} + u_r\frac{\partial u_r}{\partial r} + \frac{u_\theta}{r}\frac{\partial u_r}{\partial \theta} + \frac{u_\phi}{r\sin\theta}\frac{\partial u_r}{\partial \phi} - \frac{u_\theta^2 + u_\phi^2}{r}\right)$$
$$= \frac{1}{r^2\sin\theta}\left[\frac{\partial}{\partial r}(r^2\sin\theta T_{rr}) + \frac{\partial}{\partial\theta}(r\sin\theta T_{\theta r}) + \frac{\partial}{\partial\phi}(rT_{\phi r})\right]$$
$$- \frac{T_{\theta\theta} + T_{\phi\phi}}{r} + b_r$$

$$\theta:\rho\left(\frac{\partial u_\theta}{\partial t} + u_r\frac{\partial u_\theta}{\partial r} + \frac{u_\theta}{r}\frac{\partial u_\theta}{\partial \theta} + \frac{u_\phi}{r\sin\theta}\frac{\partial u_\theta}{\partial \phi} + \frac{u_r u_\theta}{r} - \frac{u_\phi^2\cot\theta}{r}\right)$$
$$= \frac{1}{r^2\sin\theta}\left[\frac{\partial}{\partial r}(r^2\sin\theta T_{r\theta}) + \frac{\partial}{\partial\theta}(r\sin\theta T_{\theta\theta}) + \frac{\partial}{\partial\phi}(rT_{\phi\theta})\right]$$
$$- \frac{T_{\phi\phi}\cot\theta}{r} + \frac{T_{r\theta}}{r} + b_\theta$$

$$\phi\,\rho\left(\frac{\partial u_\phi}{\partial t} + u_r\frac{\partial u_\phi}{\partial r} + \frac{u_\theta}{r}\frac{\partial u_\phi}{\partial \theta} + \frac{u_\phi}{r\sin\theta}\frac{\partial u_\phi}{\partial \phi} + \frac{u_r u_\phi}{r} + \frac{u_\theta u_\phi\cot\theta}{r}\right)$$
$$= \frac{1}{r^2\sin\theta}\left[\frac{\partial}{\partial r}(r^2\sin\theta T_{r\phi}) + \frac{\partial}{\partial\theta}(r\sin\theta T_{\theta\phi}) + \frac{\partial}{\partial\phi}(rT_{\phi\phi})\right]$$
$$+ \frac{T_{r\phi}}{r} + \frac{T_{\theta\phi}\cot\theta}{r} + b_\phi \tag{5.6.4}$$

Summary

In this chapter we have provided the physical and mathematical basis for Newton's second law, conservation of linear momentum, applied to a continuum ... any continuum, solid or fluid. The concept and role of the stress tensor, $\underline{\underline{T}}$, was the key ingredient from the insight and labors of Cauchy. However, we cannot solve any problems until the stresses in $\underline{\underline{T}}$ are expressed in terms of the deformation, through the four dependent variables, which are velocity (three components) and pressure.

Along with the conservation of mass, which is one equation, and conservation of momentum, which is three equations, we should have four equations for four unknowns. The next chapter is focused on relating the stresses to the deformation of the continuum, a general topic known as rheology, and the goal is to discover appropriate constitutive equations that can be inserted into the Cauchy momentum balance for the stress tensor.

Problems

5.1

$$\underline{\underline{A}} = \begin{bmatrix} 2 & 1 & 0 \\ 3 & 2 & 1 \\ 1 & 0 & 3 \end{bmatrix}, \quad \underline{\underline{B}} = \begin{bmatrix} 1 & 2 & 4 \\ 2 & 3 & 0 \\ 3 & 2 & 1 \end{bmatrix}$$

a. The product of two 2nd rank tensors (i.e. matrices) is given by $\underline{\underline{A}} \cdot \underline{\underline{B}} = A_{ik}B_{kj} = P_{ij}$ using the Einstein summation notation. Calculate $\underline{\underline{P}}$.
b. Does $\underline{\underline{A}} \cdot \underline{\underline{B}} = \underline{\underline{B}} \cdot \underline{\underline{A}}$?
c. Calculate the determinant of $\underline{\underline{A}}$ and the determinant of $\underline{\underline{B}}$.

d. How are the results in part (c) related to the determinant of $\underline{\underline{P}}$?

5.2

$$\underline{\underline{A}} = \begin{bmatrix} 2 & 1 & 0 \\ 3 & 2 & 1 \\ 1 & 0 & 3 \end{bmatrix}, \ \underline{c} = (2, 2, 1)$$

a. The product of a second rank tensor and a first rank tensor is given by $\underline{\underline{A}} \cdot \underline{c} = A_{ik}c_k = f_i$. Calculate $\underline{f}$.

b. The product of a first rank tensor and a second rank tensor is given by $\underline{c} \cdot \underline{\underline{A}} = c_j A_{ji} = g_i$
Calculate $\underline{g}$. Does it equal $\underline{f}$ from part (a)?

c. Show that $\underline{c} \cdot \underline{\underline{A}}^T = \underline{\underline{A}} \cdot \underline{c}$ where the superscript T indicates the transpose.

d. Parts (a), (b), (c) raise the question, under what circumstances would a vector $\underline{a}$ and a tensor $\underline{\underline{C}}$ have the same product regardless of their order? In other words, what is the property of $\underline{\underline{C}}$ such that $C_{ij}a_j = a_j C_{ji}$?

e. Calculate the magnitude of the vector $\underline{c}$.

f. Calculate the magnitudes of the vectors $\underline{f}$ and $\underline{g}$. Are they related?

5.3

$$\underline{\underline{A}} = \begin{bmatrix} 2 & 1 & 0 \\ 3 & 2 & 1 \\ 1 & 0 & 3 \end{bmatrix}, \ \underline{\underline{B}} = \begin{bmatrix} 1 & 2 & 4 \\ 2 & 3 & 0 \\ 3 & 2 & 1 \end{bmatrix}$$

a. The "inner product" of two 2nd rank tensors is given by $S = \underline{\underline{A}} : \underline{\underline{B}} = A_{ij}B_{ij}$ where we now have a double summation. The inner product results in the scalar S, which is a zero-order tensor. Calculate S.

b. For 2 × 2 matrices, $\underline{\underline{C}}, \underline{\underline{D}}$, with components C_{ij}, D_{ij}, show that $\underline{\underline{C}} : \underline{\underline{D}} = \mathrm{tr}\left(\underline{\underline{C}}^T \cdot \underline{\underline{D}}\right)$ by evaluating the terms. The notation tr is the trace of a matrix which is the sum of the terms on the diagonal.

c. Show that $\left(\underline{\underline{C}} \cdot \underline{\underline{D}}\right)^T = \underline{\underline{D}}^T \cdot \underline{\underline{C}}^T$ for the system in part (b).

5.4

a. The divergence of a vector yields a scalar, for example $\underline{\nabla} \cdot \underline{G} = \partial G_j / \partial x_j = K$. Calculate K for $\underline{G} = \left(2x_1, x_1 x_2^2, x_2 \sin x_3\right)$ and evaluate it at $\underline{x}_o = (0, 1, \pi/2)$.

b. Suppose the velocity field, $\underline{u}$, for an incompressible fluid can be represented by the gradient of a potential field, ϕ, i.e. $\underline{u} = \underline{\nabla}\phi$. What is the differential equation for ϕ when you impose the conservation of mass? Does it have a name?

c. Calculate the curl of $\underline{u}$ from part (b). What does this result tell you about the flow field?

d. From the product rule show that $\underline{\nabla} \cdot (\rho \underline{u}) = (\underline{u} \cdot \underline{\nabla}\rho) + \rho(\underline{\nabla} \cdot \underline{u})$, where ρ is variable density.

e. What does the expression $\delta_{mn}(\partial u_m / \partial x_n)$ equal for an incompressible fluid?

5.5

$$\underline{\underline{F}} = \begin{bmatrix} e^{x_1} & x_2 x_1 & \sin x_3 \\ 0 & \sin x_2 \cos x_3 & \sin x_2 \cos x_3 \\ 3x_1 & 0 & x_1 e^{x_3} \end{bmatrix}$$

a. The divergence of a tensor yields a vector such as $\underline{\nabla} \cdot \underline{\underline{F}} = \partial F_{ij} / \partial x_j = M_i$.
Calculate $\underline{M}$ and evaluate at $\underline{x}_o = (0, 1, \pi/2)$.

b. The tensor product of two vectors, also called a dyad, has the form, $\underline{\underline{H}} = \underline{c}\,\underline{d}$, where the tensor components are $H_{ij} = c_i d_j$. For $\underline{c} = (2, 2, 1), \underline{d} = (1, 2, 1)$, calculate $\underline{\underline{H}}$.

c. Dyads can also involve operators, for example $\underline{\underline{J}} = \underline{\nabla}\,\underline{h}$ where $\underline{h} = \left(2x_1, x_1 x_2^2, x_2 \sin x_3\right)$. Calculate $\underline{\underline{J}}$ and then evaluate it at $\underline{x}_o = (0, 1, \pi/2)$.

d. The velocity gradient tensor is the dyad $\underline{\underline{D}} = \underline{\nabla}\,\underline{u}$, where $\underline{u} = (u_1, u_2, u_3)$ is the velocity vector. Show that the rate of strain tensor, $\underline{\underline{E}} = \frac{1}{2}\left(\underline{\underline{D}} + \underline{\underline{D}}^T\right)$, is symmetric.

e. Show that the spin tensor, $\underline{\underline{\Omega}} = \frac{1}{2}\left(\underline{\underline{D}} - \underline{\underline{D}}^T\right)$, is skew-symmetric, i.e. $\Omega_{ij} = -\Omega_{ji}$.

5.6 For flow in two-dimensional Cartesian coordinates consider the stress tensor at the point $\underline{x}_0 = (1,0)$ is

$$T_{ij} = \begin{bmatrix} 3 & 1 \\ 1 & -2 \end{bmatrix}$$

a. Sketch the unit normal $\underline{n} = (1,1)/\sqrt{2}$ to a planar surface at $\underline{x}_0$.
b. Sketch a local portion of the planar surface at $\underline{x}_0$.
c. Calculate the unit tangent vector, $\underline{\tau}$, to the planar surface. Let its x-component be positive.
d. Calculate the stress vector $\underline{t}$ at $\underline{x}_0$.
e. Sketch $\underline{t}$ at $\underline{x}_0$.
f. Calculate the normal component of $\underline{t}$ at $\underline{x}_0$.
g. Calculate the tangential component of $\underline{t}$ at $\underline{x}_0$.

5.7 Fully developed, axisymmetric, steady flow in a circular cylinder of radius a is shown in Figure 7.1. The only velocity component is the z-velocity, $u_z(r) = (2Q/\pi a^2)(1 - r^2/a^2)$, which is parabolic in the radial coordinate, r. The volumetric flow rate through the tube is Q and the fluid viscosity is μ. It has stress tensor components

$$\underline{\underline{T}} = \begin{bmatrix} T_{rr} & 0 & T_{rz} \\ 0 & T_{\theta\theta} & 0 \\ T_{zr} & 0 & T_{zz} \end{bmatrix}$$

where $T_{rr} = T_{\theta\theta} = T_{zz} = -p$ and $T_{rz} = T_{zr} = \mu \frac{du_z}{dr}$. The pressure decreases linearly with z for flow in the +z-direction, $p = p_0 - (8\mu Q/\pi a^4)(z - z_0)$.

a. Let $p_0 = 0, z_0 = 0$. Express $\underline{\underline{T}}$ in terms of μ, Q, a, r.
b. The outward pointing unit normal to the tube wall, $\underline{n} = (n_r, n_\theta, n_z)$, must point into the fluid to calculate the fluid stress on the wall. Calculate the stress vector $\underline{t}$ at the tube wall. Call it $\underline{t}_w$.
c. What is the normal component, $\underline{t}_w \cdot \underline{e}_r$, of $\underline{t}_w$? What is its tangential or shear component, $\underline{t}_w \cdot \underline{e}_z$?
d. For a = 1 cm, $\mu = 0.01\ g \cdot cm^{-1} \cdot s^{-1}$, $Q = 50\,cm^3/s$ sketch $\underline{t}_w$ at z = −a, 0, a. What is happening to the vector $\underline{t}_w$ as z varies? Why?
e. To understand the tangential component, the wall shear, better, plot $u_z(r)$ for a = 1 cm, Q = 25, 50, 75 cm^3/s for r between 0 and a. Describe your results, including the velocity profile at the wall. How does it correlate to part (d)?
f. For arbitrary μ, Q, a calculate the total shear force, f_z, on the tube wall over the section $0 \le z \le L$. What increases/decreases f_z?
g. Find the total normal force, f_r, for the same section as in part (f). What increases/decreases f_r?

5.8 Fully developed, steady flow in a two-dimensional channel of width 2b is shown in Figure 6.9. The only velocity component is the x-velocity, $u(y) = (3Q/4bL_z)(1 - y^2/b^2)$, which is parabolic in the y coordinate. The volumetric flow rate through the channel is Q, the fluid viscosity is μ, and L_z is a unit distance in the z-direction. It has stress tensor components

$$\underline{\underline{T}} = \begin{bmatrix} T_{xx} & T_{xy} \\ T_{yx} & T_{yy} \end{bmatrix}$$

where $T_{xx} = T_{yy} = -p$ and $T_{xy} = T_{yx} = \mu \frac{du}{dy}$. The pressure decreases linearly with x for flow in the +x-direction, $p = p_0 - (3\mu Q/2b^3 L_z)(x - x_0)$.

a. Let $p_0 = 0,\ x_0 = 0$. Express $\underline{\underline{T}}$ in terms of μ, Q, b, L_z, x, y.
b. The outward pointing unit normal to the upper wall at y = b is $\underline{n} = (n_x, n_y)$ and must point into the fluid to calculate the fluid stress on the wall. Calculate the stress vector $\underline{t}$ at the channel wall. Call it $\underline{t}_w$.
c. What is the normal component, $\underline{t}_w \cdot \underline{e}_y$, of $\underline{t}_w$? What is its tangential or shear component, $\underline{t}_w \cdot \underline{e}_x$?
d. For $b = L_z = 1$ cm, $\mu = 0.01\,g \cdot cm^{-1} \cdot s^{-1}$ (water), $Q = 50\,cm^3/s$ sketch $\underline{t}_w$ at x = −b, 0, b. What is happening to the vector $\underline{t}_w$ as x varies? Why?
e. To understand the tangential component, the wall shear, better, plot u(y) for $b = L_z = 1$ cm, Q = 25, 50, 75 cm^3/s for y between -b and b. Describe your results, including the velocity profile at the wall. How does it correlate to part (d)?
f. For arbitrary μ, Q, b, L_z calculate the total shear force, f_x, on the upper wall over the section $0 \le x \le L, 0 \le z \le L_z$. What increases/decreases f_x?
g. Find the total normal force, f_y, for the same section as in part (f). What increases/decreases f_y?

6 Constitutive Equations I: Inviscid and Newtonian Viscous Fluids

Introduction

In Chapter 5 we developed the conservation of linear momentum in the form of the Cauchy momentum equation. On the left hand side of this balance is the fluid inertia, the product of the fluid density and its acceleration. The acceleration consists of time and spatial derivatives of the velocity vector, $\underline{u}$. The right hand side has the divergence of the stress tensor, $\underline{\underline{T}}$, which so far is not described in any detail other than the meaning of its components; the fact that it is symmetric, which satisfies the conservation of angular momentum locally in the fluid; and its relationship to the stress vector, $\underline{t} = \underline{\underline{T}} \cdot \underline{n}$, which was the motivation for postulating its existence in order to make use of the divergence theorem.

The stress tensor has been treated in a general way, so the Cauchy momentum equation is valid for a general continuum, i.e. either fluids or solids. It is in the details of how the stress tensor depends on the motion of the continuum that will describe a fluid as opposed to a solid. To close the system of four equations (three components of momentum conservation and one mass conservation) we are limited to solving for four unknowns. We already know that those four will be the three components of velocity, $\underline{u}$, and pressure, p. So it is essential to relate $\underline{\underline{T}}$ to pressure, p, and velocity, $\underline{u}$, and its derivatives in order to accomplish that goal. The mathematical relationship between the stress tensor and the continuum motion is called the "constitutive equation" and the study of such behavior is called "rheology." This is an elegant field which can involve a good bit of tensor analysis. For our purposes, though, we will focus on the main physical insights.

A primary issue in any constitutive relation is that it "looks" the same to different observers who are in different coordinate systems. For example, a person stretches a spring and finds that, for small stretch, the required force, F, is linearly related to the relative change in length of the spring. This is Hooke's law, a well-known constitutive equation. If, however, a second observer sees this happening while in a reference frame moving with respect to the experimenter, the same constitutive equation should be reported. That is, the constitutive equation should be independent of the observer. This concept is called the "principle of material frame indifference," and it limits our choices of what aspects of the continuum deformation are responsible for the stresses. Once we agree on a constitutive relation between fluid stresses and fluid motion, that is, a legitimate stress tensor depending on deformation, then we will insert that relation into the Cauchy momentum equation to arrive at a closed system of governing equations.

The resulting system of governing equations are partial differential equations in space and time for the variables of velocity and pressure, and accordingly, when integrated, need to have the equivalent of integration constants (or functions) specified for those variables on any boundary. Biological boundaries, such as blood vessels or airways, provide a rich array of interesting behavior. They can behave like a rigid, stationary, boundary that is common to other areas of general fluid mechanics. However, there can be prescribed movement like a swimming stroke, flexibility such as a wheezing airway for which the boundary movement is a result of the force balance between the fluid and the elastic structure, porosity which allows fluid to leak into or out of blood vessels for example, concentrated stresses such as surface tension, and others. In this chapter we will treat the simplest boundary conditions and then explore others in later chapters.

Topics Covered

We begin our exploration of constitutive equations by employing the stress tensor used for fluid statics, developed in the previous chapter. It permits only pressure forces, no viscous forces, so no dependence on fluid deformation such as velocity gradients. Inserting this simple stress tensor into the Cauchy momentum equations yields the Euler equations. When the Euler equations are integrated along a streamline, the result is the Bernoulli equation, which is discussed in reference to steady and unsteady applications. Then we will devise a constitutive equation for a Newtonian viscous fluid defined as an incompressible fluid with a constant viscosity. This is done for Cartesian, cylindrical and spherical coordinate systems. By inserting that constitutive equation into the related momentum balance, we will derive the Navier–Stokes equations which are used extensively in fluid mechanics, in general, since air and water are mostly Newtonian. Calling on our experience in Chapter 2, a sample problem related to pulsatile flow in the vasculature is scaled to demonstrate the appearance of governing dimensionless parameters, in this case the Reynolds number, the Womersley parameter and an aspect ratio. The connection between the viscous constitutive model and the kinematics of fluid deformation is examined to give us more insight into the mechanics. The Navier–Stokes equations in other coordinate systems are listed and then finally we will address some simple forms of boundary conditions for inviscid and viscous flows that are a starting point for later, more complicated, situations. The interesting aspects of biofluids which are non-Newtonian will be addressed in Chapter 10.

6.1 Inviscid Fluid: Euler Equations

When a fluid is at rest, we saw in Chapter 5 that the stress tensor is simply the pressure multiplied by the identity tensor,

$$\underline{\underline{T}} = -p\underline{\underline{I}} \quad \text{or} \quad T_{ij} = -p\delta_{ij} \tag{6.1.1}$$

While fluids do indeed have viscosity, when the Reynolds number is large, Re >> 1, regions of the flow may behave as an inviscid fluid, particularly if they are far enough away from a boundary. Equation (6.1.1) represents the simplest stress tensor for a flowing fluid, one that has no appreciable viscous properties, no shear stresses, and only fluid pressure. Inserting Eq. (6.1.1) into the Cauchy momentum balance, derived in Chapter 5 as Eq. (5.2.2) or Eq. (5.2.3), gives us

$$\rho\left(\frac{\partial u_i}{\partial t} + u_j\frac{\partial u_i}{\partial x_j}\right) = \frac{\partial T_{ik}}{\partial x_k} + b_i \quad \rightarrow \quad \rho\left(\frac{\partial u_i}{\partial t} + u_j\frac{\partial u_i}{\partial x_j}\right) = -\delta_{ik}\frac{\partial p}{\partial x_k} + b_i \tag{6.1.2}$$

Since δ_{ik} is zero for $i \neq k$ and one for $i = k$, it has the effect of substituting i for k in the pressure term

$$-\delta_{ik}\frac{\partial p}{\partial x_k} = -\frac{\partial p}{\partial x_i} \tag{6.1.3}$$

which simplifies Eq. (6.1.2) to

$$\rho\left(\frac{\partial u_i}{\partial t} + u_j\frac{\partial u_i}{\partial x_j}\right) = -\frac{\partial p}{\partial x_i} + b_i \tag{6.1.4}$$

in index form and

$$\rho\left(\frac{\partial \underline{u}}{\partial t} + (\underline{u}\cdot\underline{\nabla})\underline{u}\right) = -\underline{\nabla}p + \underline{b} \tag{6.1.5}$$

in vector form. The terminology used to discuss the various components of Eq. (6.1.5) is shown in Figure 6.1, where the unsteady acceleration, the convective acceleration, the pressure gradient force (or stress) and the body force are identified. Since Eq. (6.1.5) has first-order spatial derivatives, only one condition can be applied at a boundary. Typically that condition is one of impenetrability, i.e. that the fluid velocity normal to the boundary $\underline{u}\cdot\underline{n}$ equals the

unsteady acceleration | convective acceleration | pressure force | body force

$$\rho\left(\frac{\partial \underline{u}}{\partial t} + \left(\underline{u}\cdot\underline{\nabla}\right)\underline{u}\right) = -\underline{\nabla}\,p + \underline{b}$$

Figure 6.1 Names of the terms in the Euler equation.

normal velocity of the boundary $\underline{u}_b\cdot\underline{n}$, where $\underline{u}_b$ is the boundary velocity and $\underline{n}$ is the unit normal to the boundary surface

$$\underline{u}\cdot\underline{n} = \underline{u}_b\cdot\underline{n} \tag{6.1.6}$$

So the fluid velocity component normal to the surface must match the boundary velocity. Otherwise, the tangential velocity at a boundary is not restricted to match the boundary velocity, i.e. it is allowed to slip since there is no viscosity. A good example is the stagnation point flow, $(u, v) = (Ax, -Ay)$, where the fluid does not penetrate the wall at $y = 0$, since $v = 0$ there, but it does slip along the wall in the x-direction at speed $u = Ax$.

Equations (6.1.4) or (6.1.5) represent conservation of linear momentum for an inviscid fluid. They are known as the Euler equations, named for the Swiss mathematician and physicist Leonhard Euler (1707–1783). Along with conservation of mass, we now have four equations for four unknowns – the three components of velocity and the pressure. As an example of using the Euler equations, let's find the pressure field for stagnation point flow, where $\underline{u} = (Ax, -Ay)$. The x- and y-components of Eq. (6.1.4) are

$$\rho\left(\left(Ax\frac{\partial}{\partial x} - Ay\frac{\partial}{\partial y}\right)Ax\right) = -\frac{\partial p}{\partial x}, \quad \rho\left(\left(Ax\frac{\partial}{\partial x} - Ay\frac{\partial}{\partial y}\right)(-Ay)\right) = -\frac{\partial p}{\partial y} \tag{6.1.7}$$

which simplifies to

$$\frac{\partial p}{\partial x} = -\rho A^2 x, \quad \frac{\partial p}{\partial y} = -\rho A^2 y \tag{6.1.8}$$

Integrating the x-component gives us

$$p = -\frac{1}{2}\rho A^2 x^2 + f(y) \tag{6.1.9}$$

where we have added the function of integration $f(y)$ since the integration was partial with respect to x. Taking the partial derivative of p with respect to y in Eq. (6.1.9), and inserting it into the y-momentum balance of Eq. (6.1.8) gives us

$$\frac{\partial p}{\partial y} = -\rho A^2 y = \frac{df}{dy} \quad \rightarrow \quad f = -\frac{1}{2}\rho A^2 y^2 + c_0 \tag{6.1.10}$$

We now know $f(y)$ to within an integration constant, c_0. Inserting f into Eq. (6.1.9) leads to the form

$$p = -\frac{1}{2}\rho A^2 (x^2 + y^2) + c_0 \tag{6.1.11}$$

If we call the pressure at the stagnation point, $p(x = 0, y = 0) = p_0$, the stagnation pressure, then that sets $c_0 = p_0$ and the final form is

$$p - p_0 = -\frac{1}{2}\rho A^2 (x^2 + y^2) \tag{6.1.12}$$

Interestingly, lines of constant pressure (isobars) are circles centered at the origin where the pressure is the highest value, p_0, see Figure 6.2 where shading has been used to indicate pressure amplitude. Note that $p = p_0$ at the origin and then $p < p_0$ elsewhere.

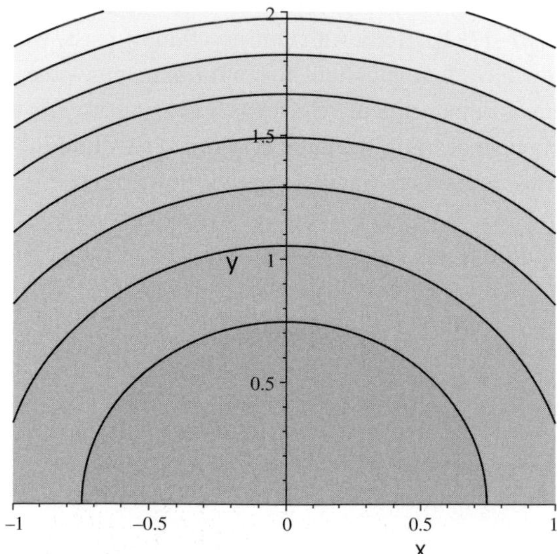

Figure 6.2 Contour and shaded plot of the pressure field for stagnation point flow.

6.2 Bernoulli Equation

Integrating the Euler equation along a streamline, at an instant of time, leads to the Bernoulli equation, named for the Dutch–Swiss mathematician Daniel Bernoulli (1700–1782), who also was a physician. Our derivation of the Bernoulli equation uses a vector identity for the vectors $\underline{A}$ and $\underline{B}$ and the gradient operator $\underline{\nabla}$

$$\underline{A} \times (\underline{\nabla} \times \underline{B}) + \underline{B} \times (\underline{\nabla} \times \underline{A}) + (\underline{A} \cdot \underline{\nabla})\underline{B} + (\underline{B} \cdot \underline{\nabla})\underline{A} = \underline{\nabla}(\underline{A} \cdot \underline{B}) \tag{6.2.1}$$

When we let $\underline{A} = \underline{B} = \underline{u}$, Eq. (6.2.1) can be rearranged to become an identity for the convective acceleration term in Eq. (6.1.4)

$$(\underline{u} \cdot \underline{\nabla})\underline{u} = \frac{1}{2}\underline{\nabla}(\underline{u} \cdot \underline{u}) - \underline{u} \times (\underline{\nabla} \times \underline{u}) \tag{6.2.2}$$

Substituting Eq. (6.2.2) into Eq. (6.1.4), and assuming a conservative body force which is represented as the gradient of a potential, $\underline{b} = -\rho\underline{\nabla}\Phi_b$, and assuming constant density, results in the form

$$\frac{\partial \underline{u}}{\partial t} + \frac{1}{2}\underline{\nabla}(\underline{u} \cdot \underline{u}) + \frac{1}{\rho}\underline{\nabla}p + \underline{\nabla}\Phi_b = \underline{u} \times (\underline{\nabla} \times \underline{u}) \tag{6.2.3}$$

At an instant of time, the velocity vector can be represented as $\underline{u} = q\underline{\sigma}$, where q is the magnitude of the velocity $q = (\underline{u} \cdot \underline{u})^{1/2}$, and $\underline{\sigma}$ is a unit vector tangent to the line. With this substitution Eq. (6.2.3) becomes

$$\frac{\partial q}{\partial t}\underline{\sigma} + q\frac{\partial \underline{\sigma}}{\partial t} + \frac{1}{2}\underline{\nabla}(\underline{u} \cdot \underline{u}) + \frac{1}{\rho}\underline{\nabla}p + \underline{\nabla}\Phi_b = \underline{u} \times (\underline{\nabla} \times \underline{u}) \tag{6.2.4}$$

The component of Eq. (6.2.4) along the streamline is found by taking the dot product with $\underline{\sigma}$

$$\underline{\sigma} \cdot \left[\frac{\partial q}{\partial t}\underline{\sigma} + q\frac{\partial \underline{\sigma}}{\partial t} + \frac{1}{2}\underline{\nabla}(\underline{u} \cdot \underline{u}) + \frac{1}{\rho}\underline{\nabla}p + \underline{\nabla}\Phi_b\right] = \underline{\sigma} \cdot [\underline{u} \times (\underline{\nabla} \times \underline{u})] \tag{6.2.5}$$

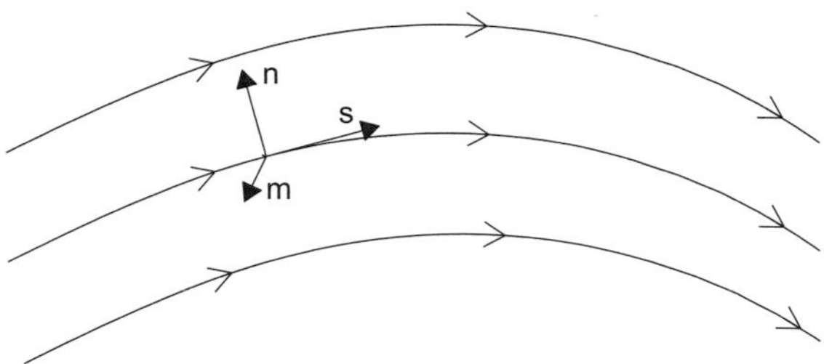

Figure 6.3 Local coordinate system (s, m, n) at a point on a streamline.

To simplify Eq. (6.2.5) further we recognize that $\underline{\sigma}\cdot\underline{\sigma}=1$, $\underline{\sigma}\cdot\partial\underline{\sigma}/\partial t=\frac{1}{2}\partial(\underline{\sigma}\cdot\underline{\sigma})/\partial t=0$ and $\underline{\sigma}\cdot\underline{\nabla}=\partial/\partial s=0$, which is the directional derivative along the streamline. Here, s is the distance along the streamline in the local coordinate system shown in Figure 6.3. Also, since the curl of the velocity, $\underline{\nabla}\times\underline{u}$, is perpendicular to the velocity, then $\underline{u}\times(\underline{\nabla}\times\underline{u})$ is a vector which is perpendicular to $\underline{\sigma}$. i.e. $\underline{\sigma}\cdot[\underline{u}\times(\underline{\nabla}\times\underline{u})]=0$ on the right hand side. These simplifications lead to

$$\frac{\partial q}{\partial t}+\frac{\partial}{\partial s}\left(\frac{q^2}{2}+\frac{1}{\rho}p+\Phi_b\right)=0 \tag{6.2.6}$$

Integrating Eq. (6.2.6) with respect to s, i.e. along the streamline, gives us the general form of the Bernoulli equation,

$$\int^{s}\frac{\partial q}{\partial t}ds+\left(\frac{q^2}{2}+\frac{1}{\rho}p+\Phi_b\right)=f(n,m,t) \tag{6.2.7}$$

Since the integration is partial with respect to the streamline coordinate, s, then the right hand side of Eq. (6.2.7) is at most a function of time and the other two local coordinate directions, measured by n, in the principal normal direction, and m, in the binormal direction, as shown in Figure 6.3. The pertinent Frenet–Serret equations for curves in space define these directions. Changing n or m changes the operating point to a different streamline and that changes the value of f.

For steady flow and two points along the same streamline, the right hand side of Eq. (6.2.7) is a constant, and the Bernoulli equation simplifies to

$$p_1+\frac{1}{2}\rho q_1^2+\rho g z_1=p_2+\frac{1}{2}\rho q_2^2+\rho g z_2=\text{constant} \tag{6.2.8}$$

where the gravity potential is $\Phi_b=gz$ and gravity is in the negative z-direction. For zero gravity the form is even simpler,

$$p_1+\frac{1}{2}\rho q_1^2=p_2+\frac{1}{2}\rho q_2^2=\text{constant} \tag{6.2.9}$$

and possesses the interpretation of the "Bernoulli effect" that regions of faster fluid velocity have lower pressures, and vice versa, since a change in pressure must accompany an opposite change in velocity.

Clinical Correlation 6.1

The Bernoulli effect is readily visualized by most anyone with a soft enough nose. With the mouth closed, try inhaling forcefully with a strong sniff. For most people, this will cause the soft tissue sides of their nose to be sucked inward, pinching the nose a bit. The reason for this is readily seen in Figure 6.4 where a streamline is drawn starting from a distance from the nose where the pressure is atmospheric, p_a, and the fluid speed is negligible. The other end of the streamline is inside the nose and there the velocity is significant $q_2 >> 0$, as shown with

Figure 6.4 Bernoulli effect for a strong sniffing inhalation through the nose. The result is a partial collapse or pinching of the soft nasal tissue since $p_2 < p_a$.

the vectors. Therefore the pressure there is less than atmospheric, $p_2 < p_a$. However, it is also p_a that surrounds the nose, so there is a net force sucking the nose into partial collapse.

Assuming that the flow is irrotational since viscosity is absent, i.e. $\underline{\nabla} \times \underline{u} = 0$, we can define the velocity as the gradient of a velocity potential, ϕ,

$$\underline{u} = \underline{\nabla}\phi \tag{6.2.10}$$

since the curl of a gradient of any function is zero, i.e. $\underline{\nabla} \times \underline{\nabla}\phi = 0$. Equation (6.2.3) then becomes

$$\underline{\nabla}\left(\frac{\partial\phi}{\partial t} + \frac{1}{2}(\underline{\nabla}\phi \cdot \underline{\nabla}\phi) + \frac{1}{\rho}p + \Phi_b\right) = 0 \tag{6.2.11}$$

Since the gradient of the term in brackets in Eq. (6.2.11) is zero, it must be independent of the spatial variables, e.g. x, y and z in rectangular coordinates. Then it is, at most, a function of time, call it $c + G(t)$, where c is a constant

$$\frac{\partial\phi}{\partial t} + \frac{1}{2}(\underline{\nabla}\phi \cdot \underline{\nabla}\phi) + \frac{1}{\rho}p + \Phi_b = c + G(t) \tag{6.2.12}$$

It is easy to show that we can let $G(t) = 0$. For example, define a new velocity potential $\hat{\phi} = \phi - \int_{t_0}^{t} G(t')dt'$ and substitute this relationship into Eq. (6.2.12)

$$\frac{\partial\hat{\phi}}{\partial t} + \frac{1}{2}(\underline{\nabla}\hat{\phi} \cdot \underline{\nabla}\hat{\phi}) + \frac{1}{\rho}p + \Phi_b = c \tag{6.2.13}$$

The cancellation of G comes from the unsteady, time derivative term. However, there is no effect on the velocity with this new potential field, since $\underline{u} = \underline{\nabla}\hat{\phi} = \underline{\nabla}\phi$. So setting $G(t) = 0$ in Eq. (6.2.12) yields the identical form as Eq. (6.2.13) while leaving ϕ as is,

$$\frac{\partial\phi}{\partial t} + \frac{1}{2}(\underline{\nabla}\phi \cdot \underline{\nabla}\phi) + \frac{1}{\rho}p + \Phi_b = c \tag{6.2.14}$$

There would be no change in the derivation if we had subtracted a constant reference pressure from p. Suppose we let $\hat{p} = p - p_0$. Clearly $\underline{\nabla}\hat{p} = \underline{\nabla}p$ and we would have $\hat{p}$ in Eq. (6.2.14) instead of p. Then we could consider $c = p_0/\rho$ and the form is consistent, including the ever popular choice of $c = 0$.

In addition to momentum conservation, keep in mind we must also conserve mass. For an incompressible fluid we have

$$\underline{\nabla} \cdot \underline{u} = \underline{\nabla} \cdot (\underline{\nabla}\phi) = \nabla^2\phi = 0 \tag{6.2.15}$$

where ∇^2 is the Laplacian, e.g. $\nabla^2 = \partial^2/\partial x^2 + \partial^2/\partial y^2 + \partial^2/\partial z^2$ in rectangular coordinates. In general we now have two equations, Eq. (6.2.14) and Eq. (6.2.15), for the two unknowns, p and ϕ. Once ϕ is solved in Eq. (6.2.15), then the three velocity components may be found from Eq. (6.2.10). It is very helpful that, in fact, Eq. (6.2.15) is independent of p. So the process would be to solve Eq. (6.2.15) first for ϕ, given the pertinent boundary conditions for the problem. Once known, the pressure field, p, may be solved using Eq. (6.2.14).

For example, the velocity potential for stagnation point flow must satisfy

$$u = \frac{\partial \phi}{\partial x} = Ax, \quad v = \frac{\partial \phi}{\partial y} = -Ay \tag{6.2.16}$$

Integrating the x-component first gives us

$$\phi = \frac{A}{2}x^2 + f(y) \tag{6.2.17}$$

which we insert into the y-component

$$\frac{\partial \phi}{\partial y} = -Ay = \frac{df}{dy} \qquad \rightarrow \qquad f = -\frac{A}{2}y^2 + c_0 \tag{6.2.18}$$

which solves for $f(y)$ to within a constant, c_0, which we can set to zero. The resulting solution for ϕ is

$$\phi = \frac{A}{2}\left(x^2 - y^2\right) \tag{6.2.19}$$

We see again that stagnation point flow is incompressible since ϕ in Eq. (6.2.19) clearly satisfies the continuity equation of Eq. (6.2.15)

$$\nabla^2\phi = \frac{\partial^2 \phi}{\partial x^2} + \frac{\partial^2 \phi}{\partial y^2} = A - A = 0 \tag{6.2.20}$$

The pressure field from the steady form of Eq. (6.2.14) and assuming no body force, $\Phi = 0$, is

$$p = -\frac{\rho}{2}\left(\underline{\nabla}\phi \cdot \underline{\nabla}\phi\right) = -\frac{\rho A^2}{2}\left(x^2 + y^2\right) \tag{6.2.21}$$

which is the same result as in Eq. (6.1.12) with $p_0 = 0$.

Recall that streamlines are lines for which the stream function, ψ, is constant and that they are parallel to the velocity field. By contrast, we will see that lines of constant ϕ are perpendicular to the velocity field, and are called potential lines. The relationship between the stream function and the potential function is given by their definitions with respect to the velocity components

$$\frac{\partial \phi}{\partial x} = \frac{\partial \psi}{\partial y} = u, \quad \frac{\partial \phi}{\partial y} = -\frac{\partial \psi}{\partial x} = v \tag{6.2.22}$$

A vector perpendicular to a streamline, i.e. a constant ψ line, is given by the gradient operator $\underline{\nabla}\psi = (\partial\psi/\partial x, \partial\psi/\partial y) = (-v, u)$; while a vector

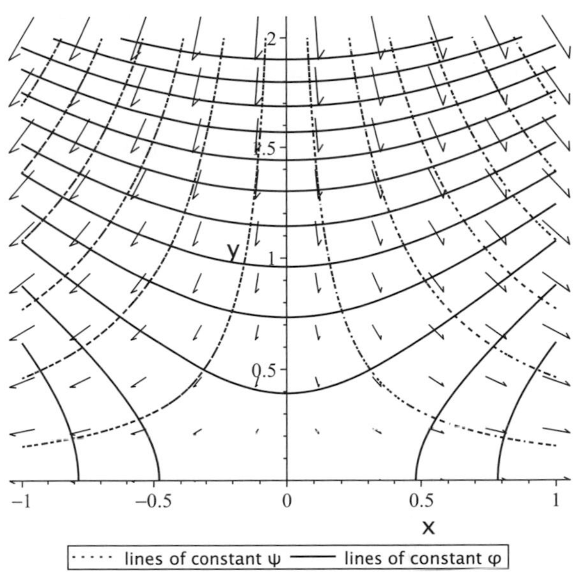

Figure 6.5 Showing for stagnation point flow the streamlines, which are lines of constant ψ, and the potential lines which are lines of constant ϕ.

perpendicular to a potential line, i.e. a constant ϕ line, is $\underline{\nabla}\phi = (u, v)$, the velocity vector. So the proof that the lines are perpendicular is that the dot product of these two vectors is zero, $\underline{\nabla}\psi \cdot \underline{\nabla}\phi = (-vu + uv) = 0$. The fact that the fluid flows from one potential line to another gives this type of fluid mechanics, inviscid and irrotational, the name "potential flow."

We learned in Chapter 3 that streamlines and particle pathlines coincide for steady flows. Since the Bernouilli equation is the integral of momentum conservation along a streamline, it is not an energy equation in general, though often mistakenly referred as such. An energy equation comes from integrating the force on a particle along its pathline. So the two concepts are completely different. It is only in steady flow that the concepts coincide, i.e. for steady flow the Bernoulli equation, Eq. (6.2.8) looks like an energy equation. There is the pressure term, the kinetic energy term and the potential term. However, for unsteady flow it does not look like an energy equation since there is no energy counterpart to $\partial\phi/\partial t$ for example in Eq. (6.2.14).

Worked Example 6.1

An example of an unsteady, incompressible, inviscid flow is given by the velocity potential

$$\phi = B\sin\left(k(x-ct)\right)e^{-ky} \tag{6.2.23}$$

which satisfies Eq. (6.2.15) and is a traveling wave form that propagates in the positive x-direction with wave speed, c, and wavelength $\lambda = 2\pi/k$. The resulting velocity field is

$$u = \frac{\partial\phi}{\partial x} = Bk\cos\left(k(x-ct)\right)e^{-ky}, \quad v = \frac{\partial\phi}{\partial y} = -Bk\sin\left(k(x-ct)\right)e^{-ky} \tag{6.2.24}$$

An example velocity field is shown in Figure 6.6 for $k = c = B = t = 1$. Both u and v components decay exponentially as y increases.

The corresponding pressure field is given by Eq. (6.2.14) with $\Phi_b = 0$, $c = 0$,

$$\begin{aligned}
p &= -\rho\left[\frac{\partial\phi}{\partial t} + \frac{1}{2}(\underline{\nabla}\phi\cdot\underline{\nabla}\phi)\right] \\
&= -\rho\left[-kcB\cos\left(k(x-ct)\right)e^{-ky} + \frac{1}{2}\left[\left(k^2B^2\cos^2(k(x-ct))e^{-2ky}\right) + \left(k^2B^2\sin^2(k(x-ct))e^{-2ky}\right)\right]\right] \\
&= -\rho\left[-kcB\cos\left(k(x-ct)\right)e^{-ky} + \frac{k^2B^2}{2}e^{-2ky}\left[\cos^2(k(x-ct)) + \sin^2(k(x-ct))\right]\right] \\
&= -\rho\left[-kcB\cos\left(k(x-ct)\right)e^{-ky} + \frac{k^2B^2}{2}e^{-2ky}\right]
\end{aligned} \tag{6.2.25}$$

and plotted in Figure 6.7, which shows lines of constant pressure (isobars) while the shading is darker for higher values and lighter for lower values.

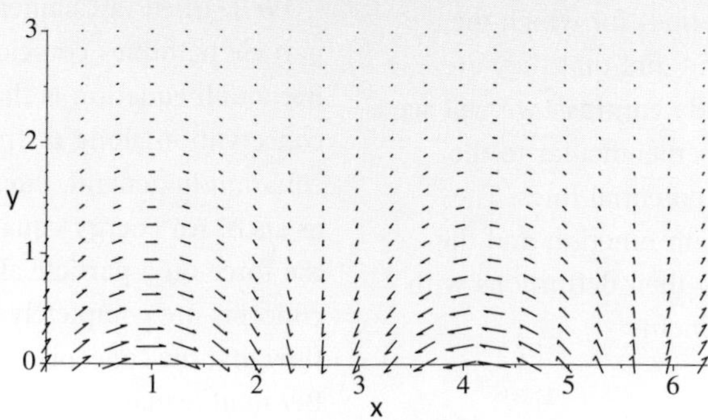

Figure 6.6 Velocity vector field for an unsteady incompressible inviscid flow corresponding to the velocity potential $\phi = \sin\left(k(x-ct)\right)e^{-ky}$ for $k = c = B = t = 1$.

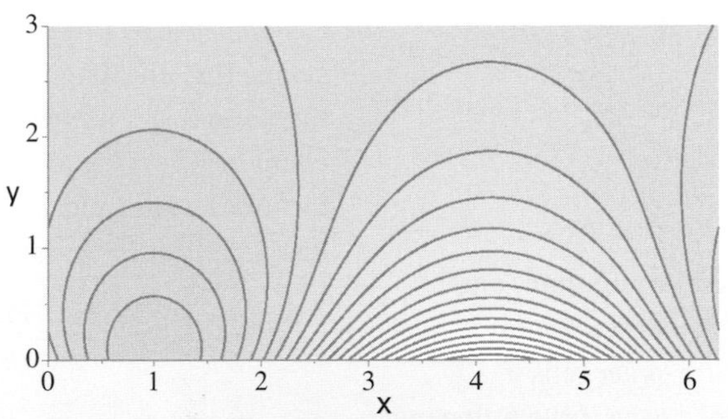

Figure 6.7 Lines of constant pressure, shaded darker for higher values, for an unsteady incompressible inviscid flow corresponding to Figure 6.6.

6.3 Newtonian Viscous Fluid

For a flowing fluid in general we reckon that there are additional stresses related to the fluid viscosity and relative motion. So an additional stress term is added to Eq. (6.1.1) called the extra stress, or viscous stress tensor, $\underline{\underline{\tau}}$, as follows

$$\underline{\underline{T}} = -p\underline{\underline{I}} + \underline{\underline{\tau}} \quad \text{or} \quad T_{ij} = -p\delta_{ij} + \tau_{ij} \tag{6.3.1}$$

The viscous stress tensor, $\underline{\underline{\tau}}$, must depend on the fluid motion in some way. If we assumed that it depends on the fluid velocity, $\underline{u}$, itself, rather than any of its derivatives, then an observer in a moving frame would see a different velocity and obtain a different constitutive equation. That would violate the principle of material frame indifference. So at the outset we assume that the stresses depend on the relative motion between fluid particles as a form of internal friction. Mathematically the dependence would be on fluid velocity gradients, i.e. spatial derivatives of the velocity field. For the 3 components of velocity, there are gradients in each of the 3 orthogonal directions, so a total of 9, call them the rate of deformation tensor, $\underline{\underline{D}}$. It is also known as the velocity gradient tensor symbolized by $\underline{\underline{D}} = \underline{\nabla u}$, and has components

$$D_{ij} = \frac{\partial u_i}{\partial x_j} = \begin{pmatrix} \frac{\partial u_1}{\partial x_1} & \frac{\partial u_1}{\partial x_2} & \frac{\partial u_1}{\partial x_3} \\ \frac{\partial u_2}{\partial x_1} & \frac{\partial u_2}{\partial x_2} & \frac{\partial u_2}{\partial x_3} \\ \frac{\partial u_3}{\partial x_1} & \frac{\partial u_3}{\partial x_2} & \frac{\partial u_3}{\partial x_3} \end{pmatrix} \quad \Leftrightarrow \quad \underline{\underline{D}} = \underline{\nabla}\, \underline{u} \tag{6.3.2}$$

$\underline{\underline{D}}$ is the Lagrangian derivative of the deformation gradient tensor formed from derivatives of the mapping, $\underline{\underline{F}} = F_{ij} = \partial X_i / \partial X_0^j$, that is

$$\underline{\underline{D}} = \frac{\partial}{\partial t}\left(\underline{\underline{F}}\right)_{\underline{X}_0} = \frac{D}{Dt}\left(\underline{\underline{F}}\right) \tag{6.3.3}$$

In general we assume that each of the nine components of τ_{ij} is a linear combination (sum) of all of the nine velocity gradients, as follows

$$\tau_{ij} = C_{ijk\ell} D_{k\ell} = C_{ijk\ell} \frac{\partial u_k}{\partial x_\ell} \tag{6.3.4}$$

where the summation for repeated indices is used. Since each index has three possible values, $C_{ijk\ell}$ is a fourth-ranked tensor with $3^4 = 81$ components which are the coefficients of each gradient term. For example, Eq. (6.3.4) says that τ_{13} is given by the sum of velocity gradient terms, each with a unique coefficient

$$\begin{aligned}\tau_{13} = C_{1311}\frac{\partial u_1}{\partial x_1} + C_{1312}\frac{\partial u_1}{\partial x_2} + C_{1313}\frac{\partial u_1}{\partial x_3} \\ + C_{1321}\frac{\partial u_2}{\partial x_1} + C_{1322}\frac{\partial u_2}{\partial x_2} + C_{1323}\frac{\partial u_2}{\partial x_3} \\ + C_{1331}\frac{\partial u_3}{\partial x_1} + C_{1332}\frac{\partial u_3}{\partial x_2} + C_{1333}\frac{\partial u_3}{\partial x_3}\end{aligned} \tag{6.3.5}$$

For an isotropic fluid or general continuum, one whose constitutive equation is independent of direction in the fluid, the fourth-order tensor, $C_{ijk\ell}$, must be isotropic. From tensor analysis the following identity holds for an isotropic fourth-order tensor,

$$C_{ijk\ell} = \lambda\delta_{ij}\delta_{k\ell} + \mu(\delta_{ik}\delta_{j\ell} + \delta_{i\ell}\delta_{jk}) \tag{6.3.6}$$

which says that the 81 coefficients are reduced to two, λ, μ. That is a great simplification. Materials like wood, which have a preferred direction with or against the grain, are not isotropic materials and require more complicated constitutive equations. Inserting Eq. (6.3.6) into Eq. (6.3.4) yields

$$\tau_{ij} = \mu\left(\frac{\partial u_i}{\partial x_j} + \frac{\partial u_j}{\partial x_i}\right) + \lambda\frac{\partial u_k}{\partial x_k}\delta_{ij} \tag{6.3.7}$$

The constant μ, is the dynamic viscosity, or just the viscosity, and multiplies the shearing and normal velocity gradients. The constant λ is the bulk viscosity which multiplies the divergence of the velocity, which is only normal velocity gradients, and is related to fluid dilatation or compressibility. The fact that they appear as constants indicates they do not depend on velocity gradients. However, they may depend on other factors such as temperature, pressure, or other substances in the fluid.

The first bracketed term of Eq. (6.3.7) is related to the velocity gradient tensor, $\underline{\underline{D}}$ in Eq. (6.3.2), through the definition of the "rate of strain tensor," $\underline{\underline{E}}$, in Cartesian coordinates is

$$\begin{aligned}E_{ij} = \frac{1}{2}(D_{ij} + D_{ji}) = \frac{1}{2}\left(\frac{\partial u_i}{\partial x_j} + \frac{\partial u_j}{\partial x_i}\right) \quad \Leftrightarrow \\ \underline{\underline{E}} = \frac{1}{2}\left(\underline{\underline{D}} + \underline{\underline{D}}^T\right) = \frac{1}{2}\left((\underline{\nabla u}) + (\underline{\nabla u})^T\right)\end{aligned} \tag{6.3.8}$$

Clearly $E_{ij} = E_{ji}$, or $\underline{\underline{E}} = \underline{\underline{E}}^T$, so the rate of strain tensor is symmetric as it must be for the stress tensor to be symmetric. The diagonal components, $i = j$, of E_{ij} are the normal strain rates and the off-diagonal components, $i \neq j$, are the shear strain rates. For an incompressible fluid, $\partial u_k/\partial x_k = 0$, and we can rewrite Eq. (6.3.7) for the extra stress tensor as

$$\tau_{ij} = 2\mu E_{ij} = \mu\left(\frac{\partial u_i}{\partial x_j} + \frac{\partial u_j}{\partial x_i}\right) \tag{6.3.9}$$

Then the full constitutive equation, Eq. (6.3.1), has the form

$$T_{ij} = -p\delta_{ij} + 2\mu E_{ij} \tag{6.3.10}$$

6.4 Constitutive Equation in Other Coordinate Systems

The stress tensor in cylindrical components has the form of Eq. (6.3.1), where the stress tensor, $\underline{\underline{T}}$, and the extra stress tensor, $\underline{\underline{\tau}}$, have the components

$$\underline{\underline{T}} = \begin{bmatrix} T_{rr} & T_{r\theta} & T_{rz} \\ T_{\theta r} & T_{\theta\theta} & T_{\theta z} \\ T_{zr} & T_{z\theta} & T_{zz}\end{bmatrix} \quad \text{and} \quad \underline{\underline{\tau}} = \begin{bmatrix} \tau_{rr} & \tau_{r\theta} & \tau_{rz} \\ \tau_{\theta r} & \tau_{\theta\theta} & \tau_{\theta z} \\ \tau_{zr} & \tau_{z\theta} & \tau_{zz}\end{bmatrix} \tag{6.4.1}$$

As in Eq. (6.3.9), the extra stress tensor is related to the rate of strain tensor, which depends on the fluid velocity vector $\underline{u} = (u_r, u_\theta, u_z)$, whose components are in the (r, θ, z)-directions, respectively. The strain rate tensor components are

$$\begin{aligned}E_{rr} &= \frac{\partial u_r}{\partial r} & E_{r\theta} = E_{\theta r} &= \frac{1}{2}\left(r\frac{\partial}{\partial r}\left(\frac{u_\theta}{r}\right) + \frac{1}{r}\frac{\partial u_r}{\partial \theta}\right) \\ E_{\theta\theta} &= \left(\frac{1}{r}\frac{\partial u_\theta}{\partial \theta} + \frac{u_r}{r}\right) & E_{rz} = E_{zr} &= \frac{1}{2}\left(\frac{\partial u_z}{\partial r} + \frac{\partial u_r}{\partial z}\right) \\ E_{zz} &= \frac{\partial u_z}{\partial z} & E_{\theta z} = E_{z\theta} &= \frac{1}{2}\left(\frac{\partial u_\theta}{\partial z} + \frac{1}{r}\frac{\partial u_z}{\partial \theta}\right)\end{aligned} \tag{6.4.2}$$

Assuming an incompressible fluid, this makes the extra stress tensor components, $\tau_{ij} = 2\mu E_{ij}$, in cylindrical coordinates

$$\tau_{rr} = 2\mu \frac{\partial u_r}{\partial r} \qquad \tau_{r\theta} = \tau_{\theta r} = \mu\left(r\frac{\partial}{\partial r}\left(\frac{u_\theta}{r}\right) + \frac{1}{r}\frac{\partial u_r}{\partial \theta}\right)$$

$$\tau_{\theta\theta} = 2\mu\left(\frac{1}{r}\frac{\partial u_\theta}{\partial \theta} + \frac{u_r}{r}\right) \qquad \tau_{rz} = \tau_{zr} = \mu\left(\frac{\partial u_z}{\partial r} + \frac{\partial u_r}{\partial z}\right)$$

$$\tau_{zz} = 2\mu \frac{\partial u_z}{\partial z} \qquad \tau_{\theta z} = \tau_{z\theta} = \mu\left(\frac{\partial u_\theta}{\partial z} + \frac{1}{r}\frac{\partial u_z}{\partial \theta}\right) \tag{6.4.3}$$

where we have used the symmetry of the stress tensor explicitly.

The Cauchy momentum equation in cylindrical coordinates has components as follows

$$r:\ \rho\left(\frac{\partial u_r}{\partial t} + u_r\frac{\partial u_r}{\partial r} + \frac{u_\theta}{r}\frac{\partial u_r}{\partial \theta} + u_z\frac{\partial u_r}{\partial z} - \frac{u_\theta^2}{r}\right)$$
$$= \left\{\frac{1}{r}\left[\frac{\partial}{\partial r}(rT_{rr}) + \frac{\partial T_{\theta r}}{\partial \theta} + \frac{\partial}{\partial z}(rT_{zr})\right] - \frac{T_{\theta\theta}}{r}\right\} + b_r$$

$$\theta:\ \rho\left(\frac{\partial u_\theta}{\partial t} + u_r\frac{\partial u_\theta}{\partial r} + \frac{u_\theta}{r}\frac{\partial u_\theta}{\partial \theta} + u_z\frac{\partial u_\theta}{\partial z} + \frac{u_r u_\theta}{r}\right)$$
$$= \left\{\frac{1}{r}\left[\frac{\partial}{\partial r}(rT_{r\theta}) + \frac{\partial T_{\theta\theta}}{\partial \theta} + \frac{\partial}{\partial z}(rT_{z\theta})\right] + \frac{T_{r\theta}}{r}\right\} + b_\theta$$

$$z:\ \rho\left(\frac{\partial u_z}{\partial t} + u_r\frac{\partial u_z}{\partial r} + \frac{u_\theta}{r}\frac{\partial u_z}{\partial \theta} + u_z\frac{\partial u_z}{\partial z}\right)$$
$$= \left\{\frac{1}{r}\left[\frac{\partial}{\partial r}(rT_{rz}) + \frac{\partial T_{\theta z}}{\partial \theta} + \frac{\partial}{\partial z}(rT_{zz})\right]\right\} + b_z \tag{6.4.4}$$

Note that the left hand side brackets contain the acceleration terms, and the curly brackets on the right hand side contain the $\nabla \cdot \underline{\underline{T}}$ contribution to the momentum balance. We can see right away that the divergence of a tensor in curvilinear coordinates involves extra terms compared to Cartesian tensors, such as $-T_{\theta\theta}/r$ and $+T_{r\theta}/r$ in the r and θ components, respectively, of Eqs. (6.4.4). The body force vector is $\underline{b} = (b_r, b_\theta, b_z)$, which for gravity translates to $\underline{b} = \rho\underline{g} = (\rho g_r, \rho g_\theta, \rho g_z)$.

The stress tensor in spherical coordinates is the following

$$\underline{\underline{T}} = \begin{bmatrix} T_{rr} & T_{r\theta} & T_{r\phi} \\ T_{\theta r} & T_{\theta\theta} & T_{\theta\phi} \\ T_{\phi r} & T_{\phi\theta} & T_{\phi\phi} \end{bmatrix} \text{ and } \underline{\underline{\tau}} = \begin{bmatrix} \tau_{rr} & \tau_{r\theta} & \tau_{r\phi} \\ \tau_{\theta r} & \tau_{\theta\theta} & \tau_{\theta\phi} \\ \tau_{\phi r} & \tau_{\phi\theta} & \tau_{\phi\phi} \end{bmatrix} \tag{6.4.5}$$

The strain rate tensor is

$$E_{rr} = \frac{\partial u_r}{\partial r} \qquad E_{r\theta} = E_{\theta r} = \frac{1}{2}\left(r\frac{\partial}{\partial r}\left(\frac{u_\theta}{r}\right) + \frac{1}{r}\frac{\partial u_r}{\partial \theta}\right)$$

$$E_{\theta\theta} = \left(\frac{1}{r}\frac{\partial u_\theta}{\partial \theta} + \frac{u_r}{r}\right)$$

$$E_{r\phi} = E_{\phi r} = \left(\frac{1}{r\sin\theta}\frac{\partial u_\phi}{\partial \phi} + \frac{u_r}{r} + \frac{u_\theta \cot\theta}{r}\right)$$

$$E_{\phi\phi} = \frac{1}{2}\left(\frac{1}{r\sin\theta}\frac{\partial u_r}{\partial \phi} + r\frac{\partial}{\partial r}\left(\frac{u_\phi}{r}\right)\right)$$

$$E_{\theta\phi} = E_{\phi\theta} = \frac{1}{2}\left(\frac{\sin\theta}{r}\frac{\partial}{\partial \theta}\left(\frac{u_\phi}{\sin\theta}\right) + \frac{1}{r\sin\theta}\frac{\partial u_\theta}{\partial \phi}\right) \tag{6.4.6}$$

where the velocity components (u_r, u_θ, u_ϕ) are in the (r, θ, ϕ)-directions, respectively. Then, since $\tau_{ij} = 2\mu E_{ij}$, the extra stress tensor components are

$$\tau_{rr} = 2\mu\frac{\partial u_r}{\partial r} \qquad \tau_{r\theta} = \tau_{\theta r} = \mu\left(r\frac{\partial}{\partial r}\left(\frac{u_\theta}{r}\right) + \frac{1}{r}\frac{\partial u_r}{\partial \theta}\right)$$

$$\tau_{\theta\theta} = 2\mu\left(\frac{1}{r}\frac{\partial u_\theta}{\partial \theta} + \frac{u_r}{r}\right)$$

$$\tau_{r\phi} = \tau_{\phi r} = \mu\left(\frac{1}{r\sin\theta}\frac{\partial u_r}{\partial \phi} + r\frac{\partial}{\partial r}\left(\frac{u_\phi}{r}\right)\right)$$

$$\tau_{\phi\phi} = 2\mu\left(\frac{1}{r\sin\theta}\frac{\partial u_\phi}{\partial \phi} + \frac{u_r}{r} + \frac{u_\theta\cot\theta}{r}\right)$$

$$\tau_{\theta\phi} = \tau_{\phi\theta} = \mu\left(\frac{\sin\theta}{r}\frac{\partial}{\partial \theta}\left(\frac{u_\phi}{\sin\theta}\right) + \frac{1}{r\sin\theta}\frac{\partial u_\theta}{\partial \phi}\right) \tag{6.4.7}$$

The Cauchy momentum equation in spherical coordinates has the following components

$$
\begin{aligned}
r:\rho\left(\frac{\partial u_r}{\partial t}+u_r\frac{\partial u_r}{\partial r}+\frac{u_\theta}{r}\frac{\partial u_r}{\partial\theta}+\frac{u_\phi}{r\sin\theta}\frac{\partial u_r}{\partial\phi}-\frac{u_\theta^2+u_\phi^2}{r}\right)\\
=\frac{1}{r^2\sin\theta}\left[\frac{\partial}{\partial r}(r^2\sin\theta T_{rr})+\frac{\partial}{\partial\theta}(r\sin\theta T_{\theta r})+\frac{\partial}{\partial\phi}(rT_{\phi r})\right]-\frac{T_{\theta\theta}+T_{\phi\phi}}{r}+b_r\\
\theta:\rho\left(\frac{\partial u_\theta}{\partial t}+u_r\frac{\partial u_\theta}{\partial r}+\frac{u_\theta}{r}\frac{\partial u_\theta}{\partial\theta}+\frac{u_\phi}{r\sin\theta}\frac{\partial u_\theta}{\partial\phi}+\frac{u_r u_\theta}{r}-\frac{u_\phi^2\cot\theta}{r}\right)\\
=\frac{1}{r^2\sin\theta}\left[\frac{\partial}{\partial r}(r^2\sin\theta T_{r\theta})+\frac{\partial}{\partial\theta}(r\sin\theta T_{\theta\theta})+\frac{\partial}{\partial\phi}(rT_{\phi\theta})\right]-\frac{T_{\phi\phi}\cot\theta}{r}+\frac{T_{r\theta}}{r}+b_\theta\\
\phi\ \rho\left(\frac{\partial u_\phi}{\partial t}+u_r\frac{\partial u_\phi}{\partial r}+\frac{u_\theta}{r}\frac{\partial u_\phi}{\partial\theta}+\frac{u_\phi}{r\sin\theta}\frac{\partial u_\phi}{\partial\phi}+\frac{u_r u_\phi}{r}+\frac{u_\theta u_\phi\cot\theta}{r}\right)\\
=\frac{1}{r^2\sin\theta}\left[\frac{\partial}{\partial r}(r^2\sin\theta T_{r\phi})+\frac{\partial}{\partial\theta}(r\sin\theta T_{\theta\phi})+\frac{\partial}{\partial\phi}(rT_{\phi\phi})\right]+\frac{T_{r\phi}}{r}+\frac{T_{\theta\phi}\cot\theta}{r}+b_\phi
\end{aligned}
\tag{6.4.8}
$$

6.5 Navier–Stokes Equations

Inserting Eq. (6.3.9) into the index form of the Cauchy momentum equation, Eq. (5.2.2), results in

$$
\begin{aligned}
\rho\left(\frac{\partial u_i}{\partial t}+u_j\frac{\partial u_i}{\partial x_j}\right)&=\frac{\partial T_{ik}}{\partial x_k}\\
&=\frac{\partial}{\partial x_k}\left[-p\delta_{ik}+\mu\left(\frac{\partial u_i}{\partial x_k}+\frac{\partial u_k}{\partial x_i}\right)\right]
\end{aligned}
\tag{6.5.1}
$$

The right hand side simplifies to

$$
-\frac{\partial p}{\partial x_i}+\mu\left(\frac{\partial^2 u_i}{\partial x_k\partial x_k}+\frac{\partial}{\partial x_k}\left(\frac{\partial u_k}{\partial x_i}\right)\right)\tag{6.5.2}
$$

And we see that the last term can be further simplified by reversing the order of differentiation

$$
\frac{\partial}{\partial x_k}\left(\frac{\partial u_k}{\partial x_i}\right)=\frac{\partial}{\partial x_i}\left(\frac{\partial u_k}{\partial x_k}\right)=0\tag{6.5.3}
$$

since we have assumed an incompressible fluid. Taking things one step further, we assume that incompressibility is based on constant density. This combination of constant density and constant viscosity is the definition of a Newtonian viscous fluid. The resulting equation is

$$
\rho\left(\frac{\partial u_i}{\partial t}+u_j\frac{\partial u_i}{\partial x_j}\right)=-\frac{\partial p}{\partial x_i}+\mu\frac{\partial^2 u_i}{\partial x_k^2}+b_i\tag{6.5.4}
$$

in index form while in vector form it is

$$
\rho\left(\frac{\partial \underline{u}}{\partial t}+(\underline{u}\cdot\underline{\nabla})\underline{u}\right)=-\underline{\nabla}p+\mu\nabla^2\underline{u}+\underline{b}\tag{6.5.5}
$$

Equations (6.5.4) or (6.5.5) are known as the Navier–Stokes equations, named after the French engineer and physicist, Claude-Louis Navier (1785–1836) and the British mathematician and physicist Sir George Gabriel Stokes (1819–1903). It is one of the most used equations in fluid mechanics and, as we saw in Figure 6.1 for the Euler equations, the Navier–Stokes equations have the same named terms but now there is also a viscous force term, shown in Figure 6.8.

The components of Eq. (6.5.5) where the velocity vector is $\underline{u} = (u, v, w)$ in the (x, y, z)-directions, respectively, are

unsteady acceleration	convective acceleration		pressure force	viscous force	body force

$$\rho\left(\frac{\partial \underline{u}}{\partial t}+\left(\underline{u}\cdot\underline{\nabla}\right)\underline{u}\right)=-\underline{\nabla}\,p+\mu\nabla^2\underline{u}+\underline{b}$$

Figure 6.8 Names of the terms in the Navier–Stokes equations.

$$\mathrm{x}:\ \rho\left(\frac{\partial u}{\partial t}+u\frac{\partial u}{\partial x}+v\frac{\partial u}{\partial y}+w\frac{\partial u}{\partial z}\right)$$
$$=-\frac{\partial p}{\partial x}+\mu\left[\frac{\partial^2 u}{\partial x^2}+\frac{\partial^2 u}{\partial y^2}+\frac{\partial^2 u}{\partial z^2}\right]+b_x$$

$$\mathrm{y}:\ \rho\left(\frac{\partial v}{\partial t}+u\frac{\partial v}{\partial x}+v\frac{\partial v}{\partial y}+w\frac{\partial v}{\partial z}\right)$$
$$=-\frac{\partial p}{\partial y}+\mu\left[\frac{\partial^2 v}{\partial x^2}+\frac{\partial^2 v}{\partial y^2}+\frac{\partial^2 v}{\partial z^2}\right]+b_y$$

$$\mathrm{z}:\ \rho\left(\frac{\partial w}{\partial t}+u\frac{\partial w}{\partial x}+v\frac{\partial w}{\partial y}+w\frac{\partial w}{\partial z}\right)$$
$$=-\frac{\partial p}{\partial z}+\mu\left[\frac{\partial^2 w}{\partial x^2}+\frac{\partial^2 w}{\partial y^2}+\frac{\partial^2 w}{\partial z^2}\right]+b_z \qquad (6.5.6)$$

Equations (6.5.5) are second order in spatial derivatives, compared to first order for an inviscid fluid, and these come from the viscous effects. Consequently two conditions need to be imposed on the boundary, and typically this is setting the fluid velocity equal to the boundary velocity on the boundary

$$\underline{u}=\underline{u}_b \qquad (6.5.7)$$

so that the boundary is both impermeable and requires no relative slip with the fluid. The nonlinear convective acceleration term $(\underline{u}\cdot\underline{\nabla})\underline{u}$ is a source of interesting fluid mechanical behavior and also a formidable challenge when seeking solutions. As we saw in Section 6.2 we can write it as $(\underline{u}\cdot\underline{\nabla})\underline{u}=(q\underline{\sigma}\cdot\underline{\nabla})\underline{u}=q\partial\underline{u}/\partial s$, where s is the streamline coordinate. Flows where the velocity does not change along the streamline have zero convective acceleration and solutions are more available. The following analysis of fully developed, steady flow in a two-dimensional channel is an example.

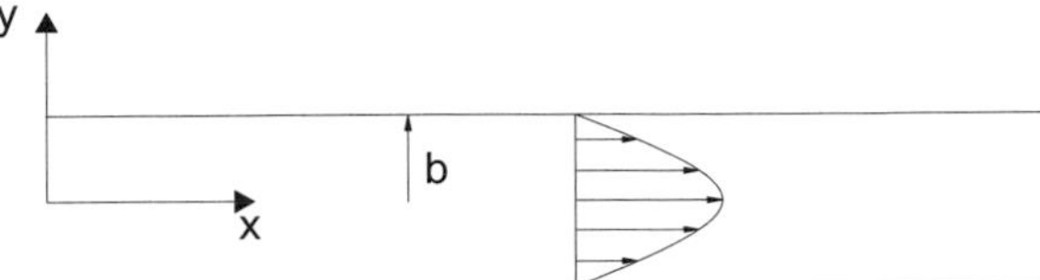

Figure 6.9 Steady Newtonian viscous flow in a channel called Hagen–Poiseuille flow.

An example of a solution to the Navier–Stokes equations is Hagen–Poiseuille flow in a two-dimensional channel. Consider Newtonian viscous flow in a channel of height 2b, as shown in Figure 6.9. The two-dimensional (x,y) Navier–Stokes and continuity equations are, respectively,

$$\text{x-momentum:}\ \frac{\partial u}{\partial t}+u\frac{\partial u}{\partial x}+v\frac{\partial u}{\partial y}=-\frac{1}{\rho}\frac{\partial p}{\partial x}+\nu\left(\frac{\partial^2 u}{\partial x^2}+\frac{\partial^2 u}{\partial y^2}\right)$$
$$\text{y-momentum:}\ \frac{\partial v}{\partial t}+u\frac{\partial v}{\partial x}+v\frac{\partial v}{\partial y}=-\frac{1}{\rho}\frac{\partial p}{\partial y}+\nu\left(\frac{\partial^2 v}{\partial x^2}+\frac{\partial^2 v}{\partial y^2}\right)$$
$$\text{continuity:}\ \frac{\partial u}{\partial x}+\frac{\partial v}{\partial y}=0 \qquad (6.5.8)$$

We can significantly simplify Eqs. (6.5.8) by imposing the assumptions:

a. Steady flow: $\partial u/\partial t=\partial v/\partial t=0$.
b. Unidirectional flow: $v=0$, which implies from continuity that $\partial u/\partial x=0$, i.e. fully developed and $u=u(y)$ only. Also from y-momentum $\partial p/\partial y=0$ so $p=p(x)$ only.
c. Then $\partial^2 u/\partial x^2=0$ in x-momentum because $\partial u/\partial x=0$ everywhere.

The simplified momentum balance is now

$$\mu\frac{d^2 u}{dy^2}=\frac{dp}{dx} \qquad (6.5.9)$$

The ordinary y-derivative is used in Eq. (6.5.9), because u only depends on y. Since $p=p(x)$ it must be that dp/dx is a constant, ensuring that $u=u(y)$. That implies the pressure is a linear function of x.

Integrating Eq. (6.5.9) twice gives the quadratic form

$$u = \frac{dp}{dx}\frac{y^2}{2\mu} + c_1 y + c_2 \tag{6.5.10}$$

and the unknown constants of integration, c_1 and c_2, are determined from the boundary conditions. For this problem the wall is at rest, so the no-slip conditions at the walls are

$$u(y = b) = 0, \quad u(y = -b) = 0 \tag{6.5.11}$$

which, because of the symmetry around the centerline, is entirely equivalent to $u = 0$ at the top wall and $du/dy = 0$ at the centerline, $y = 0$,

$$u(y = b) = 0, \quad \frac{du}{dy}(y = 0) = 0 \tag{6.5.12}$$

Imposing either Eqs. (6.5.11) or Eqs. (6.5.12) on Eq. (6.5.10) determines the integration constants as

$$c_1 = 0, \quad c_2 = -\frac{b^2}{2\mu}\frac{dp}{dx} \tag{6.5.13}$$

The solution for the velocity field is then

$$u = \frac{1}{2\mu}\left(-\frac{dp}{dx}\right)(b^2 - y^2) \tag{6.5.14}$$

For flow in the $+x$-direction, $dp/dx < 0$ so that $u > 0$. The velocity profile is parabolic in y. The maximum velocity is on the centerline (which is why $du/dy = 0$ there), and is

$$u_{max} = \frac{b^2}{2\mu}\left(-\frac{dp}{dx}\right) \tag{6.5.15}$$

The average velocity is defined as

$$u_{avg} = \frac{1}{A}\int_A u\, dA \tag{6.5.16}$$

where A is the cross-sectional area of the flow. For this two-dimensional problem $A = 2bL_z$, where L_z is some depth in the z-direction, and $dA = dy\,dz$ where $0 \leq z \leq L_z, \; -b \leq y \leq +b$

$$\begin{aligned} u_{avg} &= \frac{1}{2bL_z}\int_0^{L_z}\int_{-b}^{b} \frac{1}{2\mu}\left(-\frac{dp}{dx}\right)(b^2 - y^2)\, dy\, dz \\ &= \frac{b^2}{3\mu}\left(-\frac{dp}{dx}\right) = \frac{2}{3}u_{max} \end{aligned} \tag{6.5.17}$$

From Eq. (6.5.17) we see that the average velocity is two-thirds of the maximum velocity. The flow rate is simply

$$q = \int_0^{L_z}\int_{-b}^{b} u\; dy\, dz = 2bL_z u_{avg} = \frac{2b^3 L_z}{3\mu}\left(-\frac{dp}{dx}\right) \tag{6.5.18}$$

which increases linearly with driving pressure, i.e. if the pressure gradient is doubled, then the flow rate is doubled. Likewise, if the viscosity is doubled the flow rate decreases by half. However, the strongest effect comes from the channel width, so if b is doubled the average velocity increases by a factor of 8.

Another approach to two-dimensional flows that have a stream function is to represent the two remaining components of the Navier–Stokes equations as a single equation in terms of ψ. This is easiest seen in Eqs. (6.5.8), where $u = \partial\psi/\partial y, v = -\partial\psi/\partial x$, which satisfies conservation of mass identically. We can take the partial y-derivative of the x-component and the partial x-derivative of the y-component, and subtract to eliminate pressure. Then replacing all velocities in terms of ψ we find

$$\frac{\partial}{\partial t}(\nabla^2\psi) + \frac{\partial\psi}{\partial y}\frac{\partial}{\partial x}(\nabla^2\psi) - \frac{\partial\psi}{\partial x}\frac{\partial}{\partial y}(\nabla^2\psi) = \nu\nabla^4\psi \tag{6.5.19}$$

where $\nabla^4\psi = \nabla^2(\nabla^2\psi)$ and $\nabla^2 = \partial^2/\partial x^2 + \partial^2/\partial y^2$ is the Laplacian operator in rectangular coordinates. Equation (6.5.19) is the balance of momentum in one scalar variable, ψ, where the first term is the unsteady acceleration, the next two are the nonlinear convective acceleration, and the right hand side is the viscous term. It simplifies the analysis compared to solving the original two coupled momentum equations. Once determined, ψ can be inserted back into Eq. (6.5.8) to find the pressure.

Clinical Correlation 6.2

The liquid layer between the chest wall and lung is called the pleural fluid, and the local geometry is like a two-dimensional channel, similar to Figure 6.9. One of the channel walls can be the lung and the other the chest wall. The typical pleural liquid layer width is 40 μm in normal lungs. However, a major difference from Hagen–Poiseuille flow is that during breathing the chest wall and lung move relative to one another in the x-direction as they slide. One of the functions of the pleural fluid is to lubricate this sliding. So we can think of this problem as holding one of the walls fixed, say the upper wall, while moving the lower wall in the x-direction. For simplicity, let's say that the lower wall moves at a constant speed, U_0. The system still satisfies all of the assumptions which simplified the Navier–Stokes equations that led to Eq. (6.5.10). However, now we have a different set of boundary conditions

$$u(y = b) = 0, \quad u(y = -b) = U_0 \tag{6.5.20}$$

so the constants of integration will be different. Imposing the constraints of Eq. (6.5.20) on Eq. (6.5.10) gives us

$$c_1 = -\frac{U_0}{2b}, \quad c_2 = -\frac{b^2}{2\mu}\frac{dp}{dx} + \frac{U_0}{2} \tag{6.5.21}$$

The final form of the velocity is

$$u = -\frac{1}{2\mu}\frac{dp}{dx}\left(b^2 - y^2\right) + \frac{U_0}{2b}(b - y) \tag{6.5.22}$$

which has flow driven by the pressure gradient as well as the wall motion.

To examine the solution, define the dimensionless variables

$$Y = \frac{y}{b}, \quad U = \frac{u}{\left(-\frac{dp}{dx}\right)\frac{b^2}{\mu}} \tag{6.5.23}$$

Inserting Eqs. (6.5.23) into Eq. (6.5.22) gives the dimensionless result

$$U = \frac{1}{2}\left(1 - Y^2\right) + \frac{1}{2}\beta(1 - Y) \tag{6.5.24}$$

where the resulting dimensionless parameter is $\beta = \mu U_0 / \left(-b^2\, dp/dx\right)$, which is the ratio of the wall velocity, U_0, to a measure of the Hagen–Poiseuille velocity from a driving pressure gradient $(-dp/dx)b^2/\mu$ as we saw in Eq. (6.5.14).

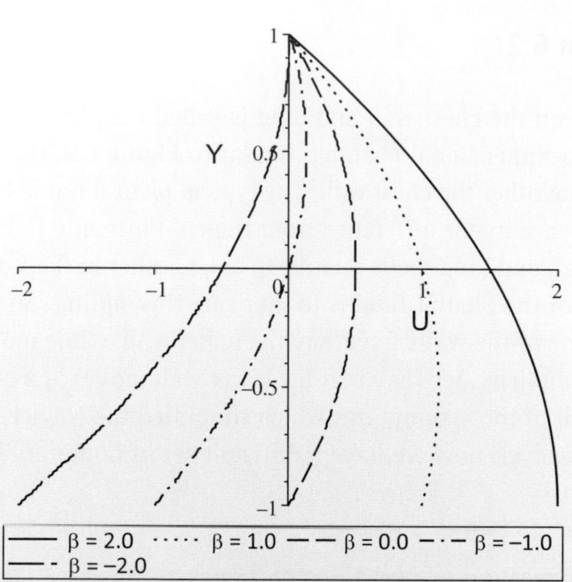

Figure 6.10 Velocity profiles U(Y) for a model of pleural liquid flow, $\beta = 0, \pm 1.0, \pm 2$

A selection of velocity profiles is shown in Figure 6.10, for $\beta = 0, \pm 1.0, \pm 2$. For $\beta = 0$ the lower wall has zero velocity and the profile is a parabola symmetric with respect to the midline, $Y = 0$. Negative and positive values of β skew the parabola accordingly.

6.6 Scaling the Navier–Stokes Equations

The Navier–Stokes equations are often solved for flow problems where one or more terms may be neglected, initially, because they are small. In order to justify such simplifications, it is essential to cast the balance in dimensionless form. For flows with a characteristic velocity scale, U_s, and length scale, L_s, define the dimensionless variables as the following

$$\underline{X} = \frac{\underline{x}}{L_s}, \quad T = \frac{t}{L_s/U_s}, \quad \underline{U} = \frac{\underline{u}}{U_s}, \quad P = \frac{p}{\rho U_s^2} \tag{6.6.1}$$

The time scale of L_s/U_s is based on convection while the pressure scale, ρU_s^2, emphasizes flows where inertial effects dominate. Inserting these forms into Eq. (6.5.5) with zero body force gives us

$$\frac{\partial \underline{U}}{\partial T} + (\underline{U} \cdot \underline{\nabla})\underline{U} = -\underline{\nabla} P + \frac{1}{Re} \nabla^2 \underline{U} \tag{6.6.2}$$

where $Re = U_s L_s / \nu$ is the Reynolds number, the ratio of inertial to viscous effects. The Reynolds number is named for Osborne Reynolds (1842–1912) who was a professor at Owens College in Manchester, now University of Manchester, in England. More information about his contributions appear in Chapter 13, Box 13.3. Equation (6.6.2) is often used for high Reynolds number flows, $Re \gg 1$. For large enough Re the viscous term is neglected to recover inviscid flow, using the Euler equations, Eq. (6.1.5), at least in flow regions away from boundaries.

If, instead, we scale pressure as $P = p/(\mu U_s/L_s)$, which emphasizes viscous shear, a $1/Re$ factor appears in front of the new pressure gradient term. Multiplying through by Re the result is

$$\mathrm{Re}\left(\frac{\partial \underline{U}}{\partial T}+(\underline{U}\cdot\underline{\nabla})\underline{U}\right)=-\underline{\nabla}P+\nabla^2\underline{U} \tag{6.6.3}$$

Equation (6.6.3) is a form often used for low Reynolds number flows, Re<<1. For small enough values the acceleration terms on the left hand side are neglected. The resulting balance of viscous and pressure gradient terms yields the Stokes equations

$$-\underline{\nabla}P+\nabla^2\underline{U}=0 \tag{6.6.4}$$

Many biofluid mechanical applications have an imposed time scale like the heartbeat, breathing frequency, ciliary beat frequency and others. In those flows the dimensionless time is better represented as $T=\omega t$, where $\omega=2\pi f$ and f is the frequency. Using this new time scale in Eq. (6.6.3) leads to

$$\alpha^2\frac{\partial \underline{U}}{\partial T}+\mathrm{Re}(\underline{U}\cdot\underline{\nabla})\underline{U}=-\underline{\nabla}P+\nabla^2\underline{U} \tag{6.6.5}$$

The new dimensionless group is the Womersley parameter, $\alpha=L_s\sqrt{\omega/\nu}$, which is the ratio of unsteady acceleration to viscous effects. For $\alpha<<1$ the unsteady acceleration term is often ignored and the flow is considered steady or quasi-steady.

Going back to an inertial pressure scale, $P=p/\rho U_s^2$, with a characteristic frequency, f, where $T=f\cdot t$, Eq. (6.6.2) becomes

$$\mathrm{St}\frac{\partial \underline{U}}{\partial T}+(\underline{U}\cdot\underline{\nabla})\underline{U}=-\underline{\nabla}P+\frac{1}{\mathrm{Re}}\nabla^2\underline{U} \tag{6.6.6}$$

where St is the Strouhal number, $\mathrm{St}=fL_s/U_s=\alpha^2/2\pi\mathrm{Re}$ and is named for the Czech physicist Vincenc Strouhal (1850–1922), who studied vortex shedding from air flow over wires in 1878 (Strouhal, 1878). This periodic vortex shedding at frequency f gives the "Aeolian tones" or "singing wire" effect that intrigued observers as telegraph wires were added to the landscape in the nineteenth century. Experiments can focus on the St(Re) dependence. More information on vortex shedding for flow over a cylinder, like a wire, is presented in Section 13.1. It is a feature of the nonlinear convective acceleration terms in the Navier–Stokes equations that an oncoming steady air flow over an object can cause time-periodic velocities.

An additional useful form is for situations where the length scale in one direction differs from that in the other direction. For example, a two-dimensional channel flow model would have a length scale for the main flow direction in x which can be the channel length, L. In contrast, a length scale across the channel in the y-direction could be its half-width, b. Often the ratio $\varepsilon=b/L<<1$ for long channels, or tubes, so some simplifications become available. Choose the following definitions of the dimensionless variables

$$U=\frac{u}{U_s},\quad V=\frac{v}{\varepsilon U_s},\quad T=\omega t,\quad P=\frac{p}{\mu U_s/\varepsilon b},\quad X=\frac{x}{L},\quad Y=\frac{y}{b} \tag{6.6.7}$$

The x-velocity is scaled on U_S. However, the y-velocity is scaled on εU_S, because it is smaller in general and leaves the continuity equation with no appearance of ε, i.e. $\partial U/\partial X+\partial V/\partial Y=0$. Now the pressure scale is chosen to be $\mu U_s/\varepsilon b$. Inserting Eqs. (6.6.7) into Eqs. (6.5.6) without body forces gives us the following dimensionless components of momentum conservation

$$\begin{aligned}
&\alpha^2\frac{\partial U}{\partial T}+\varepsilon\mathrm{Re}\left(U\frac{\partial U}{\partial X}+V\frac{\partial U}{\partial Y}\right)=-\frac{\partial P}{\partial X}+\varepsilon^2\frac{\partial^2 U}{\partial X^2}+\frac{\partial^2 U}{\partial Y^2}\\
&\varepsilon^2\alpha^2\frac{\partial V}{\partial T}+\varepsilon^3\mathrm{Re}\left(U\frac{\partial V}{\partial X}+V\frac{\partial V}{\partial Y}\right)=-\frac{\partial P}{\partial Y}+\varepsilon^2\left(\varepsilon^2\frac{\partial^2 U}{\partial X^2}+\frac{\partial^2 U}{\partial Y^2}\right)
\end{aligned} \tag{6.6.8}$$

We will use the form in Eq. (6.6.8) in Section 11.1 for steady lubrication approximations where $\varepsilon<<1$ and $\varepsilon\mathrm{Re}<<1$ and terms multiplied by either are neglected leading to the resulting forms

$$\begin{aligned}
-\frac{\partial P}{\partial X}+\frac{\partial^2 U}{\partial Y^2}&=0\\
\frac{\partial P}{\partial Y}&=0
\end{aligned} \tag{6.6.9}$$

to give a locally parabolic X-velocity profile.

6.7 Kinematics of a Fluid Element

Recalling that we formed the symmetric rate of strain tensor, $\underline{\underline{E}}$ in Eq. (6.3.8), by adding the velocity gradient tensor to its own transform, we also can form an antisymmetric (or skew symmetric) tensor, $\underline{\underline{\Omega}}$, from $\underline{\underline{D}}$ by subtracting its transpose, $\underline{\underline{\Omega}} = \frac{1}{2}\left(\underline{\underline{D}} - \underline{\underline{D}}^T\right)$, with components as follows

$$\Omega_{ij} = \frac{1}{2}(D_{ij} - D_{ji}) = \frac{1}{2}\left(\frac{\partial u_i}{\partial x_j} - \frac{\partial u_j}{\partial x_i}\right) \Leftrightarrow \underline{\underline{\Omega}} = \frac{1}{2}\left(\underline{\underline{D}} - \underline{\underline{D}}^T\right) \tag{6.7.1}$$

and we see that $\Omega_{ij} = -\Omega_{ji}$, or $\underline{\underline{\Omega}} = -\underline{\underline{\Omega}}^T$, the definition of antisymmetry. When $i = j$ the component is zero. The tensor $\underline{\underline{\Omega}}$ is called the "vorticity tensor," which reflects the local fluid rotation rate, and has only three independent components. These can be related to the vorticity vector, $\underline{\omega} = \underline{\nabla} \times \underline{u}$, whose components are

$$\omega_i = \varepsilon_{ijk}\left(\partial u_k / \partial x_j\right) \tag{6.7.2}$$

where ε_{ijk} is the permutation tensor (Levi-Civita tensor) and his values of: zero if any two indices are the same; +1 if ijk are in the pattern 1–2–3–1–2–3 etc.; and −1 if ijk are in the pattern 3–2–1–3–2–1 etc. It is a third-order isotropic tensor. The vorticity vector components relate to $\underline{\underline{\Omega}}$ as

$$\begin{aligned}
\omega_i &= \varepsilon_{ijk}\Omega_{kj} \\
\omega_1 &= \Omega_{32} - \Omega_{23} = \left(\frac{\partial u_3}{\partial x_2} - \frac{\partial u_2}{\partial x_3}\right) \\
\omega_2 &= \Omega_{13} - \Omega_{31} = \left(\frac{\partial u_1}{\partial x_3} - \frac{\partial u_3}{\partial x_1}\right) \\
\omega_3 &= \Omega_{21} - \Omega_{12} = \left(\frac{\partial u_2}{\partial x_1} - \frac{\partial u_1}{\partial x_2}\right)
\end{aligned} \tag{6.7.3}$$

From a different viewpoint, the components of $\underline{\omega}$ are in $\underline{\underline{\Omega}}$ as shown

$$\begin{aligned}
\underline{\underline{\Omega}} &= \frac{1}{2}\begin{bmatrix} 0 & \left(\frac{\partial u_1}{\partial x_2} - \frac{\partial u_2}{\partial x_1}\right) & \left(\frac{\partial u_1}{\partial x_3} - \frac{\partial u_3}{\partial x_1}\right) \\ \left(\frac{\partial u_2}{\partial x_1} - \frac{\partial u_1}{\partial x_2}\right) & 0 & \left(\frac{\partial u_2}{\partial x_3} - \frac{\partial u_3}{\partial x_2}\right) \\ \left(\frac{\partial u_3}{\partial x_1} - \frac{\partial u_1}{\partial x_3}\right) & \left(\frac{\partial u_3}{\partial x_2} - \frac{\partial u_2}{\partial x_3}\right) & 0 \end{bmatrix} \\
&= \frac{1}{2}\begin{bmatrix} 0 & -\omega_3 & \omega_2 \\ \omega_3 & 0 & -\omega_1 \\ -\omega_2 & \omega_1 & 0 \end{bmatrix}
\end{aligned} \tag{6.7.4}$$

By components, again we can use the permutation tensor to show that $\Omega_{ij} = -(1/2)\varepsilon_{ijk}\omega_k$. A useful relationship between the permutation tensor and the identity tensor (Kronecker delta) is

$$\varepsilon_{ijk}\varepsilon_{klm} = \delta_{il}\delta_{jm} - \delta_{im}\delta_{jl} \tag{6.7.5}$$

Clearly, the rate of deformation tensor, $\underline{\underline{D}}$, is the sum of $\underline{\underline{E}}$ and $\underline{\underline{\Omega}}$,

$$\underline{\underline{D}} = \underline{\underline{E}} + \underline{\underline{\Omega}} \tag{6.7.6}$$

showing that the rate of deformation can be decomposed into a strain rate and a rotational rate.

Let us look more closely at a fluid volume, say in two dimensions, to see how these velocity gradients correspond to deformation rates. Four types of fluid motion are shown in Figure 6.11. Translation and rotation do not involve deformation of the volume, so are not going to provide the internal friction or relative fluid particle motion that we seek for viscous effects projected to the stresses. However, shear strain and dilatation due to compressibility are just what we seek.

To gain more physical insight into the decomposition of these deformation rates, consider a

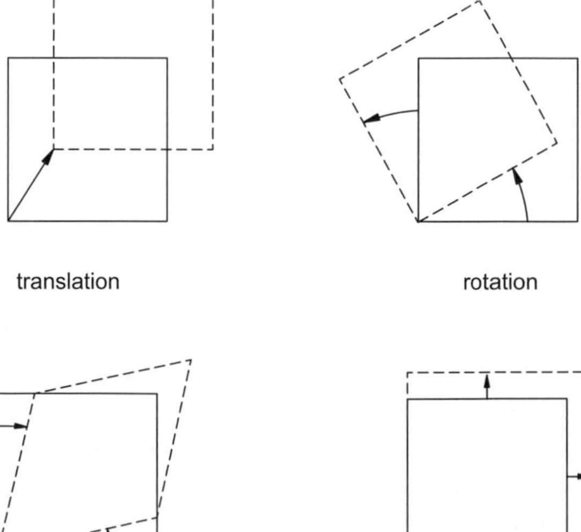

Figure 6.11 Four types of motion for a fluid volume in two dimensions.

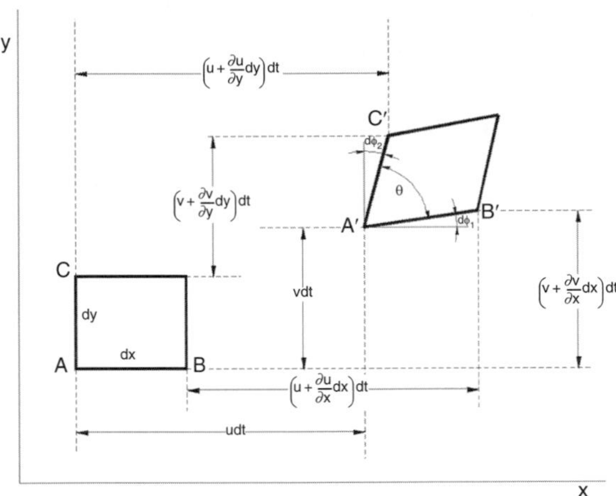

Figure 6.12 Fluid element in two dimensions undergoing a deformation.

two-dimensional fluid parcel that is undergoing translation and deformation in Figure 6.12.

Fluid point A has velocity components(u, v), while fluid point B, which starts at a distance dx away in the x-direction, has components $(u + (\partial u/\partial x)dx,\ v + (\partial v/\partial x)\,dx)$ and fluid point C, which starts at a distance dy away in the y-direction, has components $(u + (\partial u/\partial y)dy, v + (\partial v/\partial y)\,dy)$. During the time, dt, A is convected to a new location, A′, a distance $(u\,dt,\ v\,dt)$in the (x, y) directions, respectively. The velocity components of B and C come from the product of the differential distance, dx and dy, respectively, and the spatial rate at which the velocity is changing in those directions. The relative velocity of B compared to A is the difference of their respective components, which is $((\partial u/\partial x)\,dx,\ (\partial v/\partial x)\,dx)$. The relative velocity of C compared to A is the difference of their respective components, which is $((\partial u/\partial y)dy,\ (\partial v/\partial y)dy)$. Over the time duration, dt, the relative distances the points move are the product of dt and the relative velocities.

The x-component of strain for the line element AB to A′B′is the x-component of the change in length of A′B′ divided by the original length x-component of length, which was simply dx. That ratio is $(\partial u/\partial x)dx\,dt/dx = (\partial u/\partial x)dt$ and the rate of x-direction straining for AB is $\partial u/\partial x$. Likewise, the y-component of strain for the line element AC to A′C′ is derived from their relative y-velocities and found to be $(\partial v/\partial y)dy\,dt/dy = (\partial v/\partial y)dt$. So the rate of y-component straining for AC is $\partial v/\partial y$. These terms are the diagonal of the strain rate tensor, E_{ij}, and represent strain rates in the normal direction, or normal strain rates

$$E_{11} = \frac{\partial u}{\partial x},\quad E_{22} = \frac{\partial v}{\partial y} \tag{6.7.7}$$

In the third dimension we would find the z-strain rate to be $E_{33} = \partial w/\partial z$. These normal strain rates contribute to the normal stresses which are the diagonal terms of the stress tensor T_{ij}. Pure dilatation of the fluid volume, as shown in Figure 6.11, would consist only of these normal strain rates. The original fluid volume is $V_0 = dx\,dy$. The dilated fluid volume would be $V = (1 + \partial u/\partial x)dx \times (1 + \partial v/\partial y)dy$. The relative change in volume is then

$$\frac{V - V_0}{V_0} = \frac{\left(1 + \frac{\partial u}{\partial x}\right)\left(1 + \frac{\partial v}{\partial y}\right)dx\,dy - dx\,dy}{dx\,dy} = \frac{\partial u}{\partial x} + \frac{\partial v}{\partial y} + O\left(\frac{\partial u}{\partial x}\frac{\partial v}{\partial y}\right) \tag{6.7.8}$$

Immediately we see that, ignoring nonlinear terms as being small, the right hand side of Eq. (6.7.8) is the divergence of the velocity field. So we find that the normal strain rates tell us about the dilatation of the fluid volume

$$\underline{\nabla} \cdot \underline{u} = E_{kk} = E_{11} + E_{22} + E_{33} = \frac{\partial u}{\partial x} + \frac{\partial v}{\partial y} + \frac{\partial w}{\partial z} \tag{6.7.9}$$

using the summation notation, where E_{kk} is the trace of the strain rate tensor. An incompressible fluid would leave Eq. (6.7.9) equal to zero.

The off-diagonal terms of E_{ij} are the shear strain rates which we can relate to the angles $d\phi_1$ and $d\phi_2$. For small angles, the angle in radians is equal to its tangent, so $\tan\,(d\phi_1) \approx d\phi_1$. The tangent of $d\phi_1$ is the distance of the y-displacement of B to B′ divided by the new x-length of A′B′. The former is $(\partial v/\partial x)dx\,dt$ and the latter is $dx + (\partial u/\partial x)dx\,dt$. The ratio is

$$d\phi_1 = \frac{(\partial v/\partial x)dx\,dt}{dx(1+(\partial u/\partial x)dt)} = (\partial v/\partial x)dt(1-(\partial u/\partial x)dt + O(dt^2)) = (\partial v/\partial x)dt + O(dt^2) \quad (6.7.10)$$

where we have used a Taylor series for $(1+q)^{-1} = 1 - q + O(q^2)$, $q << 1$ and then neglected terms of order $(dt)^2$. A similar approach is taken for $d\phi_2$. The tangent of $d\phi_2$ is the distance of the x-displacement of C to C′ divided by the new y-length of A′B′. The former is $(\partial u/\partial y)dy\,dt$ and the latter is $dy + (\partial v/\partial y)dy\,dt$. The ratio is

$$d\phi_2 = -\frac{(\partial u/\partial y)dy\,dt}{dy(1+(\partial v/\partial y)dt)} = -(\partial u/\partial y)\,(1-(\partial v/\partial y)dt + O(dt^2))dt = -(\partial u/\partial y)\,dt + O(dt^2) \quad (6.7.11)$$

where a minus sign is introduced to maintain the assumed convention that positive angular velocities are in the counterclockwise direction.

This tells us that the shear component of the strain rate tensor is

$$E_{12} = E_{xy} = \frac{1}{2}\left(\frac{\partial u}{\partial y} + \frac{\partial v}{\partial x}\right) = \frac{1}{2}\frac{d}{dt}(\phi_1 - \phi_2) \quad (6.7.12)$$

while the rotation rate tensor is

$$\Omega_{21} = \Omega_{yx} = \frac{1}{2}\left(\frac{\partial v}{\partial x} - \frac{\partial u}{\partial y}\right) = \frac{1}{2}\frac{d}{dt}(\phi_1 + \phi_2) \quad (6.7.13)$$

The angle θ, which measures the distortion of the fluid element, is related to the other two angles by $\theta = \pi/2 - \phi_1 + \phi_2$. The time rate of decrease of θ is given by

$$\frac{d\theta}{dt} = -\frac{1}{2}\frac{d}{dt}(\phi_1 - \phi_2) = -E_{12} \quad (6.7.14)$$

So E_{12} measures the local fluid element rate of distortion, while Ω_{21} measures its local rotation. For example, the fluid element remains a rectangle if $\phi_2 = \phi_1$ and then θ remains a right angle and does not change with time, i.e. no shear strain rate, $E_{12} = 0$.

6.8 Navier–Stokes Equations in Other Coordinate Systems

Cylindrical

$$r:\ \rho\left(\frac{\partial u_r}{\partial t} + u_r\frac{\partial u_r}{\partial r} + \frac{u_\theta}{r}\frac{\partial u_r}{\partial \theta} + u_z\frac{\partial u_r}{\partial z} - \frac{u_\theta^2}{r}\right) = -\frac{\partial p}{\partial r} + \mu\left[\frac{1}{r}\frac{\partial}{\partial r}\left(r\frac{\partial u_r}{\partial r}\right) + \frac{1}{r^2}\frac{\partial^2 u_r}{\partial \theta^2} + \frac{\partial^2 u_r}{\partial z^2} - \frac{u_r}{r^2} - \frac{2}{r^2}\frac{\partial u_\theta}{\partial \theta}\right] + b_r$$

$$\theta:\ \rho\left(\frac{\partial u_\theta}{\partial t} + u_r\frac{\partial u_\theta}{\partial r} + \frac{u_\theta}{r}\frac{\partial u_\theta}{\partial \theta} + u_z\frac{\partial u_\theta}{\partial z} + \frac{u_r u_\theta}{r}\right) = -\frac{1}{r}\frac{\partial p}{\partial \theta} + \mu\left[\frac{1}{r}\frac{\partial}{\partial r}\left(r\frac{\partial u_\theta}{\partial r}\right) + \frac{1}{r^2}\frac{\partial^2 u_\theta}{\partial \theta^2} + \frac{\partial^2 u_\theta}{\partial z^2} - \frac{u_\theta}{r^2} + \frac{2}{r^2}\frac{\partial u_r}{\partial \theta}\right] + b_\theta$$

$$z:\ \rho\left(\frac{\partial u_z}{\partial t} + u_r\frac{\partial u_z}{\partial r} + \frac{u_\theta}{r}\frac{\partial u_z}{\partial \theta} + u_z\frac{\partial u_z}{\partial z}\right) = -\frac{\partial p}{\partial z} + \mu\left[\frac{1}{r}\frac{\partial}{\partial r}\left(r\frac{\partial u_z}{\partial r}\right) + \frac{1}{r^2}\frac{\partial^2 u_z}{\partial \theta^2} + \frac{\partial^2 u_z}{\partial z^2}\right] + b_z \quad (6.8.1)$$

Spherical

$$r:\ \rho\left(\frac{\partial u_r}{\partial t} + u_r\frac{\partial u_r}{\partial r} + \frac{u_\theta}{r}\frac{\partial u_r}{\partial \theta} + \frac{u_\phi}{r\sin\theta}\frac{\partial u_r}{\partial \phi} - \frac{u_\theta^2 + u_\phi^2}{r}\right) = b_r - \frac{\partial p}{\partial r} + \mu\left[\nabla^2 u_r - \frac{2u_r}{r^2} - \frac{2}{r^2}\frac{\partial u_\theta}{\partial \theta} - \frac{2u_\theta\cot\theta}{r^2} - \frac{2}{r^2\sin\theta}\frac{\partial u_\phi}{\partial \phi}\right]$$

$$\theta:\ \rho\left(\frac{\partial u_\theta}{\partial t} + u_r\frac{\partial u_\theta}{\partial r} + \frac{u_\theta}{r}\frac{\partial u_\theta}{\partial \theta} + \frac{u_\phi}{r\sin\theta}\frac{\partial u_\theta}{\partial \phi} + \frac{u_r u_\theta}{r} - \frac{u_\phi^2\cot\theta}{r}\right) = b_\theta - \frac{1}{r}\frac{\partial p}{\partial \theta} + \mu\left[\nabla^2 u_\theta + \frac{2}{r^2}\frac{\partial u_r}{\partial \theta} - \frac{u_\theta}{r^2\sin^2\theta} - \frac{2\cos\theta}{r^2\sin^2\theta}\frac{\partial u_\phi}{\partial \phi}\right]$$

$$\phi:\ \rho\left(\frac{\partial u_\phi}{\partial t} + u_r\frac{\partial u_\phi}{\partial r} + \frac{u_\theta}{r}\frac{\partial u_\phi}{\partial \theta} + \frac{u_\phi}{r\sin\theta}\frac{\partial u_\phi}{\partial \phi} + \frac{u_r u_\phi}{r} + \frac{u_\theta u_\phi\cot\theta}{r}\right) = b_\phi - \frac{1}{r\sin\theta}\frac{\partial p}{\partial \phi} + \mu\left[\nabla^2 u_\phi - \frac{u_\phi}{r^2\sin^2\theta} + \frac{2}{r^2\sin\theta}\frac{\partial u_r}{\partial \phi} + \frac{2\cos\theta}{r^2\sin^2\theta}\frac{\partial u_\theta}{\partial \phi}\right] \quad (6.8.2)$$

where the Laplace operator in spherical coordinates is

$$\nabla^2 = \frac{1}{r^2}\frac{\partial}{\partial r}\left(r^2\frac{\partial}{\partial r}\right) + \frac{1}{r^2\sin\theta}\frac{\partial}{\partial \theta}\left(\sin\theta\frac{\partial}{\partial \theta}\right) + \frac{1}{r^2\sin^2\theta}\frac{\partial^2}{\partial \phi^2} \quad (6.8.3)$$

Summary

We have examined two constitutive equations relating the stress tensor to fluid motion, one for inviscid fluids and one for Newtonian viscous fluids. For inviscid fluids the fluid stresses are only in the normal direction and depend on fluid pressure, not fluid velocity gradients. By inserting its constitutive equation for the stress tensor into the Cauchy momentum equation, we derived the Euler equations and examined some simple cases using them. We also integrated the Euler equations along a streamline to obtain different forms of the Bernoulli equation. For viscous fluids we included the pressure contribution to the stresses, as for inviscid fluids, but also an additional contribution called the viscous stress tensor. The viscous stress tensor was assumed to depend in general on the velocity gradients. We started by assuming that the stress tensor, in general, has nine components and each of them depends on a linear combination of velocity gradients of which there are also nine (three velocity components and three gradient directions). Starting with 81 constants, we simplified to isotropic fluids so that the dependence of stresses on velocity gradients decreased to two constants, the dynamic viscosity, μ, and the bulk viscosity, λ. A further simplification to incompressibility reduced the dependence to one constant, μ, which we will just call the viscosity. For a given temperature, pressure, and concentration of dissolved substances, μ and ρ are constant and this gives us a Newtonian viscous fluid. Using its constitutive equation in the Cauchy momentum equation yielded the Navier–Stokes equations. Both the Euler equations and the Navier–Stokes equations are subject to boundary conditions, the former only need match the normal velocity of the boundary while the latter must match both normal and tangential components. We have tied together our previous efforts in scaling the governing equations to find the dimensionless parameters of a model, unsteady viscous problem with two length scales. Finally we related the kinematics of a fluid volume element in two dimensions to understand how local deformation relates to components of the rate of strain tensor and rate of rotation tensor. The constitutive equations for non-Newtonian fluids and their use is explored in Chapter 10. In the next chapters, Chapters 7–9, we will focus on Newtonian viscous fluids and their many applications to biofluid mechanics.

Problems

6.1

a. For inviscid stagnation point flow the Bernoulli equation is given by Eq. (6.1.12). What is the stress tensor $\underline{\underline{T}}$ for this flow?
b. Calculate the stress vector, $\underline{t}$, on the lower wall where $y = 0$.
c. What is the normal component of $\underline{t}$? Call it t_y.
d. Calculate the force on the wall over the interval $-L \leq x \leq L,\ -z_0/2 \leq z \leq z_0/2$, where z_0 is a unit distance in the z-direction. Let $p_0 = 0$.
e. Explain your result in part (d) with regard to the direction of the force and the direction of flow.

6.2

a. The Eulerian pressure field for stagnation point flow is given in Eq. (6.1.12). Show that curves of constant pressure (isobars), say $p = p_1$, are circles centered at the origin. What is their radius?
b. Find the Lagrangian pressure, $P(X_0, Y_0, t)$ by inserting the Lagrangian velocities $\underline{U}(\underline{X}_0, t)$ into the Bernoulli equation.

c. Let $X_0 = L, Y_0 = nL, T = At$ and define a dimensionless pressure $\pi = (P - p_0)/\frac{1}{2}\rho A^2L^2$. Plot π for $n = 0.3,\ 1.0,\ 2.0,\ 3.0$ for $0 \leq T \leq 1$.
d. How do your results in part (d) compare to the Eulerian and Lagrangian speeds calculated in Eqs. (3.5.14) and (3.5.15)?

6.3

a. Figure 6.6 is a snapshot of the unsteady potential velocity field, $\underline{u} = \nabla\phi$, defined in Eq. (6.2.24), based on the velocity potential in Eq. (6.2.23). Animate the vector field plot for $0 \leq t \leq 2\pi$ over the range $0 \leq x \leq 2\pi$ setting $k = c = B = 1$.
b. Now let $c = -1,\ k = B = 1$ and describe your new animation result compared to part (a).
c. Set $c = B = 1,\ \ k = 2$ and repeat the animation. What are the changes you see?
d. Go back to $k = c = B = 1$, but now let $\rho = 1$ and animate the pressure gradient field, ∇p, that is affiliated with the flow in part (a) for $\Phi_b = 0$ in Eq. (6.2.14). How does it compare to part (a)?

6.4

a. For the viscous channel flow with a moving lower wall, the x-velocity, u, is given by Eq. (6.5.23). What is the stress tensor $\underline{\underline{T}}$ for this flow?
b. What is the stress vector, $\underline{t}$, on the upper wall?
c. Find the stress component of part (b) in the tangential, x-direction, call it t_x. This is the wall shear stress.
d. Assume $dp/dx < 0$ and make the wall shear dimensionless using the $T_x = t_x/(-b\,dp/dx)$. Express your answer in terms of the dimensionless parameter $\beta = \mu U_0/(-b^2 dp/dx)$
e. What critical value of β leads to no shear on the upper wall? How does this correlate to Figure 6.10?

6.5

a. For the viscous channel flow with a moving lower wall, the x-velocity, u, is given by Eq. (6.5.23). What is the stress tensor $\underline{\underline{T}}$ for this flow?
b. What is the stress vector, $\underline{t}$, on the lower wall?
c. Find the stress component of part (b) in the tangential, x-direction, call it t_x. This is the wall shear stress.
d. Assume $dp/dx < 0$ and make the wall shear dimensionless using the $T_x = t_x/(-b\,dp/dx)$. Express your answer in terms of the dimensionless parameter $\beta = \mu U_0/(-b^2 dp/dx)$
e. What critical value of β leads to no shear on the lower wall? How does this correlate to Figure 6.10?

6.6

a. Let $d_i = \varepsilon_{ijk} a_j b_k$ where the vectors are $\underline{a} = (2,\ 2,\ 1)$, $\underline{b} = (1,\ 2,\ 1)$. Calculate $\underline{d}$.
b. Calculate the curl of the vector $\underline{c} = \left(x_2 \sin(x_1) + x_3,\ x_3 x_2^2,\ x_1\right)$ and evaluate at $\underline{x}_0 = (\pi/2,\ 1,\ 3)$.
c. The parabolic velocity profile in Eq. (2.8.1) has the form $u = u_m\left(1 - (y/b)^2\right)$. Calculate its vorticity. What direction is the vorticity vector pointing? What is its corresponding vorticity tensor $\underline{\underline{\Omega}}$?
d. In Eq. (6.7.3) we found $\omega_i = \varepsilon_{ijk}\Omega_{kj}$. Now let's solve for $\underline{\underline{\Omega}}$ in terms of $\underline{\omega}$. First multiply both sides of $\omega_i = \varepsilon_{ijk}\Omega_{kj}$ by ε_{irs} and let $\varepsilon_{ijk} = \varepsilon_{jki}$. Now use Eq. (6.7.5) to substitute for your products of permutation tensors with the products of identity tensors. Show that $\Omega_{rs} = -\varepsilon_{irs}\omega_i/2 = -\varepsilon_{rsi}\omega_i/2$.
e. What is the vorticity, $\underline{\omega}$, for the stagnation point flow, $\underline{u} = (Ax,\ -Ay)$? What is the significance of your answer?

6.7

a. For the channel flow with a moving lower wall, the velocity is given by Eq. (6.5.3). Calculate the vorticity, ω.
b. Make your solution to part (a) dimensionless by using the scales $Y = y/b, \Omega = \mu\omega/(-b\,dp/dx)$ using the definition of the parameter $\beta = \mu U_0/\left(-b^2 dp/dx\right)$.
c. Plot your solution for Ω in part (b) across the channel for $\beta = -2,\ -1,\ 0,\ 1,\ 2$.

d. What happens to the zero vorticity location in part (c)? What is the correlation to Figure (6.10)?

6.8

a. What is the strain rate tensor, $\underline{\underline{E}}$, for the channel flow with a moving wall where u is given in Eq. (6.5.23).
b. The three invariants of $\underline{\underline{E}}$, presented in Eq. (10.1.2), do not change with rotation of the coordinate system. The first invariant is $I_1 = \mathrm{tr}(\underline{\underline{E}}) = E_{ii}$. What is its value for any incompressible fluid velocity field?
c. The third invariant is $I_3 = \det(\underline{\underline{E}})$. Calculate it for the flow in part (a).
d. The second invariant can be formulated as $I_2 = \mathrm{tr}(\underline{\underline{E}} \cdot \underline{\underline{E}}) = E_{ij}E_{ji}$. What is its form for the flow in part (a)?
e. A scalar measure of the shear rate intensity of a flow, $\dot{\gamma}$, is based on the second invariant, $\dot{\gamma} = \sqrt{2I_2} = \sqrt{2E_{ij}E_{ji}}$. It plays a central role in the derivation of non-Newtonian fluid models. What is $\dot{\gamma}$ for the flow in part (a)?
f. Scale your solution in part (d) as $\dot{\Gamma} = \mu\dot{\gamma}/(-b\,dp/dx)$, $Y = y/b$ and plot $\dot{\Gamma}$ across the channel width for $\beta = -2, -1, 0, 1, 2$. How does it correlate to Figure 6.10?

7 Steady Newtonian Viscous Flow

Introduction

This is a transitional chapter in that the previous ones were geared for a general audience with some special treatments for biofluids applications. However, the main development of the conservation of mass and momentum, the analysis of fluid deformation, the relation between stress and strain rate, and the approach of scalings and dimensionless parameters were otherwise quite general. In this chapter the full impact of our focus is on biomedical applications in both physiological and technological settings. The first six chapters have brought us to this point. Those classrooms with students having sufficient background and comfort level with the content of Chapters 1–6 can certainly start with this chapter as their entry point into biofluid mechanics. To that end, this chapter introduces some internal flows through conduits as well as external flows over spheres, cylinders and elephants. Applications are made to vascular hemodynamics, respiratory airflow, the lung's liquid lining, microfluidics and tissues. The implications range from blood pressure distribution to evolution of anatomical dimensions. A brief discussion of computational fluid dynamics (CFD) introduces this powerful approach to solving complicated flows and geometries.

Topics Covered

We start with solving the Navier–Stokes equations for flow in a circular cylinder, the so-called Hagen–Poiseuille flow. The results, Poiseuille's Law, relate the pressure drop down the tube to the volume flow rate in it, their ratio is called the resistance to flow. This simple yet elegant law (formula) is extended to blood flow in the vasculature, where we compare the pressure drop in the largest artery (the aorta) to that in the very small arteries (arterioles). The exercise teaches us that the largest drop is in those smaller vessels, and that is where the body controls much of the distribution of blood flow by modifying the diameters of arterioles. A review of the relevant vascular anatomy and physiology is provided. Then flow resulting from gravity and wall motion is presented as a means of analyzing ciliary clearance of mucus from the bronchial tree. There are several diseases in which this ciliary mechanism is overwhelmed by abnormally large amounts of mucus or mucus with abnormal constitutive properties such as its viscosity. Abnormal mucus occurs in diseases such as asthma, emphysema, chronic bronchitis, and cystic fibrosis. A review of the relevant respiratory anatomy and physiology is also provided. Core-annular flow is analyzed and one of the applications is for vascular flows where the red blood cells are found mainly in the central region of the tube cross-section (the core), while a cell-free, plasma layer is adjacent to the walls (the annulus). This leaves us with two regions of different viscosities since the red blood cells increase the apparent viscosity from their self-interactions. The same concept is applied to air flow in the liquid (mucus) coated airways as another mechanism of mucus clearance that resembles a cough. Further uses of Poiseuille's Law are its influence on the evolution of anatomy for the carrying vessels and their branches. When an airway or artery bifurcates from one parent tube to two daughter tubes, power minimization dictates a special ratio of the daughter to parent diameters that, in fact, is often found in the respiratory and vascular systems. This is called Murray's law or the Hess–Murray law. In addition, the branch angle of the bifurcation has a power-minimized optimal value. The lesson that biofluid mechanics has influenced evolution underscores the importance and scope of this field. The technology boom of microfluidics, which is especially prevalent in biomedical applications for small, hand-held analytical chips, often consists of flows in microchannels that are of various rectangular cross-sectional shapes. So 3D flow in a rectangular duct is presented and solved using a Fourier series approach. For most students, this development provides a useful introduction to the Fourier technique. Low Reynolds number flows over a sphere and cylinder are analyzed. The analysis leads to the derivation of Stokes drag for the sphere and Stokes' paradox for a cylinder with resolution. The cylinder flow is further investigated arrays of cylinders, modeling packed fibrils found in tissues. Finally there is a brief introduction to some concepts of CFD with two examples: a berry aneurysm in the brain's arterial circulation and wind blowing over an elephant.

7.1 Hagen–Poiseuille Flow in a Circular Tube

Consider Newtonian viscous flow in a cylindrical tube of radius a as shown in Figure 7.1. The Navier–Stokes Equations in cylindrical coordinates, see Chapter 5, simplify for the assumptions of unidirectional flow, $(u_r, u_\theta, u_z) = (0, 0, u_z)$, where (u_r, u_θ, u_z) are the velocity components in the (r, θ, z)-directions, respectively. Additional simplifications are steady conditions, $\partial/\partial t = 0$, fully developed velocity, $\partial u_z/\partial z = 0$, and axisymmetry, $\partial/\partial\theta = 0$, resulting in only the z-component of momentum conservation,

$$\frac{dp}{dz} = \mu\frac{1}{r}\frac{d}{dr}\left(r\frac{du_z}{dr}\right) \tag{7.1.1}$$

The pressure gradient, dp/dz, is constant in Eq. (7.1.1) and $u_z = u_z(r)$ only. Equation (7.1.1) can be integrated twice to yield

$$u_z(r) = \frac{1}{4\mu}\frac{dp}{dz}r^2 + c_1 \ln r + c_2 \tag{7.1.2}$$

where c_1 and c_2 are the constants of integration. The boundary conditions are no-slip at $r = a$ and finite velocity at $r = 0$,

$$u_z(r = a) = 0, \quad u_z(r = 0) \text{ finite} \tag{7.1.3}$$

Imposing Eq. (7.1.3) on Eq. (7.1.2) leads to the solution of c_1 and c_2. Note that we must have $c_1 = 0$ so that the solution does not become unbounded at $r = 0$ because of the $\ln(r)$ term in Eq. (7.1.2). The final form for $u_z(r)$ is the parabolic profile

$$u_z(r) = \frac{1}{4\mu}\left(-\frac{dp}{dz}\right)(a^2 - r^2) \tag{7.1.4}$$

To conform to Figure 7.1, flow in the positive z-direction requires that pressure is a decreasing function of z, so $dp/dz < 0$ and $(-dp/dz)$ is carried as a positive quantity. The maximum velocity, $u_{z\ max}$, occurs at $r = 0$ and is

$$u_{z\ max} = \frac{a^2}{4\mu}\left(-\frac{dp}{dz}\right) \tag{7.1.5}$$

while the cross-sectional average velocity, $u_{z\ avg}$, is

$$u_{z\ avg} = \frac{1}{\pi a^2}\int_0^{2\pi}\int_0^a u_z(r)\, r\, dr\, d\theta = \frac{1}{8}\frac{a^2}{\mu}\left(-\frac{dp}{dz}\right) = u_{z\ max}/2 \tag{7.1.6}$$

So the average velocity is 1/2 the maximum velocity, $u_{z\ avg} = u_{z\ max}/2$, for Poiseuille flow in a tube, as opposed to flow in a channel from Chapter 6, where the average was 2/3 of the maximum. Solving for dp/dz in terms of $u_{z\ avg}$ or $u_{z\ max}$ in Eq. (7.1.6), and substituting into the velocity solution of Eq. (7.1.4) gives us

$$u_z(r) = 2u_{z\,avg}\left(1 - \left(\frac{r}{a}\right)^2\right) = u_{z\ max}\left(1 - \left(\frac{r}{a}\right)^2\right) \tag{7.1.7}$$

The volumetric flow rate is $Q = \pi a^2 u_{z\,avg}$, resulting in the form

$$Q = \frac{\pi a^4}{8\mu}\left(-\frac{dp}{dz}\right) = \frac{\pi a^4}{8\mu}\frac{\Delta p}{L} \tag{7.1.8}$$

where $-dp/dz = (p_u - p_d)/L = \Delta p/L$ and p_u is the upstream pressure while p_d is the downstream pressure, separated by the distance L.

Equation (7.1.8) is known as Hagen–Poiseuille's Law or often simply Poiseuille's Law for Jean Léonard Marie Poiseuille (1797–1869), a French physician, physicist and physiologist, and for Gotthilf Heinrich Ludwig Hagen (1797–1884), a German physicist and hydraulic engineer. The two researchers discovered this relationship independently and at approximately the same time.

Another common expression of Eq. (7.1.8) rearranges to

$$\Delta p = QR, \quad R = \frac{8\mu L}{\pi a^4} \tag{7.1.9}$$

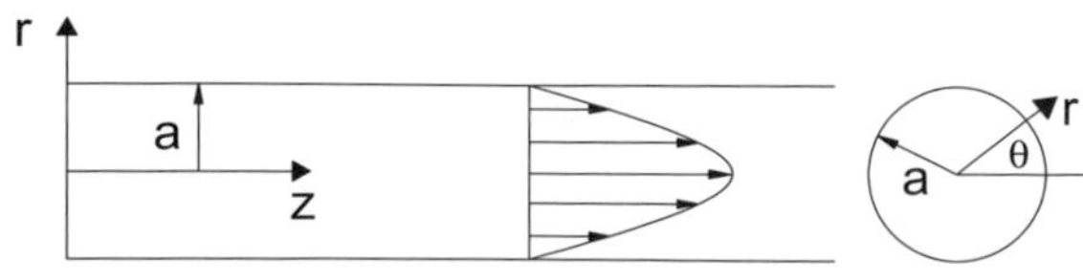

Figure 7.1 Hagen–Poiseuille flow in a cylindrical tube.

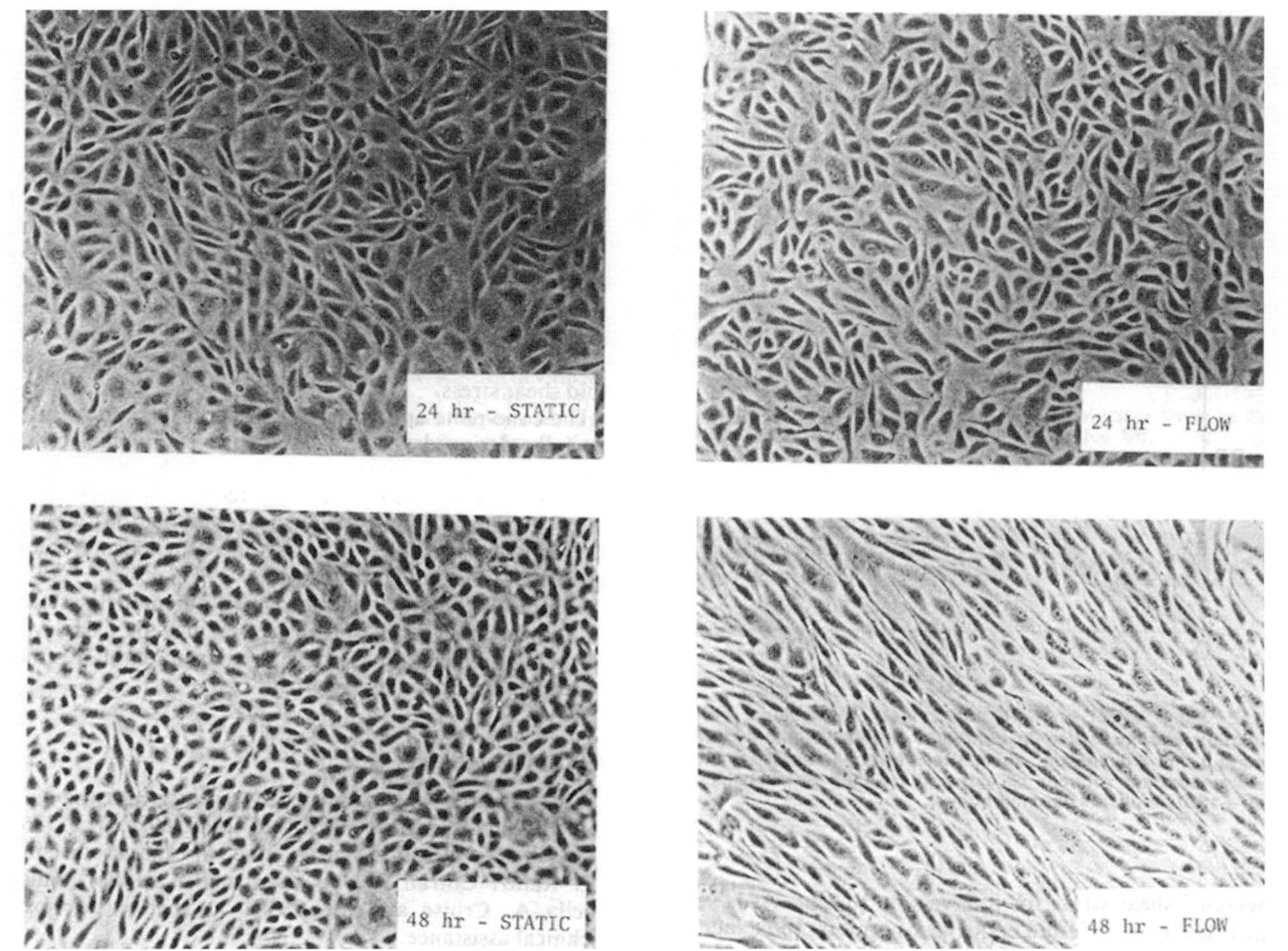

Figure 7.2 Bovine aortic endothelial cell shape and orientation for 24 and 48 hour static and flow conditions at 8 dyn/cm^2 wall shear stress. Note the cell alignment in the flow direction, especially after 48 hours.

where R is the Poiseuille flow resistance. The $\Delta p - Q$ relationship has a very strong dependence on the tube radius, to the fourth power. For fixed Δp, $Q \sim a^4$. Because of this sensitivity to the radius, the fluid mechanics of vessels which can change their radius is particularly intriguing. Poiseuille's Law in the form of either Eq. (7.1.8) or Eq. (7.1.9) is often used in a global approach toward understanding blood flow in the vasculature and air flow in the lung. We will see later that the lung application of Poiseuille's Law is not a very good one for many of the larger airway generations. Nevertheless, medical textbooks and research investigations are replete with use of Poiseuille's Law and make use of its elegant simplicity to interpret normal physiology, pathophysiology, diagnostics and responses to therapeutics.

Fluid shear stress on the tube wall is called "wall shear stress," sometimes indicated as WSS or τ_w. The cells which form the inner surface of the blood vessels, known as the endothelial cells or the endothelium, are sensitive to the level of shear stress from the overlying blood flow. Fluid shear on those cells can modify their structure and function (Dewey et al., 1981). Figure 7.2 shows bovine aortic endothelial cell shape and orientation for 24 hour static vs flow conditions, and 48 hour static vs flow conditions. The WSS for the flow is 8 dyn/cm^2. Note the cell alignment in the flow direction, especially after 48 hours. This phenomenon, where physical forces modify cell behavior, is called "mechanotransduction."

Lung airway epithelial cells are exposed to the smaller fluid shears from the air flow due to the decrease in viscosity. However, two-phase flows of liquid plugs in airways, which occurs in many diseases, like congestive heart failure and asthma, can strongly influence their function and survival (Huh et al., 2007). We discuss this in Section 16.6.

In Poiseuille flow the fluid shear stress is represented by the extra stress tensor component, τ_{rz}, whose form is

$$\tau_{rz} = \mu \frac{du_z}{dr} = -4 \frac{\mu u_{z\,avg} r}{a^2} \tag{7.1.10}$$

Evaluated at the boundary, the wall shear is given by

$$\tau_w = \tau_{rz}\big|_{r=a} = -4 \frac{\mu u_{z\,avg}}{a} = -4 \frac{\mu Q}{\pi a^3} \tag{7.1.11}$$

In addition to the effects of fluid shear on the cells, the overall shear level in the flow becomes an important aspect of its constitutive behavior, as we will explore in Chapter 10. For high enough shear stress, blood behaves as a Newtonian fluid. Since fluid shear is zero on the centerline and maximal at the wall, reasonable estimates of an "average" shear level assuming Poiseuille flow could be an algebraic average, $(\tau_w - \tau(r=0))/2 = \tau_w/2 = -2\mu u_{z\,avg}/a$, or an area averaged shear, $\frac{1}{\pi a^2}\int_0^{2\pi}\int_0^a \tau_{rz}\, r\, dr d\theta = -\frac{8}{3}\frac{\mu u_{z\,avg}}{a}$

Box 7.1 Cardiovascular Anatomy and Physiology

The cardiovascular system consists of the heart (Figure 1.3), which pumps blood through the systemic and pulmonary circulations, as shown in Figure 7.3. With regard to gas exchange, the systemic arteries, starting with the aorta, carry oxygenated (red) blood the heart's left ventricle to the tissues. The actual mass transport occurs by diffusion across the systemic capillary walls, where the blood also picks up carbon dioxide. Then the systemic veins return the deoxygenated (blue) blood to the right side of the heart, via the inferior vena cava and superior vena cava. From the right atrium blood flows to the right ventricle and then is

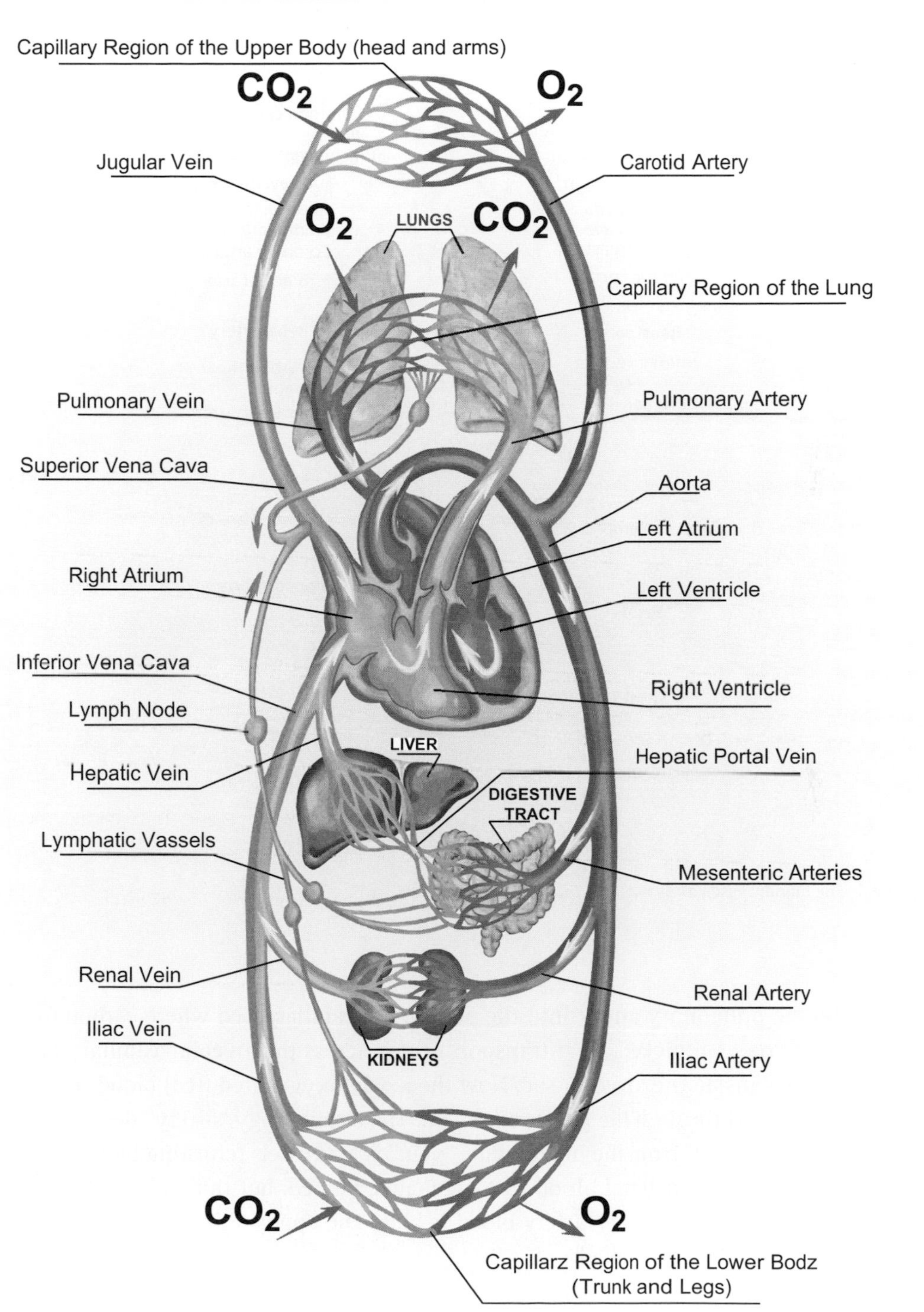

Figure 7.3 The cardiovascular system consisting of the heart, systemic circulation and pulmonary circulation with arteries, veins and capillary beds.

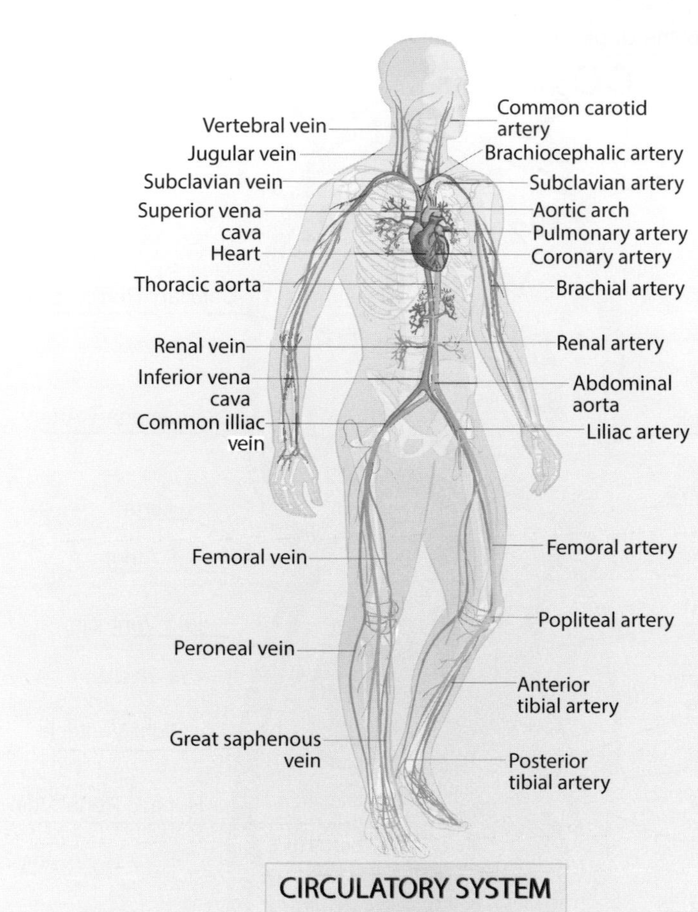

Figure 7.4 The human vascular system.

pumped via the pulmonary artery into the pulmonary capillary bed where carbon dioxide is off-loaded and oxygen picked. That transport occurs across the alveolar-capillary membrane with inhaled gas inside the alveolar sac. Now the newly oxygenated (red) blood returns to the left side of the heart through the pulmonary vein. The terminology "artery" designates vessels which carry blood away from the heart, while "vein" is for vessels returning blood to the heart. Often students consider arterial blood as oxygenated, i.e. red, but that is only true for the systemic circulation. Pulmonary artery blood is blue. The opposite is true for the pulmonary vein (red) vs systemic veins (blue).

The aorta splits into primary arteries entering the arms, legs, and head and neck which then establish their own branching patterns, see Figure 7.4. There are many arterial

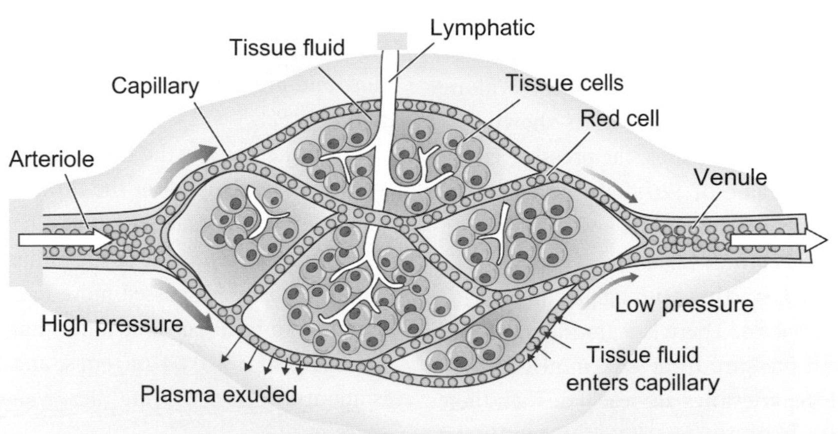

Figure 7.5 Capillary bed including arteriole, capillary and venule.

branches from the aorta which go to the head (brain), upper limbs (bone, muscle), lower limbs (bone muscle), thorax (heart muscle) and abdomen (liver, pancreas, spleen, kidneys, reproductive and gastrointestinal) as shown in Figure 7.3. The sequence of arterial branching is like a tree. There are more and more branches with smaller diameters, but the sum total at any generation of branches yields an ever increasing total cross-sectional area. Generally the arterial side of the circulation is categorized in descending diameter vessels starting with the aorta whose diameter is ~25–30 mm in an adult male, then to arteries (~4 mm), arterioles (~30 microns), and capillaries (~8 microns) see Figure 7.5. It is in the capillaries where the oxygen is delivered to the tissues and carbon dioxide is taken into the blood. Then the deoxygenated blood flows through a converging tree in the venous system with ever-decreasing cross-sectional area and increasing vessel diameters, starting with venules (~20 microns) and then veins (~5 mm) to the vena cava (~30 mm) which enters the right atrium of the heart that passes it along to the right ventricle, see Figure 7.5. In addition to oxygen delivery and carbon dioxide clearance, the porous capillaries allow flow through their walls as shown in Figure 7.5. An additional system called the lymphatics assists in clearing excess fluids. More details of this capillary filtration flow are examined in Section 11.6.

7.2 Applying Poiseuille's Law: Vascular Hemodynamics

The blood pressure in the systemic vasculature is illustrated in Figure 7.6, with the different levels along the flow route from aorta to vena cava is shown. Because of the heart contractions, the pressure is shown with its high value of 120 mmHg in the aorta during contraction (systolic) and its low value of 80 mmHg during relaxation (diastolic). The mean value is interpolated, starting with (approximately) ~100 mmHg in the aorta. There is a gradual decrease in the mean blood pressure from ~100 mmHg to ~85 mmHg until the arterioles are reached. Then there is a dramatic drop from ~85 mmHg to ~35 mmHg across the arterioles. This establishes the blood pressure at the upstream end of the capillaries at ~35 mmHg. The downstream end of the capillaries has a blood pressure of ~15 mmHg.

The pressure drop down the aorta is complicated since the flow is strongly unsteady due to the heart beat and the walls are elastic so deform carrying a pulse wave. Indeed the interaction of the wall elasticity and flow can actually increase the systolic pressure downstream, as Figure 7.6 shows with higher systolic pressures just downstream of the aorta. However, we can consider the steady, time-averaged flow and pressure and apply Poiseuille's Law to that concept.

Since the aorta is the first tube in the branching system, essentially all of the flow goes through this one tube. The cardiac output can be taken as $Q = 80\,\text{cm}^3/\text{s}$, while the aortic radius $a = 1.25\,\text{cm}$, the blood viscosity (discussed more in this chapter) is assumed to be about four times that of water in this situation, so $\mu = 0.04\,\text{gm/cm}\cdot\text{s}$, and the aortic length is about $L = 25\,\text{cm}$. Using these values in Eq. (7.1.9) gives us

$$\begin{aligned}(p_u - p_d)_{aorta} &= \frac{(80)(0.04)(8)(25)}{(3.14)(1.25)^4} \\ &= 162\ \text{dyn/cm}^2 \sim 0.12\ \text{mmHg} \qquad (7.2.1)\end{aligned}$$

since $(1\ \text{mmHg} = 1333\ \text{dyn/cm}^2)$. So the pressure drop in the aorta of 0.12 mmHg is negligible compared to the total drop from aorta to vena cava of ~100 mmHg.

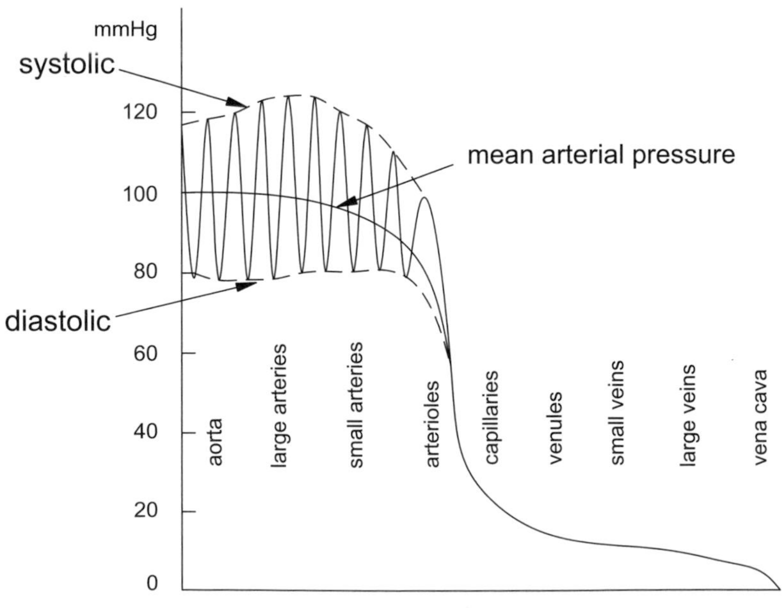

Figure 7.6 Pressure drop along the systemic vascular tree.

By comparison, the arterioles pose a significant resistance to flow with a major contribution to the pressure drop. They are specially designed with smooth muscle in their walls, so can contract in response to neural and biochemical stimuli, and consequently create smaller diameters, or larger during relaxation. Since diameter is such a strong contributor to the pressure flow relationship, it is critical for this control to be working properly. Using some representative values, suppose there are approximately 5×10^6 arterioles. These are all parallel tubes. So assuming the cardiac output is equally distributed at the arteriolar generation, the flow in each arteriole tube is $Q = 80/5 \times 10^6 = 1.6 \times 10^{-5}\,\mathrm{cm}^3/\mathrm{s}$. The arteriole radius is approximately $a = 0.0015$ cm and representative length is $L = 0.20$ cm. From these parameter values, the pressure drop across the arterioles is

$$(p_u - p_d)_{\text{arterioles}} = \frac{(1.6 \times 10^{-5})(0.04)(8)(0.2)}{(3.14)(0.0015)^4} = 64,385\,\mathrm{dyn/cm^2} \sim 48\,\mathrm{mmHg} \tag{7.2.2}$$

The pressure drop across the arterioles of 48 mmHg is very significant compared to the overall drop of 100 mmHg ... essentially half of the total. It is represented in the Figure 7.6 as a steep descent. This is powerful control of flow and pressure, and it is the arterioles that are called upon to shift blood flow from one major capillary bed to another. For example, eating and digesting food demands more blood flow to the gut, extreme heat more blood flow to the skin, and running more blood flow to the muscles. It is control of the arteriolar diameters that provide such shifts in response to physiological requirements.

Reducing the pressure to 35 mmHg at the upstream end of the capillaries is important. Capillaries are porous and fluid can exit across the capillary wall into the tissues. We address this issue in Sections 11.6 and 15.2. Collections of this fluid in the tissues are called edema, and need to be cleared primarily by the lymphatics. So keeping a constraint on the intravascular pressure of the capillaries helps to maintain normal fluid balance in the tissues. As an extreme example, spraining your ankle tears at capillaries and they leak so much you can see the swelling quickly as edema fluid accumulates. In the pulmonary circulation, the blood pressures are already quite low; the pulmonary artery pressures are 25/10 mmHg systolic/diastolic and both diminish progressing downstream into the pulmonary vasculature. Nevertheless, it is essential to keep pulmonary capillary blood pressures sufficiently low, since vascular fluid leaking into the tissues can then fill the alveolar air spaces, causing pulmonary edema, which is life threatening because of severely compromised gas exchange when there is liquid where air should be.

Equation (7.1.9) has the appearance of Ohm's Law used in electrical circuits where $\Delta V = IR_e$, the voltage drop, is analogous to the pressure drop; the current, I, is analogous to the flow, Q; and the electrical resistance, R_e, is analogous to the fluid flow resistance, R. As a consequence, two flow resistances, R_1 and R_2, which are arranged in series are calculated as $R_{tot} = R_1 + R_2$, while resistances in parallel are added as $1/R_{tot} = 1/R_1 + 1/R_2$.

Box 7.2 | Jean Léonard Marie Poiseuille

Poiseuille trained for a year, 1815–1816, as an undergraduate student at École Polytechnique in Paris with Augustin-Louis Cauchy and other noted faculty. Following many experiments on small diameter glass tubes, Poiseuille actually determined the flow relationship as

Figure 7.7 Jean Léonard Marie Poiseuille, French physician, physicist and physiologist.

$Q = Kd^4 \Delta p/L$, where Δp was the pressure drop, d the tube diameter, and L its length. Here K was a constant fitted to the data (Poiseuille, 1840; 1841). He showed that K increases with temperature. The concept of viscosity, μ, was not yet well appreciated in those days, but now we know that $K = \pi/128\mu$. That gives us Eq. (7.1.9) and μ decreases with increasing temperature, so K increases as Poiseuille observed.

One of his goals was to measure and understand blood pressure, which he did in dogs and horses (Poiseuille, 1828). As part of the experimental apparatus, Poiseuille invented the mercury manometer, which kept the pressure measurements to workable physical amplitudes that were easier to record. To this day, blood pressure is reported in millimeters of mercury by using the shorthand notation, mmHg. A typical reading is 120/80 mmHg, systolic/diastolic. Systolic is during the heart contraction (systole) and diastolic during the relaxed heart phase (diastole). Since mercury is 13.6 times the density of water, the normal systolic pressure if measured by water would be $120 \times 13.6 = 1632\ \text{mmH}_2\text{O}$, or approximately 1.6 meters. It is hard to put a measuring device of that size on the physician's desk, so we have Poiseuille to thank for that important invention. In addition, the unit of viscosity in the cgs system is the poise, $1\ \text{poise} = 1\ \text{gm/cm-sec}$, also named for Poiseuille. These highlights and additional details of the life of Poiseuille are given by Sutera and Skalak (1993).

Box 7.3 Respiratory Airway Anatomy and Physiology

The lung has both a vascular network as well as an airway network. As pictured in Figure 7.8 (a), there are two entry pathways into the upper respiratory tract, one through the mouth and the other through the nose. This redundancy has the advantage of allowing breathing when one or the other is closed off, from disease or trauma for example. Both the mouth and the nasal passages warm and humidify incoming air. The two pathways merge in the back of the throat in the pharynx. The larynx is the site of the vocal cords and is situated at the proximal end of the trachea. The trachea is the first of a branching network of tubes called the trachea-bronchial tree extending into the lower respiratory tract, Figure 7.8(b). Its diameter in an adult male is typically 2 cm while its length is 10 to 12 cm. The trachea bifurcates, one primary bronchus goes to the right lung, and one to the left. The tubes continue to go through a series of branching bifurcations with ever smaller diameter bronchi, so that in an idealized system

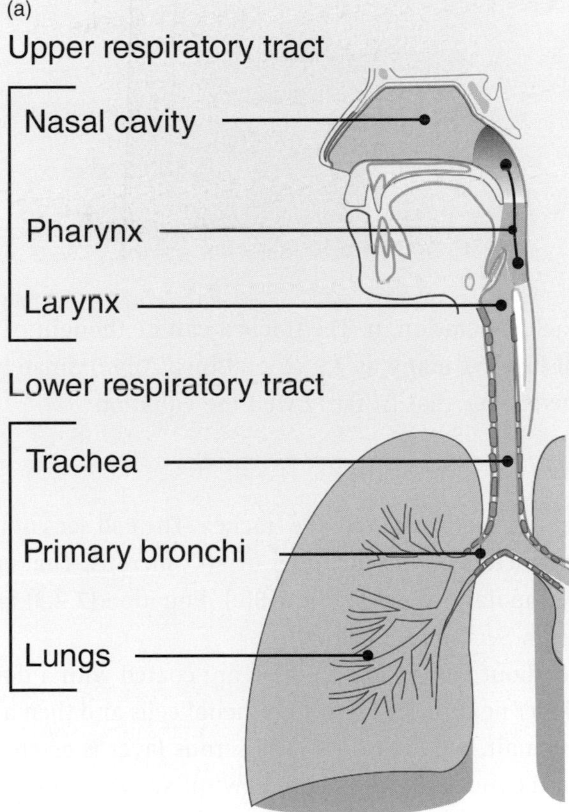

Figure 7.8 Human respiratory system: (a) upper and lower respiratory tract; (b) airways, alveoli and pulmonary circulation.

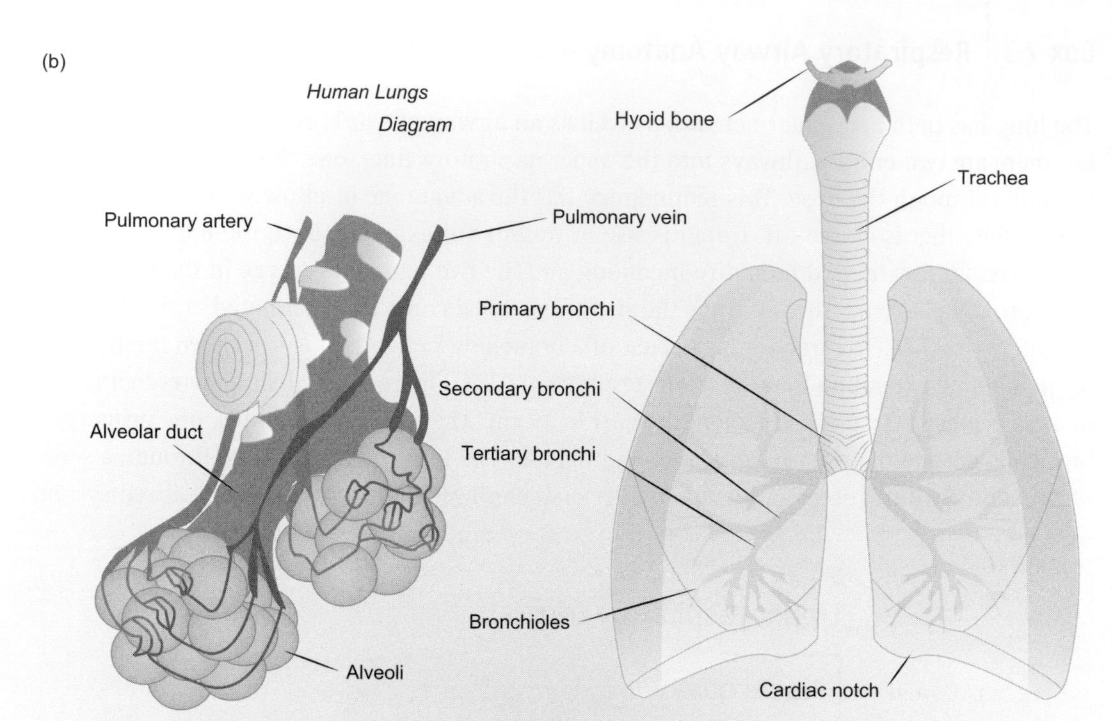

Figure 7.8 (*continued*)

there are 2^n tubes at each generation, n. The trachea can be thought of as the $n = 0$ generation. A lung will have as many as 23 generations. Approximately the first 15 airway generations have diameters, d_n, that fit fairly well the equation

$$d_n = d_0 2^{-n/3} \tag{7.2.3}$$

where d_0 is the diameter of generation zero, the trachea. The end sacs of the respiratory system are the alveoli whose diameters are on the order of 200 microns. That is where gas exchange occurs with pulmonary capillary blood Figure 7.8(b). Equation (7.2.3) is derived from energy minimization principles in Section 7.5 below.

The airways down to about generation 15 or so are coated with a double liquid layer consisting of a serous layer next to the airway epithelial cells and then a mucus layer which is in contact with the lumen air, see Figure 7.9. The serous layer is essentially Newtonian and mostly water; however, the mucus layer is non-Newtonian whose properties are discussed in Chapters 10 and 14. The airways have ciliated cells which propel the bilayer toward the mouth

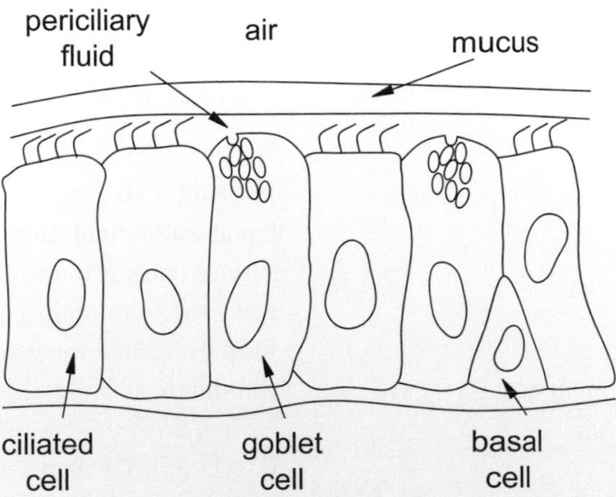

Figure 7.9 Airway epithelium showing mucus-producing goblet cells, ciliated cells and basal cells. Two liquid layers are shown, the serous or periciliary layer and the mucus layer.

where it is swallowed and eliminated through the GI tract in a process called "pulmonary toilet" (see Figure 15.18(b) for a scanning electron microscopic image of rabbit sinonasal cilia). This is an essential defense mechanism of the lung to sweep out inhaled particles, which may be infectious or otherwise harmful, see Section 17.8 for a mathematical treatment of ciliary motion. However, the lung's liquid lining is also the first barrier encountered by inhaled medications as well.

7.3 Gravity and Boundary Driven Flow: Respiratory Mucociliary Clearance

Consider thin film flow as shown in Figure 7.10. This flow could represent a simplified model of mucociliary transport in the respiratory system or an industrial coating process. For simplicity, a single Newtonian liquid layer of constant thickness, h, is assumed. The wall has a non-zero velocity, u_w, in this case directed against gravity. This wall velocity is a simple representation of the airway wall cilia which

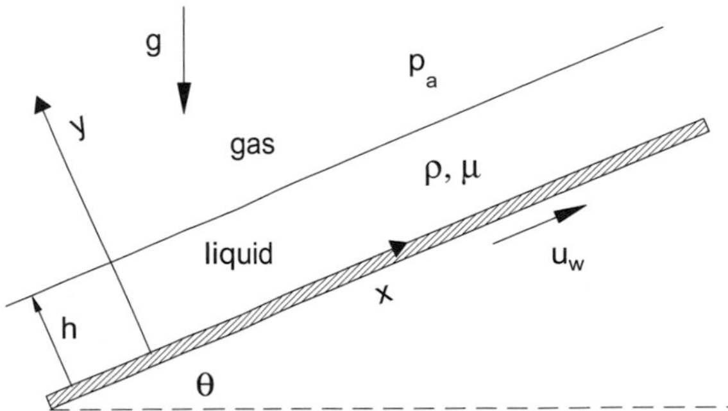

Figure 7.10 Fluid film of thickness h flows on a moving plate at angle θ.

move the liquid layer toward the mouth. See Section 17.8 for the origin of this simplification. For these steady flow assumptions, with only the x-direction velocity, u, we also ignore surface tension effects at the gas–liquid interface. The remaining terms in the Navier–Stokes equations, including the body force from gravity, are

$$\begin{aligned} 0 &= -\frac{\partial p}{\partial x} + \mu\frac{d^2u}{dy^2} - \rho g \sin\theta \\ 0 &= -\frac{\partial p}{\partial y} - \rho g \cos\theta \end{aligned} \tag{7.3.1}$$

We can integrate the y-component of Eq. (7.3.1) partially with respect to y to find

$$p = -\rho g y \cos\theta + f(x) \tag{7.3.2}$$

From Figure 7.10 and our assumptions, we see that the boundary condition on the pressure in the liquid is $p = p_a$ for $y = h$, which is shown by the relationship

$$p(y = h) = p_a = -\rho g h \cos\theta + f(x) \tag{7.3.3}$$

Since we can assume h is a constant, independent of x, then f is also independent of x and is the constant

$$f = p_a + \rho g h \cos\theta \tag{7.3.4}$$

Inserting this solution for f into Eq. (7.3.2) gives the linear pressure dependence on y

$$p = \rho g(h - y)\cos\theta + p_a \tag{7.3.5}$$

The x-component of Eq. (7.3.1) integrates twice directly, since $\partial p/\partial x = 0$, and is found to have the form

$$u = \frac{\rho g \sin\theta}{\mu}\frac{y^2}{2} + c_1 y + c_2 \tag{7.3.6}$$

The no-slip condition at the wall is

$$u(y = 0) = u_w \tag{7.3.7}$$

which gives us $c_2 = u_w$. Later, in Section 15.5, we will use a slip condition related to the porosity of the ciliary spacing, see Figure 15.18(b). Then because gas viscosity is so much smaller that liquid viscosity, and the gas is assumed to have negligible motion, we can assume no shear stress at the liquid–gas interface, $y = h$. This condition is

$$\mu\frac{du}{dy}_{y=h} = 0 \tag{7.3.8}$$

which solves for c_2, and the final form is

$$u = \frac{\rho g \sin\theta}{\mu}\left(\frac{y^2}{2} - hy\right) + u_w \tag{7.3.9}$$

In Chapter 16 a more complete treatment of an air-liquid (fluid–fluid) interface is presented, including surface tension and surface-tension gradients. We can make the solution in Eq. (7.3.9) into a dimensionless form by scaling the velocity as $U = \mu u/\rho g h^2$ and the coordinate as $Y = y/h$,

$$U = \left(\frac{Y^2}{2} - Y\right)\sin\theta + U_w \tag{7.3.10}$$

The dimensionless wall velocity, $U_w = \mu u_w/\rho g h^2$, is the ratio of the viscous shear stress $\mu u_w/h$ from the wall motion to the gravity effects through pressure, $\rho g h$. So it represents the competition between the wall drag and gravity forces. It can be cast in terms of two dimensionless groups we have already discussed, $U_w = Fr^2/Re$, for the definitions of the Reynolds number $Re = \rho u_w h/\mu$ and the Froude number $Fr = u_w/\sqrt{gh}$.

From Eq. (7.3.10) the term including $\sin\theta$ is the gravity effect, and ranges from 0 to $-(\sin\theta)/2$ as Y ranges $0 \leq Y \leq 1$. So $U_w >> 1/2$ will be wall-dominated flow.

For mucociliary clearance, parameter values for healthy individuals are fluid density $\rho = 1 g/cm^3$, fluid viscosity $\mu = 10$ poise which is 1000 times the viscosity of water, layer thickness $h = 10\,\mu m$ (Knowles and Boucher, 2002) and ciliary wall velocity $u_w = 40\,\mu m/s$ (Matsui et al., 1998; Bottier et al., 2017a; 2017b). With $g = 980\ cm/s^2$ we calculate $U_w = 40$, indicating that the system is dominated by the wall in healthy subjects. However, consider a disease process where the ciliary velocity is significantly decreased, as can occur from smoking, e.g. $u_w = 5\,\mu m/s$. The layer thickness can be larger due to accumulation of inflammation fluids, say $h = 40\,\mu m$, and the viscosity can be smaller due to this dilution, $\mu = 5$ poise. Then $U_w = 0.16$ and the system is dominated by gravity. The lung airways are a tree-like structure, see Figure 7.8, so all orientations are present. Consequently, some will not clear mucus but, instead, will accumulate it, leading to bronchitis and pneumonia. More modeling of mucociliary propulsion appears in Section 17.6.

7.4 Core-Annular Two-Phase Flow in a Circular Cylinder: Vascular and Respiratory Applications

Two-phase flow is found in biofluid mechanics of both the respiratory and vascular systems. Lung airways are lined with a liquid layer, which can be represented as fluid 2 in Figure 7.11 and exists in the region $b \le r \le a$, where a is the tube radius. Air occupies the core region $0 \le r \le b$ and is fluid 1. In vascular flows, the red blood cells congregate in a region $0 \le r \le b$, which creates a local viscosity that is several times larger than that of the plasma. The thin region $b \le r \le a$ is a cell-free fluid layer which has the viscosity of the plasma. For convenience, let's define the dimensionless geometric parameter, λ, as $\lambda = b/a$.

We will restrict ourselves to steady, axisymmetric, fully developed, unidirectional flow in the z-direction. In addition we will leave out any effects of surface tension at the fluid–fluid interface for now. In general there are three velocity components (u_r, u_θ, u_z) corresponding to the (r, θ, z)-directions. For unidirectional axisymmetric flow $(u_r, u_\theta, u_z) = (0, 0, u_z)$ and $\partial/\partial\theta = 0$.

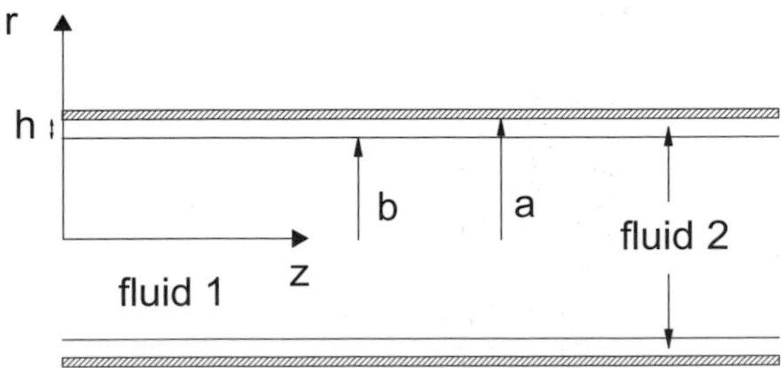

Figure 7.11 Core-annular, two-phase flow in a tube.

The governing equations for the two phases come from the z-component of the Navier–Stokes equations and, with fluids 1 and 2 denoted by subscripts, are

$$\begin{aligned} &\mu_1 \frac{1}{r}\frac{d}{dr}\left(r\frac{du_{z1}}{dr}\right) = \frac{dp_1}{dz} \qquad 0 \le r \le b \qquad \text{fluid 1} \\ &\mu_2 \frac{1}{r}\frac{d}{dr}\left(r\frac{du_{z2}}{dr}\right) = \frac{dp_2}{dz} \qquad b \le r \le a \qquad \text{fluid 2} \end{aligned} \tag{7.4.1}$$

and the pressure gradients will be constants.

The boundary conditions are no-slip at the wall, finite velocity at the centerline and matching of the velocity at the fluid–fluid interface. These are called the kinematic boundary conditions. Then also we must match the stress vector components at the fluid–fluid interface, which are called the stress boundary conditions. Stated mathematically these are

$$\begin{aligned} &u_{z2}(r=a)=0 && \text{no-slip at the wall} \\ &u_{z1}(r=0)\ \text{finite} && \text{no singularity at the centerline} \\ &u_{z1}=u_{z2}\ \text{at}\ r=b && \text{continuous velocities at the interface} \\ &(t_1-t_2)\cdot e_r=0\ \text{at}\ r=b && \text{continuous normal stress at the interface} \\ &(t_1-t_2)\cdot e_2=0\ \text{at}\ r=b && \text{continuous tangential stress at the interface} \end{aligned} \tag{7.4.2}$$

The fluid stresses for this flow are given by the remaining components of the stress tensor in cylindrical coordinates

$$\begin{bmatrix} T_{rr} & T_{r\theta} & T_{rz} \\ T_{\theta r} & T_{\theta\theta} & T_{\theta z} \\ T_{zr} & T_{z\theta} & T_{zz} \end{bmatrix} = \begin{bmatrix} -p & 0 & 0 \\ 0 & -p & 0 \\ 0 & 0 & -p \end{bmatrix} + \begin{bmatrix} \tau_{rr} & \tau_{r\theta} & \tau_{rz} \\ \tau_{\theta r} & \tau_{\theta\theta} & \tau_{\theta z} \\ \tau_{zr} & \tau_{z\theta} & \tau_{zz} \end{bmatrix} \tag{7.4.3}$$

For Poiseuille flow, the extra stress tensor, $\underline{\underline{\tau}}$, simplifies to

$$\begin{bmatrix} \tau_{rr} & \tau_{r\theta} & \tau_{rz} \\ \tau_{\theta r} & \tau_{\theta\theta} & \tau_{\theta z} \\ \tau_{zr} & \tau_{z\theta} & \tau_{zz} \end{bmatrix} = \begin{bmatrix} 0 & 0 & \mu\dfrac{\partial u_z}{\partial r} \\ 0 & 0 & 0 \\ \mu\dfrac{\partial u_z}{\partial r} & 0 & 0 \end{bmatrix} \tag{7.4.4}$$

For the stresses on the interface, the unit normal pointing into the fluid is $\underline{n} = (n_r, n_\theta, n_z) = (-1, 0, 0)$. The corresponding interfacial stress vector is for fluid 1 is $\underline{t}_1 = \underline{\underline{T}}_1 \cdot \underline{n}$, which translates to

$$\begin{bmatrix} t_{1r} \\ t_{1\theta} \\ t_{1z} \end{bmatrix} = \begin{bmatrix} -p_1 & 0 & \mu_1\dfrac{\partial u_{z1}}{\partial r} \\ 0 & -p_1 & 0 \\ \mu_1\dfrac{\partial u_{z1}}{\partial r} & 0 & -p_1 \end{bmatrix}_{r=b} \begin{bmatrix} -1 \\ 0 \\ 0 \end{bmatrix} = \begin{bmatrix} p_1 \\ 0 \\ -\mu_1\dfrac{\partial u_{z1}}{\partial r} \end{bmatrix}_{r=b} \tag{7.4.5}$$

So the normal and tangential stresses on the interface in fluid 1 are, respectively,

$$t_{1r} = p_1, \quad t_{1z} = -\mu_1\left(\frac{\partial u_{z1}}{\partial r}\right)_{r=b} \tag{7.4.6}$$

The same procedure applies for fluid 2 at the interface, resulting in

$$t_{2r} = p_2, \quad t_{2z} = -\mu_2\left(\frac{\partial u_{z2}}{\partial r}\right)_{r=b} \tag{7.4.7}$$

The two stress component boundary conditions of Eq. (7.4.2) are now simply

$$\begin{aligned} &p_1(r=b) = p_2(r=b) \\ &-\mu_1\left(\frac{du_{z1}}{dr}\right)_{r=b} = -\mu_2\left(\frac{du_{z2}}{dr}\right)_{r=b} \end{aligned} \tag{7.4.8}$$

Since the pressures in either fluid are independent of r, the first of Eqs. (7.4.8) states that the pressure at any value of z is the same in both fluids, so the constant pressure gradients in z are equal,

$$\frac{dp_1}{dz} = \frac{dp_2}{dz} = \frac{dp}{dz} = \text{constant} \tag{7.4.9}$$

Equations (7.4.1) are now

$$\begin{aligned} &\mu_1\frac{1}{r}\frac{d}{dr}\left(r\frac{du_{z1}}{dr}\right) = \frac{dp}{dz} \quad 0 \le r \le b \\ &\mu_2\frac{1}{r}\frac{d}{dr}\left(r\frac{du_{z2}}{dr}\right) = \frac{dp}{dz} \quad b \le r \le a \end{aligned} \tag{7.4.10}$$

which can be integrated directly as we did for Poiseuille flow

$$\begin{aligned} u_{z1} &= \frac{dp}{dz}\frac{r^2}{4\mu_1} + c_1\ln(r) + c_2 \quad 0 \le r \le b \\ u_{z2} &= \frac{dp}{dz}\frac{r^2}{4\mu_2} + c_3\ln(r) + c_4 \quad b \le r \le a \end{aligned} \tag{7.4.11}$$

The four boundary conditions remaining are used to determine the four constants of integration, c_1, c_2, c_3, c_4,

$$\begin{aligned} &u_{z2}(r=a) = 0 \\ &u_{z1}(r=0) = \text{finite} \\ &u_{z1}(r=b) = u_{z2}(r=b) \\ &-\mu_1\left(\frac{du_{z1}}{dr}\right)_{r=b} = -\mu_2\left(\frac{du_{z2}}{dr}\right)_{r=b} \end{aligned} \tag{7.4.12}$$

Since fluid 1 has $r = 0$ in its domain, we much choose $c_1 = 0$ as in Poiseuille flow, which satisfies the second of Eqs. (7.4.12). Then the remaining three conditions determine the rest of the constants. Using the scales

$$U_{Z1} = \frac{u_{z1}}{(-dp/dz)a^2/\mu_2}, \quad U_{Z2} = \frac{u_{z2}}{(-dp/dz)a^2/\mu_2}, \quad R = \frac{r}{a} \tag{7.4.13}$$

the dimensionless velocities are

$$\begin{aligned} U_{Z1} &= \frac{1}{4\delta}\left[\delta(1-\lambda^2) + (\lambda^2 - R^2)\right] \quad 0 \le R \le \lambda \\ U_{Z2} &= \frac{1}{4}(1 - R^2) \quad \lambda \le R \le 1 \end{aligned} \tag{7.4.14}$$

where the dimensionless parameters are the viscosity ratio $\delta = \mu_1/\mu_2$ and the length ratio $\lambda = b/a$. Note that if the viscosities of the two fluids are equal, $\delta = 1$, the two solutions are identical and together give Poiseuille flow of a single fluid.

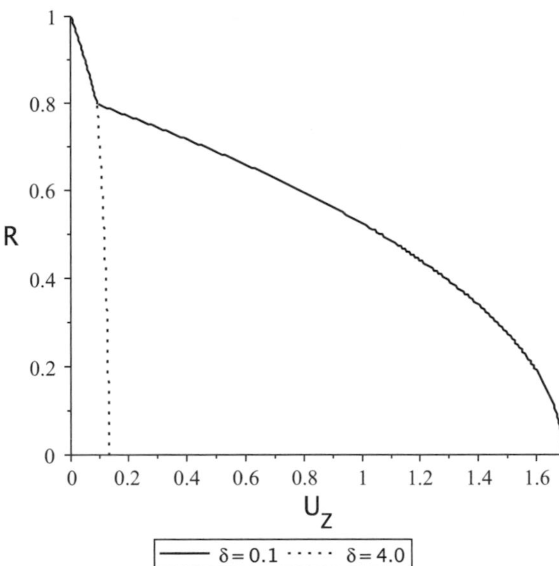

Figure 7.12 Core-annular, two-phase flow velocity profiles for $\lambda = b/a = 0.8$ and $\delta = \mu_1/\mu_2 = 0.1, 4.0$.

Both profiles are parabolas. A plot of the two velocity profiles is shown in Figure 7.12 where $\lambda = 0.8$ and $\delta = 4.0, 0.1$ to illustrate the solutions. In the actual respiratory application, δ is much smaller than 0.1. The density of air at 20 °C is $\rho_{air} = 1.205 \times 10^{-3}$ g/cm^{3}and the kinematic viscosity is $\nu_{air} = 0.15$ cm^2/s, so the bulk (dynamic) viscosity is $\mu_{air} = \nu_{air} \times \rho_{air} = 1.82 \times 10^{-4}$ g/cm · s. The viscosity of the liquid layer is at least that of water, $\mu_{H_2O} = 0.01$ g/cm · s, and because of the mucus content it can be much higher, for example $\mu_{mucus} \sim 100 - 1000 \times \mu_{H_2O} \sim 1 - 10$ g/cm · s (Denton, 1963). So $10^{-4} \le \delta \le 10^{-2}$ is a reasonable range for a lung airway. Also in reality λ is much closer to 1 in a healthy airway.

From the balance of forces on the fluid 2 layer, we know that it is being driven both by the pressure gradient and also from the shearing at the interface. Fluid 1 is "dragging" it along. In order to determine the relative effects of these two contributions, let's calculate the fluid 2 annular layer velocity field in the absence of the the fluid 1 core. The integrated z-component of Navier–Stokes is, again,

$$\hat{u}_{z2} = \frac{dp}{dz}\frac{r^2}{4\mu_2} + c_5 \ln(r) + c_6 \quad b \le r \le a \tag{7.4.15}$$

but now with different integration constants which are determined from no-slip at the wall and no-shear at the interface since fluid 1 has been removed,

$$\hat{u}_{z2}(r = a) = 0 \qquad \left(\frac{d\hat{u}_{z\,2}}{dr}\right)_{r=b} = 0 \tag{7.4.16}$$

Using the same scales as in Eq. (7.4.13), the dimensionless solution is

$$\hat{U}_{Z2} = \frac{1}{4}\left(1 - R^2\right) + \frac{1}{2}\lambda^2 \ln(R) \qquad \lambda \le R \le 1 \tag{7.4.17}$$

For another comparison, compute the dimensionless volume flow rate for both cases as in Eq. (7.4.18).

$$Q_2 = \int_\lambda^1 U_{Z2}\, R\, dR = \frac{1}{16}\left(1 - \lambda^2\right)^2$$
$$F_2 = \int_\lambda^1 \hat{U}Z2R\, dR = \frac{3}{16}\lambda^4 - \frac{1}{4}\lambda^2\left(1 + \lambda^2 \ln(\lambda)\right) + \frac{1}{16} \tag{7.4.18}$$

Comparing $\hat{U}_{Z2}$ of Eq. (7.4.17) to U_{Z2} in Eq. (7.4.14), there is still the parabolic term, but now an extra term involving the natural log. Since $\lambda \le R \le 1$, the term $\ln(R)$ is negative and the velocities are slower when fluid 1 is not present, given the same pressure gradient.

The comparison of the two volume flow rates is shown in Figure 7.13. For either case, as the fluid 2 film becomes thinner, $\lambda \to 1$, the velocities decrease. For all values of λ the volume flow rate is greater when fluid 1 is present, Q_2, compared to the absence of fluid 1, F_2. This is due to the shearing stress at the fluid–fluid interface. However, the difference between them diminishes as $\lambda \to 1$.

To appreciate more fully the relative effect, we can plot $(Q_2 - F_2)/F_2$ as shown in Figure 7.14. There we see that as the film becomes very thin, $\lambda \to 1$, the relative importance of the shear from fluid 1 increases substantially for driving the flow rate in fluid 2.

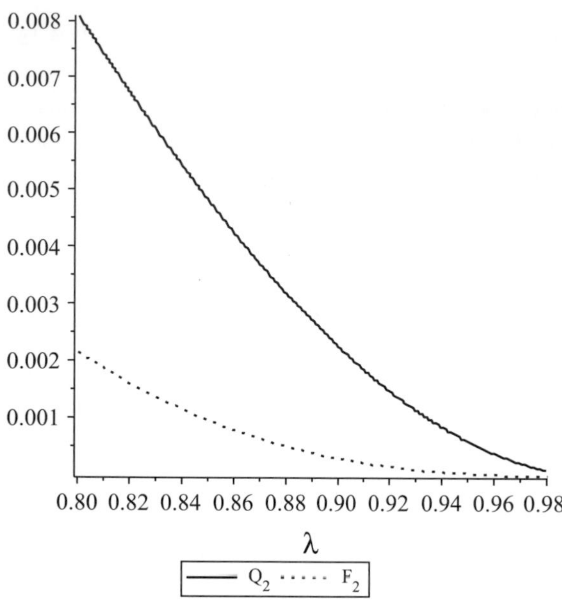

Figure 7.13 Volume flow rates for fluid 2 with fluid 1 present, Q_2, and without fluid 1 present, F_2.

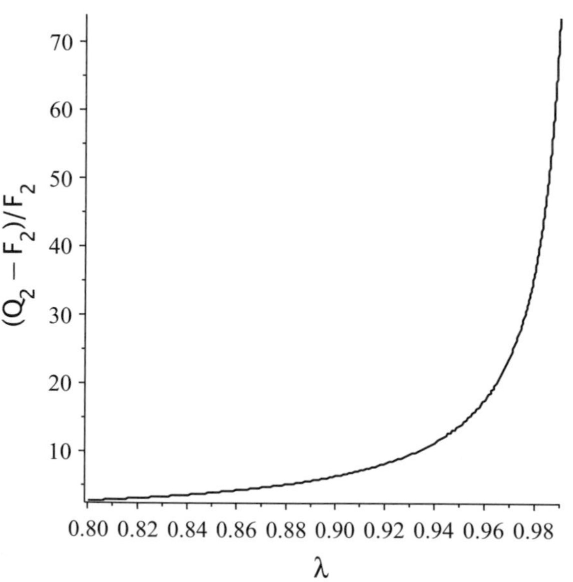

Figure 7.14 Relative fluid 2 volume flow rate with fluid 1 present, Q_2, compared to without fluid 1 present, F_2.

7.5 Flow and Anatomical Evolution: Murray's Law

Fluid mechanics can influence many biological activities. One of them is growth and form of the vessels that carry them. How branching tree structures, such as the pulmonary airways or a vascular network of blood vessels, are organized can be thought of as depending on the flow they contain. Is there a structure feature that optimizes the function while minimizing the energy, or rate of energy, cost? Does evolution respond to this need? This mechanotransduction question was investigated initially by the Swiss physician and physiologist Walter R. Hess in 1917 (Hess, 1917). However, the more general idea and clarification of the underlying principles came from the American biologist Cecil D. Murray in 1926 (Murray, 1926a; 1926b). The principles apply to plant life as well (McCulloh et al., 2003).

Consider the branching structure in Figure 7.15. The parent tube with radius r_0 is the zero-order generation of the tree. It conveys a flow which is fed into two daughter tubes, which are at generation 1. The daughter tubes have radii r_{11} and r_{12}. The question arises, what configuration of r_0, r_{11}, r_{12} results in the lowest power requirement?

The power requirement for flow in a tube, i.e. rate at which energy is needed, can be deduced by assuming Poiseuille flow. From Poiseuille's law, the flow rate is given by

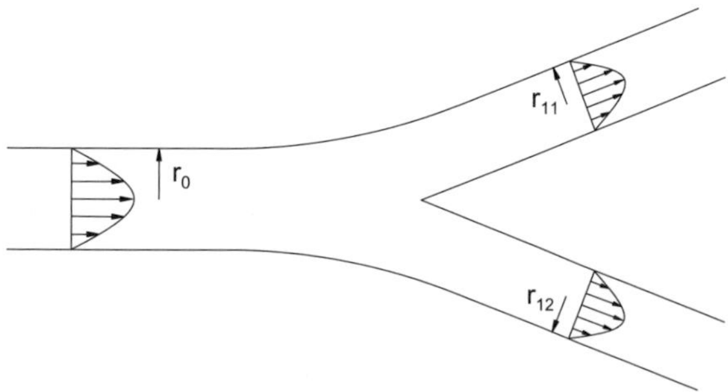

Figure 7.15 Bifurcation of a biological branching system: airways, blood vessels, which convey a flow.

$$Q = \frac{\pi r^4}{8\mu}\frac{\Delta p}{L} \tag{7.5.1}$$

The definition of the power in a flow, P_f, is

$$P_f = Q\Delta p = Q^2 r^{-4}\left(\frac{8\mu L}{\pi}\right) \tag{7.5.2}$$

The metabolic power, P_m, requirement to maintain the biomass of the tube and its fluid can be considered to scale with the tube volume, for example,

$$P_m = m\pi r^2 L \tag{7.5.3}$$

where the coefficient, m, depends on the metabolic rate. So the total (flow + metabolic) power requirement, P_t, is

$$P_t = Q^2 r^{-4}\left(\frac{8\mu L}{\pi}\right) + m\pi r^2 L \tag{7.5.4}$$

As r approaches zero or infinity, P_t, becomes unbounded for fixed flow rate, fluid viscosity, tube length and metabolic rate. A sample plot of Eq. (7.5.4), choosing three sets of parameter values for convenience, is shown in the Figure 7.16. Each curve has a local minimum.

The optimal operating point for the system is at the minimum value of the power. To locate this point, we set to zero the derivative of the power with respect to r

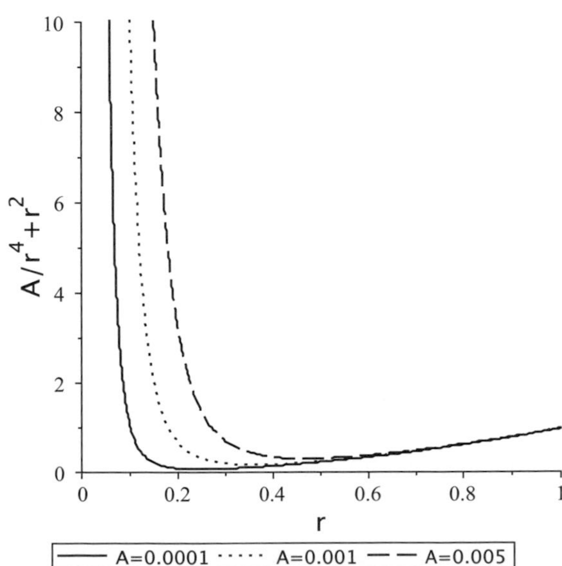

Figure 7.16 Sample plots of a power function.

$$\frac{dP_t}{dr} = -4Q^2 r^{-5}\left(\frac{8\mu L}{\pi}\right) + 2m\pi r L = 0 \tag{7.5.5}$$

which yields the relationship between flow and tube radius that is optimal

$$Q = kr^3 \qquad \text{where } k = \frac{\pi}{4}\left(\frac{m}{\mu}\right)^{1/2} \tag{7.5.6}$$

We note that the constant, k, depends only on the metabolic rate and the fluid viscosity, so is the same for all tubes in the tree. It does not depend on tube length.

For the three tubes in Figure 7.15, each has its own flow-radius criterion

$$Q_0 = kr_0^3, \quad Q_{11} = kr_{11}^3, \quad Q_{12} = kr_{12}^3 \tag{7.5.7}$$

to minimize power in each tube. However, from conservation of mass we know that

$$Q_0 = Q_{11} + Q_{12} \quad \rightarrow \quad kr_0^3 = kr_{11}^3 + kr_{12}^3 \tag{7.5.8}$$

From Eq. (7.5.8) we see that the relation amongst the tube radii is

$$r_0^3 = r_{11}^3 + r_{12}^3 \tag{7.5.9}$$

For a symmetric structure, $r_{11} = r_{12} = r_1$ and we have from Eq. (7.5.9)

$$r_0^3 = 2r_1^3 \tag{7.5.10}$$

Solving for r_1 gives us

$$r_1 = 2^{-\frac{1}{3}} r_0 \tag{7.5.11}$$

For the second generation we would have

$$r_2 = 2^{-\frac{1}{3}} r_1 = 2^{-\frac{1}{3}}\left(2^{-\frac{1}{3}} r_0\right) = 2^{-\frac{2}{3}} r_0 \tag{7.5.12}$$

Then in general for any generation number, n, we have

$$r_n = 2^{-\frac{n}{3}} r_0 \tag{7.5.13}$$

This equation works well for lung airways as was described in Eq. (7.2.3). (Weibel, 1963) and blood vessels (Murray, 1926a, 1926b) and even the xylem system in trees (McCulloh, 2003).

7.6 Murray's Law with Variable Length and Branch Angle

Suppose we have the same system as in Figure 7.15, but fix the total length of the bifurcation tubes and its width, see Figure 7.17. This would be an anatomical constraint, for example, like the size of the chest or organ. What branch angle and relative tube lengths would minimize the power consumption? Let the bifurcation be symmetric with the angle 2θ between the daughter tubes. The parent tube starts at $z = 0$ and ends at $z = \zeta$, which will be a variable. The daughter tubes end at $z = L$ and their distance from the z-axis is d. The length of a daughter is $L_1 = \sqrt{d^2 + (L-\zeta)^2}$. The inlet pressure is p_0, the outlet pressure is p_2 and the pressure at $z = \zeta$ is p_ζ. Neglecting the metabolic power for this example, the overall power is

$$P_t = Q_0\left(p_0 - p_\zeta\right) + 2Q_1\left(p_\zeta - p_2\right) \tag{7.6.1}$$

But since $Q_1 = Q_0/2$

$$P_t = Q_0(p_0 - p_2) \tag{7.6.2}$$

which is the overall pressure drop times the total flow.

Poiseuille's Law for the tubes is

$$Q_0 = \frac{\pi r_0^4}{8\mu}\frac{\left(p_0 - p_\zeta\right)}{\zeta}, \quad Q_1 = \frac{\pi r_1^4}{8\mu}\frac{\left(p_\zeta - p_2\right)}{\sqrt{d^2 + (L-\zeta)^2}} \tag{7.6.3}$$

And setting $Q_1 = Q_0/2$ and rearranging to solve for the pressure at the bifurcation, p_ζ, gives us

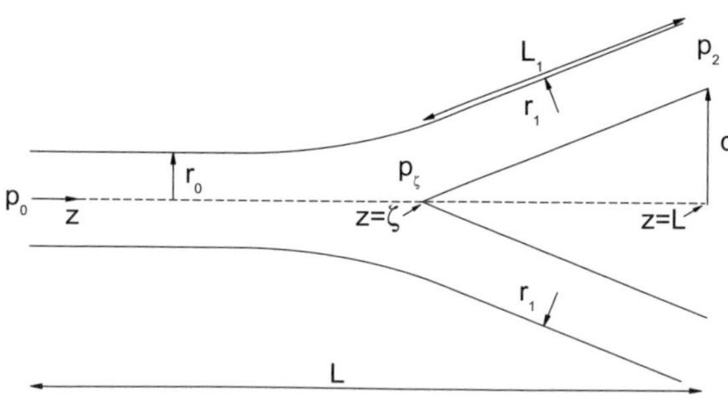

Figure 7.17 Bifurcation of a biological branching system: airways, blood vessels, which convey a flow.

$$\frac{\pi r_0^4}{\mu}\frac{\left(p_0 - p_\zeta\right)}{\zeta} = 2\frac{\pi r_1^4}{\mu}\frac{\left(p_\zeta - p_2\right)}{\sqrt{d^2 + (L-\zeta)^2}} \tag{7.6.4}$$

Equation (7.6.4) can be rearranged to solve for p_ζ

$$p_\zeta = \frac{p_0 r_0^4\sqrt{d^2 + (L-\zeta)^2} + p_2 2r_1^4\zeta}{\left(r_0^4\sqrt{d^2 + (L-\zeta)^2} + 2r_1^4\zeta\right)} \tag{7.6.5}$$

Inserting Eq. (7.6.5) into Eq. (7.6.3) to find Q_0, and then substituting that form of Q_0 into Eq. (7.6.2) yields

$$\begin{aligned} P_t &= Q_0(p_0 - p_2) \\ &= \frac{\pi r_0^4 r_1^4}{4\mu}(p_0 - p_2)^2\frac{1}{\left(r_0^4\sqrt{d^2 + (L-\zeta)^2} + 2r_1^4\zeta\right)} \end{aligned} \tag{7.6.6}$$

To minimize the power, take the derivative of Eq. (7.6.6) with respect to ζ and set it to zero, which will give us the minimum value of ζ, call it ζ_0,

$$\begin{aligned} \frac{dP_t}{d\zeta}_{\zeta=\zeta_0} &= \frac{\pi r_0^4 r_1^4}{4\mu}(p_0 - p_2)^2\frac{d}{d\zeta}\left[\frac{1}{\left(r_0^4\sqrt{d^2+(L-\zeta)^2}+2r_1^4\zeta\right)}\right] \\ &= 0 \end{aligned} \tag{7.6.7}$$

$$\zeta_0 = L - \frac{2r_1^4 d}{\sqrt{r_0^8 - 4r_1^8}} \tag{7.6.8}$$

We note that, in general,

$$d/(L-\zeta) = \tan\theta \tag{7.6.9}$$

so we can rewrite Eq. (7.6.8) as

$$\zeta_0 = L - \frac{2r_1^4}{\sqrt{r_0^8 - 4r_1^8}}(L-\zeta_0)\tan\theta \tag{7.6.10}$$

Rearranging (7.6.10) gives us

$$(L-\zeta_0)\left[1 - \frac{2r_1^4}{\sqrt{r_0^8 - 4r_1^8}}\tan\theta\right] = 0 \tag{7.6.11}$$

The solution to Eq. (7.6.11), $\zeta_0 = L$, is not physical, so we use the other solution

$$1 - \frac{2r_1^4}{\sqrt{r_0^8 - 4r_1^8}} \tan\theta = 0 \quad \rightarrow \quad \tan\theta = \sqrt{\frac{r_0^8}{2r_1^8} - 1} \tag{7.6.12}$$

Since $1 + \tan^2\theta = 1/\cos^2\theta$, we can simplify Eq. (7.6.12) to become

$$1/\cos^2\theta = \frac{r_0^8}{2r_1^8} \quad \rightarrow \quad \cos\theta = 2\left(\frac{r_1}{r_0}\right)^4 \tag{7.6.13}$$

From Murray's law, which included metabolic power, we found $r_1/r_0 = 2^{-1/3}$ so

$$\cos\theta = 2 2^{-\frac{4}{3}} = 2^{-\frac{1}{3}} = 0.79 \qquad \theta = \cos^{-1}(0.79) = 37.5^\circ \tag{7.6.14}$$

The total bifurcation angle is $2\theta = 75^\circ$. The corresponding parent tube length can be determined from Eq. (7.6.9)

$$\frac{\zeta_0}{L} = 1 - \frac{d}{L}\frac{1}{\tan\theta} = 1 - 1.3\frac{d}{L} \tag{7.6.15}$$

Since the branch angle θ is fixed, Eq. (7.6.15) states that the parent tube shortens as d increases and that for positive ζ_0 there is a constraint on d/L such that

$$\frac{d}{L} < 0.77 \tag{7.6.16}$$

7.7 Flow in a Rectangular Duct: Microfluidic Channels

Microfluidic systems are designed to handle small fluid volumes flowing in sub-millimeter channels. There are various goals including specific lab-on-chip devices to automated, high throughput bioassays for single or multiple targeted constituents. In Figure 7.18 we see a microfluidic device designed to expose human small airway epithelial cells (SAECs) to liquid plug propagation (Huh et al., 2007). (A) The microfabricated small airways are comprised of PDMS upper and lower chambers sandwiching a porous membrane. (B) SAECs are grown on the membrane with perfusion of culture media in both upper and lower chambers until they become confluent. (C) Once confluence is achieved, media are removed from the upper chamber, forming an air–liquid interface over the cells. During air–liquid interface (ALI) culture, the cells are fed basally and undergo cellular differentiation. (D) Physiologic airway closure is recreated in the microfluidic system by exposing the differentiated cells to plug flows. (E) Liquid plugs created in a plug generator progress over a monolayer of the epithelial cells and rupture in the downstream region, reopening the in vitro small airways. (F) SAECs seeded into the upper chamber attach to a membrane within 5 h after seeding in the absence of fluid flows. Continuous perfusion of media supports cell growth into monolayers with typical epithelial appearance. Confluence is reached ~6 days after seeding, at which time ~95% of the membrane surface inside the microchannel is uniformly covered with SAECs (scale bars: 150 μm).

Common to microfluidic structures is flow in rectangular channels motivating us to derive the 3D velocity field. The analysis employs Fourier series, which is often new or puzzling to many students, so the process is developed step by step. Consider steady, unidirectional, fully developed flow in a rectangular duct as shown in Figure 7.19 where the sides are of length 2b and 2c. The flow is governed by the balance of the pressure gradient and viscous forces in the Navier–Stokes equation

$$\frac{dp}{dx} = \mu\left(\frac{\partial^2 u}{\partial y^2} + \frac{\partial^2 u}{\partial z^2}\right) \tag{7.7.1}$$

while the no-slip boundary conditions are

$$u(y = \pm b) = 0, \quad u(z = \pm c) = 0 \tag{7.7.2}$$

or, equivalently, no-slip plus symmetry

$$\begin{aligned} u(y = +b) = 0, \quad \frac{\partial u}{\partial y}(y = 0) = 0 \\ u(z = +c) = 0, \quad \frac{\partial u}{\partial z}(z = 0) = 0 \end{aligned} \tag{7.7.3}$$

Choosing the following scales

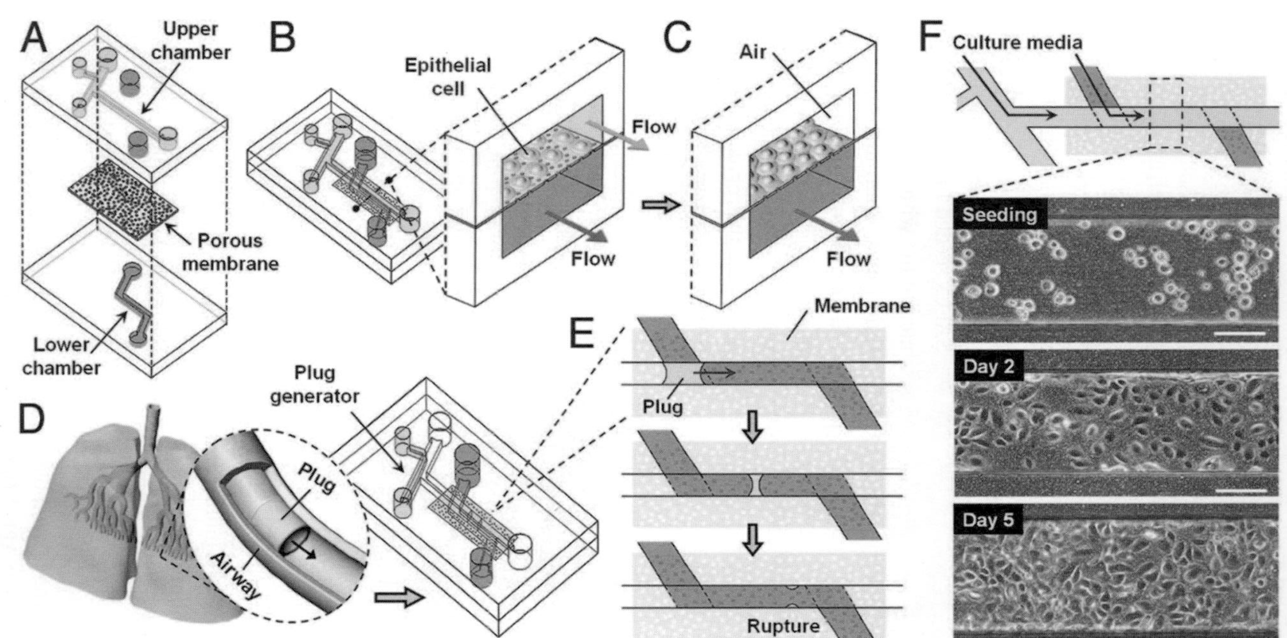

Figure 7.18 A microfluidics device for exposing respiratory airway epithelial cells to liquid plug propagation (Huh et al., 2007, Figure 1). See text for explanation.

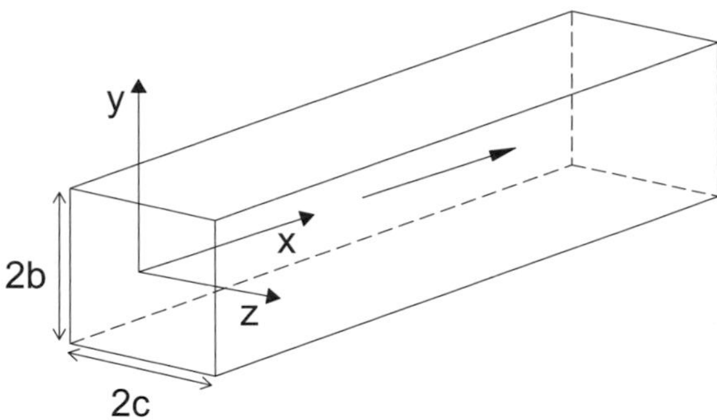

Figure 7.19 Flow in a rectangular duct.

$$U = \frac{u}{(-dp/dx)b^2/\mu}, \quad \eta = \frac{y}{b}, \quad \zeta = \frac{z}{b} \tag{7.7.4}$$

the governing equations and boundary conditions become the dimensionless form

$$\frac{\partial^2 U}{\partial \eta^2} + \frac{\partial^2 U}{\partial \zeta^2} = -1 \tag{7.7.5}$$

and the boundary conditions are

$$\begin{aligned} &U(\eta = 1) = 0, \quad \frac{\partial U}{\partial \eta}(\eta = 0) = 0, \quad \lambda = \frac{c}{b} \\ &U(\zeta - \lambda) = 0, \quad \frac{\partial U}{\partial \zeta}(\zeta = 0) = 0 \end{aligned} \tag{7.7.6}$$

The parameter $\lambda = c/b$ is the ratio of the rectangle sides.

We can approach this problem by seeking to express the structure of the solution in the η-direction as a Fourier series, while leaving the structure in the ζ-direction as unknown. This is called a partial Fourier expansion, or more generally a partial eigenfunction expansion. In this case the eigenfunctions are the sines and cosines,

$$U(\eta, \zeta) = f(\eta) + \sum_{n=1}^{\infty} a_n(\zeta)\cos(k_n\eta) + b_n(\zeta)\sin(k_n\eta) \tag{7.7.7}$$

The unknown coefficients, k_n, are called the eigenvalues. The function $f(\eta)$ will be the Hagen–Poiseuille flow in the limit as $\lambda \to \infty$ and the flow is essentially between parallel plates.

From the symmetry boundary condition $\partial U/\partial \eta = 0$ at $\eta = 0$, we can see that the sine terms will drop out of Eq. (7.7.7)

$$\frac{df}{d\eta}_{\eta=0} + \sum_{n=1}^{\infty} k_n b_n(\zeta) = 0, \quad \therefore \; b_n(\zeta) = 0, f'(0) = 0 \tag{7.7.8}$$

That leaves us with the Fourier cosine series and information about f

$$U(\eta, \zeta) = f(\eta) + \sum_{n=1}^{\infty} a_n(\zeta)\cos(k_n\eta) \tag{7.7.9}$$

Inserting Eq. (7.7.9) into the conservation of linear momentum, Eq. (7.7.5), results in

$$\frac{d^2 f}{d\eta^2} - \sum_{n=1}^{\infty} k_n^2 a_n \cos(k_n\eta) + \sum_{n=1}^{\infty} \frac{d^2 a_n}{d\zeta^2}\cos(k_n\eta) = -1 \tag{7.7.10}$$

which may be solved by collecting terms

$$\frac{d^2 f}{d\eta^2} = -1, \quad \sum_{n=1}^{\infty}\left(\frac{d^2 a_n}{d\zeta^2} - k_n^2 a_n\right)\cos(k_n\eta) = 0 \tag{7.7.11}$$

The equation for f is readily integrated to yield

$$f = -\frac{\eta^2}{2} + D\eta + E \tag{7.7.12}$$

However, we know that $f'(0) = 0 = D$, so

$$f = E - \frac{\eta^2}{2} \tag{7.7.13}$$

while the infinite series term is zero when

$$\frac{d^2 a_n}{d\zeta^2} - k_n^2 a_n = 0 \tag{7.7.14}$$

which is an ODE for the coefficient a_n. The solution for a_n is a linear combination of terms proportional to $e^{k_n\zeta}$ and $e^{-k_n\zeta}$. The $\cosh(k_n\zeta)$ and $\sinh(k_n\zeta)$ functions may be used as they are simply linear combinations of the two exponential forms $\cosh(\theta) = (e^{\theta} + e^{-\theta})/2$ and $\sinh(\theta) = (e^{\theta} - e^{-\theta})/2$. Recall that the derivative of $\cosh(\theta)$ is $\sinh(\theta)$ and the derivative of $\sinh(\theta)$ is $\cosh(\theta)$. So let a_n be

$$a_n = A_n \cosh(k_n\zeta) + B_n \sinh(k_n\zeta) \tag{7.7.15}$$

The velocity solution has the complete form

$$U(\eta,\zeta) = \left(E - \frac{\eta^2}{2}\right) + \sum_{n=1}^{\infty}(A_n \cosh(k_n\zeta) + B_n \sinh(k_n\zeta))\cos(k_n\eta) \tag{7.7.16}$$

From the symmetry condition $\partial U/\partial\zeta = 0$ at $\zeta = 0$, we find by inserting Eq. (7.7.16) that

$$\sum_{n=1}^{\infty}(k_n B_n)\cos(k_n\eta) = 0 \tag{7.7.17}$$

which is solved for

$$B_n = 0 \tag{7.7.18}$$

Now the form of our solution has reduced to

$$U(\eta,\zeta) = \left(E - \frac{\eta^2}{2}\right) + \sum_{n=1}^{\infty} A_n \cosh(k_n\zeta)\cos(k_n\eta) \tag{7.7.19}$$

The no-slip boundary condition at $\eta = 1$ applied to Eq. (7.7.19) gives us

$$U(\eta = 1,\zeta) = \left(E - \frac{1}{2}\right) + \sum_{n=1}^{\infty} A_n \cosh(k_n\zeta)\cos(k_n) = 0 \tag{7.7.20}$$

which is solved by

$$E = \frac{1}{2},\quad k_n = \left(\frac{2n-1}{2}\right)\pi \tag{7.7.21}$$

So we have found the eigenvalues, k_n, from this condition such that $\cos(k_n) = 0$.

The other no-slip boundary condition is at $\zeta = \lambda$

$$U(\eta,\ \zeta=\lambda) = \frac{1}{2}(1-\eta^2) + \sum_{n=1}^{\infty} A_n \cosh(k_n\lambda)\cos(k_n\eta) = 0 \tag{7.7.22}$$

which is an equation for the unknown coefficients, A_n. We need to represent $(1-\eta^2)/2$ in a similar Fourier cosine series so we can match the terms.

The Fourier expansion using the same eigenfunctions and eigenvalues is

$$\frac{1}{2}(1-\eta^2) = \sum_{n=1}^{\infty} F_n \cos(k_n\eta) \tag{7.7.23}$$

To determine the unknown coefficients, F_n, multiply both sides of Eq. (7.7.23) by $\cos(k_m\eta)$ and integrate from $-1 \le \eta \le 1$

$$\frac{1}{2}\int_{-1}^{1}(1-\eta^2)\cos(k_m\eta)\,d\eta = \int_{-1}^{1}\left(\sum_{n=1}^{\infty} F_n \cos(k_n\eta)\right)\cos(k_m\eta)\,d\eta \tag{7.7.24}$$

The eigenfunctions form an orthonormal set since

$$\int_{-1}^{1}\cos(k_n\eta)\ \cos(k_m\eta)\,d\eta = \delta_{mn} = \begin{cases} 0 & m \ne n \\ 1 & m = n \end{cases} \tag{7.7.25}$$

so the right hand side of Eq. (7.7.24) yields

$$\delta_{mn}F_n = F_m \tag{7.7.26}$$

The left hand side integrates to

$$\frac{1}{2}\int_{-1}^{1}(1-\eta^2)\cos(k_m\eta)\,d\eta = \frac{2\sin(k_m)}{k_m^3} - \frac{\cos(k_m)}{k_m^2} = \frac{2(-1)^{m+1}}{k_m^3} \tag{7.7.27}$$

since $\cos((2n-1)\pi/2) = 0$ and $\sin((2n-1)\pi/2) = (-1)^{(n+1)}$. So the two sides give us F_m, or equivalently F_n,

$$F_n = 2\frac{(-1)^{n+1}}{k_n^3} \tag{7.7.28}$$

Inserting the solution of Eq. (7.7.28) into Eq. (7.7.23), we see in Figure 7.20 a comparison of the exact function, $(1-\eta^2)/2$, to one-term ($N = 1$) and ten-term ($N = 10$) series solutions where $1 \le n \le N$. The ten-term series solution looks to be very accurate.

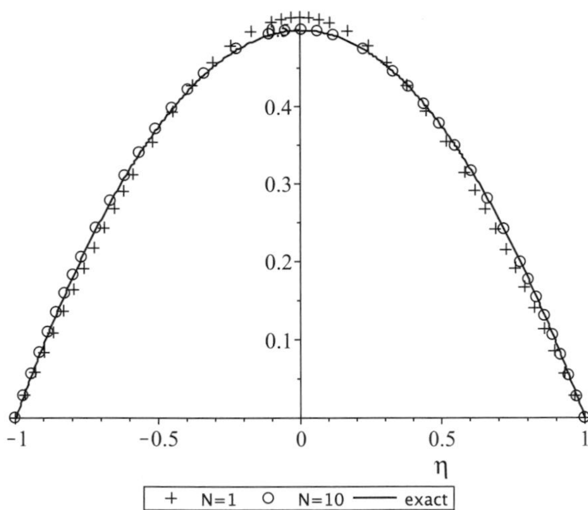

Figure 7.20 Exact function $(1-\eta^2)/2$ and series solutions for N = 1 and N = 10.

Using the resulting form of the no-slip condition in Eq. (7.7.22) yields

$$U(\eta,\zeta=\lambda)=\sum_{n=1}^{\infty}\left[2\frac{(-1)^{n+1}}{k_n^3}+A_n\cosh(k_n\lambda)\right]\cos(k_n\eta) = 0 \tag{7.7.29}$$

which solves the A_n coefficients

$$A_n=-2\frac{(-1)^{n+1}}{k_n^3\cosh(k_n\lambda)} \tag{7.7.30}$$

The final form for the velocity is

$$U(\eta,\zeta)=2\sum_{n=1}^{\infty}\frac{(-1)^{n+1}}{k_n^3}\left(1-\frac{\cosh(k_n\zeta)}{\cosh(k_n\lambda)}\right)\cos(k_n\eta)$$

$$-1\le\eta\le1,\quad -\lambda\le\zeta\le\lambda \tag{7.7.31}$$

In the limit as $\lambda\to\infty$, $\cosh(k_n\lambda)\to\infty$, so the solution in Eq. (7.7.31) approaches flow between parallel plates. Figure 7.21 shows 3D plots of the velocity $U(\eta,\zeta)$, with contour lines of constant velocity (isovelocity), for $\lambda = 1, 5$ using a 10-term series. When $\lambda = 1$ in Figure 7.21(a), the cross-section is square since $-1\le\eta\le1,\ -1\le\zeta\le1$. The maximum velocity (on the centerline) is slightly greater than the Poiseuille flow maximum of 0.25 for a circular cylinder whose radius equal to b. In addition to the shape differences of the velocity profiles, it remains that the cross-section of the tube would be πb^2, while the cross-section of the rectangular duct is $4b^2$, nearly one third larger, This makes the flow faster since the resistance is smaller assuming the same driving pressure gradient. When $\lambda = 5$ in Figure 7.21 (b), the flow is looking more 2D in the plane $\zeta = 0$, noting that now $-1\le\eta\le1,\ -5\le\zeta\le5$. The maximum velocity at the centerline approaches the Hagen–Poiseuille limit of 0.5 in a 2D channel.

7.8 Low Reynolds Number Flow over a Sphere

There are many biofluid mechanics applications where external, low Reynolds number flows around particles are significant. They include the movement of aerosols in environmental air and deposition in the lung, as well as slow flows through tissues. Aerosols can be solid or liquid, and range in size from several nanometers to 100 μm. Solid particles include dust, pollen, and products of combustion like wildfires, automobile exhaust and cigarette smoke. Liquid particles are found in clouds, fog, sprays, medical inhalants and cough-produced droplets which can carry disease. Particle shapes can be irregular, but generally are modeled with simple geometries like spheres or cylinders. For example, we have been using Stokes drag on a sphere to model aerosol particles in Sections 2.7 and 3.9. Sir George Stokes derived this formula while investigating air friction on pendulum bobs, which affected the accuracy of clocks (Stokes, 1851). Let's derive Stokes drag.

Flow around a sphere generally has velocity components (u_r, u_θ, u_ϕ) which must satisfy the general conservation of mass, Eq. (4.6.9), and Navier–Stokes equations, Eq. (6.8.2), in spherical coordinates, see Figure 7.22. The small particles and slow flows of interest have negligible Reynolds numbers,

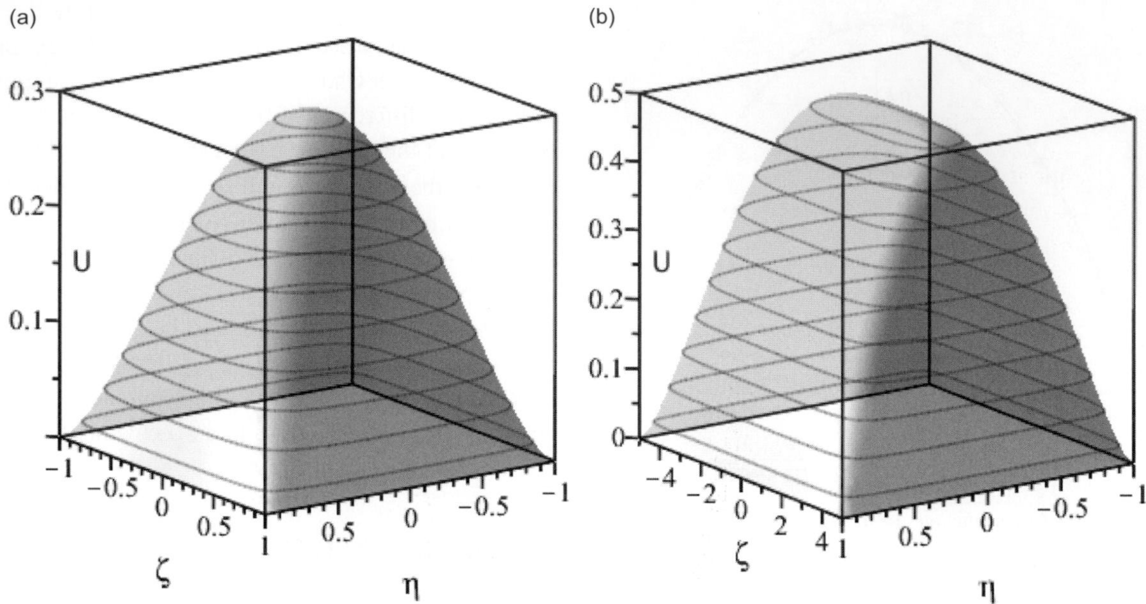

Figure 7.21 Two examples of 3D plots for the velocity profile of flow in a rectangular duct, $U(\eta, \zeta)$. A square cross-section is shown in (a) where $\lambda = c/b = 1$, and a rectangular cross-section is shown in (b) where $\lambda = c/b = 5$.

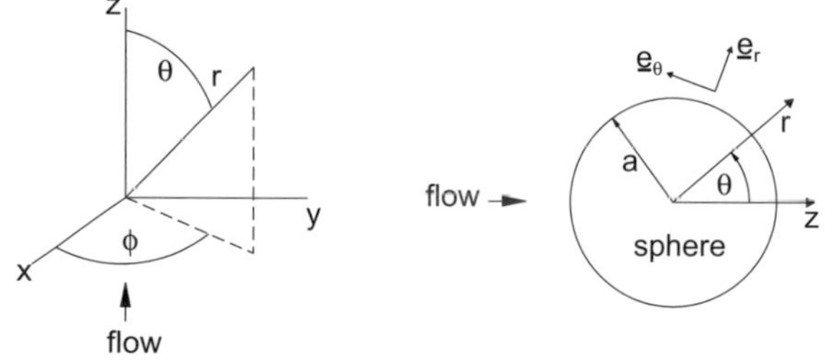

Figure 7.22. Low Reynolds number flow over a sphere of radius a in spherical coordinates.

$Re_a = aU_\infty/\nu << 1$, where the subscript indicates the length scale is the radius, a, and U_∞ is the uniform fluid velocity at infinity. We saw in Eq. (6.6.6.2) that the limit of $Re \to 0$ for a steady flow means the convective acceleration terms drop out in the Navier–Stokes equations, and we end up with the Stokes equations Eq. (6.6.4). As we will learn, however, that step omits too much.

Aligning the flow along the z-axis we see there is axisymmetry so that $u_\phi = 0, \partial/\partial\phi = 0$. This is a further simplification and allows us to use the stream function, ψ, to represent the remaining velocity components, $(u_r, u_\theta, 0)$, presented in Eq. (4.6.11)

$$u_r = \frac{1}{r^2 \sin\theta}\frac{\partial\psi}{\partial\theta}, \quad u_\theta = -\frac{1}{r\sin\theta}\frac{\partial\psi}{\partial r} \tag{7.8.1}$$

In Eq. (6.5.20) we expressed momentum conservation terms of the stream function, ψ, in rectangular coordinates. Setting the acceleration terms equal to zero in Eq. (6.5.20), the resulting form is $\nabla^4\psi = \nabla^2(\nabla^2\psi) = 0$ where $\nabla^2 = \partial^2/\partial x^2 + \partial^2/\partial y^2$. The same approach can be done in spherical coordinates and involves taking the curl of the vector form of Navier–Stokes equations and using vector identities; we can eventually reduce the momentum balance to a similar single equation for ψ

$$\nabla^2(\nabla^2\psi) = \nabla^4\psi = 0$$
$$\nabla^2 = \left(\frac{\partial^2}{\partial r^2} + \frac{\sin\theta}{r^2}\frac{\partial}{\partial\theta}\left(\frac{1}{\sin\theta}\frac{\partial}{\partial\theta}\right)\right) \tag{7.8.2}$$

The boundary conditions are no-penetration and no-slip on the sphere

$$u_r(r=a) = \left(\frac{1}{r^2 \sin\theta} \frac{\partial \psi}{\partial \theta}\right)_{r=a} = 0,$$
$$u_\theta(r=a) = \left(-\frac{1}{r \sin\theta} \frac{\partial \psi}{\partial r}\right)_{r=a} = 0 \tag{7.8.3}$$

The conditions at $r \to \infty$ are uniform z-velocity U_∞, which decomposes into $u_r(r \to \infty) = U_\infty \cos\theta$ and $u_\theta(r \to \infty) = -U_\infty \sin\theta$. For example, at $\theta = 0$ the velocity is only radial, $(u_r, u_\theta) = (U_\infty, 0)$ and aligned with z, while for $\theta = \pi/2$ the velocity is $(u_r, u_\theta) = (0, -U_\infty)$ keeping in mind that $u_\theta < 0$ for a velocity in the clockwise direction, see Figure 7.22. In terms of the stream function those conditions are

$$u_r(r \to \infty) = \left(\frac{1}{r^2 \sin\theta} \frac{\partial \psi}{\partial \theta}\right)_{r\to\infty} = U_\infty \cos\theta$$
$$u_\theta(r \to \infty) = \left(-\frac{1}{r \sin\theta} \frac{\partial \psi}{\partial r}\right)_{r\to\infty} = -U_\infty \sin\theta \tag{7.8.4}$$

Integrating Eqs. (7.8.4) gives us the form of ψ at $r \to \infty$

$$\psi_{r\to\infty} = \psi_\infty = \frac{U_\infty}{2} r^2 \sin^2\theta \tag{7.8.5}$$

Try a solution of the form

$$\psi(r, \theta) = f(r) \sin^2\theta \tag{7.8.6}$$

and insert into Eq. (7.8.2) to arrive at an equation for $f(r)$

$$\left(\frac{d^2}{dr^2} - \frac{2}{r^2}\right)^2 f = 0 \tag{7.8.7}$$

The general solution to Eq. (7.8.7) is

$$f = \frac{A}{r} + Br + Cr^2 + Dr^4 \tag{7.8.8}$$

where the integration constants A, B, C, D are chosen to satisfy the boundary conditions. For the condition at infinity in Eq. (7.8.5) we choose $D = 0$ and $C = U_\infty/2$. Then the no-slip and no-penetration conditions of Eqs. (7.8.3) give $A = U_\infty a^3/4$ and $B = -3U_\infty a/4$. The result is

$$\psi = \frac{U_\infty}{2}\left(r^2 + \frac{a^3}{2r} - \frac{3ar}{2}\right) \sin^2\theta \tag{7.8.9}$$

The first term in the parentheses corresponds to the uniform flow. The second term is called a doublet, a form that arises for inviscid flows when a source and sink are very close together. Combined they represent inviscid flow over a sphere. The third term is known as a Stokeslet and contains all of the vorticity. Note that $\psi(r = a) = 0$ on the sphere's surface and $\psi(\theta = 0, \pi) = 0$, which are the downstream and upstream stagnation streamlines.

The streamlines, actually stream surfaces or tubes, are lines of constant ψ from Eq. (7.8.9) and are shown in Figure 7.23(a) where the radial variable is $R = r/a$ and the sphere surface is located at $R = 1$. The flow is from left to right and note the symmetry fore and aft which is what we expect from neglecting the inertial terms. The velocity components come from substituting Eq. (7.8.9) into Eqs. (7.8.1)

$$u_r = U_\infty \cos\theta \left(1 - \frac{3a}{2r} + \frac{a^3}{2r^3}\right),$$
$$u_\theta = -U_\infty \sin\theta \left(1 - \frac{3a}{4r} - \frac{a^3}{4r^3}\right) \tag{7.8.10}$$

and a sample dimensionless profile is shown in Figure 7.23(a) at $\theta = \pi/2$ where $u_r = 0$ and $u_\theta/U_\infty = -(1 - 3R^{-1}/4 - R^{-3}/4)$. The profile matches zero velocity on the sphere and then gradually increases toward $u_\theta/U_\infty = -1$ as $R \to \infty$, again recalling that this velocity is in the negative $\underline{e}_\theta$-direction. The vorticity vector, $\underline{\omega}$, discussed in Section 6.7, is a measure of the local fluid rotation or shear. For spherical coordinates the general components are $\underline{\omega} = \omega_r \underline{e}_r + \omega_\theta \underline{e}_\theta + \omega_\phi \underline{e}_\phi$. However, due to the axisymmetry in this flow there is only the $\underline{e}_\phi$ component

$$\omega_\phi = \frac{1}{r}\frac{\partial(r u_\theta)}{\partial r} - \frac{1}{r}\frac{\partial u_r}{\partial \theta} = -\frac{3}{2} U_\infty a \frac{\sin(\theta)}{r^2} \tag{7.8.11}$$

The forms of ψ, u_r, and u_θ in Eq. (7.8.9) and Eqs. (7.8.10) are for zero velocity on the sphere with the fluid at infinity passing from left to right at speed U_∞. The coordinate system is attached to the sphere, where the no-slip and no-penetration conditions are satisfied by $u_r(r = a) = 0$ and $u_\theta(r = a) = 0$. However, an alternative reference frame is to consider

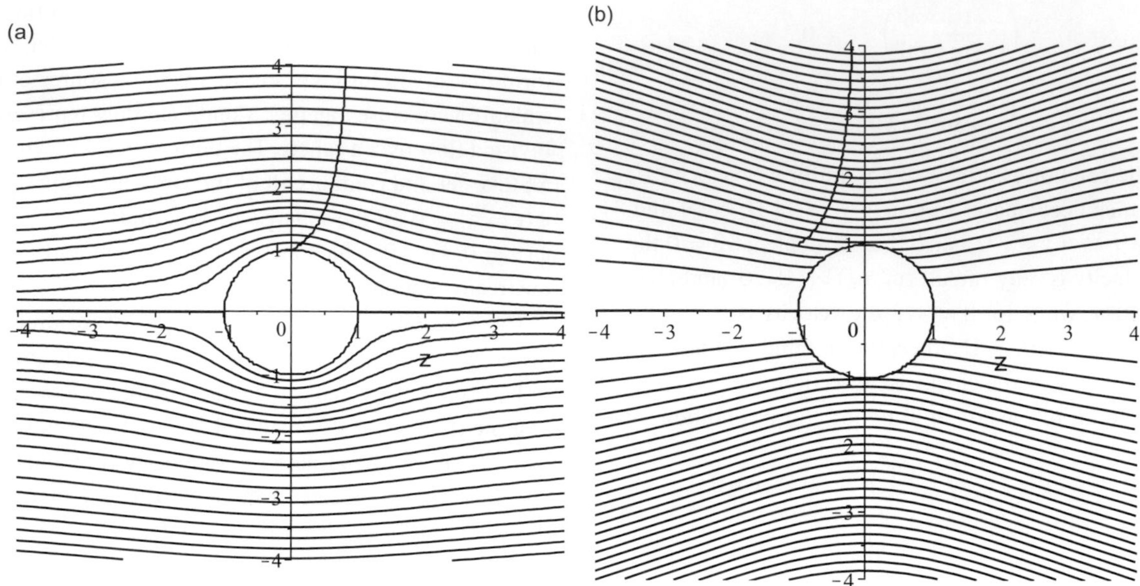

Figure 7.23 Streamlines and velocity profile for flow over a sphere at low Reynolds number: (a) coordinates attached to sphere; (b) coordinates attached to stagnant fluid.

the sphere moving right to left, at velocity $-U_\infty \underline{e}_z = -U_\infty(\cos\theta\, \underline{e}_r - \sin\theta\, \underline{e}_\theta)$, through an otherwise still fluid. Then the no-slip and no-penetration conditions on the sphere are $u_r(r = a) = -U_\infty \cos\theta$ and $u_\theta(r = a) = U_\infty \sin\theta$. The stream function for this flow, call it $\hat{\psi}$, is just our original stream function minus the stream function at infinity

$$\hat{\psi} = \psi - \psi_\infty = \frac{U_\infty}{2}\left(\frac{a^3}{2r} - \frac{3ar}{2}\right)\sin^2\theta \qquad (7.8.12)$$

and the corresponding velocities, $\hat{u}_r$, $\hat{u}_\theta$, are the originals minus $U_\infty \underline{e}_z$ so that

$$\hat{u}_r = u_r - U_\infty \cos\theta = U_\infty \cos\theta\left(-\frac{3a}{2r} + \frac{a^3}{2r^3}\right)$$
$$\hat{u}_\theta = u_\theta - (-U_\infty \sin\theta) = -U_\infty \sin\theta\left(-\frac{3a}{4r} - \frac{a^3}{4r^3}\right) \qquad (7.8.13)$$

These new streamlines are lines of constant $\hat{\psi}$ from Eq. (7.8.12) and are shown in Figure 7.23(b). In this frame the sphere is pushing fluid ahead and off to the sides to make room for itself, while entraining fluid that fills in behind. The streamlines converge and then diverge consistent with the velocity being higher when closer to the sphere. The dimensionless velocity profile for $\theta = \pi/2$ is shown and only involves $\hat{u}_\theta/U_\infty = (3/(4R) + 1/(4R^3))$ since, again, $\hat{u}_r = 0$ there. These are velocities in the positive θ-direction, starting at the sphere surface, $R = 1$, where $\hat{u}_\theta/U_\infty = 1$, and gradually decaying to zero as $R \to \infty$.

The corresponding pressure field from the Stokes equations is

$$p = p_\infty - \frac{3}{2}\frac{\mu U_\infty a}{r^2}\cos\theta \qquad (7.8.14)$$

Define a dimensionless relative pressure as $P = (p - p_\infty)/(\mu U_\infty/a) = -3\cos\theta/2R^2$ using Eq. (7.8.14). A plot of P on the sphere surface, $P(R = 1) = -3\cos\theta/2$, is shown in Figure 7.24(a), demonstrating that $P > 0$ on the upstream face $\pi/2 \le \theta \le 3\pi/2$ and $P < 0$ on the downstream face $0 \le \theta \le \pi/2 + 3\pi/2 \le \theta \le 2\pi$. The maximum surface pressure occurs at $\theta = \pi$, which is the

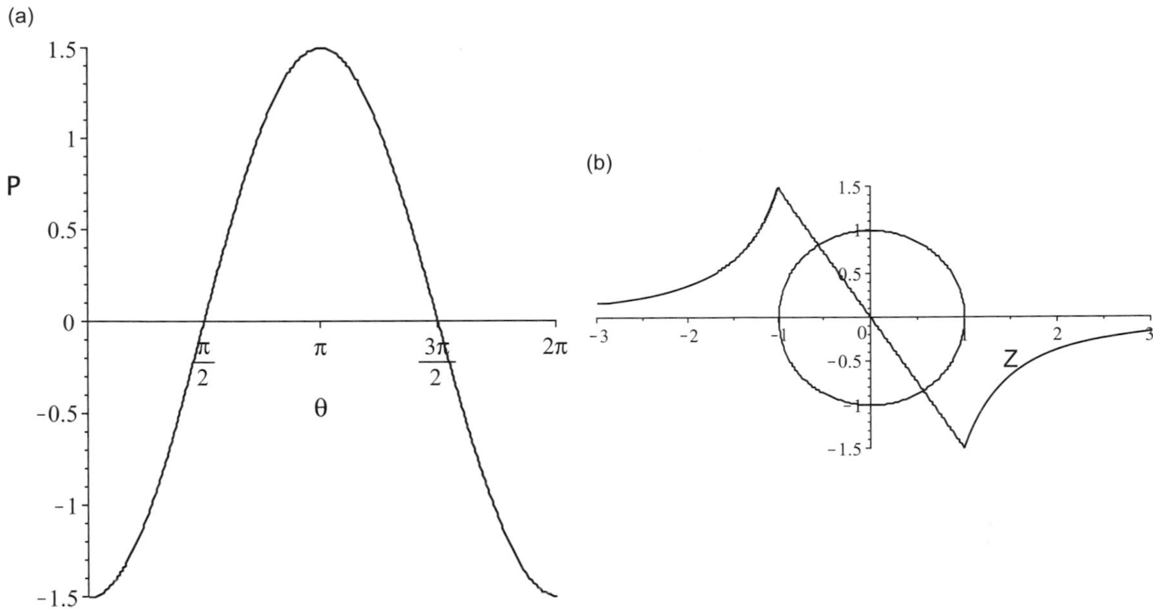

Figure 7.24 Dimensionless pressure distribution P (a) on the sphere and (b) on the streamline $\psi(X, Y) = 0$, which is the z-axis.

upstream stagnation point, while the minimum surface pressure is at $\theta = 0$, where the downstream stagnation point sits. Figure 7.24(b) shows P along the streamline $\psi(X, Y) = 0$ with segments: upstream stagnation streamline $P(X \leq -1, Y = 0)$; sphere surface $P(1 \leq X \leq -1, Y = (1 - X^2)^{1/2})$; downstream stagnation streamline $P(1 \leq X, Y = 0)$ where $X = x/a$ and $Y = y/a$.

The stress tensor components from Eq. (6.4.7) that we need are $T_{rr} = -p + 2\mu(\partial u_r/\partial r)$ containing the pressure and normal viscous stress, and $T_{r\theta} = \mu r(\partial(u_\theta/r)/\partial r) + (\mu/r)(\partial u_r/\partial\theta)$, which has the viscous tangential. The surface stresses are $\underline{t} = \underline{\underline{T}} \cdot \underline{n}$ where $\underline{n} = \underline{e}_r$, which must be evaluated at $r = a$. The z-component is $t_z = \underline{t} \cdot \underline{e}_z = -p\cos\theta - \tau_{r\theta}\sin\theta$ where the surface shear stress is $\tau_{r\theta} = \mu e_{r\theta} = -(3\mu U_\infty \sin\theta)/(2a)$ and the surface normal stress is $T_{rr} = -p + \tau_{rr} = -p$ with no viscous normal stresses on the boundary. The result is

$$t_z = -\left(p_\infty - \frac{3}{2}\frac{\mu U_\infty}{a}\cos\theta\right)\cos\theta + \frac{3\mu U_\infty \sin^2\theta}{2a} \tag{7.8.15}$$

The drag force is the integral of t_z over the sphere surface

$$F_d = \int_0^{2\pi}\int_0^{\pi} t_z a^2 \sin\theta \, d\theta \, d\phi = 6\pi\mu a U_\infty \tag{7.8.16}$$

which is Stokes drag that we used in Eq. (2.7.5) to study the erythrocyte sedimentation rate. The pressure term contributes one-third of the drag force, $2\pi\mu a U_\infty$, which is called “form drag,” “shape drag” or “pressure drag.” The surface shear contributes two-thirds, $4\pi\mu a U_\infty$, which is called “friction drag.” Almost 60 years later, in 1909, Robert A. Millikan (1868–1953), while a professor at the University of Chicago, used Stokes drag to show that ionized oil drops had quantized charges. This led to the discovery of the electron, for which he won the Nobel Prize in Physics in 1923 (Millikan, 1911; 1913). Additional biological flow applications include tiny swimming microorganisms such as algae, bacteria, protozoa, spermatozoa and other gametes. Stokes drag is used to calculate their motion, interaction with one another and energy requirements (Pedley and Kessler, 1992).

While small individually, the total biomass of all bacteria is ~15% of the total global biomass (Bar-On et al., 2018).

All is not well, however, because the Stokes solution actually violates its assumptions. Consider the large r, $r/a \to \infty$, behavior of the velocities in Eqs. (7.8.10). The dominant r-dependent terms are those proportional to $1/r$, since $1/r^3$ dies out faster. The radial velocity then looks like $u_r \sim U_\infty \cos\theta + u'$ where $u' = U_\infty \cos\theta(3a/2r)$ and $u' << U_\infty \cos\theta$ far from the sphere. A representative convective acceleration term from Eq. (6.8.2) is $\rho u_r(\partial u_r/\partial r) = \rho(U_\infty \cos\theta + u')(\partial u'/\partial r)$, which linearizes to $\rho U_\infty \cos\theta\,(\partial u'/\partial r)$ by neglecting the $u'(\partial u'/\partial r)$ term. Substituting for u′ this term becomes $\sim \left(\rho U_\infty^2/a\right) R^{-2}$. By comparison, a representative viscous term is $\mu\left(\partial^2 u_r/\partial r^2\right) \sim (\mu U_\infty/a^2)R^{-3}$. Therefore the ratio of convective acceleration to viscous forces as $r/a \to \infty$ is

$$\left.\frac{\text{convective acceleration}}{\text{viscous force}}\right|_{R\to\infty} \sim \mathrm{Re}_a R \qquad (7.8.17)$$

Equation (7.8.17) shows that, in the low Reynolds number limit $\mathrm{Re}_a \to 0$, convective acceleration can be comparable to viscous forces as $R \to \infty$, i.e. far enough from the sphere, such that $\mathrm{Re}_a R = O(1)$. That violates our assumption. For the cylinder, this contradiction is fatal at the outset.

The remedy involves the inclusion of convective acceleration and in 1910 Oseen provided the first correction to the sphere by adding a linearized term, $(\underline{U}_\infty \cdot \underline{\nabla})\underline{u} = U_\infty \partial \underline{u}/\partial z$ (Oseen, 1910),

$$F_d = 6\pi\mu a U_\infty \left(1 + \frac{3}{8}\mathrm{Re}_a + \frac{9}{40}\mathrm{Re}_a^2 \ln(\mathrm{Re}_a) + O\left(\mathrm{Re}_a^2\right)\right) \quad \text{sphere} \qquad (7.8.18)$$

In 1913 Oseen arrived at a correction to the drag (Oseen, 1913), which is the second term in Eq. (7.8.18). Proudman and Pearson added the third term in 1957 using the method of matched asymptotic expansions, which is discussed in Section 17.4 (Lagerstrom and Cole, 1955; Proudman and Pearson, 1957; Van Dyke, 1975).

7.9 Low Reynolds Number Flow over a Cylinder

Proceeding in a similar fashion as we did with the sphere, consider perpendicular flow over a cylinder, or circle, as shown in Figure 7.25. The stream function in cylindrical coordinates for $u_z = 0$ or, equivalently, polar coordinates, is defined by $u_r = (1/r)(\partial\psi/\partial\theta)$ and $u_\theta = -\partial\psi/\partial r$. Writing the momentum balance in terms of the stream function gives us the same form as Eq. (7.8.2), but with a different Laplacian operator

$$\nabla^2(\nabla^2\psi) = \nabla^4\psi = 0 \quad \nabla^2 = \frac{\partial^2}{\partial r^2} + \frac{1}{r}\frac{\partial}{\partial r} + \frac{1}{r^2}\frac{\partial^2}{\partial\theta^2} \qquad (7.9.1)$$

As $r \to \infty$ the velocity approaches the constant value $U_\infty \mathbf{e}_x$ which, in polar coordinates, is $u_r \sim U_\infty \cos\theta$ and $u_\theta \sim -U_\infty \sin\theta$. The corresponding stream function form at infinity is $\psi_{r\to\infty} \sim U_\infty r \sin\theta$, which motivates a separation of variables form. Switching to dimensionless variables, let $\hat{\psi} = f(R)\sin\theta$ where $R = r/b$ and $\hat{\psi} = \psi/bU_\infty$. Substituting this form into the dimensionless form of Eq. (7.9.1) yields an equation and solution form for f(R)

$$\left(\frac{d^2}{dR^2} + \frac{1}{R}\frac{d}{dr} - \frac{1}{R^2}\right)^2 f(R) = 0 \;\to$$
$$f(R) = A_1R^3 + B_1R\ln R + C_1R + \frac{D_1}{R} \qquad (7.9.2)$$

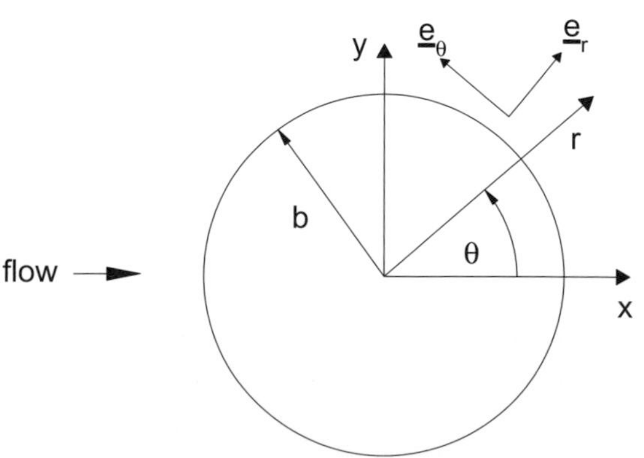

Figure 7.25 Perpendicular flow over a cylinder of radius b.

On the cylinder surface we impose no-slip $u_\theta(R=1) \sim df/dR|_{R=1} = 0$ and no-penetration, $f(R=1)=0$, i.e. the cylinder is a streamline. Further, we can set $A_1 = 0$ to suppress the R^3 dependence, using the principle of minimum singularity. That leaves us with

$$f(R) = D_1\left(2R\ln R - R + \frac{1}{R}\right) \tag{7.9.3}$$

However, no choice of D_1 permits the large distance form, $\hat{\psi}|_{R\to\infty} \sim R\sin\theta$, because of the $\ln R$ term, so there is no solution. This is Stokes' paradox.

$$F_d = 4\pi\mu U_\infty\left(\Delta_1 - 0.87\Delta_1^3 + O\left(\Delta_1^4\right)\right)$$
$$\Delta_1 = \left(\ln\left(4/Re_b\right) - \gamma + \frac{1}{2}\right)^{-1} \quad \text{cylinder} \tag{7.9.4}$$

Again, the resolution is to include a linearized form of convective acceleration. For flow over a cylinder, Lamb (1911) found a solution to the velocity field resolving Stokes' paradox in this manner. The resulting drag force per unit length of the cylinder, F_d, is the first term of Eq. (7.9.4), where $Re_b = bU_\infty/\nu$ is the Reynolds number, b is the cylinder radius and $\gamma = 0.5772$ is the Euler constant. The second term in Eq. (7.9.4) is due to Kaplun and Lagerstrom, who also used the method of matched asymptotic expansions, which we discuss in Section 17.4 (Kaplun and Lagerstrom, 1957; Van Dyke, 1975).

Flow Perpendicular to an Array of Cylinders

Though disappointing so far, we actually can find solutions to flow over a cylinder, if we replace the conditions at infinity with conditions at a finite distance. For example, consider flow through a square array of cylinders, a sample portion is shown in Figure 7.26, where we will consider perpendicular and parallel flow separately. This could be a model of packed fibers or a matrix of molecular fibrils like collagen or elastin, see Figure 15.5, discussed in Section 15.6.

For large enough spacing, the velocity field will have lines of symmetry. An example is shown for the center cylinder surrounded by a dashed box. Happel (1959) replaced that complicated situation with an equivalent radius, $r = r_0$, where the boundary conditions can be applied. Equating the hatched areas, the radius of the equivalent system is $\pi r_0^2 = L^2$. A key parameter in describing a packed bed of fibers is its porosity, ϕ, which is the volume occupied by the fluid divided by the total volume. In Figure 7.26 that would be $\phi = \left(L^2 - \pi b^2\right)/L^2 = 1 - (b/r_0)^2$. In general, the porosity has the range $0 \le \phi \le 1$. However, since this approach requires large enough spacing, the results are better for the range $0.4 - 0.5 \le \phi \le 1$.

Happel imposed no shear stress at $r = r_0$ and the radial velocity $u_r(r = r_0) = U_{avg}\cos\theta$ where U_{avg} is the average velocity through the array. Then the

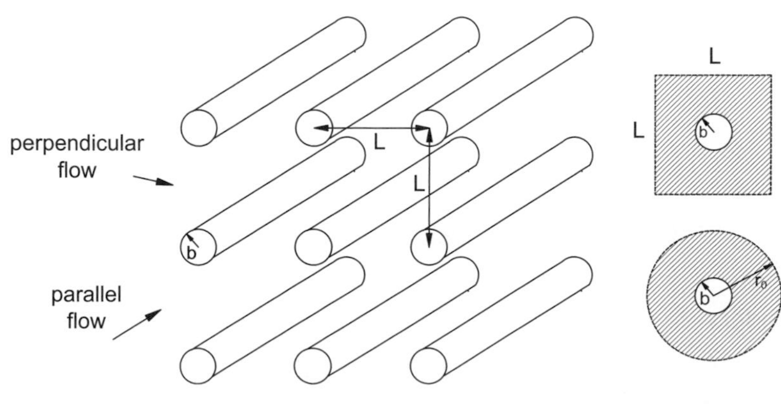

Figure 7.26 Perpendicular and parallel flow through a sample region of a square array of cylinders with radius b and spacing L. The equivalent model with radius r_0 is shown and the hatched areas are equal (Happel, 1959).

no-slip and no-penetration velocity boundary conditions on the cylinder are $u_r(r = b) = u_\theta(r = b) = 0$. The stress vector is $\underline{t} = \underline{\underline{T}} \cdot \underline{n} = \underline{\underline{T}} \cdot \underline{e}_r = \underline{\underline{T}} \cdot (1, 0, 0) = (T_{rr}, T_{r\theta}, 0)$. The θ component of the stress vector is $t_\theta = \underline{t} \cdot \underline{e}_\theta = \underline{t} \cdot (0, 1, 0) = T_{r\theta} = \tau_{r\theta} = \mu(r(\partial(u_\theta/r)/\partial r) + r^{-1}(\partial u_r/\partial\theta))$. The final condition is setting the shear stress equal to zero at $r = r_0$, i.e. $\tau_{r\theta}(r = r_0) = 0$. Imposing these four boundary conditions on Eq. (7.9.2) solves for A_1, B_1, C_1, D_1.

Going back to the Stokes equations (i.e. Navier–Stokes with zero acceleration), the pressure is found to be $p = \mu\cos\theta(C_1 r - D_1/r)$. The drag force comes from the x-component of fluid stress on the cylinder is $t_x = \underline{t} \cdot \underline{e}_x = \underline{t} \cdot (\cos\theta \underline{e}_r - \sin\theta \underline{e}_\theta) = (T_{rr}\cos\theta - \tau_{r\theta}\sin\theta)_{r=b}$ where $T_{rr} = -p + 2\mu(\partial u_r/\partial r)$. The drag force, F_x, per unit length in the x-direction is the integral

$$F_x = \int_0^{2\pi} t_x b\, d\theta = 2\pi\mu B_1 \quad B_1 = \frac{2U_{avg}}{\ln\left(\frac{r_0}{b}\right) - \frac{1}{2}\frac{(r_0^4 - b^4)}{(r_0^4 + b^4)}}$$
$$= \frac{4U_{avg}}{\ln\left(\frac{1}{1-\phi}\right) - \frac{(1-(1-\phi)^2)}{(1+(1-\phi)^2)}} \tag{7.9.5}$$

The x-direction force balance on the square control volume in Figure 7.26 is $F_x = p_{-L/2}L - p_{+L/2}L = \Delta pL$ since, due to symmetry, the convective contributions are equal on the upstream and downstream faces so subtract out. The pressure gradient in the x-direction can be sensibly estimated as $\Delta p/L = -dp/dx = F_x/L^2 = F_x/\pi r_0^2$ so that

$$-\frac{dp}{dx} = \frac{8\mu U_{avg}}{r_0^2}\frac{1}{\ln\left(\frac{1}{1-\phi}\right) - \frac{(1-(1-\phi)^2)}{(1+(1-\phi)^2)}} \tag{7.9.6}$$

Flow Parallel to an Array of Cylinders

Flow parallel to the cylinder axis obeys Eq. (7.1.2), which we used for internal flow but now can apply to external flow. Imposing no-slip on the cylinder surface, $u_z(r = b) = 0$, and no shear stress at $r = r_0$, $(du_z/dr)_{r=r_0} = 0$ determines c_1, c_2, and the resulting solution is

$$u_z(r) = -\frac{1}{4\mu}\frac{dp}{dz}\left(b^2 - r^2 + 2r_0^2\ln\left(\frac{r}{b}\right)\right) \quad b \le r \le r_0 \tag{7.9.7}$$

We can make Eq. (7.9.7) dimensionless by defining $R = r/b$ and $U_Z = u_z/(b^2(-dp/dz)/\mu)$

$$U_Z(R) = \frac{1}{4}\left(1 - R^2 + \frac{2}{(1-\phi)}\ln R\right)$$
$$1 \le R \le \frac{1}{\sqrt{1-\phi}} \tag{7.9.8}$$

Figure 7.27 shows $U_Z(R)$ for four values of the porosity, $\phi = 0.6, 0.7, 0.8, 0.9$. As ϕ increases, the cylinders are farther apart and present a smaller resistance to the flow. The velocities are higher and extend further into the fluid. The velocity averaged across the total cross-section is

$$U_{avg} = \frac{1}{\pi r_0^2}\int_0^{2\pi}\int_b^{r_0} u_z r\, dr\, d\theta$$
$$= -\frac{dp}{dz}\frac{r_0^2}{8\mu}\left(-2\phi - \phi^2 + 2\ln\left(\frac{1}{1-\phi}\right)\right) \tag{7.9.9}$$

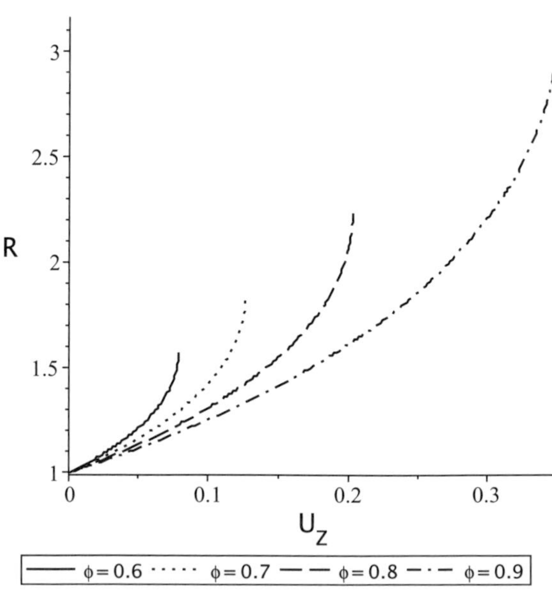

Figure 7.27 Dimensionless velocity profiles $U_Z(R)$ for four values of $\phi = 0.6, 0.7, 0.8, 0.9$.

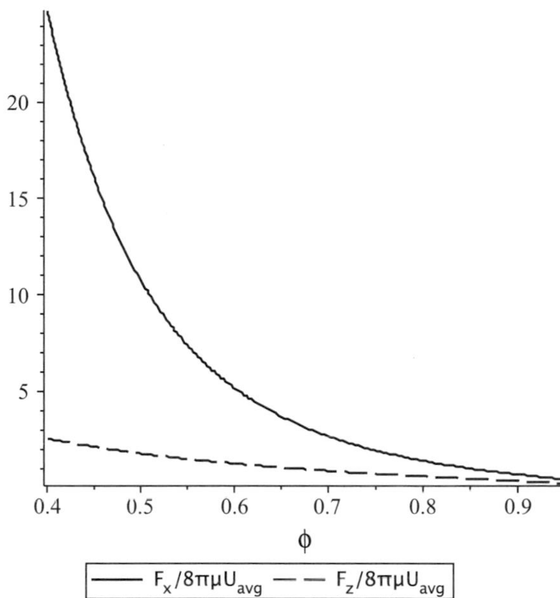

Figure 7.28 Dimensionless drag forces for flow through a square array of cylinders: perpendicular, $F_x/8\pi\mu U_{avg}$, and parallel $F_z/8\pi\mu U_{avg}$, $0.4 \le \phi \le 0.95$.

The force per unit length in the z-direction is

$$F_z = \int_0^{2\pi} \mu \frac{\partial u_z}{\partial r}_{r=b} b\, d\theta = \pi\left(-\frac{dp}{dz}\right)\left(r_0^2 - b^2\right)$$
$$= \frac{8\pi\mu U_{avg}\phi^2}{\left(-2\phi - \phi^2 + 2\ln\left(\frac{1}{1-\phi}\right)\right)} \tag{7.9.10}$$

where we have solved Eq. (7.9.9) for $(-dp/dz)$ and inserted the result into Eq. (7.9.10).

Figure 7.28 compares the dimensionless drag forces for perpendicular flow through a square array of cylinders, $F_x/8\pi\mu U_{avg}$, and parallel flow $F_z/8\pi\mu U_{avg}$ in the porosity range $0.4 \le \phi \le 0.95$. Both forces decrease as ϕ increases as one would expect. The parallel flow has lower drag, the absence of shape/ pressure drag being an important contributor.
A full range of Reynolds number behavior for flow over a cylinder, including turbulence, is explored in Section 13.1.

7.10 Computational Fluid Dynamics (CFD)

In this chapter we are solving the Navier–Stokes equations and mass conservation for simplified flows and geometries which permit analytical solutions. As the boundary shapes and flow details become more challenging, numerical methods must be employed. Recall in Section 3.10 we explored numerical methods for solving a first-order, ordinary differential equation (ODE), $dh/dx = -h$, with boundary condition $h(x = 0) = 1$ using the Euler Method. From that introduction we appreciated that the numerical solution was a list of numbers, e.g. Eq. (3.10.16), consisting of two columns: the incremental steps in the independent variable, x_i, and the corresponding dependent variable, i.e. the solution h_i. Now consider that the governing equations for fluid mechanics are coupled, nonlinear, partial differential equations with independent variables of space (x, y, z) and time (t), and dependent variables of pressure $p(x, y, z, t)$ and three components of velocity $\underline{u}(x, y, z, t)$. There are several numerical methods used to solve these problems which are increasingly available in commercial CFD software packages that are user friendly: Volume of Fluid, Finite Element, Spectral Element, Boundary Integral, Finite Difference – to name a few. Typically the static boundary shape is established, often by a computer aided design (CAD) program in engineering applications. For 2D or 3D simulations, the fluid region is divided into discrete 2D or 3D cells called the mesh. The spatial discretization of the governing equations is based on this mesh, as are the solutions which are calculated at each time step. By comparison, in our ODE example of Section 3.10 the "mesh" was simply the x_i values separated by steps of Δx, and we discretized the ODE as $h_{i+1} = h_i - h_i\Delta x$ in Eq. (3.10.14) using Euler's Method. For CFD we have steps in two or three directions and also in time. If the boundary shape also changes with time, then the approach must take that into account.

For CFD biomedical applications the boundary shapes often come from medical imaging, such as computed tomography (CT) scans, which are unique to a particular patient. Then the flow computations are used to assess function and dysfunction, as well as to simulate interventions ranging from drug delivery to surgery. With a simulation tool in hand, the clinician can test and compare the predicted results of interventions in advance, in order to choose the best strategy for that patient. This is personalized, or patient-specific medicine (Shojima et al., 2004; Pekkan et al., 2008; Taylor and Figueroa, 2009; Yin et al., 2010).

An aneurysm is a localized out-pouching of a blood vessel wall due to local weakness of the tissue. Interaction with the blood flow can affect the creation and development of aneurysms. Particularly critical locations for aneurysms are the cerebral arteries and sections of the aorta in the thorax and abdomen. In addition to the risk of wall rupture, aneurysms can provide a locus for blood clot formation. Figure 7.29 shows angiography of a cerebral artery with an aneurysm, the large balloon-like structure in the center. A ruptured cerebral artery aneurysm can cause a stroke with accompanying neurological deficits or death. Blood clots that form in the aneurysm can break off pieces which flow further downstream as emboli. Those emboli can become stuck as the vessel diameters decrease, causing a localized infarction due to flow stoppage.

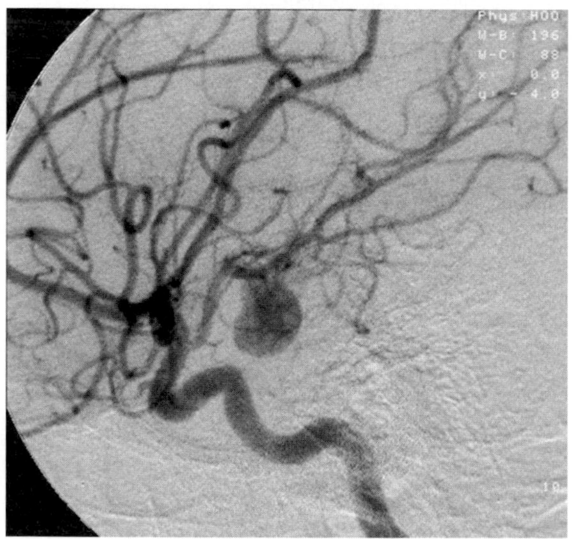

Figure 7.29 Cerebral angiography showing a berry aneurysm, the balloon-like structure in the center of the field.

Patient-specific CFD examples are shown in Figure 7.30 for intracranial berry aneurysm where Figure 7.30(a) are streamlines and Figure 7.30(b) is the wall shear stress (WSS) distribution. High WSS may initiate aneurysm growth, while low WSS may facilitate that growth leading to rupture by inducing degenerative changes in the aneurysm wall. This anterior communicating artery aneurysm was imaged with 4D flow MRI and time-of-flight MRA at Northwestern University School of Medicine. The TOF MRA data was segmented using open source software ITK-Snap to generate patient-specific vascular geometries. The numerical solution of the Navier–Stokes equations was obtained with a finite-volume solver Fluent (ANSYS). The flow was assumed to be Newtonian and the arterial walls were rigid. An unstructured, tetrahedral mesh with the nominal element size of 150 microns was generated on the domain following a mesh independence study. The 4D flow MRI measurements were used to prescribe patient-specific inlet and outlet flow waveforms. Second- and third-order schemes were used for the discretization in time and in space, respectively. Transient CFD simulations with a time step of 1.5 ms were carried out for three cardiac cycles and the last cycle was used for the flow analysis.

Another example involves airflow over an elephant as examined in the studies of Dudley et al. (2013), where both CFD and wind tunnel experiments were performed. Rather than placing a real elephant in a wind tunnel, which would weigh heavily on the budget, they used an aluminum model elephant scaled ~1:20, i.e. model height ~ 0.2 m and actual adult male elephant height ~ 4 m. Both the measured drag force and heat transfer compared well to the CFD

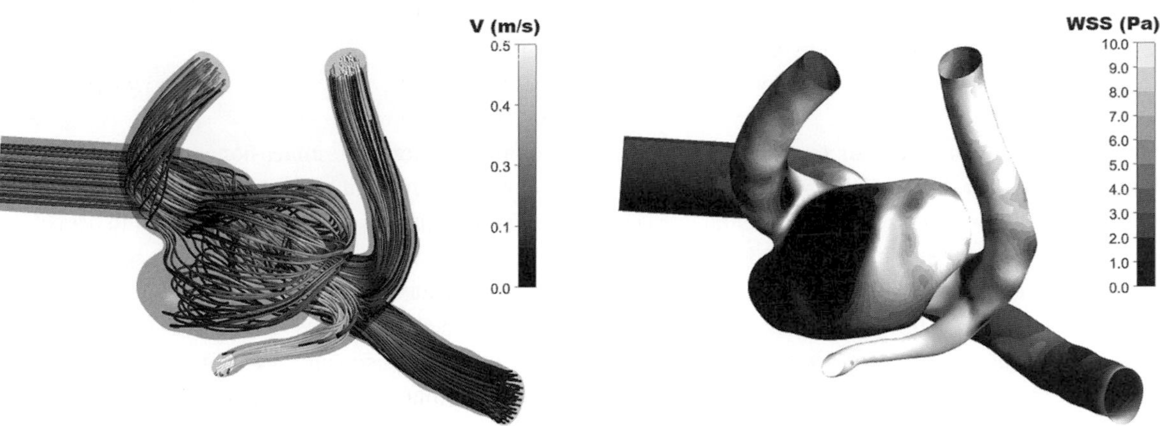

Figure 7.30 CFD results for flow through a cerebral aneurysm: (a) streamlines; (b) wall shear stress. Courtesy of Professor Vitaliy L. Rayz, Purdue University Department of Biomedical Engineering.

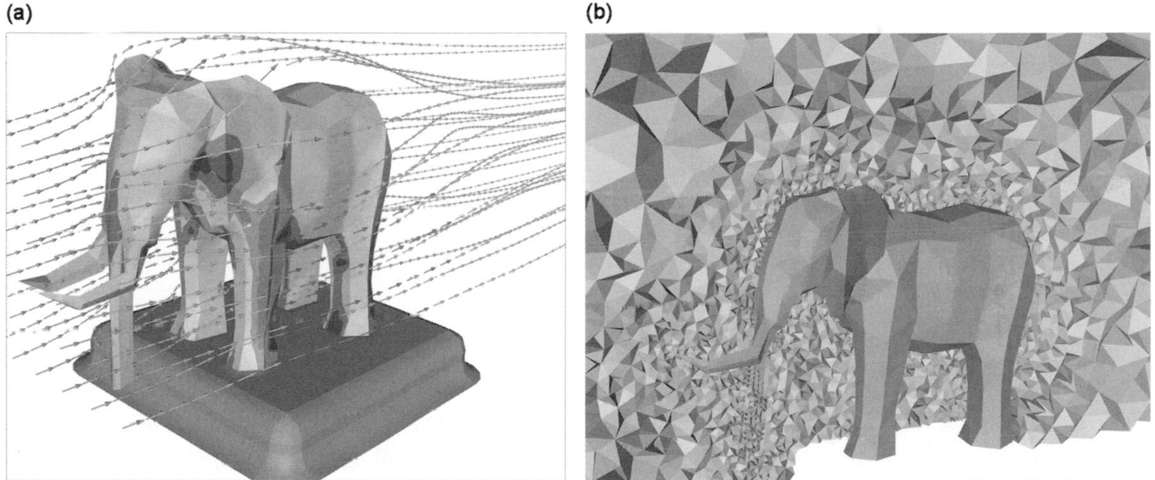

Figure 7.31 (a) CFD of flow over an elephant boundary shape showing streamlines. (b) Volume mesh for elephant CFD.

predictions. Heat transfer for animals affects their food needs to regulate body temperature and adapt, or not, to the wind and temperatures of their surroundings. In the spirit of more elephants, another CFD approach (Symscape) is shown in Figure 7.31(a) where streamlines are formed by end-to-end velocity vectors for a 5 m/s wind. Figure 7.31(b) shows the volume mesh for the elephant CFD in Figure 7.31(a). The statistics on the mesh are: 391,658 tetrahedral volume elements, 30,850 face elements and 73,694 notes using the solver OpenFOAM (incompressible with the k-omega SST turbulence model on Windows) run by Caedium software.

Summary

Steady viscous flow is a good starting point for basic fluid mechanics and especially biofluid mechanics. With an approach focusing on fundamental processes, several very important biological applications were discussed, including vascular flows, mucociliary clearance, two-phase flows in respiration which is a simple model of mucus clearance by airflow in the lung, and also blood flow with a cell-free layer near the blood vessel wall. In addition, the role of fluid mechanics in the evolution of vascular anatomy was explored and shows the impact of biofluid mechanical studies over a vast time period. We developed a solution for flow in a rectangular duct which is appropriate for three-dimensional channels as may be found in microfluidics devices, for example. Low Re, external flows over spheres, cylinders and arrays of cylinders were presented and the history of their development with increasingly sophisticated methods to create valid solutions. The power of CFD approaches was shown for cerebral berry aneurysms and elephants. Before we move on to additional flow phenomena, our next step is to introduce time-dependent flow of Newtonian fluids to develop our understanding of these important flows that we all know so well in our pulse and our breathing.

Problems

7.1

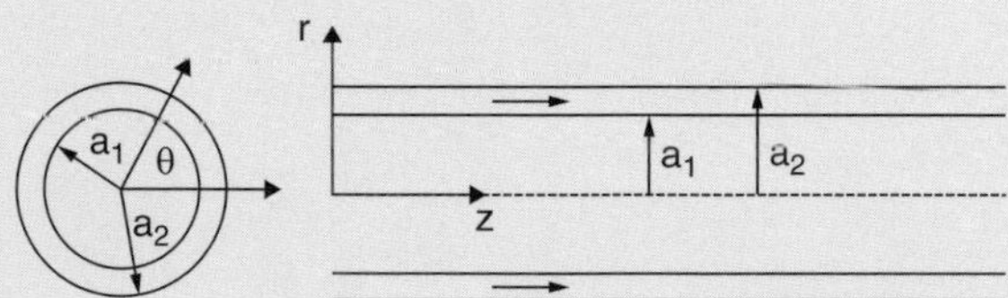

For indwelling catheters there is a flow between the catheter and the wall of the vessel. This may be an IV, or an arterial line, or even a left ventricular assist device called an intra-aortic balloon. To model the flow, consider steady, fully developed, unidirectional viscous flow between concentric cylinders. There is a constant pressure gradient, dp/dz, driving the flow for fluid which is in the gap $a_1 \le r \le a_2$, see the figure. There is no fluid for $0 \le r \le a_1$.

From the notes, a normal aortic pressure drop from Eq. (7.2.1) is 162 dyn/cm^2 over a length of 25 cm. So let $dp/dz = -162/25 = -6.5\,\text{dyn/cm}^3$. The aortic radius is $a_2 = 1.25$ cm, and the blood viscosity is $\mu = 0.04$ poise. The inner cylinder radius, a_1, will be that of the intra-aortic catheter device which can be varied.

a. Solve for the z-direction dimensional fluid velocity $u_z(r)$starting with Eq. (7.1.2) and applying no-slip at the two boundaries.
b. Scale u_z on $U_s = (-dp/dz)a_2^2/\mu$ so that $U_z = u_z/U_s$ and plot $U_z(R)$ where $R = r/a_2$ for $\lambda = a_1/a_2 = 0.2, 0.4, 0.6, 0.8$.
c. Solve for the maximum dimensional velocity, u_{max}. (Hint: use your knowledge of calculus) Now make u_{max} dimensionless using the velocity scale as in part (b), $U_{max} = u_{max}/U_s$ and plot U_{max} versus λ in the range $0.01 \le \lambda \le 0.99$.
d. Solve for the dimensional flow rate, q, and then scale it on $U_s\pi a_2^2$ so $Q = q/U_s\pi a_2^2$. Plot Q vs λ for the same range in part (c).
e. Withdrawing the catheter puts a constant z-velocity on the inner boundary, call it U_0. Repeat parts (a) and (b) except for the plot use only one value of $\lambda = a_1/a_2 = 0.4$, but three values of $\beta = U_0/U_s = 0,\ 0.3,\ 0.5$.

7.2

a. For flow due to the moving wall and gravity, Eq. (7.3.10), plot U(Y) for $U_w = -0.5, -0.25, 0, 0.25, 0.5, 0.75$ for $\theta = \pi/4$ and describe your results.

b. What is the criterion for all velocities to be positive for $0 \le \theta \le \pi/2$? Hint: what is the largest negative value of $Y^2/2 - Y$ and for what value of Y?

c. Calculate the volume flow rate, q, of the liquid layer. Let the z-axis unit length be L.

d. Calculate the dimensionless flow rate $Q = \mu q/\rho g h^3 L$. Plot Q vs $0 \le \theta \le \pi/2$ for the six values of U_w in part (a).

e. What is the criterion to have $Q > 0$ for $0 \le \theta \le \pi/2$? Is this the same or different from part (b) and why?

7.3

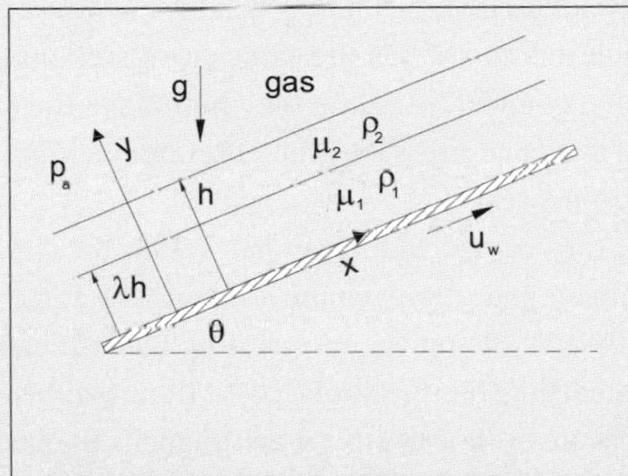

Consider a two-layer system with wall motion and gravity as shown in the figure. Fluid 1 (the serous layer) occupies $0 \le y \le \lambda h$ where $0 < \lambda < 1$. It has viscosity μ_1 and density ρ_1. Fluid 2 (the mucus layer) occupies $\lambda h \le y \le h$ and has viscosity μ_2 and density ρ_2. Repeating the initial integration of the Navier–Stokes equations, the x-direction velocities for each is

$$u_1 = \frac{\rho_1 g \sin\theta}{\mu_1}\frac{y^2}{2} + a_1 y + b_1 \quad 0 \le y \le \lambda h$$
$$u_2 = \frac{\rho_2 g \sin\theta}{\mu_2}\frac{y^2}{2} + a_2 y + b_2 \quad \lambda h \le y \le h \tag{1}$$

a. Solve u_1 and u_2 assuming $\rho_1 = \rho_2 = \rho$. You will need to impose the four boundary conditions: no-slip at the moving wall, $u_1(y=0) = u_w$; no-shear at the air–liquid interface, $\mu_2(du_2/dy)_{y=h} = 0$; continuous velocity at the liquid–liquid interface, $u_1(y=\lambda h) = u_2(y=\lambda h)$; and continuous shear stress at the liquid–liquid interface, $\mu_1(du_1/dy)_{y=\lambda h} = \mu_2(du_2/dy)_{y=\lambda h}$

b. Make your solutions dimensionless with the following scales

$$U_1 = \frac{u_1}{\rho g h^2/\mu_1}, U_2 = \frac{u_2}{\rho g h^2/\mu_1}, Y = \frac{y}{h} \tag{2}$$

and define two dimensionless parameters as $\beta = \mu_1/\mu_2$ and $U_w = u_w/(\rho g h^2/\mu_1)$. What are your dimensionless solutions?

c. For $\lambda = 1/2, \theta = \pi/4, U_w = 1$ plot the velocities $U_1(Y), 0 \le Y \le \lambda$ and $U_2(Y), \lambda \le Y \le 1$, for $\beta = 0.1, 1.0, 10$ on the same graph. Describe your results.

d. Plot the total dimensionless flow rate $Q = \int_0^\lambda U_1\,dY + \int_\lambda^1 U_2\,dY$ for $\lambda = 1/2, \theta = \pi/4$ and $\beta = 0.1, 1.0, 10$ over the wall velocity range $0 \le U_w \le 1$. Describe your results.

e. What is a typical value of β? What conditions or disease can change that value, discuss two? Repeat for λ.

7.4 For core-annular flow, see Figure 7.11, we derived the dimensionless velocity fields, U_{Z1}, U_{Z2} in Eq. (7.4.14). Let $\lambda = b/a$ and $\beta = \mu_2/\mu_1 = 1/\delta$.

a. Calculate the dimensional volumetric flow rates, q_1, q_2, for each fluid in its respective region using their definitions

$$q_1 = \int_0^{2\pi}\int_0^{b} u_{z1}\, r\,dr\,d\theta = 2\pi a^2 U_s \int_0^{\lambda} U_{Z1}\, R\,dR,$$
$$q_2 = \int_0^{2\pi}\int_b^{a} u_{z2}\, r\,dr\,d\theta = 2\pi a^2 U_s \int_\lambda^{1} U_{Z2}\, R\,dR$$

where $U_s = (-dp/dz)a^2/\mu_2$. Perform the integrations using Eq. (7.4.14).

b. Plot the ratio q_2/q_1 for $\lambda = 0.8, 0.9, 0.95$ in the range $0.1 \leq \beta \leq 10$. Place the three curves on the same graph with labels. Describe your results.

c. Calculate the clearance time, t_c, required to move the equivalent of all of fluid 2 out of a tube of length 6a. Scale t_c on $\mu_2/a(-dp/dz)$ and plot your dimensionless clearance time, $T_c = t_c a(-dp/dz)/\mu_2$, as a function of λ for $0.8 \leq \lambda \leq 0.99$. Describe what is happening to the clearance time.

d. As an experimentalist in blood flow, you decide to infer an "apparent viscosity," μ_{app}, of the total system by inverting Poiseuille's Law to solve for viscosity and measuring the total flow out of your tube apparatus, $q_{tot} = q_1 + q_2$. Define it as $\mu_{app} = \frac{\pi a^4}{8 q_{tot}}\left(-\frac{dp}{dz}\right)$. Plot μ_{app}/μ_1 for $\lambda = 0.8, 0.9, 0.95$ in the range $0.1 \leq \beta \leq 10$. Put these three curves on the same graph and label them. Do they share a common point? If so, why?

e. The cell-free layer is serum, region 2, with viscosity essentially that of water, while the whole blood in region 1 has a viscosity of about four times that of water. What is your error $(\mu_{blood} - \mu_{app})/\mu_{blood}$ as a function of λ? Plot it for $0.8 \leq \lambda \leq 0.99$.

7.5 Suppose metabolic consumption is proportional to tube surface area. Derive the relationship between the parent and daughter tube radii for symmetric branching.

7.6

a. For the velocity field $U(\eta, \zeta)$ in Eq. (7.7.31) plot contours of constant velocity for a square channel using a ten term series, $1 \leq n \leq 10$. Choose contour values $U = 0.28, 0.25, 0.20, 0.15, 0.10, 0.05, 0.001$. Describe the relative shapes of the contours.

b. Plot velocity profiles of the flow in part (a) in the η-direction for $\zeta = 0.00, 0.70, 0.95, 0.99$. What is happening to U and $\partial U/\partial \eta$ near $\eta = \pm 1$. Why?

c. A dimensionless wall shear related to part (b) is $\tau_w = (\partial U/\partial \eta)|_{\eta=1}$. Plot it as a function of ζ and describe your result. How does it compare to your answer in part (b)?

d. Equation (7.7.31) has the form $U = \sum_1^\infty u_n$. For arbitrary λ calculate the dimensionless flow rate contribution from the nth component, u_n, and call it q_n. Integrate η, ζ directly over the upper right quadrant which, from symmetry, will be 1/4th of q_n. This is best done by pencil.

e. The total dimensionless flow rate, then, is $Q = \sum_1^\infty q_n$ from part (d). For finite calculations, let $Q = \sum_1^N q_n$ and plot Q for N = 1 and N = 10 over the range $0.01 \leq \lambda \leq 2$. What is the relative accuracy of the two approximations?

f. The dimensional channel flow rate is $Q^* = ((-dp/dx)b^4/\mu)Q$. For a circular cylinder with radius *a*, the dimensional flow rate is given by Poiseuille's Law, call it Q_P^*. What is the ratio Q^*/Q_P^* when the square has the same cross-sectional area as the cylinder? Assume they both have the same viscosity and pressure gradient. Comment on your answer.

g. Repeat your calculation in part (c) for the circular cylinder and square channel having the same perimeter. Biological construction (growth and remodeling) has metabolic costs. If the perimeter is a measure of those costs, which shape is the better option? How does that compare to typical anatomy?

7.7 The Fourier series solution for flow in rectangular channel, $U(\eta, \zeta)$, is given by Eq. (7.7.31), where $U = u/U_s, \eta = y/b, \zeta = z/b$ and $U_s = (-dp/dx)b^2/\mu$. For a particle dropped at $x = 0, y = b, z = 0$, its x-equation of motion using a one-term series is

$$\frac{dx_p}{dt} = U_s \frac{(-1)^{1+1}}{k_1^3}\left(1 - \frac{1}{\cosh(k_1\lambda)}\right)\cos\left(k_1 \frac{y_p}{b}\right)$$

where we are assuming that the particle travels at the same speed as the fluid in the x-direction.

The y position of the particle is based on its terminal velocity, v_t,

$$\frac{dy_p}{dt} = -v_t$$

a. Scale time on b/v_t so that you have a dimensionless time, $T = v_t t/b$. Solve for the particle trajectories $X_p(T) = x_p/L$ and $Y_p(T) = y_p/b$. Hint: start with the y equation which is uncoupled.
b. You should end up with a dimensionless parameter, $\beta = bU_s/Lv_t$. For $\beta = 1$, plot trajectories for $\lambda = 0.5$, 1.0, 2.0, 3.0, 5.0 until they impact on the lower wall. Discuss the dependence on λ.
c. With regard to part (b), if there is added expense as λ increases, what value would you choose to give the longest axial displacement for the lowest cost?
d. Let there be a splitter plate in the x–z plane for $x \geq L, (X \geq 1)$ which divides the channel into an upper half and lower half. Consider two cell types, 1 and 2, where type 2 is twice the diameter of type 1, so the ratio of terminal velocities is $v_{t2} = 4v_{t1}$ and then $\beta_2 = \beta_1/4$. Plot the two cell trajectories to their impact on the lower wall, X_{p1} vs Y_{p1} (type 1) and X_{p2} vs Y_{p2} (type 2) for $\beta_1 = 1$ and place them on the same graph. Make sure in your plots you can "see" the entire larger channel. Let $\lambda = 1$.
e. Now adjust β_1 to get the best separation at $x = L$, so that cell type 1 enters the upper channel and cell type 2 enters the lower channel with a comfortable margin of error.
f. Repeat part (e) for $\lambda = 3$ and discuss the difference in your results.

7.8 The stream function, $\psi(r, \theta)$, the dimensionless pressure, $P(r, \theta)$, and the velocity components $\underline{u} = u_r(r, \theta)\underline{e}_r + u_\theta(r, \theta)\underline{e}_\theta$ for flow over a sphere are derived in the text.

a. Set $a = 1$, $U_\infty = 1$ and create a contour plot of ψ for $1 \leq r \leq 4, 0 \leq \theta \leq 2\pi$. Choose the contours for $\psi = 0, 0.05, 0.5, 1, 1.5, 2, 3$. You may have to call for a higher number of points calculated for good resolution. Also create a field plot for the velocity vector field over the same domain. Plot these two together on one graph. Discuss your result.
b. Create a contour plot of the speed, s, for $1 \leq r \leq 4, 0 \leq \theta \leq 2\pi$ and choose the following values: $s = 0.1, 0.3, 0.5, 0.6, 0.7, 0.75, 0.78, 0.82, 0.87$. Use color shading for s. What do you see in terms of symmetry?
c. Create a contour plot of the pressure, P, for $1 \leq r \leq 4, 0 \leq \theta \leq 2\pi$ and choose the following contour pressure values: $P = \pm 1, \pm 0.7, \pm 0.5, \pm 0.2, \pm 0.3, \pm 0.15, \pm 0.1, \pm 0.07, \pm 0.05$. Use color shading for P. Which side has the higher pressures? Why?
d. Create a contour plot of the vorticity ω_ϕ over the region $1 \leq r \leq 4, 0 \leq \theta \leq 2\pi$. Choose contour values $\omega_\phi = \pm 1, \pm 0.7, \pm 0.5, \pm 0.3, \pm 0.2, \pm 0.1, \pm 0.05$. Describe the upper half versus the lower half of your plot in terms of local fluid rotation, i.e. clockwise or counterclockwise are + or – values of ω?
e. Use the relationship between the polar unit vectors $\underline{e}_r$, $\underline{e}_\theta$ in terms of the rectangular unit vectors $\underline{e}_x$, $\underline{e}_y$ to derive the velocity field as $\underline{u}(x, y) = u_x\underline{e}_x + u_y\underline{e}_y$. You will need to substitute $r = (x^2 + y^2)^{1/2}$ and $\theta = \arctan(y/x)$ as a final step. The flow is in the positive x-direction. Create the following five curves which are shifted to make it easier to visualize: u_x for $x = 0.01, 1 \leq y \leq 5$; $u_x + 0.707$ for $x = 0.707, 0.707 \leq y \leq 5$; $u_x + 1$ for $x = 1, 0.01 \leq y \leq 5$; $u_x + 1.5$ for $x = 1.5, 0.01 \leq y \leq 5$; and $u_x + 2$ for $x = 2, 0.01 \leq y \leq 5$. Also, plot the surface of the sphere (circle) in the upper right quadrant. Plot these all on the same graph and explain the result. Why didn't we use $x = 0$, $y = 0$ in the plots?

Image Credits

Figure 7.2 Source: Figure 14 of Dewey CF, Bussolari SR, Gimbrone MA, Davies PF. The dynamic-response of vascular endothelial-cells to fluid shear-stress. *Journal of*

Biomechanical Engineering – Transactions of the ASME. 1981; 103(3):177–85.

Figure 7.3 Source: Anatomy and Physiology, Connexions website http://cnx.org/content/col11496/1.6/

Figure 7.4 Source: Ivanna Olijnyk/ Getty

Figure 7.5 Source: National Cancer Institute, National Institutes of Health http://training.seer.cancer.gov/anatomy/cardiovascular/blood/classification.html

Figure 7.8(a) Source: National Cancer Institute

Figure 7.9 Source: pablofdezr/Shutterstock

Figure 7.29 Source: Lucien Monfils, licensed under GNU Free Documentation License https://commons.wikimedia.org/wiki/File:Aneurysem.jpg

Figure 7.31(a) Source: Original Image Courtesy of Symscape www.symscape.com/blog/elephants-computational-fluid-dynamics

Figure 7.32(b) Source: Original Image Courtesy of Symscape www.symscape.com/blog/elephants-computational-fluid-dynamics

8 Unsteady Newtonian Viscous Flow

Introduction

In Chapter 7 we studied several aspects of steady flows in biological applications. However, many important biofluid mechanics situations involve time-dependent, unsteady flows. For example, blood is driven by the beating heart which imparts both a steady and time-periodic pressure and flow into the arterial tree. The human respiratory system is essentially purely time periodic, because the air goes in and out without an imposed steady contribution. Bird flapping and fish swimming are also unsteady phenomena. Whether a flow is steady or unsteady can depend on the frame of reference. For example, the falling ball/particle at low Reynolds number, which we examined in Section 2.7, can achieve a terminal velocity, v_t, that is constant. From the ball frame, the flow coming towards it becomes steady, since the weight equals the drag. A constant drag is a good indication that the velocity field, which determines the drag, is constant in that frame. However, an observation point fixed to an Eulerian (laboratory) frame, but along the trajectory of the ball, registers an unsteady velocity as it changes from zero to nonzero and back again as the ball passes by.

Flows can also become unsteady while being forced with purely steady inputs. That transition often involves some form of fluid instability, like the development of turbulence in pipe flow for large enough Reynolds number, Re. Another example is flow-induced vibrations of elastic boundaries. That occurs during expiratory airflow over the vocal cords causing phonation, i.e. speech. Wheezing breath sounds, common to asthma and emphysema patients, also fall in this category (Gavriely et al., 1984). Another type of boundary related instability is seen when a child blows through a straw into a glass of milk. The flow exiting the straw results in a string of bubbles, rhythmically pinching off the air into separate packets from a surface tension instability known as the Plateau–Rayleigh instability (Rayleigh, 1892b), which we analyze in Section 16.3. In that case, the boundary is a fluid–fluid boundary at the air–liquid interface that creates the surface tension. Great glee is derived from making the process audible.

In the Navier–Stokes equations, the effect of unsteadiness is included in the unsteady inertial term, $\rho\partial\underline{u}/\partial t$. In Section 6.6 we made the Navier–Stokes equations dimensionless, Eqs. (6.6.5) and (6.6.6), and found a dimensionless parameter which multiplies this term. For one set of scales that coefficient is the Womersley parameter $\alpha = b(\omega/\nu)^{1/2}$, the ratio of unsteady inertia to viscous stresses. The size of α helps us to gauge the overall influence of the unsteady momentum contribution. For large $\alpha >> 1$, it can be dominant, while for small $\alpha >> 1$, it can be negligible so that the phenomenon is quasi-steady. Since b is a length scale, the smaller the structures, like beating cilia which are on the order of microns in size, the smaller the value of α. For a different set of scales the key dimensionless parameter was the Strouhal number, $St = fb/\bar{u}$, the ratio of unsteady to convective effects. St is related to α through the Reynolds number, $St = \alpha^2/2\pi Re$.

Topics Covered

The first flow situation investigated is the well-known problem of a suddenly started wall from rest to a constant speed in its plane. This is known as Stokes' first problem, and permits us to understand the fundamental concepts of unsteady effects in a viscous fluid. The fluid above the wall gradually is dragged to the new speed, but it takes time for that to happen depending on the parameters of density and viscosity. To solve the problem, a similarity solution method is introduced which allows us to cast a partial differential equation into an ordinary differential equation. Adding a second wall which is stationary and parallel allows an example of using numerical methods. Then we tackle Stokes' second problem, which is the same infinite wall with fluid above as his first problem. However, in the second Stokes' problem the wall oscillates back and forth along the x-axis, introducing the frequency $\omega = 2\pi f$ as a forcing parameter. To solve this problem we use the separation of variables technique which is appropriate after transients die out.

In contrast to the two Stokes' problems, which generate unsteady flow from boundary motion, next we explore physiological applications where the flow is generated by unsteady pressure gradients: pulsatile flow, as found in arteries, and oscillatory flow, as found in lung airways. First these are addressed as flows in channels using the separation of variables approach. The advantage for this geometry is that the spatial part of the solution is readily described by functions known to the student, the hyperbolic functions sinh and cosh. One of the important clinical respiratory applications for oscillatory flow is small tidal volume, high frequency ventilation like a dog panting. The discussion includes the flow interaction with a diffusible solute, like oxygen in air, and mechanisms of enhanced delivery in neonatal mechanical ventilators are discussed. Then we repeat the oscillatory fluid mechanics analysis for a circular tube. In that case the spatial component depends on Bessel Functions, which are tabulated special functions that are likely new to the student. Hopefully there is better footing having done the channel solutions first, so the student can focus on the results, which are qualitatively similar, and get their bearings. Adding a steady background flow to the oscillatory motion gives us pulsatile tube flow, which we analyze and compare to measured velocities in arteries and its Fourier modes.

Finally we formulate a simple cough model for transporting mucus out of the lung. Coughing is a very important mechanism for cleansing the lung and the inability to cough effectively, as in neuromuscular diseases or in post-surgical patients with extreme pain of the abdomen or chest, can lead to frequent bouts of pneumonia and respiratory insufficiency.

8.1 Stokes' First Problem

Consider a fluid occupying the half-space $y \geq 0$ with a wall at $y = 0$, see Figure 8.1. For $t < 0$, the wall velocity is zero, $u_w = 0$, and fluid is motionless. However, for $t \geq 0$ the wall moves to the right at speed $u_w = u_0$. Far from the wall, $y \to \infty$, the fluid is still motionless. This unsteady flow problem was first solved by George Gabriel Stokes (1819–1903) in 1851. Stokes held the Lucasian Chair of Mathematics at the University of Cambridge, a professorship held earlier by Sir Isaac Newton. It is called Stokes' first problem, one of his several solutions to problems involving the motion of pendulums and the effects of fluid drag on them as we studied in Section 7.8 (Stokes, 1851). It is also called the Rayleigh problem. We are interested in how the shearing of the fluid from the wall motion propagates through the fluid in the y-direction.

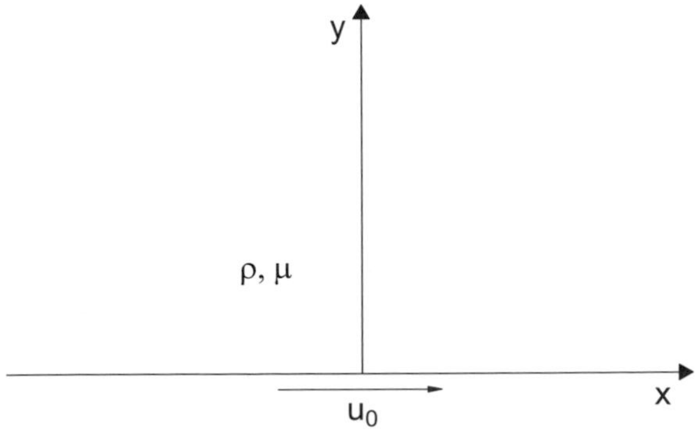

We anticipate that the resulting time-dependent flow will be unidirectional in the x-direction, $u(y, t)$, and there is no imposed pressure gradient. The x-component of the Navier–Stokes equation simplifies to

$$\frac{\partial u}{\partial t} = \nu \frac{\partial^2 u}{\partial y^2} \tag{8.1.1}$$

The boundary conditions are

$$\begin{aligned} &u(y, t = 0) = 0, \quad u(y = 0, t) = u_0 \\ &u(y \to \infty, t) = 0 \end{aligned} \tag{8.1.2}$$

The wall motion can be thought of as related to the Heaviside function, H(t), or "step function" so that the boundary condition is $u_w = u_o H(t)$ where $H(t < 0) = 0, H(t \geq 0) = 1$.

Figure 8.1 Stokes' first problem of a suddenly started wall at speed u_0 for $t \geq 0$.

Equation (8.1.1) is an unsteady diffusion equation that has the same form as an unsteady heat equation. The diffusion in our case is the diffusion of vorticity or shear from the wall into the fluid. To solve this problem, we use a method called a "similarity solution." This approach works for a number of problems where there is a boundary condition at infinity, as we have here. The idea is to define a new independent variable which is a combination of the original two independent variables, and seek to express the partial differential equation as an ordinary differential equation in the new variable. To this end, we define a new independent variable, η, as

$$\eta = k\frac{y}{t^a} \tag{8.1.3}$$

where the unknown exponent a will be determined to achieve our goal of obtaining an ODE. The idea will be to find $u(\eta)$. From the chain rule we know that the derivatives which appear in Eq. (8.1.1) can be expressed as

$$\begin{aligned}
\frac{\partial}{\partial t} &= \frac{\partial \eta}{\partial t}\frac{d}{d\eta} = -akyt^{-a-1}\frac{d}{d\eta} = -a\eta t^{-1}\frac{d}{d\eta} \\
\frac{\partial}{\partial y} &= \frac{\partial \eta}{\partial y}\frac{d}{d\eta} = kt^{-a}\frac{d}{d\eta} \\
\frac{\partial^2}{\partial y^2} &= kt^{-a}\frac{d}{d\eta}\left(kt^{-a}\frac{d}{d\eta}\right) = k^2t^{-2a}\frac{d^2}{d\eta^2}
\end{aligned} \tag{8.1.4}$$

Using these forms of the derivatives in the new variable, Eq. (8.1.1) becomes

$$-a\eta t^{-1}\frac{dU}{d\eta} = \nu k^2 t^{-2a}\frac{d^2U}{d\eta^2} \tag{8.1.5}$$

where we define the dimensionless velocity as

$$U(\eta) = \frac{u(y,t)}{u_0} \tag{8.1.6}$$

We see that choosing $a = 1/2$ cancels the t variable from Eq. (8.1.5) and leaves us with an ODE in η, which is our goal,

$$-\frac{1}{2}\eta\frac{dU}{d\eta} = \nu k^2\frac{d^2U}{d\eta^2} \tag{8.1.7}$$

It is convenient to choose $k = 1/(2\sqrt{\nu})$, which reduces Eq. (8.1.7) to

$$\frac{d^2U}{d\eta^2} + 2\eta\frac{dU}{d\eta} = 0 \tag{8.1.8}$$

where the final form of the similarity variable is

$$\eta = \frac{y}{2\sqrt{\nu t}} \tag{8.1.9}$$

Equation (8.1.8) is subject to the dimensionless boundary conditions

$$\begin{aligned}
U(\eta = 0) &= 1 \\
U(\eta \to \infty) &= 0
\end{aligned} \tag{8.1.10}$$

We have experienced this system before in Section 4.5 for the time-independent concentration field associated with inviscid stagnation point flow. The solution involves the error function Eq. (4.5.25). The major difference is that the variable η now has both y and t dependence. Equation (8.1.8) can be thought of as a first-order ODE for the variable $f = dU/d\eta$, that is,

$$\frac{df}{d\eta} + 2\eta f = 0 \tag{8.1.11}$$

Separating variables for Eq. (8.1.11) and integrating once gives us

$$\begin{aligned}
&\int\frac{df}{f} = -2\int\eta\, d\eta + c_0 \;\rightarrow\; \ln f = -\eta^2 + c_0 \;\rightarrow \\
&f = e^{(c_0-\eta^2)} = c_1 e^{-\eta^2}
\end{aligned} \tag{8.1.12}$$

where $c_1 = e^{c_0}$ is the constant of integration. We can express Eq. (8.1.12) in terms of U,

$$\frac{dU}{d\eta} = c_1 e^{-\eta^2} \tag{8.1.13}$$

and integrate again to find

$$U(\eta) = c_1\int_0^{\eta} e^{-\eta'^2}\, d\eta' + c_2 \tag{8.1.14}$$

The integral of Eq. (8.1.14) does not have an analytical solution, and is related to the error function which is tabulated. The two constants of integration, c_1 and c_2, are determined from the boundary conditions of Eq. (8.1.2). The first condition is at the wall

$$U(\eta = 0) = 1 = c_2 \tag{8.1.15}$$

while the second boundary condition is at infinity

$$U(\infty) = c_1 \int_0^\infty e^{-\eta'^2}\, d\eta' + 1 = 0 \tag{8.1.16}$$

The definite integral of Eq. (8.1.16) is known to have a solution

$$\int_0^\infty e^{-\eta'^2}\, d\eta' = \frac{\sqrt{\pi}}{2} \tag{8.1.17}$$

which solves for c_1 in Eq. (8.1.16) $c_1 = -2/\sqrt{\pi}$. The final result is

$$U(\eta) = 1 - \frac{2}{\sqrt{\pi}} \int_0^\eta e^{-\eta'^2}\, d\eta' \tag{8.1.18}$$

The error function is a tabulated special function defined in Eq. (4.5.22) and our flow solution can be represented by

$$U(\eta) = 1 - \mathrm{erf}(\eta) = \mathrm{erfc}(\eta) \tag{8.1.19}$$

where erfc is the co-error function.

A plot of $U(\eta)$ is shown in Figure 8.2. It is a compact way to show the solution, but switching back to y and t variables allows us a better vantage point to see what is going on.

To see the plot in the y coordinate for different times, choose $\nu = 1\ \mathrm{cm^2/s}$ and plot U(y) for different values of t, as shown in Figure 8.3(a). Recall that these

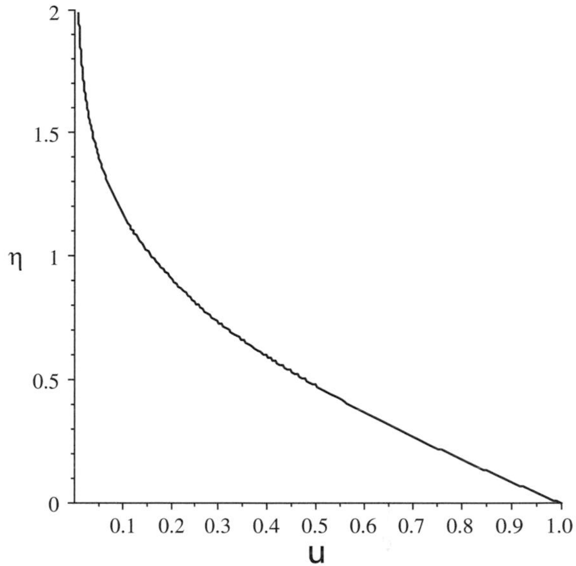

Figure 8.2 $U(\eta)$ for Stokes' first problem.

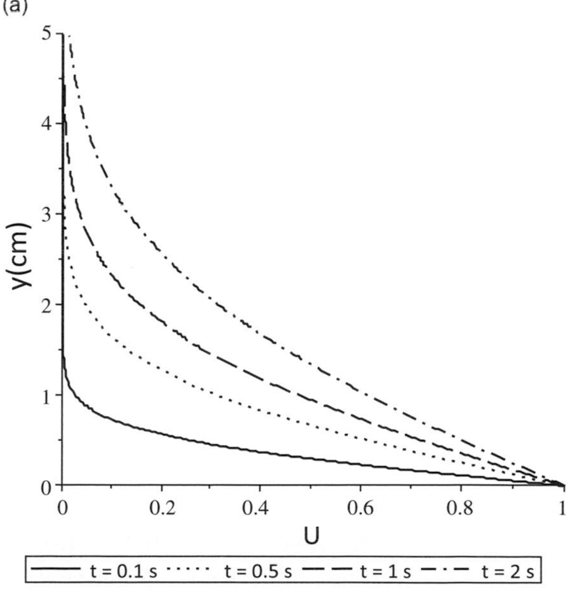

Figure 8.3 (a) U(y) for $\nu = 1\ \mathrm{cm^2/s}$ at different values of t. (b) Velocity vectors for $t = 1$ s.

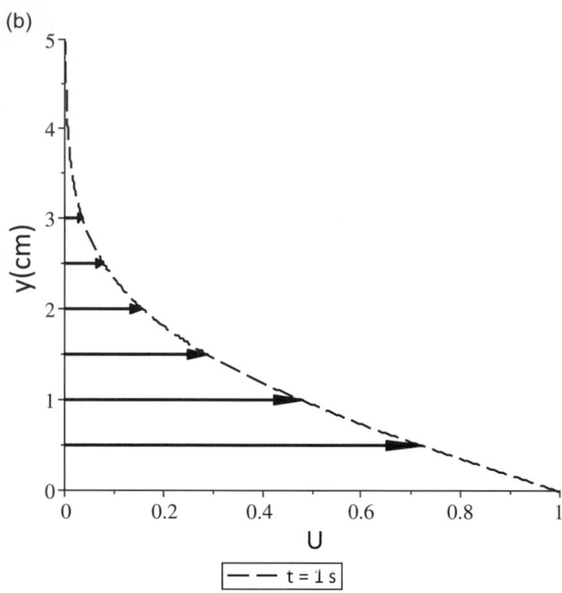

Figure 8.3 *(continued)*

curves are the velocity profile, the tips of the x-velocity vectors as they vary with y and t, see Figure 8.3(b). As time increases, the velocity profile extends further into the fluid as the shear (vorticity) propagates from the wall. Since Eq. (8.1.1) is a "diffusion" equation for the diffusion of shear, we can think of the kinematic viscosity, ν, as the related diffusion coefficient.

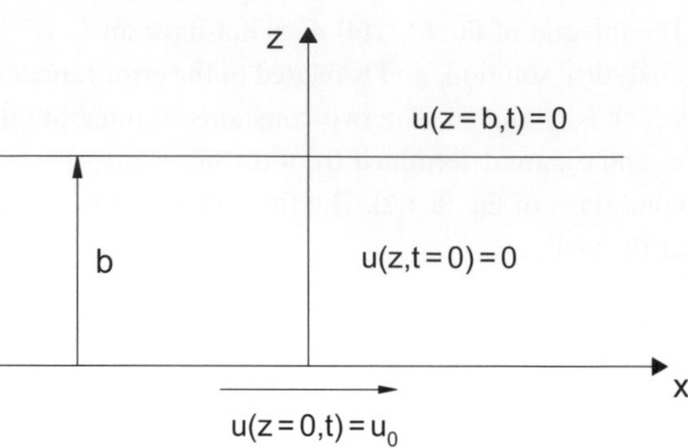

Figure 8.4 Stokes' first problem with a stationary, parallel second wall.

Because the condition at infinity is zero velocity, the effect is felt everywhere in the fluid once the wall starts to move. To get an appreciation of how deep the majority of the effect is felt, however, we select a value of η that yields a finite velocity which is close to zero. For example, $\eta = 1.819$ yields $U = 0.01$. Then using the definition of η in Eq. (8.1.9), the corresponding value of y where $U = 0.01$, or $u = 0.01\,u_0$, is $y_{0.01} = 3.638\sqrt{\nu t}$. So $y_{0.01}$ propagates into the fluid at a distance proportional to $\sim \sqrt{\nu t}$, with velocity proportional to its time derivative $\sim \frac{1}{2}\sqrt{\nu/t}$.

Box 8.1 Stokes' First Problem with a Parallel, Stationary Second Wall – Numerical Solution

Now consider Stokes' first problem with a stationary second wall which is parallel to the moving wall and adds a no-slip boundary condition at the location $z = b$. Equation (8.1.1) still governs the flow but the initial and boundary conditions are

$$\begin{aligned} &u(y, t=0) = 0, u(y=0, t) = u_0 \\ &u(y=b, t) = 0 \end{aligned} \tag{8.1.20}$$

Scaling the problem as $U = u/u_0, \eta = y/b, T = \nu t/b^2$, the dimensionless system becomes

$$\begin{aligned}
&\frac{\partial U}{\partial T} = \frac{\partial^2 U}{\partial \eta^2} \\
&U(\eta, T = 0) = 0, U(\eta = 0, T) = 1 \\
&U(\eta = 1, T) = 0
\end{aligned} \tag{8.1.21}$$

The steady state solution of Eq. (8.1.21) clearly is the linear profile $U^s = 1 - \eta$. To solve the unsteady problem, we can no longer use the similarity variable approach since the zero-velocity boundary condition is no longer at infinity. An analytical approach involves the use of Laplace transforms of the governing equations. However, computational software often includes a numerical solution method for the unsteady heat/diffusion equation. An example using Maple is shown in Code 8.1(a), where the default method is a second-order (in space and time) centered, implicit finite difference scheme. The equivalent MATLAB® code is shown in Code 8.1(b).

```
(a)
restart;
PDE := diff(U(eta, T), T) = diff(U(eta, T), eta, eta);
```

$$PDE := \frac{\partial}{\partial T} U(\eta, T) = \frac{\partial^2}{\partial \eta^2} U(\eta, T) \qquad (1)$$

```
IBC := {U(eta, 0) = 0, U(0, T) = 1, U(1, T) = 0};
```

IBC := {U(0, T) = 1, U(1, T) = 0, U(η, 0) = 0} (2)

```
sol := pdsolve(PDE, IBC, numeric, spacestep = 0.001);
```

sol := **module**() ... **end module** (3)

```
PL1 := sol[plot](T = 0.01, linestyle = 2, legend = "T = 0.01", color = black):
PL2 := sol[plot](T = 0.05, linestyle = 3, legend = "T = 0.05", color = black):
PL3 := sol[plot](T = 0.1, linestyle = 4, legend = "T = 0.1", color = black):
PL4 := sol[plot](T = 0.2, linestyle = 5, legend = "T = 0.2", color = black):
PL5 := plot(1 - eta, eta = 0..1, linestyle = 1, color = black, legend = "steady solution"):
with(plots):
display(PL1, PL2, PL3, PL4, PL5, labels = [η, "U"], labelfont = [HELVETICA, 12]);
```

```
(b)
% This MATLAB script requires that the files"Chapter8_pdeSystem8_1_21.m",
% "Chapter8_bcSystem8_1_21.m"and "Chapter8_icSystem8_1_21.m"
% is located in the same folder of this script
clear all; close all; clc
x = linspace(0, 1, 100);
t = 0:0.001:0.2;
```

Code 8.1 Code examples: (a) Maple, (b) MATLAB®.

```
m  = 0;
sol =
pdepe(m,@Chapter8_pdeSystem8_1_21,@Chapter8_icSystem8_1_21,@Chapter8_bcSystem8_1
_21,x,t);
U  = sol(:,:,1);
eta = x;
plot(eta,U(20,:),':k'), hold on;
plot(eta,U(50,:),'--k'),
plot(eta,U(100,:),'-.k'), hold on;
plot(eta,U(200,:),'-k'), hold on;
plot(eta,1-eta,'-r'), hold on;
l = legend('T=0.01','T=0.05','T=0.1','T=0.2','steady solution','location','northeast');
l.FontSize = 11;
l.FontName = 'Arial';
xlabel('U')
ylabel('\eta')
axis([0 1 0 1])
ax = gca;
ax.FontSize = 11;
ax.FontName = 'Arial';
```

The following MATLAB function must be saved as "Chapter8_pdeSystem8_1_21.m"

```
functiondUdT = Chapter8_pdeSystem8_1_21(T,eta,U)
dUdT  = diff(diff(U,eta),eta);
end
```

The following MATLAB function must be saved as "Chapter8_bcSystem8_1_21.m"

```
function [pl,ql,pr,qr] = Chapter8_bcSystem8_1_21(xl,ul,xr,ur,t)
  pl = ul-1;
ql = 0;
pr = ur;
qr = 0;
end
```

The following MATLAB function must be saved as "Chapter8_icSystem8_1_21.m"

```
function u0 = Chapter8_icSystem8_1_21(x)
  u0 = 0*x;
end
```

Code 8.1 *(continued)*

Plots of the solution are shown in Figure 8.5 showing that the profile transient approaches the steady state solution. $U(\eta)$ is plotted for $T = 0.01, 0.05, 0.1, 0.2$ and the steady solution.

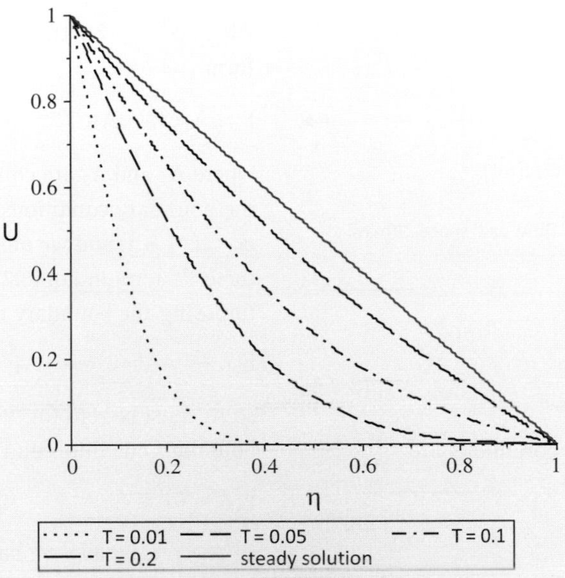

Figure 8.5 Numerical solution of Stokes' first problem with a second wall.

8.2 Stokes' Second Problem

Consider a fluid in the half-space $z \geq 0$, and the wall at $z = 0$ is oscillating in its own plane with velocity $u_w = u_0 \cos(\omega t)$, see Figure 8.6. We assume that the resulting fluid velocity only has an x-component and that there is no applied pressure gradient. The resulting form of the Navier–Stokes equation is the same as for Stokes' first problem

$$\frac{\partial u}{\partial t} = \nu \frac{\partial^2 u}{\partial z^2} \tag{8.2.1}$$

However, now the boundary conditions are

$$\begin{aligned} &u(z = 0, t) = u_0 \cos(\omega t) \\ &u(z \to \infty, t) = 0 \end{aligned} \tag{8.2.2}$$

We define the scaled variables as

$$U = \frac{u}{u_0}, T = \omega t, Z = \frac{z}{L_s} \tag{8.2.3}$$

where we will determine the length scale, L_s, to simplify the problem. The variable Z in Eq. (8.2.3) is unrelated to the one in Eq. (8.1.9). Inserting the forms of Eq. (8.2.3) into Eq. (8.2.1) leads to the dimensionless momentum equation

$$\frac{\omega L_s^2}{\nu} \frac{\partial U}{\partial T} = \frac{\partial^2 U}{\partial Z^2} \tag{8.2.4}$$

To eliminate the parameters, we choose $L_s = \sqrt{2\nu/\omega}$, which makes

$$Z = \frac{z}{\sqrt{2\nu/\omega}} \tag{8.2.5}$$

The factor of $\sqrt{2}$ in our choice of L_s at this stage simplifies the final form of the solution. The dimensionless governing equation is

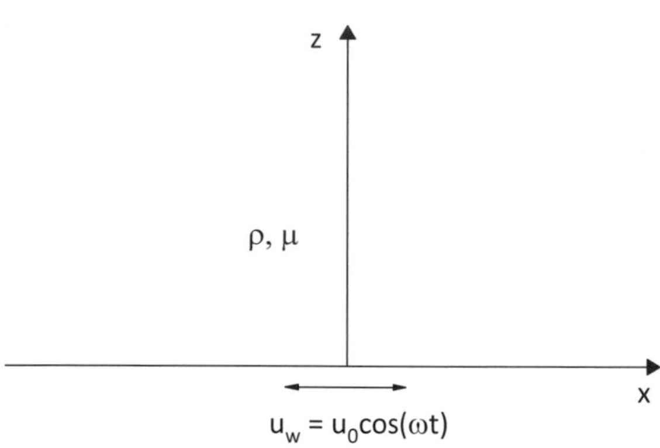

Figure 8.6 Oscillating wall with a fluid half-space, Stokes' second problem.

$$\frac{\partial^2 U}{\partial Z^2} - 2\frac{\partial U}{\partial T} = 0 \tag{8.2.6}$$

The dimensionless boundary conditions are

$$\begin{aligned} U(Z=0,T) &= \cos T = \Re\{e^{-iT}\} \\ U(Z\to\infty,T) &= 0 \end{aligned} \tag{8.2.7}$$

Recall that $e^{\pm iT} = \cos T \pm i\sin T$ and $\Re\{\}$ means the real part of the bracketed term.

To solve the partial differential equation, Eq. (8.2.6), we use the technique of separation of variables. This technique assumes that a function of two variables, say, $f(x,t)$, can be represented as the product of two functions, each depending only on one of the variables, i.e. $f(x,t) = g(x)\,h(t)$. Clearly such an assumption would not be correct for Stokes' first problem in Section 8.1, where the solution method depends on combining the two independent variables into one similarity variable. For Stokes' second problem, however, the time dependence is characterized by the oscillatory wall, so we expect that, after any transients due to startup, which are related to Stokes' first problem as we saw, the solution for the velocity field will be oscillatory with the same frequency as the wall, though the phase will be Z -dependent

Let F(Z) be a complex function, $F = F_r + iF_i$, where F_r is the real part of F and F_i is its imaginary part. We propose the solution to U as

$$U(Z,T) = \Re\{F(Z)\,e^{-iT}\} \tag{8.2.8}$$

Using Eq. (8.2.8) in Eq. (8.2.6) gives us

$$\frac{d^2F}{dZ^2} - \sigma^2 F = 0 \tag{8.2.9}$$

where $\sigma^2 = -2i$. The solution of Eq. (8.2.9) is of the form

$$F = A_1 e^{\sigma Z} + A_2 e^{-\sigma Z} \tag{8.2.10}$$

where A_1 and A_2 are constants to be determined from the boundary conditions. The square root of σ^2 is $\sigma = \pm(1-i)$ and we only need the positive root since the $e^{-\sigma Z}$ term in Eq. (8.2.10) serves the negative root. Imposing the boundary condition at infinity,

$$F(Z\to\infty) = 0 \quad \to \quad A_1 = 0 \tag{8.2.11}$$

so A_1 must be zero for the solution to be bounded. The boundary condition on the wall is

$$F(Z=0) = A_2 = 1 \tag{8.2.12}$$

so $A_2 = 1$. Finally we have

$$\begin{aligned} U(Z,T) &= \Re\{e^{-\sigma Z}e^{-iT}\} = e^{-Z}\Re\left\{e^{-i\,(T-Z)}\right\} \\ &= e^{-Z}\cos(T-Z) \end{aligned} \tag{8.2.13}$$

which is a time-periodic velocity in the x-direction, with a phase angle Z compared to the wall, and the amplitude of the velocity decreases exponentially with Z.

A plot of U for various times in the cycle is shown Figure 8.7. Most of the change in velocity occurs within the region $Z \le 1$. When $Z = 1$, the dimensional z value is $z = \sqrt{2\nu/\omega}$. This distance is often called the "Stokes layer," $\delta_S = \sqrt{2\nu/\omega}$, though the factor of $\sqrt{2}$ is sometimes omitted. We chose it to be L_s in our scalings. It decreases with smaller kinematic viscosity and larger frequency. The Stokes layer is a type of boundary layer, the region near a boundary where rapid changes occur in the velocity profile to match the boundary condition. Some authors choose $Z = 2\pi$, the distance to the next plane where the fluid is in-phase with the wall, as the definition of the Stokes layer thickness.

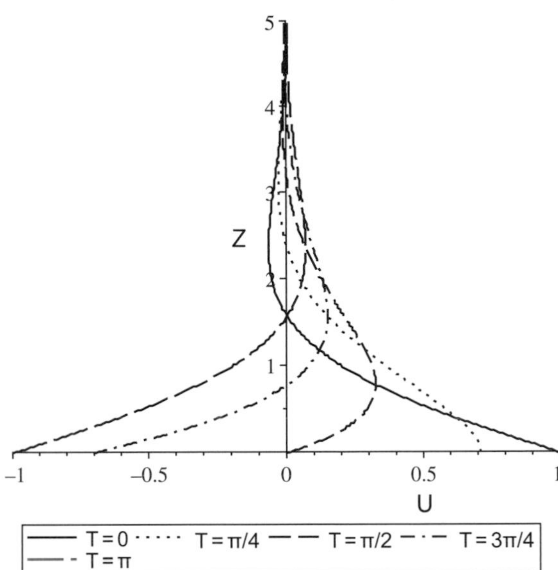

Figure 8.7 Stokes' second problem, U(Z), for five values of T.

8.3 Oscillatory Flow in a Channel

Consider oscillatory flow in a channel as shown in Figure 8.8. The x-pressure gradient is given as the sinusoidal function

$$\frac{\partial p}{\partial x} = B^p \cos(\omega t) \tag{8.3.1}$$

where B^p is the amplitude of the time-periodic pressure gradient and ω is the forcing frequency. For $v = 0$ the remaining terms in the x-component Navier–Stokes equations are

$$\frac{\partial u}{\partial t} = -\frac{1}{\rho}\frac{\partial p}{\partial x} + \nu\frac{\partial^2 u}{\partial y^2} = -\frac{1}{\rho}B^p\cos(\omega t) + \nu\frac{\partial^2 u}{\partial y^2} \tag{8.3.2}$$

The boundary conditions are no-slip at the upper and lower wall, or equivalently, no-slip at the upper wall and symmetry at the midline

$$u(y = \pm b) = 0 \quad \text{or} \quad \begin{cases} u(y = b) = 0 \\ \left.\dfrac{\partial u}{\partial y}\right|_{y=0} = 0 \end{cases} \tag{8.3.3}$$

We can make the system dimensionless choosing the scales

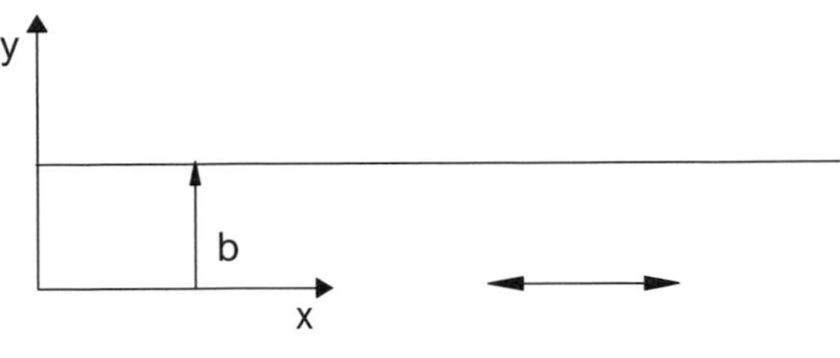

Figure 8.8 Oscillatory flow in a channel.

$$U = \frac{u}{U_s},\ Y = \frac{y}{b},\ T = \omega t \tag{8.3.4}$$

where the velocity scale is $U_s = (-B^p)b^2/\mu$ and we can assume $B^p < 0$ so that $U_s > 0$. Inserting Eq. (8.3.4) in Eq. (8.3.2) leads to the dimensionless form

$$\frac{\partial^2 U}{\partial Y^2} - \alpha^2\frac{\partial U}{\partial T} = -\cos T = -\Re\{e^{iT}\} \tag{8.3.5}$$

The boundary conditions on the flow are

$$U(Y = 1) = 0,\ \left.\frac{\partial U}{\partial Y}\right|_{Y=0} = 0 \tag{8.3.6}$$

The parameter α is the Womersley parameter where $\alpha = b\sqrt{\omega/\nu}$, which is the ratio of the unsteady inertia to viscous effects (Womersley, 1955). Womersley's work focused on arterial blood flow and built on an earlier analysis by Lambossy (1952). It is related to the Reynolds number, Re, and the Strouhal number, St, such that $\alpha^2 = \omega b^2/\nu = (2\pi f b/U)(Ub/\nu) = 2\pi \mathrm{St}\,\mathrm{Re}$. Another interpretation of the Womersley parameter is the ratio of two time scales, viscous diffusion and the period of the forced oscillation

$$\alpha^2 = \frac{b^2\omega}{\nu} = \frac{b^2/\nu}{1/\omega} = \frac{\text{viscous diffusion time}}{\text{oscillation period}} \tag{8.3.7}$$

We can try a separation of variables approach, choosing the time dependence to match the forcing frequency

$$U(Y,T) = \Re\{G(Y)e^{iT}\} \tag{8.3.8}$$

Here, G is in general complex, so we eventually want the real part of the product of G with e^{iT} as shown.

Inserting Eq. (8.3.8) into Eq. (8.3.5) gives us a resulting form in which each term is multiplied by e^{iT}. Dividing those out leads to an equation just for G

$$\frac{d^2G}{dY^2} - \sigma^2 G = -1 \tag{8.3.9}$$

where $\sigma^2 = i\alpha^2$, $\sigma = \sqrt{2}\alpha(1+i)/2$ and we will consider $\pm\sigma$ as the positive and negative roots. The boundary conditions on G are

$$G(Y=1) = 0, \left.\frac{dG}{dY}\right|_{Y=0} = 0 \tag{8.3.10}$$

First we solve the homogeneous problem of Eq. (8.3.9) by setting the right hand side to zero

$$\frac{d^2G_h}{dY^2} - \sigma^2 G_h = 0 \tag{8.3.11}$$

For the form of the homogeneous solution we assume $G_h \sim e^{qY}$, and substitute it into Eq. (8.3.11), which yields an equation for q, $q^2 - \sigma^2 = 0$so that $q = \pm\sigma$. Then the general form of G_h is

$$G_h = C_1 e^{qY} + C_2 e^{-qY} \tag{8.3.12}$$

where C_1 and C_2 are unknown constants to be determined by the boundary conditions. For our purposes, it is somewhat easier to use an equivalent form of G_h using the hyperbolic cosine and hyperbolic sine functions, cosh and sinh, which are just linear combinations of the exponentials

$$G_h = A_3 \cosh(\sigma Y) + A_4 \sinh(\sigma Y) \tag{8.3.13}$$

where $\cosh z = (e^z + e^{-z})/2$, $\sinh z = (e^z - e^{-z})/2$. The particular solution is simply the constant

$$G_p = \frac{1}{\sigma^2} \tag{8.3.14}$$

So the total solution is the sum of the homogeneous and particular solutions

$$G = A_3 \cosh(\sigma Y) + A_4 \sinh(\sigma Y) + \frac{1}{\sigma^2} \tag{8.3.15}$$

Imposing the two boundary conditions gives us the equations for A_3 and A_4

$$\left.\frac{dG}{dY}\right|_{Y=0} = A_4\sigma = 0 \quad \Rightarrow \quad A_4 = 0$$
$$G(Y=1) = A_3 \cosh\sigma + \frac{1}{\sigma^2} = 0 \quad \Rightarrow \quad A_3 = -\frac{1}{\sigma^2 \cosh\sigma} \tag{8.3.16}$$

Inserting the solutions for A_3 and A_4 in Eqs. (8.3.16) into Eq. (8.3.15) yields the solution for G

$$G(Y) = \frac{1}{\sigma^2}\left[1 - \frac{\cosh(\sigma Y)}{\cosh\sigma}\right] \tag{8.3.17}$$

Finally, the full solution for U(Y, T) is

$$U(Y,T) = \Re\{G(Y)e^{iT}\} = \Re\left\{\frac{1}{\sigma^2}\left[1 - \frac{\cosh(\sigma Y)}{\cosh\sigma}\right]e^{iT}\right\}$$
$$\sigma = \sqrt{2}\alpha(1+i)/2 \tag{8.3.18}$$

In Figure 8.9 are plotted three examples of the velocity profiles for this flow using $\alpha = 0.1, 1.0, 10.0$ at time values within a half cycle of $T = 0$, $\pi/4$, $\pi/2$, $3\pi/4$, π. For $\alpha = 0.1$ in Figure 8.9(a), the coefficient of the unsteady acceleration term in Eq. (8.3.5) is $\alpha^2 = 0.01$, and its contribution is negligible, $\partial^2U/\partial Y^2 \approx \cos T$. The result is a sequence of parabolic shapes that we would obtain, in essence, from solving the steady equations. Note that when $T = \pi/2$, the pressure gradient is zero and the velocity of the fluid is zero. At the other extreme, for $\alpha = 10$ in Figure 8.9(c), we see the appearance of the Stokes layer. The balance in Eq. (8.3.5) for large α can be viewed as $\partial U/\partial T = \left(\partial^2U/\partial Y^2 + \cos T\right)/\alpha^2$. The unsteady term is dominant away from the boundaries where $\partial^2U/\partial Y^2 << 1$, i.e. $\partial U/\partial T \approx \cos T/\alpha^2 \rightarrow U \approx \sin T/\alpha^2$. There is a core region which behaves as an inviscid fluid with a flat profile. The Stokes layer is adjacent to the wall where $\partial^2U/\partial Y^2 >> 1$ and the velocity profile matches the no-slip condition at the walls. The dimensionless Stokes layer thickness is $\delta_S/b = 1/\alpha$ when $\alpha >> 1$. So we can think of $1/\alpha$ as the ratio of the Stokes layer thickness to the channel half-width. Because the pressure gradient amplitude, B^p, is fixed, as the frequency increases the balance $U \approx -\sin T/\alpha^2$ indicates that the velocity amplitude in the core decreases as$1/\alpha^2$. We see that in the sequence of figures that the velocity amplitude is decreasing. The choice of

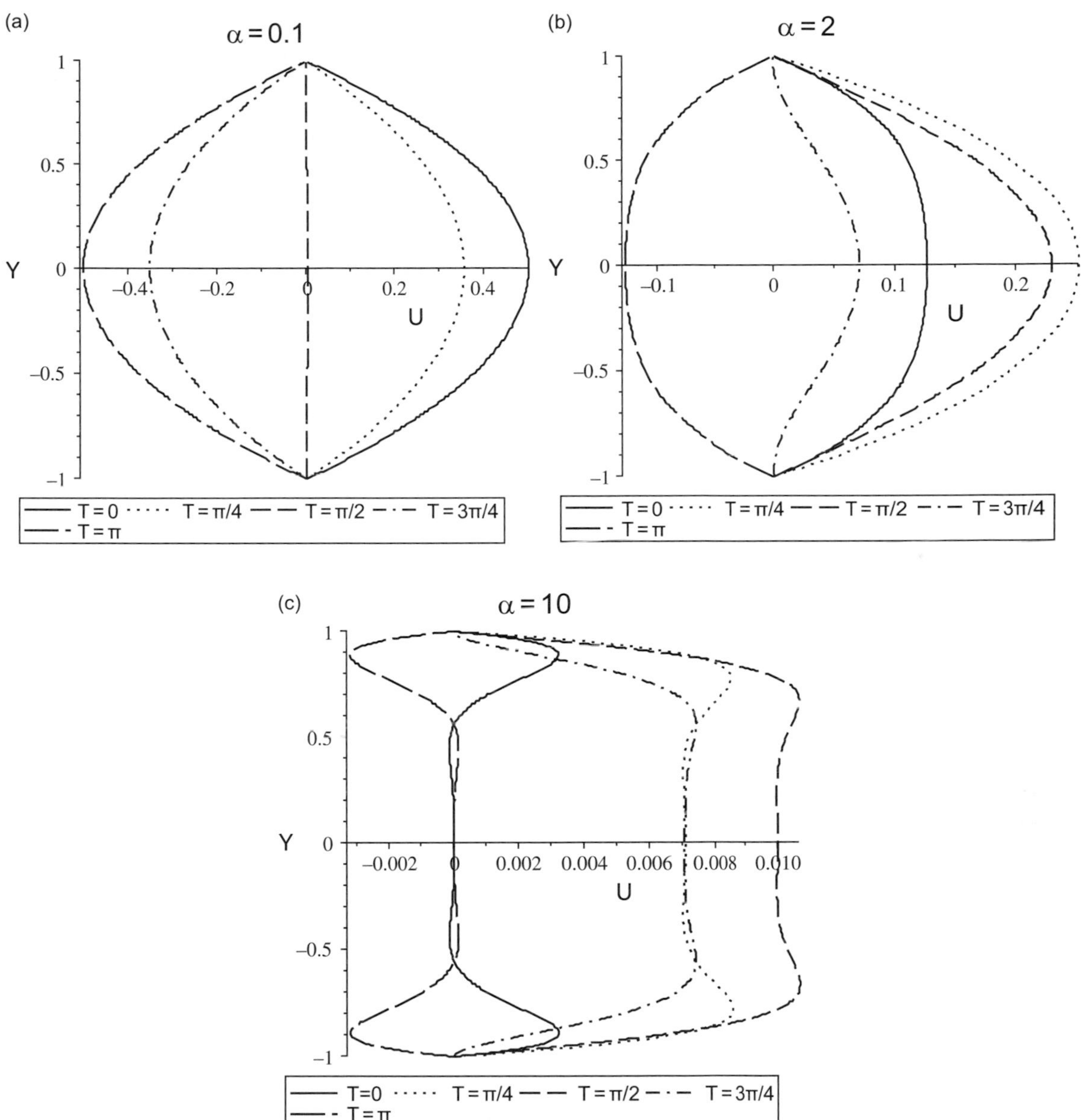

Figure 8.9 Velocity profiles at five time values during a half cycle, $T = 0,\ \pi/4,\ \pi/2,\ 3\pi/4,\ \pi$ for $\alpha = 0.1, 2, 10$.

$\alpha = 1$ in Figure 8.9(b) is an intermediate value which shows that the viscous and unsteady terms compete relatively equally.

The flow rate, q, of this oscillatory channel flow is the integral of u across the cross-section or, due to the symmetry, it is twice the integral of u from the centerline to the wall

$$q(t) = 2\int_0^{L_z}\int_0^{b} u(y,t)\,dy\,dz = 2L_z\int_0^{b} u(y,t)\,dy \qquad (8.3.19)$$

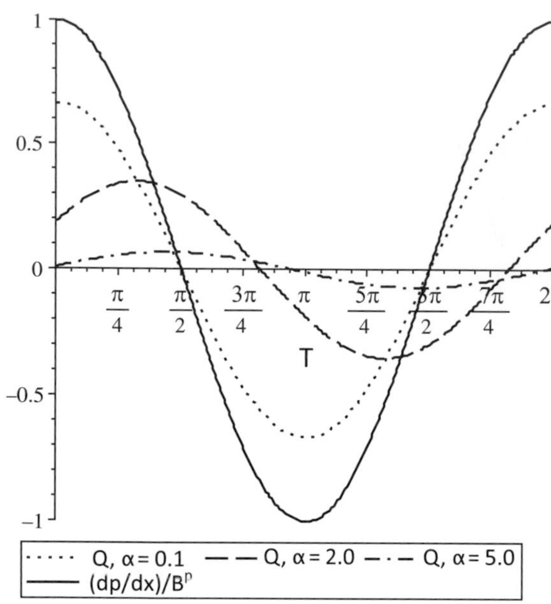

Figure 8.10 Flow rate Q(T) for three values of α and the dimensionless pressure gradient $(dp/dx)/B^p$.

where L_z is a unit length in the z-direction. Inserting our scaled variables from Eq. (8.3.4), the corresponding dimensionless flow rate, $Q(T) = q(t)/(L_z b U_s)$, is defined by

$$Q(T) = 2\int_0^1 U(Y,T)\,dY \tag{8.3.20}$$

Inserting the solution for U(Y, T) from Eq. (8.3.18) into Eq. (8.3.20) and performing the integration we find

$$Q(T) = \Re\left\{\frac{2}{\sigma^2}\left(1 - \frac{\tanh\sigma}{\sigma}\right)e^{iT}\right\} \tag{8.3.21}$$

Plots of Q(T) are shown in Figure 8.10 for three values of $\alpha = 0.1, 2.0, 5.0$ and the dimensionless pressure gradient $(dp/dx)/B^p = \cos T$. For $\alpha = 0.1$ the pressure gradient and flow are almost in phase. However, as αincreases the flow increasingly lags the pressure.

To analyze the flow, we can represent the function in the bracket of Eq. (8.3.21) as a complex number, K,

$$\frac{2}{\sigma^2}\left(1 - \frac{\tanh\sigma}{\sigma}\right) = K = K_r + iK_i \tag{8.3.22}$$

where K_r is the real part of K and K_i the imaginary part. Both K_r and K_i are real numbers and depend on α. The left hand side of Eq. (8.3.22) is complicated because $\tanh\sigma$ is the hyperbolic tangent of a complex number. In general, for a complex number $g = g_r + ig_i$, its hyperbolic tangent is

$$\tanh(g_r + ig_i) = \frac{\sinh(g_r + ig_i)}{\cosh(g_r + ig_i)} = \frac{\sinh g_r \cos g_i + i\cosh g_r \sin g_i}{\cosh g_r \cos g_i + i \sinh g_r \sin g_i} \tag{8.3.23}$$

So the process of sorting out K_r and K_i in Eq. (8.3.22) is available but tedious. However, calculation software can readily do this chore for us

$$K_r = \frac{\sqrt{2}\sinh\left(\frac{\sqrt{2}}{2}\alpha\right)\cosh\left(\frac{\sqrt{2}}{2}\alpha\right) - \sqrt{2}\sin\left(\frac{\sqrt{2}}{2}\alpha\right)\cos\left(\frac{\sqrt{2}}{2}\alpha\right)}{\alpha^3\left(\sinh^2\left(\frac{\sqrt{2}}{2}\alpha\right) + \cos^2\left(\frac{\sqrt{2}}{2}\alpha\right)\right)}$$

$$K_i = \left(\frac{\sqrt{2}\sinh\left(\frac{\sqrt{2}}{2}\alpha\right)\cosh\left(\frac{\sqrt{2}}{2}\alpha\right) + \sqrt{2}\sin\left(\frac{\sqrt{2}}{2}\alpha\right)\cos\left(\frac{\sqrt{2}}{2}\alpha\right)}{\alpha^3\left(\sinh^2\left(\frac{\sqrt{2}}{2}\alpha\right) + \cos^2\left(\frac{\sqrt{2}}{2}\alpha\right)\right)} - \frac{2}{\alpha^2}\right) \tag{8.3.24}$$

Then the flow rate, Q(T), derives from the real part of the product of K with e^{iT} in Eq. (8.3.21)

$$\begin{aligned} Q(T) &= \Re\{(K_r + iK_i)(\cos T + i\sin T)\} \\ &= K_r \cos T - K_i \sin T \end{aligned} \tag{8.3.25}$$

If we let $K_r = |K|\cos\phi,\ K_i = -|K|\sin\phi$, then Eq. (8.3.25) takes the form

$$\begin{aligned} Q(T) &= |K|(\cos\phi\cos T + \sin\phi\sin T) \\ &= |K|\cos(T - \phi) \end{aligned} \tag{8.3.26}$$

where $|K| = \left(K_r^2 + K_i^2\right)^{1/2}$ is the flow magnitude and $\phi = \tan^{-1}(-K_i/K_r)$ is the phase angle relative to the pressure gradient, which is cos(T). Both depend on α as shown in Figure 8.11, where ϕ starts at zero and increases asymptotically to $\pi/2$ as α increases. The flow amplitude decreases monotonically with an asymptote at zero as α increases.

The flow amplitude decreases with increasing α since the pressure gradient magnitude, B^p, is fixed. The momentum balance at high frequency in the inviscid core is approximately

$$B^p \sim \rho\frac{\partial u}{\partial t} \quad \rightarrow \quad B^p \sim \rho\omega^2 d \tag{8.3.27}$$

where d is the stroke distance of the fluid. It is related to the stroke volume or tidal volume, i.e. the volume of fluid displaced in half the cycle. Clearly, as frequency increases, d must decrease for fixed B^p. To keep the stroke volume constant as α increases, B^p must increase as a function of α, a situation covered in Clinical Correlation 8.1.

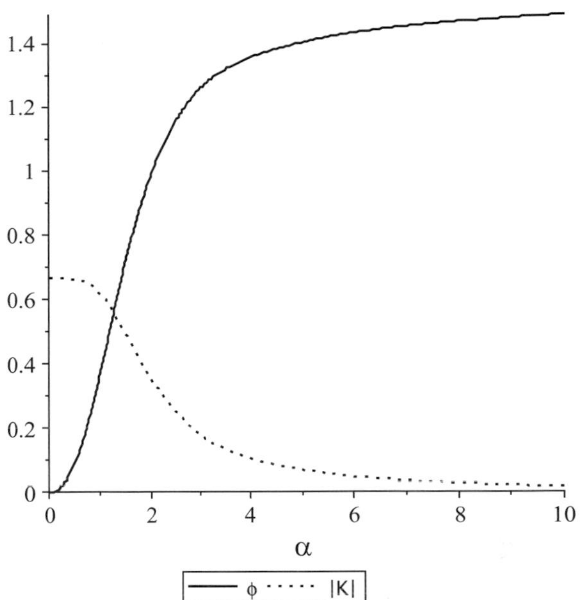

Figure 8.11 Dependence of the flow amplitude, $|K|$, and phase angle, ϕ, on α.

Clinical Correlation 8.1 | Volume-Cycled Mechanical Ventilation

For mechanical ventilators used in hospitals and some home care situations, the patient is connected to the ventilator with a tube in the trachea, an endotracheal tube. The settings for the ventilator include the breathing frequency and tidal volume, which allows volume cycling rather than the pressure cycling investigated above. For our purposes, the tidal volume is the integral of the dimensional flow, q(t), during the inspiration phase or the expiration phase. For inspiration, consider that it starts at $t = t_0$ when $q(t_0) = 0$. Then it persists for half of a cycle duration, π/ω, until the next zero crossing for flow where $q(t_0 + \pi/\omega) = 0$. The tidal volume, V_T, is calculated from integrating the flow during this half-cycle

$$V_T = 2L_z \int_{t_0}^{t_0+\frac{\pi}{\omega}} \int_0^b u(y,t)\, dy\, dt \tag{8.3.28}$$

Using the scales from Eq. (8.3.4) to replace u with U_sU, y with $b\eta$, and t with T/ω in Eq. (8.3.28) gives us

$$V_T = L_z \frac{bU_s}{\omega} \int_{T_0}^{T_0+\pi} Q(T)\, dT \tag{8.3.29}$$

where we have inserted the results of Eq. (8.3.20) for the flow rate, Q, and note that the dimensionless zero crossing is at $T_0 = \omega t_0$. Now using the results of (8.3.26) we substitute for Q(T) to find

$$V_T = L_z \frac{U_s b}{\omega} \int_{\phi-\frac{\pi}{2}}^{\phi+\frac{\pi}{2}} |K| \cos(T-\phi)\, dT = L_z \frac{U_s b}{\omega} 2|K| \tag{8.3.30}$$

In Eq. (8.3.30) we have identified $T_0 = -\phi + \pi/2$ where $\cos(T-\phi)$ crosses zero and is positive for $\phi - \pi/2 \le T \le \phi + \pi/2$. To keep the tidal volume constant, Eq. (8.3.30) yields the solution for the pressure gradient B^p within U_s as

$$U_s = \frac{\omega V_T}{2bL_z|K|} \quad \rightarrow \quad -B^p = \frac{\omega \mu V_T}{2L_z|K|b^3} \tag{8.3.31}$$

Since $|K|$ monotonically decreases as α increases, Eq. (8.3.31) shows that the pressure gradient must increase to keep the same value of V_T. We can rewrite Eq. (8.3.31) as

$$(-B^p) = \frac{\mu U_0}{|K|b^2} \qquad U_0 = \omega d, \quad d = \frac{V_T}{2L_z b} \tag{8.3.32}$$

In Eq. (8.3.32), d is a stroke length, the tidal volume divided by the cross-sectional area. The velocity associated with that oscillating stroke length is $U_0 = \omega d$. We can rescale our velocity solution from Eq. (8.3.18) on this new velocity, call it $U_1 = u/U_0$. Then the relationship is given by

$$U_1 = \frac{u}{U_0} = \frac{U_s}{U_0}\frac{u}{U_s} = \frac{U_s}{U_0} U = \frac{(-B^p)b^2}{\mu U_0} U = \frac{1}{|K|} U \tag{8.3.33}$$

Velocity profiles from the solution of $U_1(Y, T)$ in Eq. (8.3.33) are shown in Figure 8.12 for five time values during a half-cycle, $T = 0, \pi/4, \pi/2, 3\pi/4, \pi$, for $\alpha = 0.1, 2.0, 10.0$. Note that with the new velocity scale $U_0 = \omega d$, derived from keeping the tidal volume constant, the dimensionless velocity magnitudes are of similar size, independent of α.

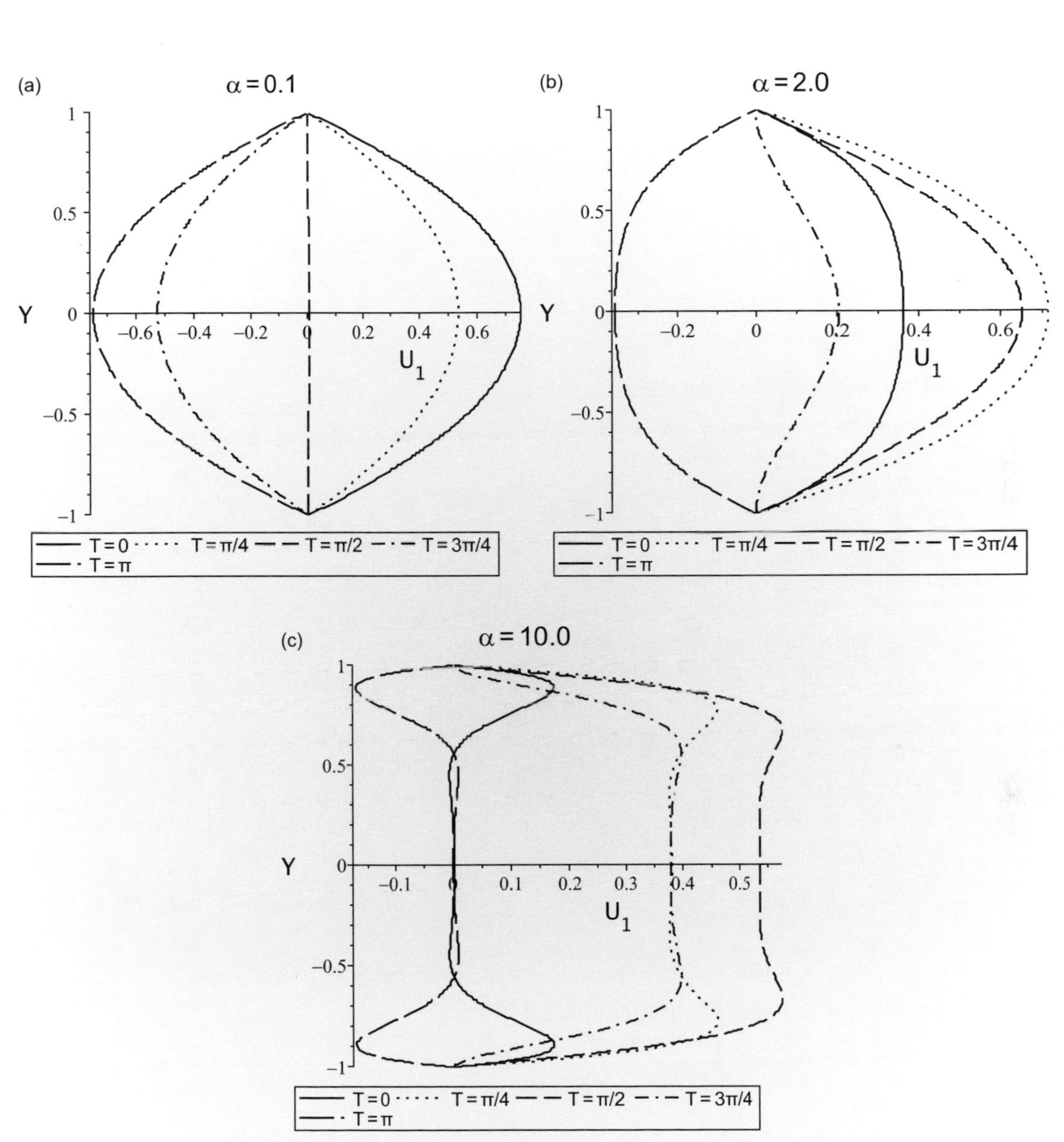

Figure 8.12 Velocity profiles $U_1(Y,T)$ for volume-cycled oscillations at five time values, $T = 0, \pi/4, \pi/2, 3\pi/4, \pi$, for $\alpha = 0.1, 2.0, 10.0$.

The Maple code for Figure 8.12b, including animation, is shown in Code 8.2(a). The equivalent code in MATLAB® is shown in Code 8.2(b).

```
(a)
restart;
with(plots):

σ_t := sqrt(2)·alpha/2 + I. sqrt(2)·alpha/2 :

F := 2/σ_t^2 · (1 - tanh(σ_t)/σ_t) :

Kr := evalc(Re)F)):
Ki := evalc(Im((f))):

G := 1/σ^2 · (1 - cosh(sigma·Y)/cosh(sigma)) :

Sigma := alpha/Sqrt(2) . (1+1):

UU := (Ki^2+Kr^2)^(-1/2) · Re(G· exp(I.T)):

PL6:=plot([subs(alpha=2.0, T=0, UU), Y, Y=-1..1], color=back, linestyle=1, legend="T=0"):

PL7:=plot([subs(alpha=2,  T=Pi/4, UU), Y, Y=-1..1], color=back, linestyle=2, legend
         ="T=π/4"):

PL8:=plot([subs(alpha=2, T=Pi/2, UU), Y, Y=-1..1], color=back, linestyle=3, legend
         ="T=π/2"):

PL9:=plot([subs(alpha=2, T=3.Pi/4, UU), Y, Y=-1..1], color=back, linestyle-4, legend
         ="T=3π/4"):

PL10:=plot([subs(alpha=2, T=Pi, UU), Y, Y=-1..1], color=back, linestyle=5, legend="T=π"):

display (Pl6, PL7, PL8, PL9, PL10, labels = [U[1] , "Y"], title ="α=2.0") :
animate (plot,  [[subs(alpha=2, UU), Y, Y=-1..1]], T = 0..2·Pi) ;
```

```
(b)
clear all, close all, clc
syms sigmasigma_talphaFYTKiKr
sigma_t  = sqrt(2)*alpha/2 + 1i*sqrt(2)*alpha/2;
F        = 2/sigma_t^2*(1-tanh(sigma_t)/sigma_t);
Kr       = real(F);
Ki       = imag(F);
sigma    = alpha/sqrt(2)*(1+1i);
G        = 1/sigma^2*(1-cosh(sigma*Y)/cosh(sigma));
UU       = (Ki^2 + Kr^2)^(-1/2)*real(G*exp(1i*T));
U_eval   = matlabFunction(UU);
```

Code 8.2 Code examples: (a) Maple, (b) MATLAB®.

```
Y_eval = linspace(-1,1,100);
alpha_eval = 2;
figure(1)
T_eval = 0;
plot(U_eval(T_eval,Y_eval,alpha_eval),Y_eval,'k'), hold on
T_eval = pi/4;
plot(U_eval(T_eval,Y_eval,alpha_eval),Y_eval,'--k')
T_eval = pi/2;
plot(U_eval(T_eval,Y_eval,alpha_eval),Y_eval,'-.k')
T_eval = 3*pi/4;
plot(U_eval(T_eval,Y_eval,alpha_eval),Y_eval,':k')
T_eval = pi;
plot(U_eval(T_eval,Y_eval,alpha_eval),Y_eval,'.k')
title('\alpha=2.0')
legend('T=0','T=\pi/4','T=\pi/2','T=3\pi/4','T=\pi','location','southeast')
xlabel('U')
ylabel('Y')
figure(2)
delta_T = 0.01;
for T_eval = 0:delta_T:2*pi
plot(U_eval(T_eval,Y_eval,alpha_eval),Y_eval,'-k')
axis([-1.1 1.1 -1 1])
  xlabel('U')
  ylabel('Y')
  drawnow
end
ax = gca;
ax.FontSize = 11;
ax.FontName = 'Arial';
```

Code 8.2 *(continued)*

8.4 High Frequency Ventilation

Mechanical ventilators are used for supporting respiration in a variety of critical care situations. They typically mimic normal respiratory frequencies and tidal volumes for the patient. At rest, adults breathe 10–20 times per minute and neonates 30–60 times per minute. Tidal volumes are typically much larger than the "dead space" of the lung, i.e. the volume of the conducting airways that lead to the end air sacs called alveoli. However, the related pressures and stretching of the lung from a normal tidal volume can be large enough to cause pulmonary injury, especially in the face of lung disease. Reducing the tidal volume has proved to be protective in a wide range of patients (Brower et al., 2000). Smaller tidal volumes require higher frequencies to maintain enough fresh air into the lungs. The well-known observation of dog panting, where respiratory rates of 300 breaths per minute at very small tidal volumes, served as a model for high frequency ventilators which are used in newborns (Henderson et al., 1915; Briscoe et al.,

1954). Those ventilators have rates up to 900 breaths per minute, 15 Hz, and tidal volumes that are smaller than the dead space.

As a model for HFV, let us solve the transport equation from Section 4.5, Eq. (4.5.15), for the concentration field c(x, y, t), where c is the concentration of a diffusible gas related to respiration, like oxygen or carbon dioxide

$$\frac{\partial c}{\partial t} + u\frac{\partial c}{\partial x} + v\frac{\partial c}{\partial y} = D_m\left(\frac{\partial^2 c}{\partial x^2} + \frac{\partial^2 c}{\partial y^2}\right) \tag{8.4.1}$$

The terms multiplied by the velocities u and v in Eq. (8.4.1) represent the convective transport, while the terms multiplied by the molecular diffusivity, D_m, represent the diffusive transport. Values of D_m depend on the constituent or solute of interest, what solvent it is diffusing through, and the temperature and pressure. For example, oxygen diffusing through air has $D_{O_2\text{-air}} \sim 0.18\,\text{cm}^2/\text{s}$ at 1 atmosphere of pressure and temperature of 25 °C. However, oxygen diffusion through water (as in blood) is much slower, $D_{O_2\text{-water}} = 2.1 \times 10^{-5}\,\text{cm}^2/\text{s}$. Note that for oscillatory flow in a channel, $v = 0$. Using the same scales as Eq. (8.3.4), with the added definitions that $C = c/C_0$, $X = x/b$. Also, we choose to keep the tidal volume constant in this example, so use the dimensionless velocity form $U_1 = u/U_0$ where $U_0 = \omega d$ and d is the stroke distance of the oscillation amplitude, see Eq. (8.3.32). Then Eq. (8.4.1) becomes

$$\alpha^2 Sc\frac{\partial C}{\partial T} + Pe_0 U_1\frac{\partial C}{\partial X} = \left(\frac{\partial^2 C}{\partial X^2} + \frac{\partial^2 C}{\partial Y^2}\right) \tag{8.4.2}$$

There are two dimensionless parameters in Eq. (8.4.2): the Womersley parameter, $\alpha = b\sqrt{\omega/\nu}$, and the Péclet number, $Pe_0 = U_0 b/D_m$, which is the ratio of convective to diffusive transport on the basis of the velocity scale U_0. We can rewrite the Péclet number as $Pe_0 = (\omega b^2/\nu)(\nu/D_m)A = \alpha^2 ScA$, which introduces the Schmidt number, $Sc = \nu/D_m$, as the ratio of shear (vorticity) diffusion to molecular diffusion and the dimensionless stroke amplitude, A, which is $A = d/b = V_T/2L_z b^2$. For oxygen transport in air, the Schmidt number is $Sc_{O_2\text{-air}} = \nu_{air}/D_{O_2\text{-air}} = 0.15\ \text{cm}^2/\text{s}/0.19\ \text{cm}^2/\text{s} = 0.78$ while for water it is $Sc_{O_2\text{-water}} = \nu_{water}/D_{O_2\text{-water}} = 0.009\ \text{cm}^2/\text{s}/2.1\times 10^{-5}\text{cm}^2/\text{s} \sim 428$. The velocity $U_1 = U/|K|$ is given in Eq. (8.3.33) with $U(Y,T) = \Re\{G(Y)e^{iT}\}$ where G is given in Eq. (8.3.17).

A solution form for the concentration field, C(X, Y, T), is given by

$$C(X,Y,T) = -\gamma X + \gamma\Re\{H(Y)e^{iT}\} \tag{8.4.3}$$

Note that Eq. (8.4.3) has a steady, linear X-dependency so that the X-gradient of concentration is the dimensionless constant, $-\gamma$, whose dimensional form is $-\gamma^* = -\gamma C_0/b$. Then it also has the oscillatory and Y-dependent term, $\Re\{H(Y)e^{iT}\}$. Inserting Eq. (8.4.3) into Eq. (8.4.2) gives us an ordinary differential equation for H in terms of G,

$$\frac{d^2H}{dY^2} - i\alpha^2 ScH = \frac{Pe_0}{|K|}G \tag{8.4.4}$$

H will be the sum of homogenous and particular solutions to Eq. (8.4.4), i.e. $H = H_h + H_p$. The homogenous solution, for the right hand side of Eq. (8.4.4) equal to zero, is

$$H_h = B_1\cosh(\beta Y) + B_2\sinh(\beta Y) \tag{8.4.5}$$

where $\beta^2 = i\alpha^2 Sc = \sigma^2 Sc$ and B_1 and B_2 are constants to be determined from the boundary conditions. The particular solution will have the same structure as G(Y), so let

$$H_p = B_3 + B_4\cosh(\sigma Y) \tag{8.4.6}$$

Inserting Eq. (8.4.6) into Eq. (8.4.4) and matching terms leads to solutions for the constants B_3 and B_4.

In terms of the boundary conditions, from symmetry $\partial H/\partial Y = 0$ at $Y = 0$, which determines that $B_2 = 0$. Then for insulated boundaries $\partial H/\partial Y = 0$ at $Y = 1$, which gives us B_1. The final form for H is

$$H = B_1\cosh(\beta Y) + B_3 + B_4\cosh(\sigma Y) \tag{8.4.7}$$

where the constants are given by

$$B_1 = \frac{Pe_0\tanh\sigma}{|K|(\beta^2 - \sigma^2)\sigma\beta\sinh\beta},\ B_3 = \frac{Pe_0}{|K|\beta^2\sigma^2},$$
$$B_4 = \frac{-Pe_0}{|K|(\beta^2 - \sigma^2)\sigma^2\cosh\sigma} \tag{8.4.8}$$

The flux of dissolved constituent in the x-direction is given by $j(x, y, t) = uc - D_m(\partial c/\partial x)$. The mass flow rate, $\dot{m}(t)$, is the integral of j across the cross-section of the channel,

$$\dot{m}(t) = \int_0^{L_z}\int_{-b}^{b}\left(uc - D_m\frac{\partial c}{\partial x}\right)dy\,dz \tag{8.4.9}$$

Using our scaled variables in Eq. (8.4.9), the dimensionless form of the mass flow rate, $\dot{M}(T)$, is

$$\dot{M}(T) = \int_0^1\left(\frac{Pe_0}{|K|}UC - \frac{\partial C}{\partial X}\right)dY \tag{8.4.10}$$

where $\dot{M} = \dot{m}/(2D_mL_zC_0)$and we have used the symmetry condition in Y for the Y-integral limits.

The integrand of Eq. (8.4.10) includes oscillatory and steady terms. Eventually we are interested in the steady mass flow rate. To start with, the purely diffusive term,$-\partial C/\partial X = \gamma$, is steady. Then it is helpful to rewrite U and C as the following

$$C = -\gamma X + \gamma\Re\{H(Y)e^{iT}\}$$
$$= -\gamma X + \gamma\frac{1}{2}\left(H(Y)e^{iT} + \bar{H}(Y)e^{-iT}\right)$$
$$U - \Re\{G(Y)e^{iT}\} = \frac{1}{2}\left(G(Y)e^{iT} + \bar{G}(Y)e^{-iT}\right) \tag{8.4.11}$$

where $\bar{G}$ is the complex conjugate of G and $\bar{H}$ is the complex conjugate of H. Clearly the real part of a complex function, $\Re\{\Gamma\}$, is ½ the sum of that function plus its complex conjugate, i.e. $\Re\{\Gamma\} = (1/2)(\Gamma + \bar{\Gamma}) = (1/2)((\Gamma_r + i\Gamma_i) + (\Gamma_r - i\Gamma_i)) = (1/2)(2\Gamma_r) = \Gamma_r$. Now by inserting Eq. (8.4.11) into Eq. (8.4.10) we see that the product UC has terms proportional to $e^{\pm iT}$, $e^{\pm 2iT}$ and e^0. The products proportional to e^0 are steady, so the steady, or time-averaged, mass flow rate, becomes

$$\langle\dot{M}\rangle = \frac{1}{2\pi}\int_0^{2\pi}\dot{M}(T)\,dT = \gamma\int_0^1\left(\frac{Pe_0}{|K|}\frac{1}{4}(G\bar{H} + \bar{G}H) + 1\right)dY \tag{8.4.12}$$

where the symbol $\langle\dot{M}\rangle$ indicates the time-average of $\dot{M}$. Equations (8.4.8) show that $H \sim Pe_0$, so the convective contribution to the mass flow rate in Eq. (8.4.12) is proportional to Pe_0^2 and, consequently, A^2.

Let us apply Eq. (8.4.12) to a two-dimensional version of HFV in a premature neonate who weighs 2.0 kg. For a cycling frequency, f = 15 Hz, the tidal volume is usually in the range $V_T = 1.5 - 3.0\,\text{cm}^3/\text{kg}$. Choosing the lower value, let $V_T = 3\ \text{cm}^3$. The tracheal diameter is 0.5 cm so set $b = L_z = 0.25$ cm. Under these circumstances $\alpha \sim 6$ and $A \sim 100$ and we assume Sc = 0.8 and let $\gamma = 1$, noting the entire mass flow rate is proportional to γ. The result for $\langle\dot{M}\rangle$ varying α is shown in Figure 8.13. Note that for $\alpha = 0$ the time-averaged mass flow rate is $\langle\dot{M}\rangle = 1$, which is purely X-diffusion without any convection. As α increases the convection–diffusion interaction increases its contribution to be $\langle\dot{M}\rangle \sim 2200$ at $\alpha = 10$, an increase which is three orders of magnitude larger than diffusion only. The curve is approximately linear for $5 \leq \alpha \leq 10$.

What is the mechanism of the enhanced mass flow rate in a system with no net, time-averaged fluid flow and tidal volumes smaller than the dead space?

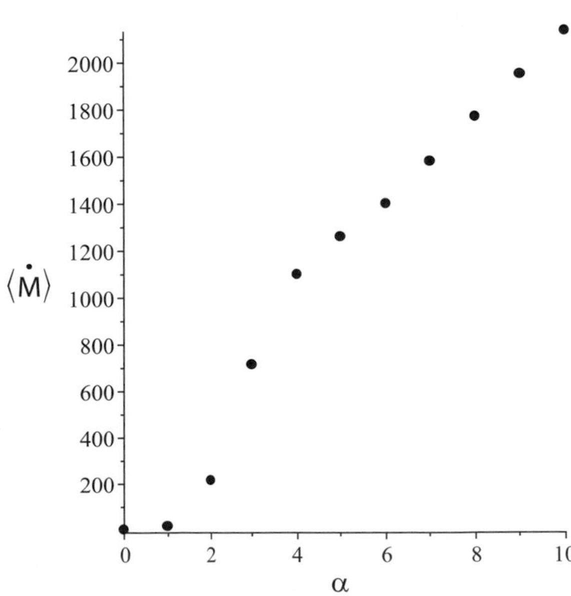

Figure 8.13 $\langle\dot{M}\rangle$ vs α for A = 10, Sc = 0.8, $\gamma = 1$.

Figure 8.14 helps to explain this phenomenon. For $\alpha = 3$, Figure 8.14(a) shows the velocity profile, $U_1(Y, T)$ for four values of T in the oscillation. In Figure 8.14(b) we have a contour plot of $C(X, Y, T)$ for $T = \pi/2$ showing lines of constant C with darker shading for higher values of C and lighter shading for lower values. The fluid velocity profile at this same value of time is shown in Figure 8.14(a) signified by a

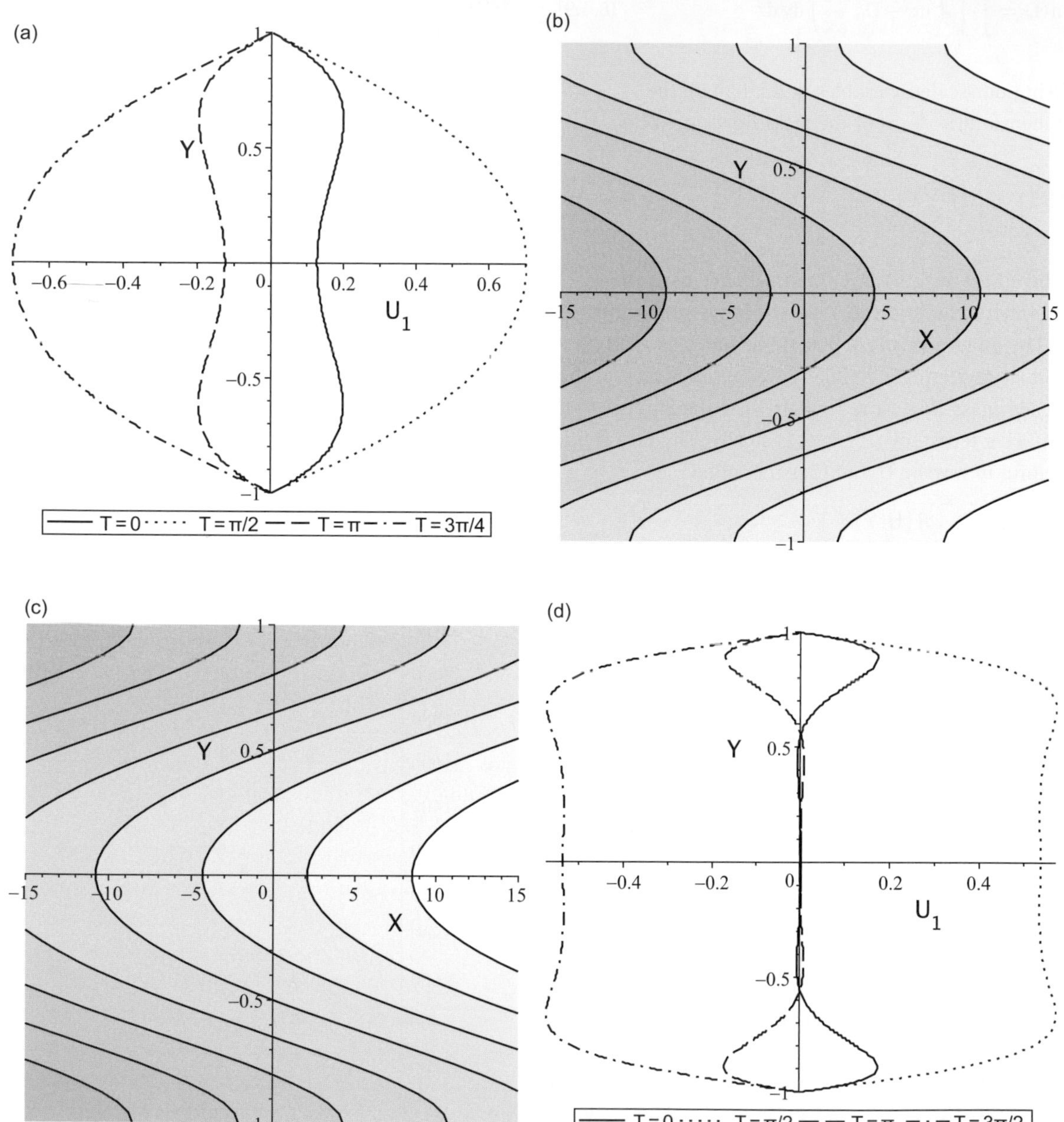

Figure 8.14 For $\alpha = 3$: (a) $U_1(Y, T)$ at four values of T; (b) $C(X, Y, T = \pi/2)$; (c) $C(X, Y, T = 3\pi/2)$. For $\alpha = 10$: (d) $U_1(Y, T)$ at four values of T; (e) $C(X, Y, T = \pi/2)$; (f) $C(X, Y, T = 3\pi/2)$. Remaining parameter values are $A = 100$, $\gamma = 1$, $Sc = 0.8$.

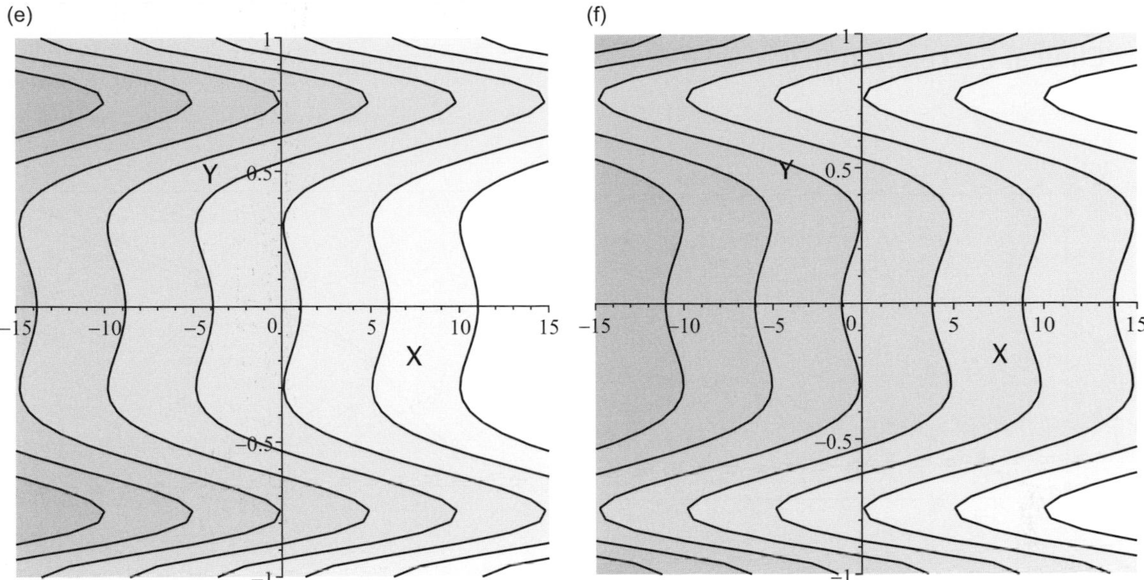

Figure 8.14 (*continued*)

dotted line. The fluid flows in the positive direction, to the right, with maximal velocity at $Y = 0$ tapering to no-slip at the walls. The fluid near the centerline in Figure 8.14(b) is darker than the fluid near the walls, so there is lateral gradient of solute. That gradient causes lateral diffusion so solute moves from the faster moving region near the center to the slower moving region near the walls. Then, during the recovery stroke, those same solute molecules will not travel as far in the negative direction, because it is in a slower moving stream. That is a net movement in the positive X-direction for the solute. Then in Figure 8.14(c) at $T = 3\pi/2$ the centerline fluid is lighter compared to the wall region. There is lateral diffusion, then, toward the centerline from the wall region, which puts more solute into an advantaged position for the next cycle when the $T = \pi/2$ condition is repeated. At a higher frequency, $\alpha = 10$, the same concept is shown in Figure 8.14(d–f), where the faster moving fluid now is near the walls, not the centerline. Consequently, the darker and lighter shadings are focused in the wall region. The sequence outlined in Figure 8.14 constitutes a net movement of mass due to the coupling of axial convection with lateral diffusion. Basically, the oscillating fluid grabs solute from the rich end and dumps it off at the poor end, a Robin Hood of gas exchange.

Enhanced solute transport due to this mechanism of axial fluid velocity coupling with lateral solute diffusion was investigated by G. I. Taylor for steady Poiseuille flow. It is known as "Taylor dispersion" (Taylor, 1953) and is also proportional to Pe^2. Often enhanced dispersion like this is viewed through the concept of an "effective diffusivity," $D_{eff} = \langle \dot{m} \rangle / (|\partial c/\partial x|)$, the dimensional ratio of the time-averaged mass flow rate to the axial concentration gradient. Our dimensionless version is simply $\langle \dot{M} \rangle / \gamma$, which equals 1 for pure diffusion.

Clinical Correlation 8.2

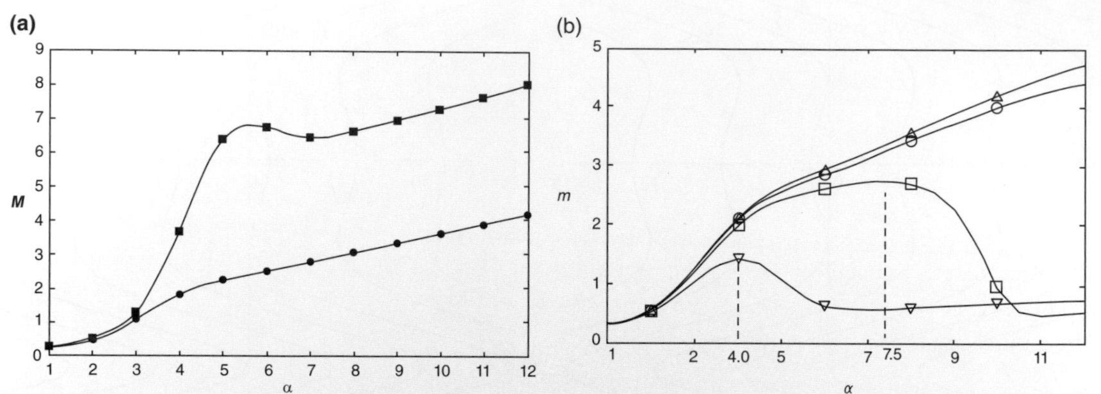

Figure 8.15 (a) Time-averaged mass flow rate (here indicated as M) vs αfor a straight tube (•) and a curved tube (■). (b) Time-averaged mass flow rate (here indicated as m) vs αfor a straight, flexible tube at four values of its stiffness parameter $\kappa = 10^3(\nabla)$, $10^4(\square)$, $10^5(\circ)$ and $10^8(\Delta)$, which is rigid. Note the local maxima in both (a) and (b).

In later chapters we discuss the fluid mechanics of flows in tubes which have flexibility, Chapter 9, and curvature, Chapter 12. These are two important features of physiologic tubes like blood vessels and airways, as is the general increase in cross-sectional area for flow into a branching system. The relationship $\langle \dot{M} \rangle$ vs α is more complicated under these three conditions (Eckmann and Grotberg, 1988; Godleski and Grotberg, 1988; Dragon and Grotberg, 1991). For example, there can be local maxima with flexibility and curvature (Eckmann and Grotberg, 1988; Dragon and Grotberg, 1991), as shown in Figure 8.15. Unlike Figure 8.13, which is monotonically increasing, these curves have distinct peaks. Local maxima indicate there can be optimal driving frequencies, which has been found in clinical applications (Bohn et al., 1980).

8.5 Oscillatory Flow in a Tube

Oscillatory flow is an important feature of biofluid mechanics. The respiratory system depends on the cycling of air into and out of the lung. Cardiovascular flow is a combination of a steady velocity due to mean arterial pressure with an overlay of an oscillatory component due to cardiac contractions.

A basic fluid mechanics model is oscillatory flow in a pipe of radius a, as shown in Figure 8.16. The flow is driven by a time-dependent pressure gradient, $dp/dz = B^p \cos \omega t$. Assuming axisymmetric flow with

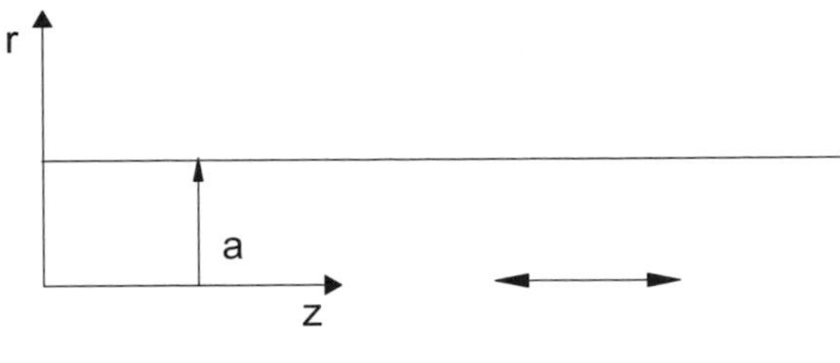

Figure 8.16 Oscillatory flow in a tube of radius a.

no radial or azimuthal velocities or pressure gradients, the axial velocity is a function of the radial position and time only, i.e. $u_z(r,t)$. The relevant form of the z-momentum equation in cylindrical coordinates is

$$\rho\frac{\partial u_z}{\partial t} = -B^p\cos(\omega t) + \mu\frac{1}{r}\frac{\partial}{\partial r}\left(r\frac{\partial u_z}{\partial r}\right) \tag{8.5.1}$$

Choosing the following scales

$$U_Z = \frac{u_z}{(-B^p)a^2/\mu}, \quad T = \omega t, \quad R = \frac{r}{a} \tag{8.5.2}$$

where we consider $B^p < 0$.

Equation (8.5.1) becomes

$$\frac{\partial^2 U_Z}{\partial R^2} + \frac{1}{R}\frac{\partial U_Z}{\partial R} - \alpha^2\frac{\partial U_Z}{\partial T} = -\cos T = -\Re\{e^{iT}\} \tag{8.5.3}$$

The dimensionless parameter α is the Womersley parameter, $\alpha = a\sqrt{\omega/\nu}$, for a cylindrical tube. The boundary conditions on the axial velocity, $U_Z(R,T)$, are no- slip at the pipe wall, $U_Z(R=1)=0$, and U_Z is finite.

The problem may be solved by assuming a separation of variables approach, let

$$U_Z(R,T) = \Re\{G_1(R)e^{iT}\} \tag{8.5.4}$$

When Eq. (8.5.4) is inserted into the partial differential equation for U_Z, Eq. (8.5.3), an ordinary differential equation for $G_1(R)$ results since all of the e^{iT} terms cancel,

$$\frac{d^2G_1}{dR^2} + \frac{1}{R}\frac{dG_1}{dR} - i\alpha^2 G_1 = -1 \tag{8.5.5}$$

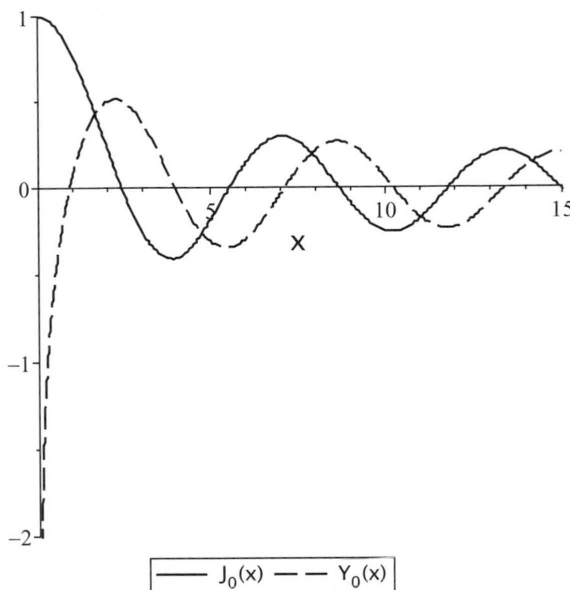

Figure 8.17 Bessel functions $J_0(x)$, $Y_0(x)$.

As we did for the oscillatory channel flow, to solve Eq. (8.5.5) we need to find its homogeneous and particular solutions. The homogeneous equation is obtained by setting the right hand side of Eq. (8.5.5) to zero,

$$\frac{d^2G_{1h}}{dR^2} + \frac{1}{R}\frac{dG_{1h}}{dR} - i\alpha^2 G_{1h} = 0 \tag{8.5.6}$$

Equation (8.5.6) is a form of Bessel's equation, in this case Bessel's equation of the first kind, zeroth-order. Solutions are linear combinations of the tabulated special functions called Bessel functions. For the form of Eq. (8.5.6), the appropriate choices are

$$G_{1h} = A_5 J_0(\lambda R) + A_6 Y_0(\lambda R) \quad \lambda = (-i)^{1/2}\alpha = \left(\frac{i-1}{\sqrt{2}}\right)\alpha \tag{8.5.7}$$

where J_0 is the zero-order Bessel function of the first kind and Y_0 is the zero-order Bessel function of the second kind. In many texts the equivalent definition of λ is $\lambda = (i)^{3/2}\alpha$ since $(-i)^{1/2} = (-1)^{1/2}(i)^{1/2} = i^{3/2}$. Note that the argument λR is complex.

To get a sense of Bessel functions, in Figure 8.17 is a plot of $J_0(x)$ and $Y_0(x)$ when they have the real

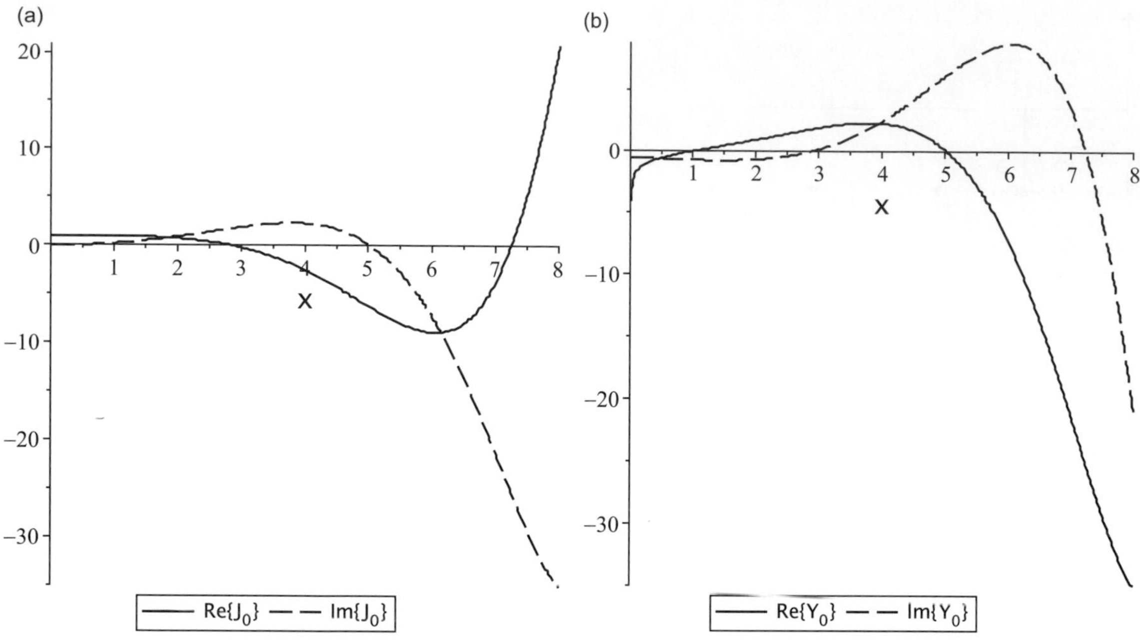

Figure 8.18 Bessel functions (a) J_o and (b) Y_o of complex argument $(-i)^{1/2}x$. Real and imaginary parts are plotted.

argument x. We can see that $Y_0(x)$ is unbounded as x approaches zero. When we examine both functions with a complex argument, a similar trend appears, see Figure 8.18. Let's choose the argument to be $(-i)^{1/2}x$. Then $J_0\left((-i)^{1/2}x\right)$ and $Y_0\left((-i)^{1/2}x\right)$ can be plotted, making sure we examine both their real and imaginary parts. Again, the magnitude of Y_o becomes unbounded as x approaches zero by virtue of its real part.

The particular solution to Eq. (8.5.5) is readily seen to be

$$G_{1p} = 1/i\alpha^2 = -i/\alpha^2 \tag{8.5.8}$$

so the total solution is the sum

$$G_1(R) = G_{1h} + G_{1p} = A_5 J_0(\lambda R) + A_6 Y_0(\lambda R) - \frac{i}{\alpha^2} \tag{8.5.9}$$

Now we can solve the unknown constants A_3 and A_4 by imposing the boundary conditions. Because W must be finite, we choose $A_6 = 0$ in Eq. (8.5.9) due to the unbounded behavior of Y_o as R approaches zero. Imposing the no-slip condition means $G_1(R = 1) = 0$, which yields a solution for the constant A_5

$$A_5 = \frac{i}{\alpha^2 J_0(\lambda)} \tag{8.5.10}$$

Inserting A_5 into Eq. (8.5.9) and combining into Eq. (8.5.4) gives the complete solution for U_Z as

$$U_Z(R,T) = \Re\left\{-\frac{i}{\alpha^2}\left(1 - \frac{J_0(\lambda R)}{J_0(\lambda)}\right)e^{iT}\right\} \tag{8.5.11}$$

The range of $\alpha = a\sqrt{\omega/\nu}$, where a is the tube radius, is fairly large in the animal kingdom. The mouse aortic diameter is $d = 0.1$ cm and a typical heart rate is 500 bpm. Assuming $\nu = 0.04\ \text{cm}^2/\text{s}$ the Womersley parameter value is $\alpha = 1.8$. The elephant aorta has $d = 10$ cm with heart rate 30 bpm, which calculates to $\alpha = 44$, whilst human $d = 3$ cm at 70 bpm results in $\alpha = 20$, i.e. between the mouse and elephant. The aorta is amongst the largest vessels within each species, so would represent a maximum value for each. Further out in the vascular network, the vessel diameters decrease causing α to decrease. Though the heart rate, i.e. the frequency, increases as species decrease in size, its contribution is a square root. The linear dependence on aortic radius dominates and α decreases.

Sample velocity profiles are shown in Figure 8.19 for $\alpha = 0.1, 2, 10$. For $\alpha = 0.1$ in Figure 8.19(a), it is easy to see that the velocity profile shape is parabolic at each value of time, as one expects from the steady solution. The entire cross-section of fluid is in-phase with the pressure gradient, so when $T = \pi/2$, the

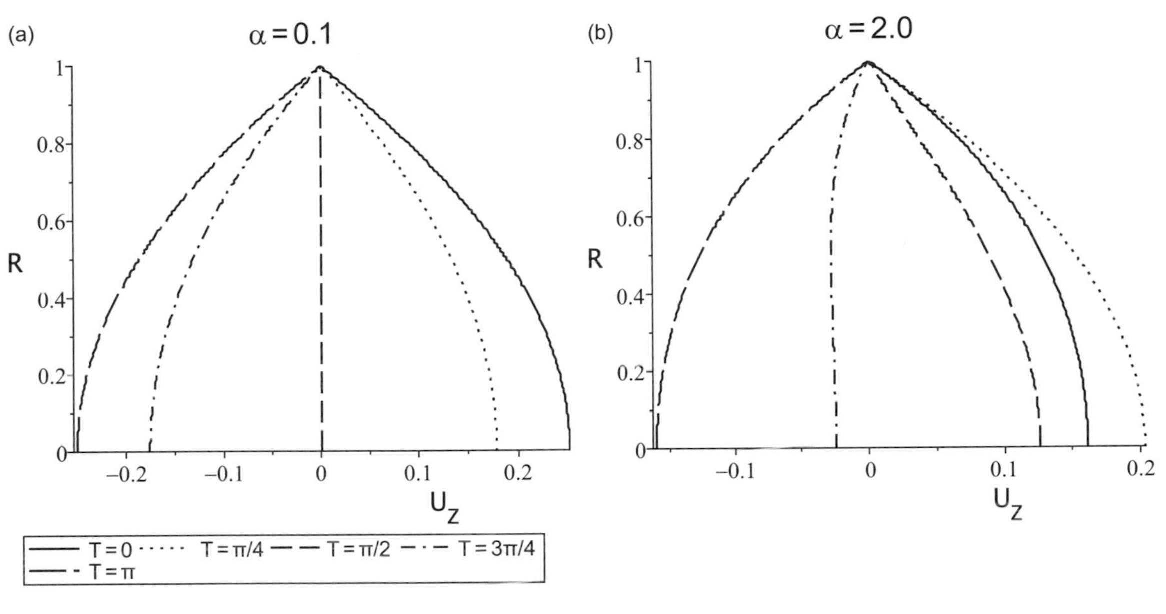

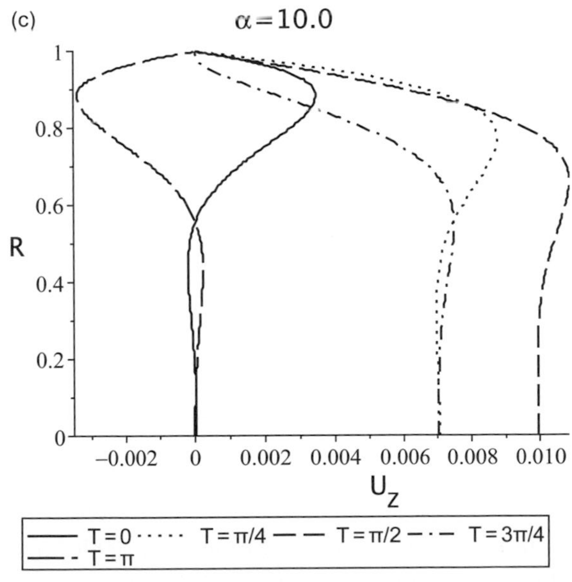

Figure 8.19 Velocity profiles $U_Z(R, T)$ for oscillatory tube flow at five values of time during the half cycle, $T = 0, \pi/4, \pi/2, 3\pi/4, \pi$, for $\alpha = 0.1, 2.0, 10.0$.

pressure gradient $dP/dz = \cos(\pi/2) = 0$, and the velocity is zero for all R. For $\alpha = 2$ in Figure 8.19(b) the fluid does not stop at $T = \pi/2$ but its inertia causes flow. For $\alpha = 10.0$ in Figure 8.19(c) the velocity profile is fairly flat in the core region. Since there is no shear for a flat profile, the core behaves as an inviscid flow region. The velocity profile then changes rapidly near the wall to satisfy the no-slip boundary condition, as we saw in the Stokes layer. For large enough values of α one can think of $1/\alpha = \sqrt{\nu/\omega}/a \sim \delta_s/a$ as the ratio of the Stokes layer thickness, $\delta_s \sim \sqrt{\nu/\omega}$, to the tube radius. As we saw in channel, the amplitude of the oscillatory velocities decreases with increasing α for constant pressure gradient amplitude, B^p.

The dimensionless flow rate is the integral of the velocity across the cross-section of the tube

$$Q(T) = \int_0^{2\pi}\int_0^1 U_Z(R,T)\, R\, dR\, d\theta$$
$$= \frac{2\pi}{\alpha^2}\Re\left\{ i e^{iT} \int_0^1 \left(1 - \frac{J_0(\lambda R)}{J_0(\lambda)}\right) R\, dR \right\} \quad (8.5.12)$$

where $Q = q/(-B^p a^4/\mu)$ and q is the dimensional flow rate. To integrate the Bessel function, we define a new variable, $\hat{R} = \lambda R$, so that

$$\int_0^1 J_0(\lambda R)\, R\, dR = \frac{1}{\lambda^2}\int_0^\lambda J_0(\hat{R})\, \hat{R}\, d\hat{R} \quad (8.5.13)$$

which has a known integral

$$\int J_0(\hat{R})\, \hat{R}\, d\hat{R} = \hat{R} J_1(\hat{R}) \quad (8.5.14)$$

where J_1 is the first-order Bessel function of the first kind. Applying the upper and lower limits gives us

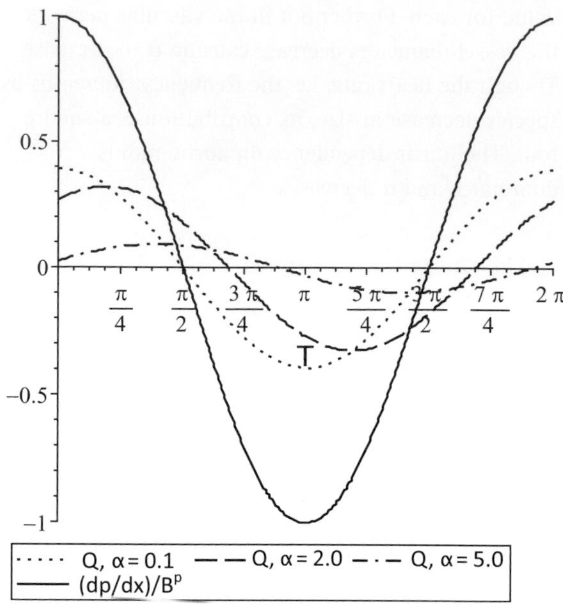

Figure 8.20 Q(T) for three choices of α and $(dp/dz)/B^p$.

$$\frac{1}{\lambda^2}\int_0^\lambda J_0(\hat{R})\, \hat{R}\, d\hat{R} = \frac{J_1(\lambda)}{\lambda} \quad (8.5.15)$$

Now Equation (8.5.12) simplifies to

$$Q(T) = \frac{2\pi}{\alpha^2}\Re\left\{ -i\left(\frac{1}{2} - \frac{J_1(\lambda)}{\lambda J_0(\lambda)}\right) e^{iT} \right\} \quad (8.5.16)$$

Plots of Q(T) for three choices of α, and $(dp/dz)/B^p = \cos(T)$ are shown in Figure 8.20, which is similar to the results we saw for oscillatory channel flow in Figure 8.10. Q increasingly lags the pressure gradient as α increases, similar to our solution in the case of oscillatory channel flow.

Worked Example 8.1

Because many problems in physics involve separation of variables in cylindrical coordinates, Bessel's Equation and Bessel Functions arise often. However, the entire concept that there are specialized functions which satisfy specific differential equations is sometimes puzzling to students. A simple example for additional insight is to consider a well-known differential equation for the function $f_1(x)$,

$$\frac{d^2 f_1}{dx^2} + f_1 = 0 \tag{8.5.17}$$

From experience we know that non-zero solutions to Eq. (8.5.17) are

$$f_1 = k_1 \sin x + k_2 \cos x \tag{8.5.18}$$

the sum of two linearly independent, tabulated, special functions as in Eq. (8.5.7). Since $\sin x$ and $\cos x$ have important trigonometric interpretations, we usually do not think of them as tabulated special functions which are defined as solutions to a specific differential equation like Eq. (8.5.17). But they are! What we found in the channel flow problem was that separation of variables led to the spatial component expressed in terms of $\sinh Y$, $\cosh Y$, or equivalently, e^{Y}, e^{-Y}, which are, again, tabulated functions. However, students generally have much more familiarity with exponential functions and trigonometric functions, compared to Bessel Functions.

8.6 Pulsatile Flow in a Tube

Pulsatile flow in a tube is the overlay of oscillatory flow with steady flow. The pressure gradient is the sum of two such terms

$$\frac{\partial p}{\partial z} = B^s + B^p \cos(\omega t) \tag{8.6.1}$$

The amplitudes are the steady amplitude, B^s, and the time-periodic amplitude, B^p. Since the governing equations are linear and the boundary conditions are zero velocity for both steady and unsteady flows, we can add the solutions we have derived for each. Then the total velocity solution is

$$u_z(r,t) = u_z^s(r) + u_z^p(r,t) \tag{8.6.2}$$

where u_z^s is the steady flow velocity field and u_z^p is the oscillatory, time-periodic solution. From Eq. (7.1.4) the steady pipe flow is

$$u_z^s = -\frac{B^s}{4\mu}\left(a^2 - r^2\right) \tag{8.6.3}$$

The oscillatory velocity field is given by Eq. (8.5.11), which dimensionally is

$$u_z^p = \frac{(-B^p)a^2}{\mu}\Re\left\{\frac{i}{\alpha^2}\left(1 - \frac{J_0(\lambda R)}{J_0(\lambda)}\right)e^{iT}\right\} \tag{8.6.4}$$

The sum of the two velocity fields, in dimensionless form, is the pulsatile flow solution

$$U_Z(R,T) = \frac{1}{4}\left(1 - R^2\right) + \beta\Re\left\{\frac{i}{\alpha^2}\left(1 - \frac{J_0(\lambda R)}{J_0(\lambda)}\right)e^{iT}\right\} \tag{8.6.5}$$

where $U_Z = u_z/(-B^s a^2/\mu)$ and $\beta = B^p/B^s$ is the ratio of the unsteady to steady pressure gradients.

Examples of the pulsatile velocity profiles are shown in Figure 8.21(a) for five values of time assuming $\alpha = 2$ and $\beta = 0.3$. Note that there is no backflow for these parameter choices. However, Figure 8.21(b) shows $\alpha = 10$ and $\beta = 35$ for eight values of time, which results in flatter profiles, a distinctive Stokes layer near the wall and flow reversal over part of the cycle. These features recreate

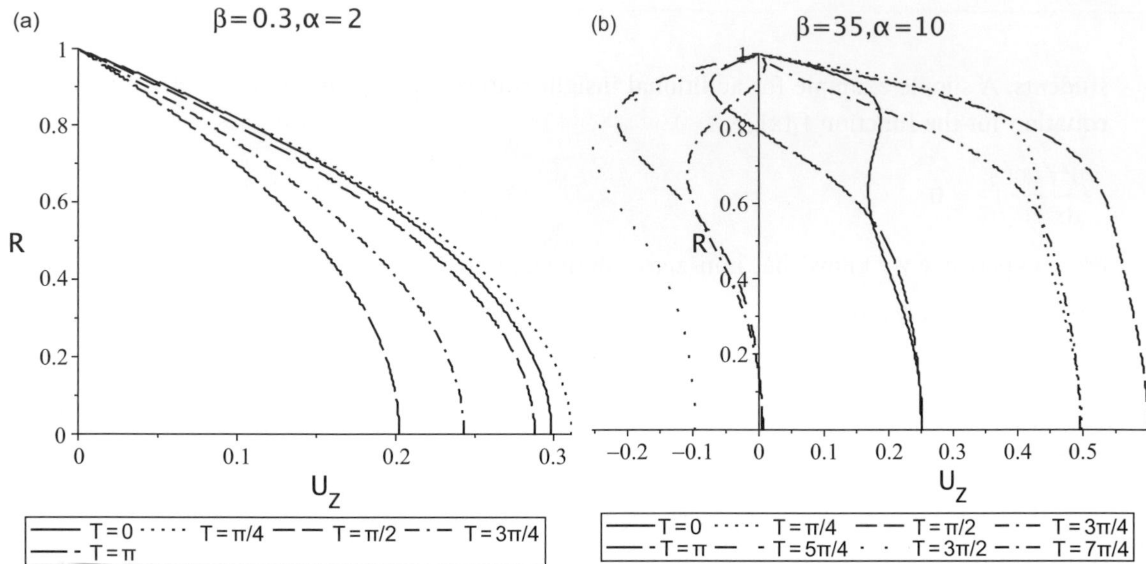

Figure 8.21 Velocity profiles $U_Z(R, T)$ for pulsatile pipe flow at several times during the cycle. (a) $\alpha = 2$, $\beta = 0.3$; (b) $\alpha = 10$, $\beta = 35$.

experimental observations in the dog descending aorta where hot-film anemometry was used to measure the centerline velocity (Ling et al., 1973). As we saw in the channel oscillatory flow, it is necessary to boost the oscillatory pressure gradient amplitude, B^p, as α increases to maintain an oscillatory stroke volume. In the present case, that means increasing β.

Box 8.2 | Pulsatile Pressure Gradient Fourier Analysis

The arterial pressure has a steady component, $p^s(z)$, and a time-periodic component, $p^p(z, t)$, so that $p = p^s + p^p$. For applications to real pulsatile arterial flows, the pressure gradient, dp/dz, must be known/measured and accurately represented. The unsteady term, $\partial p^p/\partial z$, is certainly not a simple cosine function of time as we have modeled so far. Fortunately, since the governing equations are linear, it is useful to decompose $\partial p^p/\partial z$ into its Fourier components, or harmonics, that are integer multiples of the fundamental frequency ω.

Figure 8.22(a) shows a typical unsteady pressure gradient, $\partial p^p/\partial z$, measured in the femoral artery of dogs by subtracting simultaneous pressure readings, Δp, taken at two nearby axial locations a distance, Δz, apart (McDonald, 1955). Forming their ratio for the gradient of the unsteady pressure, the following Fourier components of the resulting signal emerge

$$-\left(\frac{\partial p^p}{\partial z}\right) = D_1 \cos(T + \phi_1) + D_2 \cos(2T + \phi_2) + D_3 \cos(3T + \phi_3) + D_4 \cos(4T + \phi_4) \tag{8.6.6}$$

where $T = \omega t$ and the units are mmHg/cm. A typical pulse rate for the dogs was $f = 150$ bpm $= 2.5/\mathrm{s}$ making $\omega = 2\pi f = 15.7$ rad/s. Consequently, the dimensionless time-axis in Figure 8.22(a) corresponds to $0 \le t \le 1/f$ or $0 \le t \le 0.4\mathrm{s}$. The amplitudes are $D_1 = 0.78$, $D_2 = 1.32$, $D_3 = -0.74$, $D_4 = -0.41$. The phase angles, ϕ_n, in degrees are $0°39'$, $-82°45'$, $26°30'$, $-16°39'$ for $n = 1, 2, 3, 4$, respectively, and in radians are $\phi_1 = 0.0036\pi$, $\phi_2 = -0.4597\pi$, $\phi_3 = 0.1472\pi$, $\phi_4 = -0.0925\pi$. The individual harmonics are shown in Figure 8.22(b), where, notably, it is the second harmonic, $2\omega t$, which is dominant. The maximum positive deflection, 3.1, is larger than the maximum negative deflection, -2.1, weighting flow in the positive z-direction. The steady pressure gradient measured to be $-(dp^s/dz) = 0.13$ mmHg/cm yielding a steady flow of 1 ml/s. This value of (dp^s/dz) is ~1/25th of the maximum positive deflection, justifying our selection of $\beta = 35$ in Figure 8.21. Each Fourier component of $\partial p^p/\partial z$, then, is a forcing term in the corresponding Bessel equation to calculate its contribution to fluid velocity and flow.

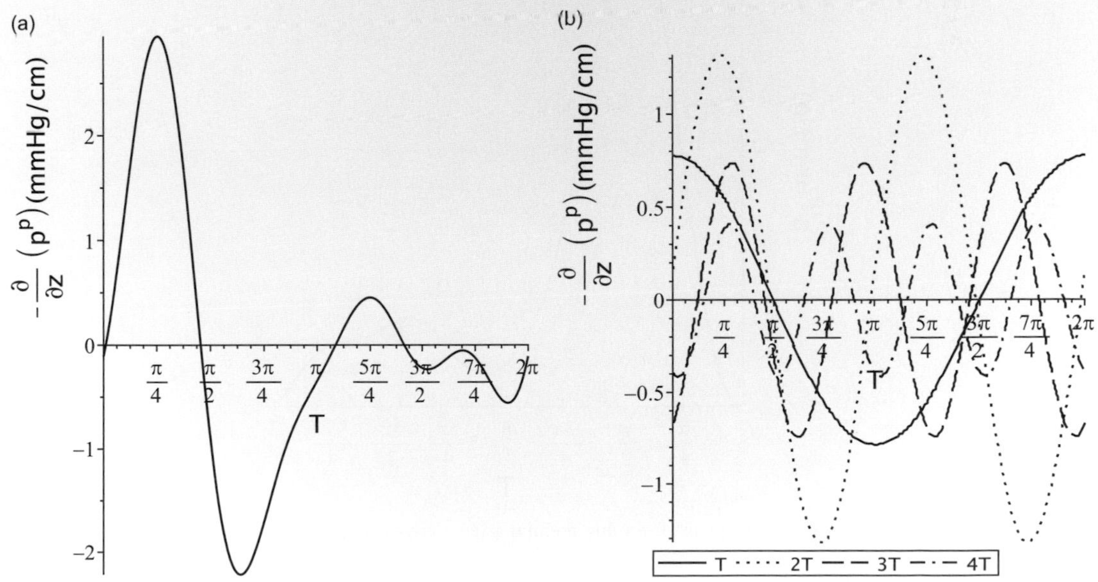

Figure 8.22 (a) Example of $\partial p^p/\partial z$ in a typical artery; (b) Fourier components of (a).

In Section 9.5 we will learn that the pressure wave propagates at speed c along an artery due to its elastic walls. For example, a simple traveling sine wave is given by $f = \sin(k(z - ct))$, which moves in the positive z-direction at speed c, wavelength $\lambda = 2\pi/k$ and frequency $\omega = kc$. The z-derivative of f is $\partial f/\partial z = k\cos(k(z - ct))$ and its time derivative is $\partial f/\partial t = -kc\cos(k(z - ct))$, telling us that $(\partial f/\partial z) = -(1/c)(\partial f/\partial t)$. Use this relationship to change variables in Eq. (8.6.6) such that $-\partial p^p/\partial z = (1/c)(\partial p^p/\partial t)$ and integrate with respect to time, t, to find the unsteady pressure at a fixed value of z

$$p^p = \frac{c}{\omega}\int^{T} D_1\cos(T + \phi_1) + D_2\cos(2T + \phi_2) + D_3\cos(3T + \phi_3) + D_4\cos(4T + \phi_4)\,dT + B_0 \tag{8.6.7}$$

where B_0 is an integration constant and we changed variables to dimensionless time, T. Choosing B_0 to set the initial pressure relative to zero, the resulting pressure wave form is shown in Figure 8.23 where we chose $c = 7$ m/s as a typical wave speed for the dog femoral artery. Note the rapid rise and fall in the first half of the cycle, again dominated by the second harmonic, while the second, smaller peak is influenced by higher harmonics.

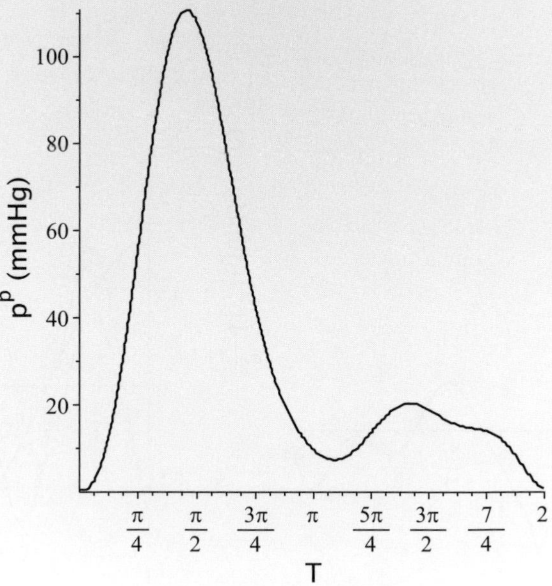

Figure 8.23 Unsteady pressure wave form, p^p, for a dog femoral artery derived from Eq. (8.6.6).

8.7 Mucus Transport from a Cough

The liquid lining of the lung's airways is an important defense barrier to inhaled particles such as dust, soot, bacteria, viruses, spores, etc. We learned in Section 3.9 about particle impaction in the lung by studying the critical Stokes number for a particle in stagnation point flow as a model of flow at a branching airway. Particle deposition occurs in these carinal regions and elsewhere along the airway surface. Since the airway surface is coated with a liquid film, as discussed in Section 7.2 with Figure 7.9, the particles stick to the film. It is the action of cilia underneath the film which beat rhythmically (10–20 beats per second) to propel the liquid toward the mouth where it is swallowed. More details of mucociliary propulsion are discussed in Section 17.8. The ciliary effectiveness can be reduced by smoking, for example, which can paralyze the cilia. It is common for smokers to accumulate an excess of mucus, which then has to be expelled by coughing. Asthmatics have similar issues of increased airway mucus, and coughing is needed to assist the transport.

Consider a simple model of a cough as shown in Figure 8.24 in which a thin viscous layer of thickness h, representing the combined serous/mucus bilayer of the airway liquid lining, is subjected to an unsteady flow from the air above it. That air flow imposes both an axial pressure gradient, dp/dx, within the liquid as well as a shear stress, τ_c, at the air–liquid interface.

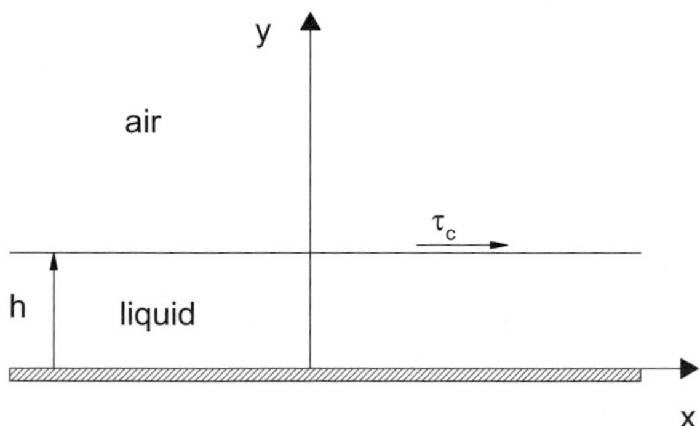

Figure 8.24 Simple model of a cough.

We start with the x-component of the Navier–Stokes equation in the layer, assuming unidirectional flow.

$$\rho\frac{\partial u}{\partial t} = -\frac{\partial p}{\partial x} + \mu\frac{\partial^2 u}{\partial y^2} \tag{8.7.1}$$

The shear stress wave form can be approximated by an exponential decay function, as shown in Figure 8.25, given by the relationship

$$\tau_c(t) = \tau_0 e^{-\frac{t}{T_0}} \tag{8.7.2}$$

with duration of the cough ~ $5T_0$ and maximum shear stress τ_0 which occurs at $t = 0$. If the length of the airway is L, then we choose the following scalings for our variables

$$X = \frac{x}{L}, \quad Y = \frac{y}{h}, \quad T = \frac{t}{T_0}, \quad U = \frac{u}{U_s}, \quad P = \frac{p}{\tau_0} \tag{8.7.3}$$

For the liquid layer we have the constitutive relationship $\tau_{12} = \mu\partial u/\partial y$. Inserting our scales, including scaling τ_{12} on τ_0, into this equation gives us the balance that $\tau_0 \sim \mu U_s/h$, from which we define U_s as $U_s = \tau_0 h/\mu$.

Inserting the scalings of Eq. (8.7.3) into Eq. (8.7.1) gives us the dimensionless form

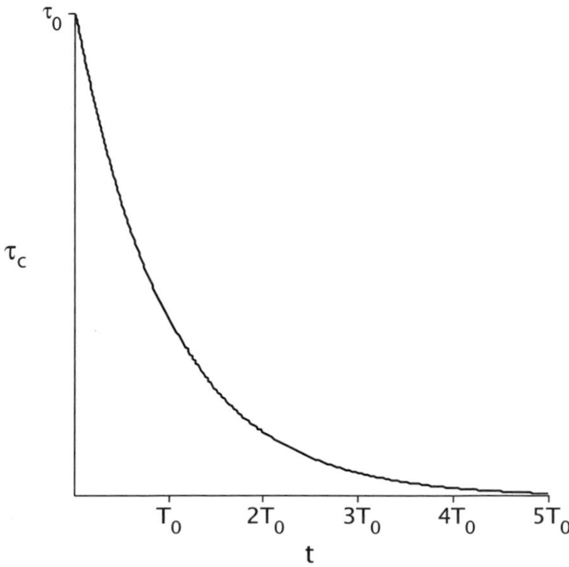

Figure 8.25 Shear stress applied to the thin viscous film.

$$\left(\frac{h^2}{\nu T_0}\right)\frac{\partial U}{\partial T} = -\left(\frac{h}{L}\right)\frac{\partial P}{\partial X} + \frac{\partial^2 U}{\partial Y^2} \tag{8.7.4}$$

Let's compute a typical value of the unsteady term coefficient, $h^2/\nu T_0$, which is a form of the Womersley parameter. The viscosity of mucus is a complicated topic discussed later in the Chapters 10 and 14. It is non-Newtonian. However, we can use the constant values from Section 7.4 as $\mu = 1 - 10\ \text{g·cm}^{-1}\text{·s}^{-1}$(poise). In disease, mucus viscosity can be even larger. In normal health, the film thickness is $h \sim 10\,\mu\text{m}$ (Song et al., 2009), but can increase in disease. Let's set it to $h = 50\ \mu\text{m} = 5 \times 10^{-3}$ cm. A typical cough duration for healthy subjects is approximately $5T_0 = 0.5\,\text{s}$, which can be estimated by personal experience. Assuming the mucus density is similar to water, $1\,\text{g/cm}^3$, then the dimensionless parameter estimate we seek is $(h^2/\nu T_0) = (5 \times 10^{-3})^2/(10/1)(0.2) = 1.25 \times 10^{-5}$. So we can readily neglect the unsteady inertial term in Eq. (8.7.4). The other dimensionless parameter, h/L, multiplying the pressure gradient term is also very small. Airway lengths are roughly $L \sim 6a$, where a is the airway radius, and the liquid linings are thin compared to the airway radius, $h/a \sim 10^{-2}$. Neglecting the pressure gradient within the thin liquid film, then, is also justified. That leaves us with

$$\frac{\partial^2 U}{\partial Y^2} = 0 \tag{8.7.5}$$

and the fluid motion in the layer comes only from the applied shear stress from the airflow, τ_c.

The boundary conditions in dimensional form are

$$u(y=0) = 0 \qquad \text{no-slip}$$
$$\mu \left.\frac{\partial u}{\partial y}\right|_{y=h} = \tau_c(t) \qquad \text{continuous shear} \tag{8.7.6}$$

Using the scales of Eq. (8.7.3) leads to the dimensionless form of Eq. (8.7.6),

$$U(Y=0) = 0, \quad \left.\frac{\partial U}{\partial Y}\right|_{Y=1} = \tau(T) \tag{8.7.7}$$

where $\tau = \tau_c/\tau_0$ is the dimensionless shear stress, which is now

$$\tau(T) = e^{-T} \tag{8.7.8}$$

Solutions to Eq. (8.7.5), after integrating partially twice with respect to Y, are

$$U(Y,T) = f_3(T)Y + f_4(T) \tag{8.7.9}$$

So the solution is a linear function of Y with time-dependent integration functions. Imposing the no-slip boundary condition makes $f_4 = 0$, and then the shear stress condition at the interface makes $f_3(T) = \tau(T)$. The final form for U, then, is

$$U(Y,T) = \tau(T)Y = e^{-T}Y \tag{8.7.10}$$

A plot of Eq. (8.7.10) for U vs Y at several values of T is shown in Figure 8.26. The linear profile decreases with time during the cough.

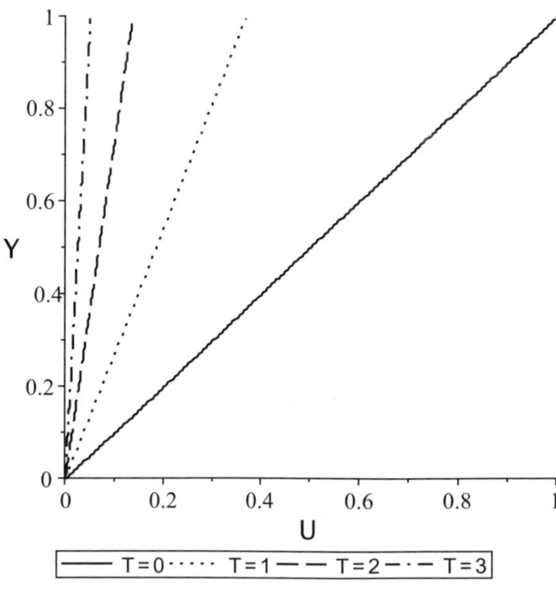

Figure 8.26 Cough flow field in the mucus layer of the lung.

Clinical Correlation 8.3 | Cough Effectiveness

The ability to cough effectively is essential for health. Individuals with compromised cough are susceptible to accumulating mucus in their lungs where inhaled contaminants can collect and cause pneumonia (bacteria, virus, fungus). Several situations reduce or prevent the ability to cough. Among these are post-surgical patients, especially abdominal or thoracic surgery, who refrain from coughing effectively due to the pain. Others are those with neuromuscular conditions such as spinal cord injury, muscular dystrophy, multiple sclerosis, amyotrophic lateral sclerosis (Lou Gehrig's Disease) and polio (Sivasothy et al., 2001). In addition, the mucus itself can become very viscous and viscoelastic in certain diseases, like cystic fibrosis, making it more difficult to shear and flow. Viscoelasticity is discussed in Chapter 14.

How can a cough be more effective? Let's examine our results. From Eq. (8.7.10) and the scalings of Eq. (8.7.3), the corresponding dimensional velocity is

$$u(y,t) = \frac{\tau_0}{\mu} e^{-\frac{t}{T_0}} y \tag{8.7.11}$$

The dimensional flow rate, q, is the integral of u across the layer and into a unit depth in the z-direction

$$q(t) = \int_0^{L_z}\int_0^h u(y,t)\, dy\, dz = \frac{\tau_0 h^2 L_z}{2\mu} e^{-\frac{t}{T_0}} \tag{8.7.12}$$

The volume of the layer, V_c, that is propelled by the cough is the integral of q(t) over the cough duration, say $0 \leq t \leq 5T_0$,

$$V_c = \int_0^{5T_0} q(t)\, dt = \frac{\tau_0 h^2 L_z T_0}{2\mu} \tag{8.7.13}$$

As Eq. (8.7.13) shows, ways to increase the effectiveness of a cough is to increase the effort level and time duration, τ_0 and T_0. If pain is restricting the cough, then treating the pain is indicated for better cough. If neuromuscular diseases are involved, then improving overall strength within the limits of the illness is important, ranging from better nutrition to breathing exercises. We also see that reducing μ increases V_c. Clinically that is achieved by providing moisture in the breathing circuit to add water to the mucus and dilute it as much as possible. Also, mucolytic agents can be applied which break some of the bonds between mucus molecular components and thereby reduce the viscosity.

Summary

In this chapter we have examined unsteady viscous flows which are important in biomedical applications. The initial concepts from Stokes' first and second problems are fundamental to all branches of fluid mechanics, including the diffusion of shear (or vorticity) from the suddenly started wall to the development of the Stokes layer from an oscillating wall. The former gave us an opportunity to use the similarity solution method while the latter yielded to the separation of variables method. Introducing a second, parallel wall gave an opportunity to utilize numerical solution methods to the suddenly started wall. In addition to unsteadiness generated by boundary motion, in biofluid mechanics there are important instances of oscillatory flows generated by oscillatory pressure gradients, vascular and respiratory, which we explored at length with clinical applications to pulsatile flow in arteries and small tidal volume, high frequency ventilation in the mechanical ventilation of humans, especially neonates.

Oscillatory flow, either in a channel or tube, is basically the same physics. However, the analysis is more challenging for the tube since the separation of variables technique leads to a Bessel equation, which has Bessel functions with a complex argument as solutions. In either case, we have the competition of pressure gradient, viscous, and unsteady inertial terms. For small Womersley parameter, $\alpha << 1$, unsteady inertia is not important and the balance is between the pressure gradient and the viscous effects. Flow and pressure gradient are, more or less, in phase. For a thin mucus film the value of α was negligible, as was the influence of the pressure gradient within the film, so we ignored both unsteady inertia and the pressure gradient. This let us simplify a cough analysis. As α increases so that $\alpha >> 1$ in the channel or tube, we see the development of a Stokes layer near the boundary where viscous effects are important, and essentially an inviscid plug flow in the core of the channel or tube where unsteady inertia dominates. Pressure gradient and flow are out of phase, the flow lag increasing as α increases. To keep the stroke volume constant, the pressure gradient amplitude needs to increase as α increases.

Problems

8.1 When we analyzed Stokes' first problem of the suddenly started wall the similarity method led to the general solution for the dimensionless velocity,

$$U(\eta) = c_1 \int_0^{\eta} e^{-\eta'^2}\, d\eta' + c_2$$

For the suddenly started wall, the boundary conditions which solved c_1 and c_2 were $U(\eta = 0) = 1$ and $U(\eta \to \infty) = 0$.

a. Now consider the fluid and wall both are moving at speed 1, but then at $t = 0$ the wall suddenly stops, while the fluid at infinity is still traveling at speed 1. What is the solution for $U(\eta)$ now? Plot it over the range $0 \leq \eta \leq 3$ and describe your result.
b. Set $\nu = 1\ \text{cm}^2/\text{s}$ and plot $U(y)$ over the range $0 \leq y \leq 3\,\text{cm}$ for $t = 0.01, 0.1, 0.3, 0.5$ s.
c. Now plot U for $y = 0.1, 0.5, 1, 5$ cm over the range $0 \leq t \leq 2\text{s}$. At a distance of 1 cm, how long does it take for the velocity to be reduced by half?

8.2

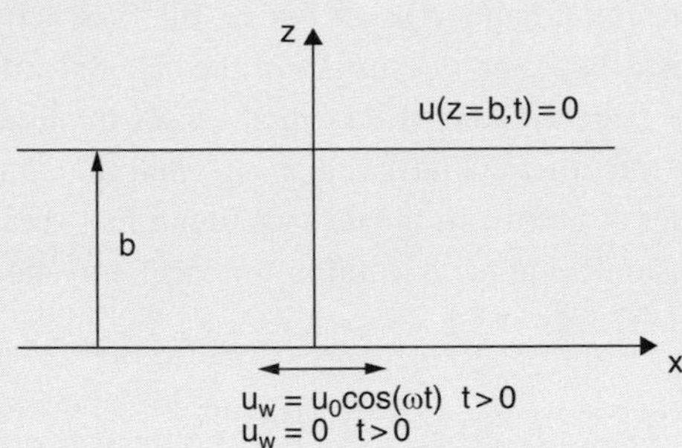

Stokes' second problem with a stationary second wall and startup.

Now consider Stokes' second problem with a stationary second wall which is parallel to the moving wall. Instead of seeking only the time-periodic solution, we will include the startup from zero fluid velocity. Your computational software will be handy for this goal.

The same governing equation applies,

$$\frac{\partial u}{\partial t} = \nu \frac{\partial^2 u}{\partial z^2} \tag{1}$$

which is a one-dimensional diffusion, or heat, equation. The fluid starts out at rest for $t = 0$, $u(z, t = 0) = 0$. Then the lower wall oscillatory motion is applied for $t \geq 0$. The boundary and initial conditions are

$$\begin{aligned} u(z = 0, t \geq 0) &= u_0 \cos(\omega t) \\ u(z = b, t) &= 0 \\ u(z, t = 0) &= 0 \end{aligned} \tag{2}$$

a. Scale the problem as follows

$$U = \frac{u}{u_0}, \zeta = \frac{z}{b}, T = \omega t \tag{3}$$

and show that the dimensionless forms of Eq. (1) and Eqs. (2) are

$$\begin{aligned} \alpha^2 \frac{\partial U}{\partial T} &= \frac{\partial^2 U}{\partial \zeta^2} \\ U(\zeta = 0, T = 0) &= \cos T \\ U(\zeta = 1, T) &= 0 \\ U(\zeta, T = 0) &= 0 \end{aligned} \tag{4}$$

where α is the Womersley parameter.

b. Set $\alpha = 0.1$ and solve the dimensionless governing equations using your software's capability for numerical solutions of the one-dimensional, unsteady heat equation. Try a spatial step size of 0.01. Plot $U(\zeta)$ over the range $0 \leq \zeta \leq 1$ for $T = T_0, T_0 + 2\pi, T_0 = 4\pi$ where $T_0 = \pi/16$. Try plotting the T_0 curve with a solid line and the other two curves using symbols, say circles and boxes, for 20 points over the domain. That will make it easier to see if the curves overlap. What do you see? Why?
c. Repeat part (c) for $\alpha = 1, 10$. Now what do you see? Why? Does the startup effect dissipate after 1, 2 or more cycles?
d. Create an animation of part (c) for the $\alpha = 10$ solution and run it for two cycles with 100 frames steps. Describe your result in terms of the Stokes layer relative to the domain width.

8.3 In the text we used computational methods to solve for Stokes' first problem, the suddenly started lower wall with a stationary, parallel second wall shown in Figure 8.4. The dimensionless governing equations are given in Eq. (8.1.21) and the solution shown in Figure 8.5, where we see the velocity profile approaches the steady solution, $U^s = 1 - \eta$.

Now consider that we have established this steady profile and suddenly stop the lower wall.

a. Show that the dimensionless governing equations are

$$\frac{\partial U}{\partial T} = \frac{\partial^2 U}{\partial \eta^2}$$

$$U(\eta, T = 0) = 1 - \eta, U(\eta = 0, T) = 0$$

$$U(\eta = 1, T) = 0$$

b. Solve the system numerically and plot velocity profiles for $T = 0.002, 0.02, 0.05, 0.1, 0.2, 0.5$.
c. Animate your solution for $0 \leq T \leq 0.25$ using 100 frames. Describe the results.

8.4 Pulsatile flow in a channel of width 2b is the overlay of oscillatory flow with steady flow. The pressure gradient is the sum of two such terms $\partial p/\partial x = B^s + B^p \cos\omega t$. The amplitudes are the steady component, $B^s < 0$, and the time-periodic component, $B^p < 0$. Since the governing equations are linear and the boundary conditions are zero velocity for both steady and unsteady flows, we can add the solutions we have derived for each. Then the total velocity solution is $u^T = u^s(y) + u^p(y,t)$, where u^s is the steady solution given in Eq. (6.5.14) and u^p is the oscillatory solution derived from the dimensional form of Eq. (8.3.18) as $u^p = \left((-B^p)b^2/\mu\right) U(Y,T)$.

a. Define the dimensionless total velocity, $U^T = u^T/(-B^s b^2/\mu)$, by adding the steady and unsteady components. Use the scaled variables, $Y = y/b, T = \omega t$ and let the ratio of pressure gradients be $\beta = B^p/B^s$.
b. Animate $U^T(Y,T)$ for $\alpha = 2.0, \beta = 0.3$.
c. Repeat part (b) for $\alpha = 10, \beta = 60$. What are two major differences compared to part (b)?
d. Define the dimensionless flow rate as $Q = \int_{-1}^{1} U^T\,dY$ and plot Q vs T for one period using the two sets of parameters in part (b) and (c). Compare the flow rates.
e. Define the dimensionless wall shear as $\tau_w(T) = (\partial U/\partial Y)_{Y=1}$. Plot $\tau_w(T)$vs T for $\alpha = 2.0, \beta = 0.3$ and $\alpha = 10, \beta = 60$ and compare the two results. Hint: for the periodic term take the Y derivative inside the brackets, $(\partial U^p/\partial Y)_{Y=1} = \Re\left((\partial G/\partial Y)_{Y=1}\, e^{iT}\right)$.
f. How does wall shear affect endothelial cells? Hint: explore the literature.

8.5

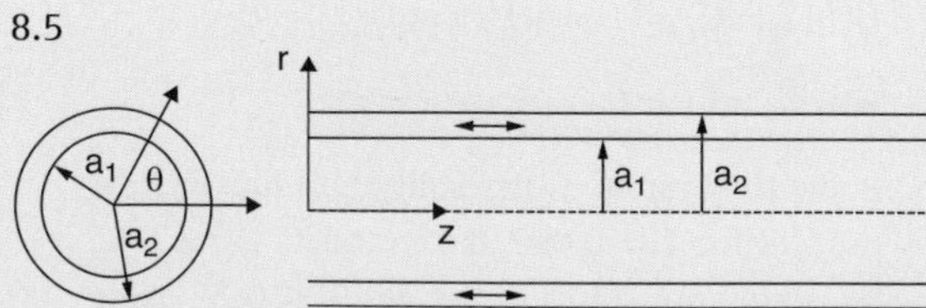

For the aortic balloon assist device, the flow between the aortic wall and the surface of the balloon catheter is flow between concentric cylinders, see the figure, where the inner cylinder is at $r = a_1$ and the outer cylinder at $r = a_2$. In the text we found that the dimensionless general solution for oscillatory flow in a cylinder was

$$U_Z(R,T) = \frac{u_z}{(-B^p)a_2^2/\mu} = \Re\{G_0(R)e^{iT}\}$$

letting $a = a_2$, where u_z is the dimensional velocity and

$$G_0(R) = A_1 J_0(\lambda R) + A_2 Y_0(\lambda R) - \frac{i}{\alpha^2} \qquad \lambda = (-i)^{1/2}$$

$$\alpha = \left(\frac{i-1}{\sqrt{2}}\right)\alpha$$

For our purposes regarding flow between concentric cylinders, consider $R = r/a_2$, so there is no-slip at the outer cylinder, $R = 1$, and the inner cylinder, $R = B = a_1/a_2$. Then the range of R is $B \le R \le 1$.

a. Solve A_1, A_2 for $G_0(R)$ imposing the no-slip boundary conditions at $R = B, 1$. Plot W(R,T) for $T = 0, \pi/4, \pi/2, 3\pi/4$ with $\alpha = 1,\ B = 0.8$. Repeat for $\alpha = 1,\ B = 0.5$. Compare your results.
b. Now solve for $G_0(R)$ imposing the boundary conditions and plot W(R,T) for $T = 0, \pi/4, \pi/2, 3\pi/4$ with $\alpha = 10, B = 0.8$. Repeat for $\alpha = 10,\ B = 0.5$. Compare your results.
c. The dimensionless flow rate is $Q = \mathrm{Re}\left\{\int_B^1 W R\,dR\ e^{iT}\right\}$. Plot Q(T) over one cycle for the parameter combinations: $\alpha = 1, B = 0.8$, $\alpha = 5, B = 0.8$, $\alpha = 1, B = 0.5$ and $\alpha = 5, B = 0.5$. Discuss your results.
d. The dimensionless shear stress is $\tau = \mathrm{Re}\{(dG_0/dR)e^{iT}\}$. Evaluated at the walls define $\tau_1 = \tau(R = 1)$, $\tau_B = \tau(R = B)$. Plot τ_1, τ_B for $\alpha = 1, B = 0.8$ and $\alpha = 10, B = 0.8$. Where is the wall shear largest?

e. Repeat part (d) for $\alpha = 1$, $B = 0.5$ and $\alpha = 10$, $B = 0.5$. Now where is the largest shear stress?

8.6 The pulsatile pressure gradient from Section 8.6 has the steady and oscillatory components as follows

$$-\frac{dp^s}{dz} = 0.13 \quad \text{mmHg/cm}$$

$$-\left(\frac{\partial p^p}{\partial z}\right) = D_1 \cos(\omega t + \phi_1) + D_2 \cos(2\omega t + \phi_2) + D_3 \cos(3\omega t + \phi_3) + D_4 \cos(4\omega t + \phi_4) \quad (1)$$

where the units of D_n are mmHg/cm. Values of D_n and ϕ_n are given in the text.

a. The steady component of Eq. (1) leads to a steady flow of $q^s = 1$ ml/s. Assuming blood viscosity is 0.04 poise, what is the radius of this artery under Poiseuille flow? Hint: change mmHg to dynes/cm^2.
b. Let's consider each Fourier component of the unsteady pressure gradient as $-\partial p_n^p/\partial z = D_n \cos(T_n)$ where $T_n = \omega_n t + \phi_n$ and eventually $\omega_n = n\omega$. Then $-\partial p^p/\partial z$is the sum of these four components, $n = 1, 2, 3, 4$, in Eq. (1). The corresponding dimensional flow rate component, from Eq. (8.5.16), is

$$q_n(T_n) = \frac{2\pi D_n a^4}{\mu \alpha_n^2} \Re\left\{-i\left(\frac{1}{2} - \frac{J_1(\lambda_n)}{\lambda_n J_0(\lambda_n)}\right)e^{iT_n}\right\} \quad (2)$$

where $\alpha_n^2 = \omega_n a^2/\nu$ and $\lambda_n = (i-1)\alpha_n/\sqrt{2}$ Plot the total flow sum, $q = q^s + q_1 + q_2 + q_3 + q_4$, vs $T = \omega t$ for one cycle, $0 \le T \le 2\pi$. Describe your results.
c. What is the maximal forward flow rate and when in the cycle does it occur? Same questions for the maximal backward flow rate.
d. Plot each unsteady Fourier flow component separately on the same graph and describe your result. Is there a dominant component? If so, which one and why?

8.7

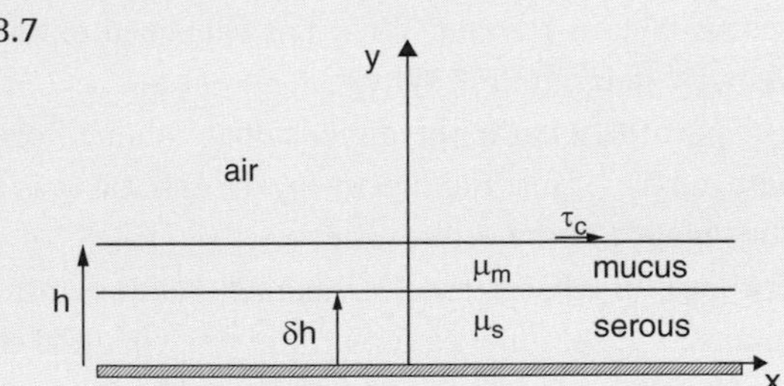

In class we discussed a single fluid layer on a wall and subjected to a shear stress on its upper surface given by

$$\tau_c(t) = \tau_0 e^{-\frac{t}{T_0}}$$

Consider the two-layer system as shown in the figure, representing the serous and mucus layers. The serous layer has thickness δh, where $0 < \delta < 1$, and viscosity, μ_s. The mucus layer has thickness, $(1-\delta)h$, and viscosity, μ_m. The full thickness of both layers is h. Typically the mucus viscosity is much larger than the serous layer viscosity, $\beta = \mu_m/\mu_s$ where $100 < \beta < 1000$ and $\mu_s \sim 0.01$ poise is similar to water.

Neglecting inertia and assuming thin films, the velocity profiles for the serous fluid layer, $u_s(y,t)$, and the mucus layer, $u_m(y,t)$, are the linear forms

$$u_s(y,t) = a_s(t) + b_s(t)y, \quad u_m(y,t) = a_m(t) + b_m(t)y$$

a. Solve for the coefficients $a_s(t)$, $b_s(t)$, $a_m(t)$, $b_m(t)$ by imposing no-slip at $y = 0$, continuous velocity and shear stress at $y = \delta h$, and the imposed cough shear stress at $y = h$.
b. Define the dimensionless variables $U_s = \mu_s u_s/\tau_0 h$, $U_m = \mu_s u_m/\tau_0 h$, $T = t/T_0$, $Y = y/h$ and the viscosity ratio $\beta = \mu_m/\mu_s$. Plot the two velocity profiles in their respective Y-ranges on the same graph for $\delta = 1/2$, $\beta = 10$ and $T = 0, 1, 2, 4$ and describe your result. What happens for $\beta = 100$?
c. Repeat part (b) for a dehydration state where $\delta = 1/4$ and the total thickness is 3/4 so only the

serous layer is thinner. What has happened to the velocity magnitudes? Why?

d. For part (b), what is the dimensional volume flow rate, $q_m(t)$, of just the mucus layer? Assume L_z is a unit length in the z-direction.

e. For part (d) what is the dimensional volume of the mucus, v_m, that moves forward in one cough? Let the duration of the cough be $0 \leq t \leq 5T_0$. You can assume $e^{-5} \sim 0$.

f. For part (e) let the dimensionless mucus volume be $V_m = \frac{\mu_s v_m}{\tau_0 h^2 T_0 L_z}$ and define the viscosity ratio as $\beta = \mu_m/\mu_s$. Plot V_m over the range $0.01 \leq \delta \leq 0.99$ for five values of $\beta = 1, 2, 5, 10, 100$ and describe your result. Dehydration can reduce the thickness of the serous layer, i.e. reduced $\delta < 1/2$. How does that affect mucus clearance.

g. For part (f) calculate the limit of $V_m(\delta)$ as $\beta \rightarrow \infty$. In this limit, what value of δ gives the maximal value of V_m ? Is that a physiologic value?

h. For part (f), now calculate the total volume, V_{tot}, which is the sum of the two layers also using the same scales as V_m. Plot your solution over the range $0.01 \leq \delta \leq 0.99$ for five values of $\beta = 1, 2, 5, 10, 100$ and describe your result.

i. For part (h) calculate the limit of $V_{tot}(\delta)$ as $\beta \rightarrow \infty$. Describe your result in relation to your velocity field plots of parts (b) and (c).

Image Credits

Figure 8.15a Source: Figure 7b from Eckmann DM, Grotberg JB. Oscillatory flow and mass transport in a curved tube. *J Fluid Mech.* 1988; 188:509–27.

Figure 8.15b Source: Figure 4 from Dragon CA, Grotberg JB. Oscillatory flow and dispersion in a flexible tube. *J Fluid Mech.* 1991; 231:135–55.

9 Flow in Flexible Tubes

Introduction

Flow through flexible tubes is a significant aspect of biofluid mechanics, especially in the vascular and respiratory systems. For rigid boundaries like a pipe, we were content to calculate fluid stresses on the pipe wall whose position was prescribed. This was important for estimating fluid shear stresses on the wall which would be imparted to cells which form its internal layer. When the wall is flexible, however, the fluid stresses change the shape of the wall, which, in turn, modifies the fluid stresses. In this respect, the location of the wall is an unknown which must be solved, and the additional equation allowing that is a solid mechanical equation for the wall internal stresses and configuration. In general fluid mechanics terminology, these are fluid–structure interactions (FSI), i.e. the wall stresses and fluid stresses influence one another to determine the shape of the boundary. The level of detail or sophistication when analyzing flow through flexible tubes can vary with the wall model, the fluid model, or both. For the tube wall some models take a global view of relating its cross-sectional area to the cross-sectionally averaged fluid pressure, regardless of the shape of the cross-section, which can be circular when inflated or more flattened when compressed. Other wall models make a point-by-point force balance along the fluid–wall interface. The fluid itself can be inviscid, viscous via a friction factor (see Chapter 13) added to the Bernoulli Equation, or fully viscous using the Navier–Stokes equations. Interesting phenomena are associated with these flows. For example, in rigid tube flows we find that increasing the driving pressure increases the volumetric flow rate. However, flexible tube flows can exhibit a maximal flow rate which becomes independent of further increases in driving pressure. This flow limitation phenomenon is an underlying mechanism in lung airway diseases like asthma and emphysema which lead to reduced gas exchange and shortness of breath. Often this flow limitation is accompanied by wheezing breath sounds which is a fluid-elastic instability. Vessel flexibility allows the sensation of an arterial pulse that is felt in the wrist and other locations, and from which the heart beat rate per minute is often ascertained. Pulse propagation in the branching arterial system brings into play the events at a bifurcation where the pulse pressure wave can have both reflected and transmitted components depending on the conditions. Heart rate is one of the most fundamental vital signs collected by health care professionals around the world, and is usually recorded for every patient contact. In addition, measuring blood pressure using an arm cuff, called a sphygmomanometer, is entirely dependent on collapsing the artery in the upper arm (brachial artery) by inflating the cuff and then deflating it, listening to Korotkoff sounds, which are another form of fluid-elastic instability.

Topics Covered

We start by analyzing steady viscous flow in a flexible tube which has a linear relationship between its radius and the pressure difference across its wall, a linear "tube law." For sufficient assumptions, inertia is ignored and conditions vary slowly in the axial flow direction. The result is locally Poiseuille flow; however, in this case the calculated parabolic profile is different at each axial position and there is no maximal flow limitation. Then we discuss a nonlinear tube law relating cross-sectional area to internal pressure, which is often used to model physiologically relevant collapsible tubes. Flows in these tubes can reach a maximum either through viscous effects or inertial effects, the latter determined by wave speed properties of the tube. Next we investigate pulse propagation in flexible tubes, related to the arterial system, and analyze what happens when a pulse wave reaches a bifurcation in the arterial tree with reflected and transmitted waves, which can lead to complex signals. Flow-induced oscillations underlying wheezing breath sounds and Korotkoff sounds heard while measure blood pressure are investigated and, finally, lumped parameter models are reviewed.

9.1 Steady Viscous Flow in a Flexible Tube

Consider viscous flow in a flexible tube as shown in Figure 9.1. Fluid enters the tube at $z = 0$ where the radius is a_0 and exits at $z = L$ where the radius is a_L. The pressure outside the tube is a constant p_{ext}. As the fluid flows down the tube the internal pressure, p, decreases with increasing z, so the radius, a(z), also decreases. The dashed lines in Figure 9.1 are parallel for the limiting case of a rigid tube where the radius is constant, $a(z) = a_0 = a_L$, and the flow is Poiseuille.

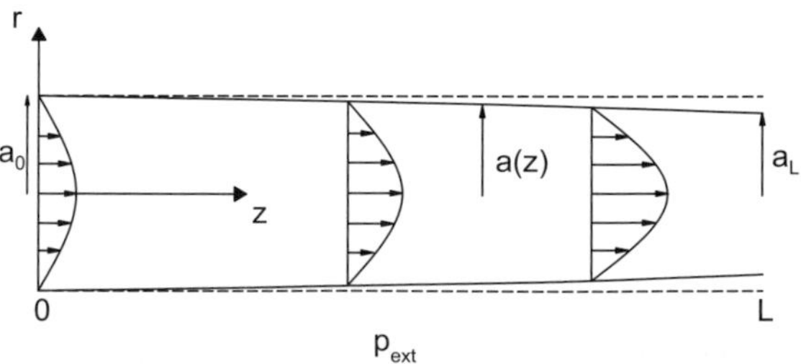

Figure 9.1 Steady viscous flow in a flexible tube.

In Section 6.6 we scaled the Navier–Stokes equations for a channel flow. Here we will scale the corresponding steady form in cylindrical coordinates, see Section 6.7, assuming axisymmetric flow, $(\partial/\partial\theta) = 0$, with no azimuthal velocities, $u_\theta = 0$. The radial and axial velocity components, u_r, u_z, satisfy the simplified Navier–Stokes equations in cylindrical coordinates,

$$\begin{aligned} r:\ \rho\left(u_r\frac{\partial u_r}{\partial r}+u_z\frac{\partial u_r}{\partial z}\right) &= -\frac{\partial p}{\partial r}+\mu\left(\frac{1}{r}\frac{\partial}{\partial r}\left(r\frac{\partial u_r}{\partial r}\right)+\frac{\partial^2 u_r}{\partial z^2}-\frac{u_r}{r^2}\right) \\ z:\ \rho\left(u_r\frac{\partial u_z}{\partial r}+u_z\frac{\partial u_z}{\partial z}\right) &= -\frac{\partial p}{\partial z}+\mu\left(\frac{1}{r}\frac{\partial}{\partial r}\left(r\frac{\partial u_z}{\partial r}\right)+\frac{\partial^2 u_z}{\partial z^2}\right) \end{aligned} \quad (9.1.1)$$

The continuity equation is

$$\frac{1}{r}\frac{\partial}{\partial r}(r u_r)+\frac{\partial u_z}{\partial z}=0 \quad (9.1.2)$$

The kinematic boundary conditions are no-penetration and no-slip at the tube wall position which is $r = a(z)$, and finite velocity throughout the domain

$$u_r(r=a(z))=0,\ u_z(r=a(z))=0 \qquad \text{finite velocities} \quad (9.1.3)$$

The position of the wall, $a(z)$, is not known beforehand. It is not prescribed, but instead is part of the solution subject to an additional equation, the stress boundary condition. We need to impose a stress boundary condition at the wall which balances the fluid stress with the elastic stress of the deformed wall defined through $a(z)$. Such problems in fluid mechanics are called free boundary problems. Here we use a simple linear relationship, or linear "tube law," where the fluid pressure relative to the external pressure balances the tube elastic stress which comes from a radial displacement

$$p(r=a(z))-p_{ext}=k\left(a(z)-a_{eq}\right) \quad (9.1.4)$$

As the internal pressure at the wall increases (decreases) then the tube radius increases (decreases). The constant k is an elastic modulus. A linear tube law assumes that the wall strain is small i.e.

$$\left|\frac{a(z)-a_{eq}}{a_{eq}}\right| << 1 \quad (9.1.5)$$

For our purposes, we can choose the equilibrium radius $a_{eq} = a_0 = a(z = 0)$ and that is consistent with defining $p_{ext} = 0$ and $p(z = 0) = 0$. Flow will be achieved by lowering the downstream pressure at $z = L$.

For this problem we choose the following scales, which are similar to those of Eq. (6.5.1),

$$U_Z=\frac{u_z}{u_s},\ U_R=\frac{u_r}{\varepsilon u_s}, P=\frac{p}{\left(\frac{\mu u_s}{\varepsilon a_0}\right)}, R=\frac{r}{a_0}, Z=\frac{z}{L},\ B=\frac{a}{a_0} \quad (9.1.6)$$

In Eq. (9.1.6) u_s is a characteristic velocity scale which we can choose to be the average velocity at the entrance. The scale for r is a_0, which is the entrance radius; however, the scale for z is L, the tube length. Inserting the scales of Eq. (9.1.6) into Eqs. (9.1.1) yields the dimensionless steady Navier–Stokes equations in cylindrical coordinates

$$\begin{aligned} R:\ \varepsilon^3 Re\left(U_R\frac{\partial U_R}{\partial R}+U_Z\frac{\partial U_R}{\partial Z}\right) &= -\frac{\partial P}{\partial R}+\varepsilon^2\left(\frac{1}{R}\frac{\partial}{\partial R}\left(R\frac{\partial U_R}{\partial R}\right)+\varepsilon^2\frac{\partial^2 U_R}{\partial Z^2}-\frac{U_R}{R^2}\right) \\ Z:\ \varepsilon Re\left(U_R\frac{\partial U_Z}{\partial R}+U_Z\frac{\partial U_Z}{\partial Z}\right) &= -\frac{\partial P}{\partial Z}+\frac{1}{R}\frac{\partial}{\partial R}\left(R\frac{\partial U_Z}{\partial R}\right)+\varepsilon^2\frac{\partial^2 U_Z}{\partial Z^2} \end{aligned} \quad (9.1.7)$$

where the aspect ratio of the tube is $\varepsilon = a_0/L$ and the Reynolds number is $Re = a_0 u_s/\nu$.

For biofluid mechanical applications (e.g. blood vessels, lymph vessels, lung airways) the tubes are generally much longer than their radius. So we can assume that ε is small, i.e. $\varepsilon << 1$, which makes ε^2 and ε^3 even smaller. In addition, for many flows of interest the Reynolds number is not too large, so we can also assume $\varepsilon Re << 1$. These scales and assumptions allow us to simplify the Navier–Stokes equations by neglecting the terms they multiply. The resulting form of Eq. (9.1.7) is

$$\begin{aligned} R:&\quad 0=-\frac{\partial P}{\partial R} \\ Z:&\quad 0=-\frac{\partial P}{\partial Z}+\frac{1}{R}\frac{\partial}{\partial R}\left(R\frac{\partial U_Z}{\partial R}\right) \end{aligned} \quad (9.1.8)$$

From the R-component of Eq. (9.1.8) we see that $\partial P/\partial R = 0$ so that $P = P(Z)$ only. The continuity equation remains intact and we note that radial velocities, scaled on εu_s, are much smaller than axial velocities, scaled on u_s,

$$\frac{1}{R}\frac{\partial}{\partial R}(RU_R) + \frac{\partial U_Z}{\partial Z} = 0 \tag{9.1.9}$$

In fact, the process of choosing these different scales for the two velocity components comes from making sure both are represented in the asymptotic form of conservation of mass, Eq. (9.1.9). The dimensionless kinematic boundary conditions are

$$U_R(R = B(Z)) = 0,\ U_Z(R = B(Z)) = 0 \quad \text{finite velocities} \tag{9.1.10}$$

while using the scales in the stress boundary condition, Eq. (9.1.4) becomes

$$B(Z) - 1 = \gamma P(Z) \tag{9.1.11}$$

In Eq. (9.1.11) we have used the result that P is the same for any value of R, i.e. $P(R = B(Z)) = P(Z)$. The dimensionless parameter γ is defined as

$$\gamma = \frac{\mu u_s/\varepsilon a_0}{k a_0} \sim \frac{\text{fluid shear stress}}{\text{elastic wall stress}} \tag{9.1.12}$$

and is the ratio of the elastic wall stress to the fluid shear stress. After neglecting ε and εRe, this fluid to solid stress ratio, γ, is our only remaining parameter.

We can integrate the Z-component of momentum in Eq. (9.1.8) to find the familiar form seen in Poiseuille flow

$$U_Z(R,\ Z) = \frac{R^2}{4}\frac{dP}{dZ} + c_1(Z)\ln(R) + c_2(Z) \tag{9.1.13}$$

which has the appearance of Eq. (7.1.2) for Poiseuille flow in a circular cylinder of constant radius. The difference is that dP/dZ is not constant and will depend on Z, and that c_1 and c_2 are functions of Z rather than constants. The kinematic boundary conditions at this order are

$$U_Z(R = B(Z)) = 0, \quad U_Z(R = 0) \quad \text{finite} \tag{9.1.14}$$

Applying these boundary conditions to Eq. (9.1.13) solves for $c_1 = 0$ and $c_2 = -B^2(dP/dZ)/4$ so that the resulting form of the Z velocity component is

$$U_Z(R, Z) = -\frac{1}{4}\frac{dP}{dZ}(B^2 - R^2) \tag{9.1.15}$$

Equation (9.1.15) reveals the results of our assumptions, that the velocity profile is locally parabolic, just a different parabola for each value of Z.

At this stage we impose the integral mass balance since the volumetric flow rate must be constant. Going back to the dimensional definition of the flow rate equaling the integral of the velocity across the cross-section, we have

$$q = \int_0^{2\pi}\int_0^{a(z)} u_z\, r\, dr\, d\theta = 2\pi a_0^2 u_s \int_0^{B(Z)} U_Z R\, dR = -\pi a_0^2 u_s \frac{B^4}{8}\frac{dP}{dZ} \tag{9.1.16}$$

Since our velocity scale, u_s, is the average velocity at the entrance, then $\pi a_0^2 u_s = q$, so Eq. (9.1.16) becomes

$$\frac{B^4}{8}\frac{dP}{dZ} = -1 \tag{9.1.17}$$

Taking the Z derivative of the stress boundary condition, Eq. (9.1.11), results in

$$\gamma\frac{dP}{dZ} = \frac{dB}{dZ} \tag{9.1.18}$$

and substituting Eq. (9.1.18) into Eq. (9.1.17) gives the form

$$B^4\frac{dB}{dZ} = -8\gamma \tag{9.1.19}$$

We can separate variables and integrate Eq. (9.1.19)

$$\int_1^B B^4\, dB = -8\gamma\int_0^Z dZ \quad \Rightarrow \quad \frac{1}{5}(B^5 - 1) = -8\gamma Z \tag{9.1.20}$$

Note that the lower limits of integration match, that is, $B = 1$ at $Z = 0$. We can rearrange Eq. (9.1.20) to solve for B,

$$B = (1 - 40\gamma Z)^{1/5} \tag{9.1.21}$$

Let's explore more about the limitation of our analysis in Eq. (9.1.5) that elastic strains for the tube are small. A way to start is to choose the largest possible strain and make sure it is kept small and under what constraints on the parameters. The largest wall displacement is at the tube exit where the radius is

$$B(Z = 1) = (1 - 40\gamma)^{1/5} = B_1 \tag{9.1.22}$$

Consequently, our constraint of small wall strain is

$$1 - B_1 << 1 \quad \rightarrow \quad 1 - (1 - 40\gamma)^{1/5} << 1 \tag{9.1.23}$$

Let $\beta_0 = 40\gamma << 1$. Then Eq. (9.1.23) becomes

$$1 - (1 - \beta_0)^{1/5} << 1 \tag{9.1.24}$$

The Taylor series for the bracketed term, assuming $\beta_0 << 1$, is

$$(1 - \beta_0)^{1/5} = 1 - \frac{1}{5}\beta_0 - \frac{2}{25}\beta_0^2 + O(\beta_0^3) \tag{9.1.25}$$

Inserting Eq. (9.1.25) into Eq. (9.1.24) and keeping only the linear term in the Taylor series yields

$$\frac{1}{5}\beta_0 << 1 \quad \rightarrow \quad \gamma << \frac{1}{8} \tag{9.1.26}$$

So the assumptions that $\varepsilon << 1$ and $\varepsilon Re << 1$ are further constrained by requiring that $\gamma << 1/8$, that is, the tube must be sufficiently stiff so that only small deformations occur. That way the flow is locally parabolic at any Z position.

Some plots of the dimensionless wall position, B(Z), from Eq. (9.1.21) are shown in Figure 9.2 for different values of γ. As γ decreases, e.g. from k increasing in its denominator, the wall is stiffer and the wall deformation is reduced. The limiting case of $\gamma \rightarrow 0$ is a rigid tube.

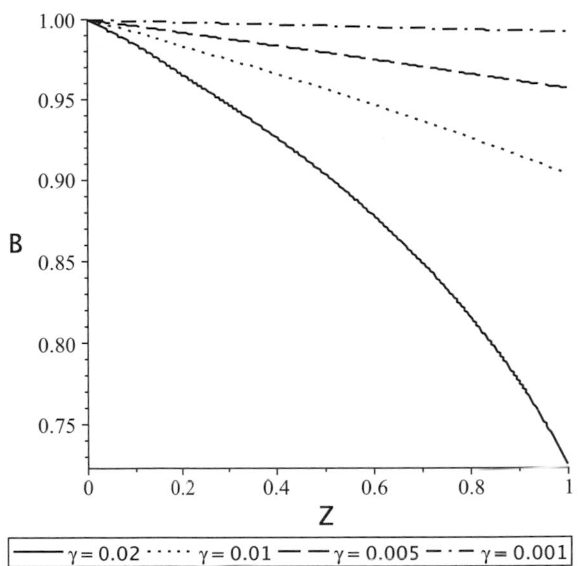

Figure 9.2 Wall position, B(Z) for four values of $\gamma = 0.02, 0.01, 0.005, 0.001$.

Figure 9.3 shows $U_Z(R, Z)$ for several values of Z and $\gamma = 0.01$. Note that as Z increases the maximum velocity at the centerline increases while the maximum radius decreases. The dimensionless centerline velocity is 2 at the entrance, as we expect, since our definition of the velocity scale, u_s, is the average velocity at the entrance. Recall from our analysis of Poiseuille flow in Chapter 7, Eqs. (7.1.15) and (7.1.16), that the centerline maximum velocity is twice the average velocity. As the tube radius tapers down the tube length, the fluid speed must increase to keep the flow rate constant.

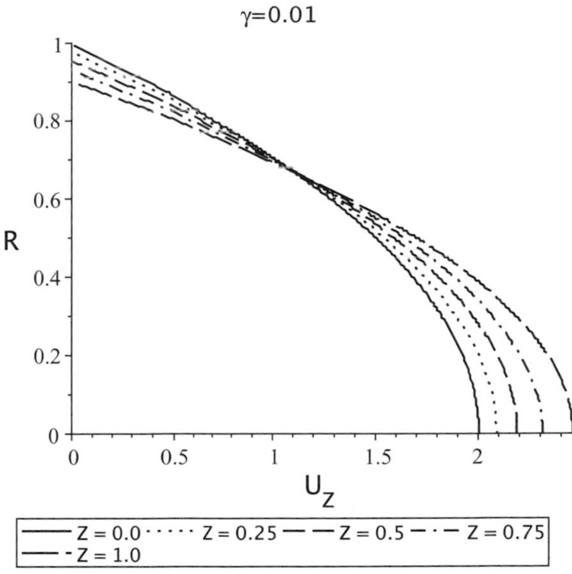

Figure 9.3 Axial velocity, U_Z vs R for Z = 0, 0.25, 0.5, 0.75, 1 along the tube for $\gamma = 0.01$.

We can now go to the conservation of mass equation, Eq. (9.1.9), to solve for the radial fluid velocity U_R. Separating the variables and integrating gives us

$$\int \partial(RU_R) = RU_R = -\int \frac{dU_Z}{dZ} R\partial R + c_3(Z) \tag{9.1.27}$$

Inserting Eq. (9.1.15) into the right hand side integrand and performing the integration leads to U_R

$$U_R = \frac{1}{4}\frac{d^2P}{dZ^2}\left(B^2\frac{R}{2} - \frac{R^3}{4}\right) + \frac{1}{4}BR\frac{dP}{dZ}\frac{dB}{dZ} + \frac{c_3}{R} \qquad c_3 = 0 \tag{9.1.28}$$

and we must choose $c_3 = 0$ for finite solutions. The no-penetration condition, $U_R(R = B(Z)) = 0$, in Eq. (9.1.10) is automatically satisfied by Eq. (9.1.28).

$U_R(R, Z)$ is plotted in Figure 9.4 for several values of Z and $\gamma = 0.01$. Note that U_R is zero at the centerline, $R = 0$, as we would expect from symmetry, and also at the tube wall, $R = B(Z)$, so there is no penetration. For all other values of R, $U_R < 0$ as the tapering tube forces the fluid toward the centerline in the negative R-direction.

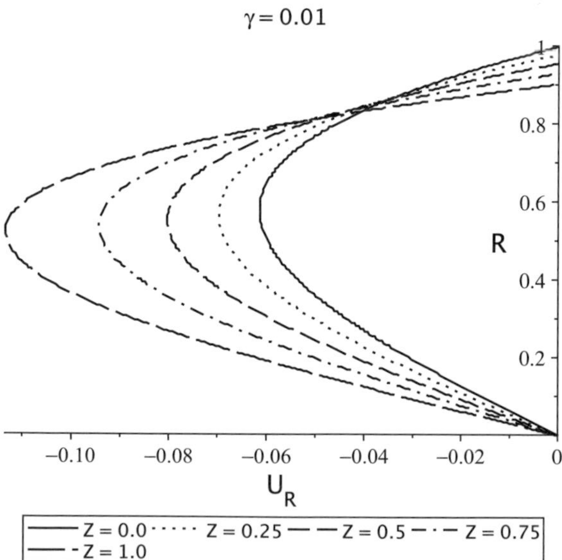

Figure 9.4 Radial velocity, U_R vs R for Z = 0, 0.25, 0.5, 0.75, 1 for $\gamma = 0.01$.

9.2 The Nonlinear Tube Law

To characterize the flexibility of a tube, a common approach in biological applications is to consider a thin-walled tube which is inflated and deflated by an internal pressure (Grotberg and Jensen, 2004). The tube cross-sectional area and shape are recorded, and the result is a graph similar to that of Figure 9.5 for thin-walled tubes (Flaherty et al., 1972).

In Figure 9.5 the dimensionless area and dimensionless pressure are given by

$$\alpha = \frac{A}{A_0}, \quad \hat{\pi} = \frac{p - p_{ext}}{\kappa} \tag{9.2.1}$$

where A_0 is the cross-sectional area when the internal pressure equals the external pressure, $\hat{\pi} = 0, \alpha = 1$. The constant $\kappa = (h/a_0)^3 Y/12(1 - \sigma)$, where Y is the Young's modulus of the elastic material, h is the tube wall thickness, a_0 is the radius, where $A_0 = \pi a_0^2$, and σ is the Poisson ratio of the material. Note that in Figure 9.5 the cross-section takes on different shapes. For inflation, $\hat{\pi} > 0$, the cross-section is circular. For deflation, $\hat{\pi} < 0$, the tube begins to collapse, first into an elliptical shape losing stability to the lowest circumferential wave mode. Then, for further decreases of the internal pressure, the cross-sectional shape has point contact between the upper and lower

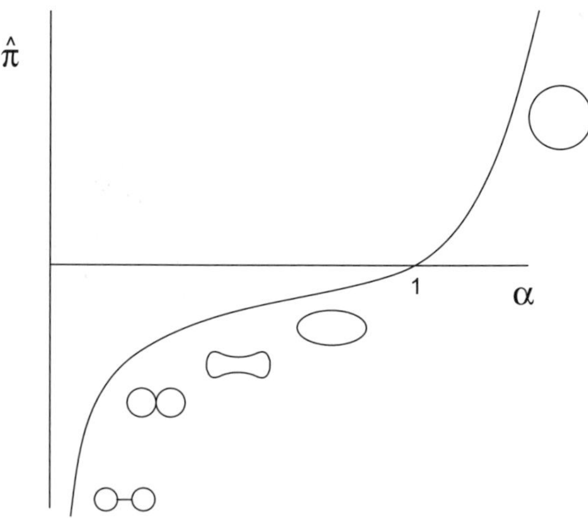

Figure 9.5 Nonlinear tube law.

wall segments, and eventually to a shape with line contact, resulting in two distinct parallel channels separated by the line. For thicker walled tubes, or tubes walls formed from concentric elastic cylinders with different material properties in the different layers, the cross-section can have more circumferential waves (Wiggs et al., 1997; Yang et al., 2007).

While some tube law models treat the entire range (Elad et al., 1988), the collapse region $\hat{\pi} < 0, \alpha < 1$ is of great interest, and a curve fit for this limb of the curve (Shapiro, 1977) is

$$\hat{\pi} = 1 - \alpha^{-n} \quad \Leftrightarrow \quad \alpha = (1 - \hat{\pi})^{-\frac{1}{n}} \tag{9.2.2}$$

The curve-fit parameter, n, depends on the tube stiffness and generally is in the range $0.5 \leq n \leq 2.5$. For example, collapse experiments on dog and sheep tracheas showed n in the range $0.3 \leq n \leq 2.1$ (Aljuri et al., 1999).

The dimensionless tube law in this collapse region is given in Figure 9.6(a) where $\hat{\pi}(\alpha)$ is shown for four values of n, while the corresponding value of the dimensionless tube compliance $d\alpha/d\hat{\pi} = (d\hat{\pi}/d\alpha)^{-1} = \alpha^{n+1}/n$ is in Figure 9.6(b). The tube is more compliant as n decreases.

The complexity of analyzing flexible tube flows grows substantially when increasing the details of the fluid modeling, and/or the elastic tube modeling and/or how they are coupled. Figure 9.7 shows experimental and computational results for flow through a flexible tube from left to right. Note how the wall tapers downstream as the internal pressure is dropping below the external pressure. The tube is flattening, forming a series of cross-sections similar to those shown in Figure 9.5. Then suddenly the tube opens to match its attachment to the downstream rigid pipe. There is very good correlation between the experiment and computations because the fluid and elastic wall models were both 3D (Heil, 1997).

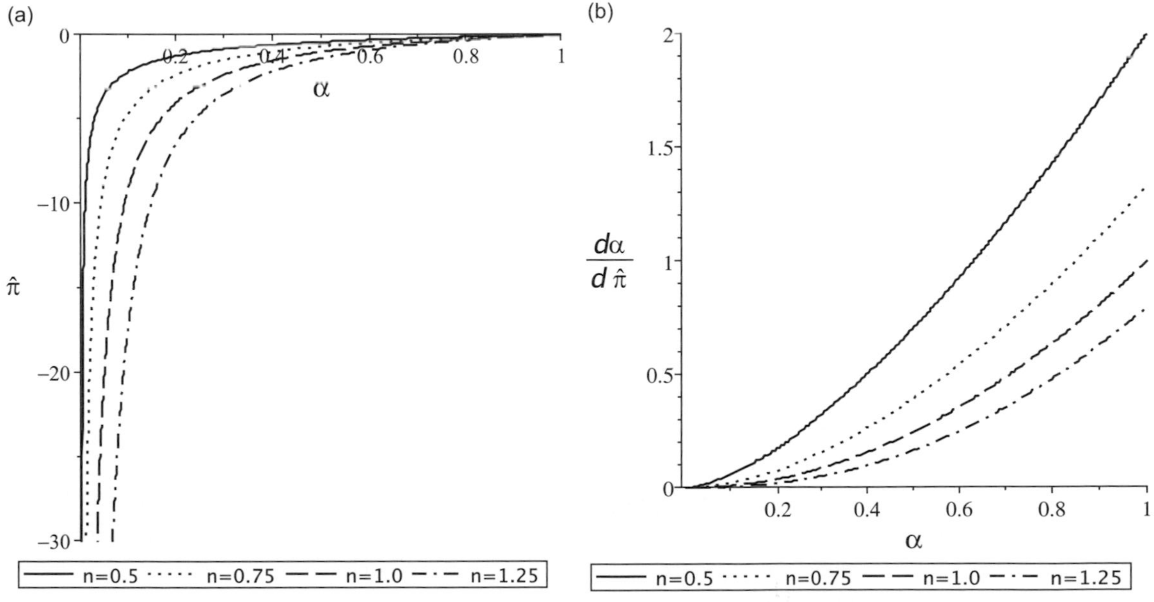

Figure 9.6 (a) Tube law, $\hat{\pi}$ vs α, in the collapsed region for four values of n, and (b) the corresponding value of tube compliance $d\alpha/d\hat{\pi}$.

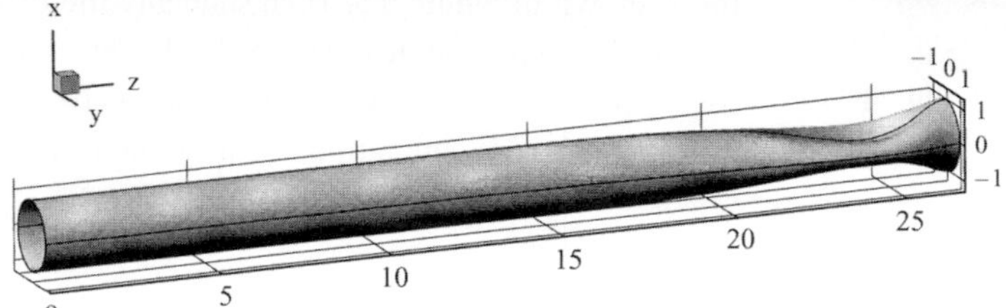

Figure 9.7 Comparison of computed and experimentally observed tube wall shapes for flow in a flexible tube from left to right.

9.3 Viscous Flow Limitation

Solving the three-dimensional velocity field for flow in an elastic tube, whose radius may, in general, depend on both the axial (z) and azimuthal (θ) directions, is a significant challenge. For thick-walled tubes, an additional complexity arises since the full three-dimensional strain field within the elastic solid itself must also be solved. The stress boundary condition where they meet is, as we learned in Section 9.1, part of a free boundary problem since the location of the fluid–solid interface is one of the unknowns. Many approaches to these complicated fluid–structure interactions seek to simplify one or both of the main components, i.e. the fluid flow and/or the elastic deformation. Here we choose to simplify both.

Without solving the velocity field in detail, we can consider the related flow vs pressure gradient relationship for these nonlinear tubes as appearing like a Poiseuille Law. That is basically the result of the scalings in Section 9.1, that the flow is locally Poiseuille. To do this, we express the flow rate, Q, which is the same constant at every axial position, as proportional to dp/dz, which will not be a constant, and to A^2, where A is the cross-sectional area that is pressure dependent. Using A^2 lends the comparison to Poiseuille Law for a tube of radius a, where Q is proportional to a^4 or the square of the cross-sectional area

$$Q = c_0 \frac{A^2}{\mu}\left(-\frac{dp}{dz}\right) \tag{9.3.1}$$

where c_o is a proportionality constant. From the definitions in Eq. (9.2.1) we have $(dp/dz) = \kappa(d\hat{\pi}/dz)$ and $A = \alpha A_o$. Using those and also defining a dimensionless axial coordinate, $Z = z/L$, we substitute into Eq. (9.3.1) to find

$$Q = c_o \frac{A_o^2 \kappa}{\mu L} \alpha^2 \left(\frac{-d\hat{\pi}}{dZ} \right) \tag{9.3.2}$$

Using $\alpha = (1 - \hat{\pi})^{-1/n}$ from Eq. (9.2.2) and defining a dimensionless flow rate as $F = Q/\left(c_o A_o^2 \kappa / \mu L\right)$ gives us a dimensionless version of Eq. (9.3.2)

$$F = (1 - \hat{\pi})^{-2/n} \left(\frac{-d\hat{\pi}}{dZ} \right) \tag{9.3.3}$$

Equation (9.3.3) can be used to find the Z-independent flow rate, F, in terms of the end pressures for the tube. Separating variables and integrating Eq. (9.3.3) for $n \neq 2$ over the length of the tube yields

$$F \int_0^1 dZ = - \int_{\hat{\pi}_o}^{\hat{\pi}_1} (1 - \hat{\pi})^{-2/n} d\hat{\pi} \quad \Rightarrow$$

$$F = \frac{n}{(n-2)} \left[(1 - \hat{\pi}_1)^{(n-2)/n} - (1 - \hat{\pi}_o)^{(n-2)/n} \right] \tag{9.3.4}$$

We can choose the entrance pressure to be $\hat{\pi}_o = 0$. Then flow is accomplished by negative values of the downstream pressure, $\hat{\pi}_1 < 0$,

$$F = \frac{n}{(n-2)} \left[(1 - \hat{\pi}_1)^{(n-2)/n} - 1 \right] \tag{9.3.5}$$

In Eq. (9.3.5) the term $(1 - \hat{\pi}_1)^{(n-2)/n}$ approaches zero as $\hat{\pi}_1 \to -\infty$ when $n < 2$. So in this limit of continuing to lower the downstream pressure, the flow rate reaches a maximum

$$\lim_{\hat{\pi}_1 \to -\infty} F = \frac{n}{(2-n)} \tag{9.3.6}$$

A plot of F vs $\hat{\pi}_1$ is shown in Figure 9.8. This figure demonstrates "flow limitation," in this case due to viscous effects. Why does this happen? It is because there are two competing effects of lowering the downstream pressure. First, it increases the driving pressure difference between the inlet and outlet of the flexible tube, which tends to increase flow. Second,

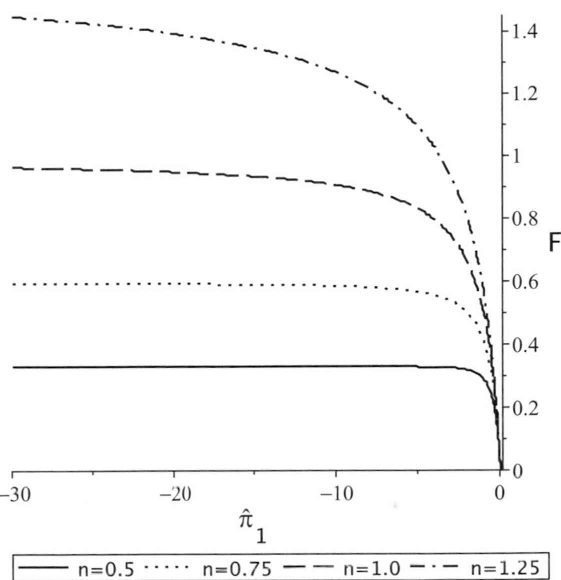

Figure 9.8 F vs $\hat{\pi}_1$ showing flow limitation due to viscosity for four values of $n = 0.5, 0.75, 1.0, 1.25$.

however, it reduces the cross-sectional area of the tube, which increases the viscous resistance to flow tending to lower the flow rate. The two effects balance as the flow reaches a maximum value.

9.4 Wave Speed and Inertial Flow Limitation

Consider inviscid flow in a tube which has a flexible section whose cross-sectional area, A, depends on the local fluid pressure, P, minus the external pressure taken to be zero in this example, Figure 9.9. As we saw in Sections 9.1 and 9.2, a "tube law" is the function P (A) or A(P). The average fluid velocity is U in the flexible region and it also depends on P through the Bernoulli Equation. In the rigid portion of the tube we have the cross-sectional area, A_o, the velocity, U_o, and the pressure, P_o, which are constants. Writing the Bernoulli Equation between the two stations we have

$$P_o + \frac{1}{2}\rho U_o^2 = P + \frac{1}{2}\rho U^2 \tag{9.4.1}$$

Recall that the $\rho U^2/2$ terms come from the convective acceleration or inertia of the Euler equation. Since

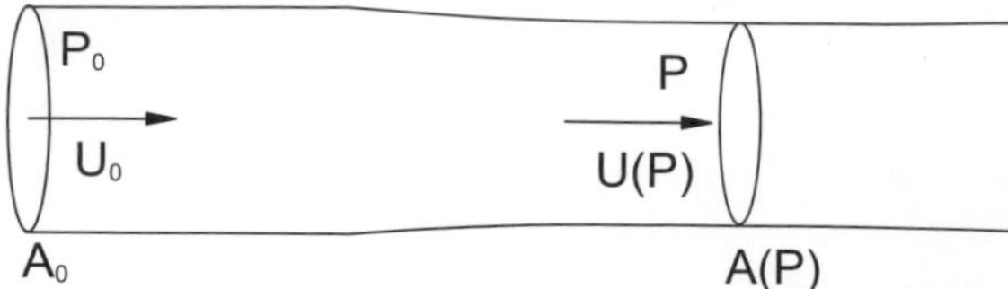

Figure 9.9 Flexible tube flow and deformation for long waves.

flow limitation is defined as the system state when flow reaches a maximum with pressure, i.e. $dQ/dP = 0$, let's investigate the implications of that criterion for the tube by taking the derivative of Eq. (9.4.1) with respect to P

$$0 = 1 + \frac{1}{2}\rho 2U \frac{dU}{dP} \tag{9.4.2}$$

However, since the average velocity can be defined as the flow rate, Q, divided by the area, A,

$$U(P) = \frac{Q(P)}{A(P)} \tag{9.4.3}$$

then Eq. (9.4.2) becomes

$$0 = 1 + \rho U \frac{d}{dP}\left(\frac{Q}{A}\right) = 1 + \rho U \left(\frac{A\frac{dQ}{dP} - Q\frac{dA}{dP}}{A^2}\right) \tag{9.4.4}$$

Imposing the criterion of flow limitation, $dQ/dP = 0$, results in

$$0 = 1 - \rho U \left(\frac{Q\frac{dA}{dP}}{A^2}\right) = 1 - \rho U^2 \frac{1}{A}\frac{dA}{dP} \tag{9.4.5}$$

Rearranging Eq. (9.4.5) gives the criterion for the fluid speed at flow limitation

$$U^2 = \frac{A}{\rho}\frac{dP}{dA} = \frac{1}{\rho}\frac{dP}{\frac{dA}{A}} \tag{9.4.6}$$

The specific elastance of the tube, E, is defined as the change in pressure for a relative change in area,

$$E = A\frac{dP}{dA} \tag{9.4.7}$$

and its inverse is the specific compliance.

Long waves on the elastic tube travel at speed, C_W, which is related to the elastance and the fluid density as

$$C_W = \sqrt{E/\rho} \tag{9.4.8}$$

So the criterion for flow limitation is that the local fluid velocity, U, equals the local fluid-elastic wave speed, C_W,

$$U = C_W \tag{9.4.9}$$

The limitation of flow is physically related to the wave speed since further changes in downstream pressure information, as communicated through the wall elasticity, cannot propagate upstream. This is a different mechanism from the viscous flow limitation of Section 9.3. Since the wave speed, C_W, depends on the tube's specific elastance, E, it will vary from one airway generation to another. The further out into the airway tree, the more compliant are the airways and E decreases. However, because of the rapidly increasing total cross-sectional area, the speeds are also decreasing. We note that in diseased airways such as emphysema, the elastance, E, of the tubes is reduced so flow limitation during expiration occurs at lower speeds and flow rates, leading to shortness of breath. Wheezing breath sounds also occur in emphysema and asthma and the connection to flow limitation is explored in Section 9.9.

The axial position where the fluid pressure is low enough to cause partial, local collapse of the tube becomes the location of a fluid-elastic shock with rapid changes in cross-sectional area and fluid speed over a short distance (Shapiro, 1977; Elad et al., 1989; Kamm and Schroter, 1989). The forms of the governing equations for this new fluid mechanical shock are identical to two other well-known shock waves: compressible gas flow and free surface waves. We will see in Section 16.6 that biofluid mechanics investigations of thin films with surface tension led to the discovery of another, the GBG shock. This is one of the benefits of studying and exploring biofluid mechanics, because it leads to new fluid physics.

Clinical Correlation 9.1 | Forced Expiration

Figure 9.10(a) shows the use of a spirometer to measure flow rate during inspiration and expiration. This is a common pulmonary function test used clinically to evaluate pulmonary health and diseases like asthma, emphysema, fibrosis and other conditions. To avoid flow through the nose, it is pinched closed with a clip. Typically the mouthpiece measures a pressure drop across a known flow resistance to infer flow rate, and that signal is integrated to yield volume. In Figure 9.10(b) there are tracings of flow vs lung volume during inspiration and four effort levels of forced expiration. For expiration, the lung volume starts fully inflated at total lung capacity (TLC) and decreases to residual volume (RV), when no more air can be expelled. Any reader can try this maneuver, just be cautious as to who, or what, is downstream. Regardless of the effort, however, the forced expiration curves all lay on top of one another in their descending limbs. That effort-independent segment is flow limitation, i.e. the flow reaches a maximum regardless of the driving pressure difference.

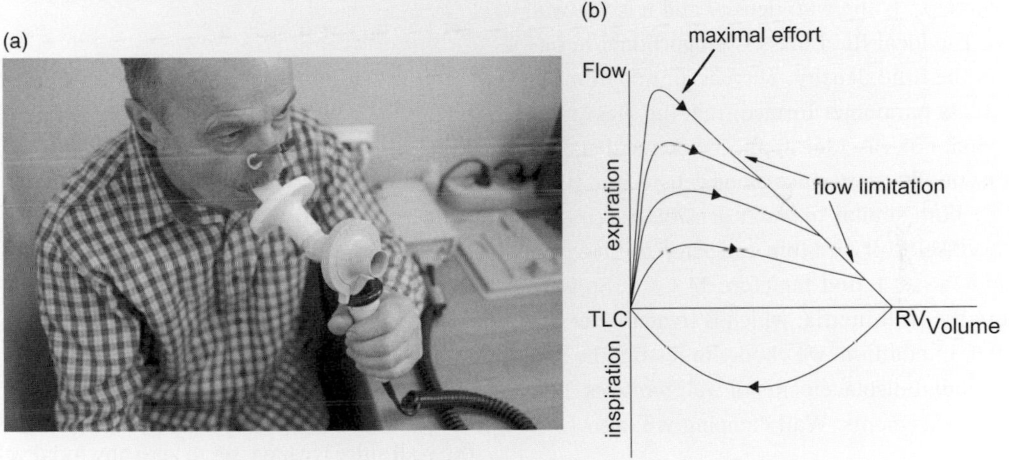

Figure 9.10 (a) Using a spirometer to measure respiratory flow; (b) flow vs lung volume at four effort levels.

9.5 Pulse Propagation in a Flexible Tube

Pulse propagation occurs primarily in the arterial system as blood is forced into it by the cardiac cycle. A propagating wave is shown in Figure 9.11.

A flexible tube with an unperturbed radius, a, carries a wave which travels at speed c. The wavelength is λ, and the radial position of the wall is a traveling wave to the right with form

$$b(z,t) = a(1 + \delta \cos(k(z - ct))) \tag{9.5.1}$$

where $k = 2\pi/\lambda$ is the wave number, $c = f\lambda$ is the wave speed where f is the frequency and a δ is the wave amplitude. We assume small amplitude

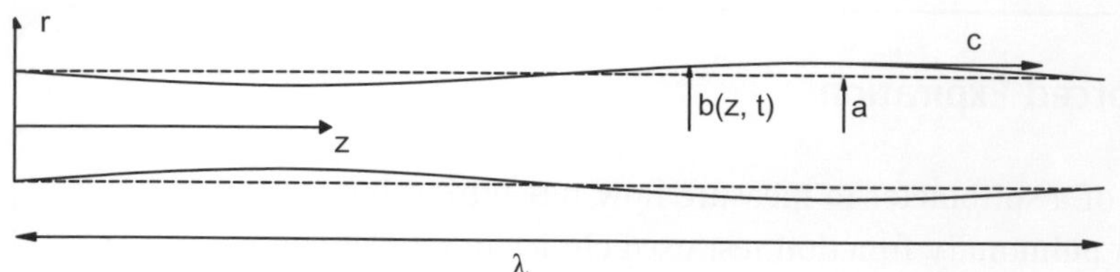

Figure 9.11 Wave propagation for a flexible tube.

disturbances so $\delta << 1$. The fluid has density, ρ, and viscosity, μ, and the velocity components in the cylindrical coordinate system are $(r, \theta, z) \sim (u_r, u_\theta, u_z)$. We will consider axisymmetric flow, so $u_\theta = 0$ and $\partial/\partial\theta = 0$. The disturbance velocities u_r, u_z will be proportional to δ so also small.

The tube is an elastic body and there are many possible levels of modeling its details. We will use a simple approach. The local wall mass is proportional to $\rho_w h$ where ρ_w is the wall density and h is the wall thickness. The local fluid mass is proportional to ρa where ρ is the fluid density. There is an important dimensionless parameter formed from the mass ratio $M = \rho_w h/\rho a$. For vascular applications, the density ratio is essentially unity since blood density and tissue density are both similar to water density, $\rho_w/\rho \sim 1$. However, vessel walls are thin compared to the vessel radius, so $h/a << 1$, and therefore, $M << 1$. So we can ignore the wall inertia, which is treated later in Section 9.9. In addition, we choose to restrict the wall motion to radial displacements of wall particles, but no axial displacements. Wall damping will also be ignored.

From these assumptions, the stress boundary condition on the wall simplifies to the use of a "tube law," such that the local cross-sectional area depends on the local fluid pressure relative to the external pressure, $A = A(p), \pi b^2 = A$.

The kinematic boundary conditions for a viscous fluid are that the fluid and wall velocity are equal at the wall. For the radial direction we have

$$\frac{\partial b}{\partial t} = u_r(r = b(z, t)) \tag{9.5.2}$$

since the radial wall velocity is just the time derivative of its position. Equation (9.5.2) shows directly that $u_r = O(\delta)$ since $\partial b/\partial t = O(\delta)$. We can expand the right hand side of Eq. (9.5.2) using the general Taylor series

$$f(x) = \sum_{n=0}^{\infty} \frac{(x - x_0)^n}{n!} \frac{d^n f}{dx^n}\bigg|_{x=x_0} = f(x_0) + (x - x_0)\frac{df}{dx}\bigg|_{x=x_0} + \frac{(x - x_0)^2}{2}\frac{d^2 f}{dx^2}\bigg|_{x=x_0} + \cdots \tag{9.5.3}$$

Let $f = u_r, x = b, x_0 = a$

$$u_r(r = b) = u_r(r = a) + (b - a)\frac{\partial u_r}{\partial r}\bigg|_{r=a} + \cdots \tag{9.5.4}$$

Note that $b - a = O(\delta a)$, so the first term on the right hand side is $O(\delta)$ while the second term is $O(\delta^2)$. Substituting Eq. (9.5.4) into Eq. (9.5.2) and dropping $O(\delta^2)$ gives the form we seek

$$\frac{\partial b}{\partial t} = u_r(r = a) \tag{9.5.5}$$

In the axial direction we have zero fluid velocity at the wall since we have set to zero any axial velocity of the wall material

$$u_z(r = b) = u_z(r = a) + \cdots = 0 \tag{9.5.6}$$

again using a truncated Taylor series of Eq. (9.5.3) with $f = u_z$. The governing fluid mechanics equations are the unsteady, axisymmetric Navier–Stokes equations, Eq. (6.5.17), and conservation of mass, Eq(4.6.5), in cylindrical coordinates.

Rather than simplify the system of equations with words, we can do it with mathematics to back up the words and show what is actually involved. Make these equations dimensionless by choosing the following definitions

$$T = \omega t,\quad R = \frac{r}{a},\quad Z = \frac{z}{\lambda},\quad U_R = \frac{u_r}{\varepsilon\delta\omega a},$$
$$U_Z = \frac{u_z}{\delta\omega a},\quad P = \frac{\varepsilon p}{\delta\rho(\omega a)^2} \tag{9.5.7}$$

where $\varepsilon = a/\lambda$. Note that the velocities are scaled on δas anticipated. Inserting the definitions of Eqs. (9.5.7) into Eq. (6.5.17) and Eq. (4.6.5) results in the dimensionless forms

$$\text{R-momentum}: \quad \varepsilon^2\left[\frac{\partial U_R}{\partial T} + \delta\varepsilon\left(U_R\frac{\partial U_R}{\partial R} + U_Z\frac{\partial U_R}{\partial Z}\right)\right]$$
$$= -\frac{\partial P}{\partial R} + \frac{\varepsilon^2}{\alpha^2}\left(\frac{\partial^2 U_R}{\partial R^2} + \frac{1}{R}\frac{\partial U_R}{\partial R} + \varepsilon^2\frac{\partial^2 U_R}{\partial Z^2} - \frac{U_R}{R^2}\right)$$
$$\text{Z-momentum}: \quad \frac{\partial U_Z}{\partial T} + \delta\varepsilon\left(U_R\frac{\partial U_Z}{\partial R} + U_Z\frac{\partial U_Z}{\partial Z}\right)$$
$$= -\frac{\partial P}{\partial Z} + \frac{1}{\alpha^2}\left(\frac{\partial^2 U_Z}{\partial R^2} + \frac{1}{R}\frac{\partial U_Z}{\partial R} + \varepsilon^2\frac{\partial^2 U_Z}{\partial Z^2}\right)$$
$$\text{continuity}: \quad \frac{1}{R}\frac{\partial}{\partial R}(RU_R) + \frac{\partial U_Z}{\partial Z} = 0 \tag{9.5.8}$$

where $\alpha^2 = \omega a^2/\nu$ and α is the Womersley parameter. We saw in Chapter 8 that $\alpha^2 >> 1$ in the major arteries, so the viscous terms in Eq. (9.5.8) multiplied by $1/\alpha^2 << 1$ can be ignored as a first approximation. In addition, we can simplify Eq. (9.5.8) by assuming long waves, $\varepsilon << 1$, with small amplitude, $\delta << 1$. The surviving terms of Eqs. (9.5.8) are

$$\text{R}: \quad 0 = -\frac{\partial P}{\partial R}$$
$$\text{Z}: \quad \frac{\partial U_Z}{\partial T} = -\frac{\partial P}{\partial Z} \tag{9.5.9}$$
$$\text{continuity}: \quad \frac{1}{R}\frac{\partial}{\partial R}(RU_R) + \frac{\partial U_Z}{\partial Z} = 0$$

Switching back to the dimensional versions of Eq. (9.5.9) gives us

$$\text{r-momentum}: \quad 0 = -\frac{\partial p}{\partial r}$$
$$\text{z-momentum}: \quad \rho\frac{\partial u_z}{\partial t} = -\frac{\partial p}{\partial z} \tag{9.5.10}$$
$$\text{continuity}: \quad \frac{1}{r}\frac{\partial}{\partial r}(ru_r) + \frac{\partial u_z}{\partial z} = 0$$

From the r-component of Eq. (9.5.10), we see that $\partial p/\partial r = 0$, which implies that p depends on z and t only, i.e. p(z, t). Equations (9.5.10), of course, assume that the velocity disturbances, u_r and u_z, and the pressure disturbance, p, are all $O(\delta)$ as was evident in their choice of scales in Eqs. (9.5.7).

Our goal is to analyze the wave behavior of this system, which involves the propagation of pressure, radial displacement and fluid velocity as a function of z and t. The first step toward this goal is to average the governing equations across the cross-section to eliminate the r-dependence of the dependent variables. The cross-sectional area of the tube is given by $A = \pi b^2$, which is also time-dependent.

First, we define cross-sectional area average of a function, f(r, z, t),

$$\bar{f}(z,t) = \frac{1}{\pi b^2}\int_0^{2\pi}\int_0^{b} f(r,z,t)\,r\,dr\,d\theta \sim \frac{1}{\pi a^2}\int_0^{2\pi}\int_0^{a} f(r,z,t)\,r\,dr\,d\theta \tag{9.5.11}$$

where Eq. (9.5.11) simplifies because f(r, z, t) is already small, of $O(\delta)$, allowing us to replace $b(z, t) \sim a$. Let's area-average the continuity equation

$$\frac{1}{\pi a^2}\int_0^{2\pi}\int_0^{a}\left(\frac{1}{r}\frac{\partial}{\partial r}(ru_r)\right) r\,dr\,d\theta + \frac{1}{\pi a^2}\int_0^{2\pi}\int_0^{a}\frac{\partial u_z}{\partial z}\,r\,dr\,d\theta = 0 \tag{9.5.12}$$

The first term can be simplified as follows

$$\frac{1}{\pi a^2}\int_0^{2\pi}\int_0^{a}\frac{1}{r}\frac{\partial(ru_r)}{\partial r}\,r\,dr\,d\theta = \frac{1}{\pi a^2}\int_0^{2\pi}\int_0^{a} d(ru_r)\,d\theta$$
$$= \frac{1}{\pi a^2}\int_0^{2\pi} ru_r\Big|_{r=0}^{r=a}\,d\theta$$
$$= \frac{1}{\pi a^2}\int_0^{2\pi}\left(au_r\big|_{r=a} - 0\right)d\theta \tag{9.5.13}$$

From Eq. (9.5.2) we know that $u_r(r = a) = \partial b/\partial t$, and then Eq. (9.5.13) becomes

$$\frac{1}{\pi a^2}\int_0^{2\pi}\int_0^a \frac{1}{r}\frac{\partial(ru_r)}{\partial r}\, r\,dr\,d\theta = \frac{2}{a}\frac{\partial b}{\partial t} \tag{9.5.14}$$

The second term of Eq. (9.5.12) is the average $\overline{\partial u_z/\partial z}$ which can be related to the z-derivative of $\bar{u}_z$ by

$$\overline{\frac{\partial u_z}{\partial z}} = \frac{1}{\pi a^2}\int_0^{2\pi}\int_0^a \frac{\partial u_z}{\partial z}\, r\,dr\,d\theta = \frac{\partial}{\partial z}\left[\frac{1}{\pi a^2}\int_0^{2\pi}\int_0^a u_z\, r\,dr\,d\theta\right] = \frac{\partial \bar{u}_z}{\partial z} \tag{9.5.15}$$

where we have simply moved the z-derivative out of the integrand. This is allowed because for small amplitude the limits of integration do not depend on z.

So the continuity equation averages to

$$\frac{2}{a}\frac{\partial b}{\partial t} + \frac{\partial \bar{u}_z}{\partial z} = 0 \tag{9.5.16}$$

Multiplying Eq. (9.5.16) by πa^2 we arrive at

$$\text{continuity}: \quad \frac{\partial A}{\partial t} + \frac{\partial Q}{\partial z} = 0 \tag{9.5.17}$$

since $Q = \pi a^2 \bar{u}_z$ and $\partial A/\partial t = \partial(\pi b^2)/\partial t = 2\pi b(\partial b/\partial t) \sim 2\pi a(\partial b/\partial t)$ for $\delta << 1$.

Turning our attention to the z-momentum balance, we area-average the component of Eq. (9.5.10)

$$\frac{1}{\pi a^2}\int_0^{2\pi}\int_0^a \rho\frac{\partial u_z}{\partial t}\, r\,dr\,d\theta - \frac{1}{\pi a^2}\int_0^{2\pi}\int_0^a \frac{\partial p}{\partial z}\, r\,dr\,d\theta = 0 \tag{9.5.18}$$

Firstly, we know from the r-component of momentum balance in Eq. (9.5.10) that p is independent of r, so the second term in Eq. (9.5.18) becomes

$$\frac{1}{\pi b^2}\int_0^{2\pi}\int_0^b \frac{\partial p}{\partial z}\, r\,dr\,d\theta = \frac{\partial p}{\partial z} \tag{9.5.19}$$

The simplification of the first term, $\overline{\partial u_z/\partial t}$, in Eq. (9.5.18) follows the same steps we took for $\overline{\partial u_z/\partial z}$ above, but in this case we bring the time derivative out of the integrand so that $\overline{\partial u_z/\partial t} = \partial \bar{u}_z/\partial t$. That leaves us with the averaged conservation of momentum

$$\frac{\partial \bar{u}_z}{\partial t} = -\frac{1}{\rho}\frac{\partial p}{\partial z} \tag{9.5.20}$$

Multiplying both sides by πa^2 and rearranging gives us

$$\frac{\partial Q}{\partial t} = -\frac{A_0}{\rho}\frac{\partial p}{\partial z} \tag{9.5.21}$$

where $A_0 = \pi a^2$. The tube law can now be imposed by expanding the z-derivative of pressure as

$$\frac{\partial p}{\partial z} = \frac{dp}{dA}\frac{\partial A}{\partial z} \tag{9.5.22}$$

which holds for small amplitude. So Eq. (9.5.21)

$$\frac{\partial Q}{\partial t} = -\frac{A_0}{\rho}\frac{dp}{dA}\frac{\partial A}{\partial z} \tag{9.5.23}$$

The specific elastance, E, of the tube is defined similarly to Eq. (9.4.7) $E = A_0(dp/dA)$, which can be assumed constant for small amplitude waves. The wave speed, c, of long waves on this tube is the same as in Eq. (9.4.8), i.e. $c = C_w = (E/\rho)^{1/2}$. This speed is related to the Moens–Korteweg wave speed of a tube, which is $C_{MK} = (Yh/2a\rho)^{1/2}$, where Y is the tube's Young's modulus, h is its wall thickness, ρ is the fluid density and a is the unperturbed tube radius. C_{MK} is derived from a more detailed stress balance on a thin-walled elastic cylinder.

The final result for z-momentum is

$$\text{z-momentum}: \quad \frac{\partial Q}{\partial t} + c^2\frac{\partial A}{\partial z} = 0 \tag{9.5.24}$$

where $c = C_w$. The system of two, coupled, first-order partial differential equations, Eq. (9.5.23) and Eq. (9.5.24), forms this wave-propagation problem,

$$\begin{aligned} &\frac{\partial A}{\partial t} + \frac{\partial Q}{\partial z} = 0 \\ &c^2\frac{\partial A}{\partial z} + \frac{\partial Q}{\partial t} = 0 \end{aligned} \tag{9.5.25}$$

If we differentiate the first equation by t and the second by z, and then subtract, we eliminate the Q dependence and arrive at

$$\frac{1}{c^2}\frac{\partial^2 A}{\partial t^2} = \frac{\partial^2 A}{\partial z^2} \tag{9.5.26}$$

which is the "wave equation" for A(z, t). Likewise, because of the linear relationships, the wave equation also governs the pressure, flow and the average velocity

$$\frac{1}{c^2}\frac{\partial^2 p}{\partial t^2} = \frac{\partial^2 p}{\partial z^2}, \quad \frac{1}{c^2}\frac{\partial^2 Q}{\partial t^2} = \frac{\partial^2 Q}{\partial z^2}, \quad \frac{1}{c^2}\frac{\partial^2 \bar{u}_z}{\partial t^2} = \frac{\partial^2 \bar{u}_z}{\partial z^2} \tag{9.5.27}$$

Clinical Correlation 9.2

The arterial pulse can be felt (palpated) not only in the wrist, but also in many other locations shown in Figure 9.12. Students can use their fingertips with gentle pressure to find them. In the respiratory system, the collapsibility of the airways is countered by the addition of structural

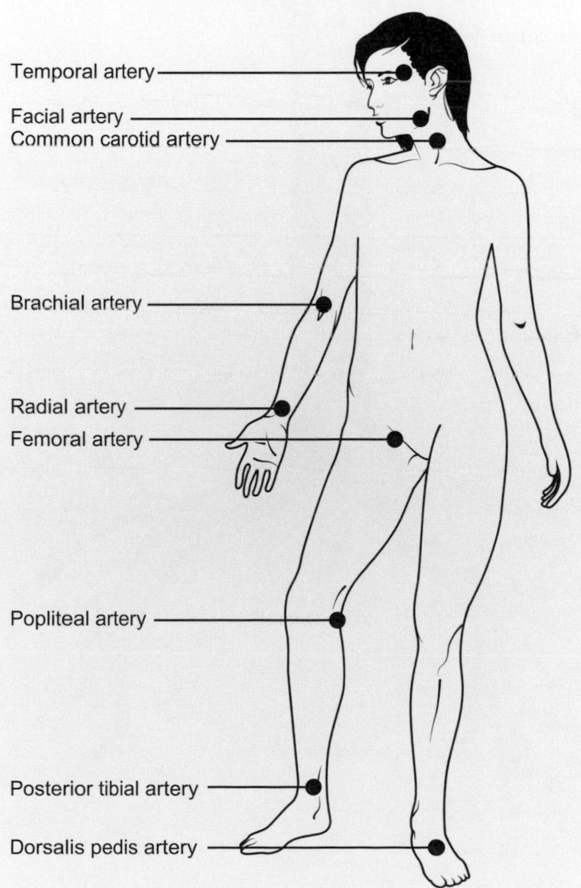

Figure 9.12 Pulse locations.

supports in their walls. Figure 9.13(a) shows the cartilage rings of the trachea and major bronchi which assist to keep the airways open. The student can palpate the cartilage rings of their own trachea which are below the thyroid cartilage (Adams apple) of the anterior neck. It should feel like a washboard, which is a bumpy surface used to wash clothes or make music, Figure 9.13(b).

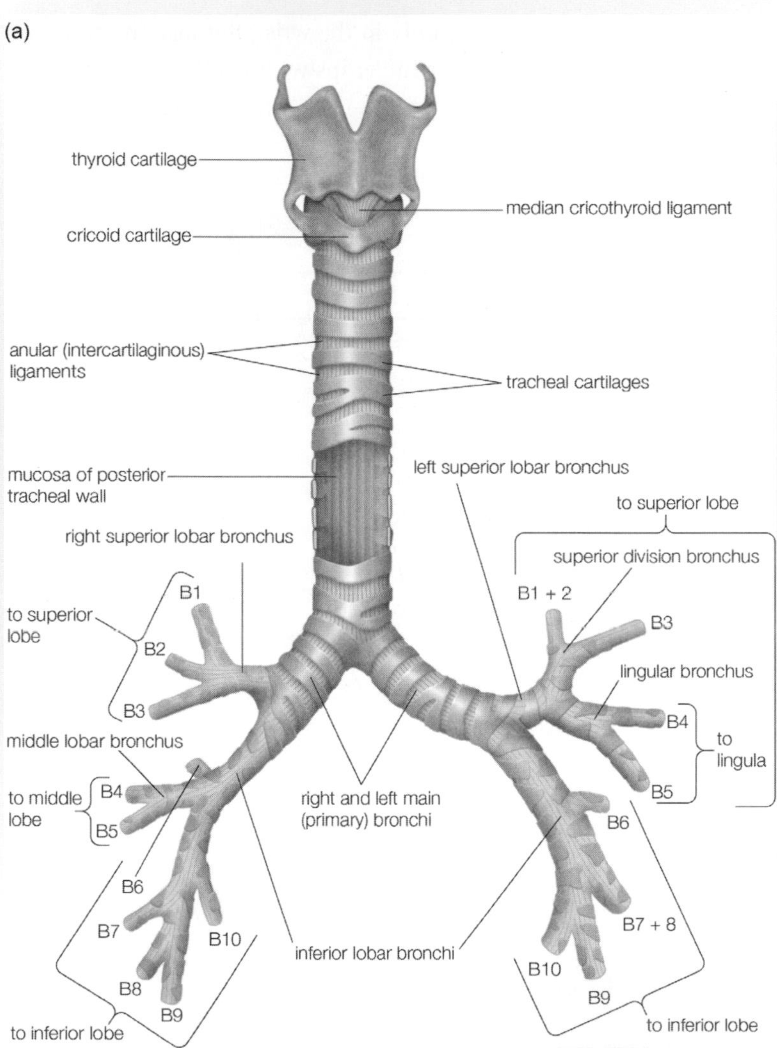

Figure 9.13 (a) Cartilage rings of the trachea and bronchi; (b) washboard used in music.

(b)

Figure 9.13 (*continued*)

The mechanical energy of fluid flow is obtained by taking the dot product of the velocity vector with the Cauchy Equation, Eq. (5.2.3) for momentum balance

$$\underline{u}\cdot\left(\rho\left(\frac{\partial \underline{u}}{\partial t}+(\underline{u}\cdot\underline{\nabla})\underline{u}\right)=\underline{\nabla}\cdot\underline{\underline{T}}+\underline{b}\right) \tag{9.5.28}$$

In Eq. (9.5.28) we replace the stress tensor with the pressure and extra stress tensor, $\underline{\underline{T}}=-p\,\underline{\underline{I}}+\underline{\underline{\tau}}$, from Eq. (6.3.1) and neglect body forces, $\underline{b}=\underline{0}$. Using our limiting simplifications of the Navier–Stokes equations shown in Eqs. (9.5.10), Eq. (9.5.28) simplifies to

$$\rho u_z\frac{\partial u_z}{\partial t}=\frac{\rho}{2}\frac{\partial u_z^2}{\partial t}=-u_z\frac{\partial p}{\partial z} \tag{9.5.29}$$

Since the pressure gradient, $\partial p/\partial z$, is independent of r, the cross-sectional area average of the equation is simply

$$\frac{\rho}{2}\frac{\partial \overline{u_z^2}}{\partial t}=-\bar{u}_z\frac{\partial p}{\partial z} \tag{5.30}$$

which says that the time rate of change of the kinetic energy per unit volume of fluid is balanced by the rate of pressure gradient work displacing the fluid.

9.6 Pulsatile Flow and Wave Propagation

Suppose there is a steady flow in the tube while the waves are propagating, like in an artery. How do we analyze that system? Let the steady average axial fluid velocity be W_0. The new velocity with a superimposed pulse wave is then $W_0 + u_z(r, z, t)$ where $u_z << W_0$. Then the z-momentum equation becomes

$$\rho\left(\frac{\partial}{\partial t}(W_0 + u_z) + (W_0 + u_z)\frac{\partial}{\partial z}(W_0 + u_z)\right) = -\frac{\partial p}{\partial z} \tag{9.6.1}$$

and mass conservation is

$$\frac{1}{r}\frac{\partial}{\partial r}(ru_r) + \frac{\partial}{\partial z}(W_0 + u_z) = 0 \tag{9.6.2}$$

Since W_0 is constant and since $u_z << W_0$, we can ignore the $u_z(\partial u_z/\partial z)$ term in Eq. (9.6.1) to arrive at

$$\rho\left(\frac{\partial u_z}{\partial t} + W_0\frac{\partial u_z}{\partial z}\right) = -\frac{dp}{dz} \tag{9.6.3}$$

In addition, the conservation of mass equation becomes

$$\frac{1}{r}\frac{\partial}{\partial r}(ru_r) + \frac{\partial u_z}{\partial z} = 0 \tag{9.6.4}$$

To analyze this flow situation, let's change to a coordinate system moving with the average fluid velocity so that the wave will appear to be similar to the situation of zero background flow. Define the primed coordinates as

$$\begin{aligned} z' &= z - W_0 t \\ t' &= t \end{aligned} \tag{9.6.5}$$

To change coordinates in the governing equations, we will need to use the chain rule for partial differentiation

$$\begin{aligned} \frac{\partial}{\partial z} &= \frac{\partial z'}{\partial z}\frac{\partial}{\partial z'} + \frac{\partial t'}{\partial z}\frac{\partial}{\partial t'} = \frac{\partial}{\partial z'} \\ \frac{\partial}{\partial t} &= \frac{\partial z'}{\partial t}\frac{\partial}{\partial z'} + \frac{\partial t'}{\partial t}\frac{\partial}{\partial t'} = -W_0\frac{\partial}{\partial z'} + \frac{\partial}{\partial t'} \end{aligned} \tag{9.6.6}$$

Then Eq. (9.6.3) becomes

$$\rho\left(\frac{\partial u_z}{\partial t'} - \cancel{W_0\frac{\partial u_z}{\partial z'}} + \cancel{W_0\frac{\partial u_z}{\partial z'}}\right) = -\frac{dp}{dz'} \tag{9.6.7}$$

The two convective acceleration terms cancel in prime space and we get

$$\rho\frac{\partial u_z}{\partial t'} = -\frac{dp}{dz'} \tag{9.6.8}$$

Equation (9.6.8) is mathematically identical to the z-momentum form of Eq. (9.5.10). All of the analysis we did for the unprimed, (z, t), system can now be done for the primed, (z′, t′), system. The resulting waves and their propagation speeds are all relative to the average fluid speed, W_0. This process of transforming the governing equations from one frame to another, where the frames are related by a constant velocity, is called a Galilean Transformation. It is explored more generally in Chapter 11, Eqs. (11.1.18)–(11.1.21), where, as here, the Navier–Stokes equations are shown to be invariant to it.

9.7 Solutions to the Wave Equation

We derived the wave equation for pressure and average axial velocity in the flexible tube as

$$\begin{aligned} \frac{1}{c^2}\frac{\partial^2 p}{\partial t^2} &= \frac{\partial^2 p}{\partial z^2} \\ \frac{1}{c^2}\frac{\partial^2 \bar{u}_z}{\partial t^2} &= \frac{\partial^2 \bar{u}_z}{\partial z^2} \end{aligned} \tag{9.7.1}$$

General solutions to these wave equations are the linear combinations $f(z - ct)$ and $g(z + ct)$,

$$\begin{aligned} p &= p_i f(z - ct) + p_r g(z + ct) \\ \bar{u}_z &= \bar{u}_{z\,i} f(z - ct) + \bar{u}_{z\,r} g(z + ct) \end{aligned} \tag{9.7.2}$$

where the subscripts i and r for the constants indicate incident or reflected waves, respectively. The function $f(z - ct)$ is a wave traveling in the positive z-direction since increasing t requires increasing z for the argument and value of f to remain the same. On the other hand, $g(z + ct)$ is a wave traveling in the negative z-direction.

To check that these are indeed solutions, let $\eta = z - ct$ and $\zeta = z + ct$. Using the chain rule for differentiation, note that

$$\frac{\partial f(\eta)}{\partial t} = \frac{df}{d\eta}\frac{\partial \eta}{\partial t} = -c\frac{df}{d\eta}, \quad \frac{\partial^2 f(\eta)}{\partial t^2} = c^2\frac{d^2 f}{d\eta^2}$$

$$\frac{\partial f(\eta)}{\partial z} = \frac{df}{d\eta}\frac{\partial \eta}{\partial z} = \frac{df}{d\eta}, \quad \frac{\partial^2 f(\eta)}{\partial z^2} = \frac{d^2 f}{d\eta^2}$$

$$\frac{\partial g(\zeta)}{\partial t} = \frac{dg}{d\zeta}\frac{\partial \zeta}{\partial t} = c\frac{dg}{d\zeta}, \quad \frac{\partial^2 g(\zeta)}{\partial t^2} = c^2\frac{d^2 g}{d\zeta^2} \tag{9.7.3}$$

$$\frac{\partial g(\zeta)}{\partial z} = \frac{dg}{d\zeta}\frac{\partial \zeta}{\partial z} = \frac{dg}{d\zeta}, \quad \frac{\partial^2 g(\zeta)}{\partial z^2} = \frac{d^2 g}{d\zeta^2}$$

Insert *p* from Eq. (9.7.2) into Eq. (9.7.1), and using the results of Eq. (9.7.3), yields

$$\frac{1}{c^2}\left(p_i c^2 \frac{d^2 f}{d\eta^2} + p_r c^2 \frac{d^2 g}{d\zeta^2}\right) = \left(p_i \frac{d^2 f}{d\eta^2} + p_r \frac{d^2 g}{d\zeta^2}\right) \tag{9.7.4}$$

which is satisfied for arbitrary *f* and *g*.

Typically one chooses sinusoidal traveling waves such as we did in Eq. (9.5.1) for $b(z, t)$

$$f(z-ct) = \cos(k(z-ct)), \quad g(z+ct) = \cos(k(z+ct)) \tag{9.7.5}$$

where the wave number, k, is inversely related to the wave length, $k = 2\pi/\lambda$, and $kc = \omega$ is the angular frequency.

Inserting the pressure and velocity wave forms of Eqs. (9.7.2) into the z-momentum equation, Eq. (9.5.20), gives us

$$\rho(-c\bar{u}_{z\,i} f' + c\bar{u}_{z\,r} g') = p_i f' + p_r g' \tag{9.7.6}$$

and collecting terms in Eq. (9.7.6) we find the relationship between incident and reflected wave amplitudes

$$p_i = \rho c \bar{u}_{z\,i}$$
$$p_r = -\rho c \bar{u}_{z\,r} \tag{9.7.7}$$

Equation (9.7.7) shows that the amplitude of the pressure wave is proportional to the product of the fluid density, the wave speed, and the fluid velocity amplitude. Pressure and velocity are "in phase" for the incident wave, but 180 degrees "out of phase" for the reflected wave because of the negative sign.

9.8 Waves at a Bifurcation: Transmission and Reflection

When a pulse wave encounters a vascular bifurcation or junction, the wave is both transmitted and reflected, see Figure 9.14. The conditions at the junction are continuous pressure

$$p_i + p_r = p_{T_1} = p_{T_2} \tag{9.8.1}$$

where the subscript "i" is for incident, "r" is for reflected, T_1 is transmitted into daughter tube 1, and T_2 is transmitted into daughter tube 2.

Also there is conservation of mass

$$Q_i + Q_r = Q_{T_1} + Q_{T_2} \tag{9.8.2}$$

where $Q = \bar{u}_z A$, the product of the average velocity and the cross-sectional area. From Eqs. (9.7.6) the incident and reflected flows are

$$Q_i = \bar{u}_{z\,i} A = +\frac{A}{\rho c} p_i, \quad Q_r = \bar{u}_{z\,r} A = -\frac{A}{\rho c} p_r \tag{9.8.3}$$

The quantity $\rho c/A$ is the characteristic impedance, Z, of the tube and is the ratio of the oscillatory pressure amplitude to the oscillatory flow amplitude

$$Z = \frac{p}{Q} = \frac{\rho c}{A} \tag{9.8.4}$$

Inserting Eqs. (9.8.3) and Eq. (9.8.4) into Eq. (9.8.2), the conservation of mass, yields

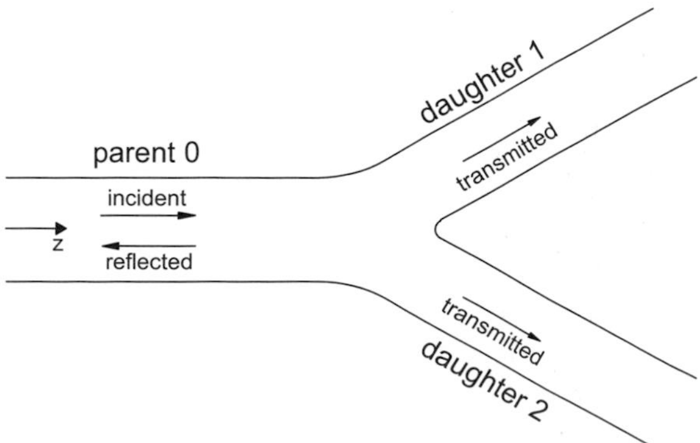

Figure 9.14 Wave reflection and transmission at a bifurcation.

$$\frac{p_i - p_r}{Z_0} = \frac{p_{T_1}}{Z_1} + \frac{p_{T_2}}{Z_2} \tag{9.8.5}$$

where the subscripts are for the parent (0) and the two daughters (1, 2) where $Z_n = p_n/Q_n = \rho c_n/A_n$ for $n = 0, 1, 2$. In general, we assume that the wave speeds are all different, i.e. $c_0 \neq c_1 \neq c_2$. Using the pressure continuity, Eq. (9.8.1), yields the dimensionless pressure amplitude ratios

$$\frac{p_r}{p_i} = \frac{Z_0^{-1} - (Z_1^{-1} + Z_2^{-1})}{Z_0^{-1} + (Z_1^{-1} + Z_2^{-1})} = R \tag{9.8.6}$$

and

$$\frac{p_{T_1}}{p_i} = \frac{p_{T_2}}{p_i} = \frac{2Z_0^{-1}}{Z_0^{-1} + (Z_1^{-1} + Z_2^{-1})} = I \tag{9.8.7}$$

I and R are selected so that we have the simpler forms

$$p_r = Rp_i, \; p_{T_1} = p_{T_2} = Ip_i \tag{9.8.8}$$

Let the incident pressure wave be traveling to in the +z-direction

$$P_i = P_m \cos(k(z - c_0 t)) \tag{9.8.9}$$

where P_m is its amplitude. According to Eq. (9.8.8), the reflected pressure wave is $P_r = RP_m \cos(k(z + c_0 t))$, which propagates in the –z-direction at speed c_0. The transmitted pressure waves are $P_{T_1} = IP_m \cos(k(z - c_1 t))$ and $P_{T_2} = IP_m \cos(k(z - c_2 t))$, which propagate in the +z-direction at speeds c_1 and c_2 into daughter tubes 1 and 2, respectively.

The pressure in the parent tube, P, is the sum of the incident and reflected pressure waves,

$$P = P_i + P_r = P_m \cos(k(z - c_0 t)) + RP_m \cos(k(z + c_0 t)) \tag{9.8.10}$$

while the flow in the parent tube, Q, is the sum of the incident and reflected flows

$$Q = Q_i + Q_r = \frac{A}{\rho c_0} P_m \cos(k(z - c_0 t)) - \frac{A}{\rho c_0} RP_m \cos(k(z + c_0 t)) \tag{9.8.11}$$

where we have substituted Eqs. (9.8.3) into Eq. (9.8.10). Note that for zero reflection, $R = 0$, in Eqs. (9.8.10) and (9.8.11) the pressure and flow wave forms are the same, i.e. traveling in the +z-direction, in phase, though with different amplitudes, P_m vs $AP_m/\rho c_0$. Once there is a reflection, $R \neq 0$, the pressure and flow are different and out of phase with one another.

The transmitted flows are

$$Q_{T_1} = \frac{p_{T_1}}{Z_1} = I\frac{p_i}{Z_1}, \; Q_{T_2} = \frac{p_{T_2}}{Z_2} = I\frac{p_i}{Z_2} \tag{9.8.12}$$

Worked Example 9.1 | Pulse Wave Reflection and Transmission

Example: Let $A_1 = A_2 = 0.6A_0$ so that $A_1 + A_2 = 1.2A_0$, a 20% increase in cross-sectional area which is a physiological value. Consider constant fluid density and let the local wave speeds all be equal, $C_w = c_0 = c_1 = c_2$. The impedances for the daughter tubes, Z_1 and Z_2, can be related to that of the parent Z_0

$$Z_1 = Z_2 = \frac{1}{0.6}\frac{\rho c_0}{A_0} = \frac{Z_0}{0.6} \tag{9.8.13}$$

Using these impedance values in Eq. (9.8.6) and Eq. (9.8.7), the values of R and I are

$$
\begin{aligned}
R &= \frac{1^{-1} - \left[\left(\frac{1}{0.6}\right)^{-1} + \left(\frac{1}{0.6}\right)^{-1}\right]}{1^{-1} + \left[\left(\frac{1}{0.6}\right)^{-1} + \left(\frac{1}{0.6}\right)^{-1}\right]} = \frac{1 - 1.2}{1 + 1.2} = -\frac{1}{11} \\
I &= \frac{2(1^{-1})}{1^{-1} + \left[\left(\frac{1}{0.6}\right)^{-1} + \left(\frac{1}{0.6}\right)^{-1}\right]} = \frac{2}{1 + 1.2} = \frac{10}{11}
\end{aligned}
\tag{9.8.14}
$$

Inserting these results for R and I into Eqs. (9.8.8) shows us that $p_r = Ip_i = -p_i/11$, so 9.09% of the incident pressure wave amplitude is reflected. Also, $p_{T_1} = p_{T_2} = Ip_i = 10p_i/11$ so 90.9% is transmitted at this bifurcation.

9.9 Flow-Induced Oscillations

In clinical medicine there are important examples of flow-related wave propagation in flexible tubes. One is wheezing breath sounds that are so prevalent in asthma and emphysema. Another is the Korotkoff sounds heard over a compressed artery, usually while measuring blood pressure. These are instances of a broad field called fluid–structure interaction, or FSI, that is important in many industrial applications like airplane wings, ship hulls, wind over buildings, etc.

Figure 9.15(a, b) shows the process of measuring blood pressure at the brachial artery using an inflatable cuff and pressure gauge known as a sphygmomanometer. The cuff is inflated until the flow in the artery stops. Then the cuff is deflated slowly. Flow-induced oscillations called Korotkoff sounds are heard when the cuff pressure reaches systolic pressure (120 mmHg) as blood starts to re-enter the partially collapsed vessel. The sounds disappear when the cuff pressure equals the diastolic pressure (80 mmHg) and the vessel is no longer collapsed.

(a)

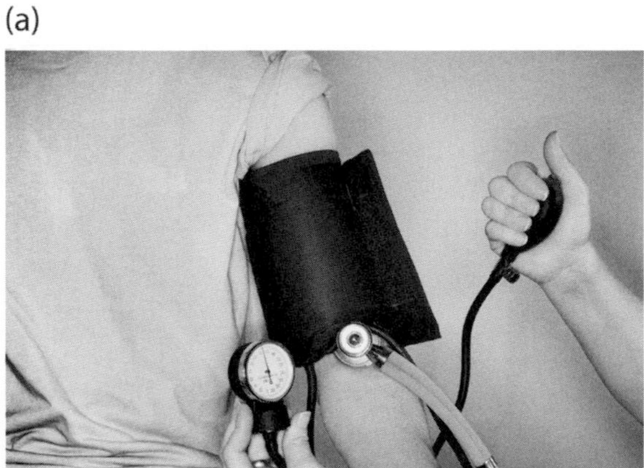

Figure 9.15 (a) Measuring blood pressure with a cuff and pressure sensor (sphygmomanometer) at the brachial artery and stethoscope over the brachial artery. (b) Blood pressure wave form and linear drop of cuff pressure vs time. The Kortkoff sounds heard through the stethoscope start at the systolic pressure (120 mmHg) and disappear at the diastolic pressure (80 mmHg). In Italian, *pressione* = pressure, *arteriosa* = artery, *manicotto* = cuff, *sanguigna* = blood.

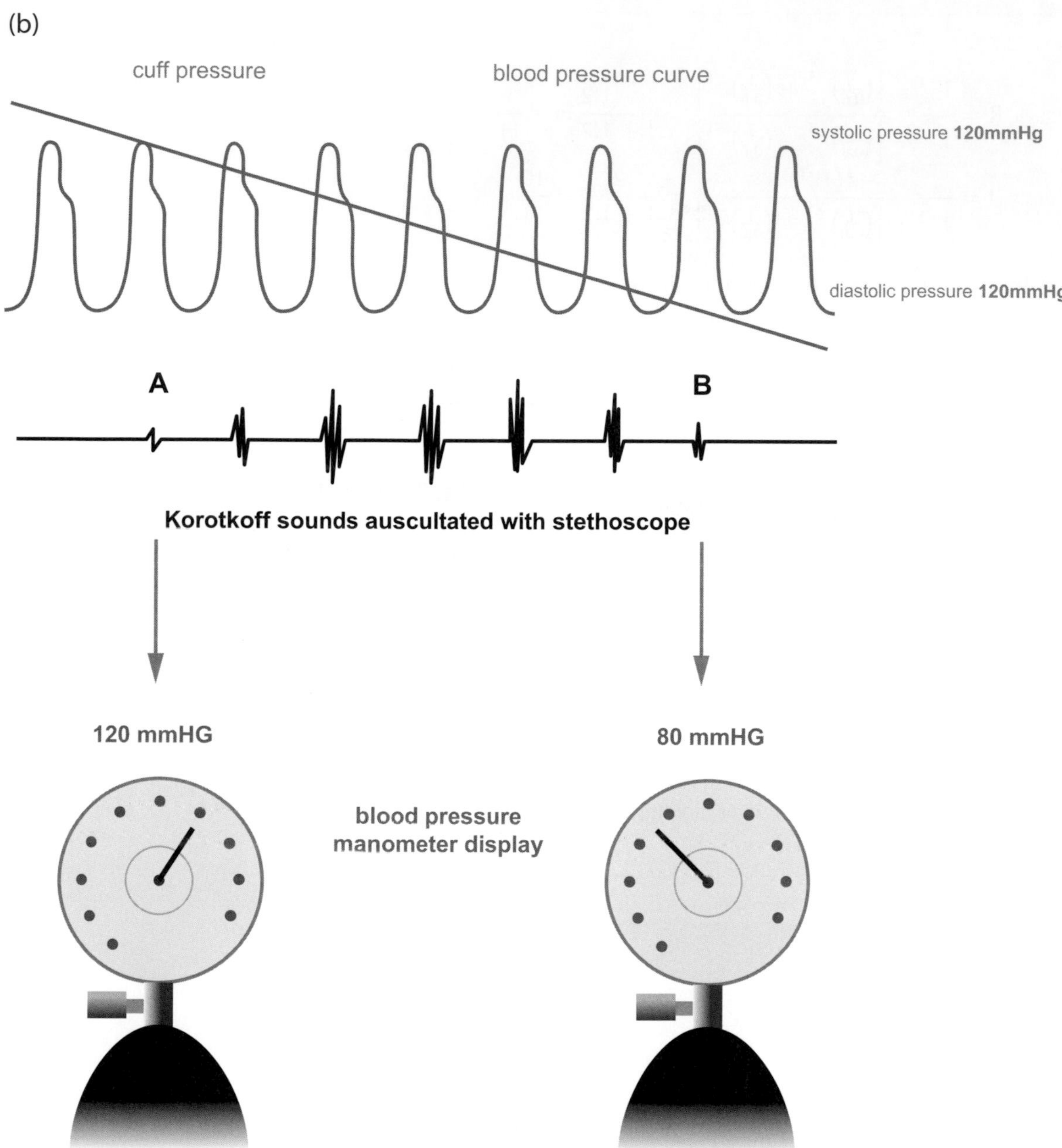

Figure 9.15 (*continued*)

Wheezing breath sounds are prevalent in asthma and emphysema, especially during expiration, and are heard during stethoscopic exams. If loud enough, they may be heard without a stethoscope. Figure 9.16(a) consists of a respiratory sound recording from an asthmatic patient where the oscillations (A) indicate wheezing (Gavriely et al., 1984). The power spectrum (B) of the 1st second of (A) is during inspiration and shows a linear relationship in a log-power vs log-frequency plot insert. The power spectrum in (C) is the

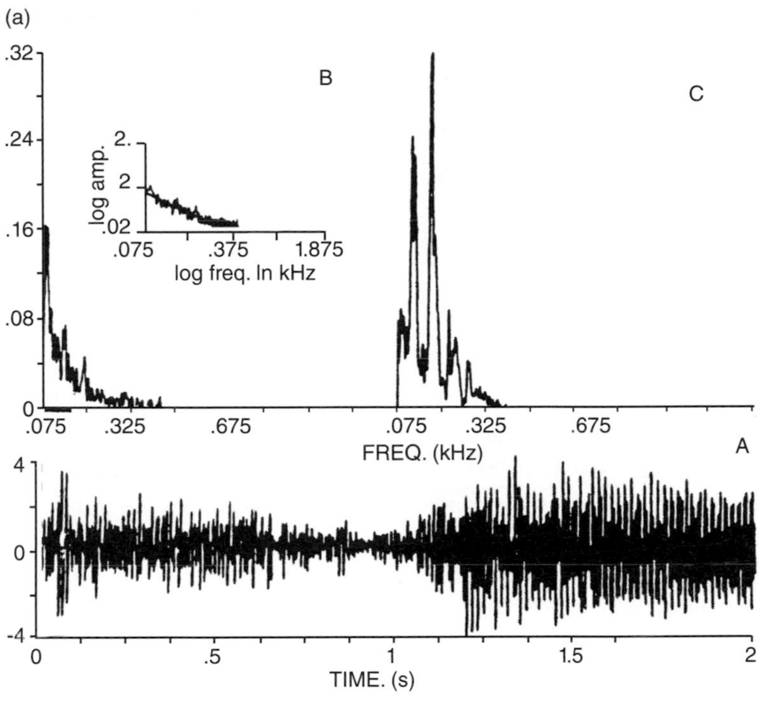

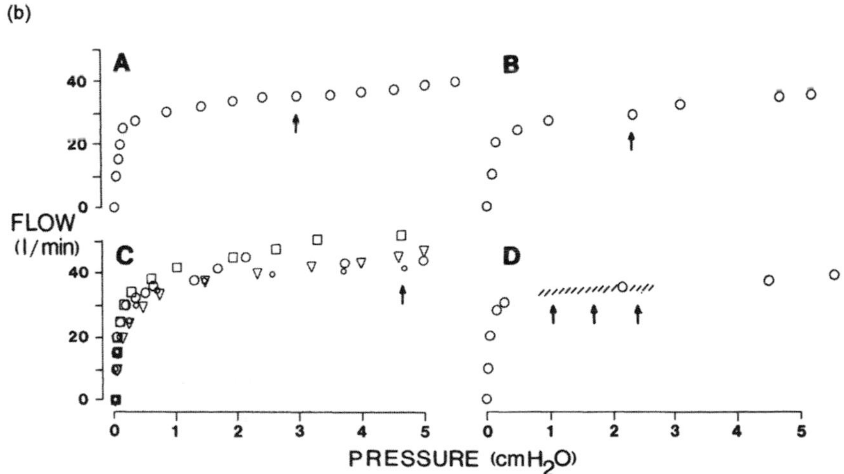

Figure 9.16 (a) (A) Wheezing respiratory sound recordings from asthmatic patient; (B) power spectrum of the 1st second of (A), which is inspiration; (C) power spectrum of the 2nd second of (A), which is expiration with a wheeze. Note sharp peaks in (C).
(b) Pressure–flow curves in a constant volume lung preparation indicating onset of wheezing sounds at the arrows where flow is limited.

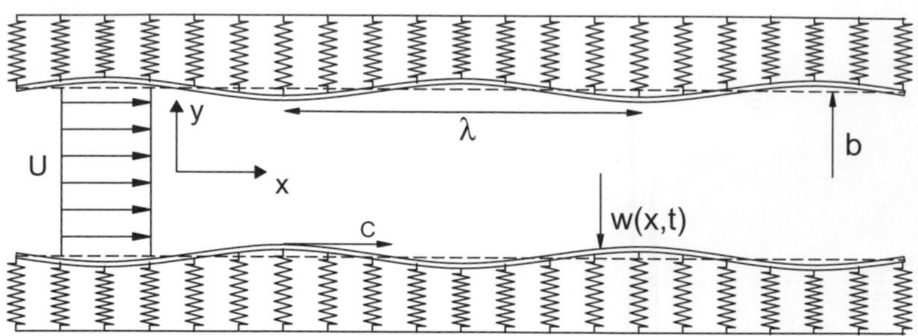

Figure 9.17 A channel model of flow-induced oscillations with uniform base state velocity U, wave speed c, wavelength, λ and lower wall position w(x, t).

2nd second of (A), which is during expiration with a pronounced wheeze giving sharp peaks in the spectrum between approximately 100 and 300 Hz. In Figure 9.16(b) flow vs driving pressure in an excised lung is shown indicating flow limitation, where the flow reaches a maximum accompanied by the appearance of wheezing (arrows) (Gavriely and Grotberg, 1988). For the lung, wheezing is an indicator of flow limitation in that airway (Gavriely et al., 1987; Grotberg and Gavriely, 1989).

In the preceding discussions of flow through flexible tubes the wall mass has been ignored. However, it becomes an essential parameter when studying flow-induced oscillations since the underlying physics of an oscillation like this is the presence of a restoring force (wall elasticity) and inertia (wall mass) which causes the system to overshoot its equilibrium state. Such a flow is shown in Figure 9.17 comprised of a two-dimensional flexible channel with base-state velocity U (Grotberg and David, 1980). We can consider the flow irrotational so that the velocities are given by $\underline{u} = \underline{\nabla}\phi$, see Eq. (6.2.10), where ϕ is known as the velocity potential. Then conservation of mass is given by the Laplace equation, Eq. (6.2.15). The corresponding conservation of momentum is the Bernoulli Equation, Eq. (6.2.14).

In addition to the kinematic boundary conditions that there is no cross-flow at the walls, we also have the stress boundary condition, here applied to the lower wall $y = w(x, t)$ assuming symmetry with respect to the midline, $y = 0$. The model uses the von Kármán plate equation which has several terms: the plate inertia, $\rho_w h \partial^2 w/\partial t^2$, where ρ_w is the wall density and h is the wall thickness; the bending stiffness, $D\partial^4 w/\partial x^4$, where $D = Yh^3/12(1-\sigma)$ and Y is the Young's modulus of the plate material while σ is its Poisson's ratio; and we add an elastic foundation term $E_0(b + w)$, where E_0 is related to the spring constants in Figure 9.17. The elastic foundation can represent surrounding tissues. The pressure, $p(\phi)$, is represented in terms of the velocity potential from the Bernoulli Equation. It is applied as a force to the plate and the resulting stress boundary condition in dimensionless form is

$$M\frac{\partial^2 W}{\partial T^2} + B\frac{\partial^4 W}{\partial X^4} + (1 + W) + P(\Phi) = 0 \qquad (9.9.1)$$

where $M = \rho_w h/\rho_f b$ is the wall to fluid mass ratio, and $B = D/(E_0 b^4)$ is the bending stiffness to spring elastance ratio. Because the density of blood is nearly the same as wall (tissue) density, i.e. $\rho_w \sim \rho_f$, the result is that $M \sim h/b$ in vascular applications. However, in lung airway applications $\rho_w \sim 1000\rho_f$ so $M \sim 1000h/b$. We will study the implications of these vastly different parameter ranges. The scales used in Eq. (9.9.1) are $\Phi = \phi/bC_0$, $W = w/b$, $T = C_0 t/b$, $X = x/b$, $Y = y/b$, $P = (p - p_{ext})/\rho_f C_0^2$ where $C_0 = \sqrt{E_0 b/\rho_f}$ is a characteristic wave speed.

The base state of the system is uniform X-velocity $S = U/C_0$ so $\Phi_0 = SX$ with parallel flat walls, $W_0 = -1$, see Figure 9.17. To explore the stability of this system we add a small perturbation in the form of a traveling wave, so the dependent variables now have the form

$$(\Phi, W) = (SX, \ -1) + \varepsilon(\Phi_1(Y), \ W_1)e^{i(K(X-CT))} \tag{9.9.2}$$

where ε is small, ε << 1. In Section 17.6 there are more background details presented for perturbation methods related to stability analysis, applied to the breakup of a liquid jet. The dimensionless wave speed is $C = c/C_0$, where c is the dimensional wave speed, while the dimensionless wave number is $K = bk$, where $k = 2\pi/\lambda$ is the dimensional wave number and λis the wavelength. The dimensional angular frequency is $\omega = 2\pi f$, where f is the frequency and the dimensionless angular frequency is $\Omega = \omega b/C_0$. We note that $c = \omega/k = f\lambda$ while $C = \Omega/K$. The form of the perturbation in Eq. (9.9.2) utilizes the separation of variables method, where the Y-dependence appears in the coefficient of the traveling wave term containing X and T. When all the equations are linearized in ε and satisfied, a characteristic equation for C results

$$C^2(1+M\tanh(K)) - 2CS - (1+BK^4)\frac{\tanh(K)}{K} + S^2 = 0 \tag{9.9.3}$$

Equation (9.9.3) is a quadratic equation with solution

$$C = \frac{S \pm G^{1/2}}{(1+M\tanh(K))} \tag{9.9.4}$$

where

$$G = -S^2 M\tanh(K) + (1+M\tanh(K))(1+BK^4)\frac{\tanh(K)}{K} \tag{9.9.5}$$

Equation (9.9.3) and its solution, Eq. (9.9.4), are called the dispersion equation since C depends on K. Different wavelengths travel at different speeds, so an initial disturbance shape, which may contain many wavelengths, sees them separate or "disperse." Not surprisingly, $\Phi_1(Y)$ depends on the hyperbolic sinh and cosh functions, as we saw in Chapter 8 for separation of variables in an unsteady channel flow. So the hyperbolic tangent, tanh, appears in Eq. (9.9.4).

There are two solutions in Eq. (9.9.4) corresponding to the positive, C^+, and negative, C^-, root. A plot of these two wave solutions is shown in Figure 9.18 for $M = 1$, $B = 1$ and $K = 1$. Note that the negative root, C^-, changes from an upstream wave, $C^- < 0$, to a downstream wave, $C^- > 0$, as S increases. The value of S where $C^- = 0$ we define as the stasis velocity, S_s, so that $C^-(S = S_s) = 0$. From Eq. (9.9.3) we see that setting $C^- = 0$ gives us S_s as

$$S_s = \left[(1+BK^4)\frac{\tanh(K)}{K}\right]^{1/2} \tag{9.9.6}$$

which is not directly dependent on M, as expected for zero frequency. We will see, however, that the value of M contributes to the selection of K, so that S_s does vary with M indirectly. This stasis velocity is the counterpart of flow limitation discussed in Section 9.4 because waves cannot propagate upstream at this fluid speed. Also in Figure 9.18 the two waves coalesce at the same speed, $C_c = 1.066$ for $S_c = 1.877$. That occurs because $G = 0$ in Eq. (9.9.4)

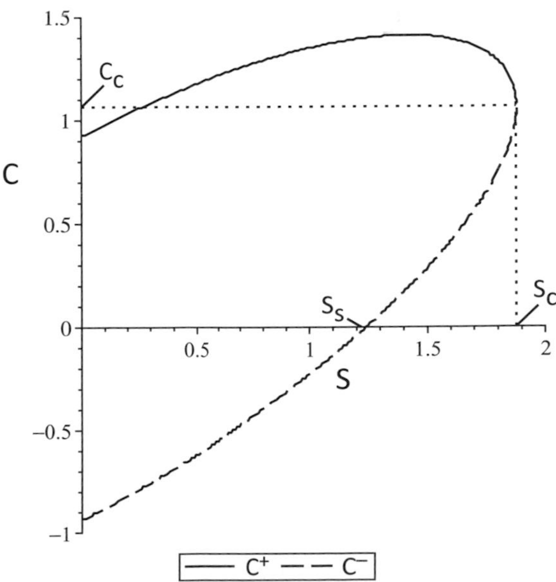

Figure 9.18 Wave solutions C^+, C^- for $M = 1$, $B = 1$, $K = 1$; S_s is the stasis velocity, S_c is the critical velocity and C_c is the critical wave speed.

when $S = S_c$. In addition, the stasis velocity is $S_s = 1.234$.

For general values of the parameters, the condition to find S_c comes from setting $G = 0$ as we just saw, which leads to its definition

$$G = 0 \quad \rightarrow \quad S_c = \left[\frac{(1 + M\tanh(K))(1 + BK^4)}{MK}\right]^{1/2} \tag{9.9.7}$$

In addition, for $G > 0$ we see that $S < S_c$ and the system is neutrally stable, that is, the perturbation does not grow or decay in this frictionless example system. However, if $S > S_c$ then $G < 0$ and $G^{1/2}$ in Eq. (9.9.4) becomes imaginary causing C to be complex. For complex C the solution looks like

$$C = \frac{S \pm i(-G)^{1/2}}{(1 + M\tanh(K))} = C_r \pm iC_i \quad \text{for } S > S_c, G < 0 \tag{9.9.8}$$

where $C_r = S/(1 + M\tanh(K)) > 0$ is the real part and $C_i = (-G)^{1/2}/(1 + M\tanh(K)) > 0$ is the imaginary part. Recall that the traveling wave dependency is $e^{iK(X-CT)} = e^{iK(X-C_rT)}e^{\pm C_iT}$. When $S > S_c$ the system grows exponentially in time due to the e^{+C_iT} dependency and the system is unstable. However, this is exactly what we are seeking, because the small perturbation grows enough in size to be observed during this instability. Nonlinear elastic effects then stabilize the wave to a new larger amplitude (Grotberg, 1984; Grotberg and Gavriely, 1989).

According to Eq. (9.9.7), the critical velocity S_c depends on K as shown in Figure 9.19. When $S < S_c$ the channel is neutrally stable and when $S > S_c$ it is unstable as the two regions are labeled. While we set $B = 1$ and $M = 1$ in the model, the perturbation is posed for any value of K. For Figure 9.18 we chose $K = 1$ just to view the initial features of the solution. However, in many fluid mechanical stability theories, including this one, K (or equivalently the wavelength λ) is selected by the physics of the problem. Notice that there is a local minimum for the $S_c(K)$ equation plotted in Figure 9.19. Since we usually think of a system starting from zero flow velocity, say the beginning of expiration in the lung, the channel reaches instability at the lowest value of $S_c(K)$, i.e. the local minimum in Figure 9.19. The minimum, then, defines the onset of the flutter instability that would be observed, and the flutter wave number $K = K_F = 0.685$ and the flutter velocity, $S_c(K = K_F) = S_F = 1.685$ are defined there. From Eq. (9.9.8) the corresponding flutter wave speed is $C(K = K_F) = C_F = 1.057$, which are different from the critical values in Figure 9.18 where we chose $K = 1$.

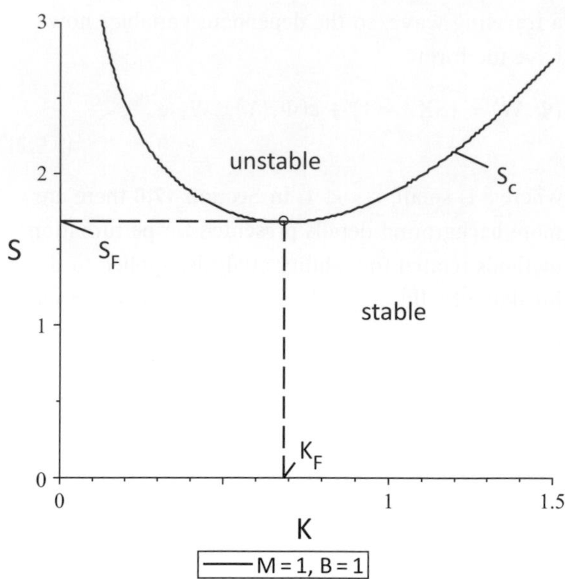

Figure 9.19 S_c vs K showing a local minimum defining K_F and S_F. Stable and unstable regions are indicated. $B = 1$, $M = 1$.

Figure 9.20(a) shows the wave propagation results for $M = 10, 50$ and 100 with $B = 1$ and $K = K_F$ for each value of M. Note that as M increases the C^+ and C^- curves become more symmetric and S_s and S_F approach one another so that flutter (S_F) and flow limitation (S_s) are essentially simultaneous, as was noted in physiological measurements of airways (Gavriely et al., 1987; Gavriely and Grotberg, 1988). Figure 9.20(b) plots S_F, S_s, K_F and C_F for $1 \leq M \leq 100$ and $B = 1$ indicating asymptotic behavior for large wall mass. S_s and S_F asymptote toward 1 while C_F approaches zero and K_F a constant.

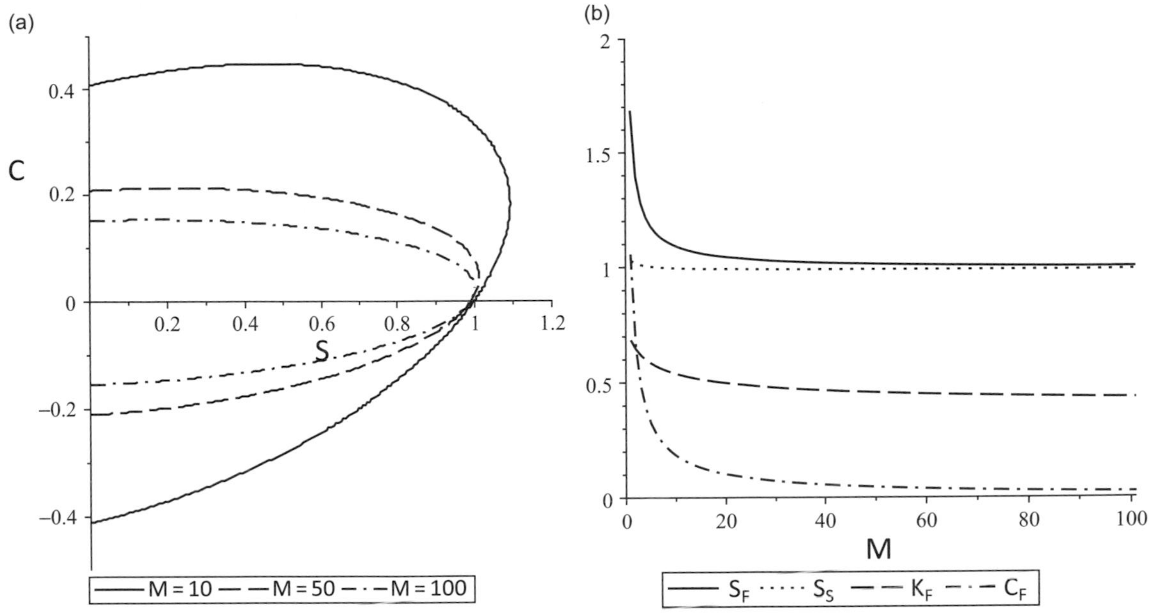

Figure 9.20 (a) Wave propagation for $M = 10, 50, 100$, with $B = 1$ and $K = K_F$ for each value of M. (b) S_F, C_F, K_F and S_s vs M for $B = 1$.

What about the dimensional frequency, f? It will be equal to $f = \Omega_F C_0/2\pi b = K_F C_F C_0/2\pi b$. For $M = 100$ the results in Figure 9.20(b) are that $K_F = 0.433$, $C_F = 0.024$. Using air density as $\rho_f = 10^{-3}\,g/cm^3$ and a very flattened airway, $b = 0.05\,cm$, we also need the wall elastance, which we take as $E_0 b = 5 \times 10^4\,dyn/cm^2$, which falls in the physiologic range for airways (Fry, 1968; Dawson and Elliott, 1977). Solving for the frequency we find $f = 234\,Hz$, which is typical for a wheeze. Further modeling includes simple versions of damping in the wall model and fluid viscosity through a friction factor term in the Bernoulli equation (Grotberg, 1984) as well as the Navier–Stokes equations (Larose and Grotberg, 1997).

In Section 9.5 we studied pulse wave propagation for a system where the wave speed depended on the wall elastance and fluid density for long waves. In that limit, the wavelength itself, did not influence the wave speed. However, Eq. (9.9.4) states that the wave speed, C (or c), depends on the wavelength of the disturbance, as represented by $K = 2\pi b/\lambda$. Such waves are called "dispersive waves." So if you apply a disturbance to the wall having many wavelengths, its shape will change going downstream as the components travel at different speeds. The disturbance disperses.

An important aspect of dispersive waves is the "group velocity," which measures the speed of energy propagation. While the phase velocity, c, is given by $c = \omega/k$, the group velocity is given by

$$c_g = \frac{\partial \omega}{\partial k} \tag{9.9.9}$$

The group velocity is readily appreciated when summing two waves of slightly different wavelengths. Let that sum be

$$h(x,t) = \text{Re}\left\{e^{i(k_1 x - \omega_1 t)} + e^{i(k_2 x - \omega_2 t)}\right\} \tag{9.9.10}$$

and define averages and differences of k, ω as $k_{avg} = (k_1 + k_2)/2$, $\omega_{avg} = (\omega_1 + \omega_2)/2$,

$\Delta k = (k_1 - k_2)/2, \Delta\omega = (\omega_1 - \omega_2)/2$. Insert these into Eq. (9.9.10) and rearrange the result,

$$\begin{aligned} h(x,t) &= \mathrm{Re}\left\{e^{i(k_{avg}x+\Delta kx-\omega_{avg}t-\Delta\omega T)} + e^{i(k_{avg}x-\Delta kx-\omega_{avg}t+\Delta\omega t)}\right\} \\ &= \mathrm{Re}\left\{e^{i(k_{avg}x-\omega_{avg}t)}\left(e^{i(\Delta kx-\Delta\omega t)} + e^{-i(\Delta kx-\Delta\omega t)}\right)\right\} \\ &= 2\mathrm{Re}\left\{e^{i(k_{avg}x-\omega_{avg}t)}\cos\left(\Delta kx - \Delta\omega t\right)\right\} \\ &= 2\cos\left(k_{avg}(x - c_{avg}t)\right)\cos\left(\Delta k(x - c_g t)\right) \end{aligned} \tag{9.9.11}$$

where $c_{avg} = \omega_{avg}/k_{avg}$, $c_g = \Delta\omega/\Delta k$.

An example of h(x, t) is shown in Figure 9.21 for $k_{avg} = 2/m$, $\Delta k = 0.1/m$, $\omega_{avg} = 10/s$, $\Delta\omega = 0.1/s$, which makes $c_{avg} = 5\,m/s$ and $c_g = 1\,m/s$ at $t = 0$. The higher frequency phase velocity travels five times faster than the group velocity envelope, both in the positive x-direction. Other important flow-induced oscillations occur in biomedical applications including vascular flows (Jensen, 1990; Bertram et al., 1991; Pedley and Luo, 1998; Grotberg and Jensen, 2004), phonation (Mittal et al., 2013) and snoring (Elliott et al., 2011), which can lead to obstructive sleep apnea (Xu et al., 2006).

The oscillating walls shown in Figure 9.17 are, in fact, made of tissue including the airway epithelial cells. Researchers have shown that oscillating airway epithelial cells causes them to secrete pro-inflammatory molecules (Almendros et al., 2007; Puig et al., 2005). So wheezes are not only a sign of disease, like asthma and emphysema, they can cause injury and inflammation which promotes more wheezing (Grotberg, 2019). How this may impact a paradigm shift in clinical medicine and the use of stethoscopes is discussed along with respiratory crackles in Section 16.6.

9.10 Lumped Parameter Models

Lumped parameter models of flows in flexible tubes are popular in medical and physiological applications. They utilize electrical circuit analogs where features of the fluid–solid system are separated into compartments. For example, airflow dynamics in the lung can be viewed as the analog RLC circuit shown in Figure 9.22. The analogs are: voltage, E, for pressure, p; current, I, for flow rate, Q, through the airway compartment; electrical resistance, R, for flow resistance, R; inductance, L, for inertance, L; and

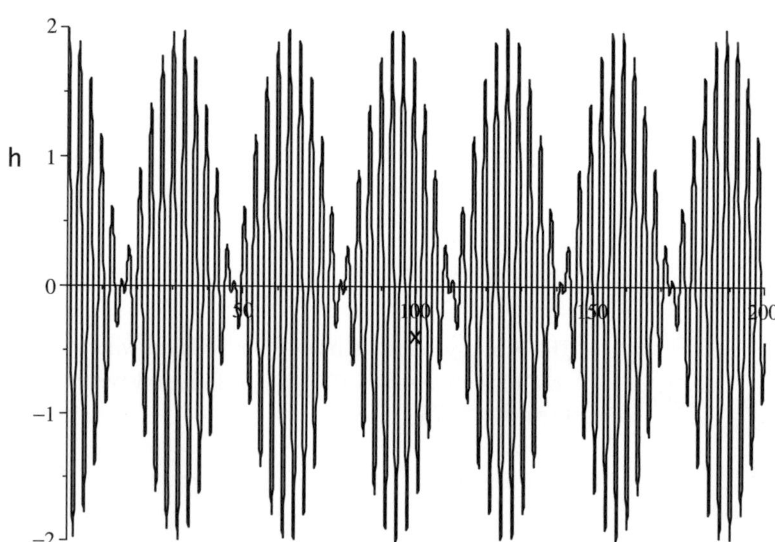

Figure 9.21 Beating phenomenon in wave propagation as an example of group velocity.

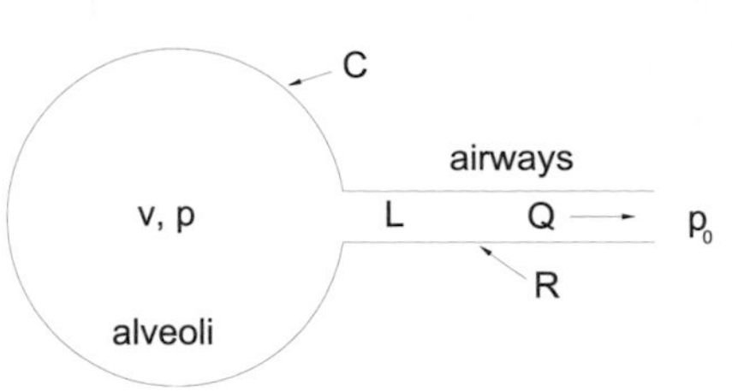

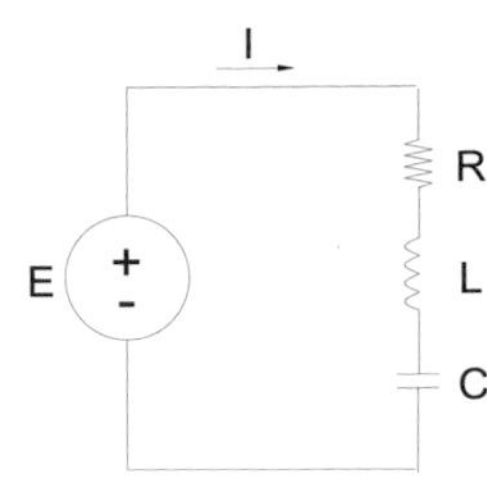

Figure 9.22 RLC electrical circuit analog of airflow in the lung.

capacitance, C, for compliance, C, of the alveolar compartment. Lung compliance depends strongly on surface tension phenomena from the alveolar liquid lining, as discussed in Section 16.6. The pressure in the alveolar compartment is $p = v/C$, while the relevant circuit elemental equations for pressure drops are $\Delta p = QR$ for the resistor and $\Delta p = L\,dQ/dt$ for the inductor. The input pressure at the mouth is p_0.

The governing equation for the lung volume comes from adding the pressure drop across each element in series $v/C - LdQ/dt - RQ = p_0$. The flow rate is $Q = -dv/dt$ from mass conservation where v(t) is the lung volume in the alveolar compartment. The negative sign sets expiration as a positive flow. Using this substitution and rearranging yields the governing equation for the RLC circuit

$$L\frac{d^2v}{dt^2} + R\frac{dv}{dt} + \frac{1}{C}v = p_0 \tag{9.10.1}$$

subject to initial conditions $v(0) = v_0, dv/dt_{t=0} = 0$. For normal subjects typical parameter values are $R = 2\ cmH_2O/(liter/s)$, $C = 0.15\ liter/cmH_2O$, $L = 0.01\ cmH_2O/(liter/s^2)$.

When $L = 0$ and $p_0 = 0$, the solution to Eq. (9.10.1) is $v \sim v_0 e^{-t/RC}$, which decays exponentially at a rate controlled by the time constant, RC.

Using small amplitude pressure oscillations at the mouth, see Figure 9.23, respiratory investigators calculate R, L and C from experimental data through models like this or ones with additional circuit elements. Normal ranges are established and deviations can be used to diagnose disease (Dubois et al., 1956; Dorkin et al., 1988; Lutchen et al., 1996; Kaczka and Dellacá, 2011).

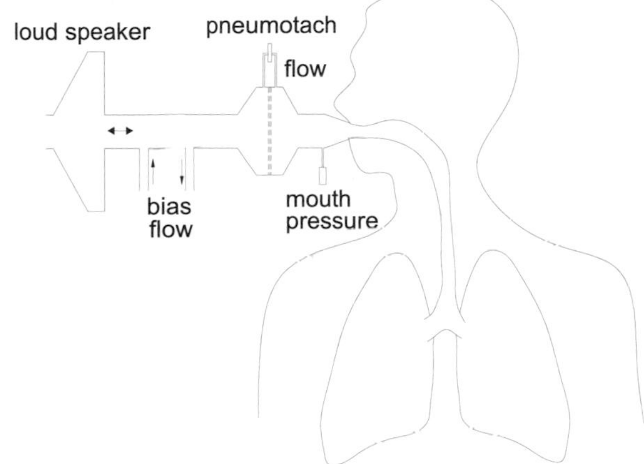

Figure 9.23 Forced oscillations of the lung.

Forced oscillations are modeled by $p_0 = \bar{p}(1 + \delta \sin(\omega t))$ where $\delta << 1$. After the homogeneous solution decays, the $L = 0$, time-periodic particular solution is

$$v = \bar{p}C\left(1 + \delta\frac{(\sin(\omega t) - \omega RC\cos(\omega t))}{1 + (\omega RC)^2}\right) \tag{9.10.2}$$

It is routine in circuit theory to express sums of sin and cos terms as a single sinusoid with a phase lag. Here the relevant general form in Eq. (9.10.2) is $A\sin x - B\cos x$. Consider the point (A, B) in rectangular coordinates and its polar equivalent, $A = R\cos(\phi), B = R\sin(\phi)$, where the radial hypotenuse is $R = (A^2 + B^2)^{1/2}$ and the polar angle has tangent $\tan\phi = B/A$. Then the sum is $R(\cos(\phi)\sin x - \sin(\phi)\cos x) = R\sin(x - \phi)$. Using this identity with $A = 1, B = \omega RC$, the dimensionless form of Eq. (9.10.2) is

$$V = 1 + \delta \frac{\sin(T - \theta)}{(1 + \Omega^2)^{1/2}} \quad (9.10.3)$$

where $T = \omega t$ and $V = v/\bar{p}C$. The dimensionless frequency is $\Omega = \omega RC$ and the phase angle is $\theta = \tan^{-1}(\Omega)$. We will learn in Section 14.3 that Ω is similar to the Deborah number, $De = \omega\lambda$, where λ is the relaxation time for a viscoelastic material. In the present application, RC is a viscoelastic relaxation time of the system: R represents the viscous loss and 1/C is the elastance. The unsteady dimensional pressure amplitude is $\Delta p = 2\delta\bar{p}$ while the dimensional volume amplitude is $\Delta v = 2\delta\bar{p}C/(1 + \Omega^2)^{1/2}$.

Sample plots are shown in Figure 9.24 for $\Omega = 1, \delta = 0.2$ involving V and the dimensionless pressure input, $P = p_0/\bar{p}$. Figure 9.24(a) presents P and V for one cycle, $0 \leq T \leq 2\pi$, where the phase lag is $\theta = \tan^{-1}(1) = \pi/4$. Figure 9.24(b) is a parametric plot of V vs P, the volume-pressure loop, for the same time interval. It has important characteristics of its overall slope, a measure of total compliance, $C_{tot} = \Delta v/\Delta p$ (Grotberg and Davis, 1980). In addition, the dimensional hysteresis area inside the loop measures the work per cycle, $w = \int_{cycle} p\, dv = \int_0^{2\pi/\omega} p(dv/dt)\, dt$. Much of the hysteresis is due to surface tension and surfactant effects discussed in Section 16.6.

The two-element Windkessel model of arterial blood flow was established by German physician and physiologist Otto Frank (1865–1944). The Frank–Starling law of the heart is named for him and Ernest Starling. The model has an electrical circuit analog of a resistor and capacitor in parallel as shown in Figure 9.25, where R is the total arterial peripheral resistance and C is the total arterial compliance. By contrast, the lung circuit elements were in series since the compliance was relegated to the alveolar compartment and resistance to the airway compartment separately. There are three-element and four-element Windkessel models that add more circuit components (Frank, 1930; Kelly et al., 1992; Stergiopulos et al., 1999; Westerhof et al., 2009).

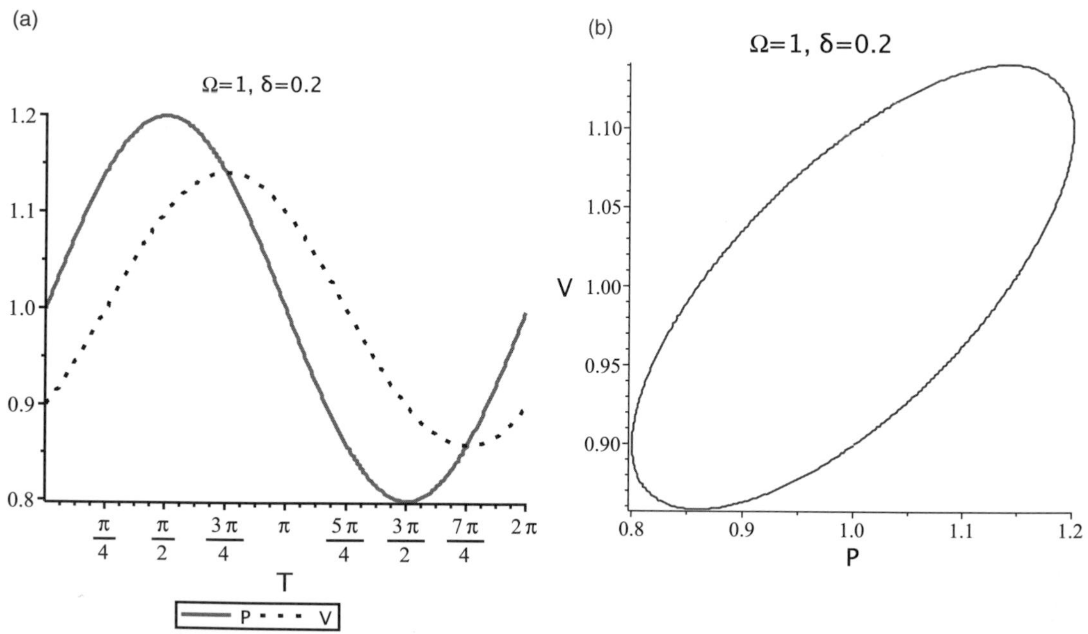

Figure 9.24 (a) V(T) and P(T) for $\Omega = 1$, $\delta = 0.2$ and (b) the volume–pressure loop is the parametric plot V(P).

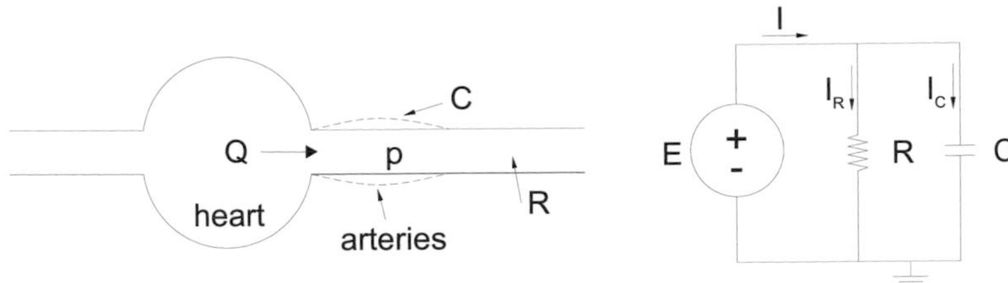

Figure 9.25 Two-element Windkessel model of arterial blood flow and pressure.

The total current I, or flow Q, is the sum of the two elemental currents, $I = I_R + I_C \rightarrow Q = Q_R + Q_C$, or

$$C\frac{dp}{dt} + \frac{1}{R}p = Q \tag{9.10.4}$$

subject to the initial condition $p(0) = p_i$, where Q is the flow rate in cm^3/s and p is the pressure in mmHg. For a given flow input, we can compute the resulting pressure using Eq. (9.10.4).

Let the total cardiac cycle have a time duration of T_c in seconds. Start with systole at $t = 0$ and model it as a sinusoidal flow input over the time period $0 \leq t \leq T_s$, where T_s is the duration of systole. Systole is followed by diastole $T_s \leq t \leq T_c$, which has no input flow.

Mathematically, Q is

$$Q = \begin{cases} Q_0 \sin(\pi t/T_S) & 0 \leq t \leq T_s \quad \text{systole} \\ 0 & T_s \leq t \leq T_c \quad \text{diastole} \end{cases} \tag{9.10.5}$$

To calculate the flow rate amplitude, Q_0, the stroke volume of the heart, V_s, is ejected during systole. The time integral yields

$$V_s = \int_0^{T_s} Q_0 \sin(\pi t/T_S)\, dt = \frac{2Q_0 T_s}{\pi} \quad \rightarrow \quad Q_0 = \frac{\pi V_s}{2T_s} \tag{9.10.6}$$

The solution to Eq. (9.10.4) for the systolic phase, including the initial condition, is

$$p_s = \left(p_i + \frac{\pi Q_0 R^2 C T_s}{\pi^2 R^2 C^2 + T_s^2}\right) e^{-\frac{t}{RC}} - \frac{Q_0 R T_s (\pi RC \cos(\pi t/T_s) - T_s \sin(\pi t/T_s))}{\pi^2 R^2 C^2 + T_s^2} \tag{9.10.7}$$

For forced oscillations of the lung, we ignored the transient term, but for arterial flow it is an important feature combined with the sinusoidal particular solution. The pressure at the end of systole is $p_{es} = p_s(t = T_s)$

$$p_{es} = p_i e^{-\frac{T_s}{RC}} + \frac{\pi Q_0 R^2 C T_s}{\pi^2 R^2 C^2 + T_s^2}\left(1 + e^{-\frac{T_s}{RC}}\right) \tag{9.10.8}$$

The diastolic phase solution is

$$p_d = p_{es} e^{-\frac{\tau}{RC}} \tag{9.10.9}$$

where $\tau = t - T_s$ and $T_s \leq t \leq T_c$. Though $Q = 0$, the elemental flows are not zero in diastole. From $Q_R + Q_C = 0$ we know that $Q_C = -Q_R < 0$, so the capacitor is discharging through the resistor. Mechanically it is the elastic arterial compartment which was distended during systole that is returning to its original volume, squeezing the blood downstream through the distal resistance.

Normal range parameter values are $R = 1\ \text{mmHg}/(\text{cm}^3/\text{s})$ and $C = 1.2\ \text{cm}^3/\text{mmHg}$. For a heart rate of 72 beats per minute, $T_c = 60/72 = 0.83\ \text{s}$, and a usual fraction of systolic vs cardiac cycle time intervals is $T_s = 0.4T_c = 0.33\ \text{s}$. The initial pressure is the end diastolic value of the previous heartbeat, let it be $p_i = 80\ \text{mmHg}$. With a stroke volume of $V_s = 80\ \text{cm}^3$, Eq. (9.10.6) gives

$Q_0 = \pi V_s/2T_s = 377\ \text{cm}^3/\text{s}$. Figure 9.26 is a plot of the systolic and diastolic arterial pressure vs time exhibiting a systolic pressure of 120 mmHg. The end diastolic pressure, $p_d(t = T_c) = 78.3\ \text{mmHg}$, is very nearly equal to p_i. A reading of 120 mmHg/80 mmHg is normal using a sphygmomanometer, see Figure 9.15.

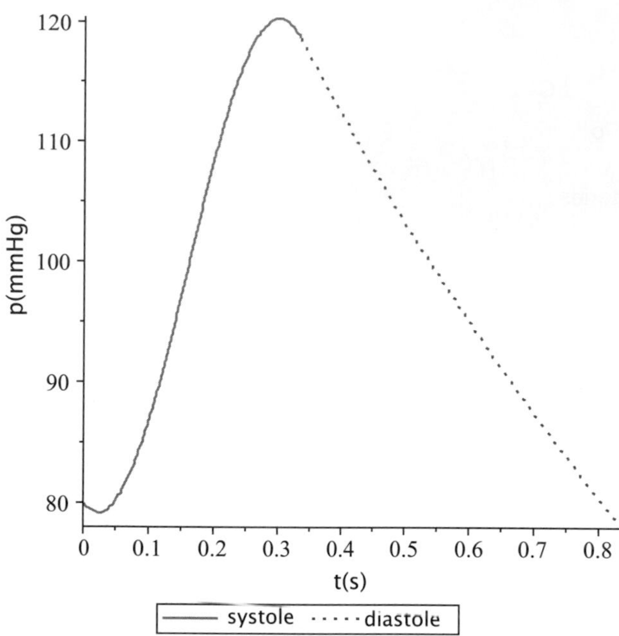

Figure 9.26 Arterial pressure for a cardiac cycle using the two-element Windkessel model.

Summary

In this chapter we touched on one of the unique properties of biological flows, that the vessel conveying that flow can be flexible. We started with the simple situation of a tube with a linear relationship between radius and transmural pressure. During flow it maintains a circular cross-section, but the radius tapers along the flow axis. We found what constraints exist on the parameter ranges and how the results line up with our intuition for locally parabolic velocity profiles. The more complicated nonlinear tube law introduced other mechanical issues such as the non-circular, azimuthal-dependent, shape of the cross-section. Rather than treating that in detail, only the tube's cross-sectional area was considered, and how it relates to a transmural pressure, a macroscopic approach. Nevertheless, we found by assuming locally Poiseuille flow versus pressure balance that flow limitation occurs due to viscous effects. Next we found flow limitation due to inertial effects and its relationship to the local wave speed along the tube wall. When the ratio of the fluid speed to the wave speed is unity, flow limitation ensues. Because the tube's specific elastance determines the wave speed, flow limitation can cause pathologically low outflows in lung diseases like asthma and emphysema. Both forms of flow limitation apply to forced expiration, the viscous limitation likely at lower lung volumes when all airway velocities fall below their local wave speed. Pulse propagation presented an interesting analysis of wave propagation, especially its application to arterial bifurcations where there are both transmitted and reflected waves. Flow through a flexible channel where wall mass was included in the analysis led to flutter waves that were related to vascular Korotkoff sounds and wheezing breath sounds at flow limitation in lung airways. Finally, lumped parameter models provided physical insight with mathematical tractability to determine overall behavior of lung and arterial flows with viscous and elastic properties.

Problems

9.1

a. For Eq. (9.1.15) plot U_Z vs R for $Z = 0, 1/3, 2/3, 1$ and let $\gamma = 0.018$. What happens to the profile as Z increases? Why?

b. Plot the profiles in part (a) offset and scaled to fit, i.e. plot $Z + (0.1)U_Z(R, Z)$, so they are distributed along the flexible tube.

c. Plot B(Z) from Eq. (9.1.21) and add it to the profiles in part (b).

d. From Eq. (9.1.28) plot U_R vs R for $Z = 0, 1/3, 2/3, 1$ and let $\gamma = 0.018$. What direction are these velocities? Why?

e. Now display the offset plots $R + (0.3)U_R$ vs Z for $R = 0.2, 0.4, 0.6$ and $0 \leq Z \leq 1$. Which curve has the largest change in velocity over the Z interval? Explain your answer.

f. Plot the velocity vector field for $\gamma = 0.018$. Add to it the wall shape B(Z). How does your result correlate to the above velocity profile plots?

9.2 For flow in a flexible tube with a linear tube law we found the dimensionless velocities U_R, U_Z in Eq. (9.1.15) and Eq. (9.1.28).

a. The stream function, ψ, in cylindrical coordinates is defined by $U_R = -(1/R)(\partial\psi/\partial Z)$, $U_Z = (1/R)(\partial\psi/\partial R)$. Find the stream function for this flow. Let $\psi(R = 0) = 0$.

b. Plot streamlines for this flow assuming $\gamma = 0.01$ and discuss your results.

c. The dimensional average axial fluid velocity, $u_{z\ avg}$, is $u_{z\ avg} = \frac{1}{\pi a(z)^2}\int_0^{2\pi}\int_0^{a(z)} u_z r\, dr\, d\theta$. Make this relationship dimensionless using the scales $B(z) = a(z)/a_0$, $U_Z = u_z/u_s$; $R = r/a_0$, $U_{Z\ avg} = u_{z\ avg}/u_s$.

d. Plot the dimensionless average axial velocity, U_{Zavg} vs Z for $\gamma = 0.003, 0.005, 0.01$ and discuss your results.

e. Plot the dimensionless wall shear stress $\tau_w = (\partial U_Z/\partial R)|_{R=B(Z)}$ vs Z for $\gamma = 0.003, 0.005, 0.01$. Where is its magnitude a maximum and why?

9.3 For flow in a flexible tube with a nonlinear tube law we found the dimensionless Poiseuille-like relationship between the flow rate F and the pressure gradient $d\hat{\pi}/dZ$ in Eq. (9.3.3).

In the text we integrated the separated form of Eq. (9.3.3) from the lower limit $Z = 0$ to the upper limit $Z = 1$ and the corresponding pressures $\hat{\pi}(Z = 0) = \hat{\pi}_0 = 0$ to $\hat{\pi}(Z = 1) = \hat{\pi}_1 < 0$, and found the flow rate, F, as a function of negative downstream pressure $\hat{\pi}_1$ and the flexibility parameter n in Eq. (9.3.5).

a. Now go back and integrate the separated form of Eq. (9.3.3) from the lower limit, $Z = 0$, to the variable upper limit, $Z = Z$, on the left hand side and the corresponding pressures on the right hand side, from $\hat{\pi}_0 = 0$ to $\hat{\pi}$. Rearrange your result to find $\hat{\pi}(Z)$ in terms of F.

b. Let $\hat{\pi}_1 = -1$ and plot $\hat{\pi}(Z)$ over the range $0 \leq Z \leq 1$ for $n = 0.75, 1.00, 1.25, 1.50$ and discuss your results.

c. Let $\hat{\pi}_1 = -1$ and plot the dimensionless cross-sectional area $\alpha(Z)$ for $n = 0.75, 1.00, 1.25, 1.50$ and discuss your results.

d. Since we do not have the details of a velocity profile in this approach, we will have to estimate the wall shear stress. In Poiseuille flow the wall shear is $\tau_w = \mu(du_z/dr)|_{r=a} = -4\mu u_{z\ avg}/a$. Assume a similar form of the wall shear based on the local value of $u_{zavg}(z)$ and a(z). Let $\hat{\pi}_1 = -1$ and plot $\tau_w(Z)/\tau_w(Z = 0)$ for $n = 0.75, 1.00, 1.25, 1.50$ and discuss your results.

9.4 We learned that the pressure and flow waves have the form Eq. (9.8.10) and Eq. (9.8.11), respectively. Consider a system of parent and daughter tubes where the daughter tubes have cross-sectional areas given by $A_1 = 0.8A_0, A_2 = 0.6A_0$ and A_0 is the parent tube cross-sectional area. Also, let the daughter wave speeds be $c_1 = c_0,\ c_2 = 0.8\,c_0$ where c_0 is the parent wave speed.

a. Calculate I and R.
b. Make the pressure and flow wave forms dimensionless with the following scales $T = kc_0t,\ Z = kz,\ \hat{P} = P/P_m,\ \hat{Q} = Q/(P_mA/\rho c_0)$. Plot the two waves $\hat{P}$, $\hat{Q}$ for $0 \le Z \le 2\pi$ at $T = 3\pi/8$ and $T = 3.5\pi/8$. Place all four curves on the same graph. Describe your results. Which direction does the wave travel for each? Which wave leads the other?
c. Repeat part (b) for $T = 7\pi/8$ and $T = 7.5\pi/8$. Compare your descriptions.
d. Animate these two waves on the same graph. For clarity try separating the two vertically by animating $1 + \hat{Q}$ and $-1 + \hat{P}$ over the Z range, $0 \le Z \le 4\pi$. Animate for the time range $0 \le T \le 2n\pi$. If your software allows continuous looping in time, just let $n = 1$. If not, let n be a large enough integer that you can observe several cycles. Again, describe what you see and relate to parts (b) and (c).
e. Using the relationships from Eqs. (9.8.3) $\bar{u}_{zi} = p_i/\rho c,\ \bar{u}_{zr} = -p_r/\rho c$, the average velocity wave is given by

$$\bar{U}_z = \bar{U}_{zi} + \bar{U}_{zr} = \frac{1}{\rho c} P_m \cos(k(z - c_0t)) - \frac{1}{\rho c} RP_m \cos(k(z + c_0t))$$

From Eq. (9.5.28) the rate of mechanical working is given by $W = -\bar{U}_z(\partial P/\partial z)$. Determine W and let the dimensionless rate of working be scaled as $\hat{W} = \rho c_0 W/kP_m^2$. Plot $\hat{W}$ for $T = 3\pi/8$ and $T = 7\pi/8$ over the Z range $0 \le Z \le 2\pi$ with the two curves labeled on the same graph. Describe your results.
f. Animate $\hat{W}$ over the same Z and T ranges in part (d). Describe your results.
g. What is the time average of $\hat{W}$ for any value of Z? Explain your answer.

9.5 The dimensionless wave speed, C, for the flutter model is given in Eq. (9.9.8) and the critical value is for $G = 0$ which determines $S = S_c$ in Eq. (9.9.7). Under those conditions

a. Plot C vs the dimensionless wave number, K. Use the parameter values $(B, M) = (1, 1), (1, 5), (5, 1), (5, 5)$ over the range $0.01 \le K \le 2$. Describe your results. What happens as $K \to 0$ and why?
b. The dimensionless group velocity is $C_g = \partial\Omega/\partial K$ where the dimensionless frequency is $\Omega = KC$. Plot C_g vs K for the same parameter choices in part a. How does C_g compare to C?
c. An example of group velocity can be seen in the propagation of two waves of similar wavelength, or wave number as they create "beating". Let $K_1 = 0.3,\ K_2 = 0.33$ and calculate the corresponding values of C_1, C_2 for $M = 1,\ B = 1,\ S = 0.5$ for neutrally stable waves. Add these two waves as $H(X,\ T) = \cos(K_1(X - C_1T)) + \cos(K_2(X - C_2T))$. Plot H over the range $0 \le X \le 300\pi$ for $T = 0$. Describe your result.
d. Animate part c for $0 \le T \le 100\pi$. What is the average phase velocity? What is the group velocity? Choose dimensional parameters as you may need them.

9.6

a. For the RLC model for forced respiratory oscillations let $\delta = 0.2$ and $\Omega = 0.5$. Plot the dimensionless V(T) and P(T) on the same graph over the time interval $0 \le T \le 2\pi$. Describe differences from the $\Omega = 1$ plots in Figure 9.24(a).
b. Repeat part (a) for $\Omega = 2.0$. How does the graph differ from part (a) and Figure 9.24(a).

c. Plot V(P) loops for $\delta = 0.2$ and $\Omega = 0.5,\ 1.0,\ 2.0$ over the interval $0 \leq T \leq 2\pi$. What is the direction of increasing time, clockwise or counterclockwise? Describe the differences in the loops.

d. Find and expression for the dimensionless total compliance, $\chi = C_{tot}/C$ for arbitrary parameter values. Plot χ vs Ω for $0 \leq \Omega \leq 10$. What happens to the total compliance as dimensionless frequency increases?

e. For arbitrary parameter values derive an expression for the dimensionless hysteresis area of the V(P) loop, i.e. work per cycle using $W = \int_0^{2\pi} P(dV/dT)dT$. Plot W vs Ω for $0 \leq \Omega \leq 10$. Describe your result. If there is a maximum, at what value of Ω does it occur? Does this system have a resonant frequency?

f. The dimensional power dissipated through the resistor is $p_R = Q^2R$. Integrate p_R over one cycle to find the work, w_R, over one period. Make the result dimensionless, $W_R = w_R/\bar{p}^2C$, and show that it is the same as W in part (e).

9.7 The RLC circuit shown in Figure 9.22 can also be used for passive, tidal expiration where the lung volume, v(t), decreases from its initial value, $v_0 = v_{FRC} + v_T$, to the final, steady state value, v_{FRC}. v_{FRC} is the functional residual capacity and v_T is the tidal volume.

a. Find the analytical solution to Eq. (9.10.1) for $L = 0$ and subject to initial conditions $v(0) = v_0$. Here we will let $p_0 = v_{FRC}/C$ for the steady state solution.

b. The product RC has units of time and is called the time constant of the system in part (a). Plot v vs t for $RC = 0.2, 0.3, 0.4$ sand choose normal values of $v_T = 0.5$ liters, $v_{FRC} = 2.5$ liters. Explain the difference in your curves.

c. From part (a), find an expression for the flow rate, Q. What is the maximum value of Q and when does it occur?

d. Using your solutions in parts (b) and (c), make parametric, flow-volume plots with v on the horizontal axis and Q on the vertical axis. Where is $t = 0$ on your three curves for the three values of RC? How do the flow rates compare?

e. Normal parameter values for R, C are given in Section 9.10 and v_T, v_{FRC} in part (b). Obstructive lung disease, like asthma, has a higher resistance due to bronchoconstriction and the patient operates at a higher lung volume, so let $R = 4\ cmH_2O/(liter/s)$, $C = 0.15\ liters/cmH_2O$, $v_T = 0.5$ liters, $v_{FRC} = 3.5$ liters. Restrictive lung disease like pulmonary fibrosis results in smaller compliance and smaller lung volumes. Let $R = 2\ cmH_2O/(liter/s)$, $C = 0.10\ liters/cmH_2O$, $v_T = 0.33$ liters, $v_{FRC} = 2.0$ liters. Plot your $L = 0$ solutions as flow–volume curves for normal, asthmatic, and fibrotic lungs. Which have the higher flow rates? How do the volumes compare? Why does the restrictive lung have a similar maximal flow rate as the normal lung?

f. Now allow for nonzero inertance with $L = 0.01\ cmH_2O/(liter/s^2)$. Solve Eq. (9.10.1) for v(t) adding the initial flow condition, $dv/dt_{t=0} = 0$, either analytically or numerically, for normal, asthmatic and fibrotic lungs. Plot the results as flow–volume curves. How does this graph compare to the one in part (e)?

g. Flip your results in part (f) by plotting −v on the horizontal axis. You can try to relabel the tick marks as positive values or just ignore their negative signs. How does your new graph compare to Figure 9.10(b)?

9.8

a. For the two-element Windkessel model of arterial blood flow calculate the pressure wave, use the normal parameter values in Section 9.10 but let the artery be less compliant. Try $C = 0.9\ cm^3/mmHg$. Calculate and plot p(t) for the cardiac cycle. How does it compare to Figure 9.26? How does the end diastolic pressure compare to the initial pressure, p_i ?

b. Use the end diastolic pressure from part (a) as a new value of p_i and plot the new waveform for this second heartbeat. You can use the same range of t. Now compare the initial to the end diastolic pressures. Are the closer?

c. Plot a third cardiac pressure cycle using the end diastolic pressure from part (b) as p_i Now compare the starting and end pressures for the cycle. What is happening as you add more and more cycles? Why?
d. Using the normal parameter values from Section 9.10, slow down the heart rate to 60 beats per minute. Starting with $p_i = 80$ mmHg in the first cycle, use the end diastolic pressure for the initial pressure of the next systole as was done in parts (a–c). Plot five cycles overlaying one another on the interval $0 \leq t \leq T_c$. Describe your results.
e. Starting with the normal parameter values in Section 9.10, reduce the stroke volume to $V_s = 60\ \text{cm}^3$. Calculate five cycles using the end diastolic pressure as the starting systolic pressure of the next cycle. Plot your curves overlaying one another. Describe your result.
f. Power in the circuit is dissipated through the resistor and is given by $p_R = Q_R^2 R$, where Q_R is the flow through the resistor, $Q_R = p/R$. For normal parameter values given in Section 9.10, plot p_R for a full cardiac cycle. Integrate the result with respect to time over the cycle to derive the work, and designate the values for systole and diastole and the total.

Image Credits

Figure 9.1 Source: Fig 18 from Heil M. Stokes flow in collapsible tubes: computation and experiment. *J Fluid Mech.* 1997; 353:285–312.

Figure 9.13a Source: Encyclopaedia Brittanica/ Contributor/ Getty

Figure 9.13b Source: Fawn-Love/ Getty

Figure 9.15a Source: John Rensten/ Getty

Figure 9.16a Source: Fig 2 of Gavriely N, Palti Y, Alroy G, Grotberg JB. Measurement and theory of wheezing breath sounds. *Journal of Applied Physiology.* 1984; 57(2):481–92.

Figure 9.16b Source: Fig 2 of Gavriely N, Palti Y, Alroy G, Grotberg JB. Measurement and theory of wheezing breath sounds. *Journal of Applied Physiology.* 1984; 57(2):481–92.

10 Constitutive Equations 2 – Generalized Newtonian Fluids

Introduction

In Chapter 6 we introduced the constitutive equations for a Newtonian viscous fluid which has constant viscosity. It is a rheological model that covers a great deal of flow situations in general fluid mechanics applications: air and water being primary examples. We developed the Newtonian fluid constitutive relationship in Eq. (6.3.9), where the extra stress tensor due to viscous effects, $\underline{\underline{\tau}}$, is linearly related to the rate of strain tensor, $\underline{\underline{E}}$, through the tensor equation $\underline{\underline{\tau}} = 2\mu\underline{\underline{E}}$. The linear proportionality comes through the constant viscosity, μ. However, biofluids are often much more complicated. For example, they can be solutions of macromolecules, like mucus, suspensions of cells, like whole blood, or mixtures of biological materials like ketchup. The resulting macroscopic viscosity is no longer constant. Fluids with non-constant viscosity are usually called "non-Newtonian fluids." Amongst other dependencies, the viscosity can vary locally with the strain rate, i.e. the flow details that are contained in $\underline{\underline{E}}$. It also can depend on the history of the strain rate, a feature of viscoelastic fluids covered later in Chapter 14.

In this chapter we describe and explore the flow of fluids whose viscosity depends on the instantaneous strain rate. A scalar measure of the strain rate intensity is defined as $\dot{\gamma} = \sqrt{2E_{ij}E_{ij}}$ ($\dot{\gamma}\sim$ "gamma-dot"), where the Einstein notation of repeated indices is assumed. $\dot{\gamma}$ is chosen because it is an invariant of $\underline{\underline{E}}$, so does not change under coordinate transformation. That is an essential feature of any constitutive relationship, that two observers in different coordinate systems measure the same constitutive properties of the material.

In this chapter we will study constitutive relationships of the form $\underline{\underline{\tau}} = 2\mu_{eff}(\dot{\gamma})\underline{\underline{E}}$, which has the appearance of a Newtonian fluid model. The main difference is that the viscosity is not constant, but is an "effective viscosity" which depends on the strain rate invariant, $\mu_{eff}(\dot{\gamma})$. Such non-Newtonian fluids are called "generalized Newtonian fluids," making for cautious sentence structures. As an example of macromolecular fluids, the viscosity depends on strain rate because the macromolecules are often long, multi-branched and tangled so resist flow. They tend to orient longitudinally with increasing shear strain rates becoming less tangled, which reduces the effective viscosity. For suspensions like red blood cells in whole blood, it is their propensity to stick together, forming chains or rouleaux as discussed in Chapter 2, that creates similar issues. Increasing the shear strain rates break up the aggregate rouleaux into shorter chains, until they disappear and the RBCs are individualized. Here, too, the effective viscosity decreases with increasing $\dot{\gamma}$.

Another significant feature of some generalized Newtonian fluids is that they do not always behave as fluids. A common example is ketchup, which, in fact, is a biofluid made of a suspension of biological particles: tomato bits. Remove the cap of a ketchup bottle then turn it upside down. Very likely there will be no flow. Do the same for a bottle of beer and your friends will ask you to buy the next round. The difference is that the static tomato bits in suspension are in contact with one another and form a structure that has solid properties. Ketchup will not flow until a critical shear stress, called the yield stress, τ_y, is applied, like tapping the bottle. Exceeding the yield stress breaks up the aggregate structure and allows the tomato bits to flow. Both blood and mucus have yield stresses, beer does not.

Topics Covered

To start this subject matter, we first describe the essentials of rheological measurement and terminology. Unraveling complex fluid behavior starts with subjecting the fluid to a very simple shear flow between parallel plates to derive the effective viscosity, $\mu_{eff}(\dot{\gamma})$. Rheological nomenclature describing the general behavior of the fluid sorts into six types. Three classifications are for fluids with no yield stress, $\tau_y = 0$, and for $\mu_{eff}(\dot{\gamma})$, which increases with $\dot{\gamma}$ (dilatant), decreases with $\dot{\gamma}$ (pseudoplastic) or is independent of $\dot{\gamma}$ (Newtonian). The other three classifications are similar to the first three, but with a finite yield stress, $\tau_y > 0$, and include Bingham fluids and, more generally, Herschel–Bulkley fluids. A new dimensionless parameter, the Bingham number, arises which is the ratio of the yield stress, τ_y, to a characteristic shear stress in the flow. Once $\mu_{eff}(\dot{\gamma})$ is established, the constitutive law, $\underline{\underline{\tau}} = 2\mu_{eff}(\dot{\gamma})\underline{\underline{E}}$, can be inserted into the Cauchy momentum balance to derive, along with conservation of mass, the equations of motion.

Specific models of generalized Newtonian fluids are explored for their features in tube flows. First is the power-law fluid which is popular for its simplicity, since $\mu_{eff}(\dot{\gamma})$ is taken to be proportional to some power of $\dot{\gamma}$. While simple, the power-law fluid model is limited to a range of $\dot{\gamma}$, however, since $\mu_{eff}(\dot{\gamma})$ can become unbounded as $\dot{\gamma} \to 0$ or $\dot{\gamma} \to \infty$, depending on the exponent. Then we examine the Casson fluid model, which has a yield stress, and is frequently used for blood flow. The behavior of both the yield stress and the high strain rate viscosity, μ_∞, depend on the hematocrit, H. Increasing H increases the oxygen delivery capacity of the blood, but also increases μ_∞, which affects the work of the heart to pump the blood. So there are important optimal values of H that balance those opposing effects. Diseases such as sickle cell anemia have increased values of the yield stress which can have a dramatic effect on flows in small vessels, even leading to local microinfarcts of tissues. We briefly examine other constitutive models which are popular both in general and also specifically for biological fluids, and which have finite values of viscosity in the limits of $\dot{\gamma} \to 0$ and $\dot{\gamma} \to \infty$, unlike the power-law model. Finally we analyze the operations of some viscometers and their related flows to measure the rheological properties of fluids.

10.1 Rheology and Constitutive Equations

The rheological properties of a fluid are tested in a viscometer. A sketch of the main principles is shown in Figure 10.1. For this planar Couette flow, the fluid is contained between parallel walls which are a distance h apart. (Maurice Marie Alfred Couette, 1858–1943, was a French physicist.) The lower wall is fixed while the upper wall moves at speed V_0 in its own plane. After a long enough time, it can be shown that any flowing fluid will obtain the linear velocity profile indicated. In practice, a Couette flow is generated by concentric cylinders of radius R_1 and $R_2 = R_1 + h$. The inner cylinder, for example, is rotated with measures of the required torque for a constant rotational frequency, while the outer cylinder is fixed. For $h/R_1 << 1$ the flow is locally parallel and looks like Figure 10.1 and the torque measurement gives the applied shear stress, τ.

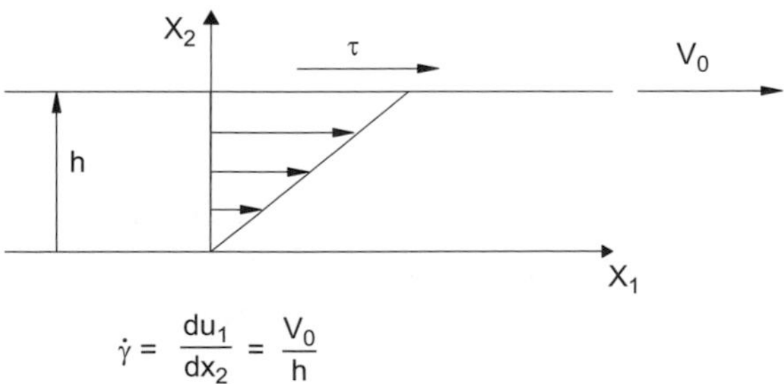

Figure 10.1 Main features of planar Couette flow which is a simple viscometric flow. The applied shear stress is τ and the resulting uniform shear strain rate is $\dot{\gamma} = du_1/dx_2 = V_0/h$.

The rate of strain tensor, E_{ij}, for a Cartesian coordinate system is symmetric and was introduced in Section 6.3, Eq. (6.3.8), as

$$E_{ij} = \frac{1}{2}\left(\frac{\partial u_i}{\partial x_j} + \frac{\partial u_j}{\partial x_i}\right) \tag{10.1.1}$$

Second-order tensors, in general, have three scalar invariants which do not change with coordinate transformation. For a tensor, $\underline{\underline{A}}$, they are defined by

$$I_1 = \mathrm{tr}\left(\underline{\underline{A}}\right) = A_{ii}$$
$$I_2 = \frac{1}{2}\left(\left(\mathrm{tr}\left(\underline{\underline{A}}\right)\right)^2 - \mathrm{tr}\left(\underline{\underline{A}}\cdot\underline{\underline{A}}\right)\right)$$
$$= \frac{1}{2}\left((A_{ii})^2 - \mathrm{tr}(A_{ik}A_{kj})\right) = \frac{1}{2}\left(A_{ii}A_{jj} - A_{ik}A_{ki}\right)$$
$$I_3 = \det\left(\underline{\underline{A}}\right) = \varepsilon_{ijk}A_{1i}A_{2j}A_{3k} \tag{10.1.2}$$

where the invariants are I_1, I_2 and I_3 and the Einstein notation for repeated indices is used. For the rate of strain tensor, $\underline{\underline{A}} = \underline{\underline{E}}$, the fact that $\underline{\underline{E}}$ is symmetric, and that we are dealing with incompressible fluids, makes for some simplifications. For an incompressible fluid, $I_1 = E_{ii} = 0$, since it is the divergence of the velocity field

$$E_{ii} = \frac{1}{2}\left(2\frac{\partial u_1}{\partial x_1} + 2\frac{\partial u_2}{\partial x_2} + 2\frac{\partial u_3}{\partial x_3}\right) = \frac{\partial u_1}{\partial x_1} + \frac{\partial u_2}{\partial x_2} + \frac{\partial u_3}{\partial x_3}$$
$$= 0 \tag{10.1.3}$$

I_3 is either zero or negligibly small, so $I_2 \sim E_{ij}E_{ji} = \underline{\underline{E}} : \underline{\underline{E}} = \mathrm{tr}(\underline{\underline{E}}\cdot\underline{\underline{E}})$ is the main invariant for the rate of strain tensor.

The relationship between stress and strain rate is called the constitutive equation for a fluid. It must be independent of the observer's reference frame, so constitutive equations that depend on I_2 fulfill that requirement. The definition of the scalar strain rate, $\dot{\gamma}$, used in rheological studies is given by I_2 in the definition

$$\dot{\gamma} = \sqrt{2E_{ij}E_{ji}} \tag{10.1.4}$$

so $\dot{\gamma}$ is invariant to coordinate transformation.

For the rheological flow in Figure 10.1, where there is only one velocity component, $u_1(x_2)$ or $u(y)$, the non-zero terms of E_{ij} in Eq. (10.1.1) are

$$E_{12} = E_{21} = \frac{1}{2}\frac{du_1}{dx_2} = \frac{1}{2}\frac{du}{dy} \tag{10.1.5}$$

Inserting Eq. (10.1.5) into Eq. (10.1.4) gives us

$$\dot{\gamma} = \sqrt{2\left(E_{12}^2 + E_{21}^2\right)} = \sqrt{4E_{12}^2} = \sqrt{\left(\frac{du_1}{dx_2}\right)^2} = \left|\frac{du_1}{dx_2}\right| = \frac{V_0}{h} \tag{10.1.6}$$

We see from the flow in Figure 10.1 that $du_1/dx_2 \geq 0$, so we can restate Eq. (10.1.6) as

$$\dot{\gamma} = \frac{du_1}{dx_2} = \frac{V_0}{h} \tag{10.1.7}$$

The general form of the stress tensor is from Section 6.3, Eq. (6.3.1),

$$\underline{\underline{T}} = -p\,\underline{\underline{I}} + \underline{\underline{\tau}} \quad \Leftrightarrow \quad T_{ij} = -p\delta_{ij} + \tau_{ij} \tag{10.1.8}$$

where the extra stress tensor due to viscous effects is $\underline{\underline{\tau}}$.

For a Newtonian fluid we discussed the constitutive equation in Section 6.3, Eq. (6.3.9). The Newtonian model is $\underline{\underline{\tau}} = 2\mu\underline{\underline{E}}$ with constant viscosity μ. In this chapter we will investigate more complicated fluid behavior where the viscosity is not constant, but instead depends on the instantaneous strain rate. In a similar approach as that for Newtonian fluids, the relationship between $\underline{\underline{\tau}}$ and $\underline{\underline{E}}$ for a "generalized Newtonian fluid" is posed using the familiar form in Cartesian coordinates

$$\underline{\underline{\tau}} = 2\mu_{eff}(\dot{\gamma})\underline{\underline{E}} \quad \Leftrightarrow \quad \tau_{ij} = 2\mu_{eff}(\dot{\gamma})E_{ij} = \mu_{eff}(\dot{\gamma})\left(\frac{\partial u_i}{\partial x_j} + \frac{\partial u_j}{\partial x_i}\right) \tag{10.1.9}$$

The function $\mu_{eff}(\dot{\gamma})$ is called an "effective viscosity" which depends in general on the shear rate scalar intensity, $\dot{\gamma}$, as defined in Eq. (10.1.4). A generalized Newtonian fluid is one for which the shear stress is related to the strain rate through the form of Eq. (10.1.9) and is restricted to being a function of the instantaneous strain rate. Fluids whose stress depends on the time history of strain rate do not fit this definition and include viscoelastic fluids, which are discussed later in Chapter 14. It is a bit confusing to call a non-Newtonian fluid by the name "generalized Newtonian fluid," but as long as our definitions are clear this is a minor obstacle. The proposition that the effective viscosity depends on $\dot{\gamma}$ means that this material property of the fluid is seen to be the same regardless of coordinate transformation since $\dot{\gamma}$ is an invariant, see Eq. (10.1.4). So the constitutive equation, Eq. (10.1.9), satisfies what is known as the "Principle of Material Frame Indifference." It means that two observers who are in different reference frames, say one rotating or translating compared to the other, measure the same constitutive property of the fluid.

Using this definition in Eq. (10.1.9) for the rheologic flow of Figure 10.1, the only non-zero values of the extra stress tensor, $\underline{\underline{\tau}}$, are $\tau_{12} = \tau_{21} = \tau$ while the only non-zero values of the strain rate tensor, $\underline{\underline{E}}$, are $E_{12} = E_{21} = (1/2)du_1/dx_2$, so that the stress–strain rate relationship in the Couette flow is

$$\tau = \mu_{eff}(\dot{\gamma})\frac{du_1}{dx_2} = \mu_{eff}(\dot{\gamma})\dot{\gamma} \tag{10.1.10}$$

where we have substituted for du_1/dx_2 from Eq. (10.1.7). By rearranging Eq. (10.1.10) we derive the definition of the effective viscosity

$$\mu_{eff}(\dot{\gamma}) = \frac{\tau}{\dot{\gamma}} \tag{10.1.11}$$

A wide range of results from a viscometric analysis is possible. The curves in Figure 10.2 are representative

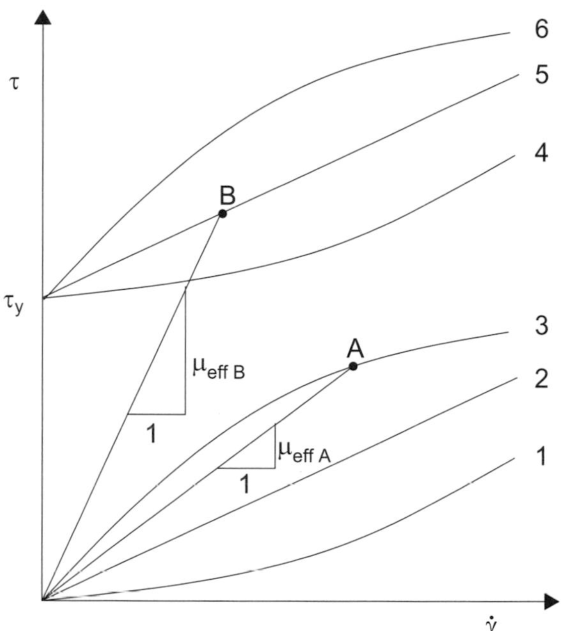

Figure 10.2 Rheological nomenclature: curve 2 Newtonian constant viscosity; curve 1 dilatant (shear thickening); curve 3 pseudoplastic (shear thinning); curve 5 Bingham yield stress with constant differential viscosity; curve 4 Herschel–Bulkley yield stress with shear thickening; and curve 6 Herschel–Bulkley yield stress with shear thinning. Examples of the effective viscosity definition $\mu_{eff} = \tau/\dot{\gamma}$ are shown as the slope of a straight line from the origin to the operating points, A and B.

of different patterns and their names. Curves 1, 2 and 3 all pass through the origin. First, curve 2 is a straight line through the origin which is a Newtonian fluid with constant viscosity. Examples are water and air. Curve 3 is a pseudoplastic fluid and its effective viscosity defined in Eq. (10.1.11) is the slope of the straight line connecting the origin to an operating point on the curve. Point A is an example with $\mu_{eff\,A}$ indicated as the slope. As $\dot{\gamma}$ increases along this curve, $\mu_{eff\,A}(\dot{\gamma})$ decreases, which is the pseudoplastic behavior, also known as "shear thinning." Biofluid examples are blood and mucus. Curve 1 is a dilatant fluid where $\mu_{eff}(\dot{\gamma})$ is increasing as $\dot{\gamma}$ increases. This increase of the apparent or effective viscosity with strain rate is called "shear thickening." Examples are cornstarch and pastes. For these systems the increase in strain rate forces fluid out from between the particles, so that they drag against one another more strongly and increase the viscosity.

Curves 4, 5 and 6 do not pass through the origin, but instead start at a yield stress, τ_y. When $\tau \leq \tau_y$ the material in the Couette channel behaves like a solid and there is no flow or strain rate, i.e. $\dot{\gamma} = 0$. It is only when $\tau > \tau_y$ that the material flows, i.e. $\dot{\gamma} > 0$. E. C. Bingham in 1916 described such a relationship (Bingham, 1916) and a "Bingham Fluid" is indicated in curve 5 in Figure 10.2. It has a yield stress which, when exceeded, continues as a straight line. In 1926, W. H. Herschel and R. Bulkley described yield stress fluids with a more general curvilinear behavior once they began to flow, i.e. shear thickening (dilatant) or shear thinning (pseudoplastic) effects (Herschel and Bulkley, 1926), see curves 4 and 6 in Figure 10.2. Curves 4, 5 and 6 are representative of Herschel–Bulkley fluids, although the Bingham fluid of curve 5 can be considered a special case. Again, μ_{eff} is the slope of the line connecting the origin to the operating point, point B on curve 5 (Bingham). For point B, clearly the slope approaches infinity as the operating point approaches τ_y at $\dot{\gamma} = 0$. An infinite viscosity results in solid behavior, which is consistent with no flow.

Often the student will think of μ_{eff} as the tangent to the $\tau(\dot{\gamma})$ curve, which is tempting but incorrect. The tangent or instantaneous slope is called the "differential viscosity," μ_{diff},

$$\mu_{diff} = d\tau/d\dot{\gamma} \tag{10.1.12}$$

see Figure 10.3. For a Newtonian fluid the effective and differential viscosities are the same, i.e. $\mu_{eff} = \mu_{diff} =$ constant. For a Herschel–Bulkley fluid which exhibits shear thinning as it flows, as in Figure 10.3, μ_{diff} decreases as $\dot{\gamma}$ increases, while one that exhibits shear thickening has μ_{diff} increasing as $\dot{\gamma}$ increases. A Bingham fluid has constant μ_{diff}.

Figure 10.4 shows how the interaction of suspended particles, macromolecules or polymers can form structures in static conditions which break up when an applied shear stress exceeds the yield stress. Further increases in the applied shear stress can then align anisotropic particles with the flow, break up

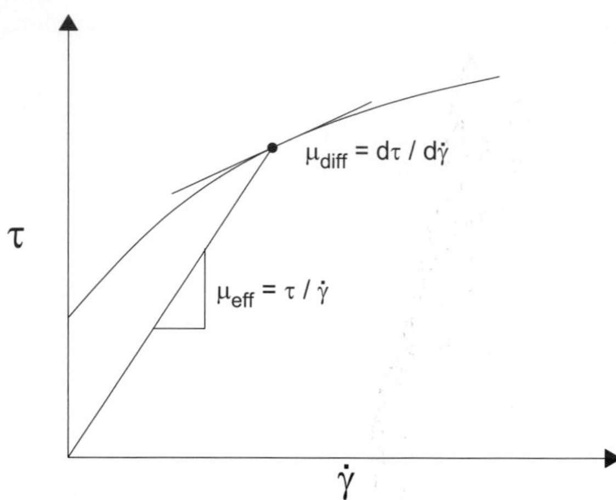

Figure 10.3 Effective viscosity, μ_{eff}, compared to the differential viscosity, μ_{diff}, for a Herschel–Bulkley fluid.

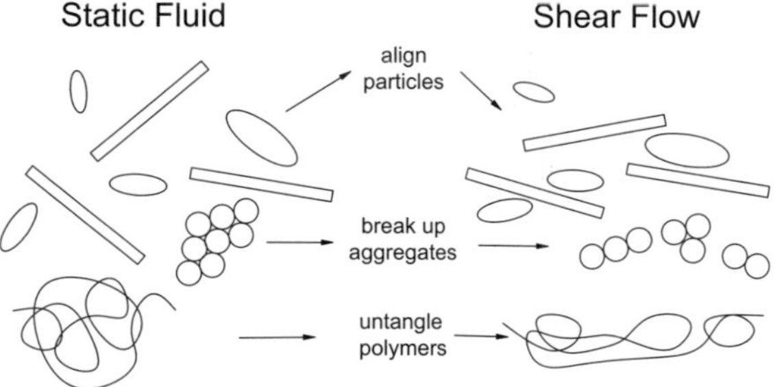

Figure 10.4 The basis of yield stress and shear thinning fluid behavior. Shear stress breaks up aggregates of particles when the yield stress is exceeded, allowing flow. Then flow to the right aligns anisotropic particles, further breaks up aggregates, and untangles polymers to reduce the effective viscosity.

aggregates into smaller ones, and untangle macromolecules to reduce the effective viscosity, the essence of shear thinning or pseudo-plastic behavior.

10.2 Cauchy Momentum Equation for Generalized Newtonian Fluids

The form of our generalized Newtonian fluid stress tensor, $\underline{\underline{T}}$, in terms of the pressure, p, and the extra stress tensor, $\underline{\underline{\tau}}$, is from Eqs. (10.1.8)

$$\underline{\underline{T}} = -p\underline{\underline{I}} + \underline{\underline{\tau}} = -p\underline{\underline{I}} + 2\mu_{eff}(\dot{\gamma})\underline{\underline{E}} \tag{10.2.1}$$

which we now can insert into the Cauchy momentum equation. For steady, fully developed, unidirectional, axisymmetric flow in a pipe using cylindrical coordinates, the remaining stress tensor components of Eq. (10.2.1), according to the details in Eqs. (5.6.3), are

$$T_{rr} = -p, T_{zz} = -p, \ T_{rz} = \tau_{rz} = 2\mu_{eff}(\dot{\gamma})E_{rz} \tag{10.2.2}$$

The shear strain rate tensor component in cylindrical components for this flow (see Section 6.4, Eq. (6.4.2)) is given by

$$E_{rz} = \frac{1}{2}\left(\frac{\partial u_z}{\partial r} + \frac{\partial u_r}{\partial z}\right) = \frac{1}{2}\frac{\partial u_z}{\partial r} \tag{10.2.3}$$

since $u_r = 0$. So the shear stress component of Eq. (10.2.2) becomes

$$\tau_{rz} = \mu_{eff}(\dot{\gamma})\frac{du_z}{dr} \tag{10.2.4}$$

and the strain rate invariant is

$$\dot{\gamma} = \sqrt{2E_{ij}E_{ji}} = \sqrt{4E_{rz}^2} = \left|\frac{du_z}{dr}\right| \tag{10.2.5}$$

For the constant pressure gradient, the z-component of the Cauchy momentum equation, Eq. (5.6.4), simplifies to

$$\frac{1}{r}\frac{d}{dr}(r\tau_{rz}) = \frac{dp}{dz} \tag{10.2.6}$$

where $p = p(z)$ only while $u_z = u_z(r)$. Separating variables in Eq. (10.2.6) and integrating gives us

$$\int d(r\tau_{rz}) = \frac{dp}{dz}\int r\,dr + c_1 \quad \rightarrow \quad \tau_{rz} = \frac{dp}{dz}\frac{r}{2} + \frac{c_1}{r} \tag{10.2.7}$$

For finite shear stress we must choose $c_1 = 0$, and the result is the simple and elegant finding that the shear stress varies linearly with r,

$$\tau_{rz} = \frac{dp}{dz}\frac{r}{2} \tag{10.2.8}$$

The magnitude of the shear stress varies from $|\tau_{rz}| = 0$ at the centerline, $r = 0$, to a maximum at the wall $r = a$, which we call the wall shear magnitude, τ_w,

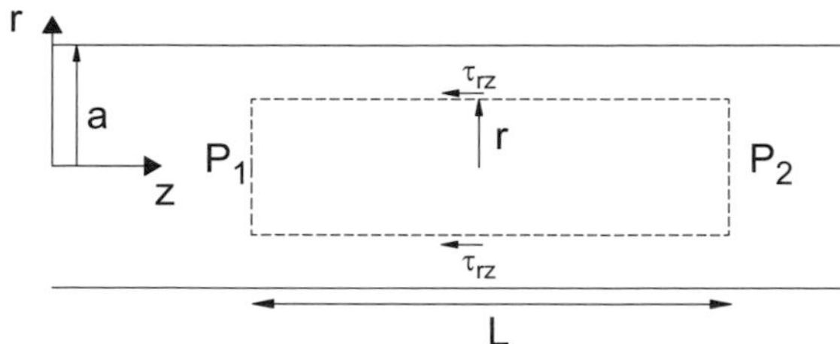

Figure 10.5 Control volume and force balance for a tube flow in the positive z-direction.

$$\tau_w = \left|\tau_{rz}\right|_{r=a} = \frac{a}{2}\left|\frac{dp}{dz}\right| \tag{10.2.9}$$

Equation (10.2.8) does not depend on the constitutive equation, so is a general result which only assumes fully developed flow with a constant pressure gradient and finite stress at the origin.

Why is Eq. (10.2.8) so simple? A sketch of the force balance answers this question. Figure 10.5 shows flow in a cylindrical tube and a control volume is indicated by a dashed line. The end pressures are P_1 and P_2, which act on the cross-sectional area πr^2, and the shear stress on the sides is τ_{rz}, which acts on the lateral surface area $2\pi rL$. So the axial force balance is

$$(P_1 - P_2)\pi r^2 + 2\pi rL\tau_{rz} = 0 \tag{10.2.10}$$

Rearranging Eq. (10.2.10) leads to

$$\tau_{rz} = \frac{(P_2 - P_1)}{L}\frac{r}{2} = \frac{dp}{dz}\frac{r}{2} \tag{10.2.11}$$

which is identical to Eq. (10.2.8). Equation (10.2.11) shows that $\tau_{rz} < 0$ for $dp/dz < 0$, so flow in the positive z-direction is a balance between the driving pressure and the viscous resistance. The flow being unidirectional and axisymmetric, finite shear stress at $r = 0$, and the axial pressure gradient a constant are all that were assumed for this balance. No details about the flow field or constitutive equation were required. Indeed, the balance holds for a solid as well, so is general to a continuum under these restrictions.

Substituting for the shear stress component, $\tau_{rz} = \mu_{eff}(\dot{\gamma})(du_z/dr)$ from Eq. (10.2.4) in Eq. (10.2.8) leaves us with the form

$$\mu_{eff}(\dot{\gamma})\frac{du_z}{dr} = \frac{dp}{dz}\frac{r}{2} \tag{10.2.12}$$

which is the first integral of the Cauchy momentum equation for this flow in a cylinder.

10.3 Power-Law Fluids

Portions of the constitutive relationship, τ vs $\dot{\gamma}$, for shear thinning and shear thickening fluids can be modeled by a power-law fit to the data

$$\tau = m\dot{\gamma}^n \tag{10.3.1}$$

where the effective viscosity is given by

$$\mu_{eff} = \frac{\tau}{\dot{\gamma}} = m\dot{\gamma}^{n-1} \tag{10.3.2}$$

Such fluids are also known as Ostwald–de Waele fluids for the first use of power-law approaches by British chemist Armand de Waele (1887–1966) (de Waele, 1923) and, popularly, for the German/Prussian physical chemist Friedrich Wilhelm Ostwald (1853 – 1932). However, it was Ostwald's son, Wolfgang Ostwald, who published the fundamental paper in 1925 (Ostwald, 1925). Friedrich Wilhelm Ostwald won the Nobel Prize in Chemistry in 1909 for his work on catalysis, chemical equilibria and reaction rates. He is considered one of the founders of the field of physical chemistry.

The exponent n is called the flow behavior index. The constant coefficient, m, is the flow consistency index, whose units are $dyn{\cdot}cm^{-2}{\cdot}s^n$, which depend on n. Figure 10.6(a) shows the shear stress vs shear strain rate $\tau(\dot{\gamma})$ for a power-law fluid for several choices of m and n. Figure 10.6(b) shows the effective viscosity vs shear strain rate, $\mu_{eff}(\dot{\gamma})$, for the same choices of m and n in Figure 10.6(a). A Newtonian fluid is recovered when $n = 1$, which makes m the constant viscosity.

When $n < 1$ the model represents a pseudoplastic, shear thinning fluid, since μ_{eff} decreases with increasing $\dot{\gamma}$. For example, human sputum, which includes mucus, fits a power law with, $n = 0.2$, $m = 105\ dyn{\cdot}cm^{-2}\cdot s^n$ over a range

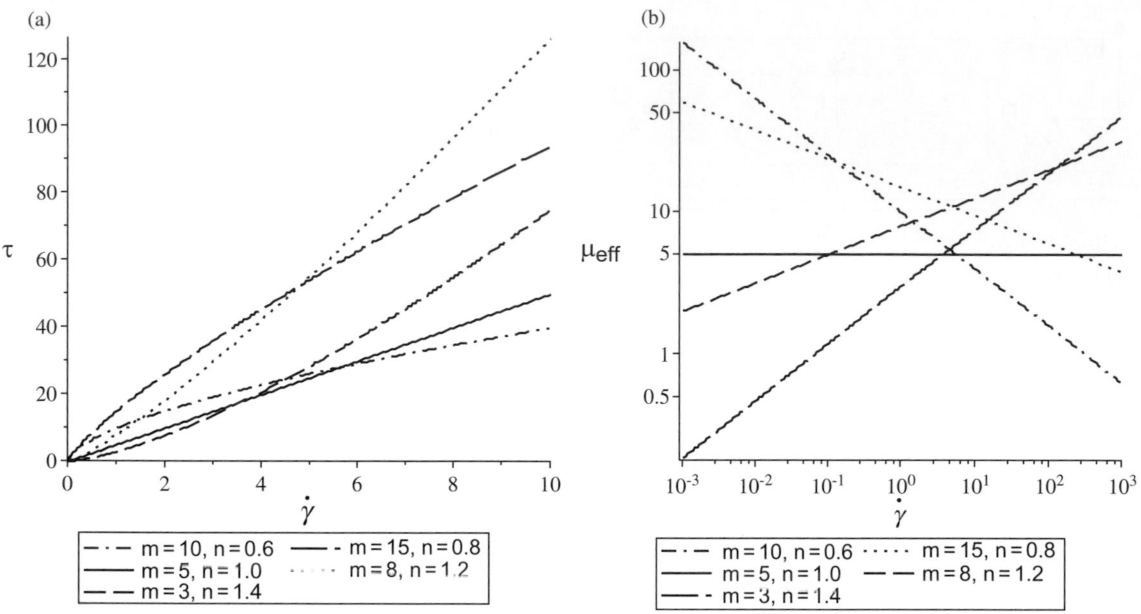

Figure 10.6 (a) Power-law fluid stress vs strain rate, $\tau(\dot{\gamma})$, for examples of Newtonian $n = 1$, shear thinning $n = 0.6, \quad 0.8$ and shear thickening $n = 1.2, 1.4$, for selected values of m. (b) Effective viscosity vs strain rate, $\mu_{eff}(\dot{\gamma})$, for the same choices of m and n in (a).

$10^{-3} \leq \dot{\gamma} \leq 10^{3}\ s^{-1}$ (Velez-Cordero and Lauga, 2013). Other investigators have found a range of $0.17 \leq n \leq 0.48$ and $3.5 \leq m \leq 155\ \text{dyn}\cdot\text{cm}^{-2}\cdot\text{s}^{n}$ in health and disease (Chatelin et al., 2017). Diseases that strongly influence mucus properties include cystic fibrosis, asthma, emphysema and infections. Blood is more uniform with $n \sim 0.7,\ \ m \sim 0.17\ \text{dyn}\cdot\text{cm}^{-2}\cdot\text{s}^{n}$ (Walburn and Schneck, 1976; Hussain et al., 1999). When $n > 1$ the power-law fluid is a dilatant, shear thickening fluid, since μ_{eff} increases with increasing $\dot{\gamma}$. For example, $n \sim 3$–4 for cornstarch (Balmforth et al., 2007).

The power-law fluid model is somewhat limited because of its unbounded effective viscosity as $\dot{\gamma} \to 0$ for $n < 1$ and $\dot{\gamma} \to \infty$ for $n > 1$. For that reason, it is typical to apply this model to a restricted range of $\dot{\gamma}$. Several rheological models with finite viscosities in both limits of $\dot{\gamma}$ are discussed in Section 10.7. Because of its simplicity, however, the power law is more amenable to analytical solutions of the governing equations than are the more complex models to follow in this chapter.

For a fully developed, unidirectional, steady, constant pressure gradient flow of a power-law fluid in a tube of radius a, we derived the resulting form of the Cauchy momentum equation, Eq. (10.2.12). Inserting the power-law model effective viscosity, Eq. (10.3.2), into Eq. (10.2.12) yields, after some rearrangement

$$m\left(-\frac{du_z}{dr}\right)^{n-1}\left(-\frac{du_z}{dr}\right) = \left(-\frac{dp}{dz}\right)\frac{r}{2} \tag{10.3.3}$$

Since u_z must be zero at the wall and positive at the centerline $r = 0$, the overall expectation is that $du_z/dr \leq 0$. With that in mind, we replaced $\dot{\gamma} = |du_z/dr| = -du_z/dr$ to derive Eq. (10.3.3), and also multiplied through by –1. Now both bracketed terms of Eq. (10.3.3) on the left hand side have the same function raised to an exponent. So adding exponents then simplifies to

$$\left(-\frac{du_z}{dr}\right)^n = \left(-\frac{dp}{dz}\right)\frac{r}{2m} \tag{10.3.4}$$

and taking the 1/n root we get

$$\frac{du_z}{dr} = -\left(\frac{1}{2m}\left(-\frac{dp}{dz}\right)\right)^{\frac{1}{n}} r^{\frac{1}{n}} \tag{10.3.5}$$

Equation (10.3.5) can be integrated with a constant of integration solved by imposing the no-slip condition at $r = a$, $u_z(r = a) = 0$. The result is

$$u_z = \frac{n}{n+1}\left(\frac{1}{2m}\left(-\frac{dp}{dz}\right)\right)^{\frac{1}{n}}\left(a^{\frac{n+1}{n}} - r^{\frac{n+1}{n}}\right) \tag{10.3.6}$$

We can scale Eq. (10.3.6) by defining dimensionless variables as

$$U_Z = \frac{u_z}{u_{z\,avg}}, \quad R = \frac{r}{a} \tag{10.3.7}$$

where $u_{z\,avg}$ is the average fluid velocity

$$u_{z\,avg} = \frac{1}{\pi a^2}\int_0^{2\pi}\int_0^{a} u_z\, r\, dr\, d\theta = \frac{n}{3n+1}\left(-\frac{1}{2m}\frac{dp}{dz}\right)^{1/n} a^{\frac{n+1}{n}} \tag{10.3.8}$$

The dimensionless form becomes

$$U_Z(R) = \frac{3n+1}{n+1}\left(1 - R^{\frac{n+1}{n}}\right) \tag{10.3.9}$$

Examples of $U_Z(R)$ are shown in Figure 10.7 for $n = 0.5, 0.7$ shear thinning, 1.0 Newtonian and 1.5 shear thickening. The profile is parabolic (Poiseuille) for $n = 1$, with the centerline velocity twice the average, $U_Z(R = 0) = 2.0$ as we showed in Eq. (7.1.6). For the other values the shape is no longer parabolic, becoming flatter as n decreases, i.e. shear thinning, and more pointed as n increases, i.e. shear thickening.

The corresponding effective viscosity, $\mu_{eff}(\dot{\gamma})$, is spatially dependent and can now be calculated since we know $u_z(r)$. Starting with the definition

$$\mu_{eff} = m\dot{\gamma}^{n-1} = m\left(-\frac{du_z}{dr}\right)^{n-1} \tag{10.3.10}$$

take the r-derivative of u_z in Eq. (10.3.6). Insert the result into Eq. (10.3.10), which gives us

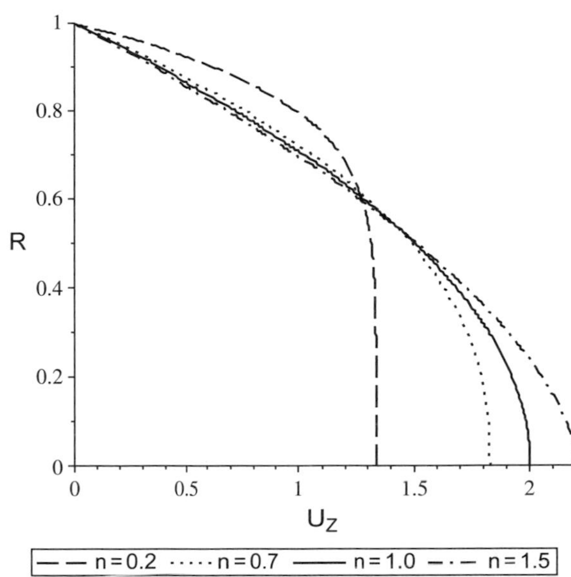

Figure 10.7 Velocity profiles of power-law fluids in a tube for n = 0.2, 0.7 (shear thinning), 1.0 (Newtonian) and 1.5 (shear thickening).

$$\mu_{eff} = m\left(\frac{1}{2m}\left(-\frac{dp}{dz}\right)\right)^{\frac{n-1}{n}} r^{\frac{n-1}{n}} \tag{10.3.11}$$

Defining a dimensionless effective viscosity as $\tilde{\mu}_{eff} = \mu_{eff}/m(-(dp/dz)b/2m)^{(n-1)/n}$ in Eq. (10.3.11) and using the dimensionless radial variable $R = r/a$ yields

$$\tilde{\mu}_{eff} = R^{\frac{n-1}{n}} \tag{10.3.12}$$

Plots of $\tilde{\mu}_{eff}(R)$ are shown in Figure 10.8. When $n = 1$ the fluid is Newtonian and has a constant viscosity, $\tilde{\mu}_{eff} = 1$, for all of R. For a shear thinning (pseudoplastic) fluid, $n = 0.5$, and $\tilde{\mu}_{eff} \to \infty$ as $R \to 0$ where $dU_Z/dR \to 0$. When the fluid is shear thickening (dilatant), $n = 1.5$ and $\tilde{\mu}_{eff} \to 0$ as $R \to 0$.

10.4 Herschel–Bulkley Fluids

Mucus, mayonnaise, blood, applesauce and ketchup are just a few examples of biofluids that exhibit a yield stress, τ_y, so are generally Herschel–Bulkley fluids. The value of τ_y varies widely depending on the

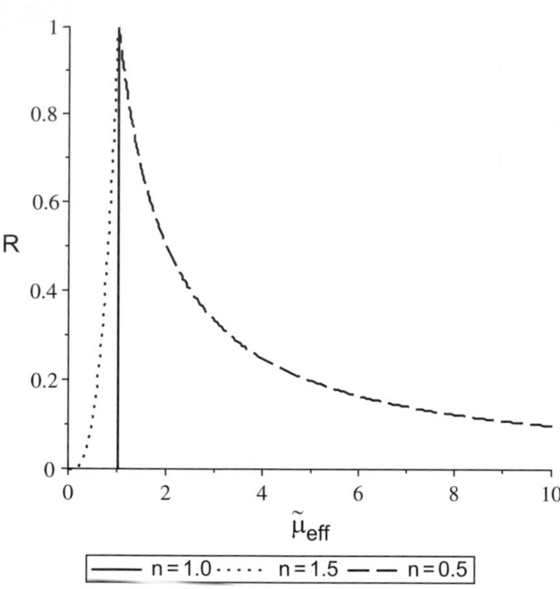

Figure 10.8 The dimensionless effective viscosity, $\tilde{\mu}_{eff}$ vs R for n = 0.5 (shear thinning), 1.0 (Newtonian) and 1.5 (shear thickening).

fluid. For example, the yield stress has typical values $\tau_y \sim 0.03, 44\ \text{dynes/cm}^2$ for blood and ketchup, respectively. Mucus yield stress has a broad range depending on the source (e.g. respiratory, cervical, gastrointestinal, snail foot), methods of collection and measurement, state of hydration and health. For normal canines, mucus collected in a tracheal pouch and measured by magnetic rheometer have a range of $\tau_y = 5.6 - 13\,\text{dyn/cm}^2$ (Edwards and Yeates, 1992). Homogenized sputum from adult cystic fibrosis patients measured at $\tau_y = 142\,\text{dyn/cm}^2$ (Broughton-Head et al., 2007). Yet a higher range, $\tau_y = 400 - 600\,\text{dyn/cm}^2$, has been measured for normal sputum using a cone-plate rheogoniometer, a version of the viscometer discussed in Section 10.8 (Davis, 1973). Because it is difficult to perform flow experiments with real mucus, a number of investigators create mucus stimulants from other compounds such as cross-linked polysaccharides (King et al., 1989), mayonnaise (Basser et al., 1989), locust bean gum (Hassan et al., 2006) and Carbopol® (Hu et al., 2015; Hu, 2019). There are many models of Herschel–Bulkley fluids (Herschel and Bulkley, 1926), but a simple approach to a constitutive equation is to combine a yield stress with a power-law relationship in the form

$$\tau = \tau_y + m'\dot{\gamma}^{n'} \qquad \tau > \tau_y \tag{10.4.1}$$

The flow consistency and behavior indexes are now m′ and n′, respectively. In fact, this form of a Herschel–Bulkley constitutive equation is also called a "yield-power-law fluid." The related effective viscosity is derived by dividing Eq. (10.4.1) by $\dot{\gamma}$,

$$\mu_{eff} = \frac{\tau}{\dot{\gamma}} = \frac{\tau_y}{\dot{\gamma}} + m'\dot{\gamma}^{n'-1} \tag{10.4.2}$$

Note that $n' = 1$ is a Bingham fluid as shown in Figure 10.2 where $\tau - \tau_y = m'\dot{\gamma}^1$ is linear in $\dot{\gamma}$ so that $\mu_{diff} = d\tau/d\dot{\gamma} = m'$ is constant in Eq. (10.4.1). Also in Figure 10.2 note that for $n' < 1$ both μ_{eff} and μ_{diff} are decreasing with increasing $\dot{\gamma}$, while $n' > 1$ is shear thickening where μ_{eff} and μ_{diff} are decreasing with increasing $\dot{\gamma}$.

For Herschel–Bulkley fluid flow in a tube of radius a, we start the analysis with the Cauchy momentum equation of momentum conservation already integrated once, Eq. (10.2.12). Set $\dot{\gamma} = |du_z/dr| = -du_z/dr \geq 0$ since we expect the velocity profile to have a non-positive slope for all values of r. Then Eq. (10.2.12) becomes

$$\left(\tau_y\left(-\frac{du_z}{dr}\right)^{-1} + m'\left(-\frac{du_z}{dr}\right)^{n'-1}\right)\frac{du_z}{dr} = \frac{dp}{dz}\frac{r}{2} \tag{10.4.3}$$

Multiplying through by −1 allows us to combine the $(-du_z/dr)$ terms

$$\tau_y + m'\left(-\frac{du_z}{dr}\right)^{n'} = \left(-\frac{dp}{dz}\right)\frac{r}{2} \tag{10.4.4}$$

Rearranging Eq. (10.4.4) to isolate du_z/dr leads to the form

$$-\frac{du_z}{dr} = \left(\frac{1}{m'}\right)^{1/n'}\left[\left(-\frac{dp}{dz}\right)\frac{r}{2} - \tau_y\right]^{1/n'} \tag{10.4.5}$$

Since the right hand side of Eq. (10.4.5) must be real and non-negative, there is a constraint on the square-bracketed term that is must be non-negative.

But this is simply the statement that the magnitude of the fluid shear stress at radial position r, from Eq. (10.2.8), must exceed the yield stress for there to be flow

$$|\tau_{rz}| = \left(-\frac{dp}{dz}\right)\frac{r}{2} \geq \tau_y \tag{10.4.6}$$

This constraint restricts the values of r to be $r \geq r_y$, where the yield surface is at $r = r_y$, defined as

$$r_y = 2\tau_y/(-dp/dz) \tag{10.4.7}$$

Also, because r cannot exceed the tube radius, a, the Herschel–Bulkley velocity field is for fluid in the radial range $r_y \leq r \leq a$. For the central core of the flow, $0 \leq r \leq r_y$, the shear stress is below τ_y and the material is unyielded. It moves like a solid plug. The speed of this plug is the same as that of the fluid where they meet, at $r = r_y$. So $u_z = u_z(r = r_y)$ for the solid plug in the radial range $0 \leq r \leq r_y$.

For example, a plot of the magnitude of $|\tau_{rz}|$ from Eq. (10.2.8) is shown in Figure 10.9 for two values of the pressure gradient, $-(1/2)(dp/dz)$. For A the shear stresses are all less than the yield stress, $\tau < \tau_y$, so there is no strain rate, $\dot{\gamma} = 0$, and consequently no flow. For B there is a range of r as shown, $r_y < r \leq a$, where the shear stress is above the yield stress, $\tau_y < \tau \leq \tau_w$. The yield value of $r = r_y$ is where the shear stress equals the yield stress, $\tau(r = r_y) = \tau_y$. Since $\dot{\gamma} = 0$ for $0 \leq r \leq r_y$, the material behaves as a solid plug in this region.

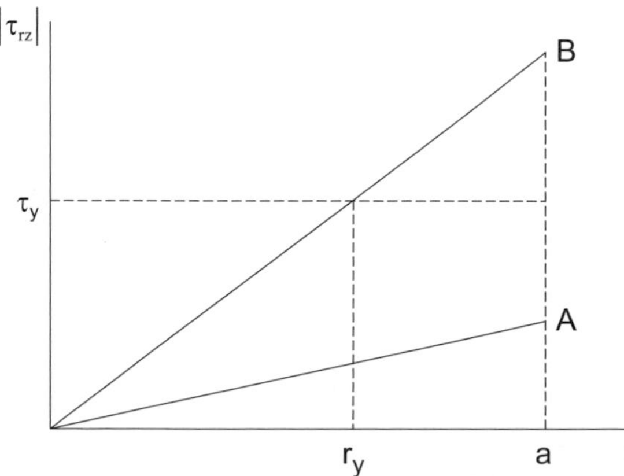

Figure 10.9 Shear stress magnitude, $|\tau_{rz}|$, dependence on radial position, for two values of the slope $-(1/2)(dp/dz)$.

Integrating Eq. (10.4.5) and solving the integration constant by enforcing no-slip, $u_z = 0$, at $r = a$, leads to the form

$$u_z = \frac{n'}{n'+1}\frac{2}{\left(-\frac{dp}{dz}\right)}\left(\frac{1}{m'}\right)^{1/n'}$$
$$\left\{\left[\left(-\frac{dp}{dz}\right)\frac{a}{2} - \tau_y\right]^{(n'+1)/n'} - \left[\left(-\frac{dp}{dz}\right)\frac{r}{2} - \tau_y\right]^{(n'+1)/n'}\right\}$$
$$r_y \leq r \leq a$$

$$u_z = u_z(r = r_y) = \frac{n'}{n'+1}\frac{2}{\left(-\frac{dp}{dz}\right)}\left(\frac{1}{m'}\right)^{1/n'}$$
$$\left[\left(-\frac{dp}{dz}\right)\frac{a}{2} - \tau_y\right]^{(n'+1)/n'} \quad 0 \leq r \leq r_y \tag{10.4.8}$$

Scale Eq. (10.4.8) by defining a dimensionless radius, $R = r/a$, and dimensionless velocity, $U_z = u_z/\left[a(-(dp/dz)(a/2m'))^{1/n'}\right]$ to find the final form of U_z

$$U_z = \frac{n'}{n'+1}\left[(1 - Bn)^{(n'+1)/n'} - (R - Bn)^{(n'+1)/n'}\right]$$
$$Bn \leq R \leq 1$$

$$U_z = U_z(R = Bn) = \frac{n'}{n'+1}(1 - Bn)^{(n'+1)/n'} \quad 0 \leq R \leq Bn \tag{10.4.9}$$

In Eq. (10.4.9) we introduce the dimensionless parameter, Bn, which is the Bingham number. In general, Bn is a ratio of the yield stress to a characteristic shear stress in a flow

$$Bn \sim \frac{\text{yield stress}}{\text{shear stress}} \tag{10.4.10}$$

For our flow a good measure of the characteristic shear stress is the wall shear from Eq. (10.2.9), $\tau_w = (a/2)(-dp/dz)$. In our tube flow, then, Bn is given by

$$Bn = \frac{\tau_y}{\tau_w} = \frac{\tau_y}{\left(-\frac{a}{2}\frac{dp}{dz}\right)} = \frac{r_y}{a} \tag{10.4.11}$$

Clearly, for there to be any flow at all the range of Bn is restricted to $0 \leq Bn < 1$ or, equivalently, $0 \leq r_y < a$. There is no flow for $Bn \geq 1$, or $r_y \geq a$, because the largest shear stress in the fluid, i.e. its wall shear, does not exceed the yield stress. In that case, the pressure gradient $(-dp/dz)$ is not large enough to create yielding shear stresses in the material.

Some examples of the flow field in Eq. (10.4.9) are shown in Figure 10.10. In Figure 10.10(a) the velocity field is plotted for $n' = 1$, which is a Bingham fluid, and

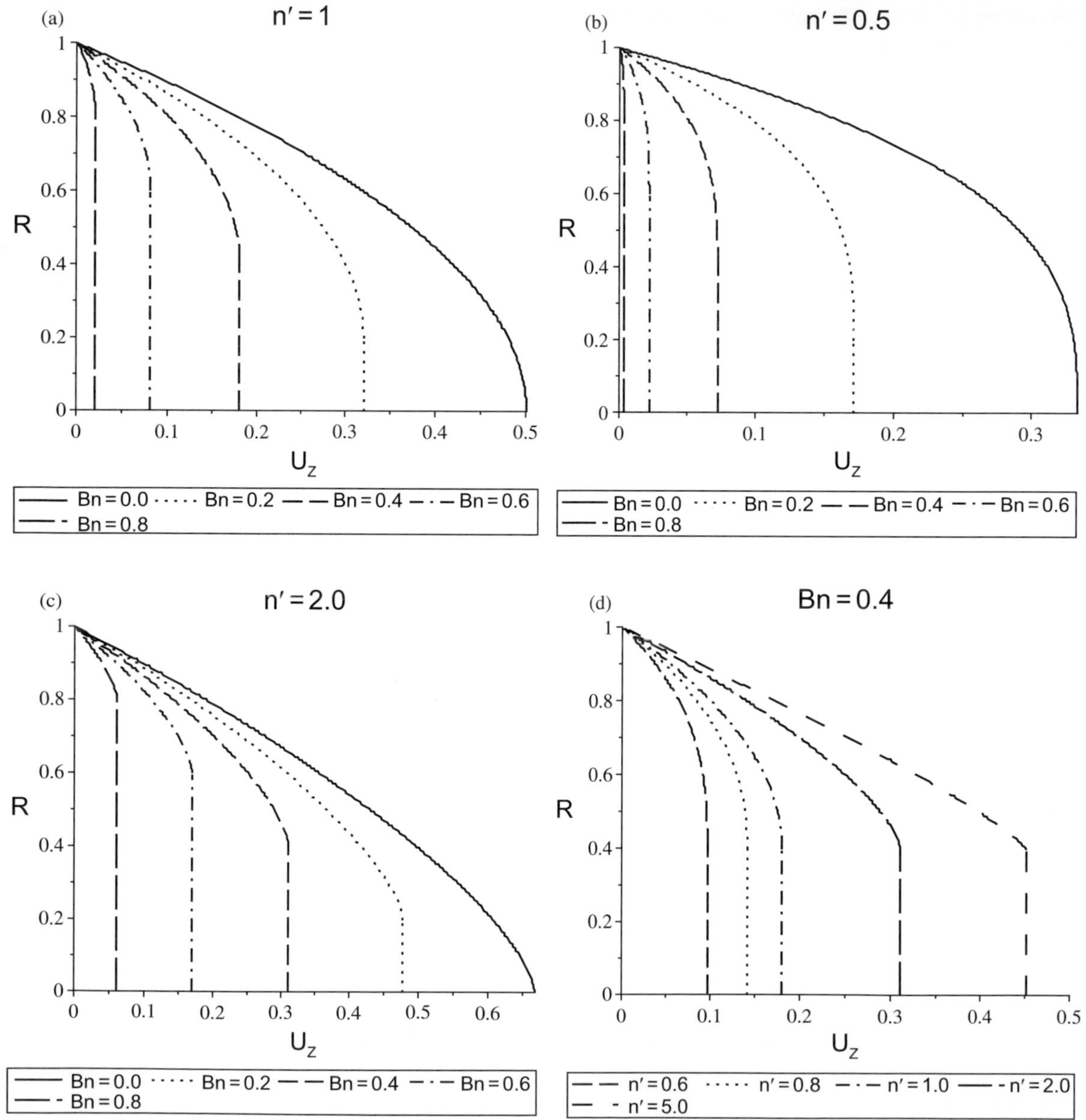

Figure 10.10 (a) $U_z(R)$ for $n' = 1$ and $Bn = 0.0, 0.2, 0.4, 0.6, 0.8$. (b) Same as (a) but for $n' = 0.5$. (c) Same as (a) but for $n' = 2$. (d) $U_z(R)$ for $Bn = 0.4$, $n' = 0.6, 0.8, 1.0, 2.0, 5.0$.

Bn = 0.0, 0.2, 0.4, 0.6, 0.8. The combination of $n' = 1$ and $bn = 0$ is a Newtonian fluid and the resulting profile is parabolic in R, which is Poiseuille flow. Then as Bn increases the solid plug core region, $0 \leq R \leq Bn$, increases while the fluid region, $Bn \leq R \leq 1$, decreases. Overall, increasing Bn slows down the fluid until the velocity is zero when $Bn = 1$, not plotted. Figure 10.10(b) is the same as Figure 10.10 (a) but for $n' = 0.5$, a shear thinning behavior with differential viscosity $\mu_{diff} = \partial\tau/\partial\dot{\gamma} = m'n'\dot{\gamma}^{n'-1}$ that decreases with increasing $\dot{\gamma}$ since $n' < 1$. For each value of Bn, the scaled velocities are smaller than the corresponding values in Figure 10.10(a). Figure 10.10(c) is the same as Figure 10.10(a) but for $n' = 20$, a shear thickening behavior with differential viscosity $\mu_{diff} = \partial\tau/\partial\dot{\gamma} = m'n'\dot{\gamma}^{n'-1}$ that increases for increasing $\dot{\gamma}$ since $n' > 1$. For each value of Bn, the scaled velocities are larger than the corresponding values in Figure 10.10(a). Because the velocity scale is $a(a(-dp/dz)/2m')^{1/n'}$, comparing dimensional values amongst Figure 10.10(a, b, c) requires specific assignments of m', dp/dz, a and n'. Figure 10.10(d) shows the flow for $Bn = 0.4$ with $n' = 0.6$, 0.8, 1.0, 2.0, 5.0. Increasing n' increases the dimensionless velocities to an asymptotic value of $U_z \sim 1 - R$ as $n' \rightarrow \infty$, which we saw for the power-law fluid. Since $R = Bn$ at the interface with the solid plug, the plug speed is $1 - Bn \tilde{} 0.6$ in this limit.

Results of rheological testing for ketchup appear in (Koocheki et al., 2009), where the data are fit either to a power-law or Herschel–Bulkley model. An example at 25 °C temperature over a shear rate range of $0 \leq \dot{\gamma} \leq 300\ s^{-1}$ for the power-law model is $m = 19.34\ Pa{\cdot}s^n$, $n = 0.228$ and for the Herschel–Bulkley model is $m' = 16.18\ Pa{\cdot}s^{n'}$, $n' = 0.25$, $\tau_y = 4.41\,Pa$. The stress unit Pa is a pascal, which is $1\ newton{\cdot}m^{-2} = 1\,kg \cdot m^{-1}{\cdot}s^{-2}$ and $1\ Pa = 10.0\,dyn \cdot cm^{-2}$.

10.5 Casson Fluid Model for Blood

As we learned in Chapter 2, blood is a complex fluid made of a liquid serum and suspended particles ranging from red blood cells (RBCs), which carry oxygen, white blood cells (WBCs), which are part of the immune system, and platelets, which are used in forming blood clots. In Chapter 2 we discussed the propensity of RBCs to stick together and form chains or rouleaux. That feature figures strongly in the rheological properties of blood at low strain rates. Data from blood rheology experiments (Merrill et al., 1963) are shown in Figure 10.11 for different values of the hematocrit, H, and temperatures. As H increases or temperature decreases, the curve is shifted upward to higher values of τ for a given $\dot{\gamma}$, i.e. more viscous. The axes are $\sqrt{\tau}$ vs $\sqrt{\dot{\gamma}}$, and the data form a linear relationship with slope $\sqrt{\mu_\infty}$ where μ_∞ is the limit of the effective viscosity as the strain rate becomes very large. The lines also have intercepts on the $\sqrt{\tau}$-axis which are the square root of the yield stress $\sqrt{\tau_y}$. For there to be flow, it is necessary that the shear stress exceeds the yield stress, i.e. $\tau > \tau_y$, so that the strain rate, $\dot{\gamma}$, is greater than zero, i.e. $\dot{\gamma} > 0$.

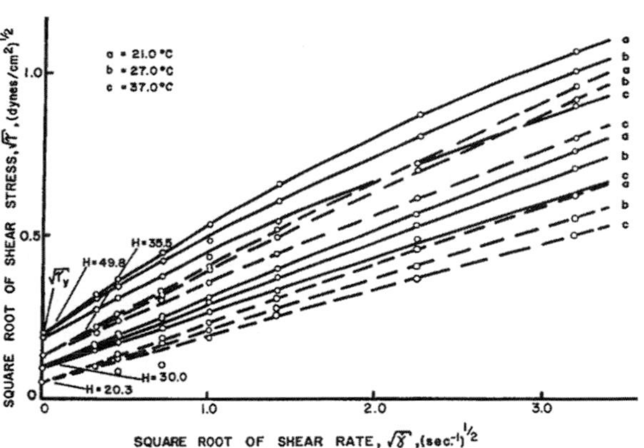

Figure 10.11 $\tau^{1/2}$ vs $\dot{\gamma}^{1/2}$ for blood at different hematocrits and temperatures.

The data over a large range of strain rate follow the Casson equation

$$\sqrt{\tau} = \sqrt{\tau_y} + \sqrt{\mu_\infty \dot{\gamma}} \qquad \tau > \tau_y \tag{10.5.1}$$

which was developed by Casson (1959) to model flow of ink with pigments suspended in oil. As an example, for a hematocrit of 39% the data look to have an intercept of $\sqrt{\tau_y} \sim 0.17$, $\tau_y \sim 0.029\ dyn/cm^2$ and a slope of $\sqrt{\mu_\infty} \sim 0.70/3.4 \sim 0.2$, $\mu_\infty \sim 0.04$ poise. This value of μ_∞ is a typical high strain rate value which is ~ 3–4 times the viscosity of water, $\mu_{H_2O} = 0.01$ poise. The Casson model is another form of a Herschel–Bulkley fluid with a yield stress and shear thinning.

Both μ_∞ and τ_y vary with the hematocrit, H. In the rheological data of Cokelet et al. (1963), viscosity was measured at varying levels of $\dot{\gamma}$ and H and interpreted through the Casson model. They found the dependence of μ_∞ on H to be

$$\mu_\infty = \mu_0(1 - H)^{-5/2} \tag{10.5.2}$$

where μ_0 is the viscosity of the fluid without suspended particles. In the case of blood, that fluid is the serum or plasma. Plasma viscosity is determined by its water-content and concentration of macromolecules. It has a normal range of 0.011–0.013 poise at 37 °C. Equation (10.5.2) shows that increasing H increases μ_∞ in a nonlinear way, but increases with μ_0 linearly. Equation (10.5.2) is known as the Brinkman–Roscoe relation (Brinkman, 1952; Roscoe, 1952) for the two authors who independently published it in 1952.

Einstein developed a formula for the high strain rate viscosity, μ_∞, of a dilute suspension of rigid spheres (Einstein, 1906). Its form is

$$\mu_\infty = \mu_0\left(1 + \frac{5}{2}\phi\right) \tag{10.5.3}$$

The volume fraction, ϕ, is the volume of the spheres divided by the volume of the suspension, which for blood would be the hematocrit, $\phi = H$. The system is considered dilute so there are no interactions between spheres, and it is valid for $\phi << 1$. In the absence of any spheres, $\phi = 0$ and the suspension viscosity is simply the fluid viscosity, μ_0. For $H << 1$, Eq. (10.5.2) from Cokelet et al. (1963) has a Taylor series expansion $\mu_\infty \sim \mu_0(1 + 5H/2) + O(H^2)$ which is equivalent to the theory of Einstein, Eq. (10.5.3).

G. I. Taylor (1932) developed an equation for the viscosity of a fluid–fluid suspension which has some appeal for modeling RBCs since they have some fluid properties. Rather than the rigid spheres of Einstein, Taylor treated spherical fluid drops whose own viscosity is μ_d. The viscous drops are suspended in a different fluid whose viscosity is μ_0. The relationship Taylor found is

$$\frac{\mu_\infty}{\mu_0} = 1 + \phi\left[\frac{\mu_0 + \frac{5}{2}\mu_d}{\mu_0 + \mu_d}\right] \tag{10.5.4}$$

Note that for zero volume fraction, $\phi = 0$, $\mu_\infty = \mu_0$, and as the drop viscosity increases toward rigidity, $\mu_d \to \infty$, the Einstein formula is recovered.

The yield stress is also H-dependent. At very low strain rates the rouleaux form interconnecting chains causing the blood to have solid, elastic structure properties. As the imposed shear stress increases, the structure is broken at the yield stress. The dependence of yield stress, τ_y, on H was found through experimentation to be

$$\tau_y = A(H - H_c)^3 \tag{10.5.5}$$

where $A \sim 0.6 - 1.2\,\text{dyn} \cdot \text{cm}^{-2}$ and $H_c = 0.05 - 0.08$ (Merrill et al., 1963; Merrill, 1969). Below a critical value of the hematocrit, $H < H_c$, there is no yield stress, since the RBCs are far enough apart to avoid rouleaux formation. For $H > H_c$, the measured yield stress fits well to Eq. (10.5.5). Increasing H increases τ_y as a cubic dependence. A typical hematocrit of $H = 0.45$, with $H_c = 0.05$, and $A = 0.6\ \text{dyn} \cdot \text{cm}^{-2}$ gives a yield stress of $\tau_y = 0.038\ \text{dyn} \cdot \text{cm}^{-2}$.

Squaring both sides of Eq. (10.5.1) and rearranging gives the relationship

$$\tau = \tau_y + 2\sqrt{\mu_\infty \tau_y}\dot{\gamma}^{1/2} + \mu_\infty \dot{\gamma} \tag{10.5.6}$$

and plots of $\tau(\dot{\gamma})$ using Eqs. (10.5.2) and (10.5.5) in Eq. (10.5.6) are shown in Figure 10.12. Clearly blood is a Herschel–Bulkley fluid which has a yield stress and shear thinning properties, Curve 6 in Figure 10.2. The two parameters, τ_y and μ_∞, provide a spectrum of behavior.

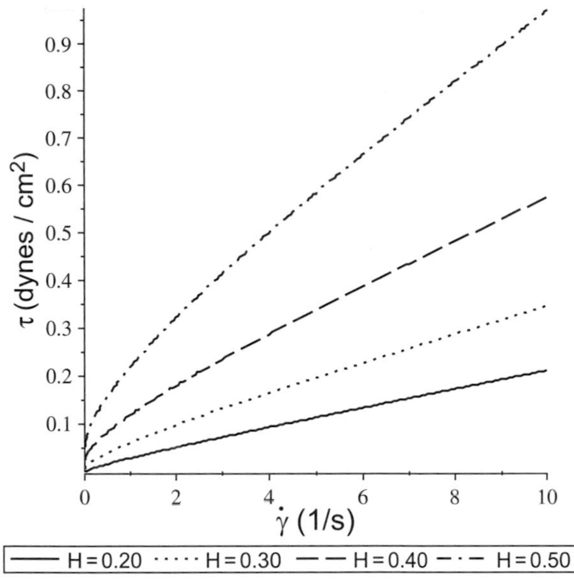

Figure 10.12 Casson fluid model $\tau(\dot{\gamma})$ for four values of the hematocrit, H.

From the definition that $\tau = \mu_{\text{eff}}(\dot{\gamma})\dot{\gamma}$, we can rearrange it to solve for $\mu_{\text{eff}}(\dot{\gamma}) = \tau/\dot{\gamma}$ to find

$$\begin{aligned}\mu_{\text{eff}}(\dot{\gamma}) &= \frac{\tau}{\dot{\gamma}} = \mu_\infty + 2\sqrt{\frac{\mu_\infty \tau_y}{\dot{\gamma}}} + \frac{\tau_y}{\dot{\gamma}} \\ &= \mu_\infty \left(1 + \sqrt{\frac{\tau_y}{\mu_\infty \dot{\gamma}}}\right)^2 \end{aligned} \tag{10.5.7}$$

and it is readily seen that $\mu_{\text{eff}}(\dot{\gamma} \to \infty) = \mu_\infty$.

Some sample plots of μ_{eff} vs $\dot{\gamma}$ for blood are given in Figure 10.13. Figure 10.13(a) is rheological data from Chien et al. (1966) showing measured viscosities for hematocrits of 90%, 45% and 0%. Figure 10.13(b) shows calculated μ_{eff} vs $\dot{\gamma}$ using Eq. (10.5.7) for four values of the hematocrit, $H = 0.20, 0.30, 0.40, 0.50$. In both figures the shear thinning behavior is evident as the viscosities decrease with increasing $\dot{\gamma}$ and asymptotically approach a constant value, μ_∞. Most of the shear dependence of μ_{eff} for $\dot{\gamma} \leq 10\ \text{s}^{-1}$.

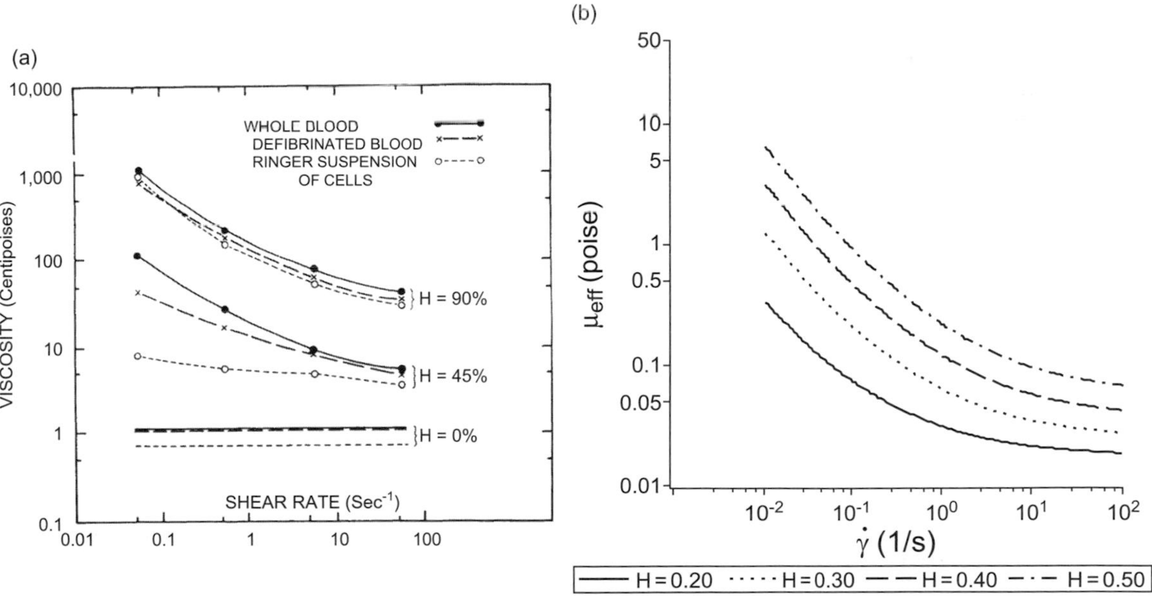

Figure 10.13 (a) Measured shear-dependent blood viscosity for hematocrits of 90%, 45% and 0%. (b) Calculated μ_{eff} vs $\dot{\gamma}$ from Eq. (10.5.7) for four values of the hematocrit, $H = 0.20, 0.30, 0.40, 0.50$.

Clinical Correlation 10.1 | Optimal Hematocrit

Increasing the hematocrit permits larger amounts of oxygen to be transported to the tissues, but there is a cost of work on the heart because the blood viscosity also increases as shown in Eq. (10.5.2). Is there an optimal value of H? Suppose we want to minimize the work of the heart while keeping oxygen delivery constant by finding an optimal hematocrit, H. Oxygen delivery is proportional to the blood flow rate, Q, multiplied by the hematocrit, H. Let $K = QH$ be constant. The work of the heart is $w = pQ$, and we can replace p with Poiseuille's Law to achieve

$$w = pQ = \left(\frac{8\mu LQ}{\pi a^4}\right)Q = \frac{8\mu L}{\pi a^4}\left(\frac{K}{H}\right)^2 \tag{10.5.8}$$

Inserting the H-dependent viscosity from Eq. (10.5.2) yields the form

$$w = \left(\frac{K}{H}\right)^2 \frac{8\mu L}{\pi a^4} = K^2 \frac{8L}{\pi a^4} \frac{\mu_0}{H^2(1-H)^{2.5}} \tag{10.5.9}$$

Clearly W in Eq. (10.5.9) becomes unbounded as $H \to 0$ and as II $\to 1$. Let the dimensionless work be W defined as

$$W = \frac{\pi a^4 w}{8\mu_0 L K^2} = \frac{1}{H^2(1-H)^{2.5}} \tag{10.5.10}$$

A plot of W vs H is shown in Figure 10.14. The optimal value of H is derived by taking the H-derivative of Eq. (10.5.10) and setting it to zero $dW/dH = 0$ and that occurs for $H = 0.44$.

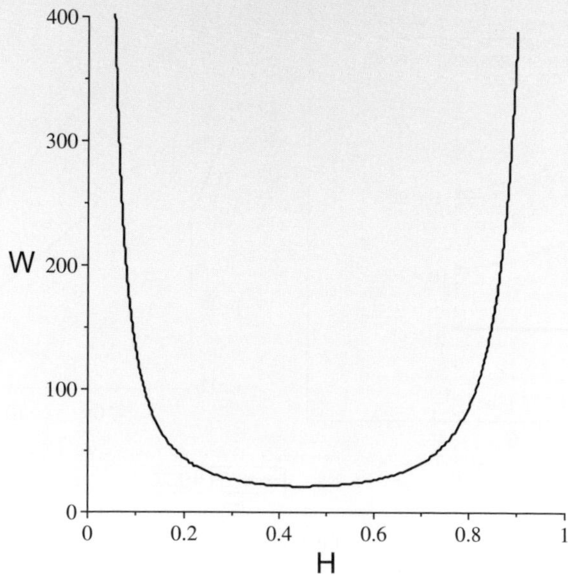

Figure 10.14 Dimensionless work vs hematocrit, H, for fixed oxygen delivery.

Interestingly, this value of the hematocrit falls within the normal range of humans. The corresponding viscosity from Eq. (10.5.2) is $\mu_\infty = 4.26\mu_0$.

Now suppose we want to keep the pressure gradient, fixed, but maximize oxygen delivery. The oxygen delivery, q_{O_2}, is proportional to the product of the cardiac output multiplied by the hematocrit,

$$q_{O_2} = Q\,x\,H = \frac{\pi a^4}{8\mu_0}\left(-\frac{dp}{dz}\right)H(1-H)^{2.5} \tag{10.5.11}$$

Let the dimensionless Q_{O_2} be defined as

$$Q_{O_2} = \frac{8\mu_0 q_{O_2}}{\pi a^4(-dp/dz)} = H(1-H)^{2.5} \tag{10.5.12}$$

A plot of Eq. (10.5.12) is shown in Figure 10.15. The maximum, where $dQ_{O_2}/dH = 0$, occurs at $H = 0.29$. The corresponding viscosity according to Eq. (10.5.2) is $\mu_\infty = 2.35\mu_0$. It is routine during open heart surgery, for example, to dilute the hematocrit of the patient in order to achieve this optimization.

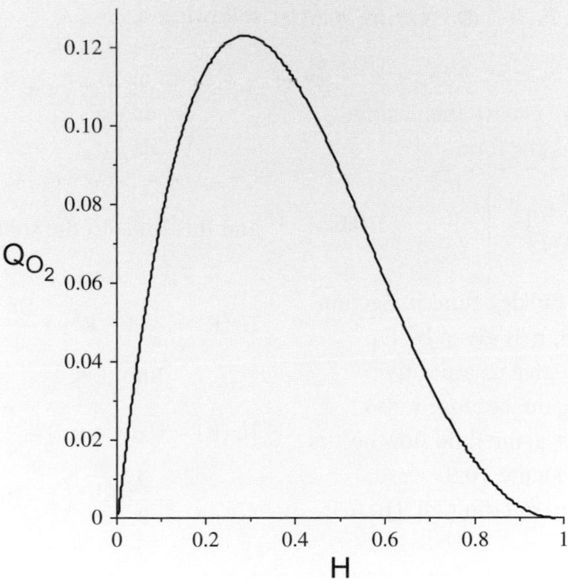

Figure 10.15 Oxygen delivery Q_{O_2} vs H.

10.6 Casson Fluid Flow in a Tube

Starting with Eq. (10.2.8), the first integral of the Cauchy momentum equation for momentum balance is

$$\mu_{eff}(\dot{\gamma})\frac{du_z}{dr}=\frac{dp}{dz}\frac{r}{2} \tag{10.6.1}$$

Substituting for the Casson effective viscosity in Eq. (10.6.1) gives us

$$\left(1+\sqrt{\frac{\tau_y}{\mu_\infty\dot{\gamma}}}\right)^2\frac{du_z}{dr}=\frac{r}{2\mu_\infty}\frac{dp}{dz} \tag{10.6.2}$$

We determined previously that in this flow $\dot{\gamma}=|du_z/dr|=-du_z/dr\geq 0$. Expanding Eq. (10.6.2) and multiplying through by –1 gives us

$$\left(-\frac{du_z}{dr}\right)+2\sqrt{\frac{\tau_y}{\mu_\infty}}\left(-\frac{du_z}{dr}\right)^{1/2}+\frac{\tau_y}{\mu_\infty}+\frac{r}{2\mu_\infty}\frac{dp}{dz}=0 \tag{10.6.3}$$

which is a quadratic equation for $(-du_z/dr)^{1/2}$. Its solution is

$$\left(-\frac{du_z}{dr}\right)^{1/2}=\sqrt{\frac{\tau_y}{\mu_\infty}}\left[-1\pm\left(\frac{r}{2\tau_y}\left(-\frac{dp}{dz}\right)\right)^{1/2}\right] \tag{10.6.4}$$

Because $(-du_z/dr)^{1/2}>0$, we choose the positive root of Eq. (10.6.4), which has the form

$$\left(-\frac{du_z}{dr}\right)^{1/2}=\sqrt{\frac{\tau_y}{\mu_\infty}}\left[-1+\left(\frac{r}{r_y}\right)^{1/2}\right] \tag{10.6.5}$$

As before with the Herschel–Bulkley fluid in Section 10.4, the yield radial position, r_y is given by Eq. (10.4.7). We require $r>r_y$ to give us a positive solution and yielded flow. Again, because r also cannot exceed the tube radius, a, the fluid flow occurs in the region $r_y\leq r\leq a$, see Figure 10.9.

Returning to our derivation, squaring Eq. (10.6.5) yields

$$-\frac{du_z}{dr}=\left(-\frac{dp}{dz}\right)\frac{r_y}{2\mu_\infty}\left[1-2\left(\frac{r}{r_y}\right)^{1/2}+\frac{r}{r_y}\right] \tag{10.6.6}$$

Equation (10.6.6) can be integrated to yield

$$u_z(r)=\frac{dp}{dz}\frac{r_y}{2\mu_\infty}\left[r-\frac{4}{3}\left(\frac{1}{r_y}\right)^{1/2}r^{3/2}+\frac{1}{2r_y}r^2\right]+c_2 \tag{10.6.7}$$

Use the no-slip condition, $u_z(r=a)=0$, to evaluate c_2, and the final dimensional form is

$$u_z(r)=\left(-\frac{dp}{dz}\right)\frac{1}{2\mu_\infty}\left[r_y(a-r)-\frac{4\sqrt{r_y}}{3}\left(a^{3/2}-r^{3/2}\right)+\frac{1}{2}(a^2-r^2)\right],\quad r_y\leq r\leq a \tag{10.6.8}$$

It is readily seen from Eq. (10.6.8) that when $r_y=0$ the solution simplifies to Poiseuille flow, i.e. parabolic with r^2 dependence. That is the case of no yield stress, $\tau_y=0$. For $\tau_y>0$ the other terms appear and have additional dependences on the radial position, r and $r^{3/2}$.

To make Eq. (10.6.8) dimensionless, choose the following scales

$$U_Z=\frac{u_z}{\left(-\frac{dp}{dz}\right)\frac{a^2}{\mu_\infty}},\quad R=\frac{r}{a} \tag{10.6.9}$$

and this leads to the solution

$$U_Z(R)=\frac{1}{4}(1-R^2)+\frac{Bn}{2}(1-R)-\frac{2}{3}Bn^{1/2}\left(1-R^{3/2}\right),\quad Bn\leq R\leq 1$$
$$U_Z(R)=U_Z(R=Bn)=\frac{1}{4}(1-Bn^2)+\frac{Bn}{2}(1-Bn)-\frac{2}{3}Bn^{1/2}\left(1-Bn^{3/2}\right),\quad 0\leq R\leq Bn \tag{10.6.10}$$

The Bingham number was introduced in Section 10.4 as $Bn=r_y/a=\tau_y/\tau_W$. We note, then, that $Bn=R_y=r_y/a$, the scaled yield radius. The first term in Eq. (10.6.10) in the yielded region, $Bn \leq R \leq 1$, is the parabolic contribution and for no yield stress, $Bn = 0$, it is the only term and the flow becomes Poiseuille. As Bn increases the other terms cause a non-parabolic shape. The unyielded region, $0 \leq R \leq Bn$ is where the shear stress is less than the yield stress, so that region is convected as a solid plug at the speed $U_z(R = Bn)$ given in Eq. (10.6.10).

Figure 10.16 shows a plot of the velocity profile for five choices of Bn. $Bn = 0$ is Poiseuille flow while the other curves show an overall decrease in the velocities as Bn increases. The velocity profiles are non-parabolic for $Bn \leq R \leq 1$, and there is solid plug flow for $0 \leq R \leq Bn$. For there to be flow, we must have $Bn < 1$, which translates to the need for a sufficient pressure gradient $-dp/dz > 2\tau_y/a$, and that critical value increases with increasing yield stress and decreasing tube radius.

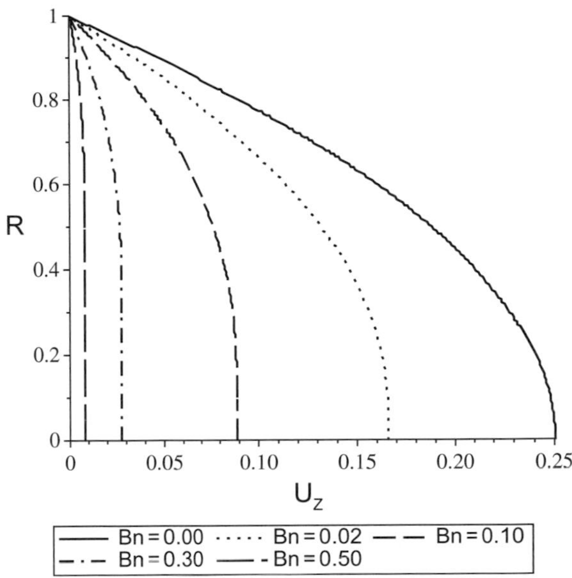

Figure 10.16 Velocity profiles, $U_Z(R)$, for Casson flow in a tube for five values of $Bn = 0, 0.02, 0.10, 0.30, 0.50$.

Clinical Correlation 10.2 | Sickle Cell Disease

Diseases can change the yield stress of blood. The red blood cells in sickle cell disease are sickle in shape, fragile, and very sticky, particularly when in low oxygen environments.

Figure 10.17 shows a normal RBC and a sickle cell anemia RBC. The sickle cell is stiffer and makes it more difficult to flow through small capillaries. Also, the surface is more sticky for the sickled RBC than for a normal RBC.

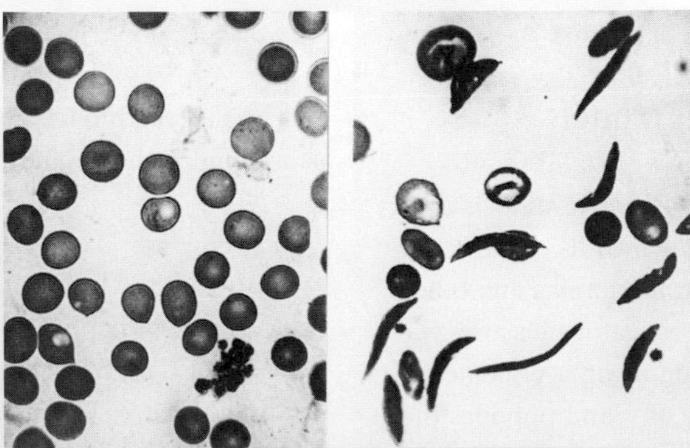

Figure 10.17 Normal and sickle cells.

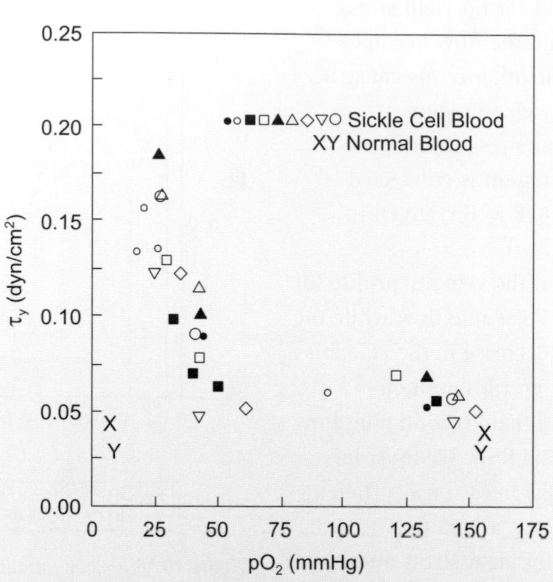

Figure 10.18 Increase of yield stress, τ_y, with lower oxygen tension for sickle cell blood compared to normal blood.

The stickiness to one another raises the yield stress as shown in Figure 10.18 adapted from Morris et al. (1993) and they also become sticky to the inside of the blood vessels, i.e. the endothelium, blocking flow. The increase in the yield stress can have dramatic consequences as we have seen in Figure 10.16. Flow rates decrease at τ_y, through R_c or Bn, increases, and may become zero if the critical pressure gradient, $2\tau_y/a$, is not exceeded. It is a particular risk for flow in smaller vessels where the radius, a, is reduced. This form of low flow state in small vessels for sickle cell anemia patients is known as sludging and can cause micro-infarcts in the downstream tissue.

10.7 Other Rheological Models

Here we present additional constitutive models of generalized Newtonian fluids, which have attractive features often used in biofluid mechanics and other applications. Typically an investigator measures the rheological properties of a fluid using a viscometer which ramps through a range of $\dot{\gamma}$ and provides the effective viscosity as a function of shear rate, $\mu_{eff}(\dot{\gamma})$. Then suitable constitutive models are tested to see how well they may, or may not, fit the $\mu_{eff}(\dot{\gamma})$ data.

The Carreau model, due to Pierre J. Carreau (1972), is a popular choice and given by the equation

$$\frac{\mu_{eff}(\dot{\gamma}) - \mu_\infty}{(\mu_0 - \mu_\infty)} = \left[1 + (\lambda\dot{\gamma})^2\right]^{\frac{n-1}{2}} \tag{10.7.1}$$

which has four parameters: the time constant, λ ; the power index, n ; the constant viscosity at low strain rate, μ_0, in the limit $\dot{\gamma} \to 0$; and the constant viscosity at high strain rate, μ_∞, in the limit $\dot{\gamma} \to \infty$. For high shear rates, $\lambda\dot{\gamma} >> 1$, the Carreau model takes the form of a power-law relationship that we discussed in Section 10.3

$$\frac{\mu_{eff}(\dot{\gamma}) - \mu_\infty}{(\mu_0 - \mu_\infty)} \sim (\lambda\dot{\gamma})^{n-1} \qquad \lambda\dot{\gamma} >> 1 \tag{10.7.2}$$

For small shear rates, $\lambda\dot{\gamma} << 1$, a Taylor series expansion of Eq. (10.7.1) is

$$\frac{\mu_{eff}(\dot{\gamma}) - \mu_\infty}{(\mu_0 - \mu_\infty)} = \left[1 + \left(\frac{n-1}{2}\right)(\lambda\dot{\gamma})^2 + O((\lambda\dot{\gamma})^4)\right] \quad \lambda\dot{\gamma} << 1 \tag{10.7.3}$$

The resulting parabolic shape of Eq. (10.7.3) governs the shape of the curve near its small shear rate asymptote.

Velez-Cordero and Lauga (2013) have collected several data sources for mucus rheological properties including human sputum (Dawson et al., 2003), human cervico-vaginal mucus (Lai et al., 2007) and pig small intestine mucus (Sellers et al., 1991). Their general power-law fit to that data yieldsm $= 105$ dyn·cm^{-2}·s^n, $n = 0.2$. They also employ a Carreau model for the data from Hwang et al. (1969) on human bronchial mucus ($n = 0.08, \lambda = 2154.6$ s, $\mu_0 = 21{,}875$ poise, $\mu_\infty \sim 0$) and human cervical mucus ($n = 0.27, \lambda = 631.04$ s, $\mu_0 = 1457$ poise, $\mu_\infty \sim 0$) for the shear range $10^{-6} \leq \dot{\gamma} \leq 10^0$.

These three curves are shown in Figure 10.19 where the viscosity units are expressed as poise for comparison to other models. Since water is approximately 10^{-2} poise, we can appreciate from Figure 10.19 that the mid-range mucus viscosities of 10^1–10^3 poise are 10^3 to 10^5 times larger. In disease states it can even be higher. Difficulty in moving mucus out of the lung airways, for example, causes significant medical problems in asthma, emphysema and cystic fibrosis.

A variation of the Carreau model is the Carreau–Yasuda model (Carreau, 1972; Yasuda and Cohen, 1981), which includes an additional fitting parameter, a, in the familiar form

$$\frac{\mu_{eff}(\dot{\gamma}) - \mu_\infty}{(\mu_0 - \mu_\infty)} = [1 + (\lambda\dot{\gamma})^a]^{\frac{n-1}{a}} \tag{10.7.4}$$

μ_0 is the constant viscosity in the limit $\dot{\gamma} \to 0$, while μ_∞ is the constant viscosity in the limit $\dot{\gamma} \to \infty$. Here λ is a time constant while a and n are exponents chosen to give dilatant or pseudoplastic behavior. The choice of $a = 2$ recovers the Carreau model. Again we obtain a power-law relationship for large shear rate but for small shear rates we have

$$\frac{\mu_{eff}(\dot{\gamma}) - \mu_\infty}{(\mu_0 - \mu_\infty)} = \left[1 + \left(\frac{n-1}{a}\right)(\lambda\dot{\gamma})^a + O((\lambda\dot{\gamma})^{2a})\right] \quad \lambda\dot{\gamma} << 1 \tag{10.7.5}$$

and the exponent, a, can be chosen to fit more closely the behavior in this shear range rather than the parabolic, $a = 2$, of the Carreau model alone.

Sample parameter values used in computations in Gijsen et al. (1999), who measured fluid velocities in a model carotid artery bifurcation using laser Doppler anemometry. Blood, itself, is unsuitable for visualizing the flow so a clear blood simulant KSCN-X was used and its non-Newtonian properties modified to resemble that of blood. Overall, the simulant and blood data from Thurston (1979) result in parameter values of the Carreau–Yasuda model of $\mu_0 = 22 \times 10^{-3}$ Pa·s, $\mu_\infty = 2.2 \times 10^{-3}$ Pa·s, $a = 0.644$, $\lambda = 0.110$ s, $n = 0.392$ over the shear range $10^0 \leq \dot{\gamma} \leq 10^3$.

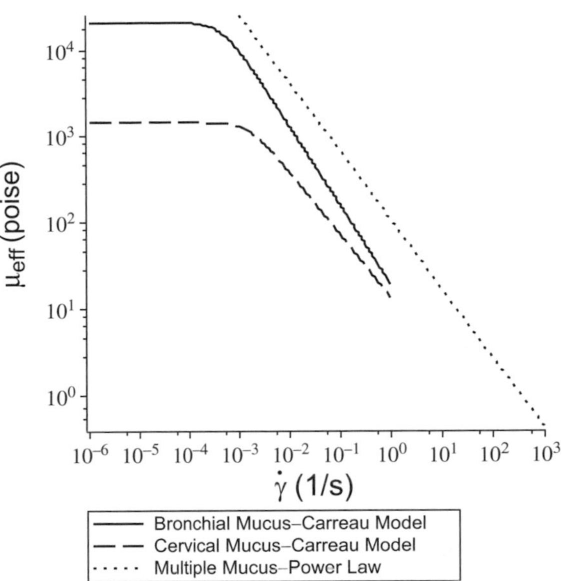

Figure 10.19 Carreau and power-law models for several mucus sources.

Cho and Kensey (1991) fit blood rheological data from several sources (Gelin, 1961; Wells et al., 1961; Rand et al., 1964; Schmid-Schönbein and Wells, 1971; Skalak et al., 1981; Guyton, 1981; Biro, 1982) using several rheological models. For the Carreau-Yasuda model he found $\mu_0 = 0.56$ poise, $\mu_\infty = 0.0345$ poise, $a = 1.25$, $\lambda = 1.902$ s, $n = 0.22$.

The modified Cross model has the form

$$\frac{\mu_{eff}(\dot{\gamma}) - \mu_\infty}{\mu_0 - \mu_\infty} = \frac{1}{\left(1 + (\lambda\dot{\gamma})^m\right)^b} \qquad (10.7.6)$$

The original Cross model had $b = 1$ and $m = 2/3$ (Cross, 1965). Cho and Kensey (1991) also used the modified Cross model to fit the reported blood data and found the parameter values to be $\lambda = 3.736\,\text{s}$, $m = 2.406$, $b = 0.254$, $\mu_0 = 0.56$ poise, $\mu_\infty = 0.0345$ poise. Walburn and Schneck (1976) fit blood rheology data to a power-law fluid model $\mu_{eff} = \mu_0\dot{\gamma}^{n-1}$. For a hematocrit of $H = 0.45$ at 37 °C typical values of the parameters were found to be $\mu_0 = 0.149$ poise·s^{n-1} and $n = 0.775$.

Figure 10.20 shows a comparison of the Carreau–Yasuda, modified Cross and power-law models for blood. The curves are quite similar for $10^0 \le \dot{\gamma} \le 10^3$ but diverge for lower values of $\dot{\gamma}$.

Quemada (1977) developed a viscosity theory for a concentrated suspension of particles from physical arguments and minimization of energy. The result is an effective viscosity

$$\mu_{eff} = \mu_p\left(1 - \frac{1}{2}k\phi\right)^{-2}, \quad k = \frac{k_0 + k_\infty\dot{\gamma}_r^{1/2}}{1 + \dot{\gamma}_r^{1/2}} \qquad (10.7.7)$$

where $\mu_p = 0.012$ poise is the plasma viscosity and $\phi = H$. The function k depends on the relative shear rate, $\dot{\gamma}_r = \dot{\gamma}/\dot{\gamma}_c$. In Quemada (1978) the model was fit to existing blood rheological data. Fitting the measurements in Schmid-Schönbein et al. (1973) for two hematocrit levels of normal blood, the model parameters are: $\phi = H = 0.40$, $k_\infty = 1.84$, $k_0 = 4.65$, $\dot{\gamma}_c = 2.23\ s^{-1}$; and $\phi = H = 0.45$, $k_\infty = 2.07$, $k_0 = 4.33$, $\dot{\gamma}_c = 1.88\ s^{-1}$. For sickle cell anemia the data in Schmid-Schönbein and Wells (1971) fit well with the corresponding model parameters $\phi = H = 0.45$, $\mu_p = 0.014$ poise, $k_\infty = 2.83$, $k_0 = 4.63$, $\dot{\gamma}_c = 4.94\ s^{-1}$. These three cases are shown in Figure 10.21.

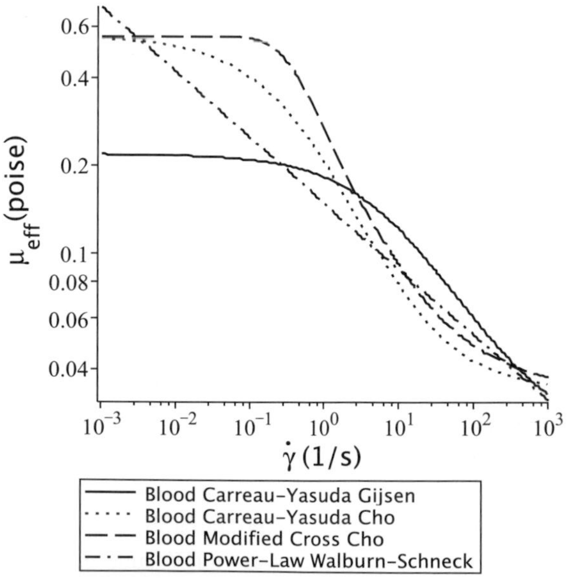

Figure 10.20 Blood viscosity vs shear rate for Carreau–Yasuda, modified Cross and power-law models.

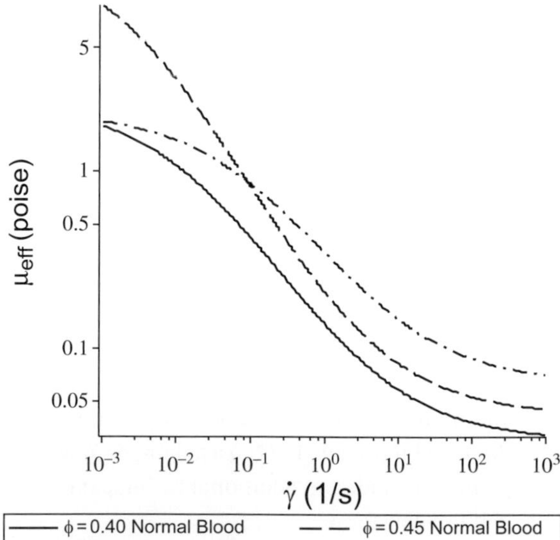

Figure 10.21 Quemada viscosity theory for normal blood at hematocrit $\phi = H = 0.40$ and 0.45 and sickle cell anemia blood at hematocrit $\phi = H = 0.40$.

Compared to mucus, blood is not very viscous, but in small vessels where the shear rate is low, it can cause reduced flows and jeopardize oxygenation or trigger clot formation. This is particularly an issue in sickle cell anemia as Figure 10.21 shows, where the sickle cell blood viscosity is elevated over normal blood for the same hematocrit of 40%. The predictions from the rheological models shown in Figure 10.20 and the normal blood of Figure 10.21 compare favorably to the data shown in Figure 10.13 (a).

10.8 Viscometers and Viscometric Flows

The planar Couette flow shown in Figure 10.1 is an idealized concept that some viscometers try to mimic. As we saw in Figure 10.8 for flow of a power-law fluid in at tube, the strain rate invariant, $\dot{\gamma}$, can vary spatially in the flow field and consequently the viscosity, $\mu(\dot{\gamma})$, also varies. For viscometric flows it is an essential, simplifying feature, that the viscosity has only one value in the deforming fluid. So $\dot{\gamma}$ needs to be uniform. Here are two examples of viscometers that are designed to achieve that goal.

The concentric-cylinder viscometer is shown in Figure 10.22. A fixed outer cylinder of radius R_2 houses a rotating inner cylinder of radius R_1. The gap between them contains the fluid. The gap width is

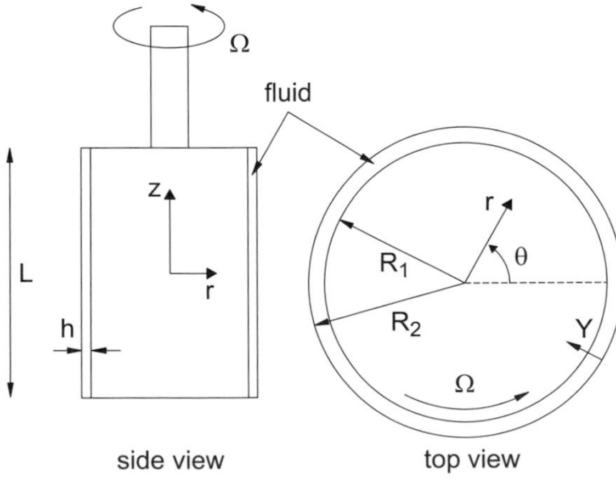

Figure 10.22 Concentric-cylinder viscometer.

$h = R_2 - R_1$ and it is small so that $h/R_2 = \varepsilon << 1$. In this flow, the fluid velocity in cylindrical coordinates is $\underline{u} = (u_r, u_\theta, u_z)$ and both $u_r = 0$ and $u_z = 0$. Flow is only in the azimuthal direction and the no-slip boundary conditions on u_θ are $u_\theta(r = R_1) = \Omega R_1,\ u_\theta(r = R_2) = 0$. The torque, T, required for the rotation is measured by the device. From Eq. 4.6.4 the simplified conservation of mass is

$$\frac{\partial u_\theta}{\partial \theta} = 0 \tag{10.8.1}$$

which tells us that $u_\theta = u_\theta(r)$ only, assuming no variations in z. For the fixed dimensions of the viscometer, we anticipate constant $\dot{\gamma}$ and effective viscosity, $\mu_{eff}(\dot{\gamma})$, everywhere in the flow field for a given rotational frequency, Ω. Under those conditions, it is appropriate to use the Navier–Stokes equations for momentum balance. From Eq. (6.8.1) in cylindrical coordinates the remaining term in the azimuthal direction

$$\frac{1}{r}\frac{d}{dr}\left(r\frac{du_\theta}{dr}\right) - \frac{u_\theta}{r^2} = 0 \tag{10.8.2}$$

To solve Eq. (10.8.2) in this flow it is convenient to make it dimensionless so we can justify further simplifications from the specific geometry of the device. Choose the scales

$$U_\theta = \frac{u_\theta}{\Omega R_1} \quad Y = \frac{R_2 - r}{h} \tag{10.8.3}$$

The new spatial variable, Y, is shown in Figure 10.22 and has the opposite directionality of r. It has the values $Y = 0$ on the outer cylinder and $Y = 1$ on the inner cylinder so $0 \leq Y \leq 1$, which is a good sign we have the right scaling. From the definitions in Eq. (10.8.3) we can use the chain rule of differentiation $d/dr = (dY/dr)d/dY = -(1/h)d/dY$ and also the fact that $r = R_2(1 - \varepsilon Y)$. Inserting these into Eq. (10.8.2) and rearranging yields the dimensionless form

$$(1 - \varepsilon Y)\frac{d^2U_\theta}{dY^2} - \varepsilon\frac{dU_\theta}{dY} - \varepsilon^2\frac{U_\theta}{(1 - \varepsilon Y)} = 0 \tag{10.8.4}$$

In the limit of $\varepsilon << 1$ Eq. (10.8.4) we can ignore terms multiplied by ε and ε^2 to arrive at the simplified equation

$$\frac{d^2U_\theta}{dY^2} = 0 \tag{10.8.5}$$

whose general solution is $U_\theta = c_1 Y + c_2$. The no-slip boundary conditions in dimensionless form are $U_\theta(Y = 0) = 0,\ U_\theta(Y = 1) = 1$, which solves for c_1 and c_2. The resulting solution in dimensionless, and dimensional, form is

$$U_\theta = Y \quad \rightarrow \quad u_\theta = \Omega R_1 \left(\frac{R_2 - r}{R_2 - R_1} \right) \tag{10.8.6}$$

The only term in the rate of strain tensor in cylindrical coordinates for this flow is given by Eq. (6.4.2), which is simplified due to the unidirectional velocity field

$$E_{r\theta} = E_{\theta r} = \frac{1}{2} \left(\frac{du_\theta}{dr} - \frac{u_\theta}{r} \right) \tag{10.8.7}$$

Expressing Eq. (10.8.7) using the scaled variables further simplifies it in the limit of $\varepsilon << 1$

$$E_{r\theta} = E_{\theta r} = \frac{1 \Omega R_1}{2h} \left(-\frac{dU_\theta}{dY} - \varepsilon \frac{U_\theta}{(1 - \varepsilon Y)} \right) \sim -\frac{1 \Omega R_1}{2h} \left(\frac{dU_\theta}{dY} \right) = -\frac{1 \Omega R_1}{2h} \tag{10.8.8}$$

From the definition of $\dot{\gamma}$ in Eq. (10.1.4) represented in cylindrical coordinates we have, $\dot{\gamma} = \sqrt{2E_{ij}E_{ij}} = \sqrt{2(E_{\theta r}^2 + E_{r\theta}^2)} = \Omega R_1/h$. Since $\dot{\gamma}$ is a constant throughout the fluid, then so is the viscosity $\mu(\dot{\gamma})$ for a given angular frequency, Ω. This flow recreates the planar Couette flow where $\dot{\gamma} = V_0/h$ in Eq. (10.1.7) with $V_0 = \Omega R_1$ and h the same gap width. The shear stress everywhere in the fluid is given by Eq. (6.4.3): $\tau_{r\theta} = \tau_{\theta r} = 2\mu_{eff} E_{r\theta} = -\mu_{eff} \Omega R_1/h$. The torque, T, must balance the integral of the moment force per unit area from the fluid, $R_1 \tau_{r\theta}$. Integrated over the surface of the inner cylinder give the magnitude of T as

$$T = \int_{-L/2}^{L/2} \int_0^{2\pi} (R_1 |\tau_{r\theta}|) R_1 \, d\theta \, dz = \frac{2\pi\mu_{eff} \Omega L R_1^3}{h} \tag{10.8.9}$$

The viscometer measures T and uses Eq. (10.8.9) to register a viscosity, $\mu_{eff} = hT/2\pi\Omega L R_1^3$.

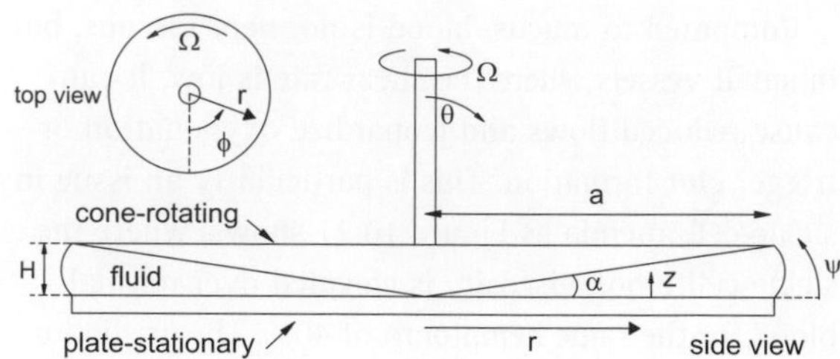

Figure 10.23 Cone-plate viscometer.

The cone-plate viscometer is a popular instrument for measuring rheological properties of fluids. It consists of a fixed flat plate and an inverted cone which contacts the plate at its apex, see Figure 10.23. The fluid is placed in the narrow space between the cone and plate. The cone is rotated at angular velocity Ω in the ϕ-direction which is part of the spherical coordinate system (r, θ, ϕ) centered at the cone apex. The torque, T, required for the rotation is measured. The angle of the cone is α which is assumed to be very small, $\alpha << 1$. Since the angle, θ, is close to $\pi/2$, it is convenient to define the angle $\psi = \pi/2 - \theta$. Then $0 \le \psi \le \alpha$ and $\psi << 1$ as well. The vertical variable z is related to the spherical coordinates by $z = r \tan\psi \sim r\psi$ for $\psi << 1$. So the cone surface is at $z = h = r \tan\alpha \sim r\alpha$ and its maximum is at $r = a$ where $h(r = a) = a\alpha = H$ as shown.

The resulting flow is unidirectional in the ϕ-direction, u_ϕ, so $u_r = 0$ and $u_\theta = 0$. The conservation of mass equation for steady incompressible flow in spherical coordinates, Eq. (4.6.9), simplifies to

$$\frac{\partial u_\phi}{\partial \phi} = 0 \tag{10.8.10}$$

Equation (10.8.10) tells us that $u_\phi = u_\phi(r, \theta)$. As with the concentric-cylinder flow, we use the Navier–Stokes equations anticipating uniform μ_{eff}. The ϕ-component of the spherical form of the Navier–Stokes equations, from Eq. (6.8.2), simplifies to

$$\frac{1}{r^2} \frac{\partial}{\partial r} \left(r^2 \frac{\partial u_\phi}{\partial r} \right) + \frac{1}{r^2 \sin\theta} \frac{\partial}{\partial \theta} \left(\sin\theta \frac{\partial u_\phi}{\partial \theta} \right) - \frac{u_\phi}{r^2 \sin^2\theta} = 0 \tag{10.8.11}$$

Using our definition of ψ we can replace $\sin\theta = \cos\psi \sim 1$ in Eq. (10.8.11). Then multiply through by r^2 to find

$$\frac{\partial}{\partial r}\left(r^2\frac{\partial u_\phi}{\partial r}\right)+\frac{\partial^2 u_\phi}{\partial\theta^2}-u_\phi=0 \tag{10.8.12}$$

The boundary conditions on Eq. (10.8.12) are no-slip at the plate and at the cone surfaces, i.e. $u_\phi(z=0)=0,\ \ u_\phi(z=h=r\alpha)=\Omega r$.

To solve Eq. (10.8.12) it will help to, again, use dimensionless variables. Let $U_\phi = u_\phi/a\Omega$ and then define dimensionless radial and vertical variables as $R=r/a, Z=z/H=R\psi/\alpha$. This choice of scales keeps the values of R bounded by $0\le R\le 1$ as well as the values of Z, $0\le Z\le 1$. With the chain rule we change from θ to Z derivative using $\partial/\partial\theta=(\partial Z/\partial\theta)\partial/\partial Z=-(R/\alpha)\partial/\partial Z$. The dimensionless form of Eq. (10.8.12) becomes

$$\frac{\partial^2 U_\phi}{\partial Z^2}+\frac{\alpha^2}{R^2}\left(\frac{\partial}{\partial R}\left(R^2\frac{\partial U_\phi}{\partial R}\right)-U_\phi\right)=0 \tag{10.8.13}$$

For $\alpha<<1$ Eq. (10.8.13) simplifies to

$$\frac{\partial^2 U_\phi}{\partial Z^2}=0 \tag{10.8.14}$$

whose solution is the linear function $U_\phi=c_1Z+c_2$. The dimensionless no-slip boundary conditions are $U_\phi(Z=0)=0$, $U_\phi(Z=R)=R$, yielding the final solution form in dimensionless, and dimensional, forms

$$U_\phi=Z\quad\rightarrow\quad u_\phi=\Omega r\left(\frac{(\pi/2)-\theta}{\alpha}\right) \tag{10.8.15}$$

The strain rate tensor component for this flow is given in Eq. (6.4.6) as

$$\begin{aligned}E_{\theta\phi}=E_{\phi\theta}&=\frac{1}{2}\left(\frac{\sin\theta}{r}\frac{\partial}{\partial\theta}\left(\frac{u_\phi}{\sin\theta}\right)+\frac{1}{r\sin\theta}\frac{\partial u_\theta}{\partial\phi}\right)\\&=\frac{1}{2}\left(\frac{1}{r}\frac{\partial u_\phi}{\partial\theta}\right)=-\frac{1}{2}\frac{\Omega}{\alpha}\end{aligned} \tag{10.8.16}$$

since $u_\theta=0$, and $\sin\theta\sim 1$. Then the strain rate invariant is

$$\dot\gamma=\sqrt{2\left(E_{\theta\phi}^2+E_{\phi\theta}^2\right)}=\frac{\Omega}{\alpha} \tag{10.8.17}$$

which is consistent with our assumption that $\dot\gamma$ is constant in the fluid for a given set of parameters α, Ω. This form of $\dot\gamma$ is identical to the plane Couette flow of Eq. (10.1.7), $\dot\gamma=V_0/h$ with upper plate velocity,$V_0=\Omega r$, and the gap width, $h=\alpha r$. The design of the cone-plate viscometer keeps V_0/h constant since they both are linearly proportional to r. Because $\dot\gamma$ is constant, the viscosity $\mu_{eff}(\dot\gamma)$ is also constant in the fluid which justifies our using the Navier–Stokes equations. The shear stress, given in Eq. (6.4.7), anywhere in the fluid is also constant

$$\tau_{\theta\phi}=2\mu_{eff}E_{\theta\phi}=-\mu_{eff}\frac{\Omega}{\alpha} \tag{10.8.18}$$

The torque, T, must balance the integral of the moment force per unit area from the fluid, $r\tau_{\theta\phi}$. The magnitude of T, then, is

$$T=\int_0^{2\pi}\int_0^a\left(r\left|\tau_{\theta\phi}\right|\right)r\,dr\,d\phi=\frac{2}{3}\pi\mu a^3\frac{\Omega}{\alpha} \tag{10.8.19}$$

Since T is measured by the viscometer, the viscosity, μ_{eff}, can be calculated from Eq. (10.8.19), $\mu_{eff}=3T\alpha/2\pi a^3\Omega$.

There are many other kinds of viscometers including falling ball, U-tube and rectangular slit with their own set of features and limitations. For generalized Newtonian fluids in either the concentric cylinder or cone-plate viscometers, we expect $\mu_{eff}(\dot\gamma)$ to change with different values of the rotational frequency, Ω, i.e. different values of $\dot\gamma$. It is typical for modern viscometers to have an automated feature so that the user selects a range of $\dot\gamma$ and chooses how many values of $\dot\gamma$ will be used within that range. The device translates that request into selections of Ω. The machine sets the first rotational frequency, Ω_1, waits for a steady state, and then reports $\mu_{eff}(\dot\gamma_1)$. Then it changes to the next frequency, Ω_2, waits for a steady state, and reports that viscosity $\mu_{eff}(\dot\gamma_2)$, and so on through the entire range. Of course there can be problems in any of the devices including edge or end effects and fluids with a yield stress or poorly mixed for which there are additional features in the device to sort things out.

Summary

Rheology of non-Newtonian fluids is a very complicated subject, so in this chapter we have taken some simple first steps and then built upon them. Initially we discussed the importance of the strain rate invariant, $\dot{\gamma}$, in the definition of a generalized Newtonian fluid whose extra stress–strain rate relationship looks just like a Newtonian fluid, except the viscosity depends on $\dot{\gamma}$, $\tau_{ij} = 2\mu_{eff}(\dot{\gamma})E_{ij}$. The various forms of rheological behavior such as yield stress, shear thinning and shear thickening were explored both qualitatively and with specific flow examples. Several fluid models were examined in simple tube flows including power-law, Herschel–Bulkley and Casson. Additional rheological models were also presented including Carreau, Carreau–Yasuda, modified Cross and Quemada. They have advantages of finite effective viscosities in the limits of zero or infinite shear rates. However, analytical solutions to flow problems using those models are usually not possible and one must use computational fluid dynamics (CFD). Some commercial CFD software packages provide a menu of rheological models from which the user can select for any flow being simulated. Rheological data from blood, mucus and ketchup were presented in the contexts of the various models used to describe their behavior. The rheology of blood and mucus plays a role in a number of diseases such as asthma, emphysema, cystic fibrosis and sickle cell anemia which were also discussed.

Problems

10.1

a. For flow of a power-law fluid in a channel, the x-component of momentum conservation in the Cauchy momentum equation $d\tau/dy = dp/dx$ where $\tau = \tau_{xy} = \tau_{12}$. Assume $dp/dx < 0$ is a constant negative pressure gradient so the fluid is flowing in the +x-direction. Integrate this equation and include the constant of integration.

b. The component of the constitutive equation we want is $\tau = \mu_{eff}\, du/dy$. Insert your solution for τ from part (a) into this equation and solve for your constant of integration by imposing the symmetry condition at the centerline, $du_1/dy = 0$ at $y = 0$.

c. Let u in the upper half of the channel, $0 \leq y \leq b$, be u_1. We expect $\dot{\gamma} = -du_1/dy > 0$ in this region, so insert $\mu_{eff} = m\dot{\gamma}^{n-1}$ into your result of part (b) and multiply through by −1 so all forms of the velocity derivative have the form $(-du_1/dy)$. Isolate du_1/dy on one side and recall $(-dp/dx) > 0$. Solve for u_1 by integrating and impose the no-slip boundary condition $u_1(y = b) = 0$.

d. Repeat your analysis for the lower half of the channel, $-b \leq y \leq 0$, where $u = u_2$ and $\dot{\gamma} = du_2/dy > 0$. It will help to let $y(dp/dx) = (-y)(-dp/dx)$ so both terms in parentheses are non-negative. Impose no-slip at the lower wall.

e. Make your two solutions dimensionless scaling y on b, so that $Y = y/b$, and u on $u_{avg} = \frac{1}{b}\int_0^b u_1\, dy$ so $U_1 = u_1/u_{avg}$, $U_2 = u_2/u_{avg}$. Now plot the two dimensionless velocity solutions vs Y, in their respective ranges of Y, for n = 0.5, 1.0, 1.5. Discuss your results.

f. Plot the dimensionless effective viscosity for the upper half of the channel, $0 \leq Y \leq 1$. Define it as $\tilde{\mu}_{eff} = \mu_{eff} / \left[m(-(bdp/dx)/m)^{\frac{n-1}{n}} \right]$ and let $n = 0.5, 1.0, 1.5$. For $n = 0.5$ try $0.2 \leq Y \leq 1$. Describe your results.

10.2 The solution u_z for fully developed flow of a power-law fluid in a pipe is given in Eq. (10.3.6), where a is the tube radius. Consider the tube is flexible, so that $a = a(z)$ and dp/dz depends on z.

a. Calculate the flow rate, q, for any z-position.
b. From your result in part (a), enforce that q is a constant, $q = \pi a_0^2 u_s$. Solve for dp/dz in terms of $a(z)$, m, n and u_s.
c. The stress boundary condition for a linearly elastic tube from Eq. (9.1.4) gives us $dp/dz = k\, da/dz$. Insert this on the left hand side of your answer to part (b) and integrate to solve for $a(z)$. The limits of the z integral are 0 to z and the corresponding $a(z)$ integral are a_0 to $a(z)$.
d. Scale your answer in part (c) so that $B(Z) = a(z)/a_0$ and $Z = z/L$. The entrance radius at $z = 0$ is a_0. Show that the solution has the form $B(Z) = (1 - \beta_n Z)^{\frac{1}{3n+2}}$, where

$$\beta_n = 2\gamma_n(3n+2)\left(\frac{3n+1}{n}\right)^n, \quad \gamma_n = \frac{m u_s^n / \varepsilon a_0^n}{k a_0}$$

and $\varepsilon = a_0/L$. Note that for $n = 1$, the dimensionless parameter $\gamma_n = \gamma$ from Eq. (9.1.12). What is its physical interpretation?
e. Set $\gamma_n = 0.01$, letting m vary with n, and plot $B(Z)$ for $n = 0.5, 1.0, 1.5$. Describe your results.
f. Now make the velocity dimensionless as $U_Z = u_z/u_s$, where the velocity scale is the average entrance speed $u_s = q/\pi a_0^2$. Let $R = r/a_0$. Hint: you will need to substitute your solution for dp/dz from part (b). For the values of n in part (e), plot $U_Z(R)$ for $Z = 0$ and $0 \leq R \leq B$ your results for $\gamma_n = 0.01$. Repeat for $Z = 1$. Discuss your results.

10.3

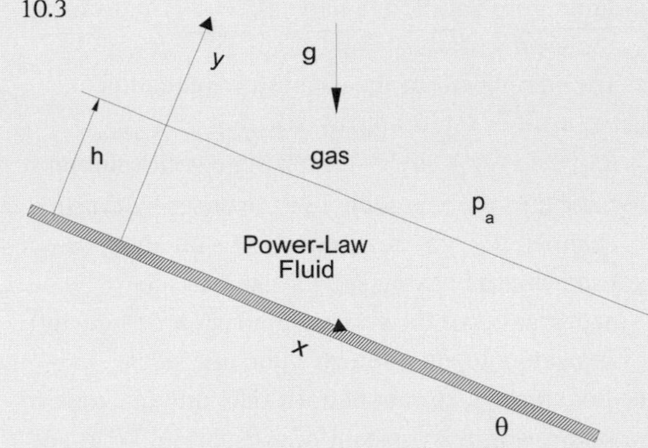

The liquid lining of the lung can behave like a power-law fluid. Clearance due to gravity is represented in the figure for flow of film of thickness h flowing down an inclined plane.

a. The momentum balance component we want is $d\tau/dy + \rho g \sin\theta = 0$ where $\tau = \tau_{xy} = \tau_{12}$. Integrate this equation and impose the boundary condition of no shear at the air–liquid interface, $\tau(y = h) = 0$, to solve your constant of integration.
b. The constitutive equation component we want is $\tau = \mu_{eff} du/dy$ where $\mu_{eff} = m\dot{\gamma}^{n-1}$. Substitute your solution for τ from part (a) and let $\dot{\gamma} = du/dy \geq 0$. Solve for u and impose the no-slip boundary condition $u(y = 0) = 0$ to solve for your constant of integration.
c. Let $\rho = 1\ \text{g/cm}^3$, $g = 980\ \text{cm/s}^2$, $\theta = \pi/4$, $h = 0.02\ \text{cm}$. Then for the effective viscosity set $\mu_{eff} = m\dot{\gamma}^{n-1} = 20\dot{\gamma}^{n-1}$ poise so $m = 20$. Plot $u(y)$ for $n = 0.2, 0.3, 0.4$ and describe your results.
d. Repeat part (c) with $h = 0.03\ \text{cm}$. What has happened and why?

10.4 Consider a Herschel–Bulkley fluid in a channel driven by a constant axial pressure gradient $dp/dx < 0$. The channel walls are at $y = \pm b$.

a. The momentum balance is $d\tau/dy - dp/dx = 0$. Integrate to find τ and impose the symmetry/no shear condition on the centerline that $\tau(y = 0) = 0$. That will determine the constant of integration.

b. From your solution in part (a), set $\tau = \tau_y$ to find the yield surface, $y = y_y$.
c. The component of the constitutive equation we want is $\tau = \mu_{eff} du/dy$ where $\mu_{eff} = \left(\tau_y/\dot{\gamma} + m\dot{\gamma}^{n-1}\right)$. Substitute your solution for τ into this equation. For the upper half of the channel, $0 \le y \le b$, assume $\dot{\gamma} = -du/dy \ge 0$. Hint: multiply through by −1 so that all appearances of the velocity gradient are $(-du/dy)$. Rearrange to get $(-du/dy)^n$ on one side.
d. Take the 1/n-th root of both sides and integrate to find u within a constant of integration. What are the limits on y for real solutions? Keep in mind $-dp/dx > 0$. To solve for the integration constant, apply the no-slip condition $u(y = b) = 0$.
e. Now make your solution dimensionless using the scales $Y = y/b$, $U = u/b((b/m)(-dp/dx))^{1/n}$. Define the Bingham number as $Bn = \tau_y/(-bdp/dx)$. Where are the unyielded and yielded regions in terms of Y and Bn? What is the velocity in the unyielded region?
f. Plot your solution for the upper half of the channel, $0 \le Y \le 1$ for $n = 0.5, Bn = 0.0, 0.3, 0.5, 0.7$. Repeat for $n = 1.0$, $Bn = 0.0, 0.3, 0.5, 0.7$ and $n = 1.5$, $Bn = 0.0, 0.3, 0.5, 0.7$. Describe your results in terms of the effects of n and Bn. At what value of Bn is there no flow?
g. Calculate the total dimensionless flow rate, Q, by integrating U with respect to Y over the yielded and unyielded regions and adding them together. Plot Q over the range $0 \le Bn \le 1$ for $n = 0.5$, 1.0, 1.5. Discuss your result.

10.5

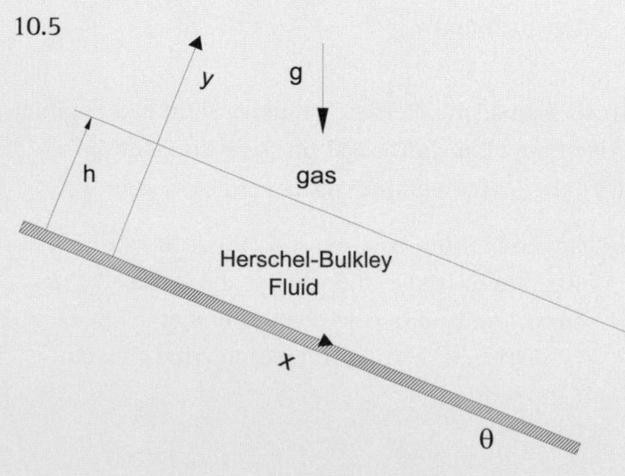

Gravity drainage and ciliary wall motion were discussed in Section 7.3 for transport of the lung's liquid lining. Let's consider it to be a single layer, with thickness h, of a Herschel–Bulkley fluid on a stationary inclined plane, as in the figure. Harnessing gravity-driven drainage is a key ingredient of chest physical therapy for cystic fibrosis, for example.

a. The x-component of momentum balance is $d\tau/dy + \rho g \sin\theta = 0$ where $\tau = \tau_{12} = \tau_{xy}$ is the shear stress. Integrate this expression to find τ. Enforce the boundary condition of no shear at the air–liquid interface, $\tau(y = h) = 0$, to solve the constant of integration.
b. Using your solution in part (a), set $\tau = \tau_y$, the yield stress, at $y = y_y$, the yield value of y. Express the dimensionless yield surface, $Y_y = y_y/h$, in terms of the Bingham number, $Bn = \tau_y/(\rho g h \sin\theta)$. What are the limits of Bn for flow?
c. From Eq. (10.1.9) the relevant component is $\tau = \mu_{eff}(du/dy)$. Use the definition of μ_{eff} in Eq. (10.4.2) with $\dot{\gamma} = du/dy > 0$ and solve for u in the yield region. Apply the no-slip boundary condition $u(y = 0) = 0$ to solve for the integration constant.
d. Make your solution dimensionless from part (c) using the scaled variables $Y = y/h$, $U = u/\left(h(\rho g h \sin\theta/m)^{1/n}\right)$. What is the range of Y for yield flow in terms of Bn? What about the unyielded range of Y?
e. What is the velocity U in the unyielded region?
f. Plot U across the entire layer for $n = 0.3, Bn = 0.3$, 0.5, 0.7. Repeat for $n = 0.7, Bn = 0.3, 0.5, 0.7$. Describe your results.
g. Calculate the dimensionless flow rate in both layers by integrating U with respect to Y in the appropriate regions. Plot the total flow rate, Q, over the range $0 \le Bn \le 1$ for $n = 0.3, 0.5, 0.7$. Describe your result. What happens to gravity clearance mechanisms when diseases increase the yield stress?
h. For $\rho = 1\,gm/cm^3, g = 980\ cm/s^2$, $h = 0.01\,cm$, $\theta = \pi/4$, $Bn = 0.7$ what is the yield stress? Is this physiologic?

10.6 Blood flow in vessels has an interesting feature that there is a cell-free layer near the vessel wall, i.e. there are no RBCs in this layer so it is Newtonian. Assume that the tube radius is *a* and that the thickness of this layer is a − b, as shown in the figure.

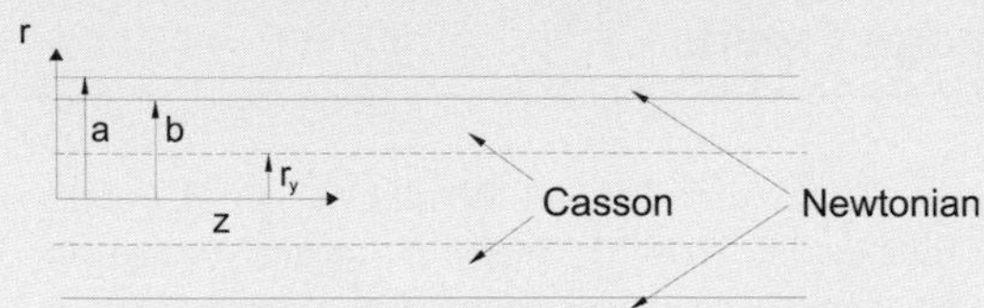

Let the constant pressure gradient be, $dp/dz < 0$ for both fluids. The Casson model velocity is given in Eq. (10.6.7) with an unknown constant of integration, c_2. It applies to the region $r_y \le r \le b$ in the figure. Call it $u_{z2}(r)$. The region around the centerline $0 \le r \le r_y$, where the shear stress is below the yield stress $\tau < \tau_y$, has the uniform velocity that matches the Casson velocity at $r = r_y$. Call it $u_{z1} = u_{z2}(r = r_y)$. The velocity for a Newtonian model in the cell-free layer, $b \le r \le a$, is given in Eq. (7.1.2). Call it $u_{z3}(r)$ and, to avoid confusion, let its two constants of integration be b_1, b_2. The constant viscosity in the Newtonian layer is μ. For the three unknown coefficients, c_2, b_1, b_2, there are three boundary condition equations: EQ1, EQ2, EQ3. Your computational software will be useful here.

a. Define the no-slip boundary condition on the Newtonian layer at the tube wall. Call this EQ1.
b. The fluid velocity must be continuous at the fluid–fluid interface. Call this EQ2.
c. The fluid shear stress must be continuous at the fluid–fluid interface. Use the original expression for the Casson shear stress, $\tau_c(r = b) = \left(\tau_y + 2\sqrt{\mu_\infty \tau_y}\,\dot{\gamma}_2^{1/2} + \mu_\infty \dot{\gamma}_2\right)_{r=b}$ where $\dot{\gamma}_2 = (-du_{z2}/dr)_{r=b}$. For the Newtonian side the shear stress is $\tau_n = \mu\dot{\gamma}_3$ $\dot{\gamma}_3 = (-du_{z3}/dr)_{r=b}$. Call this EQ3.
d. Let $dp/dz = -2$ dynes/cm^3, $\mu = 0.01$ poise, $\mu_\infty = 0.04$ poise, $a = 0.1$ cm, $b = 0.08$ cm. The cell-free layer is not this thick but this allows a better view of the velocities. Solve ODE1, ODE2, ODE3 for c_2, b_1, b_2.
e. Using your solutions in part (d), plot the velocity field for the entire range of $0 \le r \le a$ which includes the three regions $0 \le r \le r_y$ (solid plug), $r_y \le r \le b$ (Casson fluid), $b \le r \le a$ (Newtonian fluid) for $\tau_y = 0.01, 0.03, 0.06$ dyn/cm^2. Show all three cases on the same graph.
f. Repeat part (d) with $b = 0.095$ cm.
g. Discuss your results from part (e) and (f). How is the wall shear affected by the choice of τ_y?

10.7

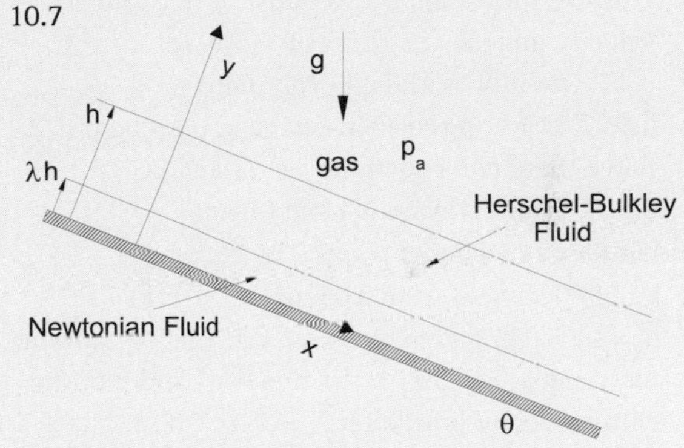

Consider a two-layer fluid system with total thickness, h, on an inclined plane. There is a Newtonian fluid next to the wall $0 \le y \le \lambda h$ and a Herschel–Bulkley fluid above, $\lambda h \le y \le h$.

a. The Newtonian layer is like the periciliary, serous layer. It satisfies the momentum balance $\mu d^2u_N/dy^2 + \rho g \sin\theta = 0$. Integrate twice to find u_N and solve one of your two constants of integration by setting $u_N(y = 0) = 0$. Call the remaining integration constant, b_1.
b. The Herschel–Bulkley layer is like the mucus layer and satisfies conservation of momentum, $d\tau_{HB}/dy + \rho g \sin\theta = 0$. Integrate this expression once and solve for the constant of integration by setting the shear stress to zero at the air–liquid interface, $\tau_{HB}(y = h) = 0$.
c. Your expression for τ_{HB} in part (b) can be used to find an expression for the yield value of y. Set $\tau_{HB} = \tau_y$, the yield stress, and solve for $y = y_y$. Defining the Bingham number as

$Bn = \tau_y/(\rho g h \sin\theta)$, what is the dimensionless version y_y/h in terms of Bn?

d. For the mucus layer, the component of the constitutive equation we want is $\tau_{HB} = \mu_{eff}\, du_{HB}/dy$ where $\mu_{eff} = \tau_y/\dot{\gamma} + m\dot{\gamma}^{n-1}$ You can assume $\dot{\gamma} = du_{HB}/dy > 0$. Insert your solution of τ_{HB} in part (b) and solve for u_{HB} by integrating. There should be one constant of integration, call it b_2. What is the constraint on the range of y/h for real solutions?

e. To solve for b_1 and b_2, we must have continuous velocity and shear stress at $y = \lambda h$. Let $u_{HB} = u_N$ at $y = \lambda h$ be EQN1 and $\mu\, du_N/dy = -\rho g \sin\theta(h - y)$ at $y = \lambda h$ be EQN2. Solve these two equations, EQN1 and EQN2, for b_1, b_2. Your software is handy here.

f. For baseline data let $\rho = 1.0\ \text{g/cm}^3$, $g = 980\ \text{cm/s}^2$, $\mu = 0.05$ poise, $\theta = \pi/4$, $h = 0.05$ cm and $m\dot{\gamma}^{n-1} = 5\dot{\gamma}^{0.35-1}$ poise. Now set the yield stress to $\tau_y = 10\ \text{dyn/cm}^2$ and plot the entire velocity profile for $\lambda = 0.02, 0.1, 0.2$. Describe your results. What is the effect of λ? Are there similarities to the shapes?

g. Repeat part (f) for a larger yield stress, $\tau_y = 20\ \text{dyn/cm}^2$. Compare your plots to the results in part (f). What has happened to the flow? Are the effects dependent on λ?

Image Credits

Figure 10.11 Source: Fig 6 of Merrill EW, Shin H, Cokelet G, Gilliland ER, Wells RE, Britten A. Rheology of human blood, near and at zero flow-effects of temperature and hematocrit level. *Biophysical Journal.* 1963; 3(3):199–213.

Figure 10.13a Source: Fig 3 of Chien S, Usami S, Taylor HM, Lundberg JL, Gregersen MI. Effects of hematocrit and plasma proteins on human blood rheology at low shear rates. *Journal of Applied Physiology.* 1966; 21(1):81–7.

Figure 10.17 Source: Bettmann/ Contributor/ Getty

Figure 10.18 Author original adapted from Fig 1 of Morris CL, Rucknagel DL, Joiner CH. Deoxygenation-induced changes in sickle cell-sickle cell adhesion. *Blood.* 1993;81(11):3138–45.

Figure 10.19 Author original adapted from Fig 2 of Velez-Cordero JR, Lauga E. Waving transport and propulsion in a generalized Newtonian fluid. *Journal of Non-Newtonian Fluid Mechanics*, Volume 199, September 2013, pages 37–50.

11 Lubrication Theory

Introduction

The flow of lubricating fluid between surfaces that slide over one another is of vital interest to both industry and biomedicine. For industry the goal is usually to reduce friction and extend the life of moving parts by cutting down on wear. In biofluids there are similar goals with additional features that affect normal physiology and pathophysiology. Red blood cell flow in capillaries and synovial fluid flow between joints are important applications. For any of these applications, the flow analysis is simplified when the characteristic gap distance, h, between the surfaces is small compared to their characteristic length, L, such that $h/L << 1$. Physically, this small ratio implies that locally the flow is nearly parallel to the surfaces which form the gap, allowing that the distance between the surfaces may change slowly in the lengthwise/streamwise direction. We have seen these circumstances in previous chapters. For example, in Section 9.1 we examined steady flow in a flexible tube where the ratio of the entrance radius, a_0, was small compared to the tube length, L, i.e. $\varepsilon = a_0/L << 1$. We made the governing equations dimensionless using a_0 to scale the radial variable, r, and L to scale the axial variable, z. The small parameter ε appears in front of several terms, Eqs. (9.1.7), and then we neglected terms multiplied by ε, ε^2, ε^3 leading to locally parabolic flow. Also, in Section 10.8 we studied viscometric flow fields which became simplified under conditions that there are two length scales in the problem whose ratio was small. For the cone-plate viscometer that ratio was the largest gap width, $H = a \tan\alpha$, compared to the largest radius, a. Their ratio is $H/a = \tan\alpha \sim \alpha$ for $\alpha << 1$ and we were able to neglect terms in the Navier–Stokes equations that were multiplied by α or powers of α, while retaining terms that were of $O(1)$. In the absence of a pressure gradient, this led to locally rectilinear flow. In general, the idea is that the viscous term involving the second spatial derivative of the lengthwise velocity in the spanwise variable is the largest, and must balance the pressure gradient in the lengthwise direction. That is the gist of a lubrication theory which we will explore more in this chapter.

Topics Covered

We start with a very traditional problem, the slide block, to bring together the basic ideas of lubrication theory and its important ramifications. Care is taken to present the problem formulation using dimensionless parameters, the pressure distribution and forces on the block, and the velocity field and streamlines. From this analysis we make our own model of a red blood cell flowing through a capillary whose diameter is similar to the RBC diameter so there is a tight fit with a thin lubrication layer between the RBC and the capillary wall. Though a somewhat primitive approach, it allows the student to calculate important features that would otherwise require numerical methods beyond the scope of this book. In addition to the problem formulation, it includes the force balance on the RBC and the resulting pressure drop across it, and the study of "leakback" flow which comes from the unexpected observation/calculation that the RBC travels faster than the average fluid speed in a capillary. A corollary to this analysis is the passage of a urinary stone through the urethra, which is a painful experience.

Then we build on those results to model the flow of a series of red blood cells in a capillary. As part of the analysis, we explore well-known phenomena in blood rheology related to the microcirculation: the Fåhræus–Lindqvist effect which is the reduction of apparent blood viscosity in tube flow as decreases in the tube diameter fall below ~ 300 μm; the inverse Fåhræus–Lindqvist effect where the apparent viscosity increases as the decreasing tube diameter falls below ~ 7 μm; and the Fåhræus effect where the hematocrit of flowing blood can be lower than its exit value into larger vessels. One of the explanations of the Fåhræus–Lindqvist effect is that there is a cell-free layer of serum near the tube wall, and that feature we examine using the results of our analysis in Section 7.4 of core-annular flow where the two regions have different viscosities. The annulus of serum has a viscosity about 1/4th of the core whole blood viscosity. We continue to explore capillary flow by introducing permeability for the walls which permit radial filtration of fluid out of, and into, the capillary. The balance of transmural hydrostatic and oncotic pressure differences is given by Starling's equation. This phenomenon is responsible for important physiological functions of normal tissues and specialized organ functions like the kidneys. It is also a source of disease symptoms and signs like kwashiorkor, which is found in severe malnutrition. Filtration flows couple with hydrostatics to create steady and transient flows in the pleural fluid, bounded by the lung and the inside surface of the thorax. Finally, shifting from the vasculature, we examine squeeze film flows such as those that occur in joints. When the two bones of a joint are pushed toward one another due to weight-bearing, the thin layer of synovial fluid between them is squeezed out laterally, supporting and distributing the weight while protecting the bones' cartilaginous surfaces from direct contact and wear. With a spirit of fancy, we also treat this flow as the closing jaws of a sea creature in a predator–prey scenario. The prey is taken to be a solid particle with Stokes drag in the squeeze film flow field, and criteria and strategies for its capture are explored. Finally, using our knowledge from Chapter 10, we repeat the squeeze-film analysis for a non-Newtonian, power-law fluid model to study the shear thinning properties of synovial fluid and the resulting effective viscosity field, $\mu_{eff}(x, y, t)$.

11.1 The Slide Block

Suppose you take a playing card, the seven of diamonds, and throw it along its flat side skimming the tabletop. With the leading edge slightly farther from the table than the trailing edge, the card is at a small angle with the table. Thrown at the right speed, the card appears to ride on a cushion of air and travel a long distance. Card dealers are good at this. If the angle or speed is too large, the card will flip onto its back stopping the motion and revealing your hand. If the angle is negative, i.e. the trailing edge farther from the wall than the leading edge, the flow above the card forces it toward the wall and motion grinds to a halt from frictional contact with the table. This problem is shown in see Figure 11.1, in what is traditionally called the "slide block problem." The slide block is moving to the left near a wall and at a small angle. Changing coordinate systems to be attached to the block, the wall moves to the right at speed U_w, see Figure 11.1. The block has length L, front gap width d_1 and a smaller rear gap width d_2. The position of the block surface is the straight line $y = h(x)$ where

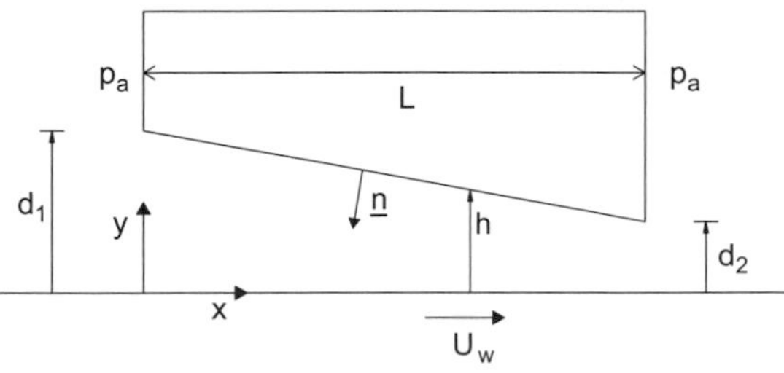

Figure 11.1 Slide block problem.

$$h(x) = d_1 - \frac{(d_1 - d_2)}{L}x \qquad (11.1.1)$$

The slope magnitude is $\varepsilon = (d_1 - d_2)/L$ and we assume that $\varepsilon << 1$. The fluid pressures at the entrance and exit, $x = 0, L$, are both p_a, so there is no applied pressure gradient to the system, though one develops in the gap between the block and wall due to the slope of the block.

We formulated a dimensionless version of the Navier–Stokes equations in Chapter 6, Eq. (6.6.8), for flow in a two-dimensional channel with L as x scale and b as the y scale. The ratio of these scales was defined as $\varepsilon = b/L$. For the slide block problem, $(d_1 - d_2)$ can be substituted for b. Equation (6.6.8) simplifies for $\varepsilon << 1$ and $\varepsilon Re << 1$, where Re is the Reynolds number, $Re = \rho U_w(d_1 - d_2)/\mu$. If we ignore terms multiplied by ε or εRe, the result is Eq. (6.6.9), which has the dimensional form

$$-\frac{\partial p}{\partial x}+\mu\frac{\partial^2 u}{\partial y^2}=0$$
$$\frac{\partial p}{\partial y}=0 \qquad (11.1.2)$$

with the pressure being only dependent on the axial position, $p = p(x)$. The full conservation of mass, or continuity equation, is retained in this approximation

$$\frac{\partial u}{\partial x}+\frac{\partial v}{\partial y}=0 \qquad (11.1.3)$$

For the present application, the boundary conditions of no-slip on both surfaces

$$u(y=0)=U_w, \quad u(y=h(x))=0$$
$$v(y=0)=0, \quad v(y=h(x))=0 \qquad (11.1.4)$$

From the y-component of Eq. (11.1.2), we find that the pressure depends only on x since $\partial p/\partial y = 0$, i.e. $p = p(x)$ only. That dependence allows us to integrate the x-component of Eq. (11.1.2) partially with respect to y and find

$$u(x,y)=\frac{1}{2\mu}\frac{dp}{dx}y^2+a(x)y+b(x) \qquad (11.1.5)$$

where a(x) and b(x) are functions of integration to be determined from the boundary conditions. We note that the velocity profile is locally parabolic with curvature dp/dx. Imposing the no-slip boundary conditions on u at $y = 0$ and $y = h(x)$ in Eq. (11.1.4) determines a(x) and b(x), and the solution for u(x, y) in Eq. (11.1.5) becomes

$$u(x,y)=\frac{1}{2\mu}\frac{dp}{dx}(y^2-hy)+U_w\left(1-\frac{y}{h}\right) \qquad (11.1.6)$$

The fluid velocity results from two contributions on the right hand side: a pressure gradient, dp/dx contribution with the related quadratic dependence in y; and the moving wall contribution with coefficient U_w and its related linear dependence in y. In this case the pressure gradient is not imposed from the outside, since we are assuming the upstream and downstream pressures at the end of the block are both equal to p_a. Instead, dp/dx is induced from the non-parallel geometry of the gap.

The two-dimensional volume flow rate, Q, is given by the integral of u across the gap

$$Q=\int_0^h u\,dy=-\frac{1}{\mu}\frac{dp}{dx}\frac{h^3}{12}+U_w\frac{h}{2} \qquad (11.1.7)$$

and it is constant, independent of x. Equation (11.1.7) is a form of the "Reynolds equation" used extensively in lubrication theory. Solving Eq. (11.1.7) for dp/dx gives us

$$\frac{dp}{dx}=\frac{12\mu}{h^3}\left(U_w\frac{h}{2}-Q\right) \qquad (11.1.8)$$

We can separate variables and integrate Eq. (11.1.8)

$$p=12\mu\int\left(\frac{U_w}{2h^2}-\frac{Q}{h^3}\right)dx+c \qquad (11.1.9)$$

Using the definition of $h(x) = d_1 - \varepsilon x$ in Eq. (11.1.9) yields

$$p=\frac{6\mu}{\varepsilon}\left(\frac{U_w}{h}-\frac{Q}{h^2}\right)+c \qquad (11.1.10)$$

We can solve for the two constants, Q and c, by imposing the two end pressure conditions on p, i.e. $p(x=0)=p_a$, $p(x=L)=p_a$, and recalling that $h(x=0)=d_1$, $h(x=L)=d_2$. The result is

$$p(x=0)=p_a=\frac{6\mu}{\varepsilon}\left(\frac{U_w}{d_1}-\frac{Q}{d_1^2}\right)+c,$$
$$p(x=L)=p_a=\frac{6\mu}{\varepsilon}\left(\frac{U_w}{d_2}-\frac{Q}{d_2^2}\right)+c \qquad (11.1.11)$$

Solving for Q and c in Eqs. (11.1.11) gives us

$$Q=U_w\left(\frac{d_1d_2}{d_1+d_2}\right), \quad c=p_a-\frac{6\mu U_w}{\varepsilon(d_1+d_2)} \qquad (11.1.12)$$

So the pressure field is solved,

$$p(x)-p_a=\frac{6\mu U_w}{\varepsilon h^2(x)}\left(\frac{(d_1-h(x))(h(x)-d_2)}{d_1+d_2}\right) \qquad (11.1.13)$$

which makes it clear that the left hand side is zero at the ends where $h(x=0)=d_1$, $h(x=L)=d_2$. From Eq. (11.1.13) we see that the pressure gradient varies linearly with the viscosity, i.e. $dp/dx \sim \mu$. This implies that the velocity field u(x, y) in Eq. (11.1.6) is independent of μ since the term $(1/\mu)(dp/dx)$ cancels the μ dependence. Substituting Eq. (11.1.8) into Eq. (11.1.6) and using the solution for Q in Eq. (11.1.12) yields u(x, y)

$$u(x,y) = \frac{6}{h^3}\left(U_w\frac{h}{2} - U_w\left(\frac{d_1 d_2}{d_1 + d_2}\right)\right)(y^2 - hy) + U_w\left(1 - \frac{y}{h}\right) \quad (11.1.14)$$

To continue in our analysis, let's make it dimensionless by defining the following scaled variables similar to Eq. (6.6.7) with $b = d_1 - d_2$ and but no ε in the pressure scale

$$U = \frac{u}{U_w},\ V = \frac{v}{\varepsilon U_w}, X = \frac{x}{L},\ Y = \frac{y}{(d_1 - d_2)},$$
$$H = \frac{h}{(d_1 - d_2)},\ P = \frac{p - p_a}{\mu U_w/(d_1 - d_2)} \quad (11.1.15)$$

Inserting Eqs. (11.1.15) into Eq. (11.1.13) the dimensionless pressure becomes

$$P(X) = \frac{6}{\varepsilon(1 + 2D)}\left(\frac{X(1 - X)}{(1 + D - X)^2}\right) \quad (11.1.16)$$

where the dimensionless wall position is $H(X) = 1 + D - X$. The ends of the gap are at $X = 0, 1$, and the dimensionless parameter D is defined as $D = d_2/(d_1 - d_2)$.

The physical meaning of D is very important. As shown in Figure 11.2 it is the ratio of the narrowest gap width at $x = L$, d_2, to the projection of the block in the x-direction to the entrance at $x = 0$, $d_1 - d_2$. Fluid entering the gap from the left "sees" the region $0 \leq y \leq d_2$ as dominated by the wall dragging it to the right. The block is not directly in the pathway. On the other hand, the region $d_2 \leq y \leq d_1$, is dominated by the block projection over the distance $d_1 - d_2$ which is, more or less, in the pathway of the incoming fluid. The block slope induces a pressure gradient which influences the flow dramatically. D is a measure of this competition between the wall and the block to influence the resulting flow.

A plot of P(X) is shown in Figure 11.3 for $\varepsilon = 0.2$ and $D = 0.25, 0.50, 0.75, 1.00$. As D increases the pressures decrease and the maximum shifts to the left. To find the value of X where the pressure is a maximum, $X = X_m$, we set the X-derivative of the pressure to zero, $dP/dX = 0$, at $X = X_m = (1 + D)/(1 + 2D)$. Inserting X_m into Eq. (11.1.16) gives us the maximum pressure, P_m,

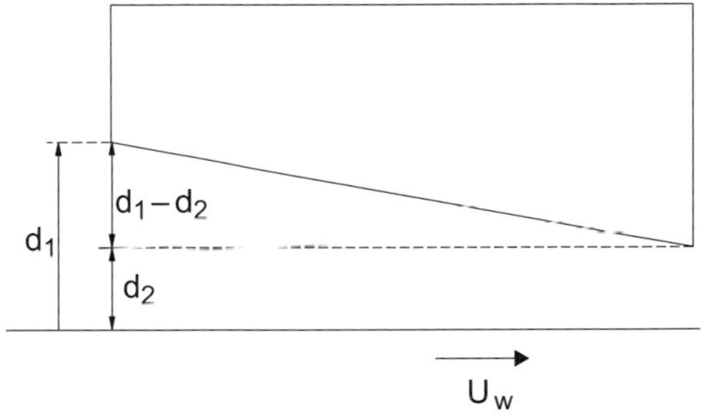

Figure 11.2 Physical interpretation of the dimensionless parameter $D = d_2/(d_1 - d_2)$.

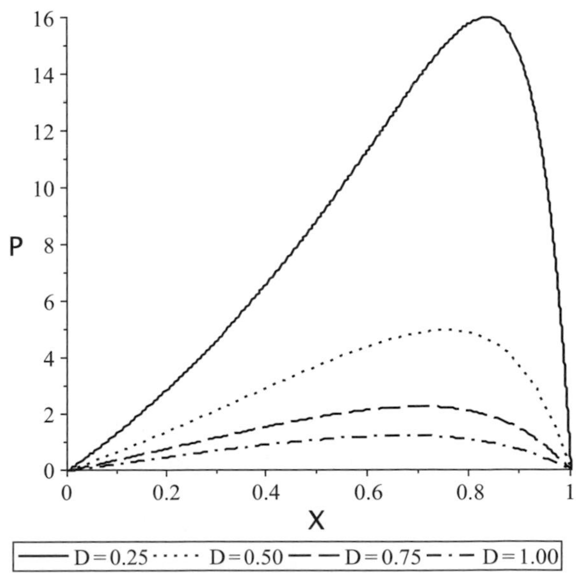

Figure 11.3 Pressure P vs X for four values of $D = 0.25, 0.50, 0.75, 1.00$.

$$P(X_m) = P_m = \frac{3}{2\varepsilon}\frac{1}{D(1+D)(1+2D)} \tag{11.1.17}$$

The high pressures developed in the gap between the slide block and wall keeps them separated, so they do not rub against one another. It is this principle which underlies the industrial application of lubricants, which motivated the field of lubrication theory. Typically a circular cross-section journal rotates inside a circular cross-section bearing housing. The gap between them is very small, and they are slightly off center from one another when a load is applied to the journal. From the rotating journal, lubricant (oil) is dragged into a converging gap geometry, creating large pressures that prevent surface-to-surface contact, see Figure 11.4.

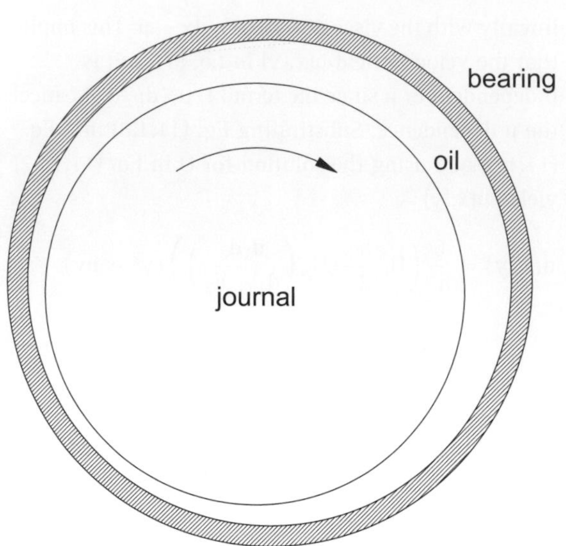

Figure 11.4 Journal bearing rotation dragging oil into a converging gap geometry.

Box 11.1 Galilean Invariance

We changed reference frames for the slide block problem from one attached to the wall, with the block translating to the left at constant velocity $-U_w$, to one attached to the block with the wall translating to the right at speed $+U_w$. Why can we do that? Consider the two reference frames shown in Figure 11.5, the primed system is translating to the right at speed U relative to the unprimed system. Starting in the (x, y, t) frame, the velocities and pressure (u, v, p) obey the Navier–Stokes and continuity equations

$$\begin{aligned}
&\frac{\partial u}{\partial t} + u\frac{\partial u}{\partial x} + v\frac{\partial u}{\partial y} = -\frac{1}{\rho}\frac{\partial p}{\partial x} + \nu\left(\frac{\partial^2 u}{\partial x^2} + \frac{\partial^2 u}{\partial y^2}\right)\\
&\frac{\partial v}{\partial t} + u\frac{\partial v}{\partial x} + v\frac{\partial v}{\partial y} = -\frac{1}{\rho}\frac{\partial p}{\partial y} + \nu\left(\frac{\partial^2 v}{\partial x^2} + \frac{\partial^2 v}{\partial y^2}\right)\\
&\frac{\partial u}{\partial x} + \frac{\partial v}{\partial y} = 0
\end{aligned} \tag{11.1.18}$$

What happens to these equations in the primed system? The relationships between the variables in the two frames are

$$\begin{aligned}
&x' = x - Ut, \quad y' = y, \quad t' = t\\
&u' = u - U, \quad v' = v, \quad p' = p
\end{aligned} \tag{11.1.19}$$

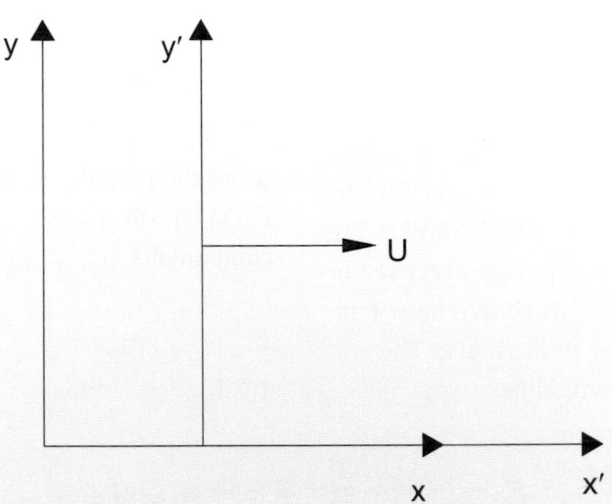

Figure 11.5 Two reference frames, primed and unprimed, related by a constant velocity, U.

Changing to the primed frame, $(x, y, t) \to (x', y', t')$, derivatives in the unprimed frame are related to derivatives in the primed frame by the chain rule of partial differentiation

$$\begin{aligned}
\frac{\partial}{\partial x} &= \cancel{\frac{\partial t'}{\partial x}}\frac{\partial}{\partial t'} + \frac{\partial x'}{\partial x}\frac{\partial}{\partial x'} + \cancel{\frac{\partial y'}{\partial x}}\frac{\partial}{\partial y'} = \frac{\partial}{\partial x'} \\
\frac{\partial}{\partial y} &= \cancel{\frac{\partial t'}{\partial y}}\frac{\partial}{\partial t'} + \cancel{\frac{\partial x'}{\partial y}}\frac{\partial}{\partial x'} + \frac{\partial y'}{\partial y}\frac{\partial}{\partial y'} = \frac{\partial}{\partial y'} \\
\frac{\partial}{\partial t} &= \frac{\partial t'}{\partial t}\frac{\partial}{\partial t'} + \frac{\partial x'}{\partial t}\frac{\partial}{\partial x'} + \cancel{\frac{\partial y'}{\partial t}}\frac{\partial}{\partial y'} = \frac{\partial}{\partial t'} - U\frac{\partial}{\partial x'}
\end{aligned} \tag{11.1.20}$$

Inserting Eqs. (11.1.20) into Eqs. (11.1.18) results in

$$\begin{aligned}
&\frac{\partial u'}{\partial t'} \boxed{-U\frac{\partial u'}{\partial x'}} + u'\frac{\partial u'}{\partial x'} \boxed{+U\frac{\partial u'}{\partial x'}} + v'\frac{\partial u'}{\partial y'} = -\frac{1}{\rho}\frac{\partial p'}{\partial x'} + \nu\left(\frac{\partial^2 u'}{\partial x'^2} + \frac{\partial^2 u'}{\partial y'^2}\right) \\
&\frac{\partial v'}{\partial t'} \boxed{-U\frac{\partial v'}{\partial x'}} + u'\frac{\partial v'}{\partial x'} \boxed{+U\frac{\partial v'}{\partial x'}} + v'\frac{\partial v'}{\partial y'} = -\frac{1}{\rho}\frac{\partial p'}{\partial y'} + \nu\left(\frac{\partial^2 v'}{\partial x'^2} + \frac{\partial^2 v'}{\partial y'^2}\right) \\
&\frac{\partial u'}{\partial x'} + \frac{\partial v'}{\partial y'} = 0
\end{aligned} \tag{11.1.21}$$

which have the same form as in the unprimed system due to the cancelation of the boxed terms. It is straightforward to prove the same in three dimensions for a constant velocity shift in an arbitrary direction. A shift of reference frames by a constant velocity like U is called a Galilean transformation. Therefore, these fluid mechanics equations are "Galilean invariant." The difference comes in the boundary conditions since $u' = u - U$. In the slide block example, the unprimed system had $u = -U_W$ on the slide block and $u = 0$ on the wall. The transformation to the primed frame would be setting $U = -U_W$, the primed system translating to the left, so that $u' = u + U_W$. Then in the primed frame we get to $u' = 0$ on the slide block and $u' = U_W$ on the wall.

11.2 Slide Block Forces and Flows

Let's find the forces on the block. The lift force is the resultant force in the y-direction and the drag force is the resultant force in the x-direction. To do this we will need the stress vector on the block, $\underline{t} = \underline{\underline{T}}\cdot\underline{n}$, where $\underline{n}$ is the unit normal to the block which points into the fluid, as shown in Figure 11.1. The details of deriving the rate of strain tensor, stress tensor and stress vector and then simplifying for $\varepsilon << 1$ are shown here. First we need the unit normal vector to the block. The upward pointing unit normal would be

$$\underline{\hat{n}} = \frac{\nabla G}{|\nabla G|} \quad \text{where } G = y - h(x) = 0 \tag{11.2.1}$$

and the downward pointing unit normal into the fluid is just $\underline{n} = -\underline{\hat{n}}$. Using the definitions of Eq. (11.2.1) and rearranging, we find the unit normal is

$$\underline{n} = \frac{\left(\frac{dh}{dx}, -1\right)}{\left(1 + \left(\frac{dh}{dx}\right)^2\right)^{1/2}} = \frac{\left(\varepsilon\frac{dH}{dX}, -1\right)}{\left(1 + \varepsilon^2\left(\frac{dH}{dX}\right)^2\right)^{1/2}}$$

$$= \left(\varepsilon\frac{dH}{dX}, -1\right) + O(\varepsilon^2) \tag{11.2.2}$$

where we have used a Taylor series expansion for the denominator for $\varepsilon << 1$,

$$\left(1 + \varepsilon^2\left(\frac{dH}{dX}\right)^2\right)^{-1/2} = 1 - \frac{\varepsilon^2}{2}\left(\frac{dH}{dX}\right)^2 + O(\varepsilon^4) \tag{11.2.3}$$

The dimensional stress tensor for the fluid is $\underline{\underline{T}} = -(p - p_a)\delta_{ij} + 2\mu E_{ij}$ where the strain rate tensor components are $E_{ij} = (1/2)(\partial u_i/\partial x_j + \partial u_j/\partial x_i)$. Using the scales from Eq. (11.1.15), the rate of strain tensor becomes

$$E_{ij} = \frac{U_w}{(d_1 - d_2)}\begin{bmatrix} \varepsilon\frac{\partial U}{\partial X} & \frac{1}{2}\left(\frac{\partial U}{\partial Y} + \varepsilon^2\frac{\partial V}{\partial X}\right) \\ \frac{1}{2}\left(\frac{\partial U}{\partial Y} + \varepsilon^2\frac{\partial V}{\partial X}\right) & \varepsilon\frac{\partial V}{\partial Y} \end{bmatrix}$$

$$= \frac{U_w}{(d_1 - d_2)}\left(\begin{bmatrix} 0 & \frac{1}{2}\frac{\partial U}{\partial Y} \\ \frac{1}{2}\frac{\partial U}{\partial Y} & 0 \end{bmatrix} + O(\varepsilon)\right) \tag{11.2.4}$$

and the stress tensor is

$$\underline{\underline{T}} = \frac{\mu U_w}{(d_1 - d_2)}\begin{bmatrix} -P & \frac{\partial U}{\partial Y} \\ \frac{\partial U}{\partial Y} & -P \end{bmatrix} + O(\varepsilon) \tag{11.2.5}$$

using the pressure scale, $\mu U_w/(d_1 - d_2)$, from Eq. (11.1.15). The stress vector, $\underline{t} = \underline{\underline{T}}\cdot\underline{n}$, has components (t_x, t_y) given by

$$\begin{bmatrix} t_x \\ t_y \end{bmatrix} = \frac{\mu U_w}{(d_1 - d_2)}\begin{bmatrix} -P & \frac{\partial U}{\partial Y} \\ \frac{\partial U}{\partial Y} & -P \end{bmatrix}\begin{bmatrix} \varepsilon\frac{dH}{dX} \\ -1 \end{bmatrix}$$

$$= \frac{\mu U_w}{(d_1 - d_2)}\begin{bmatrix} -\varepsilon P\frac{dH}{dX} - \frac{\partial U}{\partial Y} \\ P \end{bmatrix} + O(\varepsilon) \tag{11.2.6}$$

where we have kept the term $-\varepsilon P dH/dX$ as $O(1)$ since we know $P \sim 1/\varepsilon$ from Eq. (11.1.16). A more formal approach is to define a new pressure variable as $\hat{P} = \varepsilon P$ and consider $\hat{P}$ an $O(1)$ quantity in Eq. (11.2.6).

It is straightforward to define a dimensionless stress tensor, $\underline{\underline{\sigma}}$, as $\underline{\underline{\sigma}} = \underline{\underline{T}}/(\mu U_w/(d_1 - d_2))$. The related dimensionless stress vector defined as $\underline{s} = \underline{t}/\mu U_w(d_1 - d_2)$ has x- and y-components defined by $\underline{s} = \underline{\underline{\sigma}}\cdot\underline{n}$ in component form

$$\begin{bmatrix} s_x \\ s_y \end{bmatrix} = \begin{bmatrix} -P & \frac{\partial U}{\partial Y} \\ \frac{\partial U}{\partial Y} & -P \end{bmatrix}\begin{bmatrix} \varepsilon\frac{dH}{dX} \\ -1 \end{bmatrix} = \begin{bmatrix} -\varepsilon P\frac{dH}{dX} - \frac{\partial U}{\partial Y} \\ P \end{bmatrix} + O(\varepsilon) \tag{11.2.7}$$

We can use the dimensional stress vector components from Eq. (11.2.6). To find the lift force we need the stress vector on the block and then its component in the y-direction, $t_y = \underline{t}\cdot\underline{e}_y$, where

$$t_y = p - p_a \tag{11.2.8}$$

In two dimensions, the dimensional lift force per unit depth in the z-direction is f_y. It is the integral of t_y over the interval $0 \le x \le L$, assuming the rest of the block is surrounded by pressure p_a,

$$f_y = \int_0^L t_y\, dx = \int_0^L (p - p_a)\, dx \tag{11.2.9}$$

Using our scalings in Eq. (11.1.15) this simplifies to a dimensionless lift force, F_y, where the scale for the force is the pressure scale multiplied by the length, $\mu U_w L/(d_1 - d_2) = \mu U_w/\varepsilon$,

$$F_y = \frac{f_y}{\mu U_w/\varepsilon} = \int_0^1 P\,dX = \frac{1}{\varepsilon}\left[-\frac{12}{(1+2D)} + 6\ln\left(\frac{1+D}{D}\right)\right] \tag{11.2.10}$$

Now we can also calculate the drag force. The x-component of the stress vector is the dot product of the stress vector with the unit vector in the x-direction, $\underline{e}_x$, $t_x = \underline{t} \cdot \underline{e}_x$ from Eq. (11.2.6). Dimensionally it is

$$t_x = -(p - p_a)\frac{dh}{dx} - \mu\frac{\partial u}{\partial y}\bigg|_{y=h} \tag{11.2.11}$$

We see that there are two contributions to the x-direction force, or drag. One is the component of pressure acting in the x-direction on the block surface which projects in that direction due to the slope, dh/dx. This is sometimes called "pressure drag." The other is the shear stress or "skin friction," which is proportional to the viscosity. The corresponding dimensional drag force is

$$f_x = \int_0^L \left[-(p - p_a)\frac{dh}{dx} - \mu\frac{\partial u}{\partial y}\bigg|_{y=h(x)}\right] dx \tag{11.2.12}$$

and its dimensionless counterpart is

$$F_x = \frac{f_x}{\mu U_w/\varepsilon} = \int_0^1 -\varepsilon P\frac{dH}{dX} - \frac{dU}{dY}\bigg|_{Y=H(X)}$$

$$dX = -\frac{6}{(1+2D)} + 4\ln\left(\frac{1+D}{D}\right) \tag{11.2.13}$$

For $\varepsilon = 0.1$ the two forces are plotted in Figure 11.6. Notice that the vertical force (lift) increases dramatically as D becomes small.

The dimensionless x-velocity is defined in Eq. (11.1.15) as $U = u/U_w$ and the dimensionless form of Eq. (11.1.6) is

$$U(X,Y) = \frac{\varepsilon}{2}\frac{dP}{dX}(Y^2 - H(X)Y) + \left(1 - \frac{Y}{H(X)}\right) \tag{11.2.14}$$

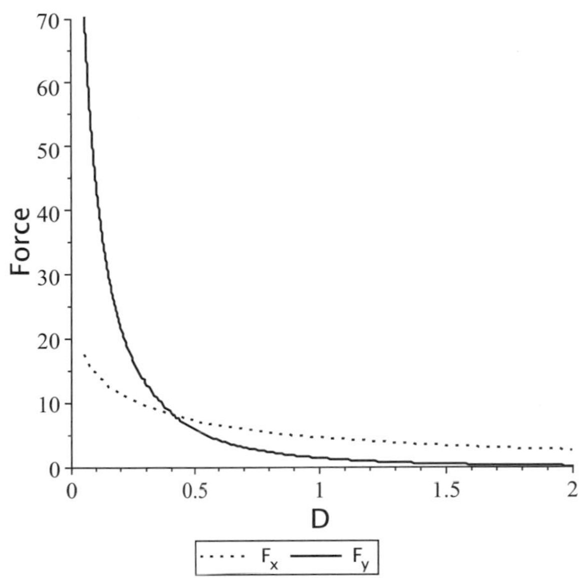

Figure 11.6 F_x and F_y vs D for $\varepsilon = 0.1$.

which is order one since $P \sim 1/\varepsilon$. The pressure gradient, dP/dX, is derived from the X-derivative of Eq. (11.1.16).

For 2D flow in rectangular coordinates, we saw from Eq. (4.3.2) that the velocity field is related to the stream function, ψ, by $u = \partial\psi/\partial y$, $v = -\partial\psi/\partial x$. Using the scales of Eq. (11.1.15) we find the dimensionless versions as $U = \partial\Psi/\partial Y$, $V = -\partial\Psi/\partial X$ for the definition of a dimensionless stream function $\Psi = \psi/U_w(d_1 - d_2)$. We can integrate the U velocity with respect to Y to find Ψ

$$\Psi(X,Y) = \int U dY + f(X) = \frac{\varepsilon\, dP}{2dX}\left(\frac{Y^3}{3} - H(X)\frac{Y^2}{2}\right) + \left(Y - \frac{Y^2}{2H(X)}\right) + f(X) \tag{11.2.15}$$

The integration function f(X) is determined by setting $\Psi = 0$ on the wall where $Y = 0$. That makes $f(X) = 0$.

Because $dP/dX \sim 1/\varepsilon$ the stream function $\Psi(X, Y)$ in Eq. (11.2.15) is independent of ε. Plotting the streamlines is helpful for understanding the flow. The streamlines are constant values of $\Psi(X, Y)$, so we can plot contours of Ψ. Different constant values yield different streamlines. But what are the possible values of Ψ?

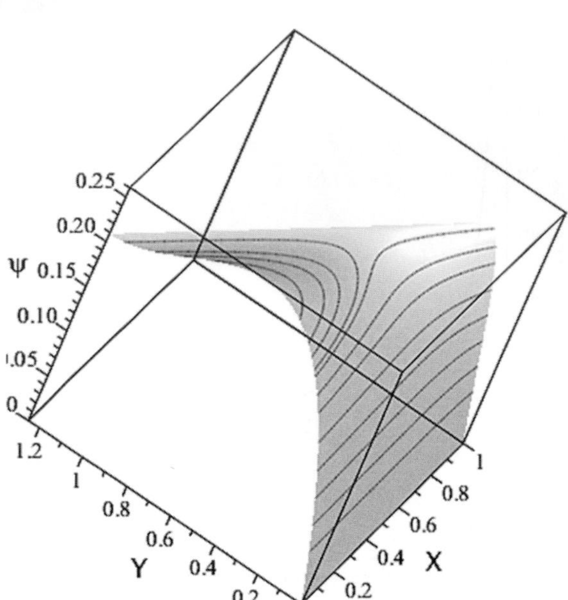

Figure 11.7 Three-dimensional plot of the stream function $\Psi(X, Y)$ giving the range of values for Ψ.

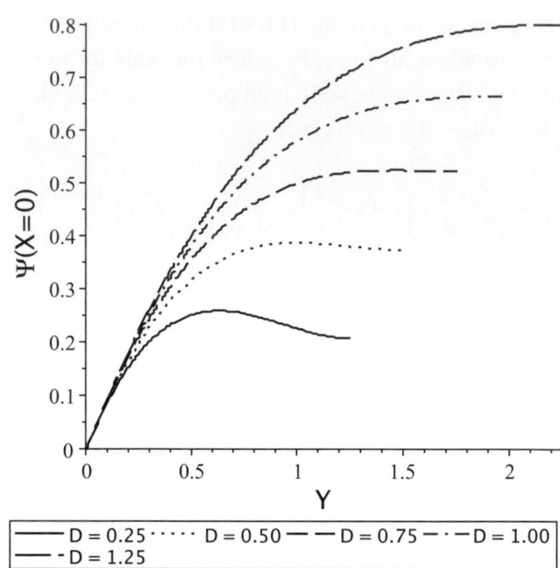

Figure 11.8 $\Psi(X = 0, Y)$ for five values of $D = 0.25, 0.50, 0.75, 1.00, 1.25$.

One way to explore the values is to make a three-dimensional plot of $\Psi(X, Y)$. Figure 11.7 shows $\Psi(X, Y)$ over the range $0 \leq X \leq 1, 0 \leq Y \leq H(X)$, which covers the gap. The range of Ψ is $0 \leq \Psi < 0.25$ approximately.

They can be learned from plotting Ψ vs Y at the entrance to the flow field, $\Psi(X = 0, Y)$. Figure 11.8 shows such plots for five values of $D = 0.25, 0.50, 0.75, 1.00, 1.25$.

The U velocity profile for $D = 0.25$ is shown in Figure 11.9(a) for four values of X which include the inlet, $X = 0$, the outlet, $X = 1$, and the point where the pressure is a maximum, $X = X_m$. For this value of D, the velocity at $X = 0$ is bi-directional, which results from the competition between the two effects driving the flow: (i) the lower wall dragging fluid to the right by virtue of its viscosity, and (ii) the induced pressure field which tends to force fluid from regions of higher to lower pressure. At $X = 0$, fluid closer to the lower wall is more influenced by the wall drag to the right, while fluid closer to the block is more influenced by the positive (adverse) pressure gradient from the slope of the block which forces it to the left. Farther downstream at $X = 0.25$ the block influence is diminished since the distance from the wall is less, and the left-directional flow is reduced. For both of these profiles the pressure gradient is positive at $X < X_m$, $dP/dX > 0$, and from Eq. (11.1.2) this defines their curvatures as positive, $\partial^2 U/\partial Y^2 > 0$. At $X = X_m = 0.833$ in Figure 11.9(a) the pressure is at its maximum where $dP/dX = 0$. Now from Eq. (11.1.2) we see that $\partial^2 U/\partial Y^2 = 0$, which implies that U is linear as shown. At $X = 1$ the pressure gradient is negative, so $\partial^2 U/\partial Y^2 < 0$. For $D = 0.25$, the velocity profile at the outlet, $X = 1$, has a region near the wall where the fluid travels faster than the wall. This feature comes from the additive effects of the wall drag and the negative pressure gradient which is steep there. So fluid can travel faster than the wall. That phenomenon is called "leakback." The velocity field in Figure 11.9(b) is for $D = 1.25$ and there is no backflow at $X = 0$ nor leakback at $X = 1$, but there is a linear profile at $X_m = 0.625$. By setting $\partial U/\partial Y = 0$ at $X = 1, Y = 0$, it is straightforward to show that leakback, i.e. $\partial U/\partial Y > 0$ there, only occurs for $0 < D < 1$. Similarly, setting $\partial U/\partial Y = 0$ at $Y = H, X = 0$ shows that backflow, i.e. $\partial U/\partial Y > 0$

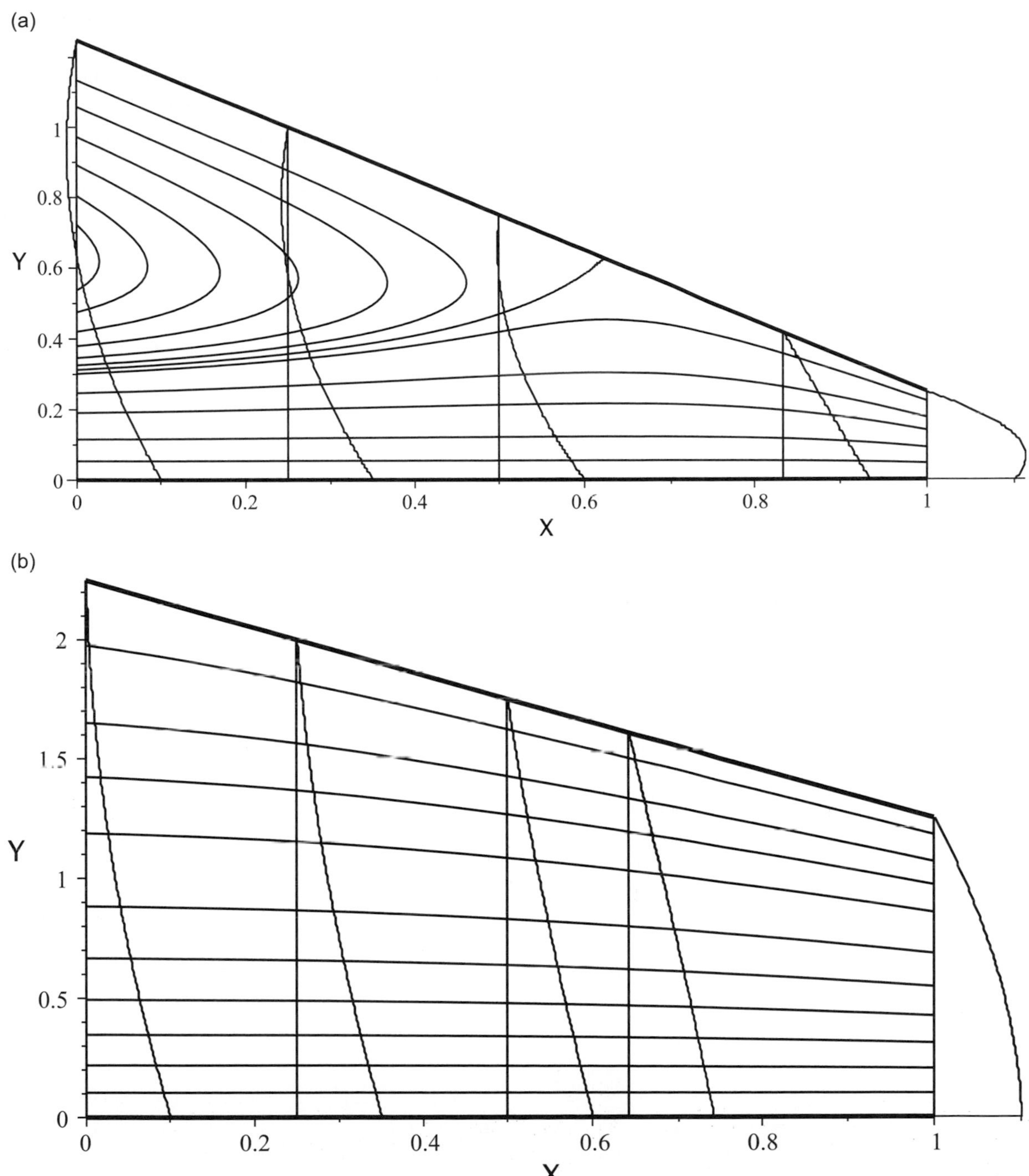

Figure 11.9 Velocity profiles of U and streamlines for (a) $D = 0.25, X_m = 0.833$ and (b) $D = 1.25, X_m = 0.625$.

there, only occurs in the same D range. This another indication of the physical interpretation of this dimensionless ratio. When $D < 1$, the system is dominated by the block.

The streamlines are also shown in Figure 11.9. For $D = 0.25$ in Figure 11.9(a) there is a stagnation streamline which intersects with the block and separates the reverse flow streamlines from the pass through streamlines. It must have the same stream function value as the block surface, call that ψ_s. We can readily find ψ_s at the entrance location of the block $\psi_s = \psi(X = 0, Y = H(X = 0)) = 0.208333333$. For $D = 1.25$ there are no reversed flow streamlines: all of them pass through the gap.

It is worthwhile to point out that the solutions for $u(x, y)$ in Eq. (11.1.14) and $p(x)$ in Eq. (11.1.13) become $-u(x, y)$ and $-p(x)$, respectively, for the wall moving in the negative x-direction, $u(y = 0) = -U_w$. The velocities are reversible since we have neglected the nonlinear acceleration terms. The fluid pressure is negative compared to p_a, and in extreme conditions can cause cavitation of the fluid. In Section 17.7 we will take into account the nonlinear accelerations terms using a perturbation approach and show that the resulting system is not reversible. Then when the wall is oscillated, so that $U_w = a\cos(\omega t)$, there is net steady streaming in the positive x-direction.

The Maple code for creating Figure 11.9(a) is shown in Code 11.1(a); the corresponding MATLAB® code is Code 11.1(b).

```
(a)
restart
with(plots):
with(plottools):
#Define the pressure, P.  Use "DD"  since "D" is a protected symbol.

             6                 (   X·(1-X)    )
P :=  ----------------------- ·( ------------ )
      epsilon ·(1+2·DD)        (  (1+DD-X)^2  )

#Find location of pressure maximum. Xm
Xm :- solve(diff(P, X) =0, X) :
H := 1+DD-X:
#Define the pressure, P.  Use "DD"  since "D" is a protected symbol.

             6                 (   X·(1-X)    )
P :=  ----------------------- ·( ------------ ) :
      epsilon ·(1+2·DD)        (  (1+DD-X)^2  )

#Find location of pressure maximum. Xm
Xm := solve(diff(P, X) =0, X) :
H := 1+DD-X:
#Define the streamfunction ψ (psi)

       epsilon               ( Y^3    H·Y^2 )         Y^2
Psi := ------- ·diff(P, X) · ( ---  - ----- )  +Y  - ----- ;
          2                  (  3       2   )        2·H
```

$$\psi := \frac{1}{2}\left(\epsilon\left(\frac{6\,(1-X)}{\epsilon(1+2DD)\,(1+DD-X)^2} - \frac{6X}{\epsilon\,(1+2DD)\,(1+DD-X)^2} + \frac{12X(1-X)}{\epsilon(1+2\,DD)\,(1+DD-X)^3}\right)\left(\frac{Y^3}{3} - \frac{(1+DD-X)\,Y^2}{2}\right)\right) + Y - \frac{Y^2}{2\,(1+DD-X)} \qquad (1)$$

Code 11.1 Code examples: (a) Maple, (b) MATLAB®.

```
U := diff(psi, Y);
```

$$U := \frac{1}{2}\left(\epsilon\left(\frac{6\ (1-X)}{\epsilon(1+2\ DD)\ (1+DD-X)^2} - \frac{6X}{\epsilon(1+2DD)\ (1+DD-X)^2} + \frac{12\ X(1-X)}{\epsilon(1+2\ DD)\ (1+DD-X)^3}\right) Y^2 - (1+DD-X)\ Y\right) + 1 - \frac{Y}{1+DD-X} \quad (2)$$

```
DD := 0.25 :
#Find stagnation streamline by solving for psi on the block surface at x = 0
psiSt := subs(X = 0, Y = subs(X = 0, H), psi);
                    psiSt := 0.208333333                    (3)
#Form the plot of the streamlines using "contourplot",...,Note the range X = 0,...,1 and Y = 0,...,H. Select
values of psi for contours.
PL1 := contourplot(psi, X = 0,..., 1, Y = 0,..., H, color= black, contours= [[0.05, 0.1, 0.15, 0.18, 0.204,
    psiSt, 0.213, 0.22, 0.23, 0.24, 0.25, 0.257], numpoints = 5000) :

#Create the slide block and wall surfaces.
LineBlock := line([0, subs(X = 0, H)], [[1, subs(X = 1, H)], color = black, thickness = 3):
LineWall := line ([[0, 0], [[1, 0], color = black, thickness = 3) :
PL4 := display (PL 1, Line Block, LineWall) :
#Let K be a scale factor for velocity to fit into the combined figure. Shift each velocity profile by its x
value. Parametric plot so horizontal flow.
K := 0.1:
PL21 := plot([subs(epsilon= 0.1, X = 0, K· U), Y, Y = 0,...,subs(X = 0, H)], color= black, linestyle
      = 1) :

PL22 := plot([[subs(epsilon= 0.1, X = 0.25, K· U + X), Y, Y = 0,...,subs(X = 0.25, H)], color= black,
      linestyle = 1) :

PL23 := plot([[subs(epsilon = 0.1, X = 0.5, K ·U + 0.5), Y, Y = 0,...,subs(X = 0.5, H)], color= black,
      linestyle = 1) :

PL24 := plot([[subs(epsilon = 0.1, X = Xm, K· U + Xm), Y, Y = 0,...,subs(X = Xm, H)], color= black,
      linestyle = 1)

PL25 := plot([[subs(epsilon= 0.1, X = 1.0, K· U + X), Y, Y = 0,...,subs(X = 1.0, H)], color= black,
      linestyle = 1) :


PL26 := display(PL21, PL22, PL23, PL24, PL25):

#Form vertical axes for each profile
Line1 := line ([[0.00, 0], [[0.00, subs(X  = 0.00, H)], color= black) :
Line2 := line ([[0.25, 0], [[0.25, subs (X = 0.25, H)], color= black) :
Line3 := line([0.50, O], [0.50, subs(X = 0.50, H)], color= black):
Line4 := line([Xm, O], [Xm, subs(X = Xm, H)], color= black) :
Line5 := line([[1.00, 0], [[1.00, subs(X = 1.00, H)], color= black) :
PL36  :=  display(Line1, Line2, Line3, Line4, Line5) :
display(PL4, PL26, PL36, labels= ["X","Y"]);
```

Code 11.1 *(continued)*

(b)

```
clear all; close all; clc;
syms X Y d epsilon p h psi u
p      = 6/(epsilon*(1+2*d))*((X*(1-X))/(1+d-X)^2);
eqn    = diff(p,X) == 0;
Xm     = solve(eqn,X)
h      = 1 + d - X;
psi = epsilon/2*diff(p,X)*(Y^3/3-h*Y^2/2)+Y-Y^2/(2*h);
u      = simplify(diff(psi,Y))
hnew   = subs(h, [X,d], [0,0.25])
psiSt  = subs(psi, [X,Y,d], [0,hnew,0.25])
PSI    = matlabFunction(psi);
U      = matlabFunction(u);
H      = matlabFunction(h);
XM     = matlabFunction(Xm);
D      = 0.25;
Epsilon = 0.1;
x      = linspace(0,1,1000);
y      = linspace(0,1,500);
[X,Y]  = meshgrid(x',y');
Y      = H(X,D).*Y;
PSI    = PSI(X,Y,D,Epsilon);
contour(X,Y,PSI,[0.05, 0.1, 0.15, 0.18, 0.204, psiSt, 0.213, 0.22, 0.23, 0.24, 0.25, 0.257],'-k');
hold on
plot(x,H(x,D),'k','linewidth',2)
plot(x,0*x,'k','linewidth',2)
K        = 0.1;
XtoPlot = 0;      YtoPlot = y'*H(XtoPlot,D); plot(K*U(XtoPlot,YtoPlot,D)+XtoPlot,YtoPlot,'-k')
XtoPlot = 0.25;   YtoPlot = y'*H(XtoPlot,D); plot(K*U(XtoPlot,YtoPlot,D)+XtoPlot,YtoPlot,'-k')
XtoPlot = 0.5;    YtoPlot = y'*H(XtoPlot,D); plot(K*U(XtoPlot,YtoPlot,D)+XtoPlot,YtoPlot,'-k')
XtoPlot = XM(D);  YtoPlot = y'*H(XtoPlot,D); plot(K*U(XtoPlot,YtoPlot,D)+XtoPlot,YtoPlot,'-k')
XtoPlot = 1;      YtoPlot = y'*H(XtoPlot,D); plot(K*U(XtoPlot,YtoPlot,D)+XtoPlot,YtoPlot,'-k')
XtoPlot = 0;      plot(XtoPlot+0*y,y'*H(XtoPlot,D),'-k')
XtoPlot = 0.25;   plot(XtoPlot+0*y,y'*H(XtoPlot,D),'-k')
XtoPlot = 0.5;    plot(XtoPlot+0*y,y'*H(XtoPlot,D),'-k')
XtoPlot = XM(D);  plot(XtoPlot+0*y,y'*H(XtoPlot,D),'-k')
XtoPlot = 1;      plot(XtoPlot+0*y,y'*H(XtoPlot,D),'-k')
xlabel('X')
ylabel('Y')
axis equal
axis([-0.05,1.15,-0.05,1.3])
```

Code 11.1 *(continued)*

Box 11.2 Lubrication Flows for Fun

Lubrication flows occur in many biofluid mechanics situations. Play is an important activity for developing self-esteem, social skills, physical strength and coordination. Figure 11.10 shows a child on a water slide using gravity and thin fluid films to full benefit.

Figure 11.10 Fun at the water park.

11.3 A Model RBC in a Capillary

We have examined the non-Newtonian properties of blood in Chapter 10 with regard to effective viscosity, $\mu_{eff}(\dot{\gamma})$, and the many rheological models used to describe it. That behavior was independent of any flow boundary influencing the mechanics. However, as blood traverses through the vasculature, it passes from large vessels to much smaller ones, including flowing through two sets of capillaries. One capillary bed is the lung and the other is in the systemic tissues. (In fact, there are three special circulations where two consecutive capillary beds occur in the systemic circulation: the kidney, gut/liver and brain. These are called "portal circulations.") While a typical RBC is a biconcave disc with a diameter of approximately 7 μm, it squeezes through capillaries which can have vessel diameters as narrow as 3–4 μm. Figure 11.11. shows RBCs flowing from right to left in a human capillary of 7 μm diameter. From the fluid flow and interaction of the fluid stresses with the cell membrane elastic stresses, the RBC is deformed into a parachute or helmet shape. At this scale, the fluid mechanics concerns the flow of plasma, whose viscosity is constant, interacting with the RBC and capillary wall.

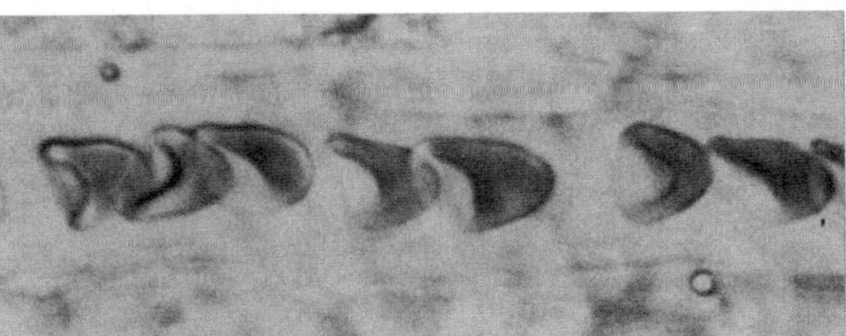

Figure 11.11 Parachute shape of RBCs as they flow from right to left in a capillary of 7 microns diameter.

Fluid dynamic investigations of the capillary flow seen in Figure 11.11 start with analyzing a single RBC passing through a narrow tube. This situation is shown in Figure 11.12 where the flow is right to left. Figure 11.12(a) are cell shapes at a cell velocity of

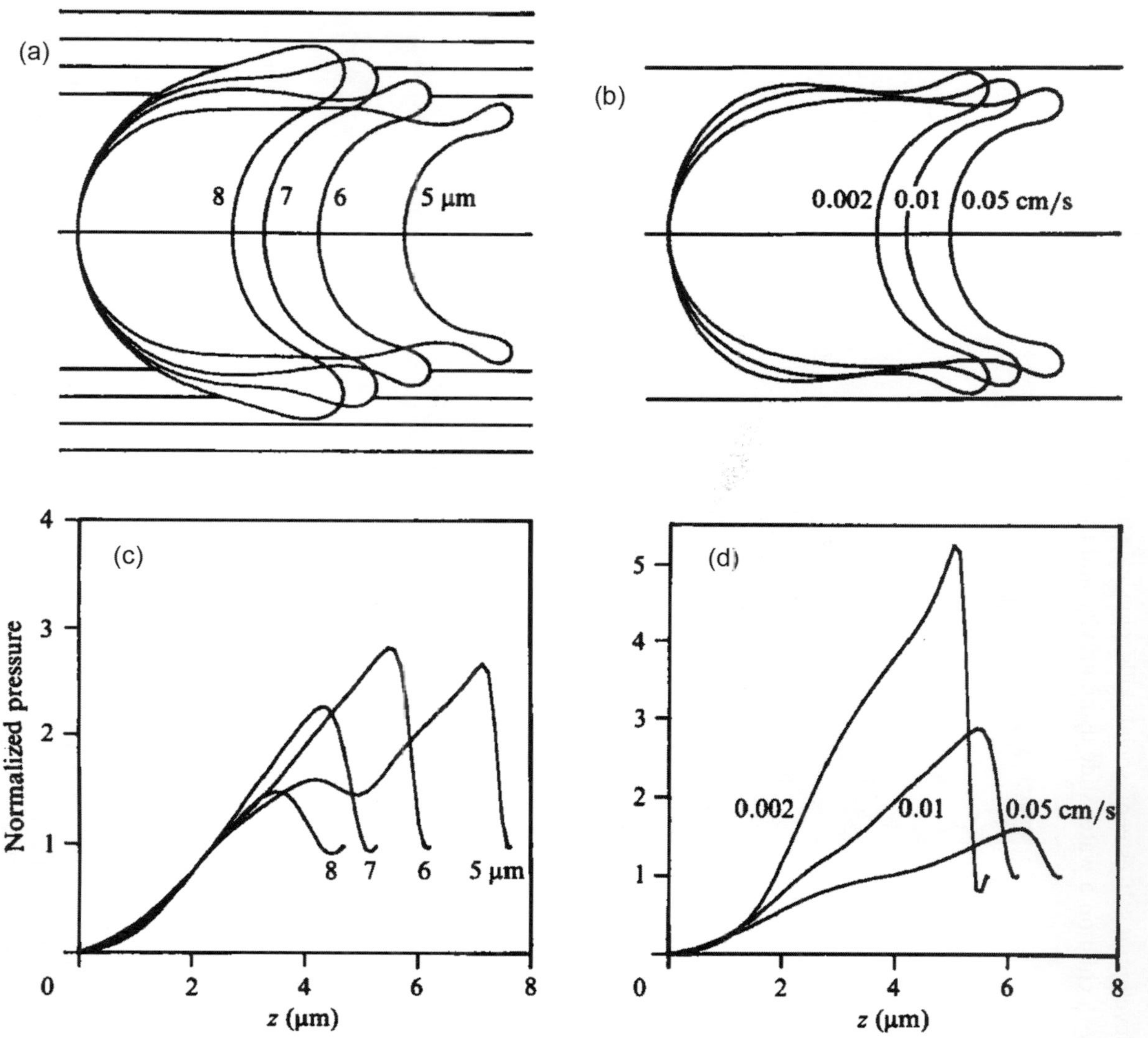

Figure 11.12 Computed cell shapes in capillary flow from right to left: (a) cell velocity of 0.01 cm/s, vessel diameters as shown; (b) 6-micron vessel, cell velocities as shown; (c, d) variation of normalized pressure in the gap corresponding to cell shapes in (a) and (b).

0.01 cm/s for different vessel diameters in μm and Figure 11.12(b) cell shapes in a 6 μm diameter vessel for different cell velocities (Secomb et al., 1986). Figure 11.12(c, d) are the corresponding normalized pressures in the gap showing the upstream end pressure great than the downstream end with a maximum in between. In this computational study the fluid mechanics was simplified using lubrication theory applied to the plasma in the narrow gap between the RBC and capillary wall. The cell membrane was modeled as an elastic thin shell structure. In both figures the general parachute shape results from the balance of fluid and elastic stresses.

Box 11.3 Capillary Blood Velocity

Let's assume there are approximately 5 billion, $n = 5 \times 10^9$, capillaries in the human body, each with a diameter of 8 μm. The entire cardiac output of $Q = 5$ liters/min must pass through those capillaries. The average blood velocity in the capillaries, V_c, then, is a straightforward conservation of mass calculation. Let the volume flow rate through one capillary be $Q_c = Q/n$. Then dividing Q_c by the cross-sectional area of a single capillary πa_c^2 gives us V_c. Typically a_c is 4 μm or $a_c = 4 \times 10^{-4}$ cm. Then the resulting velocity, $V_c = Q_c/\pi a_c^2$, in terms of cm and seconds is

$$V_c = \frac{Q_c}{\pi a_c^2} = \frac{\frac{5 \text{ liters}}{\text{min}} \times \frac{1}{n} \times \frac{1000 \text{ cm}^3}{\text{liter}} \times \frac{1 \text{ min}}{60 \text{ s}}}{\pi (4 \times 10^{-4} \text{ cm})^2} = 0.033 \text{ cm/s} \tag{11.3.1}$$

This value of V_c is an average. Blood distribution into different tissues and their capillary beds is controlled by arterioles either dilating or contracting to increase or decrease the flow. When a particular tissue is more active, it requires a greater blood flow rate and the arterioles there relax and dilate. For example, digesting food causes more flow to the gastrointestinal tract, while running sends more flow to the muscles. The control can be from a number of inputs including neurologic, hormonal, and local levels of pH, O_2, CO_2 amongst others.

What we learned in Section 11.1 for lubrication theory of the slide block problem provides enough tools to investigate our own simplified model of red blood cell (RBC) flow in capillaries. Consider a red blood cell flowing in a capillary approximated as a tapered cylindrical section, a frustum, see Figure 11.13. The capillary wall is located at $r = a$ and the RBC surface is at $r = a - h(z)$ where $h(z) = d_1 - (d_1 - d_2)z/L$, similar to $h(x)$ for Section 11.1. The definitions of d_1, d_2, L, U_w are all the same as in Section 11.1. The lubrication assumption, $\varepsilon = (d_1 - d_2)/L << 1$, remains the same so the governing momentum balance in the z-direction is

$$\mu\left(\frac{\partial^2 u_z}{\partial r^2} + \frac{1}{r}\frac{\partial u_z}{\partial r}\right) = \frac{dp}{dz} \tag{11.3.2}$$

where $p = p(z)$ only. We can make use of the thin fluid layer dimension compared to the tube radius, $\delta = (d_1 - d_2)/a << 1$, to simplify Eq. (11.3.2) even further. To show this, use the scales as in Section 11.1

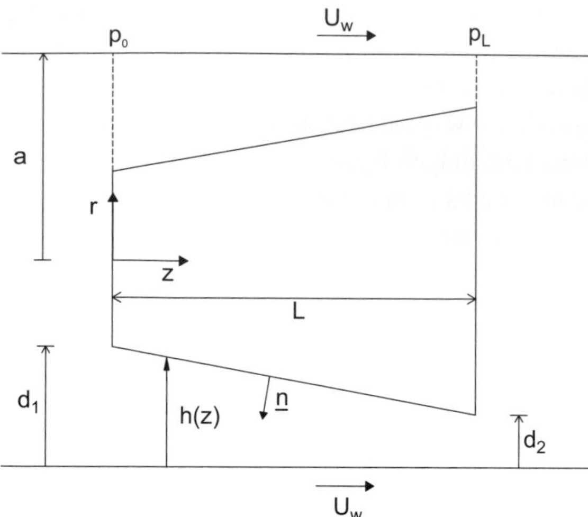

Figure 11.13 A section of a cone, a frustum, as a model for red blood cell flow in a capillary.

but define a dimensionless local coordinate system, Y, attached to the wall

$$Z = \frac{z}{L}, Y = \frac{a - r}{d_1 - d_2}, U_Z = \frac{u_z}{U_w}, \quad U_R = \frac{u_r}{\varepsilon U_w},$$
$$P = \frac{p}{\mu U_w/(d_1 - d_2)}, H = \frac{h}{d_1 - d_2} \tag{11.3.3}$$

where $H = 1 + D - Z$. The Y-coordinate is similar to Section 11.1, where $Y = 0$ on the capillary wall, $r = a$, and $Y = H(Z)$ on the RBC surface, $r = a - h(z)$. The definition of Y can be rearranged to solve for $r = a(1 - \delta Y)$ and inserted into the $1/r$ term of Eq. (11.3.2). Inserting Eqs. (11.3.3) into Eq. (11.3.2), employing the chain rule to express the z and r derivatives in the new variables, Z and R, and with some rearranging we find

$$\frac{\partial^2 U_Z}{\partial Y^2} - \frac{\delta}{1 - \delta Y}\frac{\partial U_Z}{\partial Y} = \varepsilon \frac{dP}{dZ} \tag{11.3.4}$$

where we have anticipated that $\varepsilon P \sim O(1)$.

Assuming $\delta << 1$ we can drop the curvature term it multiplies in Eq. (11.3.4) to find the same form as in Section 11.1,

$$\frac{\partial^2 U_Z}{\partial Y^2} = \varepsilon \frac{dP}{dZ} \tag{11.3.5}$$

which corresponds to Eq. (11.1.2) and shows that the flow in the thin cylindrical layer is not influenced by the curvature term and, instead, sees a flat geometry. With Eq. (11.3.5) we have fully recovered the analysis of Section 11.1 with $Z = X$ and $Y = Y$. The difference between the two analyses is that the end pressures were the same in Section 11.1 while here they are different, i.e. for the RBC problem $p = p_0$ at $z = 0$ and $p = p_L$ at $z = L$. Carrying on in dimensionless form, these end pressure conditions are

$$P(Z = 0) = P_0 = p_0/(\mu U_w/(d_1 - d_2))$$
$$P(Z = 1) = P_1 = p_L/(\mu U_w/(d_1 - d_2)) \tag{11.3.6}$$

Integrating Eq. (11.3.5) twice yields a parabolic profile in Y with two functions of integration which are solved by imposing the two no-slip conditions that $U_Z(Y = 0) = 1$ and $U_Z(Y = H(Z)) = 0$. The final result is

$$U_Z = \frac{\varepsilon}{2}\frac{dP}{dZ}\left(Y^2 - HY\right) + 1 - \frac{Y}{H} \tag{11.3.7}$$

which is the dimensionless version of Eq. (11.1.6). As before, the pressure distribution in the lubrication layer is derived from integral mass balance. In the cylindrical RBC slide block the dimensional integral mass balance for the volume flow rate Q through the gap is

$$Q = \int_0^{2\pi}\int_{a-h(z)}^{a} u_z\, r\, dr\, d\theta \tag{11.3.8}$$

Using our scales and assumption that $\delta << 1$, the dimensionless integral mass conservation is

$$\hat{Q} = \int_0^{H(Z)} U_Z\, dY = -\frac{\varepsilon H^3}{12}\frac{dP}{dZ} + \frac{H}{2} \tag{11.3.9}$$

where $\hat{Q} = Q/2\pi a^2 \delta U_w$ and we have ignored the $O(\delta^2)$ terms. Equation (11.3.9) is the dimensionless counterpart to Eq. (11.1.7) for the cylindrical setting. Rearranging Eq. (11.3.9) for the pressure gradient

$$\frac{dP}{dZ} = \frac{12}{\varepsilon H^3}\left(\frac{H}{2} - \hat{Q}\right) \tag{11.3.10}$$

Equation (11.3.10) is the dimensionless version of Eq. (11.1.8). Inserting $H = 1 + D - Z$ and integrating with respect to Z gives us

$$P = \frac{6}{\varepsilon}\left(\frac{1}{H} - \frac{\hat{Q}}{H^2}\right) + \hat{c} \quad (11.3.11)$$

which is the counterpart of Eq. (11.1.10). $\hat{Q}$ and $\hat{c}$ are solved from imposing the two pressure end conditions of Eq. (11.3.6) and the final result is

$$\begin{aligned} \hat{Q} &= \frac{D(1+D)}{(1+2D)} + \frac{\varepsilon}{6}\frac{D^2(1+D)^2}{(1+2D)}(P_0 - P_1) \\ \hat{c} &= P_0 - \frac{6}{\varepsilon(1+2D)} + \frac{D^2}{(1+2D)}(P_0 - P_1) \end{aligned} \quad (11.3.12)$$

which yields the pressure field

$$\begin{aligned} P(Z) = P_0 + (P_1 - P_0)\frac{D^2}{(1+2D)}\left[\frac{(2(1+D)-Z)Z}{(1+D-Z)^2}\right] \\ + \frac{6}{\varepsilon(1+2D)}\left(\frac{Z(1-Z)}{(1+D-Z)^2}\right) \end{aligned} \quad (11.3.13)$$

It is readily seen in Eq. (11.3.13) that $P(Z=0) = P_0$ and, with a little more effort, that $P(Z=1) = P_1$. Also, when $P_0 = P_1$ we recover Eq. (11.1.16) for the original slide block problem.

Examples of the pressure distribution P(Z) are shown in Figure 11.14 for $\varepsilon = 0.3$, $P_1 = 1$, $P_0 = 0$ and $D = 0.3, 0.4, 0.5, 1$. Note the upstream pressure is larger than the downstream pressure, and there is a local maximum as we saw in Figure 11.3 and Figure 11.12(c, d). The local maximum reduces and shifts to the left with increasing D until it is lost.

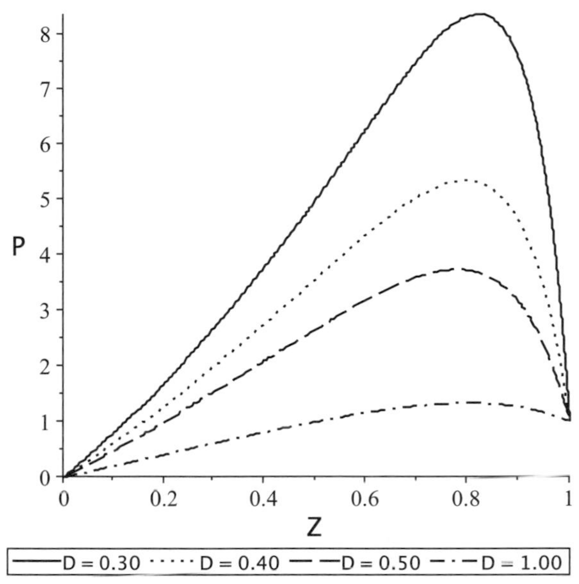

Figure 11.14 Pressure distribution P(Z) for the RBC model flow, for $P_1 = 1, P_0 = 0, \varepsilon = 0.3$ and four values of D.

Clinical Application 11.1 | **Urinary Stones**

While this analysis is focused on RBC flow in a capillary, the general approach is applicable to other objects forced through tight, liquid-filled conduits by a driving pressure. The major components of the urinary system are shown in Figure 11.15(a). Urine is made by the kidneys which filter the blood, as discussed in Section 11.6. The urine flows from the collecting system of the kidneys to the bladder via the ureters, helped along by peristaltic waves. The bladder is elastic and can expand to accommodate more urine. As its internal pressure increases, the owner feels the urge to urinate and forces the fluid out through the urethra.

However, problems arise when kidney or bladder stones are present. They form from high concentrations of minerals in the urine which crystallize, see Figure 11.15(b). Typically a condition of older individuals in developed countries, bladder stones are a public health issue

(a)

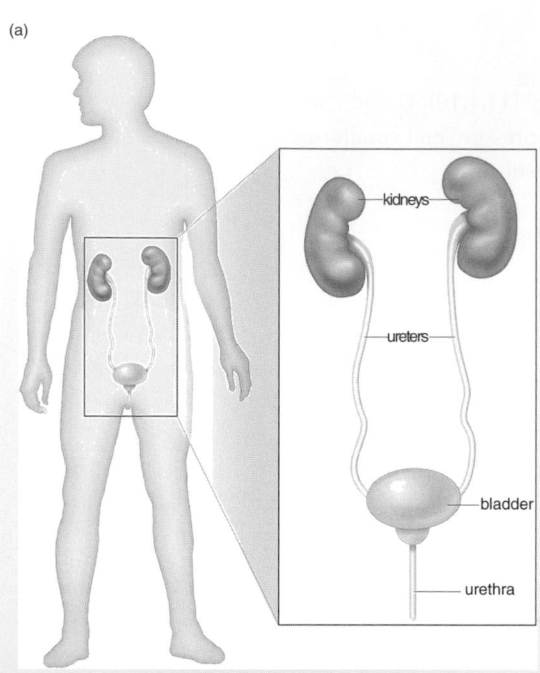

(b)

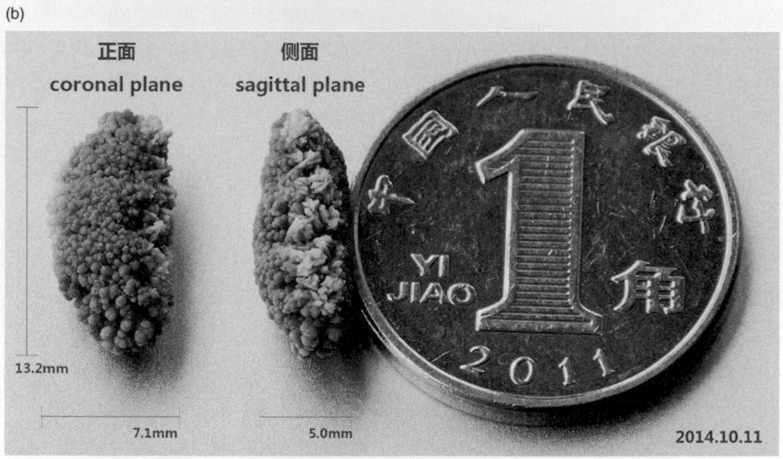

Figure 11.15 (a) The urinary system consisting of the kidneys, ureters, bladder and urethra. (b) Kidney stone passed through urethra in urine.

in developing countries, especially among children who are chronically dehydrated leaving them with highly concentrated urine. If a stone is small enough, it can pass through the urethra without difficulty. When it is only slightly smaller than the urethra diameter, the flow situation is similar to the RBC and Figure 11.13 and high driving pressure is required to pass the stone. Larger ones can be broken into smaller pieces using high energy acoustic shockwaves in a process called lithotripsy, or crushed manually through a cystoscope, which is very uncomfortable to say the least. The smaller pieces can then flow out more easily. Failing that, they can be removed surgically. In addition to humans, urinary stones are found in pets like dogs with similar challenges for treatment.

Ureters can also become blocked by abdominal contents pressing on them, for example during pregnancy. A stent can be inserted to keep the flow open, but this introduces interesting fluid mechanical challenges such as reflux from the bladder back up to the kidneys, which can then be a source of infection (Cummings et al., 2004; Siggers et al., 2009).

11.4 RBC Model Forces and Flows

The dimensional force balance on the RBC is the upstream pressure pushing on the rear of the RBC equaling the downstream pressure pushing on the front of the RBC and the x-direction fluid force on the RBC from the flow in the gap, i.e. the drag force. This balance is given dimensionally by

$$p_L\pi(a-d_2)^2 = p_O\pi(a-d_1)^2 + \int_0^L\int_0^{2\pi}\left(-p\frac{dh}{dz}+\mu\left.\frac{\partial u_z}{\partial r}\right|_{r=a-h(z)}\right) \times (a-h(z))\,d\theta\,dz \tag{11.4.1}$$

Changing to our dimensionless variables, Eq. (11.4.1) becomes

$$P_1\frac{(a-d_2)^2}{2aL} = P_0\frac{(a-d_1)^2}{2aL} + \int_0^1\left.-\varepsilon P\frac{dH}{dZ}-\frac{\partial U_Z}{\partial Y}\right|_{Y=H} dZ \tag{11.4.2}$$

Carrying out the integration in Eq. (11.4.2) by inserting our solutions for P and U_Z from Eqs. (11.3.13) and (11.3.7), the pressure drop across the RBC is

$$P_1-P_O = \frac{\left[-\frac{6}{1+2D}+4\ln\left(\frac{1+D}{D}\right)\right]}{\left[\frac{\varepsilon}{2\delta}+\frac{\varepsilon D^2}{1+2D}\right]} \sim \frac{2\delta}{\varepsilon}\left[-\frac{6}{1+2D}+4\ln\left(\frac{1+D}{D}\right)\right] \tag{11.4.3}$$

where $2\delta/\varepsilon = 2L/a$.

The pressure drop, $P_1 - P_O$, in Eq. (11.4.3) depends on the three dimensionless geometrical parameters, D, ε and δ which are formed from the four dimensional values of d_1, d_2, a and L. To explore their effect, let us constrain the geometry by insisting on a prescribed volume for the RBC. The volume of our RBC model, V_{RBC}, is that of a frustum of a right circular cone whose equation in our parameters is

$$V_{RBC} = \frac{\pi L}{3}\left((a-d_1)^2+(a-d_1)(a-d_2)+(a-d_2)^2\right) \tag{11.4.4}$$

The normal range for V_{RBC} is 80–100 μm^3. To explore the pressure drop, choose $V_{RBC} = 100\,\mu m^3$, set the capillary radius $a = 2.5\,\mu m$, and let d_2 vary from $0.1 \le d_2 \le 0.2\,\mu m$ for fixed values of d_1. L is then calculated from Eq. (11.4.4) for each choice of d_1 and d_2. A plot of the dimensional pressure drop results is

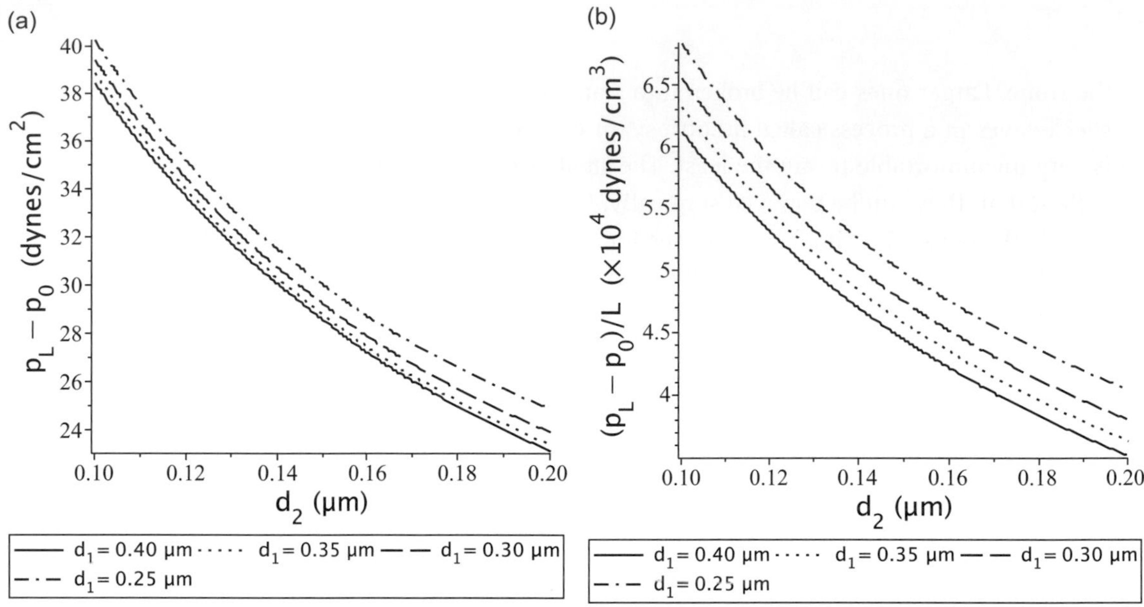

Figure 11.16 (a) RBC dimensional pressure drop $p_L - p_0$, in dyn/cm^2, vs d_2 for four values of d_1 with $a = 2.5\,\mu m$ and $V_{RBC} = 100\,\mu m^3$, $U_w = 0.01$ cm/s and $\mu = 0.012$ poise. (b) Dimensional overall pressure gradient $(p_L - p_0)/L$, in 10^4 dyn/cm^3 for the corresponding parameter values in (a).

shown in Figure 11.16(a) where we have multiplied $P_1 - P_0$ by the pressure scale $\mu U_w/(d_1 - d_2)$ and set $U_w = 0.01\,\text{cm/s}$ with $\mu = 0.012$ poise appropriate for serum. Notice how the upstream pressure increases dramatically as d_2 decreases, while less so when d_1 decreases. In a tight squeeze, the high pressure at the rear of the RBC is responsible for the inward deformation causing the parachute shape. Dividing that pressure drop by the length, L, yields an average pressure gradient across the RBC, as shown in Figure 11.16(b).

Leakback Flow and U_{RBC} Relative to U_{avg}

In the wall frame the RBC moves by at speed $U_{RBC} = U_w$ and the fluid has zero velocity on the wall. An important observation is that the RBC speed is faster than the average fluid speed. To calculate

Clinical Correlation 11.1 | RBC Volume

The volume of a red blood cell is a routine measurement for blood samples where it is called the mean corpuscular volume, or mean cell volume, MCV. It is obtained by multiplying a volume of blood by the hematocrit and dividing the result by the number red blood cells in that volume. For example, for a hematocrit of 0.44 and a RBC count of 4.4×10^6 per μL, where a microliter is $1\ \mu L = 10^{-6}$ L. Then the MCV is

$$\text{MCV} = \frac{(0.44) \times (1 \times 10^{-6}\ \text{L})}{4.4 \times 10^{6}\,\text{RBCs}} = 100 \times 10^{-15}\ \text{L} \tag{11.4.5}$$

where 10^{-15} L = 1 femtoliter (fL) = 1 μm^3. There are a number of diseases that influence MCV and usually are part of an anemia disease process where the hematocrit is lower than normal. With lowered hematocrit, patients with anemia have less oxygen carrying capacity of their blood and symptoms are fatigue, weakness, and shortness of breath.

Measuring the MCV in anemia is an important way to sort out the source of the problem. There is microcytic anemia, normocytic anemia and macrocytic anemia when the MCV is in low, normal or high ranges. The normal range of MCV is 80–100 fL and normocytic anemia occurs in the range. Microcytic anemia has MCV < 80 fL while macrocytic anemia has MCV > 100 fL. The most common cause of microcytic anemia is iron deficiency, which can be from inadequate dietary intake/absorption or iron loss from bleeding, say from the gastrointestinal tract or the uterus. In underdeveloped countries parasitic worms cause significant intestinal blood loss leading to microcytic anemia in children. In macrocytic anemia, even though the RBC is larger than normal, the RBC count is low so that the hematocrit is low. Sources of this condition include megaloblastic anemia due to DNA replication disorders causing the cells to grow too large before they divide. Another cause is alcoholism. Normocytic anemia occurs from a variety of conditions including a decreased production rate of RBCs, an increased destruction of RBCs from hemolysis, or loss of RBCs from bleeding.

this interesting feature, consider the fluid velocity at the exit, z = L, in the block frame, shown in Figure 11.12(a). The velocity is uz in the region $0 \le y \le d_2$ where $y = a - r$ and the values larger than U_w are called the "leakback." The same phenomenon observed in the wall frame is shown in Figure 11.12(b) where the velocities are now $u_z - U_w$. The observer sees the RBC moving to the left but ejecting a layer of fluid near the wall to the right. That layer "leaks back." We know from the slide block analysis that leakback is due to the negative pressure gradient region in the lubrication layer, to the right of the maximum. For the region along the backside of the RBC, $0 \le r \le a - d_2$, the fluid velocity is simply $-U_w$. The fluid is in contact with the RBC surface there and must have the same z-velocity due to no-penetration. The overall fluid volume flow rate in the tube at z = L is

$$\begin{aligned} q &= \int_0^{2\pi}\int_{a-d_2}^{a} \left(u_z|_{z=L} - U_w\right) r\,dr\,d\theta - \int_0^{2\pi}\int_0^{a-d_2} U_w\, r\,dr\,d\theta \\ &= \int_0^{2\pi}\int_{a-d_2}^{a} u_z|_{z=L}\, r\,dr\,d\theta - \pi a^2 U_w \end{aligned} \tag{11.4.6}$$

where we have lumped together the two integrals of U_w which give us $-\pi a^2 U_w$. Now we define the average fluid speed in the tube as $U_{avg} = -q/\pi a^2$, where $q < 0$ in this example, so both U_w and U_{avg} are positive values. The integral term involving u_z in Eq. (11.4.6) is our definition of Q, where $Q = 2\pi a^2 \delta U_w \hat{Q}$ in Eq. (11.3.12), and holds for any value of z, including z = L. Inserting these forms into Eq. (11.4.6) and rearranging gives us

$$\frac{U_{avg}}{U_w} = 1 - 2\delta\hat{Q} \quad < 1 \tag{11.4.7}$$

which shows that the RBC is traveling faster than the average fluid speed. This result is not intuitive and shows how fluid mechanics can be very helpful in understanding physiology. Many sources implicate the leakback flow as the cause of $U_{avg}/U_w < 1$. However, that is only part of the picture. Equation (11.4.7) shows that it is not only the leakback flow, but the entire flow between the lower wall and the RBC surface that causes $U_{avg}/U_w < 1$. This is seen in the wall frame shown in Figure 11.17(b) where all velocities in the gap are slower than the RBC, including

Clinical Application 11.2 | **Implications for Industry and Clinical Applications**

The fact that the RBC travels faster than the average fluid velocity has a long history in fluid mechanics related to a very different application. Liquid flow rates through glass capillary tubes can be measured by placing a long bubble in the tube to record its speed. The bubble diameter is nearly the same as the tube diameter, so there is a thin film between them, like our RBC model. Initially it was assumed that the bubble speed equaled the average fluid speed. However, in 1935 Fairbrother and Stubbs published an experimental study showing that assumption to be wrong (Fairbrother and Stubbs, 1935). Using 2 mm and 3 mm diameter glass tubes that were 4 feet long, they measured the bubble speed, U_b, by timing its travel over a specified distance. The fluid volume flow rate was measured by weighing the tube outflow after a specified time interval. That flow rate divided by the tube cross-sectional area formed the average fluid speed, U_{avg}. They found that the bubble velocity, U_b, traveled faster than the average fluid speed, U_{avg}. Plotting U_b vs U_{avg} they fit the data to an empirical formula

$$U_b = U_{avg} + kU_b \tag{11.4.8}$$

In Eq. (11.4.8), Ca is the capillary number, $Ca = \mu U_b/\sigma$, in this case based on the bubble velocity, U_b. The fluid viscosity is μ and σ is the surface tension between the air in the bubble and the fluid. They found $k = 1.0\,Ca^{1/2}$ fit their data and the study was restricted to very small values of Ca, $0 \leq Ca \leq 0.015$. The experimental fluid used was high viscosity glycerin and concentrated sucrose solutions. That allowed the Ca range to be explored with smaller fluid and bubble velocities that were easier to measure. Further improvements on the formula came from the analytical work of Bretherton (1961) who found $k = 1.29(3Ca)^{2/3}$, again restricted to $Ca << 1$. The capillary number, in general, is the dimensionless ratio of viscous to surface tension forces

$$Ca = \frac{\mu U}{\sigma} \sim \frac{\text{viscous forces}}{\text{surface tension forces}} \tag{11.4.9}$$

where U is a characteristic fluid velocity and appears in interfacial fluid mechanics problems, see Section 16.5.

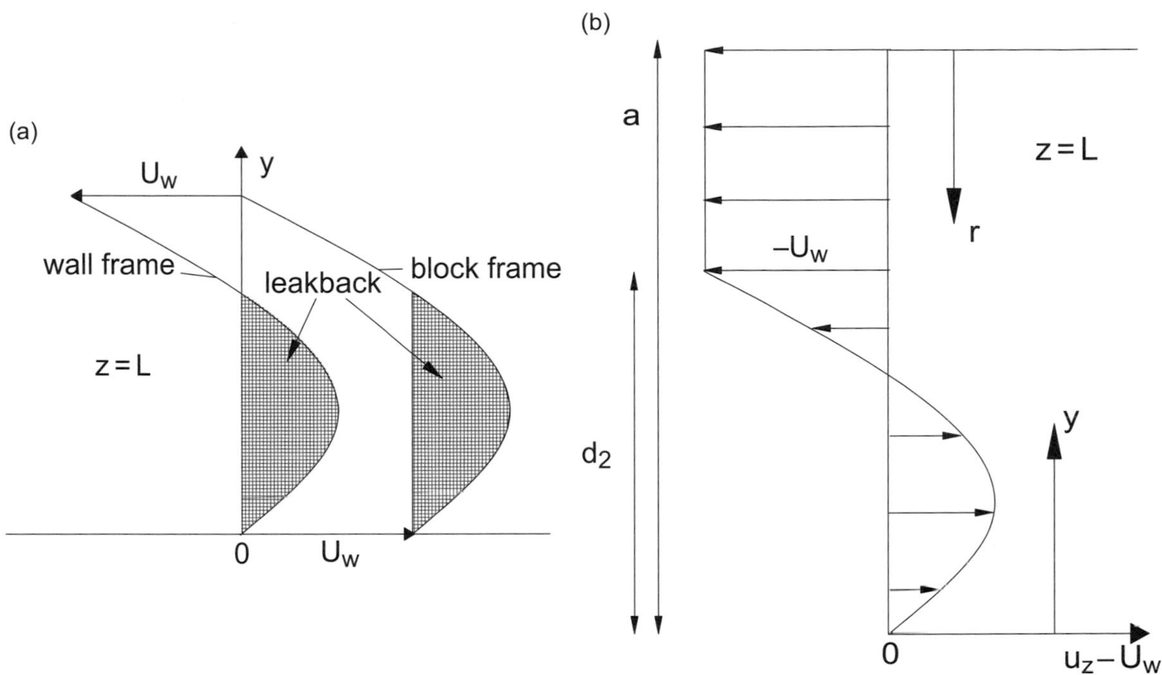

Figure 11.17 (a) Fluid velocity in the wall frame and RBC frame showing "leakback" at z = L. (b) Wall frame fluid velocity at z = L from wall to centerline.

the leakback, which is in the opposite direction. Values of $\hat{Q}$ can be calculated from Eq. (11.3.12).

11.5. Microcirculation and Apparent Blood Viscosity

The Fåhræus–Lindqvist Effect

Recall that Poiseuille Law is the linear relationship between the volumetric flow rate, Q, and pressure drop ΔP for fully developed Newtonian tube flow. From Section 7.1 in Eq. (7.1.8) we derived Poiseuille's Law as $Q = (\Delta P/L)(\pi d^4/128\mu)$ where d is the tube diameter, L is the tube length and μis the fluid viscosity. It is instructive to define an "apparent viscosity," μ_{app}, of the fluid for steady, laminar, fully developed tube flows by rearranging Poiseuille's Law

$$\mu_{app} = \frac{\Delta P}{L}\frac{\pi d^4}{128Q} \tag{11.5.1}$$

Equation (11.5.1) forms the basis of tube flow-based viscometers, for example, since all parameters on the right hand side can be measured.

Using this definition, Fåhræus and Lindqvist in 1931 studied blood flow (their own blood) in a series of glass capillary tubes of different diameters in the range 40–505 μm (Fåhræus and Lindqvist, 1931). The driving pressure was 100 mmHg and tube lengths were also changed. Sodium citrate was added to the blood to prevent clotting. They found that the apparent viscosity of blood, as defined in Eq. (11.5.1), can depend on the tube diameter, for a range of tube diameters below 300 μm. Using water as the standard viscosity, the apparent blood viscosity relative to water, $\mu_{rel/w} = \mu_{app}/\mu_{H_2O}$, was found to decrease as the tube diameter decreased. The results are in Figure 11.18, which shows $\mu_{rel/w}\sim$ 4–5 for large enough tube diameters. That finding is consistent with what we explored in Chapter 10. Shear rates were high enough to eliminate rouleaux formation, so the blood

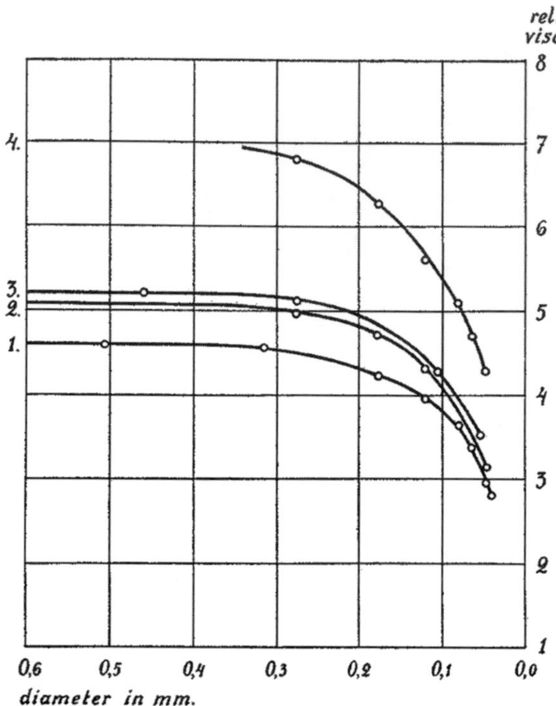

Figure 11.18 Decrease of relative blood viscosity with decreasing tube diameter, the Fåhræus–Lindqvist effect.

was in a Newtonian viscosity range of four to five times the viscosity of water. The top curve of Figure 11.18, which approaches $\mu_{rel/w} \sim 7$, is from a sample of their blood that had some serum removed to raise the hematocrit and increase the viscosity.

The main feature of Figure 11.18, however, is the reduction in $\mu_{rel/w}$ for decreasing tube diameter, down to $\mu_{rel/w} \sim$ 2.7–3.5 for a tube of 40 μm diameter. This is called the Fåhræus–Lindqvist effect and was also reported independently by Martini et al. (1930). Blood does not follow Poiseuille's Law under these circumstances. Poiseuille's Law was the accepted model of tube flows, including for blood, since it was published in 1846 (Poiseuille, 1846). As a physician, Poiseuille's work on flow in glass capillary tubes was motivated by vascular applications. However, blood itself was never used in his experiments, surely due to the lack of anticoagulants in those days to prevent clotting. He initially used distilled water at 10 °C but later branched out to other fluids, including wine, and also modified the fluid temperature (Sutera and Skalak, 1993). So, for Poiseuille's Law to fail in a blood flow apparatus was stunning.

The explanation for the reduction of μ_{app} is that a cell-free layer of thickness $d_{cf} \sim$ 2–5 μm exists adjacent to the tube wall. It results partially from fluid dynamic migration of cells away from the wall due their local spinning in a shear flow, as well as the simple fact that RBCs do not come in fractions of their volume so cannot "fit" in a uniformly dispersed arrangement. The cell-free layer only has serum viscosity, which is lower than that of whole blood. It is negligible until its ratio to the tube radius, d_{cf}/a, becomes large enough. The effect starts at a tube diameter of ~ 300 μm, where this ratio is 2/150–5/150. This ratio value range may seem small, but recall that it is the shear stress at the wall which resists the pressure gradient. Then as the tube diameter becomes even smaller the effect increases as shown in Figure 11.18. The cell-free layer is also known as the "plasma skimming layer," which refers to the non-uniform splitting of flow at a bifurcation. One daughter tube receives more of this layer and fewer RBCs than the other daughter, so the flow details affect the local hematocrit (Fung, 1973; Schmid-Schönbein et al., 1980). August Krogh was awarded the 1920 Nobel Prize in Physiology or Medicine for his studies of flow in capillaries (Krogh, 1922).

Within limits we can consider this a core-annular flow as in Section 7.4, where the core is the blood and the thin annulus is the cell-free layer of plasma. From Eq. (7.4.14) the dimensionless axial velocities of regions 1 (core) and 2 (annulus) were given by

$$U_{Z1} = \frac{1}{4\delta}\left[\delta\left(1-\lambda^2\right) + \left(\lambda^2 - R^2\right)\right] \quad 0 \le R \le \lambda$$

$$U_{Z2} = \frac{1}{4}\left(1-R^2\right) \quad \lambda \le R \le 1 \tag{11.5.2}$$

where the dimensionless velocities are related to their dimensional versions by $U_{Z1} = u_{z1}/U_s$, $U_{Z2} = u_{z2}/U_s$ and the velocity scale is $U_s = (-dp/dz)a^2/\mu_2$. The dimensionless radial position is $R = r/a$ where a is the tube radius. The dimensionless parameters are the viscosity ratio $\delta = \mu_1/\mu_2$ and the radius ratio $\lambda = b/a$ where b is the radius of the core region 1. So region 1 is in the radial range $0 \le r \le b$ or $0 \le R \le \lambda$ and region 2 is in the radial range $b \le r \le a$ or $\lambda \le R \le 1$. The overall dimensional fluid flow rate, Q, is given by integrating the dimensional velocities over the cross-section,

$$Q = \int_0^{2\pi}\int_0^{b} u_{z1}\, r\, dr\, d\theta + \int_0^{2\pi}\int_b^{a} u_{z2}\, r\, dr\, d\theta \tag{11.5.3}$$

Inserting our scaled variables, Eq. (11.5.3) becomes

$$\tilde{Q} = \int_0^{\lambda} U_{Z1} R\, dR + \int_{\lambda}^{1} U_{Z2} R\, dR - \frac{1}{16}\left(1 - \lambda^4\left(1 - \frac{1}{\delta}\right)\right) \tag{11.5.4}$$

where $\tilde{Q} = Q/(2\pi(-dp/dz)a^4/\mu_2)$. The apparent viscosity for the entire flow is derived from Poiseuille Law so that $\mu_{app} = (-dp/dz)\pi a^4/8Q$. Dividing the apparent viscosity by μ_2 gives us a relative viscosity, $\mu_{rel} = \mu_{app}/\mu_2 = 1/16\tilde{Q}$.

For the Fåhræus–Lindqvist effect let $\delta = 4$ so that the viscosity of the core blood is four times the viscosity of the cell-free layer serum. This is a normal ratio for the high strain rate Newtonian behavior of blood. Define $d_{cf} = a - b$ as the cell-free layer thickness, so that $\lambda = (a - d_{cf})/a = 1 - \eta$ where $\eta = d_{cf}/a$. The Fåhræus–Lindqvist effect is seen by letting d_{cf} be fixed and start with a much larger than d_{cf}, so that $\eta \sim 0$. Then progressively reduce a. This is shown in Figure 11.19, where μ_{rel} decreases from 4 to 1 as η increases $0 \le \eta \le 0.2$. For $d_{cf} = 4\,\mu m$, for example, reducing the tube diameter to $40\,\mu m$ ($a = 20\,\mu m$) as in the Fåhræus–Lindqvist experiment is represented by the upper limit $\eta = 4/20 = 0.2$.

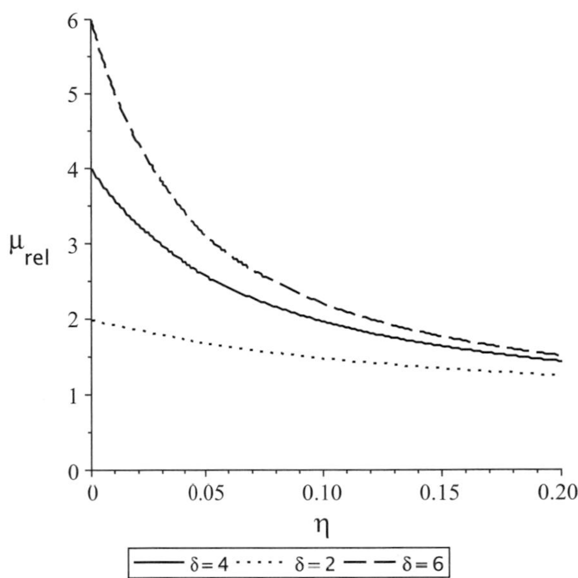

Figure 11.19 Relative apparent viscosity for core-annular flow as a function of $\eta = d_{cf}/a$ with $\delta = 2, 4, 6$.

Inverse Fåhræus–Lindqvist Effect

While Figure 11.18 shows $\mu_{rel/w}$ decreasing with decreasing tube diameter down to a diameter of 40 μm, further reductions in tube diameter reveal in interesting behavior. When the diameter falls below approximately 7 μm, roughly the diameter of an RBC, further decreases in diameter lead to increasing $\mu_{rel/w}$, i.e. the inverse Fåhræus–Lindqvist effect. Pries et al. (1992) developed an empirical fit to data from the literature involving 15 publications with tube diameters ranging from ~3 μm to 2000 μm and discharge hematocrits up to $H_D = 0.90$. (How the discharge hematocrit varies from the flow hematocrit is discussed below.) For $H_D = 0.45$ they fit the relative viscosity data, now based on the serum viscosity, μ_s, with the equation

$$\left(\mu_{rel/s}\right)_{0.45} = \left(\frac{\mu_{app}}{\mu_s}\right)_{0.45} = 220e^{-1.3\,d} + 3.2 - 2.44e^{-0.06\,d^{0.645}} \tag{11.5.5}$$

where d is the tube diameter in units of μm. The additional dependence on H_D was incorporated into the second fit equation

$$\mu_{rel/s} = \frac{\mu_{app}}{\mu_{serum}} = 1 + \left(\left(\mu_{rel/s}\right)_{0.45} - 1\right)\left(\frac{(1 - H_D)^C - 1}{(1 - 0.45)^C - 1}\right) \tag{11.5.6}$$

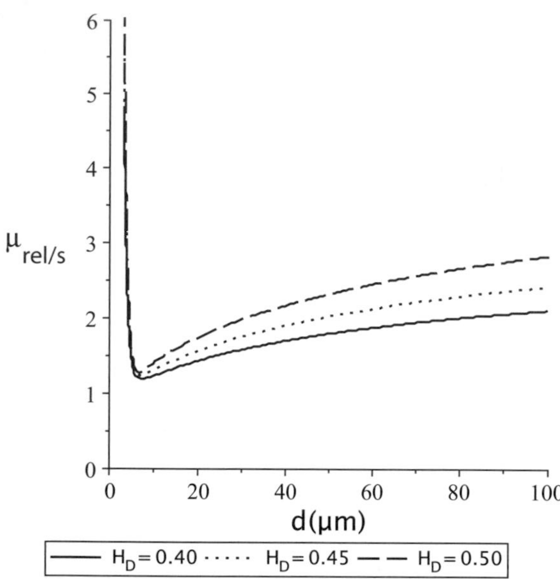

Figure 11.20 $\mu_{rel/s}$ vs d for three values of the discharge hematocrit H_D (based on Pries (1992)).

where

$$C = \left(0.8 + e^{-0.075d}\right)\left(-1 + \frac{1}{1 + 10^{-11}d^{12}}\right) + \left(\frac{1}{1 + 10^{-11}d^{12}}\right) \quad (11.5.7)$$

and the velocities were taken to be 50 tube diameters per second or higher.

Figure 11.20 shows the results from Eq. (11.5.6) for three values of the discharge hematocrit, $H_D = 0.40, 0.45, 0.50$. Starting at a diameter $d = 100\ \mu m$, all three curves show decreasing relative viscosity with decreasing diameter. The higher hematocrit has the higher $\mu_{rel/s}$, as expected. However, all three curves have a local minimum at approximately $d = 7\ \mu m$. Further decreases in diameter lead to higher relative viscosities. The reason for this inverse effect is that once the tube diameter is the same size as the RBC, or smaller, the RBCs are in a line, one after the other, as shown in Figure 11.11. Then lubrication flow dynamics take over as we examined in Section 11.2. Let's explore that further.

Calculate the Apparent Viscosity in a Capillary

With reference to Figure 11.16 let us compare the pressure drop across the RBC to the pressure drop of Poiseuille flow under the same conditions. Though the RBC travels a bit faster than the average fluid speed, just for a rough comparison let $U_{avg} = U_w = 0.01$ cm/s in a 2.5 μm radius tube. The Poiseuille pressure gradient is $dp/dz = 8\mu U_{avg}/a^2$. Substituting the parameter values for $a = 2.5\,\mu m$ and $\mu = 0.012$ poise, we find $dp/dz = 15,360\ dyn/cm^3$ for Poiseuille flow of just the serum. From Figure 11.16(b) the smallest value of $d_1 = 0.25$ and $d_2 = 0.1$ yields $\sim (p_L - p_0)/L \sim 70{,}000\ dyn/cm^3$ and $L = 5.89\ \mu m$. So, with these parameter choices, the pressure gradient across the RBC is roughly four to five times the pressure gradient of just a serum flow.

Consider the series of RBCs in capillary blood flow shown in Figure 11.21. The RBC lengths are all L and the distance from one RBC front to the next is L_H. For example, let $L_H = 2L$ so the RBCs have length L and are also a distance L apart. Under those circumstances, the overall pressure gradient is

$$\begin{aligned}\frac{(p_{02} - p_{01})}{2L} &= \frac{1}{2}\left(\frac{(p_{02} - p_{L1})}{L} + \frac{(p_{L1} - p_{01})}{L}\right)\\ &= \frac{1}{2}(15,360 + 70,000)\\ &= 42{,}680\ dyn/cm^3\end{aligned} \quad (11.5.8)$$

An overall apparent viscosity for the flow of an RBC and the fluid in between is $\mu_{app} = a^2((p_{02} - p_{01})/2L)/8U_{avg} = 0.033$ poise, which is nearly three times the serum value of 0.012 poise. This is an example of the inverse Fåhræus–Lindqvist effect. We also note that for a given V_{RBC} the hematocrit of the blood in this capillary flow is given by $H = V_{RBC}/\pi a^2 L_H$. In our example, $L_H = 2L = 11.8\ \mu m$, $a = 2.5\ \mu m$ and $V_{RBC} = 100\ \mu m^3$ so that $H = 0.43$, which is in the normal range.

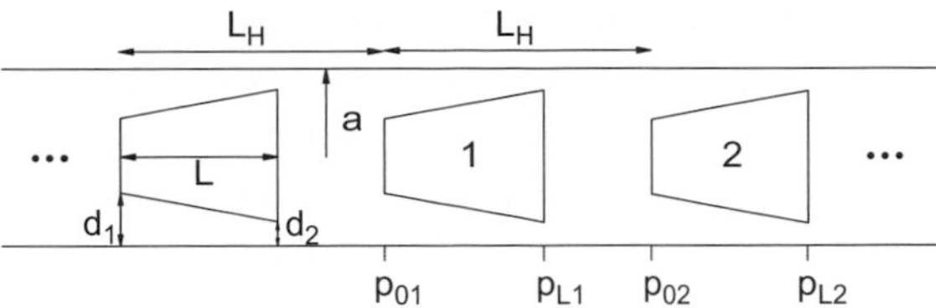

Figure 11.21 Series of RBCs flowing in capillary of radius a. Each RBC has length L and the distance from one RBC front to the next is L_H.

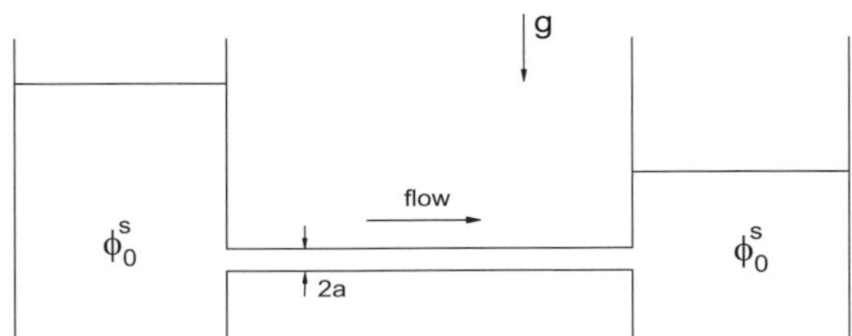

Figure 11.22 Two reservoirs of a suspension with flow from left to right.

Fåhræus Effect of Hemodilution in Capillary Flow

Another ramification of the cell-free layer is that the average hematocrit in the vessels which have a cell-free layer is lower than the hematocrit found in the large vessels. Here is an example. Consider two reservoirs connected by a tube of radius a, see Figure 11.22. The reservoirs contain a fluid with suspended particles. The solid volume fraction, ϕ^s, is the volume of the solid particles divided by the total volume. For blood $\phi^s$is the hematocrit, H. Similarly, the fluid volume fraction, ϕ^f, is the fluid volume divided by the total volume. We note that $\phi^s + \phi^f = 1$. Each reservoir has a constant $\phi^s = \phi_0^s$. There is flow from the left reservoir to the right reservoir through the tube.

The suspension velocity in the tube is radially dependent, $u_z = u_z(r)$, and the fluid volume flow rate, the rate at which fluid is delivered to the downstream reservoir, is defined as

$$Q_f = \int_0^{2\pi}\int_0^a u_z\phi^f\, r\, dr\, d\theta \tag{11.5.9}$$

Let the solid fraction be radially dependent in the tube, i.e. $\phi^s = \phi^s(r)$. Then the rate at which the solid particles are delivered to the downstream reservoir is

$$Q_s = \int_0^{2\pi}\int_0^a u_z\phi^s(r)\, r\, dr\, d\theta \tag{11.5.10}$$

For the downstream reservoir to receive an average concentration of ϕ_0^s, the rate of particle flow divided by the rate of the total flow (fluid plus particles) must be ϕ_0^s,

$$\phi_0^s = \frac{Q_s}{Q_f + Q_s} = \frac{\int_0^{2\pi}\int_0^a u_z\phi^s\, r\, dr\, d\theta}{\int_0^{2\pi}\int_0^a u_z\phi^s\, r\, dr\, d\theta + \int_0^{2\pi}\int_0^a u_z\phi^f\, r\, dr\, d\theta}$$

$$= \frac{\int_0^{2\pi}\int_0^a u_z\phi^s\, r\, dr\, d\theta}{\int_0^{2\pi}\int_0^a u_z\, r\, dr\, d\theta} \tag{11.5.11}$$

where we have used $\phi^s + \phi^f = 1$. The area-average solid fraction, $\bar{\phi}^s$, in the tube is

$$\bar{\phi}^s = \frac{1}{\pi a^2}\int_0^{2\pi}\int_0^a \phi^s\, r\, dr\, d\theta \tag{11.5.12}$$

The ratio of the average concentration in the tube to the constant concentration in the reservoirs is

$$\frac{\bar{\phi}^s}{\phi_0^s} = \frac{\left(\frac{1}{\pi a^2}\int_0^{2\pi}\int_0^a \phi^s\, r\, dr\, d\theta\right)\left(\int_0^{2\pi}\int_0^a u_z\, r\, dr\, d\theta\right)}{\int_0^{2\pi}\int_0^a u_z\phi^s\, r\, dr\, d\theta}$$

$$= \frac{2}{a^2}\frac{\left(\int_0^a \phi^s\, r\, dr\right)\left(\int_0^a u_z\, r\, dr\right)}{\int_0^a u_z\phi^s\, r\, dr} \tag{11.5.13}$$

If either u_z is constant, or ϕ^s is constant, or both, the integrals simplify to $\bar{\phi}^s/\phi_0^s = 1$. For radially dependent u_z and ϕ^s, however, the ratio is always less than one, $\bar{\phi}^s/\phi_0^s < 1$, and the average solid fraction in the tube is less than in the reservoirs. For blood, the hematocrit in the tube is less than the hematocrit of the reservoirs, as discovered by Fåhræus (1929). This phenomena is known as the Fåhræus effect.

11.6 Filtration Flow in a Capillary

In this section we examine flow not only along the axis of a capillary, but also across its wall, often called a filtration flow. Capillary walls are designed to be permeable to allow exchange of fluids and molecules at various levels. There are three categories of the permeability range as shown in Figure 11.23. Continuous capillaries only allow crossflow through the narrow space between the endothelial cells, the intercellular clefts, whose dimension is on the order of 4 nm. They only allow small molecules to cross, like water and ions. Fenestrated capillaries have larger pores that are 60–80 nm in diameter. They are large enough to permit protein molecules through, the renal capillaries of the glomerulus being a prime example. Sinusoidal capillaries have much larger openings, on the order of 30–40 μm, and occur in specialized circulations like the liver, spleen and bone marrow.

For our fluid mechanical analysis, consider flow in a tube section as a model capillary shown in Figure 11.24. Fluid flows from left to right due to an imposed pressure difference, but can also exit or enter the capillary across its membrane walls because they are permeable. The tube has radius a and a permeable section of length L, as shown. The remaining tube, upstream and downstream of the permeable section, will be impermeable. The fluid pressure internal to the capillary is p, while the constant external fluid pressure is p_i, the interstitial fluid pressure. Also the oncotic pressure, π, due to non-diffusible large molecules like proteins, influences the fluid transport across the capillary membrane. Oncotic pressure has the opposite sense of the hydrostatic pressure, i.e. flow goes from regions of lower to higher π. The interstitial oncotic pressure is π_i and the capillary oncotic pressure is π_c, and both are considered constant. The entrance pressure at $z = 0$ is p_0 and the exit pressure at $z = L$ is p_L.

Starling's equation is a well-known model of the permeability and crossflow conditions at a membrane. It is named for British physiologist Ernest Henry Starling (1866–1927), who is also recognized for the

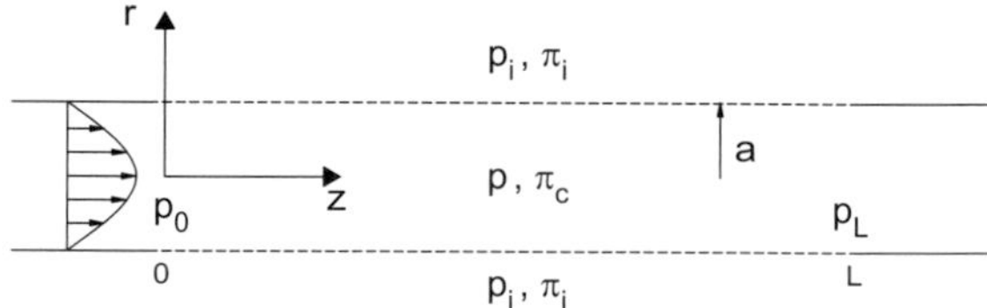

Figure 11.24 Flow through a porous tube as a model of capillary flow and fluid balance.

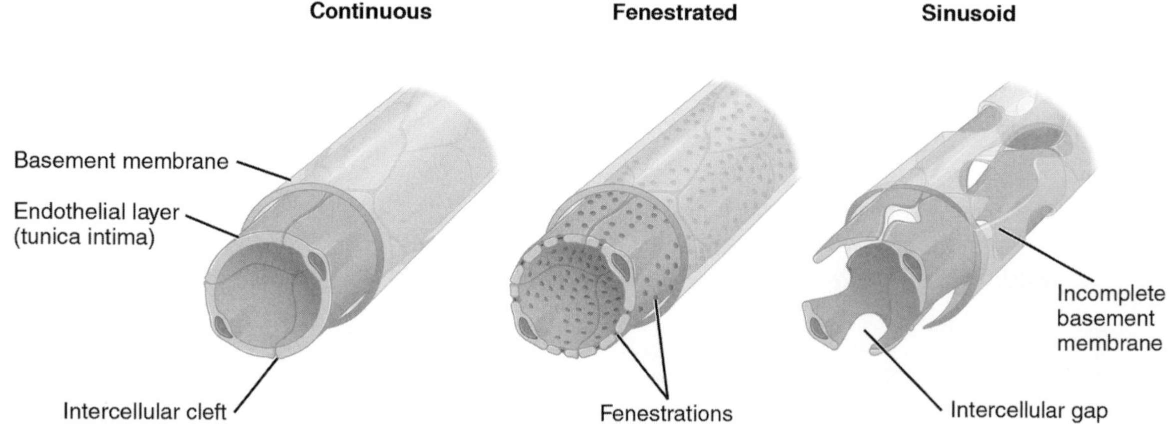

Figure 11.23 Three capillary types of permeability: continuous, fenestrated and sinusoid.

Frank–Starling law of the heart. Starling's equation is a well known model of the permeability and cross flow conditions at a membrane (Starling, 1896), and is stated as

$$J_V = L_P A[(p - p_i) - \sigma(\pi_c - \pi_i)] \tag{11.6.1}$$

where J_V is the volume per time flow rate across the membrane, L_P is the hydraulic conductivity in units of velocity (cm/s) per pressure unit (e.g. cmH_2O or mmHg), A is the surface area of the membrane and σ is the reflection coefficient. To solve a fluid mechanics problem, we need a velocity form of Eq. (11.6.1), so divide Eq. (11.6.1) by A to arrive at the cross-membrane fluid velocity, u_r,

$$u_r(r = a) = L_P[(p - p_i) - \sigma(\pi_c - \pi_i)] \tag{11.6.2}$$

For $\varepsilon = a/L << 1$ lubrication theory applies, and in this case we also assume axisymmetry. The simplified Navier–Stokes equations are, again, the balance of axial pressure gradient with the viscous shear. Let the velocity components be (u_r, u_θ, u_z) in the (r, θ, z) directions respectively, and assuming no azimuthal velocity, $u_\theta = 0$, and axisymmetry $\partial/\partial\theta = 0$. The r- and z-momentum components are

$$\begin{gathered} \frac{dp}{dr} = 0 \\ -\frac{dp}{dz} + \mu \frac{1}{r}\frac{\partial}{\partial r}\left(r\frac{\partial u_z}{\partial r}\right) = 0 \end{gathered} \tag{11.6.3}$$

Similarly, conservation of mass in cylindrical coordinates is

$$\frac{1}{r}\frac{\partial}{\partial r}(ru_r) + \frac{\partial u_z}{\partial z} = 0 \tag{11.6.4}$$

The boundary conditions on the velocities are that the axial velocity will obey no-slip at $r = a$, and be finite at $r = 0$, while the radial velocity is zero at $r = 0$ but equal to the right hand side of Eq. (11.6.2) at the wall, $r = a$. This is the permeability condition.

The boundary conditions on the fluid, then, are

$$\begin{aligned} &u_z(r=0) \quad \text{finite}, \quad u_z(r=a)=0 \\ &u_r(r=0)=0, \qquad u_r(r=a)=L_P[(p-p_i)-\sigma(\pi_c-\pi_i)] \end{aligned} \tag{11.6.5}$$

The wall is impermeable for $L_P = 0$, which would yield Poiseuille flow as our solution which is a parabolic profile. For small permeability, we might expect that the flow is locally parabolic, which we can show with our scaling.

Select the lubrication scales as the following

$$U_R = \frac{u_r}{\varepsilon w_0}, \quad U_Z = \frac{u_z}{w_0}, \quad R = \frac{r}{a}, \quad Z = \frac{z}{L}, \quad P = \frac{p}{\mu w_0/\varepsilon a} \tag{11.6.6}$$

which are similar to Eq. (6.6.7) for rectangular coordinates. Here, w_0 is a characteristic axial velocity which we choose to be the average velocity for Poiseuille flow, i.e. an impermeable tube. From Eq. (7.1.6) it is defined as $w_0 = a^2(p_0 - p_L)/8\mu L$. Inserting this value of w_0 into the pressure scale, $\mu w_0/\varepsilon a$, in Eq. (11.6.6) yields $P = 8p/(p_0 - p_L)$. When we insert the definitions of Eq. (11.6.6) into Eqs. (11.6.3) and (11.6.4), the resulting equation for Z-momentum conservation is

$$0 = -\frac{dP}{dZ} + \frac{1}{R}\frac{\partial}{\partial R}\left(R\frac{\partial U_Z}{\partial R}\right) \tag{11.6.7}$$

where $P = P(Z)$ only. The Reynolds number is $Re = \rho a w_0/\mu$ and we have assumed $\varepsilon Re << 1$ as part of our lubrication assumption. The conservation of mass, Eq. (11.6.4), simplifies to

$$\frac{1}{R}\frac{\partial}{\partial R}(RU_R) + \frac{\partial U_Z}{\partial Z} = 0 \tag{11.6.8}$$

while the boundary conditions become

$$\begin{aligned} &U_Z(R=0) \quad \text{finite}, \qquad U_Z(R=1)=0 \\ &U_R(R=0)=0, \qquad U_R=\lambda^2[(P-P_i)-\pi_m] \end{aligned} \tag{11.6.9}$$

The dimensionless permeability parameter is $\lambda^2 = (\mu L_P)/(\varepsilon^2 a)$ and the dimensionless net oncotic pressure parameter is $\pi_m = \sigma(\pi_c - \pi_i)/(\mu w_0/\varepsilon a)$. For λ^2 to be order one in size, i.e. $\lambda^2 = O(1)$, we need $\mu L_P/a \sim O(\varepsilon^2)$. So L_P cannot be too large. The term $1/L_P$ can be considered as a flow resistance, then λ^2 reflects the ratio of axial flow resistance to the leakage flow resistance when rearranged as follows

$$\lambda^2 = \frac{L_P\mu}{\varepsilon^2 a} = \frac{\dfrac{\mu L}{a^4}}{\dfrac{1}{L_P La}} \sim \frac{\text{axial flow resistance}}{\text{radial leakage flow resistance}} \tag{11.6.10}$$

The axial flow resistance in the numerator of Eq. (11.6.10) is from Poiseuille Law.

Integrating Eq. (11.6.7) twice gives us

$$U_Z(R, Z) = \frac{1}{4}\frac{dP}{dZ}R^2 + c_0(Z)\ln(R) + d_0(Z) \tag{11.6.11}$$

where $c_0(Z)$ and $d_0(Z)$ are functions of integration. The boundary conditions on U_Z are no-slip at the wall and finite velocity at the centerline

$$U_Z(R = 1) = 0, \quad U_Z(R = 0) \quad \text{finite} \tag{11.6.12}$$

The finite condition at the centerline requires $c_0(Z) = 0$, while the no-slip condition at the wall leads to the form

$$U_Z(R, Z) = -\frac{1}{4}\frac{dP}{dZ}(1 - R^2) \tag{11.6.13}$$

The solution for U_R comes from integrating the continuity equation, Eq. (11.6.8),

$$\begin{aligned}\int d(RU_R) &= -\int R\frac{\partial U_Z}{\partial Z}dR + c_1(Z)\\ &= \frac{1}{4}\frac{d^2P}{dZ^2}\int R(1 - R^2)dR + c_1(Z)\end{aligned} \tag{11.6.14}$$

where $c_1(Z)$ is a function of integration. Evaluate the integrals of Eq. (11.6.14), divide through by R and set $c_1 = 0$ to preserve finite solutions at $R = 0$. The result is

$$U_R(R, Z) = \frac{1}{4}\frac{d^2P}{dZ^2}\left(\frac{R}{2} - \frac{R^3}{4}\right) \tag{11.6.15}$$

The last condition is the leak condition that allows radial fluid flow through the wall due to the pressure difference between the fluid in the tube and the external pressure,

$$U_R(R = 1) = \frac{1}{16}\frac{d^2P}{dZ^2} = \lambda^2[(P - P_i) - \pi_m] \tag{11.6.16}$$

Equation (11.6.16) is an ordinary differential equation for P(Z) with homogeneous, P_h, and particular, P_p, solutions that sum to be

$$P = P_h + P_p = \left(a_0 e^{4\lambda Z} + b_0 e^{-4\lambda Z}\right) + (P_i + \pi_m) \tag{11.6.17}$$

We can impose the pressures at $Z = 0$ and $Z = 1$ to determine the unknown constants a_0 and b_0. For example, let $P(Z = 0) = P_0$, $P(Z = 1) = P_1$. Then the solution Eq. (11.6.17) is

$$\begin{aligned}P(Z) = &\left(\frac{P_1 - P_0e^{-4\lambda} - (P_i + \pi_m)\left(1 - e^{-4\lambda}\right)}{e^{4\lambda} - e^{-4\lambda}}\right)e^{4\lambda Z}\\ &+ \left(\frac{-P_1 + P_0e^{4\lambda} + (P_i + \pi_m)\left(1 - e^{4\lambda}\right)}{e^{4\lambda} - e^{-4\lambda}}\right)e^{-4\lambda Z}\\ &+ (P_i + \pi_m)\end{aligned} \tag{11.6.18}$$

In Sections 7.1 and 7.2 we introduced some data for blood vessels, including capillaries. Typical values for the dimensional pressure data are $p_0 = 35$ mmHg and $p_L = 15$ mmHg so $p_0 - p_L = 20$ mmHg. The interstitial hydraulic pressure can vary with the tissue and is often negative. We will let it be $p_i = -2$ mmHg. The capillary oncotic pressure is taken to be $\pi_c = 26$ mmHg, while the interstitial value will be $\pi_i = 5$ mmHg. The reflection coefficient varies with the molecule. For albumin, σ is 0.816 in frog mesenteric capillaries, while for myoglobin it is 0.348 (Michel, 1980). We will assume $\sigma = 1$ These choices make $P_0 = 8p_0/(p_0 - p_L) = 14$, $P_1 = 8p_L/(p_0 - p_L) = 6.0$, $P_i = 8p_i/(p_0 - p_L) = -0.8$, $\pi_m = 8\sigma(\pi_c - \pi_i)/(p_0 - p_L) = 8.4$, which will constitute a base state of data for our calculations.

The permeability of a capillary depends on the function of the organ or tissue. For example, the kidney is designed to allow relatively large flows across the glomerular capillaries which are quite permeable because of their fenestrations, see Figure 11.23. The net crossflow (filtration) for an adult is 125 ml/min, known as the glomerular filtration rate. For the other capillaries in the body the total net filtration is about 20 ml/min. A typical value of frog muscle capillaries is $L_P = 0.74 \times 10^{-7}$cm/(s·cm$H_2O$)

(Curry and Frokjaerjensen, 1984) and for frog mesenteric capillaries $L_P = 1.5 - 15 \times 10^{-7}$ cm/(s·cmH$_2$O) (Michel, 1980). In the presence of ANP (atrial-natriuretic-peptide) L_P increased from 6.6×10^{-7} cm/(s·cmH$_2$O) to 40×10^{-7} cm/(s·cmH$_2$O) for frog mesenteric capillaries (Huxley et al., 1987). For our example calculation, we choose $L_P = 15 \times 10^{-7}\text{cm}/(\text{s·cmH}_2\text{O})$. In cgs units, $1\ \text{cmH}_2\text{O} = \rho g h$, where ρ is water density (1 g/cm^3), g is the gravitational constant (980 cm/s^2) and $h = 1$ cm. From the definition of λ^2 we can calculate its value,

$$\begin{aligned}
\lambda^2 &= \frac{\mu L_P}{\varepsilon^2 a} \\
&= 15 \times 10^{-7} \frac{\text{cm}}{\text{s·cmH}_2\text{O}} \times \frac{1\ \text{cmH}_2\text{O}}{980\,\text{g}/(\text{cm·s}^2)} \\
&\quad \times 0.04 \frac{\text{g}}{\text{cm·s}} \times \frac{1}{0.0003\ \text{cm}} \times \frac{1}{0.003^2} \\
&= 0.0085
\end{aligned} \tag{11.6.19}$$

From Eq. (11.6.19) we see that $\lambda = 0.092$. In Eq. (11.6.19) we used $L \sim 1\ \text{mm} = 0.1\ \text{cm}$, $a \sim 3\,\mu\text{m} = 3 \times 10^{-4}\text{cm}$ so $\varepsilon = 0.003$. Also, we choose $\mu = 0.015$ poise, which falls in the range of Figure 11.20.

Figure 11.25 shows plots of P vs Z from Eq. (11.6.18) for four values of the permeability parameter, $\lambda = 0.01, 0.33, 0.66, 1.00$, whose range includes the continuous and fenestrated capillaries of Figure 11. 23. The $\lambda = 0.01$ example is a nearly impermeable wall, which results in essentially Poiseuille flow with a constant pressure gradient. Notice that as λ increases the pressure gradients are steeper at both endpoints, $Z = 0, 1$, but flatter near the middle.

An example of U_Z vs R at five values of Z is shown in Figure 11.26 for the same base state pressure data as in Figure 11.25. Notice that the maximum velocities at $R = 0$ are all close to 2. That reflects our choice of velocity scale which is the average velocity for an impermeable tube ($\lambda = 0$), i.e. Poiseuille flow. We learned earlier Section 7.1 that the maximum velocity in Poiseuille flow is twice the average, so with small permeability it is nearly twice. This is a good check on our approach. Also in Figure 11.26 you can see that the velocity is largest at $Z = 0$, then slows

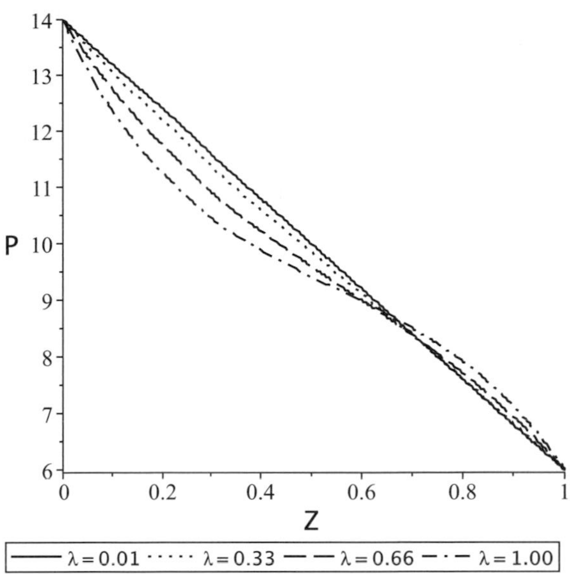

Figure 11.25 Pressure field, P(Z), for four values of the permeability parameter λ.

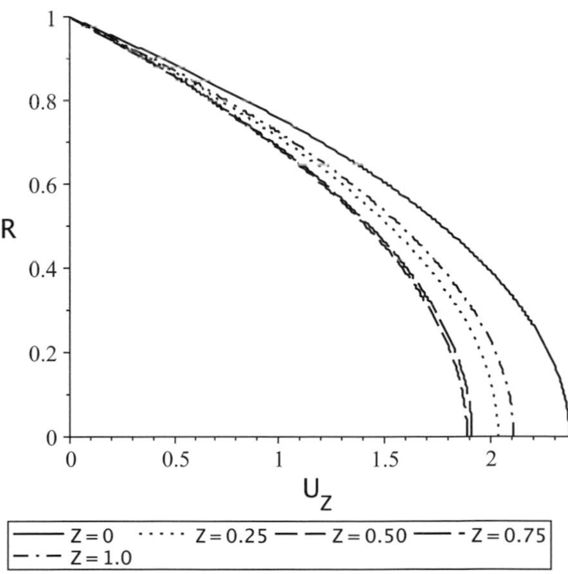

Figure 11.26 Velocities U_Z vs R at five values of Z for $\lambda = 0.3$.

down at $Z = 0.25$ and more at $Z = 0.50$ as fluid flows out of the tube across the wall. However, then it increases at $Z = 0.75$ and more at $Z = 1.00$ as fluid flows back into the tube across the wall. The maximum velocities at $R = 0$ are greater than 2 near the capillary ends, $Z = 0, 1$, where the slope dP/dZ increases as λ increases, as seen in Figure 11.25.

The radial velocity, U_R at the membrane, $R = 1$, is the key to the fluid exchange between the capillary fluid and the interstitial fluid. Figure 11.27 shows plots of $U_R(R = 1, Z)$ for four values of λ. Recall that the radial velocity is scaled on εw_0, so the dimensional values are much smaller than the axial velocities which are scaled on w_0. U_R is positive for $0 \leq Z \leq Z_0$ and negative for $Z_0 \leq Z \leq 1$ where Z_0 is the zero crossflow location, $U_Z(R = 1, Z = Z_0) = 0$. The dimensional net radial flow rate, q_r, is the integral of u_r at $r = a$ over the surface of the cylinder

$$q_r = \int_0^L \int_0^{2\pi} u_r(r = a, z)\, a\, d\theta\, dz \tag{11.6.20}$$

Using this definition and our scales, the dimensionless net crossflow is

$$Q_R = \frac{q_r}{2\pi a^2 w_0} = \int_0^1 U_R(R = 1, Z)dZ \tag{11.6.21}$$

which includes the upstream outflow segment, $0 \leq Z \leq Z_0$, and the downstream inflow segment, $Z_0 \leq Z \leq 1$. From Figure 11.27 it is obvious that there is a net loss of fluid from the capillary under normal circumstances. This fluid flows into the interstitial tissue where it is drained away by the lymphatic system. Including all capillaries other than those in the renal system, estimates are that about 20 ml/min of capillary fluid enters the interstitium. Various estimates are that there are 10 billion capillaries in an adult totaling 25,000 miles each with a length of 1 mm.

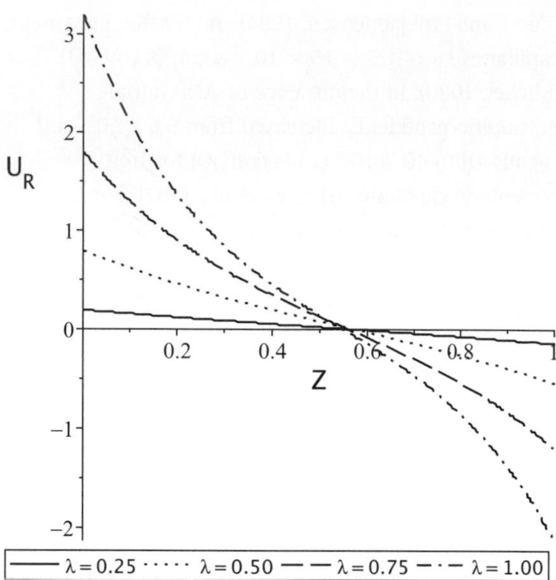

Figure 11.27 The radial velocity U_R evaluated at the membrane wall, $R = 1$, for four values of the permeability parameter λ.

The stream function, ψ, for axisymmetric flows in cylindrical coordinates was shown in Eq. (4.6.6) to be defined by $u_r = -(1/r)(\partial\psi/\partial z)$, $u_z = (1/r)(\partial\psi/\partial r)$. Using the scales in Eq. (11.6.6) the corresponding dimensionless stream function must be $\Psi = \psi/a^2 w_0$ so that

$$U_R = -\frac{1}{R}\frac{\partial\Psi}{\partial Z}, \quad U_Z = \frac{1}{R}\frac{\partial\Psi}{\partial R} \tag{11.6.22}$$

From the definition of U_R we can insert Eq. (11.6.15) and integrate to find Ψ,

$$\Psi = -R\int^Z U_R + c_4(R) = -\frac{1}{4}\frac{dP}{dZ}\left(\frac{R^2}{2} - \frac{R^4}{4}\right) \tag{11.6.23}$$

It is customary to assign $\Psi = 0$ on an impermeable boundary. For our flow that would be at the centerline, $R = 0$. Imposing $\Psi(R = 0) = 0$ on Eq. (11.6.23) determines $c_4(R) = 0$.

Figure 11.28 shows the streamlines and velocity vector field for $\lambda = 1$. The Maple code for Figure 11.28 is shown in Code 11.2(a), and the MATLAB® code is Code 11.2(b).

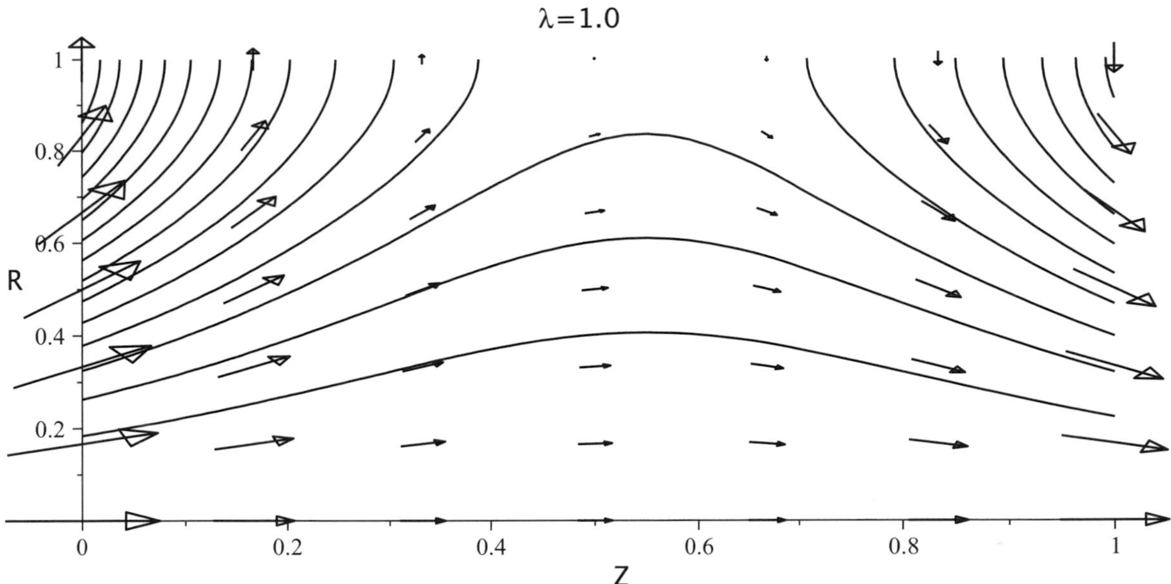

Figure 11.28 Streamlines and velocity vector field for $\lambda = 1$.

```
(a)
restart
with(plots):
#Define pressure P
P := a·exp(4·lambda·Z) + b·exp(-4·lambda·Z) + Pint + Pim;
```

$$P := a e^{4\lambda Z} + b\, e^{-4\lambda Z} + Pint + Pim \tag{1}$$

```
#solve for integration constants a,b from the boundary conditions and
assign their values
EQ1 := subs (Z=0, P) = P0 :
EQ2 := suns (Z=1, P) = P1 :
S := solve ({EQ1, EQ2}, {a, b}) :
assign(s):
#Define pressure data, recall P = 8p/(p0-pL)
```

$$Pint := -\frac{2\cdot 8}{20} : Pim := \frac{8\cdot(26-1)}{20} : P0 := \frac{8\cdot 35}{20} : P1 := \frac{8\cdot 15}{20} :$$

```
:
#Define UZ(R, Z) plot at different values of Z
```

$$UZ := -\frac{1}{4}\cdot \mathrm{diff}(P,\ Z)\cdot(1-R^2);$$

$$UZ := -\frac{\left(-\frac{4\left(\frac{24\, e^{-4\lambda}}{5}+\frac{16}{5}\right)\lambda\, e^{4\lambda Z}}{e^{4\lambda}-e^{-4\lambda}} - \frac{4\left(\frac{24\, e^{4\lambda}}{5}+\frac{16}{5}\right)\lambda\, e^{-4\lambda Z}}{e^{4\lambda}-e^{-4\lambda}}\right)(-R^2+1)}{4} \tag{2}$$

Code 11.2 Code examples: (a) Maple, (b) MATLAB®.

```
#Define UR(R, Z)
```

$$UR := -\frac{1}{4}\cdot \text{diff}(\text{diff}(P, Z), Z)\cdot\left(\frac{R}{2}-\frac{R^3}{3}\right);$$

UR :=

$$\frac{\left(-\frac{16\left(\frac{24\,e^{-4\lambda}}{5}+\frac{16}{5}\right)\lambda^2 e^{4\lambda Z}}{e^{4\lambda}-e^{-4\lambda}}+\frac{16\left(\frac{24\,e^{4\lambda}}{5}+\frac{16}{5}\right)\lambda^2 e^{-4\lambda Z}}{e^{4\lambda}-e^{-4\lambda}}\right)\left(\frac{1}{2}R-\frac{1}{3}R^3\right)}{4} \quad (3)$$

```
#Define the stream function as psi
```

$$psi := -\frac{1}{4}\cdot \text{diff}(P,\ Z)\cdot\left(\frac{R^2}{2}-\frac{R^4}{4}\right);$$

$$\psi := -\frac{\left(-\frac{4\left(\frac{24e^{-4\lambda}}{5}+\frac{16}{5}\right)\lambda e^{4\lambda Z}}{e^{4\lambda}-e^{-4\lambda}}-\frac{4\left(\frac{24e^{4\lambda}}{5}+\frac{16}{5}\right)\lambda e^{-4\lambda Z}}{e^{4\lambda}-e^{-4\lambda}}\right)\left(\frac{1}{2}R^2-\frac{1}{4}R^4\right)}{4} \quad (4)$$

```
#Create streamline plot using "contourplot" of psi
PL1 := contourplot(subs(lambda=1.0,psi),Z=0,...,1,R=0,...,1, Color=black, title="λ - 1.0", labels
     = ["Z","R"], contours =14):
#Create velocity vector field plot using "fieldplot" for (UZ,UR) vector
PL2 := fieldplot([subs(lambda=1.0,UZ),subs(lambda=1.0,UR)], Z = 0,...,1 R = 0,...,1, arrows
     = SLIM, grid = [7,7]) :
display(PL1, PL2);
```

(b)

```
clear all; close all; clc
syms a b P Z Pint Pim lambda  P0 P1 R
P              = a*exp(4*lambda*Z) + b*exp(-4*lambda*Z) + Pint + Pim;
EQ1            = subs(P,Z,0) == P0;
EQ2            = subs(P,Z,1) == P1;
EQS            = [EQ1,EQ2];
VARS           = [a b];
[aSol,bSol]    = solve(EQS,VARS);
PintEval       = 2*8/20;
PimEval        = 8*(26-1)/20;
P0Eval         = 8*35/20;
P1Eval         = 8*15/20;
aEval          = subs(aSol,[Pint,Pim,P0,P1],[PintEval,PimEval,P0Eval,P1Eval]);
bEval          = subs(bSol,[Pint,Pim,P0,P1],[PintEval,PimEval,P0Eval,P1Eval]);
PEval          = subs(P,[Pint,Pim,a,b],[PintEval,PimEval,aEval,bEval]);
UZ             = -1/4*diff(PEval,Z)*(1-R^2);
UR             = 1/4*diff(PEval,Z,Z)*(R/2-R^3/3);
PSI            = -1/4*diff(PEval,Z)*(R^2/2-R^4/4);
PSIfunction    = matlabFunction(PSI);
```

Code 11.2 *(continued)*

```
UZfunction  = matlabFunction(UZ);
URfunction  = matlabFunction(UR);
r           = linspace(0,1,1000);
z           = linspace(0,1,500);
[R,Z]       = meshgrid(r',z');
lambda      = 1;
contour(Z,R,PSIfunction(R,Z,lambda),14,'-k')
hold on;
quiver(Z(1:50:end,1:100:end),R(1:50:end,1:100:end),...
    UZfunction(R(1:50:end,1:100:end),Z(1:50:end,1:100:end),lambda),...
    URfunction(R(1:50:end,1:100:end),Z(1:50:end,1:100:end),lambda),'-k')
title('\lambda = 1')
xlabel('Z')
ylabel('R')
axis([0 1 0 1])
```

Code 11.2 *(continued)*

Clinical Correlation 11.2

Under normal circumstances there is a net outflow of fluid from the capillaries into the interstitium, which is then drained away by the lymphatics to be returned, eventually, to the circulation. However, the lymphatic drainage system has a limited flow capacity, which can be exceeded if the capillary outflow is too large. When that happens, fluid accumulates in the interstitium resulting in edema.

A number of clinical situations cause edema. Standing or sitting for long periods of time can increase the blood pressure in the capillaries of the lower extremities causing swelling of the feet and ankles. A similar sequence occurs with right heart failure which increases systemic venous blood pressure, or left heart failure increasing pulmonary capillary blood pressure. In terms of the model, the edema in these examples derives from increases in p_0 and p_L. Another mechanism involves increasing the hydraulic conductivity, L_P, so that the capillary walls are more permeable. For example, histamine is an inflammatory biochemical released by mast cells and white blood cells from a variety of circumstances including immune responses, exposure to irritants and toxins, and mechanical injury. It makes the capillary walls more permeable, creating local tissue swelling of the affected region. Finally, decreases in the blood oncotic pressure, π_c, drives more fluid into the interstitium. That can happen during severe malnutrition when blood protein levels are reduced. A well-known clinical entity in famine-stricken children called "kwashiorkor" includes not only local tissue swelling but also fluid accumulation in the abdominal cavity causing a distended abdomen, see Figure 11.29.

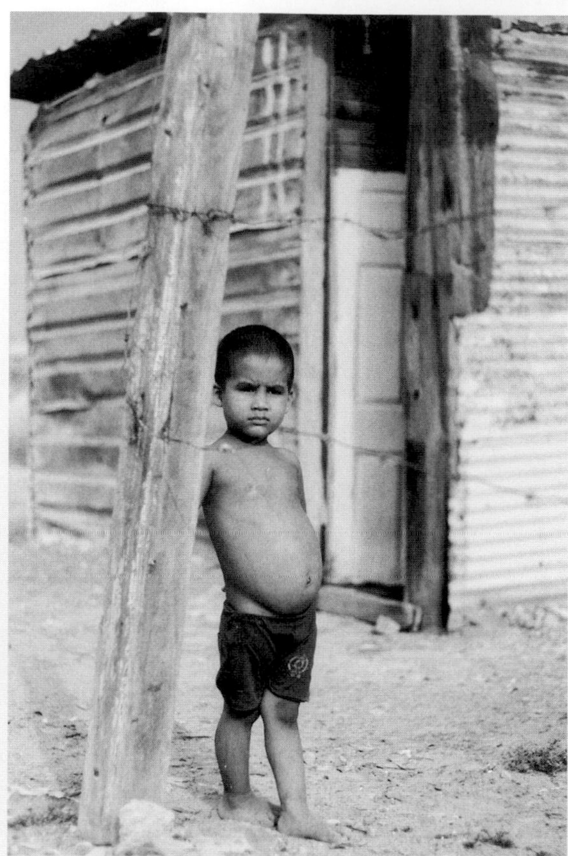

Figure 11.29 Barcelona, Venezuela June 2, 2018: child in the poor community "ciudad de los mochos" with problems of malnutrition kwashiorkor.

When edema occurs through pulmonary capillaries, overflow from the interstitium into the alveolar air sacs may result. This situation, called pulmonary edema, is potentially life threatening, because gas exchange is impaired. Drug therapy is aimed at removing excess fluid through the kidneys, lowering blood pressure generally, and improving cardiac function.

Let's use the model to gain insight into the mechanisms of edema formation. The net radial fluid transport across the membrane, Q_R, is given by Eq. (11.6.21). Figure 11.30 is a plot of Q_R vs λ for four values of the capillary pressure, π_c. The normal value of π_c is 25 mmHg due primarily to albumin in the blood. We see in Figure 11.30 that this value results in $Q_R > 0$ over the range $0 \leq \lambda \leq 1.0$. So under normal circumstances the capillaries are losing intravascular fluid to the surrounding interstitium. The lymphatic system drains the excess away. Kwashiorkor is modeled by using lower values of $\pi_c = 23, 21, 19$ mmHg because of decreased blood levels of albumin from malnutrition. The curves plotted in Figure 11.30 show larger values of Q_R. When those values exceed the lymphatic drainage capacity, edema results.

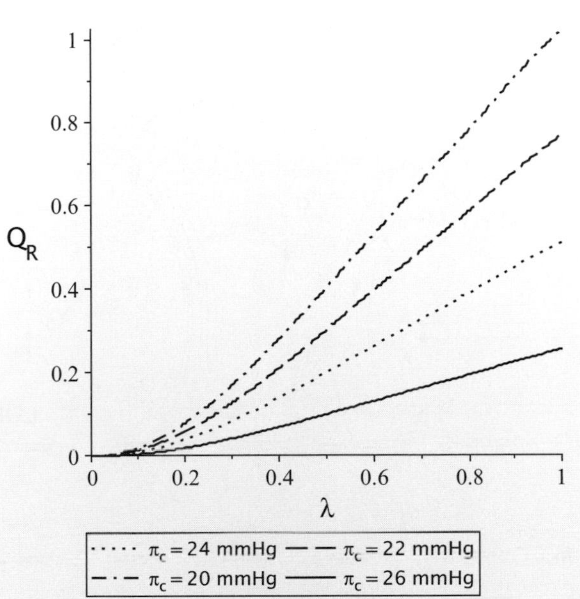

Figure 11.30 Q_R vs λ for four values of π_c.

Box 11.4 Pleural Squeeze Film and Filtration Flow

Another example of a squeeze film occurs in the intrapleural fluid of the lung. Figure 11.31(a) shows a cross-section of a supine lung including the visceral pleural membrane covering the lung, the parietal pleural membrane covering the inside of the chest cavity, and the pleural fluid in between. Typically the gap between the pleural membranes is on the order of 5–50 μm in a healthy individual. In addition to lubricating the relative motion between the lung and thoracic surfaces, as discussed in Chapter 6, Clinical Correlation 6.2, the pleural fluid pressure determines lung inflation and its distribution.

The lung is made of tissue, whose density is similar to water, ~ 1 g/cm^3, but is also inflated with air, whose density is much smaller, ~ 1.25×10^{-3} g/cm^3. Consequently it has an average density less than water, $\rho_L \sim 0.2$–0.3 g/cm^3. The surrounding pleural fluid is mostly water, $\rho_{PF} \sim 1$ g/cm^3, so the lung is buoyant and floats upward, while squeezing intrapleural fluid downward as shown. Analysis shows the transient time governing the cylinder off-set, Y(t), is ~ 5 minutes (Grotberg and Glucksberg, 1994), which compares well with experiments.

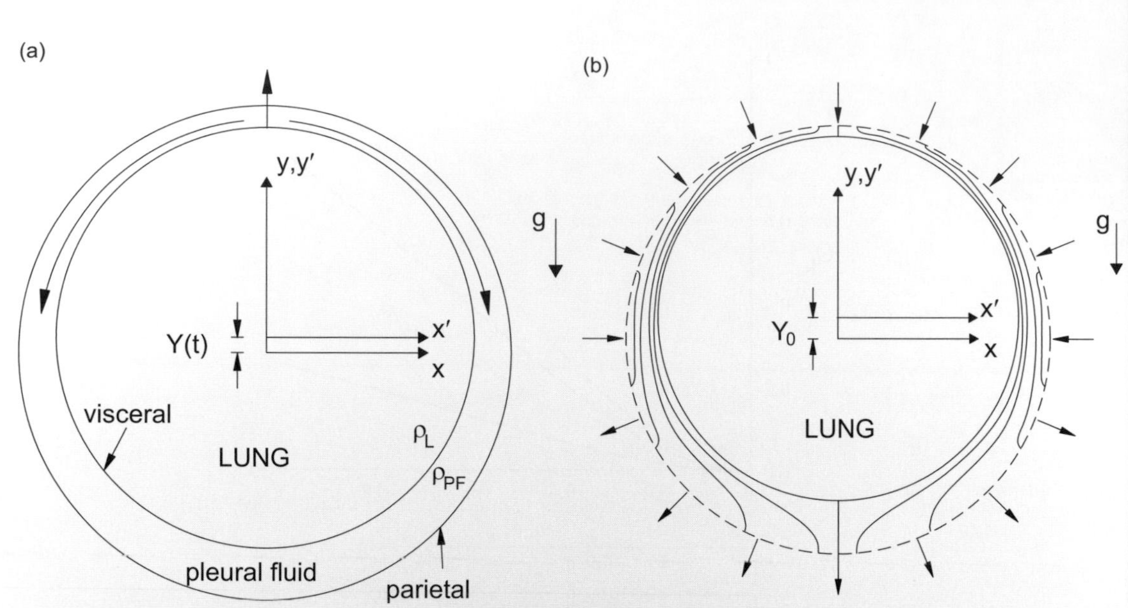

Figure 11.31 Supine lung cross-section showing pleural fluid, visceral pleural membrane and parietal pleural membrane. (a) Transient flow as buoyant lung rises and squeezes fluid from top to bottom. (b) Effect of transparietal flow Starling equation and lymphatics, creates a steady state balance of fluid entering the top and exiting the bottom. Streamlines sketched.

For steady state, $Y = Y_0$, however, we must include the transparietal flow, which depends on the Starling equation as well as lymphatics (Staub et al., 1985; Miserocchi et al., 1993; Miserocchi, 1997), see Figure 11.31(b). There is a steady state balance of fluid entering the top and exiting the bottom which applies a balancing force to the lung buoyancy, see the sketched streamlines (Haber et al., 2001). When the inflow exceeds the outflow, pleural fluid accumulates which is called a pleural effusion and lung inflation becomes compromised. Pleural mesothelial cells are sensitive to the fluid shear stresses imposed by these flows, especially their permeability and release of bioactive factors (Waters et al., 1996; 1997).

11.7 Squeeze Film: Flow in Joints

A thin layer of synovial fluid occurs in joints to help lubricate the motion between the two bone ends that make up the joint. In Figure 11.32 we see the synovial fluid, bone ends, cartilage covering the bone surface in the joint, muscle, tendon, and the synovial membrane which keeps the fluid within the joint and also secretes it.

When there is a force, say from weight bearing, in the axis of the bones, they are pushed toward one another and the synovial fluid is squeezed outward from between them. A typical example is the distal

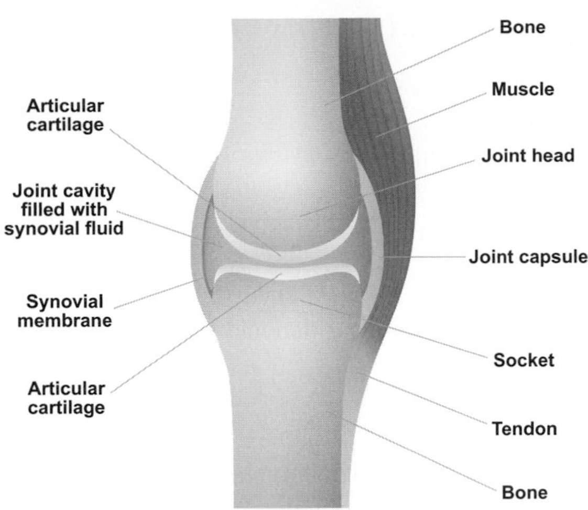

Figure 11.32 Example of a joint with synovial fluid in the "joint cavity."

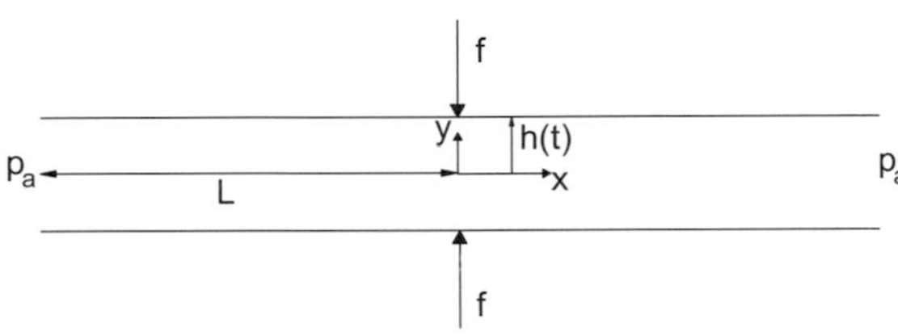

Figure 11.33 Squeeze film model of two parallel plates each under force f being pushed toward one another.

femur (thigh bone) at the knee joint where it is forced toward the proximal tibia (shin bone).

As a simple model, consider the two parallel plates of length 2L shown in Figure 11.33. A force, f, is applied to the top and bottom plates of length 2L, causing them to move toward each other. In this two-dimensional problem, f is a force per unit depth in the z-direction, so has units of force/length. The fluid between the plates is squeezed outward to the positive and negative x-directions where the end pressures are p_a. The distance to the upper plate is h(t), which is the same for the lower plate in this symmetric arrangement. Its initial position is $h(t = 0) = h_0$. The ratio of the initial height to the length of the lubrication layer is $\varepsilon = h_0/L$. We will assume $\varepsilon \ll 1$ and the lubrication approximation again yields a locally parabolic profile which now is time-dependent

$$u(x, y, t) = \frac{\partial p}{\partial x}\frac{y^2}{2\mu} + a(x, t)y + b(x, t) \tag{11.8.1}$$

where $p = p(x, t)$ and the unknown functions of integration are a(x, t) and b(x, t). The boundary conditions are symmetry along the x-axis, no-slip and no-penetration at the upper plate,

$$\left.\frac{\partial u}{\partial y}\right|_{y=0} = 0, \qquad u(y = h(t)) = 0$$
$$v(y = 0) = 0, \; v(y = h(t)) = \frac{dh}{dt} \tag{11.8.2}$$

Note that no-penetration at the upper plate means that the vertical fluid velocity evaluated there is equal to the plate velocity, dh/dt. Applying the boundary conditions on u to Eq. (11.8.1) yields

$$\frac{\partial u}{\partial y}(y=0) = 0 = a(x, t)$$

$$u(y=h) = \frac{\partial p}{\partial x}\frac{h^2}{2\mu} + b(x, t) = 0 \;\Rightarrow\; b(x, t) = -\frac{1}{2\mu}\frac{\partial p}{\partial x}h^2 \tag{11.8.3}$$

So the solution of u is

$$u = -\frac{1}{2\mu}\frac{\partial p}{\partial x}\left(h^2 - y^2\right) \qquad -L \leq x \leq L, 0 \leq y \leq h(t) \tag{11.8.4}$$

Turning to the solution of v, we calculate it from the continuity equation as follows

$$v = -\int \frac{\partial u}{\partial x}\, dy + c(x, t)$$
$$= \frac{1}{2\mu}\frac{\partial^2 p}{\partial x^2}\left(h^2 y - \frac{y^3}{3}\right) + c(x, t) \tag{11.8.5}$$

Applying the two boundary conditions on v from Eq. (11.8.2) determines the unknown function of integration, c(x, t), from $v(y = 0) = c = 0$, and poses a differential equation for the pressure, p,

$$v(y = h) = \frac{h^3}{3\mu}\frac{\partial^2 p}{\partial x^2} = \frac{dh}{dt} \tag{11.8.6}$$

Using the results of Eq. (11.8.6) to solve for $\partial^2 p/\partial x^2$ and inserting that into Eq. (11.8.5) gives us v,

$$v = \frac{3}{2h^3}\frac{dh}{dt}\left(h^2 y - \frac{y^3}{3}\right) \tag{11.8.7}$$

The differential equation for the pressure in Eq. (11.8.6) is

$$\frac{\partial^2 p}{\partial x^2} = \frac{3\mu}{h^3}\frac{dh}{dt} \tag{11.8.8}$$

where the right hand side depends only on time, t. Integrating Eq. (11.8.8) once with respect to x results in

$$\frac{\partial p}{\partial x} = \frac{3\mu}{h^3}\frac{dh}{dt}x + d(t) \tag{11.8.9}$$

Pressure is symmetric about the line $x = 0$, so $\partial p/\partial x(x = 0) = d(t) = 0$. Integrating once more with respect to x yields

$$p = \frac{3\mu}{2h^3}\frac{dh}{dt}x^2 + e(t) \tag{11.8.10}$$

We impose the end pressure at $x = L$ to solve for the unknown function of integration, e(t). The end pressure is p_a,

$$p - p_a = -\frac{3\mu}{2h^3}\frac{dh}{dt}(L^2 - x^2) \tag{11.8.11}$$

The net vertical fluid force per unit depth in the z-direction is the integral of the net pressure along the upper plate, and it must equal the imposed force, f, which we consider as a force per unit depth in the z-direction. The balance involves integrating the fluid pressure along the upper plate

$$f = \int_{-L}^{L} (p - p_a)\,dx = 2\int_{0}^{L} (p - p_a)\,dx \tag{11.8.12}$$

where we have used the symmetry of the pressure to simplify the integral over the entire plate to doubling the integral over half of the plate. Inserting Eq. (11.8.11) into Eq. (11.8.12) leads to

$$f = -2\mu\left(\frac{L^3}{h^3}\frac{dh}{dt}\right) \tag{11.8.13}$$

which is a differential equation for h(t),

$$\frac{dh}{dt} = -\frac{f}{2\mu L^3}h^3 \tag{11.8.14}$$

We can solve Eq. (11.8.14) by separating variables and integrating

$$\int_{h_0}^{h}\frac{dh}{h^3} = -\frac{f}{2\mu L^3}\int_0^t dt \quad \Rightarrow \quad \left[-\frac{1}{2}\frac{1}{h^2}\right]_{h_0}^{h}$$
$$= -\frac{1}{2}\left(\frac{1}{h^2} - \frac{1}{h_0^2}\right) = -\frac{f}{2\mu L^3}t \tag{11.8.15}$$

Rearranging Eq. (11.8.15) and defining a dimensionless plate position and time as

$$H = \frac{h}{h_0}, \quad T = \frac{t}{T_s} = \frac{\varepsilon^2 f t}{\mu L} \tag{11.8.16}$$

yields the dimensionless form of the plate position

$$H = \frac{1}{\sqrt{1+T}} \tag{11.8.17}$$

The time scale, $T_s = \mu L/\varepsilon^2 f$, can be quite large for our assumption that $\varepsilon = h_0/L << 1$, so it takes a long time for the plates to move closer. Consider the opposite problem of pulling two plates apart that have a fluid between them, say with a force –f, and the planes are very close together, so $\varepsilon << 1$. A common example known to students is trying to pull apart a microscope slide and its cover slip. It is difficult to do without breaking the cover slip. This phenomenon is called "viscous adhesion," since it gives the impression that the two surfaces are stuck to one another. In fact, it is squeeze film behavior. The student soon realizes it is easier to push the cover slip parallel to the slide and then off of it to separate the two. Note that the velocity of the plate decreases with increasing time T, $dH/dT = (-1/2)(1+T)^{-3/2}$, as the viscous resistance to outflow increases for a narrower gap.

Choose dimensionless variables consistent with Eqs. (11.8.16)

$$U = \frac{u}{U_s}, \quad V = \frac{v}{\varepsilon U_s}, P = \frac{p}{\mu U_s/\varepsilon h_0}, X = \frac{x}{L}, Y = \frac{y}{h_0} \tag{11.8.18}$$

where U_s is the velocity scale in the x-direction. Consistent with the scalings, U_s is derived from the time scale, T_s, and the flow distance, L, such that $U_s = L/T_s = \varepsilon^2 f/\mu$. Viewed in reverse, $T_s = L/U_s$ is a convective transit time. That definition of U_s simplifies the pressure scale to be $\mu U_s/\varepsilon h_0 = f/L$.

Inserting Eqs. (11.8.18) into Eq. (11.8.11) yields the dimensionless pressure. Further inserting Eq. (11.8.17) into the result yields that the pressure is independent of time

$$P - P_a = -\frac{3}{2H^3}\frac{dH}{dT}\left(1 - X^2\right) = \frac{3}{4}\left(1 - X^2\right) \quad (11.8.19)$$

which makes sense since f is constant and the plate must always support it. The dimensionless form of the x-velocity from Eq. (11.8.4) becomes

$$U = -\frac{1}{2}\frac{\partial P}{\partial X}\left(H^2 - Y^2\right) = \frac{3}{4}X\left(H^2 - Y^2\right)$$
$$-1 \leq X \leq 1, \; 0 \leq Y \leq H \quad (11.8.20)$$

The X-velocity profile, U, is parabolic in the Y-direction, consistent with the lubrication approximation. Its values increase linearly with X since at each station, X_0, the volume of fluid from $X = 0$ to X_0 is being forced out by the plates. A plot of U is shown in Figure 11.34 at $X = 0, 1/3, 2/3, 1$ for $T = 0$.

The Y-velocity is

$$V = \frac{3}{2H^3}\frac{dH}{dT}\left(H^2Y - \frac{Y^3}{3}\right) = -\frac{3}{4}\left(H^2Y - \frac{Y^3}{3}\right)$$
$$-1 \leq X \leq 1, \; 0 \leq Y \leq H \quad (11.8.21)$$

and the stream function is derived from integrating $U = \partial\psi/\partial Y$

$$\psi = -\frac{1}{2}\frac{\partial P}{\partial X}\left(H^2Y - \frac{Y^3}{3}\right) = \frac{3}{4}X\left(H^2Y - \frac{Y^3}{3}\right) \quad (11.8.22)$$

where we have set $\psi = 0$ for $Y = 0$. A selection of streamlines is also shown in Figure 11.34. At the walls they are perpendicular to the boundary, obeying no-slip and no-penetration, then curve toward the centerline becoming closer and closer together near the centerline where the axial velocities are increasing.

We are more familiar with 2D channels whose boundaries don't move. Under those conditions, the flow rate was constant along the flow axis. Here, however, the moving wall changes that concept. Start with the conservation of mass equation and integrate it with respect to y over the interval $0 \leq y \leq h(t)$ as follows

$$\int_0^{h(t)} \left(\frac{\partial u}{\partial x} + \frac{\partial v}{\partial y}\right) dy = \frac{\partial}{\partial x}\int_0^{h(t)} u\, dy + \int_{v(y=0)}^{v(y=h(t))} dv = \frac{\partial q}{\partial x} + \frac{dh}{dt} = 0 \quad (11.8.23)$$

In Eq. (11.8.23) we have substituted $v(y = 0) = 0$, $v(y = h(t)) = dh/dt$ and defined $q = \int_0^{h(t)} u\, dy$. For $dh/dt < 0$ we must have $\partial q/\partial x > 0$, as is apparent in Figure 11.34 where the velocity profile is increasing in magnitude as x increases.

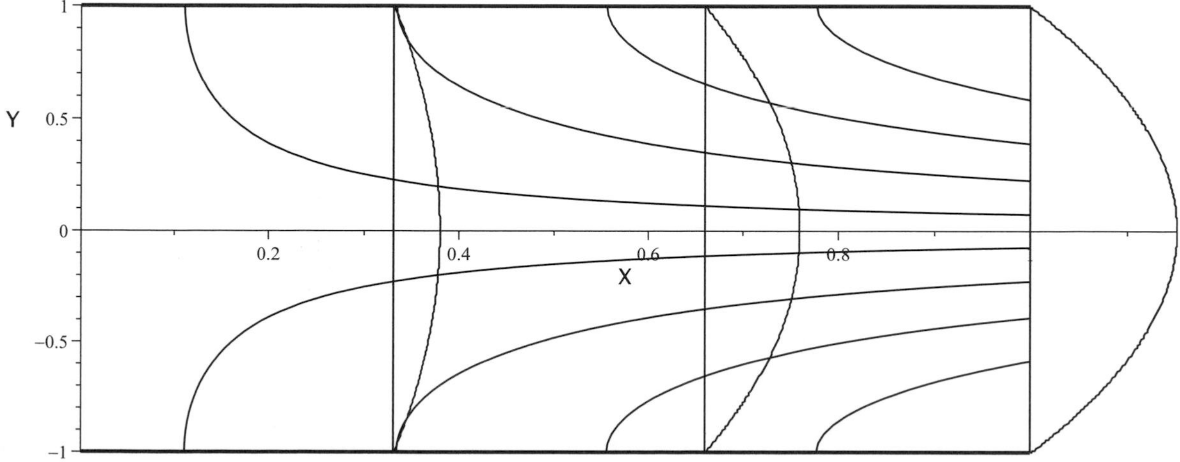

Figure 11.34 U vs Y at $X = 0, 1/3, 2/3, 1$ for $T = 0$ with streamlines.

Worked Example 11.1 Squeeze Film Predator–Prey

The dimensionless axial fluid velocity for the squeeze film problem is given by Eq. (11.8.20) with $H = (1 + T)^{-1/2}$. Now suppose this is a model of a submerged aquatic predator closing its mouth to catch some prey. Let the prey be passive, a particle of mass m and diameter d_p located at $x_p(t)$, $y_p(t)$. For a first study, let it be confined to the x-axis, that is, let $y_p = 0$. Using Stokes drag on the particle from Section 2.7 and Section 7.8, the force balance on it has the dimensional form

$$m\frac{d^2x_p}{dt^2} + 3\pi\mu d_p\left(\frac{dx_p}{dt} - u(x = x_p, y = 0, t)\right) = 0 \tag{11.8.24}$$

Using the scales $U = u/U_s$, $X = x/L$, $X_p = x_p/L$, $T = t/T_s$ where $U_s = \varepsilon^2 f/\mu$, $T_s = L/U_s$, Equation (11.8.24) has dimensionless form

$$\mathrm{Stk}\frac{d^2X_p}{dT^2} + \frac{dX_p}{dT} - \frac{3}{4}X_p\left(\frac{1}{1+T}\right) = 0 \tag{11.8.25}$$

where Stk is the Stokes number, $\mathrm{Stk} = m/(3\pi\mu d_p T_s)$, which is the ratio of the particle (prey) inertia to the drag on it. Let the particle start at $X_p = 0.2$ with zero initial velocity, $X_p(T = 0) = 0.2$, $dX_p/dT(T = 0) = 0$.

Equation (11.8.25) can be solved numerically for $X_p(T)$ given values of Stk and imposing the initial conditions. Figure 11.35 shows plots of $X_p(T)$ for $0 \leq T \leq 12$ for $\mathrm{Stk} = 0.001, 0.3, 0.75, 2.0, 5.0$. If the predator "bite" ends at $T = T_f = 12$, when

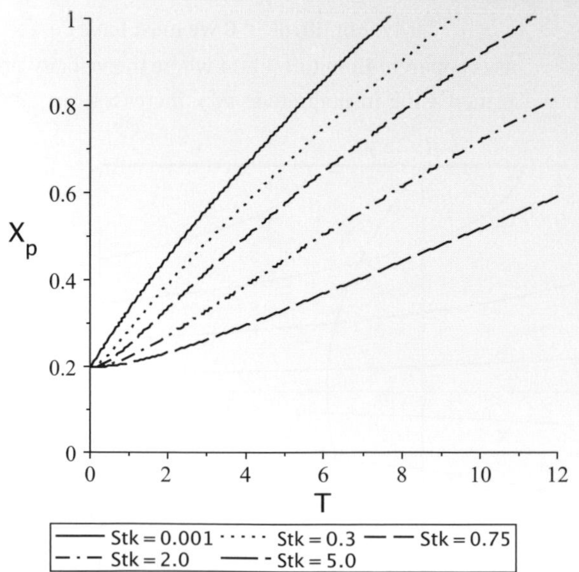

Figure 11.35 Predator–prey in a squeeze film bite.

$H = (1 + 12)^{-1/2} = 0.28$, then Figure 11.35 shows that three of the values $Stk = 0.001, 0.3, 0.75$ are small enough for the prey to "escape" by exiting the squeeze film because $X_p(T_f) > 1$. The inertia is small enough that the drag can accelerate the prey from rest. For the other two values, $Stk = 2.0, 5.0$, the prey is captured.

The Maple code to generate Figure 11.35 is shown in Code 11.3(a) and the MATLAB® code is Code 11.3(b).

(a)

```
restart;
with (plots) :
#Define the fluid velocity located at the particle position
U(T) := 3/4 ·Xp(T)·(H(T)^2 - Yp(T)) :
H(T):= 1/Sqrt (1+T)
Yp(T):= 0:
ode := Stk·diff(Xp(T), T, T) + diff(Xp(T), T) - U(T):
ics := Xp(0) = 0.2, D(Xp) (0) = 0:
#Solve ode+ics for 5 values of the Stokes number Stk and create plots for each
#Define the final time Tf
Tf := 12:

sol1 := dsolve({subs(Stk = 0.001, ode), ics}, numeric):
PL1 := odeplot( sol1, [T, Xp(T)]], T = 0 .. Tf, legend = "Stk=0.001 ", color = black, linestyle = 1) :
sol2 := dsolve({subs (Stk = 0.3, ode), ics}, numeric):
PL2  := odeplot(sol2, [T, Xp(T)]], T = 0 .. Tf, legend = "Stk=0.3", color = black, linestyle = 2)  :
sol3 := dsolve({subs(Stk = 0.75, ode), ics}, numeric):
PL3  := odeplot(sol3, [T, Xp(T)]], T = 0 .. Tf, legend = "Stk=0.75", color = black, linestyle = 3) :
sol4 := dsolve({subs (Stk = 2.0, ode), ics}, numeric):
PL4  := odeplot(sol4, [T, Xp(T)]], T = 0 .. Tf, legend = "Stk=2.0", color = black, linestyle = 4) :
sol5 := dsolve({subs (Stk = 5, ode), ics}, numeric):
PL5  := odeplot(sol5, [T, Xp(T)]], T = 0 .. Tf, legend = "Stk=5.0", color = black, linestyle = 5) :
display(PL1, PL2, PL3, PL4, PL5, view = [0 .. Tf, 0 .. 1]], labels = ["T", X[p]] );
```

(b)

```
% This MATLAB script requires that the file "Chapter11_odeEquation11_8_24.m"
% is located in the same folder of this script
clear all; close all; clc
tspan    = [0 12];
x0       = 0.2;
Dx0      = 0;
ICS      = [x0 Dx0];
Stk      = 0.001;
[t,sol]  = ode45(@(t,sol) Chapter11_odeEquation11_8_24(t,sol,Stk),tspan,ICS);
x        = sol(:,1);
Dx       = sol(:,2);
plot(t,x,'-k'), hold on;
Stk      = 0.3;
[t,sol]  = ode45(@(t,sol) Chapter11_odeEquation11_8_24(t,sol,Stk),tspan,ICS);
x        = sol(:,1);
Dx       = sol(:,2);
```

Code 11.3 Code examples: (a) Maple, (b) MATLAB®.

```
plot(t, x, ':k')
Stk      = 0.75;
[t, sol] = ode45(@(t, sol) Chapter11_odeEquation11_8_24(t, sol, Stk), tspan, ICS);
x        = sol(:, 1);
Dx       = sol(:, 2);
plot(t, x, '--k')
Stk      = 2.0;
[t, sol] = ode45(@(t, sol) Chapter11_odeEquation11_8_24(t, sol, Stk), tspan, ICS);
x        = sol(:, 1);
Dx       = sol(:, 2);
plot(t, x, '-.k')
Stk      = 5.0;
[t, sol] = ode45(@(t, sol) Chapter11_odeEquation11_8_24(t, sol, Stk), tspan, ICS);
x        = sol(:, 1);
Dx       = sol(:, 2);
plot(t, x, '.-k')
axis([0 12 0 1])
legend('Stk=0.001', 'Stk=0.3', 'Stk=0.75', 'Stk=2.0', 'Stk=5.0', 'location', 'northwest')
xlabel('T')
ylabel('X_p')
```

The following MATLAB function must be saved with the name "Chapter11_odeEquation11_8_24.m"

```
function dxdt = Chapter11_odeEquation11_8_24(t, x, Stk)
  yp       = 0;
  h        = 1/sqrt(1+t);
  u        = 3/4*x(1)*(h^2-yp);
  dxdt     = zeros(2, 1);
  dxdt(1)  = x(2);
  dxdt(2)  =(-x(2) + u)/Stk;
end
```

Code 11.3 *(continued)*

11.8 Non-Newtonian Squeeze Film

The synovial fluid in joints provides the lubrication and is non-Newtonian due to a high concentration of hyaluronic acid molecules. It exhibits shear thinning behavior (Ogston and Stanier, 1953) and has been modeled in a variety of ways. The data in (Caygill and West, 1969) were fit with a power-law fluid model which included a non-zero viscosity at infinite shear rate, $\mu_{eff} = \mu_\infty + m\dot{\gamma}^{n-1}$, where the units of m are poise·s or pascal·s^n depending on which units are used. They found that $\mu_{eff} = 0.007 + 0.99\dot{\gamma}^{-2/3}$Pa·s (N·s/m^2) fit their data well, so $n = 1/3$, which was assumed before the fit. Later, the same data was reevaluated without assuming a value for n, resulting in a similar expression, $\mu_{eff} = 0.0105 + 1.664\dot{\gamma}^{-0.793}$ Pa·s, n ~ 0.2 (Lai et al., 1978). Recall that 1 Pa · s (pascal-second) = 10 poise, so μ_∞ in their models are, respectively, 0.07 and 0.105 poise, compared to water, which is 0.01 poise. The investigators in Mazzucco et al. (2002) used a simple Cross fluid model, $\mu_{eff} = \mu_0 / \left(1 + (c\dot{\gamma})^d\right)$, with a finite viscosity, μ_0, at zero shear rate. Typical values from fitting their data were $\mu_0 = 1.3\,\mathrm{Pa \cdot s}, c = 4.2, d = 0.54$. For $c\dot{\gamma} >> 1$ the Cross model simplifies to a power law, $\mu_{eff} \sim \mu_0 (c\dot{\gamma})^{-d} \sim 0.63\dot{\gamma}^{-\frac{1}{2}}\ \mathrm{Pa \cdot s}$. So $d = 0.54$ gives us approximately $n \sim 0.5$ for a power-law fit. Several other models are discussed in Lai et al. (1978). For our purposes, we will assume a simple power-law model with $\mu_\infty \sim 0$. We will leave n as a parameter and know that for synovial fluid a range of $0.2 \lesssim n \lesssim 0.5$ is physiologic.

With the simplifications due to lubrication theory, it becomes possible to solve some additional flows of non-Newtonian fluids using analytical methods. For example, consider the squeeze film problem shown in Figure 11.33. The lubrication form of momentum conservation is given by

$$\frac{\partial}{\partial y}\left(\mu_{eff}(\dot{\gamma})\frac{\partial u}{\partial y}\right) = \frac{\partial p}{\partial x} \tag{11.8.26}$$

and $p = p(x, t)$ only. From Eq. (11.2.4) we know that the strain rate tensor components simplify in lubrication theory, which makes $\dot{\gamma}$ much easier, $\dot{\gamma} = \sqrt{2E_{ij}E_{ij}} \sim |\partial u/\partial y|$. Recall from Section 10.3 that the power-law fluid model is $\mu_{eff}(\dot{\gamma}) = \dot{\gamma}^{n-1}$. A first integral of Eq. (11.8.26) is

$$\mu_{eff}(\dot{\gamma})\frac{\partial u}{\partial y} = y\frac{\partial p}{\partial x} + c_1(x) = y\frac{\partial p}{\partial x} \tag{11.8.27}$$

The symmetry of the flow causes $\partial u/\partial y = 0$ at $y = 0$ so $c_1 = 0$. Solving the flow in the upper half of the channel, $0 \le x \le L, 0 \le y \le h(t)$, we anticipate that $\partial u/\partial y < 0$ for $\partial p/\partial x < 0$ in this region. So $\dot{\gamma} = -\partial u/\partial y$ in the power-law model applied to Eq. (11.8.27). Multiplying Eq. (11.8.27) through by −1 and rearranging gives us

$$\left(-\frac{\partial u}{\partial y}\right)^n = \left(-\frac{1}{m}\frac{\partial p}{\partial x}\right)y \quad \rightarrow \quad \frac{\partial u}{\partial y} = -\left(-\frac{1}{m}\frac{\partial p}{\partial x}\right)^{\frac{1}{n}} y^{\frac{1}{n}} \tag{11.8.28}$$

Integrating Eq. (11.8.28) results in an expression for u (x, y) with a function of integration, $c_2(x)$. Imposing the no-slip condition at the wall, $y = h(t)$, solves for $c_2(x)$ and the velocity field is

$$u(x, y, t) = \frac{n}{n+1}\left(-\frac{1}{m}\frac{\partial p}{\partial x}\right)^{1/n}\left(h^{\frac{n+1}{n}} - y^{\frac{n+1}{n}}\right) \tag{11.8.29}$$

The v velocity is derived from conservation of mass, so that

$$\begin{aligned} v &= -\int \frac{\partial u}{\partial x}dy + c_3(x,t) \\ &= -\frac{n}{n+1}\left(\frac{\partial}{\partial x}\left(-\frac{1}{m}\frac{\partial p}{\partial x}\right)^{1/n}\right)\left(h^{\frac{n+1}{n}}y - \frac{n}{2n+1}y^{\frac{2n+1}{n}}\right) \end{aligned} \tag{11.8.30}$$

where we have applied zero crossflow, $v = 0$, at $y = 0$ so that $c_3 = 0$. Now let $y = h(t)$ where $v = dh/dt$ in Eq. (11.8.30) so that

$$\frac{dh}{dt} = -\frac{n}{2n+1}\left(\frac{\partial}{\partial x}\left(-\frac{1}{m}\frac{\partial p}{\partial x}\right)^{1/n}\right)h^{\frac{2n+1}{n}} \tag{11.8.31}$$

Since h only depends on time, t, Eq. (11.8.31) can be integrated once with respect to x yielding

$$-\frac{n}{2n+1}\left(-\frac{1}{m}\frac{\partial p}{\partial x}\right)^{1/n}h^{\frac{2n+1}{n}} = \frac{dh}{dt}x + c_4(t) = \frac{dh}{dt}x \tag{11.8.32}$$

In Eq. (11.8.32) we have imposed symmetry in the x-direction at $x = 0$ where we expect $\partial p/\partial x = 0$, which solves $c_4(t) = 0$. Rearranging Eq. (11.8.32) and solving for the pressure gradient gives us

$$\frac{\partial p}{\partial x} = -F_2 x^n \quad F_2 = m\left(-\left(\frac{2n+1}{n}\right)\frac{1}{h^{\frac{2n+1}{n}}}\frac{dh}{dt}\right)^n \tag{11.8.33}$$

It appears that F_2 may depend on time, but as we found in the previous section the pressure is time independent so we will also show F_2 is constant. Integrating Eq. (11.8.33) with respect to x leads to the pressure field, and we can impose an end condition as we did before such that $p(x = L) = p_a$. Doing those steps gives us the pressure distribution

$$p(x,\ t) - p_a = \frac{1}{n+1}F_2\left(L^{n+1} - x^{n+1}\right) \tag{11.8.34}$$

As we did before, the force balance on the upper plate states that the integral of the pressure over the entire plate, $-L \le x \le L$, balances f. From symmetry that integral is twice the integral over $0 \le x \le L$ so that

$$f = 2\int_0^L (p - p_a)\,dx = \frac{2}{n+2} F_2 L^{n+2} \tag{11.8.35}$$

Rearranging Eq. (11.8.35) we want to solve for h(t), so substitute back for $F_2(t)$ as

$$\left(\frac{1}{h^{\frac{2n+1}{n}}}\frac{dh}{dt}\right) = -F_3$$
$$F_3 = \left(\frac{n}{2n+1}\right)\left(\left(\frac{n+2}{2m}\right)\frac{f}{L^{n+2}}\right)^{\frac{1}{n}} \tag{11.8.36}$$

Here, the constant $F_3 > 0$ is chosen to maintain tidiness and immediately we recognize from Eq. (11.8.33) that $F_2 = m\left(\left(\frac{2n+1}{n}\right)F_3\right)^n$ by inserting Eq. (11.8.36) showing it is independent of time. Separating variables in Eq. (11.8.36) we can integrate and correlate $h(t = 0) = h_0$ as the lower limit with variable upper limit

$$\int_{h_0}^{h} \frac{dh}{h^{\frac{2n+1}{n}}} = -F_3\int_0^t dt \quad \rightarrow \quad \frac{n}{n+1}\left(h_0^{-\frac{n+1}{n}} - h^{-\frac{n+1}{n}}\right) = -F_3 t \tag{11.8.37}$$

Rearranging Eq. (11.8.37) leaves us with

$$H = (1+T)^{-\frac{n}{n+1}} \tag{11.8.38}$$

where we have used the dimensionless height $H = h/h_0$ and a dimensionless time, $T = t/T_s$, where

$$T_s = \frac{1}{\left(\frac{n+1}{n}\right)h_0^{\frac{n+1}{n}}F_3} = \left(\frac{2n+1}{n+1}\right)\left(\left(\frac{n+2}{2}\right)\frac{f\varepsilon^{n+1}}{mL}\right)^{-\frac{1}{n}} \tag{11.8.39}$$

For n = 1, the Newtonian fluid case, we recover the solution in Eq. (11.8.17) with $T_s = \mu L/\varepsilon^2 f$.

To examine the influence of the flow index, n, let us choose $h_0 = 50$ μm, which gives a full gap distance of 100 μm, and L = 2 cm so that $\varepsilon = h_0/L = 0.0025$. Also choose the force f = 100 newtons (~25 lbs) and $m = 1$ Pa·s^n. Plotting H vs T for different values of n would be somewhat misleading since the time scale, T_s, depends on n. However, we can simply plot H vs dimensional time, t, and do that for n = 1.0, 0.8, 0.5, 0.3 and time $0 \le t \le 5$ s. The results are shown in

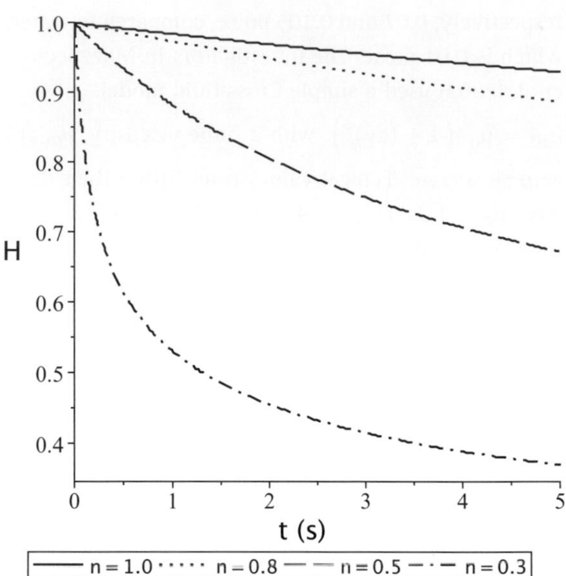

Figure 11.36 H vs t for four values of n.

Figure 11.36. H decreases with time as the force, f, is applied. It decreases more rapidly as n decreases and the shear thinning effect is stronger. The Newtonian fluid is when n = 1, and in that case m = 1 Pa·s (10 poise), which is 1000 times the viscosity of water. It would correspond to Mazzucco et al.'s (2002) value of $\mu_0 = 1.3$ Pa·s for the zero shear rate limit of the viscosity. The synovial fluid values of n are 0.3 and 0.5.

Going back to Eq. (11.8.29) we now have the full u velocity field by inserting the pressure gradient term from Eq. (11.8.33)

$$u(x, y, t) = \frac{n}{n+1}\left(\left(\frac{n+2}{2m}\right)\frac{f}{L^{n+2}}\right)^{\frac{1}{n}} x\left(h^{\frac{n+1}{n}} - y^{\frac{n+1}{n}}\right)$$
$$0 \le x \le L, \quad 0 \le y \le h(t) \tag{11.8.40}$$

We can use our scalings for the coordinates, $X = x/L$, $Y = y/h_0$, $H = h/h_0$ to simplify Eq. (11.8.40) as

$$u(x, y, t) = \frac{n}{n+1}\left(\left(\left(\frac{n+2}{2m}\right)\frac{f\varepsilon^{n+1}}{L}\right)^{\frac{1}{n}}\right) LX\left(H^{\frac{n+1}{n}} - Y^{\frac{n+1}{n}}\right) \quad 0 \le X \le 1 \quad 0 \le Y \le H(T) \tag{11.8.41}$$

With Eq. (11.8.41) it now is simple to calculate the strain rate field, $\dot{\gamma}(X, Y)$,

$$\dot{\gamma} = -\frac{\partial u}{\partial y} = -\frac{1}{h_0}\frac{\partial u}{\partial Y} = \left(\left(\frac{n+2}{2m}\right)\frac{f\,\varepsilon}{L}Y\right)^{\frac{1}{n}} X$$

$$0 \le X \le 1 \quad 0 \le Y \le H(T) \qquad (11.8.42)$$

and then the effective viscosity field, $\mu_{eff}(X, Y)$,

$$\mu_{eff} = m\dot{\gamma}^{n-1} = m\left(\left(\frac{n+2}{2m}\right)\frac{f\,\varepsilon}{L}Y\right)^{\frac{n-1}{n}} X^{n-1}$$

$$0 \le X \le 1 \quad 0 \le Y \le H(T) \qquad (11.8.43)$$

Figure 11.37 shows an example of the effective viscosity field, $\mu_{eff}(X, Y)$, for $m = 1$, $n = 1/2$, which is in the normal range of synovial fluid. Time is taken to be $T = 0$ so that the Y variable has the range $0 \le Y \le 1$. Constant viscosity curves are shown with their viscosity values indicated, and the grey shading is taken to be darker for larger values and lighter for lower values. The higher shear rates are closer to the wall and closer to the exit, so the effective viscosities are smaller there.

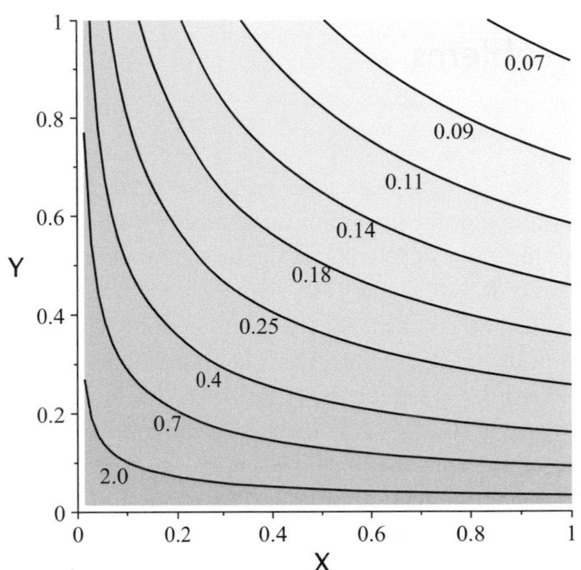

Figure 11.37 $\mu_{eff}(X, Y)$ for $m = 1$, $n = 1/2$, and $T = 0$. Constant viscosity curves are shown with their viscosity value indicated.

Summary

Lubrication theory is a robust methodology that allows us to analyze a variety of interesting flows related to biomedical applications. Just within this chapter we have studied traditional subject matter like the slide block and then expanded that analysis to single RBC flow in capillaries as well as a series of equally spaced RBCs. The inference of an apparent viscosity from the application, really the misapplication, of Poiseuille Law opens up the important concepts of the particulate nature of blood. It is the RBC dimensions relative to the tube diameter that influences the overall behavior, starting from the cell-free layer in core-annular flow to a single RBC squeezing through a tight capillary. The model is also applicable for other situations which are biologically different but fluid dynamically similar, like passing a urinary stone through the urethra. Just as important was our exploration of radial crossflow through permeable walls of the capillary and its relation to both hydrostatic pressure and oncontic pressure, the latter being a disease-related quantity that has striking repercussions in starvation leading to kwashiorkor. The squeeze film analysis, also a traditional one for lubrication theory, was extended both for a predator–prey model and then a non-Newtonian power-law model like synovial fluid which is shear thinning. So within Chapter 11 we have utilized results from Chapters 7 and 10 while going back to our first studies of particle drag presented in Chapter 2.

Problems

11.1

a. For the slide block problem let $D = 0.50, \varepsilon = 0.1$ and plot the dimensionless pressure $P(X)$. What is the value of $X = X_m$ at the pressure maximum?

b. For $D = 0.50$ plot $U(X, Y)$ for $X = 0, 0.25, 0.5, X_m, 1.0$. To display your results multiply U by 0.1 and add its X value. That should distribute the profiles along the gap. Hint: recall $0 \leq Y \leq H$ where $H = 1 + D - X$. Describe the result. Is there any reverse flow?

c. Using the stream function, plot streamlines for $D = 0.50$ including the stagnation streamline $\psi - \psi_s$. What is the value of ψ_s ? Include contours for ψ close to ψ_s to show the details of the flow. Add this streamline plot to the plot in part (b). For added visual effect, plot a line for the wall surface and a line for the block surface and add to the figure.

d. Plot $V(X, Y = 0.2)$ for $D = 0.25, 0.50, 0.75, 1.0$. Be sure that X is within the gap. What do you notice about V as D changes?

e. From your equation for U, find the critical value of D, above which there is no backflow at $X = 0$. Hint: use dU/dY evaluated at the block surface there to find your answer.

f. From your equation for U, find the critical value of D, above which there is no leakback at $X = 1$. Hint: use dU/dY evaluated at the wall surface there to find your answer.

11.2

a. The pressure distribution in the RBC gap, $P(Z)$, is given in Eq. (11.3.13) where Eq. (11.4.3) provides an expression for $(P_1 - P_0)$. Set $P_0 = 0,\ a = 2.5\,\mu m$, $d_1 = 0.4\,\mu m$, $d_2 = 0.1\ \mu m$, $V_{RBC} = 100\ \mu m^3$. What are your calculated values for $D,\ \delta,\ \varepsilon,\ L$?

b. Using the results in part (a), plot $P(Z)$. What is the location of its maximum, i.e. Z_m.

c. Plot $U_Z(Y,\ Z)$ for $Z = 0, 0.25, 0.5, Z_m, 1.0$ and $0 \leq Y \leq H(Z)$. Scale and offset your profiles so you actually plot $0.15U_Z + Z$. Now the profiles show the velocity distribution along the gap. Add lines for the RBC surface, $Y = H(Z)$, and the capillary wall, $Y = 0$, to make sure your profiles are inside the gap.

d. Compute the stream function, ψ, by integrating the relationship $U_Z = \partial\psi/\partial Y$. You can assume $\psi(Y = 0) = 0$. What is the value of the stagnation streamline? Hint: it will have the value as the streamline along the RBC surface, i.e. $\psi(Z = 0, Y = H(0)) = \psi_s$. Create a plot for the streamlines, including those near ψ_s to show the behavior, and add it to your plot from part (c).

e. How does your result compare to Figure 11.9(a) for the slide block?

11.3 From our analysis of the RBC flow we found the ratio of the average fluid speed to the RBC speed, U_w, is $U_{avg}/U_w = 1 - 2\delta\hat{Q}$. $\hat{Q}$ is given in Eq. (11.3.12) and it depends on D, ε and $(P_1 - P_0)$. Equation (11.4.3) provides an expression for $(P_1 - P_0)$. Let $P_0 = 0$, $a = 2.5\ \mu m$.

a. Set the RBC volume to $V_{RBC} = 100\ \mu m^3$. Plot U_{avg}/U_w vs d_2 over the range $0.10 \leq d_2 \leq 0.20\ \mu m$ for $d_1 = 0.4, 0.35, 0.30, 0.25\mu m$. You will need to calculate values for L, ε, D, δ accordingly. Discuss your results and their implications. What happened to the dependence on ε, L?

b. Assume the speed of the RBC is $U_w = 0.01\,cm/s$, and serum viscosity $\mu = 0.012$ poise. Let $d_1 = 0.35\ \mu m$. Plot the dimensional pressure drop across the RBC, $p_1 - 0$, vs d_2 for $0.10 \leq d_2 \leq 0.20\ \mu m$ and four values of the RBC volume, $V_{RBC} = 60, 90, 120, 150\mu m^3$. Figure 11.16 (a) is related to this calculation for

$V_{RBC} = 100\ \mu m^3$. Discuss your results including diseases which have low, normal, and high V_{RBC}.

c. For $d_1 = 0.4\ \mu m$, $V_{RBC} = 100\ \mu m^3$, $a = 2.5\ \mu m$ plot ε, δ, D over the range $0.10 \le d_2 \le 0.20\ \mu m$. Are the assumptions on the size of these parameters justified?

11.4 Consider the series of RBCs in capillary blood flow shown in Figure 11.21. The RBC lengths are all L and the distance from one RBC front to the next is L_H. Repeat the analysis of the apparent viscosity for the flow of RBCs in a capillary as appears in the text, but take account of Eq. (11.4.7). Let the RBC velocity be $U_w = 0.01\ cm/s$ in a capillary with radius $a = 2.5\ \mu m$ and serum viscosity $\mu = 0.012$ poise. Choose $d_1 = 0.35\ \mu m$, $d_2 = 0.15\ \mu m$.

a. Model a normocytic anemia by setting $V_{RBC} = 100\ \mu m^3$ and hematocrit $H = 0.30$. What are the values of L and L_H ?
b. With the parameter values in part (a), calculate δ, ε, D and the pressure scale $P_s = \mu U_w/(d_1 - d_2)$. Use Eq. (11.4.3) for the overall dimensionless pressure drop $P_1 - P_0$ and multiply by P_s to arrive at the dimensional pressure drop. Then divide by L to achieve the dimensional pressure gradient across the RBC, $(p_{L1} - p_{01})/L$.
c. From Eq. (11.4.7) calculate U_{avg}/U_w where $\hat{Q}$ is found in Eq. (11.3.12). Now rearrange to solve for U_{avg} which is the average velocity of the fluid in between the RBCs.
d. Calculate the pressure drop in the region between the RBCs, $(p_{02} - p_{L1})/(L_H - L)$, using Poiseuille law with U_{avg}.
e. Derive the apparent viscosity, μ_{app}, from the overall pressure drop $(p_{02} - p_{01})/L_H$ and compare your answer to μ.
f. Repeat steps (a) through (e) for a microcytic anemia where U_w, a, μ are the same, but $V_{RBC} = 60\,\mu m^3$, $H = 0.30$, $d_1 = 0.4\,\mu m$, $d_2 = 0.25\ \mu m$.
g. Discuss your results comparing the two situations..

11.5 The Fåhræus dilution effect is discussed in Section 11.5 where two reservoirs of RBCs, with solid fraction ϕ_0^s, are connected by a cylindrical tube whose velocity profile is $u_z(r)$. The solid fraction of RBCs in the tube is $\phi^s(r)$.

a. Calculate $\bar{\phi}^s/\phi_0^s$ for a Poiseuille velocity profile in the tube and for $\phi^s(r) = k_1(a - r)$where k_1 is a positive constant i.e. the solid fraction is a decreasing function of r and reaches zero at the wall where $r = a$. Discuss your result.
b. Repeat part (a), except now let

$$\phi^s(r) = \begin{cases} 0 & a - \delta a \le r \le a \\ k_2 & 0 \le r \le a - \delta a \end{cases}$$

where the thickness of the cell-free annular layer near the wall is δa There is a constant distribution, k_2, in the core region. Plot $\bar{\phi}^s/\phi_0^s$ as a function of δ, say for $0.5 < \delta < 1$ and discuss your result.
c. Now consider the same $\phi^s(r)$as in part (b), but a different velocity field. Let there be a core region of a constant velocity u_{z0}, where the RBCs are located, and a linear profile in the cell-free annular layer

$$u_z(r) = \begin{cases} a_1 + b_1 r & a - \delta a \le r \le a \\ u_{z0} & 0 \le r \le a - \delta a \end{cases}$$

Solve for the unknown constants a_1 and b_1 by matching velocities at the wall boundary and the edge of the core. Sketch the dimensionless velocity profile of the entire cross-section for $\delta = 0.2$, scaling velocity on u_{z0} and r on a.
d. Using the conditions in part (c), calculate $\bar{\phi}^s/\phi_0^s$ and plot as a function of δ, say for $0.05 < \delta < 0.20$ and discuss the comparison with your other results.

11.6

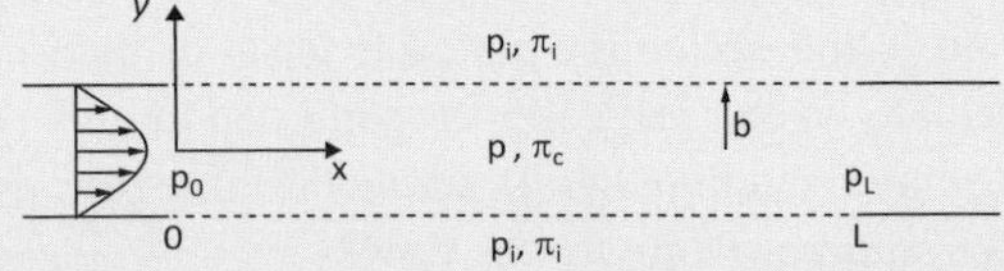

Consider flow in a 2D leaky channel section as shown in the figure. Lubrication theory yields dimensionless velocities

$$U = -\frac{1}{2}\frac{dP}{dX}\left(1 - Y^2\right), \quad V = \frac{1}{2}\frac{d^2P}{dX^2}\left(Y - \frac{Y^3}{3}\right) \tag{1}$$

where the pressure,P(X) satisfies the ODE

$$\frac{1}{3}\frac{d^2P}{dX^2} - \beta^2 P = -\beta^2(P_i + \pi_m) \tag{2}$$

In Eq. (2) $\beta^2 = \mu L_P/\varepsilon^2 b$ is the dimensionless parameter for the wall permeability and $\pi_m = \sigma(\pi_c - \pi_i)/(\mu u_0/\varepsilon b)$.

a. Solve Eq. (2) using the two boundary conditions, $P(X = 0) = P_0$ and $P(X = 1) = P_1$. Plot solutions P(X) for $0 \le X \le 1$ for $P_0 = 14$, $P_1 = 6.0$, $P_i = -0.8$, $\pi_m = 10.0$ Let $\beta = 0.01, 0.1, 1.0$. Describe your results. How are they affected by β ?
b. Plot U(X, Y) for $X = 0, 0.5, 0.8, 1.0$ and $\beta = 0.5$. Explain your results.
c. Plot the leakage velocity $V(X, Y = 1)$ for $\beta = 0.25, 0.50, 0.75, 1.00$ and discuss your results. For $\beta = 0.50$ what value of X yields $V = 0$? Call it X_0.
d. Determine the stream function $\psi(X, Y)$ by integrating the definition $U = \partial\psi/\partial Y$ and setting $\psi(Y = 0) = 0$ to solve for your X-function of integration. What is the value of the stream function that passes through the zero velocity location on the membrane, i.e. $\psi(X = X_0, Y = 1)$? Call it ψ_0, and let $\beta = 0.5$.
e. For $\beta = 0.50$ determine the value of $\psi(X, Y)$ for streamlines intersecting the following points: $(X, Y) = (0.25, 1), (0.35, 1), (0.45, 1), (0.5, 0.25), (0.5, 0.5), (0.5, 0.75)$, and call them $\psi_1, \psi_2, \psi_3, \psi_4, \psi_5, \psi_6$. Plot streamlines for the upper half of the channel flow, $0 \le X \le 1, 0 \le Y \le 1$ for your seven contour values, $\psi_0, \psi_1, \psi_2, \psi_3, \psi_4, \psi_5, \psi_6$.
f. Add to the streamline plot, a fieldplot of the velocity vector (U, V) in the same region and display them together. Describe the flow. How does kwashiorkor relate to this flow?
g. Plot the axial flow rate, Q(X) vs X for $\beta = 0.001, 0.25, 0.50, 0.75, 1.00$. Discuss your results. Is there a local minimum? If so, where and why?

11.7

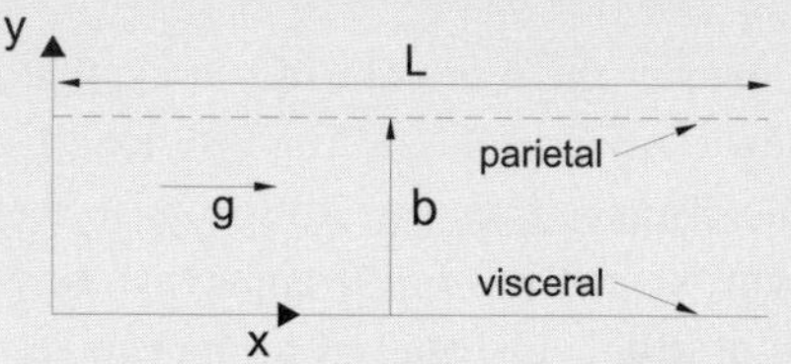

Consider the steady filtration flow system in the figure which has some features of pleural fluid flow shown in Figure (11.31). The fluid in the channel is bounded by $0 \le x \le L$ and $0 \le y \le b$. The surface $y = 0$ is impermeable, like the visceral pleural membrane, while $y = b$ is porous, like the parietal pleural membrane. Gravity acts in the +x-direction. For $\varepsilon = b/L << 1$, the governing equations are lubrication theory

$$-\frac{dp}{dx} + \mu\frac{\partial^2 u}{\partial y^2} + \rho g = 0 \tag{1}$$

where the pressure, $p = p(x)$, is independent of y. Scale Eq. (1) with the following definitions: $Y = y/b$, $X = x/L$, $U = u/u_0$, $P = p/(\mu u_0/\varepsilon b)$ and let the characteristic velocity be $u_0 = \rho g b^2/\mu$. The dimensionless form of Eq. (1) becomes

$$\frac{\partial^2 U}{\partial Y^2} = -1 + \frac{dP}{dX} \tag{2}$$

a. Integrate Eq. (2) twice with respect to Y. To solve for the two X-functions of integration, impose the boundary conditions of zero X-velocity at $Y = 0, 1$, $U(Y = 0) = 0, U(Y = 1) = 0$.
b. Now solve for V from continuity such that

$$V = -\int \frac{\partial U}{\partial X}\, dY + A_3(X) \tag{3}$$

and solve for $A_3(X)$, impose the impermeable boundary condition on the visceral pleura $V(Y=0)=0$.

c. Let the permeability condition at the parietal pleural membrane be $v(y=b)=k(p-p_0)$. The dimensionless form is

$$\frac{d^2P}{dX^2}-\beta^2P=-\beta^2P_0 \tag{4}$$

where $\beta^2=\frac{12\mu k}{\varepsilon^2 b}$. Solve Eq. (4) and solve for the two constants of integration by imposing zero crossflow at the channel ends, $U(X=0)=0,\ \ U(X=1)=0$.

d. Let $P_1(X)=P(X)-P_0$ and plot $P_1(X)$ for $\beta=0.01, 0.5, 1.0, 2.0$ and describe your results.

e. Solve for the stream function by integrating the definition $U=\partial\psi/\partial Y$. You can let $\psi(Y=0)=0$ to solve for the function of integration. Plot the streamlines for $\beta=1$.

f. Now plot U velocity profiles for $\beta=1$ at positions $X_n=0.1, 0.3, 0.5, 0.7, 0.9$ and $0\le Y\le 1$. To improve the presentation, shift each profile to its X_n position and amplify the scale so that you plot $U_{plot}=5U+X_n$. Add the streamline plot from part (e). Describe your results.

11.8 We examined the squeeze film for the force, f, pushing the plates together and found the position of the upper plate to be $H=(1+T)^{-1/2}$ in Eq. (11.8.17). However, the same approach pertains to pulling the plates apart with force –f, or $H=(1-T)^{-1/2}$. The dimensionless pressure in Eq. (11.8.19) switches sign since dH/dT is now positive. This sign change extends to dP/dX and eventually the velocities, $(U,\ V)$, and the stream function, ψ.

a. Let $P_a=0$ and plot the pressure field $P(X)$ for $-1\le X\le 1$. How does this differ from the squeezing, $dH/dT<0$, mode?

b. Plot $U(X,Y,T=0)$ for $X=0.33,\ \ 0.66,\ \ 1.00$and $-H\le Y\le H$. To place them in the gap geometry, actually plot $X+KU$ where $K=0.2$ is a scale factor to keep the curves from overlapping. How does your result differ from the squeezing, $dH/dT<0$, mode?

c. Plot the streamlines using the new stream function ψ at $T=0$ for $0\le X\le 1,\ -H\le Y\le H$ and plot it with part (b).

d. The dimensional shear stress at the upper wall is $\mu(\partial u/\partial y+\partial v/\partial x)|_{y=h(t)}$. However, the second term is negligible due to our lubrication approximation, i.e. $v<<u, \partial/\partial x<<\partial/\partial y$. So, let the dimensionless wall shear be $\tau_w=\partial U/\partial Y|_{Y=H}$ and plot it for $T=0, 0.1, 0.2, 0.3$ over the range $0\le X\le 1$. Describe your result.

e. In Eq. (11.8.23) integrate the last statement with respect to x to find $q(x,t)$. Use the condition that $q(x=0)=0$ to determine the function of integration. Make your solution dimensionless where $Q=q/U_sh_0, H=h/h_0, T=tU_s/L, X=x/L$, and plot Q over the interval $0\le X\le 1$ for $T=0, 0.1, 0.2, 0.3$. What does your plot show?

11.9 We derived the dimensionless velocities, (U, V), in Eq. (11.8.20) and Eq. (11.8.21). The dimensionless distance to the plate was found to be $H=(1+T)^{-1/2}$. For a predator-prey model, the force balance on the prey particle mass m and diameter d_p are the X_p, Y_p-components

$$Stk\frac{d^2X_p}{dT^2}+\left(\frac{dX_p}{dT}-U\big(X=X_p, Y=Y_p, T\big)\right)=0$$

$$Stk\frac{d^2Y_p}{dT^2}+\left(\frac{dY_p}{dT}-V\big(X=X_p, Y=Y_p, T\big)\right)=0$$

where Stk is the Stokes number, $Stk=m/\big(3\pi\mu d_pT_s\big)$. The X_p-component detail is shown in Eq. (11.8.25).

a. Let the particle start at $X_0=0.2, Y_0=0.2$ with zero initial velocity. Solve the governing equations numerically and plot the prey trajectory $X_p(T), Y_p(T)$ in a parametric plot for $0\le T\le 12$. Do it for $Stk=0.1, 0.5, 1, 5$ and label your curves in a legend.

b. Who escapes in part (a) and why?

c. Repeat part (a) for $X_0=0.2, Y_0=0.4$.

d. Explain your results comparing part (a) and part (c).

e. Another way for the predator to be successful is for the particle (prey) to impact the closing walls. To investigate that possibility, let's choose a starting position closer to the wall. Let $X_0 = 0.2$, $Y_0 = 0.8$ and plot two curves $Y_p(T)$ for Stk $= 0.2$ and 2 and a third curve, H(T). Let the range of time be $0 \leq T \leq 12$.
f. Is there wall-particle impact in part (e)? If so, at what value of time and for which value of Stk? Explain your result.

Image Credits

Figure 11.10 Source: Getty

Figure 11.11 Source: Fig 1 of Skalak R, Branemark PI. Deformation of red blood cells in capillaries. *Science.* 1969; 164(3880):717–9.

Figure 11.12 Source: Fig 5 of Secomb TW, Skalak R, Ozkaya N, Gross JF. Flow of axisymmetric red blood cells in narrow capillaries. *Journal of Fluid Mechanics.* 1986; 163:405–23.

Figure 11.15a Source: Encyclopaedia Brittanica/ Contributor/ Getty

Figure 11.15b Source: A lenticular kidney stone, excreted by the urine. By Wenjvn, licensed under CC-BY-SA 4.0: https://commons.wikimedia.org/wiki/File:Lenticular_kidney_stone.jpg

Figure 11.18 Source: Fig 2 of Fåhræus R, Lindqvist T. The viscosity of the blood in narrow capillary tubes. *American Journal of Physiology.* 1931; 96(3):562–8.

Figure 11.23 Source: OpenStax College, Anatomy & Physiology, Connexions website. http://cnx.org/content/col11496/1.6/

Figure 11.30 Source: Starved girl during the Nigerian-Biafran war (late 1960s). {{PD-USGov}} source http://phil.cdc.gov/ uimage ID: 6901

Figure 11.32 Source: PeterHermesFurian/ Getty

12 Laminar Boundary Layers

Introduction

Features of high Reynolds number flows relevant to our interests include viscous boundary layers and turbulence. Turbulence is discussed in Chapter 13. Here we address the concept of a boundary layer. A viscous boundary layer is a thin region near a wall or surface where the velocity parallel to the wall makes rapid changes in the cross-stream direction to meet the no-slip condition at the wall. Viscous and inertial effects balance in this boundary layer. However, farther away from the wall, outside of the boundary layer, the flow is dominated by inviscid fluid mechanics as represented by the Bernoulli equation (see Sections 6.1 and 6.2). Boundary layers can develop and grow in thickness along the axial direction at the entrance (inlet) region for vascular and airway tubes, which occur at anatomical branches and bifurcations. If the tube is long enough, the boundary layer of this "entrance flow" grows to occupy the entire cross-section, and eventually becomes fully developed, i.e. Poiseuille flow. For large enough Re the flow becomes turbulent, the topic of Chapter 13. Entrance flows can dominate the pressure–flow relationship, stresses on the cells which form the walls, as well as stresses on the blood cells. High Re phenomena also can dominate transport of soluble constituents (oxygen, carbon dioxide, anesthetics, toxins, lipids), see Section 4.5, and particulates (blood cells, cancer cells, blood clots), see Section 3.9. In general, the fluid stresses on the cells and structures which form the wall surface can influence local and regional cellular behavior, growth, and remodeling. Additional features of high Re flows include flow separation and the creation of "secondary flows." For example, when a tube is curved the axial flow experiences centripetal accelerations, which create cross-stream vortical motions which are "secondary" to the main flow direction. Axial curvature is a prominent feature of many biological tubes as they branch into their networks, and requires an additional pressure drop because of the force it imparts to deflect the fluid.

Topics Covered

We begin our foray into laminar boundary layers by examining the canonical problem of flow over a flat plate. The analysis develops the concept of a boundary layer, its thickness, and how it depends on the free stream velocity and fluid properties. By scaling the Navier–Stokes equations appropriately, they simplify to the boundary layer equations, which hold within the thin layer near the plate. Then a similarity solution method, a scheme we used in Section 8.1 for Stokes' first problem, reduces the governing partial differential equation to an ordinary differential equation in the similarity variable. The numerical solution of this resulting form is readily accomplished using computational software. Then polynomial approximations to those solutions are investigated to provide a simpler tool for investigating additional features of boundary layer flows. With this tool, the details of boundary layer velocities give an interesting example of particle deposition with applications to a clump of metastatic cancer cells or a clot in blood flow. A significant property of the boundary layer is that it grows in thickness downstream of the plate's leading edge. This feature also occurs at the entrance of tubes. For vascular flows the entrance, or inlet, region is often much shorter than the tube length. On the other hand, in respiratory applications the entrance region can be as long as the airway tube, so controls the pressure–flow relationship for the tube. This relationship is applied to the lung airway tree in a distributed resistance model to understand ventilation mechanics. Then we present the effects of axial tube curvature, which is a property of blood vessels, especially the aorta, and airway tubes. A new dimensionless parameter appears called the Dean number, $Dn = (d/2R_{ax})^{1/2}Re$, where d/2 is the tube radius, R_{ax} is the axial radius of curvature and Re is the Reynolds number. The secondary flow of cross-sectional vortices increases the resistance compared to a straight tube. The entrance boundary layer growth and tube axial curvature effects are further viewed for flow at a bifurcation, including flow separation because of fluid inertia meeting an adverse pressure gradient. Three-dimensional computational fluid dynamics and experimental fluid mechanics like particle image velocimetry (PIV), provide a visually impressive understanding of clinical situations: flow through a curved artery and lung airway flows related to aerosol deposition. Finally, a boundary layer approach to stagnation point flow, used in previous sections in its inviscid form, yields a new result satisfying no-slip at the wall. The effects of viscosity in this boundary layer will influence aerosol deposition patterns.

12.1 Boundary Layer Flow over a Flat Plate

Flows in the vascular and respiratory systems encounter bifurcations of the tubes where they branch. A model situation is shown in Figure 12.1 where flow in the upstream "parent" tube is initially uniform at its entrance and then develops a boundary layer of thickness $\delta(z)$. Within the thin boundary layer the velocity profile experiences rapid changes from the inviscid core value to meet the no-slip condition at the wall. It is within the boundary layer that viscous effects are prominent. The flow then splits at the branch and new entrance boundary layers develop in the daughter tubes. The development of a boundary layer is a well-studied phenomenon in fluid mechanics due to its importance in flow over aircraft wings, which has biological counterparts in bird wings and fish fins. The fundamentals can be appreciated by studying flow over a flat plate.

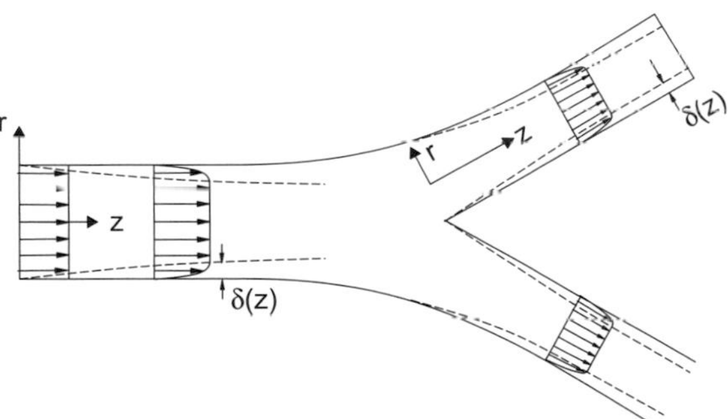

Figure 12.1 Flow at a tube entrance and into a bifurcation showing the growth of boundary layers in the parent and daughter tubes.

The most basic example of boundary layer flow is shown in Figure 12.2. A uniform upstream velocity, $u = U_\infty$, $v = 0$, encounters a flat plate at zero incidence angle, which is present for $x \geq 0$, $y = 0$. For high Reynolds number, $Re >> 1$, a thin viscous boundary layer develops on the plate. Its thickness is indicated as $y = \delta(x)$. Within the boundary layer, $0 \leq y \leq \delta(x)$, viscous effects are important as u changes rapidly in the y-direction from the no-slip condition at the wall, $u(y = 0) = 0$, to match the inviscid velocity, $u = U_\infty$ outside the boundary layer. Our goals for analyzing this flow are to calculate the velocity profiles $u(x, y), v(x, y)$ and pressure field $p(x, y)$ inside the boundary layer which match to the inviscid flow outside of it, and to determine an equation for the boundary layer thickness, $\delta(x)$.

We choose the following scales for our analysis as in Eq. (6.6.1)

$$X = \frac{x}{L}, \quad Y = \frac{y}{L}, \quad U = \frac{u}{U_\infty}, \quad V = \frac{v}{U_\infty}, \quad P = \frac{p}{\rho U_\infty^2} \tag{12.1.1}$$

where L is a length scale to be determined at the end of the analysis. Note that this pressure scale differs from lubrication theory in Chapter 11, where it was proportional to $\mu U/L$. In fact their ratio is Re. Using the definitions of the dimensionless variables of Eqs. (12.1.1) in the Navier–Stokes and continuity equations, their dimensionless forms, as in Eq. (6.6.2), are

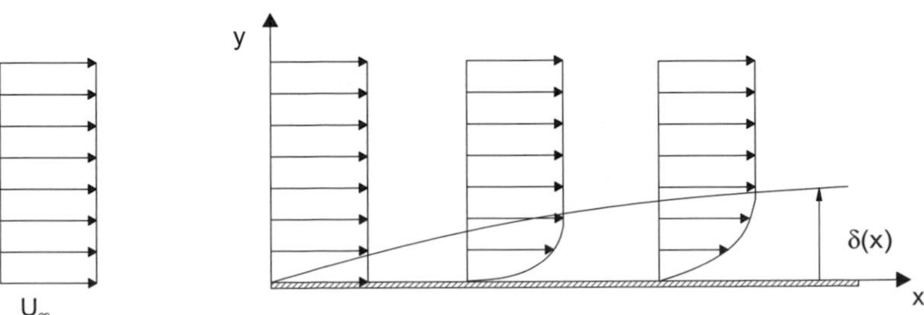

Figure 12.2 Boundary layer flow over a flat plate.

X-momentum $$U\frac{\partial U}{\partial X}+V\frac{\partial U}{\partial Y}=-\frac{\partial P}{\partial X}+\varepsilon\left(\frac{\partial^2 U}{\partial X^2}+\frac{\partial^2 U}{\partial Y^2}\right)$$

Y-momentum $$U\frac{\partial V}{\partial X}+V\frac{\partial V}{\partial Y}=-\frac{\partial P}{\partial Y}+\varepsilon\left(\frac{\partial^2 V}{\partial X^2}+\frac{\partial^2 V}{\partial Y^2}\right)$$

continuity $$\frac{\partial U}{\partial X}+\frac{\partial V}{\partial Y}=0 \tag{12.1.2}$$

where $\varepsilon = 1/\text{Re}$ and the Reynolds number is $\text{Re} = U_\infty L/\nu$. The conservation equations, Eqs. (12.1.2), are subject to the boundary conditions of no-slip and no-penetration on the plate, and matching of $U \to 1, V \to 0$ in the inviscid region $Y \to \infty$,

$$U(Y=0)=0,\ \ V(Y=0)=0,\ \ U(Y\to\infty)=1,$$

$$V(Y\to\infty)=0 \tag{12.1.3}$$

For high Reynolds number flows, $\varepsilon<<1$, we can try to ignore all terms in Eq. (12.1.2) multiplied by ε. However, we see that leads to a singularity, because dropping the $\varepsilon\partial^2U/\partial Y^2$ viscous term in the X-momentum balance reduces the equations from second order to first order in Y. In principle that situation would only allow one function of integration for U. So U cannot satisfy the two boundary conditions in Y that are required, i.e. $U(Y=0)=0,\ U(Y\to\infty)=1$. The way to preserve the highest order derivative terms in Eq. (12.1.2) in the Y-direction is to rescale Y with a function of ε. Let's define this boundary layer coordinate, η, as

$$\eta=\frac{Y}{\varepsilon^a} \tag{12.1.4}$$

where the exponent a will be determined in the analysis. In this new coordinate definition, the Y-derivatives become $\partial/\partial Y = (\partial\eta/\partial Y)\partial/\partial\eta = \varepsilon^{-a}\partial/\partial\eta$ and $\partial^2/\partial Y^2 = (\partial\eta/\partial Y)\partial/\partial\eta(\varepsilon^{-a}\partial/\partial\eta) = \varepsilon^{-2a}\partial^2/\partial\eta^2$ using the chain rule of differentiation. From Eqs. (12.1.2), the continuity equation is now

$$\frac{\partial U}{\partial X}+\frac{1}{\varepsilon^a}\frac{\partial V}{\partial\eta}=0 \tag{12.1.5}$$

To keep both terms in balance in continuity, Eq. (12.1.5), we need to rescale V in terms of ε as $V=\varepsilon^a\hat{V}$ so that the new form is

$$\frac{\partial U}{\partial X}+\frac{\partial\hat{V}}{\partial\eta}=0 \tag{12.1.6}$$

Using the boundary layer coordinate and the new velocity scale for $\hat{V}$, Eqs. (12.1.2) for momentum conservation become

$$U\frac{\partial U}{\partial X}+\frac{\varepsilon^a}{\varepsilon^a}\hat{V}\frac{\partial U}{\partial\eta}=-\frac{\partial P}{\partial X}+\varepsilon\left(\frac{\partial^2 U}{\partial X^2}+\frac{1}{\varepsilon^{2a}}\frac{\partial^2 U}{\partial\eta^2}\right)\varepsilon^a U\frac{\partial\hat{V}}{\partial X}+\frac{\varepsilon^{2a}}{\varepsilon^a}\hat{V}\frac{\partial\hat{V}}{\partial\eta}$$
$$=-\frac{1}{\varepsilon^a}\frac{\partial P}{\partial\eta}+\varepsilon\varepsilon^a\left(\frac{\partial^2\hat{V}}{\partial X^2}+\frac{1}{\varepsilon^{2a}}\frac{\partial^2\hat{V}}{\partial\eta^2}\right) \tag{12.1.7}$$

The first of Eqs. (12.1.7) is the x-momentum equation, which shows us that to preserve the $\partial^2U/\partial\eta^2$ term as $\varepsilon\to 0$ we require the exponent $a = 1/2$. That choice makes the boundary layer coordinate and the vertical velocity, respectively,

$$\eta=\frac{Y}{\varepsilon^{1/2}}=\frac{Y}{\left(\frac{1}{\text{Re}}\right)^{1/2}},\quad \hat{V}=\frac{v}{\varepsilon^{1/2}U_\infty}=\frac{V}{\left(\frac{1}{\text{Re}}\right)^{1/2}} \tag{12.1.8}$$

We see that the boundary layer thickness is proportional to $\varepsilon^{1/2}=(1/\text{Re})^{1/2}$, that is, $0\lesssim\eta\lesssim 1$ for $0\le Y\le\varepsilon^{1/2}$, and the velocity V is small, of $O(\varepsilon^{1/2})$ with $\hat{V}=O(1)$. Rewriting the governing equations, Eqs. (12.1.2), in the new variables and dropping terms multiplied by powers of ε, the new balance becomes

$$U\frac{\partial U}{\partial X}+\hat{V}\frac{\partial U}{\partial\eta}=-\frac{\partial P}{\partial X}+\frac{\partial^2 U}{\partial\eta^2}\ \ 0=-\frac{\partial P}{\partial\eta}\frac{\partial U}{\partial X}+\frac{\partial\hat{V}}{\partial\eta}=0 \tag{12.1.9}$$

We see immediately that $P = P(X)$ only, i.e. pressure does not vary across the boundary layer. Consequently the Bernoulli pressure from the inviscid flow outside of the boundary layer is felt all the way to the wall. The boundary conditions for the flow are no-slip at the wall, no cross-flow at the wall and constant X-velocity far from the plate,

$$U(\eta=0)=0,\quad \hat{V}(\eta=0)=0,\quad U(\eta\to\infty)=1 \tag{12.1.10}$$

We consider the case where $\partial P/\partial X = 0$, and cast the X-momentum equation in terms of the stream function, ψ, such that $U = \partial\psi/\partial\eta, \hat{V} = -\partial\psi/\partial X$. The choice of the stream function automatically solves continuity. Then the X-momentum equation of Eq. (12.1.9) becomes

$$\frac{\partial\psi}{\partial\eta}\frac{\partial^2\psi}{\partial\eta\partial X} - \frac{\partial\psi}{\partial X}\frac{\partial^2\psi}{\partial\eta^2} = \frac{\partial^3\psi}{\partial\eta^3} \tag{12.1.11}$$

which is now the stream function equation. Equation (12.1.11) was studied and solved by Paul Richard Heinrich Blasius (1883–1970) a German fluid dynamicist (Blasius, 1908) who was a doctoral student of Ludwig Prandtl (1875–1953) at Gottingen. Prandtl was the first physicist/engineer to identify boundary layer flow in 1905 (Prandtl, 1905).

Blasius found that a similarity solution works by defining a similarity variable, $\hat{\eta}$, and similarity stream function, $f(\hat{\eta})$, as

$$\psi = \sqrt{X}\,f(\hat{\eta}) \quad \text{where } \hat{\eta} = \eta/\sqrt{X} \tag{12.1.12}$$

Recall that we used a similarity variable to solve the suddenly started plate, Stokes' first problem in Section 8.1. The goal is to change a partial differential equation in X, Y to an ordinary differential equation in $\hat{\eta}$. Inserting Eq. (12.1.12) into Eq. (12.1.11) through the chain rule and product rule of differentiation, the result is the third-order nonlinear ordinary differential equation

$$2f''' + ff'' = 0 \tag{12.1.13}$$

Keeping track of their origins, f''' comes from the viscous term and $ff''$from the convective acceleration terms of Eq. (12.1.9). The boundary conditions for Eq. (12.1.13) are: no-penetration at the wall or, equivalently, that the wall is a streamline with a constant stream function value of $\psi(\hat{\eta} = 0) = 0$ or $f(\hat{\eta} = 0) = 0$; no-slip at the wall, $U(\hat{\eta} = 0) = 0$, which translates to $f'(\hat{\eta} = 0) = 0$; and matching of U to the inviscid velocity outside the boundary layer, $U(\hat{\eta} \to \infty) = 1$, or in terms of f, $f'(\hat{\eta} \to \infty) = 1$. So the boundary conditions on f in Eq. (12.1.13) are

$$f(\hat{\eta} = 0) = 0, f'(\hat{\eta} = 0) = 0, f'(\hat{\eta} \to \infty) = 1 \tag{12.1.14}$$

Equation (12.1.13) subject to the boundary conditions of Eq. (12.1.14) is known as the Blasius equation and can be solved numerically. In Section 17.4 we will find that our approach can be viewed formally as the first term of a perturbation series using the method of matched asymptotic expansions. The velocity components in terms of f and $\hat{\eta}$ are obtained by applying the definitions and using the chain and product rules of differentiation

$$\begin{aligned} U &= \frac{\partial\psi}{\partial\eta} = \frac{\partial\psi}{\partial\hat{\eta}}\frac{\partial\hat{\eta}}{\partial\eta} = \sqrt{X}f'(\hat{\eta})\frac{1}{\sqrt{X}} = f'(\hat{\eta}) \\ \hat{V} &= -\frac{\partial\psi}{\partial X} = -\left(\frac{1}{2\sqrt{X}}f + \sqrt{X}\frac{\partial f}{\partial\hat{\eta}}\frac{\partial\hat{\eta}}{\partial X}\right) \\ &= -\left(\frac{1}{2\sqrt{X}}f + \sqrt{X}f'\left(-\frac{\eta}{2X\sqrt{X}}\right)\right) \\ &= \frac{1}{2\sqrt{X}}(\hat{\eta}f' - f) \end{aligned} \tag{12.1.15}$$

If we let the length scale, $L = x$, then $X = x/L = 1$, and the Reynolds number is $Re = xU_\infty/\nu$, which is sometimes expressed by the symbol Re_x indicating that the length scale is based on the dimensional distance x. That makes $\varepsilon = (1/Re_x)^{1/2} = (\nu/xU_\infty)^{1/2}$. So $\hat{\eta}$ becomes

$$\hat{\eta} = \frac{Y}{\varepsilon^{1/2}\sqrt{X}} = \frac{\left(\frac{y}{x}\right)}{\left(\frac{\nu}{xU_\infty}\right)^{1/2}\sqrt{1}} = \frac{y}{\left(\frac{\nu x}{U_\infty}\right)^{1/2}} \tag{12.1.16}$$

The velocities in Eq. (12.1.15) simplify to

$$U = \frac{u}{U_\infty} = \frac{\partial\psi}{\partial\eta} = f'(\hat{\eta}),\ \hat{V} = \frac{v}{U_\infty}\left(\frac{U_\infty x}{\nu}\right)^{1/2} = \frac{1}{2}(\hat{\eta}f' - f) \tag{12.1.17}$$

Equation (12.1.13) is a third-order, nonlinear ODE subject to the boundary conditions in Eq. (12.1.14). Numerical solutions can be found using computational software. Maple, for example, recognizes the system as a boundary value problem and employs a finite difference method with Richardson extrapolation as its default. The code is shown in Code 12.1(a), where $\hat{\eta} \to \infty$ has been replaced by $\hat{\eta} = 7$ and plots of the velocities are created. The equivalent MATLAB® code is Code 12.1(b).

(a)

```
restart;
ODE := 1/2 . f(etahat)·diff(f(etahat), etahat, etahat)+diff(f(etahat), etahat, etahat, etahat)
```

$$ODE := \frac{f(etahat)\left(\frac{d^2}{detahat^2} f(etahat)\right)}{2} + \frac{d^3}{detahat^3} f(etahat) \quad (1)$$

```
#Let the boundary condition at etahat = infinity be at etahat = 7
BCS := f(0) = 0, (D(f))(0) = 0, (D(f))(7) = 1;
```

BCS := f(0) = 0, D(f)(0) = 0, D(f)(7) = 1 (2)

```
#Solve the ODE
sol := dsolve({ODE, BCS}, numeric);
```

sol := **proc**(*x_bvp*) … **end proc** (3)

```
#Create plots using odeplot.  Make them parametric plots so the velocities
are horizontal and etahat is vertical.
PL1 := plots[odeplot](sol, [diff(f(etahat), etahat), etahat], color = black, labels = [U, η̂]) :

PL2 := plots[odeplot](sol, [1/2 . etahat. diff(f(etahat), etahat) - 1/2 . f(etahat), etahat], color = black,
        labels = [V̂, η̂]) :

#Display the plots
with(plots):
display(PL1);
display(PL2);
```

(4)

(b)

```
% This MATLAB script requires that the files"Chapter12_BlasiusBL_ODESystem.m"
% "Chapter12_BlasiusBL_InitialGuess.m" and "Chapter12_BlasiusBL_BC.m"
% are located in the same folder of this script
clear all, close all, clc
etaHat_inf = 7;
etaHat     = linspace(0,etaHat_inf,1000);
solinit    = bvpinit(etaHat, @Chapter12_BlasiusBL_InitialGuess);
sol= bvp5c(@Chapter12_BlasiusBL_ODESystem,@Chapter12_BlasiusBL_BC,solinit);
f          = sol.y(1,:);
fprime     = sol.y(2,:);
u          = fprime;
v          = 1/2*(etaHat.*fprime-f);
subplot(1,2,1), plot(u,etaHat,'k'), hold on;
axis([0 1.1 0 etaHat_inf])
ylabel('$\hat{\eta}$','Interpreter','Latex')
xlabel('$U$','Interpreter','Latex')
```

Code 12.1 Code examples: (a) Maple, (b) MATLAB®.

```
subplot(1, 2, 2), plot(v,etaHat,'k'), hold on;
axis([0 1.1 0 etaHat_inf])
ylabel('$\hat{\eta}$','Interpreter','Latex')
xlabel('$\hat{V}$','Interpreter','Latex')
```

The following MATLAB function must be saved as "Chapter12_BlasiusBL_ODESystem.m"

```
function dydetaHat = Chapter12_BlasiusBL_ODESystem(etaHat,y)
dydetaHat = zeros(3,1);
dydetaHat = [ y(2)
         y(3)
        -1/2.*y(1).*y(3)];
end
```

The following MATLAB function must be saved as "Chapter12_BlasiusBL_InitialGuess.m"

```
function g = Chapter12_BlasiusBL_InitialGuess(etaHat)
g = [etaHat
   1+0*etaHat
   0*etaHat];
end
```

The following MATLAB function must be saved as "Chapter12_BlasiusBL_BC.m"

```
function BC = Chapter12_BlasiusBL_BC(ya,yb)
BC = [ya(1)
    ya(2)
    yb(2)-1];
end
```

Code 12.1 *(continued)*

Graphs of the two velocity components in the boundary layer are shown in Figure 12.3. The horizontal velocity varies rapidly within the boundary layer, and asymptotes to $U = f' = 1$ in the outer (inviscid) region. The vertical velocity is always positive, so fluid is leaving the boundary layer. In fact, for a fixed value of x, the rate at which v increases with y, i.e. $\partial v/\partial y > 0$ in the boundary layer. This is balanced in continuity by $\partial u/\partial x < 0$ as the fluid slows down due to the wall shear. The solutions have self-similar shape at each value of x, which is consistent with the assumption of a similarity solution.

The thickness of the boundary layer, $\delta(x)$, can be estimated from the U solution in Figure 12.3. Since the boundary condition away from the plate is at $\hat{\eta} \to \infty$, the solutions for the velocities are asymptotic as $\hat{\eta}$ increases and there is no finite value of $\hat{\eta}$ where $u = 1$. However, we can choose a tolerance as close as we like. For example, what is the value of $\hat{\eta}$ where

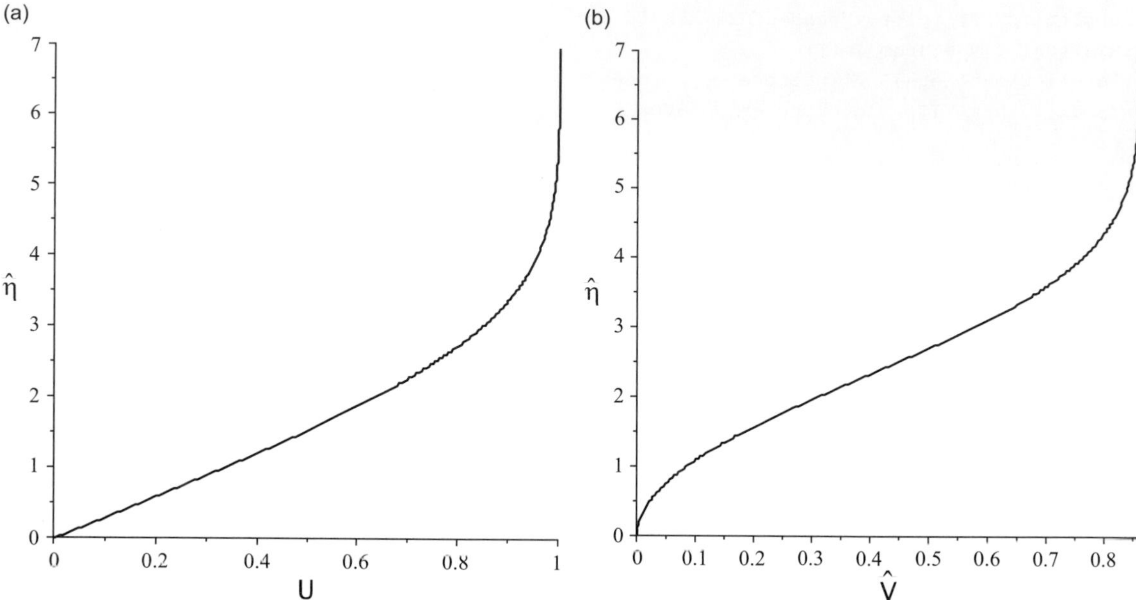

Figure 12.3 Plots of (a) U vs $\hat{\eta}$ (b) $\hat{V}$ vs $\hat{\eta}$ from the code listed.

$U = 0.99$ or 0.95? These are two levels of tolerance used commonly to define the finite boundary layer thickness.

From Eq. (12.1.16) we know that $\hat{\eta} = y/(\nu x/U_\infty)^{1/2}$. From the Blasius solution $U(\hat{\eta} = 5.0) = f'(\hat{\eta} = 5.0) = 0.99$, so within 1% error, the boundary layer thickness is $y = \delta$ where

$$\delta = 5.0\left(\frac{\nu x}{U_\infty}\right)^{1/2} \tag{12.1.18}$$

Likewise, $U(\hat{\eta} = 4.0) = f'(\hat{\eta} = 4.0) = 0.95$, so within 5% error, the boundary layer thickness is $\delta = 4.0(\nu x/U_\infty)^{1/2}$. In either case, the boundary grows as $\sqrt{x}$ along the plate. It is thinner for higher oncoming velocity, U_∞, and for smaller kinematic viscosity, ν. Note that the boundary layer equations, Eq. (12.1.9), are first order in derivatives of v ($\hat{V}$) with respect to y ($\hat{\eta}$). That means we can only satisfy one boundary condition for v, the no-penetration condition at the wall where $v(y = 0) = 0$. The condition at infinity that $v = 0$ is not satisfied and we can see that $\hat{V}(\hat{\eta} = 5) \neq 0$ in Figure 12.3(b) at the edge of the boundary layer.

The wall shear, $\tau_w = \mu(\partial u/\partial y)$ at $y = 0$, is an important quantity for boundary layer flows. Often it is cast in dimensionless form as a friction coefficient, C_f, usually called the "skin friction," where

$$C_f = \frac{\tau_w}{\frac{1}{2}\rho U_\infty^2} \tag{12.1.19}$$

The Blasius solution can be used to calculate $\partial u/\partial y$ at $y = 0$ in Eq. (12.1.19), which results in

$$C_f = \frac{0.664}{Re_x^{1/2}} \tag{12.1.20}$$

Dimensionally the wall shear depends on the parameters as

$$\tau_w = \frac{1}{2}\rho U_\infty^2 \frac{0.664}{(U_\infty x/\nu)^{1/2}} = 0.332\mu^{1/2}\rho^{1/2}U_\infty^{3/2}x^{-1/2} \tag{12.1.21}$$

Equation (12.1.21) shows that the wall shear is proportional to $U_\infty^{3/2}$, density as $\rho^{1/2}$ and viscosity as $\mu^{1/2}$. Spatially it drops off as $x^{-1/2}$ down the length of the plate.

12.2 Polynomial Approximations to Boundary Layer Flow

Prior to the arrival of computers, polynomial approximations to the Blasius solution proved to be useful for performing various calculations analytically, rather than numerically. For example, if we use the boundary layer thickness, $\delta(x)$, from Eq. (12.1.18) as the length scale in the vertical direction, then we can define the dimensionless vertical variable at

$$\zeta = \frac{y}{\delta} \qquad 0 \le \zeta \le 1 \tag{12.2.1}$$

so that $\zeta = 0$ on the plate and $\zeta = 1$ at the edge of the boundary layer (ζ is the Greek letter zeta). At this stage we do not need to specify the form of $\delta(x)$. Consider polynomial representations of u in which ζ has the range $0 \le \zeta \le 1$, in the form

$$U = \frac{u}{U_\infty} = a_0 + a_1\zeta + a_2\zeta^2 + a_3\zeta^3 + a_4\zeta^4 + \cdots \tag{12.2.2}$$

We simply need conditions on the velocity profile to solve for the unknown coefficients a_n. For a linear profile Eq. (12.2.2) is terminated after the a_1 term. Only two conditions can be imposed to solve for a_0 and a_1, and they are the no-slip condition on the wall, $U(\zeta = 0) = 0 = a_0$, and matching the constant free stream velocity at the edge of the finite boundary layer, $U(\zeta = 1) = 1 = a_1$. The result is

$$U = \zeta \qquad \text{linear} \tag{12.2.3}$$

For a quadratic approximation we add the condition that the slope of the velocity profile matches the zero slope at the edge of the boundary layer, i.e. $\partial U/\partial\zeta|_{\zeta=1} = 0$. This leads to new values of a_0 and a_1, and then a solution for a_2, giving us

$$U = 2\zeta - \zeta^2 \qquad \text{quadratic} \tag{12.2.4}$$

For a cubic approximation we see from the x-momentum balance in Eq. (12.1.9) that, with no pressure gradient, the equation yields $\partial^2 U/\partial\zeta^2|_{\zeta=0} = 0$ because U and $\hat{V}$ are both zero on the plate. This yields

$$U = \frac{3}{2}\zeta - \frac{1}{2}\zeta^3 \qquad \text{cubic} \tag{12.2.5}$$

A plot of the Blasius solution and the three polynomial approximations of linear, quadratic and cubic are shown for comparison in Figure 12.4. The quadratic polynomial (dotted curve) is a very good representation of the Blasius (solid curve) solution.

For each polynomial form of u there is a corresponding expression for v using the continuity equation. For example, the dimensional linear form is $u = U_\infty y\delta^{-1}$, which yields the dimensional v as follows

$$v = -\int_0^y \frac{\partial}{\partial x}\left(U_\infty y\delta^{-1}\right) dy = U_\infty \delta^{-2}\delta' \frac{y^2}{2} = U_\infty \delta' \frac{\zeta^2}{2} \tag{12.2.6}$$

So far we have used δ as the scale for y, but have not needed to specify its form. What is the boundary layer thickness for these different polynomial approximations? We cannot simply assume that the Blasius value, $\delta = 5.0(\nu x/U_\infty)^{1/2}$, holds for polynomial velocity profiles. In Sections 13.7 and

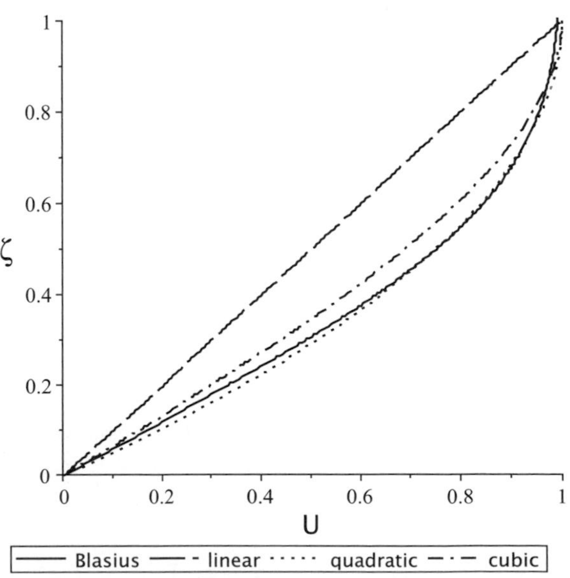

Figure 12.4 Comparison of approximate boundary layer x-velocity profiles (linear, quadratic, cubic) to the Blasius solution.

13.8 we will learn that $\delta = 3.46(\nu x/U_\infty)^{1/2}$ for the linear profile, $\delta = 5.47(\nu x/U_\infty)^{1/2}$ for the quadratic profile and $\delta = 4.64(\nu x/U_\infty)^{1/2}$ for the cubic profile. Without going through the details here, these coefficients are determined by integrating the boundary layer equations, the dimensional forms of Eq. (12.1.9) with $\partial P/\partial X = 0$, across the boundary layer $0 \le y \le \delta(x)$. So their values reflect boundary layer conservation of mass and momentum, including the balance of convective acceleration and viscous effects.

Using the linear profile boundary layer thickness, Eq. (12.2.6) becomes

$$v = 0.072\left(\frac{U_\infty^3}{\nu}\right)^{1/2}\frac{y^2}{x^{3/2}} \tag{12.2.7}$$

Some important observations of v in Eq. (12.2.7) are, firstly, that v starts at zero for $y = 0$ on the wall, but then is an increasing function of y. So not only is $v > 0$ in the boundary layer, but $\partial v/\partial y > 0$, which is balanced by $\partial u/\partial x < 0$ as the fluid slows down from the wall shear. Because $v > 0$ fluid is forced out of the boundary layer. Other observations are that increasing the free stream velocity, U_∞, increases v with a 3/2 power dependency, while farther downstream as x increases, v decreases with an $x^{-3/2}$ dependency. Defining $\hat{V} = v(x/\nu U_\infty)^{1/2}$, Eq. (12.2.7) simplifies to

$$\hat{V} = 0.87\zeta^2 \tag{12.2.8}$$

Worked Example 12.1 | Metastatic Cancer or Blood Clot Embolus

Cancer spreads by several mechanisms in the body, one of them is from blood flow. Also, blood clots can break off smaller pieces which become emboli flowing in the circulatory system. Suppose there is a clump of cancer cells or blood clot that has made its way into an arterial flow. Figure 12.5 shows this particle at the edge of the boundary layer at an arterial entrance, which we can consider planar for thin boundary layers. Let gravity be in the negative y-direction. Since the v-velocity at the edge of the boundary layer is positive, the flow tends to push the cell away from the boundary and against the gravity field. However, as the particle moves downstream the boundary layer thickens and the v-velocities all decrease so that gravity wins out. These mechanisms combine to delay its impact on the wall to a location farther downstream.

Using the linear velocity profile assumption we can make the components

$$U = \frac{u}{U_\infty} = \frac{y}{\delta},\ V = \frac{v}{U_\infty} = \delta^{-2}\delta'\frac{y^2}{2} \tag{12.2.9}$$

where $\delta = 3.46(\nu x/U_\infty)^{1/2}$ is the boundary layer thickness. If we scale x and y on the particle diameter, d_p, so $X = x/d_p$, $Y = y/d_p$, then the velocity expressions in Eq. (12.2.9), evaluated at the particle position, $X = X_p$, $Y = Y_p$, are

$$U\left(\underline{X} = \underline{X}_p\right) = \frac{1}{3.46}Re_p^{1/2}\frac{Y_p}{X_p^{1/2}},\quad V\left(\underline{X} = \underline{X}_p\right) = 0.072Re_p^{1/2}\frac{Y_p^2}{X_p^{3/2}} \tag{12.2.10}$$

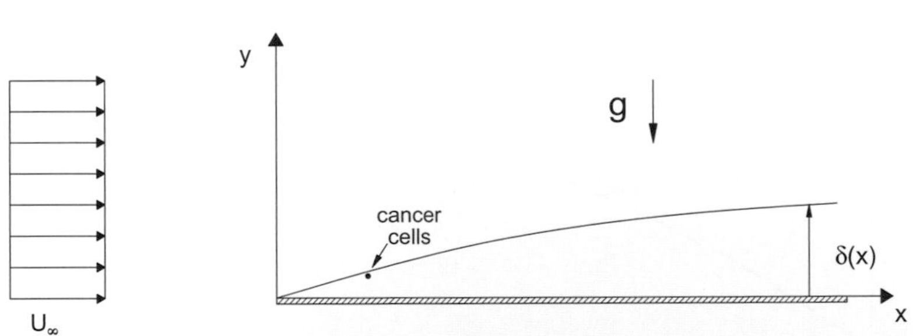

Figure 12.5 Particle flow in the boundary layer of a flat plate as a simple model of a metastatic cancer mechanism.

where $Re_p = d_p U_\infty / \nu$ is the particle Reynolds number based on its diameter. The scaled particle motion equations look similar to our previous model of aerosol deposition with gravity which we studied in a stagnation point flow, but with different U and V,

$$\begin{aligned} Stk\frac{d^2X_p}{dT^2} + \frac{dX_p}{dT} - U\left(\underline{X} = \underline{X}_p\right) &= 0 \\ Stk\frac{d^2Y_p}{dT^2} + \frac{dY_p}{dT} - V\left(\underline{X} = \underline{X}_p\right) &= -\frac{Stk}{Fr^2} \end{aligned} \tag{12.2.11}$$

Because U and V both depend on X_p and Y_p, Eq. (12.2.11) is a fully coupled pair of ODEs which can be solved numerically with computational software. The time is scaled as $T = U_\infty t / d_p$ and the dimensionless parameters are the Stokes and Froude numbers

$$Stk = \frac{\rho_p U_\infty d_p}{18\mu}, \quad Fr^2 = \frac{U_\infty^2}{g d_p \left(1 - \rho_f/\rho_p\right)} \tag{12.2.12}$$

Use the initial conditions $X_p(0) = 50$, $Y_p(0) = 0.9\delta\left(x = x_p(0)\right)/d_p = 0.9(3.46)\left(X_p(0)\right)^{1/2}/Re_p^{1/2}$ where the choice of $X_p(0)$ keeps away from the singularity at $X = 0$ and the choice of $Y_p(0)$ is 90% of the boundary layer thickness at that location for $X_p(0)$. Also let $dX_p/dT = 0$ and $dY_p/dT = 0$ at $T = 0$.

To solve this system numerically we choose the following parameter values

$$\begin{aligned} &g = 980\ cm/s^2, \mu = 0.03\ poise, \nu = 0.03\ cm^2/s \\ &\rho_f = 1.0\ g/cm^3, \rho_p = 1.3\ g/cm^3, U_\infty = 10\ cm/s, d_p = 0.02\ cm \end{aligned} \tag{12.2.13}$$

So the particle is 200 microns in diameter.

Solving the system of coupled ODEs in Eq. (12.2.11) subject to the initial conditions and parameter choices leads to Figure 12.6, which is a plot of the particle trajectory $X_p(T)$ and $Y_p(T)$, along with the boundary layer thickness. We see that the particle initially is forced away from the wall but then impacts at $X_p \sim 227$, which corresponds to a dimensional distance, multiplying $227 \times d_p = 4.53$ cm. If the vessel is shorter than 4.53 cm, the particle moves on to the next branch.

The Maple code for generating Figure 12.6 is shown in Code 12.2(a) and the MATLAB® code is Code 12.2(b).

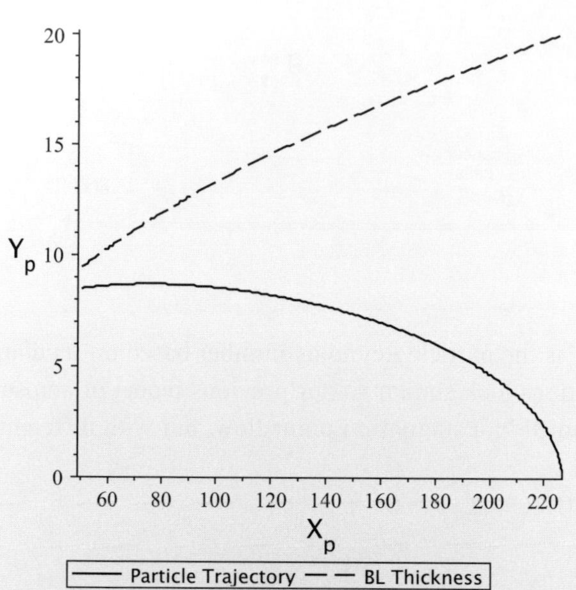

Figure 12.6 Particle trajectory to impact in a boundary layer flow and boundary layer thickness.

(a)

```
restart;
```

$$U(T) := \frac{1}{3.46} \cdot \frac{Rep^{\left(\frac{1}{2}\right)} \cdot Yp(T)}{Xp(T)^{\left(\frac{1}{2}\right)}} :$$

$$V(T) := 0.072. \frac{Rep^{\left(\frac{1}{2}\right)} \cdot Yp(T)^2}{Xp(T)^{\left(\frac{3}{2}\right)}} :$$

```
ODE1 := Stk.diff(Xp(T), T, T) + diff(Xp(T),T)-U(T) :
```

$$ODE2 := Stk.diff(Yp(T), T, T) + diff(Yp(T),T)-V(T)+\frac{Stk}{Fr^2}:$$

$$Stk := \frac{rhop \cdot Uinf \cdot dp}{18.mu}: Rep := \frac{Uinff.dp}{nu}: Fr := sqrt\left(\frac{Uinf^2}{g \cdot dp \cdot \left(1 - \frac{rhof}{rhop}\right)}\right) :$$

```
Uif := 10 : nu := 0.03 : mu := 0.03:  rhof := 1.0: rhop := 1.3: g := 980 : dp := 0.02:
#start  particle at Xp0=50, Yp=0.9 of BL thickness at Xp0
```

Code 12.2 Code examples: (a) Maple, (b) MATLAB®.

```
Xp0 := 50: Yp0 := evalf(0.9·3.46·Rep^(--)·Xp0^(1/2)) :
ICS := Xp(0) = Xp0, Yp(0) = Yp0, D(Xp)(0) = 0, D(Yp)(0) = 0 :
sol := dsolve({ODE1, ODE2, ICS}, {Xp(T), Yp(T)},numeric) :
with(plots):
Tf := 525:
Lf := sol(Tf)[2]·dp;
                         0.02Xp(T) = 4.53454770931840                    (1)
PL1 := odeplot(sol, [Xp(T), Yp(T)], 0...Tf, color=black, linestyle=1,
legend="Particle Trajectory");
PL2 := odeplot(sol,[Xp(T), 3.46·Rep^(-1/2)·Xp(T)^(1/2)], 0...Tf,color=black, linestyle=3,
legend = "BL Thickness) :
display(PL1, PL2, labels = [X[p],Y[p]],labelfont=[HELVETICA, 14]);
```

(b)

MATLAB code corresponding to Maple code reported at page 18(Chapter 12)

```
% This MATLAB script requires that the file "Chapter12_odeSystem12_2_11.m"
% is located in the same folder of this script
clear all; close all; clc
Uinf    = 10;
nu      = 0.03;
mu      = 0.03;
rhof    = 1;
rhop    = 1.3;
g       = 980;
dp      = 0.02;
Stk     = rhop*Uinf*dp/(18*mu);
Rep     = Uinf*dp/nu;
Fr      = sqrt(Uinf^2/(g*dp*(1-rhof/rhop)));
tspan   = [0 525];
x0      = 50;
Dx0     = 0;
y0      = 0.9*3.46*Rep^(-1/2)*x0^(1/2);
Dy0     = 0;
ICS     = [x0 Dx0 y0 Dy0];
[t,sol] = ode45(@(t,sol) Chapter12_odeSystem12_2_11(t,sol,Rep,Stk,Fr),tspan,ICS);
x       = sol(:,1);
Dx      = sol(:,2);
```

Code 12.2 *(continued)*

```
y        = sol(:,3);
Dy       = sol(:,4);
Lf       = dp*x(end)
plot(x,y,'-k'), hold on;
plot(x,3.46*Rep^(-1/2)*x.^(1/2),'--k')
legend('Particle Trajectory','BL Thickness','location','southwest')
xlabel('X_p')
ylabel('Y_p')
axis([50 230 0 20])
```

The following MATLAB function must be saved as "Chapter12_odeSystem12_2_11.m"

```
function dsdt= Chapter12_odeSystem12_2_11(t,s,Rep,Stk,Fr)
   U         = @(x,y) 1/3.46*Rep^(1/2)*y/(x^(1/2));
   V         = @(x,y) 0.072*Rep^(1/2)*y^2/(x^(3/2));
   dsdt      = zeros(4,1);
   dsdt(1)   = s(2);
   dsdt(2)   = -1/Stk*s(2)+1/Stk*U(s(1),s(3));
   dsdt(3)   = s(4);
   dsdt(4)   = -1/Stk*s(4)+1/Stk*V(s(1),s(3))-1/(Fr^2);
end
```

Code 12.2 (*continued*)

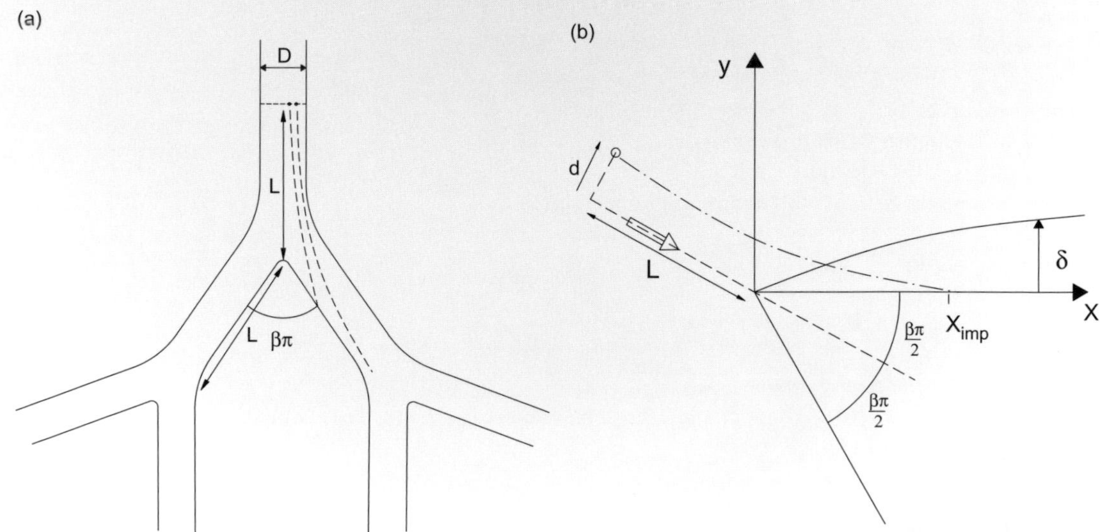

Figure 12.7 (a) Flow and particle deposition at an airway bifurcation. (b) Model of the bifurcation as flow over a wedge with boundary layer development.

This example emphasizes the "shielding" process that the positive v-velocity in the boundary layer provides by pushing the particle away from the wall. The oncoming flow in Figure 12.2 is at zero angle of incidence with the plate. However, the flow at a bifurcation, Figure 12.7(a), can be considered like flow over a wedge, where there is a non-zero incidence angle, see Figure 12.7(b). The inviscid velocity field has $v < 0$ throughout, including the boundary layer edge. Consequently, the boundary layer v can have negative values and the shielding mechanism is reduced or lost (Zierenberg et al., 2013).

12.3 Entrance Flow in a Tube: Airways and Blood Vessels

When flow enters a tube or encounters a bifurcation, as in blood vessels and airways, there is growth of a boundary layer on the tube wall along the main flow axis. Figure 12.8 shows the typical situation, a tube of radius a has entrance flow and viscous effects are confined to a boundary layer which grows in z from the wall. The velocity, U_0, is uniform in r at $z = 0$ so is also the area-averaged z-velocity, $U_0 = \bar{w}$, at every value of z. The inviscid core flow region gradually decreases as a result of the boundary layer. Far enough downstream the boundary layer joins at the centerline, which marks the end of the inviscid region. This occurs at $z = L_I$, the inviscid or inlet length. The velocity profile is not yet fully developed, however. That occurs farther downstream where $z = L_E$, which is called the entrance length. Poiseuille flow with its parabolic profile shape and whose centerline velocity is twice the average, $2\bar{w}$, is established at $z = L_E$ and persists downstream. The fluid pressure has a linear dependence on z in the Poiseuille region, but nonlinear in the entrance region, dropping much faster the closer to the entrance it is. The region $0 \leq z < L_I$ is the inviscid or inlet region while $L_I \leq z \leq L_E$ is called the "filled region" where the fully viscous flow is developing into Poiseuille flow for $z \geq L_E$.

For flow at the entrance of a tube we need an equation for the boundary layer thickness $\delta(z)$. Since this region has a non-zero pressure gradient, $dp/dz \neq 0$, and is not a flat plate, the Blasius formula

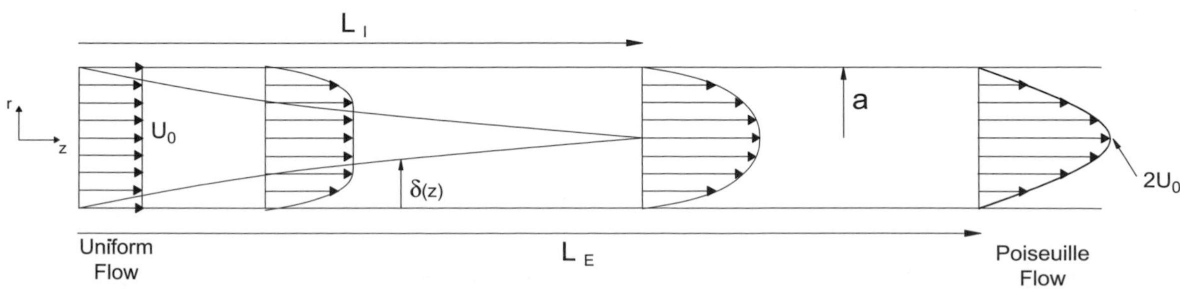

Figure 12.8 Entrance flow and pressure in a tube of radius a with boundary layer thickness, δ, and entrance length, L_E, and inviscid or inlet length, L_I.

of Eq. (12.1.18) is not correct. To incorporate the pressure gradient directly is beyond our scope at this stage. However, we can try the same functional form of the Blasius solution,

$$\delta(z) = K\sqrt{\frac{\nu z}{\bar{w}}} \tag{12.3.1}$$

for the tube. Depending on our approximations to the tube entrance boundary layer we could use $K = 5.0$(Blasius), $K = 3.46$(linear), $K = 5.47$(quadratic) and $K = 4.64$(cubic) though, again, those estimates assume zero axial pressure gradient.

A rough estimate of the inviscid length can be examined in Eq. (12.3.1) letting the boundary layer thickness equal the tube radius. That is an estimate of $z = L_I$ for $\delta = d/2$ in Eq. (12.3.1). Rearranging the result we find $L_I/d = Re/4K^2$. The analysis of entrance flow of a circular pipe performed in Mohanty and Asthana (1979) used a quartic polynomial approximation to the velocity profile dependence on the radial coordinate. They found the entrance length to be $L_E/d = 0.075\,Re$ while the inlet length was $L_I/d = 0.018\,Re$. So the inlet length is approximately one-fourth of the entrance length, $L_I \sim L_E/4$. Experiments reported in Mohanty and Asthana (1979) at Re = 1875, 2500 and 3250 corroborate the analytical results.

The entrance length, L_E, for a tube has been studied extensively using computational methods and experiments with results similar to those of Mohanty and Asthana (1979). For example, computations in Hornbeck (1964) showed $L_E/d = 0.057\,Re$ while those in Campbell and Slattery (1963) found $L_E/d = 0.068\,Re$. Experiments by Fargie and Martin (1971) found $L_E/d = 0.065\,Re$ for $760 \le Re \le 1512$. Approaches that are more inclusive for small Re include those of Atkinson et al. (1969), who found that a linear relationship between L_E/d and Re fits the calculation results well

$$\frac{L_E}{d} = 0.59 + 0.056Re \tag{12.3.2}$$

Durst et al. (2005) find, similarly, $L_E/d = 0.619 + 0.0567Re$. In addition, Durst et al. (2005) propose a nonlinear fit

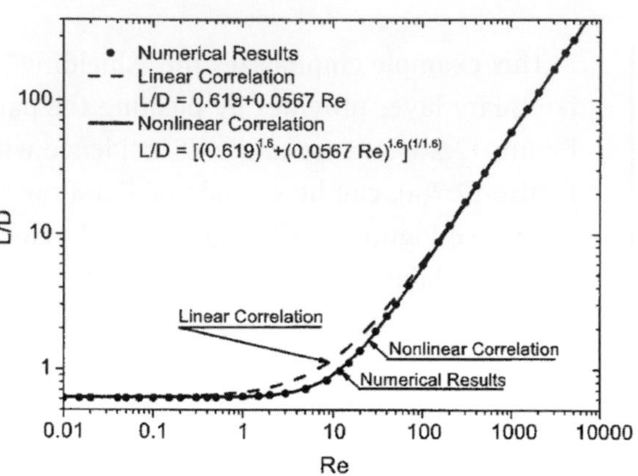

Figure 12.9 Predicted entrance length LE/d from the nonlinear correlation of Eq. (12.3.3). Comparison to linear and numerical results shown.

$$L_E/d = \left[0.619^{1.6} + (0.0567Re)^{1.6}\right]^{\frac{1}{1.6}} \tag{12.3.3}$$

which correlates better with the numerical solutions in the range $1 < Re < 100$, see Figure 12.9.

For a consistent form of our boundary layer thickness, let's continue to assume that $L_I = L_E/4$, which was found in Mohanty and Asthana (1979), and use L_E/d from Eq. (12.3.2),

$$\frac{L_I}{d} = \frac{Re}{4K^2} = \frac{1}{4}\frac{L_E}{d} = \frac{1}{4}(0.59 + 0.056Re) \tag{12.3.4}$$

Solving Eq. (12.3.4) for large Re, say Re > 100, gives us $K^2 \sim 1/0.056$ or $K \sim 4$, which is a reasonable value compared to our other choices.

The pressure drop in the inlet region can be approximated by assuming Bernoulli's equation holds in the inviscid core region

$$p_e + \frac{1}{2}\rho\bar{w}^2 = p_0(z) + \frac{1}{2}\rho w_0^2(z)$$
$$0 \le r \le a - \delta,\ 0 \le z \le L_I \tag{12.3.5}$$

where w is the z-component of velocity and we neglect any radial velocity in the core. p_e is the pressure at $z = 0$ and $\bar{w} = U_0$ is the uniform z-velocity at $z = 0$ which equals the area-averaged velocity, $\bar{w}$, at all z locations. The downstream pressure, $p_0(z)$, and inviscid core velocity, $w_0(z)$, are at an arbitrary axial position, z, as long as $z < L_I$. An alternative approach is to employ a flow-averaged energy balance (Pedley et al., 1970; 1971).

The volume flow rate through the tube is constant, Q, and equal to the integral of w(r, z), the z-velocity component, across the cross-section. The velocity field is defined in the core and boundary layer regions as

$$w = \begin{cases} w_0(z) & 0 \le r \le a - \delta \\ w_b(r, z) & a - \delta \le r \le a \end{cases} \quad (12.3.6)$$

The average axial fluid velocity is $\bar{w} = w_0(z = 0)$ where $\delta = 0$.

For the boundary layer region we introduce the local radial coordinate

$$\xi = \frac{a - r}{\delta} \quad (12.3.7)$$

whose origin is on the wall, $\xi = 0$, and increases to the edge of the boundary layer where $\xi = 1$ (ξ is the Greek letter pronounced xi). Then we scale Eqs. (12.3.6) on the core velocity,

$$W = \frac{w(r, z)}{w_0(z)} \quad (12.3.8)$$

so that, using the quadratic approximation to the boundary layer velocity

$$W = \begin{cases} 1 & 0 \le r \le a - \delta \\ W_b = 2\xi - \xi^2 & 0 \le \xi \le 1 \end{cases} \quad (12.3.9)$$

The volume flow rate at any axial location is the same

$$Q = \int_0^{2\pi}\int_0^{a-\delta} w_0\, r\, dr\, d\theta + \int_0^{2\pi}\int_{a-\delta}^{a} w_b\, r\, dr\, d\theta \quad (12.3.10)$$

We can recast Eq. (12.3.10) using the boundary layer coordinate ξ

$$Q = \pi w_0 (a - \delta)^2 + 2\pi a \delta w_0 \int_0^1 W_b \left(1 - \frac{\delta}{a}\xi\right) d\xi \quad (12.3.11)$$

Dividing Eq. (12.3.11) through by $Q = \pi a^2 \bar{w}$ and inserting the quadratic approximation to the boundary layer velocity profile, Eq. (12.3.9), into the integral of Eq. (12.3.11) gives us, after rearrangement,

$$w_0 = \bar{w}\left[\left(1 - \frac{\delta}{a}\right)^2 + 2\frac{\delta}{a}\left(\frac{2}{3} - \frac{5}{12}\frac{\delta}{a}\right)\right]^{-1} \quad (12.3.12)$$

which can be expanded in a Taylor series for $\delta/a << 1$. Neglecting terms of $(\delta/a)^2$ or smaller, the solution simplifies to

$$w_0 = \bar{w}\left(1 + \frac{2}{3}\frac{\delta}{a} + O\left(\frac{\delta}{a}\right)^2\right) \quad (12.3.13)$$

The Taylor series can be done by hand or using calculation software. The Maple code for the Taylor series is shown in Code 12.3(a) with $\hat{\delta} = \delta/a$ and the MATLAB® code is Code 12.3(b).

```
(a)
restart
w := w0.((1-deltahat)^2 + 2·deltahat·(2/3 - 5/12 .deltahat))^(-1) ;

                                  w0
        w := ------------------------------------------------------         (1)
              (1 - deltahat)^2 + 2 deltahatt (2/3 - 5 deltahat/12)

#Use the Taylor expansion command requesting 3 terms.  We will only use
the first 2 terms taylor(w, deltahat, 3);

        w0 + 2/3 w0 deltahat + 5/18 w0 deltahat^2 + O(deltahat^3)           (2)
```

```
(b)
clear all; close all; clc;
syms deltahat w0
w = w0*((1-deltahat)^2+2*deltahat*(2/3 -5/12*deltahat))^(-1);
taylor(w, deltahat, 'Order', 3)
```

Code 12.3 Code examples: (a) Maple, (b) MATLAB®.

In Eq. (12.3.12) we are neglecting terms of $O(\delta/a)^2$. Note that w_0 is an increasing function of z, since $\delta(z)$ is increasing. At $z = 0$, $\delta = 0$ and $w_0 = \bar{w}$, but then w_0 increases like $\sim z^{1/2}$. Since the volumetric flow rate is a constant, independent of z, the core fluid must speed up as the fluid in the boundary layer is slowing down due to viscosity.

Inserting this solution of w_0 into the Bernoulli equation, Eq. (12.3.5), results in the inlet flow pressure drop, Δp_I, as a function of the boundary layer viscous and inertial effects

$$p_e - p_0(z) = \Delta p_I = \frac{2}{3}\rho\bar{w}^2\frac{\delta}{a} = \frac{4K}{3}\rho\bar{w}^{3/2}\left(\frac{\nu z}{d^2}\right)^{1/2} \tag{12.3.14}$$

where we have substituted for $\delta(z)$ from Eq. (12.3.1) in Eq. (12.3.14). The pressure drop depends on contributions from inertia, ρ, and viscosity, μ.

By comparison, the Poiseuille pressure drop, Δp_P, assuming fully developed flow, over the distance z is

$$\Delta p_P = \frac{32\mu\bar{w}z}{d^2} \qquad \text{Poiseuille} \tag{12.3.15}$$

where we have substituted $Q = \pi a^2\bar{w}$. Poiseuille pressure drops only depend on viscous effects, not inertia, i.e. not density ρ. Notice that the dependence on $\bar{w}$ is stronger for entrance flow, with a 3/2 exponent in Eq. (12.3.14), compared to the linear term for Poiseuille flow in Eq. (12.3.15). The viscous shearing in the boundary layer has a strong influence on the pressure drop at the inlet.

Define a dimensionless pressure scaled on a characteristic viscous shear stress $P = p/(\mu\bar{w}/a)$ so that the dimensionless version of Eq. (12.3.14) is

$$P_e - P_0 = \Delta P_I = \frac{2K}{3}Re^{1/2}Z^{1/2} = 3.65Re^{1/2}Z^{1/2} \tag{12.3.16}$$

where $Z = z/d$ and $Re = \bar{w}d/\nu$. With this scaling, the dimensionless Poiseuille pressure drop becomes $\Delta P_P = 16Z$. The corresponding dimensionless boundary layer thickness, $\hat{\delta} = \delta/a$, is derived from Eq. (12.3.1) by dividing through by the tube radius and rearranging so that $\hat{\delta} = 2K\sqrt{Z/Re}$.

Examples of P_0 and $\hat{\delta}$ vs Z in the inlet region are shown in Figure 12.10 assuming $P_e = 0$ for four values of the Reynolds number, $Re = 200, 500, 1000, 1500$. Here we choose $K = 5.47$ for our quadratic approximation to the boundary layer profile, Eq. (12.3.9). Note that increasing Re has the effect of making the downstream pressures more negative, Figure 12.10(a) and the boundary layers thinner, Figure 12.10(b). Also, for Re = 200 we see that $\hat{\delta}(Z = 3) \sim 1$, which is its maximum allowed value but certainly outside of our restrictions for $\hat{\delta} << 1$. The linear Poiseuille pressure drop is shown for comparison in Figure 12.10(a), emphasizing the significant difference between entrance and fully developed flows.

Another common pressure scale is $\frac{1}{2}\rho\bar{w}^2$, focusing on the fluid inertia that dominates in high Reynolds number flow. With that scale the entrance flow pressure drop, over a distance L, is

$$C_{F_E} = \frac{\Delta p_I}{\frac{1}{2}\rho\bar{w}^2} = \frac{8K}{3}Re^{-1/2}\left(\frac{L}{d}\right)^{1/2} \tag{12.3.17}$$

where C_F is the coefficient of friction and the subscript E indicates entrance flow.

We would like to know the wall shear along the inlet region. The wall is made of endothelial cells for blood vessels and epithelial cells in airways. The fluid flow stresses can modify their growth, remodeling and function. The wall shear, τ_w, is defined as

$$\tau_w = \mu\left.\frac{\partial w_b}{\partial r}\right|_{r=a} \tag{12.3.18}$$

Since $\partial/\partial r = (\partial\eta/\partial r)d/d\eta = -(1/\delta)d/d\eta$ by the chain rule, we can calculate the wall shear as

$$\tau_w = -\frac{\mu w_0(z)}{\delta(z)}\left(\frac{d}{d\tilde{\eta}}(2\tilde{\eta} - \tilde{\eta}^2)\right)_{\eta=0} = -\frac{2\mu w_0(z)}{\delta(z)} \tag{12.3.19}$$

Inserting w_0 from Eq. (12.3.12) and δ from Eq. (12.3.1) into Eq. (12.3.19) gives the wall shear, which is proportional to $1/\delta$ as the largest term

$$\tau_w = -\frac{2\mu\bar{w}}{\delta} = -\frac{2}{K}(\rho\mu)^{1/2}\bar{w}^{3/2}z^{-1/2} \tag{12.3.20}$$

So the shear stress is infinite at $z = 0$, which is where the profile is uniform. However, the length-averaged shear stress is finite and useful to calculate

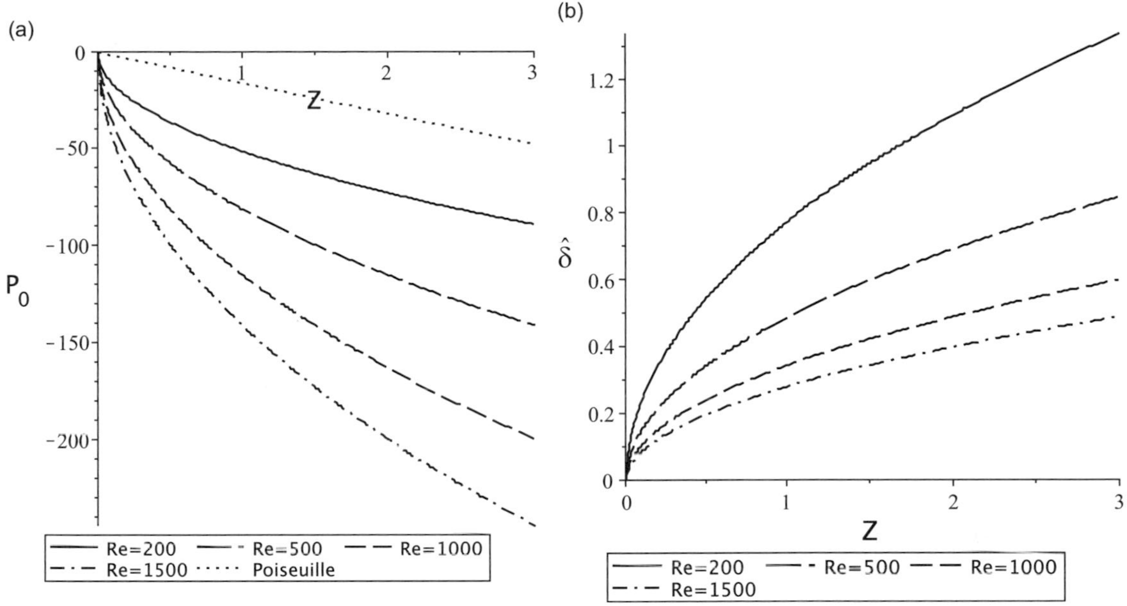

Figure 12.10 (a) Dimensionless pressure P_0 vs Z assuming the entrance pressure is $P_e = 0$, and $K = 5.47$. A linear Poiseuille pressure drop is shown for comparison. (b) Dimensionless boundary layer thickness $\hat{\delta}$ vs Z. Both (a) and (b) are shown for $Re = 200, 500, 1000, 1500$.

$$\bar{\tau}_w = \frac{1}{L}\int_0^L \tau_w \, dz = -\frac{4}{K}\left(\frac{\rho\mu}{L}\right)^{\frac{1}{2}} \bar{w}^{\frac{3}{2}} \quad (12.3.21)$$

12.4 Inspiratory Flow in the Tracheobronchial Tree

Using our quadratic boundary layer profile solution for inlet flow, the dimensional pressure drop in Eq. (12.3.14) down a uniform tube of length L is

$$p(z = 0) - p(z = L) = \Delta p_I(L) = \frac{2K}{3}\bar{w}^{3/2}\left(\frac{\rho\mu L}{a^2}\right)^{1/2} \quad (12.4.1)$$

as long as L is shorter than the inlet length, $L < L_I$. We can view the ratio of the pressure drop to the flow rate, Q, as a resistance, R_I for the inlet flow, such that

$$R_I = \frac{\Delta p_I(L)}{Q} = \frac{2K}{3}\frac{\bar{w}^{3/2}}{Q}\left(\frac{\rho\mu L}{a^2}\right)^{1/2} \quad (12.4.2)$$

Box 12.1 Lung Airway Anatomy and Inlet Flow

The lung is a branching network of tubes as shown in Figure 7.5 and sketched in Figure 12.11 as a symmetric bifurcating system. The airway generations start with the trachea at generation $n = 0$, which branches to the two main bronchi, generation $n = 1$, one to the left lung and one to the right lung. We assume that the tubes bifurcate at subsequent generations, so at any generation level there are 2^n tubes. For the mature adult lung, $0 \leq n \leq 23$. For neonates there

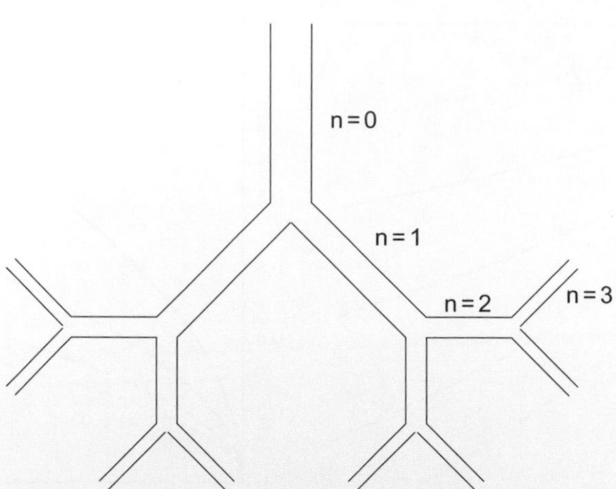

Figure 12.11 First four generations of the lung airway tree shown as symmetric bifurcations.

are fewer generations, say $0 \le n \le 16$, and the lung develops the remaining seven generations as part of its growth and development over the first several years of life.

From Sections 7.2 and 7.5 we used $d_n = d_0 2^{-n/3}$ where d_n is the diameter of an airway in generation n, and d_0 is the tracheal diameter. For an adult, $d_0 = 1.8$ cm, while for a neonate $d_0 = 0.4$ cm is typical, see Figure 12.12 from Haefeli-Bleuer and Weibel (1988) for a comparison of the measured diameter data and $d_n = d_0 2^{-n/3}$.

While the diameter fit is good, the equation $d_n = d_0 2^{-n/3}$ misses an important aspect of the airway tree that the measured total cross-sectional area has a local minimum at generation $n = 3$. So instead we can use the actual measurements of airway geometries, as shown in Figure 12.13. Airway anatomical data are shown in Figure 12.13 for the first 15 of 23 generations for an adult lung using the symmetric model (Weibel, 1963). Included are: generation number, n; number of airways at generation n (2^n); airway diameter d_n; airway length L_n; and total cross-sectional area, A_n, at generation n.

Given this geometry, which airways are dominated by inlet flow? To answer that question we need to calculate the inlet length $L_{I\,n}$ at generation n, and compare it to the airway length, L_n, at generation n. We can employ Eq. (12.3.2), $L_{E\,n}/d_n = 0.59 + 0.056\mathrm{Re}_n$, and then use the result of Mohanty and Asthana (1979) that $L_I \sim L_E/4$. First we need to calculate the Reynolds number, Re_n, at generation n. For a flow rate Q (cc/s) through the lung, we can define an average velocity at each generation $\bar{w}_n = Q/A_n$ where A_n is the total cross-sectional area at generation n shown in Figure 12.13. The corresponding Reynolds number, Re_n, is $\mathrm{Re}_n = d_n \bar{w}_n/\nu$ where ν is the kinematic viscosity of the air. A typical value is $\nu = 0.15\ \mathrm{cm}^2/\mathrm{s}$.

Using those relationships, Figure 12.14 shows how $L_n/L_{E\,n}$ and $L_n/L_{I\,n}$ varies with airway generation, n, for $Q = 1000$ cc/s. Inlet flow would occupy the entire length of the tube when $L_n/L_{I\,n} < 1$ in gross terms, which occurs for $0 \le n \le 7$ at this value of Q.

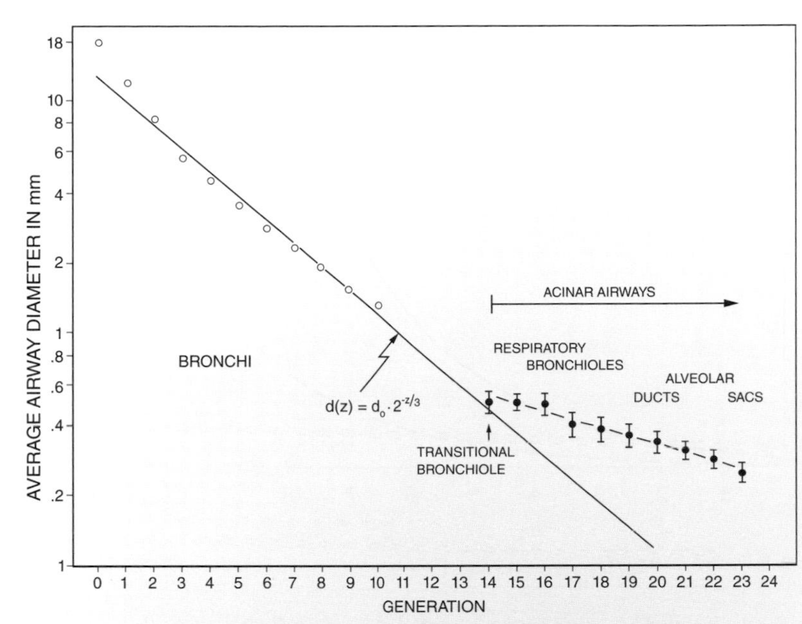

Figure 12.12 Regularized model of human airways (model A). Open circles are data taken from Weibel (1963).

Generation	Number of airways	Diameter, d_n (cm)	Length, L_n (cm)	Total cross-sectional area, A_n (cm^2)
0	1	1.800	12.00	2.54
1	2	1.220	4.76	2.33
2	4	0.830	1.90	2.13
3	8	0.560	0.76	2.00
4	16	0.450	1.27	2.48
5	32	0.350	1.07	3.11
6	64	0.280	0.90	3.96
7	128	0.230	0.76	5.10
8	256	0.186	0.64	6.95
9	512	0.154	0.54	9.56
10	1024	0.130	0.46	13.40
11	2048	0.109	0.39	19.60
12	4096	0.095	0.33	28.80
13	8192	0.082	0.27	44.50
14	16384	0.074	0.23	69.40
15	32768	0.066	0.20	113.00

Figure 12.13 Airway dimensions for symmetrical model adapted from Weibel (1963).

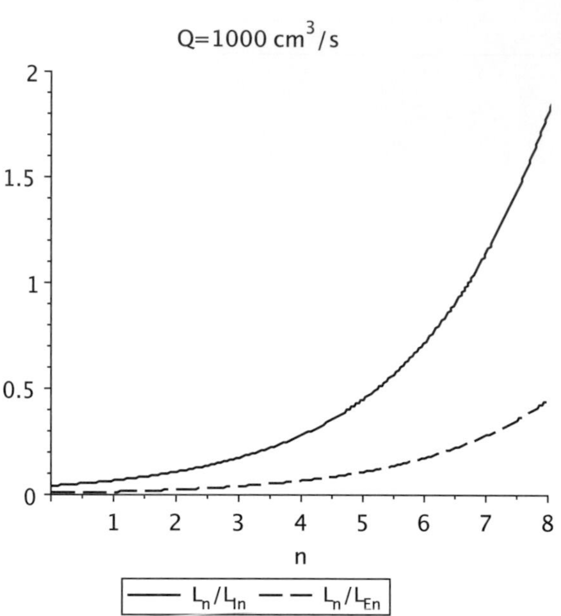

Figure 12.14 Ratio of airway length to its inlet length, $L_n/L_{I\,n}$ and entrance length $L_n/L_{E\,n}$, vs n for Q = 1000 cc/s.

For the purpose of calculating the pressure drop down the airway tree, the lung can be represented as a network of flow resistances as shown in Figure 12.15. Here we assume a symmetric system so that the pressure and flow resistance at each generation is the same for all of the airways in that generation. We can apply the Bernoulli equation for the pressure change from generation n to generation n + 1. There will be an inertial component since the average fluid velocity changes from one generation to the next due to the change in total cross-sectional area, A_n. Also, there will be a viscous pressure drop term, $\Delta p_{V\,n}$, which for Poiseuille flow is Eq. (12.3.15) and for inlet flow is Eq. (12.3.14),

$$\Delta p_n = p_n - p_{n+1} = \Delta p_{KE} + \Delta p_{V\,n} \tag{12.4.3}$$

where $\Delta p_{KE} = \frac{1}{2}\rho\left(\bar{w}_{n+1}^2 - \bar{w}_n^2\right)$ is the inertial or kinetic energy contribution to the pressure change.

For example, consider Δp_{V1}, which is the viscous pressure drop across generation n = 1. The single inlet pressure for both n = 1 airways is p_1, while the exit pressures are both assumed to be p_2. Under those circumstances we can use the circuit theory analogy that the pressure drop equals the flow times the resistance for the individual flows. So $\Delta p_{V1} = q_1 r_1$ where $q_1 = Q/2^1$ is the flow in each of the two tubes, while Q is the total flow rate, which is the same at each generation. Their individual resistances are r_1. It is useful to treat the flow at generation 1 as having an equivalent single resistance, R_1, for the total flow, Q, so that $\Delta p_{V1} = QR_1$. Equating these pressure drops gives us $q_1 r_1 = QR_1$. The total flow at n = 1 is Q, so by conservation of mass $q_1 + q_1 = Q$. Using this balance and inserting solutions for $q_1 = QR_1/r_1$ gives us the familiar relationship for the equivalent total resistance when two resistors are in parallel

$$\frac{QR_1}{r_1} + \frac{QR_1}{r_1} = Q \quad \rightarrow \quad \frac{1}{R_1} = \frac{1}{r_1} + \frac{1}{r_1} = \frac{2^1}{r_1} \tag{12.4.4}$$

In general, the total resistance at generation n, where the individual resistances are r_n and the individual flows are $q_n = Q/2^n$, will be

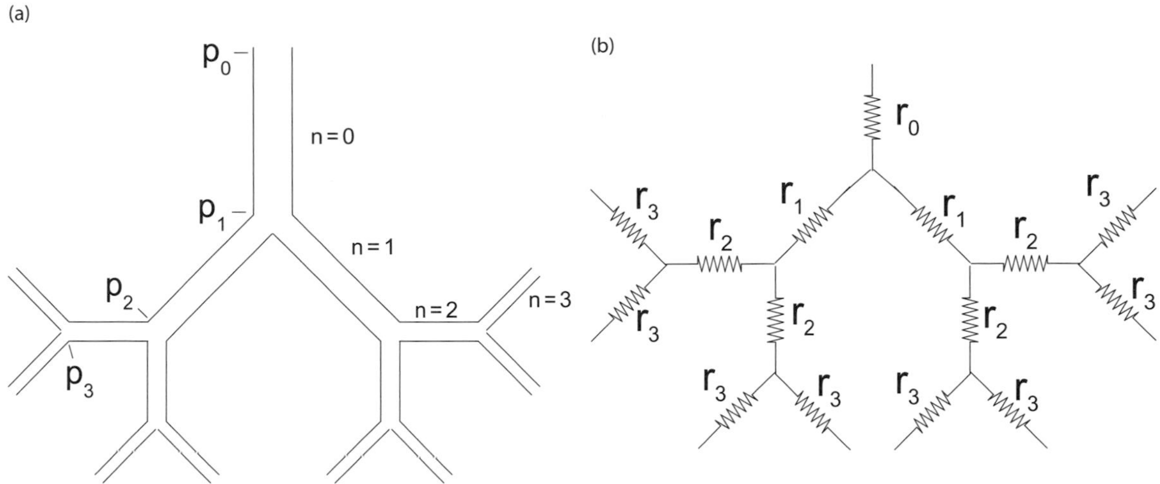

Figure 12.15 First four generations of the tracheobronchial tree showing (a) pressures and (b) equivalent flow resistances.

$$\frac{1}{R_n} = \frac{2^n}{r_n} \quad \rightarrow \quad R_n = \frac{r_n}{2^n} \tag{12.4.5}$$

The viscous pressure drop across generation n for inlet flow is

$$r_n = \frac{\Delta p_{Vn}}{q_n} = \frac{2K}{3} \frac{2^{-n/2}}{\pi^{3/2} a_n^4} (\rho \mu L_n Q)^{1/2} \quad \text{inlet flow} \tag{12.4.6}$$

where Δp_{Vn} comes from Eq. (12.4.1), which is flow dependent, while for Poiseuille flow it is

$$r_n = \frac{\Delta p_{Vn}}{q_n} = \frac{8 \mu L_n}{\pi a_n^4} \quad \text{Poiseuille flow} \tag{12.4.7}$$

which is independent of flow.

Box 12.2 Airway Flow Resistance

Let's calculate airway resistance to flow comparing Poiseuille vs inlet flow assumptions.

$$R_n = \frac{r_n}{2^n} = \frac{8 \mu L_n}{\pi a_n^4 2^n} \quad \text{Poiseuille} \tag{12.4.8}$$

For inlet flow the resistances are

$$R_n = \frac{r_n}{2^n} = \frac{2K}{3} \frac{2^{-3n/2}}{\pi^{3/2} a_n^4} (\rho \mu L_n Q)^{1/2} \quad \text{inlet flow} \tag{12.4.9}$$

Figure 12.16(a) shows the resistances, R_n, in units of $cmH_2O/(L/s)$. A 1 cmH_2O pressure unit is $1\, cmH_2O = \rho g h$ where $\rho = 1\, g/cm^3$, $g = 980\, cm/s^2$ and $h = 1\, cm$. So $1\, cmH_2O = 980\, dyn/cm^2$. Note that the resistance, R_n, has a maximum at $n = 4$.

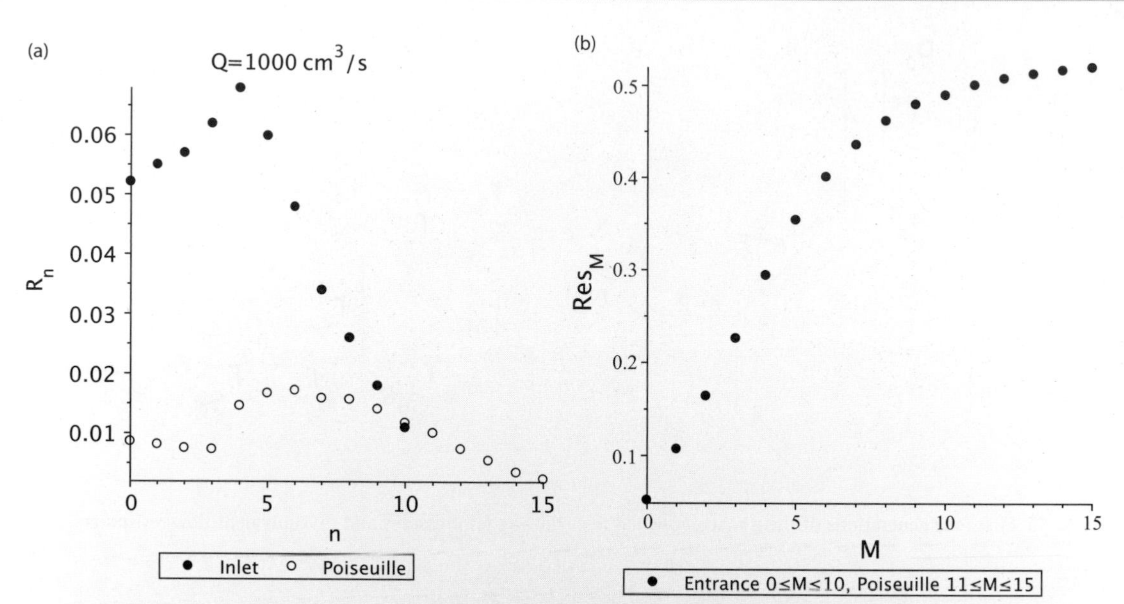

Figure 12.16 (a) Viscous resistance Rn vs airway generation (n) using entrance and Poiseuille flow models. (b) Cumulative entrance flow resistance ResM ($0 \leq n \leq M$) using entrance flow for r, $0 \leq M \leq 10$, and Poiseuille flow resistance for $11 \leq M \leq 15$ vs airway generation M. The resistance units are $cmH_2O/(L/s)$. The flow rate is 1 L/s while $\mu = 0.00018$ poise and $\rho = 0.0012\ g/cm^3$ for air.

This occurs because the total cross-sectional area has a local minimum ($2\ cm^2$) at $n = 4$, so the velocities are maximal. The major region of viscous flow resistance is in the first few generations while airways past $n = 10$ offer relatively little resistance. This is called the "quiet zone" of the lung airways, because it is difficult to diagnose airway diseases that occur in this small airways region using normal pulmonary function tests that rely on overall flow and pressure measurements. Overall resistance to flow is measured as a pulmonary function test where the overall pressure drop, from the mouth to the alveoli, $P_0 - P_{23}$, is inferred using a body plethysmograph, for example, while the total flow rate, Q, is also measured. One of the most frequent small airways diseases is inflammation from smoking cigarettes. Figure 12.16 (b) shows the cumulative resistance, $Res_M = \sum_{n=0}^{M} R_n$, using inlet flow resistances for $0 \leq M \leq 10$, where we have seen from Figure 12.14 that inlet flow conditions occur in these generations. Then Poiseuille flow resistances are used for $11 \leq M \leq 15$, since that is the region where $L_n/L_{En} > 1$. For $M = 15$, $Res_M = 0.52\ cmH_2O/(L/s)$, which is essentially the total resistance to flow offered by the lung. By $M = 6$ approximately 80% of the $M = 15$ value has been reached.

Box 12.3 Liquid Ventilation

While we naturally think of air flow in the lung, there are therapies that involve cycling the lung with perfluorocarbon (PFC) liquids which have a large oxygen and carbon dioxide carrying capacity. A typical range for PFC density is $\rho \sim 1.5 - 2.0\,\text{g/cm}^3$ and kinematic viscosity $\nu = \mu/\rho = 0.8\text{–}8.0\,\text{cm}^2/\text{s}$. This technique can involve filling the lung partially with PFC and cycling with gas, called partial liquid ventilation, or completely filling the lung with PFC, called total liquid ventilation. The goal is to open up closed airways and alveoli, reducing surface tension forces that tend to collapse them (Shaffer et al., 1992; Hirschl et al., 1996; Baba et al., 2004; Bull et al., 2007; 2009). Figure 12.17 shows a mouse submerged in PFC and breathing it. PFC was initially investigated as a blood substitute (Clark and Gollan, 1966).

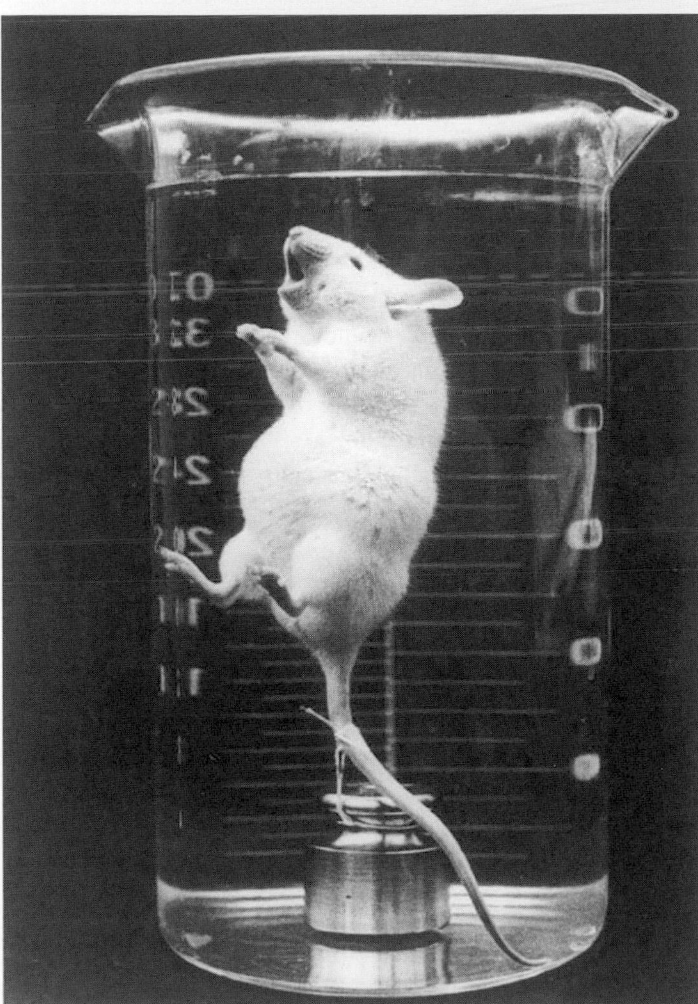

Figure 12.17 Mouse submerged in perfluorocarbon liquid and breathing it.

12.5 Flow in a Curved Tube

Airway and vascular tubes can have significant axial curvature along the flow path. Figure 12.1 shows this for a typical bifurcation while Figure 12.18 exhibits the distinctive aortic arch anatomy, both prime examples of this geometric feature.

Flow in a curved tube is shown in Figure 12.19. It is a single curved tube with axial radius of curvature R_{ax} and diameter, d. As the fluid rounds the curve, it experiences a centripetal force which is proportional to the axial velocity squared, $\sim \rho w^2/R_{ax}$. So, this force is larger for the faster moving fluid in the midline of the tube. Because of that, it is preferentially thrown towards the outer (O) wall of the curve. This skews the axial velocity profile (solid line) toward the outer wall compared to an equivalent, symmetric Poiseuille flow (dashed curve). Balancing the centripetal force is a radial pressure gradient with higher pressure at the outer wall and lower pressure at the inner (I) wall. This induced pressure gradient then causes a return flow, toward the center of axial curvature, along the top and bottom wall. Consequently, a pair of vortices is created in the cross-section with opposite rotations, as shown. Individual fluid particles, then, move in a helical pattern down the tube. While the main flow is along the axis, the vortices in the cross-section are called a "secondary flow."

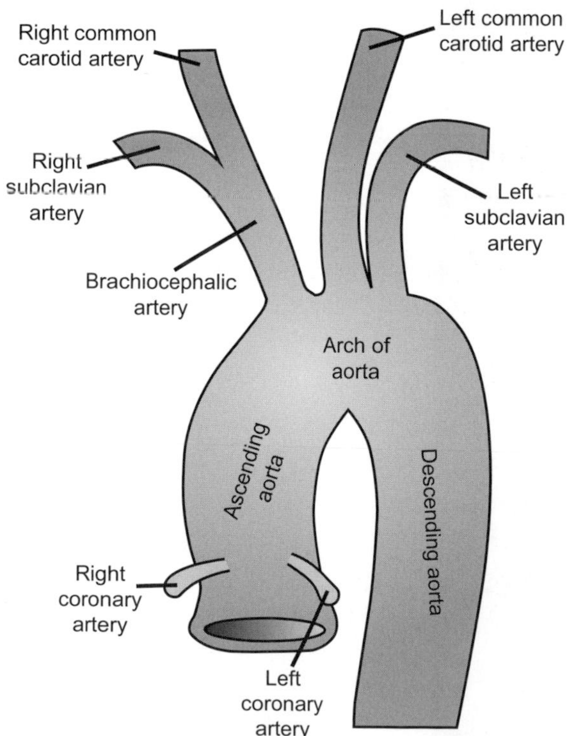

Figure 12.18 Aortic arch as an example of axial curvature.

The viscous pressure drop for fully developed flow in curved tubes depends on the Dean number, $Dn = (d/2R_{ax})^{1/2}Re$, which involves the ratio of the tube radius, d/2, to the axial radius of curvature, R_{ax}, and the Reynolds number. The friction factor for a tube, f, defined in Section 13.3, Eq. (13.3.1), is a dimensionless pressure drop. Curved tube flow, f_C, has been studied by several investigators reviewed in Ali (2001). A popular form was derived by Barua (1963) as a ratio to the Poiseuille flow friction factor, f_P, from Eq. (13.3.2)

$$\frac{f_C}{f_P} = 0.509 + 0.0918\,Dn^{1/2} \tag{12.5.1}$$

This formula fit well the data from earlier experiments (White, 1929; Adler, 1934). So there is additional friction due to the axial curvature, requiring a higher driving pressure for the same flow rate. One source of this increased flow resistance is that the skewed axial velocity profile has much steeper gradients at the

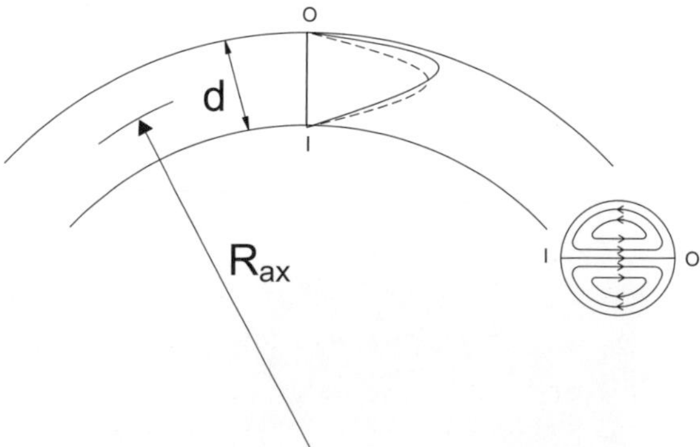

Figure 12.19 Flow in a curved tube showing skewing of velocity profile to the outer wall (solid line) compared to Poiseuille flow (dashed line). Cross-sectional view shows secondary swirling stream lines.

outer wall, so that the wall shear is increased overall. Another reason is that the vortical flow in the cross-section adds to the dissipation. We note that, although it does not affect the ratio in Eq. (12.5.1), Barua uses Fanning friction factors.

Values of $d/2R_{ax}$ are wide ranging depending on the generation of the airway. From the data in Horsfield et al. (1971) it ranges from 1/2 to 1/6 for 1.5 mm diameter airways, 1/4 to 1/10 for diameters between 1.5 and 3 mm, and 1/7 to 1/14 for diameters greater than 3 mm, except for the large central airways (Wang, 2005). For a fourth-generation airway, Figure 12.13 tells us that an individual airway diameter is $d_4 = 0.45$ cm and the total cross-sectional area is$A_4 = 2.48\ \text{cm}^2$. With an overall flow rate of $Q = 500\ \text{cm}^3/\text{s}$ the velocity in a fourth-generation airway is $U_4 = Q/A_4 \sim 400\ \text{cm/s}$. Then the Reynolds number is $Re_4 = d_4U_4/\nu \sim 1200$ assuming $\nu = 0.15\ \text{cm}^2/\text{s}$. If $d/2R_{ax} = 0.1$, then Dn for this airway is $Dn \sim 400$ and Eq. (12.5.1) yields $f_C/f_P = 2.3$. The frictional resistance is roughly double that of a straight tube. Likewise, the aortic arch has a range of curvatures, but on the average a reasonable value is $R_{ax} = 3.3\ \text{cm}$ (Redheuil et al., 2011). With a diameter of the ascending aorta at 3 cm, that makes $d/2R_{ax} \sim 1/2$. So, for an aortic flow with $Re = 4000$, Eq. (12.5.1) yields $f_C/f_P = 5.39$, which is a big effect, particularly since it is the heart which must supply this larger driving pressure.

The entry length for a curved tube (Yao and Berger, 1975) is found as a ratio to entrance length in a straight tube

$$\frac{L_{EC}}{L_{ES}} = 8e_1 Dn^{-1/2} \tag{12.5.2}$$

for large Dn where the constant was found to be $2 < e_1 < 4$. The authors used the straight tube entrance length as $L_{ES}/d = 0.06Re$, similar to Eq. (12.3.2). For example, let $Re = 2000$ and $d/2R_{ax} = 0.1$. Then the Dean number is $Dn = 632$ and Eq. (12.5.2) yields the ratio $L_{EC}/L_{ES} = 0.64$ using $e_1 = 2$. So the entrance length of a curved tube is shorter than that of the equivalent straight tube. Part of the explanation comes from the vortices, see Figure 12.19, which help to mix the shearing across the cross-section.

Clinical Correlation 12.1 | Pulsatile Flow in a Curved Tube – Newtonian and Non-Newtonian Fluids

Figure 12.20 shows the results both of experiments, using particle image velocimetry (PIV) and CFD for steady flow through a 180° curved section of a tube connecting two parallel straight tubes. It is representative of the flow found in the femoral artery or branches off the aortic arch like the brachiocephalic, left carotid and subclavian arteries. The dimensions of the physical model, made from acrylic, are $R_{ax} = 44.45\ \text{mm}$ and $d/2 = 12.7/2$ mm so that $\delta = 1/7$ so $\delta^{1/2} \sim 0.38$. Figure 12.20(a)–(c) have both the measurements and the computational results for the streamwise velocity profiles at three values of $Re = 220,\ 430$ and 640, viewing left to right. The corresponding Dean numbers are $Dn = 83,\ 163$ and 242, respectively. Clearly the velocities are larger for the higher Re value. Figure 12.20(d)–(f) show the secondary flow velocity fields in the cross-section measured in the experiments for the corresponding Re values in (a)–(c). Note the pair of counter-rotating, Dean-type vortices at $Re = 220$. They become more distorted as Re increases and a second pair of vortices appears, four in all. The new pair are called Lyne-type vortices (Lyne, 1971). Figure 12.20(g)–(i) are the CFD calculations for the same parameter values as in (d)–(f). Note the good correlation between experiments and CFD, and the axial location is 90° into the curve.

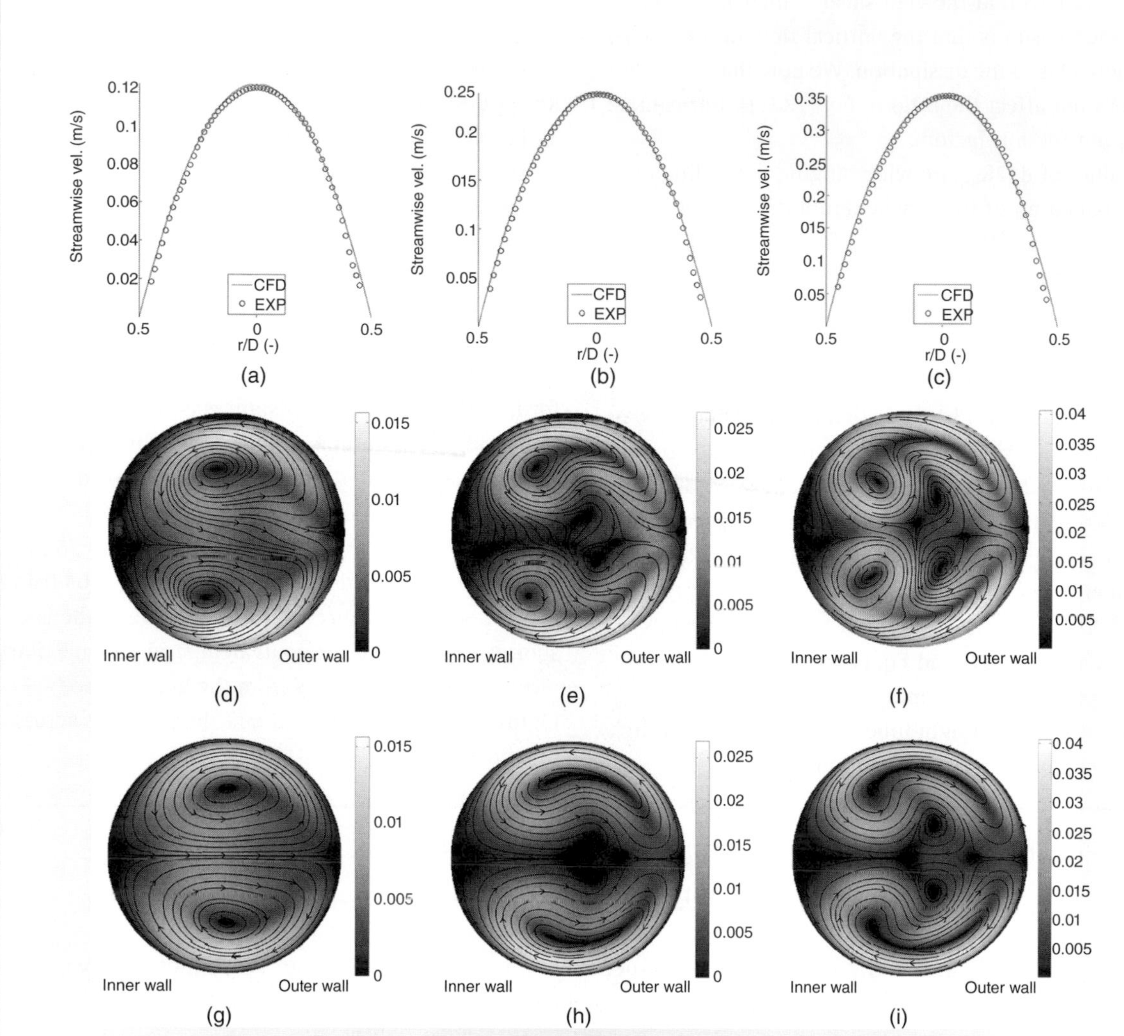

Figure 12.20 Steady flow through a curved tube section. (a)–(c) Steady flow mean streamwise velocity flow profile experimental measurements vs CFD comparisons for, left to right, Re = 220, 430 and 640 at the entrance to the curved tube. (d)–(f) Secondary flow velocity fields from experiments and (g)–(i) secondary flow velocity fields from CFD.

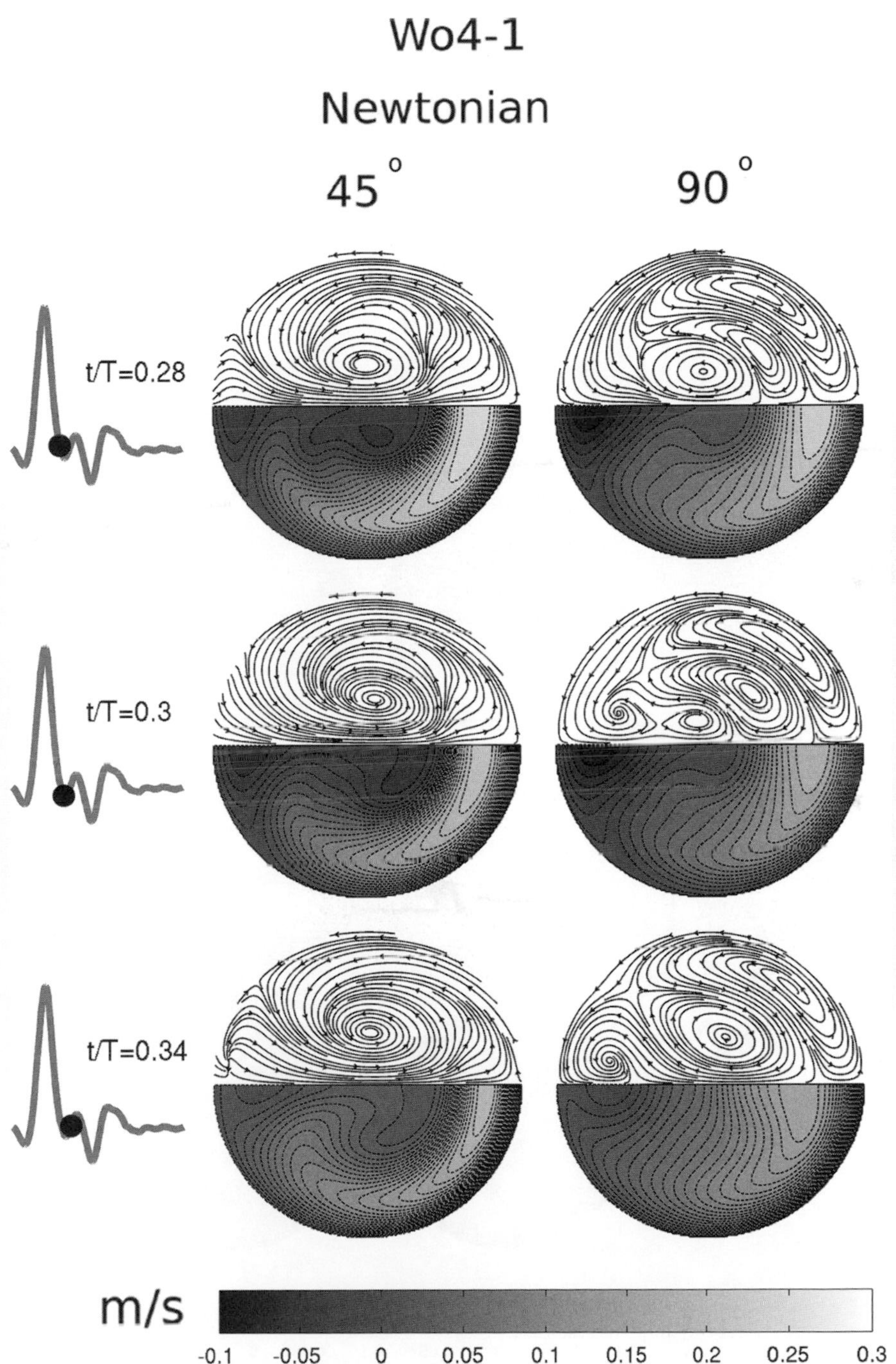

Figure 12.21 CFD results for pulsatile flow through a curved tube section at 45° (left column) and 90° (right column) into the curve at $t/T = 0.3$. (a) Newtonian fluid, contours of streamwise iso-velocity lines (bottom half) and secondary flow streamlines (top half). (b) Non-Newtonian fluid secondary flow streamlines (top) and streamwise iso-velocity lines (bottom).

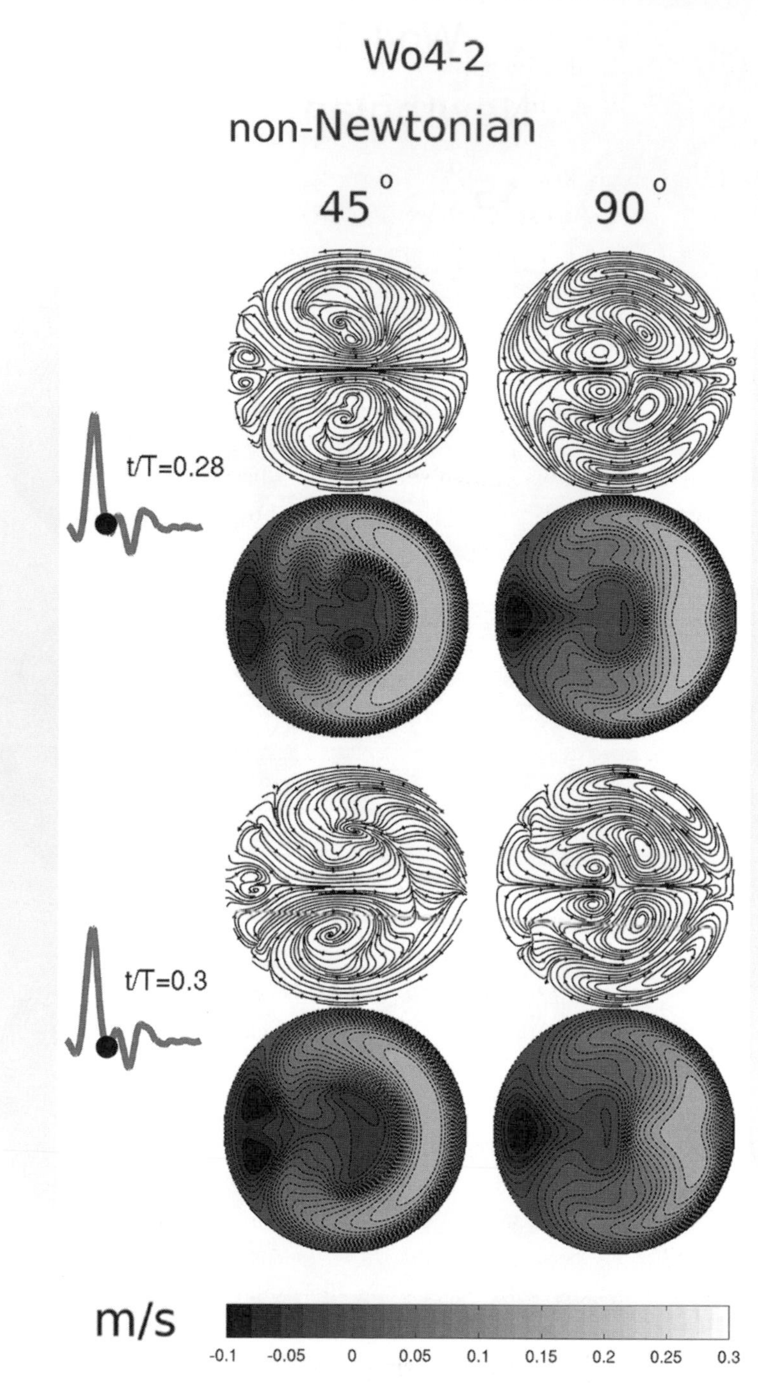

Figure 12.21 (*continued*)

Figure 12.21 shows CFD results for pulsatile flow through the curved tube. The figures are for the cross-section taken at 45° into the curve (left column) and 90° (right column). The Womersley parameter value is $\alpha = 4.2$, the average Reynolds number is $Re_{avg} = 360$ and maximum Reynolds number is $Re_0 = 1650$. As discussed in Chapter 8, the heart creates a volumetric flow-rate waveform as shown based on physiologic measurements (Holdsworth et al., 1999). The results are for $t/T = 0.3$ in the deceleration phase of the cycle. Figure 12.21 (a) is a Newtonian fluid with contours of streamwise iso-velocity lines in the bottom half and secondary flow streamlines in the top half. The 45° pattern has the single pair of Dean-type vortices, but the 90° pattern now has four pairs of vortices, two more added to the Dean-type and Lyne-type. Figure 12.21(b) is a non-Newtonian fluid (Quemada model, see Chapter 11) which captures the behavior of blood. There are secondary flow streamlines and streamwise iso-velocity lines, however, are now shown separately to discern top–bottom asymmetries across the equator. Again there are four pairs of vortices but asymmetry top to bottom, especially at 45°.

Box 12.4 Particle Image Velocimetry

The fluid velocity measurements in Figure 12.20 were obtained using particle image velocimetry (PIV). A typical experimental apparatus is shown in Figure 12.22. The fluid is seeded with reflective particles that are small enough to follow the flow field, i.e. the Stokes number is small, $Stk << 1$. Typically they are spherical in shape with a diameter range ~10–100 μm. In liquid flows these microspheres are often made from glass or polystyrene, and chosen to have the same density of the fluid to avoid settling (or rising) due to gravity. Gas flows, on the other hand, are seeded with droplets of water or oil. The particles pass through a laser sheet created by sending the beam through a cylindrical lens and bouncing it off of a mirror, as shown. The sheet illuminates a narrow slice of the flow domain, indicated as z_1. The laser is pulsed at time interval Δt, and the particle images are recorded synchronously with a CCD camera. Two sequential frames, at t_1 and $t_2 = t_1 + \Delta t$, are selected for comparison. Figure 12.22 shows three particles which maintain their relative spacing from one another in the two frames, but as a group they are displaced. Their corresponding displacements, $\Delta x, \Delta y$, yield the velocities, $u = \Delta x/\Delta t, v = \Delta y/\Delta t$. Using cross-correlation image processing, this calculation is computerized and the velocity vector field calculated as a function of space and time.

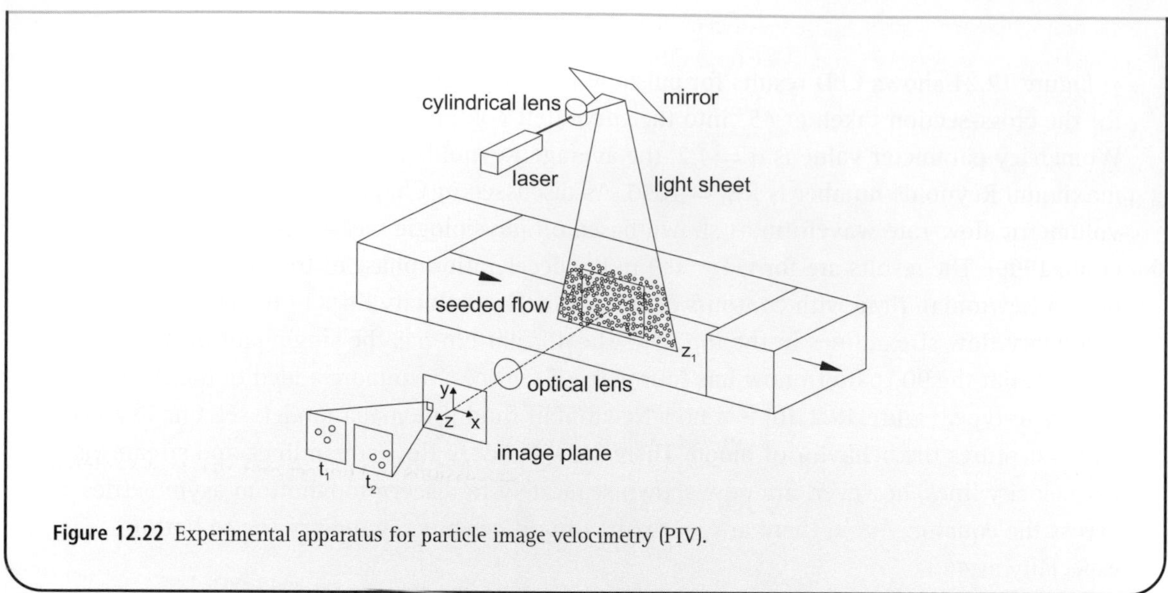

Figure 12.22 Experimental apparatus for particle image velocimetry (PIV).

12.6 Flow Through a Bifurcation

Putting our results together, so far, Figure 12.23 shows flow through a bifurcation or branch. The flow in the parent tube splits at the carina and a complicated flow in the daughter tubes results. The velocity profile in the daughters is a developing flow with an entrance boundary indicated by the dashed lines. Also, because of the axial curvature there are secondary swirling flows. Detailed studies of flow visualization (Schreck and Mockros, 1970; Olson,

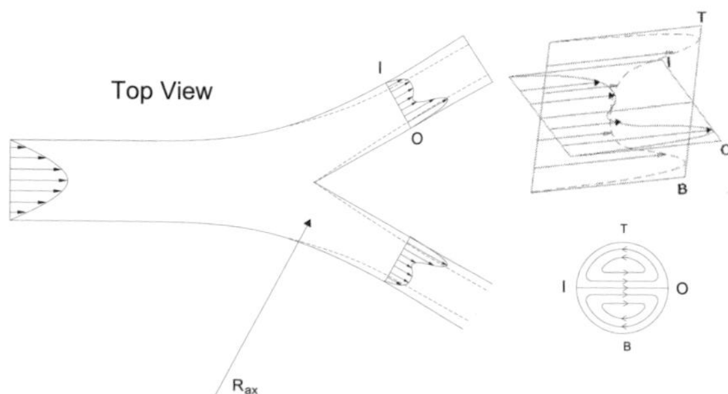

Figure 12.23 Inflow at a bifurcation or branch. Top insert shows axial velocity profiles for the I–O plane and the T–B plane. Bottom insert shows secondary swirling flow.

1971) and pressure drop measurements (Pedley et al., 1970a, 1970b), in airway models give this background picture. The results are expressed as a ratio to the pressure drop of Poiseuille flow in the same diameter and length tube from Eq. (12.3.15), $\Delta p_p = 32\mu\bar{w}L/d^2$. Calling this ratio, Z, their results were in the form

$$Z = \frac{\Delta p}{\Delta p_p} = \frac{C}{4\sqrt{2}}\left(\frac{d}{L}\right)^{1/2} Re^{1/2} \tag{12.6.1}$$

where C was a constant determined from the measurements. Its average value turned out to be $C = 1.85$ so the coefficient is $1.85/4\sqrt{2} = 0.33$. For simple entrance flow in a single long tube we would have $Z_E = (K/24)(d/L)^{1/2}Re^{1/2}$ using Eqs. (12.3.14) and (12.3.15), which is of similar form to Eq. (12.6.1) except for the coefficient. For $K = 5.0$ the coefficient is $5/24 = 0.21$. So the real system has higher pressure drops than that of a single long tube. Using the definition of the friction factor, f, from Eq. (13.3.1), the single tube entrance flow pressure drop has $f_E = f_P Z = (8/3)K(d/L)^{1/2}Re^{-1/2}$ where $f_P = 64/Re$ is given in Eq. (13.3.2).

The reason for the higher pressure drop in a real airway system is seen in Figure 12.23 where the velocity profiles in the plane of the bifurcation (inside, I, to outside, O) and the perpendicular plane (top, T, to bottom, B) are shown. In the I–O plane the peak velocity is skewed toward the O wall, as we saw in curved tube flow of Figure 12.19, so the wall shear is very high there. In the T–B plane the velocity profile is a symmetric M-shape, and consequently has boundary layers and also losses in the core flow which is no longer flat. The cross-sectional insert shows the secondary swirling flows which add to the dissipation. Of course, the carina of the branch creates an obstacle to the parent tube flow, which splits and diverts the stream, hence convective acceleration becomes an important aspect of the pressure drop.

Figure 12.24 shows CFD results for flow through two bifurcations including velocity vectors, contours of constant velocity (isovelocity) magnitude (speed), and in-plane streamtraces (streamlines). Figure 12.24 (a) is general entrance flow into a straight tube, Figure 12.24(b) is flow through generations $n = 3 - 5$ (upper airways) and Figure 12.24(c) shows generations $n = 7 - 9$ (central airways), with $n = 0$ being the trachea. Note the stronger skewed profiles in the upper airways where the entrance Reynolds number is larger, $Re = 894$, compared to the central airways entrance, $Re = 125$, at a total flow rate into the lung of 30 L/s. However, even at generation 9 the skew is evident. See also discussions in Longest and Vinchurkar (2007).

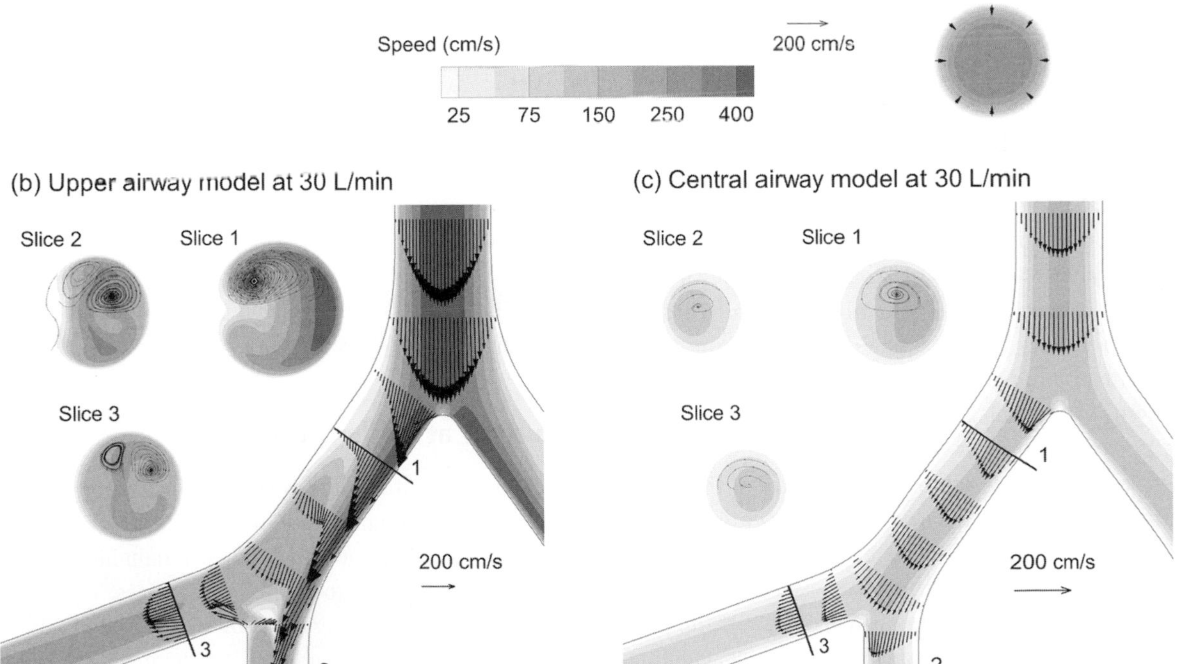

Figure 12.24 CFD results for flow through a single tube (top), the upper airway generations 3-5 and the central airways generations $n = 7$–9.

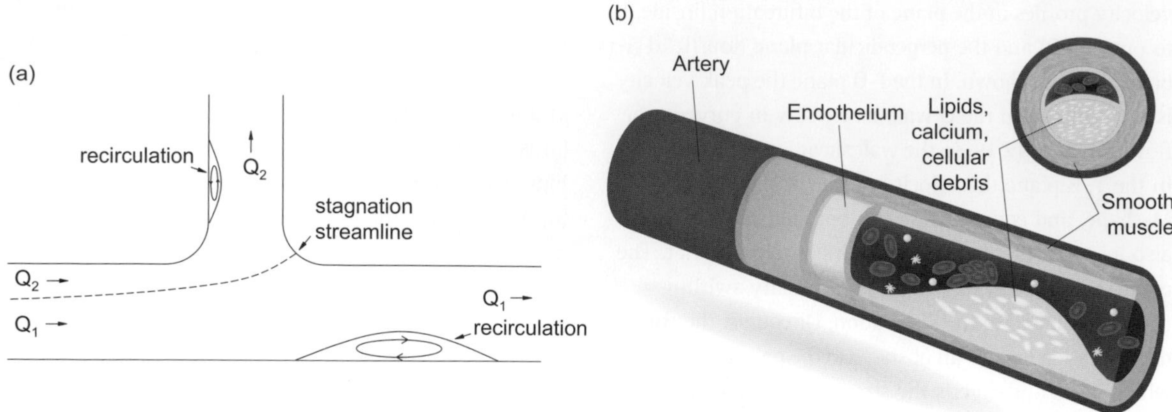

Figure 12.25 (a) Flow separation at a branch. (b) atherosclerotic plaque in the wall of an artery.

Another feature at a bifurcation is flow separation, which has significant implications in the vascular system. An example is shown in Figure 12.25(a). The total flow rate $Q_1 + Q_2$ enters from the parent tube. The bifurcation is a sharp turn of 90° into a smaller vessel. Q_1 continues in the main vessel while the lower flow rate, Q_2, goes into the smaller branch. In both vessels, flow separates from the wall leaving recirculation regions. These regions have relatively low shear which has been implicated in enhancing the production of atherosclerotic plaques on the underlying wall (Caro et al., 1971). In addition to the geometric effects, generally the total cross-section is increasing as one progresses from the parent to daughter tubes, so the velocity slows down, presenting an adverse pressure gradient from a Bernoulli perspective. The pressure locally is increasing downstream. The geometry of the carotid artery bifurcation is particularly susceptible to flow separation. The development of atherosclerotic plaques there can lead to a stenosed vessel with lower flow rates reducing oxygenation of the brain, Figure 12.25(b). Plaques consist of lipids, calcium and cellular debris in the arterial wall and can become sites of blood clot formation. Fragments of the clot can be swept along with the flow, as we saw in Section 12.2. In this case, the embolus travels into the brain vasculature and can occlude a smaller vessel, which causes a stroke.

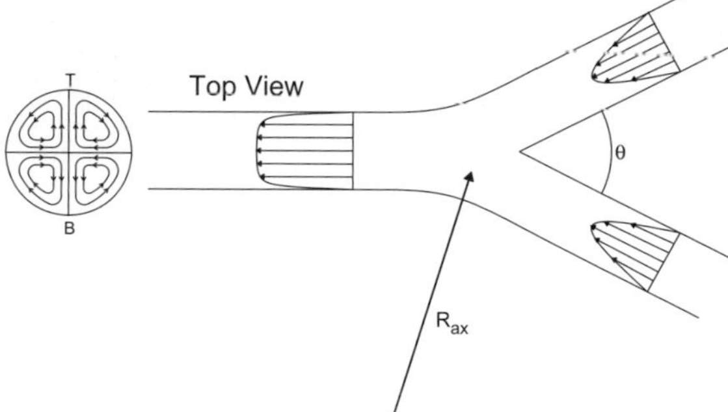

Figure 12.26 Outflow at a bifurcation showing four vortices in the parent tube made from the vortex pair in each daughter tube.

Outflow at a Bifurcation

Outflow at a bifurcation is sketched in Figure 12.26. The cross-sectional view shows the results of having two vortex pairs, one pair from each daughter, flow into the parent tube to create four vortices. The top view shows that converging flows have a more blunt velocity profile than parabolic. So there will be a higher level of viscous dissipation in this flow compared to the Poiseuille equivalent due to the effects of convergence and swirling (Schreck and Mockros, 1970).

Box 12.5 Bifurcation and Curved Tube Flow Visualization and CFD

In Pedley et al. (1971) a model of a respiratory bifurcation was created from transparent acrylic. The range of Reynolds numbers based on the daughter tube diameter was $50 < Re_D < 4500$, and smoke was used for flow visualization. Figure 12.27(a) shows the single pair of vortices during inspiration, similar to the sketch in Figure 12.23, but a double pair during expiration, Figure 12.27(b), as depicted in Figure 12.26.

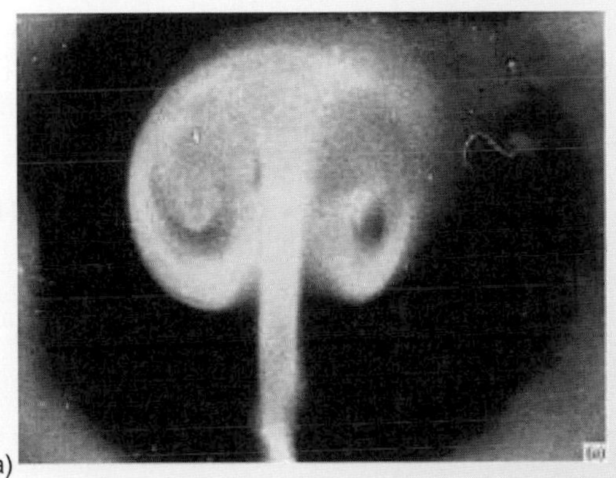
(a)

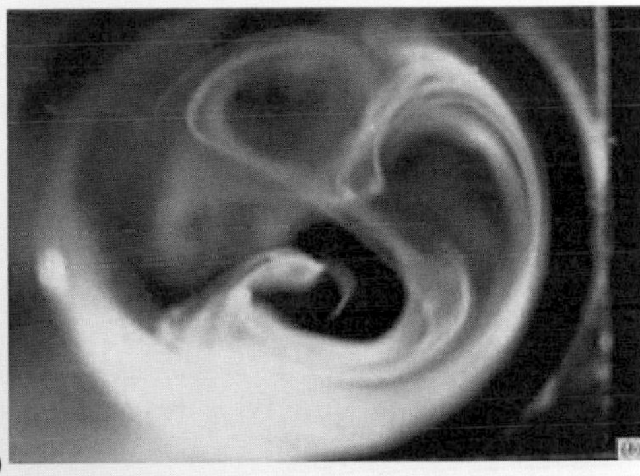
(b)

Figure 12.27 (a) Inspiratory flow at a bifurcation constructed from transparent acrylic. A pair of secondary flow vortices were visualized using smoke. (b) Expiratory flow showing the combination of two sets of vortex pairs.

12.7 Viscous Stagnation Point Flow

The inviscid Eulerian velocity field for 2D stagnation point flow is

$$U = Ax, \; V = -Ay \tag{12.7.1}$$

and we expect that far enough away from the wall, where $y \rightarrow \infty$, this description is accurate. The pressure field associated with the inviscid flow comes from the Bernoulli equation

$$P + \frac{1}{2}\rho\left(U^2 + V^2\right) = P + \frac{1}{2}\rho A^2\left(x^2 + y^2\right) = P_0 \tag{12.7.2}$$

where P_0 is the pressure at the origin, the stagnation pressure.

For a real fluid, though, this solution violates the viscous no-slip condition on the wall ($y = 0$), since $U(y = 0) = Ax$ instead of zero. So in the vicinity of the wall, we need to formulate the problem with viscosity using the Navier–Stokes equations

$$\begin{aligned} \text{x-momentum}: \quad & \rho\left(u\frac{\partial u}{\partial x} + v\frac{\partial u}{\partial y}\right) = -\frac{\partial p}{\partial x} + \mu\left(\frac{\partial^2 u}{\partial x^2} + \frac{\partial^2 u}{\partial y^2}\right) \\ \text{y-momentum}: \quad & \rho\left(u\frac{\partial v}{\partial x} + v\frac{\partial v}{\partial y}\right) = -\frac{\partial p}{\partial y} + \mu\left(\frac{\partial^2 v}{\partial x^2} + \frac{\partial^2 v}{\partial y^2}\right) \end{aligned} \tag{12.7.3}$$

and continuity

$$\text{continuity}: \quad \frac{\partial u}{\partial x} + \frac{\partial v}{\partial y} = 0 \tag{12.7.4}$$

We will want these solutions to satisfy no-slip and no-penetration at the wall,

$$u(y=0) = 0, \quad v(y=0) = 0 \tag{12.7.5}$$

and then also match to the inviscid velocity and pressure far from the wall

$$(u,\ v,\ p) \to (U,\ V,\ P), \quad y \to \infty \tag{12.7.6}$$

The approach to solving Eqs. (12.7.3) and (12.7.4) subject to the boundary conditions Eqs. (12.7.5) and (12.7.6) is to propose forms of the solutions which turn out to simplify the analysis. Let the velocities be given by

$$\begin{aligned} & u = xf'(y), \quad v = -f(y) \\ & p - p_0 = -\frac{1}{2}\rho A^2 (x^2 + F(y)) \end{aligned} \tag{12.7.7}$$

where the unknown functions f(y) and F(y) will be determined. Inserting the forms of Eq. (12.7.7) into the continuity equation, Eq. (12.7.4), gives us

$$f'(y) - f'(y) = 0 \tag{12.7.8}$$

so continuity is satisfied identically. Inserting Eq. (12.7.7) into the momentum conservation, Eqs. (12.7.3) gives us

$$\begin{aligned} & \text{x-momentum}: \quad \rho\left(xf'^2 - xff''\right) = \rho A^2 x + \mu\left(xf'''\right) \\ & \text{y-momentum}: \quad \rho(ff') = \frac{1}{2}\rho A^2 F' - \mu\left(f''\right) \end{aligned} \tag{12.7.9}$$

Dividing out the common factor of x in the x-momentum equation, and rearranging the constants for both equations, gives us

$$\begin{aligned} & \text{x-momentum}: \quad \nu f''' + ff'' - f'^2 + A^2 = 0 \\ & \text{y-momentum}: \quad \frac{1}{2}A^2 F' = \nu f'' + ff' \end{aligned} \tag{12.7.10}$$

where $\nu = \mu/\rho$ is the kinematic viscosity. The boundary conditions on the velocities are

$$f(0) = 0, f'(0) = 0, f'(\infty) = A \tag{12.7.11}$$

and the boundary condition on the pressure is

$$F(0) = 0 \tag{12.7.12}$$

so that $P = P_0$ at the origin.

To make this system dimensionless, and scale out the dimensional parameters, let

$$\eta = \frac{y}{\sqrt{\nu/A}}, \quad \varphi(\eta) = \frac{f(y)}{\sqrt{A\nu}}, \quad \Phi = \frac{AF(y)}{\nu} \tag{12.7.13}$$

Using the forms of Eq. (12.7.13) in the x-momentum of Eq. (12.7.10) and its boundary conditions of Eq. (12.7.11) leads to the system for x-momentum conservation

$$\begin{aligned} & \varphi''' + \varphi\varphi'' - \varphi'^2 + 1 = 0 \\ & \varphi(0) = 0, \ \varphi'(0) = 0, \ \varphi'(\infty) = 1 \end{aligned} \tag{12.7.14}$$

The y-momentum equation of Eq. (12.7.10) can be integrated, giving us

$$\frac{1}{2}A^2(F - F_0) = \nu f' + \frac{1}{2}f^2 \tag{12.7.15}$$

where F_0 is a constant of integration. Using the dimensionless variables of Eq. (12.7.13) in Eq. (12.7.15) results in

$$\begin{aligned} & \Phi - \Phi_0 = 2\varphi' + \varphi^2 \\ & \Phi(0) = 0 \end{aligned} \tag{12.7.16}$$

The boundary condition on Φ tells us that $\Phi_0 = 0$ since both $\varphi'(0) = 0$ and $\varphi(0) = 0$. So the equation for Φ is

$$\Phi = 2\varphi' + \varphi^2 \tag{12.7.17}$$

The system of Eqs. (12.7.14) is independent of Φ, and can be solved separately. Then the solution of φ can be used to solve for Φ in Eq. (12.7.17).

The x- and y-velocity components, scaled on their inviscid counterparts, are

$$\begin{aligned} & \frac{u}{U} = \frac{xf'(y)}{Ax} = \varphi'(\eta) \\ & \frac{v}{V} = \frac{-f(y)}{-Ay} = \frac{\varphi(\eta)}{\eta} \end{aligned} \tag{12.7.18}$$

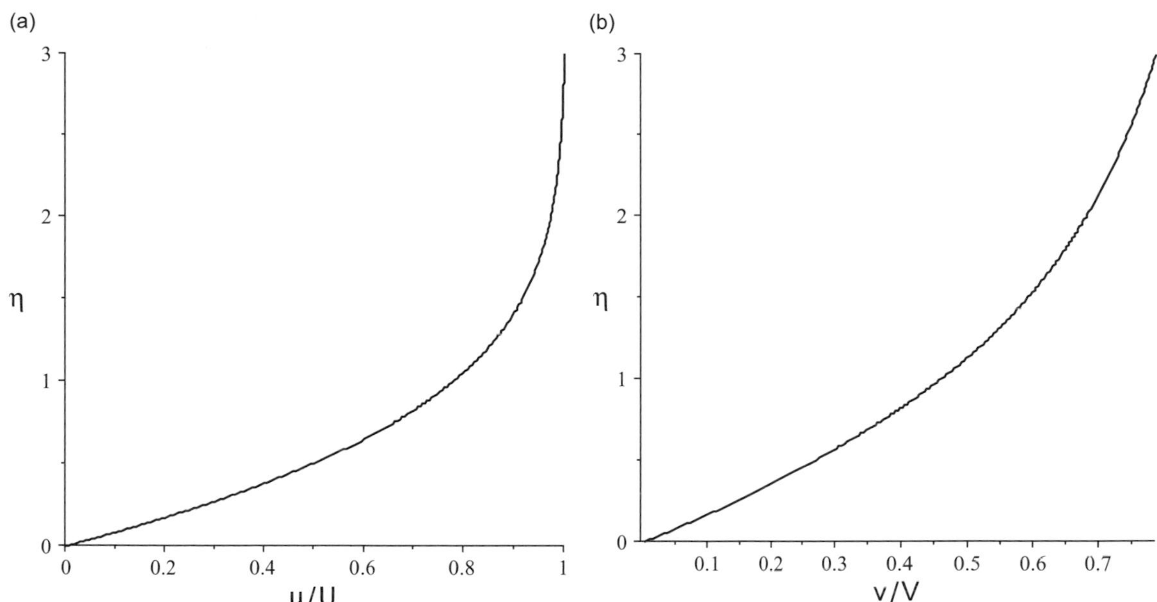

Figure 12.28 Numerical solutions for (a) u/U vs η and (b) v/V vs η v/V vs η.

Numerical solutions to Eq. (12.7.14) for $\varphi(\eta)$, inserted into Eqs. (12.7.18) yield the dimensionless velocities, u/U and v/V, in Figure 12.28. The condition at $\varphi \to \infty$ was taken to be $\varphi = 3$. While these two velocity components look similar to the boundary layer flow over a flat plate from Section 12.1, there are some very important differences. First, the numerical solution in Figure 12.28(a) shows that for $u/U = 0.99$, $\eta = \tilde{\delta} = 2.4$, which is the finite thickness of this viscous "boundary layer." Unlike flow over a flat plate, the boundary layer thickness, $\tilde{\delta}$, here is independent of x. The dimensional boundary layer thickness is $\delta = \tilde{\delta}\sqrt{\nu/A} = 2.4\sqrt{\nu/A}$, which increases with increasing kinematic viscosity, ν, and decreasing external velocity, A.

Second, since $U = Ax$, $u \sim Ax$, the x-velocity is increasing with x, rather than decreasing. Finally, in Figure 12.28(b) we see that $v/V > 0$ and, since $V = -Ay$, this implies that $v < 0$. As a consequence, fluid is flowing into the boundary layer rather than out of it.

Figure 12.29 gives an overall view of this flow. For $y > \delta$ the flow is inviscid, while for $0 \le y \le \delta$ the boundary layer profile matches no-slip at the wall and the inviscid velocity at $y = \delta$. The x-velocity at $x = \pm 2$ in the inviscid region is twice that at $x = \pm 1$, while the magnitude of the inviscid y-velocity at $y = 2$ equals the magnitude of the inviscid x-velocity at $x = \pm 2$. Streamlines in the inviscid region are the hyperbolas we saw in Chapter 3. What about the streamlines in the boundary layer?

To work more efficiently with this flow, we can employ polynomial solution forms as we did for entrance flow in Section 12.2. Let us try a parabolic approximation

$$\frac{u}{U} = a_0 + a_1\zeta + a_2\zeta^2 = 2\zeta - \zeta^2 = 2\left(\eta\tilde{\delta}\right) - \left(\eta\tilde{\delta}\right)^2 \tag{12.7.19}$$

where $\zeta = y/\delta = \eta/\tilde{\delta}$. This form satisfies the three boundary conditions: $u/U = 0$ at $\zeta = \eta = 0$; $u/U = 1$ at $\zeta = 1, \eta = \tilde{\delta}$; and $\partial(u/U)/\partial\zeta = 0$ at $\zeta = 1, \eta = \tilde{\delta}$.

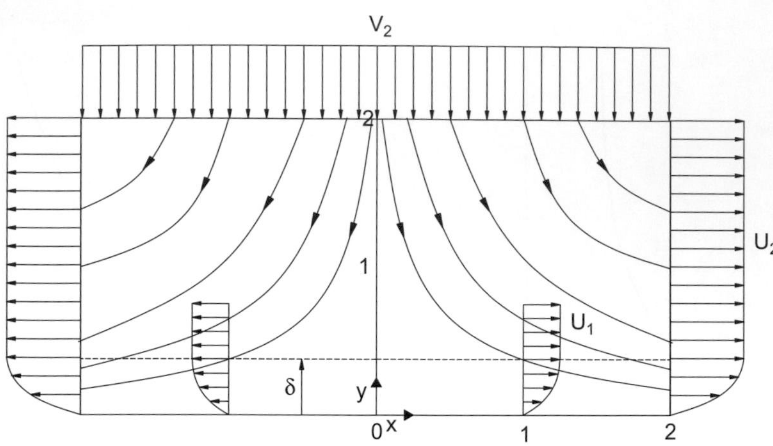

Figure 12.29 Viscous stagnation point flow showing representative velocities and streamlines.

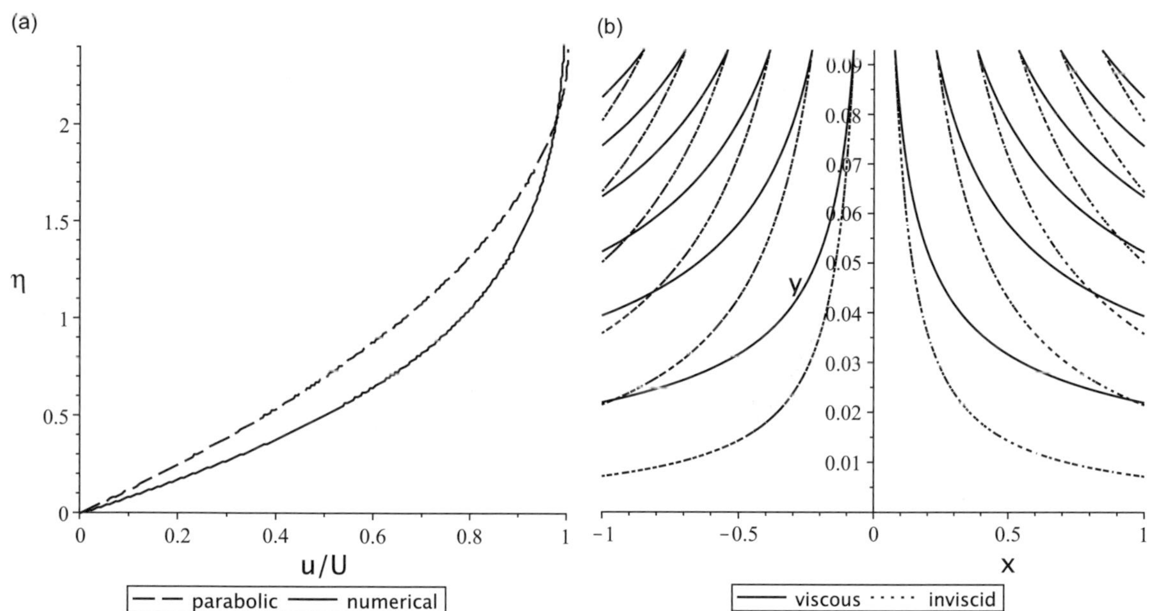

Figure 12.30 (a) Comparison of u/U for numerical vs parabolic polynomial approximation. (b) Streamlines compared for inviscid vs viscous flow in the boundary layer region $0 \leq y \leq \delta$ for $A = 100\ \text{s}^{-1}$, $\nu = 0.15\ \text{cm}^2\text{s}^{-1}$. Note the viscous streamlines are flatter.

Figure 12.30(a) compares the numerical solution of u/U, from Figure 12.28(a), to the parabolic polynomial solution if Eq. (12.7.19) in the η variable. The parabolic form runs slower in the midrange of η but does better near the wall and edge of the boundary layer.

From the polynomial form of u in Eq. (12.7.19) written as $u = Ax\left(2y/\delta - y^2/\delta^2\right)$ we can derive a polynomial form of v from continuity

$$v = -\int \frac{\partial u}{\partial x}\, dy + g(x) = -Ay\left(\frac{y}{\delta} - \frac{1}{3}\frac{y^2}{\delta^2}\right) \qquad (12.7.20)$$

where we have enforced the boundary condition that $v(y=0)=0$ so $g=0$. We note that $u(y=\delta)=Ax$, which is the same value as the inviscid flow $U=Ax$, as we expect from the matching. However, $v(y=\delta)=-(2/3)A\delta$, which is 2/3 the inviscid value of $V(y=\delta)=-A\delta$. So we anticipate the streamlines to be flatter than their inviscid counterparts in the boundary layer. Figure 12.30(b) demonstrates exactly that result, the overlay of the inviscid, hyperbola, streamlines drawn in the boundary layer region versus the viscous streamlines for airflow with $A=100\,s^{-1}, \nu=0.15\,cm^2s^{-1}$ so $\delta=2.4\sqrt{\nu/A}=0.093\,cm$.

With regard to the pressure field, Eq. (12.7.7) can be made dimensionless as

$$P=\frac{p-p_0}{\mu A}=-\frac{1}{2}(\xi^2+\Phi(\eta)) \tag{12.7.21}$$

where the dimensionless x variable is now $\xi=x/\sqrt{\nu/A}$, which has the same scaling as η and we have substituted $F(y)=(\nu/A)\Phi$ from Eq. (12.7.13). From Eq. (12.7.17) and Eq. (12.7.18) we know that

$$\Phi=2\varphi'+\varphi^2=2\left(\frac{u}{U}\right)+\left(\eta\frac{v}{V}\right)^2 \tag{12.7.22}$$

We already have u/U in terms of η in Eq. (12.7.19). Since $y=\delta\eta/\tilde{\delta}$ we can rewrite Eq. (12.7.20) to find v/V as a function of η

$$\frac{v}{V}=\left(\frac{\eta}{\tilde{\delta}}-\frac{1}{3}\left(\frac{\eta}{\tilde{\delta}}\right)^2\right) \tag{12.7.23}$$

Substituting Eqs. (12.7.19) and (12.7.23) into Eq. (12.7.22) gives us the form of Φ we need to insert into Eq. (12.7.21). The result is

$$P=-\frac{1}{2}\left(\xi^2+2\left(2\left(\frac{\eta}{\tilde{\delta}}\right)-\left(\frac{\eta}{\tilde{\delta}}\right)^2\right)+\left(\eta\left(\frac{\eta}{\tilde{\delta}}-\frac{1}{3}\left(\frac{\eta}{\tilde{\delta}}\right)^2\right)\right)^2\right) \tag{12.7.24}$$

In Figure 12.31 there is a comparison of lines of constant pressure, isobars, for viscous stagnation point flow, Figure 12.31(a), and inviscid stagnation point flow, Figure 12.31(b). We studied inviscid stagnation point flow in Section 6.1. To within a multiplicative constant, the isobar contours for the inviscid flow are circles such that $P_{inviscid}\sim-(\xi^2+\eta^2)$, see Eq. 6.1.12. Those are shown in Figure 12.31(b), which is the same as Figure 6.2. For both figures, the dimensionless pressure is zero at the origin and negative elsewhere. Note that viscosity distorts the isobars considerably compared to the inviscid solution.

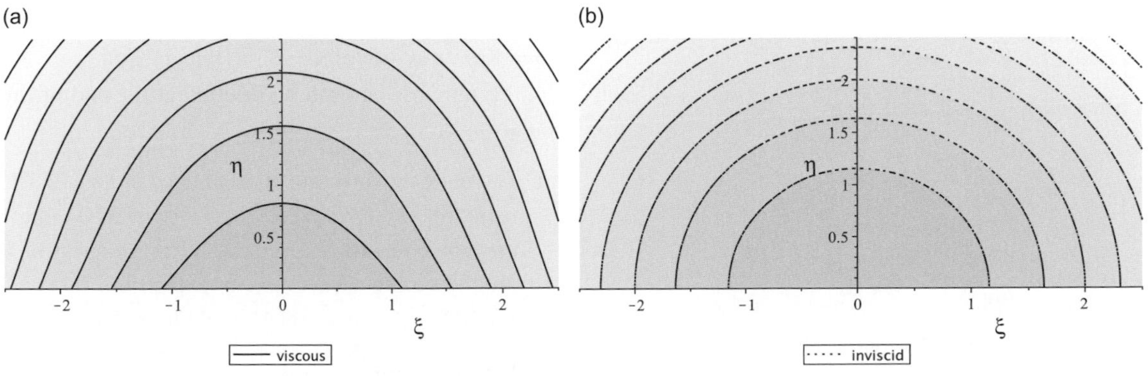

Figure 12.31 Comparison of lines of constant pressure, isobars, for (a) viscous stagnation point flow vs (b) inviscid stagnation point flow, both in the region of the viscous boundary layer.

Summary

In this chapter our interests were focused on high Reynolds number, laminar phenomena which involve boundaries. The fundamental problem of flow over a flat plate, at zero incidence angle, gave us our initial insight into how boundary layers behave and why it is so critical to analyze them. It gave us a chance to use scaling skills and revisit similarity solutions. In this case, however, the answer needed to be solved numerically which brought in use of computational software available to the student. This, then, carried over to the topic of inlet, or entrance, flows into tubes which has a relatively large impact on respiratory fluid mechanics. In addition to boundary layers, we also explored secondary vortical flows in the cross-section of a tube due to axial curvature. Particularly for vascular flows, such curvature effects can induce complicated systems of vortices which number up to four pairs, with asymmetries stemming from the non-Newtonian fluid behavior of blood. The power of experimental and 3D computational fluid dynamics techniques showed us these fascinating patterns. Then for airway flows there are both entrance and curvature effects which CFD reveals in 3D. Lastly, stagnation point flow also has a boundary layer when viscous effects are included. Those solutions were compared to the inviscid solution to give us an appreciation of how fluid motion changes near a boundary when viscosity is added. In the next chapter, Chapter 13, the Reynolds number will be increased enough so that boundary layers, as well as other regions of the fluid, become turbulent.

Problems

12.1 To employ the definitions of the similarity stream function, $\psi = \sqrt{X} f(\hat{\eta})$, and similarity coordinate, $\hat{\eta} = \eta/\sqrt{X}$, in Eq. (12.1.11), we need to use the chain and product rules for derivatives.

a. Show that $\dfrac{\partial \psi}{\partial X} = \dfrac{1}{2\sqrt{X}}(f - \hat{\eta} f')$ where $f' = df/d\hat{\eta}$.

b. Show that $\dfrac{\partial \psi}{\partial \eta} = f'$ and then

$$\frac{\partial^2 \psi}{\partial \eta^2} \frac{1}{\sqrt{X}} f'', \quad \frac{\partial^3 \psi}{\partial \eta^3} = \frac{1}{X} f'''$$

c. Show that $\dfrac{\partial^2 \psi}{\partial \eta \partial X} = -\dfrac{1}{2X} \hat{\eta} f''$

d. Insert your results for parts (a)–(c) into Eq. (12.1.11) and show that

$$f'\left(-\frac{1}{2X}\hat{\eta} f''\right) - \frac{1}{2\sqrt{X}}(f - \hat{\eta} f')\frac{1}{\sqrt{X}} f'' = \frac{1}{X} f'''$$

e. Simplify part (d) to arrive at Eq. (12.1.13).

12.2 In addition to the quadratic boundary layer profile, the entrance region of a tube can also be modeled with linear, $w_b(r,z) = w_0(z)\,\xi$, or cubic boundary, $w_b(r,z) = w_0(z)\left(3\xi/2 - \xi^3/2\right)$, profiles, where $\xi = (a - r)/\delta$. The boundary layer thickness is $\delta = K\sqrt{\nu z/\bar{w}}$ where $K = 3.46\,(\text{linear})$ and $4.64\,(\text{cubic})$, which can be inserted at the end of the analysis.

a. Calculate the flow rate, Q, as defined in Eq. (12.3.10) using the two profiles. Set your answer to $Q = \pi a^2 \bar{w}$ and solve for the two versions of w_0 in terms of $\bar{w}$ and δ/a using a Taylor series expansion. Drop the $O(\delta/a)^2$ terms. They should look like $w_0 = \bar{w}(1 + k_0 \delta/a)$ where k_0 has different values for the two profiles. What are they?

b. Using your solutions in part (a) for w_0, insert into the Bernoulli equation, Eq. (12.3.5), and solve for $p_e - p_0$ using a Taylor series, neglecting $O(\delta/a)^2$.

c. Scale your pressure drop solutions using the dimensionless variables $P = p/(\mu\bar{w}/a)$, $Z = z/d$, $Re = \bar{w}d/\nu$. Organize them to look like $P_e - P_0 = k_1\sqrt{Re}\sqrt{Z}$ where k_1 depends on K and k_0 for the two profiles. What are your values of k_1 after inserting for K? How do your solutions compare to Eq. (12.3.16).
d. Set $P_e = 0$ and plot the dimensionless P_0 vs $Z = z/d$ for $0 \leq Z \leq 3$ and Re = 200, 500, 1000, 1500 for the cubic approximation. Describe your result.
e. Calculate the corresponding length-averaged wall shear $\bar{\tau}_w = -\frac{1}{L}\int_0^L \mu\left(\frac{\partial w}{\partial r}\right)_{r=a} dz$ for both profiles. Keep K unspecified. Make your solution dimensionless, $\bar{T}_w = \bar{\tau}_w/(\mu\bar{w}/a)$, in terms of $Re = \bar{w}d/\nu$, K and $\lambda = L/d$.
f. For the cubic profile, plot T_w vs Re for $\lambda = 1, 2, 3$ over the Reynolds number range $500 \leq Re \leq 1500$. Discuss your results.
g. As a check on part (f), form the ratio of the tube length, L, to the inviscid length, L_I, using Eq. (12.3.4). Plot $L/L_I = \lambda/(L_I/d)$ for $\lambda = 1,\ 2,\ 3$ and $500 \leq Re \leq 1500$. Is $L/L_I < 1$ for the graph in part (e)?
h. Both endothelial cells of the blood vessels and airway epithelial cells of the lung are subject to the shear stress from entrance flow. Discuss how one of them responds to shear stress.

12.3 An alternative approach to determine the pressure drop in the entrance region of tube flow is the energy balance,

$$\iint_{A_e} pw\,dA - \iint_{A_0} pw\,dA = \iint_{A_0} \frac{1}{2}\rho q^2 w\,dA - \iint_{A_e} \frac{1}{2}\rho q^2 w\,dA + D \quad (1)$$

where p is the pressure, w is the z-component of velocity, $q = (\underline{u}\cdot\underline{u})^{1/2} = (w^2)^{1/2}$ is the speed, and $dA = r\,dr\,d\theta$ The first two terms are the net rate at which pressure does work on the flow, the second two terms are the change in kinetic energy, and the last term is the viscous dissipation rate, D. Divide the balance by the volume flow rate, $Q = \pi a^2\bar{w}$, where $\bar{w}$ is the cross-sectional average velocity. The result is the flow-averaged energy equation

$$\hat{p}_e - \hat{p}_0 = \frac{1}{2}\rho\left(\hat{q}_0^2 - \hat{q}_e^2\right) + \frac{D}{Q} \quad (2)$$

where the terms are defined by

$$\hat{p} = \frac{1}{Q}\iint_A pw\,dA,\ \hat{q}^2 = \frac{1}{Q}\iint_A q^2 w\,dA,\ D = \int_0^z D_0(z)\,dz$$

and $D_0(z) = \mu\iint_A \left(\frac{\partial w}{\partial r}\right)^2 + \left(\frac{1}{r}\frac{\partial w}{\partial \theta}\right)^2 dA$.

Assume station e is the tube entrance at z = 0, where the velocity is uniform and equal to the average $\bar{w}$ and the pressure is p_e.

a. Calculate $\frac{1}{2}\rho\hat{q}_e^2$ at the entrance $z = 0$.
b. Calculate $\hat{p}_e$ at the entrance $z = 0$.
c. To find the remaining terms in the entrance region, use the core and quadratic boundary layer forms of Eq. (12.3.6) where $w_b = w_0(2((a-r)/\delta) - ((a-r)/\delta)^2)$. The integral conservation of mass still yields $w_0 = \bar{w}\left(1 + 2\delta/3a + O(\delta/a)^2\right)$ as derived in Eq. (12.3.13). Calculate $\frac{1}{2}\rho\hat{q}_0^2$ in terms of $\bar{w}$ keeping $O(\delta/a)$ in your Taylor expansion.
d. Calculate $D_0(z)$ in terms of $\bar{w}$ and note that it has no contribution from the core region, since $\partial w_0/\partial r = 0$. Keep $O(1/\delta)$ as the largest term.
e. For the quadratic profile $\delta(z) = K\sqrt{\nu z/\bar{w}}$ where $K = 5.47$ can be inserted at the end of the analysis. Insert this form into $D_0(z)$ from part (d) and integrate with respect to z to find D. Then divide by Q.
f. Insert your answers into Eq. (2) and make it dimensionless with the scaled variables $\hat{P} = \hat{p}/(\mu\bar{w}/a)$, $Z = z/d$, $Re = \bar{w}d/\nu$ where $d = 2a$ is the tube diameter. Organize the result to look like $\hat{P}_e - \hat{P}_0 = k_1\sqrt{Re}\sqrt{Z}$ where k_1 depends on K. What is your value of k_1 after you insert $K = 5.47$?
g. What are the percent contributions from the kinetic energy versus the viscous dissipation terms? How does your answer compare to Eq. (12.3.16), which was derived using the Bernoulli equation for the center streamline in the inviscid core? Calculate the relative difference of their

coefficients, which multiply $\sqrt{Re}\sqrt{Z}$, by subtracting their values and dividing by the larger. Is it of $O(\delta/a)^2$, i.e. the terms we dropped?

f. What diameter arteries have $U_\infty = U_{avg}$ equal to 10 cm/s? Are they long enough for the cancer cells to impact?

12.4

a. Repeat the metastatic cancer/blood clot model for particle deposition in boundary layer flow over a flat plate. Use the quadratic x-velocity profile from Eq. (12.2.4). What is the corresponding form of v? For the quadratic profile recall that $\delta = 5.47(\nu x/U_\infty)^{1/2}$.
b. Following the development of Eq. (12.2.10) for the linear profile, scale the quadratic profile so $X = x/d_p,\ Y = y/d_p$ and determine its forms of U and V at $X = X_p, Y = Y_p$ in terms of X_p, Y_p, Re_p. What are the governing dimensionless equations in terms of Stk, Re_p and Fr?
c. Insert U and V from part (b) into Eqs. (12.2.11). Use the initial conditions $X_p(0) = 50,\ Y_p(0) = (0.9)\delta(x = x_p(0))/d_p$ where the choice of $X_p(0)$ keeps away from the singularity at $x = 0$ and the choice of $Y_p(0)$ is 90% of the boundary layer thickness at that location for $X_p(0)$. Also let dX_p/dT and dY_p/dT equal zero at $T = 0$, where $T = U_\infty t/d_p$ is the dimensionless time variable. Use the parameter values of Eq. (12.2.13) We want to find out what size of a clump of cancer cells, what value of d_p, will lead to a solution where the particle lands on the wall.

   Start with $d_p = 0.02$ cm, which would be a clump of cells. Plot the particle position until it hits the wall – this should take about 900 dimensionless time units. What is the dimensional downstream distance where it impacts, and at what dimensional time?
d. Now increase d_p to 0.03 cm and repeat part (c). How many time units until impact? What is the dimensional downstream distance where it impacts?
e. Now decrease d_p to 0.015 cm and repeat part (c). How many time units until impact? What is the dimensional downstream distance where it impacts?

12.5 The flow resistance, R_n, across airway generation n is given for Poiseuille flow in Eq. (12.4.8) and entrance flow in Eq. (12.4.9). For this exercise, let the airway diameter be $d_n = d_0 2^{-n/3}$ at generation n, where $n = 0$ is the trachea. Assume a bifurcating system so there are 2^n identical tubes at generation n.

a. What is the equation for the total cross-sectional area, A_n, for all of the tubes at generation n?
b. What is the equation for the average velocity, U_n, at generation n?
c. The Reynolds number at generation n is $Re_n = U_n d_n/\nu$. Let $\nu = 0.15\ cm^2/s$, $d_0 = 2$ cm, $Q = 750\ cm^3/s$and plot Re_n for $0 \le n \le 15$, treating n as a continuous variable. Describe your result.
d. Let the airway lengths be $L_n = 3d_n$. We want to know how L_n compares to the inlet length, L_{In}. Plot the ratio $L_n/L_{In} = (L_n/d_n)/(L_{In}/d_n)$ for $0 \le n \le 15$ where $L_{In}/d_n = (0.59 + 0.056Re_n)/4$ from Eq. (12.3.4). For $L_n/L_{In} < 1$, our model is appropriate. To be safe, what range of $0 \le n \le N$ is the ratio $L_n/L_{In} < 0.8$? Make a note of N.
e. The resistance, R_n, at generation n is given by Eq. (12.4.9) for the inlet flow, call it R_{nI}, and Eq. (12.4.8) for Poiseuille flow, call it R_{nP}. For our choice of airway network geometry, it will turn out that R_{nP} is a constant independent of n. Let $\mu = 0.00018$ poise, $\rho = 0.0012\ g/cm^3$, $K = 5.47$ and note that $a_n = d_n/2$. On the same graph plot R_{nI} over the range $0 \le n \le N$ and R_{nP}over the range $N \le n \le 15$.
f. The cumulative resistance corresponding to your figure in part (e) is related to the integrals of R_{nI} and R_{nP}. Let $F_n = \int_0^n R_{nI}\,dn$, which will be for generations $0 \le n \le N$. Then define $G_n = F_n(n = N) + R_{nP} \times (n - N)$, which will be for generations $N \le n \le 15$. Plot F_n and G_n on the same graph over their respective regions.

g. Compare your results to Figure 12.16(a, b). In particular, how does your formulation for total cross-sectional area, A_n, compare to the measured values in Figure 12.13?

12.6

a. In Eq. (12.7.19) we derived a parabolic approximation to the u velocity in the boundary layer of viscous stagnation point flow, see Eq. (12.7.19). From that we obtained the corresponding v velocity, Eq. (12.7.20). Find u(x, y) for a linear profile assumption and then its corresponding v(x, y). What are they?
b. Using Stokes drag the governing equations for a particle released in the viscous stagnation point boundary layer are

$$\rho_p \pi \left(\frac{d_p}{2}\right)^3 \frac{d^2 x_p}{dt^2} + 3\pi\mu d_p \left(\frac{dx_p}{dt} - u\left(x_p, y_p\right)\right) = 0$$

$$\rho_p \pi \left(\frac{d_p}{2}\right)^3 \frac{d^2 y_p}{dt^2} + 3\pi\mu d_p \left(\frac{dy_p}{dt} - v\left(x_p, y_p\right)\right) = 0$$

where the particle diameter is d_p with density ρ_p. For initial conditions let the particle start at $x = 0.5\delta, y = 0.8\delta$ where $\delta = 2.4\sqrt{\nu/A}$ is the constant boundary layer thickness. Also let the initial velocity equal the fluid velocity where it is starting. Write the full dimensional governing equations and initial conditions.
c. Choose these scales to make your system in part (b) dimensionless, $X_p = x_p/\delta, Y_p = y_p/\delta, T = At$ and let $Stk = \rho_p\left(d_p^2 A/18\mu\right)$ be the Stokes number. Rewrite the dimensionless forms of the governing equations and initial conditions.
d. Solve the system in part (c) plotting your results for Stokes number $Stk = 0.1, 0.5$ for the time range $0 \le T \le 10$. Discuss your results.
e. Solve for the stream function by integrating $u = \partial\psi/\partial y$ and imposing $\psi(y = 0) = 0$ so that the wall is a streamline with no cross-flow. Now make the stream function dimensionless as $\Psi = \psi/A\delta^2$. Plot a streamline through the starting point of the two particles in part (d) and compare to the trajectories in one graph. What do you see?
f. The particle deposition governing equations for inviscid stagnation point flow were derived in Eq. (3.9.5). There the length scale was L, but the form is the same if the length scale is δ and time is scaled the same. Use the same initial conditions as in part (c) and solve the inviscid trajectories for $Stk = 0.1, 0.5$ for the time range $0 \le T \le 2$. It may be convenient to do this numerically as well. Now plot all four solutions, the two viscous solutions from part (d) and the two inviscid solutions here.
g. Discuss your results in part (f). Form a ratio of the y-component of fluid velocity with the viscous form, from part (a), in the numerator and the inviscid form, $-Ay$, in the denominator. How does this ratio influence the relative time it takes for the particle to approach the wall? Keep in mind $0 \le y/\delta \le 1$ in the boundary layer region.

Image Credits

Figure 12.7a Source: Figure 1 in Zierenberg, Halpern, Filoche, Sapoval, Grotberg. An asymptotic model of particle deposition at an airway bifurcation. *Mathematical Medicine and Biology*. 2013; 30:131–56.

Figure 12.7b Source: Figure 2 in Zierenberg, Halpern, Filoche, Sapoval, Grotberg. An asymptotic model of particle deposition at an airway bifurcation. *Mathematical Medicine and Biology*. 2013; 30:131–56.

Figure 12.9 Source: Fig 5 of Durst F, Ray S, Unsal B, Bayoumi OA. The development lengths of laminar pipe and channel flows. *Journal of Fluids Engineering – Transactions of the ASME*. 2005; 127(6):1154–60.

Figure 12.12 Source: Figure 15 from Haefeli-Bleuer, H, Weibel ER. Morphometry of the human pulmonary acinus. *The Anatomical Record*. 1988; 220:401–14.

Figure 12.17 Source: Fig 3 from Lowe KC. Blood substitutes: from chemistry to clinic. *Journal of Materials Chemistry*. 2006; 16:4189–96.

Figure 12.18 Source: public domain.

Figure 12.20 Source: based on Fig 7 in Wyk S, Wittberg LP, Bulusu KV, Fuchs L, Plesniak MW. Non-Newtonian perspectives on pulsatile blood-analog flows in a 180 degrees curved artery model. *Physics of Fluids*.

2015; 27(7). Monotonic greyscale courtesy of M. Plesniak.

Figure 12.21a Source: based on Fig 14 in Wyk S, Wittberg LP, Bulusu KV, Fuchs L, Plesniak MW. Non-Newtonian perspectives on pulsatile blood-analog flows in a 180 degrees curved artery model. *Physics of Fluids*. 2015; 27(7). Monotonic greyscale courtesy of M. Plesniak.

Figure 12.21b Source: based on Fig 18 in Wyk S, Wittberg LP, Bulusu KV, Fuchs L, Plesniak MW. Non-Newtonian perspectives on pulsatile blood-analog flows in a 180 degrees curved artery model. *Physics of Fluids*. 2015; 27(7). Monotonic greyscale courtesy of M. Plesniak.

Figure 12.24 Source: Courtesy of Dr. P.W. Longest specifically for this book.

Figure 12.25 Source: ttsz/ Getty

Figure 12.27a Source: Fig 3a in Pedley TJ, Schroter RC, Sudlow MF. Flow and pressure drop in systems of repeatedly branching tubes. *Journal of Fluid Mechanics*. 1971; 46(2):365–83.

Figure 12.27b Source: Fig 3b in Pedley TJ, Schroter RC, Sudlow MF. Flow and pressure drop in systems of repeatedly branching tubes. *Journal of Fluid Mechanics*. 1971; 46(2):365–83.

13 Turbulence

Introduction

Turbulent flow is characterized by random and chaotic variations in pressure and velocity. It involves the appearance of unsteady vortices, or eddies, of many different sizes which interact with one another. Typically the energy of the larger eddies feeds into the smaller eddies where it dissipates from viscous friction. Once turbulence appears, it can have a dramatic effect on the nature of the flow. General engineering applications such as delivery of water or oil through pipes, over long distances, focus on the pressure–flow relationship and the increased frictional losses from turbulence, which has implications for choosing pumping strategies. On the same basis, turbulent blood flow exiting the heart into the aorta puts additional demands on cardiac function. In biomedical applications, such turbulent flows can produce sounds which are the source of heart murmurs or lung sounds that are heard with a stethoscope. Turbulent flows can also dominate stresses on the cells which form bounding walls, as well as stresses on living components of the fluid, like blood cells. Turbulent mixing can strongly influence the transport of soluble constituents and particulates as well as heat.

Although turbulence arises in many types of biological flows, for our purposes we will focus on four basic situations: flow over bluff bodies, jets, boundary layer flows, and pipe flows. The criterion for the appearance of turbulence usually is when the flow exceeds a critical Reynolds number, $Re = UL/\nu$. The particular value of the critical Re depends on the flow situation. Examples for flow over bluff bodies include external flow past living structures, both animal and plant life, in either air or water. A blue whale can hit $Re \sim 10^8$, which vastly surpasses the critical value, without expending excessive effort. In these flows the resultant forces and transport phenomena for dissolved gaseous and nutrient species, as well as heat, are focal points for investigation. Frostbite is a risk on a frigid, windy day. A jet is a flow emanating from a smaller orifice into a larger, quiescent fluid domain, like a jet engine. A cough is an example of a jet which has gas and liquid droplet components. It has implications for the spread of respiratory diseases, including epidemics like influenza or tuberculosis. Flow through an artificial heart valve is another important application of jets, representative of medical devices which interact with the blood: blood pumps, intra-aortic balloons and vascular stents. They can damage red blood cells and trigger clotting. Boundary layer flow over a flat plate, as we saw in Chapter 12, relates to aerodynamic and hydrodynamic external flows over wings and fins. The boundary layer can start at the leading edge, $x = 0$, where it is laminar. However, further downstream it can transition to a turbulent boundary layer when a critical Reynolds number is achieved. As with laminar flow, there is also an entrance length, at the inlet of a pipe, associated with the growth of a turbulent boundary layer, which can evolve to fully developed, turbulent conditions downstream. Once achieved, fully turbulent pipe flow is well represented in biofluid mechanics. The respiratory and cardiovascular systems are prime examples of internal pipe flows which can easily reach a turbulent range.

Turbulent flow is a large and complex field. Underlying principles that provided background for the treatment here can be found in Schlichting (1968), White (1974), (1986) and Pope (2000) as well as further extensions.

Topics Covered

The chapter starts with the well-studied topic of flow over a cylinder of diameter D, in a fluid of kinematic viscosity ν, with free stream velocity, U. The flow characteristics such as the its drag, streamline patterns and transition from laminar to turbulent flow undergo various qualitative and quantitative changes as the Reynolds number, $Re_D = UD/\nu$, is increased. This introduces us to rich fluid flow behaviors including laminar vortex shedding and turbulent vortex shedding. These have a major impact on what we observe in nature and how we tame technology. Then we cover the traditional analysis of turbulent flows through manipulations of the Navier–Stokes equations into a form that shifts the turbulent convective acceleration to be viewed as a turbulent shear stress. These concepts are applied to three remaining applications which include: fully developed turbulent flow in a pipe with applications to blood flow and respiratory air flow; wall bounded turbulent flows including developing flow over a flat plate which are applied to entrance flows; and turbulent jet flow that occurs in aquatic propulsion and cough, for example. Finally the last two sections are devoted to the traditional analysis of boundary layer flows using integral methods, including laminar and turbulent regimes.

13.1 Examples of Turbulent Flows in Biofluid Mechanics
13.2 Analysis of Turbulent Flow
13.3 Fully Developed Turbulent Pipe Flow
13.4 Wall Bounded Turbulent Flow
13.5 Turbulent Boundary Layer on a Flat Plate
13.6 Turbulent Jets
13.7 Integral Balances for Boundary Layer Flow over a Flat Plate
13.8 Solutions to the Momentum Integral Equation

13.1 Examples of Turbulent Flows in Biofluid Mechanics

Turbulence is encountered in a variety of flow situations that occur in industrial as well as biological applications. Visualization of turbulent flows is an essential aspect of characterizing and understanding the underlying dynamics. While there are a large variety of flow situations that involve turbulence, there are four basic ones that we will consider: flow over bluff bodies, pipe flows, boundary layer flows and jets. In this introductory section let's view the primary characteristics of flow over a bluff body (a cylinder) as a sequence of insights into the complexity of this system. The other three topics are examined with more detail in the following sections.

Figure 13.1(a) shows flow over a cylinder for $Re_D = UD/\nu < 5$ where D is the cylinder diameter, U is the free stream velocity and ν is the kinematic viscosity. For this small inertia range the streamlines are nearly symmetric fore and aft. Figure 13.1(b) is in the range $5 < Re_D < 40$ and we see a pair of vortices attached to the cylinder in the wake. The bounding streamline is shown, and the outer streamlines are more asymmetric fore and aft. The point where the bounding streamline intersects the cylinder surface is the separation point for the flow. Figure 13.1(c) is for $40 < Re_D < 150$ and the vortices are now shedding periodically in an alternating pattern forming the Kármán vortex street (von Kármán, 1911; 1912). This flow is laminar but with a time-periodic instability. Figure 13.1(d) shows $150 < Re_D < 3 \times 10^5$ where the upstream boundary layer is laminar until its point of separation, and the vortex street is turbulent. Figure 13.1(e) is in the range $3 \times 10^5 < Re_D < 3.5 \times 10^6$ where the boundary layer becomes turbulent and separates from the cylinder further downstream along the cylinder surface. The wake is fully turbulent and the vortex street has disappeared. The scale of the turbulence, i.e. the eddy sizes, is smaller now. Figure 13.1(f) is in the range $Re_D > 3.5 \times 10^6$ where the turbulent vortex street returns, as in Figure 13.1 (d), but the wake is narrower.

Figure 13.2 shows flow visualization of regimes related to Figure 13.1. Figure 13.2(a) from Coutanceau and Bouard (1977) at $Re_D = 24.3$ is an oil flow with magnesium cuttings for visualization. It corresponds to the attached vortex pair sketched in Figure 13.1(b). Figure 13.2(b) is airflow with smoke at $Re_D = 200$ and corresponds to the laminar Kármán vortex street sketched in Figure 13.1(c). The vortex shedding frequency is 28 Hz. Figure 13.2(c) is the turbulent wake behind a cylinder sketched in Figure 13.1(d). It is airflow over a 2 inch diameter cylinder using helium-filled soap bubbles (neutrally buoyant) for imaging. Figure 13.2(d) is a satellite image of air flow over the Juan Fernandez Islands off the Chilean coast with $Re_D \sim 10^9$ from the high volcanic mountains, and corresponds to the turbulent Kármán vortex street sketched in Figure 13.1(f). Clouds provide the flow visualization.

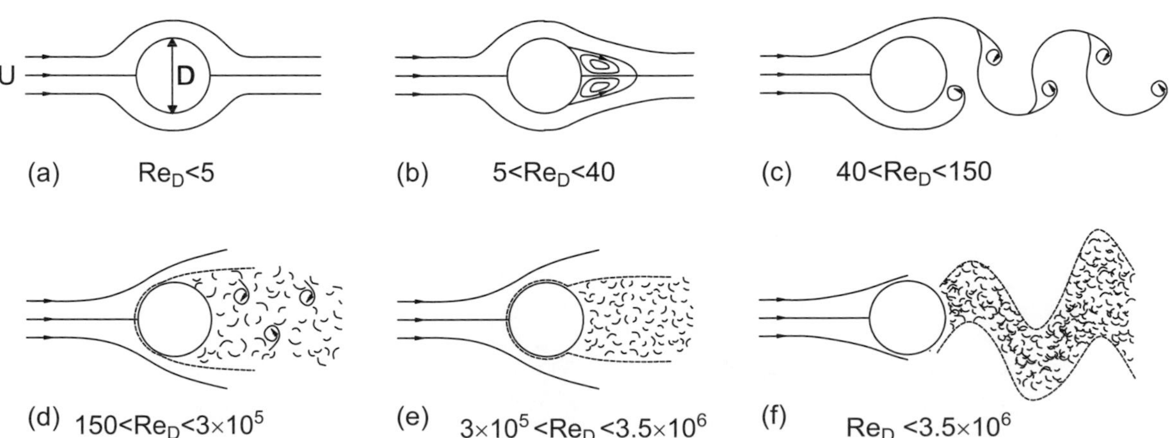

Figure 13.1 Flow over a cylinder for various ranges of $Re_D = UD/\nu$, see text for discussion.

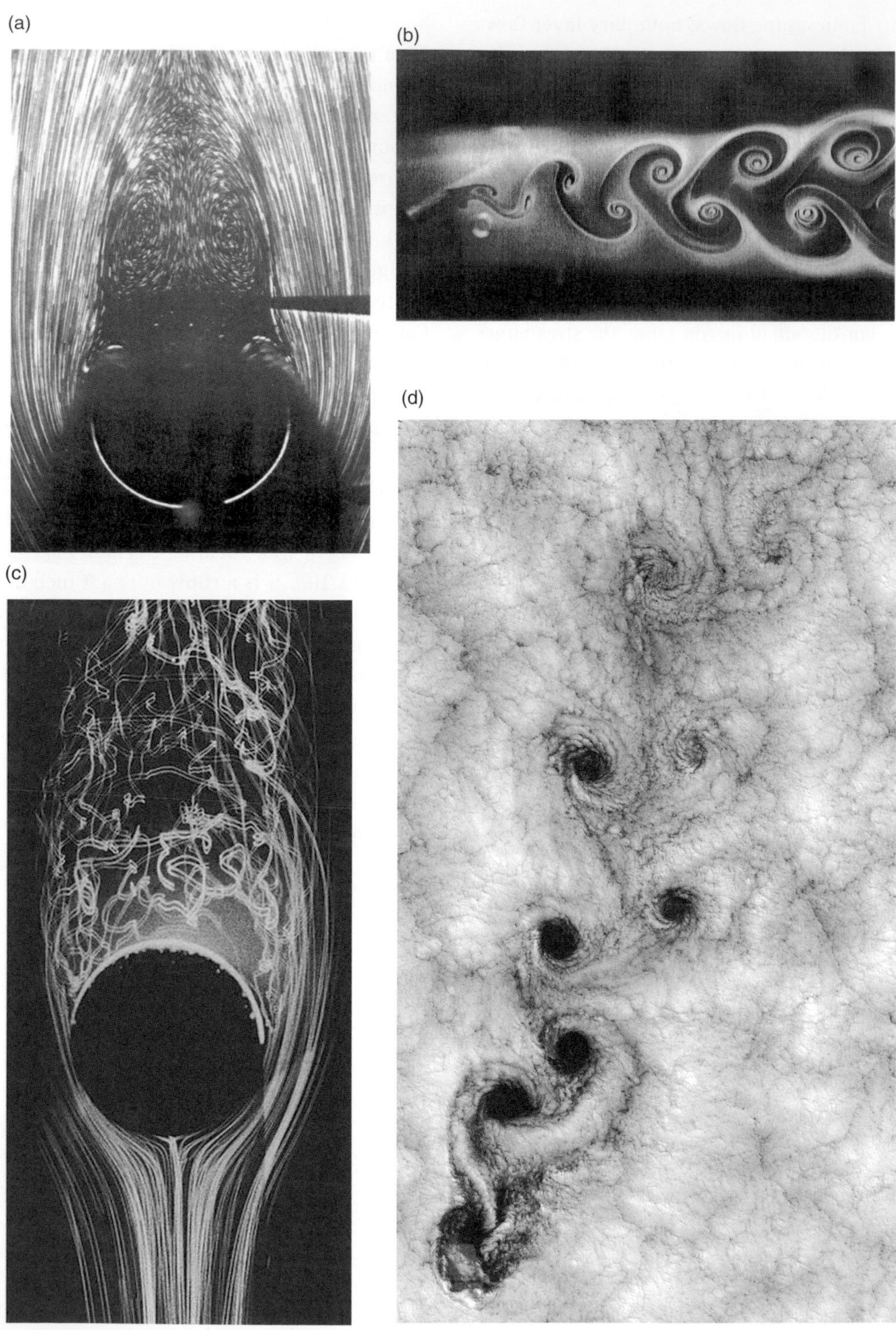

Figure 13.2 (a) Attached vortex pair $Re_D = 24.3$; (b) laminar Kármán vortex street $Re_D = 200$, frequency = 28 Hz; (c) turbulent airflow wake; and (d) Landsat 7 image of clouds flowing over the Juan Fernandez Islands showing a turbulent Kármán vortex street.

In Figure 13.3 the drag coefficient, C_D, of the cylinder is plotted vs the Reynolds number based on its diameter, $Re_D = UD/\nu$. The dimensionless drag coefficient is defined as $C_D = (F_d/A)/(\frac{1}{2}\rho U^2)$ where F_d is the drag force and A is a reference area, such as the area projected perpendicular to the flow. Drag force has two contributions, the viscous shear or "skin friction" and the pressure drop across the cylinder called "form drag" or "pressure drag" due to the shape. The range $4.2 < Re_D < 800{,}000$ in the graph is based on the measurements of Wieselsberger (1921; 1922) who used nine different cylinders with diameters ranging from 0.05 to 300 mm in air flows ranging from 1.2 to 36 m/s. The range $10^6 < 10^6 < Re_D < 10^7$ comes from the data in Roshko (1961), who used an 18 inch diameter cylinder with air flow velocities kept under 0.25 of the sound speed to avoid fluid compressibility effects. The range $\sim 0 < Re_D < 4.2$ matches to the small Re_D work of Lamb (1911). For much of the range $Re_D < 100$ we find $C_D \sim Re_D^{-1} \rightarrow F_d \sim (\mu AU)/2D$. The drag force is independent of inertial effects, ρ. It is linearly proportional to viscous effects, the product μU. See Section 17.5 for the details of $Re_D << 1$. When $100 < Re_D < 2 \times 10^5$ there is not much dependence on Re_D, so that $C_D \sim 1 \rightarrow F_d \sim \rho U^2$. The drag force depends on inertial effects and is quadratic in velocity, i.e. ρU^2. In the range

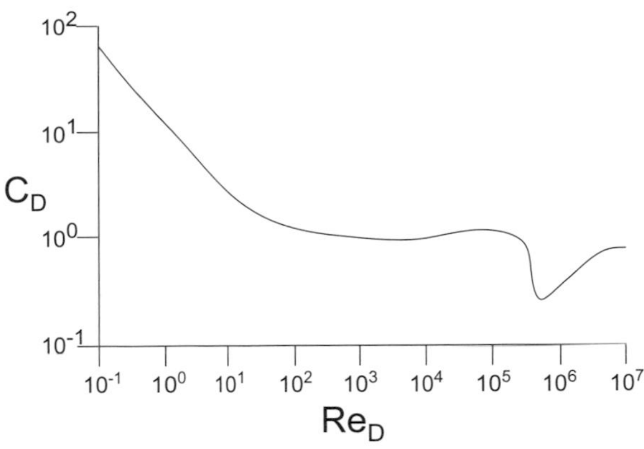

Figure 13.3 Drag coefficient, C_D, for flow over a cylinder versus the Reynolds number, Re_D.

$2 \times 10^5 < Re_D < 5 \times 10^5$ there is a sudden dip to $C_D \sim 0.3$ which comes back for higher Re_D to ~ 0.7 at $Re_D \sim 1.0 \times 10^6 - 3.5 \times 10^6$. This dip corresponds to the onset of turbulence in the boundary layer which is seen in Figure 13.1(e), causing the point of separation to be further downstream along the cylinder surface, which narrows the wake and reduces the form drag. Similar behavior is seen for flow over a sphere. This is why golf balls are dimpled to achieve drag reduction at lower Re_D by shifting the C_D vs Re_D curve to the left. Dimpling promotes boundary layer turbulence.

Box 13.1 | Homeland Security Application

Biofluid mechanics has many examples of turbulent flow over a cylinder. Consider the cylinder to be a tree trunk or limb in the breeze, or submerged plant life in flowing oceans, lakes or rivers. A cylinder can be a model of a leg or arm for analyzing swimming mechanics and evaluating efficient stroke styles. A cylinder is also the shape of the human body. In Figure 13.4 turbulent flow over a human is visualized using the schlieren photography technique (Craven et al., 2014). The goal is to develop instrumentation for sampling chemical contamination related to homeland security. At 0 m/s the upward vertical flow is induced by body heat causing a buoyant thermal plume. As the walking velocity increases, the turbulent flow becomes more horizontal.

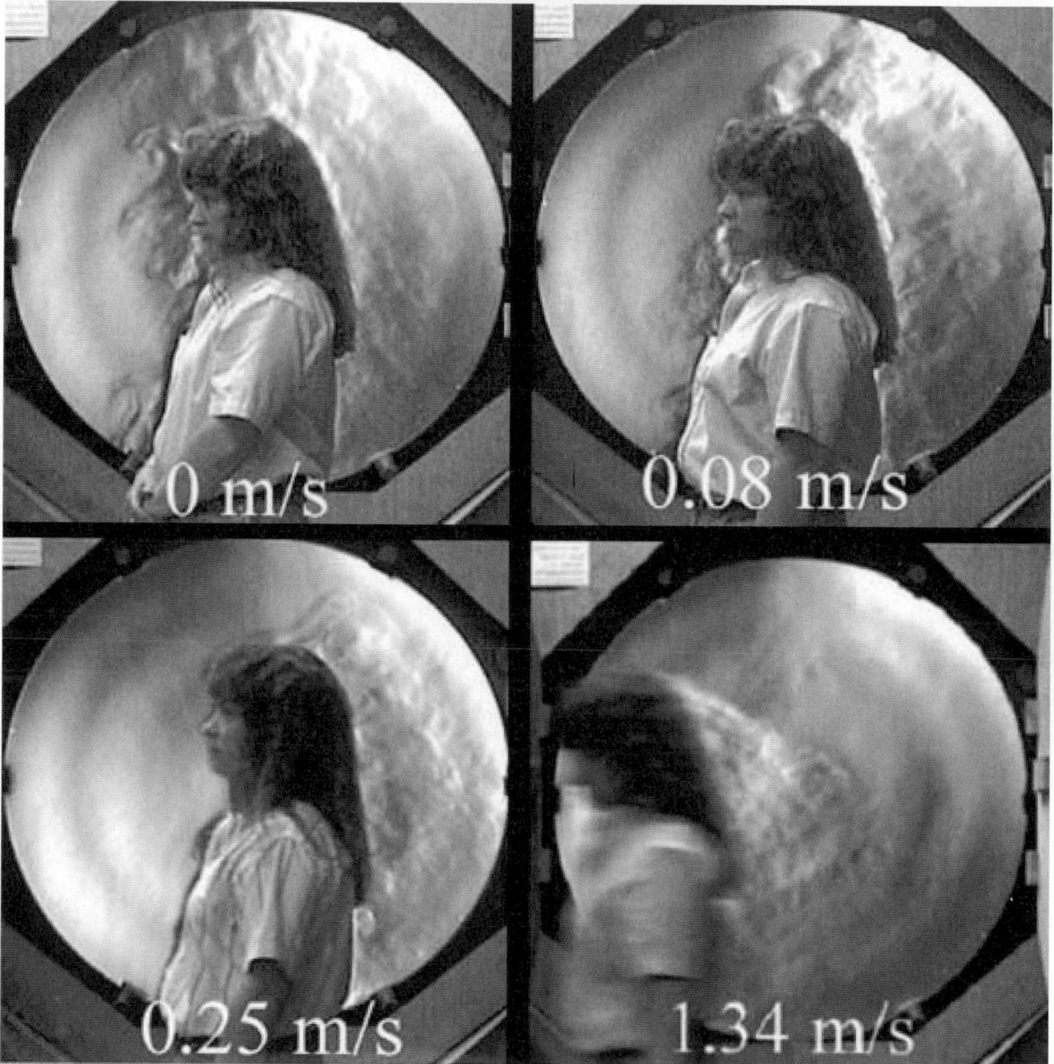

Figure 13.4 Schlieren images of a human walking at different speeds showing turbulent air flow. For standing still, 0 m/s, the flow is vertical due to a buoyant thermal plume from the body temperature. As the walking speed increases the flow becomes more horizontal.

Box 13.2 Explosive Seed Dispersal

Some plants distribute their seeds by an explosive process. As the seed pod dries, its asymmetric architecture makes it spring-loaded, ready to launch. *Impatiens* (family Balsaminaceae) can be provoked to explode by physical contact, hence the common name "touch-me-not." The sandbox tree (*Hura crepitans*) can grow to a height of 60 m and eject their flat 2 cm diameter seeds at 70 m/s, landing as far as 100 m away. The process is so loud it is given the nickname "dynamite tree." The seed, or particle, Reynolds numbers are quite large, in the turbulent range, so Stokes drag is not applicable.

There is a similar curve to Figure 13.3 for flow over a spherical particle, where the particle Reynolds number is $Re_p = d_p|\underline{u}_p|/\nu$. It has a flat region where $C_D = 0.44$ for $500 < Re_p < 2 \times 10^5$due to a turbulent wake. This C_D yields a drag force of magnitude $F_d = 0.44\pi\rho d_p^2|\underline{u}_p|2/8$, which is very different from Stokes drag. It is a function of the inertia involving density and the square of the velocity, while Stokes drag depends on viscosity and is linear in velocity. The vector form is $\underline{F}_d = -0.44\pi\rho d_p^2\underline{u}_p|\underline{u}_p|/8$, which maintains the correct directionality opposing the velocity.

Using the scales $T = t/(V_0/g)$, $X_p = x_p/(V_0^2/g)$, $Y_p = y_p/(V_0^2/g)$, the equations of motion for a particle given an initial velocity, V_0, at angle θ, from a height, h, in a negative Y-direction gravity field are

$$\frac{d^2X_p}{dT^2} + \beta\frac{dX_p}{dT}\left(\left(\frac{dX_p}{dT}\right)^2 + \left(\frac{dY_p}{dT}\right)^2\right)^{1/2} = 0$$

$$\frac{d^2Y_p}{dT^2} + \beta\frac{dY_p}{dT}\left(\left(\frac{dX_p}{dT}\right)^2 + \left(\frac{dY_p}{dT}\right)^2\right)^{1/2} + 1 = 0 \qquad (13.1.1)$$

where $\beta = 0.44\rho V_0^2/\rho_p g d_p$ is the ratio of the initial turbulent drag to the weight. The corresponding initial conditions are

$$X_p(0) = 0, Y_p(0) = \frac{1}{Fr^2}, \left.\frac{dX_p}{dT}\right|_{T=0} = \cos\theta, \left.\frac{dY_p}{dT}\right|_{T=0} = \sin\theta \qquad (13.1.2)$$

where $Fr^2 = V_0^2/gh$ and Fr is the Froude number from Section 2.4.

Figure 13.5 shows trajectories for explosive seed dispersal comparing inviscid, $\beta = 0$, to turbulent, $\beta = 1$, results from two heights, $Fr = 1$, $Fr = 1/\sqrt{2}$, and with a launch angle $\theta = \pi/4$. The familiar parabolic shape for $\beta = 0$ is significantly modified with the turbulent drag, reducing the maximum height and the impact X-distance, the dispersal range. For this choice of parameters, the height advantage for dispersal range is apparent, but turbulent drag has reduced it from the inviscid case.

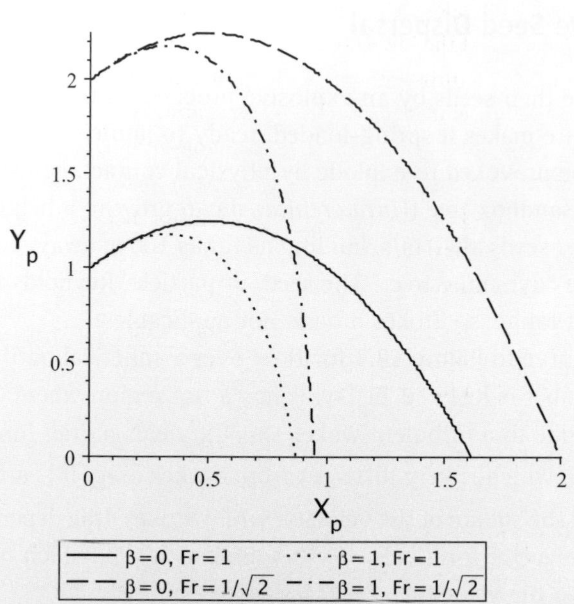

Figure 13.5 Trajectories for explosive seed dispersal at two different plant heights comparing inviscid, $\beta = 0$, to turbulent drag, $\beta = 1$, with launch angle $\theta = \pi/4$.

13.2 Analysis of Turbulent Flow

Reynolds' approach (Reynolds, 1895) to analyzing turbulent flows starts by representing the velocity vector and pressure as the sum of time-averaged and time-fluctuating components. Let's start with a flow which, other than the turbulent fluctuations, is steady. Then these dependent variables are given by

$$\underline{u}(\underline{x},t) = \bar{\underline{u}}(\underline{x}) + \underline{u}'(\underline{x},t) \quad p(\underline{x},t) = \bar{p}(\underline{x}) + p'(\underline{x},t) \tag{13.2.1}$$

Time averaging for a representative function, $f(\underline{x}, t)$, is defined as

$$\bar{f}(\underline{x}) = \frac{1}{T}\int_0^T f(\underline{x},t)\,dt \tag{13.2.2}$$

where the time interval for the averaging, T, is much longer than any oscillatory period of the turbulent fluctuations. Here, f is a velocity component or pressure. Clearly the time-average of the fluctuations are zero, i.e. $\overline{\underline{u}'} = 0$, $\overline{p'} = 0$.

The turbulent intensity, or level, is gauged by the dimensionless ratio of the root mean square of the velocity fluctuation to the steady velocity magnitude,

$$I = \left(\left(\overline{u'^2} + \overline{v'^2} + \overline{w'^2}\right)/3\right)^{1/2} / \left(\bar{\underline{u}} \cdot \bar{\underline{u}}\right)^{1/2} \tag{13.2.3}$$

For flow over an airplane or car, where the fluid (air) is otherwise stationary, the intensity is $I < 1\%$. This is viewed as the low turbulence level range. Though low intensity, the magnitude of the fluctuations can be considerable since the denominator of Eq. (13.2.3) is the average speed. The average speed of a commercial airplane is 600 mph. On the other hand, biological flows may have $1\% < I < 5\%$, which is the medium turbulence range. As an example, turbulent blood flow in the aorta is reasonably modeled by pipe flow, which is investigated in more detail below in following sections. A good approximation for I in turbulent pipe flow is given by $I = 0.16\mathrm{Re}_D^{-1/8}$. When $\mathrm{Re}_D = 4,000$, a value we will learn is at the low end of turbulent pipe Reynolds numbers, the turbulent intensity is $I \sim 0.05 = 5\%$ and the level decreases as Re_D is increased further.

Figure 13.6(a) shows an example of a single velocity component measurement u vs t at a fixed spatial location in a turbulent flow. There is a time-averaged value, $\bar{u}$, and an unsteady fluctuation, $u'(t)$. The sum is the measured velocity $u(t) = \bar{u} + u'(t)$. The scale indicates $\bar{u} = 100$ and $u' \leq 5$ so the turbulence intensity is $I < 5\%$. In Figure 13.6(b) a turbulent flow near a boundary demonstrates two instantaneous velocity profiles and the time-averaged velocity profile $\bar{u}(y)$.

Consider the Navier–Stokes equations in two dimensions,

$$\rho\left(\frac{\partial u}{\partial t}+u\frac{\partial u}{\partial x}+v\frac{\partial u}{\partial y}\right) = -\frac{\partial p}{\partial x}+\mu\left(\frac{\partial^2 u}{\partial x^2}+\frac{\partial^2 u}{\partial y^2}\right)$$
$$\rho\left(\frac{\partial v}{\partial t}+u\frac{\partial v}{\partial x}+v\frac{\partial v}{\partial y}\right) = -\frac{\partial p}{\partial y}+\mu\left(\frac{\partial^2 v}{\partial x^2}+\frac{\partial^2 v}{\partial y^2}\right) \tag{13.2.4}$$

where the subscripts indicate partial differentiation with respect to the named variable. Now insert Eqs. (13.2.1) into Eqs. (13.2.4) and take the time-average, noting that the time-average of the fluctuations are zero, and the time–average of products of fluctuations are non-zero in general, i.e. $\overline{u'} = 0$ and $\overline{v'} = 0$ but $\overline{u'v'} \neq 0$. A form like $\overline{u'v'}$ is the mathematical correlation between the two terms u' and v'. The result is

$$\rho\left(\bar{u}\frac{\partial \bar{u}}{\partial x}+\bar{v}\frac{\partial \bar{u}}{\partial y}\right) = -\frac{\partial \bar{p}}{\partial x}+\mu\left(\frac{\partial^2 \bar{u}}{\partial x^2}+\frac{\partial^2 \bar{u}}{\partial y^2}\right)-\rho\left(\overline{u'\frac{\partial u'}{\partial x}}+\overline{v'\frac{\partial u'}{\partial y}}\right)$$
$$\rho\left(\bar{u}\frac{\partial \bar{v}}{\partial x}+\bar{v}\frac{\partial \bar{v}}{\partial y}\right) = -\frac{\partial \bar{p}}{\partial y}+\mu\left(\frac{\partial^2 \bar{v}}{\partial x^2}+\frac{\partial^2 \bar{v}}{\partial y^2}\right)-\rho\left(\overline{u'\frac{\partial v'}{\partial x}}+\overline{v'\frac{\partial v'}{\partial y}}\right) \tag{13.2.5}$$

where we have employed the identity that the time-average of a spatial derivative is equal to the spatial derivative of the time-average, e.g. $\overline{\partial f/\partial x} = \partial \bar{f}/\partial x$. Note that we have shifted the averages of the fluctuation products to the right hand side. Performing the same process on the conservation of mass equation gives us

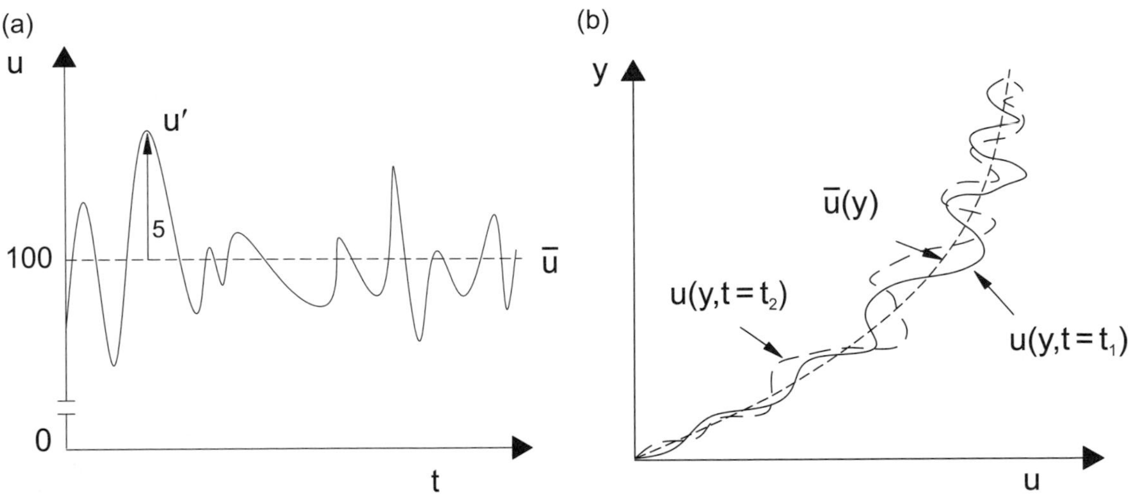

Figure 13.6 (a) Velocity measurement u vs t at a fixed location in a turbulent flow. Note that $u(t) = \bar{u} + u'(t)$ where $\bar{u}$ is the time-averaged velocity. (b) Turbulent flow near a boundary showing two instantaneous velocity samples, $u(y, t = t_1)$ and $u(y, t = t_2)$, and the time-averaged velocity profile $\bar{u}(y)$.

$$\frac{\partial \bar{u}}{\partial x}+\frac{\partial \bar{v}}{\partial y}+\frac{\partial u'}{\partial x}+\frac{\partial v'}{\partial y}=0 \tag{13.2.6}$$

If we time-average Eq. (13.2.6) we find that

$$\frac{\partial \bar{u}}{\partial x}+\frac{\partial \bar{v}}{\partial y}=0 \tag{13.2.7}$$

Inserting Eq. (13.2.7) into Eq. (13.2.6) tells us that the fluctuations satisfy the same form of mass conservation as the steady components

$$\frac{\partial u'}{\partial x}+\frac{\partial v'}{\partial y}=0 \tag{13.2.8}$$

Equations (13.2.5) can be rewritten by starting with the following identities before averaging

$$\begin{aligned}
u'\frac{\partial u'}{\partial x} &= \frac{\partial}{\partial x}(u'u') - u'\frac{\partial u'}{\partial x}\\
v'\frac{\partial u'}{\partial y} &= \frac{\partial}{\partial y}(u'v') - \frac{\partial v'}{\partial y}u'\\
u'\frac{\partial v'}{\partial x} &= \frac{\partial}{\partial x}(u'v') - v'\frac{\partial u'}{\partial x}\\
v'\frac{\partial v'}{\partial y} &= \frac{\partial}{\partial y}(v'v') - v'\frac{\partial v'}{\partial y}
\end{aligned} \tag{13.2.9}$$

Collecting the forms in Eq. (13.2.9) to correspond to the right hand side of Eq. (13.2.5) we find a simplification due to Eq. (13.2.8)

$$\begin{aligned}
u'\frac{\partial u'}{\partial x}+v'\frac{\partial u'}{\partial y} &= \frac{\partial}{\partial x}(u'u')+\frac{\partial}{\partial y}(u'v')-u'\cancel{\left(\frac{\partial u'}{\partial x}+\frac{\partial v'}{\partial y}\right)}\\
u'\frac{\partial v'}{\partial x}+v'\frac{\partial v'}{\partial y} &= \frac{\partial}{\partial x}(u'v')+\frac{\partial}{\partial y}(v'v')-v'\cancel{\left(\frac{\partial u'}{\partial x}+\frac{\partial v'}{\partial y}\right)}
\end{aligned} \tag{13.2.10}$$

Now take the time-averages of Eqs. (13.2.10) and insert into Eq. (13.2.5) to find the forms

$$\begin{aligned}
\rho\left(\bar{u}\frac{\partial \bar{u}}{\partial x}+\bar{v}\frac{\partial \bar{u}}{\partial y}\right) &= -\frac{\partial \bar{p}}{\partial x}+\frac{\partial}{\partial x}\left(\mu\frac{\partial \bar{u}}{\partial x}-\rho\overline{u'u'}\right)+\frac{\partial}{\partial y}\left(\mu\frac{\partial \bar{u}}{\partial y}-\rho\overline{u'v'}\right)\\
\rho\left(\bar{u}\frac{\partial \bar{v}}{\partial x}+\bar{v}\frac{\partial \bar{v}}{\partial y}\right) &= -\frac{\partial \bar{p}}{\partial y}+\frac{\partial}{\partial x}\left(\mu\frac{\partial \bar{v}}{\partial x}-\rho\overline{u'v'}\right)+\frac{\partial}{\partial y}\left(\mu\frac{\partial \bar{v}}{\partial y}-\rho\overline{v'v'}\right)
\end{aligned} \tag{13.2.11}$$

where we recognize that in Eq. (13.2.5) $\bar{u}_t = \bar{v}_t = 0$ by definition for a steady base flow. The nonlinear convective acceleration terms, multiplied by ρ, appear on the right hand side of Eq. (13.2.11) along with the viscous shear terms. A more compact form of Eq. (13.2.11) is

$$\rho\left(\underline{\bar{u}}\cdot\underline{\nabla}\right)\underline{\bar{u}} = -\bar{p}+\underline{\nabla}\cdot\underline{\underline{\tau}} \tag{13.2.12}$$

which has the form of the Cauchy momentum conservation balance, Eqs. (5.2.2) and (5.2.3). In the turbulent flow case, however, the extra stress tensor, $\underline{\underline{\tau}}$, is given by

$$\tau_{ij} = \bar{\tau}_{ij}+\tau'_{ij} \quad \bar{\tau}_{ij} = 2\mu\bar{E}_{ij} \quad \tau'_{ij} = -\rho\overline{u'_i u'_j} \tag{13.2.13}$$

where the first term is viscous stress tensor, $\bar{\tau}_{ij} = 2\mu\bar{E}_{ij}$, which has the same form for laminar flows. Here, μ is the viscosity and the average strain rate tensor is $\bar{E}_{ij} = (1/2)\left(\partial\bar{u}_i/\partial x_j + \partial\bar{u}_j/\partial x_i\right)$. The second term is $\tau'_{ij} = -\rho\overline{u'_i u'_j}$ and its components are known as the Reynolds stresses, due to the convective acceleration inertial terms. The system of averaged equations for conservation of mass (one component) and momentum (three components) provide four equations as we saw in laminar flow where we could then solve for the four unknowns, pressure and three components of velocity. However, now we have additional unknowns in the form of $\rho\overline{u'_i u'_j}$, which are called the "Reynolds stresses" or turbulent shear stresses, even though they originate from convective acceleration. Consequently the turbulence equations, as developed, do not have closure.

An example of a typical Reynolds stress, τ'_{ij} for a simple shear flow is shown in Figure 13.7, where the average velocity, $\bar{u}(y)$, has a positive slope, $d\bar{u}/dy > 0$. When a small parcel of fluid jumps vertically upward, $v' > 0$, it is coming from a slower fluid stream so reduces the velocity in its new location, $u' < 0$. However, if it jumps vertically downward, $v' < 0$, it will speed up the fluid there, $u' > 0$. In either case, we see that $u'v' < 0$, which also occurs when $d\bar{u}/dy < 0$. Clearly, then, $-\rho\overline{u'v'} > 0$ in Eq. (13.2.13), so the Reynolds stresses have the same sign as the viscous stresses.

Viewing this process in the context of the viscous terms has led to attempts at closure for the equations

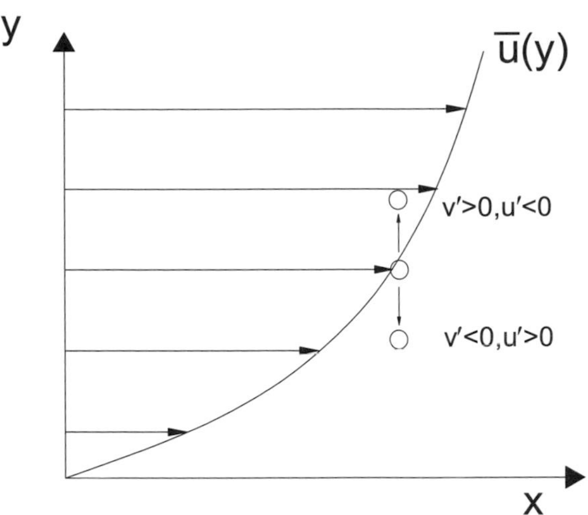

Figure 13.7 Example of Reynolds stress in a turbulent flow with average velocity profile $\bar{u}(y)$.

by introducing an "eddy viscosity," μ_t. To see its use, we separate out shear stresses into laminar and turbulent components for our simple flow,

$$\tau_{12} = \bar{\tau}_{12} + \tau'_{12} \tag{13.2.14}$$

The laminar stress is given by the Newtonian fluid assumption,

$$\bar{\tau}_{12} = \mu(d\bar{u}/dy) \tag{13.2.15}$$

However, the Reynolds stress is now represented by the assumed equivalence

$$\tau'_{12} = -\rho\overline{u'v'} = \mu_t(d\bar{u}/dy) \tag{13.2.16}$$

Unlike the viscosity, μ, which is a property of the fluid material, the turbulent or "eddy viscosity," μ_t, depends on the flow structure, so varies with position. For general turbulent flows, μ_t is much larger than μ, often several thousand times larger.

To model the eddy viscosity, Prandtl proposed his "mixing length theory" for turbulence, relying on an analogy to the mean free path in gas flow kinetics. The mixing length, ℓ, is the transverse distance a fluid parcel has to move so that its momentum changes to that of its new location. In our simple example ℓ is a differential distance in the y-direction, and the relationships for the velocity fluctuation components are

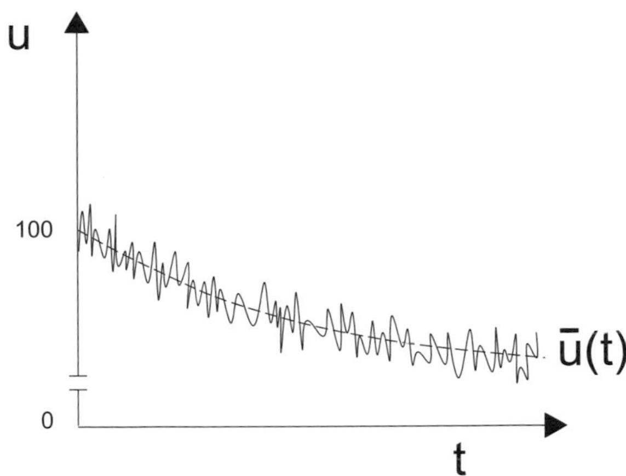

Figure 13.8 Turbulence for an unsteady background flow.

$$u' = \ell\frac{d\bar{u}}{dy}, v' = \ell\frac{d\bar{v}}{dy} \tag{13.2.17}$$

Now the turbulent shear stress is

$$\tau'_{12} = -\rho\overline{u'v'} = \rho\ell^2\left|\frac{d\bar{v}}{dy}\right|\frac{d\bar{u}}{dy} \tag{13.2.18}$$

which keeps the sign of the stress the same as the sign of the $d\bar{u}/dy$. Then for specific problems one must relate ℓ to characteristic length scales of the flow.

What about the case where the background flow is, itself, unsteady? In biofluid mechanics this is very much the normal situation. For example, blood flow is pulsatile with an underlying oscillation period of ~ 1 s in adults. The process is the same as above for steady flows, but the definition of time-averaging is broader,

$$\bar{f}(\underline{x}, t) = \frac{1}{T}\int_0^T f(\underline{x}, t + s)\, ds \tag{13.2.19}$$

Now the averaging interval, T, must be long compared to the turbulent fluctuation period but short compared to a characteristic time of the unsteady background flow. Under these circumstances, the governing equations in Eq. (13.2.12) become unsteady

$$\rho\left(\underline{\bar{u}}_t + (\underline{\bar{u}}\cdot\underline{\nabla})\underline{\bar{u}}\right) = -\underline{\nabla}\bar{p} + \underline{\nabla}\cdot\underline{\underline{\tau}} \tag{13.2.20}$$

An example of an unsteady background flow which is turbulent is shown in Figure 13.8. Equations (13.2.11) are called the Reynolds-averaged Navier–Stokes (RANS) equations. There are a number of computational methods for solving turbulent flows. The $k - \omega$ method (Wilcox, 2008) is based on RANS, but provides closure through equations that determine the eddy viscosity. As computer power has increased, it has become possible to solve the Navier–Stokes equations directly for turbulent flows, including large eddy simulation (LES) (Deardorff, 1970; Lesieur and Metais, 1996) and direct numerical simulation (DNS) (Kim et al., 1987).

13.3 Fully Developed Turbulent Pipe Flow

Figure 13.9 shows flow visualization of turbulent pipe flow for $2,000 \leq \mathrm{Re_D} \leq 20{,}000$ using a commercial product, Mearlmaid AA Pearlessence, which is used in the manufacture of glossy lipstick and eye shadow. It breaks up into small platelets when mixed with water into roughly $35 \times 10 \times 1$-micron-thick ellipsoids. A vertical 2 mm thick light sheet illuminates the flow in the 20 mm diameter pipe. These images were obtained by Dr. Jorge Peixinho as a postdoctoral student of Professor Dr. Tom Mullin at the University of Manchester where Osborne Reynolds began turbulent pipe flow studies (Personal communication, Tom Mullin). It is readily seen from the images that the scale of the eddies decreases as $\mathrm{Re_D}$ increases.

In the human body turbulent pipe flow occurs in the largest blood vessels, like the ascending aorta, which exits directly from the heart's left ventricle. For an adult ascending aorta diameter assume a value of 3.0 cm (Redheuil et al., 2011). A typical cardiac output at rest is 5 L/min. Assuming blood viscosity to be 0.03 poise, the Reynolds number based on cardiac output is $\mathrm{Re_D} = 1200$. This is an average value over time. However, since the flow is pulsatile, the Reynolds number varies with the unsteady velocity component. The peak value can be in the range $5,000 \leq \mathrm{Re_D} \leq 10{,}000$ (Stein and Sabbah, 1976) at normal resting conditions. Turbulence influences the behavior and metabolism of the endothelial cells lining the blood vessel wall (Davies et al., 1986). Vascular diseases of the aorta include aneurysm, which is a dilation of the vessel diameter from weakened walls, stenosis and coarctation, which are narrowings, and dissection, which is a slow rupture process, see Figure 13.24. These situations involve the interplay of the biology and fluid mechanics (Lasheras, 2007; Tse et al., 2011), whether turbulent or not. In addition, turbulence can be a source of damage to red blood cells due to the large turbulent shear

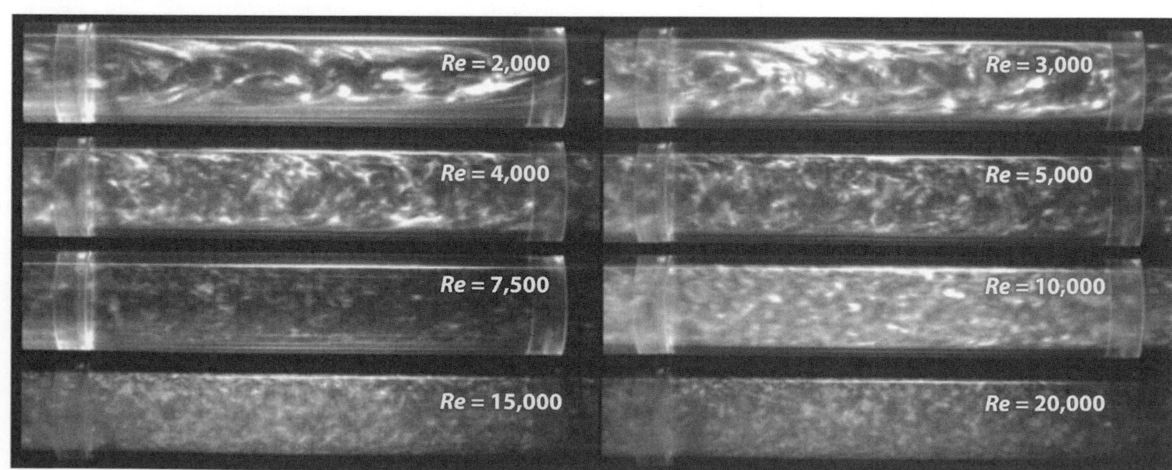

Figure 13.9 Turbulent pipe flow for increasing Reynolds number, $2,000 \leq \mathrm{Re_D} \leq 20,000$. Note that the turbulent eddy scale becomes finer as Re increases.

stresses causing them to rupture or "lyse" (Sallam and Hwang, 1984). Hemolysis from turbulence can also occur when blood flows through the valves of the heart, especially if they are artificial valves. Whenever red blood cells rupture, there are signals to the clotting mechanisms of the blood, which is an additional hazard. In large mammals the range of Re_D for vascular internal flows can extend much higher. For example, an elephant has an aortic diameter of 9 cm (Milnor, 1989) with a cardiac output of 67 L/min (Li, 1996). So its aortic Reynolds number based on the cardiac output is $Re_D \sim 5{,}000$. Since the human peak value is approximately three times the average, we could expect the elephant peak Reynolds number to easily reach $Re_D \sim 15{,}000$ at rest.

Turbulence can also occur in the large airways of the lung, say the trachea. Assuming a tracheal diameter of 1.8 cm, air kinematic viscosity of 0.15 poise and a flow rate of 1 L/s gives $Re_D \sim 4{,}500$. Exercise increases both cardiac output and respiratory ventilation so Re increases for both settings increasing the turbulence level. The critical value of Re that separates laminar from turbulent flow in a long, smooth tube is $Re_D = 2{,}300$. This is an idealized situation, however, with external disturbances minimized. In general, fluid dynamicists think of laminar flows for $Re_D < 2{,}000$ and turbulent flow for $Re_D > 4{,}000$ and transitional flow for $2{,}000 < Re_D < 4{,}000$. In the transitional region the flow can switch back and forth between laminar and turbulent. In biological applications, for example, the onset of turbulence can at even lower Re since the tubes have varying geometry and are seldom long and straight without external disturbances. In addition to respiration, there are instances of very high flow rates during coughs where flows can approach 10 L/s (Leiner et al., 1966). So a transient tracheal Reynolds number would be $Re_D = 45{,}000$ during a cough.

In terms of fully developed flows, we have already discussed the well-known Poiseuille law, which relates the pressure drop Δp_P to flow rate Q as $\Delta p_P = \left(128\mu L/\pi d^4\right)Q$ where μ is the fluid viscosity, L is the distance in z over which the pressure drop pertains, and d is the tube diameter. The dimensionless Darcy (or Darcy–Weisbach) friction factor, f, is a common way to describe fully developed tube flows ranging from laminar to turbulent regimes. It is defined as a dimensionless coefficient that relates the viscous pressure drop per pipe length, $\Delta p/L$, to the fluid kinetic energy in the form

Box 13.3 The Experiments of Osborne Reynolds

Figure 13.10(a) shows the famous experiments by Osborne Reynolds on transition to turbulence for pipe flow in 1883 (Reynolds, 1883). The apparatus consists of a tank of water which surrounds and flows into a clear tube due to gravity and exits below the level of the floor. There is a downstream valve to control the flow speed. An upstream source of dye from a bottle feeds into the tube entrance to tag the fluid particles in a streakline, which we studied in Section 3.7. There is a float on the water surface attached to a dial which indicates changes in its height. Knowing the cross-sectional area of the tank, the dial tracks changes in fluid volume. Comparing two height readings and the time interval between gives an overall flow rate and hence an average tube flow velocity.

The flow visualization results from the dye are sketched in Figure 13.10(b). For values of Re_D below a critical value, the top figure shows the dye streak as a thin line extending down the tube. Recall that for steady flow the streakline from a stationary source is a streamline. So

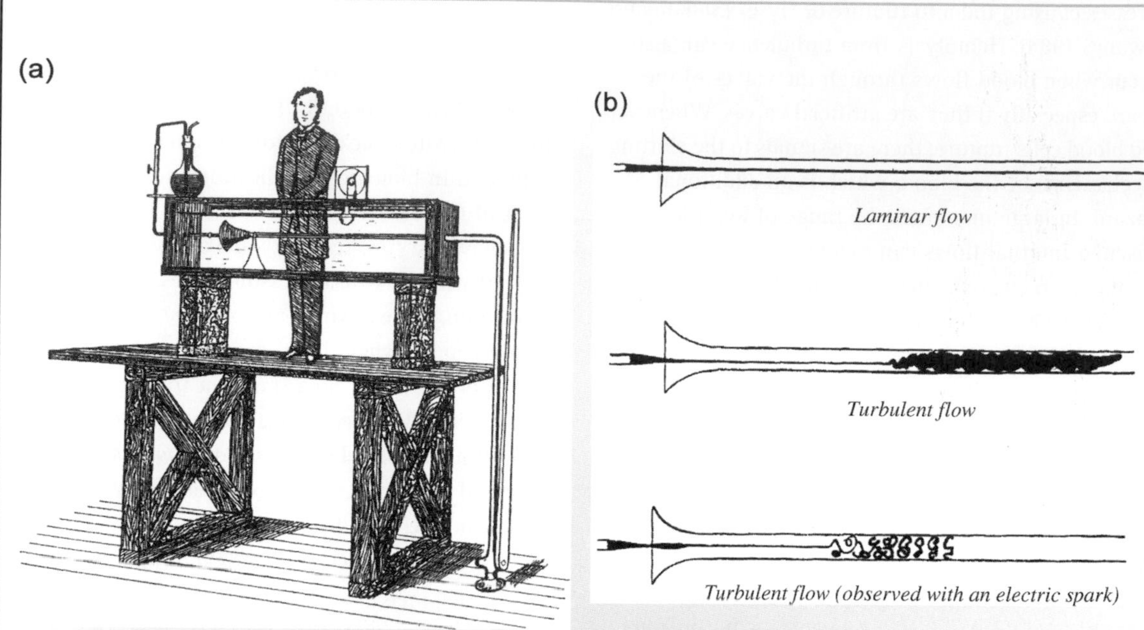

Figure 13.10 Osborne Reynolds' famous experiment on transition to turbulence for flow in a tube: (a) experimental apparatus; (b) observation results for laminar and turbulent flow. Illustration is taken from fluid dynamicist Osborne Reynolds' influential paper from 1883, "An experimental investigation of the circumstances which determine whether the motion of water in parallel channels shall be direct or sinuous and of the law of resistance in parallel channels." Water flows from the tank near the experimenter down to below the ground, through a transparent tube; and dye is injected in the middle of the flow. The turbulent or laminar nature of the flow can therefore be observed precisely.

the middle figure is above a critical value of Re and, at a distance downstream of the entrance, the dye streak is mixed vigorously into what appears to be a cloud. This is turbulent flow. To gain more visual detail, Reynolds used a spark to illuminate the cloud at a time instant, visually freezing the motion, which is sketched in the third figure showing sinuous oscillations.

Reynolds viewed Re_D as a ratio of the average fluid speed, U, to the quantity ν/D, which is a shear-related speed. It is clear from his publication (Reynolds, 1883) that the then-known Navier–Stokes equations were a helpful guide in his estimating the size and relationship of the dimensional parameters to one another. The insight to form the dimensionless parameter Re_D in such a complex setting as fluid mechanics led to studying "dynamic similitude," which involves the ratio of governing forces. A generally accepted result is that tube flow becomes turbulent for $Re_D > Re_{D\ crit}$ where $Re_{D\ crit} = 2,300$; however, this critical value is sensitive to the level of disturbances in the inflow fluid. When those disturbances are carefully reduced, as in Barnes and Coker (1904) and Schiller (1922), the critical Re_D can be much larger. Indeed, values of $Re_{D\ crit} \sim 20{,}000$ were found in Schiller (1922), for example. Reynolds' experiments on the "law of resistance" also had "geometric similitude" since he used long, 16 ft, circular pipes of diameter 1/4 and 1/2 inch. This made their length to diameter ratios equal to 768 and 384 respectively, enough to minimize any entrance effects on the overall pressure drop and flow results.

$$f = \frac{\Delta p}{\frac{1}{2}\rho U^2}\frac{D}{L} \qquad (13.3.1)$$

For fully developed flow the assumption is that $\Delta p/L$ is constant along the tube length. However, f, in general, depends on Re_D, and also the roughness of the tube surface given as a ratio to the roughness height, ε, to tube diameter, ε/D.

For example, Poiseuille flow, which is laminar, has the friction factor, f_P,

$$f_P = \frac{\Delta p_P}{\frac{1}{2}\rho U^2}\frac{D}{L} = \frac{64}{Re_D} \qquad (13.3.2)$$

where we have used the Poiseuille law, $\Delta p_p = QL128\mu/\pi D^4$, and D is the tube diameter. Since the flow rate, Q, is related to the average velocity, U, by $Q = \pi D^2 U/4$, Eq. (13.3.2) is readily obtained. While convenient, the form in Eq. (13.3.2) is somewhat misleading since the "dependence" on density, ρ, has been introduced by the scaling only. Poiseuille flow does not depend on ρ. We now turn our focus to turbulent pipe flows, a subject with a long history of contributions by distinguished figures in the field of fluid mechanics. Getting water into town and, later, oil to the refinery, drove much of turbulent pipe flow investigations.

For turbulent flow in a straight tube, the pressure drop depends on whether the tube walls are smooth or have roughness. Roughness is gauged by the height of bumps on the inner surface, ε. The classic presentation of experimental data for fully developed pipe flows is called the Moody diagram after Lewis Ferry Moody (1880–1953) (Moody, 1944) of Princeton University, who built on the earlier work of Hunter Rouse (1906–1996) (Rouse, 1943) and R. J. S. Pigott (Pigott, 1933).

An example is shown in Figure 13.11, a log–log plot where the horizontal axis is the Reynolds number, Re_D, and the vertical axis is the friction factor, f. This is the famous Moody diagram (Moody,

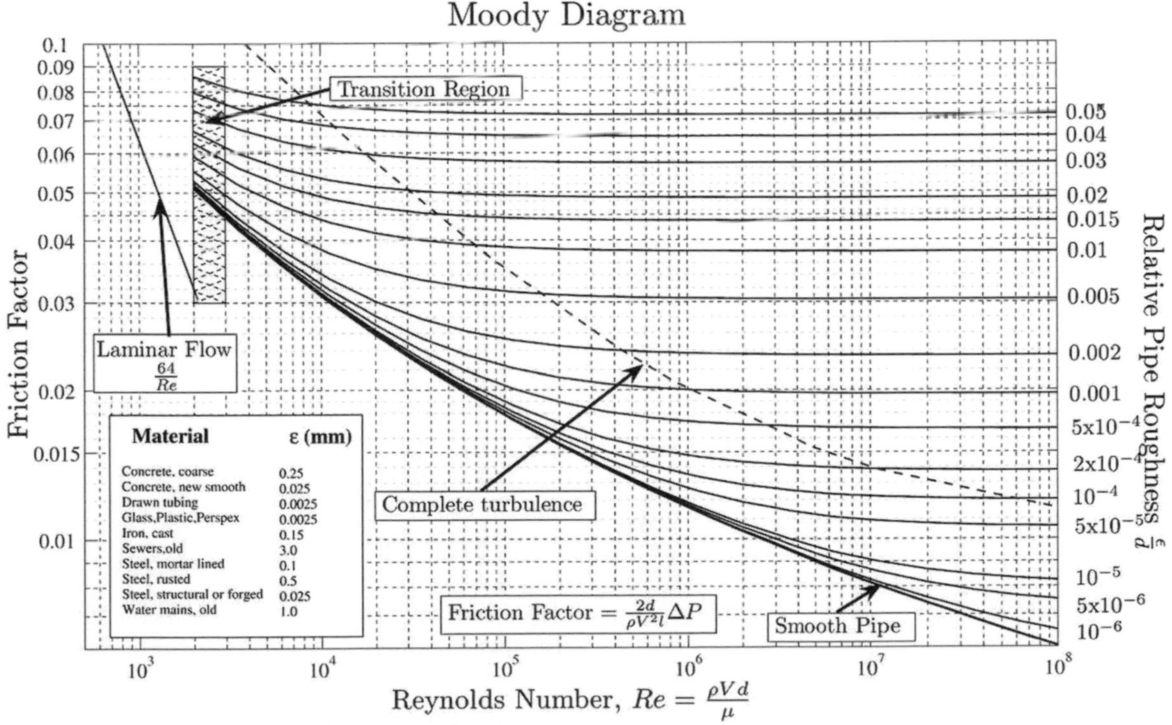

Figure 13.11 Moody diagram.

1944). Curves of constant relative roughness, ε/D, are drawn and labeled for values as small as 10^{-6}, which is very smooth, to 0.05, which is quite rough. The insert gives values of ε in mm for different pipe materials ranging from 0.0025 mm for glass to 3.0 mm for old sewers. For $Re_D < 2{,}000$ note the laminar Poiseuille flow result $f = 64/Re_D$ which has a straight slope of −1. When Re_D reaches the transition region just above 2,000, there is a dramatic increase in the value of f. This is the well-known effect of turbulence in pipe flow, that it significantly increases the required driving pressure for the same flow rate.

For large enough Re_D the curves of the Moody Diagram, Figure 13.11, level off to have zero slope, so that f asymptotes to a constant value. In Figure 13.11 that behavior is demarcated by the dashed line labeled "complete turbulence" where the pressure gradient is $\Delta p/L \sim \rho U^2/D$, which loses its dependence on viscosity, μ.

The Moody diagram is a succinct presentation of pipe flow behavior; however, a mathematical fit to that data is especially valuable. One of the most popular is the Colebrook or Colebrook–White equation (Colebrook and White, 1937; Colebrook, 1939)

$$\frac{1}{\sqrt{f}} = -2.0 \log_{10}\left(\frac{1}{3.7}\frac{\varepsilon}{D} + \frac{2.51}{Re_D\sqrt{f}}\right) \quad Re_D > 4{,}000 \tag{13.3.3}$$

Equation (13.3.3) is an implicit equation for f given a turbulent regime where $Re_D > 4{,}000$ with relative roughness ε/D. We cannot isolate f on one side of the equation, so it must be solved numerically. For a smooth pipe $\varepsilon = 0$ and Eq. (13.3.3) simplifies to

$$\frac{1}{\sqrt{f}} = -2.0 \log_{10}\left(\frac{2.51}{Re_D\sqrt{f}}\right) = 2\log_{10}\left(Re_D\sqrt{f}\right) - 0.8 \quad Re_D > 4{,}000 \text{ smooth} \tag{13.3.4}$$

which recovers Prandtl's universal law of friction for smooth pipes (Prandtl, 1933) shown on the right hand side.

An explicit equation which is also a good fit is the Swamee–Jain equation (Swamee and Jain, 1976)

$$f = 0.25\left[\log_{10}\left(\frac{1}{3.7}\frac{\varepsilon}{D} + \frac{5.74}{Re_D^{0.9}}\right)\right]^{-2} \tag{13.3.5}$$

Now f is isolated to the left hand side. For a smooth pipe, $\varepsilon = 0$, Eq. (13.3.5) simplifies to

$$f = 0.25\left[\log_{10}\left(\frac{5.74}{Re_D^{0.9}}\right)\right]^{-2} \quad \text{smooth} \tag{13.3.6}$$

The smooth pipe asymptote is shown in the Moody diagram of Figure 13.11. The slope is non-constant; however, in an earlier attempt to fit experimental data, Blasius proposed that there is a constant slope of −1/4 in his formula

$$f = 0.3164\, Re_D^{-1/4} \tag{13.3.7}$$

Equation (13.3.7) is known as the "Blasius formula" and it holds for $Re_D < 100{,}000$. Using Eq. (13.3.6) for a smooth pipe, a straight line in the log–log plot connecting the two points $Re_D = 4{,}000$, $f(Re_D = 4{,}000) = 0.04$ and $Re_D = 30{,}000$, $f(Re_D = 30{,}000) = 0.023$, has the slope

$$\frac{\log_{10}(f(\varepsilon=0, Re_D=4{,}000)) - \log_{10}(f(\varepsilon=0, Re_D=30{,}000))}{\log_{10}(Re_D=4{,}000) - \log_{10}(Re_D=30{,}000)} = -0.25 \tag{13.3.8}$$

So the Blasius formula covers an important Re_D range for biomedical applications.

The pressure drop is related to the wall shear as shown in Figure 13.12 where a control volume is used to balance the z-direction forces of a section of pipe, similar to Figure 10.5. Under those circumstances the force balance is $(P_1 - P_2)\pi a^2 = \tau_w 2\pi a L$ so that we can relate τ_w to f as

$$f = \frac{(P_1 - P_2)}{\frac{1}{2}\rho U^2}\frac{D}{L} = \frac{8\tau_w}{\rho U^2} \tag{13.3.9}$$

We note here that another popular friction factor is the Fanning friction factor, defined as the ratio $f_{Fanning} = \tau_w/\frac{1}{2}\rho U^2$, which we see from Eq. (13.3.9) is ¼ the value of the Darcy friction factor, $f_{Fanning} = f/4$.

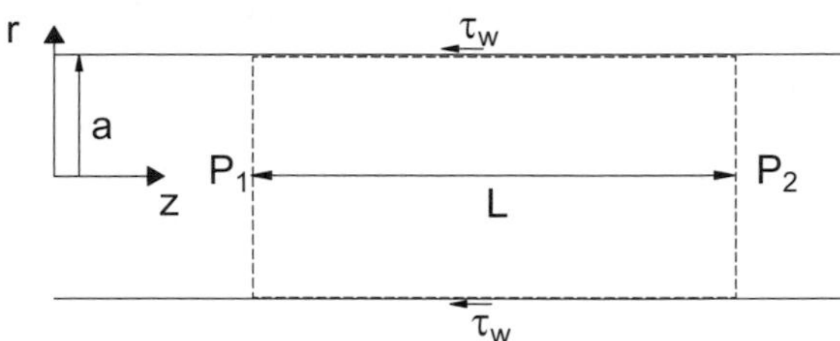

Figure 13.12 Control volume showing the balance forces in the z-direction including the pressures and wall shear τ_w.

One of the reasons for the sudden increase in pressure drop is that the lateral mixing of momentum, the turbulent shear stresses of Eq. (13.2.13), flattens the time-averaged velocity profile from parabolic to one that is blunt in the central core with thin layers near the wall where viscous dissipation is much higher. We can see this by using a popular curve fit to the average velocity profile for turbulent pipe flow called the power-law expression

$$\frac{u}{u_{max}} = \left(1 - \frac{r}{a}\right)^{1/n} \tag{13.3.10}$$

where n = 7 is a good fit for many applications and a is the tube radius. The average flow velocity of the profile in Eq. (13.3.10) is given by

$$\begin{aligned} u_{avg} &= \frac{1}{\pi a^2}\int_0^{2\pi}\int_0^{a} u_{max}\left(1 - \frac{r}{a}\right)^{1/n} r\,dr\,d\theta \\ &= 2u_{max}\frac{n^2}{(1+n)(1+2n)} \end{aligned} \tag{13.3.11}$$

Plots of u/u_{avg} for the power-law profiles are shown in Figure 13.13 along with Poiseuille flow. Notice that the profile becomes much more blunt with turbulence compared to the parabolic profile of laminar Poiseuille flow. As we learned in Section 7.1 the maximum velocity at r = 0 is twice the average velocity in Poiseuille flow, i.e. $u_{max} = 2\,u_{avg}$. For fully turbulent pipe flow that ratio is 1.22 for n = 7, i.e. $u_{max} = 1.22\,u_{avg}$. The thin layers near the wall are where high shear stresses increase the flow resistance. Equation (13.3.10), while useful, does not model the flow that is adjacent to the wall.

As we explored in Chapter 12, there are additional effects of axial pipe curvature for fully developed

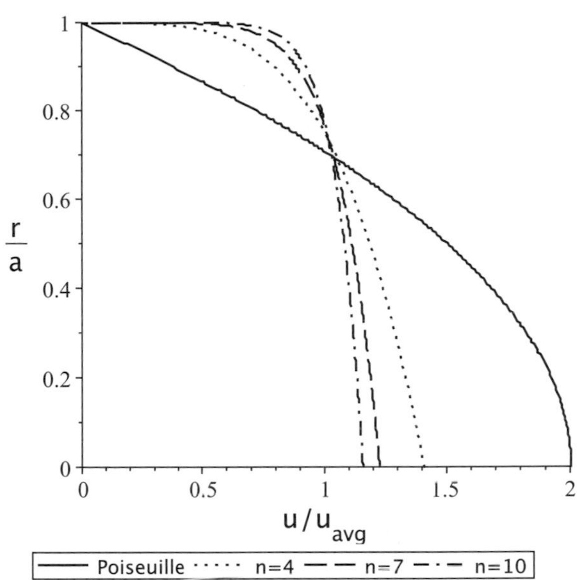

Figure 13.13 u/u_{avg} vs r/a for Poiseuille flow and for turbulent flow using power-law velocity profiles.

turbulent flow in curved tubes. The friction factor for turbulent flow in a curved tube was found by White (1932) and given as a dimensionless ratio

$$\frac{f_C}{f_0} = 1 + 0.075\text{Re}^{1/4}\left(\frac{d}{2R_{ax}}\right)^{1/2} \tag{13.3.12}$$

Here, f_0 is the friction factor for turbulent flow in a straight smooth tube given by Eq. (13.3.4). Using values of Re = 4,000 and $d/2R_{ax} = 0.5$ for the adult aorta during systole, Eq. (13.3.12) yields $f_C/f_0 \sim 1.4$. So for turbulent flow the curve increases the required driving pressure by about 40%. Turbulent flow also has entrance effects which, as we saw in Chapter 12, can be initially explored by studying flow over a flat plate.

In biomedical applications the base flow is often unsteady, for example in respiration where oscillatory flow in a tube was central to the discussions in Section 8.5 and included normal ventilation as well as high frequency ventilation and gas exchange, see Section 8.4. As the fluid oscillates back and forth it goes through acceleration and deceleration phases, and turbulence can be generated, especially during deceleration. Figure 13.14 shows Laser Doppler Velocimetry measurements of oscillatory tube flow.

Figure 13.14(a) shows laminar flow during the full cycle; however, Figure 13.14(b) shows turbulence during the deceleration phases. The two figures have the same dimensionless frequency, $\alpha = 28$, where α is the Womersley parameter discussed in Sections 8.3–8.5. They also have the same dimensionless amplitude, $A = V_T/\pi a^3$, where V_T is the tidal volume. Alternatively, two independent dimensionless groups are A and the Reynolds number based on the Stokes layer thickness, $Re_\delta = U\delta_s/\nu$, where U is the peak cross-sectional averaged velocity. The Stokes layer, $\delta_s = (2\nu/\omega)^{1/2}$, is discussed in Section 8.2, and $Re_\delta = 854$ for both Figure 13.14(a, b). The difference is that Figure 13.14(a) is measured at dimensionless radial position $r/a = 0.39$, while Figure 13.14(b) is much closer to the wall at $r/a = 0.88$, where a is the tube radius (Eckmann and Grotberg, 1991). As we will see in Sections 13.4 and 13.5, the proximity of the wall promotes turbulence.

13.4 Wall Bounded Turbulent Flow

For fully developed turbulent pipe flow, the $n = 1/7$ power-law velocity profile often fits well far enough from the pipe wall, see Eq. (13.3.10). Here we examine this near-wall region for fully developed pipe flow and gain some insights for developing turbulent boundary layer flow (discussed in Section 13.5).

Important features of the velocity profile structure near a wall take place over very short distances perpendicular to it. It is easier to see the details by switching from the linear spatial scale to a logarithmic scale. Figure 13.15 shows three regions near the wall: the viscous sublayer, the buffer layer and the logarithmic layer. The viscous sublayer remains laminar since the flow there is dominated by molecular viscosity. It is considered linear in the variable y where $y = 0$ on the wall. Then the equation for the wall shear stress, τ_w, in this local coordinate system is

$$\tau_w = \mu\frac{du}{dy} = \mu\frac{u}{y} \quad \text{or} \quad \frac{\tau_w}{\rho} = \nu\frac{u}{y} \tag{13.4.1}$$

The term τ_w/ρ in Eq. (13.4.1) appears often in turbulence analysis. Its square root has the dimensions of velocity and that form is called the "friction velocity," u_*, where $u_* = \sqrt{\tau_w/\rho}$. The wall shear is often represented in dimensionless form as a friction coefficient, c_f, as we saw in Eq. (12.1.19) when discussing laminar boundary layers. It is defined by

$$c_f = \frac{\tau_w}{\frac{1}{2}\rho U^2} = 2\frac{u_*^2}{U^2} \tag{13.4.2}$$

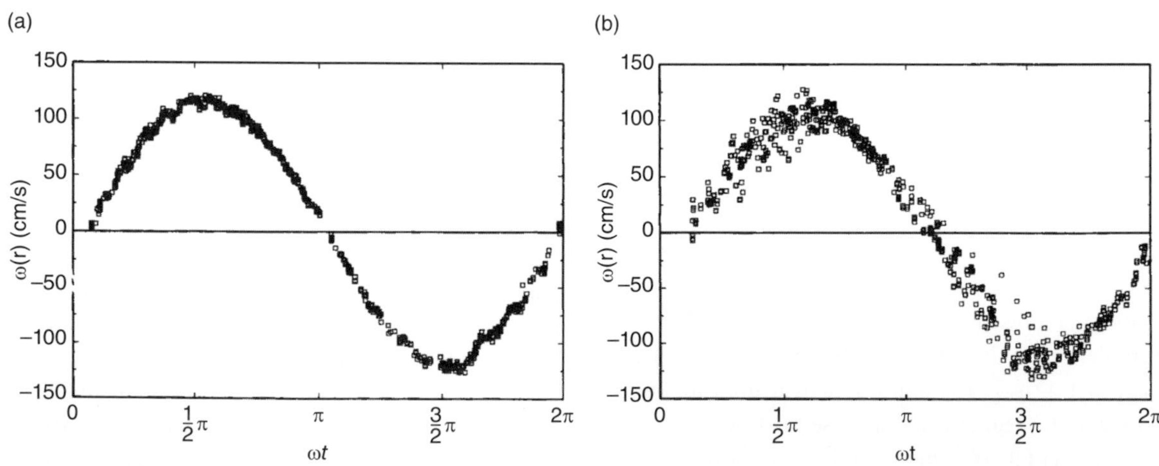

Figure 13.14 Velocity measurements for oscillatory tube flow using Laser Doppler Velocimetry for $\alpha = 28$, $A = 21.6$ and $Re_\delta = 854$. (a) Laminar flow at dimensionless radial position $r/a = 0.39$. (b) Turbulent flow during deceleration at dimensionless radial position $r/a = 0.88$. Dimensionless Stokes layer thickness is $\delta/a = 0.05$ for both (Eckmann and Grotberg, 1991).

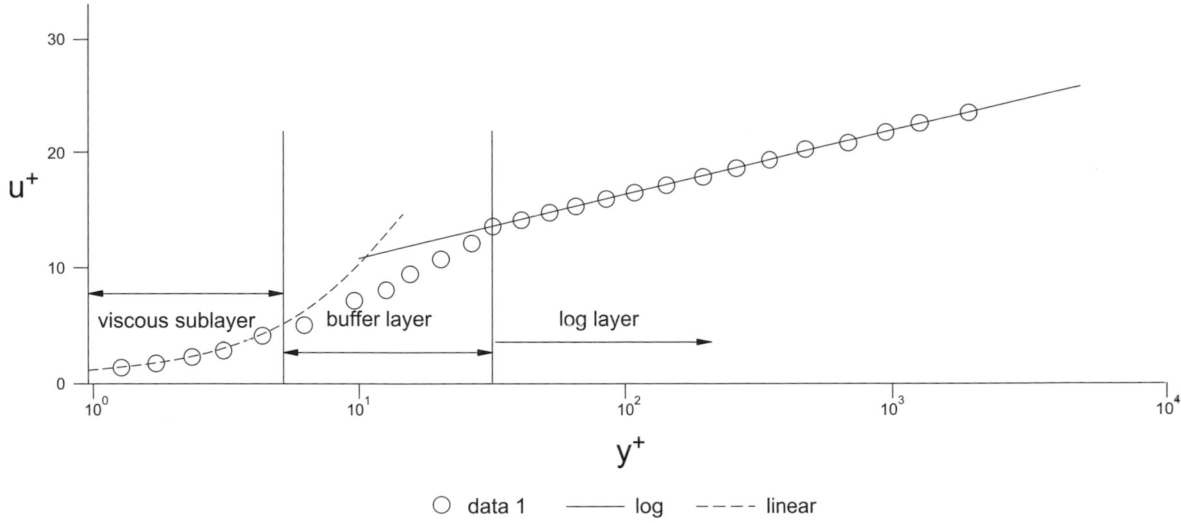

Figure 13.15 Semilog plot of representative data for fully developed turbulent pipe flow near the wall. Velocities for a pipe show the viscous sublayer, $0 \leq y^+ \leq 5$, the buffer layer, $5 \leq y^+ \leq 30$, and the logarithmic layer, $y^+ \geq 30$.

where U is a measure of the free stream or average velocity. Using this definition of u_* allows us to present the wall shear equation of Eq. (13.4.1) in dimensionless form as

$$\frac{u}{u_*} = \frac{yu_*}{\nu} \tag{13.4.3}$$

Equation (13.4.3) is dimensionless and we note that the right hand side has the form of a Reynolds number with u_* as the velocity and y as the length scale. It holds in the range $0 \leq yu_*/\nu \leq 5$ for smooth surfaces. The ratio ν/u_* in Eq. (13.4.3) has units of length and is called the "viscous length." We use it here as a scale for y, such that $y^+ = yu_*/\nu$. The dimensionless velocity ratio is symbolized as $u^+ = u/u_*$. A shorthand version of Eq. (13.4.3) is

$$u^+ = y^+ \quad 0 \leq y^+ \leq 5 \tag{13.4.4}$$

In Figure 13.15, Eq. (13.4.4) appears as a curved line since it is plotted on a semilog axis. The representative data fit well to this curve.

Equation (13.4.4) is the manifestation of dimensional analysis that we studied in Chapter 2. We expect the dimensional velocity u, to depend on the dimensional parameters τ_w, ρ, μ, y, such that $u = f_1(\tau_w, \rho, \mu, y)$. From the scalings this translates to the dimensionless relationship $u^+ = \phi_1(y^+)$ where $\phi_1(y^+) = y^+$, as in Eq. (13.4.4).

For the region of the flow where, approximately, $y^+ \geq 30$, a different equation fits the shown representative data. The dimensional analysis persists, telling us that $u^+ = \phi_2(y^+)$; however, it turns out that the best fit to data in this region is when ϕ_2 is a logarithmic relationship

$$u^+ = \frac{1}{\kappa}\ln(y^+) + B = 5.6\log_{10}(y^+) + 5.0 \tag{13.4.5}$$

where the conversion factor to $\log_{10}$ is $\ln(10) = 2.303$. A "universal" velocity law like Eq. (13.4.5) was first derived by Prandtl (1927) and further developed by von Kármán (1930). It is known as the "logarithmic law." The constant κ is known as the von Kármán constant in his honor. The fitting constants κ and B have been evaluated from experimental data and are approximately $\kappa = 0.41$ and $B = 5.0$. Slightly different values are also used. Figure 13.15 shows this log layer as a straight line for the semilog plot.

Figure 13.15 shows that between the linear sublayer and the log layer is the "buffer layer" for $5 \leq y^+ \leq 30$ where neither the law of the wall nor the logarithmic law fit the data.

Using Eq. (13.3.7) in Eq. (13.3.9) makes the wall shear stress for fully developed, turbulent pipe flow

$$\tau_w = \frac{f}{8}\rho U^2 = 0.03955\,\rho U^2 Re_D^{-1/4} \tag{13.4.6}$$

This value of τ_w makes the friction velocity

$$u_* = 0.2URe_D^{-1/8} \tag{13.4.7}$$

With this expression for τ_w we are now able to estimate the viscous sublayer thickness

$$\delta_{sublayer} = 5\nu/u_* = 25\,DRe_D^{-7/8} \tag{13.4.8}$$

For example, choose $Re_D = 5{,}000$ in Eq. (13.4.8). The ratio of the sublayer thickness to the tube diameter is $\delta_{sublayer}/D = 25/(5{,}000)^{7/8} = 0.0145$. For a 2 cm diameter tube, say the aorta or trachea, the sublayer thickness is $\delta_{sublayer} = 0.029$ cm.

Expanding Eq. (13.4.6) and using the definition of u_* gives us $\tau_w = 0.03955\rho U^{7/4}\nu^{1/4}D^{-1/4} = \rho u_*^2$. Let the pipe diameter be twice its radius, $D = 2a$, and separate $u_*^2 = u_*^{1/4}u_*^{7/4}$ so that the expression becomes $0.03325\,U^{7/4}\nu^{1/4}a^{-1/4} = u_*^{1/4}u_*^{7/4}$. Then collect terms to yield $(U/u_*)^{7/4} = (u_* a/\nu)^{1/4}/0.03325$, or take the 4/7th power of both sides, to give the form

$$\frac{U}{u_*} = 6.99\left(\frac{u_* a}{\nu}\right)^{1/7} \quad \rightarrow \quad \frac{u(y=a)}{u_*} = 8.53\left(\frac{u_* a}{\nu}\right)^{1/7} \tag{13.4.9}$$

where we have used our calculation for $n = 7$ that $u_{max} = u(y = a) = 1.22U$. Assuming Eq. (13.4.9) is true for any value of $y = a - r$, not just for $y = a$ at $r = 0$, then

$$\frac{u(r)}{u_*} = 8.53\left(\frac{u_* a}{\nu}\right)^{1/7}\left(1-\frac{r}{a}\right)^{1/7} \tag{13.4.10}$$

which recovers the form of the 1/7th power-law velocity profile from Eq. (13.3.10). So the Blasius formula for f, Eq. (13.3.7), is consistent with the 1/7th power-law profile.

We can rewrite Eq. (13.4.5) in new dimensionless variables to see better how the log law performs when extended across the entire cross-section of the pipe. Changing variables and letting $y = a - r$, gives us y^+, u^+ in terms of r and u,

$$y^+ = \frac{yu_*}{\nu} = 0.1(1-r/a)Re_D^{7/8},\; u^+ = \frac{u}{u_*} = 5\frac{u}{U}Re_D^{1/8} \tag{13.4.11}$$

Multiplying Eq. (13.4.5) by u_*/u_{max} and substituting for u_* from Eq. (13.4.7) and y^+ from Eqs. (13.4.11) gives us

$$\frac{u}{u_{max}} = \frac{0.2URe^{-1/8}}{u_{max}}\left(5.6\log_{10}\left((1-r/a)0.1Re^{7/8}\right)+5.0\right) \tag{13.4.12}$$

where the maximum velocity is $u_{max} = u(r = 0)$ and varies with the Reynolds number. Now the dimensionless spatial variable is r/a, rather than y^+, and the dimensionless velocity is u/u_{max}, rather than u^+. Equation (13.4.12) is graphed in Figure 13.16 for $Re_D = 5{,}000$ and 10,000 along with the power-law profile for $n = 7$ from Eq. (13.3.10). Note that the profiles are very similar and often the log-law profile is extended across the entire pipe cross-section for a reasonable representation of the velocities.

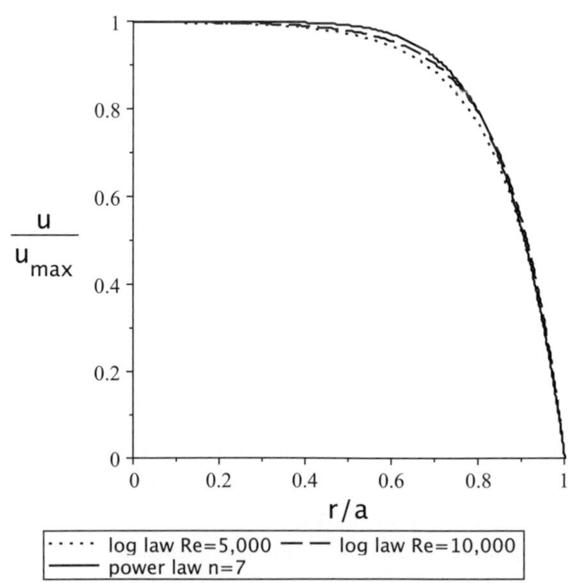

Figure 13.16 Velocity profile u/u_{max} vs r/a comparing log law to $n = 7$ power law for $Re_D = 5{,}000$ and 10,000.

13.5 Turbulent Boundary Layer on a Flat Plate

Turbulence in a developing boundary layer imposes considerable influence in biofluid mechanics, such as the fluid drag on wings and fins and the pressure drop of developing flows in tubes like blood vessels and airways. In Chapter 12 we investigated laminar flow over a flat plate and the development of the boundary layer thickness from the leading edge ($x = 0$) as $\delta(x) = 5.0(\nu x/U)^{1/2}$, see Figure 12.2 Now we consider that the incoming uniform flow with speed U is turbulent, so that U is the time-averaged uniform velocity.

Figure 13.17 compares representative laminar and turbulent velocity profiles. The vertical distance is in units of y/δ and the velocities are given as ratios u/U. The laminar profile is parabolic, which is a good approximation to the Blasius profile. Note the very different structure of the turbulent boundary layer by comparison. The slope of the velocity profile, $d(u/U)/d(y/\delta)$, at the wall is much larger, or $d(y/\delta)/d(u/U)$ is much smaller, for the turbulent boundary layer creating significantly larger wall shear

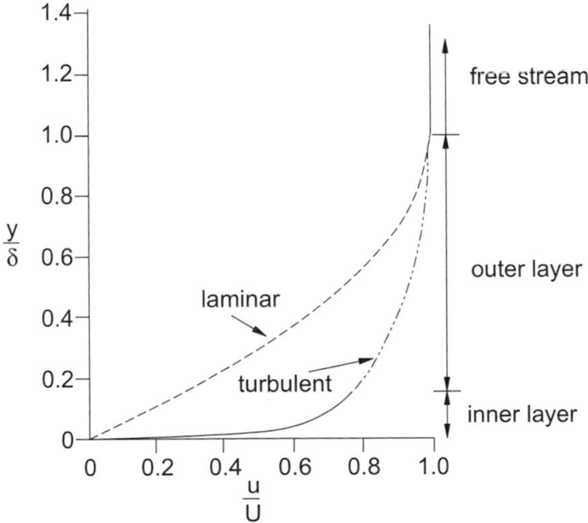

Figure 13.17 Comparing laminar and turbulent boundary layers. The turbulent boundary layer has both inner and outer layers.

stresses, τ_w. There are two major layers, the "inner layer" where $0 \leq y/\delta \leq 0.15$, and the "outer layer" where $0.15 \leq y/\delta \leq 1$, approximately. As in the pipe flow, we need to change to a logarithmic spatial scale to see the structure within the inner layer.

Figure 13.18 is a semilog plot showing the general structure of a turbulent boundary layer on a flat plate with the inner and outer layers of Figure 13.17. Within the inner layer are the three regions: the viscous sublayer, buffer layer, and logarithmic layer which we saw in Figure 13.15. As shown, It would be typical in a turbulent boundary layer of thickness δ that the inner layer exists between $0 \leq y \leq 0.15\delta$ and the outer layer between $0.15\delta \leq y \leq \delta$. However, now there is a free stream velocity outside the boundary layer, past $y \geq \delta$, where the velocity is uniform and the data level off with a constant velocity as shown. Representative data are shown in Figure 13.18. Many investigators have published such measurements, a very short list being Nikuradse (1932), Wieghardt (1944), Laufer (1952), Clauser (1956) and Wei and Willmarth, (1989).

The upper limit of the logarithmic layer is shown as $y^+ = 200$; however, it can range from 200 to 1000 depending on the flow situation and Reynolds number range, and is an active area of research for increasingly higher Reynolds numbers (Vallikivi et al., 2015). The inner layer is dominated by viscous shear, the outer layer by turbulent shear, and in the overlap (logarithmic) layer both types of shear are important. A turbulent boundary layer can develop in an initially laminar flow. Under that circumstance, the laminar boundary layer on the flat plate, starting at $x = 0$, transitions to turbulent far enough downstream. That situation is covered in Section 13.6.

We can investigate turbulent boundary layers using the momentum integral equation which is presented in Section 13.7. For $dU/dx = 0$, Eq. (13.7.29) becomes

$$\tau_w = \rho U^2 \frac{d\theta}{dx} \tag{13.5.1}$$

where θ is the momentum thickness given in Eq. (13.7.14). In terms of a friction coefficient defined from Eq. (13.4.2), we rewrite Eq. (13.5.1) as

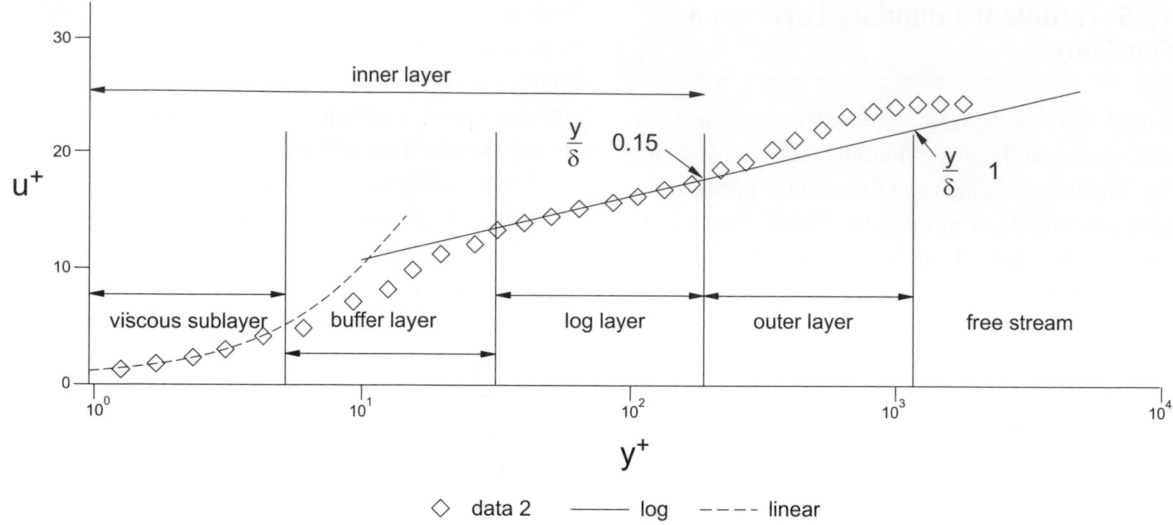

Figure 13.18 Semilog plot of representative data for flow in a turbulent boundary layer on a flat plate. Two main layers are shown: inner and outer. The inner layer includes the viscous sublayer, $0 \leq y^+ \leq 5$, the buffer layer, $5 \leq y^+ \leq 30$, and the logarithmic (overlap) layer, $30 \leq y^+ \leq 200$. The outer layer is $200 \leq y^+ \leq 1200$.

$$c_f = \frac{\tau_w}{\frac{1}{2}\rho U^2} = 2\frac{d\theta}{dx} \tag{13.5.2}$$

It turns out the logarithmic law we used for turbulent pipe flow, Eq. (13.4.5), holds well for flow over a flat plate, but cannot be extended as far into the fluid as we did for the pipe, which worked all the way to the pipe centerline. At the edge of the boundary layer, where the velocity profile is a time-averaged constant, U, we can set $u = U$ and $y = \delta$ to find

$$\frac{U}{u_*} = \frac{1}{\kappa} \ln\left(\frac{\delta u_*}{\nu}\right) + B \tag{13.5.3}$$

where, again, $\kappa = 0.41$ and $B = 5.0$. Recall that $u_* = \sqrt{\tau_w/\rho}$ so substituting for τ_w in terms of c_f, Eq. (13.5.3) becomes

$$\left(\frac{2}{c_f}\right)^{1/2} = \frac{1}{0.41} \ln\left(Re_\delta \left(\frac{c_f}{2}\right)^{1/2}\right) + 5.0 \tag{13.5.4}$$

where $Re_\delta = \delta U/\nu$ is the Reynolds number based on the boundary layer thickness. As we saw in Eq. (13.3.3), this similar expression in Eq. (13.5.4) is an implicit function of c_f for given values of Re_δ. To make the solution to Eq. (13.5.2) a bit easier, an approximation to Eq. (13.5.4) that fits fairly well was given by Prandtl,

$$c_f \sim 0.02 Re_\delta^{-1/6} \tag{13.5.5}$$

This form has ~15% error for $Re_\delta = 4 \times 10^3$ but ~10% for $Re_\delta = 2 \times 10^4$ and even better for larger Re_δ.

For the rest of Eq. (13.5.2) we can evaluate the momentum thickness, θ, defined in Eq. (13.7.14) using the power-law velocity profile of Eq. (13.3.10) with $n = 7$ as an approximation. That makes $u/U = (y/\delta)^{1/7} = g^{1/7}$, which can be inserted into Eq. (13.8.5) where $\theta = \alpha_2 \delta$ and δ is the boundary layer thickness while

$$\alpha_2 = \int_0^1 \zeta^{1/7}\left(1 - \zeta^{1/7}\right) d\zeta = \frac{7}{72} \tag{13.5.6}$$

So we have simplified Eq. (13.5.2) to become

$$0.103\left(\frac{U}{\nu}\right)^{-1/6} dx = \delta^{1/6} d\delta \tag{13.5.7}$$

Equation (13.5.7) is readily integrated to give $\delta(x)$ assuming $\delta(x = 0) = 0$ to solve for the constant of integration. The final result is

$$\delta = 0.16\left(\frac{U}{\nu}\right)^{-1/7} x^{6/7} \rightarrow \frac{\delta}{x} = 0.16 Re_x^{-1/7} \qquad (13.5.8)$$

We see immediately that the growth of the turbulent boundary layer along x goes to the 6/7th power, compared to the 1/2th power for a laminar boundary layer. So, turbulent boundary layers grow much more rapidly. Inserting Eq. (13.5.8) for δ into the definition of $Re_\delta = U\delta/\nu$, the friction coefficient of Eq. (13.5.5) becomes

$$c_f = 0.027 Re_x^{-1/7} \qquad (13.5.9)$$

The dimensional wall shear obtained from Eq. (13.5.9) is

$$\tau_w = 0.0135 \rho^{6/7} \mu^{1/7} U^{13/7} x^{-1/7} \qquad (13.5.10)$$

The growth of the wall shear stress in x goes like $x^{-1/7}$, which is much slower than the $x^{-1/2}$ dependence we found in laminar boundary layers, Eq. (12.1.21). The density dependence of $\rho^{6/7}$ is nearly linear and much larger than the laminar value of $\rho^{1/2}$. This makes sense since turbulence theory interprets part of the convective acceleration term, which is linear in ρ, as a shear stress. The dependence on viscosity, $\mu^{1/7}$, is remarkably weak, much smaller than $\mu^{1/2}$ for laminar flow, while the velocity dependence of $U^{13/7}$ is nearly U^2 and larger than the laminar value of $U^{3/2}$. With slightly different approximations, another often used formula for the turbulent boundary layer thickness is $\delta/x = 0.37 Re_x^{-1/5}$ and the wall shear friction coefficient is $c_f = 0.072 Re_x^{-1/5}$ (Schlichting, 1968).

The shape of the inner layer is dominated by the log law, Eq. (13.4.5). Let's change to new variables, u/U and y/δ, by using the relationships

$$u^+ = \frac{u}{u_*} = \frac{u}{U}\frac{U}{u_*} = \frac{u}{U}\frac{1}{\sqrt{c_f/2}},$$
$$y^+ = \frac{y u_*}{\nu} = \frac{y}{\delta}\frac{\delta U}{\nu}\sqrt{c_f/2} \qquad (13.5.11)$$

Using Eq. (13.5.5), where $c_f \sim 0.02 Re_\delta^{-1/6}$, Eqs. (13.5.11) simplify to

$$u^+ = 10\frac{u}{U} Re_\delta^{1/12}, \quad y^+ = 0.1\frac{y}{\delta} Re_\delta^{11/12} \qquad (13.5.12)$$

The forms in Eq. (13.5.12) can be inserted into the logarithmic law of Eq. (13.4.5). The result is plotted in Figure 13.17 for $Re_\delta = 30{,}000$ where the inner layer is dominated by the logarithmic law as identified. The inner layer ends at $y/\delta = 0.15$ where $y^+ \sim 190$ and the boundary layer ends at $y/\delta = 1$ where $y^+ = 1270$, which are fairly represented in Figure 13.18. The shape of the outer layer in Figure 13.17 is more complicated (Coles, 1956; Purtell et al., 1981) and we will not venture into the details. The curve provided is an estimate of those solutions in Coles (1956) and Purtell et al. (1981) to show the general idea.

We can estimate the entrance length for the turbulent boundary layer, call it L_t, in a pipe by setting $\delta = D/2$ and $x = L_t$ in Eq. (13.5.8) to find $L_t/D = 3.78 Re_D^{1/6}$, which is a very weak function of the Reynolds number based on the tube diameter, D. The scaled entrance length for turbulent flow is popularly given as

$$\frac{L_t}{D} = 4.4 Re_D^{1/6} \qquad (13.5.13)$$

in multiple and various sources, so we are very close. There are many other forms of L_t/D in the literature that depend on the range of Re and the specific assumptions. Another popular version is $L_t/D = 1.359 Re_D^{1/4}$ (Wang, 1982), which is also a weak function of Re_D. Since the turbulent boundary layer grows so rapidly, the turbulent entrance length is much shorter than laminar entrance lengths. For example, using the laminar entrance length equation, Eq. (12.3.2), at Re = 2,300 we find $L_E/D = 0.59 + 0.056 \times 2{,}300 \sim 130$, or 130 diameter lengths downstream. However, using Eq. (13.5.13), $L_t/D = 4.4 \times 2{,}300^{1/6} = 15.9$, or ~10% of the laminar value. The vigorous lateral mixing of momentum hastens the development of fully turbulent flow across the cross-section of the pipe. Other approaches include Zagarola and Smits (1998) whose results can be fit by the series

$$L_E/D \sim 2.72 Re_D^{0.2} + 11.7 Re_D^{0.1} \quad \text{turbulent} \qquad (13.5.14)$$

which is useful for a wide range of high Re_D values up to $Re_D \sim 10^6$ (personal communication, AS). Additional sources for this discussion include Schlichting, (1968) and White (1986).

Clinical Correlation 13.1

A good example of turbulent flow in human airways is seen in the work of Isabey and Chang (1981), see Figures 13.19 and 13.20.

The pressure–flow relationships from experiments of inspiratory and expiratory flow in a five-generation, hollow cast of the central airways are shown in Figure 13.20. It is organized as a Moody plot, though the d/L ratio is not included. The coefficient of friction, $C_F = \Delta p / \frac{1}{2}\rho U^2$, is plotted versus the tracheal Reynolds number, Re. Three different gases were used at various flow rates to achieve a wide range of Re: air, HeO_2 and SF_6O_2. While the kinematic viscosity of air is 0.15 cm^2/s, the value for HeO_2 is 0.54 cm^2/s and SF_6O_2 is 0.03 cm^2/s, due primarily to their density differences. HeO_2 and SF_6O_2 are used in physiological studies of the human respiratory system, so oxygen is part of the mixture.

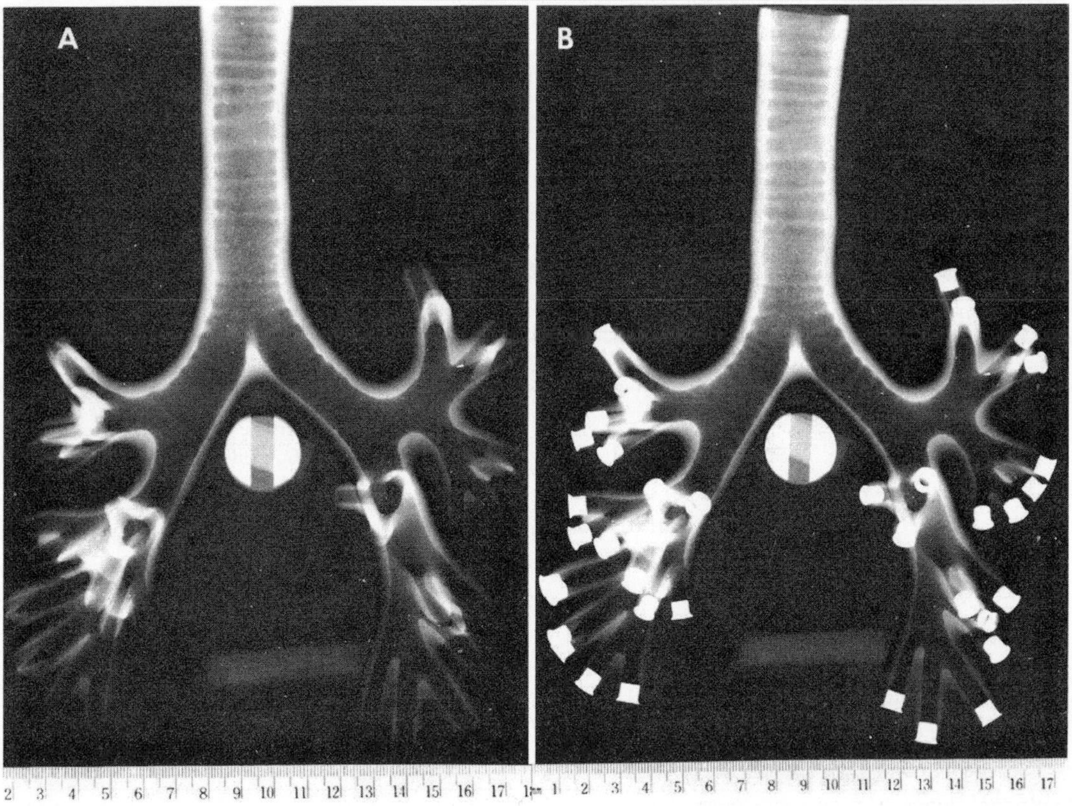

A

B

2 3 4 5 6 7 8 9 10 11 12 13 14 15 16 17 1 2 3 4 5 6 7 8 9 10 11 12 13 14 15 16 17

Figure 13.19 Flow through a five-generation hollow model cast of human central airways.

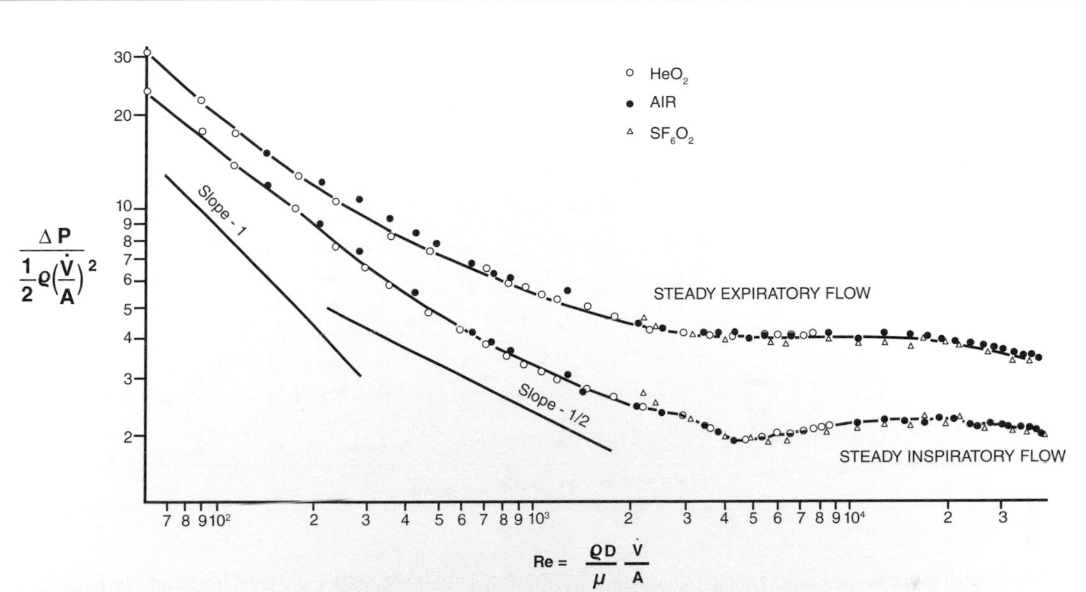

Figure 13.20 Moody diagram of flow through the hollow model of human central airways. Both inspiratory flow and expiratory flow are shown. Three gases were used to achieve the full range of Re: air, HeO_2 and SF_6O_2.

The slopes of different Re ranges are noted. In the region of Re < 200 has a slope of −1 in the log–log plot. This is the Re^{-1} dependence from Poiseuille flow we saw in Eq. (13.3.2). Then in the range 500 < Re < 1,500 the slope is −1/2, as expected when entrance flow dominates for inspiratory flow, see Eq. (12.3.17). For Re > 4,000, the slopes of both inspiratory and expiratory curves are essentially zero, as is found in rough-walled turbulence in the Moody diagram for fully developed pipe flow, Figure 13.11. We note that the pressure drop at any Reynolds number is larger during expiration than during inspiration. This is due to the converging flow, shown in Figure 12.23, with the thin boundary layers and relatively blunt velocity profile.

When the uniform flow approaching a flat plate is laminar, the boundary layer starts out as laminar at $x = 0$, where x is the distance along the plate from the leading edge. However, it can transition to become a turbulent boundary layer further downstream. In this flow the Reynolds number is defined as $Re_x = Ux/\nu$. Generally, the boundary layer becomes turbulent at $x = x_{cr}$ where $Re_x = Ux_{cr}/\nu \sim 5x10^5$. Figure 13.21 shows a sketch of this transition noting that the boundary layer is thicker and grows faster when it is turbulent. Figure 13.22 is flow visualization of this transition (Matsubara and Alfredsson, 2001) using smoke in airflow and a very low level of free stream turbulence.

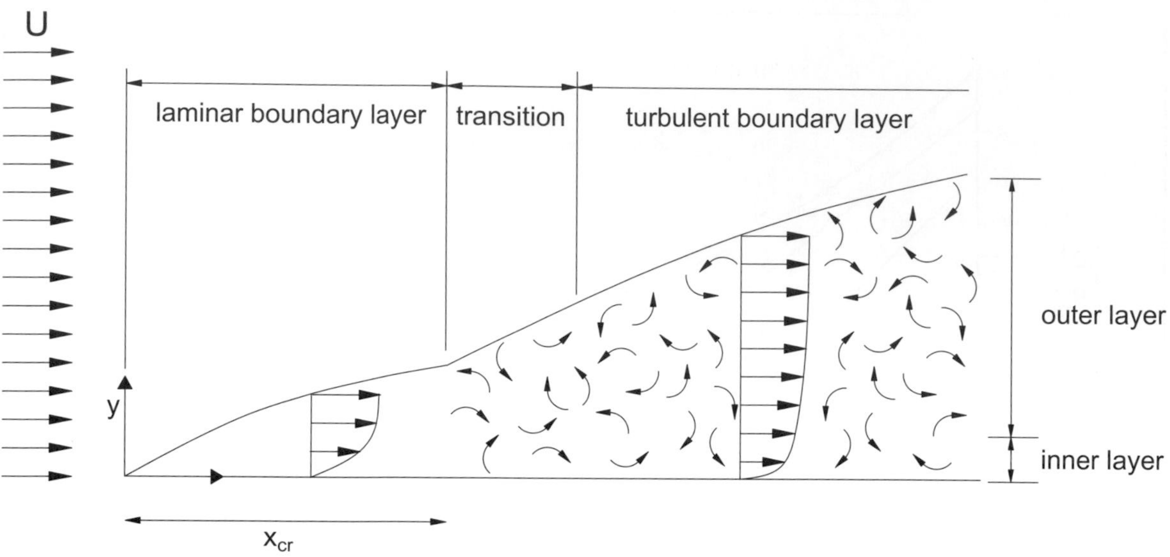

Figure 13.21 Transition from laminar to turbulent boundary layer flow over a flat plate. The upstream uniform velocity, U, is laminar.

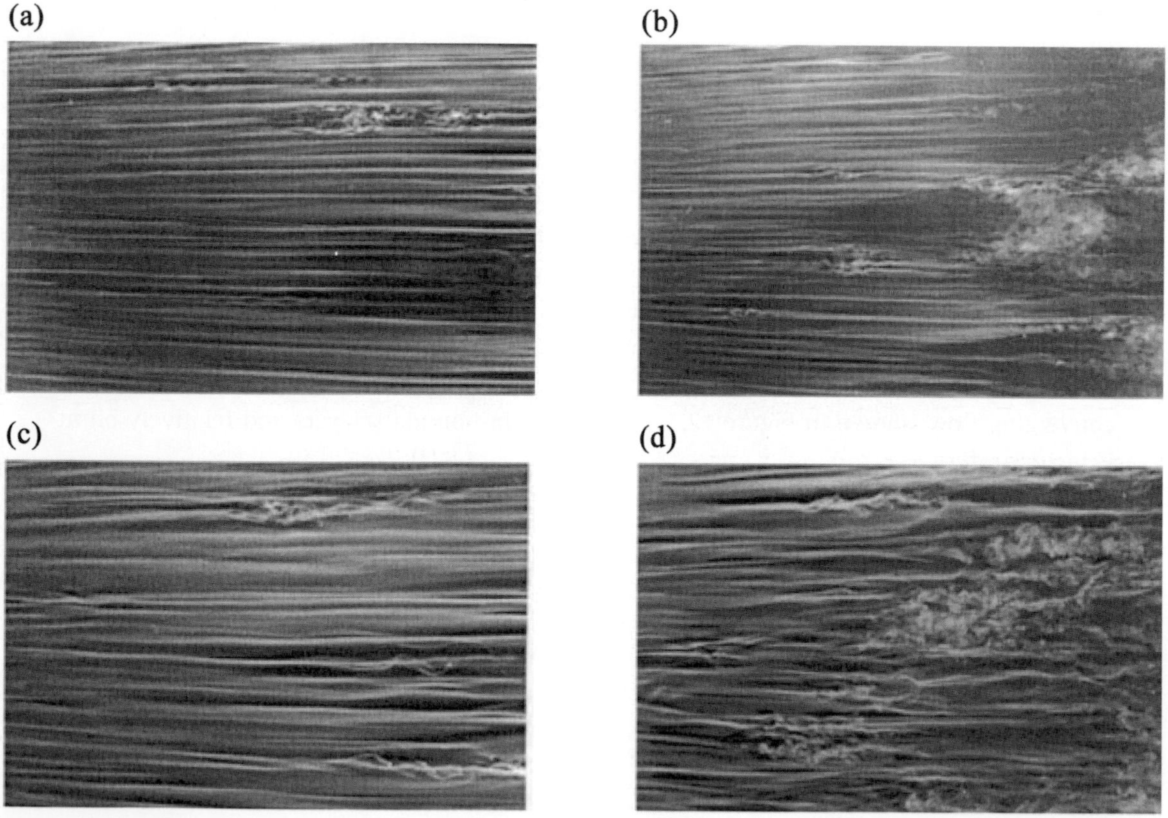

(a)

(b)

(c)

(d)

Figure 13.22 Transition from laminar to turbulent flow in a boundary layer.

Box 13.4 Shark Skin Biomimetics

A fascinating biological approach to reducing turbulence in a boundary layer, and hence the skin friction contribution to drag, is present in real shark skin as shown in Figure 13.23 from Martin and Bhushan (2014). The skin of this Mako shark has a pattern of structures called "denticles." The size of shark denticles varies over a range roughly 0.1–1 mm depending on the species and location (Domel et al., 2018). The drag reductions come from redistributing the shear stress to be more concentrated on the raised surfaces as well as confining the lateral turbulent velocity components to smaller amplitudes (Bechert et al., 1997). In addition, vortices are created in the tiny cavities which lower the drag on the overlying main flow stream. Clearly sharks benefit from being faster than their prey. Efforts to pattern surfaces for reducing drag on airplane wings, for example, is an example of "biomimetics." Biomimetics is a field that explores the advantages to engineering design by mimicking what is observed in nature.

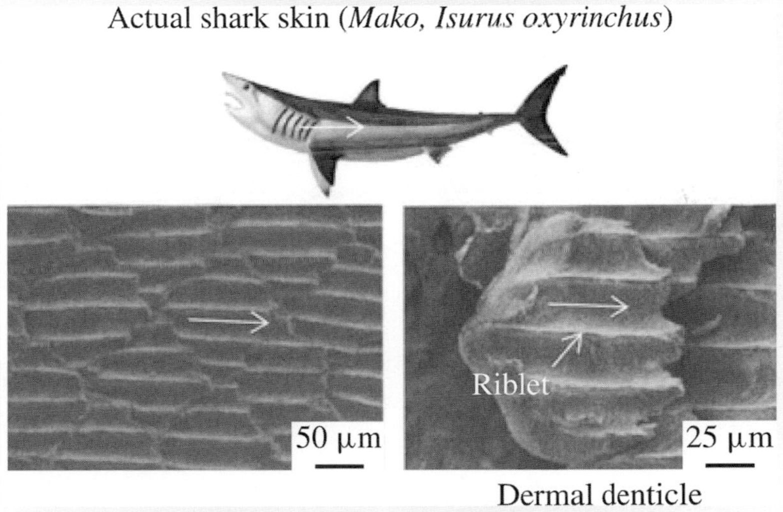

Actual shark skin (*Mako, Isurus oxyrinchus*)

Riblet

50 μm

25 μm

Dermal denticle

Figure 13.23 Skin with dermal denticles of a Mako shark viewed through a scanning electron microscope.

Box 13.5 Flow Through a Stenosis, Coarctation and Aneurysm

The above discussions outlining turbulence concepts for tubes/pipes are limited to vessels whose diameter does not change along the flow axis. For arteries and airways, a significant deviation occurs when there is a locally narrowed region or stenosis. Arterial stenoses, Figure 13.24(a), often result from atheromatous plaques as shown in Figure 12.25(b). Airway stenoses can derive from damaged walls or tumor impingement. Two features of a stenosis contribute to turbulence occurring well below the critical Reynolds number for a normal tube: the narrowing increases the blood velocity, and hence Re, while the exit from the narrowed region produces a separated flow and a confined jet which promotes flow instability.
A coarctation, usually of the aorta, is a congenital narrowing as shown in Figure 13.24(b). There can also be locally expanded or dilated regions of the tube. For blood vessels these are called aneurysms, while for the airways the condition is called bronchiectasis. An example of an aneurysm appears in Figure 7.29 for a cerebral artery. A sketch is in Figure 13.24(c). When the flow stream enters the dilation, it also can separate, forming a jet which triggers turbulence.

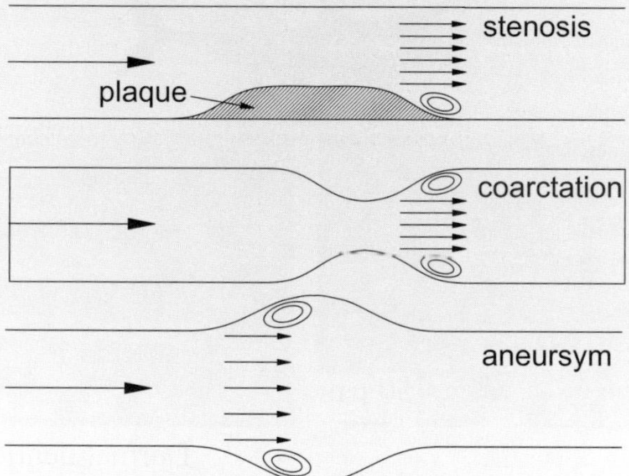

Figure 13.24 Flow through (a) stenosis, (b) coarctation and (c) aneurysm.

13.6 Turbulent Jets

Turbulent jets are found in many industrial and biofluid mechanics settings and are an example of "free turbulent flow," i.e. not confined by a wall boundary. Two industrial-type turbulent jets are shown in Figure 13.25 using laser-induced fluorescence of the incoming fluid (Dimotakis, 2000). The jet Reynolds number is defined as $Re_j = Ud/\nu$ where U is the average speed of the jet as it exits the orifice whose diameter is d. The image on the left has $Re_j \cong 2.5 \times 10^3$ for $0 < x/d < 35$ where x is the

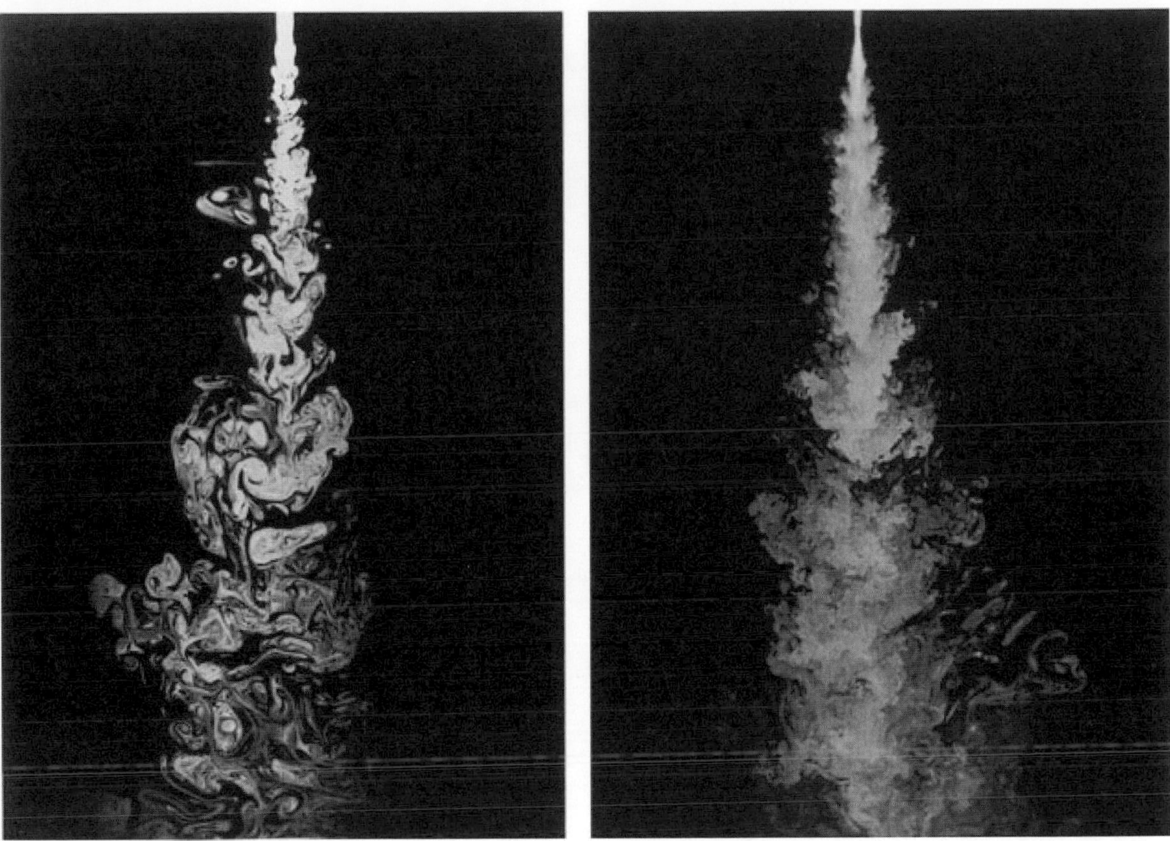

Figure 13.25 Laser induced fluorescence imaging of turbulent jets. Left: $Re_j \cong 2.5 \times 10^3$ $(0 < x/d < 35)$ with $d = 0.75$ cm Right: $Re_j \cong 10^4$ $(0 < x/d < 200)$ with $d = 0.25$ cm. Note similar shape but deeper penetration and finer scale structure for the higher Re_j. Data from Dimotakis et al. (1983).

distance from the orifice exit in the jet direction and $d = 0.75$ cm. The image on the right has a larger jet Reynolds number, $Re_j \cong 10^4$, for $0 < x/d < 200$ and $d = 0.25$ cm. The jet shapes are self-similar, widening with the same slope, but the higher Re_j jet penetrates deeper and its turbulent scale is finer.

The expiratory phase of respiration, including cough and sneeze, are biofluid turbulent jets. Short bursts of a jet are also called "puffs." Figure 13.26(a) shows a frame from a high speed schlieren video of a cough (Tang et al., 2013). Note the turbulent jet structure and the penetration distance. Figure 13.26(b) is the velocity field associated with the cough at the same instance. It is derived by comparing frames 300 to 301 using schlieren particle image velocimetry (PIV). The highest velocities, ~ 8 m/s, are in the center of the jet. Because cough usually contains many thousands of liquid droplets, both the air and droplet phase can contain communicable diseases like influenza and tuberculosis, see Section 16.7.

The details of the turbulent jet velocity field can be examined initially through the structure of a circular laminar jet, which satisfies the Navier–Stokes equations and continuity in cylindrical coordinates (Schlichting, 1968). The steady conservation equations are

$$u_r \frac{\partial u_z}{\partial r} + u_z \frac{\partial u_z}{\partial z} = \nu \frac{1}{r} \frac{\partial}{\partial r}\left(r \frac{\partial u_z}{\partial r} \right)$$
$$\frac{\partial u_r}{\partial r} + \frac{u_r}{r} + \frac{\partial u_z}{\partial z} = 0 \qquad (13.6.1)$$

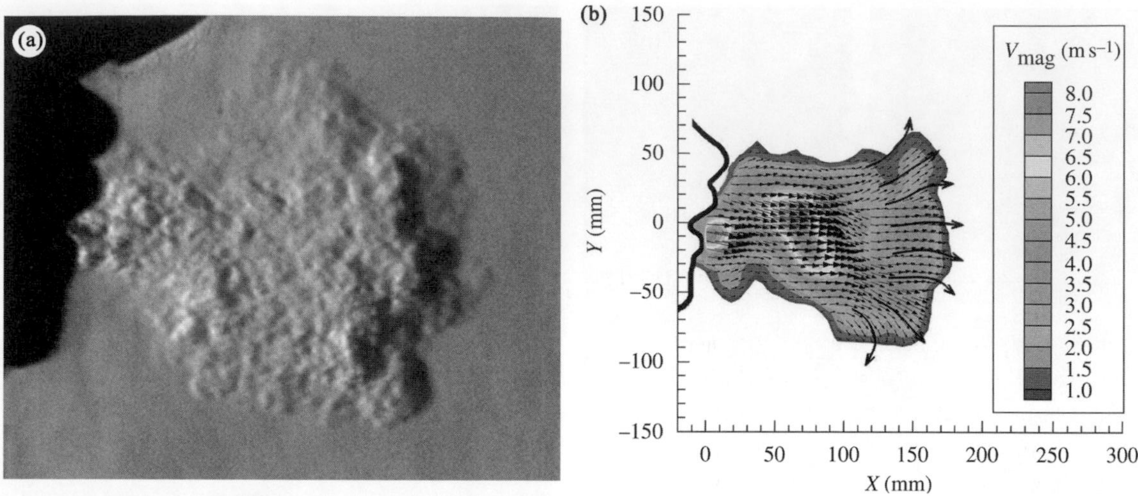

Figure 13.26 (a) Sample frame (no. 300) from high speed schlieren video of a cough showing a turbulent jet structure. (b) Instantaneous velocity–magnitude contours and vectors of a cough with highest velocities in the center and lowest at the outer edges derived from PIV using frames 300 and 301.

which assume no pressure gradient and a thin layer for the jet, i.e. $u_r/u_z<<1$ and $\partial/\partial z<<\partial/\partial r$ as we saw in Section 12.1 for the boundary layer equations. Then only the z-component of momentum is used and z-gradients of u_z do not contribute to the viscous effects. Equations (13.6.1) are subject to the boundary conditions of no cross flow and symmetry at $r = 0$ and zero velocity at infinity in r,

$$u_r(r = 0) = 0,\ \left.\frac{\partial u_z}{\partial r}\right|_{r=0} = 0,\ u_z(r \to \infty) = 0 \quad (13.6.2)$$

The stream function in cylindrical coordinates is taken to be of the form

$$\psi = \nu z\,\phi(\eta) \quad (13.6.3)$$

where $\eta = \gamma r/z$, so that the flow is viewed to be "self-similar" in the similarity variable η. Using the definition of the stream function in cylindrical coordinates for axisymmetric flow we find u_z and u_r are given by

$$u_z = \frac{1}{r}\frac{\partial\psi}{\partial r} = \gamma^2\frac{\nu}{z}\frac{\phi'}{\eta},\quad u_r = -\frac{1}{r}\frac{\partial\psi}{\partial z} = \gamma\frac{\nu}{z\eta}(\eta\phi' - \phi) \quad (13.6.4)$$

where the prime symbol means derivative with respect to η. Since the stream function form automatically satisfies continuity, we only need to insert the forms in Eqs. (13.6.4) into the Navier–Stokes equation, Eq. (13.6.1), to obtain

$$\frac{\phi\phi'}{\eta^2} - \frac{\phi'^2}{\eta} - \frac{\phi\phi''}{\eta} = \frac{d}{d\eta}\left(\phi'' - \frac{\phi'}{\eta}\right)$$
$$\rightarrow \quad \frac{d}{d\eta}\left(-\frac{\phi\phi'}{\eta}\right) = \frac{d}{d\eta}\left(\phi'' - \frac{\phi'}{\eta}\right) \quad (13.6.5)$$

which can be integrated once to yield

$$\phi\phi' = \phi' - \eta\phi'' \quad (13.6.6)$$

The solution to Eq. (13.6.6) with boundary conditions Eqs. (13.6.2) is

$$\phi = \frac{\eta^2}{1 + \frac{1}{4}\eta^2} \quad (13.6.7)$$

Inserting Eq. (13.6.7) into Eq. (13.6.4) gives us the solution for the velocities

$$u_z = \gamma^2 \frac{\nu}{z} \frac{2}{\left(1+\frac{1}{4}\eta^2\right)^2}, \quad u_r = \gamma \frac{\nu}{z} \frac{\eta - \frac{1}{4}\eta^3}{\left(1+\frac{1}{4}\eta^2\right)^2} \qquad (13.6.8)$$

The momentum flux is J where

$$J = 2\pi\rho \int_0^\infty u_z^2 r dr = \frac{16}{3}\pi\rho\gamma^2\nu^2 \quad \rightarrow \quad \gamma^2 = \frac{J}{\rho}\frac{3}{16}\frac{1}{\pi\nu^2}$$
$$= K\frac{3}{16}\frac{1}{\pi\nu^2} \qquad (13.6.9)$$

where $K = J/\rho$ is called the kinematic momentum and measures the jet intensity in units of L^4/T^2.

For a turbulent jet a "virtual viscosity," ε_0, is proposed and replaces the kinematic viscosity, i.e. $\nu = \varepsilon_0$. In addition, we now consider the velocities u_z and u_r to be the time-averaged values of the turbulent velocities. Part of our challenge is to find how ε_0 depends on the flow. Substituting for γ and replacing $\nu = \varepsilon_0$we find u_z, u_r and η to be

$$u_z = \frac{3}{8\pi}\frac{K}{\varepsilon_0 z}\frac{1}{\left(1+\frac{1}{4}\eta^2\right)^2}, \quad u_r = \frac{1}{4}\left(\frac{3}{\pi}\right)^{1/2}\frac{K^{1/2}}{z}\frac{\eta - \frac{1}{4}\eta^3}{\left(1+\frac{1}{4}\eta^2\right)^2},$$
$$\eta = \frac{1}{4}\left(\frac{3}{\pi}\right)^{1/2}\frac{K^{1/2}}{\varepsilon_0}\frac{r}{z} \qquad (13.6.10)$$

To solve for $\varepsilon_0$requires another relationship amongst the variables, in this case from velocity profile measurements. The maximum of u_z is on the centerline where $\eta = 0$, i.e. $u_z(\eta = 0) = u_{z\ max} = (3/8\pi)(K/\varepsilon_0 z)$. Since u_z tends to zero as r increases to infinity, it is easier to choose a finite measure of the width of the u_z distribution. One choice is to determine the r value where $u_z = u_{z\ max}/2$. That value of r is called $r = b_{1/2}$. Measurements (Reichardt, 1942) show that $b_{1/2} = 0.0848\,z$, linear in z, for a wide range of jet intensities. Inserting $\eta = 1.286 = \gamma b_{1/2}/z$ into u_z of Eq. (13.6.10) yields $u_z(\eta = 1.286) = u_{z\ max}/2$. Using the definition of η in Eq. (13.6.10), we find $b_{1/2} = 4(\pi/3)^{1/2} 1.286 z\varepsilon_0/K^{1/2} = 5.27 z\varepsilon_0/K^{1/2}$. Equating these two expressions for $b_{1/2}$ gives us

$$\frac{\varepsilon_0}{K^{1/2}} = \frac{0.0848}{5.27} = 0.0161 \qquad (13.6.11)$$

So we can substitute for ε_0 in Eqs. (13.6.10) in terms of K,

$$u_z = \frac{3}{8\pi}\frac{K^{1/2}}{(0.0161)z}\frac{1}{\left(1+\frac{1}{4}\eta^2\right)^2},$$
$$u_r = \frac{1}{4}\left(\frac{3}{\pi}\right)^{1/2}\frac{K^{1/2}}{z}\frac{\eta - \frac{1}{4}\eta^3}{\left(1+\frac{1}{4}\eta^2\right)^2},$$
$$\eta = \frac{1}{4}\left(\frac{3}{\pi}\right)^{1/2}\frac{1}{(0.0161)}\frac{r}{z} \qquad (13.6.12)$$

and the stream function is $\psi = \varepsilon_0 z\phi(\eta) = (0.0161)K^{1/2}z\phi(\eta)$. So $K^{1/2}$ is simply a scale factor for ψ and does not affect the shape of the streamlines, $\psi =$ constant, just their absolute values and the relative difference between them, i.e. the volume flow rate between streamlines.

Figure 13.27 shows the streamlines and u_z velocity profiles for the turbulent jet flow where $K = 10,000$. Note that the maximum velocity along the centerline decreases with z; however, the profile broadens as new fluid is entrained from the lateral flow toward the centerline. The momentum flux entering the system through a narrow tube in the wall centered at $z = 0$ would have $J = \rho U^2\pi d^2/4$ where d is the tube diameter. Without external forces or pressure gradients, this will equal J in Eq. (13.6.9) so that

$$J = \rho U^2\pi d^2/4 = \frac{16}{3}\pi\rho\gamma^2\varepsilon_0^2 = \rho K \qquad (13.6.13)$$

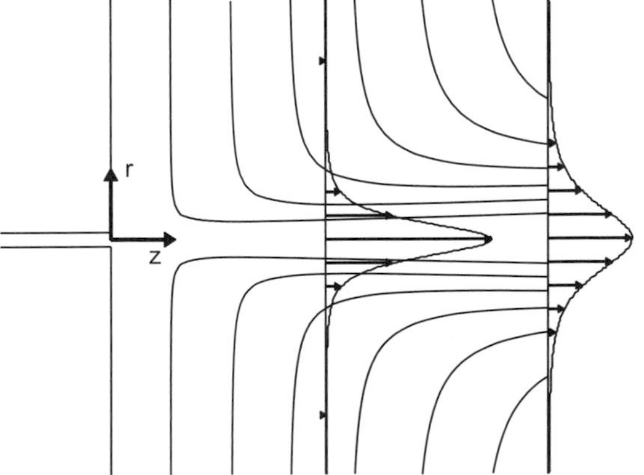

Figure 13.27 Streamlines and u_z velocity profiles for turbulent jet flow where $K = 10,000$.

where we have used $\gamma^2 = 3K/(16\pi\varepsilon_0^2)$ after substituting $\nu = \varepsilon_0$ in Eq. (13.6.9). Dividing Eq. (13.6.13) through by $\rho\nu_1^2$, where ν_1 is the actual kinematic viscosity, and defining $Re_j = Ud/\nu_1$, gives us

$$Re_j = \frac{2K^{1/2}}{\sqrt{\pi}\nu_1} \tag{13.6.14}$$

where $\nu_1 = 0.01 \ \text{cm}^2/\text{s}$ for water and $K = 500 \ \text{cm}^4/\text{s}^2$, and $Re_j \sim 2{,}500$ as in Figure 13.25. The volume flow rate comes from integrating u_z across any cross-section and is found to be

$$Q = 2\pi \int_0^\infty u_z r dr = 8\pi\varepsilon_0 z = 0.404K^{1/2}z \tag{13.6.15}$$

so that the strength of the jet, K, is directly related to the flow rate, Q. Interestingly, that is not the case for the laminar circular jet, where Eq. (13.6.15) translates to $Q = 8\pi\nu_1 z$, independent of K. In that case, at a given value of z, more intense jets with higher K (larger$u_{z\ max}$) are narrower and therefore entrain smaller additional fluid, while less intense jets, lower K (smaller $u_{z\ max}$), are broader and entrain larger additional fluid. The balance is that Q is the same.

The penetration of the jet is related to $u_{z\ max}$, which from Eq. (13.6.12) is $u_z(\eta = 0)$, or

$$u_{z\ max} = \frac{3}{8\pi}\frac{K^{1/2}}{(0.0161)z} \tag{13.6.16}$$

Clearly larger K, which implies larger Re_j, causes further penetration of the jet as seen in Figure 13.28 and Figure 13.25. The model presented does not deal with the flow near an actual jet orifice, so is more representative of conditions outside of that near-orifice region.

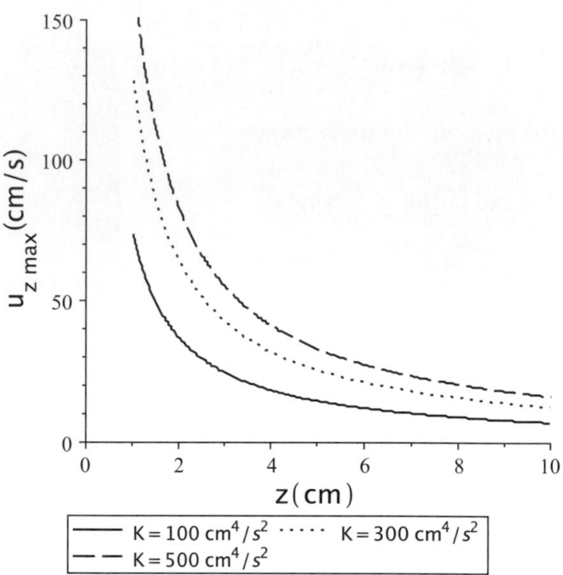

Figure 13.28 $u_{z\ max}(z)$ for three values of $K = 100, 300, 500\ \text{cm}^4/\text{s}^2$.

Artificial heart valves present a significant intersection of medical care and fluid mechanics. A typical situation is the failure of the patient's natural aortic valve, which allows blood flow out of the left ventricle of the heart into the aorta during systole, and prevents leakage back into that chamber during diastole. Failure of that valve can involve restriction of the outlet cross-section creating a high resistance to flow as well as dangerous levels of leakage back into the heart, reducing the efficiency of pumping. The design and development of artificial aortic valves has an interesting history starting in the 1950s. The first valve was designed and used by Charles Hufnagel in 1952. It was a ball valve inserted in the descending aorta, and helped to reduce leakage backflow, see Figure 13.29(a). The next technological advance in 1960 was the caged ball, which replaced the actual diseased aortic valve, see Figure 13.29(b). Initially made by Starr-Edwards, during systole the ball is forced against the cage allowing blood to flow. In diastole the ball moves back against the base and prevents backflow. Later, in 1969, came the tilting disc design from Björk-Shiley, see Figure 13.29(c), which provided less obstruction to the flow compared to the caged ball. In 1979 came the bileaflet valve in Figure 13.29(d) from St. Jude Medical. The leaflets swivel on hinges to allow forward flow and stop backflow.

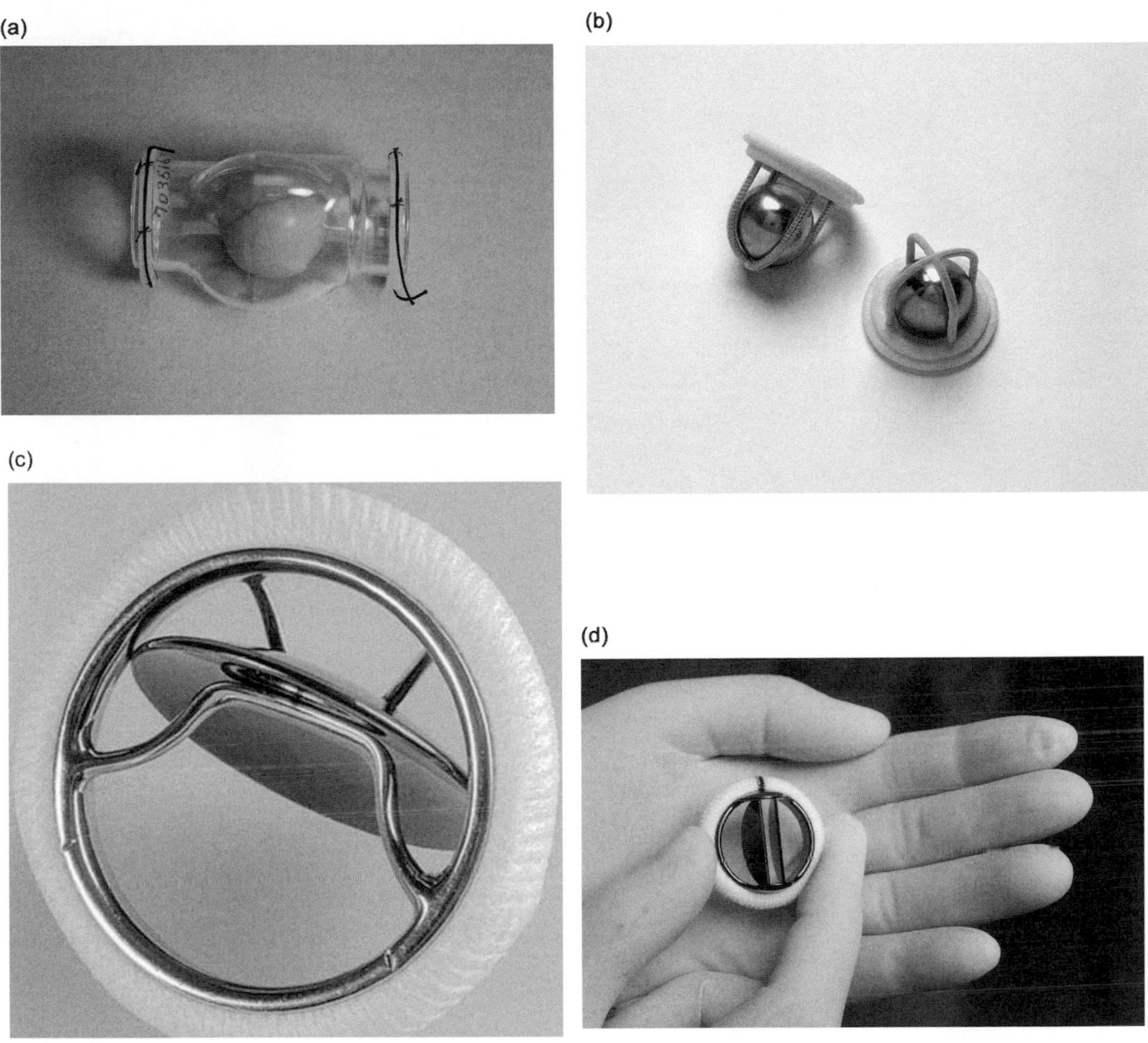

Figure 13.29 (a) Hufnagel heart valve. (b) Two caged ball valves (Starr-Edwards). (c) Tilting disc valve (Björk-Shiley) (Abu-Omar and Dunning, 2012). (d) Bileaflet heart valve (St. Jude Medical).

Figure 13.30 shows aspects of an experimental investigation (Dasi et al., 2007) of pulsatile flow through a St. Jude bileaflet aortic valve. Figure 13.30(a) is the flow circuit with PIV measuring apparatus for the valve region. The top row of Figure 13.30(b) shows the peak-flow average axial velocity profile, u(r), at four values of the axial position, $z = 10, 20, 30, 40$ mm, downstream of the valve. Note the three jets which correspond to the central orifice and two side orifices of the valve, see Figure 13.29(d). They are dominant at $z = 10$ mm and diminish downstream. The axial turbulent velocity component, $u'(r)$, is shown in the middle row where the maximal values are at $z = 20$ mm. Then the Reynolds shear stress, $\mathrm{RSS}(r) = \rho\overline{u'v'}$ is the bottom row, again highest at $z = 20$ mm. Potential damage to RBCs and triggering of clot formation is related to these flow features.

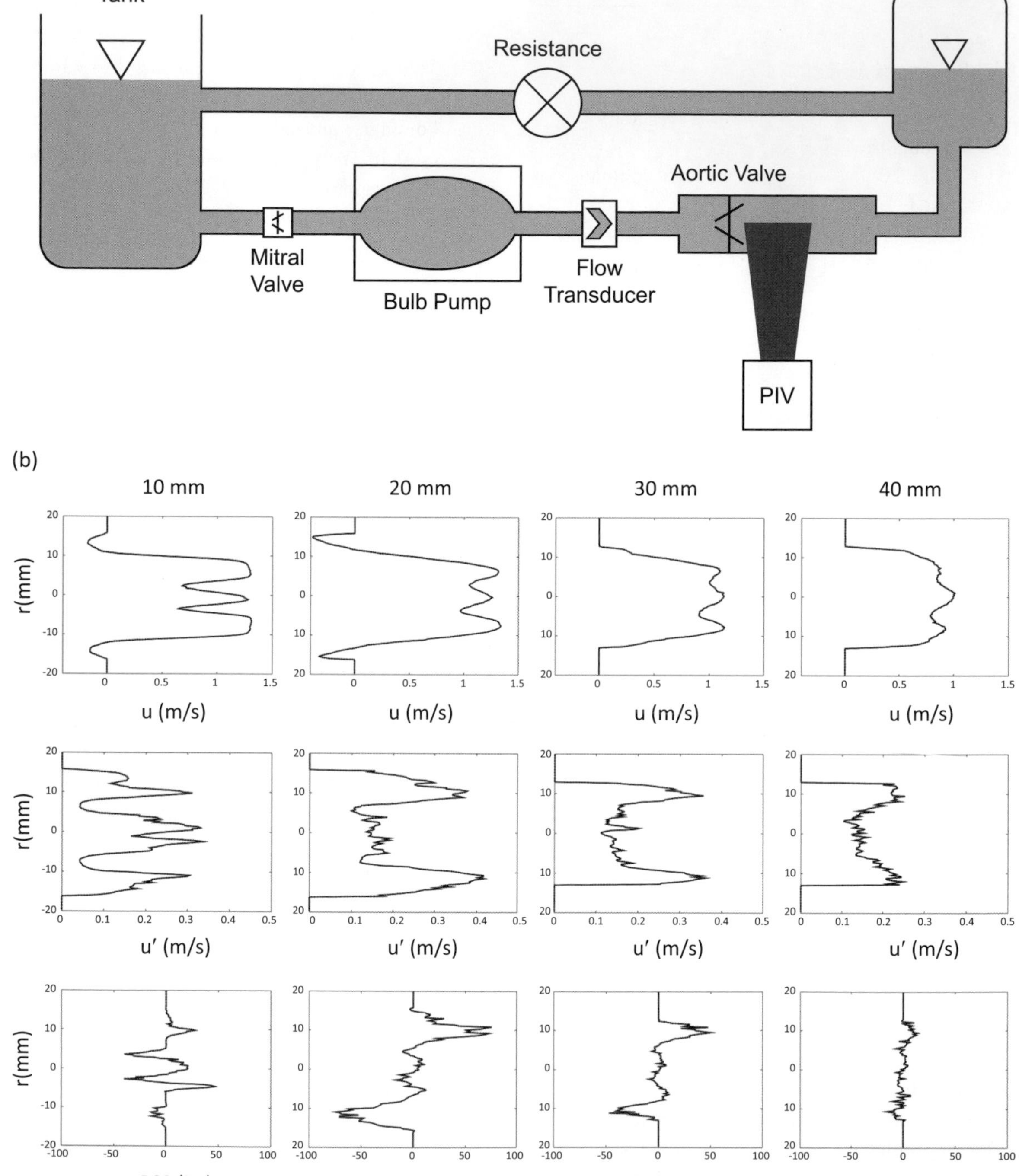

Figure 13.30 PIV measurements of pulsatile flow through a St. Jude Medical bileaflet heart valve in a straight, axisymmetric model aorta: (a) flow setup; (b) peak-flow velocity profile, u, its turbulent component u′, and Reynolds shear stress (RSS) at 10, 20, 30 and 40 mm downstream of the valve outlet.

13.7 Integral Balances for Boundary Layer Flow over a Flat Plate

Integral Mass Balance: The Displacement Thickness

The momentum integral method for boundary layers was first published by von Kármán (1881–1963) in 1921 with an adjoining paper by Ernst Pohlhausen (1890–1964). They were both students of Ludwig Prandtl at the University of Göttingen (von Kármán, 1921; Pohlhausen, 1921). In this approach the boundary layer equations are solved approximately by integrating them across the thickness of the layer with simplifying assumptions. The approach may be used for either laminar or turbulent flow. An excellent presentation can be found in the classic book by Hermann Schlichting (1968), who was also a student of Prandtl. This section serves as input to Section 12.2 for laminar boundary layers, so both laminar and turbulent flow will be examined.

Consider conservation of mass for flow between the plate and a streamline which is a distance h_0 from the wall at $x = 0$, see Figure 13.31. The oncoming uniform, dimensional velocity is U and we retain the possibility that its value, though independent of y, could be a function of x for $x > 0$. One way for that to happen is for the oncoming velocity to be at a non-zero angle of incidence with the x-axis. That is unlike the Blasius assumption of Section 12.1 where $U = U_\infty$, a constant for all x. The streamline is selected to be just outside of the boundary layer. Further downstream it is still outside of the boundary layer, but displaced by a distance δ^*, known as the displacement thickness. A control volume is shown by dashed lines, and we note the upper boundary, labeled 3, is a streamline, so has no flow across it. The lower boundary, labeled 1, is the wall and is impermeable. The integral mass balance equates the mass flow across the left hand boundary, labeled 4, to that across the right hand boundary, labeled 2,

$$\underbrace{\int_0^{h_0} (\rho U b)\, dy}_{4} = \underbrace{\int_0^{h_0+\delta^*} (\rho u b)\, dy}_{2} \tag{13.7.1}$$

where b is a depth in the z-direction. Some rearrangement of Eq. (13.7.1) leaves us with

$$U h_0 = \int_0^{h_0+\delta^*} (u - U + U)\, dy$$

$$= U(h_0 + \delta^*) + \int_0^{h_0+\delta^*} (u - U)\, dy \tag{13.7.2}$$

which simplifies to

$$\delta^* = \int_0^{h_0+\delta^*} \left(1 - \frac{u}{U}\right) dy \tag{13.7.3}$$

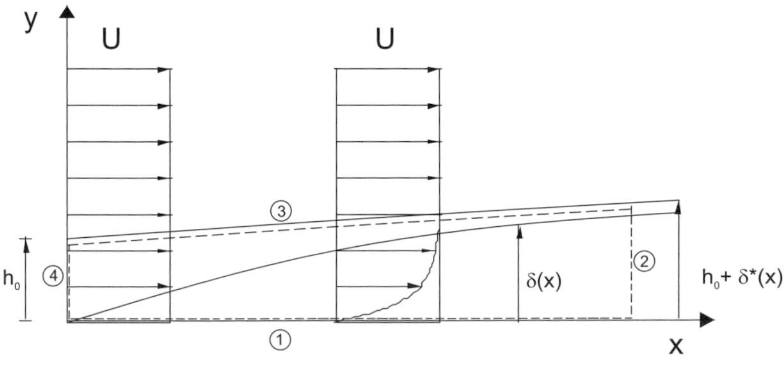

Figure 13.31 Boundary layer flow with boundary layer thickness, δ, and displacement thickness δ*.

For laminar flow, with the Blasius solution as our guide, let the dimensionless boundary layer coordinate and velocity be, respectively,

$$\hat{\eta} = \frac{y}{\left(\frac{\nu x}{U}\right)^{1/2}}, \quad f'(\hat{\eta}) = \frac{u}{U} \tag{13.7.4}$$

and rewrite Eq. (13.7.3) as

$$\begin{aligned} \delta^* &= \sqrt{\frac{\nu x}{U}} \int_0^{\hat{\eta}_1} (1 - f'(\hat{\eta}))\, d\hat{\eta} \\ &= \sqrt{\frac{\nu x}{U}} [(\hat{\eta}_1 - f(\hat{\eta}_1)) - (0 - f(0))] \\ &= \sqrt{\frac{\nu x}{U}} [(\hat{\eta}_1 - f(\hat{\eta}_1))] \end{aligned} \tag{13.7.5}$$

where the upper limit of the integration is $\hat{\eta} = \hat{\eta}_1$, which corresponds to $y = h_0 + \delta^*(x)$. In the last step of simplifying Eq. (13.7.5) we have used the no-slip boundary condition $f(\hat{\eta} = 0) = 0$.

Here $\hat{\eta}_1$ is just outside the boundary layer and a plot of $\hat{\eta} - f(\hat{\eta})$ vs $\hat{\eta}$ from the Blasius solution is shown in Figure 13.32. We are interested in the region $\hat{\eta} = \hat{\eta}_1 > 5$. The plot shows an asymptote in this region and it can be evaluated from the numerical solution to be $\hat{\eta}_1 - f(\hat{\eta}_1) = 1.721$. Inserting this value into Eq. (13.7.5) gives us the displacement thickness

$$\delta^* = 1.721\sqrt{\frac{\nu x}{U}} \sim \frac{\delta}{3} \tag{13.7.6}$$

We note that the displacement thickness is smaller than the boundary layer thickness.

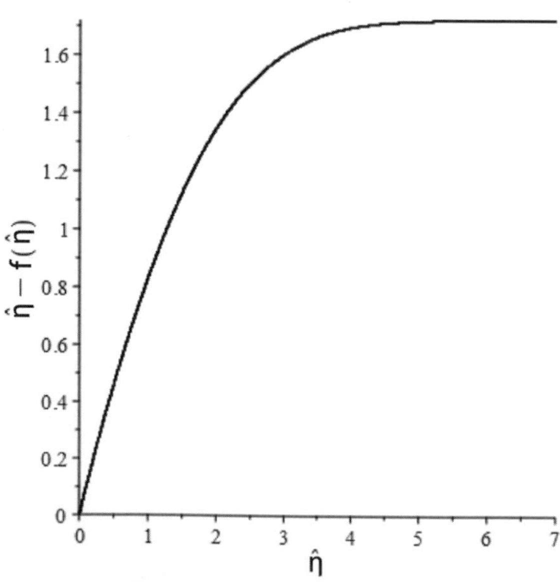

Figure 13.32 $\hat{\eta} - f(\hat{\eta})$ vs $\hat{\eta}$.

Integral Momentum Balance: The Momentum Thickness

Figure 13.33 shows boundary layer flow with a slightly different control volume. First we need to conserve mass for this control volume. The streamline that forms side 3 starts at $(0, h_0)$ and ends at (L, δ). Since sides 1 and 3 have no cross flow, the mass balance is between sides 4 and 2,

$$h_0 U = \int_0^{\delta} u\, dy \tag{13.7.7}$$

The integral balance for linear momentum for the boundary layer is

$$-\iint \rho \underline{u} \left(\underline{u} \cdot \underline{n}\right) dA + \iint \underline{t}\, dA = 0 \tag{13.7.8}$$

The x-component of Eq. (13.7.8) is

$$-\iint \rho u \left(\underline{u} \cdot \underline{n}\right) dA + \iint t_x\, dA = 0 \tag{13.7.9}$$

Again, there is no flux of mass or momentum across boundaries 1 and 3, and there is no pressure gradient, so t_x only depends on shear. There is no shear stress at boundary 3 since it is in the inviscid region. That leaves only the shear stress on boundary 1, call it τ_w, the wall shear. For laminar flow

$$\tau_w = \mu \frac{\partial u}{\partial y}_{y=0} \tag{13.7.10}$$

so the balance is

$$\underbrace{\int_0^{h_0} (-\rho\, U^2\, b)\, dy}_{4} + \underbrace{\int_0^{\delta} (\rho\, u^2\, b)\, dy}_{2} - \underbrace{\mu \int_0^{L} \frac{\partial u}{\partial y}_{y=0} b\, dx}_{1} = 0 \tag{13.7.11}$$

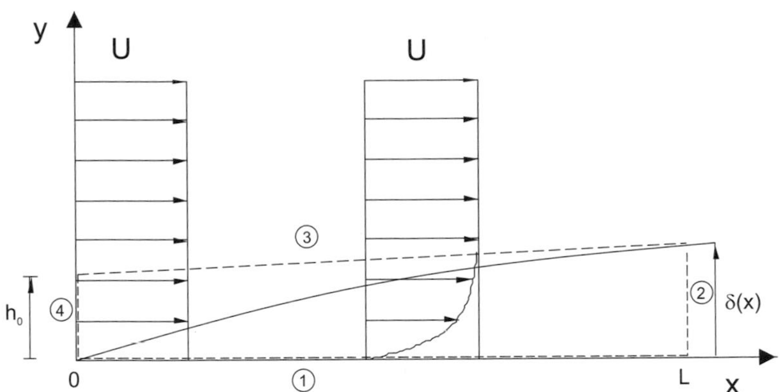

Figure 13.33 Control volume of boundary layer flow for momentum conservation.

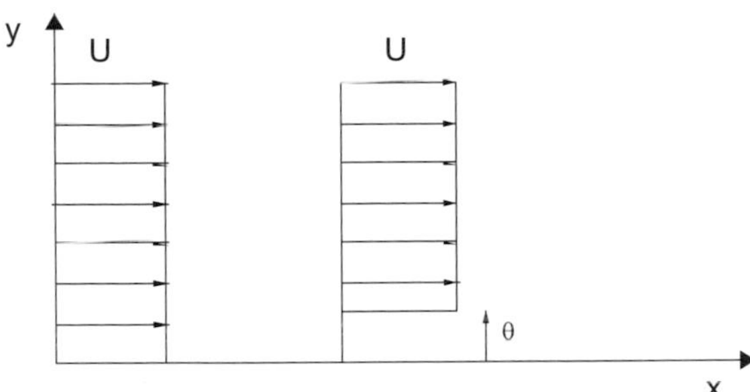

Figure 13.34 Momentum deficit of uniform flow and the momentum thickness, θ.

We can call the integrated wall shear the drag, D,

$$D = \mu \int_0^L \frac{\partial u}{\partial y}_{y=0} b\,dx \qquad (13.7.12)$$

and define the "momentum thickness," θ, as a conceptual thickness of a momentum "deficit" from the uniform inviscid flow of speed U.

An example is shown in Figure 13.34. The concept is to envision this deficit as equivalent to the drag force, D, and the relationship is

$$D = \rho U^2 b\theta \qquad (13.7.13)$$

Using this definition, we rewrite Eq. (13.7.11) and incorporate Eq. (13.7.7),

$$\theta = h_0 - \int_0^\delta \left(\frac{u^2}{U^2}\right) dy = \int_0^\delta \left[\frac{u}{U}\left(1 - \frac{u}{U}\right)\right] dy \qquad (13.7.14)$$

For laminar flow, using the Blasius variables, we rewrite Eq. (13.7.14) as

$$\theta = \left(\frac{x\nu}{U}\right)^{1/2} \int_0^{\hat{\eta}_2} (f'(1 - f')\,)\,d\hat{\eta} \qquad (13.7.15)$$

where the upper limit of the integral is taken to be the location of the boundary layer, with $\hat{\eta}_2 = 5$ as 1% tolerance. Using the Blasius solution, the integral computes to

$$\int_0^{\hat{\eta}_2} (f'(1 - f')\,)\,d\hat{\eta} = 0.664 \qquad (13.7.16)$$

so the momentum thickness is

$$\theta = 0.664 \left(\frac{x\nu}{U}\right)^{1/2} \sim \frac{\delta}{8} \qquad (13.7.17)$$

and we note that the momentum thickness is even smaller than the displacement thickness.

Integration of the Boundary Layer Equations: The Momentum Integral Equation

Shifting from a control volume approach, here the momentum equations of the boundary layer are integrated across it, using the concepts of the displacement, δ^*, and momentum, θ, thicknesses. Recall the laminar boundary layer equations developed in Section 12.1,

$$u\frac{\partial u}{\partial x}+v\frac{\partial u}{\partial y}=-\frac{1}{\rho}\frac{\partial p}{\partial x}+\nu\frac{\partial^2 u}{\partial y^2}$$
$$\frac{\partial u}{\partial x}+\frac{\partial v}{\partial y}=0 \tag{13.7.18}$$

Since the boundary layer pressure does not vary in the vertical direction, it is equal to the inviscid pressure in the outer flow, so from the Bernoulli Equation we have

$$\frac{\partial p}{\partial x}=-\rho U\frac{\partial U}{\partial x} \tag{13.7.19}$$

where U is the inviscid velocity which can be x-dependent unlike U_∞ which we treated as a constant. To develop the momentum integral equation, we integrate the momentum equation from $y=0$ to $y=h$, where h is outside the boundary layer

$$\int_0^h\left(u\frac{\partial u}{\partial x}+v\frac{\partial u}{\partial y}-U\frac{\partial U}{\partial x}\right)dy=\nu\int_0^h\frac{\partial^2 u}{\partial y^2}\,dy \tag{13.7.20}$$

The right hand side integrates directly

$$\nu\int_0^h\frac{\partial^2 u}{\partial y^2}\,dy=\nu\left.\frac{\partial u}{\partial y}\right|_0^h=\frac{\mu}{\rho}\left[0-\left.\frac{\partial u}{\partial y}\right|_{y=0}\right]=-\frac{\tau_w}{\rho} \tag{13.7.21}$$

where we have used the fact that $\partial u/\partial y=0$ at $y=h$, and τ_w is the wall shear. On the left hand side, substitute from continuity that

$$v=-\int_0^y\frac{\partial u}{\partial x}\,dy \tag{13.7.22}$$

and address the second term

$$-\int_0^h\left(\frac{\partial u}{\partial y}\int_0^y\frac{\partial u}{\partial x}\,dy\right)dy \tag{13.7.23}$$

This term can be integrated by parts, by letting

$$\tilde{u}=-\int_0^y\frac{\partial u}{\partial x}\,dy,\quad d\tilde{v}=\frac{\partial u}{\partial y}dy\quad\rightarrow$$
$$d\tilde{u}=-\frac{\partial u}{\partial x}\,dy,\quad \tilde{v}=u \tag{13.7.24}$$

so the integral is

$$\int_0^h\tilde{u}\,d\tilde{v}=\left[\tilde{u}\,\tilde{v}\right]\Big|_{y=0}^{y=h}-\int_{y=0}^{y=h}\tilde{v}\,d\tilde{u} \tag{13.7.25}$$

Inserting the forms in Eq. (13.7.24) leads to

$$-\left[u\Big|_{y=0}^{y=h}\int_0^h\frac{\partial u}{\partial x}\,dy\right]+\int_{y=0}^{y=h}u\frac{\partial u}{\partial x}\,dy$$
$$=-U\int_0^h\frac{\partial u}{\partial x}\,dy+\int_{y=0}^{y=h}u\frac{\partial u}{\partial x}\,dy \tag{13.7.26}$$

Inserting the result of Eq. (13.7.25) into Eq. (13.7.20)

$$\int_0^h\left(2u\frac{\partial u}{\partial x}-U\frac{\partial u}{\partial x}-U\frac{\partial U}{\partial x}\right)dy=-\frac{\tau_w}{\rho} \tag{13.7.27}$$

or with some rearrangement

$$\int_0^h\frac{\partial}{\partial x}\left[u(U-u)\right]dy+\frac{\partial U}{\partial x}\int_0^h(U-u)\,dy=\frac{\tau_w}{\rho} \tag{13.7.28}$$

Now we can use our definitions of the displacement thickness, Eq. (13.7.3), and the momentum thickness, Eq. (13.7.14),

$$\frac{\tau_w}{\rho}=\frac{d}{dx}\left(U^2\theta\right)+\delta^* U\frac{dU}{dx} \tag{13.7.29}$$

which is the "momentum integral equation" for 2D, incompressible, boundary layer flow over a flat plate. In general, it holds for either laminar or turbulent flows, because we are leaving the form of τ_w unspecified. Under laminar conditions, of course, Eq. (13.7.10) defines the wall shear. For turbulence, the velocities in Eq. (13.7.28) and Eq. (13.7.29) are their time-averages. Turbulent forms of τ_w are developed in Section 13.5 from data and then used to derive the corresponding boundary layer thickness, Eq. (13.5.8), using Eq. (13.7.29).

13.8 Solutions to the Momentum Integral Equation

One solution of Eq. (13.7.29) is to assume $dp/dx = 0$ as in the Blasius equations. That means $dU/dx = 0$ and the laminar equation becomes

$$\frac{\tau_w}{\rho U^2} = \frac{d\theta}{dx} = 0.664\frac{d}{dx}\left[\left(\frac{\nu x}{U}\right)^{1/2}\right] = 0.332(\mathrm{Re}_x)^{-1/2} \tag{13.8.1}$$

where we used Eq. (13.7.17) for the momentum thickness, θ, and define $\mathrm{Re}_x = Ux/\nu$. To check this answer, compute the wall shear stress from the Blasius solution,

$$\tau_w = \mu\frac{\partial u}{\partial y}\bigg|_{y=0} \tag{13.8.2}$$

and using the Blasius variables τ_w becomes

$$\tau_w = \mu U\left(\frac{U}{\nu x}\right)^{1/2} f''(0) \tag{13.8.3}$$

The Blasius solution yields $f''(0) = 0.332$ and then Eq. (13.8.3) is identical to Eq. (13.8.1). It is not surprising that the system of equations is consistent with the Blasius solution.

An advantage of the momentum integral approach is that we can apply it using approximations to the boundary layer velocity profile that are analytic rather than numeric. We explored this in Section 12.2. A general approach to solving Eq. (13.7.29) is to consider the dimensionless velocity, u, scaled on the inviscid value, U, and the boundary layer coordinate scaled on the unknown boundary layer thickness

$$g(\zeta) = \frac{u}{U}, \quad \zeta = \frac{y}{\delta(x)} \tag{13.8.4}$$

similar to Section 12.2. Using these scales, the displacement and momentum thicknesses are, respectively,

$$\delta^* = \alpha_1\delta, \quad \alpha_1 = \int_0^1 (1 - g)\, d\zeta$$
$$\theta = \alpha_2\delta, \quad \alpha_2 = \int_0^1 g(1 - g)\, d\zeta \tag{13.8.5}$$

Likewise, the wall shear is

$$\frac{\tau_w}{\rho} = \frac{\mu}{\rho}\frac{\partial u}{\partial y}\bigg|_{y=0} = \beta_1\frac{\nu U}{\delta}, \qquad \beta_1 = g'(0) \tag{13.8.6}$$

In the absence of a pressure gradient, the momentum integral equation is

$$\frac{\tau_w}{\rho} = U^2\frac{d\theta}{dx} \tag{13.8.7}$$

and substituting Eq. (13.8.6) and Eq. (13.8.5) results in

$$\frac{\nu U}{\delta}\beta_1 = \alpha_2 U^2\frac{d\delta}{dx} \tag{13.8.8}$$

which rearranged is

$$\frac{\nu\beta_1}{U\alpha_2} = \frac{1}{2}\frac{d}{dx}(\delta^2) \tag{13.8.9}$$

We can integrate Eq. (13.8.9) to find

$$\frac{\nu\beta_1}{U\alpha_2}\int_0^x dx = \frac{1}{2}\int_0^\delta d(\delta^2) \quad \rightarrow \quad \delta^2 = \frac{\nu\beta_1}{U\alpha_2}x \tag{13.8.10}$$

so the boundary layer thickness is

$$\delta = \left(\frac{2\beta_1}{\alpha_2}\right)^{1/2}\left(\frac{\nu x}{U}\right)^{1/2} \tag{13.8.11}$$

The form in Eq. (13.8.11) permits us to readily adopt polynomial approximations to $g(\zeta)$, which can be substituted into Eqs. (13.8.5) and Eq. (13.8.6) to solve for α_2 and β_1. The linear, quadratic and cubic forms for laminar boundary layers are

$$g = \zeta \text{ (linear)}, \quad 2\zeta - \zeta^2 \text{ (quadratic)}, \quad \frac{3}{2}\zeta - \frac{1}{2}\zeta^3 \text{ (cubic)} \tag{13.8.12}$$

When these profiles are inserted into Eqs. (13.8.5) and Eq. (13.8.6) to find α_2 and β_1 for Eq. (13.8.11), the result is $\delta = 3.46(\nu x/U_\infty)^{1/2}$ for the linear profile, $\delta = 5.47(\nu x/U_\infty)^{1/2}$ for the quadratic profile and

$\delta = 4.64(\nu x/U_\infty)^{1/2}$ for the cubic profile. This information was reported in Section 12.2.

Going a step further, what is known as the Kármán–Pohlhausen approximation employs a quartic polynomial, $g(\zeta) = a_0 + a_1\zeta + a_2\zeta^2 + a_3\zeta^3 + a_4\zeta^4$. To find the five constants we again have no slip at the wall, $g(0) = 0$, matching the inviscid velocity at the edge of the boundary layer, $g(1) = 1$, with the slope zero there, $g'(1) = 0$. For the cubic polynomial we assumed no pressure gradient, but now we can include it by evaluating the x-momentum equation, Eq. (13.7.18), at the wall boundary

$$\mu \left.\frac{\partial^2 u}{\partial y^2}\right|_{y=0} = -\rho U(x)\frac{dU(x)}{dx} \rightarrow g''(0) = -\frac{\delta^2}{\nu}\frac{dU(x)}{dx} = -\Lambda(x) \tag{13.8.13}$$

since u and v are zero there. We have replaced the pressure gradient with the Bernoulli term in Eq. (13.7.19). $\Lambda(x)$ is a dimensionless measure of the pressure gradient from the inviscid flow. The final condition for the quartic is zero curvature at the boundary layer edge $g''(1) = 0$. Further development of this approach leads to conditions of flow separation along the plate, the location where $(\partial u/\partial x)|_{y=0} = 0$.

Summary

Turbulent flow is a complex topic, and here we have presented an outline of how it relates to biofluid mechanics for a limited range of biological examples. Turbulence in an industrial application is not usually damaging to the fluid, say air or water. However, biofluids can have living components, like blood. So turbulence not only affects the pressure–flow relationship, but can damage the blood or trigger biochemical reactions like clotting which alter the fluid flow properties. In this chapter we examined several applications of turbulent flows including flow over bluff bodies such as a human leg or torso, fully developed turbulent pipe flow related to respiratory airflow and blood flow, developing turbulent flow over a flat plate to gauge turbulent entrance lengths in a tube such as a blood vessel or airway, and jet turbulence encountered in cough and squid propulsion. Our approach has been to focus on specific applications in a field which is highly complex.

Problems

13.1 A time-decaying, cellular flow which can be used to model homogeneous turbulence has the Eulerian velocity field

$$u = u_0 e^{-at} \sin(bx)\cos(by)$$
$$v = -u_0 e^{-at} \cos(bx)\sin(by)$$

a. What is the stream function, $\psi(x, y, t)$, for this flow? Let $\psi(y = 0) = 0$.
b. Make a 3D plot of the stream function on the domain $-2\pi \le x \le 2\pi$, $-2\pi \le y \le 2\pi$ at $t = 0$ for $u_0 = 1$, $b = 1$, $a = 0.2$.
c. Plot streamlines at $t = 0$ for $u_0 = 1$, $b = 1$, $a = 0.2$ over the region $0 \le x \le 2\pi$, $0 \le y \le 2\pi$.
d. Plot the velocity vector field corresponding to part (c) and describe your results.

e. For $u_0 = 1$, $b = 1$, $a = 0.2$ use a numerical approach to plot the pathlines $X(t), Y(t)$, i.e. the mapping, of fluid particles starting at $(X_0, Y_0) = (\pi/2, \pi/8), (\pi/2, \pi/4), (\pi/2, 3\pi/8)$ for $0 \le t \le 25$. Discuss your result.

f. Now plot the Lagrangian speed of the particle starting at $(X_0, Y_0) = (\pi/2, \pi/4)$ for the same parameter values as in part (e). Describe your result.

13.2 The aorta is the large artery coming directly out of the heart's left ventricle, see Figure 1.3 and Figure 12.18. The proximal segment, before any of the major branches go to the head and upper extremities, is called the ascending aorta. It has a typical diameter of 3 cm in an adult male. In Section 10.5 we found the high strain rate viscosity of blood is three to four times that of water. Let $\mu = 0.03$ poise, and blood density $\rho = 1.06\,\text{g/cm}^3$, which is slightly larger than water. The heart rate is HR in beats per minute and for each heart beat cycle period is divided into the systolic time, T_s, and the diastolic time, T_d. During systole the heart ejects the stroke volume, SV. Under resting conditions SV $= 70$ ml, HR $= 60$ beats/ min and $T_s/T_d = 1/2$. However, during exercise SV $= 110$ ml, HR $= 160$ beats / min and $T_s/T_d = 1/1$. For both resting and exercise conditions do the following.

a. Calculate the cardiac output, CO, in L/min.
b. Calculate the flow rate, Q, during one systolic period in units of ml/s.
c. Calculate the aortic Reynolds number for systole.
d. Calculate the wall shear τ_w assuming fully developed, steady conditions for a straight tube. For the resting condition, give τ_w for both laminar and turbulent models. Discuss your answers.
e. In Section 7.5 we discussed the power related to flow in a pipe as $P_f = Q\Delta P$. Calculate the power per axial length, P_f/L, for the rest condition and the turbulent exercise condition. Is there a benefit in maintaining laminar flow at rest?

13.3 A normal, resting inspiratory tidal volume is $V_T = 500\ \text{cm}^3$ for an adult at a breathing frequency of $f = 12$ breaths/ min. For each 5 second breathing period, $T_b = 5$ s, the inspiration time, T_i, is about half of the expiratory time, i.e. the I : E ratio is 1 : 2. So $T_i = T_b/3$ and the average inspiratory flow rate is $Q = V_T/T_i$.

a. What is Q for resting conditions? For heavy exercise let $V_T = 2,500\ \text{cm}^3$ at a breathing frequency of 30 breaths/ min and an I : E ratio of 1 : 1. What is Q for exercise?
b. Let the airway tree diameters obey the equation $d_n = d_0 2^{-n/3}$ where d_0 is the tracheal diameter and n is the airway generation. Calculate the total cross-sectional area of the airway tree as a function of n, call it A_n. Assume a perfectly bifurcating system so there are 2^n identical tubes at each generation. Plot A_n for $d_0 = 1.5$ cm and $0 \le n \le 15$. You can treat n as a continuous variable.
c. The average velocity during inspiration is $U_n = Q/A_n$. Plot U_n for $d_0 = 1.6$ cm and $0 \le n \le 15$, comparing the resting and exercise flow rates.
d. The Reynolds number at generation n can be defined as $Re_n = d_n U_n/\nu$. Plot Re_n for $0 \le n \le 15$ comparing rest to exercise. Where would you expect turbulent flow?
e. Plot the wall shear, τ_w, assuming fully turbulent flow. Plot it for the values of n where $Re_n > 2,000$ as you see in part (d).

13.4 In Section 7.5 we examined Murray's law for laminar flow. Suppose the involved vessels have turbulent flow where the pressure drop is given by

$$\Delta p = \lambda \frac{L}{2r}\frac{\rho}{2}U^2 = \lambda \frac{L}{2r}\frac{\rho}{2}\left(\frac{Q}{\pi r^2}\right)^2$$

and λ is a pipe friction coefficient. Using the metabolic power consumption proportional to tube volume, what is the relationship between the parent and daughter radii?

13.5

a. Find numerical solutions to the system in Eq. (13.1.1) and Eq. (13.1.2), plotting the trajectories for $\theta = \pi/3$ and two heights, $Fr^2 = 1$, $Fr^2 = 1/2$. For the two heights, plot three trajectories each for $\beta = 0, 1, 3$. Plot them to impact at $Y_p = 0$ by adjusting the time duration.
b. For $\beta = 1$, which height loses more dispersal distance, compared to an inviscid fluid, due to turbulent drag? Why?
c. For $\beta = 3$ what has happened to the height advantage for dispersal? Why?
d. For $\beta = 3$ at the height $Fr^2 = 1$ compare trajectories to impact for $\theta = \pi/6$, $\theta = \pi/4$, $\theta = \pi/3$. Describe your results.

13.6 A skydiver drops from an airplane and is in freefall until opening the parachute. During freefall the skydiver's velocity continues to accelerate until reaching terminal velocity, when the air drag resistance balances the weight. The flow is turbulent, so a force balance is

$$m\frac{dv}{dt} = mg - kv^2$$

where k includes the drag coefficient from the shape and projected area.

a. Make the governing equation dimensionless using the definitions $V = v/V_s$, $T = t/T_s$ where the scales are $V_s = (mg/k)^{1/2}$, $T_s = (m/gk)^{1/2}$
b. Solve the ODE from part (a) with initial condition $V(0) = 0$ and plot your solution.
c. What is the value of V at steady state, i.e. the dimensionless terminal velocity V_t? At what value of T does the velocity reach 99% terminal?
d. A typical dimensional terminal velocity is $v_t = 50$ m/s, which equals 112 miles/hr. Using this information, calculate k for a 70 kg skydiver.
e. At what dimensional value of t does the velocity reach $0.99\,v_t$ for the above conditions?
f. The recommended parachute deployment altitude is 1 km for beginners. For an initial altitude of 4 km, how much time is in freefall? Hint: consider the dimensionless position, $Y(T)$, positive downward with initial value $Y(0) = 0$ at the airplane.
g. How does this compare to the erythrocyte sedimentation rate example in Section 2.7?

13.7

a. According to the solutions for (u_r, u_z) given in Eq. (13.6.10), u_z is positive, but u_r changes sign as η increases from zero. At what positive value of $\eta = \eta_c$ does this happen?
b. Make a plot of the vector field (u_z, u_r) for $K = 100,000$ cm^4/s^2 so the u_r-component is vertical and the u_z-component is horizontal. Plot the field for $0 \leq z \leq 15$ cm, $0 \leq r \leq 2$ cm. You may have to modify the relative vector length scheme so that they are all easily visible, for example a log function. Add to this plot the straight line $r_c(z)$ corresponding to your solution of η_c. What does this line demarcate for the vector field? What is the jet Reynolds number value, Re_j?
c. Plot $u_z(r = 0)$ for $3 \leq z \leq 15$ cm and let $K = 100,000;\ 50,000;\ 10,000$ cm^4/s^2. Which ones compare favorably to Figure 13.24, assuming $u_z >> u_r$? Also plot the flow rate, Q, for the same range of z at the three values of K. What is the effect of K?
d. Plot u_r for the flow in part (b) at $z = 3, 5, 10, 15$ for $0 \leq r \leq 5$. Describe your results. How does this match up with your vector field plot of part (b)?
e. Make a contour plot of speed, s, for the flow in part (b) with z on the horizontal axis, $3 \leq z \leq 15$, and r vertical. $0 \leq r \leq 4$.

   Plot the contours $s = 5, 10, 15, 50, 100, 200, 400, 600$ cm/s. Choose an option for grayscale shading and compare your results to Figure 13.24. How might differences be due to anatomy?
f. Coughs have thousands of liquid droplet aerosols which can spread disease. For small enough particles we can assume Stokes drag, so the

governing equations for the aerosol particle trajectory are

$$m\frac{\partial^2 z_p(t)}{\partial t^2} + 3\pi\mu d_p\left(\frac{\partial z_p(t)}{\partial t} - u_z\left(r_p(t), z_p(t)\right)\right) = 0$$

$$m\frac{\partial^2 r_p(t)}{\partial t^2} + 3\pi\mu d_p\left(\frac{\partial r_p(t)}{\partial t} - u_r\left(r_p(t), z_p(t)\right)\right) = -mg$$

The particle mass is $m = \rho_p(4/3)\pi\left(d_p/2\right)^3$, where d_p is the particle diameter and ρ_p is the particle density. Plot aerosol particle trajectories for $d_p = 5$, 10, 20 μm starting at $r(0) = 2$ cm, $z(0) = 3$ cm. Let the initial particle velocity equal the local fluid velocity at the initial position. Try the time interval $0 \le t \le 1$ s. Describe your results. Which particle is being cleared by gravity better than the others? Which particle is the most dangerous for spreading disease? Why?

13.8 Using the momentum integral approach, the boundary layer thickness $\delta(x)$ is given by Eq. (13.8.11).

a. Show that the linear, quadratic, and cubic forms of the boundary layer velocity profile lead to $\delta = 3.46(\nu x/U_\infty)^{1/2}$, $\delta = 5.47(\nu x/U_\infty)^{1/2}$ and $\delta = 4.64(\nu x/U_\infty)^{1/2}$.
b. Let the boundary layer velocity profile be approximated by $g = \sin(\pi\zeta/2)$. What is its corresponding$\delta(x)$?

Image Credits

Figure 13.2a Source: Figure 4 from Coutanceau, M. and R. Bouard (1977). "Experimental determination of the main features of the viscous flow in the wake of a circular cylinder in uniform translation. Part 1. Steady flow." *Journal of Fluid Mechanics* 79(2): 231–256.

Figure 13.2b Source: Gary Koopman

Figure 13.2c Source: Courtesy of Sage Action, inc. "Helium Filled Soap Bubbles" were used for visualising the flow around the cylinder.

Figure 13.2d Source: Bob Cahalan, NASA GSFC

Figure 13.4 Source: Fig 1a from Craven BA, Hargather MJ, Volpe JA, Frymire SP, Settles GS. Design of a high-throughput chemical trace detection portal that samples the aerodynamic wake of a walking person. *IEEE Sensors Journal.* 2014; 14(6):1852–66.

Figure 13.9 Source: Fig 3 of Mullin T. Experimental studies of transition to turbulence in a pipe. In: Davis SH, Moin P, editors. *Annual Review of Fluid Mechanics*, Vol 432011. p. 1–24.

Figure 13.10a Source: Osborne Reynolds (public domain)

Figure 13.10b Source: Osborne Reynolds (public domain)

Figure 13.11 Source: Moody diagram reprinted with permission from Moody (1944).

Figure 13.14a Source: Fig 5 in Eckmann DM, Grotberg JB. Experiments on transition to turbulence in oscillatory pipe flow. *J Fluid Mech.* 1991; 222:329–50.

Figure 13.14b Source: Fig 6 in Eckmann DM, Grotberg JB. Experiments on transition to turbulence in oscillatory pipe flow. *J Fluid Mech.* 1991; 222:329–50.

Figure 13.19 Source: Fig 1 in Isabey D, Chang HK. Steady and unsteady pressure-flow relationships in central airways. *Journal of Applied Physiology.* 1981; 51(5):1338–48.

Figure 13.20 Source: Fig 3 in Isabey D, Chang HK. Steady and unsteady pressure-flow relationships in central airways. *Journal of Applied Physiology.* 1981; 51(5):1338–48.

Figure 13.22 Source: Fig 4 of Matsubara M, Alfredsson PH. Disturbance growth in boundary layers subjected to free-stream turbulence. *J. Fluid Mech.* 2001; 430:149–68.

Figure 13.25 Source: Figure 5 from Dimotakis PE. The mixing transition in turbulent flows. *J. Fluid Mech.* 2000; 409:69–98.

Figure 13.26a Source: Fig 4a in Tang JLW, Nicolle AD, Klettner CA, Pantelic J, Wang LD, Bin Suhaimi A, et al. Airflow dynamics of human jets: sneezing and breathing – potential sources of infectious aerosols. *PLoS One.* 2013; 8(4).

Figure 13.26b Source: Fig 4b in Tang JLW, Nicolle AD, Klettner CA, Pantelic J, Wang LD, Bin Suhaimi A, et al. Airflow dynamics of human jets: sneezing and breathing – potential sources of infectious aerosols. *PLoS One.* 2013; 8(4).

Figure 13.29a Source: Alan Hawk, Historical Collections, National Museum of Health and Medicine (public domain).

Figure 13.29b Source: Science & Society Picture Library/Getty

Figure 13.29c Source: Image of the Chitra valve, by Phulwari28, licensed under CC BY-SA 3.0 https://commons.wikimedia.org/wiki/File:Chitra_Valve.jpg

Figure 13.29d Source: Layne Kennedy/ Getty

Figure 13.30 Source: Author original adapted from Dasi, L. P., L. Ge, et al. (2007). "Vorticity dynamics of a bileaflet mechanical heart valve in an axisymmetric aorta." *Physics of Fluids* 19: 067105, with gratitude to M. Heitkemper, L. P. Dasi and A. P. Yoganathan.

14 Constitutive Equations 3: Viscoelastic Fluids

Introduction

This is our final foray into constitutive equations for biofluids. In Chapter 6 we discussed inviscid fluids whose stress tensor is comprised only of the pressure, $T_{ij} = -p\delta_{ij}$. Inserting this form into the Cauchy momentum equation, Eq. (5.2.2), gave us Euler's equation, Eq. (6.1.5), and its integral along a streamline, the Bernoulli equation, see Eqs. (6.2.7–6.2.9). Next we added viscous effects through the extra-stress tensor, τ_{ij}, to give the full stress tensor as $T_{ij} = -p\delta_{ij} + \tau_{ij}$. For a Newtonian fluid the extra-stress tensor is proportional to the strain rate tensor through the constant viscosity, μ, as $\tau_{ij} = 2\mu E_{ij}$. Then in Chapter 10 we analyzed generalized Newtonian fluids whose constitutive relationship looks like that of a Newtonian fluid, except the effective viscosity, μ_{eff}, depends on the instantaneous value of the invariant strain rate scalar $\dot{\gamma} = \sqrt{2E_{ij}E_{ij}}$, so that $\tau_{ij} = 2\mu_{eff}(\dot{\gamma})E_{ij}$. The nomenclature describing generalized Newtonian fluids includes shear-thinning (pseudoplastic) and shear-thickening (dilatants) fluids as well as fluids with a yield stress. We explored various yielded models including power-law, Carreau, Carreau–Yasuda, modified Cross and Quemada. For the broad topic of fluids with a yield stress, Herschel–Bulkley fluids, we examined yield-stress + power-law and Casson models. The shear strain dependence of the viscosity, as well as the yield behavior, comes from the interaction of suspended particles and entangled macromolecules, see Figure 10.4, that form these complex biofluids like apple sauce, mayonnaise, blood and mucus.

In the present chapter, we extend further into complex fluid behavior by allowing the fluid to have "memory." What do we mean by memory? When a stress is applied to an elastic solid it creates a deformation, and when that stress is removed the solid returns to its original configuration. The elastic solid has "memory" of its initial state. When a fluid has memory it is said to be viscoelastic. So instead of the viscous effect depending on the instantaneous strain rate, it can incorporate the history of the strain rate. To that end, the constitutive equation for a viscoelastic fluid is, itself, a partial differential equation. These may include time and spatial derivatives of the stress tensor and velocity field. As a consequence, it is coupled to the Cauchy momentum equation, so the two partial differential equations must be solved simultaneously. As we saw with other non-Newtonian fluids, viscoelastic behavior also arises from the entangled macromolecules within the fluid. They exhibit elastic properties like resistance to bending and stretching. It turns out that blood, mucus and synovial fluid are all viscoelastic to varying degrees, which can influence their function and vary with disease. A familiar personal experience is the increase of viscosity and elastic behavior of nasal mucus during an upper respiratory infection.

Topics Covered

We begin the chapter by introducing basic concepts of viscoelastic materials and the well-known Maxwell fluid model which consists of a spring and linear dashpot in series. From this configuration, step inputs of displacement or force are examined, revealing the concept of a stress-relaxation time, $\lambda = \mu/k$, where μ is the dashpot viscosity and k is the spring constant. This is the time it takes for the force to relax to a steady (lower) state following a step change in displacement. Then we examine oscillatory shear forcing, with k replaced by the shear modulus, G, displacement replaced by shear strain, γ, and force by shear stress, τ. Assuming the forms $\tau = \tau_0 e^{i\omega t}$, $\gamma = \gamma_0 e^{i\omega t}$, the response is treated first as a viscoelastic solid, $\tau_0 = G^* \gamma_0$, relating stress amplitude, τ_0, to strain amplitude, γ_0. The complex shear modulus is $G^* = G' + iG''$ where the shear storage modulus, G', and the loss modulus, G'', are functions of ω, G, μ and examined in detail. From the analysis, an important dimensionless parameter arises, $De = \omega\lambda$, the Deborah number. $De << 1$ indicates low frequency oscillations where the relaxation time is short compared to the oscillation period and viscous effects dominate. For $De >> 1$ the system is at high frequency with the oscillation period short compared to the relaxation time and elastic effects dominate. Then the deformation is treated as a viscoelastic fluid, $\tau_0 = \mu^*(i\omega\gamma_0)$, relating stress amplitude to the strain rate amplitude, $i\omega\gamma_0$. The complex viscosity is $\mu^* = \mu' - i\mu''$, which is easily shown to be related to $\mu' = G''/\omega$ and $\mu'' = G'/\omega$. We revisit Stokes' second problem of Section 8.2, the semi-infinite fluid bounded by a wall oscillating in its own plane, and study the influence of De on the velocity profiles compared to the Newtonian $De = 0$ case. Similarly, we analyze oscillatory channel flow, as in Section 8.3. Because many biofluids are mixtures of long chain polymers of differing lengths and properties, a single relaxation time does not model the system well, so a generalized Maxwell model is explored which consists of multiple Maxwell elements in parallel, each with a different spring constant and viscosity. Trying to apply the Maxwell model directly to a flow problem shows an important defect, that it does not transform properly under a Galilean coordinate transformation. So more sophisticated models are explored, including the upper convected Maxwell and Oldroyd B. Data for G', G'', and $|\mu^*|$ as functions of oscillatory frequency, ω, and μ_{eff} as a function of steady shear rate, $\dot{\gamma}$, are discussed for blood, mucus and synovial fluid. A very important relationship is the Cox–Merz rule (Cox and Merz, 1958), which states that the steady effective viscosity is equal to the magnitude of the complex viscosity, when $\dot{\gamma} = \omega$, i.e. $\mu_{eff}(\dot{\gamma}) = |\mu^*(\omega)|_{\dot{\gamma}=\omega}$. We show it holds fairly well for these biofluids. To gain more insight into synovial fluid and joint space mechanics, we revisit the squeeze film model of Chapter 11, but use an oscillatory force as one might experience in walking or running. For a Maxwell fluid model, we examine how the displacement amplitude of the plates depends on the parameters and its implications for joint mechanics. Then a unique property of viscoelastic fluids under a simple shearing flow is their ability to generate both a shear stress in the direction of flow, as well as a normal stress perpendicular to flow. These normal stresses account for fluid climbing a spinning rod, like cake mix on a beater shaft. For completeness we show the equivalent approach for viscoelastic fluids using memory functions, which turns out to be an integral of the Maxwell model differential equations, and then a viscoelastic solid model called the Kelvin–Voigt model.

14.1 Viscoelastic Materials

Shear deformation of a material is shown in Figure 14.1 where the displacement ΔU_1 results from the shear stress τ. The shear strain is $\gamma = \Delta U_1/\Delta x_2 = \partial U_1/\partial x_2$. Viewed as an elastic solid, the constituive equation is

$$\tau = G\gamma = G\frac{\partial U_1}{\partial x_2} \tag{14.1.1}$$

where G is the elastic shear modulus and τ is the shear stress.

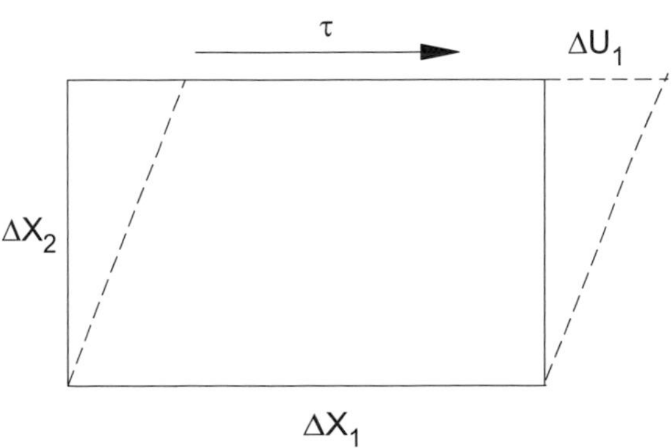

Figure 14.1 Shear deformation of a material.

To model viscoelastic materials, we are interested in combining this elastic term into the constitutive equation for a fluid, where the fluid velocity and its gradients are the working dependent variables. So, taking the time derivative of Eq. (14.1.1) gives us

$$\frac{\partial \tau}{\partial t} = G\dot{\gamma} = G\frac{\partial u_1}{\partial x_2} \tag{14.1.2}$$

where $u_1 = \dot{U}_1$ is the velocity. This relationship is in contrast to Newtonian viscous fluid behavior in which

$$\tau = \mu\frac{\partial u_1}{\partial x_2} = \mu\dot{\gamma} \tag{14.1.3}$$

where μ is the constant "shear viscosity" or "dynamic viscosity." Note that the Greek letter η is sometimes used for viscosity.

A viscoelastic fluid can be described by a combination of the ideas in Eqs. (14.1.2) and (14.1.3). The simplest example is a linear combination of the effects, known as a Maxwell fluid

$$\tau + \frac{\mu}{G}\frac{\partial \tau}{\partial t} = \mu\dot{\gamma} \tag{14.1.4}$$

In Eq. (14.1.4) we see that the $\partial\tau/\partial t$ term is multiplied by the shear stress-relaxation time, $\lambda = \mu/G$. In the limit of large G, the relaxation time is short and the material behaves as a viscous fluid with the familiar balance between $\tau$and the viscous shear rate on the right hand side as in Eq. (14.1.3). However, in the limit of large relaxation time, $(\mu/G)\partial\tau/\partial t$ dominates the left hand side of Eq. (14.1.4). Then the viscosity, μ, cancels as the behavior is dominated by elastic forces resulting in Eq. (14.1.2). The combination of terms captures the behavior in between these two limiting cases.

The Maxwell model is based on viscous (dashpot) and elastic (spring) components in series as shown in Figure 14.2. So let us use terminology from that setting as it is usually taught in an introductory physics course common to a wide spectrum of student backgrounds. Along the way we will point out the equivalence for shear deformation. The spring length is U_s and the dashpot position is U_d. For shear deformation these are representative of shear strain γ_s and γ_d, respectively. The spring constant is k, which represents the shear modulus, G, and the dashpot has

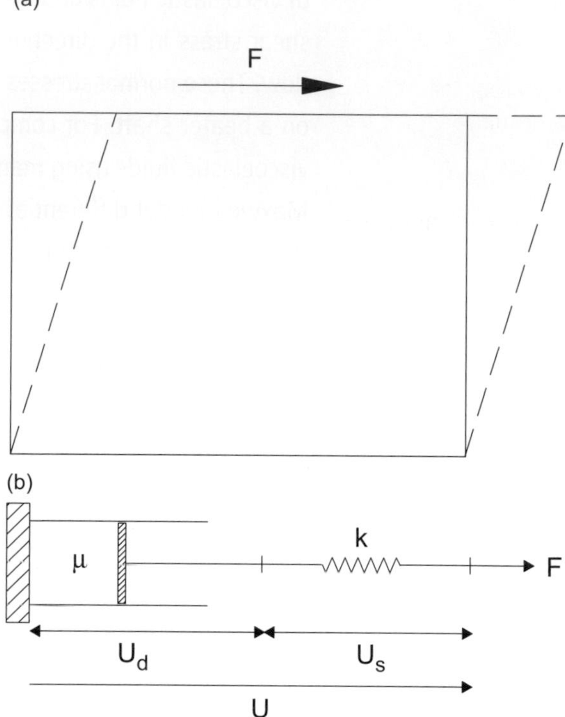

Figure 14.2 Maxwell fluid model for (a) shear deformation of a material by force F is (b) a spring and dashpot in series.

constant fluid viscosity, μ, the same quantity in both settings. The force, F, is representative of the shear stress, τ.

The force of the dashpot, F_d, is linearly related to its velocity as the piston moves through the viscous fluid filling the cylinder. The force of the spring, F_s, is linearly related to its change in length or displacement

$$F_d = \mu\dot{U}_d, \quad F_s = kU_s \tag{14.1.5}$$

The distance U is the sum of U_s and U_d,

$$U = U_d + U_s \tag{14.1.6}$$

so that U represents the total shear strain, $\gamma = \gamma_d + \gamma_s$. The time derivative of Eq. (14.1.6) is

$$\dot{U} = \dot{U}_d + \dot{U}_s \tag{14.1.7}$$

Using Eq. (14.1.5) in Eq. (14.1.7) we get

$$\dot{U} = \frac{F_d}{\mu} + \frac{\dot{F}_s}{k} \tag{14.1.8}$$

We know that in the absence of a mass, the forces are all equal,

$$F_d = F_s = F \tag{14.1.9}$$

so Eq. (14.1.8) becomes

$$\dot{U} = \frac{F}{\mu} + \frac{\dot{F}}{k} \tag{14.1.10}$$

which is equivalent to Eq. (14.1.4) for the Maxwell fluid shear model.

14.2 Maxwell Fluid Response to Step Inputs

Suppose we impose a step change in U, such that

$$U(t) = U_0 H(t) \tag{14.2.1}$$

where U_0 is a constant and H(t) is the Heaviside function, a unit step at $t = 0$,

$$H(t) = \begin{cases} 0 & t < 0 \\ 1 & t \geq 0 \end{cases} \tag{14.2.2}$$

Since U is the constant value U_0 for $t > 0$,

$$\dot{U}(t > 0) = 0 \tag{14.2.3}$$

then its time derivative is zero. So the form of Eq. (14.1.10) becomes

$$F + \frac{\mu}{k}\dot{F} = 0 \tag{14.2.4}$$

which is satisfied by

$$F = F_0 e^{-\frac{t}{\mu/k}} H(t) \tag{14.2.5}$$

We solve for the integration constant, F_0, by imposing the initial condition that assumes the initial displacement is instantaneously provided by only the spring, there being insufficient time for flow to develop in the dashpot

$$F(0) = kU(0) \quad \Rightarrow \quad F_0 = kU_0 \tag{14.2.6}$$

So the final result is

$$F = kU_0 e^{-\frac{t}{\mu/k}} H(t) \tag{14.2.7}$$

So the step change in strain gives us a step change in stress by the spring instantly lengthening. This is followed by a decay in stress as the spring shortens and pulls the dashpot through the fluid. The ratio $\mu/k = \lambda$ is the relaxation time constant for the Maxwell model. In takes longer for the spring to return to its unstressed length if the fluid is more viscous or if the spring constant is smaller, i.e. less stiff. This phenomenon is called "stress relaxation," and is shown in Figure 14.3(a, b).

We may also consider that there is a step change in F with a resultant displacement U. Let

$$F = F_0 H(t) \tag{14.2.8}$$

Inserting Eq. (14.2.8) into Eq. (14.1.10) yields

$$\dot{U} = F_0\left(\frac{H(t)}{\mu} + \frac{\dot{H}(t)}{k}\right) \tag{14.2.9}$$

which can be integrated as

$$U = F_0\left(\frac{t}{\mu} + \frac{1}{k}\right) + c \tag{14.2.10}$$

The constant of integration, c, is solved from the initial condition

$$U(0) = \frac{F(0)}{k} = \frac{F_0}{k} \tag{14.2.11}$$

which we saw in Eq. (14.2.6). This condition renders $c = 0$, and the solution for U(t) is

$$U(t) = F_0\left(\frac{1}{k} + \frac{t}{\mu}\right) H(t) \tag{14.2.12}$$

The strain continues as a linear function of time once the constant force is applied at $t = 0$, see Figure 14.4. This is called "creep" for a solid. For a fluid it is simply flow. The slope of the linear time function is F_0/μ so the creep is slower when the viscosity is larger or the force is smaller.

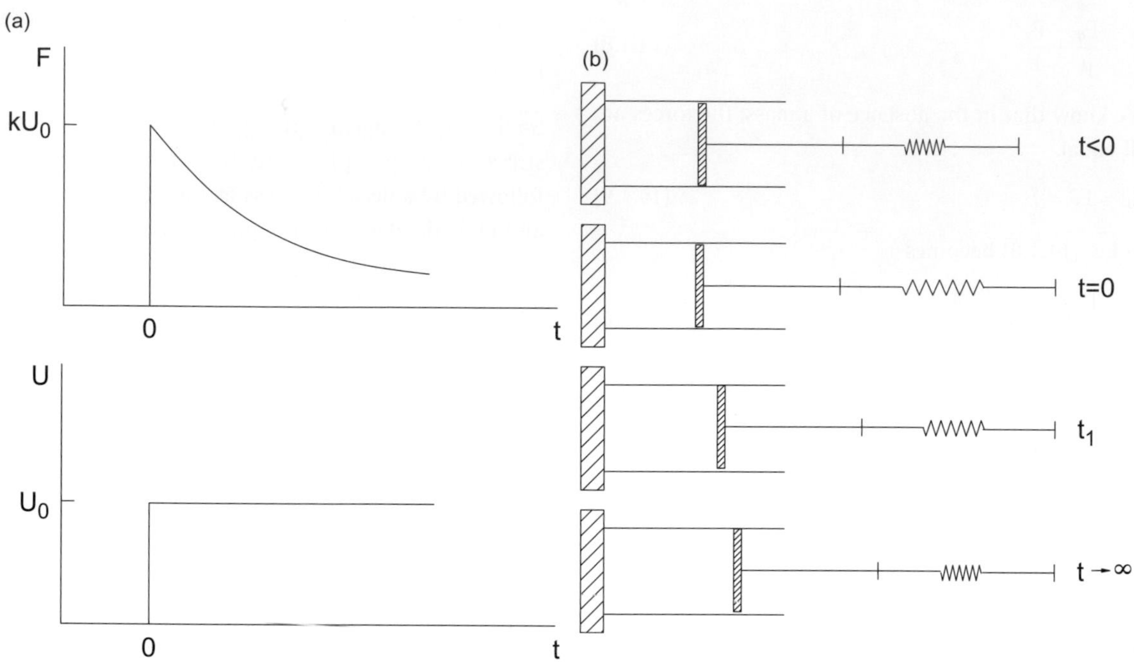

Figure 14.3 Stress relaxation for step change at t = 0. (a) Force response to step change in displacement. (b) Time sequence of dashpot and spring for stress relaxation showing spring shortening to equilibrium length pulling the piston through the viscous dashpot fluid.

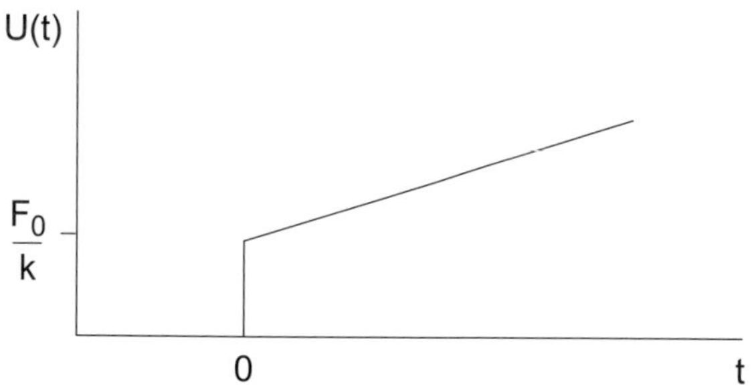

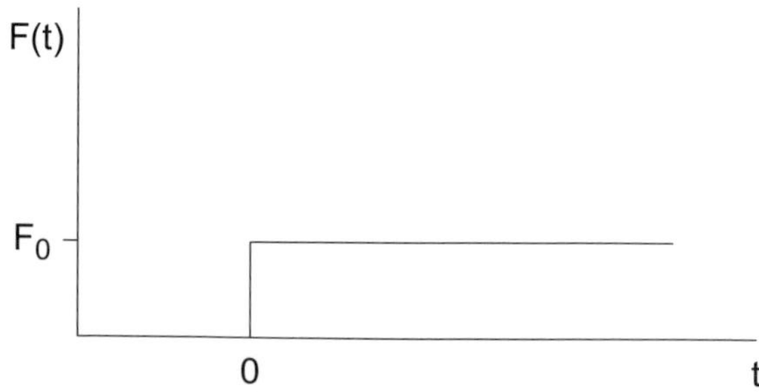

Figure 14.4 Creep or flow for step change at t = 0. Displacement response to step force input.

14.3 Maxwell Fluid Response to Oscillatory Forcing

Consider that the upper plate is oscillated by the shear stress, τ, as in Figure 14.5 so that

$$\tau = \tau_0\, e^{i\omega t}, \quad \gamma = \gamma_0\, e^{i\omega t} \tag{14.3.1}$$

where the angular frequency is $\omega = 2\pi f$ and f is the oscillation frequency. The shear stress amplitude is τ_0 while the shear strain amplitude is γ_0. Note that $e^{i\omega t} = \cos(\omega t) + i \sin(\omega t)$ and both τ_0, γ_0 can be complex. So we seek the real part of the products $\tau_0\, e^{i\omega t}$, $\gamma_0\, e^{i\omega t}$ in Eq. (14.3.1). Inserting these forms into the Maxwell model of Eq. (14.1.4) we get

$$\left(\frac{1}{\mu} + \frac{i\omega}{G}\right)\tau_0 = i\omega\gamma_0 \tag{14.3.2}$$

The limiting behavior of the system can be appreciated in Eq. (14.3.2). For $\mu \to \infty$ we have $\tau_0 = G\gamma_0$, which is purely elastic. The resistance in the

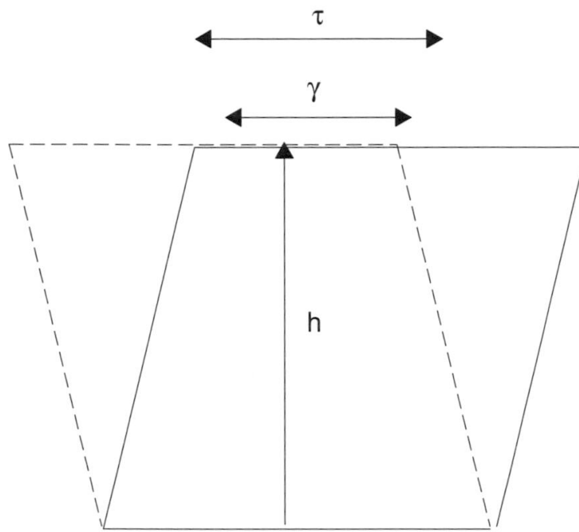

Figure 14.5 Oscillatory shear forcing of a viscoelastic fluid.

dashpot is very high so the dashpot strain, $\gamma_d \sim U_d$, goes to zero. In that limit, all of the deformation is relegated to the spring strain, $\gamma_s \sim U_s$, and the system reaches an elastic limit. As $\omega \to \infty$ the balance is also the elastic limit $\tau_0 = G\gamma_0$. Since the dashpot strain is $\gamma_d \sim 1/\mu\omega$ it cannot keep up with the faster forcing. For $G \to \infty$ the stress becomes $\tau_0 = i\omega\mu\gamma_0$, which is purely a viscous fluid. The spring is rigid so its strain decreases to zero and all of the strain is taken up by the dashpot.

Equation (14.3.2) can be rearranged to represent the stress amplitude τ_0 as a linear function of the strain amplitude, γ_0,

$$\tau_0 = G^* \gamma_0, \quad G^* = \left(\frac{(\omega\mu)^2 G + i\omega\mu G^2}{G^2 + (\omega\mu)^2} \right) \tag{14.3.3}$$

where the real and imaginary parts of the complex shear modulus $G^* = G' + iG''$ are readily found to be

$$G' = \frac{(\omega\mu)^2 G}{G^2 + (\omega\mu)^2}, \quad G'' = \frac{\omega\mu G^2}{G^2 + (\omega\mu)^2} \tag{14.3.4}$$

In Eq. (14.3.4), G' is the shear storage modulus which is in-phase with the oscillation since $e^{i0} = \cos(0) + i\sin(0) =$ and G'' is the loss modulus which is $\pi/2 = 90°$ out of phase with the oscillation since $e^{i\frac{\pi}{2}} = \cos(\pi/2) + i\sin(\pi/2) = i$.

We can define a dimensionless frequency using the Deborah number,

$$\mathrm{De} = \omega\lambda = \frac{\mu\omega}{G} = \frac{\mu/G}{1/2\pi f} \sim \frac{\text{relaxation time}}{\text{observation time (i.e. oscillation period)}} \tag{14.3.5}$$

where a system is dominated by viscosity for $\mathrm{De} << 1$ and by elasticity for $\mathrm{De} >> 1$. Rewriting Eqs. (14.3.4) gives us

$$\frac{G'}{G} = \frac{\mathrm{De}^2}{1 + \mathrm{De}^2}, \quad \frac{G''}{G} = \frac{\mathrm{De}}{1 + \mathrm{De}^2} \tag{14.3.6}$$

Note that $G'/G'' = \mathrm{De} = \omega\lambda$ or, equivalently, $\lambda = G'/\omega G''$.

Plots of these two components as functions of De are shown in Figure 14.6(a, b) in linear and log–log formats. G'/G starts at zero and asymptotes to 1. Since this is the real part of the modulus, it relates the displacement component to the in-phase force component and tells us about the elastic nature of the system under oscillation. As De increases, the strain rate amplitude of the dashpot decreases and the system approaches an elastic limit. G''/G also starts from zero but passes through a local maximum when $\mathrm{De} = 1$ where viscous and elastic effects are equal. At that value, both components are equal, $G' = G''$, called the "crossover point." Then G''/G asymptotes to zero when De increases without bound as the system is dominated by its elasticity. For $\mathrm{De} < 1$ we see that $G'' > G'$ and the system behaves more like a viscous fluid. On the other hand, when $\mathrm{De} > 1$, $G'' < G'$ and the system behaves more like an elastic solid.

The local maximum of G''/G is derived by setting to zero its derivative with respect to De and solving for the value of $\mathrm{De} = \mathrm{De_m}$ at the maximum

$$\frac{d}{\mathrm{De}}\left(\frac{G''}{G}\right)\bigg|_{\mathrm{De}=\mathrm{De_m}} = \frac{1 + \mathrm{De_m^2} - 2\mathrm{De_m^2}}{\left(1 + \mathrm{De_m^2}\right)^2} = 0 \to \mathrm{De_m} = 1 \tag{14.3.7}$$

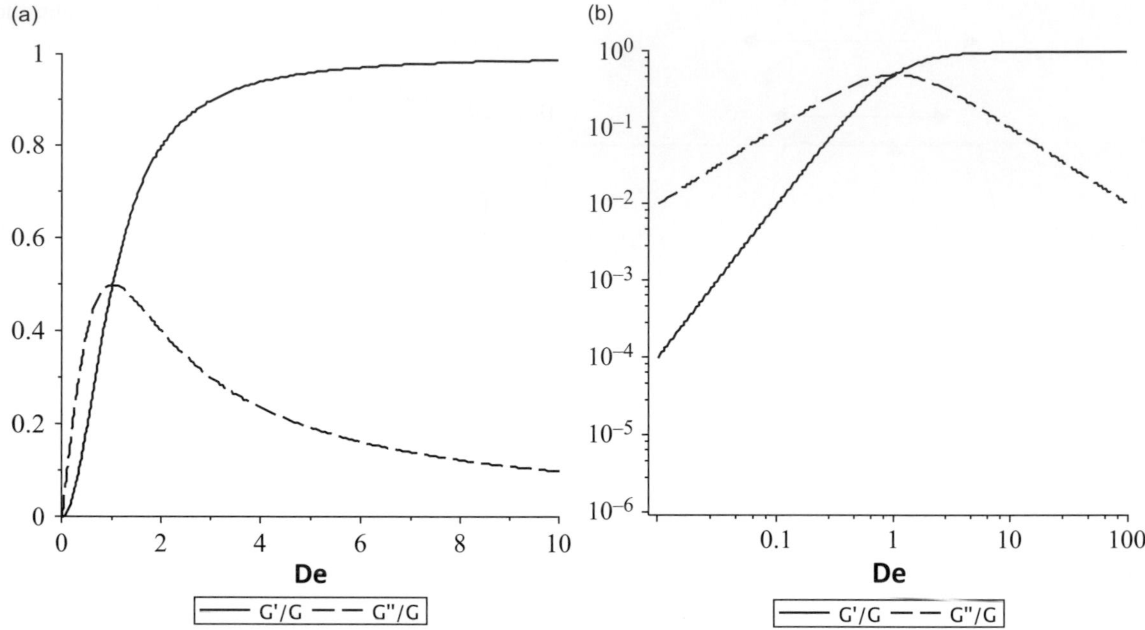

Figure 14.6 G'/G and G''/G vs De in (a) linear and (b) log–log plots.

The local maximum occurs at $De_m = 1$. The crossover value of De, call it De_c, is found by setting $G'/G = G''/G$ and solving for $De = De_c$

$$\frac{G'}{G} = \frac{G''}{G} \rightarrow \frac{De_c^2}{1+De_c^2} = \frac{De_c}{1+De_c^2} \rightarrow De_c = 1 \qquad (14.3.8)$$

So $De_m = De_c = 1$ where the values of the scaled moduli are $G'/G = G''/G = 1/2$.

When interpreting data of a Maxwell fluid where G' and G'' are plotted vs ω, the crossover frequency at $\omega = \omega_c$ identifies the relaxation time. There we know that $G'(\omega_c) = G''(\omega_c)$ and it occurs when $De_c = \omega_c\lambda = 1$. Then the relaxation time is simply $\lambda = 1/\omega_c = \lambda = 1/2\pi f_c$ where f_c is the crossover frequency. A second dimensionless group defined for oscillatory viscoelastic flows is the Weissenberg number, $Wi = \omega\gamma_0\lambda$, which differs from De by including the strain amplitude γ_0. Karl Weissenberg (1893–1976) was an Austrian physicist who made many contributions to rheology and crystallography.

Box 14.1 The Deborah Number

The Deborah number was a name created by Markus Reiner from The Technion (Israel) who also worked in the US with Eugene C. Bingham at Lafayette College (Easton, PA) in the early days of rheology (Reiner, 1964). The reference is to the prophetess Deborah who was a leader (Judge) of the Jewish people in the twelfth century BCE, see Figure 14.7. Her song after victory in battle over the Philistines included the phrase in the Bible (Judges 5:5): "The mountains

Figure 14.7 The prophetess Deborah singing in triumph over Jabin, by Gustave Doré (1865).

flowed before the Lord." Reiner saw this as an opportunity to connect "rheology" to "theology," humored by his incoming surface mail often addressed to a theology department. The idea is that only the Lord can view a mountain over a long enough time period to observe its flow. So De is the relaxation time divided by the observation time.

Representing G^* in complex polar form gives us

$$G^* = |G^*|e^{i\delta} = |G^*|(\cos\delta + i\sin\delta) \tag{14.3.9}$$

so that

$$G' = |G^*|\cos\delta, \quad G'' = |G^*|\sin\delta \tag{14.3.10}$$

The magnitude and angle are related to G' and G'' as

$$|G^*| = \left(G'^2 + G''^2\right)^{1/2}, \quad \tan\delta = \frac{G''}{G'} \tag{14.3.11}$$

where $|G^*|$ is the dynamic modulus of elasticity and $\tan\delta$ is the internal damping. Equation (14.3.11) shows us that the crossover point, where $G' = G''$, occurs where $\tan\delta = 1$ or $\delta = 45°$. Dividing $|G^*|$ by G in Eq. (14.3.11) and using the forms of Eq. (14.3.6), the dependence on De is found to be

$$\frac{|G^*|}{k} = \frac{De}{(1 + De^2)^{1/2}}, \quad \tan\delta = \frac{1}{De} \tag{14.3.12}$$

The two properties of Eq. (14.3.12) are plotted in Figure 14.8. As De increases from zero, the Maxwell fluid behaves increasingly like an elastic solid, with internal damping decreasing and the dynamic elastic modulus increases to equal G.

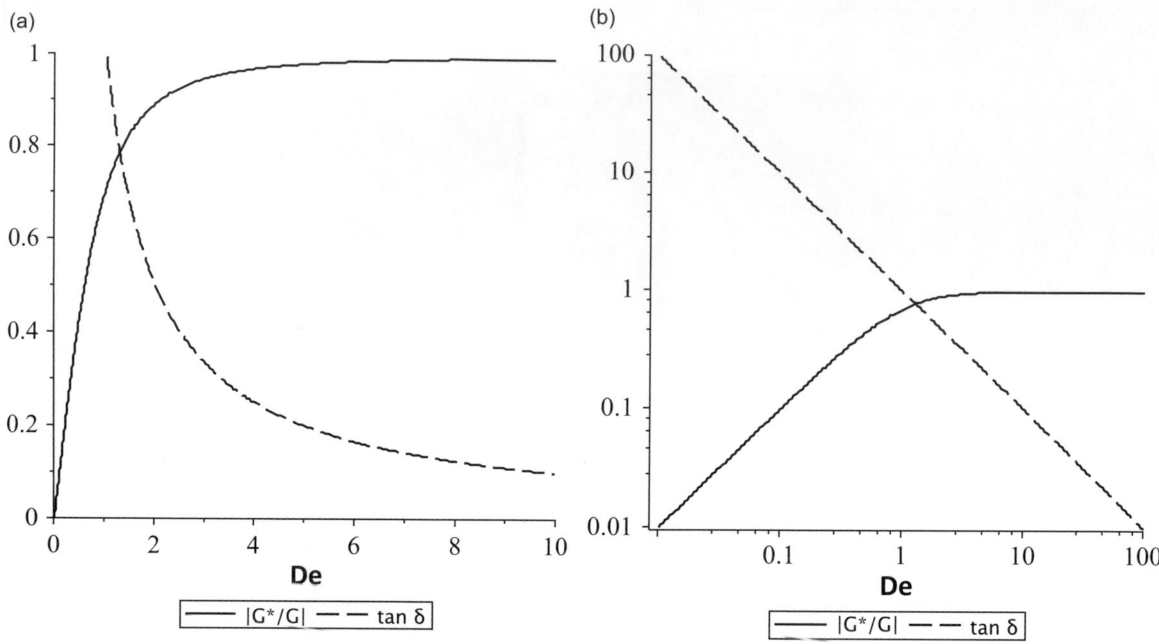

Figure 14.8 Dynamic elastic modulus scaled on G, $|G^*/G|$, and internal damping $\tan\delta$ vs De as (a) linear and (b) log–log plots.

14.4 Complex Fluid Viscosity

An alternative view is to consider the material as a fluid model, where the shear stress is linearly proportional to the strain rate. With this approach we can rearrange Eq. (14.3.2) to keep $i\omega\gamma_0 = \dot{\gamma}$ as a group, and solve for τ_0,

$$\tau_0 = i\omega\gamma_0\left(\frac{1}{\mu}+\frac{i\omega}{G}\right)^{-1} \tag{14.4.1}$$

Rearranging Eq. (14.4.1), and using the definition of $\mathrm{De} = \mu\omega/G$ yields

$$\tau_0 = \mu\frac{(1-i\mathrm{De})}{(1+\mathrm{De}^2)}(i\omega\gamma_0) = \mu^*(i\omega\gamma_0) \tag{14.4.2}$$

where the complex viscosity, μ^*, is defined as $\mu^* = (\mu' - i\mu'')$ with real and imaginary components. According to Eq. (14.4.2) those components are

$$\mu' = \mu\frac{1}{(1+\mathrm{De}^2)} = \frac{G''}{\omega}, \quad \mu'' = \mu\frac{\mathrm{De}}{(1+\mathrm{De}^2)} = \frac{G'}{\omega} \tag{14.4.3}$$

where we have included the relationships between the components of μ^* and G^*; μ' is called the dynamic viscosity and μ'' is the out-of-phase viscosity. Note that $\mu''/\mu' = \mathrm{De} = \omega\lambda$ or, equivalently, $\lambda = \mu''/\omega\mu'$. The magnitude of Eq. (14.4.2) is

$$|\tau_0| = |\mu^*|\omega\gamma_0 \tag{14.4.4}$$

where the magnitude of μ^* is given by

$$|\mu^*| = \left(\mu'^2 + \mu''^2\right)^{1/2}.$$

In dimensionless form the two components of the complex viscosity, μ'/μ and μ''/μ, and the magnitude $|\mu^*|/\mu$ are plotted vs De in Figure 14.9 in linear and log–log formats

$$\frac{\mu'}{\mu} = \frac{1}{1+\mathrm{De}^2}, \quad \frac{\mu''}{\mu} = \frac{\mathrm{De}}{1+\mathrm{De}^2}, \quad \frac{|\mu^*|}{\mu} = \frac{1}{(1+\mathrm{De}^2)^{1/2}} \tag{14.4.5}$$

For increasing De the viscous behavior μ'/μ diminishes. As we have learned, whenever the frequency is high, i.e. the oscillation period is small compared to the relaxation time and the system is not

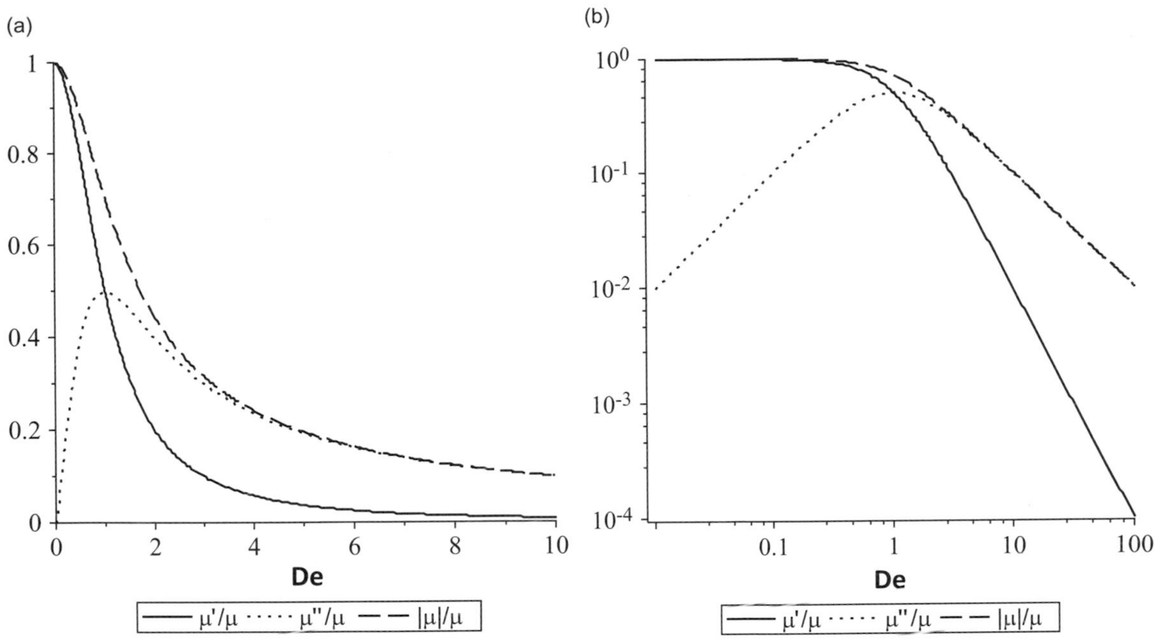

Figure 14.9 The real and imaginary components of the complex viscosity, μ'/μ and μ''/μ, and the magnitude $|\mu^*/\mu|$ vs De displayed in (a) linear and (b) log–log plots.

allowed to "relax" as the flow direction changes too quickly. For $De \to 0$ we see that $\mu'/\mu \to 1$, which is purely viscous behavior as elastic effects disappear. There is a local maximum at $De = 1$ for μ''/μ and $\mu'/\mu = \mu''/\mu = 1/2$ there. When evaluating data through the Maxwell model viscosities, the frequency at the maximum, call it ω_m, gives us the relaxation time $\lambda = 1/\omega_m$. We see that $\omega_m = \omega_c$, the crossover frequency for G' and G'' in Figure 14.6. The magnitude of the complex viscosity, $|\mu^*|/\mu$, decreases monotonically with De, indicating that the Maxwell fluid is shear thinning.

14.5 Stokes' Second Problem for a Maxwell Fluid

In Section 8.2 we studied Stokes' second problem for a wall oscillating in its own plane with a Newtonian viscous fluid above it. Let's repeat this analysis for a viscoelastic fluid. Recall there is no pressure gradient and only one component of velocity, $u_1(x_2, t) = u(y, t)$. With the stress tensor $T_{ij} = -p\delta_{ij} + \tau_{ij}$, the $i = 1$, $j = 2$ component is all that remains in the Cauchy momentum equation, Eq. (5.2.2). That component is

$$\rho \frac{\partial u}{\partial t} = \frac{\partial \tau_{12}}{\partial y} \tag{14.5.1}$$

We will see in Section 14.7 that the Maxwell model extended into continuum mechanics suffers from important deficiencies because, in general, it is not invariant to coordinate transformations. However, for this simple flow those issues disappear. The only component is

$$\tau_{12} + \lambda \frac{\partial \tau_{12}}{\partial t} = \mu \frac{\partial u}{\partial y} \tag{14.5.2}$$

Again, as $G \to \infty, \lambda \to 0$ and the balance in Eq. (14.5.2) is a Newtonian viscous fluid $\tau_{12} = \mu \partial u/\partial y$, where we are reminded that this "u" is a velocity. For $\mu \to \infty$ the time constant, $\lambda = \mu/G \to \infty$, the balance approaches $\partial \tau_{12}/\partial t \sim G\partial u/\partial y$, or when integrated $\tau_{12} \sim G\partial U/\partial y = G\gamma$, which is the elastic shear limit.

The kinematic boundary conditions are the wall velocity at $y = 0$, and zero flow far from the wall

$$u(y=0) = u_0 \cos \omega t, \quad u(y \to \infty) = 0 \tag{14.5.3}$$

where we can consider the velocity amplitude to be linearly related to the frequency $u_0 = a\omega$.

For scales, let

$$U = \frac{u}{\omega a}, \quad T = \omega t, \quad Y = \frac{y}{a}, \sigma = \frac{\tau_{12}}{G} \tag{14.5.4}$$

Insert Eqs. (14.5.4) into Eqs. (14.5.1) and (14.5.2) to find the Cauchy momentum equation

$$\alpha^2 De \frac{\partial U}{\partial T} = \frac{\partial \sigma}{\partial Y} \tag{14.5.5}$$

where $\alpha^2 = \omega a^2/\nu$ and α is the Womersley parameter while De is the Deborah number, $De = \omega\lambda$. The Maxwell constitutive equation becomes

$$\sigma + De \frac{\partial \sigma}{\partial T} = De \frac{\partial U}{\partial Y} \tag{14.5.6}$$

Now insert Eqs. (14.5.4) into the boundary conditions, Eqs. (14.5.3), to find

$$\begin{aligned} U(Y=0) &= \cos T = \Re\left(e^{-iT}\right) \\ U(Y \to \infty) &= 0 \end{aligned} \tag{14.5.7}$$

In Eq. (14.5.7), $\Re$ indicates the real part. Try a separation of variables approach using the sinusoidal time dependency from the wall,

$$\begin{aligned} U(Y, T) &= \Re\left(\hat{U}(Y) e^{-iT}\right) \\ \sigma(Y, T) &= \Re\left(\hat{\sigma}(Y) e^{-iT}\right) \end{aligned} \tag{14.5.8}$$

Inserting Eqs. (14.5.8) into Eq. (14.5.5) and Eq. (14.5.6) we arrive at coupled ODEs for $\hat{\sigma}$, and $\hat{U}$,

$$\begin{aligned} -i\alpha^2 De \hat{U} &= \frac{d\hat{\sigma}}{dY} \\ \hat{\sigma}(1 - iDe) &= De \frac{d\hat{U}}{dY} \end{aligned} \tag{14.5.9}$$

Substituting the derivative of the second equation of Eqs. (14.5.9) into the first leads to a second-order ODE for $\hat{U}$,

$$\frac{d^2\hat{U}}{dY^2} - \beta^2 \hat{U} = 0 \tag{14.5.10}$$

where $\beta^2 = -\alpha^2(De + i)$. Solutions of Eq. (14.5.10) are of the form

$$\hat{U}(Y) = c_1 e^{\beta Y} + c_2 e^{-\beta Y} \tag{14.5.11}$$

where the complex number $\beta = \beta_r + i\beta_i$ has the real and imaginary components

$$\begin{aligned} \beta_r &= \frac{\sqrt{2}\alpha}{2}\left(-De + \left(1 + De^2\right)^{1/2}\right)^{1/2} > 0 \\ \beta_i &= -\frac{\sqrt{2}\alpha}{2}\left(De + \left(1 + De^2\right)^{1/2}\right)^{1/2} < 0 \end{aligned} \tag{14.5.12}$$

The boundary conditions of Eq. (14.5.7) become

$$\begin{aligned} \hat{U}(Y \to \infty) = 0 \quad &\Rightarrow \quad c_1 = 0 \\ \hat{U}(Y = 0) = 1 \quad &\Rightarrow \quad c_2 = 1 \end{aligned} \tag{14.5.13}$$

The solution for $\hat{U}(Y)$ is

$$\hat{U}(Y) = e^{-\beta Y} \tag{14.5.14}$$

Inserting the solution for $\hat{U}(Y)$ into Eq. (14.5.9) for $\hat{\sigma}(Y)$ yields

$$\hat{\sigma} = -\frac{\beta De(1 + iDe)}{(1 + De^2)} e^{-\beta Y} \tag{14.5.15}$$

and the total solution is

$$\begin{aligned} U(Y, T) &= \Re\left(e^{-\beta Y} e^{-iT}\right) \\ \sigma(Y, T) &= \Re\left(-\frac{\beta De(1 + iDe)}{(1 + De^2)} e^{-\beta Y} e^{-iT}\right) \end{aligned} \tag{14.5.16}$$

We can simplify the velocity solution by combining the exponential terms

$$U(Y, T) = \Re\left(e^{-\beta Y} e^{-iT}\right) = e^{-\beta_r Y} \cos\left(T - (-\beta_i Y)\right) \tag{14.5.17}$$

Since the wall velocity is $\cos T$, the phase lag for the fluid velocity in Eq. (14.5.17) is $-\beta_i Y > 0$, similar to what we found for the Newtonian system in Section 8.2, Eq. (8.2.17), except now the effects of viscoelasticity influence the motion.

Results of the model are shown in Figure 14.10 for three values of $De = 0.1, 1, 10$. As De increases the fluid becomes more elastic and the amplitude of the velocity increases, while the spatial frequency in the Y-direction increases and the rate of decay decreases.

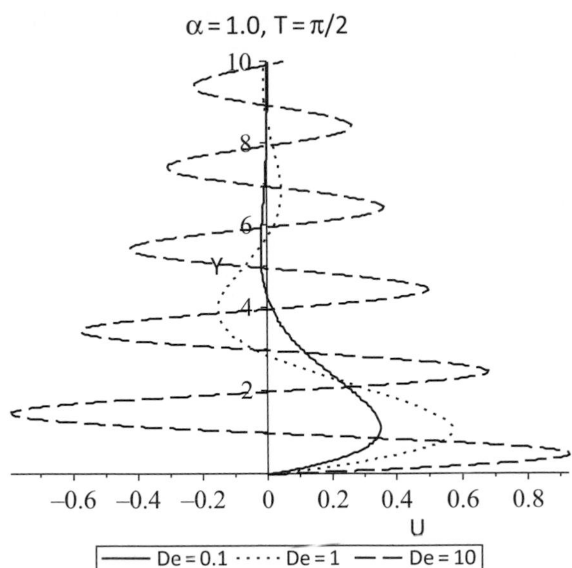

Figure 14.10 Stokes' second problem for a Maxwell fluid for De = 0.1, 1, 10.

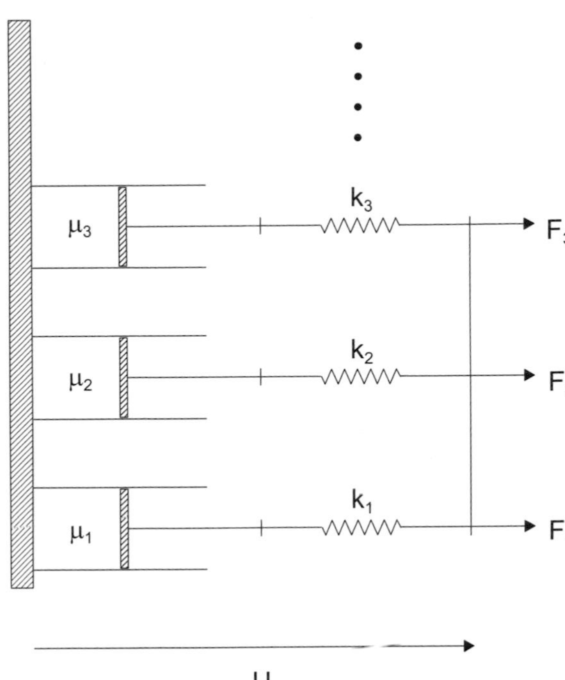

Figure 14.11 Generalized Maxwell model with multiple Maxwell elements.

14.6 Generalized Maxwell Fluid

Biofluids are often polydisperse, that is, the long chain molecules which make up the viscoelastic fluid have more than one type of molecule and more than one length. Under these circumstances, the stress relaxation, which comes from the untangling of these molecules, is going to have more than one relaxation time. The stress relaxation of this generalized system is in response to a step change in U which is experienced by all of the spring–dashpot components, see Figure 14.11. $U = U_0 H(t)$ as in Eq. (14.2.1) and the individual forces are

$$F_i = k_i U_0 e^{-\frac{t}{\lambda_i}} \tag{14.6.1}$$

where $\lambda_i = \mu_i / k_i$ is the time constant for the Maxwell element. The sum of the forces is

$$F = \sum_{i=1}^{n} F_i = U_0 \sum_{i=1}^{n} k_i e^{-\frac{t}{\lambda_i}} \tag{14.6.2}$$

or rearranged as

$$k(t) = \frac{F}{U_0} = \sum_{i=1}^{n} k_i e^{-\frac{t}{\lambda_i}} \tag{14.6.3}$$

where k(t) is called the stress-relaxation modulus and $k_i = F_i / U_0$. For a continuous distribution of relaxation times the summation in Eq. (14.6.3) can be replaced by the integral

$$k(t) = \int_0^{\infty} k(\lambda) e^{-\frac{t}{\lambda}} d\lambda \tag{14.6.4}$$

Let us consider a two-element generalized Maxwell fluid. The constitutive equations are given by

$$\begin{aligned} \tau_1 + \lambda_1 \dot{\tau}_1 &= \mu_1 \dot{\gamma}, \quad \lambda_1 = \frac{\mu_1}{G_1} \\ \tau_2 + \lambda_2 \dot{\tau}_2 &= \mu_2 \dot{\gamma}, \quad \lambda_2 = \frac{\mu_2}{G_2} \end{aligned} \tag{14.6.5}$$

and we know that the total shear stress is the sum of the two Maxwell elements

$$\tau = \tau_1 + \tau_2 \tag{14.6.6}$$

We can resolve Eqs. (14.6.5) into a single equation, which is readily done by using the Laplace transform, F(s), defined as

$$F(s) = \int_0^\infty f(t)e^{-st}\,dt \tag{14.6.7}$$

The transform of a function f(t) is F(s), while transforms of its time derivatives are

$$\begin{aligned}\int_0^\infty \frac{df}{dt}(t)e^{-st}\,dt &= sF(s) - f(0)\\ \int_0^\infty \frac{d^2f}{dt^2}(t)e^{-st}\,dt &= s^2F(s) - sf(0) - \dot{f}(0)\end{aligned} \tag{14.6.8}$$

The Laplace transforms of Eqs. (14.6.5) and (14.6.6) are

$$\begin{aligned}(1+\lambda_1 s)T_1 &= \mu_1 s\Gamma\\ (1+\lambda_2 s)T_2 &= \mu_2 s\Gamma\\ T = T_1 + T_2 &= \frac{\mu_1 s\Gamma}{(1+\lambda_1 s)} + \frac{\mu_2 s\Gamma}{(1+\lambda_2 s)}\end{aligned} \tag{14.6.9}$$

where

$$\begin{aligned}T(s) &= \int_0^\infty \tau(t)e^{-st}\,dt\\ \Gamma(s) &= \int_0^\infty \gamma(t)e^{-st}\,dt\end{aligned} \tag{14.6.10}$$

and we assume $\tau(0), \dot{\tau}(0), \gamma(0), \dot{\gamma}(0)$ are all zero. The last of Eqs. (14.6.9) can be rearrange to yield

$$\begin{aligned}&\left[1+(\lambda_1+\lambda_2)s+\lambda_1\lambda_2 s^2\right]T =\\ &\qquad\left[(\mu_1+\mu_2)s+(\mu_1\lambda_2+\mu_2\lambda_1)s^2\right]\Gamma\end{aligned} \tag{14.6.11}$$

Then the inverse Laplace transform of Eq. (14.6.11) gives us a single, second-order differential equation for this generalized Maxwell fluid

$$\begin{aligned}\tau + (\lambda_1+\lambda_2)\dot{\tau} + \lambda_1\lambda_2\ddot{\tau} &= (\mu_1+\mu_2)\dot{\gamma}\\ &+ (\mu_1\lambda_2+\mu_2\lambda_1)\ddot{\gamma}\end{aligned} \tag{14.6.12}$$

Equation (14.6.12) has the steady solution $\tau = (\mu_1+\mu_2)\dot{\gamma}$ so that the effective viscosity is the sum of the two viscous elements

$$\mu_{eff} = \tau/\dot{\gamma} = \mu_1 + \mu_2 \tag{14.6.13}$$

The general form of Eq. (14.6.12) is

$$\tau + p_1\dot{\tau} + p_2\ddot{\tau} = q_1\dot{\gamma} + q_2\ddot{\gamma} \tag{14.6.14}$$

For n Maxwell elements a similar approach yields the form of the differential equation as

$$\sum_{j=0}^{n} p_j \frac{d^j}{dt^j}\tau = \sum_{j=0}^{n} q_j \frac{d^j}{dt^j}\gamma \tag{14.6.15}$$

where $q_0 = 0$ while p_j and q_j depend on the viscoelastic parameters of the elements. Equation (14.6.15) forms the basis of the generalized Maxwell model.

Worked Example 14.1 | Viscoelastic Behavior of a Two-Element Generalized Maxwell Fluid

To find the relationship between shear stress and shear strain amplitudes for oscillatory forcing, insert Eqs. (14.3.1) into Eq. (14.6.12), to give

$$\left[1+(\lambda_1+\lambda_2)i\omega - \lambda_1\lambda_2\omega^2\right]\tau_0 = \left[(\mu_1+\mu_2)i\omega - (\mu_1\lambda_2+\mu_2\lambda_1)\omega^2\right]\gamma_0 \tag{14.6.16}$$

Rearranging Eq. (14.6.16) leaves us with the familiar form of a complex shear modulus

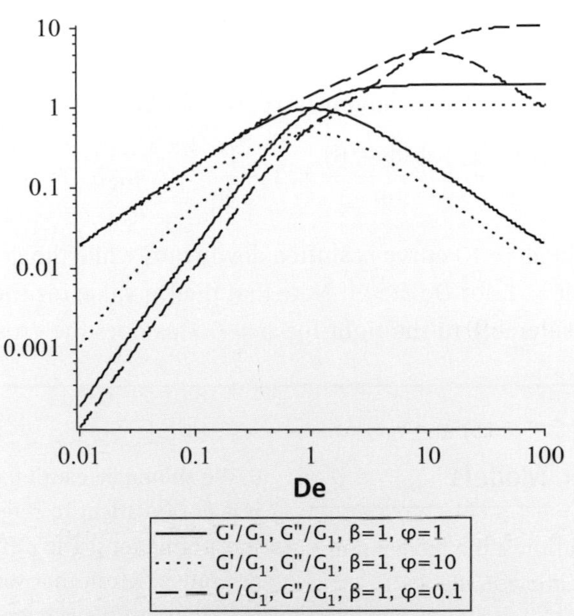

Figure 14.12 G'/G_1 and G''/G_1 vs De for a two-element Maxwell fluid model with $\beta = 1$ and $\phi = 0.1, 1, 10$. G''/G_1 curves exhibit a local maximum.

$$\tau_0 = G^*\gamma_0 = \left(G' + iG''\right)\gamma_0 \tag{14.6.17}$$

where $G^* = G' + iG''$, and G' is the shear storage modulus while G'' is the loss modulus given by

$$\begin{aligned} G' &= \frac{\left[-(\mu_1\lambda_2 + \mu_2\lambda_1)(1 - \lambda_1\lambda_2\omega^2)\omega^2 + (\mu_1 + \mu_2)(\lambda_1 + \lambda_2)\omega^2\right]}{\left[(1 - \lambda_1\lambda_2\omega^2)^2 + (\lambda_1 + \lambda_2)^2\omega^2\right]} \\ G'' &= \frac{(\mu_1 + \mu_2)(1 - \lambda_1\lambda_2\omega^2)\omega + (\mu_1\lambda_2 + \mu_2\lambda_1)(\lambda_1 + \lambda_2)\omega^3}{\left[(1 - \lambda_1\lambda_2\omega^2)^2 + (\lambda_1 + \lambda_2)^2\omega^2\right]} \end{aligned} \tag{14.6.18}$$

If we define the following dimensionless parameters for the ratios of the viscosities, $\beta = \mu_2/\mu_1$, and time constants, $\phi = \lambda_2/\lambda_1$, and let the Deborah number be $De = \omega\lambda_1$ then Eqs. (14.6.18) can be simplified to the dimensionless forms

$$\begin{aligned} \frac{G'}{G_1} &= \frac{(1 + \beta\phi + \phi(\phi + \beta)De^2)De^2}{(1 + De^2)(1 + \phi^2 De^2)} \\ \frac{G''}{G_1} &= \frac{(1 + \beta + (\beta + \phi^2)De^2)De}{(1 + De^2)(1 + \phi^2 De^2)} \end{aligned} \tag{14.6.19}$$

Sample plots of Eqs (14.6.19) are shown in Figure 14.12. G'/G_1 and G''/G_1 are plotted vs De for $\beta = 1$ and $\phi = 0.1, 1, 10$. In all cases G'/G_1 asymptotes to a constant as De >> 1 while G''/G_1 has a local maximum. The behavior of G'/G_1 can be seen in Eq. (14.6.19) where

$$\left(\frac{G'}{G_1}\right)_{De>>1} \sim \frac{(\phi(\phi+\beta)De^4)}{(\phi^2 De^4)} = \frac{\phi(\phi+\beta)}{\phi^2} = 1 + \frac{\beta}{\phi} \tag{14.6.20}$$

which explains why the $\phi = 10$ curve is shifted downward while the $\phi = 0.1$ curve is shifted upward compared to $\phi = 1$, for De >> 1. Note that the De value for the maximum of G''/G_1, call it De_m, shifts considerably to the right for $\phi = 0.1$ as does the crossover point.

14.7 More Viscoelastic Models

The Cauchy momentum equation is the momentum conservation we studied in Chapter 5, Eq. (5.2.2),

$$\rho\left(\frac{\partial u_i}{\partial t} + u_k\frac{\partial u_i}{\partial x_k}\right) = \frac{\partial T_{ij}}{\partial x_j} + b_i \tag{14.7.1}$$

where the stress tensor is the sum of the isotropic pressure contribution multiplying the Kronecker delta, or identity tensor, $-p\delta_{ij}$, and the extra, or viscous, stress tensor, τ_{ij}, is

$$T_{ij} = -p\delta_{ij} + \tau_{ij} \tag{14.7.2}$$

Substituting Eq. (14.7.2) into Eq. (14.7.1) leads us to

$$\rho\left(\frac{\partial u_i}{\partial t} + u_k\frac{\partial u_i}{\partial x_k}\right) = -\frac{\partial p}{\partial x_i} + \frac{\partial \tau_{ij}}{\partial x_j} \tag{14.7.3}$$

The constitutive equation for the Maxwell fluid is the equivalent of Eq. (14.1.10) for each component of the stress tensor

$$\tau_{ij} + \lambda\frac{\partial \tau_{ij}}{\partial t} = 2\mu E_{ij} = \mu\left(\frac{\partial u_i}{\partial x_j} + \frac{\partial u_j}{\partial x_i}\right) \tag{14.7.4}$$

where $\lambda = \mu/G$ is the time constant. Equations (14.7.3) and (14.7.4) are coupled PDEs. Recall that for a Newtonian fluid, $G \to \infty, \lambda \to 0$ and we can substitute $\tau_{ij} = \mu(\partial u_i/\partial x_j + \partial u_j/\partial x_i)$ into Eq. (14.7.3) to obtain the Navier–Stokes equation. For $\lambda > 0$ we have the coupled system, however.

We should be cautious about Eq. (14.7.4) because it is not invariant to rigid body coordinate transformations. The difficulty rests with the time derivative. Recall that we examined the Galilean invariance of the Navier–Stokes equations in Section 11.1, Eqs. (11.1.18)–(11.1.21). The prime coordinate system travels in the +x-direction at speed U compared to the unprimed system, Figure 11.4, and the relationships for the variables in the two systems are $x' = x - Ut$, $y = y'$, $u' = u - U$, $v' = v$. From the chain rule we found that the time derivatives in the two systems are related by $\partial/\partial t = \partial/\partial t' - U\partial/\partial x'$. Applied to the acceleration term, which is the material derivative of the velocity, $D\underline{u}/Dt = \partial\underline{u}/\partial t + (\underline{u}\cdot\underline{\nabla})\underline{u}$, we showed that it is invariant, here for the x-component

$$\begin{aligned}\frac{\partial u}{\partial t} + u\frac{\partial u}{\partial x} + v\frac{\partial u}{\partial y} &= \left(\frac{\partial u'}{\partial t'} - U\frac{\partial u'}{\partial x'}\right) + \left(U\frac{\partial u'}{\partial x'} + u'\frac{\partial u'}{\partial x'}\right) + v'\frac{\partial u'}{\partial y'} \\ &= \frac{\partial u'}{\partial t'} + u'\frac{\partial u'}{\partial x'} + v'\frac{\partial u'}{\partial y'}\end{aligned} \tag{14.7.5}$$

The key point is that the convective acceleration term produces $U\partial u'/\partial x'$, which cancels the $-U\partial u'/\partial x'$ term arising from the time derivative. The full Maxwell model of Eq. (14.7.4) does not have a convective acceleration term to cancel the time derivative term. So Eq. (14.7.4) is not Galilean invariant. To deal with this difficulty, a number of models for viscoelastic fluids have been developed which include convective terms to accompany the time derivative on the left hand side of Eq. (14.7.4) so that the new model is invariant. In fact, often they use the material derivative.

One example is the "upper convected Maxwell model" or UCM,

$$\tau_{ij} + \lambda \overset{\nabla}{\tau}_{ij} = 2\mu E_{ij} \tag{14.7.6}$$

where the "upper convected time derivative" is defined by

$$\overset{\nabla}{\tau}_{ij} = \frac{\partial \tau_{ij}}{\partial t} + u_k \frac{\partial \tau_{ij}}{\partial x_k} - \frac{\partial u_i}{\partial x_k}\tau_{kj} - \tau_{ik}\frac{\partial u_j}{\partial x_k} \tag{14.7.7}$$

Note that the first two terms on the right hand side of Eq. (14.7.7) form the material derivative of τ_{ij}, i.e. $D\tau_{ij}/Dt = \partial\tau_{ij}/\partial t + u_k \partial\tau_{ij}/\partial x_k$, which we know is Galilean invariant, which guarantees the same for this model. Another popular choice is the Oldroyd B model, which is an extension of the UCM

$$\tau_{ij} + \lambda_1 \tau \nabla_{ij} = 2\mu\left(E_{ij} + \lambda_2 E \nabla_{ij}\right) \tag{14.7.8}$$

where λ_1 is the relaxation time and λ_2 is the retardation time. The material derivatives of both τ_{ij} and E_{ij} are now involved. Still more models start with the generalized Maxwell approach, but replace the linear dashpots with nonlinear fluids such as power-law fluids, and linear springs with nonlinear springs (Monsia, 2011).

14.8 Rheological Measurements of Biofluids: The Cox–Merz Rule

Measuring Viscoelastic Behavior of Fluids

In Section 10.8 we discussed measurement of viscosity for generalized Newtonian fluids with examples of a concentric-cylinder viscometer (Figure 10.22), and a cone-plate viscometer (Figure 10.23). For that presentation, the viscometers were applying steady, unidirectional flow at rotational frequency Ω. The fluid response is explored for a wide range of Ω. For viscoelastic fluids, similar devices are driven in an oscillatory mode at frequency $\omega = 2\pi f$ rad/s, so that the fluid is forced to cyclically reverse direction while the torque is measured. A wide range of ω is also explored, such as several decades like $0.01 \le \omega \le 100$ rad/s. The force (torque) data are separated into the component in-phase with the oscillation, G' the shear storage modulus, and the component 90 degrees out-of-phase with the oscillation, G'' the loss modulus. Both of these are related to the complex viscosity $\mu^* = \mu' - i\mu''$ shown in Eq.(14.4.3). In this mode, these devices are measuring more than viscosity, so are called "rheometers." Additional technologies for measuring rheology of materials utilize smaller scale approaches where magnetic beads with diameters less than ~ 5 microns are placed into the material, i.e. on or inside a cell (Bausch et al., 1998). Then an oscillatory magnetic fields is applied and the measured forces and motion are decoupled to yield G' and G''.

In the steady flow viscometer the effective viscosity was measured as $\mu_{eff}(\dot{\gamma}) = \tau/\dot{\gamma}$. Can this viscosity be related to the complex viscosity, $\mu^*(\omega)$, of an oscillatory rheometer? The short answer is yes, but not always. In a fundamental 1958 paper by Cox and Merz, the authors proposed that the relationship is

$$\mu_{eff}(\dot{\gamma}) = |\mu^*(\omega)|_{\dot{\gamma}=\omega} \tag{14.8.1}$$

Equation (14.8.1) says that $\mu_{eff}(\dot{\gamma})$ is equal to the magnitude of the complex viscosity $|\mu^*(\omega)|$ for the strain rate $\dot{\gamma}(s^{-1})$ equal to the oscillation frequency $\omega = 2\pi f$ (rad/s), i.e. $\dot{\gamma} = \omega$. This equivalence in Eq. (14.8.1) is known as the Cox–Merz rule (Cox and Merz, 1958) and holds well for numerous polymeric systems. Since its publication, many viscoelastic materials have been tested in steady and oscillatory viscometers to see if the left hand side really equals the right hand side. It does not always hold, especially as the material is increasingly complicated in microstructure like foods (Bistany and Kokini, 1983; Vernon-Carter et al., 2016). For example, salad dressing emulsions do not follow the Cox–Merz rule (Riscardo et al., 2005). The equivalence is often restricted to low frequency oscillations on the right hand side. However, for blood

(Beissinger and Williams, 1985), mucus and synovial fluid it is fairly accurate.

We computed the right hand side of Eq. (14.8.1) for the Maxwell model already in Eq. (14.4.5)

$$\frac{|\mu^*(\omega)|}{\mu} = \frac{1}{(1+\mathrm{De}^2)^{1/2}} \tag{14.8.2}$$

Of course the left hand side of Eq. (14.8.1) is simply $\mu_{\mathrm{eff}} = \mu$ for a Maxwell fluid, i.e. from Eq. (14.1.3) $\mu_{\mathrm{eff}} = \tau/\dot{\gamma} = \mu$. So according to Eq. (14.8.2), the Cox–Merz rule is only accurate for the Maxwell unit when $\mathrm{De}^2 << 1$, i.e. low frequency. A Taylor series of $|\mu^*(\omega)|/\mu$ in Eq. (14.8.2) in this limit is

$$\begin{aligned}\frac{|\mu^*(\omega)|}{\mu} &= (1+\mathrm{De}^2)^{-1/2} \\ &= \left(1 - \frac{1}{2}\mathrm{De}^2 + O(\mathrm{De}^4)\right) \qquad \mathrm{De}^2 << 1\end{aligned} \tag{14.8.3}$$

where we see that $|\mu^*(\omega)|/\mu \sim 1$ for $\mathrm{De}^2 << 1$. The other asymptotic limit of Eq. (14.8.2) is for $\mathrm{De} >> 1$, which yields

$$\frac{|\mu^*(\omega)|}{\mu} \sim \frac{1}{\mathrm{De}} \qquad \mathrm{De}^2 >> 1 \tag{14.8.4}$$

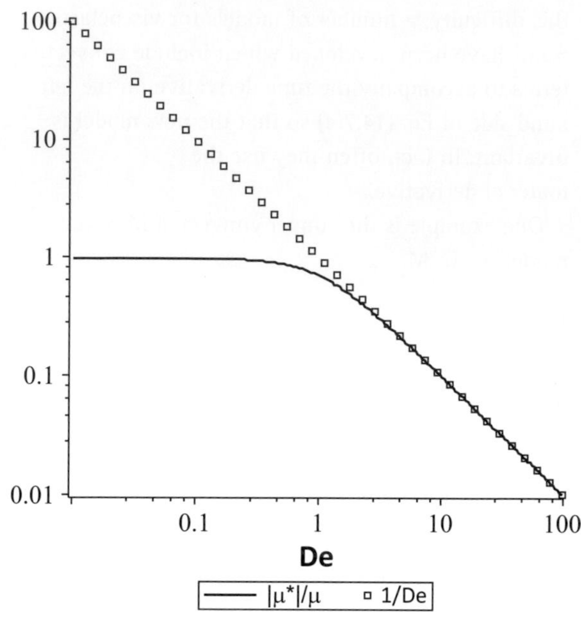

Figure 14.13 Maxwell fluid $|\mu^*|/\mu$ (solid line) and 1/De vs De (boxes).

From Eq. (14.8.2) $|\mu^*|/\mu$ vs De is shown in Figure 14.13 in a log–log plot along with 1/De, which is the $\mathrm{De} >> 1$ solution.

Worked Example 14.2

For the two-element Maxwell model we found G'/G_1 and G''/G_1 in Eq. (14.6.19). Then to calculate $|\mu^*|$ yields

$$|\mu^*| = \frac{\left[(G')^2 + (G'')^2\right]^{1/2}}{\omega} \tag{14.8.5}$$

and scaling it on μ_1 leads to

$$\frac{|\mu^*|}{\mu_1} = \frac{\left[(G'/G_1)^2 + (G''/G_1)^2\right]^{1/2}}{\mathrm{De}} \tag{14.8.6}$$

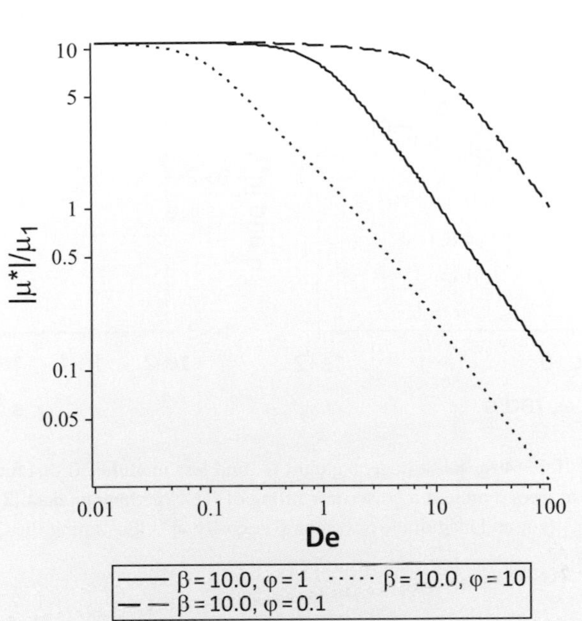

Figure 14.14 $|\mu^*|/\mu_1$ vs De for $\beta = 10, \phi = 0.1, 1, 10$.

A plot of $|\mu^*|/\mu_1$ from Eq. (14.8.6) is shown in Figure 14.14 for $\beta = 10$ and $\phi = 0.1, 1.0, 10.0$, the same parameter set used in Figure 14.12. A Taylor series expansion of $|\mu^*|/\mu_1$ for De << 1 is

$$|\mu^*|/\mu_1 = (1+\beta) + O(\mathrm{De}^2) \quad \mathrm{De} << 1 \tag{14.8.7}$$

The asymptote for De → 0 is $|\mu^*|/\mu_1 = 1+\beta$, which only involves the viscosities of the Maxwell elements. Dimensionally this limit is $|\mu^*| = \mu_1(1+\beta) = \mu_1 + \mu_2$, the sum of the two viscosities, which recovers the steady solution in Eq. (14.6.13) as we expect when the frequency, ω, approaches zero.

Blood

Figure 14.15 shows measurements of blood rheology using both oscillatory and steady shear rates (Tomaiuolo, 2014; D'Apolito et al., 2015; Tomaiuolo et al., 2016). In Figure 14.15(a) the storage modulus, G', and loss modulus, G'', are plotted versus the angular frequency, ω (in rad/s). As ω increases both moduli increase as a power law $\sim\omega^{0.7}$ with a slope of ~0.7 for the straight part of the data in this log–log plot. G'' is larger than G' indicating what we all know from experience – that blood is more of a viscous liquid than an elastic solid. The elasticity derives from interaction of the red blood cells, so is dependent on hematocrit among other parameters. Figure 14.15(b) shows the steady viscosity $\eta(\dot{\gamma})$ and the magnitude of the complex viscosity $|\eta^*(\omega)|$ plotted on the same axis where $\dot{\gamma} = \omega$, demonstrating that the Cox–Merz rule holds well for this shear-thinning fluid.

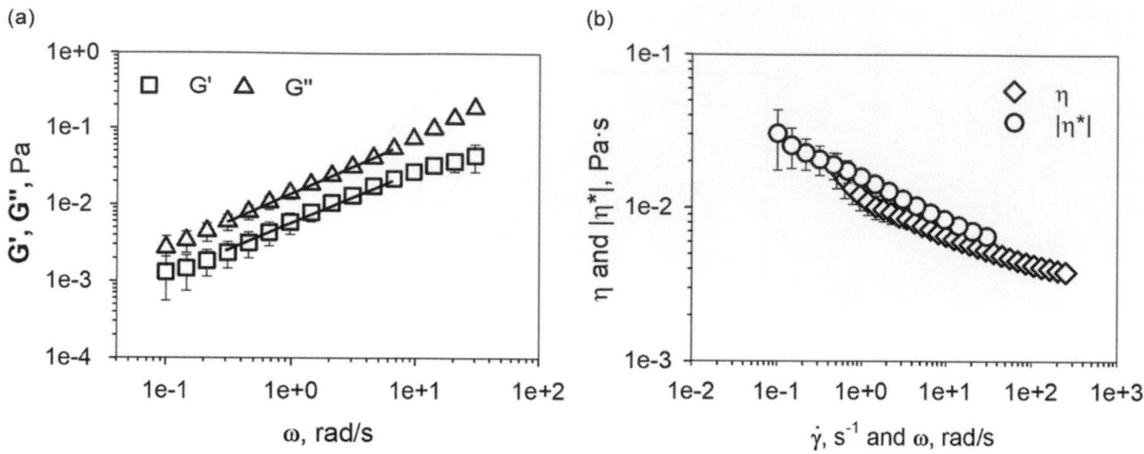

Figure 14.15 Blood rheology for Hct = 45%. (a) Storage modulus, G', and loss modulus, G'', as functions of angular frequency, ω (rad/s). The gray dashed straight lines represent a power-law fitting of the experimental data ($R^2 = 0.999$) with slope ~ 0.7. (b) Comparison of steady shear viscosity η and magnitude of complex viscosity $|\eta^*|$ illustrating the Cox–Merz rule. Here η is our μ.

Rheological literature often uses η as viscosity, as opposed to our μ. They are the same here.

Using a Maxwell model, the stress-relaxation time, λ, was measured to be ~0.03 s over the frequency range 1–3 Hz for normal hematocrits (Long et al., 2005). Adult and pediatric blood had similar behavior. Over a wider range of frequencies, however, λ did not remain constant indicating a more complicated behavior than a single Maxwell unit. For human physiology fetal heart rates are typically 150 bpm (2.5 Hz) while the adult range is 60–100 bpm (1–1.66 Hz). These values place the range of the Deborah number, $De = 2\pi f \lambda$, as approximately $0.2 \leq De \leq 0.5$. However, higher frequencies can occur in blood processing equipment, for example, extending the upper range of De. Within the animal kingdom, mouse heart rates are in the range of 500 bpm (~8 Hz), so larger De values are possible.

Mucus

Mucus is a lubricating and protective secretion which coats mucous membranes. In humans that includes the epithelial cells of the gastrointestinal, genitourinary and respiratory tracts as well as the auditory and visual systems. One of its major constituents is mucin, which is produced in specialized cells called goblet cells that can secrete onto the epithelial surface. Mucin is a large and bushy glycoprotein molecule as shown in Figure 14.16. Typically it forms long chain oligomers of repeated mucin molecules which can aggregate.

In Figure 14.17 shows a compilation of steady mucus viscosity measurements. The slopes in the log–log plots of both are in the range −1 to −0.5 with an average of −0.86 for the steady shearing (Yeates et al., 1997). Oscillatory values of $\mu' \sim G''/\omega$ are also presented in (Lai et al., 2009) but not $|\eta^*|$ which we would need to test the Cox–Merz rule. The horizontal line is water viscosity, 10^{-3} Pa·s, so the range of mucus viscosities range from tears near water to 10,000–100,000 times water for chronic bronchitis which, in addition to mucus, has inflammatory cells and debris added to the mix.

Figure 14.18 shows rheological data from oscillatory forcing of mucus (Hill et al., 2014). In Figure 14.18(c, d) G' and G'' are plotted on log–log axes vs oscillation frequency, f, (Hz) for five different mucus solids concentrations: 1.5, 2, 3, 4, 5 wt%. For Figure 14.18(c) with 1.5, 2.0 and 3.0 wt%, we see that

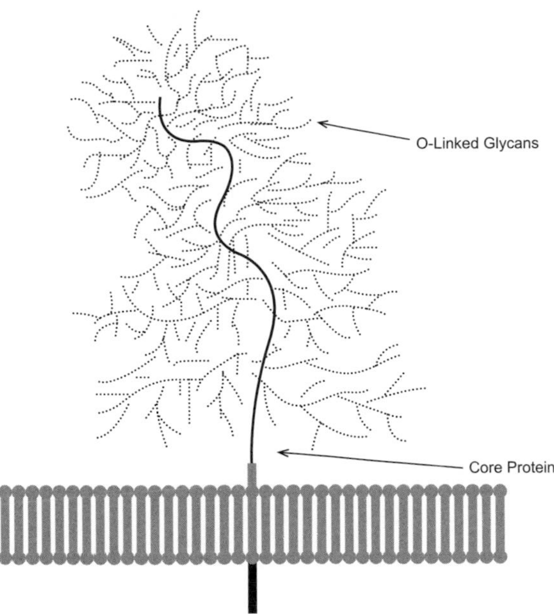

Figure 14.16 Mucin glycoprotein molecular structure showing a core protein with many glycan branches.

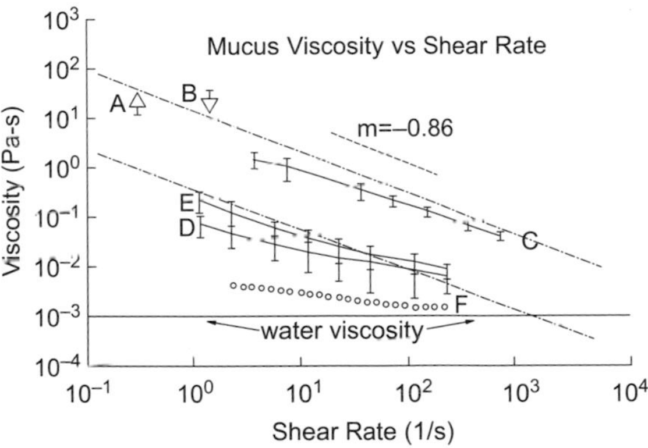

Figure 14.17 Steady viscosity of various types of human mucus. Steady shear viscosity as a function of shear rate for (a, b) chronic bronchitis mucus circle (Puchelle et al., 1983) and square (Dulfano et al., 1971); (c) nonovulatory cervical mucus (Lai et al., 2007); (d) normal gastric mucus (Curt and Pringle, 1969); (e) duodenal ulcer mucus (Curt and Pringle, 1969); and (f) tears (Tiffany, 1991). Thin dashed lines indicate the typical range of viscosity values for human mucus suggested by (Cone, 1999). The thin solid line represents the viscosity of water (10^{-3} Pa·s).

$G' < G''$ and the mucus behaves more like a viscous fluid than an elastic solid. However, for Figure 14.18 (d) with 4 and 5 wt%, $G' > G''$ and the mucus is more elastic than viscous. Figure 14.18(a) shows the complex viscosity magnitude $|\eta^*(\omega)|$ vs frequency f (Hz). Figure 14.18(b) plots the slope of a power-law curve fit to the data in Figure 14.18(a). Assuming a form $|\eta^*(\omega)| \sim \omega^{-s} = (2\pi)^{-s} f^{-s}$ such that

$$\log\left(|\eta^*(\omega)|\right) = \log\left((2\pi)^{-s}\right) - s\log(f) \tag{14.8.8}$$

and s is the slope to the data in Figure 14.18(a). The 5 wt%, for example, has $s = 0.77$ in Figure 14.18(a), which compares well to the steady shear viscosity of Figure 14.17 values where $m = -0.86$. So the Cox–Merz rule appears well intact for mucus in this range of parameter values. Recall the power-law fluid model of Section 10.3 and Eq. (10.3.2) where $\mu_{eff} = \tau/\dot{\gamma} = m\dot{\gamma}^{n-1}$. There we presented additional mucus data, Figure 10.19 reviewed in (Velez-Cordero and Lauga, 2013) to show that $n = 0.2$ gives a similar slope of −0.8 compared to −0.77 of Figure 14.18(a) and −0.86 of Figure 14.17.

We expect a spectrum of relaxation times for mucus; however, investigators look for the dominant value. In Hwang et al. (1969) the crossover frequency yielded $\lambda \sim 100$ s for lung mucus while Gilboa and Silberberg (1976) find $\lambda \sim 30$ s. Cervical mucus relaxation times are in the range 1–10 seconds (Eliezer, 1974; Litt et al., 1976; Wolf et al., 1977; Tam et al., 1980). Note that typical driving frequencies of respiratory cilia are $f = 10$ Hz and sperm swimming by the beating flagella at $f = 20 - 50$ Hz. So $De = 2\pi f\lambda \sim O(10^3)$ for these situations and the fluid behaves elastically. However, for gravity driven flows in the lung a characteristic time scale is $T_s = \mu/\rho g h$, where h is the mucus film thickness. Let $h = 10\,\mu m\ (10^{-3} cm)$, $g = 980\,cm/s^2$ and density $\rho = 1.0\ g/cm^3$. Choose the viscosity to be $\mu = 10\ g/cm{\cdot}s = 1\,Pa{\cdot}s$, which is a typical value in Figure 14.17. The result is $T_s \sim 10s$, which is close to the relaxation time.

Synovial Fluid

We introduced synovial fluid in Chapter 11 when studying the lubrication theory of a squeeze film, see Figure 11.28. There we treated it as either a Newtonian fluid (Section 11.7), or a power-law fluid (Section

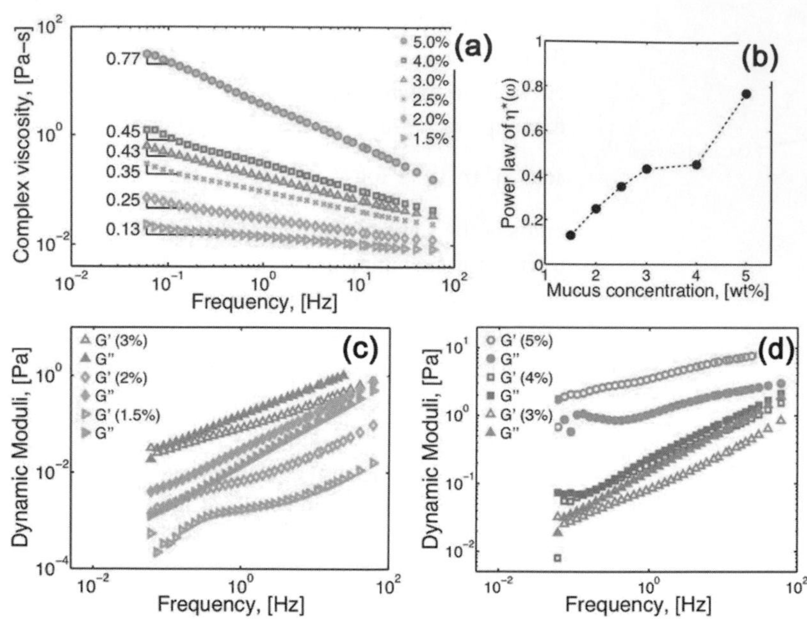

Figure 14.18 Viscoelastic properties of mucus. (a) Complex viscosity magnitude $|\eta^*(\omega)|$ vs frequency f (Hz). (b) The slope of a power-law curve fit, $|\eta^*(\omega)| \sim \omega^{-s}$ to the data in (a). G' and G'' vs f (Hz) for (c) mucus solids concentrations of 1.5, 2, 3 %wt and (d) mucus solids concentrations of 3, 4, 5 %wt.

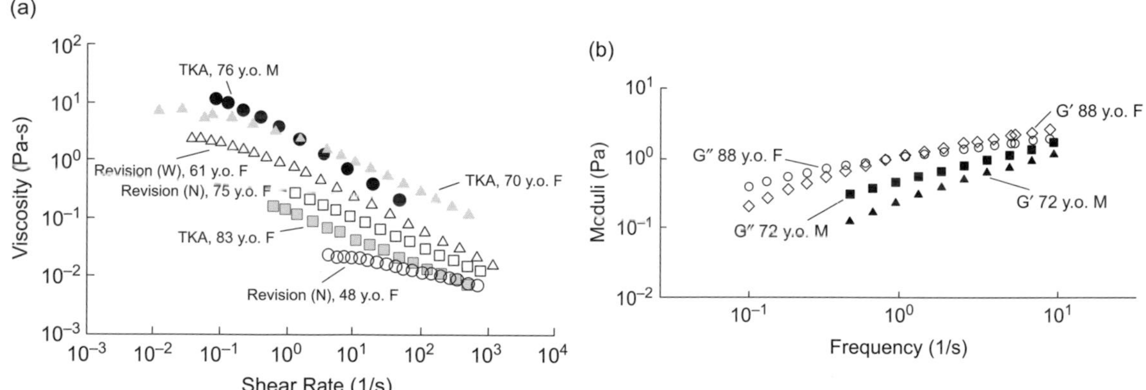

Figure 14.19 (a) Viscosity (Pa·s) vs steady shear rate (s^{-1}) for samples of synovial fluid in patients undergoing total knee arthroplasty (TKA) and revision TKA. (b) $G'$and G'' vs frequency f (Hz) for two sample patients with one showing crossover.

11.8). Now we explore its viscoelastic behavior. Figure 14.19(a) shows the viscosity (Pa·s) vs steady shear rate (s^{-1}) for samples of synovial fluid in patients undergoing total knee arthroplasty (TKA) and revision TKA. All exhibit shear thinning and most also have a low frequency plateau which we learned is a feature of the Carreau and Cross models, Section 10.7. Recall water viscosity $\sim 10^{-3}$ Pa · s, which forms the horizontal axis, so the range is 10 to 10,000 times water viscosity. Figure 14.19(b) shows G' and G'' vs frequency f (Hz) for two sample patients, one showing crossover at 0.87 Hz (for a 88-year-old female) and the other no crossover (a 72-year-old male). Walking and running frequencies are indicated.

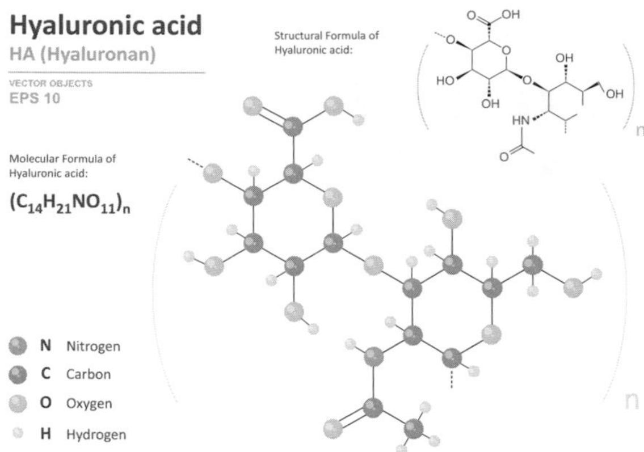

Figure 14.20 Hyaluronic acid is long chains of this disaccharide molecule where n can be 25,000.

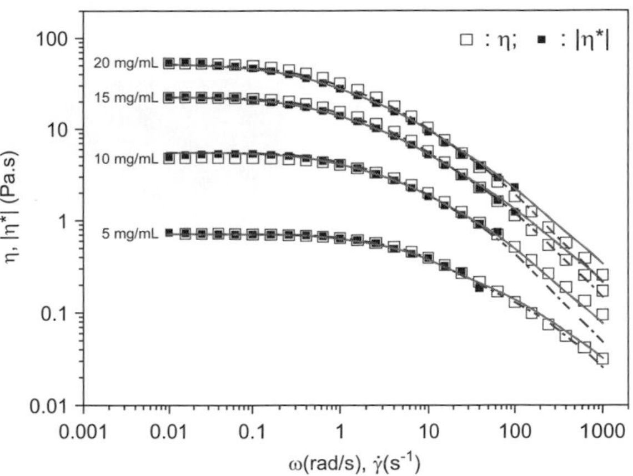

Figure 14.21 Cox–Merz plot of HA-1 solutions at 25 °C showing experimental data (symbols) for steady viscosity, η, and the magnitude of the complex viscosity |η*|. Lines are predictive models. Note the strong evidence of the Cox–Merz rule.

The experimental results are consistent with the well-known fact that synovial fluid not only acts as a viscous liquid in low frequency regions, responding to slowly moving joints, but also acts as an elastic behavior in high frequency regions, responding to rapidly moving joints.

Synovial fluid is composed primarily of hyaluronic acid (HA) molecules. HA is a disaccharide polymer and can form long chains, see Figure 14.20, where n can be 25,000, for example.

Figure 14.21 shows experimental data (symbols) for the steady viscosity, η, and the magnitude of the complex viscosity, $|\eta^*|$, vs $\omega = \dot{\gamma}$ for HA solutions. Note their overlap as strong evidence of the Cox–Merz rule holding for this fluid. The lines are for the predictive models Phan-Thien–Tanner (solid line) (Thien and Tanner, 1977) and Giesekus (dashed line) (Giesekus, 1982), which are additional, more complex, constitutive models for viscoelastic fluids.

Synovial fluid relaxation times vary depending on the state of disease for the joint and its fluids. Schurz and Ribitsch (1987) found a range 40–100 s for normal, 8–20 s for degenerative and 0.02–1.0 s for inflammatory conditions. They inferred a relaxation time as the reciprocal of the steady shear rate which corresponded to the onset of shear thinning. Other investigators used the crossover frequency $\lambda = 1/\omega_c = 1/2\pi f_c$ and found a normal range of 0.65–5.82 s (Safari et al., 1990). Still another study showed ~10 s (Lai et al., 1977). In terms of HA alone, an average relaxation time constant can be derived from the multimode approaches in Yu et al. (2014) and a value of $\lambda = 3.31$ s is reported there.

Box 14.2 Lifting, Walking and Running

Athletics exposes the joints, and the synovial fluid in them, to a wide range of frequencies. For example, basketball and running, see Figure 14.22(a, b), push the upper and lower extremities to frequencies in the range of 2 Hz, while slow-lift weight training can be at 0.1 Hz, see Figure 14.22(c).

(a)

(b)

(c)

Figure 14.22 Running and jumping are typically in the frequency range of 2 Hz: (a) John Grotberg, Grinnell College Basketball and (b) Anna Grotberg, Castle Triathlon Series. (c) Slow-lift weight training is more like 0.1 Hz, for example bicep curls.

14.9 Oscillatory Squeeze Film: Synovial Fluid

To obtain more experience with a viscoelastic fluid, consider a squeeze film as shown in Figure 14.23. The plates are of length L and average distance h_0 apart. The walls oscillate toward and away from one another at frequency, f, or angular frequency $\omega = 2\pi f$. The upper wall position is $h = h_0(1 + \delta \sin \omega t)$, while the lower wall is 180° out of phase. This makes the centerline of the film, $y = 0$, a line of symmetry for the flow where $\partial u/\partial y = 0$ and $v = 0$. This system is similar to the squeeze film we studied in the context of lubrication theory, Section 11.7, except now the driving force is oscillatory

$$f = f_0 \cos(\omega t) = \Re\left(f_0 e^{i\omega t}\right) \tag{14.9.1}$$

In the lubrication limit of $\varepsilon = h_0/L << 1$ and negligible inertia, $\alpha^2 = \omega h_0^2/\nu << 1$, the Cauchy momentum equation for this system simplifies to

$$\frac{\partial \tau_{12}}{\partial y} = \frac{\partial p}{\partial x} \qquad \frac{\partial p}{\partial y} = 0 \tag{14.9.2}$$

Since $\partial p/\partial y = 0$, we know that $p = p(x,t)$ and Eq. (14.9.2) can be integrated directly

$$\tau_{12} = \frac{\partial p}{\partial x} y \tag{14.9.3}$$

where the constant of integration has been chosen to be zero so that there is zero shear at $y = 0$.

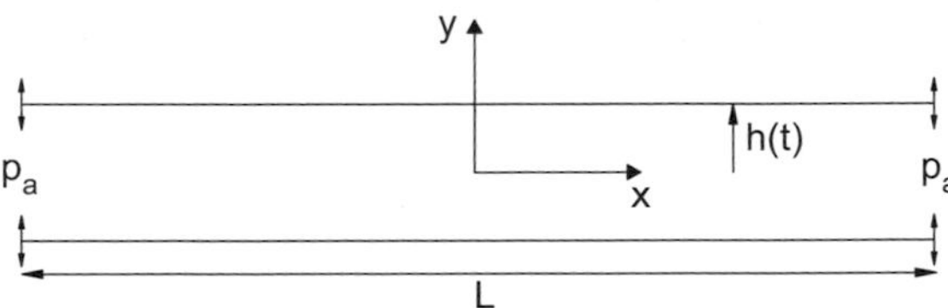

Figure 14.23 Oscillatory squeeze film.

For simplicity we choose the constitutive equation for the Maxwell fluid, which reduces to the τ_{12}-component

$$\tau_{12} + \beta \frac{\partial \tau_{12}}{\partial t} = \mu \frac{\partial u}{\partial y} \tag{14.9.4}$$

The boundary conditions are no-slip and no-penetration at the plates, which is equivalently symmetry at the midline and no-slip and no-penetration at the upper wall

$$\frac{\partial u}{\partial y}(y = 0) = 0, \quad v(y = 0) = 0 \qquad u(y = h(t)) = 0, \quad v(y = h(t)) = \frac{dh}{dt} \tag{14.9.5}$$

The continuity equation is

$$\frac{\partial u}{\partial x} + \frac{\partial v}{\partial y} = 0 \tag{14.9.6}$$

Our approach is to separate variables according to the oscillatory driving force

$$u = \hat{u}(x, y)e^{i\omega t} \qquad v = \hat{v}(x, y)e^{i\omega t} \qquad p = \hat{p}(x)e^{i\omega t} \tag{14.9.7}$$

$$(1 + iDe)\frac{dp}{dx} y = \mu \frac{\partial \hat{u}}{\partial y} \tag{14.9.8}$$

which can be directly integrated as

$$\hat{u}(x, y) = (1 + iDe)\frac{d\hat{p}}{dx}\frac{y^2}{2\mu} + c_1(x) \tag{14.9.9}$$

The remaining kinematic boundary condition on $\hat{u}$ is no-slip at $y = h(t)$. Of course, $\hat{u}$ cannot depend on t, so to satisfy no-slip we let the wall position be $h(t) = h_0(1 + \delta e^{i\omega t})$ and assume small amplitude oscillations so that $|\delta| << 1$. Then we can apply the no-slip condition at $y = h_0$, $\hat{u}(y = h_0) = 0$, as the first term of a Taylor series in δ, which solves for $c_1(x)$ in Eq. (14.9.9). The result is

$$\hat{u}(x, y) = (1 + iDe)\frac{d\hat{p}}{dx}\frac{(y^2 - h_0^2)}{2\mu} \tag{14.9.10}$$

We can derive $\hat{v}$ from the continuity equation,

$$\hat{v} = -\int \frac{\partial \hat{u}}{\partial x} dy + c_2(x)$$
$$= -(1 + iDe)\frac{d^2\hat{p}}{dx^2}\frac{(y^3/3 - h_0^2 y)}{2\mu} + c_2(x) \quad (14.9.11)$$

From symmetry we have no cross-flow along the centerline, $\hat{v}(y = 0) = 0$, which determines $c_2(x) = 0$. Finally, the vertical fluid velocity at the upper plate, $v(y = h(t))$, must equal the plate velocity, $dh/dt = i\omega h_0 \delta e^{i\omega t}$. Again using a Taylor series this condition can be applied at $y = h_0$ so that $\hat{v}(y = h_0) = i\omega h_0 \delta$, giving

$$(1 + iDe)\frac{d^2\hat{p}}{dx^2}\frac{h_0^3}{3\mu} = i\omega h_0 \delta \quad (14.9.12)$$

Equation (14.9.12) can be solved for the pressure as

$$\hat{p} = -\frac{3\mu\omega\delta(De + i)}{h_0^2(1 + De^2)}\frac{x^2}{2} + c_3 x + c_4 \quad (14.9.13)$$

Imposing symmetry in x at $x = 0$ on the pressure is $d\hat{p}/dx = 0$, which yields $c_3 = 0$. Then let the average pressure at $x = L$ be zero. That solves for c_4 and the result is

$$\hat{p} = \frac{3\mu\omega\delta(De + i)}{h_0^2(1 + De^2)}\frac{(L^2 - x^2)}{2} \quad (14.9.14)$$

The force balance on the upper plate is

$$f_0 = L_z \int_0^L \hat{p}\, dx = \frac{\mu\omega\delta L^3 L_z(De + i)}{h_0^2(1 + De^2)} \quad (14.9.15)$$

Equation (14.9.15) can be rearranged by defining a dimensionless force $F_0 = f_0/GLL_z$ and recalling that $\varepsilon = h_0/L$, so that

$$F_0 = \frac{De(De + i)\delta}{\varepsilon^2(1 + De^2)} \quad (14.9.16)$$

We can assume that F_0 is the real force magnitude input and our goal is to understand how the viscoelasticity of the synovial fluid modifies the displacement amplitude $|\delta|$, which can be obtained by rearranging Eq. (14.9.16) to solve for δ and taking the magnitude of the result

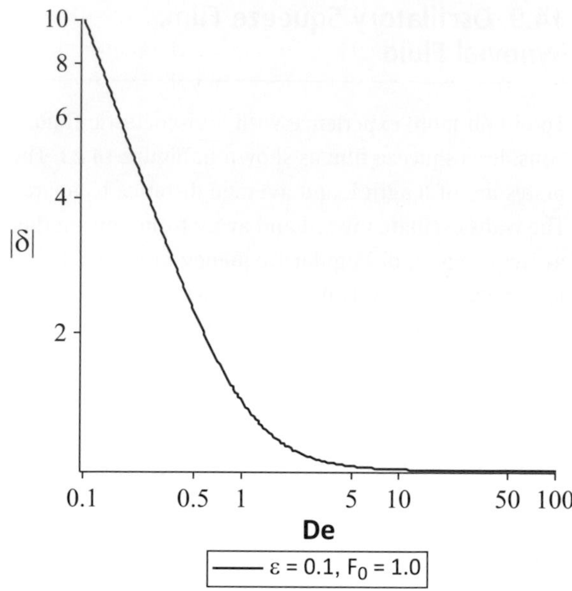

Figure 14.24 $|\delta|$ vs De for $\varepsilon = 0.1, F_0 = 1$.

$$|\delta| = \frac{\varepsilon^2 F_0}{De}\left(1 + De^2\right)^{1/2} \quad (14.9.17)$$

To ensure $|\delta| << 1$ in Eq. (14.9.17) the range of De is limited by the relationship $(1 + De^2)^{1/2}/De << 1/\varepsilon^2 F_0$.

Figure 14.24 shows the result that as the synovial fluid becomes more elastic, with De increasing, the amplitude of the motion is reduced. This is of great benefit to the knee joint, for example, so that the surfaces are kept further apart during walking (2 Hz) and running (4 Hz), which protects them from contact damage. For a relaxation time of $\lambda = 5$ s those frequencies correspond to $De = (2\pi)(2)(5) = 62.8$ and $De = (2\pi)(4)(5) = 125.6$.

14.10 Oscillatory Channel Flow

In Section 8.3 we analyzed oscillatory flow in a channel for Newtonian fluids, see Figure 8.8. For a Maxwell fluid the governing equations are

$$\rho\frac{\partial u}{\partial t} = -\frac{\partial p}{\partial x} + \frac{\partial \tau_{12}}{\partial y}$$
$$\mu\frac{\partial u}{\partial y} = \lambda\frac{\partial \tau_{12}}{\partial t} + \tau_{12} \quad (14.10.1)$$

where $\partial p/\partial x = B^p e^{i\omega t} = B^p \cos(\omega t)$. Use the following scales to make Eqs. (14.10.1) dimensionless: $U = u/U_s, \sigma = \tau_{12}/\sigma_s, T = \omega t, Y = y/b$, where the velocity and shear stress scales are $U_s = -B^p b^2/\mu, \sigma_s = -B^p b$. The resulting system is

$$\alpha^2 \frac{\partial U}{\partial T} = e^{iT} + \frac{\partial \sigma}{\partial Y}$$
$$\frac{\partial U}{\partial Y} = De\frac{\partial \sigma}{\partial T} + \sigma \qquad (14.10.2)$$

with boundary conditions of no-slip at the upper wall, $U(Y = 1) = 0$, and symmetry at the midline $\partial U/\partial Y|_{Y=0} = 0$. The two dimensionless parameters are the Deborah number, $De = \omega\lambda$, and the Womersley parameter, $\alpha = b\sqrt{\omega/\nu}$. To solve the coupled PDE system in Eqs. (14.10.2), separate variables by defining $U(Y,T) = \Re\{\hat{U}(Y)e^{iT}\}$, $\sigma(Y,T) = \Re\{\hat{\sigma}(Y)e^{iT}\}$ and substitute

$$i\alpha^2\hat{U} = 1 + \frac{d\hat{\sigma}}{dY} \qquad \frac{d\hat{U}}{dY} = (1 + iDe)\hat{\sigma} \qquad (14.10.3)$$

Taking the Y-derivative of second equation and inserting into the first leads to a single ODE for $\hat{U}$,

$$\frac{d^2\hat{U}}{dY^2} - \beta^2\hat{U} = -(1 + iDe) \qquad (14.10.4)$$

where $\beta^2 = i\alpha^2(1 + iDe)$. Taking the complex square root shows that $\beta = \alpha\sqrt{i - De}$ is

$$\beta = \frac{\alpha\sqrt{2}}{2}\sqrt{\sqrt{De^2 + 1} - De} + i\frac{\alpha\sqrt{2}}{2}\sqrt{\sqrt{De^2 + 1} + De} \qquad (14.10.5)$$

The general solution form for Eq. (14.10.4) is the sum of homogeneous and particular contributions

$$\hat{U} = A\cosh(\beta Y) + B\sinh(\beta Y) + \frac{(1 + iDe)}{\beta^2} \qquad (14.10.6)$$

Imposing the two boundary conditions solves for A and B in Eq. (14.10.6), the final form being

$$\hat{U} = \frac{(1 + iDe)}{\beta^2}\left(1 - \frac{\cosh(\beta Y)}{\cosh(\beta)}\right) \qquad (14.10.7)$$

Now that $\hat{U}$ is solved, $\hat{\sigma}$ becomes

$$\hat{\sigma} = \frac{1}{(1 + iDe)}\frac{d\hat{U}}{dY} = -\frac{1}{\beta}\frac{\sinh(\beta Y)}{\cosh(\beta)} \qquad (14.10.8)$$

Examples of viscoelastic effects on U are shown in Figure 14.25 at time $T = \pi/4$ for $\alpha = 0.1, 1, 5, 10$ and $De = 0, 0.5, 1, 2$. Figure 14.25(a, b) exhibit the phase and amplitude changes due to increasing De from Newtonian, $De = 0$, to larger elastic contributions. As α becomes larger, $\alpha >> 1$, U exhibits more oscillatory layers Figure 14.25(c, d). This is what we saw in Figure 14.10 for the oscillatory wall.

14.11 Normal Stresses

Figure 14.26 shows the Weissenberg effect, which is also called "rod climbing." For the Newtonian fluid, the rotating rod creates a centrifugal force that throws the fluid radially outward, lowering the fluid level at the rod and raising it at the container wall. By contrast, the viscoelastic fluid develops normal stresses which are stronger than the centrifugal force. These normal stresses push the fluid radially inward, forcing it to the climb the rod. The normal stresses create tension along the circular streamlines which act like a hoop stress. Rod climbing is a common experience when using an electric mixer for cake batter, as well as frosting.

Two examples of simple shear flows are shown in Figure 14.27 with velocity field $\underline{u} = (u_1(x_2), 0, 0)$ where $u_1(x_2) = \dot{\gamma}x_2$ and $\dot{\gamma} = V_0/h$ is the uniform shear strain rate. On the left is a Newtonian fluid where the fluid exerts a shear stress on the upper moving wall felt as a drag force. On the right is a viscoelastic fluid which also exerts a drag force, but in addition it generates a thrust force perpendicular to the moving wall and streamlines. This feature of viscoelastic fluids derives from the tangled chains of its

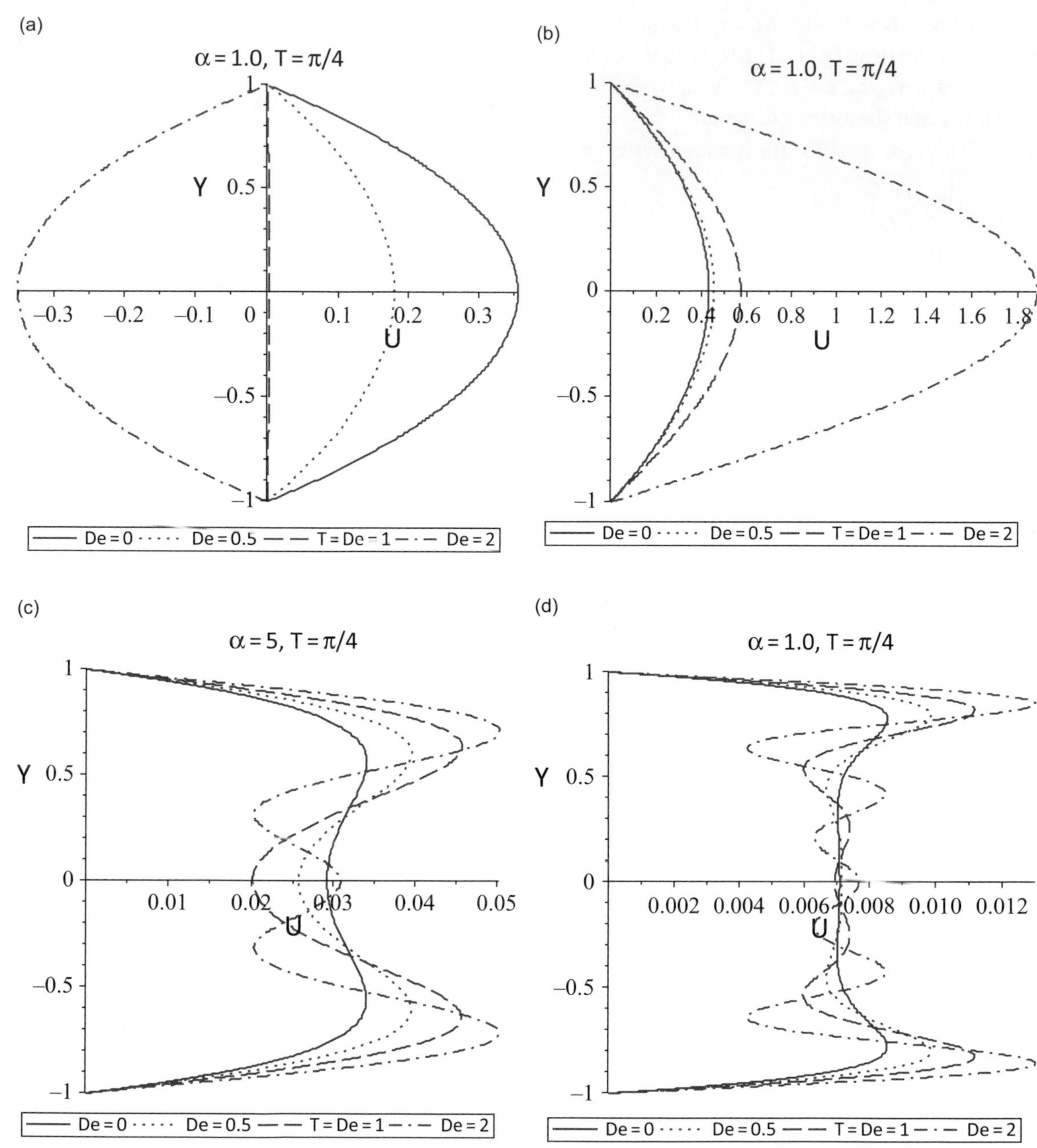

Figure 14.25 U(Y) for $T = \pi/4$ comparing De = 0, 0.5, 1.0, 2.0 for (a) $\alpha = 0.1$, (b) $\alpha = 1$, (c) $\alpha = 5$ and (d) $\alpha = 10$.

components. The design of a cone-plate viscometer (rheometer), see Figure 10.23, can include a measure of vertical thrust to derive the normal stress in addition to the torque which gives the shear stress.

The average normal stress in terms of the stress tensor, T_{ij}, is $T_m = \frac{1}{3}(T_{11} + T_{22} + T_{33})$. It was our choice to relate this average to the pressure as $T_m = -p$, consistent with the definition $T_{ij} = T_m\,\delta_{ij} + \tau_{ij}$, which separates the stress tensor into

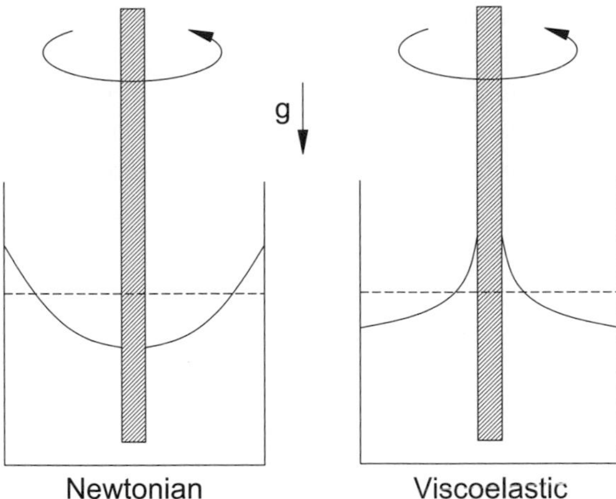

Figure 14.26 Rod climbing, the Weissenberg effect of viscoelastic fluids.

an isotropic part, $T_m \delta_{ij}$, and deviatoric part, τ_{ij}. The trace of T_{ij} is $T_{kk} = T_{11} + T_{22} + T_{33}$ where the repeated index implies the summation

$$T_{kk} = \frac{1}{3}(T_{11} + T_{22} + T_{33})\delta_{kk} + \tau_{kk}$$
$$= (T_{11} + T_{22} + T_{33}) + \tau_{kk} \quad (14.11.1)$$

Noting that $\delta_{kk} = 1 + 1 + 1 = 3$, subtract T_{kk} from both sides of Eq. (14.11.1), and the result is the trace of the extra stress tensor equals zero,

$$\tau_{kk} = \tau_{11} + \tau_{22} + \tau_{33} = 0 \quad (14.11.2)$$

Because τ_{ij} is symmetric, i.e. $\tau_{ij} = \tau_{ji}$, its nine components reduce to six independent components. In addition, Eq. (14.11.2) further reduces the extra stress tensor to five independent components, the three off-diagonal shear stresses and two normal stresses or their linear combinations, for example. With regard to viscoelastic fluids, it is standard to represent the normal stresses in terms of the first and second normal stress differences, N_1, N_2, respectively

$$N_1 = \tau_{11} - \tau_{22}, \quad N_2 = \tau_{33} - \tau_{22} \quad (14.11.3)$$

where N_1 and N_2 emphasize that it is not the absolute values of normal stresses that affect flow, but rather it is their differences. They are usually represented as the first and second normal stress coefficients, ψ_1, ψ_2, respectively

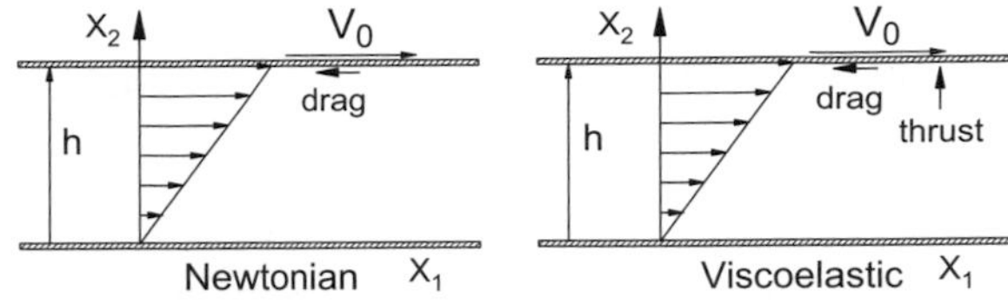

Figure 14.27 Simple shearing of Newtonian and viscoelastic fluids showing drag force in both, but also a normal thrust force in the viscoelastic fluid.

$$\psi_1 = \frac{\tau_{11} - \tau_{22}}{\dot{\gamma}^2} = \frac{N_1}{\dot{\gamma}^2}, \quad \psi_2 = \frac{\tau_{33} - \tau_{22}}{\dot{\gamma}^2} = \frac{N_2}{\dot{\gamma}^2} \quad (14.11.4)$$

In Figure 14.27 it is $\psi_1 < 0$ from the relative size of the normal stress in the vertical direction, τ_{22}. For the Newtonian fluid in Figure 14.27, $\tau_{11} = \mu \partial u_1 / \partial x_1 = 0$, $\tau_{22} = \mu \partial u_2 / \partial x_2 = 0$, $\tau_{33} = \mu \partial u_3 / \partial x_3 = 0$ so $\psi_1 = \psi_2 = 0$ as we expect. Normal stresses are potentially important in synovial fluid flow, where shearing can create thrust to keep joint surfaces apart and reduce wear.

14.12 The Equivalent Approach for a Fading Memory Fluid

Another approach to the Maxwell model is to multiply Eq. (14.7.4) by $(1/\lambda)e^{t/\lambda}$ to obtain

$$\frac{1}{\lambda} e^{\frac{t}{\lambda}} \left[\tau_{ij} + \lambda \frac{\partial \tau_{ij}}{\partial t} = 2\mu E_{ij} \right] \quad (14.12.1)$$

and note that the left hand side simplifies to

$$\frac{\partial}{\partial t}\left(e^{\frac{t}{\lambda}} \tau_{ij} \right) = 2\frac{\mu}{\lambda} e^{\frac{t}{\lambda}} E_{ij} \quad (14.12.2)$$

Multiplying through by dt and integrating from all previous times to the present time, t, via the dummy variable t'

$$\int_{-\infty}^{t} d\left(e^{\frac{t'}{\lambda}} \tau_{ij}(t') \right) = 2 \int_{-\infty}^{t} \frac{\mu}{\lambda} e^{\frac{t'}{\lambda}} E_{ij}(t')\, dt' \quad (14.12.3)$$

yields the result

$$e^{\frac{t'}{\lambda}}\tau_{ij}(t')\Big|_{-\infty}^{t} = 2\int_{-\infty}^{t} \frac{\mu}{\lambda} e^{\frac{t'}{\lambda}} E_{ij}(t')\,dt' \tag{14.12.4}$$

We assume that the stress is finite at $t' = -\infty$ so only the upper time limit, $t' = t$, contributes to the left hand side, and rearranging gives us

$$\tau_{ij}(t) = 2e^{-\frac{t}{\lambda}}\int_{-\infty}^{t} \frac{\mu}{\lambda} e^{\frac{t'}{\lambda}} E_{ij}(t')\,dt' \tag{14.12.5}$$

Since t can be treated as a constant within the t' integral, we can move terms to get

$$\tau_{ij}(t) = 2\int_{-\infty}^{t} \left(\frac{\mu}{\lambda} e^{-\frac{(t-t')}{\lambda}}\right) E_{ij}(t')\,dt' \tag{14.12.6}$$

Equation (14.12.6) says that the current value of the stress at time t depends on the history of the strain rate at previous values at times t'. The weighting term $G\exp(-(t-t')/\lambda)$ is called the relaxation modulus, which incorporates the memory of the deformation rate. The further back in time, the greater is the positive difference $t - t'$, but its effect is less and less important due to the negative sign in the exponential, i.e. it is an exponentially fading memory. Viscoelastic fluids are often called memory fluids. An additional view is to let the rate of strain tensor, E_{ij}, be defined directly as the time derivative of the strain tensor, γ_{ij}, so that $E_{ij} = \dot{\gamma}_{ij}$,

$$\tau_{ij}(t) = 2\int_{-\infty}^{t} \left(\frac{\mu}{\lambda} e^{-\frac{(t-t')}{\lambda}}\right) \dot{\gamma}_{ij}(t')\,dt' \tag{14.12.7}$$

Integrating the right hand side of Eq. (14.12.6) by parts gives us

$$\begin{aligned}\tau_{ij}(t) &= 2\int_{-\infty}^{t} \left(\frac{\mu}{\lambda} e^{-\frac{(t-t')}{\lambda}}\right) \dot{\gamma}_{ij}(t')\,dt' \\ &= 2\left(\frac{\mu}{\lambda} e^{-\frac{(t-t')}{\lambda}}\right)\gamma_{ij}\Big|_{-\infty}^{t} \\ &\quad - 2\int_{-\infty}^{t} \left(-\frac{\mu}{\lambda^2} e^{-\frac{(t-t')}{\lambda}}\right) \gamma_{ij}(t')\,dt' \end{aligned} \tag{14.12.8}$$

The first term on the right hand side is zero for the lower limit for $t' = -\infty$ assuming that $\gamma_{ij}(t' = -\infty)$ is finite. It is also zero for the upper limit when $t' = t$ since $\gamma_{ij}(t' = t)$ because the deformation is measured relative to the configuration at time t. The result is

$$\tau_{ij}(t) = -2\int_{-\infty}^{t} \left(-\frac{\mu}{\lambda^2} e^{-\frac{(t-t')}{\lambda}}\right) \gamma_{ij}(t')\,dt' \tag{14.12.9}$$

which now has the appearance of Hooke's law for a solid. The function in brackets is called the memory function of this viscoelastic material.

14.13 Kelvin–Voigt Viscoelastic Solid: Biological Tissues

A simple model for a viscoelastic solid also involves a spring and dashpot, but instead of being arranged in series like the Maxwell fluid, they are in parallel as shown in the Kelvin–Voigt model of Figure 14.28. The total force is the sum of the two contributions

$$F = kU + \mu\dot{U} \tag{14.13.1}$$

The corresponding equation for shear deformation is

$$\tau = G\gamma + \mu\dot{\gamma} \tag{14.13.2}$$

For an applied constant stress F_0, the solution for the displacement U in Eq. (14.13.1) is

$$U = \frac{F_0}{k}\left(1 - e^{-\frac{t}{\mu/k}}\right) \tag{14.13.3}$$

which is the creep of a solid that has a relaxation behavior with decay time constant $\lambda = \mu/k$, resulting in the steady state displacement F_0/k. For an applied constant displacement, U_0, there is only the elastic component of force,

$$F = kU_0 \tag{14.13.4}$$

For shear oscillations of a Kelvin–Voigt model we use the approach in Eq. (14.3.1), with $\tau = \tau_0 e^{i\omega t}, \gamma = \gamma_0 e^{i\omega t}$ inserted into Eq. (14.13.2) as

$$\tau_0 = (G + i\omega\mu)\gamma_0 \tag{14.13.5}$$

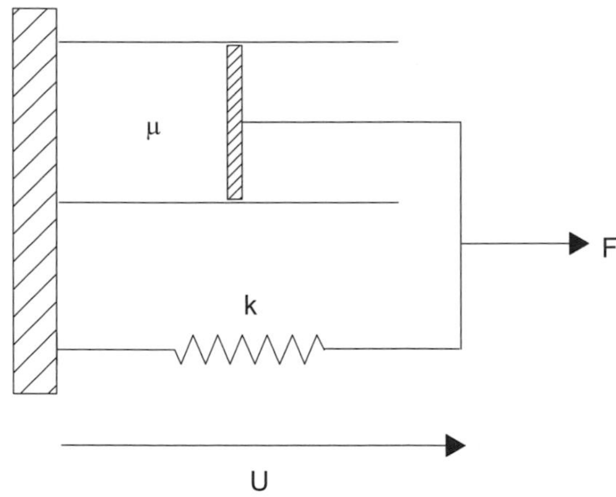

Figure 14.28 The Kelvin–Voigt model of a viscoelastic solid.

From Eq. (14.13.5) we see the complex shear modulus of a Kelvin–Voigt solid is $G^* = G' + iG''$ where $G' = G,\ G'' = \mu\omega$ so that their dimensionless versions are

$$\frac{G'}{G} = 1, \quad \frac{G''}{G} = \frac{\mu\omega}{G} = De \qquad (14.13.6)$$

The Kelvin–Voigt model behaves as an elastic solid in the limit $\mu \rightarrow 0$ or $De << 1$, and as a viscous fluid for $G \rightarrow 0$ or $De >> 1$. These are the opposite limits for the Maxwell fluid where fluid behavior occurs for $De << 1$ and elastic behavior for $De >> 1$. Extending this to a generalized Maxwell model, with n Maxwell elements in parallel with the single elastic spring, is called the Wiechert model or Maxwell–Wiechert model.

Summary

Viscoelastic fluids cover a wide variety of biological situations and their interesting behaviors characterized by complex shear modulus, components G' and G'', and complex viscosity, components μ' and μ''. In this chapter we have dealt with mucus, blood and synovial fluid as primary examples, noting how the elasticity and viscosity vary with the parameters. Mucus, in particular, is wide ranging in nature including respiratory, cervical and gastrointestinal, to which we can add the slime of creatures like slugs which assists their locomotion. Key features included identifying any crossover point on the frequency axis, $\omega = \omega_c$, where $G'(\omega_c) = G''(\omega_c)$, and the system behavior is changing from viscous dominated $G' < G''$ for $\omega < \omega_c$ to elastic dominated $G' > G''$ for $\omega > \omega_c$. In addition, the Cox–Merz rule proved to hold up well for these biofluids so that the steady shear viscosity could be equated to the magnitude of the complex viscosity, $\mu_{eff}(\dot{\gamma}) = |\mu^*(\omega)|_{\dot{\gamma}=\omega}$.

The important concepts of stress relaxation and a relaxation time were discussed and tested by forcing familiar systems over a range of frequencies: Stokes' second problem, the squeeze film of a model joint with synovial fluid, and oscillatory channel flow. The Deborah number, $De = \omega\lambda$, told us when the system was at low frequency, viscous fluid dominated, and high frequency, elastic dominated. While the Maxwell model was illuminating, it had only one relaxation time. Generalized Maxwell approaches, with multiple Maxwell units, on the other hand gave us more options, and we explored a two-unit system to see how multiple relaxation times can influence the overall system behavior. During a more rigorous examination of the Maxwell model we were able to draw upon our discussions in Chapter 11 on Galilean invariance to demonstrate its flaws, and the need for more sophisticated models like the upper convected Maxwell and Oldroyd B. Finally, we discussed the unique property of viscoelastic fluids to generate normal stresses from shear flow, which leads to the Weissenberg effect.

Problems

14.1

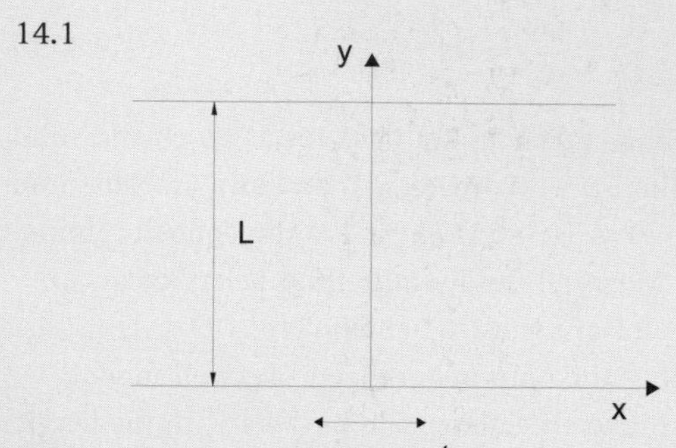

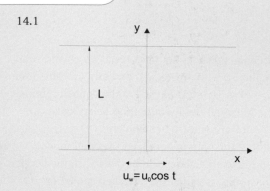

In Section 14.5 we formulated Stokes' second problem for a Maxwell Fluid, see Eq. (14.5.17) and Figure 14.10. Now we will modify that problem by imposing a second wall at $y = L$ parallel to the oscillating wall which is at $y = 0$, see figure. This second wall is stationary so the fluid velocity is zero there. This is a model of flow in the joint space since synovial fluid has viscoelastic properties. The dimensionless velocity, $U(\eta, T) = \Re(\hat{U}(\eta)e^{iT})$, has the same form as Eq. (14.5.8), except that now $\eta = y/L$. The form of $\hat{U}(\eta)$ is the same as Eq. (14.5.11).

a. The boundary conditions to determine c_1 and c_2 are now $\hat{U}(\eta = 0) = 1, \hat{U}(\eta = 1) = 0$ where, due to the scaling, the upper stationary wall is at $\eta = 1$. What are the values of c_1 and c_2 in terms of β?
b. Let $T = 0$, $\alpha = 1$ and plot the velocity profile $U(\eta)$ for $De = 0, 2, 5$. Be sure you only plot where there is fluid.
c. Repeat part (b) for $\alpha = 5$ and compare your results.
d. We found in our analysis that the shear stress is given by $\sigma(\eta, T) = \Re(\hat{\sigma}(\eta)e^{iT})$ and that $\hat{\sigma}(1 + iDe) = d\hat{U}/d\eta$. Plot the dimensionless lower wall shear $T_0 = \Re(\hat{\sigma}(\eta = 0)\, e^{iT})$ and the upper wall shear $T_1 = \Re(\hat{\sigma}(\eta = 1)\, e^{iT})$ vs T for one cycle with $De = 0, 2, 5$. Let $\alpha = 1$. Discuss your results.
e. Repeat part (d) with $\alpha = 5$ and compare your results.

14.2 Try a simplistic cough model by considering an applied shear stress $\tau(t) = \tau_0 e^{-t/T_0}$ to the surface of a single fluid layer as in Figure 8.24, but with a viscoelastic fluid. Ignoring inertia and assuming $De << 1$, the Cauchy momentum equation for conservation of momentum is $\partial\tau_{12}/\partial y = 0$, which implies that $\tau_{12} = \tau_{12}(t)$. The stress boundary condition is $\tau_{12}(t) = \tau(t)$ at $y = h$ and, therefore, for all values of y.

a. Insert for $\tau_{12}(t)$ into Eq. (14.5.2) and then integrate with respect to y to find $u(y, t)$. Use the no-slip boundary condition that $u(y = 0) = 0$.
b. Scale your result in part (a) using the dimensionless variables $U = u/(\tau_0 b/\mu)$, $Y = y/h$, $T = t/T_0$ and define $De = \lambda/T_0$. Plot U vs Y at times $T = 0, 1, 2, 3$ for $De = 0.1$.
c. Plot $U(Y, T = 1)$ for $De = 0, 0.25, 0.5$. What is the effect of De on the velocity profile?
d. Now calculate the dimensional flow rate $q(t) = L_z \int_0^b u\, dy$ for a unit length L_z in the z-direction. Using the scales in part (b), define the dimensionless flow rate as $Q = q/(\tau_0 h^2 L_z/\mu)$ and plot Q over the time range $0 \leq T \leq 5$ for $De = 0, 0.25, 0.5$. What is the effect of viscoelasticity?
e. The dimensional volume that moves during the cough is $v = \int_0^{5T_0} q\, dt$ where you can ignore terms multiplied by e^{-5}. What is your answer? What makes the volume greater? How does viscoelasticity, with $De<<1$, affect it?

14.3

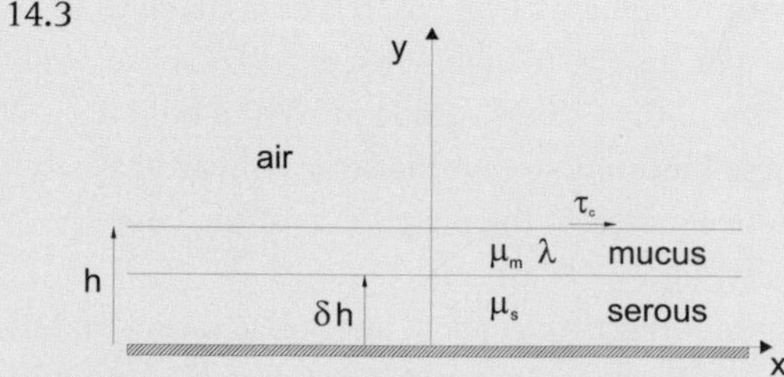

Consider a two-layer airway liquid lining as shown in the figure. The lower layer adjacent to the wall, $0 \le y \le \delta h$, is serous and Newtonian. The upper layer in contact with the air, $\delta h \le y \le h$, is viscoelastic mucus with viscosity μ_m and Maxwell stress-relaxation time λ. Neglecting inertia for thin layers and assuming De << 1, the governing equations are

$$\mu_s \frac{\partial^2 u_s}{\partial^2 y} = 0 \quad 0 \le y \le \delta h \text{ serous}$$
$$\mu_m \frac{\partial u_m}{\partial y} = \tau_{12} + \lambda \frac{\partial \tau_{12}}{\partial t} \quad \delta h \le y \le h \text{ mucus}$$

and the shear stress for all y is the applied cough shear, $\tau_c = \tau_0 e^{-t/T_0}$.

a. Integrate the serous equation twice and call the integration functions $a_s(t)$, $b_s(t)$. Apply the no-slip condition $u_s(y = 0) = 0$ and the stress condition, $\mu_s \partial u_s/\partial y = \tau_c(t)$, to find $a_s(t)$, $b_s(t)$. What is your solution for $u_s(y, t)$?
b. Insert $\tau_{12} = \tau_c$ into the mucus equation and integrate once to find $u_m(y, t)$ and let your integration function be $a_m(t)$. Solve for $a_m(t)$ by equating the mucus and serous velocities at $y = \delta h$. What is your solution for $u_m(y, t)$?
c. Use the following scales to make your velocity solutions in parts (a) and (b) dimensionless $U_s = u_s/(\tau_0 h/\mu_s)$, $U_m = u_m/(\tau_0 h/\mu_s)$, $Y = y/h$, $T = t/T_0$. Define two more dimensionless parameters, $De = \lambda/T_0$, $\beta = \mu_m/\mu_s$ to go along with δ. Show $U_s(Y, T)$, $U_m(Y, T)$ and their respective ranges of Y. Why is there no dimensionless parameter involving τ_0 ?
d. Plot the dimensionless velocities, U_s, U_m vs Y in their respective regions. Let $\beta = 10$, $\delta = 1/2$, De = 0.25 and T = 0, 1, 2, 3. Describe your results.
e. To explore the effect of viscoelasticity, plot U_s, U_m for $\beta = 10$, $\delta = 1/2$, T = 1 and De = 0, 0.25, 0.5. How does viscoelasticity affect the velocities?
f. Dehydration can reduce the serous layer thickness. To simulate this situation for comparison, set the serous layer to be $0 \le y \le h/4$ and the mucus layer $h/4 \le y \le 3h/4$ so the mucus layer thickness remains $h/2$. Plot $U_s\ (0 \le Y \le 1/4)$ and $U_m(1/4 \le Y \le 3/4)$, and let $\beta = 10$, T = 1 and De = 0, 0.25, 0.5. Display on the same plot from part (e). How does a reduction of the serous layer thickness influence the velocities? Is the De effect modified?
g. Let the dimensionless mucus flow rate for equal layer thicknesses be defined as $Q_{m2}(T) = \int_{1/2}^{1} U_m\, dY$ and the dehydrated flow rate as $Q_{m4}(T) = \int_{1/4}^{3/4} U_m\, dY$. Plot them on the same graph for $\beta = 5$, De = 0, 0.25, 0.5 over the time range $0 \le T \le 5$. What are your conclusions?

14.4 For pulsatile flow of a Maxwell fluid the applied pressure gradient has steady and time-periodic contributions, $\partial p/\partial x = B^s + B^p \cos(\omega t)$. The velocity solution in dimensionless form is the sum of the steady parabolic profile from Eq. (6.5.14) and the oscillatory profile from Section 14.10

$$U(Y,T) = U^s(Y) + U^p(Y,T) = \frac{1}{2}(1 - Y^2) + f\,\mathrm{Re}\{\hat{U}(Y)\, e^{iT}\}$$

$f = B^p/B^s$ is the ratio of the periodic to steady pressure gradient magnitudes. The dimensionless variables are $U = u/(-B^s b^2/\mu)$, $T = \omega t$, $Y = y/b$ and $\hat{U}(Y)$ is given in Eq. (14.10.7).

a. Plot $U(Y, T = \pi/2)$ for $\alpha = 5$, $f = 20$ and De = 0, 0.5, 1, 2. What is the effect of De on the profile? Which value of De is representative of blood in human physiology?
b. Repeat part (a) at time $T = \pi$. Is there any region of reverse flow? If so, does De modify it?
c. Repeat part (a) for $\alpha = 1$, $f = 0.2$. Compare your results.
d. Repeat part (b) for $\alpha = 1$, $f = 0.2$. Compare your results.
e. The dimensionless upper wall shear is $\tau_w = \tau_w^s + \tau_w^p$ where $\tau_w^s = -dU^s/dY|_{Y=1}$ and $\tau_w^p = -f\,\mathrm{Re}\{\hat{\sigma}(Y = 1)\, e^{iT}\}$ where $\hat{\sigma}(Y)$ is given in Eq. (14.10.8). Plot τ_w for $\alpha = 5$, $f = 20$ and De = 0, 0.5, 1, 2. How does viscoelasticity affect wall shear? Why?

f. The dimensionless flow rate is $Q = 2\int_0^1 U\,dY$. Plot Q for $\alpha = 5, f = 20$ and $De = 0, 0.5, 1, 2$. How does viscoelasticity affect Q? Why?
g. Repeat part (f) for $\alpha = 5, f = 20$. Compare your results.

14.5

a. For the two-element Maxwell model the dimensionless equations, Eqs. (14.6.19) describe the dependence of G'/G_1 and G''/G_1 as functions of the parameters $\beta = \mu_2/\mu_1$ and $\phi = \lambda_2/\lambda_1$. Calculate the values of $De = De_m$ where G''/G_1 is a maximum for $\beta = 1$ and $\phi = 0.1, 1.0, 10.0$.
b. Then calculate the crossover values of $De = De_c$ where $G'/G_1 = G''/G_1$ for the same parameter choices.
c. For the single component Maxwell model, we found $De_c = De_m = 1$ at both the crossover and the maximum, Eqs. (14.3.7) and (14.3.8). Is $De_m = De_c$ for these parameter choices in a two-component Maxwell system?
d. Plot G'/G_1 and G''/G_1 vs $0.01 \le De \le 100$ for $\beta = 10$ and $\phi = 0.1, 1.0, 10.0$.
e. Describe your graph in part (d) and the behavior of the different parameter choices like the presence of maxima, crossovers and asymptotic behavior for large and small De and how these change with the parameters.
f. What are the crossover values De_c and maximum points De_m for part (d). Does $De_c = De_m$ for the same parameter set?

14.6 In the notes for the oscillatory squeeze film of a viscoelastic fluid we derived the x-velocity as the real part of $u = \hat{u}(x,y)e^{i\omega t}$ and the pressure as the real part of $p = \hat{p}(x)e^{i\omega t}$, see Eq. (14.9.10) and Eq. (14.9.14). Rearranging Eq. (14.9.16) the displacement amplitude is $\delta = F_0\varepsilon^2(De - i)/De$.

a. Make the velocity dimensionless with the scales $U = u/\omega L, T = \omega t, Y = y/h_0, X = x/L$. What is $U(X,Y,T) = \hat{U}(X,Y)e^{iT}$?
b. Plot $U(X = 0.5, Y, T = \pi/8)$ for $De = 0.1, 1, 10$ over the interval $-1 \le Y \le 1$ and discuss your results. Let $F_0 = 1, \varepsilon = 0.1$.
c. Animate the three plots in part (b) on the same graph. Describe what you see.
d. Plot the flow rate $Q = \mathrm{Re}\left\{e^{iT}\int_{-1}^{1}\hat{U}\,dY\right\}$ at $X = 0.5$ for $De = 0.1, 1, 10$. How does De influence the flow rate?
e. Find the dimensionless stream function, ψ, from your solution of U in part (a) assuming $\psi(Y = 0) = 0$. Plot the streamlines for $T = 0$ with $F_0 = 1$, $\varepsilon = 0.1$ and $De = 1$. Describe your results.

14.7 In ventilatory modalities like high frequency oscillatory ventilation the liquid lining of the lung is exposed to a time-periodic shear stress, $\tau = \tau_0\cos(\omega t)$, well beyond normal breathing rates, e.g. 3–10 Hz. For a thin viscoelastic film occupying $0 \le y \le h$, with a wall at $y = 0$ and the air–liquid interface at $y = h$, the dimensionless governing equations are shown in Eq. (1)

$$\frac{\partial\sigma}{\partial Y} = 0$$
$$\sigma + De\frac{\partial\sigma}{\partial T} = \frac{\partial U}{\partial Y} \tag{1}$$

$$\frac{d\hat{\sigma}}{dY} = 0$$
$$\hat{\sigma}(1 + iDe) = \frac{d\hat{U}}{dY} \tag{2}$$

where $De = \omega\lambda$. We have ignored inertia since $\alpha = h\sqrt{\omega/\nu} << 1$ due to $h \sim$ 10–20 μm. The scaled variables are $T = \omega t,\ Y = y/h,\ U = u/(\tau_0 h/\mu)$, $\sigma = \tau_{12}/\tau_0$. The boundary conditions are $U(Y = 0) = 0$ and $\sigma(Y = 1) = e^{iT}$. Let the solutions be separated into Y and T dependent terms, $U = \mathrm{Re}\{\hat{U}(Y)e^{iT}\}, \sigma = \mathrm{Re}\{\hat{\sigma}(Y)e^{iT}\}$, and insert into Eq. (1). The result is Eq. (2). Eliminating $\hat{\sigma}$ yields an ODE for $\hat{U}$

$$\frac{d^2\hat{U}}{dY^2} = 0 \qquad (3)$$

a. Solutions to Eq. (3) are of the form $\hat{U} = A + BY$. Solve for A and B using the boundary conditions $\hat{U}(Y = 0) = 0,\ \hat{\sigma}(Y = 1) = \frac{1}{(1+iDe)} \frac{d\hat{U}}{dY}\Big|_{Y=1} = 1$.
b. With your solution for $\hat{U}$ in part (a), substitute into the second of Eqs. (2) to solve for $\hat{\sigma}$.
c. For $\lambda = 30$ s and $0.25\ \text{Hz} \leq f \leq 3\ \text{Hz}$, show that the range of De is approximately $50 \leq De \leq 500$. Plot $U(Y, T)$ at times $T = 0, \pi/2, \pi, 3\pi/2$for $De = 0$, i.e. Newtonian. Is there symmetry to this flow with respect to time? Why?
d. Repeat part (c) for $De = 50$. How does your result differ from part (c)? What about the symmetry? Recall that within the coefficient $(1 + iDe)$ the i-term is $i = e^{i\frac{\pi}{2}} = \cos(\pi/2) + i\sin(\pi/2)$.
e. To see the effect of viscoelasticity on the phase of the velocity, plot $U(Y, T = 3\pi/2)$ for $De = 0, 1, 5, 10, 50$. What has happened to the phase and magnitude of the profile? Relate your answer to Figure 14.9(b), where the behavior of $|\mu^*/\mu|$ is shown as $De >> 1$.
f. The flow rate is $Q = Re\left\{\int_0^1 U\,dY \times e^{iT}\right\}$. Plot it for one cycle with $\alpha = 0.1$ and $De = 0, 0.5, 1, 1.5, 2$. What happens to the magnitude and phase of the maximal Q?

Image Credits

Figure 14.7 Source: photos.com/ Getty

Figure 14.15a and b Source: author original adapted from Tomaiuolo, G., A. Carciati, et al. (2016). "Blood linear viscoelasticity by small amplitude oscillatory flow." *Rheol Acta* 55: 485–495, with gratitutude to G. Tomaiuolo.

Figure 14.17 Source: author original adapted from Fig 4B of Lai SK, Wang Y-Y, Wirtz D, Hanes J. Micro- and macrorheology of mucus. *Advanced Drug Delivery Reviews*. 2009; 61(2):86–100, with gratitude to J. Hanes.

Figure 14.18 Source: Fig 14.18abcd is Fig 5 from Hill DB, Vasquez PA, Mellnik J, McKinley SA, Vose A, Mu F, et al. A biophysical basis for mucus solids concentration as a candidate biomarker for airways disease. *PLoS One*. 2014; 9(2).

Figure 14.19a Source: author original adapted from Fig 1 of Mazzucco D, McKinley G, Scott RD, Spector M. Rheology of joint fluid in total knee arthroplasty patients. *Journal of Orthopaedic Research*. 2002; 20(6):1157–63. With gratitude to M. Spector.

Figure 14.19b Source: author original adapted from Fig 3 of Mazzucco D, McKinley G, Scott RD, Spector M. Rheology of joint fluid in total knee arthroplasty patients. *Journal of Orthopaedic Research*. 2002; 20(6):1157–63. With gratitude to M. Spector.

Figure 14.20 Source: Iv_design/ Getty

Figure 14.21 Source: author original adapted from Fig 7 of Yu FY, Zhang F, Luan T, Zhang ZN, Zhang HB. Rheological studies of hyaluronan solutions based on the scaling law and constitutive models. *Polymer*. 2014; 55(1):295–301. With gratitude to HB Zhang.

Figure 14.22a Source: courtesy of Grinnell College Sports Information.

Figure 14.22b Source: courtesy of peachysnaps.

Figure 14.22c Source: gawrav/ Getty

15 Porous Media

Introduction

A porous material is made of a solid structure, or matrix, and interconnected voids, or pores. A sponge is a good example. This chapter assumes the pores are filled with a fluid which is flowing through the media, like squeezing a fully soaked sponge. Early investigations of porous media stemmed from geological applications like rain-drenched hills and their penchant for mudslides, and civil engineering construction concerns of leakage flows through and under dams and dikes as well as sinking structures, some being quite famous, for example the Leaning Tower of Pisa or Kansai Airport. These flows also are critical to supplying clean water to growing urban societies. Later, the interest expanded to drilling for oil. In the biomedical arena porous media concepts are relevant to devices such as chromatography columns, artificial lungs/ membrane oxygenators and gas masks. They also underlie physiological flow through various tissues and the interstitium. These include bone, cartilage, tendon, ligaments, muscle (skeletal, cardiac, smooth), skin, lung, eye, blood vessel walls, placenta, intervertebral discs and tumors amongst others. Two key features for porous media are: its porosity, $\phi = V_V/(V_V + V_S)$, where V_S is the solid (matrix) volume and V_V is the pore (void) volume which is fluid-filled (i.e. the void volume fraction); and its permeability, K, which depends on the pore structure. The ratio $K' = K/\mu$, where μ is the fluid viscosity, is called the hydraulic conductivity and characterizes the ease at which fluid flows through the media for a given driving pressure gradient. K certainly depends on ϕ and they both vary with tissue type and disease.

Topics Covered

To begin, we examine Darcy's Law, which since 1856 has been the founding principle driving this field. We apply it to flow of plasma through an arterial wall which, though very slow, can create shear stresses on the smooth muscle cells of the wall and trigger release of bioactive molecules. Then we examine flow from a cancerous tumor where the interstitial fluid pressure is so large it creates a distributed field of fluid source. One of the consequences is that delivery of anti-cancer medications through the vasculature is impeded from this flow. In many flow situations a more accurate viscous stress contribution is required, and this is provided in the Brinkman equation. An important application to bone flow is explored. Often a porous media is in contact with a freely flowing fluid, and the boundary between them is a complicated situation to model. We examine some approaches to this tricky problem, including applications to the glycocalyx layer coating vascular endothelium, effectiveness of mucociliary pumping and airflow over a tree canopy. Once the Darcy law became established, there was a strong drive to predict K based on ϕ, in the Kozeny–Carman equation. It includes features of the tortuosity of the pathways through the pores and measures of the solid surface area per unit media volume and can be quite accurate. Further extensions involve mass transport, like the uptake of toxic gases in gas masks and the transport of oxygen and carbon dioxide in a membrane oxygenator. Finally, the more complex system of both a deformable solid matrix with a flowing fluid is addressed. Such systems are called poroelastic media and include many important examples such as skin, intervertebral discs and heart muscle.

15.1 Darcy's Law

One of the important characteristics of a porous media is its porosity, ϕ, or void volume fraction. As a simple example, consider a bucket of wet sand. The void volume, V_V, consists of the spaces between sand particles, which are filled with water, while the solid volume, V_S, consists of the particles themselves. The total volume is $V_V + V_S$. The porosity, or void volume fraction, is the ratio of the void volume to the total volume

$$\phi = \frac{V_V}{V_V + V_S} \tag{15.1.1}$$

Likewise, the solid volume fraction is $V_S/(V_V + V_S) = 1 - \phi$. The range of ϕ is $0 \leq \phi \leq 1$, where the two limits are the absence of voids, $\phi = 0$, and the absence of solid matrix, $\phi = 1$.

Figure 15.1 shows X-ray computed tomography of a packed bed of microspheres in a chromatography column. The microspheres were either ceramic or cellulose with average diameter 50 μm, 80 μm and average porosity $\phi = 0.34, 0.36$, respectively. Figure 15.1(a) is a magnified 2D slice of the ceramic packed column and Figure 15.1(b) is a 3D rendering of the void network in a cellulose column with dark regions indicating larger voids (Johnson et al., 2017). The porosity of human tissues is widely varying. Ligaments and tendons have the range $0.32 \leq \phi \leq 0.42$ (Frank et al., 1987; Chen et al., 1998). The total porosity of bone is much smaller at $\phi = 0.0881 \pm 0.02$ (Martin and Ishida, 1989) and

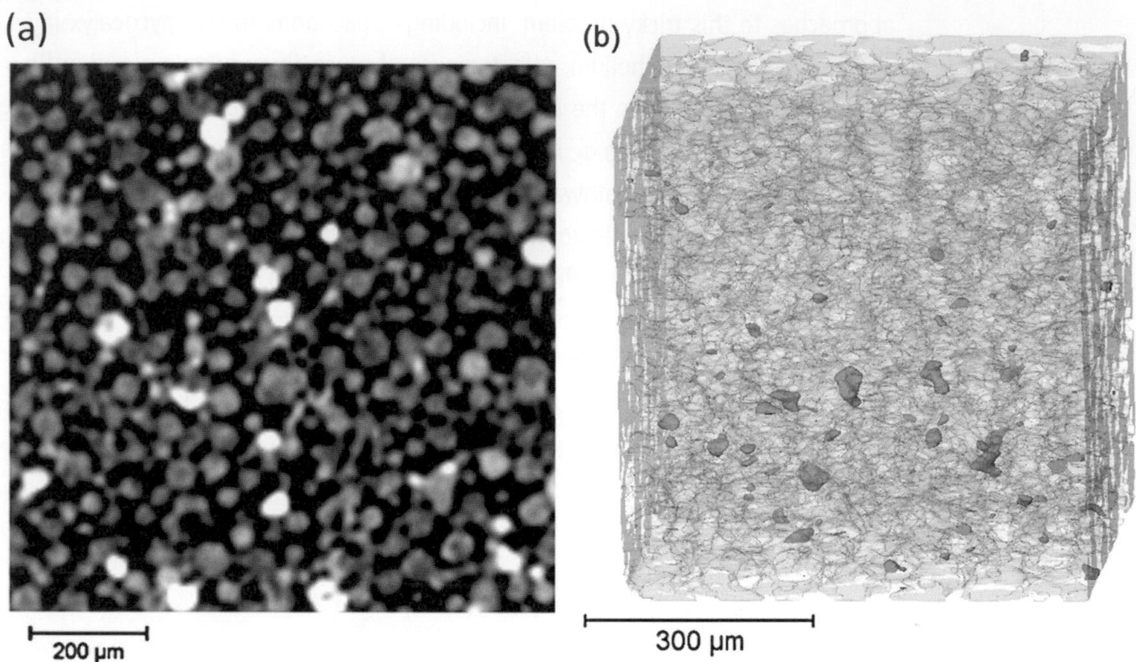

Figure 15.1 X-ray computed tomography of packed columns of microspheres: (a) ceramic in a 2D slice and (b) cellulose in a 3D rendering of the void network.

articular cartilage is much higher, $\phi = 0.85$ (Lipshitz et al., 1976).

Darcy's law was established by Henry Philibert Gaspard Darcy (1803–1858). One of his goals was to filter water by having it flow through a sand-packed pipe section, so needed to understand how much pressure drop that would take and what flow rate to expect, since he was designing the municipal water supply for Dijon, France. He posed the law as a simple proportional relationship between the instantaneous discharge rate, Q, in volume per time through the pipe, and the driving pressure $p_1 - p_2 > 0$ over a distance, $z_2 - z_1 = L$, see Figure 15.2. The cross-sectional area of the pipe is A and the fluid viscosity is μ, and his equation has the form

$$Q = -\frac{KA}{\mu}\frac{(p_2 - p_1)}{z_2 - z_1} \tag{15.1.2}$$

where K, in units of cm^2, is the permeability of the medium and depends on its structure. Dividing both sides of the equation by A, and using more general vector notation leads to

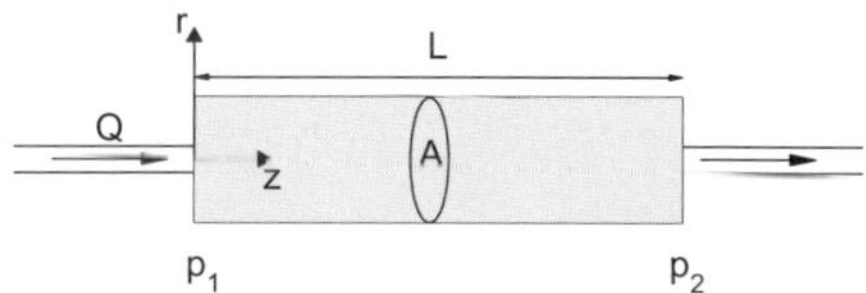

Figure 15.2 Darcy flow through a sand-filled pipe section.

$$\underline{u} = -\frac{K}{\mu}\underline{\nabla}p \tag{15.1.3}$$

where $\underline{u}$ is the velocity vector field and $\underline{\nabla}p$ is the corresponding pressure gradient vector. Equation (15.1.3) is the differential form of Darcy's law for an isotropic media. The ratio of the permeability, K, to the viscosity, μ, is called the hydraulic conductivity, $K' = K/\mu$. In biological flows the units of K' are expressed as a velocity per pressure gradient given in terms of $cm^2/mmHg \cdot s$ or $cm^2/cmH_2O \cdot s$. This is appealing since historically physiologists used water or mercury manometers to measure pressure. The hydraulic conductivity of tissues has a wide range of values. For comparison, in units of $cm^2/mmHg \cdot s$, rat abdominal muscle is $K' = 15 - 78 \times 10^{-8}$ (Zakaria et al., 1997), while cartilage is less permeable, $K' = 0.3 - 0.7 \times 10^{-8}$ (Levick, 1987), and the mesentery is more permeable, $K' = 41 - 253 \times 10^{-8}$ (Levick, 1987).

Note that, although the fluid only occupies part of the overall space, the void, Eq. (15.1.3) assumes $\underline{u}$ and p are spatially distributed everywhere in the media. Is that physically reasonable? Well, it turns out Darcy's law is derivable from first principles using volume averaging techniques for the Navier–Stokes equations (Whitaker, 1986). The averaging volume is small compared to the macroscopic length scale, say a pipe diameter, but large compared to the microscopic scale, say the distance between solid particles. We see immediately that Eq. (15.1.3) is missing inertial terms, which fits well with the low Reynolds number assumption, $Re << 1$. However, it is also missing the viscous term of the Navier–Stokes equation, $\mu\nabla^2\underline{u}$, so that it is of lower spatial order. Consequently, $\underline{u}$ cannot satisfy the no-slip condition on boundaries, for example. For more complex systems the permeability may depend on direction so the media is anisotropic. Then the permeability is represented by a tensor, $\underline{\underline{K}}$, with components K_{ij},

$$\underline{u} = -\frac{1}{\mu}\underline{\underline{K}} \cdot \underline{\nabla} p \tag{15.1.4}$$

Since the fluid is incompressible, conservation of mass is taken as

$$\underline{\nabla} \cdot \underline{u} = 0 \tag{15.1.5}$$

For constant K and μ, inserting Eq. (15.1.3) into Eq. (15.1.5) gives us

$$\underline{\nabla} \cdot \underline{u} = -\frac{K}{\mu}\underline{\nabla} \cdot (\underline{\nabla} p) = 0 \quad \rightarrow \quad \nabla^2 p = 0 \tag{15.1.6}$$

The sequence of solving a problem using the Darcy model is first to find a solution to the Laplace equation for the pressure field in Eq. (15.1.6) and then substitute it into Eq. (15.1.3) to find the velocities.

Box 15.1 | Henry Philibert Gaspard Darcy

The study of fluid flow in porous media began in the field of hydrology in the nineteenth century, to understand how water percolates through sand or soil. In 1856, Henry Philibert Gaspard Darcy (1803–1858), Figure 15.3(a), published his famous report on the public water supply of the French city of Dijon (of mustard fame), "Les fontaines publiques de la ville de Dijon" ("The public fountains of the city of Dijon") (Darcy, 1856). As part of his work to engineer the distribution of fresh water to the city's fountains, where the public could retrieve water, he determined flow resistance in pipes, which we explored in Section 13.3 as the Darcy friction factor. He also used sand-filled pipe sections to purify the water, with contaminants depositing on the sand particles. That portion of his report was called "Détermination des lois d'écoulement de l'eau á travers le sable" ("Determination of the laws of water flow through sand"). The experiments and theory presented for flow in his sand-filled pipes forms the basis of "Darcy's law," which laid the foundation for the field of hydrogeology (Simmons, 2008). Figure 15.3(b) is his original sketch of the flow column showing gravity-driven water entering the top and exiting the bottom after being filtered by the sand.

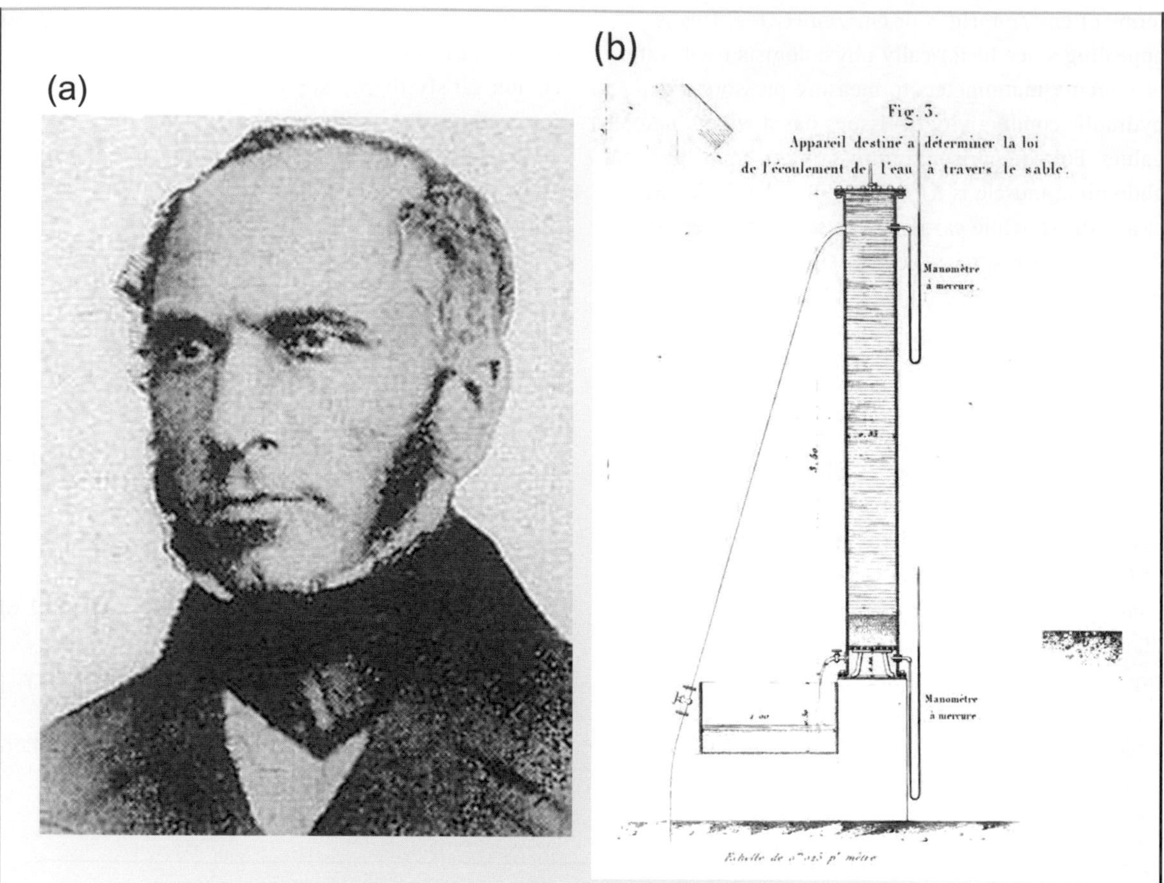

Figure 15.3 (a) Henry Philibert Gaspard Darcy (1803–1858), French hydraulic engineer. (b) Darcy's flow experiment through a sand-filled pipe section.

Box 15.2 The Interstitium

Porous media concepts are important in flows through biological tissues. As seen in Figure 15.4, the interstitium or extracellular matrix (ECM) occupies the region between blood vessels and the epithelial cells which do the work of that organ. It provides support and organization. The ECM consists of a network of fibrous materials: collagen, elastin, laminins,

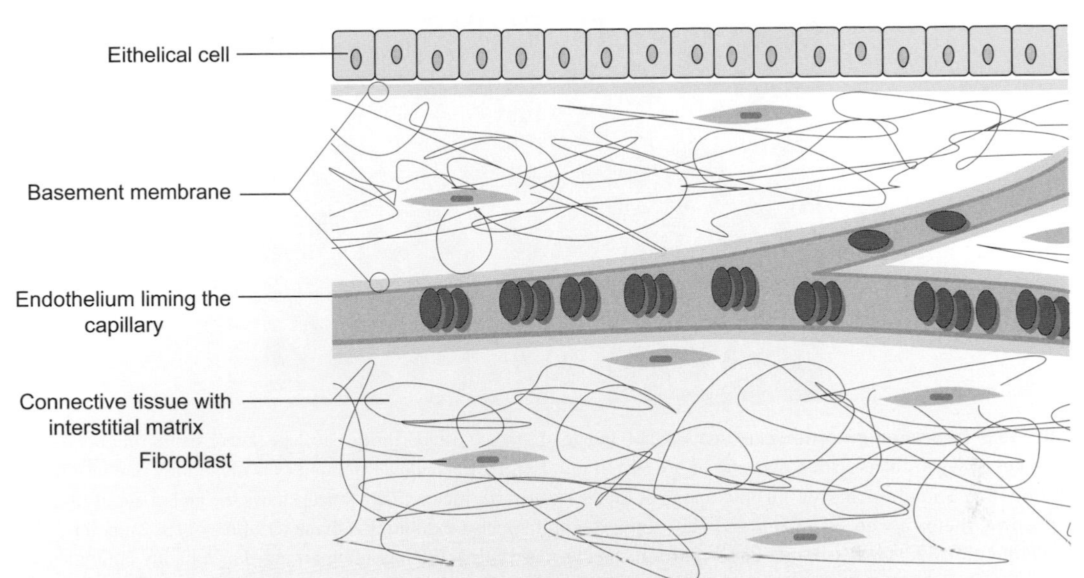

Figure 15.4 Schematic of the interstitium showing major features and components typical of extracellular matrix (ECM). Collagen fibers are intertwined with proteoglycans and interstitial cells (e.g. fibroblasts). A capillary with red blood cells is shown.

proteoglycans and fibrin. Interstitial fluid fills the pores between these matrix elements. An adult has about 10 liters of interstitial fluid, at a density of ~ 1 kg/liter that amounts to $\sim 15\%$ of body weight for a 70 kg person, and it is essentially always slowly flowing. Recently, the interstitium has been viewed as an entire organ system itself (Benias et al., 2018). Examples of soft tissues which have this kind of flow percolating through them are muscle, tendon, ligaments, cartilage, skin, blood vessel walls, lung parenchyma, cancerous tumors, intervertebral discs and the vitreous humor of the eye.

Figure 15.5 is a scanning electron microscope view of collagen fibers from a human knee (Gottardi et al., 2016). How the ECM influences cell behavior is a central theme of investigators who grow tissues outside the body on scaffolds designed to mimic the *in vivo* setting or engineered to study or promote particular cell responses. One aim of these endeavors is to regenerate damaged tissues under controlled conditions and then reinsert them into the body (Drury and Mooney, 2003; Lutolf and Hubbell, 2005). Collagen is a common biomaterial used in scaffolds.

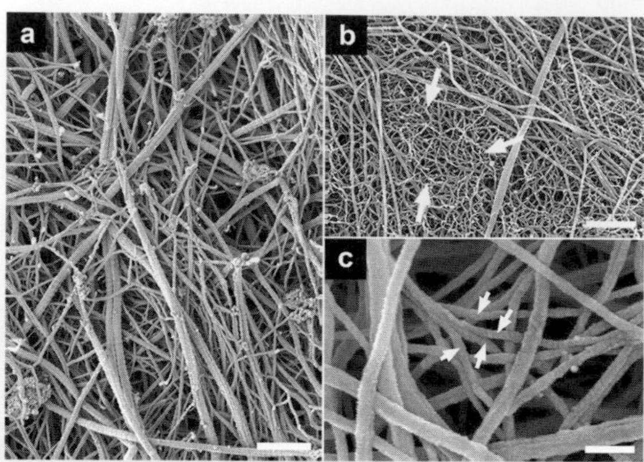

Figure 15.5 (a) Scanning electron microscope (SEM) image of grade 3 osteoarthritic cartilage (knee), exhibiting breakdown of thicker collagen fibers with a diameter of 40–60 nm into thinner fibers down to bundles made of only one prototypic fibril of 18 ± 5 nm in diameter. (b) SEM image of grade 3 osteoarthritic cartilage (knee) shows the end-stage of fiber breakdown, that is a wool-like structure (white arrows) with filaments exhibiting a diameter of $d = 13 \pm 2$ nm. (c) Degrading articular cartilage larger fibers split into smaller sized fibrils that are often arranged as a highly entangled fibrillar meshwork (white arrows). Scale bars, 500 nm (a and c); 100 nm (b).

15.2 Arterial Wall

Suppose we have a blood vessel with a porous wall as shown in Figure 15.6. The tissue annulus has an inner radius, r_i, outer radius, r_o, and corresponding pressure p_i and p_o where $(p_i - p_o) > 0$. To model the outward flow through the wall in cylindrical coordinates, assume axisymmetry, so that fluid velocity vector, $\underline{u}$, and pressure, p, do not depend on θ. Also assume that the characteristic measure of the radial pressure gradient, $(p_i - p_o)/(r_o - r_i)$, is much larger than the axial pressure gradient, $\partial p/\partial z$, which we ignore. Under these conditions, the Laplace equation, Eq. (15.1.6), simplifies to

$$\nabla^2 p = \frac{1}{r}\frac{\partial}{\partial r}\left(r\frac{\partial p}{\partial r}\right) + \cancel{\frac{1}{r^2}\frac{\partial^2 p}{\partial \theta^2}} + \cancel{\frac{\partial^2 p}{\partial z^2}} = 0$$
$$\rightarrow \frac{1}{r}\frac{d}{dr}\left(r\frac{dp}{dr}\right) = 0 \qquad (15.2.1)$$

Equation (15.2.1) implies that $r\,dp/dr = c_1$, which can be integrated to

$$p = c_1 \ln r + c_2 \qquad (15.2.2)$$

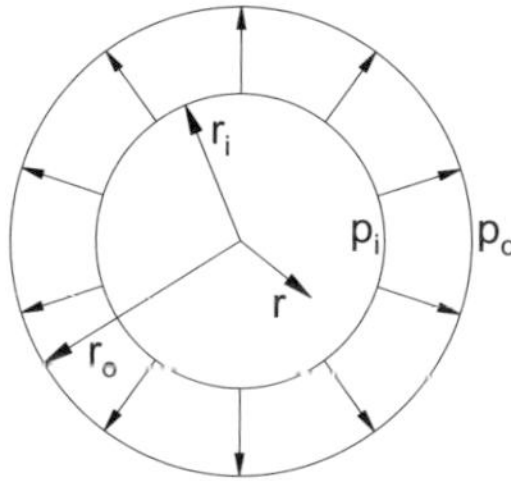

Figure 15.6 Darcy flow across a vascular wall.

Imposing pressure boundary conditions $p(r = r_i) = p_i$ and $p(r = r_o) = p_o$ will permit us to solve for the integration constants, c_1 and c_2, giving

$$p - p_o = -\frac{(p_i - p_o)}{\ln(r_o/r_i)}\ln(r/r_o) \qquad (15.2.3)$$

We can scale the solution in Eq. (15.2.3) by defining the following dimensionless variables

$$P = \frac{p - p_o}{p_i - p_o}, \quad R = \frac{r - r_i}{r_i} \qquad (15.2.4)$$

As r varies between the limits $r_i \leq r \leq r_o$, the dimensionless radial variable has the range

$0 \leq R \leq \delta$. The dimensionless parameter $\delta = (r_o - r_i)/r_i$ is the ratio of the wall thickness, $r_o - r_i$, to the inner radius. Also, as p varies as $p_i \leq p \leq p_o$, the dimensionless pressure, P, varies as $0 \leq P \leq 1$.

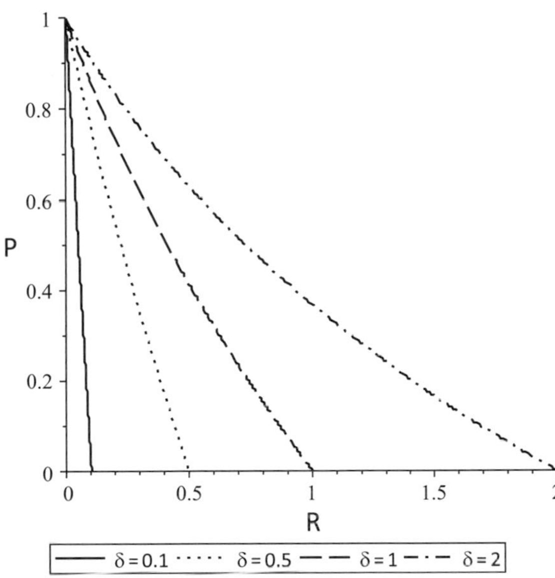

Figure 15.7 Distribution of pressure for Darcy flow through an arterial wall for $\delta = 0.1, 0.5, 1.0, 2.0$.

The pressure in dimensionless form is

$$P = 1 - \frac{\ln(1+R)}{\ln(1+\delta)} \tag{15.2.5}$$

Figure 15.7 shows the pressure from Eq. (15.2.5) for four values of $\delta = 0.1, 0.5, 1.0, 2.0$ which are in the range from Table 15.1 for human blood vessels. For $R << 1$, Eq. (15.2.5) has the Taylor series expansion

$$P = 1 - \frac{\ln(1+R)}{\ln(1+\delta)} = 1 - \frac{1}{\ln(1+\delta)}\left(R - O(R^2)\right) \tag{15.2.6}$$

which simplifies to a linear profile seen for the entire $\delta = 0.1$ solution.

The radial fluid velocity, $u_r(r)$, is given by the r-component of Eq. (15.1.3)

$$u_r = -\frac{K}{\mu}\frac{dp}{dr} = \frac{K}{\mu}\frac{(p_i - p_o)}{\ln(r_o/r_i)}\frac{1}{r} \tag{15.2.7}$$

where we have calculated dp/dr from Eq. (15.2.3). The characteristic pressure gradient is $(p_i - p_o)/(r_i - r_o)$ and we can multiply it by K/μ to obtain a characteristic velocity scale, $U_s = K(p_i - p_o)/\mu(r_o - r_i)$. However, since we want to isolate the effect of increasing wall thickness, $r_o - r_i$, we want to avoid having $r_o - r_i$ in the velocity scale to learn the effects of increasing δ.

Table 15.1 **Human vascular internal diameters and wall thicknesses with calculation of δ.**

Blood vessel	Internal diameter		Wall thickness		δ
	$2r_i$		$r_o - r_i$		$(r_o - r_i)/r_i$
Aorta	2.5	cm	0.2	cm	0.16
Medium artery	0.4	cm	0.1	cm	0.50
Arteriole	30	microns	20	microns	1.33
Precapillary	35	microns	30	microns	1.71
Capillary	8	microns	1	microns	0.25
Venule	20	microns	2	microns	0.20
Vein	0.5	cm	0.05	cm	0.20
Vena cava	3	cm	0.15	cm	0.10

Source: Data from Burton (1954).

So let us scale the velocity, u_r, on $\delta U_s = K(p_i - p_o)/\mu r_i$, and define the dimensionless radial velocity, U_R, as

$$U_R = \frac{u_r}{\delta U_s} = \frac{1}{(1+R)\ln(1+\delta)} \tag{15.2.8}$$

Solutions to the velocity field, $U_R(R)$, are shown in Figure 15.8 for the same selections of the parameter δ used in Figure 15.7. For a given value of δ the velocity, U_R, is larger closer to the inner wall where dP/dR has a larger magnitude, and then decreases with increasing R as the pressure gradient is weaker. It is the conservation of mass causing the slower velocities and weaker pressure gradients with increasing R, since larger R means smaller U_R to keep the volume flow rate the same. Increasing δ for thicker wall reduces the velocities consistent with the overall pressure gradient being reduced in magnitude.

Since the source of the fluid for this transmural flow is blood serum from the arterial lumen, the velocity at $R = 0$ (the inner wall boundary) causes a radially outward flow in the lumen. This creates a convective mechanism for molecules, like lipids, to reach the inner wall boundary which is lined with endothelium (Stangeby amd Ethier, 2002). Lipid deposition there is considered to be a building block for the development of atherosclerotic plaques.

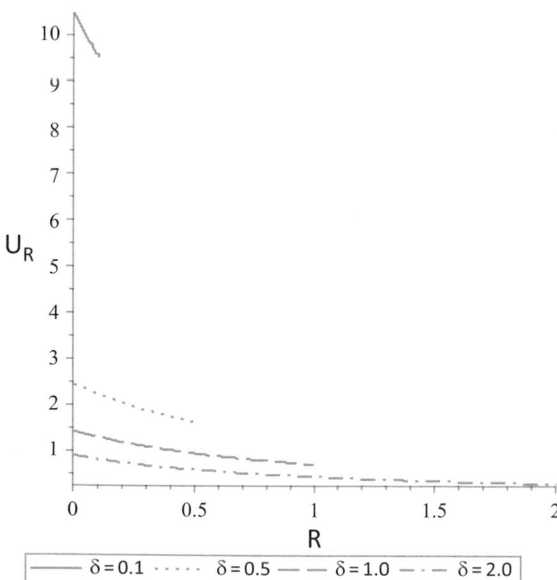

Figure 15.8 Dimensionless radial velocity, U_R, vs dimensionless radius, R, for Darcy flow through a vascular wall at four values of $\delta = 0.1, 0.5, 1.0, 2.0$.

We encountered permeability in Section 11.6 using the Starling equation, Eqs. (11.6.1) and (11.6.2), for flow across a porous membrane such as the capillary wall. There it was called L_P, the hydraulic conductivity, which is common physiological terminology. It is given in units of $cm/cmH_2O \cdot s$ as the velocity per pressure difference across the vascular wall, i.e. $u = L_P(p_i - p_o)$, in the absence of oncotic pressures. However, the phrase "hydraulic conductivity," now in the context of Darcy flow literature, is reserved for K'. Darcy's law is the velocity per pressure gradient, so the wall thickness, ℓ, needs to be introduced as

$$u = L_P \ell \frac{(p_1 - p_2)}{\ell} \tag{15.2.9}$$

and we see that $K' = L_P \ell$. In this context, Levick (1987) explains a terminology triad: K (permeability) is the specific hydraulic conductivity; K' is the hydraulic conductivity; and $K'' = L_p$ is the hydraulic conductance.

The value of L_P for the total thoracic aorta of rabbits has been measured to be $L_{P_T} = 4.07 \pm 1.3 \times 10^{-8}\ cm/(s \cdot cmH_2O)$ for a wall thickness $\ell = 0.02\ cm$ yielding $K' = L_{P_T} \cdot \ell = 8.14 \pm 2.6 \times 10^{-10} cm^2/(s \cdot cmH_2O)$ (Vargas et al., 1979). Multiplying by the serum viscosity $\mu = 0.01235$ poise put in consistent units where $1\ \text{poise} = 1\ dyn \cdot s/cm^2 = (1/980)\, cmH_2O \cdot s$ yields $K = \mu K' \tilde{\ } 1.026 \times 10^{-14} cm^2$. By comparison, K measured in human aortas was found to be $K = 1.7 \pm 0.2 \times 10^{-14} cm^2$ in a cylindrical geometry and $K = 2.9 \pm 0.3 \times 10^{-14} cm^2$ in a flat geometry (Whale et al., 1996). We can estimate the characteristic velocity for the fluid filtering through the rabbit aorta from the scale $U_s = K'(p_i - p_o)/\ell = L_{P_T}(p_i - p_o)$. For $p_i - p_o = 100\ mmHg = 136\ cmH_2O$, which is in the range of typical rabbit blood pressures, the characteristic velocity is $U_s = 5.53 \times 10^{-6} cm/s = 5.53 \times 10^{-2}\ \mu m/s$.

Multi-Layer Blood Vessel Wall

Figure 15.9 represents a cross-section of a typical artery showing three layers of tissue called the tunica externa, tunica media and tunica intima (*tunica* is Latin for "coat"). The tunica externa, also called the tunica adventitia, consists mostly of collagen, while the tunica media contains smooth muscle and elastic tissue. The tunica intima includes the endothelial cell layer, which is the inner lining of the vessel in contact with the blood, and sits on a supportive elastic layer. Porosity can vary in a tissue, for example across the arterial wall layers. Prosi et al. (2005) has computed the coupled flow through the intima using $\phi = 0.474$ and then through the media $\phi = 0.15$.

To represent more accurately the anatomy of an arterial wall, let us model it as two concentric cylindrical regions. The inner region will be the intima while the outer will be the media and externa combined together. Using Eq. (15.2.2) in each layer we obtain

$$\begin{aligned} p_1 &= c_3 \ln r + c_4, \quad r_i \le r \le r_m \\ p_2 &= c_5 \ln r + c_6, \quad r_m \le r \le r_o \end{aligned} \qquad (15.2.10)$$

where r_m divides the wall thickness into the two layers $r_i \le r \le r_m$ and $r_m \le r \le r_o$. As before, we impose pressures at the inner and outer walls for two boundary conditions. However, we have four integration constants, c_3, c_4, c_5, c_6, so two more boundary conditions are needed. Those additional

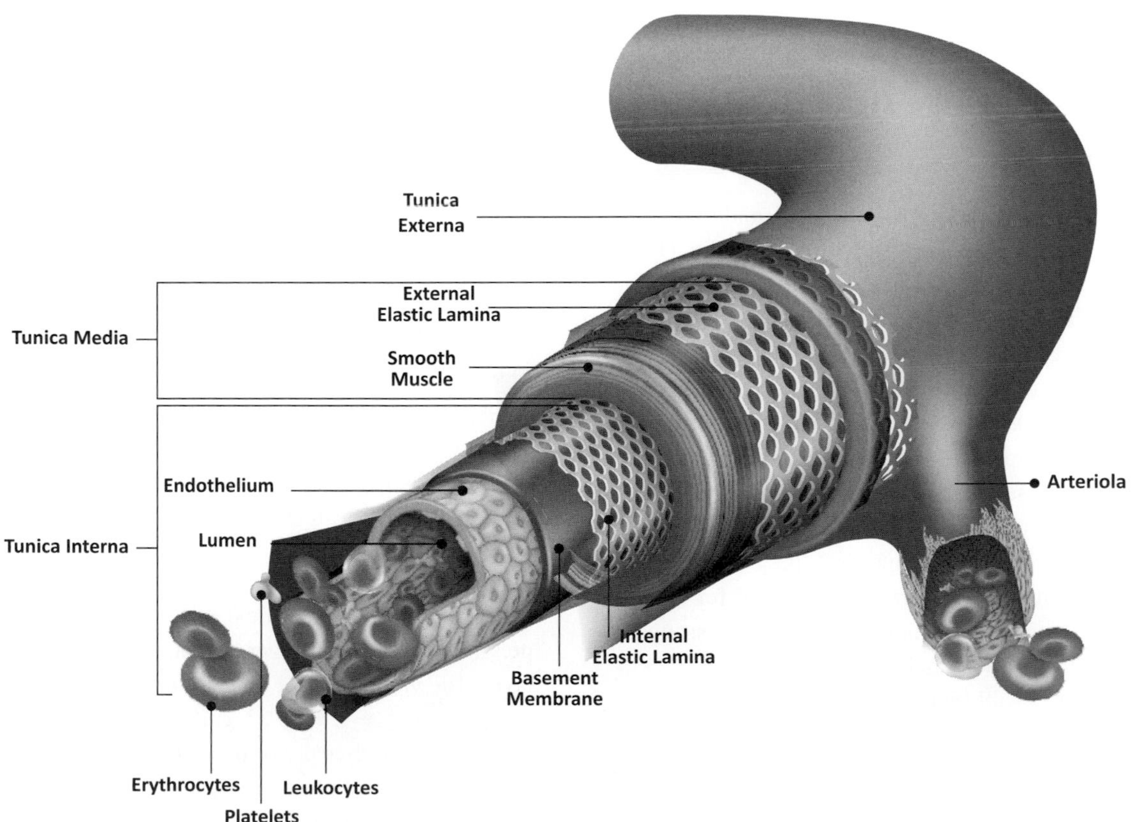

Figure 15.9 Cross-section of typical artery showing three distinct wall layers: tunica externa, tunica media and tunica intima.

conditions are matching of stress and volume flow rate at $r = r_m$. The four conditions are

$$p_1(r = r_i) = p_i, p_2(r = r_o) = p_o, p_1(r = r_m) = p_2(r = r_m), K_1 \frac{dp_1}{dr}_{r=r_m} = K_2 \frac{dp_2}{dr}_{r=r_m} \quad (15.2.11)$$

Solving for the integration constants we cast the solution in dimensionless form as

$$P_1 = \frac{-\ln(1+R) + \beta \ln(1+\delta) + (1-\beta)\ln(1+R_m)}{\beta \ln(1+\delta) + (1-\beta)\ln(1+R_m)},$$

$$0 \le R \le R_m$$

$$P_2 = \frac{\beta(-\ln(1+R) + \ln(1+\delta))}{\beta \ln(1+\delta) + (1-\beta)\ln(1+R_m)}, \quad R_m \le R \le \delta \quad (15.2.12)$$

where the dimensionless variables and parameters are

$$R = \frac{r - r_i}{r_i}, P_1 = \frac{p_1 - p_o}{p_i - p_o}, P_2 = \frac{p_2 - p_o}{p_i - p_o}, \delta = \frac{r_o - r_i}{r_i},$$

$$\beta = \frac{K_1}{K_2}, R_m = \frac{r_m - r_i}{r_i} \quad (15.2.13)$$

First note that setting $\beta = 1$ is the condition that $K_1 = K_2$ and both P_1 and P_2 simplify to Eq. (15.2.5), since the two regions have the same permeability.

The P_1 and P_2 solutions of Eq. (15.2.12) are plotted in Figure 15.10 for $\delta = 0.2$, $R_m = 0.02$, so the intima is 10% of the total thickness. There are four values of $\beta = 1, 0.5, 0.11, 0.05$ selected. Note that for $\beta = 1$ the two curves merge into one continuous solution. Then as β decreases the less permeable intima layer creates a greater and greater pressure drop and the ratio of the region 1 to region 2 slopes is $1/\beta$.

The values of L_p have been explored for these two layers. If the endothelial layer (intima) is stripped away, L_p for the remaining outer layers of the arterial wall increases to $L_{P_{OL}} = 7.73 \pm 2.8 \times 10^{-8}$ cm/(s · cmH$_2$O)(Vargas et al., 1979). Because the resistance to flow is the inverse of L_p, the series of resistances encountered by this flow consist of the endothelium, $1/L_{P_E}$, and outer layer (media + externa) without endothelium, $1/L_{P_{OL}}$. The total resistance, $1/L_{P_T}$, is given by

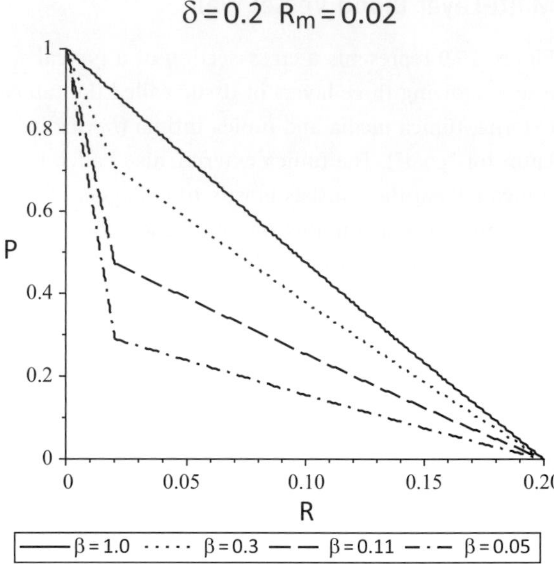

Figure 15.10 P vs R for $\delta = 0.2$, $R_m = 0.02$ and four values of $\beta = 1, 0.5, 0.11, 0.05$.

$$\frac{1}{L_{P_T}} = \frac{1}{L_{P_{OL}}} + \frac{1}{L_{P_E}} \quad \rightarrow \quad L_{P_E} = \frac{L_{P_{OL}} L_{P_T}}{L_{P_{OL}} - L_{P_T}} \quad (15.2.14)$$

Inserting the values for $L_{P_{OL}}$ and L_{P_T} into Eq. (15.2.14), the aortic endothelial L_p calculates to be $L_{P_E} = 8.6 \times 10^{-8}$ cm/(s · cmH$_2$O), which is nearly the same value as the much thicker outer layer, $L_{P_{OL}}$. So Eq. (15.2.14) tells us the approximately half of the total resistance to transmural flow is due to the endothelial layer. To obtain a representative value of β use the equivalence $\beta = K_1/K_2 = K'_1/K'_2 = \ell_E L_{P_E}/\ell_{OL} L_{P_{OL}}$, where ℓ_E is the thickness of the intima (endothelium) layer and ℓ_{OL} is the outer layer thickness for the combined media and externa. Since $L_{P_E}/L_{P_{OL}} \sim 1$ then $\beta \sim \ell_E/\ell_{OL} = R_m/(\delta - R_m)$. In Figure 15.10 where $\delta = 0.2, R_m = 0.02$, the value of $\beta = 1/9$ corresponds to the $\beta = 0.11$ curve where the pressure drops by half over the much thinner intima layer. The value of K for human aorta has been measured as $K = 2 \times 10^{-14}$ cm^2 (Whale et al., 1996). Using a serum viscosity of $\mu = 0.01235$ poise and recognizing that 1 poise = 1 dyn · s/cm^2 and 1 cm H$_2$O = 980 dyn/cm^2, the corresponding value of K′ is $K' = K/\mu = 1.587 \times 10^{-9}$(cm^2/s)/cm H$_2$O, which is about twice the rabbit value of

$K' = 8.14 \pm 2.6 \times 10^{-10}\ cm^2/(s \cdot cmH_2O)$ (Vargas et al., 1979).

While these flows are very slow, it turns out that they create $\sim 1\,dyn/cm^2$ shear stress on the smooth muscle cells in the tunica media which is large enough to activate cellular expression of bioactive molecules (Wang and Tarbell, 1995; 2000). It will take the Brinkman model, discussed in Section 15.4, for shear velocity gradients and shear stresses to appear.

15.3 Cancerous Tumors

Cancerous tumors are highly vascularized since cell growth and division are both abnormally increased. We learned in Section 11.6 that capillaries leak serum through their walls into the interstitium, and in tumors that results in a large interstitial pressure. Typical values for breast and colorectal cancer vary from 4 to 50 mmHg with an average of 20 mmHg (Less et al., 1992). Normal interstitial fluid pressure is near zero, for example in Section 11.6 we set it equal to 1 mmHg just outside of a capillary modeling filtration flow. Consequently, there is a slow radial flow of interstitial fluid out of the tumor. Various cancer treatments, however, rely on the diffusion of therapeutic molecules radially inward, opposite to the convection. For successful delivery, the inward diffusion must be faster than the outward convection (Jain, 1987). Tumors have a wide range of K' depending on the tissue source: human colon adenocarcinoma (LS174T), $K' = 45 \times 10^{-8}\ cm^2/mmHg \cdot s$; human glioblastoma (U87), $K' = 65 \times 10^{-8}\ cm^2/mmHg \cdot s$; human soft tissue sarcoma (HSTS 26T), $K' = 9.2 \times 10^{-8}\ cm^2/mmHg \cdot s$; and mouse mammary carcinoma (MCaIV), $K' = 248 \times 10^{-8}\ cm^2/mmHg \cdot s$ (Netti et al., 2000). These are significantly higher values than arterial K'.

The tumor interstitial fluid dynamics can be modeled as Darcy flow radially outward in spherical coordinates. Following Baxter and Jain (1989), we assume the pressure depends only on the radial variable, $p(r)$, so it is the same in all directions for any value of r. Consider that the tumor tissue contains leaky capillaries, discussed in Section 11.6, which can provide a source of fluid, while lymphatics, which can drain it, provide a sink. Because capillary and lymphatic diameters are very small compared to the tumor diameter, they can be treated as continuously distributed. Conservation of mass with a continuous source, σ_ρ, was presented in Section 4.1, Eq. (4.1.5); where $\sigma_\rho > 0$ is a source and $\sigma_\rho < 0$ is a sink. The steady form for constant density, ρ, in spherical coordinates is

$$\underline{\nabla} \cdot \underline{u} = \frac{1}{r^2}\frac{d}{dr}\left(r^2 u_r\right) = \frac{\sigma_\rho}{\rho} = s \qquad (15.3.1)$$

where the net source, s, may vary spatially and has units s^{-1}. For a set point pressure, p_s, reflecting the capillary/lymphatic balance, let $s = \gamma^2(p_s - p)$. For the interstitial pressure, p, below p_s, the capillary source overcomes the lymphatic sink and s is positive, a net source.

Darcy's law of Eq. (15.1.3) simplifies for spherical coordinates when there is only r-dependence,

$$u_r = -\frac{K}{\mu}\frac{dp}{dr} \qquad (15.3.2)$$

Inserting Eq. (15.3.2) into Eq. (15.3.1) leads to

$$\frac{1}{r^2}\frac{d}{dr}\left(r^2\frac{dp}{dr}\right) - B^2 p = -B^2 p_s \qquad (15.3.3)$$

where $B^2 = \mu\gamma^2/K$.

The general solution to Eq. (15.3.3) is

$$p = c_1\frac{\cosh(Br)}{Br} + c_2\frac{\sinh(Br)}{Br} + p_s \qquad (15.3.4)$$

Note that the first term is unbounded at $r = 0$, so we need to choose $c_1 = 0$, which also guarantees $dp/dr = 0$ at $r = 0$ as a symmetry condition. To solve for c_2 define the pressure at the edge of the spherical tumor, $p(r = r_o) = p_o$. Making the resulting solution dimensionless we have

$$P = 1 - \frac{\sinh(\sigma R)}{R\sinh(\sigma)} \qquad (15.3.5)$$

where the dimensionless pressure is $P = (p - p_o)/(p_s - p_o)$, the dimensionless radial position is $R = r/r_o$, and the only dimensionless parameter is $\sigma = Br_o = (\mu/K)^{1/2}\gamma r_o$, which is a ratio

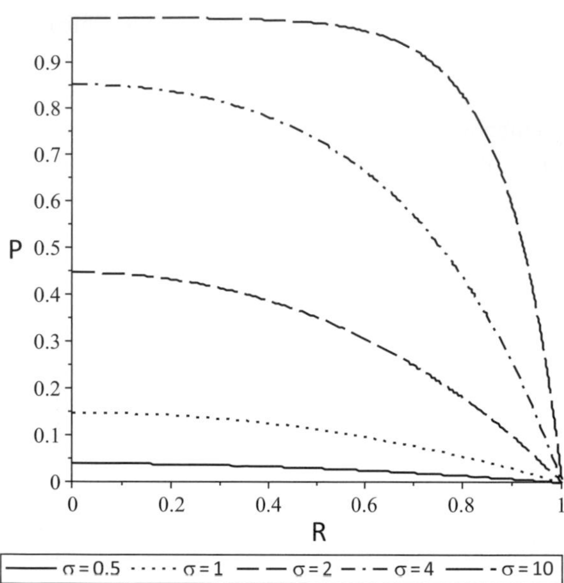

Figure 15.11 P vs R for five values of $\sigma = 0.5, 1, 2, 4, 10$.

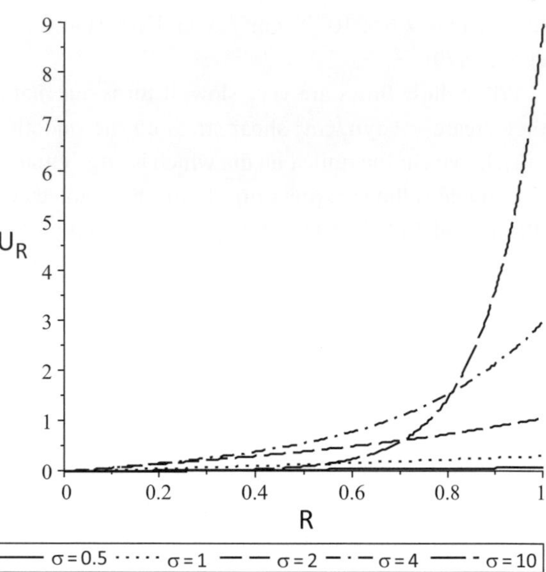

Figure 15.12 U_R vs R for five values of $\sigma = 0.5, 1, 2, 4, 10$.

of a characteristic source velocity to a characteristic Darcy's law velocity.

Figure 15.11 shows plots of the solution for P(R) from Eq. (15.3.5) for five values of $\sigma = 0.5, 1, 2, 4, 10$. As σ increases P increases for all values of R and is flatter near $R = 0$ with a more rapid dropoff toward $P = 0$ at $R = 1$. The corresponding dimensionless radial velocity field is

$$U_R = -\frac{dP}{dR} = \frac{\sigma R \cosh(\sigma R) - \sinh(\sigma R)}{R^2 \sinh(\sigma)} \tag{15.3.6}$$

where $U_R = u_r/U_s$ and $U_s = (K/\mu)(p_s - p_o)/r_o$ is the characteristic velocity scale.

Figure 15.12 shows the dimensionless radial velocity, $U_R(R)$, for five values of $\sigma = 0.5, 1, 2, 4, 10$ corresponding to the pressure distribution in Figure 15.11. The velocities all start at zero from the origin since we imposed $dP/dR = 0$ there. They increase with increasing R because the distributed net source is adding more and more flow to the system. Higher values of σ generally lead to higher velocities, though $\sigma = 10$ has those higher velocities shifted toward $R = 1$ because its pressure distribution is so flat initially. This is unlike the radial velocities for the arterial model, Figure 15.8, which decreased with increasing radius. The important difference is that the arterial model did not have a source term within the media. For an estimate of U_s choose $K' = K/\mu = 30 \times 10^{-8}\ \text{cm}^2/\text{mmHg}\cdot\text{s}$, tumor radius $r_o = 1$ cm, and pressure difference $p_s - p_o = 20$ mmHg we obtain $U_s = 0.06\ \mu\text{m/s} = 6 \times 10^{-6}$ cm/s. At the edge of the tumor, $R = 1$, and assuming $\sigma = 4$, Figure 15.12 shows $U_R \sim 3$. The dimensional velocity there is $u_r \sim 3 \times U_s = 0.18\ \mu\text{m/s}$, which falls in the reported range for tumors of 0.1–0.2 μm/s (Jain, 1987).

15.4 Brinkman Equation

In the case that ε is sufficiently large, say $\phi > 0.6$, a viscous term similar to that of the Navier–Stokes equations may be added,

$$\frac{\mu}{K}\underline{u} = -\underline{\nabla} p + \tilde{\mu}\nabla^2\underline{u} \tag{15.4.1}$$

Equation (15.4.1) is the Brinkman equation (Brinkman, 1947). Here $\tilde{\mu}$ is called the effective viscosity and can vary depending on the porous media structure, but is often taken to be equal to the fluid viscosity. With the higher order spatial derivatives in the Laplacian operator of Eq. (15.4.1), the velocity solutions can now satisfy no-slip, unlike Darcy's law.

Using the Brinkman equation, Eq. (15.4.1), with $\tilde{\mu} = \mu$, let us solve for fully developed flow in a channel of width 2b which is filled with a porous media

$$\frac{\mu}{K}u = -\frac{dp}{dx} + \mu\frac{d^2u}{dy^2} \tag{15.4.2}$$

For this flow in x-direction the y variable is in the range $-b \leq y \leq b$ and $y = 0$ is the midline. We can gain insight into the expected velocity profiles by examining the terms of Eq. (15.4.2). For K large the Darcy term is less important and the left hand side is negligible. Then the balance is between the pressure gradient and the Navier–Stokes viscous term, a system that tends to a parabolic profile. For K small the Darcy term dominates and is balanced by the pressure gradient, recovering Darcy's law which is a flat profile. With the presence of the Navier–Stokes viscous term, however, all profiles must match no-slip on the walls.

Let the pressure gradient be constant and scale velocity and the vertical variable as follows

$$U = \frac{u}{\left(-\frac{dp}{dx}\right)\frac{b^2}{\mu}}, \quad Y = \frac{y}{b} \tag{15.4.3}$$

The dimensionless momentum balance becomes

$$\frac{d^2U}{dY^2} - \alpha^2 U = -1 \tag{15.4.4}$$

where the dimensionless parameter is $\alpha^2 = 1/Da$ and $Da = K/b^2$ is the Darcy number. Da is different from the other dimensionless parameter carrying his name, the Darcy friction factor we studied in Section 13.3. The sum of the homogeneous and particular solutions is

$$U = \frac{1}{\alpha^2}\left[c_1 \cosh(\alpha Y) + c_2 \sinh(\alpha Y) + 1\right] \tag{15.4.5}$$

where c_1 and c_2 are unknown constants determined from the boundary conditions of no-slip at $Y = 1$ and

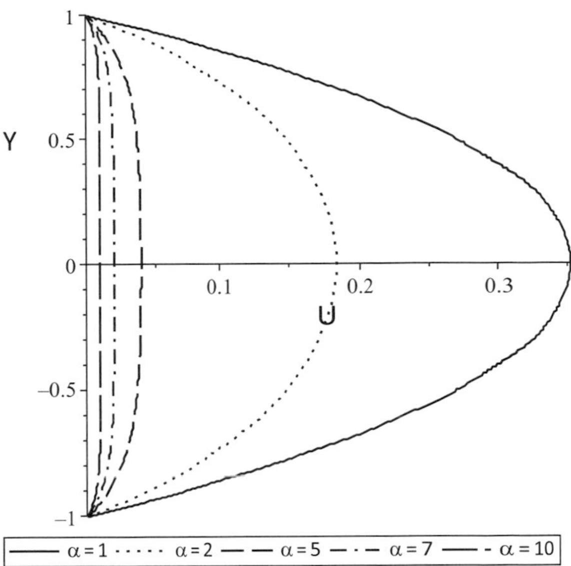

Figure 15.13 Velocity profiles in a channel using the Brinkman equation for five values of $\alpha = 1, 2, 5, 7, 10$.

symmetry condition at $Y = 0$, i.e. $U(Y = 1) = 0$, $dU/dY\,|_{Y=0} = 0$. The final form is

$$U = \frac{1}{\alpha^2}\left(1 - \frac{\cosh(\alpha Y)}{\cosh(\alpha)}\right) \tag{15.4.6}$$

Figure 15.13 shows velocity profiles for five values of $\alpha = 1, 2, 5, 7, 10$ and, as we predicted, the profile flattens as α increases, i.e. K decreases. The flat profiles develop curvature in a layer near the wall to meet the no-slip condition. The thickness of this layer is proportional to $1/\alpha$. The profiles also show that the volume flow rate decreases, as we would expect with smaller K, and that dU/dY at the walls decreases, so the wall shear is smaller.

Bone Flow

Another interesting porous media flow occurs in bone. Blood vessels, including capillaries, are contained in Haversian and Volkmann canals of bone, see Figure 15.14(a). Plasma filters across the capillaries and becomes interstitial fluid in the space between the canal wall and blood vessel. This "bone fluid" continues to flow into the lacuna-canalicular

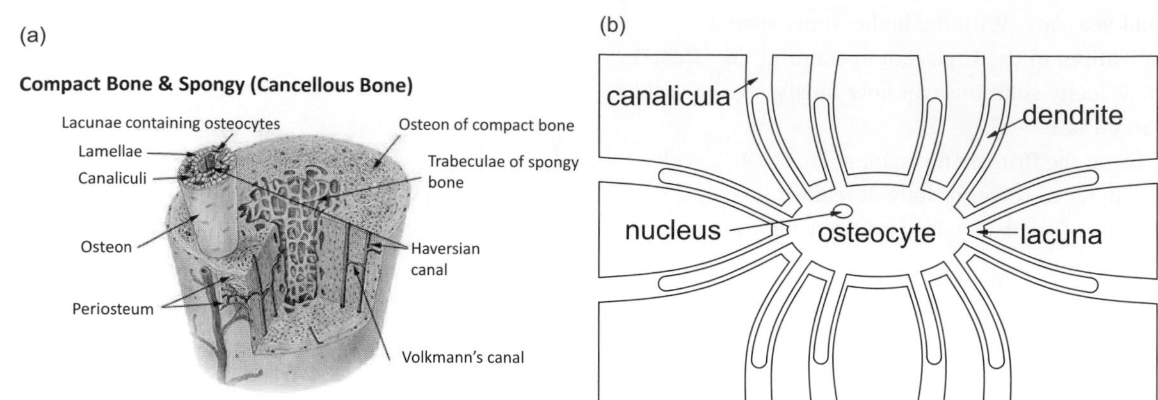

Figure 15.14 (a) Section of bone showing compact and spongy (cancellous) regions and important substructures. (b) Schematic of osteocyte, lacuna, canalicula and dendrite structures.

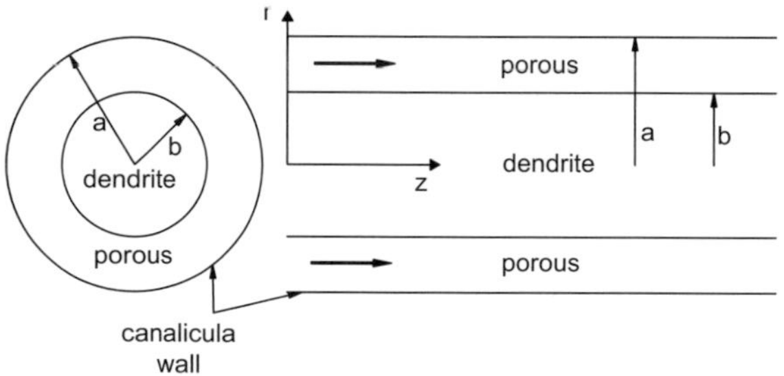

Figure 15.15 Axial annular flow between concentric cylinders containing a porous media as a model of canalicular flow with an osteocyte dendrite forming the inner cylinder and the canalicular wall forming the outer cylinder.

structure from pressure gradients and stresses created through load bearing, Figure 15.14(b).

To analyze this flow, Weinbaum et al. (1994) employed the Brinkman equation, Eq. (15.4.1), in cylindrical coordinates. Axial flow in the z-direction is driven by a pressure gradient, dp/dz, subject to no-slip boundary conditions at the outer cylinder, $r = a$, and the inner cylinder, $r = b$, see Figure 15.15. The porous media in the annulus, $b \leq r \leq a$, is comprised of proteoglycan matrix. An osteocyte cell dendrite occupies the region $0 \leq r \leq b$ and the canalicular wall forms the outer cylinder at $r = a$. The general solution to the Brinkman equation in cylindrical coordinates is

$$U_Z = \frac{1}{\alpha^2}\left[c_1 I_0(\alpha R) + c_2 K_0(\alpha R) + 1\right] \tag{15.4.7}$$

where the dimensionless radial coordinate is $R = r/a$. The dimensionless velocity is $U_Z = u_z/U_s$ where $U_s = (-dp/dz)a^2/\mu$ and, again, $\alpha^2 = 1/\text{Da}$. I_0 and K_0 are modified Bessel functions of zero order. No-slip is enforced at the outer cylinder wall, $R = 1$, and the inner cylinder wall, $R = b/a = \beta$, to solve for the two integration constants c_1 and c_2, which are

$$c_1 = \frac{K_0(\alpha\gamma) - K_0(\alpha)}{I_0(\alpha)K_0(\alpha\gamma) - I_0(\alpha\gamma)K_0(\alpha)},$$

$$c_2 = \frac{I_0(\alpha) - I_0(\alpha\gamma)}{I_0(\alpha)K_0(\alpha\gamma) - I_0(\alpha\gamma)K_0(\alpha)} \tag{15.4.8}$$

Figure 15.16 displays the velocity profiles in the annulus for $\alpha = 1, 5, 10, 20, 30$ where we see, as in Figure 15.13, that increasing α (decreasing K) reduces the flow rate and flattens the profile. Again, there is a thin layer, $\sim 1/\alpha$, near the wall where the velocity rapidly changes to meet the no-slip condition. The dimensional wall shear is found to be in the range 8–30 dyn/cm^2, which is large enough to influence cellular growth and remodeling (Weinbaum et al., 1994).

For situations where the fluid velocity is large enough the Darcy equations are usually modified to include a quadratic inertial term, the so-called Forchheimer equations,

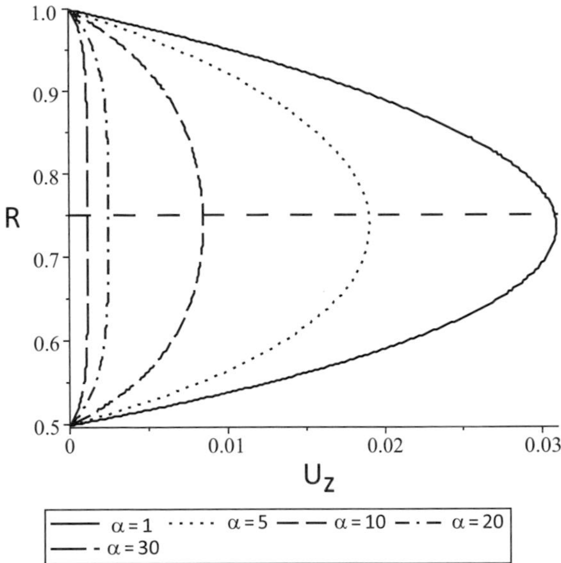

Figure 15.16 Axial velocity $U_Z(R)$ through a porous annulus at five values of $\alpha = 1, 5, 10, 20, 30$.

$$\frac{\mu}{K}\underline{u} + \beta\rho_f|\underline{u}|\underline{u} = -\underline{\nabla}p \tag{15.4.9}$$

where β is the Forchheimer coefficient. More complex flows require a combination of Brinkman and Forchheimer models,

$$\frac{\mu}{K}\underline{u} + \beta\rho_f|\underline{u}|\underline{u} = -\underline{\nabla}p + \lambda\nabla^2\underline{u} \tag{15.4.10}$$

Additional features of more realistic properties include pressure-dependent porosity and permeability. This is especially true in soft tissues where increasing pressures stretch the porous media, increasing both porosity and permeability (Guyton, 1965; Guyton et al., 1966). Flow-dependent permeability is another option explored for placental blood flow in Erian et al. (1977), see Figure 15.17. Oxygenated maternal arterial blood enters the intervillous space and leaves by maternal venous blood, which has become deoxygenated because of transfer to the fetal circulation in the chorionic villus. The maternal blood bathes the chorionic villus which contains the fetal artery (deoxygenated blood) and fetal veins (oxygenated blood) with their capillary network.

15.5 Porous Media Boundary

There are a number of examples in biofluid mechanics where a free-flowing fluid encounters porous media as a boundary. The squeeze film from Section 11.7 was viewed as a model of a load-bearing joint space like the knee. The two plates in the model, however, are actually coated with a collagen layer which is porous. Blood vessel endothelial cells are coated with a thin porous glycoprotein layer called the glycocalyx, which is in contact with the flowing blood. Measurements of the glycocalyx layer thickness are as high as 11 μm (Ebong et al., 2011), as shown in Figure 15.18(a). Flow driven by wall motion and gravity (see Section 7.3) was a simple model that we discussed for airway mucus clearance in the lung. The moving wall accounted for ciliary action, which can be considered a porous layer over the epithelium, see Figure 15.18(b). In Section 17.6 we analyze further the ciliary motion leading to propulsion of the overlying mucus-serous layer.

Airflow over a forest, the trees forming a porous media layer, and water flow over bottom vegetation are examples at a larger scale where the Reynolds numbers are turbulent. Figure 15.19(a) shows a sketch of flow over a tree canopy next to a lake. Figure 15.19 (b) is the wind tunnel model using pipe cleaners as trees with a porosity of $\phi = 0.78$, (Markfort et al., 2010). For large enough distances the trees can be viewed as surface roughness for turbulent flow.

The subject of boundary conditions at the interface between free-flowing fluid and porous media is complicated. There have been many proposed approaches and it is an active research area. A small sample of references can be found in Beavers and Joseph (1967), Larson and Higdon (1987), Hou et al. (1989), Alazmi and Vafai (2001) and Minale (2014) and the literature reviews therein.

An early approach to formulating the boundary condition at the interface between a free-flowing fluid and porous media obeying Darcy's law is found in Beavers and Joseph (1967). Consider a channel of width b conveying a fully developed flow by means of a uniform axial pressure gradient dp/dx.

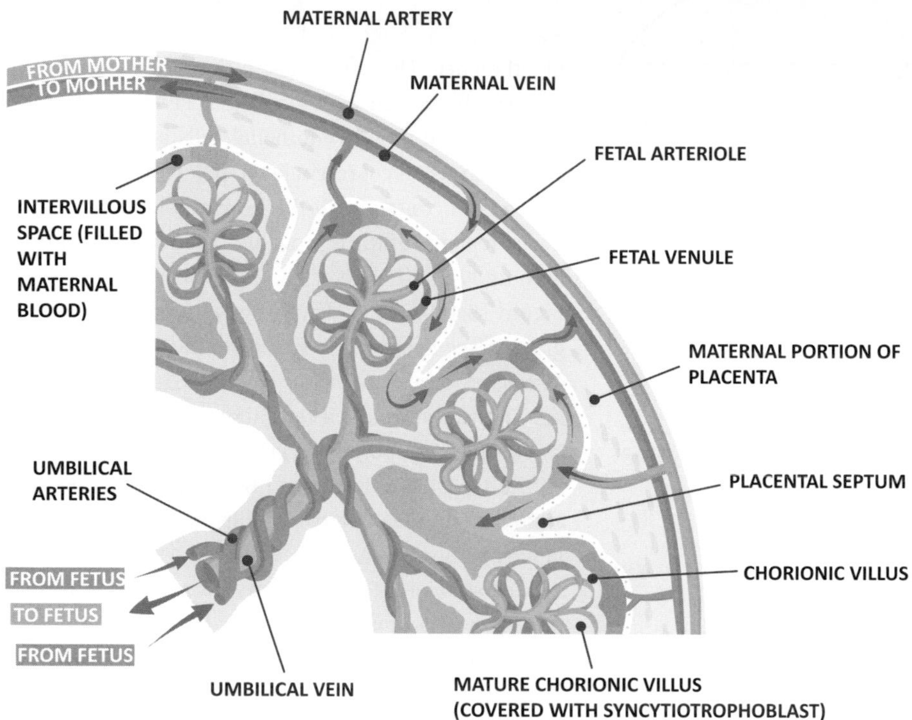

Figure 15.17 Placental blood flow showing maternal and fetal circulations and the tree-like structure called the chorionic villus.

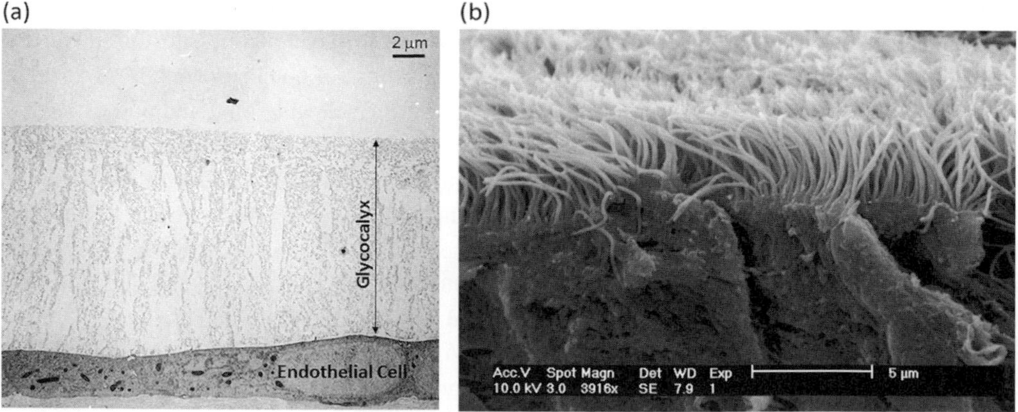

Figure 15.18 (a) Glycocalyx layer (GCX) and endothelial cell at 2500× magnification. (b) Scanning electron microscopy at nearly 4000× magnification of normal rabbit sinonasal epithelium.

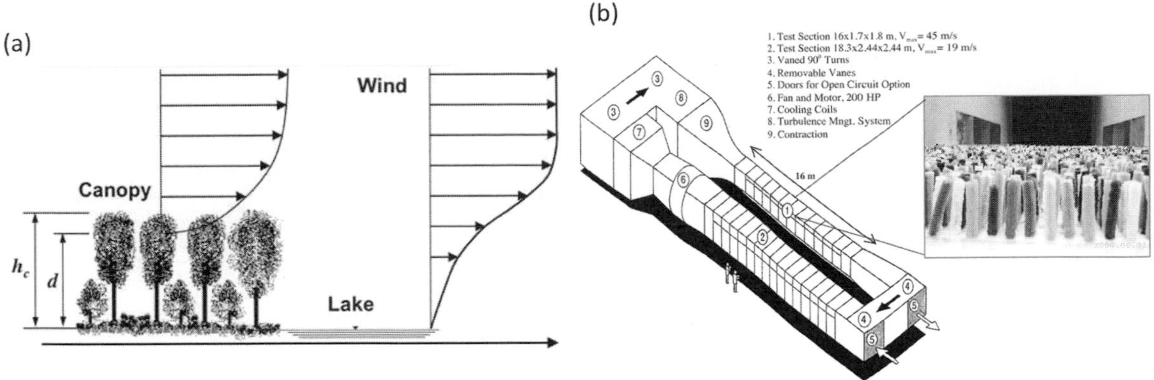

Figure 15.19 (a) Wind blowing over a tree canopy at the edge of a lake. (b) wind tunnel model of flow over trees (pipe cleaners) with porosity $\phi = 0.78$.

The governing equation and general solution for the x-velocity, u_1, are

$$\mu \frac{d^2 u_1}{dy^2} = \frac{dp}{dx} \quad \rightarrow \quad u_1 = \frac{dp}{dx}\frac{y^2}{2\mu} + c_1 y + c_2 \tag{15.5.1}$$

where $0 \le y < b$. Let the upper boundary at $y = b$ be a rigid, impermeable wall where the fluid obeys the no-slip condition, $u_1(y = b) = 0$. However, the lower wall of the channel will be the upper boundary of a porous media obeying Darcy's law for its x-velocity, u_2,

$$u_2 = -\frac{K}{\mu}\frac{dp}{dx} \tag{15.5.2}$$

which is constant for $y < 0$. Then Beavers and Joseph (1967) proposed an *ad hoc* formulation for the boundary condition at the interface $y = 0$,

$$u_1(y = 0) - u_2 = \frac{1}{\beta}\frac{du_1}{dy}_{y=0} \tag{15.5.3}$$

where β is a positive constant characterizing the interaction. Clearly, as $\beta \to \infty$ in Eq. (15.5.3) the condition asymptotes to no-slip. Otherwise, Eq. (15.5.3) is a slip condition and β is a slip coefficient. Since we expect $du_1/dy > 0$ at the lower wall for channel flow, then $u_1(y = 0) > u_2$. This slip model is motivated by the observation that the free-flowing fluid does not experience a completely solid lower boundary, but instead sees an intermittent solid separated by parts of pores which are fluid-filled, offering less drag on the flow. This approach was successfully used to model mucociliary propulsion with good data fit (Bottier et al., 2017a; 2017b).

Solving for c_1 and c_2 in Eq. (15.5.1) from the boundary conditions, let the dimensionless velocity forms be $U_1 = u_1/((-dp/dx)b^2/\mu)$ and $U_2 = u_2/((-dp/dx)b^2/\mu)$. Define the Darcy number as $Da = K/b^2$ and form a dimensionless slip coefficient $B = \beta\sqrt{K}$, so that B depends only on the characteristics properties of the porous media. The solutions are

$$U_1 = \frac{1}{2}\left[\frac{\sqrt{Da} + 2BDa}{\sqrt{Da} + B} + \frac{B(1 - 2Da)}{\sqrt{Da} + B}Y - Y^2\right]$$

$$U_2 = Da \tag{15.5.4}$$

An impermeable wall occurs when $Da = 0$ and Eq. (15.5.4) yields $U_1 = (Y - Y^2)/2$ and $U_2 = 0$. That form of U_1 is our familiar symmetric parabola around $Y = 1/2$ with no-slip at the lower and upper wall, $Y = 0, 1$. Also, for $Da > 0$, as $B \to \infty$ in Eq. (15.5.4), the channel profile asymptotes to $U_1 = (2Da + Y - Y^2)/2$, which matches $U_2 = Da$ at $Y = 0$ and there is no slip.

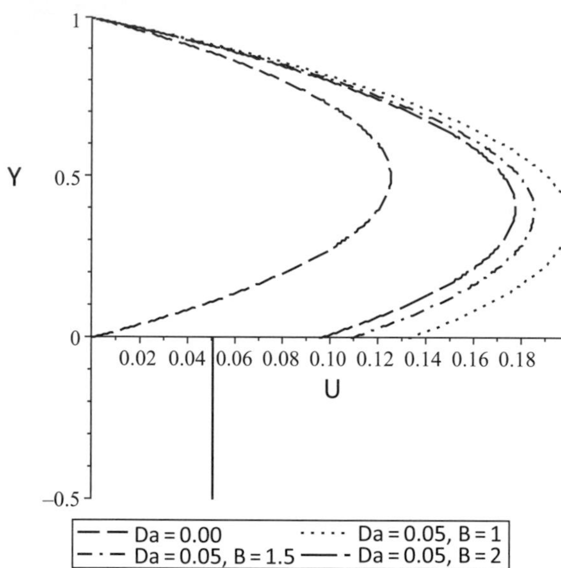

Figure 15.20 Channel flow over porous media: $U_1(0 \leq Y \leq 1)$ for Da $= 0.05$ with B $= 1.0, 1.5, 2.0$ and corresponding $U_2(Y < 0) = 0.05$; for comparison U_1 for an impermeable wall, Da $= 0$, is shown.

Figure 15.20 shows the channel velocity profile, $U_1(0 \leq Y \leq 1)$, for Da $= 0.05$ and B $= 1.0, 1.5, 2.0$. The corresponding porous media velocity is $U_2(Y < 0) = 0.05$. These U_1 profiles are asymmetric parabolic sections with slip at the lower boundary, $U_1(Y = 0) > U_2$. As B increases the flow approaches no-slip. For comparison the impermeable wall, Da $= 0.0$, solution for U_1 is shown, a symmetric parabola. Compared to the impermeable wall solution, the volume flow rate in the channel is greater over a permeable wall.

In contrast to a slip condition, Neale and Nader (1974) treat the core and porous velocity and stress fields as continuous using the Brinkman model. We can model a general situation of channel flow with a porous layer coating the walls. The flow is driven by a constant axial pressure gradient, dp/dx, which is the same for both the core and coating regions. The walls are at $y = \pm b$ and the core-coating interface is at $y = \pm\gamma b$, where $\gamma < 1$, so the coating has thickness $(1-\gamma)b$. We will assume symmetry at $y = 0$, and so can solve just the upper half flow for $0 \leq y \leq \gamma b$ in the core and $\gamma b \leq y \leq b$ in the coating.

The governing equations are the Navier–Stokes equations in the core region and the Brinkman equation, and their general solutions are

$$\text{core}: \ \mu_1 \frac{d^2 u_1}{dy^2} = \frac{dp}{dx} \ \rightarrow \ u_1 = \frac{y^2}{2\mu_1}\frac{dp}{dx} + a_1 y + b_1, \quad 0 \leq y \leq \gamma b$$

$$\text{layer}: \ \mu_2 \frac{d^2 u_2}{dy^2} - \frac{\mu_2}{K} u_2 = \frac{dp}{dx} \ \rightarrow \ u_2 = a_2 e^{\frac{y}{\sqrt{K}}} + b_2 e^{-\frac{y}{\sqrt{K}}} - \frac{K}{\mu_2}\frac{dp}{dx}, \quad \gamma b \leq y \leq b \tag{15.5.5}$$

where the core fluid viscosity is μ_1 and the coating layer fluid viscosity is μ_2. The four constants of integration are solved by symmetry at the midline, no-slip at the wall, and conditions on velocity and stress at the core–layer interface. The corresponding boundary condition equations using Neale and Nader (1974) are

$$\left.\frac{du_1}{dy}\right|_{y=0} = 0, \ u_2(y = b) = 0,$$
$$u_1(y = \gamma b) = u_2(y = \gamma b), \ \mu_1 \left.\frac{du_1}{dy}\right|_{y=\gamma b} = \mu_2 \left.\frac{du_2}{dy}\right|_{y=\gamma b} \tag{15.5.6}$$

Solving for a_1, b_1, a_2, b_2 from the boundary conditions of Eq. (15.5.6), the resulting velocities can then be made dimensionless using the following scales

$$U_1 = \frac{u_1}{\left(-\frac{dp}{dx}\right)\frac{b^2}{\mu_1}}, \ U_2 = \frac{u_2}{\left(-\frac{dp}{dx}\right)\frac{b^2}{\mu_1}}, \ Y = \frac{y}{b} \tag{15.5.7}$$

In Figure 15.21(a) we plot the velocity solutions U_1, U_2 for $\gamma = 0.8$ and $\mu_1 = \mu_2$. A wide range of Darcy number values, $10^{-5} \leq \text{Da} \leq 10$, is used to plot seven curves. Notice that as Da approaches the upper limit, the profile asymptotes to a parabola filling the entire channel. On the other hand, as Da approaches the lower limit, the velocity profile asymptotes to a parabola filling only the core so that, functionally, the core fluid feels the no-slip condition at $Y = 0.8$. Within the porous layer, the curvature of the profiles change sign for small enough Da. Figure 15.21(b) is

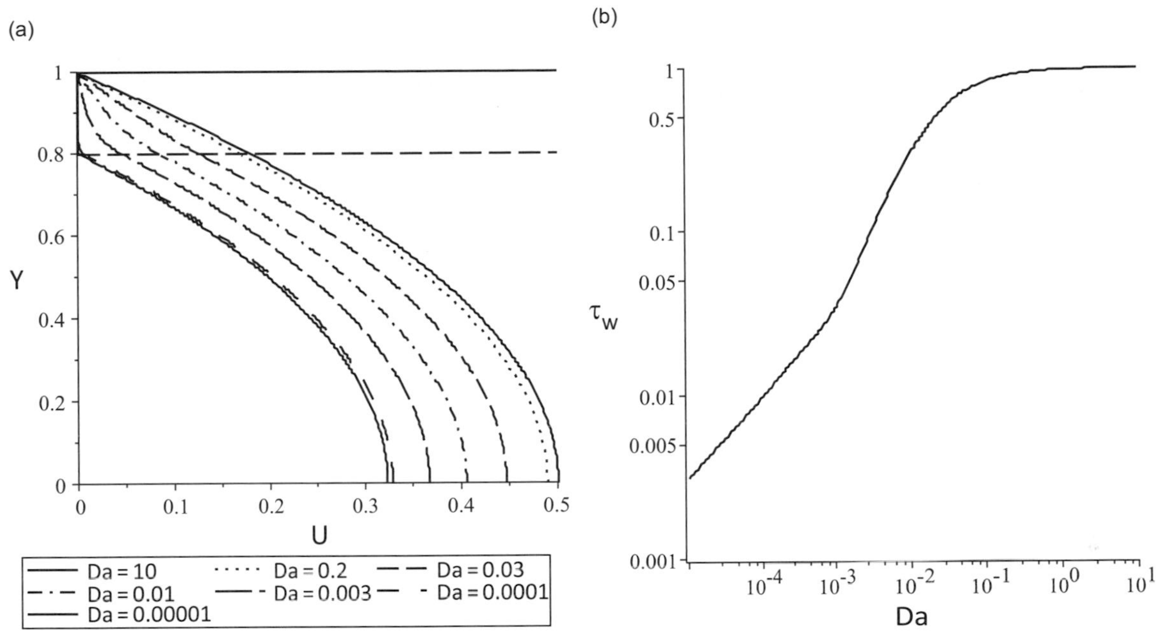

Figure 15.21 (a) Velocity profiles for $0 \leq Y \leq 1$ in a channel coated with a porous layer $0.8 \leq Y \leq 1$. The Darcy number, Da, varies in the range $10^{-5} \leq Da \leq 10$. (b) Wall shear, τ_w, vs Da.

the dimensionless wall shear, τ_w, vs Da where τ_w is defined as

$$\tau_w = -\frac{\mu \frac{du_2}{dy}}{\left(-\frac{dp}{dx}b\right)} = -\frac{dU_2}{dY} \tag{15.5.8}$$

As Da decreases the porous layer is less permeable, which significantly reduces τ_w. Since the wall is generally made of cells, the coating has a protective role keeping the wall shear low.

The endothelium shown in Figure 15.18(a) is a wavy surface coated with the glycocalyx layer. A schematic is shown in Figure 15.22(a) showing the endothelial cells (EC), their nuclei (N), the junctions (J) between cells, and the glycocalyx layer (GCX). The coupled flow problem for the core and porous layer using biphasic theory was solved in Wei et al., 2003 #424). The solution streamlines are shown in Figure 15.22(b) and show a band of streamlines which enter and exit the porous layer (upper). Then (lower) there is an enlarged view of a vortex over the junction which can enhance transport through it.

15.6 Kozeny–Carman Equation

A basic question in porous media flows is how to relate K to the porosity, ϕ, and specific geometric features of its microstructure. In biofluid mechanics, there are applications related to both biotechnology and physiological flows. For a technology example, flow through packed beds of microspheres is used for high performance liquid chromatography (HPLC) to analyze urine samples for drug screening (Sundström et al., 2013). The microspheres are silica with diameters on the order of 5 microns. For a physiologic example we saw above flows through tissues like tumors and arterial walls where the underlying microstructure of the solid phase is long fibers of collagen and other tissue components.

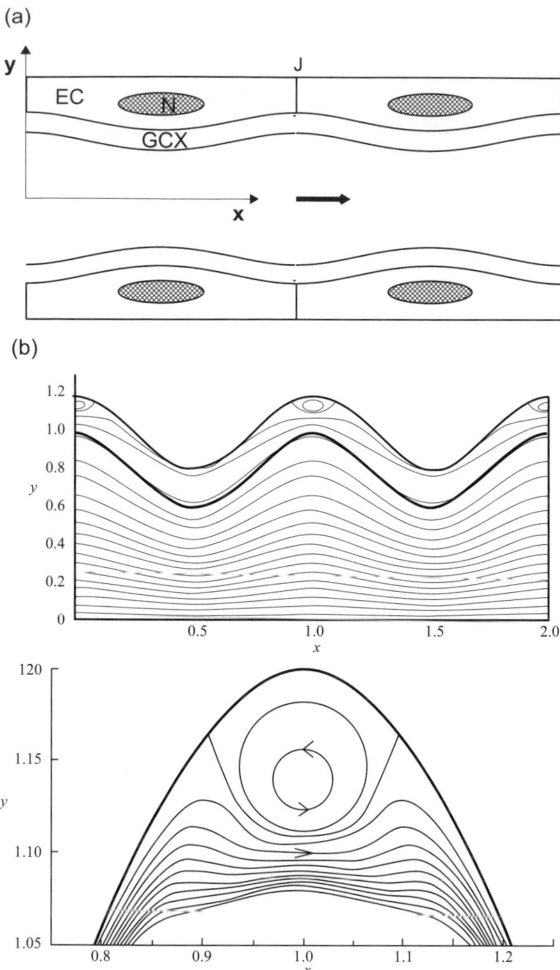

Figure 15.22 (a) Model of flow over endothelial cells (EC) with a wavy surface due to cell nuclei (N) and glycocalyx coating (GCX). (b) (Upper) streamlines for y > 0 (lower) enlarged view of streamlines and vortex at junction (J).

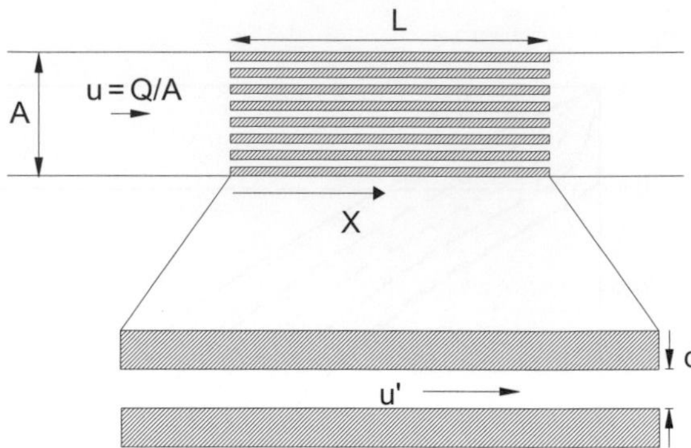

Figure 15.23 Flow through a pipe with a section of porous media idealized as identical parallel cylindrical tubes.

The Austrian hydraulic engineer and physicist, Josef Kozeny (1889–1967), tackled this problem in 1927 by viewing flow through the pores as equivalent flow through capillary tubes (Kozeny, 1927). In 1937 Philip C. Carman, a British chemical engineer, extended and modified this approach (Carman, 1937). Consider flow in a pipe of cross-sectional area A with a section of porous media of length L, as shown in Figure 15.23. The flow rate, Q, is the same everywhere, defining an average velocity in the upstream pipe, $u = Q/A$, where there is no porous media. The velocity u is known as the superficial or empty-tube velocity. The section of porous media is idealized as a system of equally spaced, identical circular tubes of diameter d. Of course, u must also be the porous media fluid velocity averaged across the tube cross-section. Let the cross-sectional area of the voids (pores) be A_V, so the local average fluid velocity within a pore is $u' = Q/A_V$. From these definitions we readily see that

$$u' = \frac{A}{A_V}u - \frac{1}{\phi}u \tag{15.6.1}$$

where ϕ is the ratio of the fluid volume, $A_V L$, to the total volume, AL, i.e. the porosity given by $\phi = A_V/A$. From Eq. (15.6.1) we see that $u' > u$ since the fluid, at constant Q, must speed up to go through a smaller cross-sectional area, $A_V < A$. Flow in these parallel straight tubes satisfies Poiseuille law where the pressure drop, $\Delta p = p(x = 0) - p(x = L) > 0$, is given by

$$\frac{\Delta p}{L} = \frac{32\mu u'}{d^2} = \frac{32\mu u}{d^2\phi} \tag{15.6.2}$$

which can be rearranged to look like Darcy's law,

$$u = -\frac{K}{\mu}\frac{dp}{dx}, \quad K = \frac{\phi d^2}{32} \tag{15.6.3}$$

since $\Delta p/L = -dp/dx$.

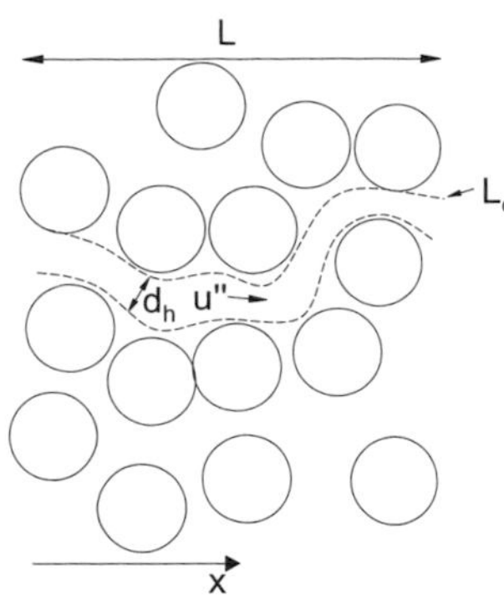

Figure 15.24 A tortuous stream tube pathway of flow through a porous material comprised of spherical particles.

In real situations, however, the pathway length of the fluid flow is not a straight tube. Rather, it is a tortuous stream tube path of length L_c and average diameter d_h, known as the hydraulic diameter, see Figure 15.24. Call the fluid velocity in the stream tube u''. For the same flow rate, Q, equal amounts of fluid volume have the same transit time, T, such that $T = L/u' = L_c/u''$ so

$$u'' = \frac{L_c}{L}u' = \frac{L_c}{L}\frac{u}{\phi} \tag{15.6.4}$$

The ratio $L_c/L > 1$ is a relative measure of the longer distance the fluid must travel, due to the tortuous path it takes. The tortuosity of the medium is defined as $\tau = (L_c/L)^2$, though some authors call the tortuosity L_c/L. For example, the tortuosity of the packed columns of microspheres seen in Figure 15.1 were calculated to be $L_c/L = 1.40$ (ceramic) and $L_c/L = 1.79$ (cellulose). Because the route is longer, we see that $u'' > u'$, and this was a major contribution from Carman (1937) to the formulation. From Poiseuille law in the tortuous stream tube, using the flow rate as $q = u''\pi d_h^2/4$, and Eq. (15.6.4), we find

$$\frac{\Delta p}{L_c} = -\frac{128\mu q}{\pi d_h^4} = -32\frac{\mu}{d_h^2}\frac{L_c}{L}\frac{u}{\phi} \tag{15.6.5}$$

Multiplying Eq. (15.6.5) by L_c/L gives us the pressure gradient in the x-direction

$$\frac{\Delta p}{L} = 32\frac{\mu\tau}{d_h^2}\frac{u}{\phi} = -\frac{dp}{dx} \tag{15.6.6}$$

with $\tau = (L_c/L)^2$. The total cross-section is the sum of the void, A_V, and solid, A_S, cross-sections, $A = A_V + A_S$. Using the definition of ϕ we see that

$$\phi = \left(\frac{A_V}{A_V + A_S}\right)\left(\frac{L}{L}\right) = \frac{V_V}{V_V + V_S} \tag{15.6.7}$$

In Eq. (15.6.7), $V_V = A_V L$ is the volume of the fluid-filled voids and $V_S = A_S L$ is the volume of the solid. The total volume is the sum $V = AL = V_V + V_S$. Rearranging Eq. (15.6.7) we obtain $V_V = \phi(V_V + V_S)$. Then since $V_S = V - V_V = V(1 - V_V/V) = V(1 - \phi)$ we can state that

$$V = \frac{V_S}{(1-\phi)} = V_V + V_S\left(\frac{1-\phi}{1-\phi}\right) \quad \rightarrow \quad V_V = \frac{\phi V_S}{1-\phi} \tag{15.6.8}$$

Let S be the surface area of the voids and divide Eq. (15.6.8) by S so that

$$\frac{V_V}{S} = \frac{\phi V_S}{(1-\phi)S} \tag{15.6.9}$$

The assumption of this approach is that the total void volume is made of n identical tortuous stream tubes, so that $V_V = n\pi d_h^2 L_c/4$, while the total wetted surface area of those stream tubes is $S = n\pi d_h L_c$. Inserting these forms into the left hand side of Eq. (15.6.9) we obtain $V_V/S = d_h/4$. This is a key step in the Kozeny–Carman strategy: define the hydraulic diameter, d_h, as one which yields identical stream tubes whose total void volume to wetted surface area ratio reflects that of the porous media. Now Eq. (15.6.9) takes the form

$$d_h = \frac{4\phi}{(1-\phi)}\frac{V_S}{S} = \frac{4\phi}{(1-\phi)S_S} \tag{15.6.10}$$

where $S_S = S/V_S$ is the specific surface area per volume of the solid. Inserting this value of d_h from Eq. (15.6.10) into Eq. (15.6.5) we find

$$\frac{dp}{dx} = -2\mu u S_S^2 \tau \frac{(1-\phi)^2}{\phi^3} \tag{15.6.11}$$

Equation (15.6.11) can be rearranged to look like Darcy's law, $u = -(K/\mu)(dp/dx)$, where K is given by

$$u = -\frac{1}{\mu}\frac{1}{2\tau S_S^2}\frac{\phi^3}{(1-\phi)^2}\frac{dp}{dx} \quad \rightarrow \quad K = \frac{1}{GS_S^2}\frac{\phi^3}{(1-\phi)^2} \tag{15.6.12}$$

Equation (15.6.12) is the Kozeny–Carman equation in a general form relating the permeability, K, to characteristic parameters of the porous media: its porosity, specific surface area per volume of solid, and the tortuosity: ϕ, S_S, τ, respectively (Kozeny, 1927; Carman, 1937). The product 2τ is called the Kozeny constant, $G = 2\tau$. The value of G depends on the shapes and sizes of the solid particles and partly on ϕ, but the typical range is $4 \leq G \leq 6$ determined from measurements, and $G = 5$ fits many applications.

For a porous media made of a packed column of spherical particles the solid volume is $V_S = \pi d_s^3/6$ while the surface area of a sphere is $S = \pi d_s^2$. Then $S_S = 6/d_s$ and for $G = 5$ we find K from Eq. (15.6.12) to be

$$K = \frac{d_s^2}{180}\frac{\phi^3}{(1-\phi)^2} \tag{15.6.13}$$

Applying Eq. (15.6.13) to the chromatography column of Figure 15.1 using $d_p = 50\ \mu m$ and $\phi = 0.34$ produces a permeability of $K = 1.25 \times 10^{-8}\,cm^2$.

Biomedical applications for interstitial tissue flows involve a random array of fibers like collagen, GAG and PGP, see Figure 15.5. In this setting, the form of the Kozeny–Carman equation is often rearranged as follows. In Eq. (15.6.9) add V_S/S to both sides to obtain

$$\frac{V_V + V_S}{S} = \frac{\phi V_S + (1-\phi)V_S}{(1-\phi)S} = \frac{V_S}{(1-\phi)S} \tag{15.6.14}$$

Now define $S_T = S/(V_V + V_S)$ as the surface area per total volume and insert Eq. (15.6.14) into Eq. (15.6.10) to find

$$d_h = \frac{4\phi}{(1-\phi)}\frac{V_S}{S} = \frac{4\phi}{S_T} \tag{15.6.15}$$

This form of d_h substituted into Eq. (15.6.6) yields the Kozeny–Carman equation for K in modified form as

$$u = -\frac{\phi^3}{2\mu\tau S_T^2}\frac{dp}{dx} \quad \rightarrow \quad K = \frac{\phi^3}{GS_T^2} \tag{15.6.16}$$

Another step often taken is to define a mean hydraulic radius as $\bar{r} = \phi/S_T$ so that K becomes

$$K = \frac{\phi\bar{r}^2}{G} \tag{15.6.17}$$

For example, in Section 7.9 we examined both perpendicular, Figure 7.26, and parallel flow, Figure 7.27, through an array of cylinders and arrived at relationships between their average velocity and pressure gradients in Eqs. (7.9.5) and (7.9.12). We can substitute those into Darcy's law, such that $U_{avg} = -(K_x/\mu)(dp/dx)$ for the perpendicular flow and $U_{avg} = -(K_z/\mu)(dp/dz)$for the parallel flow. Solving for K in each, and recognizing that $S_T = 2\pi b/\pi r_0^2$, the respective values of G compute to be

$$G_x = \frac{2\phi^3}{(1-\phi)\left[\ln\left(\frac{1}{1-\phi}\right) - \frac{(1-(1-\phi)^2)}{(1+(1-\phi)^2)}\right]},$$
$$G_z = \frac{2\phi^3}{(1-\phi)\left(-2\phi - \phi^2 + 2\ln\left(\frac{1}{1-\phi}\right)\right)} \tag{15.6.18}$$

Figure 15.25 shows a comparison of G_x for perpendicular flow and G_z for parallel flow through a square array of cylinders which are the results of

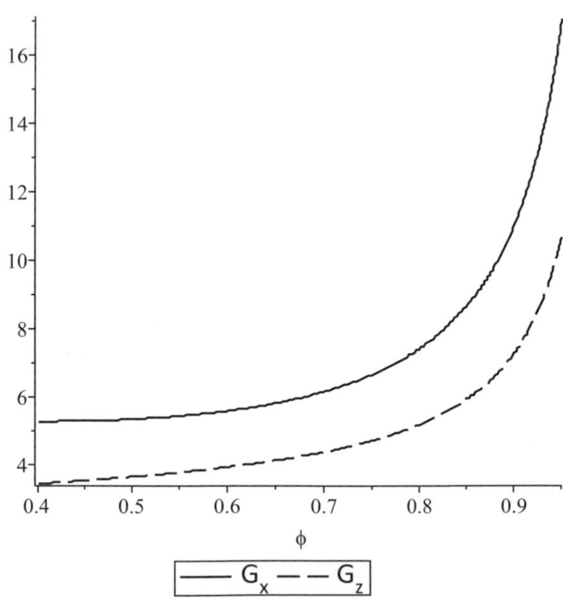

Figure 15.25 Comparison of G_x, G_z for the range $0.4 \leq \phi \leq 0.95$.

Happel (1959), except for a typographical error in G_x. For an array of cylinders oriented randomly, Happel and Brenner (1965) propose a weighted sum for the resulting G_{xz} as

$$G_{xz} = \frac{2}{3}G_x + \frac{1}{3}G_z \tag{15.6.19}$$

15.7 Mass Transport

Recall that we studied conservation of a dissolved species through the convection–diffusion equation, Eq. (4.5.14). The same balance applies to mass transport in porous media where the Péclet number, $Pe = UL/D_m$ from Eq. (4.5.26), represents the ratio of convective to diffusive transport. Often in tissues, because flows are so slow, the transport is dominated by diffusion, i.e. $Pe << 1$. For example, for flow through the rabbit aorta wall we calculated a characteristic velocity of $U = 5.53 \times 10^{-6}$ cm/s using a characteristic length scale of $L = 0.02$ cm. For a tumor we estimated $U \sim 2 \times 10^{-5}$ cm/s and $L = 1$ cm. $D_m \sim 2 \times 10^{-5}$ cm^2 s^{-1} for oxygen diffusing in water at 25 °C, so $Pe << 1$ for oxygen transport in the arterial wall but $Pe \sim 1$ for the tumor for transport over a 1 cm distance. However, the diffusivity can be slowed by the size of the molecule, and steric effects related to pore size and interaction with the matrix lead to lowered D_m, by as much as 18 to 93% from its value in a free solution (Berk et al., 1993; Pluen et al., 1999; Ramanujan et al., 2002; Wiig et al., 2003; Swartz and Fleury, 2007). For example, $D_m \sim 1 - 10 \times 10^{-7}$ cm^2/s for large molecules (dextran, immunoglobulins) diffusing through various tissues (Leddy and Guiliak, 2003; Brown et al., 2004; Leddy, 2004; Evans and Quinn, 2005; Swartz and Fleury, 2007). Consequently there are many interstitial flow processes where $Pe \geq 1$.

Medical devices that require rapid mass transport, however, are usually in the regime of $Pe >> 1$. Examples include blood oxygenators and artificial lungs where many designs involve blood flow around fiber bundles see Figure 15.26(a). The fibers are hollow and convey an internal gas flow at very high oxygen concentrations. The oxygen diffuses across the fiber wall into the blood, while carbon dioxide diffuses in the opposite direction. The fiber cylinders may be arranged in a number of ways: square arrays, staggered arrays, cross arrays, parallel to the flow and others, see Figure 15.26 (b). The blood convection can be treated like Darcy flow where the array of fibers is the matrix. The Darcy permeability, K, that the cylinders pose has been measured for several fiber sizes and arrangements in Pacella et al. (2011). The measured values differed significantly from the predictions of Eq. (15.6.13), and a better match to the

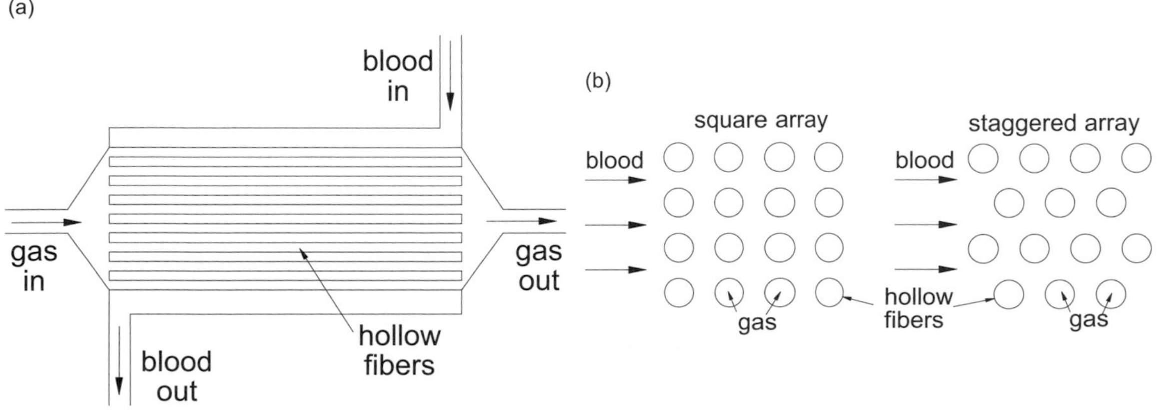

Figure 15.26 (a) Example of an oxygenator design with gas flow inside hollow fibers and blood flow around them. (b) Cross-section of (a) showing blood flow around hollow fiber arrays, square vs staggered.

data was achieved by replacing the constant term, 180, with a porosity-dependent term, $542\phi - 128$.

The actual mass transport for such complicated flows requires careful experiments and computational approaches (Mockros and Gaylor, 1975; Baker et al., 1991; Dierickx et al., 2000; 2001; Chan et al., 2006). Figure 15.27 shows the CFD results for pulsatile flow through square, Figure 15.27(a), or staggered, Figure 15.27(b), cylinder arrays. Both have a porosity, void fraction, of $\phi = (A_{tot} - A_{cyl})/A_{tot} = 0.8036$ where A_{tot} is the total area and A_{cyl} is the area of the cylinders. They also have the same Reynolds number, Womersley number and amplitude parameter values and are a snapshot at the instant of peak flow, $t = t_0$. The upper curves are constant dimensionless concentration contours, $C(x, y, t = t_0) = \text{constant}$, with $C = 1$ on the cylinder surfaces to represent the oxygen coming from the hollow fiber gas flow. Then $C = 0$ for the incoming upstream fluid, so the range is $0 \le C \le 1$. The lower curves are streamlines. Comparing Figure 15.27(a) to Figure 15.27(b) it is easy to see why the staggered array provides better oxygen transport than the square array. The square array has a trapped vortex between the cylinders which reduces the mixing that the staggered array provides.

Another example of mass transport in porous media is a gas mask which is worn to protect an individual from inhaling toxic chemicals. Potential exposures may occur in many situations ranging from job-related activities to military conflict, see Figure 15.28. In Figure 15.28(a) a military daughter learns about gas masks. The porous media component can be designed as replaceable, shown in Figure 15.28 (b). The inhaled environmental air passes through a sequence of filters which first removes aerosol particulates, then absorbs toxic gases onto activated charcoal, and then removes dust from that charcoal. Capsules and powder of activated charcoal are shown in Figure 15.28(c). The fractal-like particle shape is the source of their enormous surface area. Although ~0.1 mm wide, each particle has a surface area of several square meters. Indeed, one gram of activated charcoal can have a surface area of 3,000 m^2. That is what makes it so effective at absorbing toxic gases and serve as a strong boundary term imposed on the convection–diffusion equation, Eq. (4.5.14). It is also used in the gastrointestinal tract to absorb ingested poisons. Prior to the development of activated charcoal, the masks in Figure 15.28(d) used packed asbestos fibers to protect British civilians during World War 2. As was eventually learned, inhaled asbestos particles lodge in the lung setting up a chronic inflammatory process there (asbestosis) which can progress to a particular type of cancer called

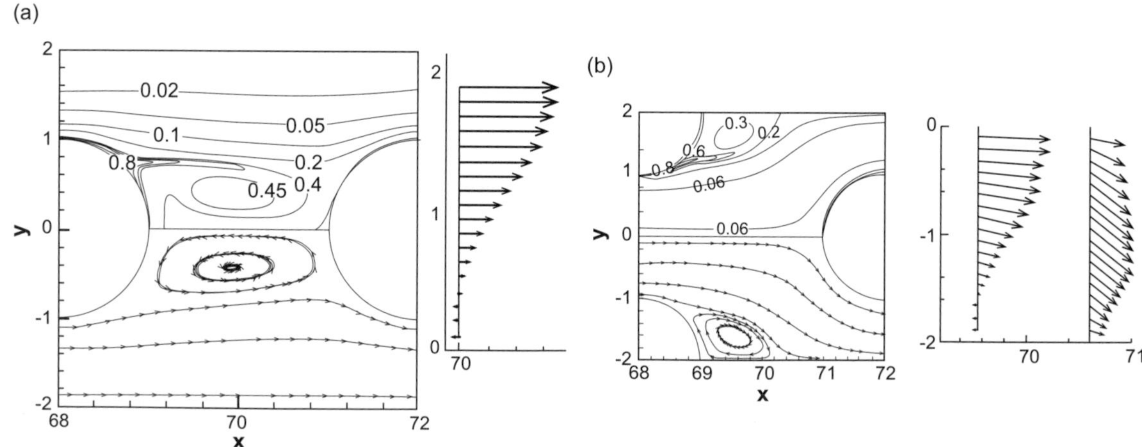

Figure 15.27 (a) Concentration contours (upper), streamlines (lower) and velocity vectors for pulsatile flow through a square array with $\phi = 0.8036$ at peak flow. Concentration on the cylinder boundary is $C = 1$. (b) Same as (a) for a staggered array.

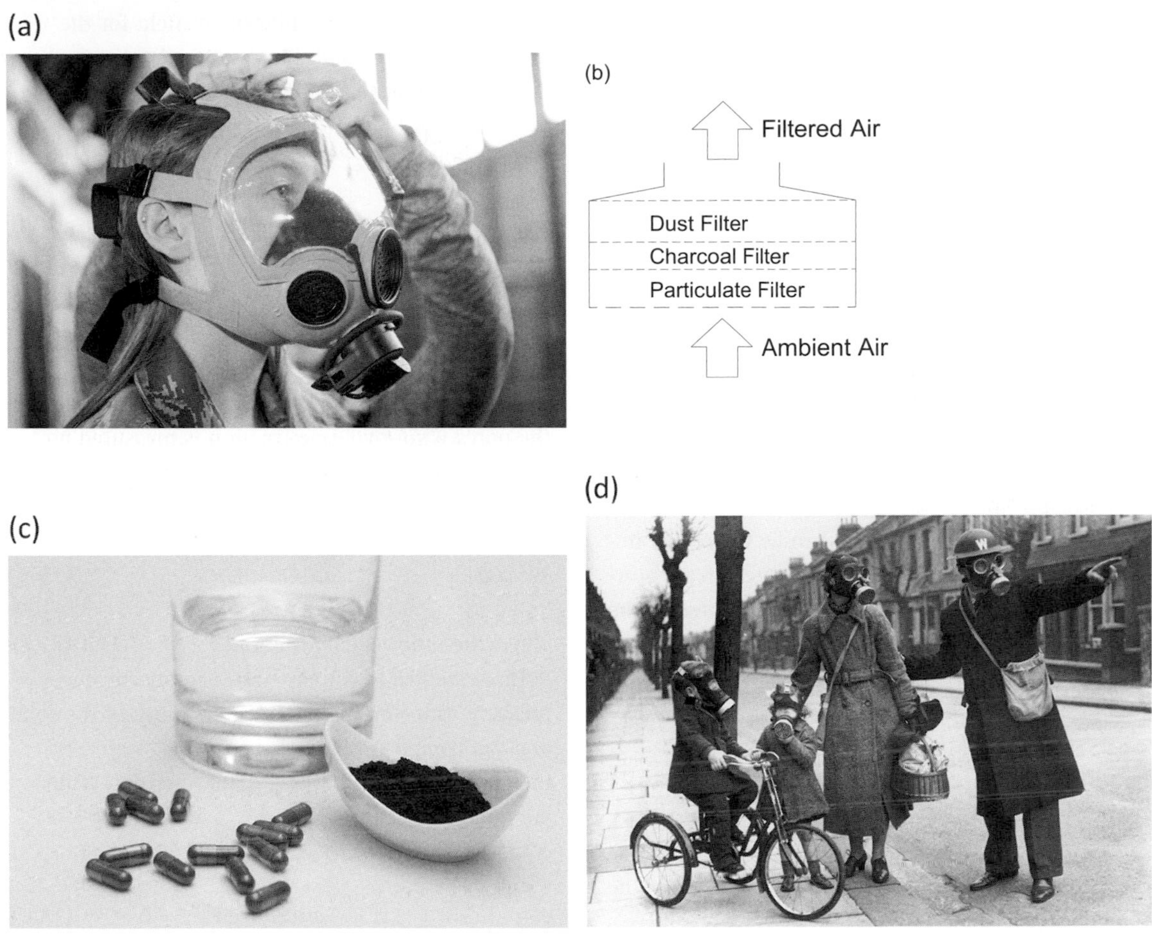

Figure 15.28 (a) A military daughter dons a gas mask at Ellsworth Air Force Base, SD, April 9, 2016. (b) Cross-sectional view of a typical air filter used in gas masks with three layers to (1) remove aerosol particulates, (2) absorb toxic gases onto activated charcoal particles and (3) remove charcoal dust. (c) Activated charcoal capsules and powder. (d) A warden gives directions to a mother and her two children during a World War II gas drill in Southend, England, March 29 1941.

mesothelioma. The design of a gas mask must keep K large enough so that the wearer can adequately ventilate at normal respiratory pressures.

15.8 Poroelastic Media

In many applications the solid phase of the porous media is deformable. This is especially true in biomedical situations, such as flow through tissues which are soft materials. Analysis of this topic started with the poroelastic theory of Biot in 1941 for soil under loading (Biot, 1941) and the slow adjustment known as consolidation. The elastic behavior for soils is attributed to surface tension forces holding the particles together, a concept proposed earlier in a theory by Terzaghi (1925) as molecular forces binding the particles. Additional analytical approaches include mixture theory (Bowen, 1980) and biphasic theory, the latter motivated by flow through cartilage (Mow et al., 1980). The biphasic and poroelastic theories are compared and contrasted by Simon

(1992) showing that the linear formulations are equivalent. We will use poroelastic theory in this section.

For rigid porous media we were only interested in the motion of the fluid. However, for poroelastic media the matrix also moves. We have to account for both. For a brief review of solid mechanics, consider a linearly elastic solid which is homogeneous and isotropic. The constitutive relationship between the Cauchy stress tensor, σ_{ij}, and the strain tensor, ε_{ij}, in Cartesian tensor notation is

$$\sigma_{ij} = \lambda\varepsilon\delta_{ij} + 2G\varepsilon_{ij} \tag{15.8.1}$$

where λ is the first Lamé constant and G is the second Lamé constant, which is also called the shear modulus as employed in Section 14.1. λ can be expressed in terms of G and the Poisson ratio, ν, such that $\lambda = 2G\nu/(1-2\nu)$. The strain tensor is related to the solid displacements, U_i, by

$$\varepsilon_{ij} = \frac{1}{2}\left(\frac{\partial U_i}{\partial x_j} + \frac{\partial U_j}{\partial x_i}\right) \tag{15.8.2}$$

The term ε in Eq. (15.8.1) is the dilatation,

$$\varepsilon = \varepsilon_{kk} = \varepsilon_{11} + \varepsilon_{22} + \varepsilon_{33} = \underline{\nabla}\cdot\underline{U} \tag{15.8.3}$$

where the repeated index implies summation per the Einstein notation.

For a poroelastic material, we must incorporate the fluid pressure, p, into the stress–strain relationship which models the combined fluid and solid behavior. Biot (1941) accomplished this by introducing the fluid pressure into Eq. (15.8.1),

$$\tau_{ij} = \lambda e\delta_{ij} + 2Ge_{ij} - \alpha p\delta_{ij} \tag{15.8.4}$$

where τ_{ij} is now the average stress tensor, and e_{ij} is the average strain tensor for the poroelastic solid now filled with a fluid. The negative sign in front of the pressure term yields tension (negative stress) for positive pressure. The coefficient multiplying the pressure term, α, is known as the Biot–Willis parameter. The strain tensor is defined as

$$e_{ij} = \frac{1}{2}\left(\frac{\partial U_i^s}{\partial x_j} + \frac{\partial U_j^s}{\partial x_i}\right) \tag{15.8.5}$$

where U_i^s is the solid displacement field for the deformation. The solid dilatation is the divergence of the solid displacement field,

$$e = e_{kk} = \frac{\partial U_k^s}{\partial x_k} = \frac{\partial U_1^s}{\partial x_1} + \frac{\partial U_2^s}{\partial x_2} + \frac{\partial U_3^s}{\partial x_3} \tag{15.8.6}$$

Defining the fluid displacement vector field as U_i^f, the fluid displacement relative to the solid displacement, W_i, is the difference

$$W_i = \phi\left(U_i^f - U_i^s\right) \tag{15.8.7}$$

where we are reminded that the fluid is confined to the pores whose volume fraction is measured by the porosity, ϕ. The relative fluid velocity is defined by the Lagrangian time derivative of Eq. (15.8.7),

$$\underline{\dot{W}} = \phi\left(\underline{\dot{U}}^f - \underline{\dot{U}}^s\right) = \phi\left(\underline{u} - \frac{\partial \underline{U}^s}{\partial t}\right) \tag{15.8.8}$$

where the fluid velocity is defined as $\underline{u} = \underline{\dot{U}}^f = \partial\underline{U}^f/\partial t$.

Biot established an equation of state for the pressure as a sum of the compressibility of the solid and the compressibility of the fluid. For an incompressible combined system, the Biot–Willis parameter takes the value $\alpha = 1$ and the state equation simplifies to

$$e + \frac{\partial W_k}{\partial x_k} = \frac{\partial}{\partial x_k}\left((1-\phi)U_k^s + \phi U_k^f\right) = 0 \tag{15.8.9}$$

Equation (15.8.9) states that the divergence of the sum of the solid and fluid displacement fields, weighted by their respective volume fractions, ϕ for the fluid and $1-\phi$ for the solid, is zero. Equation (15.8.9) states that a change in volume of the porous material is balanced by a fluid volume shift. We assume incompressibility for our applications which is typical in biological poroelastic flows.

The time derivative of Eq. (15.8.9) gives us the incompressible conservation of mass for the combined solid and fluid system

$$\frac{\partial}{\partial x_k}\left((1-\phi)\frac{\partial U_k^s}{\partial t} + \phi u_k\right) \leftrightarrow \underline{\nabla}\cdot\left((1-\phi)\frac{\partial \underline{U}^s}{\partial t} + \phi\underline{u}\right) = 0 \tag{15.8.10}$$

given in index and vector notation.

Turning to conservation of momentum for the fluid brings in Darcy's law. Since it is the fluid velocity relative to the solid velocity which creates the resistance to flow, we use $\dot{W}_i$ or $\underline{\dot{W}}$ from Eq. (15.8.8) in Darcy's law. The result is

$$\phi\left(u_i - \frac{\partial U_i^s}{\partial t}\right) = -K'\frac{\partial p}{\partial x_i} \leftrightarrow \phi\left(\underline{u} - \frac{\partial \underline{U}^s}{\partial t}\right) = -K'\underline{\nabla} p \tag{15.8.11}$$

also in index and vector notation.

Finally, the conservation of momentum for the solid, ignoring acceleration and body force terms, is the divergence of the stress tensor set to zero as required by Cauchy's momentum equation, Eq. (5.2.2),

$$\frac{\partial \tau_{ij}}{\partial x_j} = 0 \tag{15.8.12}$$

Inserting Eq. (15.8.4) into Eq. (15.8.12) leads to

$$(\lambda + G)\frac{\partial e}{\partial x_i} - \frac{\partial p}{\partial x_i} + G\frac{\partial^2 U_i^s}{\partial x_j^2} = 0 \tag{15.8.13}$$

Taking the divergence, $\partial/\partial x_i$, of Eq. (15.8.13), we obtain

$$(\lambda + G)\frac{\partial^2 e}{\partial x_i^2} - \frac{\partial^2 p}{\partial x_i^2} + G\frac{\partial^2}{\partial x_j^2}\left(\frac{\partial U_i^s}{\partial x_i}\right) = 0 \tag{15.8.14}$$

Recalling that the divergence of the solid displacement field is $e = \partial U_i^s/\partial x_i$, Eq. (15.8.14) rearranges to

$$(\lambda + 2G)\frac{\partial^2 e}{\partial x_i^2} - \frac{\partial^2 p}{\partial x_i^2} = \frac{\partial^2}{\partial x_i^2}[(\lambda + 2G)e - p] = 0 \tag{15.8.15}$$

Equation (15.8.15) tells us that, within an arbitrary function, f,

$$(\lambda + 2G)e_{kk} = p + f \tag{15.8.16}$$

as long as $\nabla^2 f = 0$. We can choose $f = 0$, so that

$$(\lambda + 2G)e = p \tag{15.8.17}$$

Solving Eq. (15.8.17) for e and substituting into Eq. (15.8.13) gives us

$$-\frac{\partial p}{\partial x_i} + (\lambda + 2G)\frac{\partial^2 U_i^s}{\partial x_k^2} = 0 \quad \leftrightarrow$$

$$-\underline{\nabla} p + (\lambda + 2G)\nabla^2 \underline{U}^s = \underline{0} \tag{15.8.18}$$

in index and vector notation.

The governing system of equations consists of conservation of fluid momentum, Eq. (15.8.11), conservation of solid momentum, Eq. (15.8.18), and conservation of total mass, Eq. (15.8.10). The first two are vector equations with three components each, while the last is a scalar equation. for a total of seven independent equations in the seven unknowns, $U_1^s, U_2^s, U_3^s, u_1, u_2, u_3, p$. If we substitute the relative velocity from Darcy's law of Eq. (15.8.11) into the conservation of mass equation, Eq. (15.8.10), we find

$$\underline{\nabla}\cdot\left(\frac{\partial \underline{U}^s}{\partial t} - K'\underline{\nabla} p\right) = \frac{\partial}{\partial t}(\underline{\nabla}\cdot\underline{U}^s) - K'\nabla^2 p = 0 \tag{15.8.19}$$

However, $\underline{\nabla}\cdot\underline{U}^s = e = p/(\lambda + 2G)$ from Eq. (15.8.17), so Eq. (15.8.19) becomes an unsteady diffusion equation for the pressure

$$\frac{\partial p}{\partial t} = D\nabla^2 p \tag{15.8.20}$$

where $D = (\lambda + 2G)K'$ is a diffusion coefficient in cm/s^2, for example. The pressure disturbance diffuses through the poroelastic media more rapidly for a stiffer matrix (larger $\lambda + 2G$) or larger K'. For a given length scale, L_s, in a problem definition, the relevant time scale is $T_s = L_s^2/D$, which has the form of an RC (resistance x capacitance) time constant in an electrical circuit. The resistance comes from $1/K'$ and the capacitance, or compliance, from $1/(\lambda + 2G)$.

Consider a one-dimensional version of Eq. (15.8.20)

$$\frac{\partial p}{\partial t} = D\frac{\partial^2 p}{\partial y^2} \tag{15.8.21}$$

and recall we have solved this diffusion problem form in Sections 8.1 and 8.2. There we studied Stokes' first and second problem of shear diffusion from a wall at $y = 0$ into a semi-infinite fluid at $y \to \infty$. The diffusion coefficient was the kinematic viscosity, ν. The first problem (Section 8.1) was for a suddenly started wall, while the second (Section 8.2) was for an oscillating wall. In both cases the fluid velocity and shear were zero at $y \to \infty$. Here we are following the diffusion of pressure.

For the suddenly imposed pressure, p_0, at $y = 0$, the boundary conditions are

$$p(y=0) = \begin{cases} 0 & t < 0 \\ p_0 & t \geq 0 \end{cases} \qquad p(y \to \infty) = 0 \tag{15.8.22}$$

From Section 8.1 we showed the similarity solution to this system is

$$\frac{p}{p_0} = 1 - \mathrm{erf}(\eta) = \mathrm{erfc}(\eta) \tag{15.8.23}$$

where $\eta = y/2\sqrt{Dt}$ is the similarity variable and erf is the error function, Eq. (8.1.20), while erfc is the co-error function.

A clinical application of this solution would be an infiltrated intravenous (IV) site, say in the back of the hand. IV fluid mistakenly enters the subcutaneous space driven by gravity and the patient gets a local swelling under the skin. How long was this going on? The skin of a mouse tail has $\lambda + 2G = 110$ mmHg and $K' = 1.5 \pm 0.2 \times 10^{-6}\ \mathrm{cm^2/mmHg \cdot s}$ so $D = 1.65 \times 10^{-4}\ \mathrm{cm^2/s}$ (Swartz et al., 1999). For a swelling characterized by $L_s = 1$ cm, the time scale is $T_s = L_s^2/D = 6060\ \mathrm{s} = 1.68\ \mathrm{h}$.

The oscillatory pressure boundary conditions on Eq. (15.8.21) are

$$p(y=0) = p_0 \cos \omega t \qquad p(y \to \infty) = 0 \tag{15.8.24}$$

Choose the scalings $T = \omega t$, $P = p/p_0$, $\eta = y/L_s$ and rewrite the governing equations in dimensionless form as

$$\frac{\partial^2 P}{\partial \eta^2} - 2\frac{\partial P}{\partial T} = 0 \tag{15.8.25}$$

with $L_s = (2D/\omega)^{1/2}$. Equation (15.8.25) is the same form as Eq. (8.2.6) and the solution is the same separated variable form as Eq. (8.2.17),

$$P(\eta, T) = \Re\{e^{-\sigma\eta} e^{-iT}\} = e^{-\eta}\cos(T - \eta) \tag{15.8.26}$$

where $\sigma = 1 - i$.

If we use the Brinkman model there is an additional term in Eq. (15.8.20),

$$\frac{\partial p}{\partial t} = \left(\frac{K(\lambda + 2G)}{\mu}\right)\nabla^2 p + K\nabla^2\left(\frac{\partial p}{\partial t}\right) \tag{15.8.27}$$

Again assume the pressure boundary conditions of Eq. (15.8.24) and choose the same scalings used in Eq. (15.8.25). If we again let

$$L_s^2 = 2\left(\frac{K(\lambda + 2G)}{\mu\omega}\right) = \frac{2D}{\omega} \tag{15.8.28}$$

then Eq. (15.8.27) becomes

$$\frac{\partial^2 P}{\partial \eta^2} - 2\frac{\partial P}{\partial T} + \beta\frac{\partial^3 P}{\partial \eta^2 \partial T} = 0 \tag{15.8.29}$$

where $\beta = \mu\omega/(\lambda + 2G)$ is the ratio of a viscous fluid stress scale to the elastic stress scale.

Assuming separation of variables, try a solution of the form

$$P(\eta, T) = \Re\{\hat{P}(\eta)e^{-iT}\} \tag{15.8.30}$$

and insert Eq. (15.8.30) into Eq. (15.8.29) to find

$$\frac{d^2\hat{P}}{d\eta^2} - \hat{\sigma}^2\hat{P} = 0 \tag{15.8.31}$$

where $\hat{\sigma}^2 = -2i/(1 - i\beta) = 2(-i + \beta)/(1 + \beta^2)$. Solutions to Eq. (15.8.31) are of the form

$$\hat{P} = Ae^{\hat{\sigma}\eta} + Be^{-\hat{\sigma}\eta} \tag{15.8.32}$$

For $\hat{\sigma} = \hat{\sigma}_R + i\hat{\sigma}_I$ we find the real and imaginary parts to be

$$\hat{\sigma}_R = \frac{\left(1 + \beta^2 + \beta(1 + \beta^2)^{1/2}\right)^{1/2}}{(1 + \beta^2)^{3/4}}, \qquad \hat{\sigma}_I = -\frac{\left(1 + \beta^2 - \beta(1 + \beta^2)^{1/2}\right)^{1/2}}{(1 + \beta^2)^{3/4}} \tag{15.8.33}$$

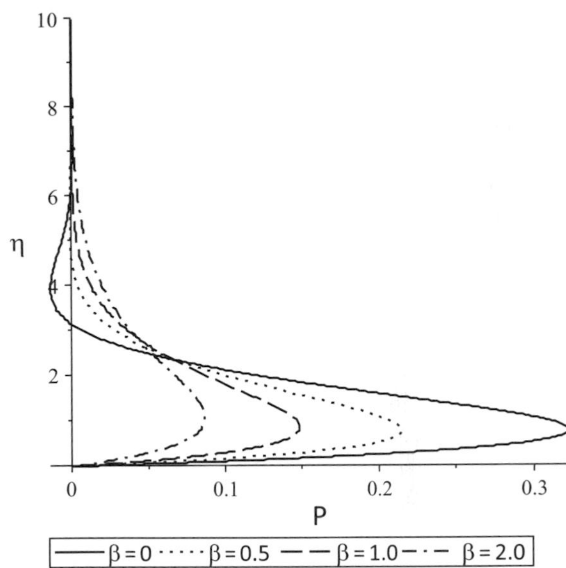

Figure 15.29 Oscillatory poroelastic pressure field using the Brinkman equation for $T = \pi/2$ for four values of $\beta = 0, 0.5, 1.0, 2.0$.

To satisfy zero pressure at $\eta \to \infty$ we must have $A = 0$ and at $\eta = 0$ we need $B = 1$, see Eq. (15.8.24). The result is

$$P = \Re\{e^{-\hat{\sigma}\eta}e^{-iT}\} = \Re\left\{e_R^{-\hat{\sigma}\eta}e^{-i(T+\hat{\sigma}_I\eta)}\right\} = e_R^{-\hat{\sigma}\eta}\cos(T+\hat{\sigma}_I\eta) \qquad (15.8.34)$$

Figure 15.29 shows the oscillatory pressure field for four values of $\beta = 0, 0.5, 1.0, 2.0$ at $T = \pi/2$. Note, for $\beta = 0$ we regain the previous result of the Darcy model, Eq. (15.8.26), since $\hat{\sigma}_R = 1, \hat{\sigma}_I = -1$. Clearly as β increases the pressure amplitude decays more rapidly as viscous effects are increasing. An example of cycled poroelastic media is cardiac tissue and hydrogel scaffolds for cardiac cells designed to mimic the strain oscillation driven by cardiac contractions (Vaughan et al., 2013). For the heart, the cycling frequency is on the order ~ 1/s. In contrast, the cycling of intervertebral discs is 1/day.

Box 15.3 Intervertebral Discs

An important example of cyclic forcing of a poroelastic material occurs in the vertebral spinal column, Figure 15.30(a), where the intervertebral discs, Figure 15.30(b), are poroelastic. These discs separate the spinal vertebrae providing cushioning and flexibility. A disc has two main structural regions: the annulus fibrosus made of fibrocartilage surrounds the gel-like nucleus pulposus. The disc is loaded during 16 hours that a person is sitting, standing, walking, and unloaded for 8 hours of sleep in a reclined position. During loading fluid is squeezed out of the disc, exiting primarily through the vertebral endplates rather than the lateral boundary of the annulus fibrosus (Nachemson et al., 1970; Ayotte et al., 2001). Then during unloading the fluid flows back in.

Figure 15.31(a) shows cake slice of a finite-element CFD domain for solving this poroelastic flow problem (Ferguson et al., 2004). To approximate the geometry of a typical lumbar segment, the disc has a diameter of 45 mm and an initial height of 12.5 mm. The central nucleus pulposus is surrounded by the annulus fibrosus and the vertebral endplates form the top and bottom boundaries. The endplates have cartilaginous and bony layers. The 3D mesh

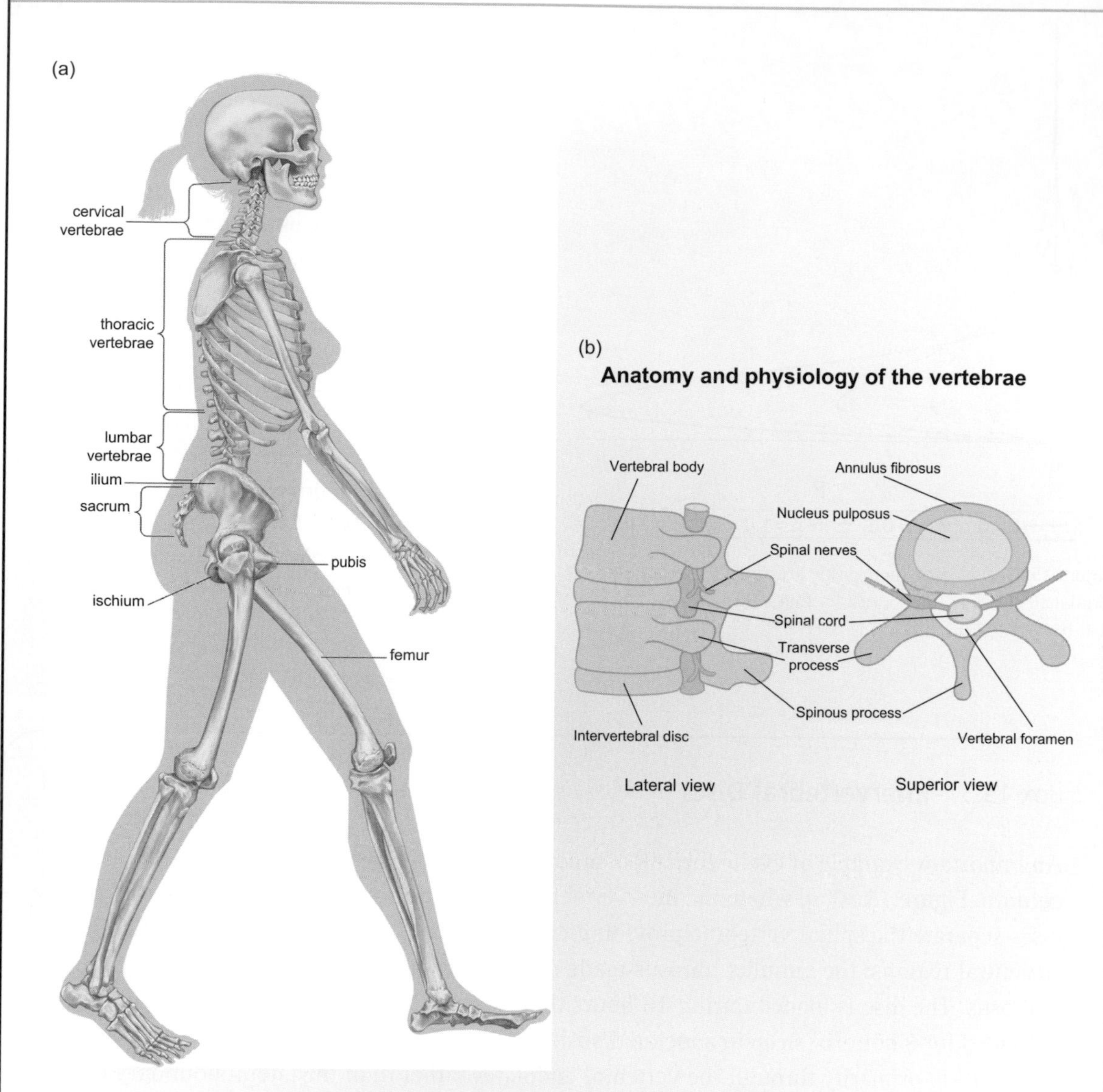

Figure 15.30 (a) Human vertebral column showing seven cervical, 12 thoracic and five lumbar vertebrae. (b) Lateral and superior views of an intervertebral disc with its annulus fibrosus and nucleus pulposus regions.

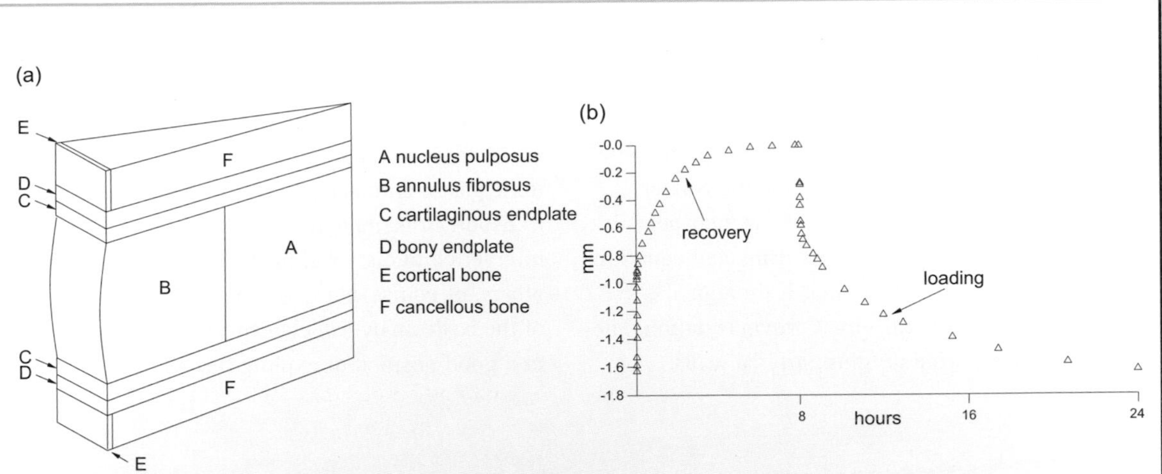

Figure 15.31 (a) A cake slice of a finite-element model of a lumbar intervertebral disc with the central nucleus pulposus surrounded by the annulus fibrosus, The vertebral endplates form the top and bottom boundaries. (b) CFD model disc height over one daily loading cycle. During the 8 hour resting period, the disc regains fully the height lost during 16 hours of loading.

consists of ~ 4,500 elements using commercial software (ABAQUS) which included a strain-dependent permeability. Figure 15.31(b) shows computed results for the 24 hour unloading–loading cycle. Approximately 11% of the total disc height is lost during loading and regained during rest. Elastic deformation of the disc accounts for approximately 37% of this change, while fluid motion accounts for the remaining 63%. Most of the height loss occurs in the first 2–3 hours.

The vertebrae are stacked, seven cervical (neck), 12 thoracic (attached to ribs) and five lumbar (lower back), Figure 15.30(a). Consequently the shrinkage is additive and accounts for our loss of height during the day which is on the order of 1 cm. This is a source of variability and error that can be clinically significant when following growth and development of children (Tillmann and Clayton, 2001). Because the disc is mostly avascular, this daily fluid oscillation has been proposed as an important mechanism for transport of slow-diffusing, large molecules (Gullbrand et al., 2015). Manipulations of the spine, like sustained extensions and flexions, can temporarily recover the spine height decrease (Owens et al., 2009), apparently promoting fluid flow back into the disc. Degeneration of the intervertebral discs is a significant clinical problem that adversely affects this fluid flow (Urban and Roberts, 2003).

Summary

In this chapter we explored a wide variety of applications for flow through porous and poroelastic media. The complexity of solving the full Navier–Stokes equations in the expanse of irregular pores is simplified by Darcy's law and the Brinkman equation along with approaches to finding K through measurements and the Kozeny–Carman equation. The applications were wide ranging: arterial walls, cancerous tumors, osteocytes, glycocalyx layer, mucociliary pumping, tree canopies, absorption of toxic gases, chromatography columns, blood-gas transport in membrane oxygenators, skin, intervertebral discs and heart muscle. There are many others, of course, but this sampling and construction of the basic analytical elements will stand the student in a good position to explore further.

Problems

15.1 Consider a tumor which exists on the semi-infinite strip $-L \le x \le L,\ -\infty \le y \le \infty,\ -\infty \le z \le \infty$. The distributed source term is given by $s = \gamma^2(p_s - p) > 0$ where p_s is a constant set point pressure. The divergence of the velocity field yields the following ODE for the pressure, p,

$$-\frac{K}{\mu}\frac{d^2p}{dx^2} = \gamma^2(p_s - p) \ \rightarrow \ \frac{K}{\mu}\frac{d^2p}{dx^2} - \gamma^2 p = -\gamma^2 p_s \tag{1}$$

a. Find the analytical solution to Eq. (1). It should consist of the homogeneous solution and the particular solution with two constants of integration.

b. Impose the following symmetric boundary conditions to solve for your two integration constants

$$\frac{dp}{dx}_{x=0} = 0,\ p(x = L) = p_L \tag{2}$$

c. Make your pressure solution dimensionless with the following scales

$$X = \frac{x}{L},\quad P = \frac{p - p_L}{p_s - p_L} \tag{3}$$

and plot P(X) for $0 \le X \le 1$ X for $\lambda = 0.5, 1, 2, 5, 10$ where $\lambda = \alpha L$ and $\alpha^2 = \mu\gamma^2/K$. Describe your results.

d. From your solution to the dimensional pressure field, find the dimensional velocity, u, in the x-direction.

e. Make your u velocity dimensionless with the following scale

$$U = \frac{u}{\frac{K}{\mu}\frac{(p_s - p_L)}{L}} \tag{4}$$

and plot U(X) for $0 \le X \le 1$ and the five values of λ from part (d). Describe your results.

15.2

a. The general solution to Brinkman flow in a cylinder is given in Eq. (15.4.7). Apply no-slip at $R = 1$ and finite velocity at $R = 0$ to solve for the constants c_1 and c_2.

b. Plot U(R) for $\alpha = 1, 2, 5, 7, 10$ and describe your results.

c. Plot the dimensionless flow rate $Q = \int_0^1 UR\,dR$ for $0.01 \le \alpha \le 100$ and describe your results. Use a log–log plot. What happens for $\alpha \to 0$? Why?

d. Plot the dimensionless wall shear, $\tau_w = (-dU/dR)_{R=1}$ wall shear for $0.01 \le \alpha \le 100$

and describe your results. Use a log–log plot. What happens for $\alpha \to 0$? Why?

15.3 Consider a spherical shell tumor which occupies the region $r_i \le r \le r_o$ with pressure boundary conditions $p(r = r_i) = p_i,\ p(r = r_o) = p_o$.

a. If there is no local source, i.e. $s = 0$, then Eq. (15.3.1) states that $r^2 u_r$ is a constant, call it c_3. Solve for u_r.
b. Insert your solution for u_r into Eq. (15.3.2) and solve for p(r). Call your integration constant c_4.
c. Impose the pressure boundary conditions to solve for c_3 and c_4. What is your solution for p(r)?
d. Use the scales $P = (p - p_o)/(p_i - p_o)$, $R = (r - r_i)/(r_o - r_i)$ and make your solution dimensionless. Let $\lambda = r_i/r_o$. What are the ranges of P, R, λ?
e. Plot P(R) for $\lambda = 0.1, 0.3, 0.5, 0.9$. Describe your result. What is the effect of λ?
f. The corresponding dimensionless velocity is $U_R = (\mu(r_o - r_i)u_r)/(K(p_i - p_o)) = -dP/dR$. Plot U_R for the same conditions in part (e). Explain your results.

15.4 For poroelastic flow using Darcy's law, the oscillatory pressure boundary condition gives $P(\eta, T) = \Re\{e^{-\sigma\eta} e^{-iT}\}$ from Eq. (15.8.26), where $\sigma = 1 - i$.

a. Plot P for $T = 0, \pi/2, \pi, 3\pi/2$ over the range $0 \le \eta \le 5$. What physical mechanisms cause the decay as $\eta \to \infty$?
b. Let the fluid velocity relative to the solid velocity be $\underline{v} = \underline{u} - (\partial \underline{U}^s/\partial t)$. Solve for the y-component, v_y, starting with Eq. (15.8.11),

$$\phi v_y = -\frac{K}{\mu}\frac{\partial p}{\partial y}$$

Scale the equation using $P = p/p_o$, $V_\eta = v_y/(p_o L_s/\mu\phi), \eta = y/L_s$ where $L_s = (2D/\omega)^{1/2}$. Let $Da = K/L_s^2$ be the Darcy number. What is V_η in terms of P?

c. Substitute P into your scaled equation in part (a) and assume separation of variables for $V_\eta = \Re\{\hat{V}_\eta(\eta) e^{-iT}\}$. What is your form for $\hat{V}_\eta(\eta)$?
d. Plot V_η at $T = \pi/2$ for $Da = 0.5, 1.0, 2.0$ over the range $0 \le \eta \le 5$. What is the effect of Da and why?

15.5

a. From Eq. (15.5.4) calculate the dimensionless flow rate in the channel, $Q = \int_0^1 U_1\, dY$. What is your form for Q?
b. What is the form of Q for Da = 0, i.e. a nonporous region in $Y < 0$? Call this Q_O. Does it depend on B?
c. Plot the relative increase in flow $(Q - Q_O)/Q_O$ vs $3 \le 1/\sqrt{Da} \le 20$ for $B = 0.8, 0.9, 1.0, 1.1, 1.2$. This is the range of parameter values which fit the experimental data in Beavers and Joseph (1967) for flow over porous blocks where the range of K was $K \sim 10^{-5} - 10^{-4}\ \text{in}^2$ and b was varied to achieve the Da range.
d. What is the overall effect of Da on your result in part (c)? What about B?

15.6 In Section 7.3 we studied a model of airway liquid clearance due to gravity drainage and ciliary motion simplified to be a tangential velocity u_w of an wall inclined at angle θ, see Figure 7.10. The velocity field was found to be $u = \rho g \sin\theta y^2/2\mu + c_1 y + c_2$ $(0 \le y \le h)$ before any boundary conditions were applied.

a. Solve u for a no shear stress at $y = h$ and then treat the wall as a slip velocity boundary condition from Eq. (15.5.3), $u(y = 0) - u_w = (1/\beta)(du/dy)_{y=0}$.
b. Make your solution in part (a) dimensionless using the scales $Y = y/h$, $U = \mu u/\rho g h^2$ and the parameter definitions $U_w = \mu u_w/\rho g h^2, B = \beta h$. What is the final form for U?
c. Plot U over the range $0 \le Y \le 1$ for $\theta = \pi/4$ and $U_w = 0.5$. Choose $\beta = 1, 3, 100$. Explain your results.
d. Calculate the dimensionless flow rate $Q = \int_0^1 U\, dY$. What is the final form?

e. Plot Q over the range $0 \le U_w \le 3$ for $B = 0.5, 1, 3, 100$. Let $\theta = \pi/4$.
f. How does B influence the efficiency of the cilia?

Image Credits

Figure 15.1a Source: Fig 3B in Johnson TF, Levison PR, Shearing PR, Bracewell DG. X-ray computed tomography of packed bed chromatography columns for three dimensional imaging and analysis. *Journal of Chromatography A*. 2017; 1487(3):108–15.

Figure 15.1b Source: Fig 5B in Johnson TF, Levison PR, Shearing PR, Bracewell DG. X-ray computed tomography of packed bed chromatography columns for three dimensional imaging and analysis. *Journal of Chromatography A*. 2017; 1487(3):108–15.

Figure 15.3a Source: Original from Darcy HPG. Les fontaines publiques de la ville de Dijon. Dalmont V, editor. Paris 1856 Plate 24, Fig. 3.

Figure 15.3b Source Simmons CT. Henry Darcy (1803–1858): Immortalised by his scientific legacy. *Hydrogeol J.* 2008; 16:1023–38. Fig. 5.

Figure 15.5 Source: Gottardi R, Hansen U, Raiteri R, Loparic M, Duggelin M, Mathys D, et al. Supramolecular organization of collagen fibrils in healthy and osteoarthritic human knee and hip joint cartilage. *PLoS One*. 2016; 11(10). Copyright: © 2016 Gottardi et al.

Figure 15.9 Source: urfinguss/ Getty

Figure 15.14a Source: U.S. National Cancer Institute's Surveillance, Epidemiology and End Results (SEER) Program (http://training.seer.cancer.gov/index.html).

Figure 15.17 Source: logika600/ Shutterstock

Figure 15.18a Source: Courtesy of Dr. E. Ebong

Figure 15.18b Source: Fig 1 from Gudis D, Zhao K-Q, Cohen NA. Acquired cilia dysfunction in chronic rhinosinusitis. *Am J Rhinol Allergy*. 2012; 26:1–6.

Figure 15.19a Source: Markfort C, Pérez ALS, Thill JW, Jaster DA, Porté-Agel F, Stefan H. Wind sheltering of a lake by a tree canopy or bluff topography. *Water Resources Research*. 2010; 46(3):W03530.

Figure 15.19b Source: Markfort C, Pérez ALS, Thill JW, Jaster DA, Porté-Agel F, Stefan H. Wind sheltering of a lake by a tree canopy or bluff topography. *Water Resources Research*. 2010; 46(3):W03530.

Figure 15.22b Source: Fig 2 of Wei HH, Waters SL, Liu SQ, Grotberg JB. Flow in a wavy-walled channel lined with a poroelastic layer. *J Fluid Mech*. 2003; 492:23–45.

Figure 15.27a Source: Fig 8a in Chan KY, Fujioka H, Bartlett RH, Hirschl RB, Grotberg JB. Pulsatile flow and mass transport over an array of cylinders: gas transfer in a cardiac-driven artificial lung. *Journal of Biomechanical Engineering – Transactions of the ASME*. 2006; 128(1):85–96.

Figure 15.27b Source: Fig 9a in Chan KY, Fujioka H, Bartlett RH, Hirschl RB, Grotberg JB. Pulsatile flow and mass transport over an array of cylinders: gas transfer in a cardiac-driven artificial lung. *Journal of Biomechanical Engineering – Transactions of the ASME*. 2006; 128(1):85–96.

Figure 15.28a Source: public domain. www.dvidshub.net/image/2527045/ellsworth-celebrates-month-military-child

Figure 15.28c Source: Sandy Aknine/ Getty

Figure 15.28d Source: Eric Harlow / Stringer/ Getty

Figure 15.30a Source: Encyclopaedia Brittanica/ Contributor/ Getty

Figure 15.30b Source: Olga Bolbot/ Shutterstock

Figure 15.31b Source: adapted from Ferguson SJ, Ito K, Nolte L-P. Fluid flow and convective transport of solutes within the intervertebral disc. *Journal of Biomechanics*. 2004; 37:213–21. With gratitude to S.J. Ferguson.

16 Interfacial Phenomena

Introduction

Many important biomedical fluid mechanics applications involve two-phase flows, e.g. a liquid and a gas or two immiscible liquids. A key feature of these flows is the boundary between the phases, called the interface. Surface tension, σ, is a force per unit length that develops at the interface and acts tangentially to it. When the interface is curved, a pressure jump arises across it which balances the forces. Drops and bubbles are static examples of this pressure jump, higher on the inside and lower on the outside. Surface tension forces do not care which fluid surrounds the other. For moving fluids, the dynamic bulk fluid momentum conservation is coupled to the interfacial mechanics through this pressure–curvature dependency. As in Chapter 9, where we dealt with flow through elastic tubes, the current chapter also focuses on a stress boundary condition, but in this case between two fluids. The coupling of surface and bulk mechanics leads to new dimensionless parameters: the capillary number, $Ca = \mu U/\sigma$, which is the ratio of viscous to surface tension effects, and the Bond number, $Bo = \rho g L^2/\sigma$, which is the ratio of gravity to surface tension effects. Since L is a characteristic length scale, the smaller the structure, the larger the influence of surface tension. We have generally ignored gravity effects in previous chapters, since the fluids were of one phase and density. An exception was Section 7.3 for a liquid layer on a moving, tilted plane to model mucociliary transport. The layer had a free surface with the atmospheric gas. For two-phase, or multiphase, flows the density differences of the involved fluids bring gravity into play.

The strength of σ varies with the chemistry of the two fluids, usually gas–liquid surface tensions are larger than liquid–liquid values. It also depends on temperature and the presence of surface active molecules called surfactants which lower σ. Soluble surfactants have a bulk concentration, C, measured in mass/volume, but also an interfacial concentration, Γ, measured in mass/area. These two concentrations affect one another through adsorption and desorption processes between the bulk and interface. Since σ generally decreases with increasing Γ, i.e. $\partial\sigma/\partial\Gamma < 0$, surfactants can significantly influence flow behavior. The strength of the surfactant to reduce surface tension brings in an additional dimensionless group, the elasticity number $E = -(\Gamma_\infty/\sigma_0)(\partial\sigma/\partial\Gamma)$, which is a dimensionless derivative of the $\sigma(\Gamma)$ relationship scaled on reference values of Γ_∞, σ_0. When Γ is distributed unevenly, an interfacial region of higher σ (lower Γ) pulls a region of lower σ (higher Γ) towards itself, dragging the underlying viscous fluid. This is called a Marangoni flow and the Marangoni number is the dimensionless group, $Ma = ECa$. Such surface tension gradients create a wide range of interesting behavior in the lung and eye, for example, including the formation of a shock wave.

Topics Covered

The chapter begins with a description of surface tension, its physical origins, contact angle with solid surfaces, and measurement. Applications include sap rising in plants, insects which walk on water, blood sampling, ocular tear drainage, and microneedles fashioned on mosquitoes sucking blood. Next we form a simple force balance for constant surface tension and static fluids, which allows us to describe drops and bubbles. We calculate the shape of drops which sit on, or hang from, a planar surface. The latter is found to have a stability limit, beyond which the drop will fall. Beads of sweat, eye drops for medicine delivery are examples, as are 3D cell cultures. Bubbles in blood and tissues can cause "the bends," while whales blow "bubble nets" to corral fish for eating. Ultrasound creates microbubbles which entrain flows potentially useful for trapping and destroying blood clots. Early work on surface tension addressed the breakup of a liquid jet into drops, which is readily seen in a dripping faucet or male urination. We perform the well-known energy calculation of this Plateau–Rayleigh instability to describe its features. Then we deal with the effects of surfactants first by describing their physic-chemical nature through adsorption isotherm models, $\Gamma(C)$, and surface equation of states $\sigma(\Gamma)$, and then through a more general interfacial stress balance with variable surface tension that surfactants create. That balance readily isolates the balance of surface tension gradients with fluid shear at the interface, the origin of Marangoni flows. A variety of applications are described for the lung, including airway closure and reopening, surfactant spreading with development a shock wave, alveolar oscillation flows and surfactant replacement therapy. In addition, infectious respiratory epidemics, involving droplets from cough and sneeze, are modeled as a timely introduction to the 2019–2020 coronavirus pandemic. Finally, we derive the kinematic boundary condition for an evolving interface using examples of a draining liquid film with applications to lung airways and capillary-gravity wave propagation at an interface.

16.1 Surface Tension
16.2 Fluid Statics: Drops and Bubbles
16.3 Breakup of a Liquid Jet
16.4 Variable Surface Tension: Surfactants
16.5 Interfacial Stress Boundary Condition for Variable Surface Tension
16.6 Respiratory Interfacial Phenomena
16.7 Respiratory Epidemics
16.8 Interfacial Kinematic Boundary Condition

16.1 Surface Tension

Surface tension exists at the interface between two immiscible fluids. For example, consider the air–water system shown in Figure 16.1(a). An interior water molecule has equal attractive (cohesive) forces acting on it from the surrounding water molecules. These result from a network of hydrogen bonding. The sum of the forces on it is zero, $\sum F = 0$. However, the surface water molecule sees an imbalance of cohesive forces, because the air above provides very little attraction compared to the water below. The resulting sum of forces is non-zero, causing a net inward force creating surface tension, σ. Surface tension behaves like an elastic membrane at the interface resisting deformation. The units of σ are force/length as shown in Figure 16.1(b), with the force vectors perpendicular to a line of length L. An alternative interpretation of σ is the surface energy per unit surface area, which explains why systems tend to seek the smallest surface area, lowest energy state.

Figure 16.2 shows a soap film attached to a frame and massless slide bar. A mass, m, is suspended from the slide bar with weight, mg, and supported by the surface tension. Both a side view and cross-sectional view are shown. Since there are two interfaces for the soap film, the force balance is $mg = 2\sigma L$ where L is the slide bar width.

The value of σ depends on the physicochemical properties of the two fluids involved and their temperature. Generally the surface tension decreases when temperature is increased. For water, $\sigma \sim 73\,\text{dyn/cm}$ at 20 °C but $\sigma \sim 70\,\text{dyn/cm}$ at body temperature, 37 °C. Blood surface tension is $\sim 58\,\text{dyn/cm}$ at body temperature. Liquid metals have much higher values, for example mercury–air is $\sim 480\,\text{dyn/cm}$. The addition of surface active molecules into the interface can modify the value of σ, generally reducing it. A very common surfactant is soap. In biofluid mechanics a major application of surface tension phenomena is the lung. The airways and alveoli of the lung are coated internally with a thin liquid layer, so there is surface tension between the resident air and this coating at their interface. Lung surfactants are produced by the Type II alveolar cells and released into the alveolar liquid lining, eventually arriving at the air–liquid interface where they can reduce σ from the air–water value to make the lungs more compliant. This is discussed later in the chapter.

When an interface is in contact with a solid surface, there is a three-phase line with a contact angle θ. Figure 16.3 shows two examples of liquid drops on a solid surface forming a contact angle. In the top case, the drop is "wetting" because $\theta < 90°$, while in the bottom case the drop is "non-wetting," $\theta > 90°$.

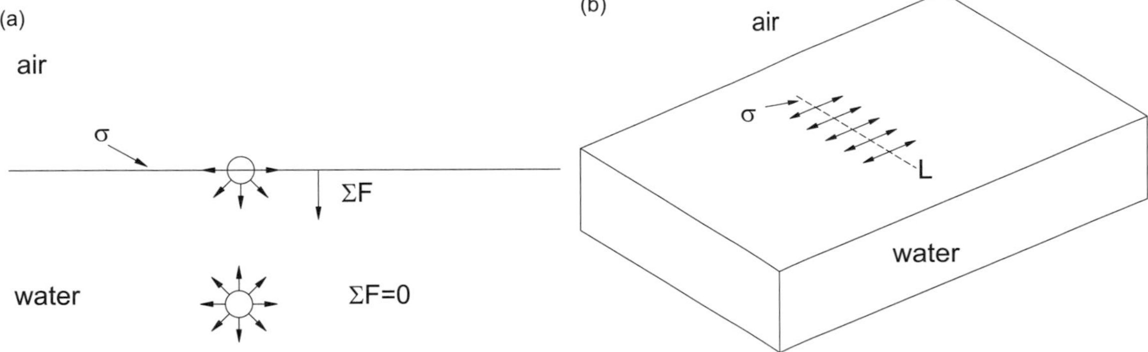

Figure 16.1 (a) Attraction forces on interior compared to interfacial water molecules. (b) Surface tension as an interfacial force per unit length, L, perpendicular to the force direction.

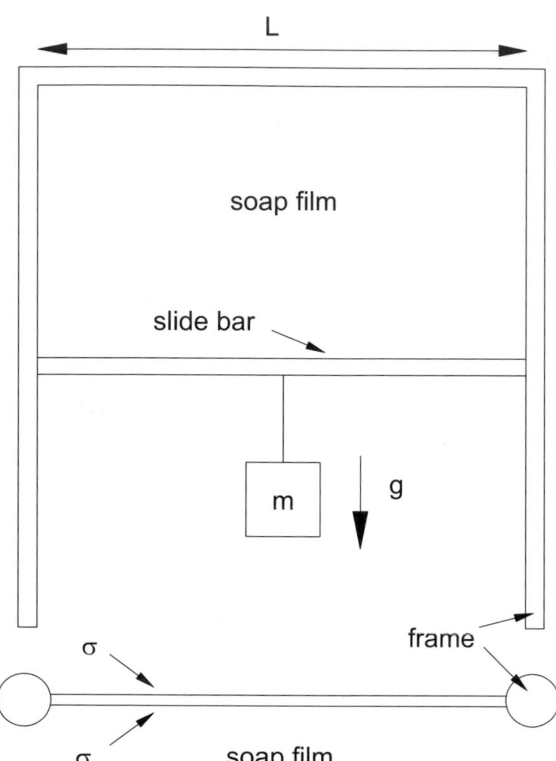

Figure 16.2 A rectangular frame with a soap film supporting a mass, viewed from the side and cross-section.

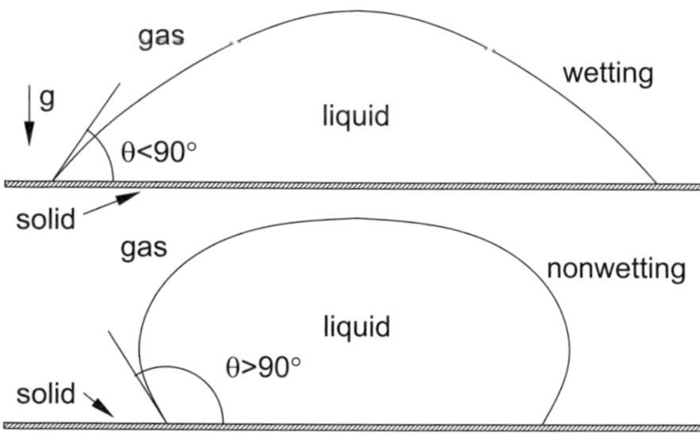

Figure 16.3 Contact angle for liquid drops on a surface, wetting vs non-wetting.

The value of θ depends on the properties of the three phases, and for any one combination, can vary within a range. For water, a surface material which wets is called hydrophilic, while one that is non-wetting is hydrophobic. Figure 16.4 shows the force balance at a contact line, which is Young's equation

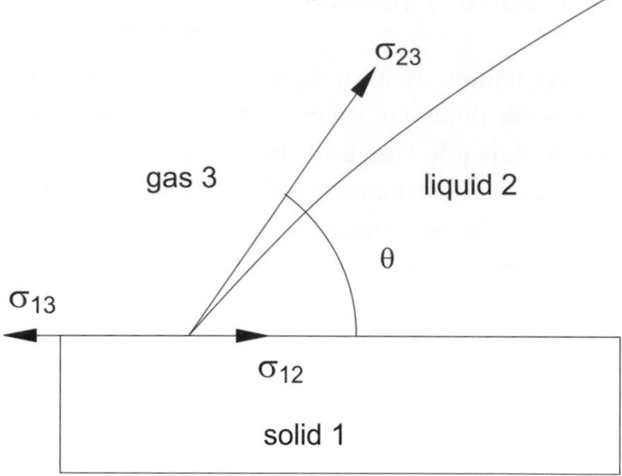

Figure 16.4 Force balance at a contact line.

$$\sigma_{13} = \sigma_{12} + \sigma_{23}\cos\theta \qquad (16.1.1)$$

where σ_{12} is the surface tension between the solid and gas, σ_{13} is the surface tension between the solid and liquid and σ_{23} is the surface tension between the air and liquid. Of course there also must be a vertical reaction at the contact line to balance $\sigma_{23}\sin\theta$.

The contact angle is an important feature of capillarity, which is the tendency for liquid to rise inside a thin tube against gravity. Figure 16.5 shows two tubes supporting liquid columns of the same density, ρ, with surface tension, σ, and contact angle, θ. The weight of a liquid column is the product of its mass and gravity, $(\rho h\pi d^2/4)g$, where h is the column height and d is the tube diameter. This weight must be supported by the vertical component of surface tension along the perimeter contact line, $\pi d\sigma\cos(\theta)$. Equating these two terms, the height, h, can be solved in terms of the remaining parameters

$$h = \frac{4\sigma\cos(\theta)}{\rho g d} \qquad (16.1.2)$$

In Figure 16.5 the diameters shown have the relationship $d_2 = d_1/2$, which causes the column

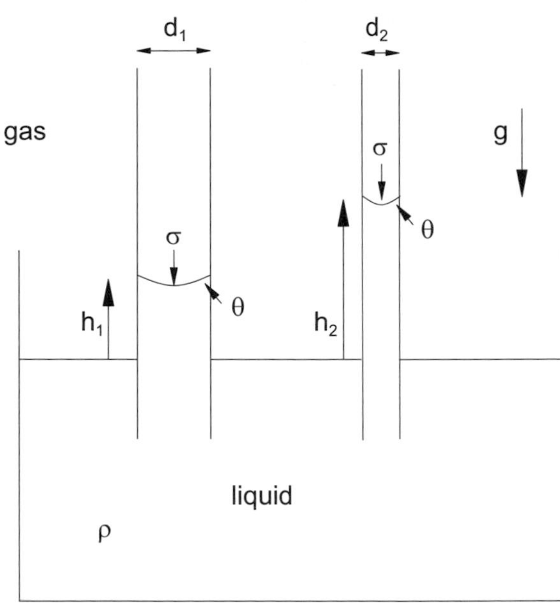

Figure 16.5 Capillarity: surface tension pulling liquid up capillary tubes against gravity.

heights to be related by $h_2 = 2h_1$. For water on clean glass, the contact angle is $\theta \sim 0$, water density is $1\ \text{g/cm}^3$, and the surface tension is $\sigma = 72\,\text{dyn/cm}$. Equation (16.1.2) then becomes $h \sim 0.3/d$ in cm using these parameter values. For example, $h = 1, 10, 100, 1000$ cm for $d = 0.3, 0.03, 0.003, 0.0003$ cm, respectively. Capillarity is an everyday occurrence in biological applications. It contributes to the rise of water (sap) in plants, from roots to leaves. Ten-metre trees are commonplace and redwoods grow to ~ 100 m. Equation (16.1.2) is also used to measure the surface tension from the column height and remaining parameter values.

Figure 16.6(a) shows a Wilhelmy balance and Figure 16.6(b) a Du Noüy ring balance. These are commonly used devices for measuring surface tension. Typically they are constructed of platinum, which is highly wettable, making the contact angle near zero, $\theta \sim 0$. In both cases $\sigma = F/L_w \cos(\theta)$ where L_w is the wetted perimeter and the force, F, is measured. Corrections are made for weight and buoyancy effects.

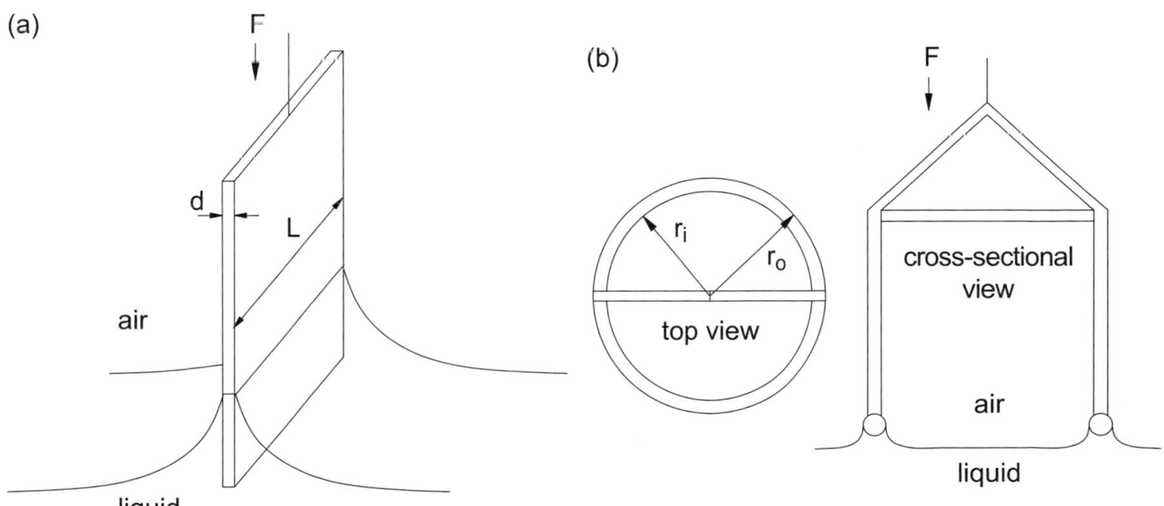

Figure 16.6 (a) Wilhelmy balance and (b) Du Noüy ring balance used for measuring surface tension.

Box 16.1 Biological Applications of Surface Tension

Figure 16.7(a) shows blood being drawn up into a thin tube by capillarity during clinical sampling. Tests run on such blood samples include glucose levels, to monitor diabetics, and hematocrit values by spinning the tube in a centrifuge. Figure 16.7(b) shows the lacrimal punctum, an opening in the lower eyelid which drains (sucks) tears from the eye surface.

Figure 16.8(a) shows a female mosquito biting a human arm. The proboscis penetrates the skin and initial blood flows can be due to capillarity, followed by active pumping. A number of diseases are spread by mosquitoes during this biting process including malaria, dengue, yellow fever, West Nile and Zika viruses, amongst others, all significant challenges to world health. Figure 16.8(b) shows a scanning electron micrograph of a titanium microneedle designed with similar dimensions as the mosquito proboscis, 25 μm inner diameter (Chakraborty and Tsuchiya, 2008). A conventional needle with 900 μm outer diameter is shown for comparison. The goal of the biomimetic microneedle is to make sampling and injections painless.

Many insects, called striders or skippers, can walk on water, see Figure 16.9(a). This is because their feet indent the interface, but do not rupture it, and surface tension plus buoyancy forces balance their weight (Bush and Hu, 2006). In many cases surface tension dominates. Figure 16.9 (b) shows that the interface meets the insect foot at angle θ. If L is the length of the foot, then the supporting force from foot one of six is $F_1 = 2\sigma L_1 \sin\theta$ since there are two contact lines. For the three pairs of feet, each pair consisting of two identical feet, the total supporting force is $2 \times (F_1 + F_2 + F_3)$. To balance the weight, mg, of the insect we must have

$$mg = 2 \times (F_1 + F_2 + F_3) = 4\sigma \sin\theta (L_1 + L_2 + L_3) \tag{16.1.3}$$

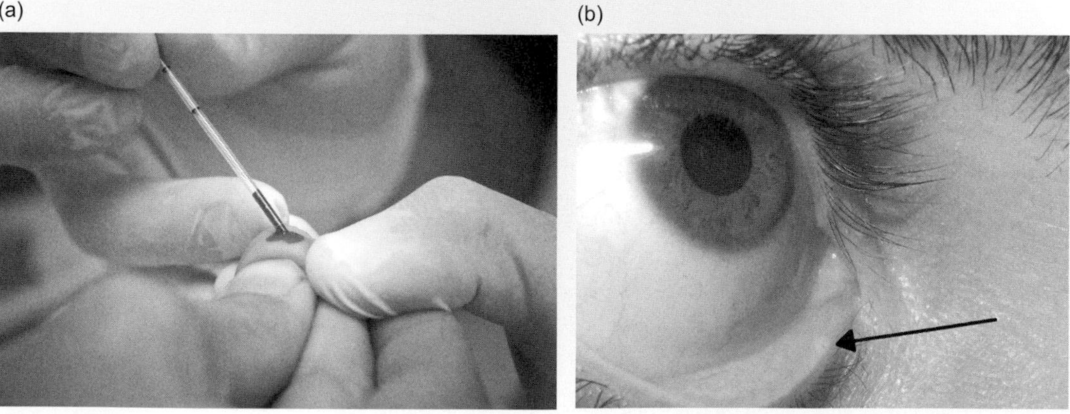

Figure 16.7 (a) Blood sampling using a capillary tube. (b) Lacrimal punctum which drains tears from the eye

(a)

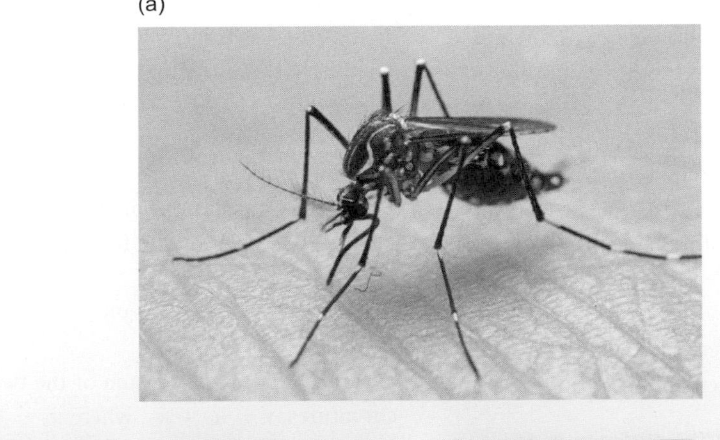

(b)

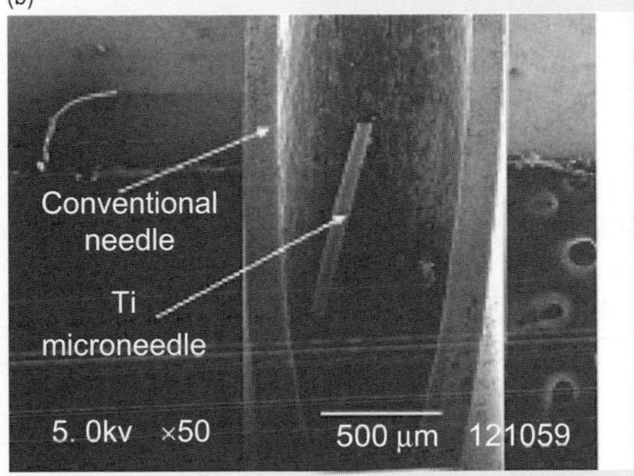

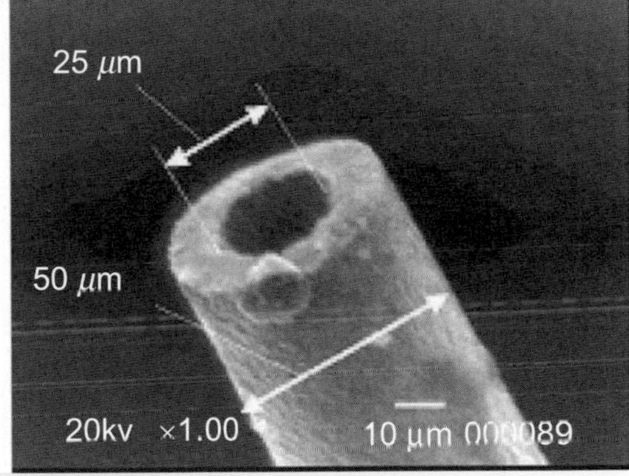

Figure 16.8 Biomimetic design: (a) mosquito (*Aedes aegypti*) sucking blood through its proboscis; (b) scanning electron microscope image of a titanium microneedle compared to a conventional needle.

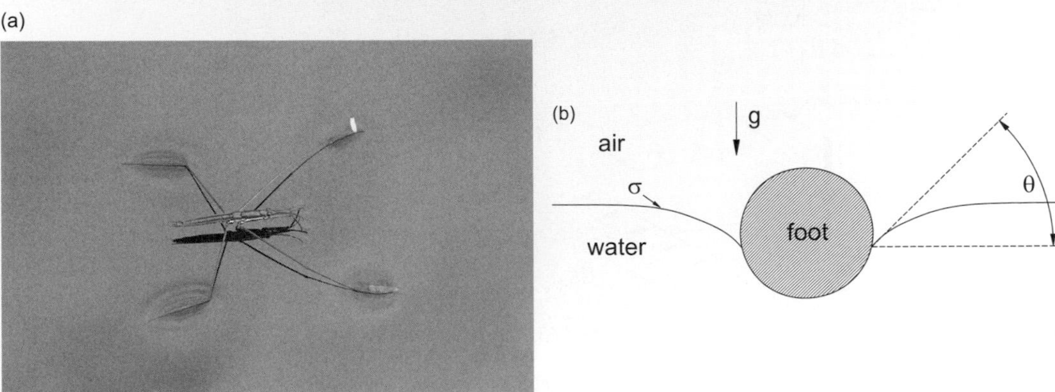

Figure 16.9 (a) A water skipper, Gerridae, Hemiptera, Sanjay Gandhi National Park, Mumbai, Maharashtra, India. (b) Surface tension supporting water strider foot in cross-section.

16.2 Fluid Statics: Drops and Bubbles

Consider the differential surface element shown in Figure 16.10(a). In the 1-direction it has a radius of curvature R_1 over the angle $d\theta_1$ with a surface curve of length $dS_1 = R_1 d\theta_1$. In the 2-direction, which is perpendicular to 1-direction, there is a radius of curvature R_2 over the angle $d\theta_2$ with a surface curve of length $dS_2 = R_2 d\theta_2$. The force balance for a static fluid consists of the pressure difference acting on the differential surface area, $(p - p_0)dS_1 dS_2$, balanced by the net surface tension effect. The vertical (normal) components of the surface tension force, in the 1- and 2-directions, are

$$\begin{aligned} 2\sigma dS_1 \sin\left(\frac{d\theta_2}{2}\right) &\sim \sigma dS_1 d\theta_2 = \sigma dS_1 \frac{dS_2}{R_2} \\ 2\sigma dS_2 \sin\left(\frac{d\theta_1}{2}\right) &\sim \sigma dS_2 d\theta_1 = \sigma dS_2 \frac{dS_1}{R_1} \end{aligned} \quad (16.2.1)$$

where we assume small angles so the sine terms are replaced by the first term of their Taylor series. The sum of the terms in Eq. (16.2.1) must balance the pressure difference

$$(p - p_0)dS_1 dS_2 = \sigma dS_1 dS_2 \left(\frac{1}{R_1} + \frac{1}{R_2}\right) \quad (16.2.2)$$

Dividing both sides of Eq. (16.2.2) by the differential area $dS_1 dS_2$ gives us the formula we are seeking

$$p - p_0 = \sigma\left(\frac{1}{R_1} + \frac{1}{R_2}\right) = \sigma\kappa \quad (16.2.3)$$

The curvature, κ, is the sum of the two principal curvatures, $\kappa = \kappa_1 + \kappa_2$, where $\kappa_1 = 1/R_1$ and $\kappa_2 = 1/R_2$. It is also twice the "mean curvature" of the surface and is equal to the divergence of the unit normal

$$\kappa = \underline{\nabla} \cdot \underline{n} \quad (16.2.4)$$

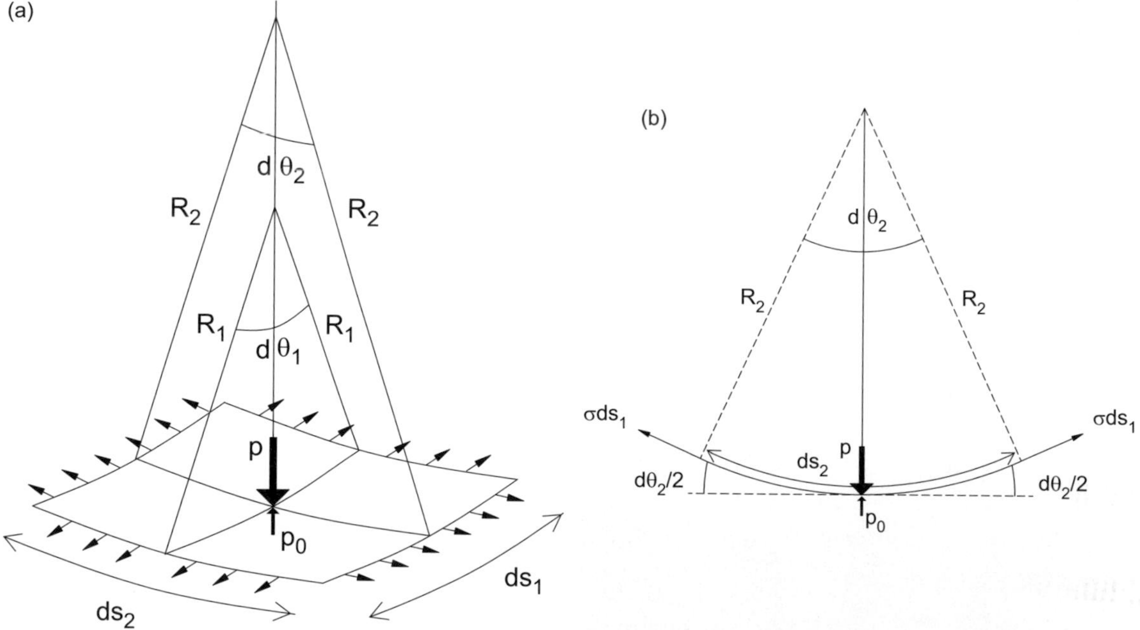

Figure 16.10 (a) Forces on a differential surface element showing the pressure difference $p - p_0$ balanced by the surface tension, σ, and the interfacial curvature. (b) Details of resolving the surface tension vertical contribution.

For a spherical bubble or drop of radius a, $R_1 = R_2 = a$, and the pressure jump across the interface, inside minus outside, is $p_i - p_o = 2\sigma/a$. For a soap bubble there is gas on the inside and outside, separated by a thin liquid film. The two interfaces have essentially the same radius of curvature, so $p_i - p_o = 4\sigma/a$. The smaller the radius, the larger the pressure jump. Equation (16.2.3) is called the Young–Laplace equation after Thomas Young (1773–1829), an English physicist, mathematician and physician, and for Pierre-Simon Laplace (1749–1827), a French mathematician, physicist and astronomer. Often it is represented as

$$\Delta p = -\sigma \underline{\nabla} \cdot \underline{n} \tag{16.2.5}$$

with the convention that the difference operator is $\Delta p = p_{+n} - p_{-n}$. In Eq. (16.2.3) this convention would be $\Delta p = p_o - p$.

Sessile Drop

Liquid drops sitting on a planar surface are important aspects of biofluid mechanics. In technical applications, the sessile drop can be imaged and the contact angle measured. Whether it is a leaf or the skin, drops are involved in transport of heat and mass. For example, the skin provides an essential cooling function through beads of sweat and its evaporation, see Figure 16.11(a). Other sources of liquid drops on the skin range from rain to fog to mist cooling stations and humidifiers. Toxic drops are encountered in a variety of situations ranging from fires involving petroleum products to military and homeland security exposures such as blistering agents (e.g. sulfur mustard, nitrogen mustard) and nerve agents (e.g. sarin, VX) amongst others. While respiratory intake is a considerable risk, the skin can also be a route of toxin delivery.

(a)

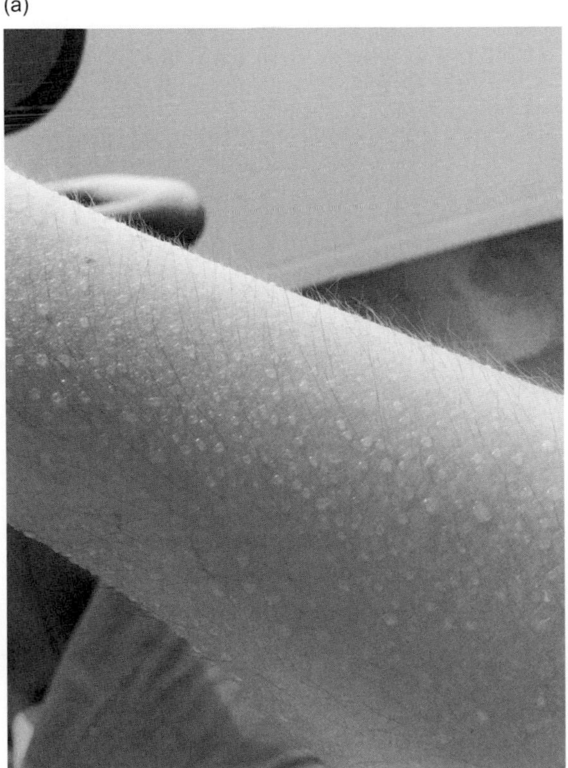

(b)

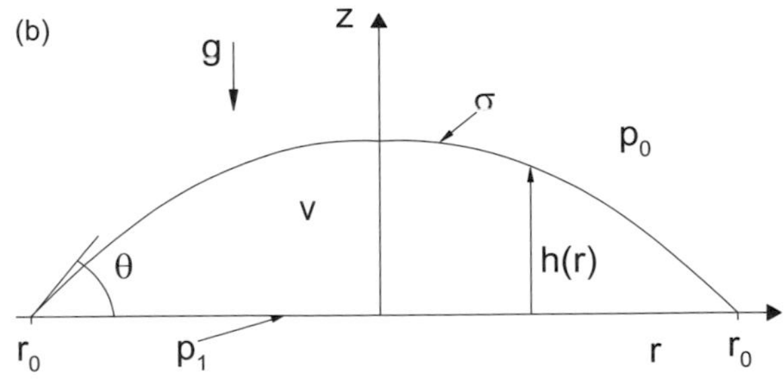

Figure 16.11 (a) Perspiration on the arm. (b) Axisymmetric drop sitting on a horizontal plane surface.

The mechanisms and rates of the drop effects include its volume and also its contact area. The shape of a drop can be calculated from Eq. (16.2.5). For an axisymmetric drop of height h(r), see Figure 16.11(b), the location of the interface is $F = z - h(r) = 0$ in cylindrical coordinates. As we saw in Eq. (11.2.1) the unit normal is given by

$$\underline{n} = (n_r, n_\theta, n_z) = \frac{\nabla F}{|\nabla F|} = \frac{\left(-\frac{dh}{dr}, 0, 1\right)}{\left(1 + \left(\frac{dh}{dr}\right)^2\right)^{1/2}} \tag{16.2.6}$$

The curvature is calculated from $\kappa = \underline{\nabla} \cdot \underline{n}$ and inserting Eq. (16.2.6) gives us

$$\kappa = \underline{\nabla} \cdot \underline{n} = \frac{1}{r}\frac{d}{dr}(r n_r) = \frac{-\frac{d^2h}{dr^2} - \frac{1}{r}\frac{dh}{dr}\left(1 + \left(\frac{dh}{dr}\right)^2\right)}{\left(1 + \left(\frac{dh}{dr}\right)^2\right)^{3/2}} \tag{16.2.7}$$

Assuming small slope, $|dh/dr| << 1$, we can ignore terms of $O\left((dh/dr)^2\right)$ or smaller. Then Eq. (16.2.7) becomes

$$\kappa = \underline{\nabla} \cdot \underline{n} = -\frac{d^2h}{dr^2} - \frac{1}{r}\frac{dh}{dr} \tag{16.2.8}$$

The hydrostatic fluid pressure in the drop is $p(z) = p_1 - \rho g z$ where p_1 is the constant pressure at $z = 0$. Evaluating this form at the interface, $z = h(r)$, Eq. (16.2.5) becomes

$$\frac{d^2h}{dr^2} + \frac{1}{r}\frac{dh}{dr} - \frac{\rho g}{\sigma}h = -\frac{(p_1 - p_0)}{\sigma} \tag{16.2.9}$$

where p_0 is the uniform air pressure. The homogeneous part of Eq. (16.2.9) is a modified Bessel's equation of zero order whose solutions are the linear combination of Bessel functions, while the particular solution is clearly $h_p = (p_1 - p_0)/\rho g$. The total solution, then, is

$$h(r) = c_1 I_0(r/\ell_c) + c_2 K_0(r/\ell_c) + h_p \tag{16.2.10}$$

where $\ell_c = (\sigma/\rho g)^{1/2}$ is the capillary length scale. For a water drop, $\ell_c = (72/1 \cdot 980)^{1/2} = 0.27\,\text{cm}$.

In Eq. (16.2.10), I_0 and K_0 are the zero-order modified Bessel functions of the first and second kind, respectively. Their derivatives arise in the analysis. They are $dI_0/dx = I_1(x)$ and $dK_0/dx = -K_1(x)$ where I_1 and K_1 are the first-order modified Bessel functions of the first and second kind, respectively. Figure 16.12 shows $I_0(x)$, $K_0(x)$, $I_1(x)$, $K_1(x)$. We will want to impose the symmetry condition that $(dh/dr)_{r=0} = 0$. That forces the constant $c_2 = 0$, since both $K_0(x), K_1(x) \to \infty$ as $x \to 0$. We can impose a second boundary condition that $h(r = r_0) = 0$, which solves for c_1. The resulting form is

$$h(r) = h_p\left(1 - \frac{I_0(r/\ell_c)}{I_0(r_0/\ell_c)}\right) \tag{16.2.11}$$

In a typical problem, we would know ρ, σ, g and can assume the surrounding atmospheric gas pressure is zero, $p_0 = 0$. The remaining dimensional parameters, p_1 and r_0, are not fundamental to the problem design, however. Instead, the contact angle, θ, and drop volume, v, would be experimental inputs. So we would like to solve p_1, r_0 in terms of θ, v.

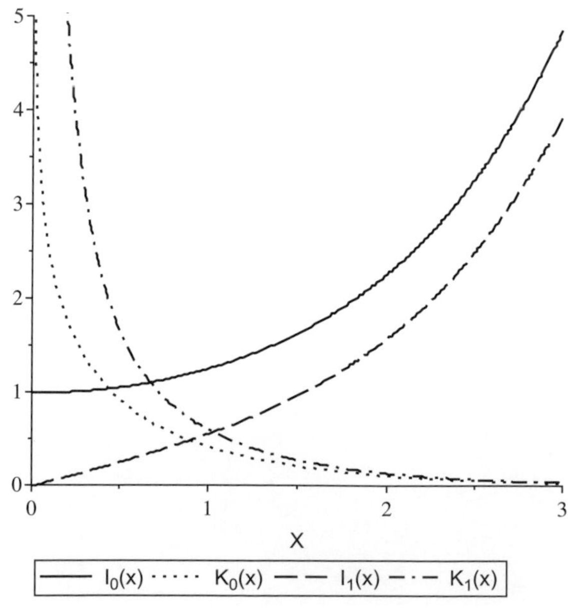

Figure 16.12 Modified Bessel functions $I_0(x)$, $K_0(x)$, $I_1(x)$), $K_1(x)$.

The contact angle condition is

$$(dh/dr)_{r=r_0} = -\tan(\theta) \quad \rightarrow \quad h_p = \ell_c \tan(\theta)\frac{I_0(R_0)}{I_1(R_0)} \tag{16.2.12}$$

which solves for h_p in terms of the dimensionless drop radius, $R_0 = r_0/\ell_c$. The assumption of small slope starts with Eq. (16.2.12) where $\tan(\theta) << 1$, which implies that $\theta << 1$. Specifying a drop volume, v, it is calculated from the integral

$$v = 2\pi \int_0^{r_0} h(r)\, r\, dr \tag{16.2.13}$$

A property of Bessel functions is that

$$\frac{d}{dx}(x^n I_n(x)) = x^n I_{n-1}(x) \quad \rightarrow \quad \int x I_0(x)\, dx = x I_1(x) \tag{16.2.14}$$

Applying Eq. (16.2.14) to Eq. (16.2.13) we obtain the expression for v

$$v = 2\pi \int_0^{r_0} h(r)\, r\, dr = \pi R_0^2 \ell_c^2 h_p \left(1 - \left(\frac{2}{R_0}\right)\frac{I_1(R_0)}{I_0(R_0)}\right) \tag{16.2.15}$$

Substituting for h_p from Eq. (16.2.12) and defining a dimensionless volume as $V = v/\ell_c^3$, Eq. (16.2.15) becomes

$$V = \pi R_0 \tan(\theta)\left(\frac{R_0 I_0(R_0)}{I_1(R_0)} - 2\right) \tag{16.2.16}$$

Figure 16.13(a) shows plots of Eq. (16.2.16), V vs R_0, for three choices of small contact angle, $\theta = 0.1, 0.2, 0.3$. Figure 16.13(b) shows the relationship between V and the dimensionless surface area defined as $A = \pi r_0^2/\ell_c^2 = \pi R_0^2$, also for $\theta = 0.1, 0.2, 0.3$. For any volume, V, the surface area covered is larger when θ is smaller.

The Bond number is the ratio of gravity to surface tension effects, defined as $Bo = \rho g L_s^2/\sigma = L_s^2/\ell_c^2$ where L_s is a characteristic length scale. For the drop problem, we can choose $L_s = v^{1/3}$, the cube root of the drop volume, so $Bo = v^{2/3}/\ell_c^2$, which makes $V = Bo^{3/2}$. Systems with $Bo << 1$ are dominated by surface tension. For air–water interfaces that means $L_s << 0.27$ cm, a size range which is prevalent in the liquid-coated small airways and alveoli of the lung. Indeed, surface tension has a strong influence on lung

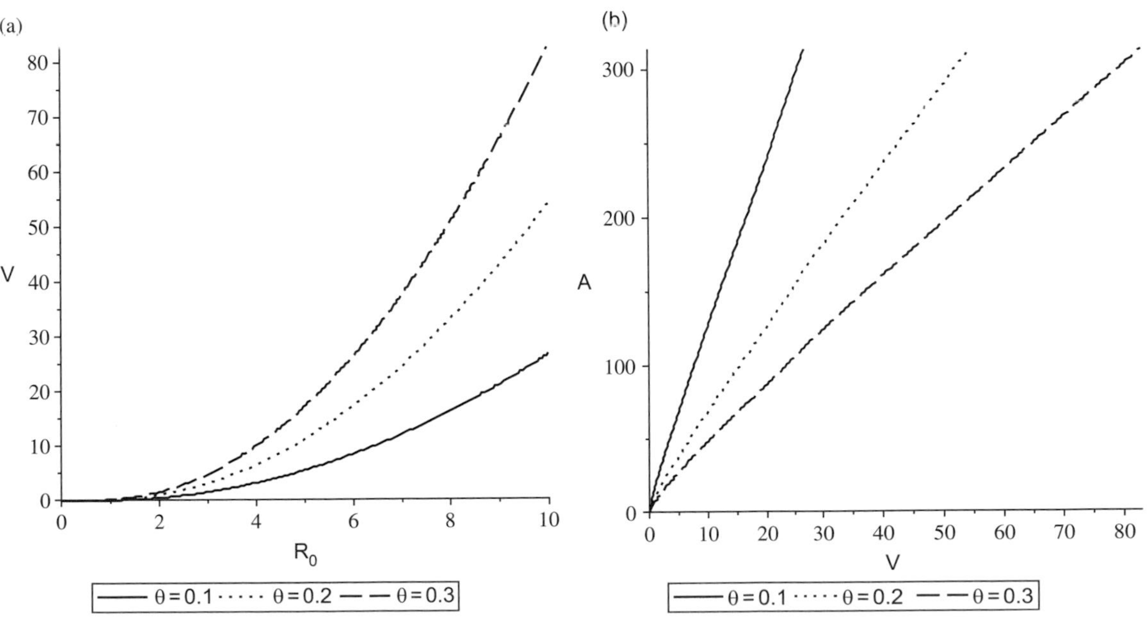

Figure 16.13 (a) V vs R_0 and (b) A vs V for three values of $\theta = 0.1, 0.2, 0.3$.

mechanical function. On the other hand, gravity dominates for Bo>>1. The features of this drop problem are controlled by the two dimensionless parameters, Bo and θ.

Scaling the interface position as $H = h/\ell_c$, and the radial variable as $R = r/\ell_c$, in Eq. (16.2.11), with h_p substituted from Eq. (16.2.12), the dimensionless height is

$$H(R) = \frac{\tan(\theta)}{I_1(R_O)}(I_0(R_O) - I_0(R)) \tag{16.2.17}$$

As we see in Figure 16.12, the denominator $I_1(R_O)$ has no zeroes for $R_O > 0$, so H(R) is a well-behaved function of parameters θ, R_O.

Figure 16.14(a) shows H(R) for $\theta = 0.2$ and $V = 0.2, 1, 3, 15, 50$, where we note the difference in the axes ranges, $0 \leq H \leq 0.22$, $0 \leq R \leq 10$. The corresponding values of $R_O = 1.1, 1.93, 2.91, 5.61, 9.62$, respectively, were determined from the root of Eq. (16.2.16). The Bond numbers are $Bo = V^{2/3} = 0.34, 1, 2.08, 4.64, 13.57$, respectively. Increasing the volume for the lower values of V is accomplished by increases in the height and width of the drop. For larger V, however, additional volume primarily widens the drop. Figure 16.14(b) shows H(R) for $V = 10$, $Bo = 4.64$, and $\theta = 0.1, 0.2, 0.3$ where, again, the values of $R_O = 6.39, 4.72, 3.96$ are the roots of Eq. (16.2.16). As θ increases, the height increases and the width decreases, as we saw in Figure 16.13.

Hanging Drop

Now let's turn our attention to a hanging or pendant drop. The setup is the same as in Figure 16.11, except g is replaced by −g and points in the +z-direction. With that sign change, Eq. (16.2.9) becomes Bessel's equation of zero order, no longer the modified Bessel's equation. The solution for h is

$$h = c_3 J_0(r/\ell_c) + c_4 Y_0(r/\ell_c) - h_p \tag{16.2.18}$$

where J_0, Y_0 are the zero-order Bessel functions of the first and second kind, respectively. Their derivatives are $dJ_0(x)/dx = -J_1(x)$ and $dY_0(x)/dx = -Y_1(x)$ where J_1, Y_1 are the first-order Bessel functions of the first and second kind, respectively, see Figure 16.15.

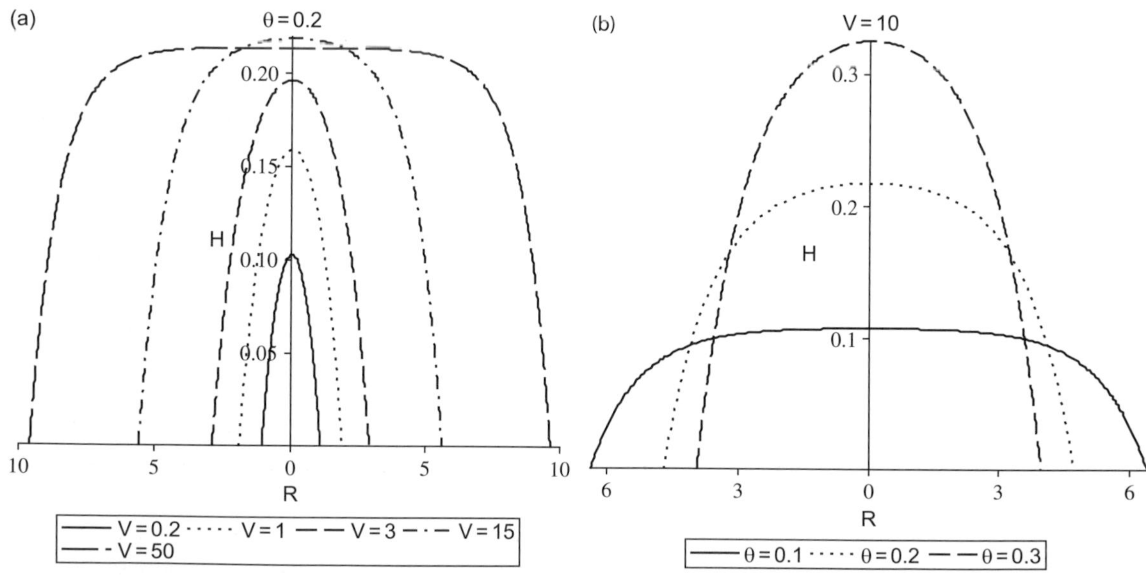

Figure 16.14 (a) H(R) for $\theta = 0.2$ and $V = 0.2, 1, 3, 15, 50$; (b) H(R) for $V = 10$ and $\theta = 0.1, 0.2, 0.3$.

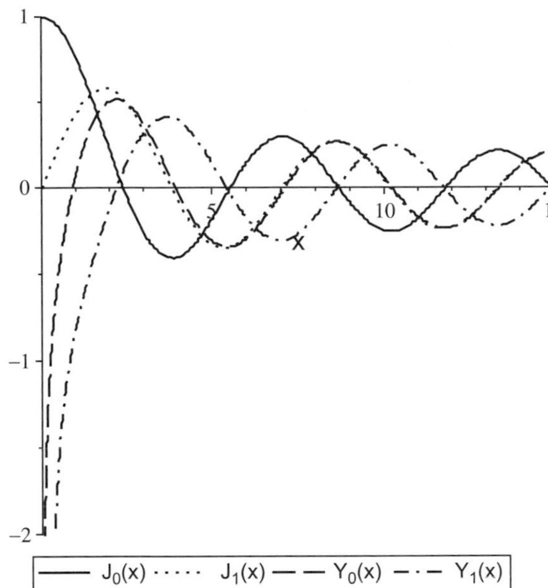

Figure 16.15 Bessel functions $J_0(x)$, $Y_0(x)$, $J_1(x)$, $Y_1(x)$.

Symmetry of $h_r(r = 0) = 0$ requires $c_4 = 0$ since both Y_0, Y_1 are unbounded at r = 0. Imposing $h(r = r_0) = 0$ solves for c_3 and the resulting form is

$$h = h_p\left(\frac{J_0(r/\ell_c)}{J_0(r_0/\ell_c)} - 1\right) \tag{16.2.19}$$

Imposing the contact angle, Eq. (16.2.12) and the volume constraint, Eq. (16.2.13), leads to the solution for h_p,

$$h_p = \ell_c \tan(\theta)\frac{J_0(R_0)}{J_1(R_0)} \tag{16.2.20}$$

and the dimensionless volume

$$V = \pi R_0^2 \tan(\theta)\left(\frac{2}{R_0} - \frac{J_0(R_0)}{J_1(R_0)}\right) \tag{16.2.21}$$

Using the same scales, H(R) is

$$H(R) = \frac{\tan(\theta)}{J_1(R_0)}(J_0(R) - J_0(R_0)) \tag{16.2.22}$$

Compared to the sitting drop, the new feature for the hanging drop is that the solution is based on oscillatory functions, as demonstrated in Figure 16.15. In the numerator, $J_0(R) - J_0(R_0) > 0$ for $0 \leq R \leq R_0$ until the volume, V, is large enough so that R_0 exceeds the value at the first local minimum, where $dJ_0(R)/dR = 0$. Since $dJ_0(R)/dR = -J_1(R)$, this minimum occurs at first zero of J_1, $R_0 = 3.83$. The denominator is zero at this value so H becomes unbounded. So there is a critical V where there is no solution to H, and the drop is unstable. Before we reach that value of V, however, H becomes very large compared to R_0 and we violate our small slope assumption, $|dH/dR| << 1$. Nonlinear curvature terms must be retained in that limit.

In Figure 16.16 we see H(R) for the hanging drop plotted with varying V and θ. Figure 16.16(a) shows H for $\theta = 0.2$ and $V = 0.1, 1, 5, 10$, which corresponds to $R_0 = 0.85, 1.76, 2.71, 3.11$ and $Bo = V^{2/3} = 0.22, 1, 2.92, 4.64$, respectively. Increasing V increases both H and R_0. For the largest value shown, V = 10, the aspect ratio is just under 1/3, which is essentially at the limit of our small slope assumption. In this R–Z plane profile, the curvature $d^2H/dR^2 = 0$ at R = 1.84. So the V = 0.1, 1 curves show only one curvature while the V = 5, 10 curves show the change in curvature, more evident in the V = 10 case. Figure 16.16 (b) shows H(R) for V = 1 and $\theta = 0.1, 0.2, 0.3$, which yields $R_0 = 2.15, 1.76, 1.54$ respectively, and $Bo = V^{2/3} = 1$. As in the sessile drop, increasing θ decreases R_0 while increasing H.

What is the critical drop volume which causes it, or part of it, to fall? That question has been studied since Tate (1864) performed experiments on fluids slowly dripping from the bottom end of a vertical tube. Such a fluid system with modernized equipment is now called a stalagmometer, see Figure 16.17. He found a linear relationship, Tate's law, between the weight of a fallen drop, w, and the tube diameter, d. Typically many drops are collected and their average weight calculated. His equation, w = 5.5 d, reasonably fit the data, where w is in units of grains and d in units of inches. The constant 5.5, therefore, is in units of grains/in where 1 grain/in = (0.065 g × 980 cm·s^{-2})/(2.54 cm) = 25.1 dyn/cm. For the purpose of measuring surface tension, Tate's law assumes that, at the moment the drop detaches, the only forces are its weight and the surface tension from the interface attached to the end of the tube acting vertically.

That assumption leads to $w = \rho g v = \pi d\sigma$, making Tate's coefficient $\pi\sigma = 5.5 \times 25.1 = 138$, or $\sigma = 44$ dyn/cm. That value is far from the expected range ~70 dyn/cm range.

As shown in Figure 16.17(a), the evolution of the falling drop, a–d, leads to detachment at e. Clearly this is much more complicated than Tate's law could encompass, given the technology of 1864. So, in

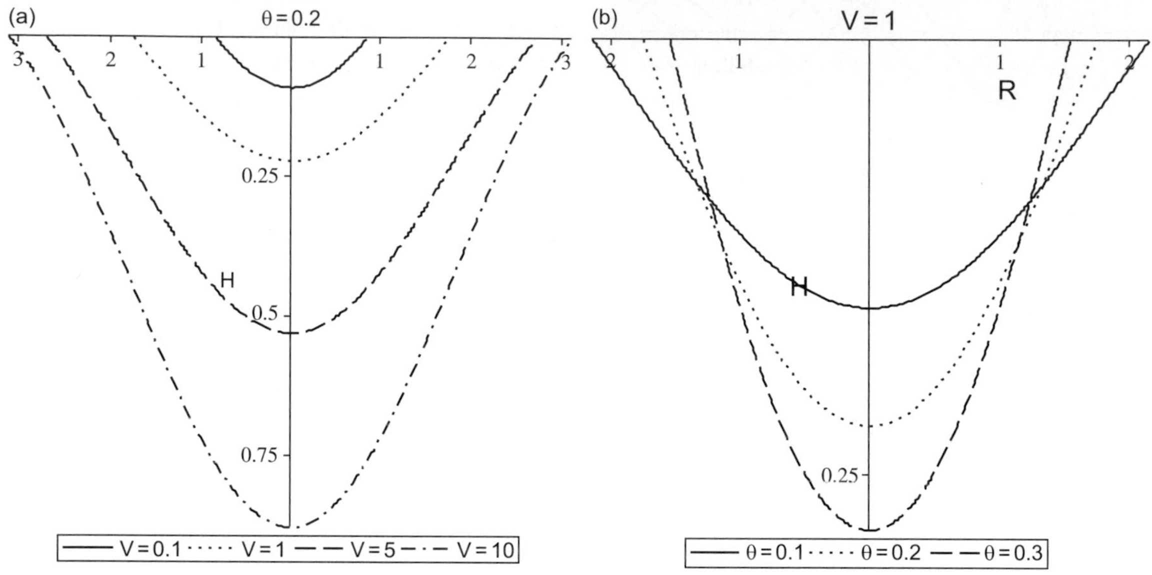

Figure 16.16 Hanging drop shapes for (a) $\theta = 0.2$ and $V = 0.1$, 1, 5, 10; and (b) $V = 1$ and $\theta = 0.1$, 0.2, 0.3.

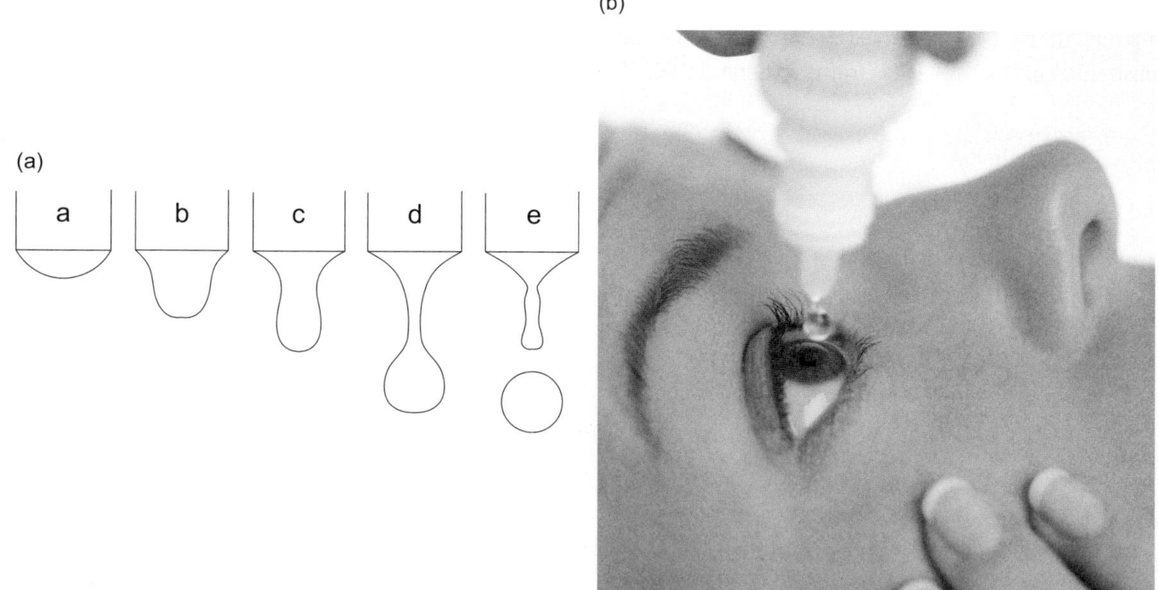

Figure 16.17 (a) Stalagmometer showing evolution of a falling drop, a–d, to the detached drop e. (b) Topical application of medicine by eye drops.

1919 Harkins and Brown sought a correction to Tate's law (Harkins and Brown, 1919). They used the capillary rise method to establish the surface tension of various fluids, and then applied Tate's law with a correction factor, $w = 2\pi r\sigma\psi\left(r/v^{1/3}\right)$. The correction factor ψ depends on $r/v^{1/3}$, the dimensionless ratio of the tube radius to the cube root of the fallen drop volume, which is a shape factor. They determined ψ experimentally over a range $0 \leq r/v^{1/3} \leq 1.6$ in steps of 0.05. Their recommendation for best accuracy is the range $0.759 \leq r/v^{1/3} \leq 0.95$, where the results were $\psi = 0.6032, 0.6000, 0.5992, 0.5998, 0.6034$ for $r/v^{1/3} = 0.75, 0.80, 0.85, 0.90, 0.95$, respectively. A sample data point from Tate (1864) is $d = 0.25$ inch and $w = 1.36$ grains. For those values, $r/v^{1/3} = 0.698$ and $\psi(0.70) = 0.61$ in Harkins and Brown (1919).

Now Tate's data can be corrected to $\sigma = 138/\pi{\cdot}0.61 = 72$ dyn/cm, which is spot on. There are more modern investigations of the falling drop as a way to measure surface tension. The analysis in Chesters (1977) forms a fundamental understanding, while computational methods used to determine σ from the shape of the pendant drop, e.g. c in Figure 16.17(a), are popular (Rotenberg et al., 1983). Figure 16.17(b) shows a common use of drops to deliver topical medication to the eye. The delivered dose depends on the drug concentration, drop volume, number of drops per treatment and treatment frequency. Typical uses include antibiotics, anti-inflammatories, lubricants for wearing contact lenses and glaucoma medications.

Box 16.2 3D Cell Culture in Hanging Drops

A hanging drop can be used to provide a 3D cell culture environment. Such a system can provide advantages over 2D cell cultures on plates/dishes or those sandwiched into layers of extracellular matrix for 3D effects. Figure 16.18 shows various stages of such a system designed for high throughput (Tung et al., 2011). Figure 16.18(a) is an illustration of the 384 hanging drop spheroid culture array plate, and its cross-sectional view. Figure 16.18(b) shows the key dimensions of the array plate. Figure 16.18(c) shows the hanging drop formation process in the array plate. The pipette tip is first inserted through the access hole to the bottom surface of the plate, and cell suspension is subsequently dispensed. The cell suspension is quickly attracted to the hydrophilic plate surface and a hanging drop is formed and confined within the plateau. Within hours, individual cells start to aggregate and eventually form into a single spheroid after around 1 day. Figure 16.18(d) is a photo of the 384 hanging drop array plate operated with a liquid handling robot capable of simultaneously pipetting 96 cell culture sites. Figure 16.18(e) is a drawing of the final humidification chamber used to culture 3D spheroids in the hanging drop array plate. The 384 hanging drop array plate is sandwiched between a 96-well plate filled with distilled water and a standard-sized plate lid. Distilled water from the bottom 96-well plate and the peripheral water reservoir prevent serious evaporation of the small volume hanging drops. Using this platform, the investigators found that drugs with different modes of action produce different responses in the physiological 3D cell spheroids compared to conventional 2D cell monolayers. To analyze

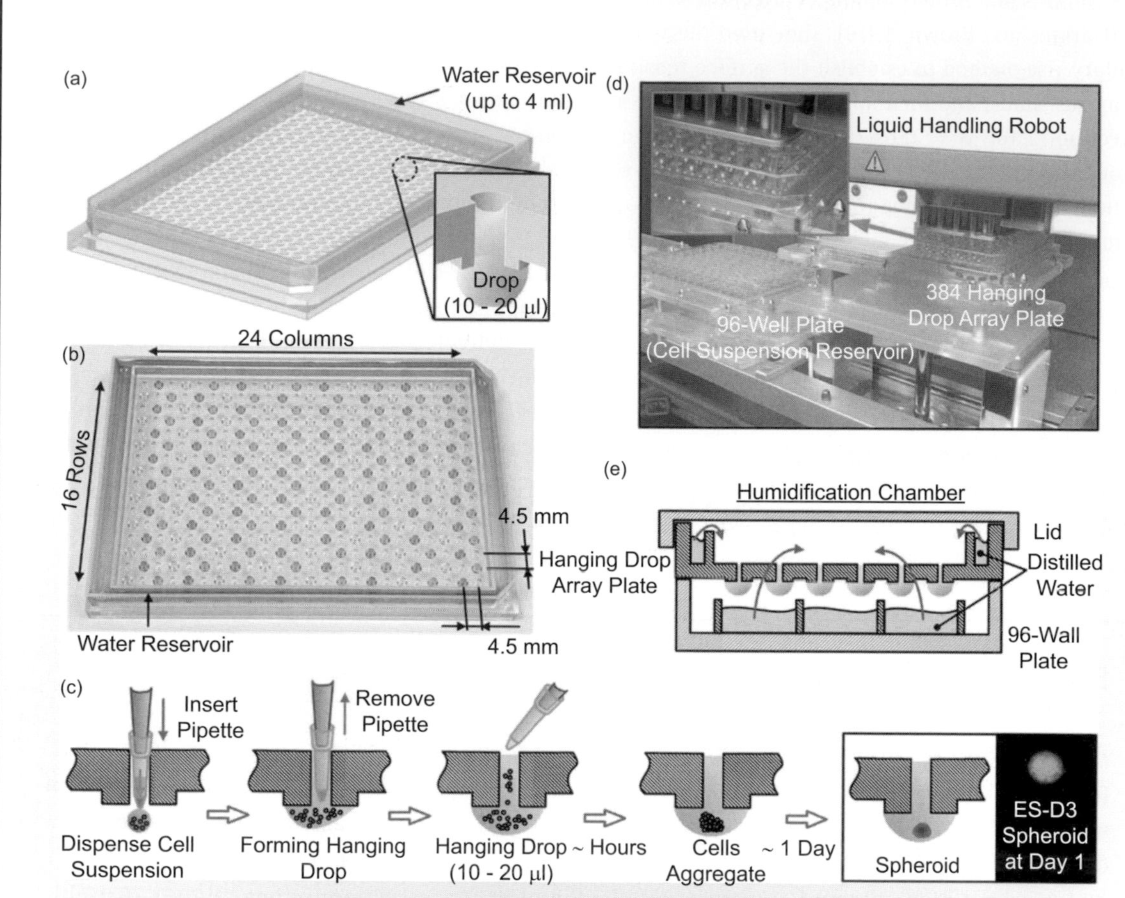

Figure 16.18 Hanging drops for cell 3D cell cultures.

the diffusive transport of gases, nutrients, signaling and other bioactive molecules, the drop shapes calculated in Eq. (16.2.22) form the boundary for the contained cells and conditions on diffusive transport.

Bubbles

Bubbles in biological flows can be a problem. For example, air embolism in the blood stream can damage the endothelium and also create a blockage when the air bubble wedges into a vessel of similar size. However, bubbles can also be harnessed for therapy as purposeful emboli to block blood flow locally in a cancerous tumor and destroy it, a process known as gas embolotherapy (Bull, 2007; Qamar, 2012).

Decompression sickness occurs when dissolved gases come out of solution to form bubbles inside the body. Decompression occurs most often when divers ascend from the high pressures underwater back to atmospheric pressures, giving it the name "divers disease." Bubbles in tissue can cause the diver to bend over in pain, called "the bends." Construction or repair of a bridge, pier, dam or ship can involve the use of a caisson, which is a structure that seals off the surrounding water using a pressurized air supply to prevent it flowing in from below, see Figure 16.19(a). These workers can also experience decompression sickness, giving it the name "caisson disease." Preventive measures include decompression in controlled smaller steps and treatment can include use of a hyperbaric chamber where the air pressure is

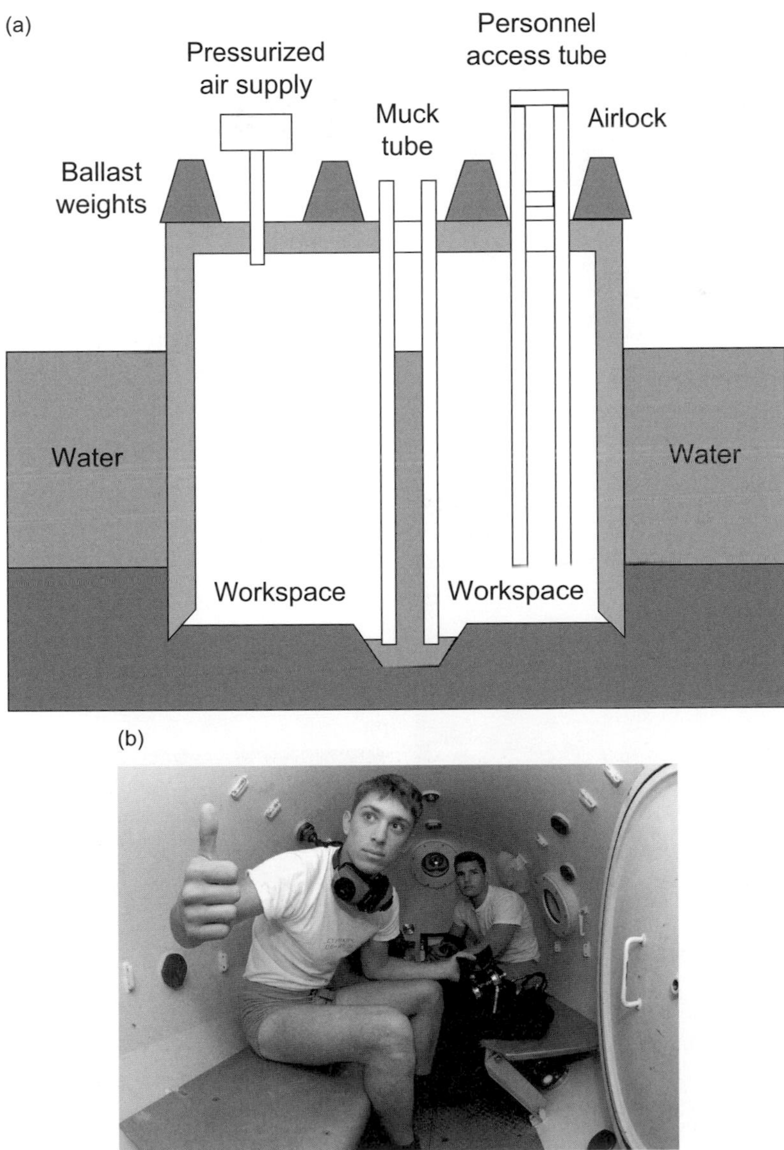

Figure 16.19 (a) Structure of a caisson showing pressurized air to keep the surrounding water from entering from below. (b) A hyperbaric chamber for slow decompression. Two United States Navy sailors prepare for training inside a decompression chamber.

increased above atmospheric and then slowly reduced, see Figure 16.19(b). The fluid mechanics and dissolution of these bubbles is investigated in Hlastala and Farhi (1973) and Hlastala and van Liew (1975).

Humpback whales use "bubble-nets" to herd prey fish, like herring, into crowded groups. The whale starts deep and swims toward the surface in a spiral blowing bubbles around the prey, see Figure 16.20. The herring will not swim through the bubbles which surround them. Then the whale takes a big mouthful, rich in protein.When a liquid experiences sufficiently low pressure, it can develop cavitation bubbles. These bubbles often occur on ship propellers, significantly damaging their surface. However, this process can be created purposefully in biomedical applications. High intensity focused ultrasound directed into tissues is used to create cavitation bubbles and destroy the local structure. This process is known as histotripsy, and applications in animal preparations include treatment of benign prostatic hyperplasia, and tumors of the liver and kidney (Khokhlova et al., 2015). When directed into fluids, they create a flow from the motion of the cavitation bubbles entraining the fluid. By controlling the flow, useful velocity fields have been explored to trap blood clots in an artery or vein by local vortical flow (Maxwell et al., 2014). Figure 16.21 shows a visualization of flow in a blood vessel phantom from crossflow (right to

Figure 16.20 Humpback whale creating a "bubble-net" to herd fish for eating.

(a)

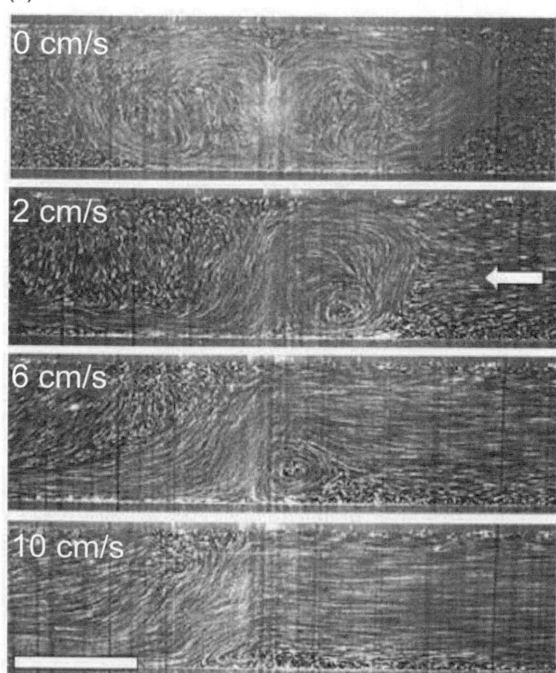

(b)

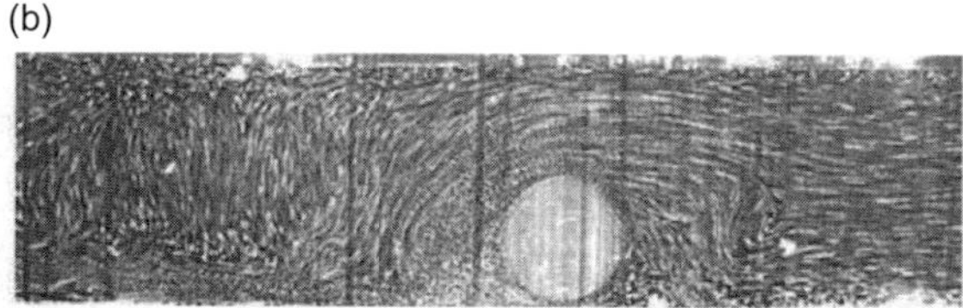

Figure 16.21 Visualization of flow in a blood vessel phantom from ultrasound-induced vortex with crossflow (right to left). (a) Crossflow at 0, 2, 6, 10 cm/s. (b) trapping of a 3 mm particle in the upstream vortex at 8 cm/s crossflow.

left) and ultrasound-induced vortical flow with the ultrasound focus at the center of top wall. In Figure 16.21(a) the crossflow is at 0, 2, 6, 10 cm/s and the scale bar is 5 mm. Note the symmetric pair of vortices at 0 crossflow become more asymmetric as the crossflow is increased. In Figure 16.21(b) a 3 mm particle is trapped in upstream vortex at 8 cm/s crossflow. This particle can represent a blood clot which would otherwise flow downstream and become a dangerous embolus.

16.3 Breakup of a Liquid Jet

A common surface tension phenomenon observed in daily life is the breakup of a liquid jet into drops. Figure 16.22(a) shows a dripping faucet where a liquid thread is stable upstream, but then develops sinusoidal perturbations of its air–liquid interface, eventually creating drops. Drop e in Figure 16.17(a) hints at this behavior. Figure 16.22(b) shows the same phenomena in a biofluid mechanics application of male micturition (urination) in the famous statue Mannekin Pis in Brussels. Additional applications include "lab on chip" microfluidic systems, ink jet printers and coating processes such as optical fibers.

This problem was originally explored by Félix Savart (1791–1841) in 1833 who was a French mathematician, physicist and medical doctor (Savart, 1833). He is well known as a co-founder of the Biot–Savart law of electromagnetism along with Jean-Baptiste Biot (1774–1862), a French physicist, astronomer and mathematician. Further advances came from the Belgian physicist Joseph Antoine Ferdinand Plateau (1801–1883) in 1873 (Plateau, 1873). The most comprehensive early treatment and analysis is due to Rayleigh (1842–1919) in 1878, which we will explore here (Rayleigh, 1878a). The instability is called the Rayleigh instability, the Plateau–Rayleigh instability or the Savart–Plateau–Rayleigh instability.

Consider a jet of fluid 1 passing through, and surrounded by, an immiscible fluid 2 as shown in Figure 16.23. For this analysis we ignore gravity and use a coordinate system that moves with the average jet speed, U. The interfacial position is $r = h(z)$. Let $h(z)$ be the constant radius, a, plus a small sinusoidal perturbation of the form

$$h(z) = a(1 + \varepsilon \cos(kz)) \tag{16.3.1}$$

(a)

(b)

Figure 16.22 (a) liquid jet breakup of a dripping faucet. (b) Liquid jet breakup of the Mannekin Pis fountain by Duquesnoy in Brussels.

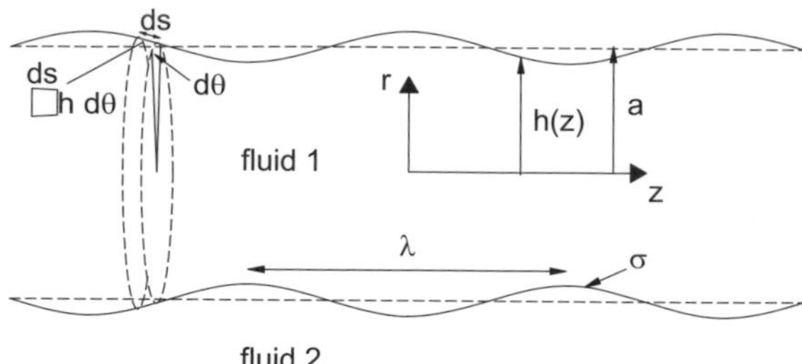

Figure 16.23 A jet of fluid 1 surrounded by fluid 2, viewed in a frame moving with fluid 1. The surface has a small amplitude sinusoidal perturbation.

The wave number is $k = 2\pi/\lambda$ where λ is the wavelength and we assume a small wave amplitude $\varepsilon << 1$. A differential surface area element is the product of $h\,d\theta$ and the arc length, ds, as shown in Figure 16.23. We can represent ds as $ds = \left(dh^2 + dz^2\right)^{1/2} = (1 + h'^2)1/2dz$ where $h' = dh/dz$. The surface area for one full wavelength, $\lambda = 2\pi/k$, is S, given by

$$S = \int_0^{2\pi/k}\int_0^{2\pi} h\left(1 + h'^2\right)^{1/2} d\theta\, dz \tag{16.3.2}$$

For $\varepsilon<<1$ we can expand the square root term of the integrand using the general Taylor series form

$$(1 + x)^n = 1 + nx + n(n-1)x^2/2 + O\left(x^3\right) \quad x<<1 \tag{16.3.3}$$

Letting $x = h'^2$ and $n = 1/2$ in Eq. (16.3.3), the integrand of Eq. (16.3.2) becomes

$$h\left(1 + h'^2\right)^{1/2} = a + \varepsilon a \cos(kz) + \frac{1}{2}\varepsilon^2 k^2 a^3 \sin^2(kz) + O\left(\varepsilon^3\right) \tag{16.3.4}$$

Then performing the integration of Eq. (16.3.2) using Eq. (16.3.4) for the integrand, we obtain

$$S = \frac{\pi^2 a}{k}(4 + \varepsilon^2 a^2 k^2) + O(\varepsilon^3) \tag{16.3.5}$$

The fluid volume, V, for one wavelength is given by

$$V = \int_0^{2\pi/k}\int_0^{2\pi}\int_0^{h} r\,dr\,d\theta\,dz = \frac{\pi^2 a^2}{k}\left(2 + \varepsilon^2\right) \tag{16.3.6}$$

Solving Eq. (16.3.6) for the radius, a, and expanding using the previous Taylor series of Eq. (16.3.3) with $x = \varepsilon^2/2$ and $n = -1/2$, gives us

$$a = \left(\frac{kV}{2\pi^2}\right)^{1/2}\left(1 + \frac{1}{2}\varepsilon^2\right)^{-1/2} = \frac{(2kV)^{1/2}}{2\pi}\left(1 - \frac{\varepsilon^2}{4}\right) + O(\varepsilon^4) \tag{16.3.7}$$

Now substitute a from Eq. (16.3.7) into S of Eq. (16.3.5) to find

$$S = 2\pi\left(\frac{2V}{k}\right)^{1/2}\left[1 + \frac{\varepsilon^2}{4}\left(\frac{k^3V}{2\pi^2} - 1\right)\right] + O(\varepsilon^3) \tag{16.3.8}$$

From Eq. (16.3.8) the undisturbed, $\varepsilon = 0$, value of S is $S_O = 2\pi(2V/k)^{1/2}$. Using this notation, rearrange Eq. (16.3.8) to reveal the relative change in surface area,

$$\frac{S - S_O}{S_O} = \left[\frac{\varepsilon^2}{4}\left(\frac{k^3V}{2\pi^2} - 1\right)\right] \tag{16.3.9}$$

Finally substitute V from Eq. (16.3.6) into Eq. (16.3.9) using only the leading-order term, $2\pi^2a^2/k$, since V is already multiplied by ε^2in Eq. (16.3.9). Then replace $k = 2\pi/\lambda$ to find

$$\frac{S - S_O}{S_O} = \left[\frac{\varepsilon^2}{4}\left(\left(\frac{2\pi a}{\lambda}\right)^2 - 1\right)\right] \tag{16.3.10}$$

Equation (16.3.10) shows that when the wavelength is greater than the circumference, $\lambda > 2\pi a$, the relative change in surface area due to the perturbation is negative, $(S - S_O)/S_O < 0$. That range of wavelengths reduces the surface area and hence the surface energy. The system prefers the lower energy state, which leads to the pinch-off instability. For $\lambda < 2\pi a$ the system is stable. This analysis depends only on the concept of interfacial surface energy, we did not need to know the value of the surface tension nor the identity of either fluid, as long as they are immiscible. So the unstable range of wavelengths could be for a liquid jet

Box 16.3 John William Strutt, 3rd Baron Rayleigh

Figure 16.24 John William Strutt, 3rd Baron Rayleigh (1842–1919), British physicist.

Lord Rayleigh is John William Strutt, 3rd Baron Rayleigh (1842–1919), a British physicist who made many seminal contributions in acoustics, optics, wave theory, electrodynamics, light scattering, electromagnetism, photography, elasticity and fluid dynamics. He received undergraduate and further training at the University of Cambridge starting in 1861 with faculty including physicist and fluid dynamicist George Gabriel Stokes, whom we discussed in Sections 8.1 and 8.2. Eventually Rayleigh became the second Cavendish Professor of Physics at Cambridge (1879–1884), a chaired professorship previously held by James Clerk Maxwell whose equations of electromagnetism revolutionized the field. Rayleigh left Cambridge to continue his scientific pursuits at his estate with a self-funded laboratory, the Baron business being good. He became a professor at the Royal Institution of Great Britain (1887–1905), a time period which included his work with William Ramsey discovering Argon in 1894. They shared the Nobel Prize in Physics awarded in 1904 for their efforts. Rayleigh would later return to be the Chancellor at University of Cambridge (1908–1919), a position first filled in the year c.1215. His 1877 book *The Theory of Sound* (Rayleigh, 1877, 1878) is the foundation of acoustics while his now-called "Rayleigh scattering" of light was the first explanation of why the sky is blue.

surrounded by gas, a gas jet surrounded by liquid, or two immiscible liquids like oil and water. In Section 17.6 a dynamic linear stability analysis of this flow reveals that the fastest growing wavelength is $\lambda = 9.015\,a$.

16.4 Variable Surface Tension: Surfactants

Variable surface tension can occur from the effects of temperature and surfactants. Surfactants are molecules which assimilate at the fluid–fluid interface and reduce the surface tension. Typical examples are detergents, soaps, foaming agents and many others. The key features of a surfactant molecule are shown in Figure 16.25(a), case (i), as the hydrophilic head, which remains in the water, and the hydrophobic tail, which sticks out of the interface. That orientation interferes with the surface tension and reduces it. Also shown are the bulk concentration, C (mass/volume); the surface concentration, Γ (mass/area); the surface tension, σ ; adsorption of bulk molecules to the interface; and desorption from the interface into the bulk. Figure 16.25(b), case (ii), is a higher C and Γ which reduce σ. Figure 16.25(c), case (iii), is the critical micelle concentration, C_{cmc}, when the interface is fully crowded and additional surfactant forms micelles in the bulk and possibly additional layers on the interface. Figure 16.25(d) is the σ vs C relationship showing cases (i), (ii) and (iii) and the typical decrease of σ leading to an asymptotic minimum surface tension for $C \geq C_{cmc}$.

A well-known relationship describing the adsorption of surfactant to the interface is the Langmuir (1918) adsorption equation

$$\Gamma(C) = \Gamma_\infty \frac{KC}{1+KC} \quad \rightarrow \quad KC = \frac{\Gamma}{\Gamma_\infty - \Gamma} \tag{16.4.1}$$

In Eq. (16.4.1) the static system reaches a saturated interface $\Gamma \rightarrow \Gamma_\infty$ as C increases and the adsorption parameter, K, measure the activity of the surfactant. Three examples of Eq. (16.4.1) are shown in Figure 16.26 for $K = K_1, \Gamma_\infty = \Gamma_{\infty 1}$; $K = K_1$, $\Gamma_\infty = 2\Gamma_{\infty 1}$; and $K = 2K_1, \Gamma_\infty = \Gamma_{\infty 1}$. The higher value of Γ_∞ yields a higher asymptote, while increasing K makes the system reach the asymptote more rapidly in C. These curves are called Langmuir isotherms. His work on this and related topics earned him the Nobel Prize in Chemistry in 1932 while working at the General Electric research laboratory in Schenectady, NY.

Using the Langmuir adsorption model, we want to relate the surface concentration, Γ, to the surface tension, σ. This is accomplished from the Gibbs adsorption equation

$$d\sigma = -k_B T \Gamma d(\ln C) \tag{16.4.2}$$

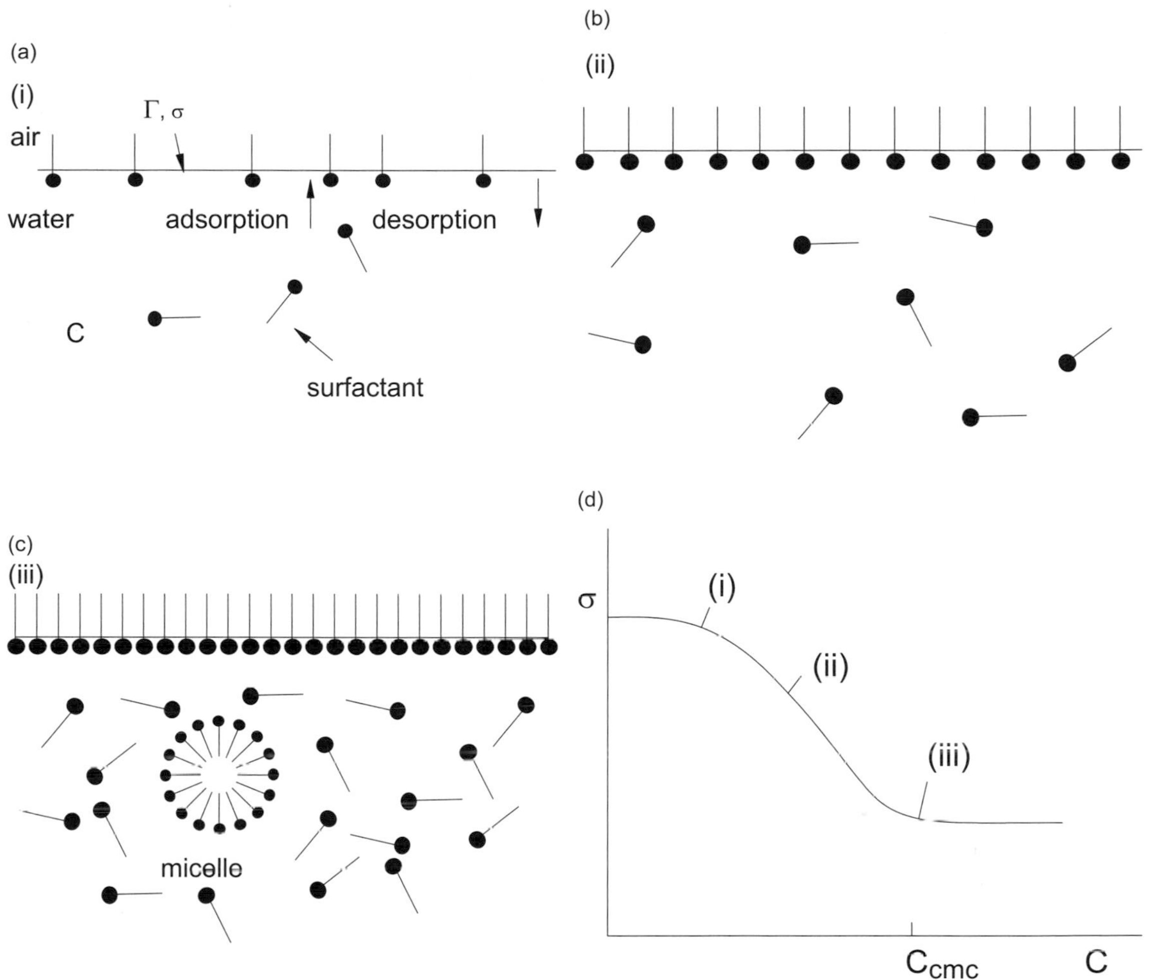

Figure 16.25 (a) Case (i): surfactant molecules in water at bulk concentration, c, and interfacial concentration, Γ; (b) case (ii): higher value of C and Γ; (c) case (iii) $C = C_{cmc}$ the critical micelle concentration with formation of a micelle; (d) σ vs Cshowing cases (i), (ii) and (iii).

where k_B is the Boltzmann constant and T is the absolute temperature. Substituting for $\Gamma(C)$ from Eq. (16.4.1) into Eq. (16.4.2) and integrating yields

$$\sigma - \sigma_0 = -k_B T \int_0^C \frac{K\Gamma_\infty}{1 + KC} dC = -k_B T\Gamma_\infty \ln(1 + KC) \tag{16.4.3}$$

where σ_0 is the surfactant-free, $C = 0$, surface tension. Equation (16.4.1) can be rearranged to give $1 + KC = \Gamma_\infty/(\Gamma_\infty - \Gamma)$ and inserting this into Eq. (16.4.3) we arrive at the $\sigma(\Gamma)$ relationship for the Langmuir model

$$\sigma(\Gamma) = \sigma_0 - k_B T\Gamma_\infty \ln\left(\frac{\Gamma_\infty}{\Gamma_\infty - \Gamma}\right) \tag{16.4.4}$$

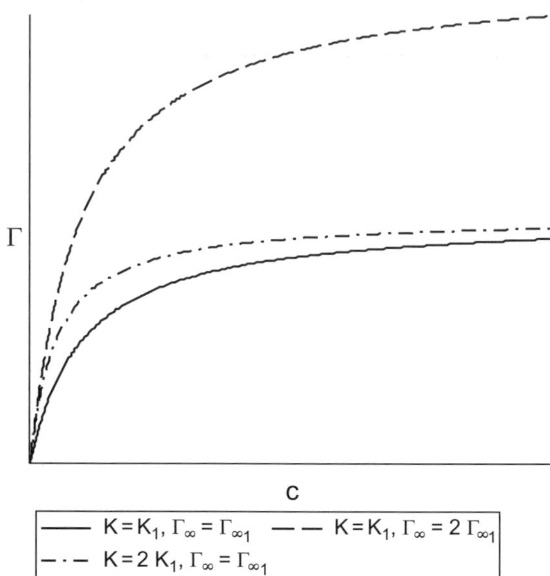

Figure 16.26 Γvs C for three combinations of Γ_∞, K.

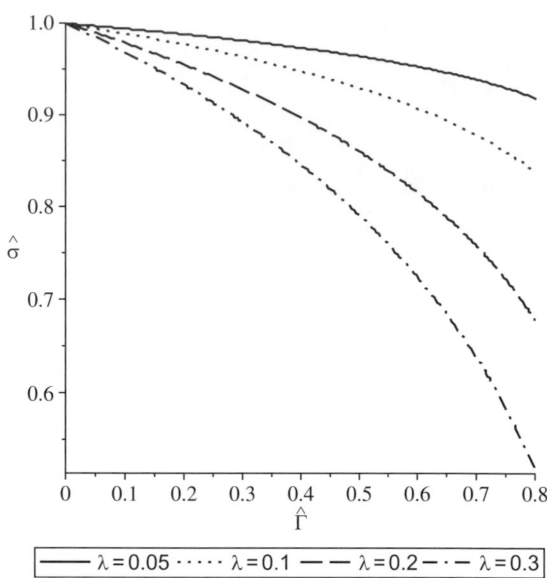

Figure 16.27 Langmuir surface equation of state for four values of $\lambda = 0.05, 0.1, 0.2, 0.3$.

	Adsorption isotherm	Surface equation of state
Henry	$KC = \dfrac{\Gamma}{\Gamma_\infty}$	$\sigma = \sigma_0 - k_B T \Gamma$
Freundlich	$KC = \left(\frac{\Gamma}{\Gamma_\infty}\right)^{1/m}$	$\sigma = \sigma_0 - k_B T \dfrac{\Gamma}{m}$
Langmuir	$KC = \dfrac{\Gamma}{\Gamma_\infty - \Gamma}$	$\sigma = \sigma_0 + k_B T \Gamma_\infty \ln\left(1 - \dfrac{\Gamma}{\Gamma_\infty}\right)$
Volmer	$KC = \dfrac{\Gamma}{\Gamma_\infty - \Gamma} e^{\left(\frac{\Gamma}{\Gamma_\infty - \Gamma}\right)}$	$\sigma = \sigma_0 - k_B T \dfrac{\Gamma_\infty \Gamma}{\Gamma_\infty - \Gamma}$
Frumkin	$KC = \dfrac{\Gamma}{\Gamma_\infty - \Gamma} e^{-\frac{2\beta\Gamma}{k_B T}}$	$\sigma = \sigma_0 + k_B T \Gamma_\infty \ln\left(1 - \dfrac{\Gamma}{\Gamma_\infty}\right) + \beta \Gamma^2$

van der Waals

$$KC = \frac{\Gamma}{\Gamma_\infty - \Gamma} e^{\left(\frac{\Gamma}{\Gamma_\infty - \Gamma} - \frac{2\beta\Gamma}{k_B T}\right)} \qquad \sigma = \sigma_0 - k_B T \frac{\Gamma_\infty \Gamma}{\Gamma_\infty - \Gamma} + \beta \Gamma^2 \tag{16.4.5}$$

There are several other models available which focus on different types of molecular behavior. A sampling of these are listed in Eq. (16.4.5), where the surface tension equations of state $\sigma(\Gamma)$ are derived from the adsorption isotherms through Eq. (16.4.2). They include models from: British chemist William Henry (1774–1836) (Henry, 1803), whose model is linear in Γ/Γ_∞, as is the leading term of a Taylor series expansion of the Langmuir model, Eq. (16.4.1) for $\Gamma/\Gamma_\infty << 1$; German chemist Herbert Freundlich (1880–1941) (Freundlich, 1909), whose model has an exponent $1/m$ where $m = 1$ recovers the Henry model; and forms with exponential dependence such as German physical chemist Max Volmer (1885–1965). Additional complexity appears in an interaction parameter, β, found in the work of Alexander Frumkin (1895–1976), who was a Russian/Soviet electrochemist (Frumkin, 1925) and Johannes Diderik van der Waals (1837–1923), a Dutch theoretical physicist. Further models not discussed are those of Fowler–Guggenheim and Hill–deBoer.

Figure 16.27 shows a plot of the Langmuir surface equation of state in dimensionless form

$$\hat{\sigma} = 1 + \lambda \ln\left(1 - \hat{\Gamma}\right) \tag{16.4.6}$$

where $\hat{\sigma} = \sigma/\sigma_0, \hat{\Gamma} = \Gamma/\Gamma_\infty, \lambda = k_B T \Gamma_\infty/\sigma_0$. Four values of $\lambda = 0.05$, 0.1, 0.2, 0.3 are shown and $\hat{\Gamma}$ is in the range $0 \le \hat{\Gamma} \le 0.8$. These are typical of regions (i) and (ii) in Figure 16.25(d) in concentrations below the critical micelle concentration, C_{CMC}.

When surface tension varies along an interface, for example from non-uniform surfactant or a temperature distribution, the gradient pulls the lower surface tension region toward the higher surface tension region. Because the underlying fluid is viscous, it is dragged along creating bulk flows. This

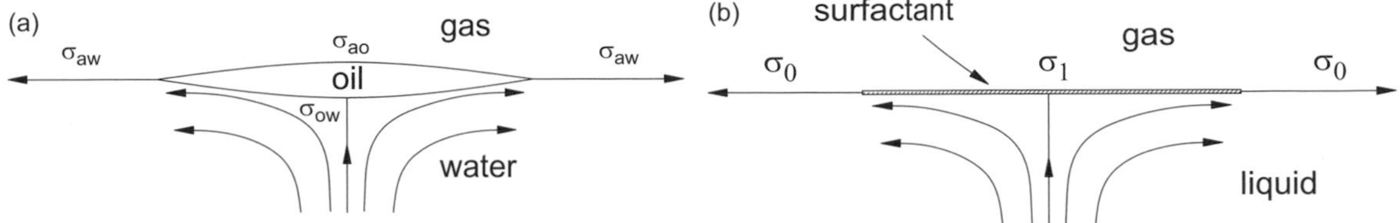

Figure 16.28 Surface spreading due to surface tension variation: (a) oil drop on water and (b) localized surfactant.

phenomenon is called Marangoni flow, named for Carlo Marangoni who studied the spreading of a liquid drop floating on the surface of another liquid and published the results in his PhD thesis at the University of Paiva, Italy in 1865 (Marangoni, 1865) and later in a journal (Marangoni, 1869). Figure 16.28 shows two surface spreading situations. Figure 16.28 (a) is an oil drop on water. Assuming that the oil and water meet at zero angle, there is spreading when $\sigma_{aw} > \sigma_{ow} + \sigma_{ao}$ where σ_{aw} is the surface tension between air and water, σ_{ow} between oil and water, and σ_{ao} between oil and air. Likewise in Figure 16.28(a) there is spreading due to a localized surfactant on the interface. Surface pressure, defined as $\Pi_s = \sigma_0 - \sigma(\Gamma)$, is commonly used in this field where clearly $\Pi_s \geq 0$. For a localized surfactant with interfacial concentration, Γ_1, the surface tension there is $\sigma(\Gamma_1) < \sigma_0$. However, it is surrounded by the surfactant-free interface with surface tension σ_0.

Prior to Marangoni's work, James Thomson (1822–1892) reported observations of wine and liquor interfaces. He noted that alcohol reduces surface tension and that this explains the interfacial spreading from regions of higher to lower alcoholic surface concentration, or lower to higher surface tension. Included are observations of "wine tears" or "wine legs," where the thin film coating a wine glass quickly evaporates the alcohol, pulling wine up the side of the glass, until its weight causes it to fall down the side as drops (Thomson, 1855), see Figure 16.29. Thomson's father, also James (1786–1849), was a famous Irish mathematician, and his brother, William, later became Lord Kelvin (1824–1907), a famous mathematical physicist and engineer. All three eventually held positions at the University of Glasgow.

Figure 16.29 Wine tears (or legs).

16.5 Interfacial Stress Boundary Condition for Variable Surface Tension

In Section 16.2 we derived the stress boundary condition at an interface for constant surface tension, Eq. (16.2.3), the Young–Laplace equation. With the introduction of surfactants (or heat), surface tension can vary in space and time, so we need a more general approach. Figure 16.30 shows the balance of stresses at an interface between two fluids. The interfacial surface, A, is bounded by a closed surface contour curve, C. The local coordinate system along C consists of the unit vectors $\underline{s}$, $\underline{n}$, $\underline{m}$ where $\underline{s}$ is tangent to C, $\underline{n}$ is normal to the surface, and $\underline{m} = \underline{s} \times \underline{n}$ is their cross product. $\underline{s}$ and $\underline{m}$ are in a local tangent plane to A, and the surface tension, σ, acts in the $\underline{m}$-direction.

The force balance involves the surface tension and the stress vectors $\underline{t}_1$, $\underline{t}_2$ acting on A. Recall that the stress vectors are related to the stress tensors through

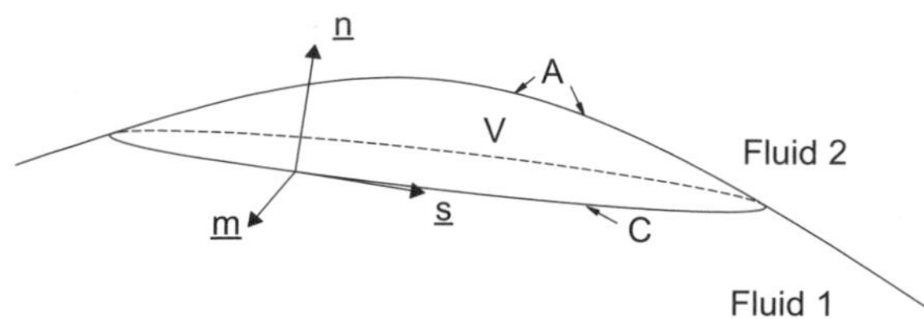

Figure 16.30 Control volume, V, with surface, A, and curve, C, for balance of stresses at a fluid–fluid interface.

the local unit normal to the surface, $\underline{n}$. In fluid 1 we have $\underline{t}_1 = \underline{\underline{T}}_1 \cdot \underline{n}$, while in fluid 2 the unit normal is the opposite direction, so $\underline{t}_2 = \underline{\underline{T}}_2 \cdot (-\underline{n})$.

$$\iint_A \underline{t}_1 + \underline{t}_2 \, dA + \oint_C \sigma \underline{m} \, ds = \iint_A \left[\underline{\underline{T}}_1 - \underline{\underline{T}}_2\right] \cdot \underline{n} \, dA + \oint_C \sigma \underline{s} \times \underline{n} \, ds = \underline{f} \qquad (16.5.1)$$

The remaining force, $\underline{f}$, is due to fluid acceleration and body forces in the bulk fluid volume, V. Since we will shrink our contour down to a point, the characteristic length scale of the volume decreases more rapidly than that of the surface, so $\underline{f} \to 0$.

We want to change the line integral in Eq. (16.5.1) to a surface integral. Consider Stokes' theorem for a vector field, $\underline{F}$,

$$\oint_C \underline{F} \cdot d\underline{s} = \iint_A (\underline{\nabla} \times \underline{F}) \cdot d\underline{A} \qquad (16.5.2)$$

where $d\underline{A} = \underline{n} \, dA$ and $d\underline{s} = \underline{s} \, ds$. Let $\underline{F} = (\underline{a} \times \underline{b})$ where $\underline{b}$ is an arbitrary constant vector. Use the vector identity $(\underline{a} \times \underline{b}) \cdot \underline{s} = \underline{b} \cdot (\underline{s} \times \underline{a})$ so that the integral along C becomes

$$\oint_C (\underline{a} \times \underline{b}) \cdot \underline{s} \, ds = \underline{b} \cdot \oint_C (\underline{s} \times \underline{a}) \, ds \qquad (16.5.3)$$

Then consider the vector identity $\underline{\nabla} \times (\underline{a} \times \underline{b}) = (\underline{\nabla} \cdot \underline{b})\underline{a} - (\underline{\nabla} \cdot \underline{a})\underline{b} + (\underline{b} \cdot \underline{\nabla})\underline{a} - (\underline{a} \cdot \underline{\nabla})\underline{b}$, which simplifies for constant $\underline{b}$ to $\underline{\nabla} \times (\underline{a} \times \underline{b}) = -(\underline{\nabla} \cdot \underline{a})\underline{b} + (\underline{b} \cdot \underline{\nabla})\underline{a}$. Using this result, insert it into the right hand side of Eq. (16.5.2)

$$\iint_A (\underline{\nabla} \times \underline{F}) \cdot \underline{n} \, dA = \iint_A \left(-(\underline{\nabla} \cdot \underline{a})\underline{b} + (\underline{b} \cdot \underline{\nabla})\underline{a}\right) \cdot \underline{n} \, dA = \underline{b} \cdot \iint_A \left(-(\underline{\nabla} \cdot \underline{a})\underline{n} + (\underline{\nabla}\underline{a}) \cdot \underline{n}\right) dA \qquad (16.5.4)$$

where $(\underline{\nabla}\underline{a}) = (\partial a_i / \partial x_j)$ and $(\underline{\nabla}\underline{a}) \cdot \underline{n} = (\partial a_i / \partial x_j) n_j$ in index notation. Equation (16.5.2) now takes the form

$$\underline{b} \cdot \oint_C (\underline{s} \times \underline{a}) \, ds = \underline{b} \cdot \iint_A \left(-(\underline{\nabla} \cdot \underline{a})\underline{n} + (\underline{\nabla}\underline{a}) \cdot \underline{n}\right) dA \qquad (16.5.5)$$

Since $\underline{b}$ is arbitrary, it must be that

$$\oint_C (\underline{s} \times \underline{a}) \, ds = \iint_A \left(-(\underline{\nabla} \cdot \underline{a})\underline{n} + (\underline{\nabla}\underline{a}) \cdot \underline{n}\right) dA \qquad (16.5.6)$$

Substitute $\underline{a} = \sigma \underline{n}$ into Eq. (16.5.6) to find

$$\oint_C (\underline{s} \times (\sigma \underline{n})) \, ds = \iint_A \left(-(\underline{\nabla} \cdot (\sigma \underline{n}))\underline{n} + (\underline{\nabla}(\sigma \underline{n})) \cdot \underline{n}\right) dA = \iint_A \left(-(\underline{n} \cdot \underline{\nabla}\sigma)\underline{n} - \sigma(\underline{\nabla} \cdot \underline{n})\underline{n} + (\underline{n} \cdot \underline{n})\underline{\nabla}\sigma + \sigma(\underline{\nabla}\underline{n}) \cdot \underline{n}\right) dA \qquad (16.5.7)$$

We note that in Eq. (16.5.7) $\underline{n} \cdot \underline{n} = 1$ and $(\underline{\nabla}\underline{n}) \cdot \underline{n} = \frac{1}{2}\underline{\nabla}(\underline{n} \cdot \underline{n}) = \frac{1}{2}\underline{\nabla}(1) = 0$. Furthermore, $\underline{n} \cdot \underline{\nabla}\sigma = 0$, since $\underline{\nabla}\sigma$ must be tangent to the surface. This feature is reflected in the notation $\underline{\nabla}\sigma = \underline{\nabla}_s \sigma$ where $\underline{\nabla}_s = (\underline{\underline{I}} - \underline{n}\underline{n}) \cdot \underline{\nabla}$ is the surface gradient obtained from the general gradient operator subtracting off the normal contribution. Consequently, Eq. (16.5.7) becomes

$$\oint_C (\underline{s} \times (\sigma \underline{n})) \, ds = \iint_A \left(-\sigma(\underline{\nabla} \cdot \underline{n})\underline{n} + \underline{\nabla}_s \sigma\right) dA \qquad (16.5.8)$$

Substituting Eq. (16.5.8) into Eq. (16.5.1) yields

$$\iint_A \left[\left(\underline{\underline{T}}_1 - \underline{\underline{T}}_2\right) \cdot \underline{n} - \sigma(\underline{\nabla} \cdot \underline{n})\underline{n} + \underline{\nabla}_s \sigma\right] dA = 0 \qquad (16.5.9)$$

Since A is arbitrary, the integrand of Eq. (16.5.9) must be zero. Substituting from Eq. (16.2.4) we find

$$\left(\underline{\underline{T}}_1 - \underline{\underline{T}}_2\right)\cdot\underline{n} = \sigma\kappa\underline{n} - \underline{\nabla}_s\sigma \tag{16.5.10}$$

The normal component is derived by taking the inner product of Eq. (16.5.10) with the unit normal $\underline{n}$

$$\left(\underline{\underline{T}}_1\cdot\underline{n}\right)\cdot\underline{n} - \left(\underline{\underline{T}}_2\cdot\underline{n}\right)\cdot\underline{n} = \sigma\kappa \rightarrow T_{1ij}n_jn_i - T_{2ij}n_jn_i = \sigma\kappa \tag{16.5.11}$$

Equation (16.5.11) states that the normal stress jump across the interface is balanced by the product of the curvature and surface tension. For an inviscid or static fluid, it becomes the Young–Laplace equation, Eq. (16.2.3). In general, a flowing viscous fluid has pressure and viscous normal stresses at a boundary, which appear in the $T_{ij}n_in_j$ terms.

The tangential component is the inner product of Eq. (16.5.10) with a unit tangent vector, $\underline{\tau}$,

$$\left(\underline{\underline{T}}_1\cdot\underline{n}\right)\cdot\underline{\tau} - \left(\underline{\underline{T}}_2\cdot\underline{n}\right)\cdot\underline{\tau} = -\underline{\tau}\cdot\underline{\nabla}_s\sigma \tag{16.5.12}$$

where we have imposed that $\underline{n}\cdot\underline{\tau} = 0$. Equation (16.5.12) states that the jump in shear stress at an interface is balanced by the surface tension gradient, $\underline{\nabla}_s\sigma$, which are Marangoni flows, first introduced in Section 16.4. Often an interface of interest is between a liquid and gas, and it is routine to treat the gas as a passive region of constant pressure without internal flow, e.g. $\underline{\underline{T}}_2 = -p_a\underline{I}$. For a relatively flat interface between a liquid and gas, the Marangoni balance simplifies to $\mu(du/dy) = d\sigma/dx$ for a unidirectional flow where $\underline{u} \sim (u(y), 0)$. The kinematic boundary at the interface is that fluid does not pass through it,

$$\underline{u}\cdot\underline{n} = 0 \tag{16.5.13}$$

Conservation of the surfactant species must be enforced both in the bulk and in the interface. The bulk conservation is the convection–diffusion equation, Eq. (4.5.14),

$$\frac{\partial C}{\partial t} + \nabla\cdot(\underline{u}C) = D_m\nabla^2 C \tag{16.5.14}$$

The interfacial conservation of mass is given by

$$\frac{\partial\Gamma}{\partial t} + \underline{\nabla}_s\cdot(\Gamma\underline{u}_s) + \Gamma(\underline{\nabla}_s\cdot\underline{n})(\underline{u}\cdot\underline{n}) = D_s\nabla_s^2\Gamma + j_n \tag{16.5.15}$$

where $\underline{u}_s$ is the surface fluid velocity, which is the fluid velocity vector minus the normal component, $\underline{u}_s = (\underline{\underline{I}} - \underline{nn})\cdot\underline{u}$ (Stone, 1990). The first two terms on the left are the familiar unsteady and convective terms, respectively. We saw that structure in the bulk transport equation, Eq. (4.5.11), for the bulk velocity and concentration. The third term on the left results from stretching (contracting) of the interface and contains the surface curvature term $\underline{\nabla}_s\cdot\underline{n}$.

The right hand side has surface diffusion, with surface molecular diffusivity, D_s, and a net source term, j_n, which is the balance of surfactant adsorption from the bulk to the interface, and desorption from the interface to the bulk

$$-D_m(\underline{n}\cdot\underline{\nabla}c) = j_n \tag{16.5.16}$$

An example of the stretching term is to consider an expanding spherical bubble with radius $R(t)$, an insoluble surfactant with no bulk concentration, making $j_n = 0$, which is uniformly distributed on the interface, so that $\nabla_s^2\Gamma = 0$. The surface mass of surfactant is $M = 4\pi R^2\Gamma$ and does not change with time, making its time derivative zero,

$$\frac{dM}{dt} = 4\pi\left(R^2\frac{d\Gamma}{dt} + 2R\Gamma\frac{dR}{dt}\right) = 0 \rightarrow \frac{d\Gamma}{dt} + 2\frac{\Gamma}{R}\frac{dR}{dt} = 0 \tag{16.5.17}$$

From Eq. (16.5.15) $\Gamma(\underline{\nabla}_s\cdot\underline{n})(\underline{u}\cdot\underline{n}) = \Gamma(2/R)(dR/dt)$, which is added to $d\Gamma/dt$, the same balance as Eq. (16.5.17). Equation (16.5.17) shows that the interfacial concentration is decreasing as the radius is increasing, i.e. $d\Gamma/dt < 0$ when $dR/dt > 0$.

Now that we have a moving viscous fluid, an important new dimensionless parameter arises called the capillary number, Ca,

$$Ca = \frac{\mu U}{\sigma} = \frac{\text{viscous forces}}{\text{surface tension forces}} \tag{16.5.18}$$

where U is a characteristic fluid velocity and μ is its viscosity. Ca is the ratio of viscous to surface tension forces.

16.6 Respiratory Interfacial Phenomena

Surfactant Models

Lung surfactants play a major role in respiratory dynamics, so details of modeling their behavior is critical. The balance of adsorption and desorption kinetics are represented in Eq. (16.5.16) as j_n. Surfactant flux from the bulk to the interface, adsorption, and from the interface to the bulk, desorption, can be modeled using Langmuir–Hinshelwood kinetics (Cyril Hinshelwood (1897–1967) received the 1956 Nobel Prize in Chemistry),

$$j_n = \begin{cases} k_a C_s(\Gamma_\infty - \Gamma) - k_d\Gamma & (\Gamma < \Gamma_\infty) \\ -k_d\Gamma & (\Gamma \geq \Gamma_\infty) \end{cases} \tag{16.6.1}$$

where k_a is the adsorption constant, k_d is the desorption constant and C_s is the bulk concentration evaluated at the interface (Chang and Franses, 1995; Ghadiali and Gaver, 2003).

Equation (16.6.1) considers two cases. When $\Gamma < \Gamma_\infty$ both adsorption and desorption are active. When $\Gamma \geq \Gamma_\infty$ only desorption occurs. But, how can $\Gamma > \Gamma_\infty$? The concept of an upper limit, Γ_∞, comes from static equilibrium considerations. However, in a dynamic system Γ can exceed Γ_∞, since surfactant may accumulate locally due to interfacial flows and deformation. As Γ increases beyond Γ_∞, it can reach a maximum value for surfactant monolayers, Γ_{max}. Adding more surfactant to the interface results in buckling or collapse of the monolayer into multiple layers. In the range $\Gamma_\infty \leq \Gamma < \Gamma_{max}$, increasing Γ continues to reduce surface tension, reaching a constant minimum σ at Γ_{max} (Otis et al., 1994; Krueger and Gaver, 2000),

$$\hat{\sigma} = \begin{cases} 1 - E\hat{\Gamma} \; (\hat{\Gamma} < 1) \\ (1 - E)e^{\frac{E}{1-E}(1-\hat{\Gamma})} \, (\hat{\Gamma} \geq 1) \end{cases} \tag{16.6.2}$$

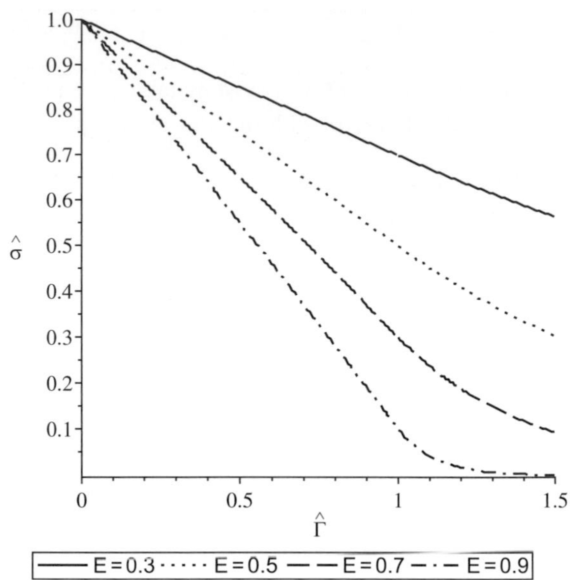

Figure 16.31 Modified Frumkin model of surface tension equation of state for four values of $E = 0.3, 0.5, 0.7, 0.9$.

To account for the full range of $0 \leq \Gamma < \Gamma_{max}$, the surface tension equation of state for lung surfactant can be modeled with a modified Frumkin approach as shown in Eq. (16.6.2) where $\hat{\Gamma} = \Gamma/\Gamma_\infty$ and $\hat{\sigma} = \sigma/\sigma_0$ (Fujioka and Grotberg, 2005). For $\hat{\Gamma} < 1$ the surface tension decreases linearly, as in a Henry model. However, once $\hat{\Gamma} \geq 1$ an exponential decay model is used. E is the dimensionless elasticity number defined as $E = -(\Gamma_\infty/\sigma_0)(\partial\sigma/\partial\Gamma)_{\Gamma\leq\Gamma_\infty}$ where $(\partial\sigma/\partial\Gamma) < 0$ is the constant negative slope. It represents the surfactant's ability to reduce surface tension, i.e. its strength. The dimensionless Marangoni number is given by $Ma = ECa$ where $Ca = \mu U/\sigma_0$ is the capillary number, based on the clean interface surface tension.

Figure 16.31 shows plots of Eq. (16.6.2) for four values of $E = 0.3, 0.5, 0.7, 0.9$. Note the linearity for $\hat{\Gamma} < 1$ and nonlinearity for $\hat{\Gamma} \geq 1$. $E = 0.7$ is a reasonable value for lung surfactant behavior (Fujioka and Grotberg, 2005).

Box 16.4 Lung Surfactants: Physiology and Disease

The fetal lung is deflated since gas exchange occurs through the placental circulation from the mother. At birth, the lungs first take in air. Because lungs are coated internally with a thin liquid film, the surface tension between that film and the resident air becomes a critical issue. If it is too high, the lung is stiff and makes breathing difficult for the newborn. For its own protection, the lung makes surfactant in the alveolar Type II cells, see Figure 16.32, which is secreted into the liquid lining and oriented at the interface to reduce surface tension. It is primarily made of dipalmitoylphosphatidylcholine (DPPC) with important related protein constituents. To appreciate its effect, Figure 16.33(a) shows a typical measurement system. Surfactant is distributed on the interface whose surface area is cycled while a Wilhelmy balance measures the surface tension as in Figure 16.6(a). The resulting plot of percent relative area versus surface tension for lung surfactant is shown in Figure 16.33(b). It exhibits a hysteresis loop, with surface compression behaving differently from surface expansion. Recall that air–water surface tension is ~ 70 dyn/cm, independent of the surface area. The loop goes to a very low value, in the range 1–5 dyn/cm. These low values protect the lung from becoming stiff from high surface tension. Another popular method for measuring the surface tension vs area relationship is the bubble surfactometer (Enhorning, 1977; Notter, 2000).

To appreciate the overall influence of surface tension and surfactants on lung mechanics, Figure 16.34 shows the volume–pressure relationship of an excised lung, which is cycled either with air or saline. Note that the air-cycled lung has a hysteresis area, like that of Figure 16.33(b). However, with saline cycling there is essentially no surface tension, so the

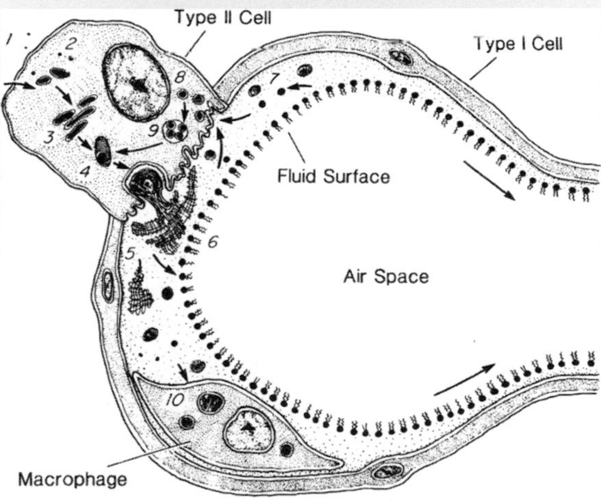

Figure 16.32 Surfactant system of the lung.

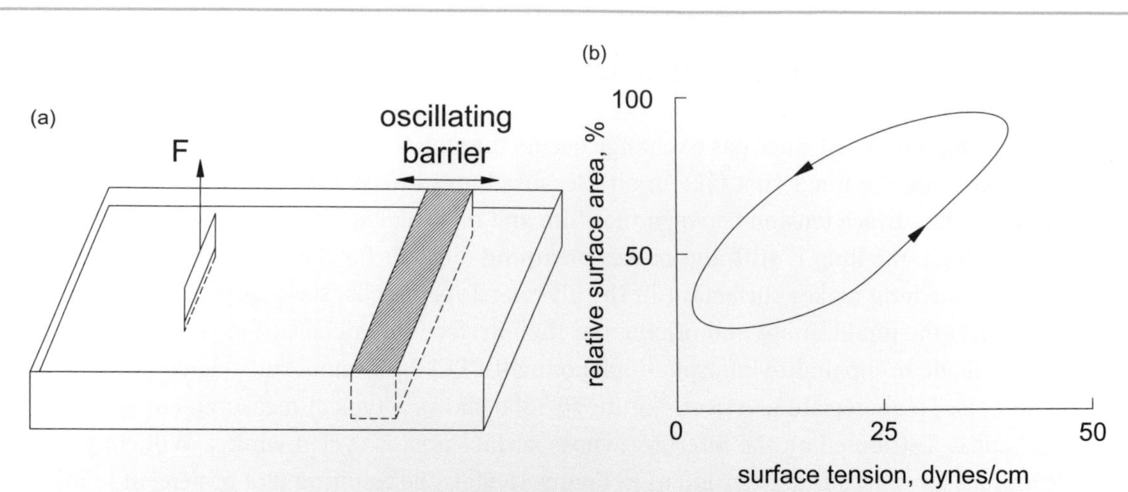

Figure 16.33 (a) Oscillating surface area compression of an interface with lung surfactant. (b) Surface tension vs relative surface area.

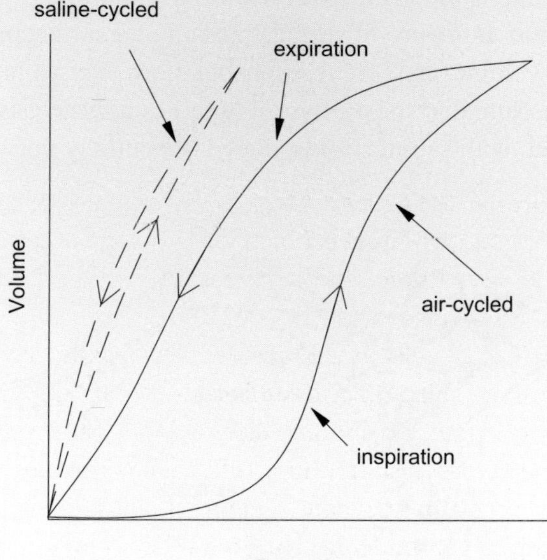

Figure 16.34 Volume–pressure curves for a lung cycled with air (solid curve) and with saline (dashed curve).

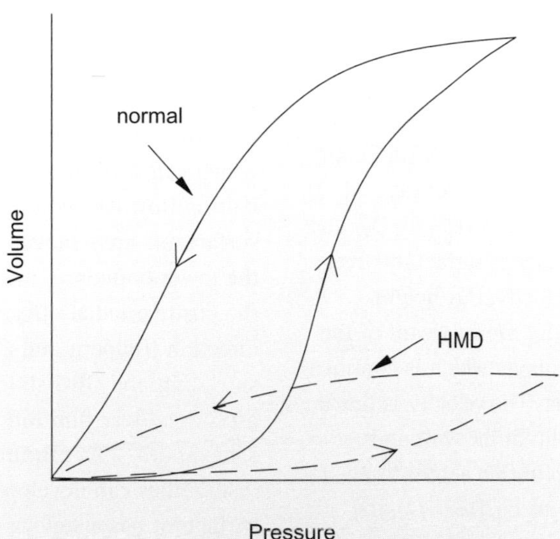

Figure 16.35 Hyaline membrane disease.

very small hysteresis area is due primarily to viscous tissue effects. The average slope is higher with saline cycling, evidence of a more compliant system. This phenomenon was first studied by Swiss physician and investigator, Kurt von Neergaard (1887–1947) in 1929 (von Neergaard, 1929). The discovery of surfactants in lungs was a pivotal point in medicine (Pattle, 1955; Clements, 1957). Clearly the effects of surface tension and surfactant are very significant for overall lung mechanics and function. This should not be surprising, since surface tension effects are proportional to the inverse of the alveolar radius, see Eq. (16.2.3), which is very small, $r_{alv} \sim 100\ \mu m$.

The maturation of the Type II cells to manufacture surfactant occurs late in gestation, so premature birth can cause breathing difficulties from stiff lungs. This is known as respiratory distress syndrome (RDS) or hyaline membrane disease (Avery and Mead, 1959). Figure 16.35 shows hyaline membrane disease resulting in a stiffer lung. Therapy can involve surfactant replacement to assist the premature neonate while lung maturation continues. This is accomplished by instilling surfactant, collected from other animal sources (pig, cow, calf) or synthetic, directly into the trachea-bronchial tree, discussed later in this section (Kendig et al., 1988; Grotberg, 1994; 2001; 2011).

Surfactant Spreading and the Shock Wave

Exploration of surfactant spreading related to respiratory applications led to the discovery of a new fluid mechanical shock wave, the GBG shock (Grotberg, 2011). Under conditions of thin film, lubrication flow with vanishing surface diffusion and no bulk surfactant, Figure 16.36(a) shows a standing shock (Borgas and Grotberg, 1988). The wall at $y = 0$ has velocity U_W and upstream of the shock the interface is surfactant-free, $\Gamma = 0$, the fluid has uniform velocity, U_W, height $h = h_0$, and mass flow rate $U_W h_0$. Downstream of the shock there is a surfactant monolayer which is confined to be motionless due to a barrier. The velocity is linear, $u = U_w(1 - y/2h_0)$, with no-slip at the wall and interface, which requires a jump in film height to $2h_0$ to conserve mass flow. According to Eq. (16.5.12), the tangential stress balance at the interface is $\mu(du/dy) = d\sigma/dx$ or $d\sigma/dx = -\mu U_w/2h_0$, so the surface tension decreases linearly with x. For a Henry's law equation of state, Γ will increase linearly with x as indicated.

Figure 16.36(b) shows an axisymmetric shock evolution where a disc of surfactant on the surface, $H = 1$, where $H = h/h_0$, starts at $t = 0$ in the region $0 \leq R \leq 1$, where $R = r/r_0$. It spreads radially, and for lubrication theory, $h_0/r_0 << 1$, and vanishing surface diffusion with no bulk surfactant, the interface evolves toward a 2:1 film height like that of Figure 16.36(a) (Gaver and Grotberg, 1990; Jensen and Grotberg, 1992). There is a minimum height which can become zero, rupturing the film and halting flow (Gaver and Grotberg, 1992). If surfactants are removed from the system, say through the lower boundary, then the spreading starts out in the positive radial direction, but then reverses direction (Halpern and Grotberg, 1992a). Surfactant spreading and GBG shocks also occur in the precorneal tear film following a blink (Braun and King-Smith, 2007; Braun, 2012). Fingering instabilities can develop during spreading so that the surfactant has a stellate pattern (Matar and Troian, 1999; Matar et al., 2002).

Surface Tension and Airway Flows

The thin liquid coating of the airways is involved in important flows which affect lung function. If the film is thick enough, it can undergo Plateau–Rayleigh

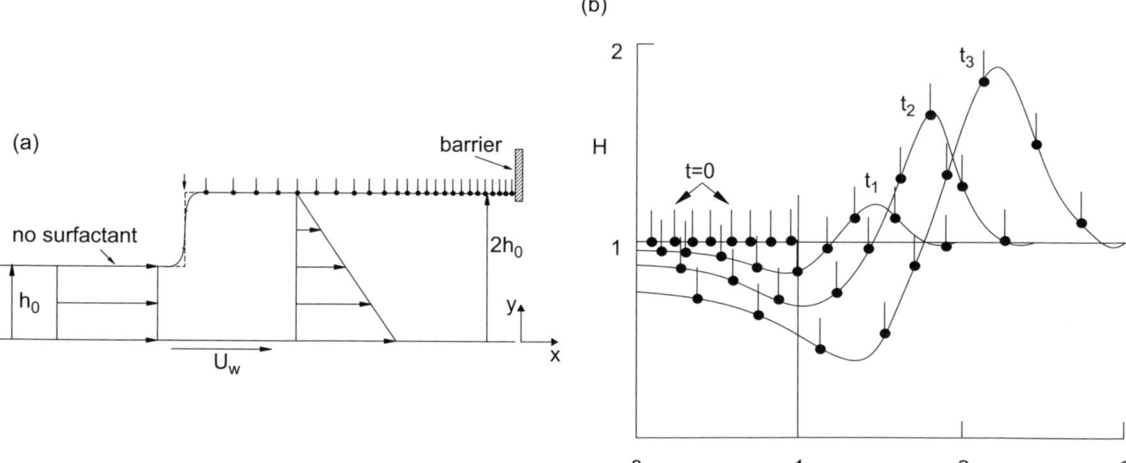

Figure 16.36 (a) Standing shock; (b) axisymmetric shock evolution.

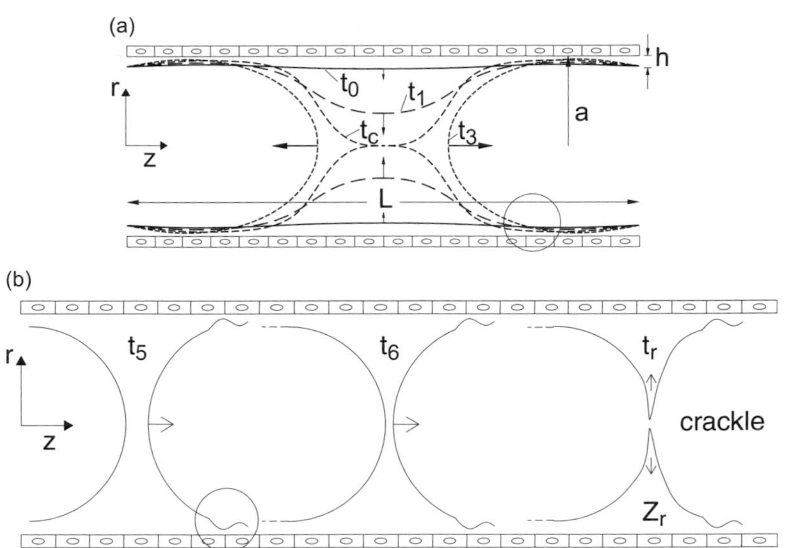

Figure 16.37 (a) Sketch of airway closure instability. (b) Airway reopening. Regions of largest fluid stresses and gradients are circled.

instability and form a liquid plug (Everett and Haynes, 1972; Hammond, 1983; Gauglitz and Radke, 1988; Kamm and Schroter, 1989). This plug closes the airway to gas exchange, as shown in Figure 16.37(a), a process called airway closure which tends to occur toward the end of expiration (Macklem et al., 1970; Greaves et al., 1986). The lung volume at which this occurs is called the closing volume. It is an important pulmonary function test which varies with age, gender and disease (Bode et al., 1976; Drummond and Milic-Emili, 2007). Heterogeneous airway closure contributes to heterogeneous ventilation distribution. At time t_0 an initial wave disturbance grows slowly; at time t_1 there is faster growth of the instability with a minimum radius, $R_{min} \sim 0.4$; at time t_c the film coalesces, $R_{min} = 0$; and for $t_c < t < t_3$ there is post-coalescence filling flow. Airway walls are flexible, which is destabilizing (Halpern and Grotberg, 1992b), while surfactants have a stabilizing effect (Cassidy et al., 1999; Halpern and Grotberg, 1993; 2003; Campana et al., 2004) and non-Newtonian effects such as viscoelastic fluids modify the stability and stress levels (Halpern et al., 2010). The airway walls are made of epithelial cells, as indicated, which experience fluid stresses from the closure process, enough to damage them (Bian et al., 2010; Tai et al., 2011a #4;Romano et al., 2019).

Following closure, inspiration forces the plug to propagate distally into the lung as shown in Figure 16.37(b). Typically the plug deposits more liquid into its trailing film than it picks up from the film ahead, causing it to shorten and eventually rupture, a process called airway reopening. At times t_5 and t_6 the plug is propagating, but getting shorter. Airway epithelial cells in this region of $0 \leq z < z_r$ experience large fluid stresses and axial stress gradients, both normal and tangential. The levels are much higher near the front interface, circled, compared to the rear interface, because the front layer is thinner, and it develops a capillary wave with a trough that reduces the local thickness even further (Howell et al., 2000; Fujioka and Grotberg, 2004; 2005; Fujioka et al., 2008; Ubal et al., 2008; Olgac and Moradoglu, 2013). It is this same reason that airway closure flow creates the largest fluid stresses on the wall during the post-coalescence flow at time t_3, as circled in Figure 16.37(a), since this is a "front" interface propagating forward.

The large stresses and gradients damage the cells (Bilek et al., 2003; Huh et al., 2007; Tavana et al.,

2010; Viola et al., 2019). Non-Newtonian fluid effects can modify the stresses (Zamankhan et al., 2012; 2018; Hu et al., 2019) as can wall flexibility (Howell et al., 2000; Zheng et al., 2009). At time $t = t_r$ the plug ruptures, producing a crackle sound which physicians hear with a stethoscope. The stresses on the cells in the vicinity of $z = z_r$ are much larger than the region $0 \leq z < z_r$, due to the interface snapping back toward the wall driving a locally high fluid stress environment also primarily at the front meniscus (Hassan et al., 2011; Yildiran et al., 2017). The cell damage there is also much higher (Huh et al., 2007). In general, the addition of surfactants reduces the fluid stresses during both propagation and rupture (Waters and Grotberg, 2002; Fujioka and Grotberg, 2005; Zheng et al., 2007; Tavana et al., 2010; 2011; Yildiran et al., 2017; Muradoglu et al., 2019).

This finding of crackle-related airway epithelial damage can provide a major change in clinical thinking. Since the invention of the stethoscope by French physician René Laennec (1781–1826) in 1816, crackle sounds were a sign of lung disease. Typically in that era, it was tuberculosis, and he died from this disease which ravaged his patients: a dedicated physician. Other causes of crackles include pulmonary edema, pulmonary fibrosis and acute respiratory distress syndrome (ARDS), which occurs from a variety of lung injuries like aspiration, infections and trauma. However, if the fluid mechanical event creating the sound is, itself, damaging the airway epithelium, then crackles can also cause disease by contributing to injury and inflammation. We met a similar issue regarding wheezing, see Section 9.9, where oscillations of airway epithelial cells promotes inflammation. So biofluid mechanics provides the basis for a major paradigm shift in the clinical use of the stethoscope, that crackles and wheezes are not only signs of disease, they also cause disease (Grotberg, 2019).

A semi-infinite plug is also a model for airway reopening and has a rear interface with an infinite, fluid filled tube ahead (Gaver et al., 1990; Naureckas et al., 1994; Perun and Gaver, 1995; Gaver et al., 1996). In addition, it pertains to the first breaths of a newborn where the lung is initially filled with a small amount of liquid. Possible airway epithelial damage during these breaths can be minimized by adding pulsations and surfactants to the inflow cycle (Gaver et al., 1996; Zimmer et al., 2005). Reviews of these two-phase airway flows are found in Grotberg (1994, 2001, 2011), Grotberg and Jensen (2004) and Ghadiali and Gaver (2008).

The effects of surface tension on mucociliary pumping is studied in Manolidis et al. (2016) for a three-dimensional model of an airway bifurcation. As in Section 7.3, the ciliary pumping is modeled as an imposed wall velocity. It is found that a minimal surface tension is necessary for efficiently removing the mucus while maintaining the mucus film thickness at physiological levels. Other approaches to modeling the ciliary motion range from the envelope model of Ross and Corrsin (1974), which replaces the ciliary tips with a continuous surface to the motion of individual cilia immersed in fluid and arranged in patterned arrays (Ding et al., 2014).

We studied a cough model of mucus transport in Section 8.7 for a single layer. In (Moriarty and Grotberg, 1999) a more thorough investigation treats a viscoelastic solid mucus slab resting on a Newtonian serous liquid layer which is adjacent to the airway wall. Surface tension competes with the elastic shear modulus in the dimensionless parameter σ/dG' where d is a characteristic layer thickness. Stability analysis reveals waves which have combined effects of both restoring forces from surface tension and elasticity.

Alveolar Flows

Figure 16.38(a) shows a two-dimensional lung alveolus with a thick fluid layer, which can occur in diseases such as acute respiratory distress syndrome (ARDS) (Ware and Matthay, 2000; Fan et al., 2018). Oscillating the radius mimics breathing, and in the presence of surfactants the flow during alveolar expansion (inspiration) and alveolar contraction (expiration) are not reversible, creating a time-averaged bulk flow. Figure 16.38(b, c) show the time-averaged streamlines which can vary as a single

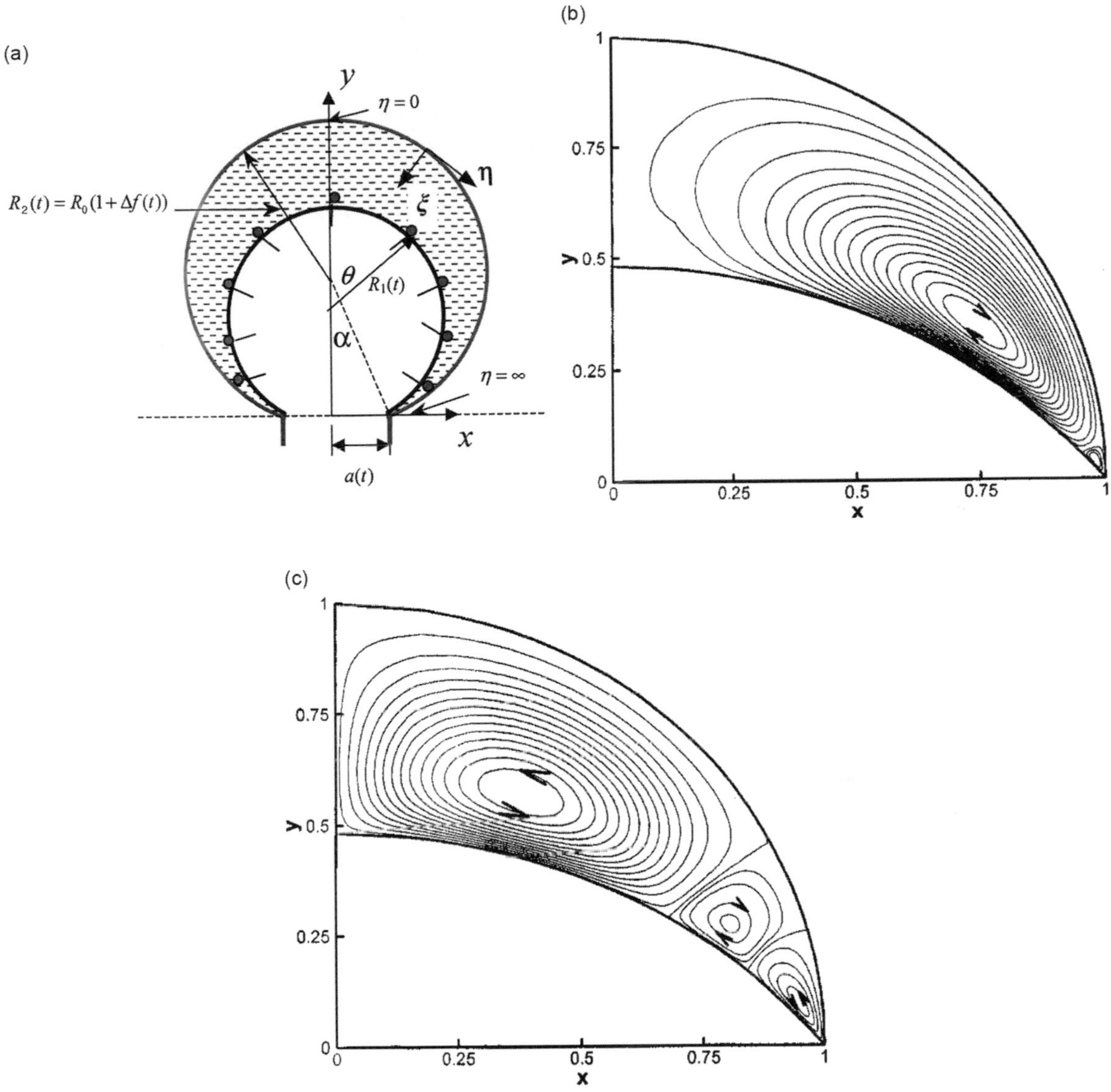

Figure 16.38 (a) Geometry for an oscillatory, two-dimensional alveolus with a thick fluid layer and surfactants. (b, c) Time-averaged streamlines for two values of the sorption parameter.

vortex to multiple vortices depending on the surfactant sorption properties. Such flows can provide a route for cell-cell signaling within the alveolus. This steady streaming from imposed oscillatory forcing is a feature of nonlinear systems and is further explored in Section 17.7.

Surfactant Replacement Therapy

A treatment for neonatal RDS (hyaline membrane disease) is to instill replacement surfactants directly into the trachea in a process called surfactant replacement therapy (SRT) (Kendig et al., 1988). The instillation is followed by applied breaths which force the liquid into the tracheobronchial tree where it forms plugs that propagate in airways and split at bifurcations. A two-step process approach is to consider the plug propgation separately from the plug splitting. During propagation the plug leaves a trailing film which coats the airway wall. This surfactant is lost to the delivery, not available to the alveoli in a timely manner. The splitting step divides the plug uneventLy based on geometric and gravity effects. Using one-dimensional conservation of mass and momentum, but applied to a three-dimensional airway tree, SRT distributions are simulated in Tai et al. (2011b #112), Filoche et al. (2015), Grotberg et al. (2017) and Kazemi et al. (2019) based on earlier work in Zheng et al. (2005, 2006), Cassidy et al. (2001a, 2001b) and Halpern et al. (1998).

In Figure 16.39 we see the simulation results for SRT in a neonate compared to an adult at two different dose volumes per kilogram: $V_D/kg = 1\,ml/kg$, $3\,ml/kg$. Half of the dose is given in the left lateral decubitus (laying on the left side) position and half in the right lateral decubitus (laying on the right side) position. In Figure 16.39(a) the neonatal airway tree has a trachea and eight generations of airways while the adult tree, Figure 16.39(d), has 12. Both are sized appropriately with 1 cm and 10 cm lengths indicated. The SRT simulation is represented by shaded spheres at the airway terminations, lighter for less and darker for more delivery as indicated by the shade bars. For the 1 kg neonate, there is good homogeneity of the distribution for both dose volumes and clearly more reaches the tree terminations for the 3 ml/kg dose volume, Figure 16.39(c), compared to 1 ml/kg, Figure 16.39(b). However, for the 70 kg adult there are significant problems at 1 ml/kg, Figure 16.39(e), where very few of the tree terminations receive surfactant and the homogeneity is poor. The gravity effects favor the two decubitus positions where delivery to the tree terminations is very focused in the most gravity-dependent regions. At 3 ml/kg the delivery is much better, Figure 16.39(f), with much more reaching the terminations and far greater homogeneity.

The neonatal clinical field of SRT successfully shifted from 3 to 1 ml/kg, while increasing the surfactant concentration of the mixture. Based on that experience, the adult SRT field did the same, but met with failure. For SRT the model showed that more surfactant mixture was being lost to airway coating, the coating cost, in adults compared to neonates per kg, primarily from the lower number of airway generations in a neonate. The surface area of neonatal conducting airways is $\sim 40\,cm^2$ while an adult is $\sim 4{,}500\,cm^2$. The adult/neonate airway surface area ratio in cm^2 is ~ 110:1, while the kg weight ratio for the patient population is ~ 70:1–3. So effective dose volumes do not scale with weight. Biofluid mechanics has a central role in creating meaningful models for basic medical science and clinical applications.

16.7 Respiratory Epidemics

During an epidemic, there is an outbreak of disease that involves many people at the same time, spreading rapidly over a defined area due to a common source, oral-fecal, respiratory or other mechanisms. When that area enlarges enough, e.g. crossing national boundaries, it is often called a pandemic. Cholera, smallpox, bubonic plague, yellow fever, typhoid fever, diphtheria, polio, measles, whooping cough, plague, malaria and influenza have all appeared as epidemics and/or pandemics. A famous respiratory virus example is the Spanish flu pandemic of 1918–1920 which infected 500 million people globally, one-third of the world population, with estimates of 20–50 million deaths, see Figure 16.40(b). The 2019–2021 coronavirus global pandemic is another example.

Fundamental to respiratory transmission is the biofluid mechanics of cough and sneeze, see Section 13.6 (Settles, 2006; Bourouiba et al., 2014; Craven et al., 2014). The entire respiratory system (lung, nose, pharynx) is coated with a thin liquid film which can be aerosolized from the explosive turbulent flow over it due to a coupled instability between the airflow and the film (Moriarty and Grotberg, 1999; Edwards et al., 2004; Seminara et al., 2020). Thousands of virus-

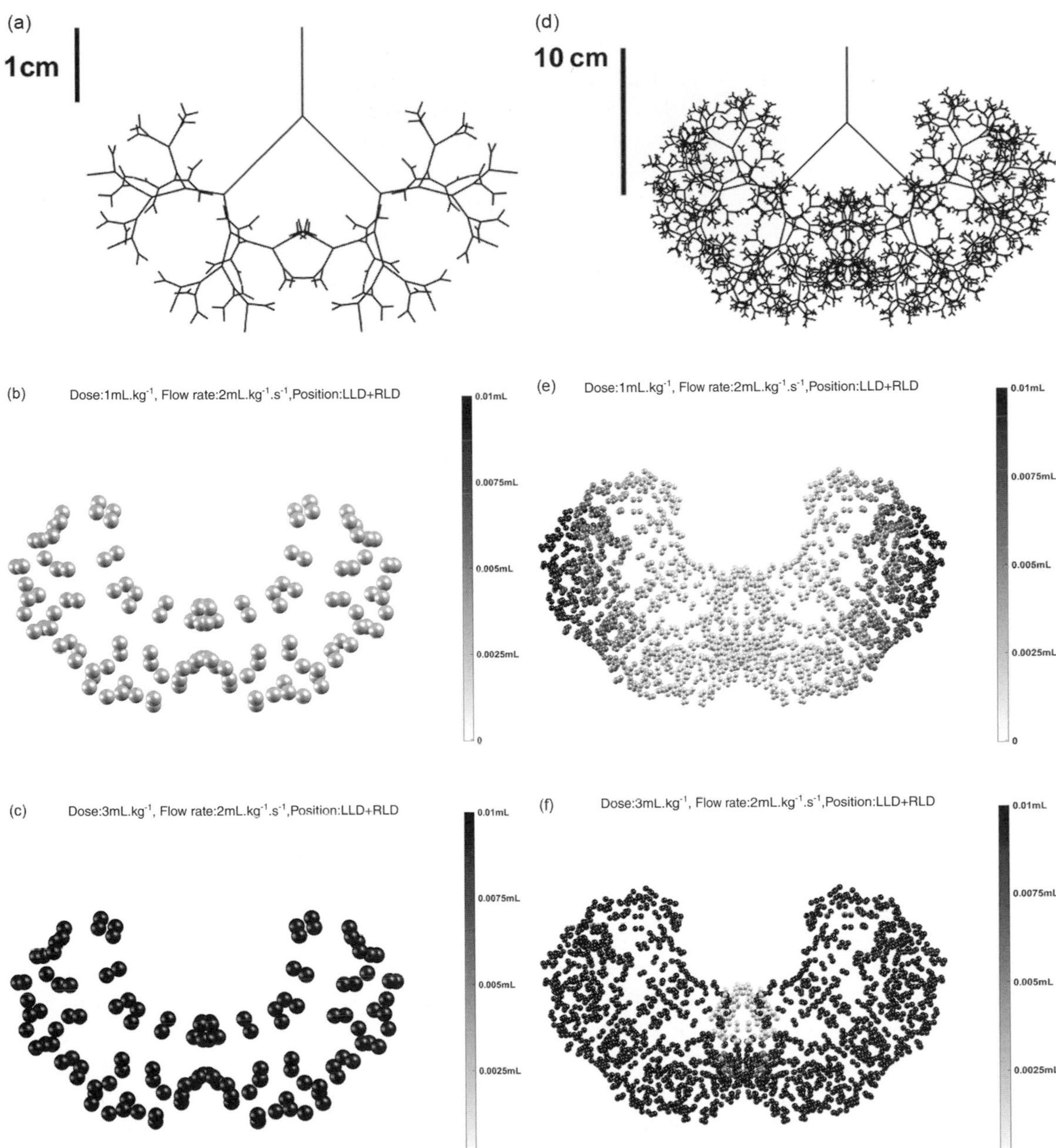

Figure 16.39 (a) Neonatal airway tree showing eight generations; (b) neonatal SRT at $V_D/kg = 1$ ml/kg; (c) neonatal SRT at $V_D/kg = 3$ ml/kg; (d) adult airway tree with 12 generations; (e) adult SRT at $V_D/kg = 1$ ml/kg; and (f) adult SRT at $V_D/kg = 3$ ml/kg.

(a)

(b)

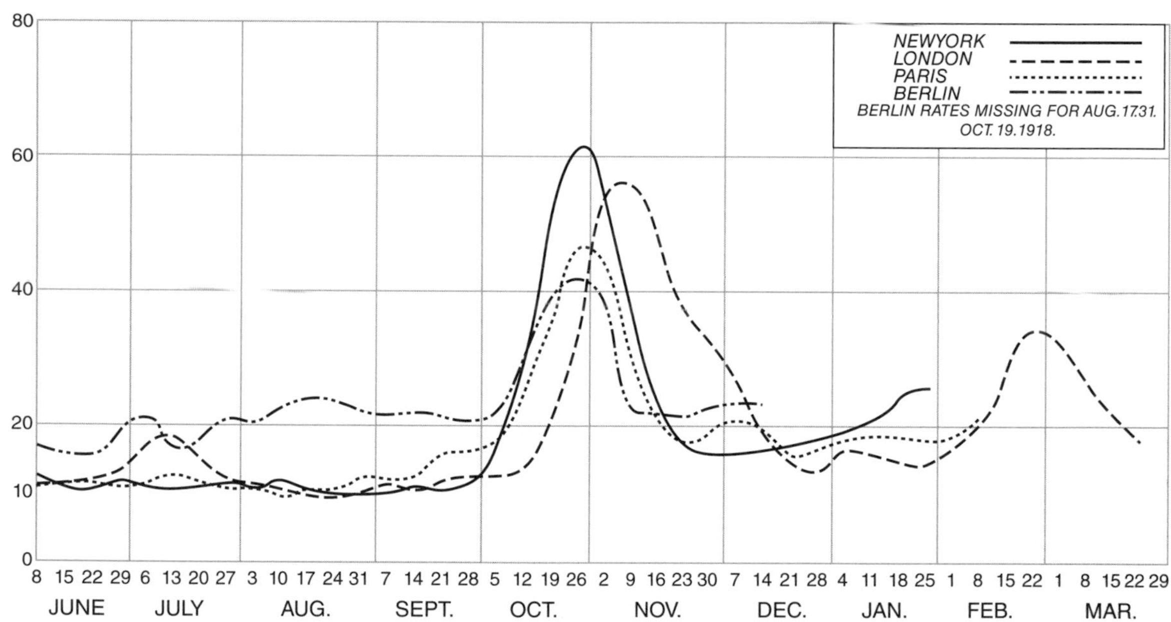

Figure 16.40 (a) Liquid droplets propelled into the air from a sneeze. (b) Spanish flu deaths in major cities.

laden droplets are propelled into the air from the infected person, see Figure 16.40(a). Those droplets can be inhaled by the susceptible person or land on surfaces which become contaminated and then picked up. Eventually, over time, the infected person recovers or dies.

We examined the terminal settling velocity, v_t, of a spherical particle in a fluid in Section 2.7 regarding the erythrocyte sedimentation rate, deriving Eq. (2.7.10): $v_t = (\rho_p - \rho_f)d_p^2 g/18\mu$. The studies in Bourouiba et al. (2014) focus on the wide range of droplet diameters, $1\ \mu m < d_p < 100 + \mu m$, their shrinkage due to evaporation, and buoyancy of the body-temperature cough jet into the cooler atmosphere. That combination can extend the travel distance of the infectious cloud before it lands or is inhaled.

A standard model for epidemics is the Kermack–McKendrick model. It relates the three categories of individuals whose total is N: the susceptible S(t), the infected I(t) and the recovered R(t) by the following coupled ODEs

$$\frac{dS}{dt} = -\beta\frac{SI}{N}$$
$$\frac{dI}{dt} = \beta\frac{SI}{N} - \gamma I \qquad (16.7.1)$$
$$\frac{dR}{dt} = \gamma I$$

The transmission term is represented by $\beta SI/N$ and the coefficient β is a rate measured in d^{-1}, where d = days. The average time between transmission contacts by the infected individual with susceptibles is $T_c = 1/\beta$ d. This term appears with a negative sign for dS/dt, since it reduces that group, and a positive sign for dI/dt, since it increases that group. The term γI represents transition from infected to recovered. γ is also a rate measured in d^{-1} and gauges how rapidly an infected individual recovers to be noninfectious. The time to recovery is $T_r = 1/\gamma$ d. This term appears with a negative sign for dI/dt, since it decreases that group, and with a positive sign for the for dR/dt, since it increases that group.

The system of N individuals starts at t = 0 with n being infected, N − n susceptible, and none recovered yet. Typically n/N<<1 so there are a comparatively small number of seed infections. Mathematically, the initial conditions on Eqs. (16.7.1) are $I(0) = n$, $S(0) = N - n$, $R(0) = 0$. At all times the sum of categories is N, $I(t) + S(t) + R(t) = N$, and we note that $dS/dt + dI/dt + dR/dt = 0$.

In addition to n/N, the system is also governed by the dimensionless ratio $R_0 = \beta/\gamma = T_r/T_c$, which is called the basic reproduction number. It is the average number of people that a single infectious person will infect over the course of their infection. Rewriting the rate equation for infections gives us one of the important roles of R_0 at t = 0

$$\frac{dI}{dt} = \left(R_0\frac{S}{N} - 1\right)\gamma I \qquad (16.7.2)$$

The initial sign of $dX_I(0)/dt$ depends on the sign of $R_0 S(0)/N - 1$. If $R_0 < N/S(0) \sim 1$ then $dI(0)/dt < 0$ and the infected category immediately shrinks and there is no epidemic. If $R_0 > N/S(0) \sim 1$ then $dI(0)/dt > 0$ and the infected category immediately grows and the epidemic begins. Once started, R_0 controls the dynamics of the ensuing process.

Examples of the model predictions are shown in Figure 16.41 for N = 1000, n = 1. Figure 16.41(a) shows the results for $\beta = 0.4/d$, $\gamma = 0.2/d$, $R_0 = 2.0$. The epidemic duration is ~ 80 d with the maximum $I_{max} \sim 150$ at $t \sim 35$ d. Both numbers impact health care capabilities, particularly I_{max}, which can cause a surge of hospitalizations and need for critical care and mechanical ventilators. Approximately 800 people recovered or, stated differently, 800 people were infected. To "flatten the curve" with mitigation there is social distancing, washing hands, coughing into your upper sleeve, disinfecting surfaces, wearing masks which filter out the droplets and avoiding touching your face. All can be effective to reduce β, say to $\beta = 0.3$. Figure 16.41(b) shows the result reducing $I_{max} \sim 60$ at $t \sim 60$ d and total duration ~120 d. The surge strain on hospitals and healthcare workers is less and fewer people, ~600, were infected, presumably saving lives.

16.8 Interfacial Kinematic Boundary Condition

In Section 7.3 we discussed clearance of the airway liquid lining by cilia, modeled as a simple moving

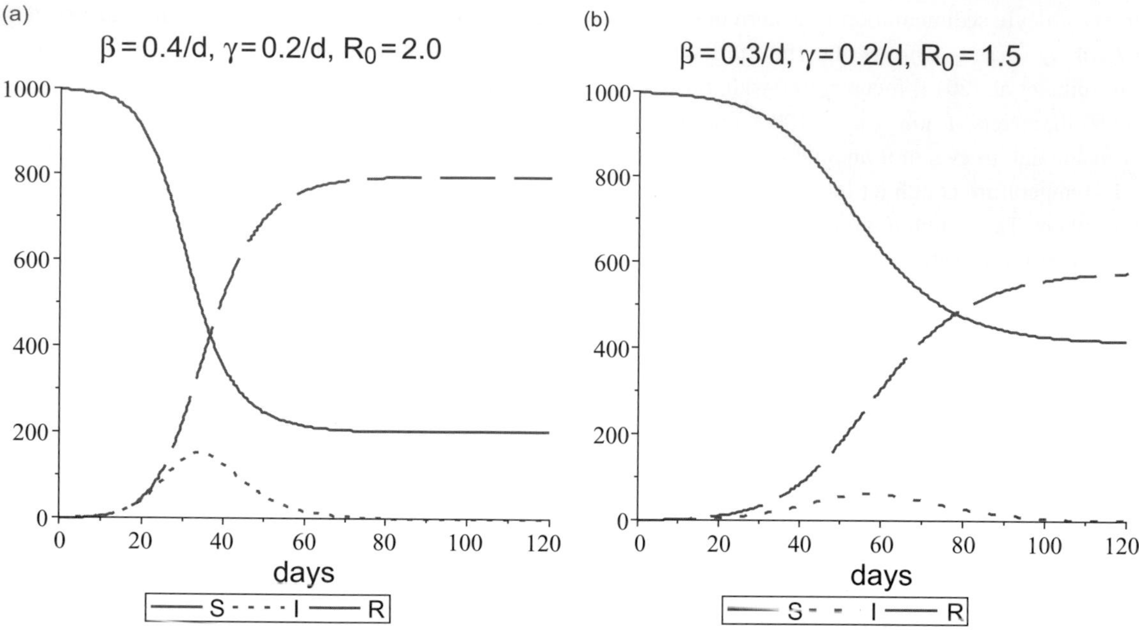

Figure 16.41 Epidemic model results for $n = 1$, $N = 1000$: (a) $\beta = 0.4/d$, $\gamma = 0.2/d$, $R_0 = 2.0$ and (b) $\beta = 0.3/d$, $\gamma = 0.2/d$, $R_0 = 1.5$.

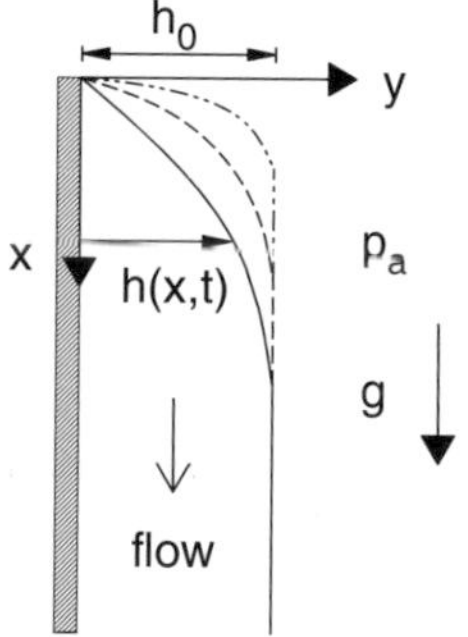

Figure 16.42 Unsteady draining film as an example of a deforming, unsteady interface.

wall, in the presence of gravity. The liquid film had uniform thickness for its infinite length. Here we analyze the non-uniform, evolving film thickness of a draining system shown in Figure 16.42 where the film length is semi-infinite. The film starts out with uniform thickness h_0, but without new fluid being introduced from above. It drains creating an interface shape $h(x, t)$ which is sketched at three time values.

For small Reynolds number and a layer that is thin compared to its length, lubrication theory gives us the x and y momentum balances and conservation of mass, respectively

$$-\frac{\partial p}{\partial x} + \mu \frac{\partial^2 u}{\partial y^2} + \rho g = 0$$

$$-\frac{\partial p}{\partial y} = 0 \tag{16.8.1}$$

$$\frac{\partial u}{\partial x} + \frac{\partial v}{\partial y} = 0$$

Pressure of the passive atmospheric gas, p_a, is constant, and we can simplify the analysis by neglecting surface tension. Then the liquid pressure at $y = hx, t$ is always p_a and, since $\partial p/\partial y = 0$, we have $p = p_a$ everywhere in the liquid. Consequently, there is no pressure gradient in the x-direction either, i.e. $\partial p/\partial x = 0$. Integrating the simplified x-component of momentum balance in Eq. (16.8.1) gives us

$$u(x, y, t) = -\frac{\rho g}{\mu}\frac{y^2}{2} + f_1(x, t)y + f_2(x, t) \quad (16.8.2)$$

The boundary conditions for Eq. (16.8.2) are no-slip at the wall, $u(y = 0) = 0$, and no-shear at the interface, $\mu(\partial u/\partial y)_{y=h(x,t)} = 0$ for a passive gas. Imposing those on Eq. (16.8.2) solves f_1, f_2 and leads to the form

$$u = \frac{\rho g}{\mu}\left(hy - \frac{y^2}{2}\right) \quad (16.8.3)$$

From conservation of mass in Eq. (16.8.1), the form for v is determined as

$$v = -\int y\frac{\partial u}{\partial x}dy + f_3(x, t) = -\frac{\rho g}{\mu}\frac{\partial h}{\partial x}\frac{y^2}{2} \quad (16.8.4)$$

where $f_3 = 0$ by imposing no penetration at the wall, $v(y = 0) = 0$. At this stage, h(x, t) is still unknown, but we can determine it from the kinematic boundary condition for the interface. Let's derive that condition.

A fluid particle on the interface whose position is (x_p, y_p) must remain on the interface, i.e. it does not leave making a hole. Clearly the relationship between x_p and y_p is

$$y_p = h(x_p, t) \quad (16.8.5)$$

For a small displacement of the surface the new position of the particle is

$$y_p + \Delta y_p = h(x_p + \Delta x_p, t + \Delta t) = h(x_p, t) + \frac{\partial h}{\partial x}\Delta x_p + \frac{\partial h}{\partial t}\Delta t \quad (16.8.6)$$

where we have used a Taylor series expansion in Eq. (16.8.6). Subtract Eq. (16.8.5) from Eq. (16.8.6) and divide through by Δt. The result is

$$\frac{\Delta y_p}{\Delta t} = \frac{\partial h}{\partial x}\frac{\Delta x_p}{\Delta t} + \frac{\partial h}{\partial t} \quad (16.8.7)$$

Taking the limit of Eq. (16.8.7) for $\Delta x_p \to 0$, $\Delta y_p \to 0$, $\Delta t \to 0$, we find $\Delta x_p/\Delta t$ converges to the x-component of the particle's Lagrangian velocity, U_p. Through the inverse mapping the Lagrangian form is readily transformed to the Eulerian velocity, u(x, t) at $x = x_p$. The same can be done for $\Delta y_p/\Delta t$, which becomes v(x, t). The final form of Eq. (16.8.7) is

$$\frac{\partial h}{\partial t} + u\frac{\partial h}{\partial x} = v \quad \text{at} \quad y = h(x, t) \quad (16.8.8)$$

Equation (16.8.8) can be expressed more generally in terms $F = y - h(x, t)$ where the interface is at $F = 0$

$$\frac{\partial F}{\partial t} + \underline{u} \cdot \underline{\nabla} F = 0 \text{ at } F = 0 \quad (16.8.9)$$

A more general form of Eq. (16.8.9) for a surface in three-dimensional space, $S = z - H(x, y, t) = 0$, is

$$\frac{\partial S}{\partial t} + \underline{u} \cdot \underline{\nabla} S = 0 \quad \text{at} \quad S = 0 \quad (16.8.10)$$

Recognize Eq. (16.8.10) as the Lagrangian time derivative of S, which is zero. If there are two fluids separated by the interface, call them fluids 1 and 2 as in Eq. (16.8.10) can be applied for each one in terms of their respective velocity fields, $\underline{u}_1$, $\underline{u}_2$,

$$\frac{\partial S}{\partial t} + \underline{u}_1 \cdot \underline{\nabla} S = 0 \text{ at } S = 0 \quad \frac{\partial S}{\partial t} + \underline{u}_2 \cdot \underline{\nabla} S = 0 \text{ at } S = 0 \quad (16.8.11)$$

Subtracting the two equations and dividing through by $|\nabla S|$ gives us

$$\underline{u}_1 \cdot \underline{n} = \underline{u}_2 \cdot \underline{n} \quad (16.8.12)$$

since $\underline{n} = \underline{\nabla} S/|\underline{\nabla} S|$. Equation (16.8.12) states that the normal fluid velocities must be equal, which makes the interface a streamline.

Substituting Eqs. (16.8.3) and (16.8.4) into Eq. (16.8.9) results in

$$\frac{\partial h}{\partial t} + \frac{\rho g}{\mu}h^2\frac{\partial h}{\partial x} = 0 \quad (16.8.13)$$

Equation (16.8.13) can be solved using a similarity variable $\xi = x/t$, see Section 8.1. The final form is

$$\frac{dh}{d\xi}\left(\frac{\rho g}{\mu}h^2 - \xi\right) = 0 \quad (16.8.14)$$

Equation (16.8.14) has two solutions. One comes from setting the bracketed term of Eq. (16.8.14) to zero so that

$$h = \left(\frac{x}{t}\frac{\mu}{\rho g}\right)^{1/2} \quad \text{for } 0 \le \frac{x}{t} \le \frac{\rho g h_0^2}{\mu} \tag{16.8.15}$$

The range of x/t gives $0 \le h \le h_0$. For fixed time, h increases with x, $h \sim x^{1/2}$, until $h = h_0$. For larger x, the other solution to Eq. (16.8.14) applies, $dh/d\xi = 0$, or h is the constant h_0

$$h = h_0 \text{ for } \frac{x}{t} > \frac{\rho g h_0^2}{\mu} \tag{16.8.16}$$

Using the scales $T = tU_s/h_0, X = x/h_0, \ H = h/h_0$, where $U_s = \rho g h_0^2/\mu$, the solution is represented more compactly as

$$H = \left(\frac{X}{T}\right)^{1/2} \quad \text{for } \ X \le T, \quad H = 1 \ \text{ for } X > T \tag{16.8.17}$$

Figure 16.43 displays H(X, T) for three values of T = 0.25, 1.0, 2.0. Note the curvilinear portions are for $0 \le X \le T$ according to Eq. (16.8.17), and H = 1 for X > T. The two solutions intersect at X = T.

Another example of the kinematic boundary condition is surface wave propagation. Consider a semi-infinite liquid (water) occupying the region $0 \ge y \ge -\infty$ whose undisturbed, flat interface with the overlying air is at y = 0, see Figure 16.44. The interface has surface tension, σ, gravity, g, acts in the negative y-direction, and the constant air pressure is p_a. Now perturb the interface with a small amplitude traveling wave whose position is $y = h(x, t) = h_0 e^{i(kx-\omega t)}$. The wavenumber is $k = 2\pi/\lambda$ where λ is the wavelength and $\omega = 2\pi f$, where f is the frequency. The wave speed or phase velocity is $c = f\lambda = \omega/k$. Waves need a restoring force and inertia to overshoot the equilibrium state. In this case there are two restoring forces, gravity and surface tension, while inertia comes from the water density. This is a dispersive system where capillary-gravity waves of different wavelengths travel at different speeds. We studied dispersion in Section 9.9 with regard to flutter waves in a flexible channel as a model of wheezing.

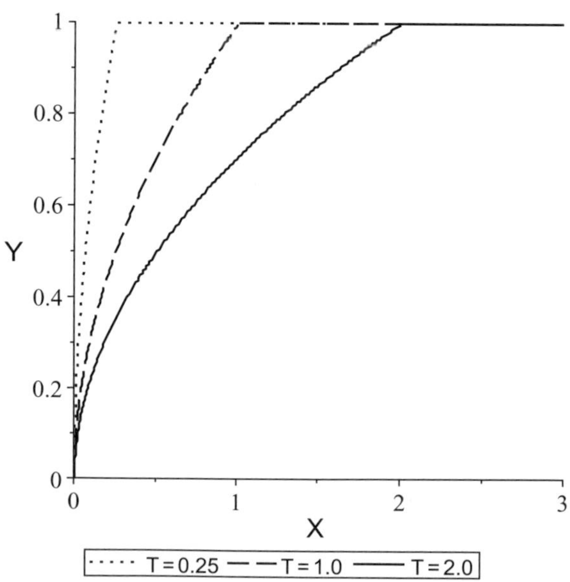

Figure 16.43 Interface evolution of a draining film.

In Section 6.2 we showed that the velocity vector can be written as the gradient of a potential, $\underline{u} = \underline{\nabla}\phi$, for inviscid flows which is appropriate here. Conservation of mass for this flow appears in Eq. (6.2.15), the Laplace's equation $\nabla^2\phi = 0$. Let $\phi = f(y)e^{i(kx-\omega t)}$ and insert to find

$$\frac{d^2f}{dy^2} - k^2 f = 0 \tag{16.8.18}$$

Using the form $f \sim e^{sy}$ we insert into Eq. (16.8.18) to find $s = \pm k$. These two roots give two linearly independent solutions

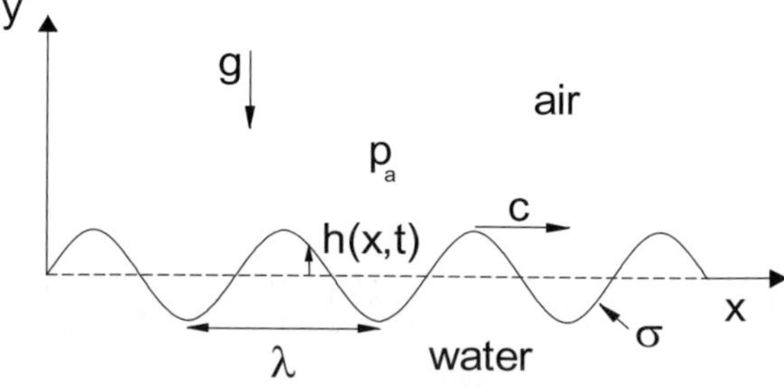

Figure 16.44 Traveling wave at an air–water interface with gravity and surface tension.

$$f = f_0 e^{ky} + g_0 e^{-ky} \qquad (16.8.19)$$

where f_0 and g_0 are constants to be determined from boundary conditions. For this deep water example, we want velocities to vanish as $y \to -\infty$, which sets $g_0 = 0$.

The stress boundary condition is $p_a - p = \sigma \partial^2 h/\partial x^2$ at $y = h$. Pressure at $y = h$comes from the unsteady Bernoulli Equation, Eq. (6.2.14), and the relevant form is $\partial\phi/\partial t|_{y=0} + \cancel{\nabla\phi\cdot\nabla\phi}/2 + p/\rho + gh = p_a/\rho$. We neglect the nonlinear term due to the small perturbation and the boundary condition has been linearized for $\partial\phi/\partial t$ by evaluating it at $y = 0$. Inserting for the pressure, the stress boundary condition becomes

$$\rho \frac{\partial\phi}{\partial t}_{y-0} + \rho g h = \sigma \frac{\partial^2 h}{\partial x^2} \qquad (16.8.20)$$

Substituting for ϕ, h solves f_0 in terms of h_0, i.e. $f_0 = -ih_0(k^2\sigma + \rho g)/\rho\omega$.

The final equation is the kinematic boundary condition

$$\frac{\partial h}{\partial t} + \cancel{\left(\frac{\partial\phi}{\partial x}\right)\left(\frac{\partial h}{\partial x}\right)} = \left.\frac{\partial\phi}{\partial y}\right|_{y=0} \qquad (16.8.21)$$

where we neglect the nonlinear term and have linearized for $\partial\phi/\partial y$ to be evaluated at $y = 0$. Inserting the forms of h, ϕ into this condition, the constant h_0 cancels out and the result is the dispersion relation $\omega(k)$ and the corresponding wave speed $c = \omega/k$,

$$\omega = \left(k^3 \frac{\sigma}{\rho} + kg\right)^{1/2}, \quad c = \frac{\omega}{k} = \left(k\frac{\sigma}{\rho} + \frac{g}{k}\right)^{1/2} \qquad (16.8.22)$$

Short waves, k large, are dominated by surface tension while long waves, k small, are dominated by gravity.

Figure 16.45 is a plot of the phase velocity, c vs λ for air and water with $g = 980\,\text{cm/s}^2$, $\rho = 1\ \text{g/cm}^3$ and $\sigma = 70, 50, 30$ dyn/cm. There is a minimum value of $c = c_0$ and at $\lambda = \lambda_0$, Figure 16.45(a). For $\lambda >> \lambda_0$ gravity dominates as the three curves for different values of surface tension merge. For $\lambda << \lambda_0$ surface tension dominates and the distinction is easier to see in the expanded region, see Figure 16.45(b).

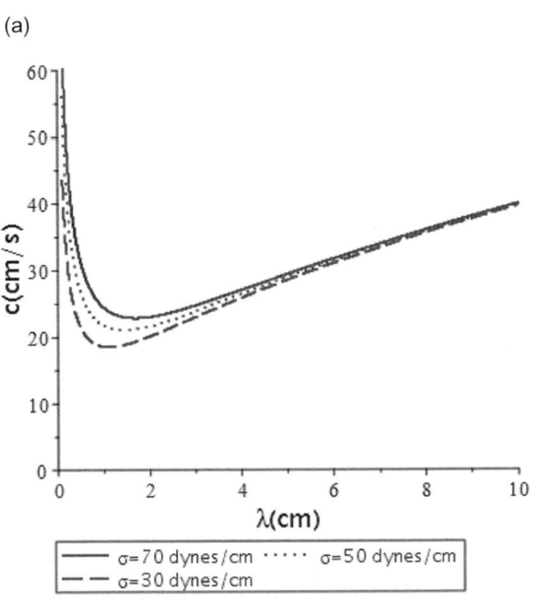

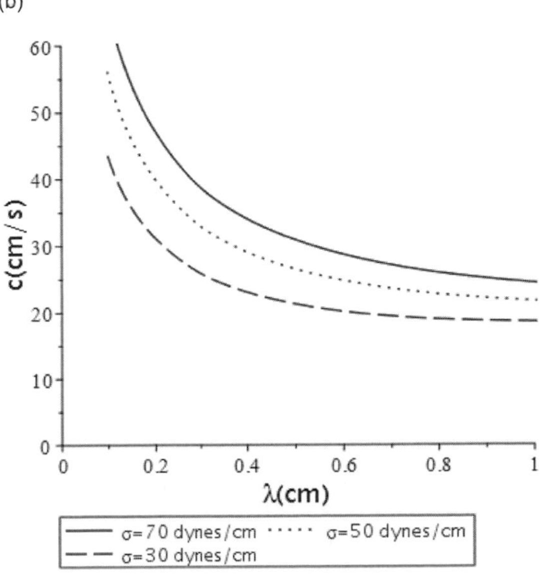

Figure 16.45 Wave speed, phase velocity and dispersion for capillary-gravity waves.

Summary

In this chapter we considered two-phase flows in biological applications. Key concepts were the description of the fluid–fluid interface, the physics and chemistry of surface tension, as well as surfactants. The derivations of the full stress boundary condition and kinematic boundary condition were complicated, but gave us an understanding of pressure jumps across curved interfaces, Marangoni effects from surface tension gradients, and the evolution of an interface.

Applications ranged from mosquitoes to whales, both seeking a meal; Caisson workers to health care professionals, both doing their job; and sweat to urination, both a relief. The organ system which is quite strongly affected by surface tension is the lung, and several flows were studied which underlie its normal function and disease states, as well as drug delivery to treat surfactant deficiency or dysfunction. Finally, the application to epidemics and pandemics was explored as a timely topic.

Problems

16.1

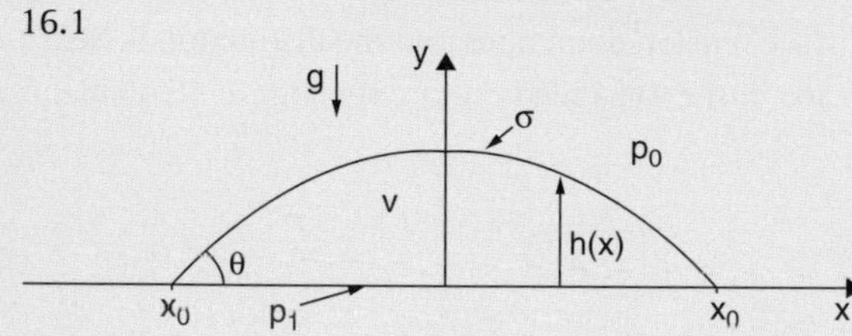

a. Let the shape of a 2D hanging drop be given by $F = y - h(x) = 0$. See the figure. What is the unit normal, $\underline{n}$, for this surface?

b. From your solution for $\underline{n}$ derive the curvature, $\kappa = \underline{\nabla} \cdot \underline{n}$. Simplify your result for small slope by neglecting terms of $(dh/dx)^2$.

c. The hydrostatic pressure is $p = p_1 - \rho g y$ and the stress boundary condition is $p_0 - p(y = h) = -\sigma\kappa$. Derive the result

$$\frac{d^2h}{dx^2} - \frac{\rho g}{\sigma}h = -\frac{(p_1 - p_0)}{\sigma} \quad (1)$$

Show that the general solution to Eq. (1) is $h(x) = c_3 \cosh(x/\ell_c) + c_4 \sinh(x/\ell_c) + h_p$ where $\ell_c = (\sigma/\rho g)^{1/2}$ is the capillary length scale and $h_p = (p_1 - p_0)/\rho g$.

d. Solve for the two constants of integration, c_3 and c_4, by imposing symmetry at the origin, $dh/dx_{x=0} = 0$, and intersection with the plane at $x = x_0$, $h(x = x_0) = 0$. What is your resulting form of $h(x)$?

e. To obtain the solution in terms of the contact angle and drop volume, first impose the contact angle condition $dh/dx_{x=x_0} = -\tan(\theta)$ which solves for h_p.

f. Now apply the drop volume, v, condition to find the relationship between x_0 and v. Let $v = 2L\int_0^{x_0} h(x)\,dx$where L is a unit depth in the z-direction.

g. Make your solutions dimensionless with the scalings $X = x/\ell_c$, $X_0 = x_0/\ell_c$, $H = h/\ell_c$ and $V = v/L\ell_c^2$. Show that H and V are

h. Plot V vs X_0 for $\theta = 0.1, 0.2, 0.3$ and $0.01 \leq X_0 \leq 30$. In a particular problem, typically V is given and X_0 is determined as a root of the V equation.

i. Plot H(X) for $V = 0.5,\ 1,\ 3$ and $\theta = 0.2$. Describe your results.

j. Plot H(X) for $V = 1$ and $\theta = 0.1,\ 0.2,\ 0.3$. Describe your results.

16.2

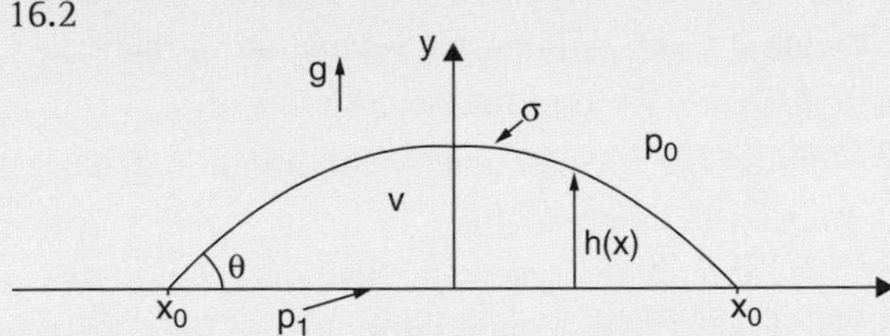

a. Let the shape of a 2D hanging drop be given by $F = y - h(x) = 0$, see the figure. Note that gravity is pointing in the positive y-direction, indicating a hanging drop. What is the unit normal, $\underline{n}$, for this surface?

b. From your solution for $\underline{n}$derive the curvature, $\kappa = \underline{\nabla}\cdot\underline{n}$. Simplify your result for small slope by neglecting terms of $(dh/dx)^2$.

c. The hydrostatic pressure is $p = p_1 - \rho g y$ and the stress boundary condition is $p_0 - p(y = h) = -\sigma\kappa$. Derive the result

$$\frac{d^2h}{dx^2} + \frac{\rho g}{\sigma}h = -\frac{(p_1 - p_0)}{\sigma} \qquad (1)$$

where $-x_0 \le x \le x_0$. Show that the general solution to Eq. (1) is $h(x) = c_3 \cos(x/\ell_c) + c_4 \sin(x/\ell_c) - h_p$ where $\ell_c = (\sigma/\rho g)^{1/2}$ and $h_p = (p_1 - p_0)/\rho g$.

d. Solve for the two constants of integration, c_3 and c_4, by imposing symmetry at the origin, $dh/dx_{x=0} = 0$, and intersection with the plane at $x = x_0$, $h(x = x_0) = 0$. What is your resulting form of h(x)?

e. To obtain the solution in terms of the contact angle and drop volume, first impose the contact angle condition $dh/dx_{x=x_0} = -\tan(\theta)$ which solves for h_p.

f. Now apply the drop volume, v, condition to find the relationship between x_0 and v. Let $v = 2L\int_0^{x_0} h(x)\,dx$ where L is a unit depth in the z-direction.

g. Make your solutions dimensionless with the scalings $X = x/\ell_c$, $X_0 = x_0/\ell_c$, $H = h/\ell_c$ and $V = v/L\ell_c^2$. Show that H and V are

$$H = \frac{\tan(\theta)}{\sin(X_0)}(\cos(X) - \cos(X_0)),\ V = 2\tan(\theta)\left(1 - \frac{X_0}{\tan(X_0)}\right)$$

What happens for $X_0 \ge \pi$? Why?

h. Plot V vs X_0 for $\theta = 0.1,\ 0.2,\ 0.3$ and $0.01 \le X_0 \le 0.9\pi$. In a particular problem, typically V is given and X_0 is determined as a root of the V equation.

i. Plot H(X) for $V = 0.2, 1.0, 3.0$ and $\theta = 0.2$. Describe your results. Try plotting as –H(X) to see the "hanging drop." Ignore the negative signs for your tick marks or replace them with positive values.

j. Plot H(X) for $V = 1$and $\theta = 0.1, 0.2, 0.3$. Describe your results.

16.3

a. In the analysis of a draining vertical film we used a similarity solution approach to solve Eq. (16.8.13). Assume the similarity variable is $\xi = x/t^a$, and apply the chain rule to $\partial/\partial x$, $\partial/\partial t$ to transform the equation, replacing x with $x = \xi t^a$ in the coefficients. Show that $a = 1$ leaves the equation only in terms of ξ.

b. The dimensionless interface position is H(X, T) as $H = (X/T)^{1/2}$ for $X \le T$ and $H = 1$ for $X > T$. For $U_s = \rho g h_0^2/\mu$, use the scales $T = tU_s/h_0$, $X = x/h_0$, $Y = y/h_0$, $U = u/U_s$, $V = v/U_s$ and $H = h/h_0$ to derive the dimensionless forms of the velocity components U and V. There will be two forms of each, one for $X \le T$ and one for $X > T$. From your solutions, what sign is U, understanding that $0 \le Y \le H$? How about V?

c. Now derive the dimensionless stream function $\psi(X, Y, T)$ in two forms associated your two forms of U and V. Assume $\psi(Y = 0) = 0$. Plot streamlines for $\psi = 0.0025, 0.02, 0.05, 0.10, 0.17, 0.25$ for $T = 2$ and $0 \leq X \leq 4$. Both of your forms for ψ will be needed. Include a plot of H as a boundary for your streamlines. What do you note about the streamlines?

d. Plot the X-velocity profile, $U(X, Y, T)$, at $T = 1$ for $X = 0.25, 0.50, 1.0, 2.0$ and include a plot of H as a boundary in Y for your profiles. Offset the profiles by adding their X value to U. Also include a vertical line $0 \leq Y \leq H$ for each X-value as a zero-line for the corresponding velocity profile. Describe your results. How does it compare to the analysis in Section 7.3?

e. Integrate your two forms of U over the range $0 \leq Y \leq H$ to find the X-direction flow rate, Q in its two forms. Plot Q at $X = 0.25, 0.50, 0.75$ for $0.01 \leq T \leq 4$ and describe your result.

16.4

a. For the plots in Figure 16.45, calculate the minimum wave speed values $c_0(\lambda_0)$. What is the effect of surface tension?

b. To see the effects of eliminating either restoring force, plot the wave speed solution as in Figure 16.45 for the case of $\sigma = 70$ dyn/cm, $g = 980$ cm^2/s. Add to your plot a curve for no surface tension, $\sigma = 0$, $g = 980$ cm^2/s, and another for no gravity, $\sigma = 70$ dyn/cm, $g = 0$. Describe the results.

c. Now plot the group velocity, $c_g = \partial\omega/\partial k$, for the same conditions in part (a) and calculate the minima. How do they compare to c?

d. Plot the velocity vector field. Make the variables dimensionless with $\xi = kx, \eta = ky, T = \omega t$, $U = \omega u / h_0 k (k^2\sigma + g), V = \omega v / h_0 k (k^2\sigma + g)$ where u, v are the dimensional velocities derived from the real parts of $u = \partial\phi/\partial x, v = \partial\phi/\partial y$. Choose $T = 0, 0 \leq \xi \leq 2\pi, 0 \geq \eta \geq -1$.

e. Now animate your plot in part (g) for $0 \leq T \leq 2\pi$. What direction is the wave propagation? What is happening to the velocity vector at a particular value of ξ, η.

16.5

a. During the development of capillary-gravity wave propagation in Section 16.8 the velocity potential was found to be $\phi = f(y)e^{i(kx-\omega t)}$ with $f = f_0 e^{ky} + g_0 e^{-ky}$ given by Eq. (16.8.19). Instead of zero y-velocity at infinity, suppose there is a wall at $y = -d$. Impose no-penetration there, i.e. $\partial\phi/\partial y|_{y=-d} = 0$ to find g_0 in terms of f_0.

b. Impose the stress boundary condition Eq. (16.8.20) to find f_0 in terms of h_0.

c. Now impose the kinematic boundary condition Eq. (16.8.21). h_0 should drop out and the result is the dispersion relationship $\omega(k)$. What is it?

d. Because there is a new parameter, d, compared to the semi-infinite fluid, we have a dimensionless parameter the Bond number $Bo = \rho g d^2/\sigma$. Additionally, it is convenient to define the dimensionless wave number, $K = kd$. Let the dimensionless frequency squared be $\Omega^2 = \omega^2 \rho d^3/\sigma$. Show that the result is $\Omega^2 = K \tanh(K)(Bo + K^2)$.

e. Calculate the dimensionless wave speed, $C = \Omega/K$. Plot C vs K for $Bo = 0.1, 1, 10$ and $0.1 \leq K \leq 10$. What does your plot show for the effect of Bo? Do waves travel faster or slower as the depth, d, increases? Why?

f. Calculate the dimensionless group velocity $C_g = \partial\Omega/\partial K$. Plot C_g vs K for $Bo = 0.1, 1, 10$ and $0.1 \leq K \leq 10$. What does your plot show for the effect of Bo?

16.6

a. The fluid velocity field for the deep water capillary-gravity waves, Section 16.8, is $\underline{u} = \nabla\phi$ where u and v are the real parts. Show that $u = \omega h_0 B e^{ky} \cos(kx - \omega t)$ and $v = \omega h_0 B e^{ky} \sin(kx - \omega t)$ where B is the dimensionless ratio $B = (k^3\sigma + k\rho g)/\rho\omega^2$.

b. The fluid particle trajectories are solved from the Lagrangian time derivative of the positions $\underline{X}(\underline{X}_0, t)$

$$\frac{\partial X}{\partial t} = \omega h_0 B e^{kY} \cos(kX - \omega t) = \omega h_0 B e^{ky} \cos(kx - \omega t) + O(h_0^2)$$

$$\frac{\partial Y}{\partial t} = \omega h_0 B e^{kY} \sin(kX - \omega t) = \omega h_0 B e^{ky} \sin(kx - \omega t) + O(h_0^2) \tag{1}$$

Since the wave disturbance is small, of $O(h_0)$, we can replace $\underline{X}$ with the Eulerian $\underline{x}$ on the right hand side of Eqs. (1) as shown. Integrate Eqs. (1) with respect to time to find $\underline{X}(\underline{x}, t)$ to within functions of integration which depend at most on x, y.

c. To choose the functions of integration we recognize that $X = x, Y = y$ when there is no disturbance, i.e. $h_0 = 0$ with our approximations. What are they?

d. Using your results from part (c) show that $\frac{(X-x)^2}{(R(y))^2} + \frac{(Y-y)^2}{(R(y))^2} = 1$. What is the radius of these circular trajectories? Where is the center?

e. For the finite-depth capillary-gravity waves discussed in Problem 16.5, where there is a wall at $y = -d$, the velocity potential was found to be $\phi = 2f_0 e^{-kd} \cosh(k(y+d)) e^{i(kx-\omega t)}$. From the analysis we know that $f_0 = (-ih_0(\rho g + \sigma k^2)e^{kd})/2\rho\omega \cosh(kd)$. What are the dimensional velocities $u = \partial\phi/\partial x, v = \partial\phi/\partial y$. Be sure to take the real parts.

f. Follow the procedure in part (b) to find the particle trajectories, $(X(t), Y(t))$, by replacing $(u, \ v)$ with $(\partial X/\partial t, \partial Y/\partial t)$ and integrating with respect to time.

g. From your results in part (f), express the form $\frac{(X-x)^2}{R_x^2(y)} + \frac{(Y-y)^2}{R_y^2(y)} = 1$. What are your results for $R_x(y), \ R_y(y)$? What shape are the trajectories? How do they compare to part (d)? What happens as $-d \to -\infty$?

16.7 Let's examine capillary-gravity wave propagation as in Section 16.8 with a viscous fluid. Let there be a wall at $y = -d$ and restrict wavelengths so that $d/\lambda << 1$ and we can use lubrication theory discussed in Chapter 11 where y-velocities are much smaller than x-velocities, $v << u$, and variations are smaller in the x-direction $\partial/\partial x << \partial/\partial y$.

a. The y-momentum equation is $\partial p/\partial y = -\rho g$ which integrates to $p = -\rho g y + p_a + f(x, t)$. Insert this into the x-momentum equation, $\mu \partial^2 u/\partial y^2 = \partial p/\partial x$, to find u(x, y, t) by integrating with respect to y twice.

b. In part (a) your solution for u should have two functions of integration, $c_1(x, t), c_2(x, t)$. Determine them from enforcing two boundary conditions of no-slip at the wall the$u(y = -d) = 0$ and no shear at interface, $y = h(x, t)$, linearized at y=0, $\partial u/\partial y|_{y=0} = 0$. What is your form of u?

c. We will need to derive v from continuity. Perform the integration $v = -\int_{-d}^{y} \frac{\partial u}{\partial x} dy$ which enforces$v(y = -d) = 0$. Your solution should involve $\partial^2 f/\partial x^2$.

d. Apply the stress boundary condition, $p_a - p_{y=h} = \sigma \partial^2 h/\partial x^2$ using your solution for p. What is the form of this equation. It should involve f and h.

e. The remaining constraint is the kinematic boundary condition linearized at y=0 and for small disturbances $v|_{y=0} = \partial h/\partial t + u\partial h/\partial x$. Insert for v and show your resulting condition.

f. For the small disturbance travelling wave let $h = h_0 e^{i(kx-\omega t)}, u = u_0(y) e^{i(kx-\omega t)}$, $v = v_0(y) e^{i(kx-\omega t)} f = f_0 e^{i(kx-\omega t)}$. Insert into your expression for u to find $u_0(y)$ in terms of f_0 and y. Then use the stress boundary condition to determine $f_0$in terms of h_0. Finally apply the kinematic boundary condition where h_0 should cancel leaving the dispersion relation $\omega(k)$. Show that $\omega = -ik^2 d^3(\rho g + k^2\sigma)/3\mu$.

g. For $\sigma = 0$ the dispersion relation simplifies to $\omega = -ik^2 d^3 \rho g/3\mu$. Let k be complex, $k = k_r + ik_i$ where $k_r, \ k_i$ are real. Determine both components of k in terms of ω and the parameters and assume $\omega$is real. You will need to satisfy both the real and imaginary parts of the dispersion equation. Show that $k_r = k_i = ((3\mu\omega)/(2d^3\rho g))^{1/2}$.

h. Insert the solution for k from part (g) into $e^{i(kx-\omega t)}$. What is the spatial decay of the wave as it propagates in the x-direction? Why does that happen? What is the wave speed $c = \omega/k_r$ and what is the effect of fluid viscosity?

i. Repeat the analysis of part (g) including surface tension. What do you find?

16.8

a. Using the governing equations for epidemic modeling, Eqs. (16.7.1), solve the system numerically for a population of 1000 starting with 100 infected people: the initial conditions $S(0) = 900, I(0) = 100, R(0) = 0$ with $\beta = 0.4$, $\gamma = 0.2$, $R_0 = 2$. Plot your results on the same graph for S(t), I(t), R(t) vs t. How does your result compare to Figure 16.41(a) in terms of the epidemic duration, maximum value of I(t), total recovered, R?
b. Repeat part (a) for $\beta = 0.3$, $\gamma = 0.2$, $R_0 = 1.5$. How does your result compare to Figure 16.41(b)?
c. Repeat the analysis in Figure 16.41(a) and calculate the values of R, I, S at $t = 20$ d.
d. Suppose the population lowers the value of R_0 during the epidemic through wearing masks, social distancing, washing hands and other active measures. Use the values from part (c) as initial conditions for the governing equations, but now set $\beta = 0.18$, $\gamma = 0.2$, $R_0 = 0.9$. Plot your solution and compare to Figure 16.41(a). What happened?
e. What are your recommendations for the best outcomes of an epidemic?

Image Credits

Figure 16.7a Source: Jody Amiet/ Getty

Figure 16.7b Source: Picture of an human eye with the lacrimal punctum (or lacrimal point) in evidence, by Diogo Melo Rocha, licensed under CC BY 2.5, https://commons.wikimedia.org/wiki/File:Lacrimal_punctum.jpg

Figure 16.8a Source: Jao Paulo Burini/ Getty

Figure 16.8b Source: Fig 2 from Chakraborty S, Tsuchiya K. Development and fluidic simulation of microneedles for painless pathological interfacing with living systems. *J Appl Phys*. 2008; 103:114701-1–9.

Figure 16.9a Source: ePhotocorp/ Getty

Figure 16.11 Source: Rachata Teyparsit/ EyeEm/ Getty

Figure 16.17b Source: Fuse/ Getty

Figure 16.18 Source: Fig 1 from Tung YC, Hsiao AY, Allen SG, Torisawa YS, Ho M, Takayama S. High-throughput 3D spheroid culture and drug testing using a 384 hanging drop array. *Analyst*. 2011;136(3):473–8.

Figure 16.19b Source: Panama City, FL (November 9, 2006) – Seaman Ethan Cypher gives the thumbs up to instructors as he and Seaman Jeff Finkel prepare to undergo decompression chamber training. US Navy photo by Mass Communication Specialist 2nd Class Jayme Pastoric (public domain), https://web.archive.org/web/20061122151509/http://www.news.navy.mil/view_single.asp?id=41011

Figure 16.20 Source: Christin Khan/NOAA/NEFSC, www.nmfs.noaa.gov/pr/species/mammals/cetaceans/humpbackwhale.htm

Figure 16.21a Source: Fig 5 in Maxwell AD, Park S, Vaughan BL, Cain CA, Grotberg JB, Xu Z. Trapping of embolic particles in a vessel phantom by cavitation-enhanced acoustic streaming. *Physics in Medicine and Biology*. 2014; 59(17):4927–43. © Institute of Physics and Engineering in Medicine. Reproduced by permission of IOP Publishing. All rights reserved

Figure 16.21b Source: Fig 7a in Maxwell AD, Park S, Vaughan BL, Cain CA, Grotberg JB, Xu Z. Trapping of embolic particles in a vessel phantom by cavitation-enhanced acoustic streaming. *Physics in Medicine and Biology*. 2014; 59(17):4927–43. © Institute of Physics and Engineering in Medicine. Reproduced by permission of IOP Publishing. All rights reserved

Figure 16.22a Source: Janine Schmitz/ Contributor/ Getty

Figure 16.22b Source: Kevin Alexander George/ Getty

Figure 16.24 Source: ilbusca/ Getty

Figure 16.29 Source: Georgia Court/ Getty

Figure 16.32 Source: Fig 1 from Hawgood S, Clements JA. Pulmonary surfactant and its apoproteins. *Journal of Clinical Investigation*. 1990; 86:1–6.

Figure 16.38a Source: Fig 2 in Wei HH, Fujioka H, Hirschl RB, Grotberg JB. A model of flow and transport in an oscillatory alveolus partially filled with liquid. *Physics of Fluids*. 2005; 17:Art. No. 031510.

Figure 16.38b, c Source: Fig 15 in Wei HH, Fujioka H, Hirschl RB, Grotberg JB. A model of flow and transport in an oscillatory alveolus partially filled with liquid. *Physics of Fluids*. 2005; 17:Art. No. 031510.

Figure 16.39a–f Source: courtesy Dr. Alireza Kazemi

Figure 16.40a Source: Smith Collection/ Gado/ Contributor/ Getty

Figure 16.40b Source: Chart showing mortality from the 1918 influenza pandemic in the US and Europe, courtesy of the National Museum of Health and Medicine.

17 Perturbation Methods

Introduction

The challenges of fluid mechanics often center on solving the set of coupled, nonlinear, partial differential equations that form conservation of momentum and mass. To simplify, recall that we neglected terms in the Navier–Stokes equations, for example, in the limit of small inertia, $\mathrm{Re} << 1$ (Sections 7.8 and 7.9); small viscosity, $1/\mathrm{Re} << 1$ (Section 12.1); and small boundary slope, $\varepsilon << 1$ (Chapter 11). Also for a fluid–fluid interface we neglected nonlinear curvature terms for small interfacial slope, $|dh/dr| << 1$ (Section 16.2), and assumed small amplitude disturbances for surface waves of a liquid jet (Section 16.3). The overall goal was to find exact, analytical solutions of the simplest version of the governing equations, without throwing out the underlying physics. However, neglecting terms can lead to inconsistencies, as we saw for $\mathrm{Re} << 1$ flow over a cylinder, which had no solution in the limit $\mathrm{Re} = 0$. Also, including the omitted terms can introduce additional physical interactions leading to qualitatively different results.

To confront these challenges, applied mathematicians and fluid dynamicists developed methods based on "perturbation theory," an approach first used for the physics of planetary motion by Lagrange, Laplace and others in the eighteenth and nineteenth centuries, i.e. long before electronic computers. Perturbation methods provide an orderly way of seeking approximate solutions that are near to the exact analytical solution of the simplified system which omits the small terms (Van Dyke, 1975; Nayfeh, 1993). If we have a system of equations containing a small dimensionless parameter, call it $\delta << 1$, and can find the exact solution for $\delta = 0$, then perturbation methods seek solutions proposed as a power series, or perturbation series, in δ. For example, the power series $\underline{u} = \underline{u}_0 + \delta \underline{u}_1 + \delta^2 \underline{u}_2 + O(\delta^3)$ could represent the velocity field where $\underline{u} = \underline{u}_0$ is the exact solution to the $\delta = 0$ problem. Typically, the next term, $\underline{u}_1$, depends on $\underline{u}_0$, and generally the solution at any order depends on known solutions at lower orders. In this chapter we will go through the perturbation approach, starting with model equations to learn the concepts and techniques. With those under our belt, we move on to some important applications for small and large Re, dynamic stability of a liquid jet, and steady streaming from convective acceleration and from nonlinear boundary motion.

Topics Covered

The first step is to learn the perturbation method approach and terminology by analyzing a quadratic equation containing a small parameter. Since the reader will be very familiar with this equation, it is easier to focus on the method. Next comes the perturbation method for a simple differential equation, which has similarities to the algebraic system. In both of these sections, the examples chosen avoid difficulties, like lost solutions. That is not the case for the next two sections, where singular algebraic and differential equations have missing solutions. Curing this difficulty involves not only rescaling variables, but in the case of a differential equation includes matching solutions valid in different parts of the domain to arrive at a composite result. From this method of matched asymptotic expansions, we finally reach a reckoning for flow over a cylinder at $Re << 1$ (Section 7.9). Also, we will discover our approach to boundary layer flow over a flat plate (Section 12.1), where it was necessary to rescale the cross-stream coordinate as $\eta = Y/(1/Re)^{1/2}$ to keep the highest-order derivative term, was prescient for matching solutions. An important topic in fluid mechanics, generally, is stability analysis. When a system is perturbed from its base state, does it return to that state, i.e. is it stable? If not, what state does it go to? We revisit the liquid jet problem from Section 16.3, but now apply a time-dependent wave disturbance. The analysis gives us the fastest growing wavelength and which compares very well with experiments. The nonlinear convective acceleration terms are a source of rich fluid behavior. As we saw in Section 13.1, a steady flow input over a cylinder can lead to time-periodic oscillation output, the Kármán vortex street. Here we will study the opposite, a time-periodic input can lead to a steady output called steady streaming. The model problem is an oscillatory slide block which builds on our earlier experience in Section 11.1 where the block motion was steady. Finally, mucus clearance depends on cilia lining the airways, and the model we explore derives steady flow from purely oscillatory cilia motion, this time from boundary condition nonlinearities.

17.1 Perturbation Method for an Algebraic Equation

Let's solve the following quadratic equation for $\varepsilon << 1$,

$$x^2 + \varepsilon x - 1 = 0 \tag{17.1.1}$$

using a regular perturbation method. First we propose a solution form for x as a power series in ε,

$$x(\varepsilon) = \sum_{n=0}^{\infty} \frac{\varepsilon^n x_n}{n!} = x_0 + \varepsilon x_1 + \frac{\varepsilon^2}{2} x_2 + O(\varepsilon^3) \tag{17.1.2}$$

where x_n are functions, in this case constants, we will determine. The notation $O(\varepsilon^3)$, with an upper case "O," means that the next term in the series is of "order" ε^3. Note that derivatives of $x(\varepsilon)$ with respect to ε, evaluated at $\varepsilon = 0$, are given by

$$\frac{d^k x}{d\varepsilon^k}_{\varepsilon=0} = x_k \tag{17.1.3}$$

To proceed, we need to expand Eq. (17.1.1) in a Taylor series for $\varepsilon << 1$. Define $F(\varepsilon)$ as

$$F(\varepsilon) = x^2 + \varepsilon x - 1 = 0 \tag{17.1.4}$$

The Taylor series for $F(\varepsilon)$ is

$$F(\varepsilon) = F(0) + \varepsilon F'(0) + \frac{\varepsilon^2}{2} F''(0) + O(\varepsilon^3) = 0 \tag{17.1.5}$$

where $' = \partial/\partial\varepsilon$. Since Eq. (17.1.5) must hold for all $\varepsilon << 1$, including $\varepsilon = 0$, setting $\varepsilon = 0$ implies $F(0) = 0$. Now we can subtract F(0) from Eq. (17.1.5), divide the remainder by ε, and again set $\varepsilon = 0$. The result is that $F'(0) = 0$, and so on. So each coefficient of ε^n in Eq. (17.1.5) is zero, and we can solve them successively for each order of ε.

At O(1) or, equivalently, $O(\varepsilon^0)$, the first term of Eq. (17.1.5) is zero,

$$F(0) = x_0^2 - 1 = 0 \tag{17.1.6}$$

where we have inserted the expansion $x(\varepsilon)$ from Eq. (17.1.2). The solution is

$$x_0 = \pm 1 \tag{17.1.7}$$

At $O(\varepsilon)$ we need to take the derivative of F with respect to ε and then set $\varepsilon = 0$

$$\begin{aligned} F'(\varepsilon) &= 2xx' + x + \varepsilon x' \\ F'(0) &= 2x_0 x_1 + x_0 = 0 \end{aligned} \tag{17.1.8}$$

The value of x_0 cancels out of Eq. (17.1.8) and we find x_1

$$x_1 = -\frac{1}{2} \tag{17.1.9}$$

At $O(\varepsilon^2)$ take the second derivative of F with respect to ε and then set $\varepsilon = 0$

$$\begin{aligned} F''(\varepsilon) &= 2x'^2 + 2xx'' + x' + x' + \varepsilon x'' \\ F''(0) &= 2x_1^2 + 2x_0 x_2 + 2x_1 = 0 \end{aligned} \tag{17.1.10}$$

Substituting for both x_0 and x_1 we find x_2 to be

$$x_2 = \pm\frac{1}{4} \tag{17.1.11}$$

So far our solution is

$$x = \pm 1 - \frac{1}{2}\varepsilon \pm \frac{1}{8}\varepsilon^2 + O(\varepsilon^3) \tag{17.1.12}$$

Let's compare the perturbation solution to the exact solution, call it $\hat{x}$, which we know from Eq. (17.1.1) as a quadratic equation is

$$\hat{x}(\varepsilon) = \frac{-\varepsilon \pm (\varepsilon^2 + 4)^{1/2}}{2} \tag{17.1.13}$$

For comparison to the perturbation solution, we need to expand $\hat{x}(\varepsilon)$ of Eq. (17.1.13) in a Taylor series

$$\begin{aligned} \hat{x}(\varepsilon) &= \hat{x}(0) + \varepsilon\hat{x}'(0) + \frac{\varepsilon^2}{2}\hat{x}''(0) + O(\varepsilon^3) \\ &= \pm 1 - \frac{1}{2}\varepsilon \pm \frac{1}{8}\varepsilon^2 + O(\varepsilon^3) \end{aligned} \tag{17.1.14}$$

which can be done by hand or by using computational software. We see that the perturbation solution and the expansion of the exact solution are equivalent term by term.

For a next step in complexity, consider the equation

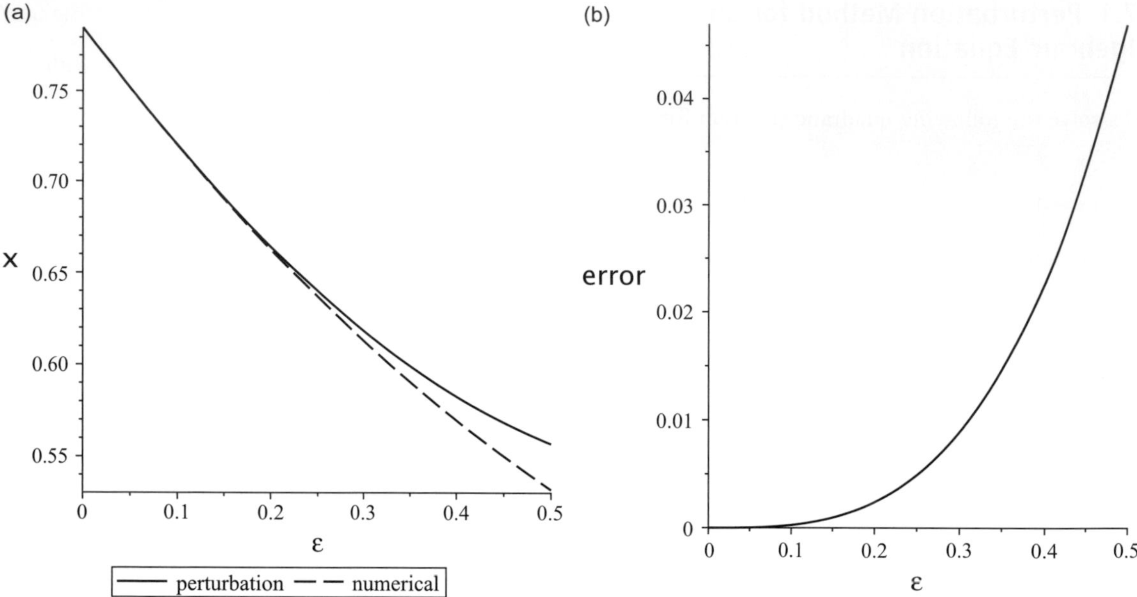

Figure 17.1 (a) Comparison of numerical and perturbation solutions of Eq. (17.1.15). (b) Error calculation.

$$x + \varepsilon \sin(x) - \frac{\pi}{4} = 0 \tag{17.1.15}$$

and seek a perturbation solution, $x(\varepsilon)$, using the expansion form of Eq. (17.1.2)

$$x = x_0 + \varepsilon x_1 + \frac{1}{2}\varepsilon^2 x_2 + O(\varepsilon^3) \tag{17.1.16}$$

keeping the first three terms. Proceeding as above, the results ordered by powers of ε give us the x_n values

$O(1)$

$$\begin{aligned} F(\varepsilon) &= x + \varepsilon \sin(x) - \frac{\pi}{4} = 0 \\ F(0) &= x_0 - \frac{\pi}{4} = 0 \quad \rightarrow \quad x_0 = \frac{\pi}{4} \end{aligned} \tag{17.1.17}$$

$O(\varepsilon)$

$$\begin{aligned} F'(\varepsilon) &= x' + \sin(x) + \varepsilon \cos(x) x' = 0 \\ F'(0) &= x_1 + \sin(x_0) = 0 \rightarrow x_1 = -\sin\left(\frac{\pi}{4}\right) = -\frac{\sqrt{2}}{2} \end{aligned} \tag{17.1.18}$$

$O(\varepsilon^2)$

$$\begin{aligned} F''(\varepsilon) &= x'' + 2\cos(x)x' + \varepsilon(-\sin(x)x'^2 + \cos(x)x'') = 0 \\ F''(0) &= x_2 + 2\cos(x_0)x_1 = 0 \rightarrow x_2 \\ &= -2\cos\left(\frac{\pi}{4}\right)\left(-\frac{\sqrt{2}}{2}\right) = 1 \end{aligned} \tag{17.1.19}$$

so that the perturbation solution is

$$x = \frac{\pi}{4} - \frac{\sqrt{2}}{2}\varepsilon + \frac{1}{2}\varepsilon^2 \tag{17.1.20}$$

We don't have an exact solution to expand for comparison, as we did in the previous example for the quadratic equation Eq. (17.1.1). However, roots of Eq. (17.1.15) are readily found numerically using computational software. Figure 17.1(a) is a comparison of the perturbation and numerical solutions which overlap for ε small enough. Figure 17.1(b) is an error calculation defined $(x_{pert} - x_{num})/x_{num}$, which shows 1% error at $x \sim 0.3$.

The Maple code for Figure 17.1 is given in Code 17.1(a) and the corresponding MATLAB® code is Code 17.1(b).

(a)

```
restart;
epsilonmax := 0.5:
N := 100:
```

$$\text{delta} := \frac{\text{epsilonmax}}{N};$$

$$\delta := 0.005000000000 \tag{1}$$

```
for i from 1 to N +1 do
epsilon[i] := (i-1)·delta:
```

$$gg[i] := yy[i] + epsilon[i]\cdot \sin(yy[i]) - \frac{Pi}{4}:$$

```
yy[i] := fsolve(gg[i] = 0, yy[i]):
```

$$xx[i] := \frac{\pi}{4} - \frac{epsilon[i]\sqrt{2}}{2} + \frac{epsilon[i]^2}{2}:$$

$$err[i] := \frac{(xx[i]-yy[i])}{yy[i]}:$$

```
end do:

with (plots)
PL1 := listplot([seq([ epsilon[i], xx[i]],i = 1..1+ N)], linestyle = 1, legend = "perturbation"):

PL2 := listplot([seq([epsilon[i], yy[i]],i = 1..1+ N)], linestyle = 3, legend = "numerical"):
display(PL1, PL2, labels = [ε. "x"], labelfont = [HELVETICA, 14], legendstyle = [font = [HELVETICA,
    10]]):
PL3 := listplot([seq([epsilon[i], err[i]], i = 1..1+ N)], labels = [ε,"error"], labelfont = [HELVETICA,
    14]):
display(PL3) :
```

(b)

```
clear all, close all, clc
syms y
epsilonmax = 0.5;
N          = 100;
delta      = epsilonmax/N
for i = 1:N+1
epsilon(i) = (i-1)*delta;
gg(i)      = y + epsilon(i)*sin(y) - pi/4 == 0;
yy(i)      = solve(gg(i));
xx(i)      = pi/4 - epsilon(i)*sqrt(2)/2 + epsilon(i)^2/2;
err(i)     = (xx(i)-yy(i))/yy(i);
end
figure(1)
plot(epsilon,xx,'k'); hold on
plot(epsilon,yy,'--k');
xlabel('\epsilon')
ylabel('x')
```

Code 17.1 Code examples: (a) Maple, (b) MATLAB®.

```
ax = gca;
ax.FontSize = 11;
ax.FontName = 'Arial';
l = legend('perturbation','numerical','location','northeast');
l.FontSize = 11;
l.FontName = 'Arial';
figure(2)
plot(epsilon,err,'-k');
xlabel('\epsilon')
ylabel('error')
ax = gca;
ax.FontSize = 11;
ax.FontName = 'Arial';
```

Code 17.1 (*continued*)

17.2 Perturbation Method for a Differential Equation

Let's solve the following nonlinear ODE for X(T) assuming $\varepsilon << 1$,

$$\frac{d^2X}{dT^2} + \varepsilon X^2 - 1 = 0 \tag{17.2.1}$$

subject to the initial conditions

$$X(T=0) = 0, \quad \left.\frac{dX}{dT}\right|_{T=0} = 1 \tag{17.2.2}$$

First we expand the dependent variable, X, in powers of ε,

$$X(T;\varepsilon) = X_0(T) + \varepsilon X_1(T) + \frac{\varepsilon^2}{2} X_2(T) \tag{17.2.3}$$

We need to expand Eqs. (17.2.1) and (17.2.2) in Taylor series for $\varepsilon << 1$. Define F, G and H as

$$\begin{aligned} F(T;\varepsilon) &= \frac{d^2X}{dT^2} + \varepsilon X^2 - 1 = 0 \\ G(\varepsilon) &= X|_{T=0} = 0 \\ H(\varepsilon) &= \left.\frac{dX}{dT}\right|_{T=0} - 1 = 0 \end{aligned} \tag{17.2.4}$$

The Taylor series for F, G and H are

$$\begin{aligned} F(T,\varepsilon) &= F(T,0) + \varepsilon F'(T,0) + \frac{\varepsilon^2}{2} F''(T,0) + O(\varepsilon^3) = 0 \\ G(\varepsilon) &= G(0) + \varepsilon G'(0) + \frac{\varepsilon^2}{2} G''(0) + O(\varepsilon^3) = 0 \\ H(\varepsilon) &= H(0) + \varepsilon H'(0) + \frac{\varepsilon^2}{2} H''(0) + O(\varepsilon^3) = 0 \end{aligned} \tag{17.2.5}$$

where $' = \partial/\partial\varepsilon$. We solve the system of Eqs (17.2.5) successively at each order of ε, since each term must equal zero as we saw for the algebraic system of Eq. (17.1.5). While the functions x_n in the expansion Eq. (17.1.2) for an algebraic equation were constants, here the $X_n(T)$ functions for a differential equations depend on time, T.

The O(1) system of Eq. (17.2.5) is

$$\begin{aligned} F(T,0) &= \frac{d^2X}{dT^2}(0) - 1 = \frac{d^2X_0}{dT^2} - 1 = 0 \\ G(0) &= X(0)|_{T=0} = X_0|_{T=0} = 0 \\ H(0) &= \left.\frac{dX}{dT}(0)\right|_{T=0} - 1 = \left.\frac{dX_0}{dT}\right|_{T=0} - 1 = 0 \end{aligned} \tag{17.2.6}$$

Integrating $d^2X_0/dT^2 = 1$ twice in the first of Eqs. (17.2.6) yields the form

$$X_0 = \frac{T^2}{2} + A_0T + B_0 \tag{17.2.7}$$

and imposing the O(1) initial conditions from Eq. (17.2.6) gives the final result

$$X_0 = \frac{T^2}{2} + T \tag{17.2.8}$$

At O(ε) we need to take the derivative of F with respect to ε and then set ε = 0

$$\begin{aligned} F' &= \frac{d^2X'}{dT^2} + X^2 + \varepsilon 2XX' = 0 \\ F'(0) &= \frac{d^2X_1}{dT^2} + X_0{}^2 = 0 \end{aligned} \tag{17.2.9}$$

Since we know X_0 from Eq. (17.2.8), the differential equation for X_1 is

$$\frac{d^2X_1}{dT^2} = -X_0^2 = -\left(\frac{T^2}{2} + T\right)^2 \tag{17.2.10}$$

Equation (17.1.9) can be integrated twice to give

$$X_1 = -\left(\frac{T^6}{120} + \frac{T^5}{20} + \frac{T^4}{12}\right) + A_1T + B_1 \tag{17.2.11}$$

The initial conditions involving G and H at this order are

$$\begin{aligned} X'(0)|_{T=0} &= X_1|_{T=0} = 0 \\ \frac{dX'}{dT}(0)\bigg|_{T=0} &= \dot{X}_1\big|_{T=0} = 0 \end{aligned} \tag{17.2.12}$$

Imposing Eqs. (17.2.12) on Eq. (17.2.11) gives the final answer

$$X_1 = -\left(\frac{T^6}{120} + \frac{T^5}{20} + \frac{T^4}{12}\right) \tag{17.2.13}$$

Clearly, in order to maintain the perturbation assumptions, we cannot let T to become too large.

Symbolic manipulation in computational software can do most of this work for you. The Maple code is shown in Code 17.2(a) and the equivalent MATLAB® file is Code 17.2(b).

(a)

```
restart;
ODE := diff(x(t),t,t) + epsilon·x(t)^2 - 1;
```

$$ODE := \frac{d^2}{dt^2}x(t) + \epsilon x(t)^2 - 1 \qquad (1)$$

```
IC1 := eval(x(t), t=0);
```

$$IC1 := x(0) \qquad (2)$$

```
IC2 := eval(diff(x(t),t=0) -1;
```

$$IC2 := \left(\frac{d}{dt}x(t)\right)\bigg|_{\{t=0\}} - 1 \qquad (3)$$

```
x(t) :=x0(t)+ epsilon·x1(t);
```

$$x := t \rightarrow x0(t) + \epsilon x1(t) \qquad (4)$$

```
taylor(ODE, epsilon, 2);
```

$$\frac{d^2}{dt^2}x0(t) - 1 + \left(\frac{d^2}{dt^2}x1(t) + x0(t)^2\right)\epsilon + O(\epsilon^2) \qquad (5)$$

```
taylor(IC1, epsilon, 2);
```

$$x0(0) + x1(0)\ \epsilon \qquad (6)$$

```
taylor(IC2, epsilon, 2);
```

$$\left(\frac{d}{dt}x0(t)\right)\bigg|_{\{t=0\}} - 1 + \left(\left(\frac{d}{dt}x1(t)\right)\bigg|_{\{t=0\}}\right)\epsilon \qquad (7)$$

Code 17.2 Code examples: (a) Maple, (b) MATLAB®.

```
(b)
clear all; close all; clc;
syms epsilon tx(t) x0(t) x1(t)
x         = x0 + epsilon*x1
dx_dt     = diff(x,t);
d2x_dt2   = diff(dx_dt,t);
ODE       = d2x_dt2 + epsilon*x^2-1
IC1       = x(0)
IC2       = dx_dt(0)-1
taylor(ODE,epsilon,'Order',2)
taylor(IC1,epsilon,'Order',2)
taylor(IC2,epsilon,'Order',2)
```

Code 17.2 (*continued*)

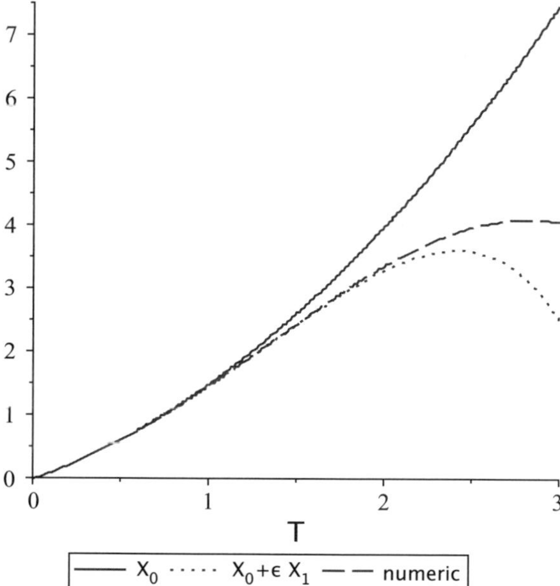

Figure 17.2 Comparison of perturbation and numerical solutions for Eq. (17.2.1).

A plot of the results is shown in Figure 17.2 comparing the one-term and two-term perturbation solutions to the numerical solution. The two-term solution is accurate just past $T = 2$.

17.3 Singular Algebraic Equation

Suppose we modify the quadratic equation Eq. (17.1.1) by shifting the position of ε to multiply the quadratic term

$$\varepsilon x^2 + x - 1 = 0 \tag{17.3.1}$$

Using the perturbation solution of Eq. (17.1.2) in Eq. (17.3.1), we find to leading order that $x_0 = 1$, which is only one root for a quadratic equation. This is a singular result since there should be two solutions. The difficulty is that ε multiplies the highest power term, which we lose to leading order when we assume $\varepsilon x^2 << 1$. The leading-order equation, then, is linear in x, which, indeed, has a single solution. To deal with this singularity, entertain the possibility that εx^2 does not remain small by rescaling the variable as

$$\xi = \delta(\varepsilon)x \tag{17.3.2}$$

Inserting Eq. (17.3.2) into Eq. (17.3.1) the resulting form is

$$\varepsilon\frac{\xi^2}{\delta^2} + \frac{\xi}{\delta} - 1 = 0 \tag{17.3.3}$$

Of the three terms in Eq. (17.3.3), consider a balance of the first and second terms such that

$$\frac{\varepsilon}{\delta^2} \sim \frac{1}{\delta} \tag{17.3.4}$$

The balance in Eq. (17.3.4) implies $\delta = \varepsilon$ and $\xi = \varepsilon x$, which means that Eq. (17.3.3) becomes

$$\xi^2 + \xi - \delta = 0 \tag{17.3.5}$$

Assume a perturbation solution in powers of δ

$$\xi = \xi_0 + \delta\xi_1 + \frac{\delta^2}{2}\xi_2 + O(\delta^3) \tag{17.3.6}$$

Proceeding as we did in Section 17.1, the steps are listed as

$$
\begin{aligned}
&F(\delta) = \xi^2 + \xi - \delta = 0 \\
&F(0) = \xi_0^2 + \xi_0 = 0 \quad \xi_0 = 0,\ -1 \\
&F'(\delta) = 2\xi\xi' + \xi' - 1 = 0 \\
&F'(0) = 2\xi_0\xi_1 + \xi_1 - 1 = 0 \quad \xi_1 = 1,\ -1 \\
&F''(\delta) = 2\xi'^2 + 2\xi\xi'' + \xi'' = 0 \\
&F''(0) = 2\xi_1^2 + 2\xi_0\xi_2 + \xi_2 = 0 \quad \xi_2 = -2, 2
\end{aligned}
\tag{17.3.7}
$$

so that the expansion in Eq. (17.3.6) has the two roots

$$
\begin{aligned}
&\xi = \delta - \delta^2 + O(\delta^3) \\
&\xi = -1 - \delta + \delta^2 + O(\delta^3)
\end{aligned}
\tag{17.3.9}
$$

Rewriting Eqs. (17.3.9) in terms of x and ε, the solutions become

$$
\begin{aligned}
&x = 1 - \varepsilon + O(\varepsilon^2) \\
&x = -\frac{1}{\varepsilon} - 1 + \varepsilon + O(\varepsilon^2)
\end{aligned}
\tag{17.3.10}
$$

The expressions in Eq. (17.3.10) match the Taylor expansion of the exact solution to Eq. (17.3.1)

$$
\begin{aligned}
x &= \frac{-1 \pm (1+4\varepsilon)^{1/2}}{2\varepsilon} = \frac{-1 \pm (1 + 2\varepsilon - 2\varepsilon^2 + \cdots)}{2\varepsilon} \\
&= 1 - \varepsilon, \quad -\frac{1}{\varepsilon} - 1 + \varepsilon
\end{aligned}
\tag{17.3.11}
$$

17.4 Singular Differential Equation: Method of Matched Asymptotic Expansions

Consider the linear, second-order ODE

$$\varepsilon \frac{d^2u}{dx^2} + \frac{du}{dx} + u = 0 \tag{17.4.1}$$

subject to the boundary conditions

$$u(0) = 0, \quad u(1) = 1 \tag{17.4.2}$$

Equation (17.4.1) with the boundary conditions will turn out to be a "singular perturbation problem," and the method outlined here for solution is called the "method of matched asymptotic expansions."

For $\varepsilon << 1$ we propose a perturbation solution, which we will call an "outer" solution to the "outer" equation, Eq. (17.4.1), denoted by a superscript "o,"

$$u^o = u_0^o + \varepsilon u_1^o + O(\varepsilon^2) \tag{17.4.3}$$

Inserting Eq. (17.4.3) into the governing equation and boundary conditions, and ordering in powers of ε, gives us at O(1)

$$\frac{du_0^o}{dx} + u_0^o = 0 \tag{17.4.4}$$

The resulting form, Eq. (17.4.4), is a first-order ODE, of lower order than Eq. (17.4.1). This reduction is the origin of the singularity in the problem, since the lower-order equation does not have a solution which can solve both boundary conditions. The general solution to Eq. (17.4.4) is

$$u_0^o(x) = Ae^{-x} \tag{17.4.5}$$

with the unknown constant of integration, A. Since Eq. (17.4.4) is first order in x, we cannot solve both boundary conditions. We choose the solution, Eq. (17.4.5), to satisfy the condition at $x = 1$ and call this the "outer" boundary condition. This condition makes $A = e$, so the final form is

$$u_0^o(x) = e^{1-x} \tag{17.4.6}$$

We are not done with the analysis, since our naïve perturbation expansion left us with a lower-order ODE. That is because the small parameter, ε, multiplies the highest-order derivative. The assumption for the expansion was that d^2u/dx^2 remains O(1) as $\varepsilon << 1$ is small, so that the size of their product is $O(\varepsilon)$. However, that assumption breaks down if d^2u/dx^2 becomes large in some region, say near the "inner" boundary at $x = 0$ where there is a boundary layer.

In order to capture that idea of arranging the variables so that we retain the highest-order derivative in the limit of $\varepsilon << 1$ we rescale x as

$$\xi = \frac{x}{\varepsilon^a} \tag{17.4.7}$$

and refer to ξ as the "inner" variable or "boundary layer" variable. Using the chain rule of differentiation, the x-derivatives become

$$\frac{d}{dx} = \frac{d\xi}{dx}\frac{d}{d\xi} \tag{17.4.8}$$

and rewriting Eq. (17.4.1) in terms of ξ gives us a differential equation for the inner velocity, $u^i(\xi)$,

$$\frac{\varepsilon}{\varepsilon^{2a}}\frac{d^2u^i}{d\xi^2}, + \frac{1}{\varepsilon^a}\frac{du^i}{d\xi} + u^i = 0 \tag{17.4.9}$$

As we saw in the singular algebraic example of Section 17.3, our goal is to balance the first term with either the second or third term. To balance with the third term we must have $a = 1/2$ and rearranging Eq. (17.4.9) gives us

$$\varepsilon^{1/2}\frac{d^2u^i}{d\xi^2}, + \frac{du^i}{d\xi} + \varepsilon^{1/2}u^i = 0 \tag{17.4.10}$$

where again we have a small parameter in front of the $d^2u/d\xi^2$ term. Now try the other balance, with the second term, which requires $a = 1$, and rewriting Eq. (17.4.9) gives us

$$\frac{d^2u^i}{d\xi^2} + \frac{du^i}{d\xi} + \varepsilon u^i = 0 \tag{17.4.11}$$

which is the desired form. Equation (17.4.11) is called the "inner" equation and $\varepsilon^a = \varepsilon$ is the boundary layer thickness.

Propose a perturbation solution to this "inner" problem,

$$u^i(\xi) = u^i_0(\xi) + \varepsilon u^i_0(\xi) + O(\varepsilon^2) \tag{17.4.12}$$

Inserting Eq. (17.4.12) into Eq. (17.4.11) leads to the O(1) problem

$$\frac{d^2u^i_0}{d\xi^2} + u^i_0\xi = 0 \tag{17.4.13}$$

The general solution to Eq. (17.4.13) is

$$u^i_0 = Be^{-\xi} + C \tag{17.4.14}$$

where B and C are constants of integration. Equation (17.4.14) must satisfy the "inner" boundary condition at $\xi = 0$, which sets $B = -C$,

$$u^i_0 = C(1 - e^{-\xi}) \tag{17.4.15}$$

To solve for C, we "match" the inner and outer solutions by setting

outer limit (inner solution) =
inner limit (outer solution) (17.4.16)

which has the mathematical equivalence

$$\lim_{x \text{ fixed}, \varepsilon\to 0}\left[u^i_0\right] = \lim_{\xi \text{ fixed}, \varepsilon\to 0}\left[u^o_0\right] \tag{17.4.17}$$

In order to take the outer limit, which is in terms of x, of the inner solution, which is terms of ξ, we use $\xi = x/\varepsilon$, and vice versa for the inner limit term,

$$\lim_{x \text{ fixed}, \varepsilon\to 0}\left[C(1 - e^{-\frac{x}{\varepsilon}})\right] = \lim_{\xi \text{ fixed}, \varepsilon\to 0}\left[e^{1-\varepsilon\xi}\right] \tag{17.4.18}$$

Taking the indicated limits leaves us with $C = e$, which is the "common part" to both expansions. This equivalence completes the inner solution

$$u^i_0 = e(1 - e^{-\xi}) \tag{17.4.19}$$

A plot of the inner and outer solutions for $\varepsilon = 0.1$ are shown in Figure 17.3, where we see the inner solution satisfies the inner boundary condition at

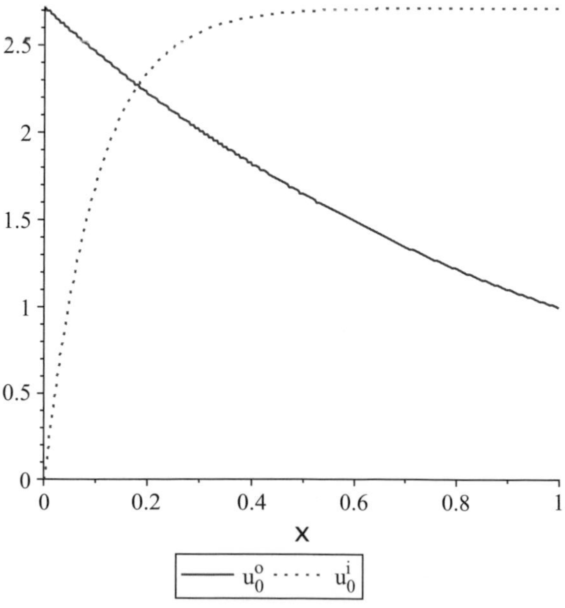

Figure 17.3 Inner and outer solutions for $\varepsilon = 0.1$.

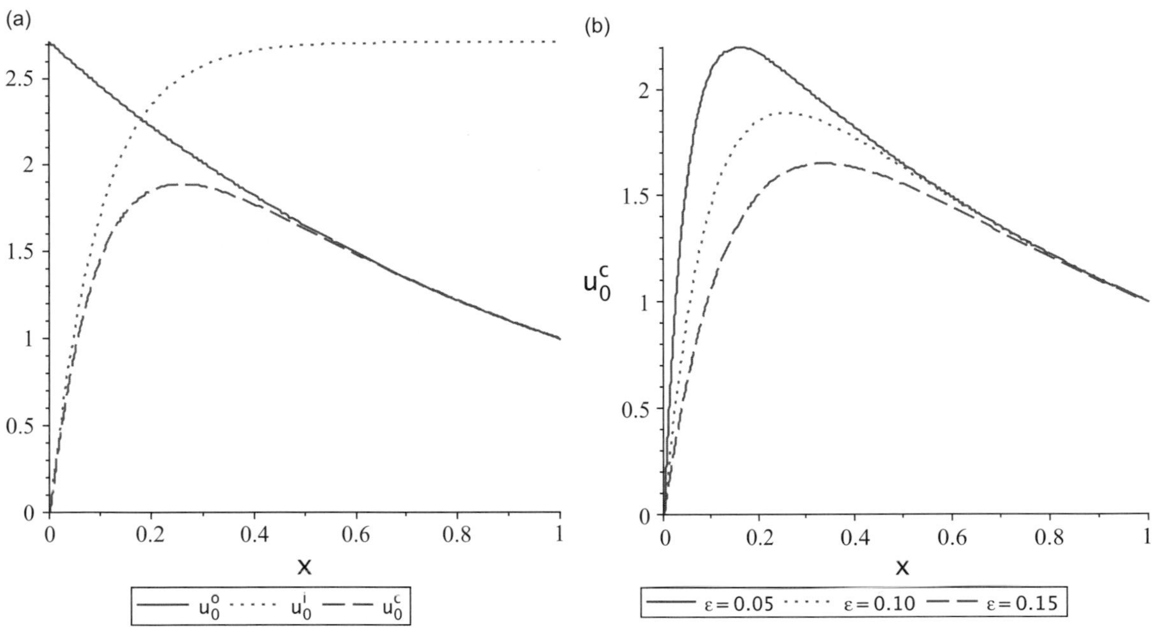

Figure 17.4 (a) Inner, outer and composite solutions for $\varepsilon = 0.1$ and (b) composite solution for $\varepsilon = 0.05, 0.10, 0.15$.

$x = 0$, while the outer solution satisfies the outer boundary condition at $x = 1$.

To create a total solution, we want contributions from both the inner and outer forms. Define the "composite solution," u_0^c, to be the sum of the inner and outer solutions and subtract the common part once, since it appears twice from the addition

$$u_0^c = u_0^o + u_0^i - \text{c.p.} \quad \rightarrow \quad u_0^c = e^{1-x} - e^{1-x/\varepsilon} \tag{17.4.20}$$

A plot of the composite solution is shown in Figure 17.4(a), along with the outer and inner solutions for $\varepsilon = 0.1$. It shows that the composite solution satisfies both boundary conditions at $x = 0, 1$ as it asymptotes to the inner and outer solutions in those regions, respectively. Figure 17.4(b) shows u_0^c for three values of $\varepsilon = 0.05, 0.10, 0.15$, illustrating the boundary layer thickness, measured by ε, increases as ε increases. The boundary layer thickness is $O(\varepsilon)$, where rapid changes in u occur to meet the boundary condition at $x = 0$. It is in the boundary layer that d^2u/dx^2 is large, as can be seen. In the boundary layer, however, $d^2u/d\xi^2$ is $O(1)$ since $\xi = O(1)$ in the boundary layer where x and ε are both small and their ratio is $O(1)$ in the limit $\varepsilon \rightarrow 0$.

How does our perturbation solution compare to the exact solution to Eq. (17.4.1)? Let $u = e^{rx}$ and insert it into Eq. (17.4.1), which gives us a quadratic for r,

$$\varepsilon r^2 + r + 1 = 0 \tag{17.4.21}$$

with two solutions

$$r^+ = \frac{-1 + (1 - 4\varepsilon)^{1/2}}{2\varepsilon}, \quad r^- = \frac{-1 - (1 - 4\varepsilon)^{1/2}}{2\varepsilon} \tag{17.4.22}$$

The general solution is

$$u(x) = De^{r^+x} + Ee^{r^-x} \tag{17.4.23}$$

Applying the two boundary conditions, Eq. (17.4.2), solves for the integration constants, D and E, and the final form is

$$u(x) = \frac{e^{r^+x} - e^{r^-x}}{e^{r^+} - e^{r^-}} \tag{17.4.24}$$

Plots of the exact and composite solutions are shown in Figure 17.5 indicating a good agreement

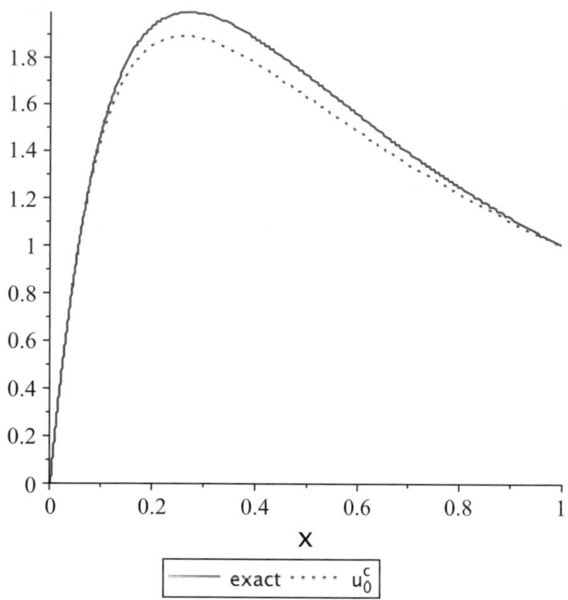

Figure 17.5 Composite and exact solutions for $\varepsilon = 0.1$.

over most of x. Improvement to the composite solution can be achieved by going to next order to find u_1^c.

17.5 Singular Flows at Low and High Reynolds Numbers

In Section 12.1 we examined flow over a flat plate, and encountered a singularity where $\varepsilon = 1/Re$ multiplied the highest-order spatial derivatives found in the viscous terms of the Navier–Stokes equations. Without specific reference to the method of matched asymptotic expansions, we actually employed it by introducing an "inner variable" in the Y-direction, $\eta = Y/\varepsilon^{1/2}$, and arriving at the inner equation, Eq. (12.1.13), the Blasius equation. The three boundary conditions included no-slip and no-penetration at the plate, but also matching to the inviscid free stream velocity far from the plate and parallel to it, i.e. the outer solution to the outer equations which are the Bernoulli equation. While the x-velocity matches at the edge of the boundary layer, the y-velocity does not. Going to next order is required for that matching to be smooth, and, for example, allow calculations of aerosol deposition for particles starting in the outer region and crossing over into the boundary layer to land on the wall (Zierenberg et al., 2013).

In Section 7.9 we also encountered a singularity involving low Reynolds number flow over a cylinder of radius b, i.e. "Stokes' paradox." The convective acceleration terms of the Navier–Stokes equations were entirely ignored, since they are multiplied by $Re_b = bU_\infty/\nu$. Under the assumption that the solution is a perturbation series in powers of $Re_b << 1$, the leading-order system is for $Re_b = 0$. Rather than the singularity occurring on the cylinder boundary, $r = b$, however, it happens far from the cylinder as $r \to \infty$. The issue is similar to that shown in Eq. (7.8.16) for a sphere, where the convective acceleration terms do not remain negligible compared to the viscous terms. For the cylinder the ratio of those two terms is proportional to the product $Re_b R \ln R$ where $R = r/b$ is the dimensionless radial coordinate. This form is different from Eq. (7.8.16) by the appearance of the $\ln R$ term, related to the two dimensionality the cylinder, or circle. Clearly, for any choice of $Re_b << 1$ a value of $R \to \infty$ can be found so that $Re_b R \ln R = 0(1)$ and is not negligible.

The resolution is to include a linearized form of convective acceleration, the Oseen approximation where Eq. (7.8.2), $\nabla^4\psi = 0$, is replaced by the following in dimensionless form

$$\left(\nabla^2 - Re_b \frac{\partial}{\partial x}\right)\nabla^2\psi = 0 \tag{17.5.1}$$

which keeps the inertial term in the main flow direction (Oseen, 1910). Here, spatial variables are scaled on b and ψ is scaled on $U_\infty b$. Lamb (1911) found a solution to Eq. (17.5.1), and the y-component of velocity is

$$-\frac{\partial\psi}{\partial x} = \Delta_1\left\{\frac{\sin(2\theta)}{2R^2} + \frac{2}{Re_b}\frac{\partial}{\partial y} \times \left[\ln(Re_b R) + e^{\frac{1}{2}Re_b R\cos\theta} K_0\left(\frac{1}{2}Re_b R\right)\right]\right\} \tag{17.5.2}$$

where $\Delta_1 = \left(\ln(4/Re_b) - \gamma + \frac{1}{2}\right)^{-1}$, K_0 is the Bessel function and $\gamma = 0.5772$, Euler's constant. It is this form which yields the drag force equation Eq. (7.9.4).

Note the appearance of the term $Re_b R$, which is equivalent to scaling the radial position, r, on the viscous length scale, ν/U_∞, rather than the cylinder radius. Calling $\hat{R} = Re_b R$, formally $\hat{R}$ is an inner variable as viewed through matched asymptotics, which developed as a mathematical method decades later and can be used to derive Eq. (17.5.2) (Lagerstrom and Cole, 1955; Kaplun and Lagerstrom, 1957; Van Dyke, 1975). For flow over a flat plate we defined the inner variable as $\eta = Y/\varepsilon^{1/2}$, and in the example system of Section 17.4 it was $\xi = x/\varepsilon$, both for $\varepsilon << 1$. These definitions "stretch" the coordinate so that small values of Y or x become $O(1)$ in the new variable. By contrast, for flow over a cylinder, the inner variable "compresses" the distance far away, so that for large R, $R >> 1$, multiplied by small Re_b, $Re_b << 1$, the variable $\hat{R} = O(1)$.

Drag calculations for the cylinder are compared to experimental data in Figure 17.6. The drag coefficient here is defined as $C_D = F_d/\rho U_\infty^2 b$ where F_d is the drag force per unit length of the cylinder. The circles represent data points selected from Tritton (1959) for $0.2 \le Re_b \le 1$ and the squares are data from Finn (1953), selected for $0.03 \le Re_b \le 1$. The "one-term" curve uses the first term of Eq. (7.9.4) for F_d. The "two-term" curve is two terms of Eq. (7.9.4) for F_d. Note that the two-term curve correlates well with the Tritton data for $0.2 < Re_b < 0.5$, approximately, and both converge to fit the Finn data for $Re_b < 0.2$.

From Eq. (7.9.4), dividing both sides by $b\rho U_\infty^2$ yields $C_D = 4\pi(\Delta_1 - 0.87\Delta_1^3)/Re_b$, which is plotted in Figure 17.7 along with Re_b^{-1} for comparison in a log–log plot. As $Re_b \to 0$ the slope of C_D approaches that of Re_b^{-1}, i.e. –1, which is reflected in Figure 13.3.

Stokes flow over a sphere gave a solution for $Re_a = 0$, Eq. (7.8.9), which is the first approximation for a perturbation series in powers of $Re_a << 1$. That was lucky, because the next-order correction term does not have a solution, a finding called the "Whitehead paradox." Oseen found a solution, which viewed through matched asymptotics is equivalent to a two-term approximation (Oseen, 1910),

$$\psi = \frac{1}{4}\left(2R^2 + \frac{1}{R}\right)\sin^2\theta - \frac{3}{2}\frac{1}{Re_a}(1+\cos\theta)\left[1 - e^{-\frac{1}{2}Re_a R(1-\cos\theta)}\right] \tag{17.5.3}$$

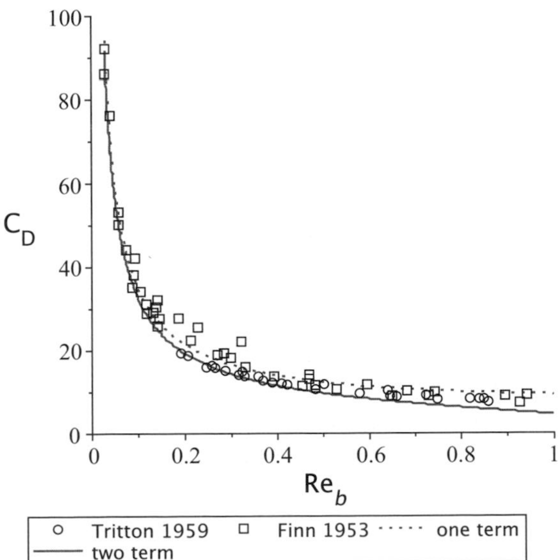

Figure 17.6 Drag coefficient C_D vs Re_b for flow over a cylinder, see text for details.

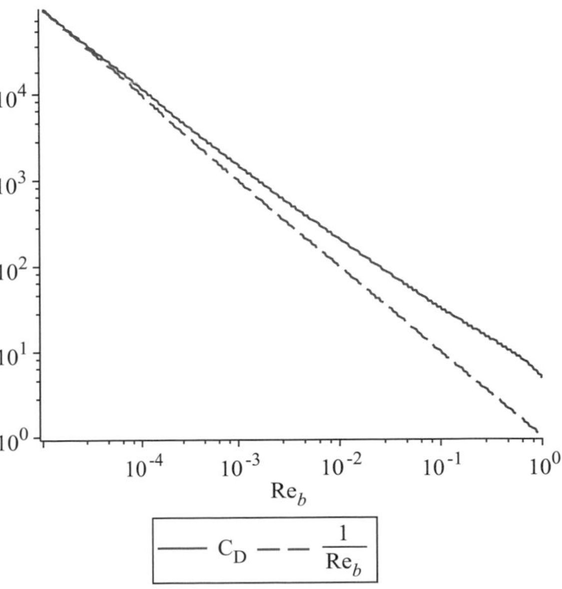

Figure 17.7 C_D and Re_b^{-1} vs Re_b in a log–log plot.

Again, note the appearance of the product $Re_a R$, which is the inner variable. The corresponding corrected drag force on the sphere appears in Eq. (7.8.17) and was found by Oseen (1913).

17.6 Stability Analysis of Jet Breakup: Fastest Growing Wavelength

Within the unstable range of wavelengths, $\lambda > 2\pi a$, determined by the static energy analysis in Section 16.3, there will be one which grows faster than the others. This suspicion comes from the observation by Plateau (1873) that the jet consistently starts to break up at a reproducible distance from the jet exit. He found that this length is approximately 6.26a. We can discover the most dangerous wave length by the following dynamic stability analysis of the liquid jet.

Let the perturbed interface position now be a sinusoidal standing wave, $\cos(kz) = \Re(e^{ikz})$, with a time-dependent amplitude e^{qt}

$$b(z,t) = a\left(1 + \varepsilon e^{ikz} e^{qt}\right) \tag{17.6.1}$$

as shown in Figure 17.8. Because of the form for b(z, t), we use a similar separation of variables for the pressure and velocities

$$p = p_0 + \varepsilon\hat{p}(r)e^{ikz}e^{qt}, \quad u_r = \varepsilon\hat{u}(r)e^{ikz}e^{qt}, \quad u_z = \varepsilon\hat{w}(r)e^{ikz}e^{qt} \tag{17.6.2}$$

Our goal will be to find the growth rate, q, as a function of the system parameters and determine which value of λ produces the maximum value of $q > 0$, which is the unstable range.

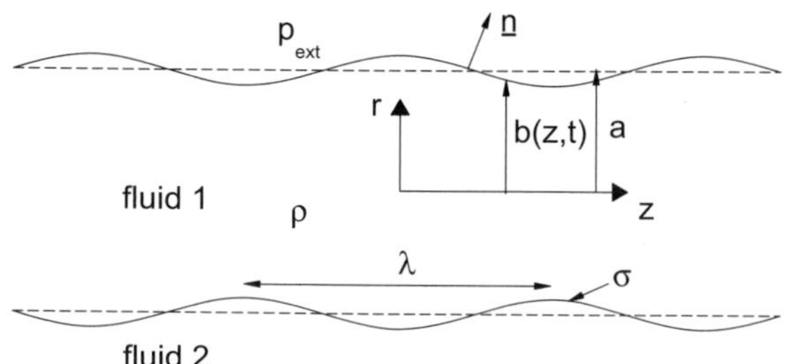

Figure 17.8 Stability analysis of the breakup of a fluid 1 jet surrounded by immiscible fluid 2.

Because $\varepsilon << 1$, the Navier–Stokes equations simplify since the nonlinear convective acceleration terms are of $O(\varepsilon^2)$ while the other terms are of $O(1)$. Also, the fluid viscosity and gravity may be ignored and we are moving in a coordinate system at the constant jet speed, U. The resulting Navier–Stokes equations are the r- and z-components

$$\begin{aligned} r: \ \rho\frac{\partial u_r}{\partial t} &= -\frac{\partial p}{\partial r} \\ z: \ \rho\frac{\partial u_z}{\partial t} &= -\frac{\partial p}{\partial z} \end{aligned} \tag{17.6.3}$$

and continuity is

$$\frac{\partial u_r}{\partial r} + \frac{u_r}{r} + \frac{\partial u_z}{\partial z} = 0 \tag{17.6.4}$$

We have the kinematic boundary condition at the interface, that the radial fluid velocity matches the interfacial radial velocity

$$\frac{\partial b}{\partial t} = u_r(r = b(z,t)) \tag{17.6.5}$$

It is vital that we expand the right hand side of Eq. (17.6.5) in a Taylor series for $b(z,t) - a = O(\varepsilon) << 1$, because we have assumed separation of variables (r, z, t) which is violated if we set $r = b(z,t)$. So $u_r(r = b(z,t)) = u_r(r = a) + (b - a)\partial u_r/\partial r|_{r=a} \cdots$ and we only keep the first term. This process of linearizing the boundary condition is similar to the analysis of wave propagation on a flexible tube in Section 9.5, Eqs. (9.5.2)–(9.5.5). These are called "free boundary" problems because the position of the boundary is part of the solution. In Section 9.5 the restoring force was the tube elastance, while here it is the surface tension. Finally we have the linearized version of Eq. (17.6.5)

$$\frac{\partial b}{\partial t} = u_r(r = a) \tag{17.6.6}$$

Inserting the forms for b and u_r from Eqs. (17.6.2) and canceling the common factor of $\varepsilon e^{ikz}e^{qt}$ we obtain

$$\hat{u}(r = a) = aq \tag{17.6.7}$$

The other kinematic condition is that we must have finite velocities.

With the presence of surface tension we also have a stress boundary condition at the interface given by the Young–Laplace equation, see Eq. (16.2.3). This is appropriate since, in the absence of viscosity, the only normal fluid stress is the pressure

$$p(r = b(z,t)) - p_{ext} = \sigma\kappa \tag{17.6.8}$$

Again we linearize this boundary condition using the Taylor expansion process which took us from Eq. (17.6.5) to Eq. (17.6.6). So Eq. (17.6.8) becomes

$$p(r = a) - p_{ext} = \sigma\kappa \tag{17.6.9}$$

The curvature κ is defined by the divergence of the outward pointing unit normal $\underline{n}$

$$\kappa = \underline{\nabla}\cdot\underline{n} = \frac{1}{r}\frac{\partial}{\partial r}(r n_r) + \frac{1}{r}\frac{\partial}{\partial \theta}(n_\theta) + \frac{\partial}{\partial z}(n_z) \tag{17.6.10}$$

The unit normal is related to the location of the interface given by $F = r - b(z,t) = 0$

$$\begin{aligned}\underline{n} = \frac{\underline{\nabla}F}{|\underline{\nabla}F|_{F=0}} &= (1, 0, -\partial b/\partial z)\left(1+(\partial b/\partial z)^2\right)^{-1/2} \\ &= (1, 0, -\partial b/\partial z) + \cdots\end{aligned} \tag{17.6.11}$$

as shown in Figure 17.8. We have used the Taylor series for $\left(1+(\partial b/\partial z)^2\right)^{-1/2} = 1-(\partial b/\partial z)^2/2+\cdots = 1+O(\varepsilon^2)+\cdots$ because $\partial b/\partial z = O(\varepsilon)$. Inserting Eq. (17.6.11) into Eq. (17.6.10) gives us the curvature

$$\kappa = \left[\frac{1}{r}\frac{\partial}{\partial r}(r)\right]_{r=b} + \frac{\partial}{\partial z}\left(-\frac{\partial b}{\partial z}\right) = \frac{1}{b} - \frac{\partial^2 b}{\partial z^2} + O(\varepsilon^2) \tag{17.6.12}$$

Substitute for κ from Eq. (17.6.12) into the stress boundary condition Eq. (17.6.9) to obtain

$$p(r = a) - p_{ext} = \sigma\left(\frac{1}{b} - \frac{\partial^2 b}{\partial z^2}\right) \tag{17.6.13}$$

Now we can substitute for b(z, t) from Eq. (17.6.1) and expand $1/b$ in Eq. (17.6.13) in a Taylor series

$$\begin{aligned}p(r = a) - p_{ext} &= \sigma\left(\frac{1}{a}\left(1+\varepsilon e^{ikz}e^{qt}\right)^{-1} + \varepsilon a k^2 e^{ikz}e^{qt}\right) \\ &= \sigma\left(\frac{1}{a} + \varepsilon\left(ak^2 - \frac{1}{a}\right)e^{ikz}e^{qt} + O(\varepsilon^2)\right)\end{aligned} \tag{17.6.14}$$

Inserting the pressure from Eq. (17.6.2) into Eq. (17.6.14) we obtain

$$\begin{aligned}&\left(p_0 + \varepsilon\hat{p}(r=a)e^{ikz}e^{qt}\right) - p_{ext} \\ &\quad = \sigma\left(\frac{1}{a} + \varepsilon\left(ak^2 - \frac{1}{a}\right)e^{ikz}e^{qt}\right)\end{aligned} \tag{17.6.15}$$

We can solve for the base state pressure, p_0, by setting $\varepsilon = 0$ in Eq. (17.6.15). That gives us

$$p_0 - p_{ext} = \frac{\sigma}{a} \tag{17.6.16}$$

Subtracting Eq. (17.6.16) from Eq. (17.6.15) leaves us with the $O(\varepsilon)$ stress boundary condition where the common term $\varepsilon e^{ikz}e^{qt}$ cancels out so that

$$\hat{p}(r = a) = \sigma\left(ak^2 - \frac{1}{a}\right) \tag{17.6.17}$$

Insert the solution forms of Eq. (17.6.2) into Eqs. (17.6.3) and Eq. (17.6.4). This leads to three coupled ODEs in r

$$\rho q\hat{u} = -\frac{d\hat{p}}{dr}, \quad \rho q\hat{v} = -ik\hat{p}, \quad \frac{d\hat{u}}{dr} + \frac{\hat{u}}{r} + ik\hat{v} = 0 \tag{17.6.18}$$

To solve this system we take the r-derivative of the second, z-momentum, equation in Eq. (17.6.18) to get $\rho q(d\hat{v}/dr) = -ik(d\hat{p}/dr)$ and eliminate $d\hat{p}/dr$ by substituting into the first, r-momentum, equation to obtain $d\hat{v}/dr = ik\hat{u}$. We can use this expression by taking the r-derivative of the third, continuity, equation in Eq. (17.6.18), and substituting for $d\hat{v}/dr$, which yields

$$\frac{d^2\hat{u}}{dr^2} + \frac{1}{r}\frac{d\hat{u}}{dr} - \left(\frac{1}{r^2} + k^2\right)\hat{u} = 0 \tag{17.6.19}$$

Equation (17.6.19) is a modified Bessel equation of order 1. We saw in Section 8.5, Eq. (8.5.6), that Bessel's equation typically must be satisfied by an r-dependent function when using separation of variables in cylindrical coordinates. Solutions to Eq. (17.6.19) are the sum of the linearly independent modified Bessel functions of order 1, I_1 and K_1,

$$\hat{u} = c_1 I_1(kr) + c_2 K_1(kr) \tag{17.6.20}$$

Because $K_1(kr) \rightarrow \infty$ as $r \rightarrow 0$ we must choose $c_2 = 0$. That leaves us with

$$\hat{u} = c_1 I_1(kr) \tag{17.6.21}$$

From the first of Eqs. (17.6.18) we know that

$$-\frac{d\hat{p}}{dr} = \rho q c_1 I_1(kr) \tag{17.6.22}$$

There is an identity for Bessel functions that $I_1(x) = dI_0(x)/dx$. Using this identity for $x = kr$ and $dx = kdr$ in Eq. (17.6.22) we get

$$-\frac{d\hat{p}}{dr} = \frac{\rho q c_1}{k}\frac{dI_0(kr)}{dr} \tag{17.6.23}$$

which can be integrated to yield

$$\hat{p} = -\frac{\rho q c_1}{k} I_0(kr) \tag{17.6.24}$$

with a zero integration constant. The kinematic boundary condition of Eq. (17.6.7) at the interface solves for c_1

$$c_1 = \frac{aq}{I_1(ka)} \tag{17.6.25}$$

Inserting Eq. (17.6.24) into the stress boundary condition, Eq. (17.6.17), and substituting c_1 from Eq. (17.6.25) yields

$$q^2 = \frac{\sigma}{\rho a^3}\frac{(ka)I_1(ka)}{I_0(ka)}\left(1-(ka)^2\right) \tag{17.6.26}$$

Taking the square root of Eq. (17.6.26) yields q in the form

$$q = \pm F^{1/2} \quad F = \frac{\sigma}{\rho a^3}\frac{(ka)I_1(ka)}{I_0(ka)}\left(1-(ka)^2\right) > 0 \quad ka < 1 \tag{17.6.27}$$

There are two roots and Eq. (17.6.27). One is stable, $q < 0$, and one unstable, $q > 0$. It is the unstable root that is observed since it grows exponentially in time. It occurs when $ka < 1$, $\lambda > 2\pi a$, which is the same criterion we found for instability in the surface energy approach of Section 16.3.

Equation (17.6.26) is called the characteristic equation or dispersion equation. We encountered a previous example of a dispersion equation in Section 9.9, Eq. (9.9.3), for the flutter instability of wheezing. In both cases the wave speed varies with the wavelength of the disturbance.

We can plot the dimensionless version of Eq. (17.6.27) by letting $Q = (\rho a^3/\sigma)^{1/2} q$ and $K = ka$, yielding

$$Q = \left[K\left(1-K^2\right)\frac{I_1(K)}{I_0(K)}\right]^{1/2} \tag{17.6.28}$$

A plot of Q(K) is shown in Figure 17.9 and here we reach our goal of identifying which wavelength grows the fastest. The value of K at the maximum computes to be $K_m = 0.697$ or $\lambda = 9.015\,a$. Since this unstable value of λ grows the fastest, it is what we observe. The physics of the system selects a preferred wavelength as we saw in the flutter problem of Section 9.9. Inserting K_m into Eq. (17.6.28) and taking the square root leads to the dimensionless growth rate for the most dangerous wavelength $Q_m = Q(K_m) = 0.343$. The dimensional growth rate is $q_m = (\sigma/\rho a^3)^{1/2} Q_m$. For a 1 cm diameter water jet in air the surface tension is $\sigma = 72\,\text{dyn/cm}$, and the density is $\rho = 1\,\text{g/cm}^3$, which gives $q_m = (\sigma/\rho a^3)^{1/2} Q_m = 8.24/\text{s}$.

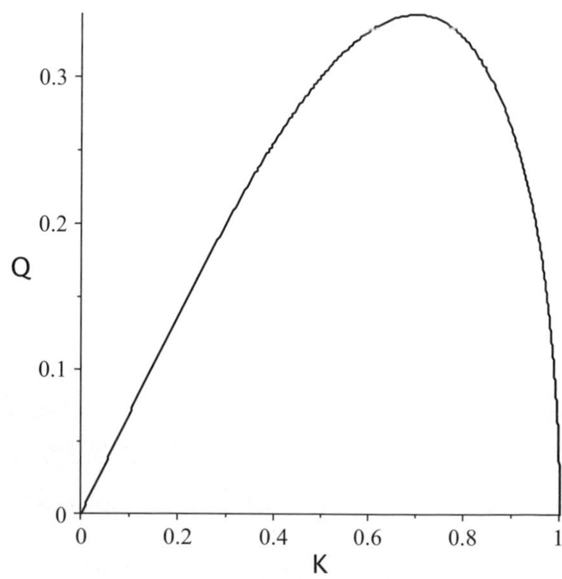

Figure 17.9 Dimensionless growth rate Q vs dimensionless wave number K from Eq. (17.6.28).

A characteristic breakup time would be $t_b = 1/q_m = 1/8.24$ s.

Figure 17.10 shows an experimental investigation of a water jet breakup, where sinusoidal perturbations were imposed using a loudspeaker driven by an audio oscillator of selected frequencies. The result is to impose a specific wavelength perturbation, thereby setting K for the perturbation. In Figure 17.10(a) there is slow growth for $K = 0.148$, in Figure 17.10(b) the fastest growth is for $K = 0.678$, which is nearly K_m, and in Figure 17.10(c) $K = 1.07$, which is a stable wavelength $\lambda = 2\pi a/1.07 < 2\pi a$, so no breakup.

(a)

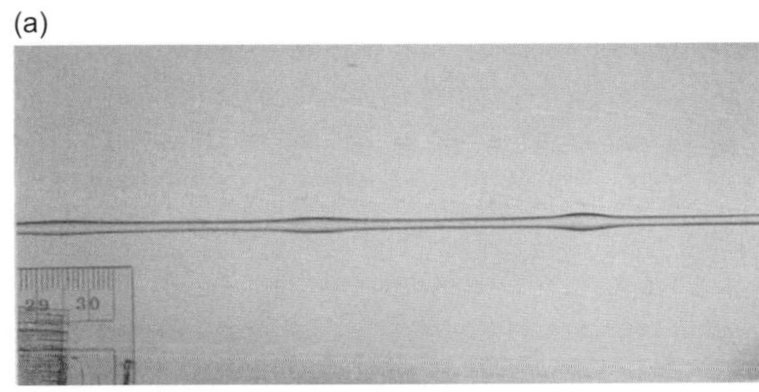

(b)

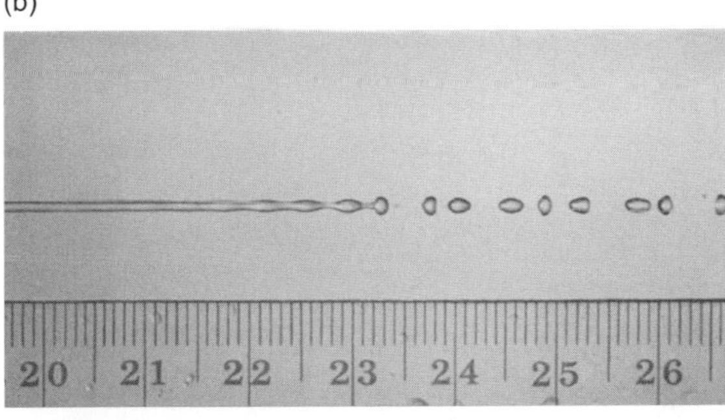

(c)

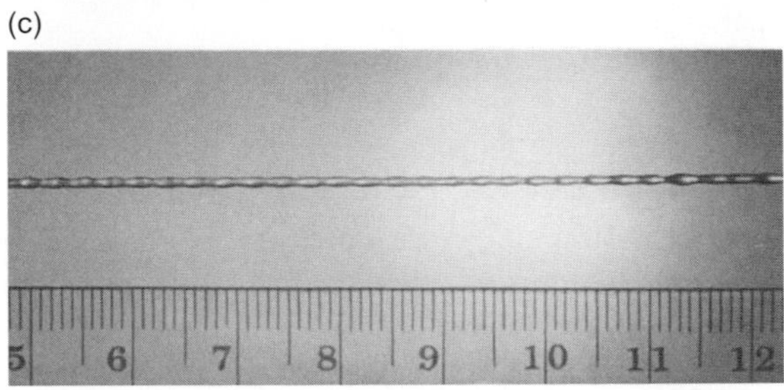

Figure 17.10 Perturbed water jet at three wavelengths: (a) K = ka = 0.148, slow growth; (b) K = 0.678, fastest growth; and (c) K = 1.07, stable.

17.7 Steady Streaming: Oscillating Slide Block

We considered the slide block problem introduced in Section 11.1 as a fundamental model of lubrication theory. That flow was steady with the wall dragging the fluid into a converging channel, see Figure 11.1. The solution to the x-velocity in Eq. (11.1.14), call it $\hat{u}(x, y)$, is reversible, since replacing U_w with $-U_w$ simply changes $\hat{u}(x, y)$ to $-\hat{u}(x, y)$. However, considered as a perturbation expansion $u(x, y) = u_0(x, y) + \varepsilon u_1(x, y) + \cdots$, where $\varepsilon = (d_1 - d_2)/L << 1$ is the channel slope, $\hat{u}(x, y)$ is actually the $O(1)$ term, i.e. $\hat{u}(x, y) = u_0(x, y)$. Extending to $O(\varepsilon)$ will bring in contributions from the nonlinear convective acceleration terms so that $u_1(x, y)$ is not reversible.

With this in mind, now we allow the lower wall to oscillate in its own plane, see Figure 17.11. Let the wall velocity be

$$U_w = a\omega \cos(\omega t) \tag{17.7.1}$$

where a is the amplitude and $\omega = 2\pi f$ is the angular frequency. This oscillation drags fluid back and forth through a relative converging, flow to the right, and diverging, flow to the left, channel geometry. The irreversibility at $O(\varepsilon)$ will cause fluid displacement per cycle, i.e. steady flow or steady streaming. Here is the analysis.

Define the Reynolds number as $Re = u_0(d_1 - d_2)/\nu$ and the Womersley parameter as $\alpha^2 = \omega(d_1 - d_2)^2/\nu$. The unsteady, dimensionless Navier–Stokes equation in stream function form, see Eq. (6.5.20), is

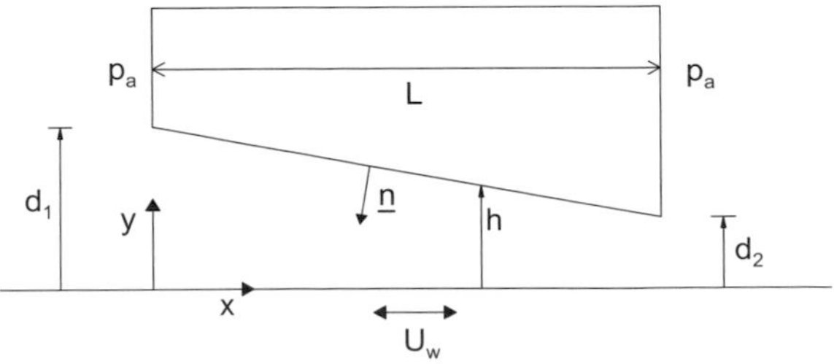

Figure 17.11 Oscillating slide block.

$$\alpha^2 \frac{\partial^3 \psi}{\partial T \partial \eta^2} + \varepsilon \mathrm{Re}\left(\frac{\partial \psi}{\partial \eta}\frac{\partial^3 \psi}{\partial \eta^2 \partial \xi} - \frac{\partial \psi}{\partial \xi}\frac{\partial^3 \psi}{\partial \eta^3}\right) = \frac{\partial^4 \psi}{\partial \eta^4} + O(\varepsilon^2) \quad (17.7.2)$$

where $\xi = x/L$, $\eta = y/(d_1 - d_2)$, $T = \omega t$, $U = u/u_0$, $V = v/\varepsilon u_0$, $P = p/P_s$, $P_S = \mu u_0/\varepsilon(d_1 - d_2)$, $u_0 = \omega(d_1 - d_2)$ and $\psi = \psi^*/u_0(d_1 - d_2)$, where ψ^* is dimensional $U = \partial\psi/\partial\eta$, $V = -\partial\psi/\partial\xi$. Note that the $O(\varepsilon^2)$ quantities, which arise from the $\partial^2 u/\partial x^2$, $\partial^2 v/\partial x^2$ viscous terms, are omitted and we assume $\mathrm{Re} = O(1)$.

The boundary conditions are no cross flow and no slip at the wall and block surfaces

$$\psi(\eta = 0) = 0, \quad \left.\frac{\partial \psi}{\partial \eta}\right|_{\eta=0} = A\cos T, \quad \left.\frac{\partial \psi}{\partial \xi}\right|_{\eta=H} = 0, \quad \left.\frac{\partial \psi}{\partial \eta}\right|_{\eta=H} = 0 \quad (17.7.3)$$

where $A = a/(d_1 - d_2)$ and $D = d_2/(d_1 - d_2)$, and the surface of the slide block is at $H = 1 + D - \xi$.

Assume a perturbation solution

$$\psi = \psi_0 + \varepsilon\psi_1 + \cdots \quad (17.7.4)$$

To simplify let $\alpha^2 = \alpha_0^2\varepsilon^2$ where $\alpha_0^2 = O(1)$, which allows us to ignore unsteady acceleration.

At $O(1)$ we have

$$\frac{\partial^4 \psi_0}{\partial \eta^4} = 0 \quad (17.7.5)$$

Separate variables and let

$$\psi_0 = \hat{\psi}_0(\xi, \eta)\cos T \quad (17.7.6)$$

Inserting Eq. (17.7.6) into Eq. (17.7.5) gives us

$$\frac{\partial^4 \hat{\psi}_0}{\partial \eta^4} = 0 \quad (17.7.7)$$

whose solution is obtained by integrating four times with respect to η to obtain

$$\hat{\psi}_0 = a_0(\xi)\eta^3 + b_0(\xi)\eta^2 + c_0(\xi)\eta + d_0(\xi) \quad (17.7.8)$$

The boundary conditions of Eqs. (17.7.3) at $O(1)$ are

$$\hat{\psi}_0(\eta = 0) = 0, \quad \left.\frac{\partial \hat{\psi}_0}{\partial \eta}\right|_{\eta=0} = A, \quad \left.\frac{\partial \hat{\psi}_0}{\partial \xi}\right|_{\eta=H} = 0, \quad \left.\frac{\partial \hat{\psi}_0}{\partial \eta}\right|_{\eta=H} = 0 \quad (17.7.9)$$

and imposing them on Eq. (17.7.8) solves for the unknown coefficients $a_0(\xi)$, $b_0(\xi)$, $c_0(\xi)$, $d_0(\xi)$, resulting in the form

$$\hat{\psi}_0 = \left(\frac{A}{H^2} + \frac{e_0}{H^3}\right)\eta^3 - \left(\frac{2A}{H} + \frac{3}{2}\frac{e_0}{H^2}\right)\eta^2 + A\eta \quad (17.7.10)$$

Equation (17.7.10) has a remaining constant, e_0, which can be solved from matching to the end pressures. The ξ-component of the $O(1)$ Navier–Stokes equations is

$$\frac{\partial^3 \psi_0}{\partial \eta^3} = \frac{\partial P_0}{\partial \xi} = \frac{\partial \hat{P}_0(\xi)}{\partial \xi}\cos T \quad (17.7.11)$$

where we let $\partial P_0/\partial\xi = \left(\partial\hat{P}_0(\xi)/\partial\xi\right)\cos T$, since the $O(1)$ pressure is not dependent on η. Assuming $\hat{P}_0$ at both ends are the same, $\hat{P}_0(\xi = 0) = \hat{P}_0(\xi = 1)$, the solution for e_0 is

$$e_0 = -\frac{2AD(1 + D)}{1 + 2D} \quad (17.7.12)$$

Velocity profiles $\hat{U}_0(\xi, \eta)$ are graphed in Figure 17.12 for $\xi = 0, 0.25, 0.5, 0.75, 1$, and $A = 1$, $D = 0.25$. They show the same behavior as the steady slide block, see Figure 11.9(a), including flow reversal and leakback, because the two systems have the same spatial equations at this order.

For next order, $O(\varepsilon)$, the governing stream function equation is

$$\frac{\partial^4 \psi_1}{\partial \eta^4} = \alpha_0^2 \frac{\partial^3 \psi_0}{\partial T \partial \eta^2} + \mathrm{Re}\left(\frac{\partial \psi_0}{\partial \eta}\frac{\partial^3 \psi_0}{\partial \eta^2 \partial \xi} - \frac{\partial \psi_0}{\partial \xi}\frac{\partial^3 \psi_0}{\partial \eta^3}\right) \quad (17.7.13)$$

The right hand side of Eq. (17.7.13) has contributions from the unsteady acceleration and the convective acceleration consisting of products of time-periodic, $\sim \cos T$, terms involving ψ_0 and its spatial derivatives. The products of those terms yield

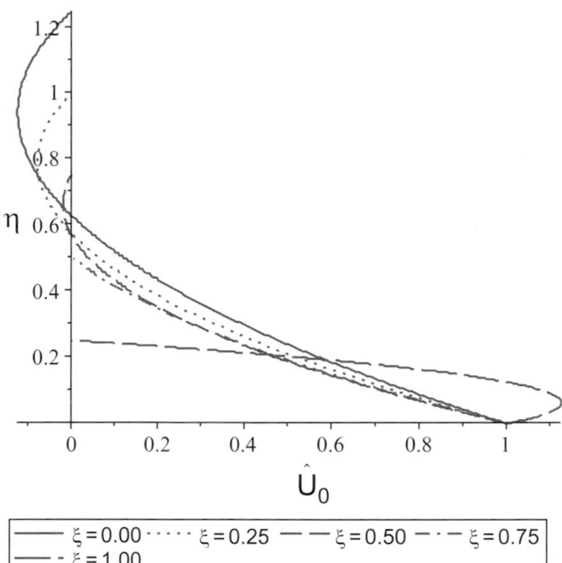

Figure 17.12 $\hat{U}_0(\xi, \eta)$ for $\xi = 0, 0.25, 0.5, 0.75, 1$, and $A = 1$, $D = 0.25$.

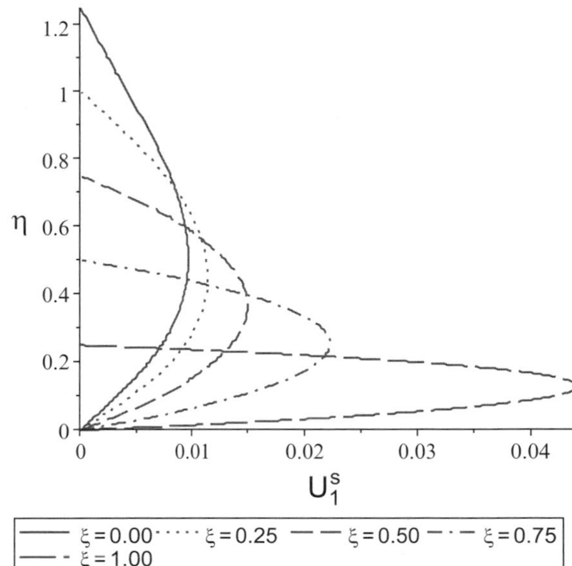

Figure 17.13 $U_1^s(\xi, \psi)$ for five values of $\xi = 0, 0.25, 0.5, 0.75, 1$, for $A = 1$, $D = 0.25$ and $Re = 1$.

$\cos^2 T = (1 + \cos 2T)/2$, the sum of a steady component, 1/2, and a doubly periodic term, $\cos 2T/2$. It is the steady contribution on the right hand side which drives steady streaming in the solution of ψ_1.

Let's look for the steady solution, first substituting the expression for ψ_0 into the two product terms of Eq. (17.7.13)

$$\frac{\partial\psi_0}{\partial\eta}\frac{\partial^3\psi_0}{\partial\eta^2\partial\xi} = \frac{\partial\hat{\psi}_0}{\partial\eta}\frac{\partial^3\hat{\psi}_0}{\partial\eta^2\partial\xi}\cos^2 T = \frac{\partial\hat{\psi}_0}{\partial\eta}\frac{\partial^3\hat{\psi}_0}{\partial\eta^2\partial\xi}\frac{1}{2}(1 + \cos 2T) \tag{17.7.14}$$

and

$$\frac{\partial\psi_0}{\partial\xi}\frac{\partial^3\psi_0}{\partial\eta^3} = \frac{\partial\hat{\psi}_0}{\partial\xi}\frac{\partial^3\hat{\psi}_0}{\partial\eta^3}\cos^2 T = \frac{\partial\hat{\psi}_0}{\partial\xi}\frac{\partial^3\hat{\psi}_0}{\partial\eta^3}\frac{1}{2}(1 + \cos 2T) \tag{17.7.15}$$

The steady component of ψ_1, call it $\psi_1^{(s)}$, satisfies

$$\frac{\partial^4\psi_1^{(S)}}{\partial\eta^4} = \frac{Re}{2}\left(\frac{\partial\hat{\psi}_0}{\partial\eta}\frac{\partial^3\hat{\psi}_0}{\partial\eta^2\partial\xi} - \frac{\partial\hat{\psi}_0}{\partial\xi}\frac{\partial^3\hat{\psi}_0}{\partial\eta^3}\right) \tag{17.7.16}$$

Inserting $\hat{\psi}_0$ from Eq. (17.7.10) and Eq. (17.7.12) into the right hand side of Eq. (17.7.16), we can integrate Eq. (17.7.16) three times to determine the ξ-direction steady streaming velocity $U_1^s = \partial\psi_1^s/\partial\eta$. Figure 17.13 shows plots of $U_1^s(\xi, \eta)$ at five values of $\xi = 0, 0.25, 0.5, 0.75, 1$, for $A = 1$, $D = 0.25$ and $Re = 1$, revealing unidirectional flow toward the narrow end of the gap.

This product of two time-periodic variables yielding a steady result was the basis of the mass transport we examined for oscillatory flow in a straight channel, Section 8.4, related to high frequency ventilation. In that case, it is the convective flux term, $u \times c$, which generates steady transport, because the x-velocity, u, is time-periodic, and the concentration, c, has both steady and time-periodic components. If the channel is tapered, however, it is also possible for the steady streaming velocity of u to combine with the steady part of c to add to the transport (Godleski and Grotberg, 1988).

In Section 11.6 we learned that axial velocity profile for inflow, inspiration, at an airway bifurcation is different from the outflow, expiration. This strongly depends on the increase in total cross-sectional area, from parent to the two daughter tubes, past generation

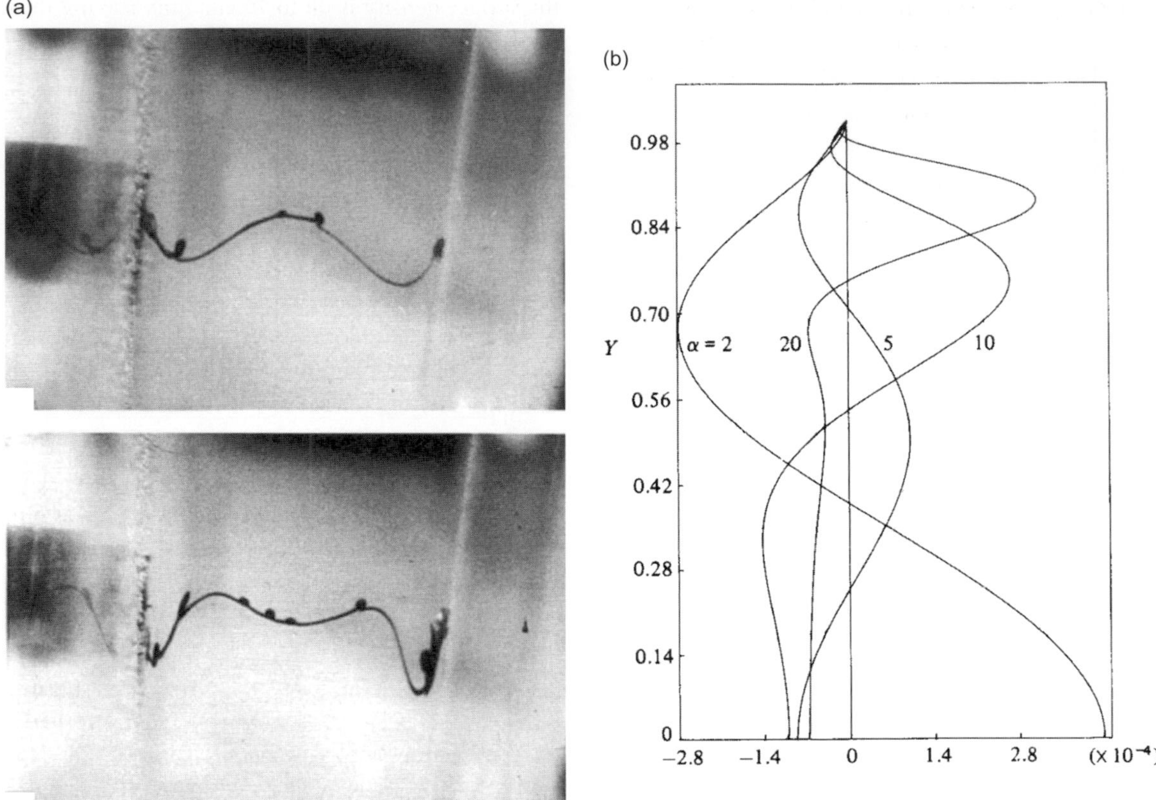

Figure 17.14 Steady streaming axial velocity for oscillatory flow in a tapered channel: (a) experimental flow visualization for $\alpha = 3.8$(upper) and $\alpha = 17.4$ (lower); (b) perturbation solution, $\alpha = 2, 5, 10, 20$.

3 or so, i.e. the cross-sectional area tapers with the wider end at the daughter's. Studies of oscillatory flow at a bifurcation show that there is bi-directional streaming, as we expect for a volume-cycled system with zero time-averaged flow (Haselton and Scherer, 1982; Scherer and Haselton, 1982). A model isolating the area expansion feature was examined both in theory and experiment for a tapered channel with volume-cycled flow (Grotberg, 1984; Gaver and Grotberg, 1986). The experimental results are shown in Figure 17.14(a) for flow visualization of an initially straight dye streak across the tapered channel, which is wider at the top. Results for two values of α are shown, $\alpha = 3.8$ (upper) and $\alpha = 17.4$ (lower). In both cases, the streaming is bi-directional, depending on the value of the Y-coordinate, with $Y = 0$ the centerline. Note that the centerline streaming is toward the wide end for $\alpha = 3.8$, but toward the narrow end for $\alpha = 17.4$. Figure 17.14(b) is the steady streaming perturbation solution, similar to the above analysis for an oscillatory slide block. The wide end is on the right, and the streaming direction depends on Y and $\alpha = 2, 5, 10, 20$. Again, the direction of streaming for $Y = 0$ reverses for larger α. The streaming, or drift, velocity forms a boundary layer, δ_D, as α becomes larger. It is thicker than the Stokes layer, δ_s, as $\delta_D = \delta_s/\varepsilon A$, where $\varepsilon A << 1$, $\varepsilon << 1$ is the small slope of the taper, and A is a dimensionless stroke amplitude.

17.8 Mucociliary Pumping

Steady streaming can also result from nonlinearities in the boundary conditions. An important example occurs in mucociliary propulsion. Ciliated epithelial cells populate the upper airways, consisting of the nose and pharyngeal regions, and the lower airways from the trachea distally to approximately generation 15. They provide an essential defense mechanism by moving the serous-mucus liquid lining to be swallowed and safely eliminated, along with inhaled particles (e.g. bacteria, viruses, pollen, vehicle exhaust) that have landed on its surface. The cilia diameters are 0.2–0.3 μm and their lengths are 5–7 μm. Each ciliated cell can have up to 200 cilia and the surface density is up to 10 cilia/μm, leaving the distance between cilia ~ 1–2 times their diameters (Sleigh et al., 1988). See the scanning electron microscopic image of cilia from rabbit sinonasal epithelium in Figure 15.8(b).

Models of ciliary propulsion include analysis of a discrete cilium beating with prescribed or resultant motion from two-way force coupling, as well as equivalent distributed force singularities to mimic the

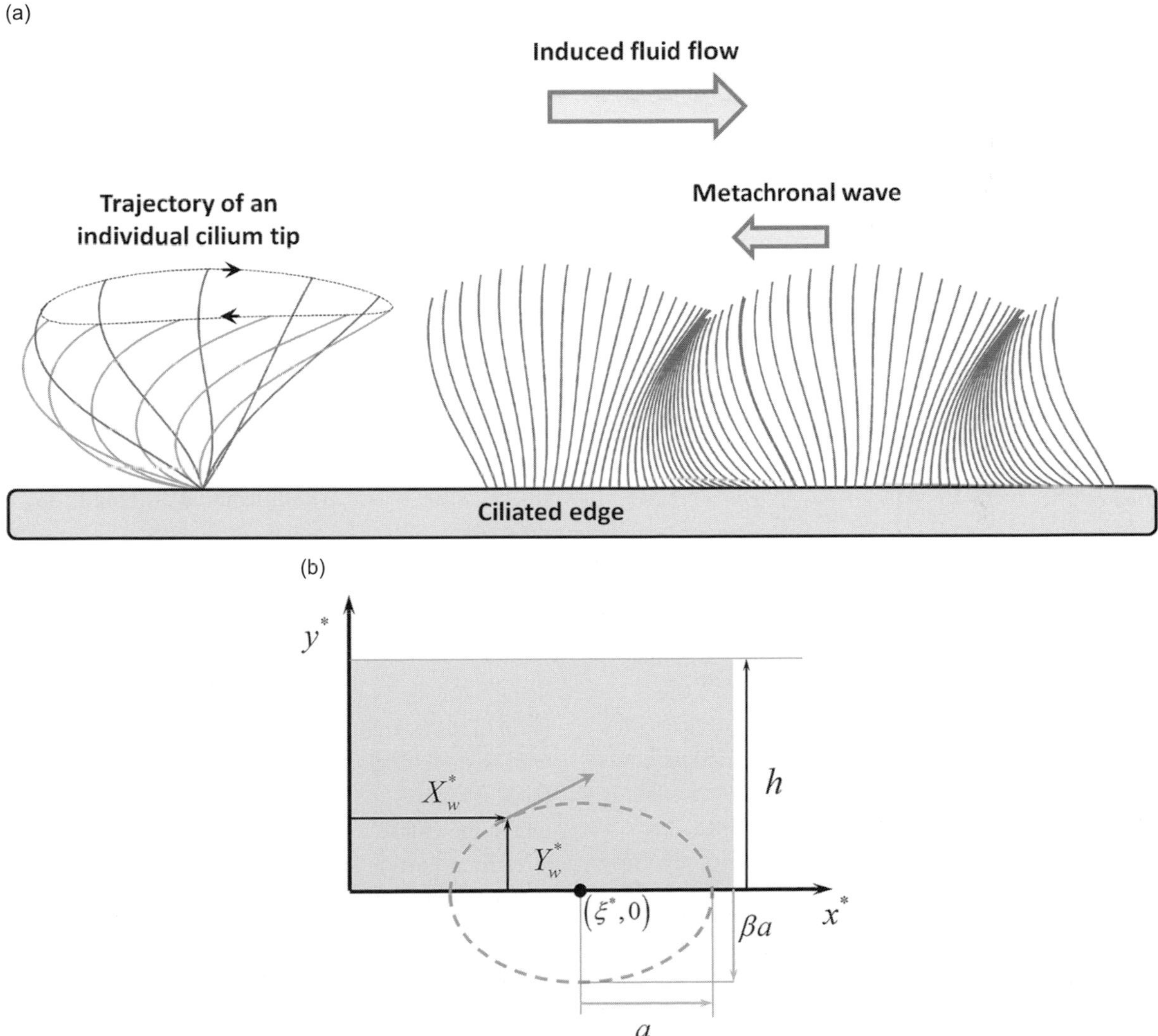

Figure 17.15 (a) Ciliary motion driving fluid flow to the right with metachronal wave to the left. (b) Elliptical trajectory of an individual cilia tip in continuous boundary model.

cilia (Smith et al., 2008). Because the cilia are close to one another, the Ross–Corrsin, or envelope, model of mucociliary propulsion treats the cilia tips as a continuous, deforming boundary (Ross and Corrsin, 1974). The tips perform an elliptical trajectory in the clockwise direction with the power stroke to the right and the recovery stroke to the left. The relative phase between cilia creates a metachronal wave propagating to the left, while the fluid layer above is "pumped" to the right, see Figure 17.15(a). This model has been successfully applied to experiments measuring fluid velocities over ciliated human nasal epithelial tissue strips obtained by brushing (Bottier et al., 2017a; 2017b).

Let a wall particle have the Lagrangian position (X_w^*, Y_w^*), Figure 17.15(b), with a fluid layer of thickness h above. We force it to execute an elliptical trajectory to mimic a ciliary tip. For an ellipse centered at a fixed position $(\xi^*, 0)$, the position of the wall particle is given by

$$X_w^* = \xi^* - a\cos(\omega t^*)$$
$$Y_w^* = a\beta\sin(\omega t^*) \tag{17.8.1}$$

Clearly we have

$$\left(\frac{X_w^* - \xi^*}{a}\right)^2 + \left(\frac{Y_w^* - 0}{a\beta}\right)^2 = \cos^2(\omega t^*) + \sin^2(\omega t^*) = 1 \tag{17.8.2}$$

which is the equation for an ellipse centered at $(\xi^*, 0)$with minor axis $2\beta a$ in the y^*-direction and major axis 2a in the x^*-direction. For $\beta = 1$ the trajectory is a circle. For $\beta = 0$ the wall particles only move in the x^*-direction. For $\beta > 0$ the particle orbits in the clockwise direction, while for $\beta < 0$ it orbits in the counterclockwise direction. As shown in Figure 17.15(a), typically the tip trajectories are fairly flat, i.e. $0 < \beta << 1$.

To model the metachronal wave of ciliary propulsion, employ a traveling wave of elliptical wall trajectories. Let ξ^* to be a continuous variable along the x^*-axis, i.e. a continuum of orbital centers, each with a corresponding wall particle executing an ellipse. The equations of the Lagrangian wall particles become

$$X_w^* = \xi^* - a\cos\left(\frac{2\pi\xi^*}{\lambda} + \omega t^*\right)$$
$$Y_w^* = a\beta\sin\left(\frac{2\pi\xi^*}{\lambda} + \omega t^*\right) \tag{17.8.3}$$

These motions are a traveling wave in the negative x^*-direction with wavelength λ and angular frequency $\omega = 2\pi f$, where f is the frequency. The wave speed is $c = f\lambda$. If we assume $\beta > 0$, the power stroke is in the positive x^*-direction, which is opposite to the wave direction, as is observed for airway cilia.

Scale these equations using

$$X_w = \frac{X_w^*}{h}, Y_w = \frac{Y_w^*}{h}, x = \frac{x^*}{h}, y = \frac{y^*}{h}, \xi = \frac{\xi^*}{h}, t = \omega t^* \tag{17.8.4}$$

to arrive at the following dimensionless expressions for the wall particle position

$$X_w = \xi - \varepsilon\cos(k\xi + t)$$
$$Y_w = \varepsilon\beta\sin(k\xi + t) \tag{17.8.5}$$

where the dimensionless amplitude is $\varepsilon = a/h$ and the dimensionless wave number is $k = 2\pi h/\lambda$. For the purposes of the analysis, we will use the fact that the amplitude of the wave is much smaller than the film thickness, i.e. $\varepsilon << 1$. The differential arc length, dS_w is defined by

$$dS_w = \left((\partial X_w/\partial\xi)^2 + (\partial Y_w/\partial\xi)^2\right)^{1/2} d\xi = \left(1 + \varepsilon k\sin(k\xi + t) + O(\varepsilon^2)\right)d\xi \tag{17.8.6}$$

using a Taylor expansion for $\varepsilon << 1$. Equation (17.8.6) shows that dS_w is time-periodic, so clearly the wall is stretching and contracting locally. Looking back at Figure 17.15(a), the contracting region is where the ciliary tips are closer together, while the stretching portion is where they are farther apart.

An example of the wall shape is shown in Figure 17.16 for $\varepsilon = 0.2$, $k = 1$, $\beta = 0.1$, $t = 0$ compared to $\sin\xi$, exhibiting the distortion of the wave shape from a sinusoid.

Now we need to calculate the Eulerian wall velocity to learn how this boundary will drive the overlying fluid layer. The Lagrangian dimensionless wall particle velocities are

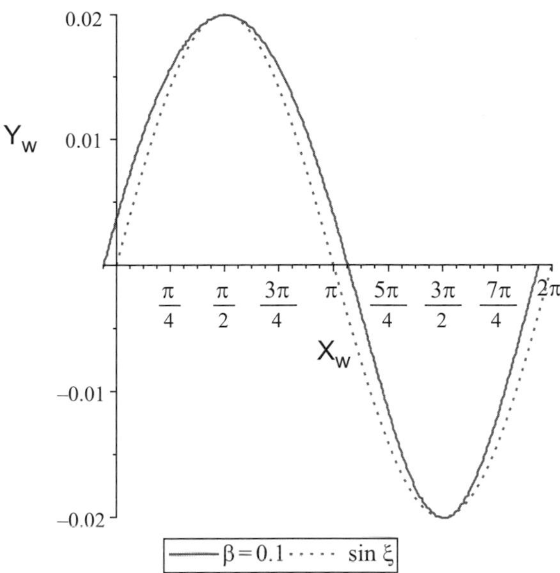

Figure 17.16 Ciliary boundary for $\varepsilon = 0.2$, $k = 1$, $\beta = 0.1$, $t = 0$ compared to $\sin\xi$.

$$
\begin{aligned}
U_w(\xi, t) - \left(\frac{\partial X_w}{\partial t}\right)_\xi &= \varepsilon \sin(k\xi + t) \\
V_w(\xi, t) = \left(\frac{\partial Y_w}{\partial t}\right)_\xi &= \varepsilon\beta \cos(k\xi + t)
\end{aligned}
\tag{17.8.7}
$$

These expressions for the wall velocities are in the Lagrangian frame of reference which has the independent variables (ξ, t). We need to express them in the Eulerian frame of reference (x, t) to use as boundary conditions in the conservation of momentum equation. First we recognize that the two frames coincide when $x = X_w$, $y = Y_w$, so from Eq. (17.8.5) we have

$$
x = \xi - \varepsilon \cos(k\xi + t) \tag{17.8.8}
$$

Then we employ a Taylor series expansion for x near ξ, which we know is true since in Eq. (17.8.8) $x - \xi$ is proportional to ε. For U_w we have

$$
\begin{aligned}
U_w(\xi, t) &= U_w(\xi, t)|_{\xi=x} + (\xi - x)\frac{\partial U_w}{\partial \xi}\Big|_{\xi=x} + \cdots \\
&= \varepsilon \sin(kx + t) + \big(-\varepsilon \cos(kx + t)\big) \\
&\quad [-\varepsilon k \cos(kx + t) + \cdots] \\
&= \varepsilon \sin(kx + t) + \varepsilon^2 k \cos^2(kx + t) + O(\varepsilon^3) \\
&= u_w(x, t)
\end{aligned}
\tag{17.8.9}
$$

Likewise for V_w we have

$$
\begin{aligned}
V_w(\xi, t) &= V_w(\xi, t)|_{\xi=x} + (\xi - x)\frac{\partial V_w}{\partial \xi}\Big|_{\xi=x} + \cdots \\
&= \varepsilon\beta \cos(kx + t) + (-\varepsilon \cos(kx + t)) \\
&\quad \left[-\varepsilon\beta k \sin(kx + t) + \cdots\right] \\
&= \varepsilon\beta \cos(kx + t) + \varepsilon^2 \beta k \cos(kx + t)\sin(kx + t) \\
&\quad + O(\varepsilon^3) = v_w(x, t)
\end{aligned}
\tag{17.8.10}
$$

The right hand side of Eqs. (17.8.9) and (17.8.10) are in the Eulerian independent variables, x, t, so these are our Eulerian wall velocities, $u_w(x, t), v_w(x, t)$ as indicated. Using trigonometric identities these simplify to

$$
\begin{aligned}
u_w(x, t) &= \varepsilon \sin(kx + t) + \frac{\varepsilon^2 k}{2}\left[\boxed{1} + \cos(2(kx + t))\right] + O(\varepsilon^3) \\
v_w(x, t) &= \varepsilon\beta \cos(kx + t) + \frac{\varepsilon^2 \beta^2 k}{2} \sin(2(kx + t)) + O(\varepsilon^3)
\end{aligned}
\tag{17.8.11}
$$

The striking result is that u_w has a steady component, emphasized in a square around the term in Eq. (17.8.11). It is proportional to $\varepsilon^2 k/2$ in the positive x-direction, and dimensionally is $u_s^* = \omega h \varepsilon^2 k/2 = 2\pi^2 f a^2/\lambda$, while v_w has no steady component, as expected. Consequently, the overlying fluid layer is subjected to a steady wall velocity driving it to the right. A steady wall velocity was used in Section 7.3 to evaluate the effectiveness of ciliary propulsion in a gravity field. We now see the origins of this assumption. Typical parameter values $f = 10\ s^{-1}$, $\lambda = 20\ \mu m$, $a = 2\ \mu m$ yield the steady wall velocity $u_s^* \sim 40\ \mu m/s$, which is the value used in Section 7.3. Smoking reduces ciliary function by decreasing f, and therefore u_s^*. Eventually the ciliated airway cells do not survive and "smoker's cough" in the morning is a well-known clinical finding where propulsion by air flow is used to compensate for ciliary function loss. A disease called primary ciliary dyskinesia is a rare, inherited disease of ciliary malfunction, resulting in chronic sinus, ear, and lung infections, since mucus clearance is impaired.

Summary

Previous chapters introduced mathematical techniques for solving problems, like separation of variables (Sections 8.2–8.7), Fourier series (Section 7.7), similarity solutions (Section 8.1) and rescaled coordinates (Section 12.1). Here we dedicate an entire chapter to a mathematical technique, because it has its own set of concepts and rules. While perturbation methods seek to find solutions near an exact one, computational fluid dynamics often is not limited to small deviations. As a consequence, there is important interplay between these two approaches. Learning the rules started with algebraic equations, both regular and singular, whose exact solutions we knew and could compare to the perturbation approach when expanded in the small parameter. For differential equations the regular problem did not have an analytic solution, while the singular problem did and led us to the method of matched asymptotic expansions. The richness of that method gave the insight needed to solve flow over a cylinder at low Reynolds number and flow over a flat plate at high Reynolds number. Both have boundary layers which require rescaling of the spatial variable to maintain the proper ordering. The plate-flow boundary layer is next to the plate, while the cylinder-flow boundary layer is at infinite distance from it. Perturbation methods are often applied to study stability of a system to small disturbances, since it gives initial insight into the physics behind the maintenance or loss of stability. It may not determine the final state, which can involve large amplitudes. Here we build on the static stability analysis of a liquid jet from Section 16.3 to find the fastest growing wavelength, which is the one observed in experiments. Finally, steady streaming resulted from purely time-periodic inputs for the oscillating slide block, through convective acceleration, and the beating cilia, through the nonlinear transformation of the boundary condition from Lagrangian to Eulerian variables. Both are qualitatively different from the leading-order solutions, giving physical insight to the mechanisms at work.

Problems

17.1 Consider the singular cubic equation $\varepsilon x^3 - x + 1 = 0$ where $\varepsilon << 1$.

a. Assume a scaled variable, $\xi = \delta(\varepsilon)x$, and substitute into the equation.
b. To determine $\delta(\varepsilon)$ balance the cubic and linear terms in ξ. What is $\delta(\varepsilon)$?
c. Substitute $\delta(\varepsilon)$ into your equation of part (a) and then simplify so there is no coefficient of ξ^3 and it is preserved. Any remaining forms of ε should be represented in terms of δ. You should have $F(\delta) = \xi^3 - \xi + \delta = 0$.
d. Let the solution be expanded as $\xi = \xi_0 + \delta\xi_1 + O(\delta^2)$. Solve $F(0) = 0$ for three roots of ξ_0.
e. Calculate $F'(\delta)$ and then set $F'(0) = 0$. Solve for ξ_1 using each value of ξ_0.
f. Express the two-term solutions for each of the three roots in terms of ξ, δ.
g. Now express the solutions in part (f) in terms of x, ε. Plot your three solutions over the range $0.05 \leq \varepsilon \leq 0.2$.
h. Try a different balance in part (b). Balance the cubic term with the constant 1. What is the new $\delta(\varepsilon)$? Does this choice preserve the cubic term as $\varepsilon \to 0$?

17.2 Consider the mixed algebraic-transcendental equation $F(\varepsilon) = \varepsilon e^{x/2} + x - 1 = 0$.

Solve this equation using a regular perturbation method for $\varepsilon << 1$, neglecting $O(\varepsilon^3)$ and smaller terms. Let $x = x_0 + \varepsilon x_1 + \frac{\varepsilon^2}{2} x_2 + O(\varepsilon^3)$.

a. Set $F(0) = 0$and solve for x_0.
b. Calculate $F'(\varepsilon)$ and then set $F'(0) = 0$ to find x_1.
c. Calculate $F''(\varepsilon)$ and then set $F''(0) = 0$ to find x_2.
d. What is your three-term solution for x?
e. There is no exact solution for x, so compare your perturbation solution to a numerical solution using your software. Plot your two solutions for $0 \leq \varepsilon \leq 0.3$. Be sure to use the real roots.
f. Plot the error over the same range of ε defined as the difference between the numerical and perturbation solutions divided by the numerical solution. At what value of ε is the error 2%?

17.3 Consider the following ODE for X(T) where $\varepsilon << 1$

$$F(T; \varepsilon) = \frac{dX}{dT} + \varepsilon X^3 - \cos(T) = 0$$

subject to the initial condition $X(0) = 0$. Solve using a perturbation expansion for keeping the first two terms, $X(T; \varepsilon) = X_0(T) + \varepsilon X_1(T) + O(\varepsilon^2)$. Be sure to expand the initial condition too.

a. Set $F(T; \varepsilon = 0)$ to solve for $X_0(T)$.
b. Calculate $F'(T; \varepsilon)$ and then set $F'(T; 0) = 0$ and solve for X_1.
c. Solve $F(T; \varepsilon) = 0$ numerically and plot X(T) for $\varepsilon = 0.2$ over the time range $0 \leq T \leq 2\pi$. Add your perturbation solution to the graph. What are the features of this comparison?
d. Add a third curve to your graph in part (d) which is just $X_0(T)$. What is the effect of $\varepsilon X_1(T)$?

17.4 Consider the dimensionless Eulerian velocity field

$$u = 1 + \varepsilon \cos(x)$$
$$v = \varepsilon y \sin(x)$$

a. Is this velocity field incompressible?
b. Using a perturbation methods for $\varepsilon << 1$, find the mapping $(X(t), Y(t))$. As covered in Chapter 3, start by replacing the Eulerian variables u, v, x and y with their Lagrangian counterparts. Show your resulting ODE system.
c. Expand the dependent variables as

$$X(t) = X_0(t) + \varepsilon X_1(t) + O(\varepsilon^2)$$
$$Y(t) = Y_0(t) + \varepsilon Y_1(t) + O(\varepsilon^2)$$

and solve for the first two terms of each assuming the initial positions are $(X(0), Y(0)) = (X_{00}, Y_{00})$.
d. Plot your solution for $\varepsilon = 0.3$ and three initial positions $(X_{00}, Y_{00}) = (1, 1), (1, 3), (1, 5)$ for $0 \leq t \leq 2\pi$.
e. Consider the timeline that starts at $t = 0$ for the locus of fluid particles $1 \leq X_{00} \leq 5, Y_{00} = 1$. Plot it at times $t = 0, \frac{\pi}{2}, \pi, \frac{3\pi}{2}, 2\pi$. Describe your results. What has happened after one full oscillatory cycle?
f. For our purposes, the differential length of the timeline in part (e) is

$$dS = \left(dX^2 + dY^2\right)^{1/2} = \left(\left(\frac{\partial X}{\partial X_{00}}\right)^2 + \left(\frac{\partial Y}{\partial X_{00}}\right)^2\right)^{1/2} dX_{00}$$

Using your perturbation solutions, calculate the length of the timeline, L(t), by integrating dS over the range of $1 \leq X_{00} \leq 5$. Plot L(t) for $\varepsilon = 0.1, 0.2, 0.3$ over one cycle and describe your findings. Do they match up with your results in part (e)?
g. Go back to your system of differential equations in part (b). Solve them numerically for the mapping of the fluid particle starting at $X_{00} = 1, Y_{00} = 3$ and $\varepsilon = 0.1, 0.2, 0.3$. Compare to your perturbation solutions and describe the results and any trends.

17.5 Consider the singular linear, second-order ODE for $\varepsilon << 1$

$$\varepsilon \frac{d^2u}{dx^2} - \frac{du}{dx} + u = 0$$

It is the same as in the text, Eq. (17.4.1), but with a sign change of the middle term. Let the boundary conditions be the same: $u(0) = 0, u(1) = 1$.

a. Let the original ODE be the outer equation $\varepsilon \frac{d^2u^o}{dx^2} - \frac{du^o}{dx} + u^o = 0$. For the expansion $u^o = u_0^o + \varepsilon u_1^o + O(\varepsilon^2)$, solve for u_0^o and evaluate the constant of integration by imposing the boundary condition at $x = 0$, i.e. $u_0^o(x = 0) = 0$.
b. Let the inner variable be $\xi = (1 - x)/\varepsilon$ so that the inner equation is $\frac{d^2u^i}{d\xi^2} + \frac{du^i}{d\xi} + \varepsilon u^i = 0$. The inner expansion is $u^i = u_0^i + \varepsilon u_1^i + O(\varepsilon^2)$. Solve for u_0^i and find the relationship between the two integration constants by imposing the condition $u_0^i(\xi = 0) = 1$.
c. To solve for the remaining constant in u_0^i, perform the match where $\lim_{\xi\to\infty} u_0^i = \lim_{x\to 1} u_0^o$.
d. Now form the composite solution $u_0^c = u_0^o + u_0^i - \text{c.p.}$ Plot u_0^c on the interval $0 \le x \le 1$ for $\varepsilon = 0.05, 0.10, 0.15$.
e. Find the exact, analytic solution by hand using the form $u \sim e^{px}$. Inserting this into the ODE leads to a quadratic equation for $p(\varepsilon)$ which has two roots, p^+, p^-. The solution is a linear combination $u = c_1 e^{p^+x} + c_2 e^{p^-x}$, where c_1, c_2 are solved from the two boundary conditions. Alternatively, use your computational software to solve for the exact solution.
f. Plot your composite solution versus the analytic solution for $\varepsilon = 0.1$. How is the comparison?

17.6 From Section 8.3 on oscillatory channel flow, the governing x-momentum equation is Eq. (8.3.2). For this exercise, let the dimensionless variables be $U = u/U_s, T = \omega t, Y = y/b$ where $U_s = -B^p b^2/\mu\alpha^2$ and let $B^p < 0$. This velocity scale is different from the one in Eq. (8.3.4) by including α^2 in the denominator, where α is the Womersley parameter, $\alpha = b\sqrt{\omega/\nu}$. The resulting dimensionless equation is

$$\frac{\partial U}{\partial T} = \cos T + \varepsilon \frac{\partial^2 U}{\partial Y^2} \tag{1}$$

where $\varepsilon = \alpha^{-2}$. Let us investigate the large $\alpha >> 1$, high frequency behavior of Eq. (1) so that $\varepsilon << 1$. Because ε multiplies the highest-order derivative in Y, this is a singular perturbation problem. The boundary conditions are no-slip on the walls, or no-slip at the upper wall and symmetry at the midline.

a. Equation (1) is the outer equation. Using the outer expansion is $U^o = U_0^o + \varepsilon U_1^o + \cdots$, show that the $O(1)$ outer solution is $U_0^o = -ie^{iT}$ which satisfies the symmetry condition $(\partial U_0^o/\partial Y)_{Y=0} = 0$.
b. The inner region boundary layer is at the wall. To retain the highest-order derivatives in the inner region, we need to rescale Y as $\eta = \varepsilon^a(1 - Y)$. What does the value of a have to be? Let the inner expansion be $U^i(\eta, T) = U_0^i + \varepsilon U_1^i + \cdots$. Show that the leading-order inner equation is

$$\frac{\partial U_0^i}{\partial T} - \frac{\partial^2 U_0^i}{\partial \eta^2} = e^{iT} \tag{2}$$

c. Solve the homogeneous problem of Eq. (2) by letting the right hand side be zero and assume a separation of variables approach, $U_0^i h(\eta, T) = F(\eta)e^{iT}$. Show that F must satisfy the equation

$$\frac{d^2F}{d\eta^2} - \beta^2 F = 0 \tag{3}$$

where $\beta^2 = i, \beta = \sqrt{2}(1 + i)/2$.
d. Solve for F by using our previous approach in Eq. (8.3.12) or your software. There should be two unknown constant coefficients of your two exponentials, call them A and B.
e. The particular solution is obviously $U_{0p}^i = -ie^{iT}$. Add this to the homogeneous solution to form U_0^i and apply the no-slip boundary condition $U_0^i(\eta = 0) = 0$. That will express B in terms of A.
f. The match in this case requires $\lim_{\eta\to\infty} U_0^i = \lim_{Y\to 1} U_0^o$, but U_0^o does not depend on Y, which makes things easier. To prevent the inner solution from becoming infinitely large in this limit, choose A appropriately. The common part remains on both sides, what is it?

g. What is the composite solution, U_0^c, i.e. the sum of the outer and inner solutions subtracting off the common part? Express it in terms of Y and T.
h. Now compare U_0^c to the $\alpha >> 1$ expansion of the exact solution in Eq. (8.3.18). Recall that the velocity scale for the exact solution is $-B^p b^2/\mu$, while the perturbation solution has $-B^p b^2/\mu\alpha^2$. So account for that difference in your comparison. Also, $\cosh(z) = (e^z + e^{-z})/2 \sim e^z/2$ for $z >> 1$. Are they the same?

17.7 In Section 7.9 we sought solutions for Stokes flow over a cylinder assuming $\hat{\psi} = f(R)\sin\theta$ where $\nabla^4\hat{\psi} = 0$ conserves momentum. The resulting equation, Eq. (7.9.2), gave $f(R) = A_1R^3 + B_1R\ln R + C_1R + \frac{D_1}{R}$. Our approach was to satisfy no-slip and no-penetration on the cylinder is a streamline and show we could not match to the large distance form $\hat{\psi}|_{R\to\infty} \sim R\sin\theta$. We now know that the conditions on the cylinder are in the outer region, while those at infinity are the inner region. Try solving the other way around, i.e. first satisfy $\hat{\psi}|_{R\to\infty} \sim R\sin\theta$ and then try to apply the cylinder conditions. What do you find?

Image Credits

Figure 17.10a Source: Fig 6 in Donnelly RJ, Glaberson W. Experiments on capillary instability of a liquid jet. *Proceedings of the Royal Society of London Series A – Mathematical and Physical Sciences.* 1966; 290 (1423):547–556.

Figure 17.10b Source: Fig 8 in Donnelly RJ, Glaberson W. Experiments on capillary instability of a liquid jet. *Proceedings of the Royal Society of London Series A – Mathematical and Physical Sciences.* 1966; 290 (1423):547–556.

Figure 17.10c Source: Fig 10 in Donnelly RJ, Glaberson W. Experiments on capillary instability of a liquid jet. *Proceedings of the Royal Society of London Series A – Mathematical and Physical Sciences.* 1966; 290 (1423):547–556.

Figure 17.14a Source: Figure 2 in Gaver, D. P. and J. B. Grotberg (1986). "An experimental investigation of oscillating flow in a tapered channel." *Journal of Fluid Mechanics* 172: 47–61.

Figure 17.14b Source: Figure 1 in Grotberg, J. B. (1984). "Volume-cycled oscillatory flow in a tapered channel." *Journal of Fluid Mechanics* 141: 249–264.

Figure 17.15a Source: Fig 1 in Bottier M, Fernandez MP, Pelle G, Isabey D, Louis B, Grotberg JB, Filoche M. A new index for characterizing micro-bead motion in a flow induced by ciliary beating: Part II, modeling. *PLoS Computational Biology.* 2017; 13(7):1005552. doi: 10.1371/journal.pcbi.1005552.

Figure 17.15b Source: Fig 2 in Bottier M, Fernandez MP, Pelle G, Isabey D, Louis B, Grotberg JB, Filoche M. A new index for characterizing micro-bead motion in a flow induced by ciliary beating: Part II, modeling. *PLoS Computational Biology.* 2017; 13(7):1005552. doi: 10.1371/journal.pcbi.1005552.

REFERENCES

Abu-Omar, Y. and J. Dunning (2012). "Prosthetic heart valves," in *Core Topics in Cardiac Anesthesia*. J. Mackay and J. Arrowsmith, eds. Cambridge, Cambridge University Press: 124–132.

Adler, M. (1934). "Strömung in gekrümmten Rohren." *Zeitschrift für Angewandte Mathematik und Mechanik* **14**(5): 257–275.

Alazmi, B. and K. Vafai (2001). "Analysis of fluid flow and heat transfer interfacial conditions between a porous medium and a fluid layer." *International Journal of Heat and Mass Transfer* **44**: 1735–1749.

Ali, S. (2001). "Pressure drop correlations for flow through regular helical coil tubes." *Fluid Dynamics Research* **28**: 295–310.

Aljuri, N., et al. (1999). "Modeling expiratory flow from excised tracheal tube laws." *Journal of Applied Physiology* **87**(5): 1973–1980.

Almendros, I., et al. (2007). "Upper-airway inflammation triggered by vibration in a rat model of snoring." *Sleep* **30**(2): 225–227.

Atkinson, B., et al. (1969). "Low Reynolds number developing flows." *AIChE Journal* **15**(4): 548–553.

Avery, M. E. and J. Mead (1959). "Surface properties in relation to atelectasis and hyaline membrane disease." *American Journal of Diseases of Children.* **97**: 517–523.

Ayotte, D. C., et al. (2001). "Direction-dependent resistance to flow in the endplate of the intervertebral disc: an ex vivo study." *Journal of Orthopaedic Research* **19**: 1073–1077.

Baba, Y., et al. (2004). "Assessment of the development of choked flow during total liquid ventilation." *Critical Care Medicine* 32(1): 201–208.

Baker, D. A., et al. (1991). "Computationally two-dimensional finite-difference model for hollow-fiber blood-gas exchange device." *Medical & Biological Engineering & Computing* **29**: 482–488.

Balashazy, I., et al. (2003). "Local particle deposition patterns may play a key role in the development of lung cancer." *Journal of Applied Physiology* **94**(5): 1719–1725.

Balmforth, N. J., et al. (2007). "Viscoplastic dam breaks and the Bostwick consistometer." *Journal of Non-Newtonian Fluid Mechanics* **142**(1–3): 63–78.

Barnes, H. T. and E. G. Coker (1904). "The flow of water through pipes." *Proceedings of the Royal Society of London* **74**: 341.

Bar-On, Y. M., et al. (2018). "The biomass distribution on Earth." *Proceedings of the National Academy of Sciences* **115**(25): 6506–6511.

Barua, S. N. (1963). "On secondary flow in stationary curved pipes." *Quarterly Journal of Mechanics and Applied Mathematics* **16**(1): 61–77.

Basser, P. J., et al. (1989). "The mechanism of mucus clearance in cough." *Journal of Biomechanical Engineering* **111**: 288–297.

Bausch, A. R., et al. (1998). "Local measurements of viscoelastic parameters of adherent cell surfaces by magnetic bead microrheometry." *Biophysical Journal* **75**(4): 2038–2049.

Baxter, L. T. and R. K. Jain (1989). "Transport of fluid and macromolecules in tumors. 1. Role of interstitial pressure and convection." *Microvascular Research* **37**(1): 77–104.

Beavers, G. S. and D. D. Joseph (1967). "Boundary conditions at a naturally permeable wall." *Journal of Fluid Mechanics* **30**: 197–&.

Bechert, D. W., et al. (1997). "Experiments on drag-reducing surfaces and their optimization with an adjustable geometry." *Journal of Fluid Mechanics* **338**: 59–87.

Beissinger, R. and M. Williams (1985). "Effects of blood storage on rheology and damage in low-stress shear flow." *Biorheology* **22**(6): 477–493.

Benias, P. C., et al. (2018). "Structure and distribution of an unrecognized interstitium in human tissues." *Scientific Reports* **8**(1): 4947.

Berk, D. A., et al. (1993). "Fluorescence photobleaching with spatial fourier analysis: measurement of diffusion in light-scattering media." *Biophysical Journal* **65**: 2428–2436.

Bertram, C. D., et al. (1991). "Application of nonlinear dynamics concepts to the analysis of self-excited oscillations of a collapsible tube conveying a fluid." *Journal of Fluids and Structures* **5**(4): 391–426.

Bian, S., et al. (2010). "Experimental study of flow fields in an airway closure model." *Journal of Fluid Mechanics* **647**: 391–402.

Bilek, A. M., et al. (2003). "Mechanisms of surface-tension-induced epithelial cell damage in a model of pulmonary airway reopening." *Journal of Applied Physiology* **94**(2): 770–783.

Bingham, E. C. (1916). "An investigation of the laws of plastic flow." *US Bureau of Standards Bulletin* **13**: 309–353.

Biot, M. A. (1941). "General theory of three-dimensional consolidation." *Journal of Applied Physics* **12**: 155–164.

Biro, G. P. (1982). "Comparison of acute cardiovascular effects and oxygen supply following hemodilution with dextran, stroma-free hemoglobin solution and fluorocarbon suspension." *Cardiovascular Research* **16**(4): 194–204.

Bistany, K. L. and J. L. Kokini (1983). "Comparison of steady shear rheological properties and small amplitude dynamic viscoelastic properties of fluid food materials." *Journal of Texture Studies* **14**(2): 113–124.

Blasius, H. (1908). "Grenrschichten in Flüssigkeiten mit kleiner Reibung." *Zeitschrift für Angewandte Mathematik und Physik* **56**: 1–37.

Bode, F. R., et al. (1976). "Age and sex differences in lung elasticity, and in closing capacity in nonsmokers." *Journal of Applied Physiology* **41**(2): 129–135.

Bohn, D. J., et al. (1980). "Ventilation by high-frequency oscillation." *Journal of Applied Physiology* **48**: 710–716.

Borgas, M. S. and J. B. Grotberg (1988). "Monolayer flow on a thin film." *Journal of Fluid Mechanics* **193**: 151–170.

Bottier, M., et al. (2017a). "A new index for characterizing micro-bead motion in a flow induced by ciliary beating: part I, experimental analysis." *Plos Computational Biology* **13**(7): 1005605.

Bottier, M., et al. (2017b). "A new index for characterizing micro-bead motion in a flow induced by ciliary beating: part II, modeling." *Plos Computational Biology* **13**(7): 1005552.

Böttiger, L. and C. Svedberg (1967). "Normal erythrocyte sedimentation rate and age." *British Medical Journal* **2**: 85–87.

Bourouiba, L., et al. (2014). "Violent expiratory events: on coughing and sneezing." *Journal of Fluid Mechanics* **745**: 537–563.

Bowen, R. M. (1980). "Incompressible porous media models by use of the theory of mixtures." *International Journal of Engineering Science* **18**(9): 1129–1148.

Braun, R. J. (2012). "Dynamics of the tear film." *Annual Review of Fluid Mechanics*. 44: 267–297.

Braun, R. J. and P. E. King-Smith (2007). "Model problems for the tear film in a blink cycle: single-equation models." *Journal of Fluid Mechanics* **586**: 465–490.

Bretherton, F. P. (1961). "The motion of long bubbles in tubes." *Journal of Fluid Mechanics* **10**: 166–188.

Brinkman, H. C. (1947). "A calculation of the viscous force exerted by a flowing fluid in a dense swarm of particles." *Applied Science Research* A **1**: 27–34.

Brinkman, H. C. (1952). "The viscosity of concentrated suspensions and solutions." *Journal of Chemical Physics* **20**: 571.

Briscoe, W. A., et al. (1954). "Alveolar ventilation at very low tidal volumes." *Journal of Applied Physiology* **7**(1): 27–30.

Broughton-Head, V. J., et al. (2007). "Unfractionated heparin reduces the elasticity of sputum from patients with cystic fibrosis." *American Journal of Physiology - Lung Cellular and Molecular Physiology* **293**: L1240–L1249.

Brower, R. G., et al. (2000). "Ventilation with lower tidal volumes as compared with traditional tidal volumes for acute lung injury and the acute respiratory distress syndrome." *New England Journal of Medicine* **342**(18): 1301–1308.

Brown, E. B., et al. (2004). "Measurement of macromolecular diffusion coefficients in human tumors." *Microvascular Research* **67**: 231–236.

Buckingham, E. (1914). "On physically similar systems; illustrations of the use of dimensional equations." *Physical Review* **4**(4): 345–376.

Buckingham, E. (1915). "The principle of similitude." *Nature* **96**(2406): 396–397.

Bull, J. L. (2007). "The application of microbubbles for targeted drug delivery." *Expert Opinion on Drug Delivery* **4**(5): 475–493.

Bull, J. L., et al. (2007). "The effects of respiratory rate and tidal volume on gas exchange during total liquid ventilation." *Critical Care Medicine* **35**: A222–A222.

Bull, J. L., et al. (2009). "Effects of respiratory rate and tidal volume on gas exchange in total liquid ventilation." *ASAIO Journal* **55**(4): 373–381.

Burton, A. C. (1954). "Relation of structure to function of the tissues of the wall of blood vessels." *Physiological Reviews* **34**(4): 619–642.

Bush, J. W. M. and D. L. Hu (2006). "Walking on water: biolocomotion at the Interface." *Annual Review of Fluid Mechanics*. **38**: 339–369.

Campana, D., et al. (2004). "A 2-D model of Rayleigh instability in capillary tubes – surfactant effects." *International Journal of Multiphase Flow* **30**(5): 431–454.

Campbell, W. D. and J. C. Slattery (1963). "Flow in the entrance of a tube." *ASME Journal of Basic Engineering* **85**: 41–45.

Carman, P. C. (1937). "Fluid flow through granular beds." *Transactions, Institution of Chemical Engineers, London* **15**: 150–166.

Caro, C. G., et al. (1971). "Atheroma and arterial wall shear: observation, correlation and proposal of a shear dependent mass transfer mechanism for atherogenesis." *Proceedings of the Royal Society Series B – Biological Sciences* **177**(1046): 109–&.

Carreau, P. J. (1972). "Rheological equations from molecular network theories." *Transactions of the Society of Rheology* **16**(1): 99–&.

Cassidy, K. J., et al. (1999). "Surfactant effects in model airway closure experiments." *Journal of Applied Physiology* **87**(1): 415–427.

Cassidy, K. J., et al. (2001a). "A rat lung model of instilled liquid transport in the pulmonary airways." *Journal of Applied Physiology* **90**: 1955–1967.

Cassidy, K. J., et al. (2001b). "Liquid plug flow in straight and bifurcating tubes." *Journal of Biomechanical Engineering* 123(6): 580–589.

Casson, N. (1959). "A flow equation for pigment oil suspensions of the printing ink type," in *Rheology of Disperse Systems*. C. Mill, ed. New York, NY, Pergamon Press.

Caygill, J. C. and G. H. West (1969). "The rheological behavior of synovial fluid and its possible relation to joint lubrication." *Med. & Biol. Engng.* **7**: 507.

Chakraborty, S. and K. Tsuchiya (2008). "Development and fluidic simulation of microneedles for painless pathological interfacing with living systems." *Journal of Applied Physics* **103**: 114701–114709.

Chan, K. Y., et al. (2006). "Pulsatile flow and mass transport over an array of cylinders: Gas transfer in a cardiac-driven artificial lung." *Journal of Biomechanical Engineering – Transactions of the ASME* **128**(1): 85–96.

Chang, C. H. and E. I. Franses (1995). "Adsorption dynamics of surfactants at the air/water interface – a critical-review of mathematical-models, data, and mechanisms." *Colloids and Surfaces A – Physicochemical and Engineering Aspects* **100**: 1–45.

Chatelin, R., et al. (2017). "Numerical and experimental investigation of mucociliary clearance breakdown in cystic fibrosis." *Journal of Biomechanics* **53**: 56–63.

Chen, C. T., et al. (1998). "A fiber matrix model for interstitial fluid flow and permeability in ligaments and tendons." *Biorheology* **35**(2): 103–118.

Chesters, A. K. (1977). "An analytical solution for the profile and volume of a small drop or bubble symmetrical about a vertical axis." *Journal of Fluid Mechanics* **81**(4): 609–624.

Chien, S., et al. (1966). "Effects of hematocrit and plasma proteins on human blood rheology at low shear rates." *Journal of Applied Physiology* **21**(1): 81–87.

Cho, Y. I. and K. R. Kensey (1991). "Effects of the non-Newtonian viscosity of blood on flows in a diseased arterial vessel. 1. Steady flows." *Biorheology* **28**(3–4): 241–262.

Clark, L. C. and F. Gollan (1966). "Survival of mammals breathing organic liquids equilibrated with oxygen at atmospheric pressure." *Science* **152**(3730): 1755–&.

Clauser, F. H. (1956). "The turbulent boundary layer." *Advances in Applied Mechanics* **4**: 1–51.

Clements, J. A. (1957). "Surface tension of lung extracts." *Proceedings of the Society for Experimental Biology and Medicine* **95**: 170–172.

Cokelet, G. R., et al. (1963). "The rheology of human blood – measurement near and at zero shear rate." *Transactions of the Society of Rheology* **7**: 303–317.

Colebrook, C. F. (1939). "Turbulent flow in pipes, with particular reference to the transition region between smooth and rough pipe laws." *Journal of the Institution of Civil Engineers* **11**(4): 133–156.

Colebrook, C. F. and C. M. White (1937). "Experiments with fluid friction in roughened pipes." *Proceedings of the Royal Society of London. Series A, Mathematical and Physical Sciences* **161**(A906): 367–381.

Coles, D. (1956). "The law of the wake in the turbulent boundary layer." *Journal of Fluid Mechanics* **1**(2): 191–226.

Cone, R. A. (1999). "Mucus," in *Mucosal Immunology*. P. L. Ogra, ed. San Diego, Academic Press: 43–64.

Coutanceau, M. and R. Bouard (1977). "Experimental determination of the main features of the viscous flow in the wake of a circular cylinder in uniform translation. Part 1. Steady flow " *Journal of Fluid Mechanics* **79**(2): 231–256.

Cox, W. P. and E. H. Merz (1958). "Correlation of dynamic and steady flow viscosities." *Journal of Polymer Science* **28**(118): 619–622.

Craven, B. A., et al. (2014). "Design of a high-throughput chemical trace detection portal that samples the aerodynamic wake of a walking person." *IEEE Sensors Journal* **14**(6): 1852–1866.

Cross, M. M. (1965). "Rheology of non-Newtonian fluids: a new flow equation for pseudo-plastic systems." *Journal of Colloid Science* **20**(5): 417–437.

Cummings, L. J., et al. (2004). "The effect of ureteric stents on urine flow: reflux." *Journal of mathematical biology* **49**(1): 56–82.

Curry, F. E. and J. Frokjaerjensen (1984). "Water-flow across the walls of single muscle capillaries in the frog,

Rana-pipiens." *Journal of Physiology – London* 350 (May): 293–307.

Curt, J. R. N. and R. Pringle (1969). "Viscosity of gastric mucus in duodenal ulceration." *Gut* 10(11): 931–&.

D'Apolito, R., et al. (2015). "Red blood cells affect the margination of microparticles in synthetic microcapillaries and intravital microcirculation as a function of their size and shape." *Journal of Controlled Release* 217: 263–272.

Darcy, H. P. G. (1856). *Les fontaines publiques de la ville de Dijon*. Paris.

Dasi, L. P., et al. (2007). "Vorticity dynamics of a bileaflet mechanical heart valve in an axisymmetric aorta." *Physics of Fluids* 19: 067105.

Davies, P. F., et al. (1986). "Turbulent fluid shear stress induces vascular endothelial cell turnover in vitro." *Proceedings of the National Academy of Sciences* 83: 2114–2117.

Davis, S. S. (1973). "Rheological examination of sputum and saliva and the effect of drugs," in *Rheology of Biological Systems*. H. L. Gabelnick and M. Litt, eds. Springfield, Charles C. Thomas: 157–194.

Dawson, M., et al. (2003). "Enhanced viscoelasticity of human cystic fibrotic sputum correlates with increasing microheterogeneity in particle transport." *Journal of Biological Chemistry* 278(50): 50393–50401.

Dawson, S. V. and E. A. Elliott (1977). "Wave-speed limitation on expiratory flow: a unifying concept." *Journal of Applied Physiology* 43: 498–515.

de Waele, A. (1923). "Viscometry and plastometry." *Journal of the Oil & Colour Chemists Association* 6: 33–39.

Deardorff, J. (1970). "A numerical study of three-dimensional turbulent channel flow at large Reynolds numbers." *Journal of Fluid Mechanics* 41(2): 453–480.

Denton, R. (1963). "The rheology of human mucus." *Annals of the NY Academy of Science* 106: 746–754.

Dewey, C. F., et al. (1981). "The dynamic-response of vascular endothelial-cells to fluid shear-stress." *Journal of Biomechanical Engineering – Transactions of the ASME* 103(3): 177–185.

Dierickx, P. W. T., et al. (2000). "Blood flow around hollow fibers." *International Journal of Artificial Organs* 23 (9): 610–617.

Dierickx, P. W., et al. (2001). "Two-dimensional finite element model for oxygen transfer in cross-flow hollow fiber membrane artificial lungs." *International Journal of Artificial Organs* 24(9): 628–635.

Dimotakis, P. E. (2000). "The mixing transition in turbulent flows." *Journal of Fluid Mechanics* 409: 69–98.

Dimotakis, P. E., et al. (1983). "Structure and dynamics of round turbulent jets." *Physics of Fluids* 26(11): 3185–3192.

Ding, Y., et al. (2014). "Mixing and transport by ciliary carpets: a numerical study." *Journal of Fluid Mechanics* 743: 124–140.

Domel, A. G., et al. (2018). "Hydrodynamic properties of biomimetic shark skin: effect of denticle size and swimming speed." *Bioinspiration & Biomimetics* 13: 056014.

Donnelly, R. J. and W. Glaberson (1966). "Experiments on capillary instability of a liquid jet." *Proceedings of the Royal Society of London SeriesA – Mathematical and Physical Sciences* 290(1423): 547–556.

Dorkin, H. L., et al. (1988). "Human respiratory input impedance from 4 to 200 Hz: physiological and modeling considerations." *Journal of Applied Physiology* 64(2): 823–831.

Dragon, C. A. and J. B. Grotberg (1991). "Oscillatory flow and dispersion in a flexible tube." *Journal of Fluid Mechanics* 231: 135–155.

Drummond, G. B. and J. Milic-Emili (2007). "Forty years of closing volume." *British Journal of Anaesthesia* 99 (6): 772–774.

Drury, J. L. and D. J. Mooney (2003). "Hydrogels for tissue engineering: scaffold design variables and applications." *Biomaterials* 24(24): 4337–4351.

Dubois, A. B., et al. (1956). "Oscillation mechanics of lungs and chest in man." *Journal of Applied Physiology* 8 (6): 587–594.

Dudley, P. N., et al. (2013). "Consider a non-spherical elephant: computational fluid dynamics simulations of heat transfer coefficients and drag verified using wind tunnel experiments." *J. Exp. Zool.* 319A: 319–327.

Dulfano, M. J., et al. (1971). "Sputum viscoelasticity in chronic bronchitis." *American Review of Respiratory Disease* 104(1): 88–98.

Durst, F., et al. (2005). "The development lengths of laminar pipe and channel flows." *Journal of Fluids Engineering – Transactions of the ASME* 127(6): 1154–1160.

Ebong, E. E., et al. (2011). "Imaging the endothelial glycocalyx in vitro by rapid freezing/freeze substitution transmission electron microscopy." *Arteriosclerosis, Thrombosis, and Vascular Biology* 31: 1908–1915.

Eckmann, D. M. and J. B. Grotberg (1988). "Oscillatory flow and mass transport in a curved tube." *Journal of Fluid Mechanics* 188: 509–527.

Eckmann, D. M. and J. B. Grotberg (1991). "Experiments on transition to turbulence in oscillatory pipe flow." *Journal of Fluid Mechanics* 222: 329–350.

Edwards, D. A., et al. (2004). "Inhaling to mitigate exhaled bioaerosols." *Proceedings of the National Academy of Sciences* **101**(50): 17383–17388.

Edwards, P. A. and D. B. Yeates (1992). "Magnetic rheometry of bronchial mucus." *ACS Symposium Series* **489**: 249–267.

Einstein, A. (1906). "Eine neue Bestimmung der Moleküldimensionen." *Annalen der Physik* **19**: 289–306.

Elad, D., et al. (1988). "Tube law for the intrapulmonary airway." *Journal of Applied Physiology* **65**(1): 7–13.

Elad, D., et al. (1989). "Steady compressible flow in collapsible tubes – application to forced expiration." *Journal of Fluid Mechanics* **203**: 401–418.

Eliezer, N. (1974). "Viscoelastic properties of mucus." *Biorheology* **11**: 61–68.

Elliott, N. S. J., et al. (2011). "Modelling and simulation of fluid-structure interactions in human snoring." *19th International Congress on Modelling and Simulation (Modsim2011)*: 530–536.

Enhorning, G. (1977). "Pulsating bubble technique for evaluating pulmonary surfactant." *Journal of Applied Physiology* **43**(2): 198–203.

Erian, F. F., et al. (1977). "Maternal, placental blood-flow – model with velocity-dependent permeability." *Journal of Biomechanics* **10**(11–1): 807–814.

Ericksen, J. L. (1960). "Anisotropic fluids." *Archives for Rational Mechanics and Analysis* **4**: 231–237.

Evans, R. C. and T. M. Quinn (2005). "Solute diffusivity correlates with mechanical properties and matrix density of compressed articular cartilage." *Archives of Biochemistry and Biophysics* **442**: 1–10.

Everett, D. H. and J. M. Haynes (1972). "Model studies of capillary condensation 1. Cylindrical pore model with zero contact angle." *Journal of Colloid and Interface Science* **38**: 125–137.

Fåhræus, R. (1929). "The suspension stability of the blood." *Physiological Reviews* **9**(2): 241–274.

Fåhræus, R. and T. Lindqvist (1931). "The viscosity of the blood in narrow capillary tubes " *American Journal of Physiology* **96**(3): 562–568.

Fairbrother, F. and A. E. Stubbs (1935). "Studies in electro-endosmosis. Part VI. The "bubble-tube" method of measurement." *Journal of the Chemical Society*: 527–529.

Fan, E., et al. (2018). "Acute respiratory distress syndrome: advances in diagnosis and treatment." *JAMA* **319**(7): 698–710.

Fargie, D. and B. W. Martin (1971). "Developing laminar flow in a pipe of circular cross-section." *Proceedings of the Royal Society of London Series A – Mathematical and Physical Sciences* **321**(1547): 461–&.

Ferguson, S. J., et al. (2004). "Fluid flow and convective transport of solutes within the intervertebral disc." *Jounal of Biomechanics* **37**: 213–221.

Filoche, M., et al. (2015). "Three-dimensional model of surfactant replacement therapy." *Proceedings of the National Academy of Sciences of the United States of America* **112**(30): 9287–9292.

Finn, R. K. (1953). "Determination of the drag on a cylinder at low Reynolds numbers." *Journal of Applied Physics* **24**: 771–773.

Flaherty, J. E., et al. (1972). "Post buckling behavior of elastic tubes and rings with opposite sides in contact." *SIAM Journal of Applied Mathematics* **23**(4): 446–455.

Frank, C., et al. (1987). "Normal ligament: structure, function, and composition," in *Injury and Repair of the Musculoskeletal Soft Tissues*. S. L.-Y. Woo and J. A. Buckwalker, eds. Savannah, GA, AAOS Press: 45–101.

Frank, O. (1930). "Schätzung des Schlagvolumens des menschlichen Herzens auf Grund der Wellen-und Windkesseltheorie." *Zeitschriftfür Biologie* **90**: 405–409.

Freundlich, H. (1909). *Kapillarchemie, eine Darstellung der Chemie der Kolloide und verwandter Gebiete*. Leipzig, Verlag der Akademischen Verlagsgesellschaft.

Frumkin, A. (1925). "Die Kapillarkurve der höheren Fettsäuren und die Zustandsgleichung der Oberflächenschicht." *Zeitschrift für Physikalische Chemie (Leipzig)* **116**(1): 466–484.

Fry, D. L. (1968). "A preliminary lung model for simulating aerodynamics of bronchial tree." *Computers and Biomedical Research* **2**(2): 111–134.

Fujioka, H. and J. B. Grotberg (2004). "Steady propagation of a liquid plug in a two-dimensional channel." *Journal of Biomechanical Engineering* **126**(5): 567–577.

Fujioka, H. and J. B. Grotberg (2005). "The steady propagation of a surfactant-laden liquid plug in a two dimensional channel." *Physics of Fluids* **17**(8): Art. No: 082102.

Fujioka, H., et al. (2008). "Unsteady propagation of a liquid plug in a liquid-lined straight tube." *Physics of Fluids* **20**(6): Art. No: 062104.

Fung, Y. C. (1973). "Stochastic flow in capillary blood vessels." *Microvascular Research* **5**: 34–48.

Gauglitz, P. A. and C. J. Radke (1988). "An extended evolution equation for liquid film breakup in cylindrical capillaries." *Chemical Engineering Science* **43**: 1457–1465.

Gaver, D. P. and J. B. Grotberg (1986). "An experimental investigation of oscillating flow in a tapered channel." *Journal of Fluid Mechanics* **172**: 47–61.

Gaver, D. P. and J. B. Grotberg (1990). "The dynamics of a localized surfactant on a thin film." *Journal of Fluid Mechanics* **213**: 127–148.

Gaver, D. P. and J. B. Grotberg (1992). "Droplet spreading on a thin viscous film." *Journal of Fluid Mechanics* **235**: 399–414.

Gaver, D. P., et al. (1990). "Effects of surface tension and viscosity on airway reopening." *Journal of Applied Physiology* **69**(1): 74–85.

Gaver, D. P., et al. (1996). "The steady motion of a semi-infinite bubble through a flexible-walled channel." *Journal of Fluid Mechanics* **319**: 25–65.

Gavriely, N. and J. B. Grotberg (1988). "Flow limitation and wheezes in a constant flow and volume lung preparation." *Journal of Applied Physiology* **64**(1): 17–20.

Gavriely, N., et al. (1984). "Measurement and theory of wheezing breath sounds." *Journal of Applied Physiology* **57**(2): 481–492.

Gavriely, N., et al. (1987). "Forced expiratory wheezes are a manifestation of airway flow limitation." *Journal of Applied Physiology* **62**(6): 2398–2403.

Gelin, L. E. (1961). "Disturbance of the flow properties of blood and its counteraction in surgery." *Acta Chirurgica Scandinavica* **122**: 287–293.

Ghadiali, S. N. and D. P. Gaver (2003). "The influence of non-equilibrium surfactant dynamics on the flow of a semi-infinite bubble in a rigid cylindrical capillary tube." *Journal of Fluid Mechanics* **478**: 165–196.

Ghadiali, S. N. and D. P. Gaver (2008). "Biomechanics of liquid-epithelium interactions in pulmonary airways." *Respiratory Physiology & Neurobiology* **163** (1–3): 232–243.

Giesekus, H. (1982). "A simple constitutive equation for polymer fluids based on the concept of deformation-dependent tensorial mobility." *Journal of Non-Newtonian Fluid Mech* **11**: 69–109.

Gijsen, F. J. H., et al. (1999). "The influence of the non-Newtonian properties of blood on the flow in large arteries: steady flow in a carotid bifurcation model." *Journal of Biomechanics* **32**(6): 601–608.

Gilboa, A. and A. Silberberg (1976). "In-situ rheological characterization of epithelial mucus." *Biorheology* **13**: 59–65.

Godleski, D. A. and J. B. Grotberg (1988). "Convection-diffusion interaction for oscillatory flow in a tapered tube." *Journal of Biomechanical Engineering* **110**(4): 283–291.

Gottardi, R., et al. (2016). "Supramolecular organization of collagen fibrils in healthy and osteoarthritic human knee and hip joint cartilage." *Plos One* **11**(10). e0163552.

Greaves, I. A., et al. (1986). "Micromechanics of the lung," in *Handbook of Physiology*. Bethesda, MD, American Physiological Society.

Grotberg, J. B. (1984). "Volume-cycled oscillatory flow in a tapered channel." *Journal of Fluid Mechanics* **141**: 249–264.

Grotberg, J. B. (1994). "Pulmonary flow and transport phenomena." *Annual Review of Fluid Mechanics* **26**: 529–571.

Grotberg, J. B. (2001). "Respiratory fluid mechanics and transport processes." *Annual Review of Biomedical Engineering* **3**: 421–457.

Grotberg, J. B. (2002). "Respiratory mechanics and gas exchange," in *Standard Handbook of Biomedical Engineering and Design*. M. Kutz, ed. London, McGraw-Hill.

Grotberg, J. B. (2011). "Respiratory fluid mechanics." *Physics of Fluids* **23**(2): 021301–021315.

Grotberg, J. B. (2019). "Crackles and wheezes: agents of injury?" *Annals of the American Thoracic Society*. doi: 10.1513/AnnalsATS.201901-022IP.

Grotberg, J. B. and S. H. Davis (1980). "Fluid-dynamic flapping of a collapsible channel: sound generation and flow limitation." *Journal of Biomechanics* **13**(3): 219–230.

Grotberg, J. B. and N. Gavriely (1989). "Flutter in collapsible tubes: a theoretical model of wheezes." *Journal of Applied Physiology* **66**: 2262–2273.

Grotberg, J. B. and M. R. Glucksberg (1994). "A buoyancy-driven squeeze-film model of intrapleural fluid dynamics: basic concepts." *Journal of Applied Physiology* **77**(3): 1555–1561.

Grotberg, J. B. and O. E. Jensen (2004). "Biofluid mechanics in flexible tubes." *Annual Review of Fluid Mechanics* **36**: 121–147.

Grotberg, J. B. and E. L. Reiss (1982). "A subsonic flutter anomaly." *Journal of Sound and Vibration* **80**(3): 444–446.

Grotberg, J. B. and E. L. Reiss (1984). "Subsonic flapping flutter." *Journal of Sound and Vibration* **92**(3): 349–361.

Grotberg, J. B., et al. (1980). "Frequency dependence of pressure-volume loops in isolated dog lobes." *J. Biomech* **13**(11): 905–912.

Grotberg, J. B., et al. (2017). "Did reduced alveolar delivery of surfactant contribute to negative results in adults with Acute Respiratory Distress Syndrome?"

American Journal of Respiratory and Critical Care Medicine 195(4): 538 -540.

Grzybowski, A. and J. Sak (2011). "Edmund Biernacki (1866–1911): discoverer of the erythrocyte sedimentation rate. On the 100th anniversary of his death." *Clinics in Dermatology* 29(6): 697–703.

Gudis, D., et al. (2012). "Acquired cilia dysfunction in chronic rhinosinusitis." *Americal Journal of Rhinology & Allergy* 26: 1–6.

Gullbrand, S. E., et al. (2015). "ISSLS prize winner: dynamic loading-induced convective transport enhances intervertebral disc nutrition." *Spine* 40: 1158–1164.

Guyton, A. C. (1965). "Interstitial fluid pressure. II. Pressure-volume curves of interstitial space." *Circ. Res.* 16: 452–460.

Guyton, A. C. (1981). *Medical Physiology*. Philadelphia, PA, W.B. Saunders Company.

Guyton, A. C., et al. (1966). "Interstitial fluid pressure. III. Its effects on resistance to tissue fluid mobility." *Circ. Res.* 19(412–419).

Haber, R., et al. (2001). "Steady-state pleural fluid flow and pressure and the effects of lung buoyancy." *Journal of Biomechanical Engineering* 123: 485–492.

Haefeli-Bleuer, B. and E. R. Weibel (1988). "Morphometry of the human pulmonary acinus." *The Anatomical Record* 220: 401–414.

Hahn, I., et al. (1993). "Velocity profiles measured for air-flow through a large-scale model of the human nasal cavity." *Journal of Applied Physiology* 75(5): 2273–2287.

Halpern, D., et al. (2010). "The effect of viscoelasticity on the stability of a pulmonary airway liquid layer." *Physics of Fluids* 22(1).

Halpern, D. and J. B. Grotberg (1992a). "Dynamics and transport of a localized soluble surfactant on a thin film." *Journal of Fluid Mechanics* 237: 1–11.

Halpern, D. and J. B. Grotberg (1992b). "Fluid-elastic instabilities of liquid-lined flexible tubes." *Journal of Fluid Mechanics* 244: 615–632.

Halpern, D. and J. B. Grotberg (1993). "Surfactant effects on fluid-elastic instabilities of liquid-lined flexible tubes: a model of airway closure." *Journal of Biomechanical Engineering* 115(3): 271–277.

Halpern, D. and J. B. Grotberg (2003). "Nonlinear saturation of the Rayleigh instability due to oscillatory flow in a liquid-lined tube." *Journal of Fluid Mechanics* 492: 251–270.

Halpern, D., et al. (1998). "A theoretical study of surfactant and liquid delivery into the lung." *Journal of Applied Physiology* 85(1): 333–352.

Hammond, P. S. (1983). "Nonlinear adjustment of a thin annular film of viscous fluid surrounding a thread of another within a circular pipe." *Journal of Fluid Mechanics* 137: 363–384.

Happel, J. (1959). "Viscous flow relative to arrays of cylinders." *AIChE Journal* 5(2): 174–177.

Happel, J. and H. Brenner (1965). *Low Reynolds Number Hydrodynamics*. Englewood Cliffs, NJ, Prentice-Hall.

Harkins, W. D. and F. E. Brown (1919). "The determination of surface tension (free surface energy), and the weight of falling drops – The surface tension of water and benzene by the capillary height method." *Journal of the American Chemical Society* 41: 499–524.

Haselton, F. R. and P. W. Scherer (1982). "Flow visualization of steady streaming in oscillatory flow through a bifurcating tube." *Journal of Fluid Mechanics* 123 (Oct): 315–333.

Hassan, A. A., et al. (2006). "Clearance of viscoelastic mucus simulant with airflow in a rectangular channel, an experimental study." *Technology and Health Care* 14(1): 1–11.

Hassan, E. A., et al. (2011). "Adaptive Lagrangian-Eulerian computation of propagation and rupture of a liquid plug in a tube." *International Journal for Numerical Methods in Fluids* 67(11): 1373–1392.

Hawgood, S. and J. A. Clements (1990). "Pulmonary surfactant and its apoproteins." *Journal of Clinical Investigation* 86: 1–6.

Heil, M. (1997). "Stokes flow in collapsible tubes: computation and experiment." *Journal of Fluid Mechanics* 353: 285–312.

Henderson, Y., et al. (1915). "The respiratory dead space." *American Journal of Physiology* 38(1): 1–19.

Henry, W. (1803). "Experiments on the quantity of gases absorbed by water, at different temperatures, and under different pressures." *Philosophical Transactions of the Royal Society of London* 93: 29–274.

Herschel, W. H. and R. Bulkley (1926). "Konsistenzmessungen von Gummi-Benzollösungen." *Kolloid Zeitschrift* 39(291–300).

Hess, W. R. (1917). "Uber die periphere Regulierung der Blutzirkulation." *Pflügers Archiv für die gesamte Physiologie des Menschen und der Tiere* 168: 439–490.

Hill, D. B., et al. (2014). "A biophysical basis for mucus solids concentration as a candidate biomarker for airways disease." *Plos One* 9(2).

Hirschl, R. B., et al. (1996). "Evaluation of gas exchange, pulmonary compliance, and lung injury during total and partial liquid ventilation in the acute respiratory distress syndrome." *Critical Care Medicine* **24**(6): 1001–1008.

Hlastala, M. P. and L. E. Farhi (1973). "Absorption of gas bubbles in flowing blood." *Journal of Applied Physiology* **35**(3): 311–316.

Hlastala, M. P. and H. D. van Liew (1975). "Absorption of *in vivo* inert gas bubbles." *Respiration Physiology* **24**: 147–158.

Holdsworth, D., et al. (1999). "Characterization of common carotid artery blood-flow waveforms in normal human subjects." *Physiol. Meas.* **20**(3): 219–240.

Hornbeck, R. W. (1964). "Laminar flow in entrance region of a pipe." *Applied Scientific Research Section A – Mechanics Heat Chemical Engineering Mathematical Methods* **13**(2–3): 224–232.

Horsfield, K., et al. (1971). "Models of the human bronchial tree." *Journal of Applied Physiology* **31**: 207–217.

Hou, J. S., et al. (1989). "Boundary conditions at the cartilage–synovial fluid interface for joint lubrication and theoretical verifications." *Journal of Biomechanical Engineering* **111**: 78–87.

Howell, P. D., et al. (2000). "The propagation of a liquid bolus along a liquid-lined flexible tube." *Journal of Fluid Mechanics* **406**: 309–335.

Hu, Y., et al. (2015). "A microfluidic model to study fluid dynamics of mucus plug rupture in small lung airways." *Biomicrofluidics* **9**(4): 044119.

Hu, Y., et al. (2019). "Effects of surface tension and yield stress on mucus plug rupture: a numerical study." *Journal of Biomechanical Engineering*. To appear.

Huh, D., et al. (2005). "Microfluidics for flow cytometric analysis of cells and particles." *Physiological Measurement* **26**(3): R73–R98.

Huh, D., et al. (2007). "Acoustically detectable cellular-level lung injury induced by fluid mechanical stresses in microfluidic airway systems." *Proceedings of the National Academy of Sciences of the United States of America* **104**: 18886–18891.

Hussain, M., et al. (1999). "Relationship between power law coefficients and major blood constituents affecting the whole blood viscosity " *Journal of Biosciences* **24** (3): 329–337.

Huxley, V. H., et al. (1987). "Increased capillary hydraulic conductivity induced by atrial-natriuretic-peptide." *Circulation Research* **60**(2): 304–307.

Hwang, S. H., et al. (1969). "Rheological properties of mucus." *Rheologica Acta* **8**: 438–448.

Isabey, D. and H. K. Chang (1981). "Steady and unsteady pressure-flow relationships in central airways." *Journal of Applied Physiology* **51**(5): 1338–1348.

Jain, R. K. (1987). "Transport of molecules across tumor vasculature." *Cancer and Metastasis Reviews* **6**(4): 559–593.

Jensen, O. E. (1990). "Instabilities of flow in a collapsed tube." *Journal of Fluid Mechanics* **220**: 623–659.

Jensen, O. E. and J. B. Grotberg (1992). "Insoluble surfactant spreading on a thin viscous film: shock evolution and film rupture." *Journal of Fluid Mechanics* **240**: 259–288.

Johnson, T. F., et al. (2017). "X-ray computed tomography of packed bed chromatography columns for three dimensional imaging and analysis." *Journal of Chromatography A* **1487**(3): 108–115.

Kaczka, D. W. and R. L. Dellacá (2011). "Oscillation mechanics of the respiratorh system: applications to lung disease." *Critical Reviews in Biomedical Engineering* **39**(4): 337–359.

Kamm, R. D. and R. C. Schroter (1989). "Is airway closure caused by a thin liquid instability?" *Respiration Physiology* **75**: 141–156.

Kaplun, S. and P. A. Lagerstrom (1957). " Asymptotic expansions of Navier-Stokes solutions for small Reynolds numbers." *Journal of Mathematics and Mechanics* **6** (5): 585–593.

Kazemi, A., et al. (2019). "Surfactant delivery in rat lungs: comparing 3D geometrical simulation model with experimental instillation." *PLOS Computational Biology* **15**(10). e107408.

Kelly, R. P., et al. (1992). "Effective arterial elastance as index of arterial vascular load in humans." *Circulation* **86**: 513–521.

Kendig, J. W., et al. (1988). "Surfactant replacement therapy at birth: final analysis of a clinical trial and comparisons with similar trials." *Pediatrics* **82**(5): 756–762.

Keyhani, K., et al. (1995). "Numerical simulation of airflow in the human nasal cavity." *Journal of Biomechanical Engineering – Transactions of the ASME* **117**(4): 429–441.

Khokhlova, V. A., et al. (2015). "Histotripsy methods in mechanical disintegration of tissue: toward clinical applications." *International Journal of Hyperthermia* **31**(2): 145–162.

Kim, J., et al. (1987). "Turbulence statistics in fully-developed channel flow at low Reynolds-number." *Journal of Fluid Mechanics* **177**: 133–166.

King, M., et al. (1989). "The role of mucus gel viscosity, spinnability, and adhesive properties in clearance by simulated cough." *Biorheology* **26**(4): 737–745.

Knowles, M. and R. C. Boucher (2002). "Mucus clearance as a primary innate defense mechanism for mammalian airways." *Journal of Clinical Investigation* **109**(5): 571–577.

Koocheki, A., et al. (2009). "The rheological properties of ketchup as a function of different hydrocolloids and temperature." *International Journal of Food Science and Technology* **44**: 596–602.

Koopmann, G. H. (1967). "Vortex wakes of vibrating cylinders at low Reynolds numbers." *Journal of Fluid Mechanics* **28**: 501–512.

Kozeny, J. (1927). "Ueber kapillare Leitung des Wassers im Boden." *Sitzungsber Akad. Wiss., Wien* **136(2a)**: 271–306.

Krogh, A. (1922). *Anatomy and Physiology of Capillaries.* New Haven, CT, Yale University Press.

Krueger, M. A. and D. P. Gaver (2000). "A theoretical model of pulmonary surfactant multilayer collapse under oscillating area conditions." *Journal of Colloid and Interface Science* **229**: 353–364.

Lagerstrom, P. A. and J. D. Cole (1955). "Examples illustrating sxpansion procedures for the Navier–Stokes Equations." *Journal of Rational Mechanics and Analysis* **4**: 817–882.

Lai, S. K., et al. (2007). "Rapid transport of large polymeric nanoparticles in fresh undiluted human mucus." *Proceedings of the National Academy of Sciences of the United States of America* **104**(5): 1482–1487.

Lai, S. K., et al. (2009). "Micro- and macrorheology of mucus." *Advanced Drug Delivery Reviews* **61**(2): 86–100.

Lai, W. M., et al. (1977). "Computation of stress relaxation function and apparent viscosity from dynamic data of synovial fluid." *Biorheology* **14**: 229–236.

Lai, W. M., et al. (1978). "Rheological equations for synovial fluid." *Journal of Biomechanical Engineering – Transactions of the ASME* **100**(4): 169–186.

Lamb, H. (1911). "On the uniform motion of a sphere through a viscous fluid." *Philosophical Magazine* **21**: 112–121.

Lambossy, P. (1952). "Oscillations forcées d'un liquide incompressible et visqueux dans un tube rigide et horizontal : calcul de la force de frottement." *Helvetica Physica Acta* **25**: 371–386.

Langmuir, I. (1918). "The adsorption of gases on plane surfaces of glass, mica and platinum." *J. Am. Chem. Soc.* **40**(9): 1361–1403.

Larose, P. G. and J. B. Grotberg (1997). "Flutter and long wave instabilities in compliant channels conveying developing flows." *Journal of Fluid Mechanics* **331**: 37–58.

Larson, R. E. and J. J. L. Higdon (1987). "Microscopic flow near the surface of two-dimensional porous media. Part 2. Transverse flow." *Journal of Fluid Mechanics* **178**: 119–136.

Lasheras, J. C. (2007). "The Biomechanics of Arterial Aneurysms." *Annual Review of Fluid Mechanics* **39**: 293–319.

Laufer, J. (1952). *The Structure of Turbulence in Fully Developed Pipe Flow.* National Advisory Committee for Aeronautics, Report 1174.

Leach, C. L., et al. (1996). "Partial liquid ventilation with perflubron in premature infants with severe respiratory distress syndrome." *New England Journal of Medicine* **335**(11): 761–767.

Leddy, H. A. (2004). "Molecular diffusion in tissue-engineered cartilage constructs: effects of scaffold material, time, and culture conditions." *Journal of Biomedical Materials Research B* **70B**: 397–406.

Leddy, H. A. and F. Guilak (2003). "Site-specific molecular diffusion in articular cartilage measured using fluorescence recovery after photobleaching." *Annals of Biomedical Engineering* **31**: 753–760.

Leiner, G. C., et al. (1966). "Cough peak flow rate." *American Journal of the Medical Sciences* **251**(2): 211–214.

Lesieur, M. and O. Metais (1996). "New trends in large-eddy simulations of turbulence." *Annual Review of Fluid Mechanics* **28**(1): 45–82.

Leslie, F. M. (1968). "Some constitutive equations for liquid crystals." *Archives for Rational Mechanics and Analysis* **28**(4): 265–283.

Less, J. R., et al. (1992). "Interstitial hypertension in human breast and colorectal tumors." *Cancer Research*: 6371–6374.

Levick, J. R. (1987). "Flow through interstitium and other fibrous matrices." *Quarterly Journal of Experimental Physiology* **72**: 409 – 438.

Li, J. K.-J. (1996). *Comparative Cardiovascular Dynamics of Mammals.* London, CRC Press.

Ling, S. C., et al. (1973). "Nonlinear analysis of aortic flow in living dogs." *Circulation Research* **33**: 198–212.

Lipshitz, H., et al. (1976). "Changes in hexosamine content and swelling ratio of articular cartilage as a function of depth from the surface." *Journal of Bone and Joint Surgery* **58A**: 1149–1153.

Litt, M., et al. (1976). "Mucus rheology – relation to structure and function." *Biorheology* **13**(1): 37–48.

Long, J.A, et al. (2005). "Viscoelasticity of pediatric blood and its implications for thetesting of a pulsatile pediatric blood pump." *ASAIO Journal* 51(5): 563–566.

Longest, P. W. and S. Vinchurkar (2007). "Validating CFD predictions of respiratory aerosol deposition: effects of upstream transition and turbulence." *Journal of Biomechanics* 40(2): 305–316.

Longest, P. W. and J. X. Xi (2007). "Computational investigation of particle inertia effects on submicron aerosol deposition in the respiratory tract." *Journal of Aerosol Science* 38(1): 111–130.

Lowe, K. C. (2006). "Blood substitutes: from chemistry to clinic." *Journal of Materials Chemistry* 16: 4189–4196.

Lutchen, K. R., et al. (1996). "How inhomogeneities and airway walls affect frequency dependence and separation of airway and tissue properties." *Journal of Applied Physiology* 80(5): 1696–1707.

Lutolf, M. P. and J. A. Hubbell (2005). "Synthetic biomaterials as instructive extracellular microenvironments for morphogenesis in tissue engineering." *Nature Biotechnology* 23(1): 47–55.

Lyne, W. H. (1971). "Unsteady viscous flow in a curved pipe." *Journal of Fluid Mechanics* 45(1): 13–&.

Macklem, P. T., et al. (1970). "The stability of peripheral airways." *Respiration Physiology* 8: 191–203.

Manolidis, M., et al. (2016). "A macroscopic model for simulating the mucociliary clearance in a bronchial bifurcation: the role of surface tension." *Journal of Biomechanical Engineering – Transactions of the ASME* 138(12).

Marangoni, C. (1865). "Sull'espansione delle goccie d'un liquido galleggianti sulla superfice di altro liquido." *PhD Thesis, Pavia, Italy.*

Marangoni, C. (1869). *Sull'espansione delle goccie d'un liquido galleggianti sulla superfice di altro liquido.* Pavia, Italy, Fratelli Fusi.

Markfort, C., et al. (2010). "Wind sheltering of a lake by a tree canopy or bluff topography." *Water Resources Research* 46(3): W03530.

Martin, R. B. and J. Ishida (1989). "The relative effects of collagen fiber orientation, porosity, density, and mineralization on bone strength." *Journal of Biomechanics* 22(5): 419–426.

Martin, S. and B. Bhushan (2014). "Fluid flow analysis of a shark-inspired microstructure." *Journal of Fluid Mechanics* 756: 5–29.

Martini, P., et al. (1930). "Die Stromung des Blutes in eigen Gefassen. Eine Abweichung vom Poiseuille'schne." *Gesetz. Deutsches Archiv fur klinische Medizin* 169: 212–222.

Matar, O. K., et al. (2002). "Surfactant transport on highly viscous surface films." *Journal of Fluid Mechanics* 466: 85–111.

Matar, O. K. and S. M. Troian (1999). "The development of transient fingering patterns during the spreading of surfactant coated films." *Physics of Fluids* 11(11): 3232–3246.

Matsubara, M. and P. H. Alfredsson (2001). "Disturbance growth in boundary layers subjected to free-stream turbulence." *Journal of Fluid Mechanics* 430: 149–168.

Matsui, H., et al. (1998). "Coordinated clearance of periciliary liquid and mucus from airway surfaces." *Journal of Clinical Investigation* 102(6): 1125–1131.

Maxwell, A. D., et al. (2014). "Trapping of embolic particles in a vessel phantom by cavitation-enhanced acoustic streaming." *Physics in Medicine and Biology* 59(17): 4927–4943.

Mazzucco, D., et al. (2002). "Rheology of joint fluid in total knee arthroplasty patients." *Journal of Orthopaedic Research* 20(6): 1157–1163.

McCulloh, K., et al. (2003). "Water transport in plants obeys Murray's law." *Nature* 421: 939–942.

McDonald, D. A. (1955). "The relation of pulsatile pressure to flow in arteries." *Journal of Physiology* 127(3): 533–552.

Merrill, E. W. (1969). "Rheology of blood." *Physiological Reviews* 49(4): 863–888.

Michel, C. C. (1980). "Filtration coefficients and osmotic reflection coefficients of the walls of single frog mesenteric capillaries." *Journal of Physiology – London* 309(Dec): 341–355.

Millikan, R. A. (1911). "The isolation of an ion, a precision measurement of its charge, and the correction of Stokes's Law." *Physical Review* 32(4): 349–397.

Millikan, R. A. (1913). "On the elementary electrical charge and the Avogadro Constant." *Physical Review* 2(2): 109–143.

Milnor, W. R. (1989). *Hemodynamics*. Baltimore, OH, Williams & Wilkins.

Minale, M. (2014). "Momentum transfer within a porous medium. II. Stress boundary condition." *Physics of Fluids* 26: 123102.

Miserocchi, G. (1997). "Physiology and pathophysiology of pleural fluid turnover." *European Respiratory Journal* 10(1): 219–225.

Miserocchi, G., et al. (1993). "Model of pleural fluid turnover." *Journal of Applied Physiology* 75: 1798–1806.

Mittal, R., et al. (2013). "Fluid dynamics of human phonation and speech." *Annual Review of Fluid Mechanics* **45**: 437–467.

Mockros, L. F. and J. D. S. Gaylor (1975). "Artificial lung design – tubular membrane units." *Med. Biol. Engin.* **13**(2): 171–181.

Mohanty, A. K. and S. B. L. Asthana (1979). "Laminar-flow in the entrance region of a smooth pipe." *Journal of Fluid Mechanics* **90**(Feb): 433–447.

Monsia, M. D. (2011). "A simplified nonlinear generalized Maxwell model for predicting the time dependent behavior of viscoelastic materials " *World Journal of Mechanics* **1**: 158–167.

Moody, L. F. (1944). "Friction factors for pipe flow." *Transactions of the ASME* **66**(8): 671–684.

Moriarty, J. A. and J. B. Grotberg (1999). "Flow-induced instabilities of a mucus-serous bilayer." *Journal of Fluid Mechanics* **397**: 1–22.

Morris, C. L., et al. (1993). "Deoxygenation-induced changes in sickle cell-sickle cell adhesion." *Blood* **81**(11): 3138–3145.

Mow, V. C., et al. (1980). "Biphasic creep and stress relaxation of articular cartilage in compression: theory and experiments." *Journal of Biomechanical Engineering* **102**: 73–84.

Mullin, T. (2011). "Experimental studies of transition to turbulence in a pipe." *Annual Review of Fluid Mechanics* **43**: 1–24.

Muradoglu, M., et al. (2019). "Effects of surfactant on propagation and rupture of a liquid plug in a tube." *Journal of Fluid Mechanics* **872**: 407–437.

Murray, C. D. (1926a). "The physiological principle of minimum work. I. The vascular system and the cost of blood volume." *Proceedings of the National Academy of Sciences USA* **12**: 207–214.

Murray, C. D. (1926b). "The physiological principle of minimum work applied to the angle of branching of arteries." *Journal of General Physiology* **9**: 835–841.

Nachemson, A., et al. (1970). "In vitro diffusion of dye through the end-plates and the annulus fibrosus of human lumbar inter-vertebral discs." *Acta Orthopaedica Scandinavica* **41**: 589–607.

Naureckas, E. T., et al. (1994). "Airway reopening pressure in isolated rat lungs." *Journal of Applied Physiology* **76**(3): 1372–1377.

Nayfeh, A. H. (1993). *Introduction to Perturbation Techniques*. New York, John Wiley and Son.

Neale, G. and W. Nader (1974). "Practical significance of Brinkman's extension of Darcy's law: coupled parallel flows within a channel and a bounding porous medium." *Canadian Journal of Chemical Engineering* 52(4): 475–478.

Netti, P. A., et al. (2000). "Role of extracellular matrix assembly in interstitial transport in solid tumors." *Cancer Research* **60**(9): 2497–2503.

Nikuradse, J. (1932). "Gesetzmäßigkeit der turbulenten Strömung in glatten Rohren." *Forsch. Arb. Ing.-Wes.* **356**.

Norton, F. H., et al. (1944). "Fundamental study of clay: VI, flow properties of kaolinite-water suspensions." *Journal of the American Ceramic Society* **27**(5): 149–164.

Notter, R. H. (2000). *Lung Surfactants: Basic Science and Clinical Applications*. Marcel Dekker, Inc.

Ogston, A. G. and J. E. Stanier (1953). "The physiological function of hyaluronic acid in synovial fluid – viscous, elastic and lubricant properties." *Journal of Physiology – London* **119**(2–3): 244–252.

Olgac, U. and M. Muradoglu (2013). "Computational modeling of unsteady surfactant-laden liquid plug propagation in neonatal airways." *Physics of Fluids* **25**(7): 071901.

Olson, D. E. (1971). "Fluid mechanics relevant to respiration: flow within curved or elliptical tubes and bifurcating systems." PhD thesis, Imperial College, London.

Oseen, C. W. (1910). "Über die Stokes'sche formel, und über eine verwandte Aufgabe in der Hydrodynamik." *Arkiv för Matematik, Astronomi och Fysik* **6**: 29.

Oseen, C. W. (1913). "Über den Gültigkeitsbereich der Stokesschen Widerstandsformel." *Arkiv för Matematik, Astronomi och Fysik* **9**(17): 1–15.

Ostwald, W. (1925). "Uber die Geschwindigkeitsfunktion der Viskositat disperser Systeme. I." *Kolloid Zeitschrift* **36**: 99–117.

Otis, D. R., Jr., et al. (1994). "Dynamic surface tension of surfactant TA: experiments and theory." *Journal of Applied Physiology* **77**(6): 2681–2688.

Owens, S. C., et al. (2009). "Changes in spinal height following sustained lumbar flexion and extension postures: a clinical measure of intervertebral disc hydration using stadiometry." *Journal of Manipulative and Physiological Therapeutics* **32**(5): 358–363.

Pacella, H. E., et al. (2011). "Darcy permeability of hollow fiber bundles used in blood oxygenation devices." *Journal of Membrane Science* **382**(1–2): 238–242.

Pattle, R. E. (1955). "Properties, function, and origin of the alveolar lining layer." *Nature* **175**: 1125–1126.

Pedley, T. J. and J. O. Kessler (1992). "Hydrodynamic phenomena in suspensions of swimming microorganisms." *Annual Review of Fluid Mechanics* **24**: 313–358

Pedley, T. J. and X. Y. Luo (1998). "Modelling flow and oscillations in collapsible tubes." *Theoretical and Computational Fluid Dynamics* **10**(1–4): 277–294.

Pedley, T. J., et al. (1970a). "Energy losses and pressure drops in models of human airways." *Respiration Physiology* **9**: 371–386.

Pedley, T. J., et al. (1970b). "The prediction of pressure drop and variation of resistance within the human bronchial airways." *Respiration Physiology* **9**: 387–405.

Pedley, T. J., et al. (1971). "Flow and pressure drop in systems of repeatedly branching tubes." *Journal of Fluid Mechanics* **46**(2): 365–383.

Pekkan, K., et al. (2008). "Patient-specific surgical planning and hemodynamic computational fluid dynamics optimization through free-form haptic anatomy editing tool (SURGEM)." *Medical & Biological Engineering & Computing* **46**(11): 1139–1152.

Perun, M. L. and D. P. Gaver (1995). "An experimental-model investigation of the opening of a collapsed untethered pulmonary airway." *Journal of Biomechanical Engineering – Transactions of the ASME* **117**(3): 245–253.

Pigott, R. J. S. (1933). "The flow of fluids in closed conduits." *Mechanical Engineering* **55**: 497–501, 515.

Plateau, J. (1873). *Statique experimentale et theorique des liquides soumis aux seules forces moleculaires*. Paris, Gauthier-Villars.

Pluen, A., et al. (1999). "Diffusion of macromolecules in agarose gels: comparison of linear and globular configurations." *Biophysical Journal* **77**: 542–552.

Pohlhausen, E. (1921). "Zur näherungsweisen Integration der Differentialgleichung der Iaminaren Grenzschicht." *Zeitschrift für Angewandte Mathematik und Mechanik* 4: 252–268.

Poiseuille, J. (1840). "Recherches expérimentales sur le mouvement des liquides dans les tubes de très petits diamètres." *Comptes rendus de l'Académie des Sciences* **11**: 961–967, 1041–1048.

Poiseuille, J. (1841). "Recherches expérimentales sur le mouvement des liquides dans les tubes de très petits diamètres." *Comptes rendus de l'Académie des Sciences* **12**: 112–115.

Poiseuille, J. L. M. (1828). "Recherehes sur la force du coeur aortique." *Archives générales de médecine* **8**: 550–554.

Poiseuille, J. L. M. (1846). "Recherches expérimentales sur Ie mouvement des liquides dans les tubes de très petits diamètres. ." *Mémoires presentés par divers savants à l'Académie Royale des Sciences de l'Institut de France* IX: 433–544.

Pope, S. B. (2000). *Turbulent Flows*. Cambridge, Cambridge University Press.

Prandtl, L. (1905). "Über Flüssigkeitsbewegung bei sehr kleiner Reibung," in *Verh. III Intern. Math. Kongr., Heidelberg 1904*. A. Krazer, ed. Leipzig, Teubner: 484–491.

Prandtl, L. (1925). "Bericht uber Untersuchungen zur ausgebildeten Turbulenz." *Z. Angew. Math. und Mech* **5**: 136–139.

Prandtl, L. (1927). "Über den Reibungswiderstand strömender Luft." *Ergeb. AVA Göttingen, III. Lfg.*: 1–5.

Prandtl, L. (1933). "Neuere Ergebnisse der Turbulenzforschung." *Zeitschrift des Vereines Deutscher Ingenieure* **77**: 105–114.

Pries, A. R., et al. (1992). "Blood viscosity in tube flow: dependence on diameter and hematocrit." *Am. J. Physiol.* **263**: H1770–H1778.

Prosi, M., et al. (2005). "Mathematical and numerical models for transfer of low-density lipoproteins through the arterial walls: a new methodology for the model set up with applications to the study of disturbed lumenal flow." *Journal of Biomechanics* **38**(4): 903–917.

Proudman, I. and J. R. A. Pearson (1957). "Expansions at small Reynolds numbers for the flow past a sphere and a circular cylinder." *Journal of Fluid Mechanics* **2** (3): 237–262.

Puchelle, E., et al. (1983). "Spinability of bronchial mucus - relationship with viscoelasticity and mucous transport properties." *Biorheology* **20**(2): 239–249.

Puig, F., et al. (2005). "Vibration enhances interleukin-8 release in a cell model of snoring-induced airway inflammation." *Sleep* 28(10): 1312–1316.

Purtell, L. P., et al. (1981). "Turbulent boundary-layer at low Reynolds-number." *Physics of Fluids* **24**(5): 802–811.

Qamar, A., et al. (2012). "Evolution of acoustically vaporized microdroplets in gas embolotherapy." *Journal of Biomechanical Engineering* **134**(3): 31010–31011–31013.

Quemada, D. (1977). "Rheology of concentrated disperse systems and minimum energy dissipation principle I. Viscosity-concentration relationship." *Rheologica Acta* **16**: 82–94.

Quemada, D. (1978). "Rheology of concentrated disperse systems III. General features of the proposed non-Newtonian model, Comparison with experimental data." *Rheologica Acta* **17**(6): 643–653.

Ramanujan, S., et al. (2002). "Diffusion and convection in collagen gels: implications for transport in the tumor interstitium." *Biophysical Journal* **83**: 1650–1660.

Rand, P. W., et al. (1964). "Viscosity of normal human blood under normothermic + hypothermic conditions." *Journal of Applied Physiology* 19(1): 117–122.

Rayleigh, J. W. S. (1877). *The Theory of Sound: Volume 1.* London, Macmillan.

Rayleigh, J. W. S. (1878a). "On the instability of jets." *Proceedings of the London Mathematical Society* 10: 4–13.

Rayleigh, J. W. S. (1878b). *The Theory of Sound: Volume 2.* London, Macmillan.

Rayleigh, J. W. S. (1892a). "On the question of the stability of the flow of liquids." *Philosophical Magazine* 34: 59–70.

Rayleigh, J. W. S. (1892b). "On the instability of a cylinder of viscous liquid under capillary force." *Philosophical Magazine Series 5* 34(207): 145–154.

Redheuil, A., et al. (2011). "Age-related changes in aortic arch geometry relationship with proximal aortic function and left ventricular mass and remodeling." *Journal of the American College of Cardiology* 58(12): 1262–1270.

Reichardt, H. (1942). "Gesetzmäßigkeiten der freien Turbulenz." *VDI-Forschungsheft* 414.

Reiner, M. (1964). "The Deborah number." *Physics Today* 17(1): 62.

Reynolds, O. (1883). "An experimental investigation of the circumstances which determine whether the motion of water shall be direct or sinuous, and of the law of resistance in parallel channels." *Proceedings of the Royal Society of London* 35: 84–99.

Reynolds, O. (1895). "On the dynamical theory of incompressible viscous fluids and the determination of the criterion." *Philosophical Transactions of the Royal Society of London A* 186: 123–164.

Riscardo, M. A., Moros, J.E., Franco, J.M., Gallegos, C., (2005). "Rheological characterisation of salad-dressing-type emulsions stabilised by egg yolk/sucrose distearate blends." *European Food Research and Technology* 220(3–4): 380–388.

Romano, F., et al. (2019). "Liquid plug formation in an airway closure model." *Physical Review Fluids* 4: 093103.

Roscoe, R. (1952). "The viscosity of suspensions of rigid spheres." *British Journal of Applied Physics* 3: 267–269.

Roshko, A. (1961). "Experiments on the flow past a circular cylinder at very high Reynolds number." *Journal of Fluid Mechanics* 10(3): 345–356.

Ross, S. M. and S. Corrsin (1974). "Results of an analytical model of mucociliary pumping." *Journal of Applied Physiology* 37: 333–340.

Rotenberg, Y., et al. (1983). "Determination of surface-tension and contact-angle from the shapes of axisymmetric fluid interfaces." *Journal of Colloid and Interface Science* 93(1): 169–183.

Rouse, H. (1943). "Evaluation of boundary roughness." *Proceedings, Second Hydraulic Conference*, University of Iowa Bulletin 27.

Safari, M., et al. (1990). "Clinical assessment of rheumatic diseases using viscoelastic parameters for synovial fluid." *Biorheology* 27: 650–674.

Sallam, A. M. and N. H. Hwang (1984). "Human red blood cell hemolysis in a turbulent shear flow: contribution of Reynolds shear stresses." *Biorheology* 21(6): 783–797.

Savart, F. (1833). "Mémoire sur la constitution des veines liquides lancées par des orifices circulaires en mince paroi." *Annali di Chimica* 53: 337–398.

Scherer, P. W. and F. R. Haselton (1982). "Convective exchange in oscillatory flow through bronchial-tree models." *Journal of Applied Physiology* 53(4): 1023–1033.

Schiller, L. (1922). "Untersuchungen über laminare und turbulente Strömung." *Forschg. Ing.-Wes. Heft 248 (or ZAMM, Bd. 2, 96–106, or Physikal. Z., Bd. 33, 14).*

Schlichting, H. (1968). *Boundary Layer Theory.* New York, McGraw-Hill.

Schmid-Schönbein, G. W., et al. (1980). "Cell distribution in capillary networks." *Microvascular Research* 19(1): 18–44.

Schmid-Schönbein, H. and R. E. Wells (1971). *Red Cell Deformation and Red Cell Aggregation: Their Influence on Blood Rheology in Health and Disease.* Heidelberg, Springer-Verlag.

Schmid-Schönbein, H., et al. (1971). "Fluid drop like behavior of erythrocytes-disturbance in pathology and its quantification." *Biorheology* 7(4): 227–234.

Schmid-Schönbein, H., et al. (1973). "Microrheology and protein chemistry of pathological red cell aggregation (blood sludge) studied *in vitro*." *Biorheology* 10: 213–227.

Schreck, R. M. and L. F. Mockros (1970). "Fluid dynamics in the upper pulmonary airways," in *AIAA 3rd Fluid and Plasma Dynamics Conference*, Los Angeles. AIAA.

Schurz, J. and V. Ribitsch (1987). "Rheology of synovial fluid." *Biorheology* 24: 385–399.

Secomb, T. W., et al. (1986). "Flow of axisymmetric red blood cells in narrow capillaries." *Journal of Fluid Mechanics* 163: 405–423.

Sellers, L. A., et al. (1991). "The rheology of pig small intestinal and colonic mucus - weakening of gel structure by non-mucin components." *Biochimica et Biophysica Acta* **1115**(2): 174–179.

Seminara, G., et al. (2020). "Biological fluid dynamics of airborne COVID-19 infection." *Rendiconti Lincei. Scienze Fisiche e Naturali*: 1–33.

Settles, G. S. (2006). "Fluid mechanics and homeland security." *Annual Review of Fluid Mechanics*. **38**: 87–110.

Shaffer, T. H., et al. (1992). "Liquid ventilation." *Pediatric Pulmonology* **14**(2): 102–109.

Shapiro, A. H. (1977). "Steady flow in collapsible tubes." *Journal of Biomechanical Engineering* **99**: 126–147.

Shapiro, A. H. and M. Y. Jaffrin (1971). "Reflux in peristaltic pumping - is it determined by Eulerian or Lagrangian mean velocity." *Journal of Applied Mechanics* **38**(4): 1060–&.

Shojima, M., et al. (2004). "Magnitude and role of wall shear stress on cerebral aneurysm - Computational fluid dynamic study of 20 middle cerebral artery aneurysms." *Stroke* **35**(11): 2500–2505.

Siggers, J. H., et al. (2009). "Flow dynamics in a stented ureter." *Math. Med. Biol.* **26**(1): 1–24.

Simmons, C. T. (2008). "Henry Darcy (1803–1858): Immortalised by his scientific legacy." *Hydrogeol. J.* **16**: 1023–1038.

Simon, B. R. (1992). "Multiphase poroelastic finite element models for soft tissue structures." *Appl. Mech. Rev.* **45**(6): 191–218.

Sivasothy, P., et al. (2001). "Effect of manually assisted cough and mechanical insufflation on cough flow of normal subjects, patients with chronic obstructive pulmonary disease (COPD), and patients with respiratory muscle weakness." *Thorax* **56**(6): 438–444.

Skalak, R. and P. I. Branemark (1969). "Deformation of red blood cells in capillaries." *Science* **164**(3880): 717–719.

Skalak, R., et al. (1981). "Mechanics of blood flow." *Journal of Biomechanical Engineering - Transactions of the ASME* **103**(2): 102–115.

Sleigh, M. A., et al. (1988). "The propulsion of mucus by cilia." *Am. Rev. Respir. Dis.* **137**: 726–741.

Smith, D. J., et al. (2008). "Modelling mucociliary clearance." *Respiratory Physiology & Neurobiology* **163**(1–3): 178–188.

Song, Y. L., et al. (2009). "Airway surface liquid depth measured in ex vivo fragments of pig and human trachea: dependence on Na+ and Cl- channel function." *American Journal of Physiology - Lung Cellular and Molecular Physiology* **297**(6): L1131–L1140.

Stangeby, D. K. and C. R. Ethier (2002). "Computational analysis of coupled blood-wall arterial LDL transport." *Journal of Biomechanical Engineering* **124**(1): 1–8.

Starling, E. H. (1896). "On the absorption of fluids from the connective tissue spaces." *Journal of Physiology* 19 (4): 312–326.

Staub, N. C., et al. (1985). "Transport through the pleura," in *The Pleura in Health and Disease*. J. Chrétien, J. Bignon and A. Hirsch, eds. New York, Dekker: 169–193.

Stein, P. D. and H. N. Sabbah (1976). "Turbulent blood flow in the ascending aorta of humans with normal and diseased aortic valves." *Circulation Research* **39**(1): 58–65.

Stergiopulos, N., et al. (1999). "Total arterial inertance as the fourth element of the windkessel model." *American Journal of Physiology - Heart and Circulatory Physiology* **276**(1): H81–H88.

Stokes, G. G. (1851). "On the effect of the internal friction of fluids on the motion of pendulums." *Transactions Cambridge Philosophical Society* **9**: 8–106.

Stone, H. A. (1990). "A simple derivation of the time-dependent convective-diffusion equations for surfactant transport along a deforming interface." *Physics of Fluids A* **2**: 111–112.

Strouhal, V. (1878). "Uber eine besondere Art der Tonerregung." *Annalen der Physik und Chemie* **5**(10). 216–251.

Sundström, M., et al. (2013). "A high-sensitivity ultra-high performance liquid chromatography/high-resolution time-of-flight mass spectrometry (UHPLC-HR-TOFMS) method for screening synthetic cannabinoids and other drugs of abuse in urine." *Analytical and Bioanalytical Chemistry* **405**(26): 8463–8474.

Sutera, S. P. and R. Skalak (1993). "The history of Poiseuille's law." *Annual Review of Fluid Mechanics* **25**: 1–19.

Swamee, P. K. and A. K. Jain (1976). "Explicit equations for pipeflow problems." *Journal of the Hydraulics Division - Proceedings of ASCE* **102**(5): 657–664.

Swartz, M. A. and M. E. Fleury (2007). Interstitial flow and its effects in soft tissues. *Annual Review of Biomedical Engineering*. **9**: 229–256.

Swartz, M. A., et al. (1999). "Mechanics of interstitial-lymphatic fluid transport: theoretical foundation and experimental validation." *Journal of Biomechanics* **32**(12): 1297–1307.

Tai, C. F., et al. (2011a). "Numerical study of flow fields in an airway closure model." *Journal of Fluid Mechanics* **677**: 483–502.

Tai, C. F., et al. (2011b). "Liquid and drug delivery into a 3D lung tree." *American Journal of Respiratory and Critical Care Medicine* **183**: A2298.

Tam, P. Y., et al. (1980). "Non-linear viscoelastic properties of cervical mucus." *Biorheology* **17**(5–6): 465–478.

Tang, J. L. W., et al. (2013). "Airflow dynamics of human jets: sneezing and breathing – potential sources of infectious aerosols." *Plos One* **8**(4). e59970.

Tate, T. (1864). "On the magnitude of a drop of liquid formed under different circumstances." *The London, Edinburgh, and Dublin Philosophical Magazine and Journal of Science* **27**(181): 176–180.

Tavana, H., et al. (2009). "Microfluidics, lung surfactant, and respiratory disorders." *Lab Medicine* **40**: 203–209.

Tavana, H., et al. (2010). "Dynamics of liquid plugs of buffer and surfactant solutions in a micro-engineered. pulmonary airway model." *Langmuir* **26**(5): 3744–3752.

Tavana, H., et al. (2011). "Epithelium damage and protection during reopening of occluded airways in a physiologic microfluidic pulmonary airway model." *Biomedical Microdevices* **13**(4): 731–742.

Taylor, C. A. and C. A. Figueroa (2009). "Patient-specific modeling of cardiovascular mechanics." *Annual Review of Biomedical Engineering* **11**: 109–134.

Taylor, G. I. (1932). "The viscosity of a fluid containing small drops of another fluid." *Proceedings of the Royal Society of London A* **138**: 41–48.

Taylor, G. I. (1953). "Dispersion of soluble matter in solvent flowing slowly through a tube." *Proceedings of the Royal Society of London A* **291**(1137): 186–203.

Terzaghi, K. (1925). *Erdbaumechanik auf bodenphysikalischer grundlage*. Leipzig, Wien, F. Deuticke.

Thien, N. P. and R. I. Tanner (1977). " A new constitutive equation derived from network theory." *Journal of Non-Newtonian Fluid Mechanics* **2**: 353–365.

Thomson, J. (1855). "On certain curious motions observable on the surfaces of wine and other alcoholic liquours." *Philosophical Magazine* **10**: 330–333.

Thurston, G. B. (1979). "Rheological parameters for the viscosity, viscoelasticity and thixotropy of blood." *Biorheology* **16**: 149–162.

Tiffany, J. M. (1991). "The viscosity of human tears." *International Ophthalmology* **15**(6): 371–376.

Tillmann, V. and P. E. Clayton (2001). "Diurnal variation in height and the reliability of height measurements using stretched and unstretched techniques in the evaluation of short-term growth." *Annals of Human Biology* **28**(2): 195–206.

Tingbeall, H. P., et al. (1993). "Volume and osmotic properties of human neutrophils." *Blood* **81**(10): 2774–2780.

Tomaiuolo, G. (2014). "Biomechanical properties of red blood cells in health and disease towards microfluidics." *Biomicrofluidics* **8**: 051501.

Tomaiuolo, G., et al. (2016). "Blood linear viscoelasticity by small amplitude oscillatory flow." *Rheologica Acta* **55**: 485–495.

Tritton, D. J. (1959). "Experiments on the flow past a circular cylinder at low Reynolds numbers " *Jounal of Fluid Mechanics* **6**(4): 547–567.

Tse, K. M., et al. (2011). "Investigation of hemodynamics in the development of dissecting aneurysm within patient-specific dissecting aneurismal aortas using computational fluid dynamics (CFD) simulations." *Journal of Biomechanics* **44**(5): 827–836.

Tung, Y. C., et al. (2011). "High-throughput 3D spheroid culture and drug testing using a 384 hanging drop array." *Analyst* **136**(3): 473–478.

Ubal, S., et al. (2008). "Stability of the steady-state displacement of a liquid plug driven by a constant pressure difference along a prewetted capillary tube." *Industrial & Engineering Chemistry Research* **47**(16): 6307–6315.

Urban, J. P. G. and S. Roberts (2003). "Degeneration of the intervertebral disc." *Arthritis Research & Therapy* **5**(3): 120–130.

Vallikivi, M., et al. (2015). "Spectral scaling in boundary layers and pipes at very high Reynolds number." *Journal of Fluid Mechanics* **771**: 303–326.

Van Dyke, M. (1975). *Perturbation Methods in Fluid Mechanics*. Stanford, CA, The Parabolic Press.

van Wyk, S., et al. (2015). "Non-Newtonian perspectives on pulsatile blood-analog flows in a 180 degrees curved artery model." *Physics of Fluids* **27**(7): 071901.

Vargas, C. B., et al. (1979). "Hydraulic conductivity of the endothelial and outer layers of the rabbit aorta." *American Journal of Physiology* **236**(1): H53–H60.

Vaughan, B. L., et al. (2013). "A poroelastic model describing nutrient transport and cell stresses within a cyclically strained collagen hydrogel." *Biophysical Journal* **105** (9): 2188–2198.

Velez-Cordero, J. R. and E. Lauga (2013). "Waving transport and propulsion in a generalized Newtonian fluid." *Journal of Non-Newtonian Fluid Mechanics* **199**: 37–50.

Vernon-Carter, E. J., et al. (2016). "Cox–Merz rules from phenomenological Kelvin–Voigt and Maxwell models." *Journal of Food Engineering* **169**: 18–26.

Viola, H., et al. (2019). "Microphysiological systems modeling acute respiratory distress syndrome that capture mechanical force-induced injury-inflammation-repair." *APL Bioengineering* 3(4): 041503.

von Kármán, T. (1911). "Über den Mechanismus des Widerstandes, den ein bewegter Körper in einer Flüssigkeit erfährt." *Nachr. Ges. Wiss. Göttingen, Math.-Phys. Kl.*: 509–517.

von Kármán, T. (1912). "Über den Mechanismus des Widerstandes, den ein bewegter Körper in einer Flüssigkeit erfährt." *Nachr. Ges. Wiss. Göttingen, Math.-Phys. Kl.*: 547–556.

von Kármán, T. (1921). "Über laminare und turbulente Reibung." *Zeitschrift für Angewandte Mathematik und Mechanik* 1(4): 233–252.

von Kármán, T. (1930). "Mechanische Ähnlichkeit und Turbulenz." *Nachrichten von der Gesellschaft der Wissenschaften zu Göttingen* 5(58–76).

von Neergaard, K. (1929). "Neue Auffassungen über einen Grundbegriff der Atemmechanik. Die Retraktionskraft der Lunge, abhängig von der Oberflächenspannung in den Alveolen." *Zeitschrift fur Gesampte Expcrimentalle Medizin* **66**: 373–394.

Walburn, F. J. and D. J. Schneck (1976). "Constitutive equation for whole human-blood." *Biorheology* 13 (3): 201–210.

Wang, C. (2005). *Inhaled Particles*. Oxford, Elsevier.

Wang, D. M. and J. M. Tarbell (1995). "Modeling interstitial flow in an artery wall allows estimation of wall shear stress on smooth muscle cells." *Journal of Biomechanical Engineering* **117**(3): 358–363.

Wang, S. and J. M. Tarbell (2000). "Effect of fluid flow on smooth muscle cells in a 3-dimensional collagen gel model." *Arteriosclerosis Thrombosis and Vascular Biology* **20**(10): 2220–2225.

Wang, Z.-Q. (1982). "Study on correction coefficients of laminar and turbulent entrance region effect in round pipe." *Applied Mathematics and Mechanics* 3(3): 433–446.

Ware, L. B. and M. A. Matthay (2000). "The acute respiratory distress syndrome." *New England Journal of Medicine* **342**(18): 1334–1349.

Waters, C. M., et al. (1996). "Shear stress alters pleural mesothelial cell permeability in culture." *Journal of Applied Physiology* **81**(1): 448–458.

Waters, C. M., et al. (1997). "Mechanical forces alter growth factor release by pleural mesothelial cells." *American Journal of Physiology-Lung Cellular and Molecular Physiology* **16**(3): L552–L557.

Waters, S. L. and J. B. Grotberg (2002). "The propagation of a surfactant laden liquid plug in a capillary tube." *Physics of Fluids* **14**(2): 471–480.

Wei, H. H., et al. (2003). "Flow in a wavy-walled channel lined with a poroelastic layer." *Journal of Fluid Mechanics* **492**: 23–45.

Wei, H. H., et al. (2005). "A model of flow and transport in an oscillatory alveolus partially filled with liquid." *Physics of Fluids* **17**: Art. No. 031510.

Wei, T. and W. W. Willmarth (1989). "Reynolds-number effects on the structure of a turbulent channel flow." *Journal of Fluid Mechanics* **204**: 57–95.

Weibel, E. R. (1963). *Morphometry of the Human Lung*. New York, Academic Press.

Weinbaum, S., et al. (1994). "A model for the excitation of osteocytes by mechanical loading-induced bone fluid shear stresses." *Journal of Biomechanics* **27**(3): 339–360.

Wells, R. E., et al. (1961). "Measurement of viscosity of biological fluids by cone plate viscometer." *Journal of Laboratory and Clinical Medicine* **57**(4): 646–&.

Westerhof, N., et al. (2009). "The arterial Windkessel." *Medical & Biological Engineering & Computing* **47**: 131–141.

Whale, M. D., et al. (1996). "The effect of aging and pressure on the specific hydraulic conductivity of the aortic wall." *Biorheology* **33**: 17–44.

Whitaker, S. (1986). "Flow in porous media. 1. A theoretical derivation of Darcy's Law." *Transport in Porous Media* **1**(1): 3–25.

White, C. M. (1929). "Streamline flow through curved pipes." *Proceedings of the Royal Society of London A* **123**: 645–663.

White, C. M. (1932). "Fluid friction and its relation to heat transfer." *Transactions of the Institute of Chemical Engineers (London)* **10**: 66–86.

White, F. M. (1974). *Viscous Fluid Flow*. London, McGraw-Hill, Inc.

White, F. M. (1986). *Fluid Mechanics*. London, McGraw-Hill.

Whitesides, G. M. (2006). "The origins and the future of microfluidics." *Nature* **442**(7101): 368–373.

Wieghardt, K. (1944). "Über die turbulente Strömung im Rohr und längs einer Platte." *Zeitschrift für Angewandte Mathematik und Mechanik* **24**(5–6): 294–296.

Wieselsberger, C. (1921). "Neuere Feststellungen über die Gesetze des Flüssigkeits und Luftwiderstandes." *Physikalische Zeitschrift* **22**: 321–328.

Wieselsberger, C. (1922). "New data on the laws of fluid resistance." *National Advisory Committee for Aeronautics* TN **84**.

Wiggs, B. R., et al. (1997). "On the mechanism of mucosal folding in normal and asthmatic airways." *Journal of Applied Physiology* **83**(6): 1814–1821.

Wiig, H., et al. (2003). "Effect of charge on interstitial distribution of albumin in rat dermis in vitro." *Journal of Physiology (London)* **550**: 505–514.

Wilcox, D. C. (2008). "Formulation of the k–ω turbulence model revisited." *AIAA Journal* **46**: 2823–2838.

Withers, P. C. (1981). "An aerodynamic analysis of bird wings as fixed aerofoils." *J. Exp. Biol.* **90**: 143–162.

Wolf, D. P., et al. (1977). "Human cervical mucus. III. Isolation and characterization of rheologically active mucin." *Fertility and Sterility* **28**: 53–58.

Womersley, J. R. (1955). "Method for the calculation of velocity, rate of flow and viscous drag in arteries when the pressure gradient is known." *J. Physiol.* **127** (3): 553–563.

Xu, C., et al. (2006). "Computational fluid dynamics modeling of the upper airway of children with obstructive sleep apnea syndrome in steady flow." *Journal of Biomechanics* **39**(11): 2043–2054.

Yang, W., et al. (2007). "Instability of the two-layered thick-walled esophageal model under the external pressure and circular outer boundary condition." *Journal of Biomechanics* **40**(3): 481–490.

Yao, L. S. and S. A. Berger (1975). "Entry flow in a curved pipe." *Journal of Fluid Mechanics* **67**(1): 177–196.

Yasuda, R. A. K. and R. Cohen (1981). "Shear-flow properties of concentrated-solutions of linear and star branched polystyrenes." *Rheologica Acta* **20**: 163.

Yeates, D. B., et al. (1997). "Physicochemical properties of mucus and its propulsion," in *The Lung: Scientific Foundations*. R. G. Crystal, et al., eds. Philadelphia, PA, Lippincott-Raven: 487–503.

Yildiran, N., et al. (2017). "Effects of surfactant on propagation and rupture of a liquid plug in a tube." *Bulletin of the American Physical Society* **62**(14). E4.1.

Yin, Y. B., et al. (2010). "Simulation of pulmonary air flow with a subject-specific boundary condition." *Journal of Biomechanics* **43**(11): 2159–2163.

Yu, F. Y., et al. (2014). "Rheological studies of hyaluronan solutions based on the scaling law and constitutive models." *Polymer* **55**(1): 295–301.

Zagarola, M. V. and A. J. Smits (1998). "Mean-flow scaling of turbulent pipe flow." *Journal of Fluid Mechanics* **373**: 33–79.

Zakaria, E. R., et al. (1997). "In vivo hydraulic conductivity of muscle: effects of hydrostatic pressure." *American Journal of Physiology-Heart and Circulatory Physiology* **273**(6): H2774–H2782.

Zamankhan, P., et al. (2012). "Steady motion of Bingham liquid plugs in two-dimensional channels." *Journal of Fluid Mechanics* **705**: 258–279.

Zamankhan, P., et al. (2018). "Steady displacement of long gas bubbles in channels and tubes filled by a Bingham fluid." *Physical Review Fluids* **3**(1).

Zheng, Y., et al. (2005). "Effect of gravity on liquid plug transport through an airway bifurcation model." *Journal of Biomechanical Engineering* **127**(5): 798–806.

Zheng, Y., et al. (2006). "Effects of inertia and gravity on liquid plug splitting at a bifurcation." *Journal of Biomechanical Engineering* **128**(5): 707–716.

Zheng, Y., et al. (2007). "Effects of gravity, inertia, and surfactant on steady plug propagation in a two-dimensional channel." *Physics of Fluids* **19**(8): Art. No.: 082107.

Zheng, Y., et al. (2009). "Liquid plug propagation in flexible microchannels: a small airway model." *Physics of Fluids* **21**(7).

Zierenberg, J. R., et al. (2013). "An asymptotic model of particle deposition at an airway bifurcation." *Mathematical Medicine and Biology - A Journal of the IMA* **30**(2): 131–156.

Zimmer, M. E., et al. (2005). "The pulsatile motion of a semi-infinite bubble in a channel: flow fields, and transport of an inactive surface-associated contaminant." *Journal of Fluid Mechanics* **537**: 1–33.

INDEX